SELECTED APPLICATIONS

TABLE 1 EXPONENTIAL FUNCTIONS

x	e^x	e^{-x}	x	e^x	e^{-x}	x	e^x	e^{-x}
0.010	1.0101	0.9901	0.300	1.3499	0.7408	0.590	1.8040	0.5543
0.020	1.0202	0.9802	0.310	1.3634	0.7334	0.600	1.8221	0.5488
0.030	1.0305	0.9704	0.320	1.3771	0.7261	0.610	1.8404	0.5434
0.040	1.0408	0.9608	0.330	1.3910	0.7189	0.620	1.8589	0.5379
0.050	1.0513	0.9512	0.340	1.4049	0.7118	0.630	1.8776	0.5326
0.060	1.0618	0.9418	0.350	1.4191	0.7047	0.640	1.8965	0.5273
0.070	1.0725	0.9324	0.360	1.4333	0.6977	0.650	1.9155	0.5220
0.080	1.0833	0.9231	0.370	1.4477	0.6907	0.660	1.9348	0.5169
0.090	1.0942	0.9139	0.380	1.4623	0.6839	0.670	1.9542	0.5117
0.100	1.1052	0.9048	0.390	1.4770	0.6771	0.680	1.9739	0.5066
0.110	1.1163	0.8958	0.400	1.4918	0.6703	0.690	1.9937	0.5016
0.120	1.1275	0.8869	0.410	1.5068	0.6637	0.700	2.0138	0.4966
0.130	1.1388	0.8781	0.420	1.5220	0.6570	0.710	2.0340	0.4916
0.140	1.1503	0.8694	0.430	1.5373	0.6505	0.720	2.0544	0.4868
0.150	1.1618	0.8607	0.440	1.5527	0.6440	0.730	2.0751	0.4819
0.160	1.1735	0.8521	0.450	1.5683	0.6376	0.740	2.0959	0.4771
0.170	1.1853	0.8437	0.460	1.5841	0.6313	0.750	2.1170	0.4724
0.180	1.1972	0.8353	0.470	1.6000	0.6250	0.760	2.1383	0.4677
0.190	1.2092	0.8270	0.480	1.6161	0.6188	0.770	2.1598	0.4630
0.200	1.2214	0.8187	0.490	1.6323	0.6126	0.780	2.1815	0.4584
0.210	1.2337	0.8106	0.500	1.6487	0.6065	0.790	2.2034	0.4538
0.220	1.2461	0.8025	0.510	1.6653	0.6005	0.800	2.2255	0.4493
0.230	1.2586	0.7945	0.520	1.6820	0.5945	0.810	2.2479	0.4449
0.240	1.2712	0.7866	0.530	1.6989	0.5886	0.820	2.2705	0.4404
0.250	1.2840	0.7788	0.540	1.7160	0.5827	0.830	2.2933	0.4360
0.260	1.2969	0.7711	0.550	1.7333	0.5770	0.840	2.3164	0.4317
0.270	1.3100	0.7634	0.560	1.7507	0.5712	0.850	2.3396	0.4274
0.280	1.3231	0.7558	0.570	1.7683	0.5655	0.860	2.3632	0.4232
0.290	1.3364	0.7483	0.580	1.7860	0.5599	0.870	2.3869	0.4190

(Continued on inside back cover)

APPLIED MATHEMATICS FOR BUSINESS, ECONOMICS, AND THE SOCIAL SCIENCES

FOURTH EDITION

APPLIED MATHEMATICS FOR BUSINESS, ECONOMICS, AND THE SOCIAL SCIENCES

Frank S. Budnick

University of Rhode Island

McGRAW-HILL, INC.

New York   St. Louis   San Francisco   Auckland   Bogotá   Caracas   Lisbon
London   Madrid   Mexico   Milan   Montreal   New Delhi   Paris   San Juan
Singapore   Sydney   Tokyo   Toronto

**APPLIED MATHEMATICS FOR BUSINESS,
ECONOMICS, AND THE SOCIAL SCIENCES**
International Editions 1993

Exclusive rights by McGraw-Hill Book Co. - Singapore for manufacture and export. This book cannot be re-exported from the country to which it is consigned by McGraw-Hill.

5 6 7 8 9 0 CMO FC 9 8 7 6

This book was set in New Century Schoolbook by Progressive Typographers, Inc.
The editors were Michael Johnson, Karen M. Minette, Margery Luhrs, and David A. Damstra; the designer was Leon Bolognese; the production supervisor was Kathryn Porzio.
New drawings were done by J & R Services, Inc.
Cover and chapter-opener illustrations were done by Dick Krepel.

Library of Congress Cataloging-in-Publication Data

Budnick, Frank.
    Applied mathematics for business, economics, and the social sciences / Frank
Budnick. - 4th ed.
        p. cm.
    Includes index.
    ISBN 0-07-008902-7
    1. Mathematics.   2. Business mathematics.   3. Social sciences - Mathematics.
I. Title.
QA37.2.B83   1993
510-dc20   92-30009

When ordering this title, use ISBN 0-07-112580-9

Printed in Singapore

ABOUT THE AUTHOR

FRANK S. BUDNICK received his B.S. degree in Industrial Engineering from Rutgers, The State University of New Jersey. His master's and doctoral degrees were earned at the University of Maryland. He is currently a Full Professor at the University of Rhode Island, where he has been teaching since 1971. He has worked in private industry as well as with the federal government. He has conducted federally funded research in the criminal justice area and in the transfer of technology between universities and industry. He is coauthor of the text *Principles of Operations Research for Management*, second edition, published by Richard D. Irwin, Inc. He also has authored *Finite Mathematics with Applications*, a McGraw-Hill text.

To my wife, Deb,
and my children, Chris, Scott, and Kerry,
I LOVE YOU!

CONTENTS

EQUATIONS AND FUNCTIONS

THE CALCULUS

PREFACE

INTRODUCTION

Mathematics is an integral part of the education of students in business, economics, and the social sciences. There is increasingly a desire to improve the level of quantitative sophistication possessed by graduates in these types of programs. The objective is not to make mathematicians of these students, but to make them as comfortable as possible in an environment which increasingly makes use of quantitative analysis and the computer. Students are discovering that they must integrate mathematics, statistical analysis, and the computer in both required and elective courses within their programs. Furthermore, organizations are becoming more effective users of quantitative tools and the computer. Decision makers will be better equipped to operate within this type of environment if they are familiar with the more commonly used types of quantitative analyses and the technology of the computer. Such familiarity can assist them in being better "critics" and "users" of these tools, and, hopefully, better decision makers.

Applied Mathematics for Business, Economics, and the Social Sciences, fourth edition, continues to provide an informal, non-intimidating presentation of the mathematical principles, techniques, and applications most useful for students in business, economics, management, and the life and social sciences. Designed primarily for a two-term course in applied mathematics (the book can be adapted easily for a one-term course) it provides a comprehensive treatment of selected topics in finite mathematics and calculus. It is appropriate for use in both two-year schools and four-year schools, as well as the "foundation" level for graduate programs having prerequisite mathematics requirements. M.B.A. and M.P.A. programs are typical graduate programs having this type of requirement.

FEATURES

The following features from the previous edition have been retained:

- ❏ A level of presentation which carefully develops and reinforces topics.
- ❏ A style which appeals to the *intuition* of students and provides a great deal of *visual reinforcement.*
- ❏ An *applied orientation* which motivates students and provides a sense of purpose for studying mathematics.
- ❏ An approach which first develops the mathematical concept and then reinforces with applications.
- ❏ An approach which minimizes the use of rigorous mathematical proofs.

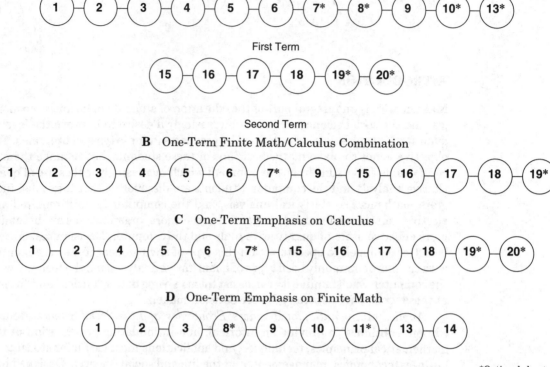

A   Two-Term Finite Math/Calculus Combination

1 — 2 — 3 — 4 — 5 — 6 — 7* — 8* — 9 — 10* — 13*

First Term

15 — 16 — 17 — 18 — 19* — 20*

Second Term

B   One-Term Finite Math/Calculus Combination

1 — 2 — 3 — 4 — 5 — 6 — 7* — 9 — 15 — 16 — 17 — 18 — 19*

C   One-Term Emphasis on Calculus

1 — 2 — 4 — 5 — 6 — 7* — 15 — 16 — 17 — 18 — 19* — 20*

D   One-Term Emphasis on Finite Math

1 — 2 — 3 — 8* — 9 — 10 — 11* — 13 — 14

*Optional chapter

Some Suggested Course Structures

Specific Features

- ❏ An increased orientation toward using the COMPUTER AS A TOOL for mathematical analysis.
- ❏ ALGEBRA FLASHBACKS are used throughut the book to assist the student in the recall of key rules or concepts. The flashback usually consists of a restatement of a rule or concept with a reference to appropriate sections of algebra review in the text.
- ❏ NOTES TO THE STUDENT which provide special insights about a mathematical concept or application.
- ❏ "POINTS FOR THOUGHT AND DISCUSSION" which allow students to pause for a moment and reconsider a concept or example from a different perspective. Their purpose is to reinforce and extend the student's understanding inducing critical thinking.

PEDAGOGICAL FEATURES

❑ COMPUTER-BASED PROBLEMS, identified in the exercise set by a computer icon, offer the student and instructor an opportunity to solve larger scale problems.

❑ MINICASES allow students to analyze and interpret a more complex, realistic application. These can provide the basis for stimulating classroom discussion.

❑ A multitude of other learning aids, including *chapter objectives, numerous solved examples, a wealth of exercises, chapter tests, lists of key terms and concepts, and summary lists of important formulas.*

NEW FEATURES AND CHANGES

The major changes in the fourth edition are organizational. First, the book has been organized into three major subsections:

   I.   *Equations and Functions*
  II.   *Finite Mathematics*
 III.   *The Calculus*

Other significant changes include the following:
Chapter 1: **Some Preliminaires** is a new chapter which discusses some fundamental concepts (beyond the review of basic principles of algebra in Appendix A) which are prerequisite to the material which follows.

The material on linear equations and systems of equations has been moved so as to precede the discussion of mathematical functions.

The material on mathematical functions has been consolidated into four chapters early in the text. Chapter 4 introduces the concept and notation of **Mathematical Functions.** Chapter 5 focuses upon **Linear Functions and Applications.** Chapter 6 discusses **Quadratic and Polynomial Functions** with applications and Chapter 7 presents **Exponential and Logarithmic Functions** with applications.

The treatment of linear programming in Chapters 10 and 11 has been reorganized slightly. Whereas applications were introduced first in the previous edition, Chapter 10 focuses upon graphical solution methods first, followed by selected applications. The section discussing computer solution methods has been moved to the end of Chapter 11, which presents the simplex method.

The material on integer-programming and goal programming has been eliminated in this edition. Although interesting extensions of linear programming do exist, these topics were determined to be of relatively minor importance.

In the discussion of calculus, the material on optimization has been split into two separate chapters (as in the second edition). Chapter 16 introduces the methodology of optimization and Chapter 17 is devoted exclusively to applications of optimization. The rationale for the separation of these topics is that there is too much material for one chapter. The pedagogy is to have students learn the mathematical methodology in Chapter 16, followed by selected applications in Chapter 17.

The material on optimization of functions of several variables has been moved to the last chapter in the text. This topic is optional for many schools and the new placement is compatible with competing texts.

Other significant changes include:

The material on differentiation (Chapter 15) has been organized into shorter sections.

Chapter 11 from the third edition (Selected Applications of Probability) has been eliminated, although some applications have been transferred to the chapter on matrix algebra.

The number of Practice Exercises has been expanded significantly to give more opportunities for reinforcement of new concepts.

In addition to these changes, the author has incorporated a signifiant number of applications (either as examples or exercises) which contain "live data". Also, the author makes a conscientious attempt to make students aware of the nature of *estimation* in applying mathematics. That is, the application of mathematical analysis in the "real world" involves approximating relationships among variables. It is important for students to understand the strengths and weaknesses of mathematical analysis.

The book includes a large number of diverse applications. The intent is that instructors cover as many applications in these chapters as they feel appropriate for their students.

Some exercises in the book are considered to be a higher level of difficulty than most others. These are preceded by an asterisk (*).

SUPPLEMENTS

*The **Instructor's Resource Manual** contains solutions to all even-numbered exercises in the text as well as all of the solutions to the chapter tests and minicases.*

*The **Student's Solutions Manual** contains solutions to the odd-numbered exercises in the text.*

*The **Professor's Assistant** is a computerized test generator that allows the instructor to create tests using questions from a standard test bank. This testing system enables the instructor to choose questions by section, question type, and other criteria. This system is available for IBM, IBM compatible, and Macintosh computers.*

*The **Print Test Bank** is a printed and bound copy of the questions found in the computerized test generator.*

MATHPACK is a software package that relieves students from tedious calculation, and includes numerous routines cross referenced with relevant sections/pages from the text.

For further information about these supplements, please contact your local college division sales representative.

ACKNOWLEDGMENTS

I wish to express my sincere appreciation to those persons who have contributed either directly or indirectly to this project. I wish to thank: Thomas Arbutiski, Community College of Allegheny County; Helen B. Chun, Community College of Allegheny County; Benjamin Eichorn, Rider College; Joseph Fadyn, Southern College of Technology; Odene Forsythe, Westark Community College; Gary Grimes, Mount Hood Community College; Anne Hughes, St. John's University; Harry Hutchins, Southern Illinois University; Harlan Koca, Washburn University of Topeka; Joyce Longman, Villanova University; Daniel J. Madden, University of Arizona; Victor McGee, Dartmouth College; Michael Mogavero, Alfred University; Dean Morrow, Robert Morris College; Richard Semmler, Northern Virginia Community College, Richard Witt, University of Wisconsin, Eau Claire; and Cathleen Zucco, Le Moyne College, for their very helpful comments during the development of the manuscript. A very special note of thanks goes to Thomas Arbutiski for his conscientious and extremely detailed reviews and suggestions.

I also wish to thank the people at McGraw-Hill with whom I worked directly. These persons include Michael Johnson, Margery Luhrs, and David Damstra. Thanks also to Karen Minette for coordinating the supplements package and Leon Bolognese for his work on the design of the book.

I am also grateful for the efforts of Shaochi Xu who assisted in developing solution sets for exercises. Special thanks also go to my QBA 520 students who served as "guinea pigs" for class testing parts of the manuscript. I would also like to acknowledge the helpful suggestions of Sandra Quinn, Kathy Bowser, and the late Elizabeth Flaherty, as well as their efforts in developing the *Instructor's Resource Manual* and *Student's Solutions Manual*.

Finally, I want to thank my wife, Deb, for her support through this ordeal, as well as the others we have shared together.

Frank S. Budnick

APPLIED MATHEMATICS FOR BUSINESS, ECONOMICS, AND THE SOCIAL SCIENCES

CHAPTER 1

SOME PRELIMINARIES

CHAPTER OBJECTIVES

❏ Discuss equations and methods of solution

❏ Present properties of inequalities and methods of solution

❏ Illustrate absolute value relationships

❏ Introduce properties of rectangular coordinate systems

This chapter provides a discussion of selected algebraic concepts. An understanding of these concepts, as well as the fundamental concepts reviewed in Appendix A, is prerequisite to the successful study of the material which follows in this text. As with the material in Appendix A, many of you will find that the material in this chapter is a review of topics which you have studied in previous mathematics courses.

1.1 SOLVING FIRST-DEGREE EQUATIONS IN ONE VARIABLE

We work continually with equations in this book. It is absolutely essential that you understand the meaning of equations and their algebraic properties.

Equations and Their Properties

An **equation** states the equality of two algebraic expressions. The algebraic expressions may be stated in terms of one or more *variables*. The following are examples of equations.

$$3x - 10 = 22 - 5x \tag{1}$$

$$\frac{2r - 5s + 8t}{3} = 100 \tag{2}$$

$$w^2 - 5w = -16 \tag{3}$$

In Eqs. (1) and (3), the variables are x and w, respectively. In Eq. (2) there are three variables, r, s, and t. The term *variable* is used because various numbers can be substituted for the letters.

The **solution** of an equation consists of those numbers which, when substituted for the variables, make the equation true. The numbers, or values of the variables, which make the equation true are referred to as the **roots** of the equation. We say that the roots are the values of the variable(s) which *satisfy the equation*. In Eq. (1), substitution of the number 0 for the variable x results in

$$-10 = 22$$

which is untrue. The value $x = 0$ is not a root of the equation. However, substitution of the number 4 for the variable x results in

$$3(4) - 10 = 22 - 5(4)$$

or
$$2 = 2$$

The value $x = 4$ is considered to be a root of the equation.

We can distinguish three types of equations. An ***identity*** is an equation which is true for all values of the variables. An example of an identity is the equation

$$6x + 12 = \frac{12x + 24}{2}$$

Another example is

$$5(x + y) = 5x + 5y$$

In each of these equations, any values that are assigned to the variables will make both sides of the equation equal.

A *conditional equation* is true for only a limited number of values of the variables. For example, the equation

$$x + 3 = 5$$

is true only when x equals 2.

A *false statement,* or *contradiction,* is an equation which is never true. That is, there are no values of the variables which make the two sides of the equation equal. An example is the equation

$$x = x + 5$$

We indicate that the two sides are not equal by using the symbol $\neq$; for this example,

$$x \neq x + 5$$

Solving an equation refers to the process of finding the roots of an equation, if any roots exist. In order to solve equations, we normally must manipulate or rearrange them. The following rules indicate allowable operations.

SELECTED RULES FOR MANIPULATING EQUATIONS

I *Real-valued expressions which are equal can be added to or subtracted from both sides of an equation.*

II *Both sides of an equation may be multiplied or divided by any nonzero constant.*

III *Both sides of an equation may be multiplied by a quantity which involves variables.*

IV *Both sides of an equation may be squared.*

V *Both sides of an equation may be divided by an expression which involves variables provided the expression is not equal to 0.*

Rules I and II lead to the creation of **equivalent equations.** *Equivalent equations are equations which have the same roots.* Rules III and IV can result in roots which are not roots of the original equation. These roots are called **extraneous roots.** Applying rule V can lead to equations which do not have all the roots contained in the original equation, or equations which are not equivalent to the original equations.

In Appendix A, Sec. A.2, the **degree of a polynomial** is defined as the degree of the highest-degree term in the polynomial. If an equation can be written in the form

$$Polynomial\ expression = 0$$

then the degree of the polynomial expression is the ***degree of the equation.*** Thus, the equation $2x - 4 = 0$ is an equation of degree 1. The equation $4r^2 - r + 10 = 0$ is an equation of degree 2. The equation $n^4 - 3n^2 + 9 = 0$ is an equation of degree 4.

Solving First-Degree Equations in One Variable

The procedure used for solving equations depends upon the nature of the equation. Let's consider first-degree equations which involve one variable. The following equations are examples.

$$3x = 2x - 5$$

$$5x - 4 = 12 + x$$

Solving equations of this form is relatively easy. By using appropriate rules of manipulation, the approach is simply to isolate the variable on one side of the equation and all constants on the other side of the equation.

EXAMPLE 1   Solve the two first-degree equations given previously.

SOLUTION

For the equation $3x = 2x - 5$, we can add $-2x$ to both sides to get

$$3x + (-2x) = 2x - 5 + (-2x)$$

or $$x = -5$$

The only value of x which satisfies this equation is -5.

For the equation $5x - 4 = 12 + x$, we can add $-x$ and 4 to both sides, getting

$$5x - 4 + 4 + (-x) = 12 + x + 4 + (-x)$$

$$5x - x = 12 + 4$$

or $$4x = 16$$

Dividing both sides by 4 (or multiplying by $\frac{1}{4}$) gives us the root of the equation:

$$x = 4$$

EXAMPLE 2   To solve the equation

$$2x + 5 = 10 + 2x$$

we can subtract $2x$ from both sides, resulting in

$$2x + 5 - 2x = 10 + 2x - 2x$$

or
$$5 = 10$$

This result is a false statement, or contradiction, indicating that there are no roots to the original equation.

EXAMPLE 3 To solve the equation

$$x - 3 = \frac{2x - 6}{2}$$

we can multiply both sides of the equation by 2, resulting in

$$2(x - 3) = 2x - 6$$

$$2x - 6 = 2x - 6$$

The two sides of the equation are *identical*, suggesting that any value can be assigned to x and the equation will be satisfied. If we attempt to isolate x on the left side of the equation, subtracting $2x$ from both sides results in

$$-6 = -6$$

This is an identity, also signaling that any value can be assigned to the variable x.

❏

PRACTICE EXERCISE
Solve the following first-degree equations:
(*a*) $4x - 10 = 8 - 2x$
(*b*) $x - 5 = -\dfrac{(-2x + 10)}{2}$
(*c*) $3x + 3 = 3x - 5$
Answer: (*a*) 3, (*b*) any real number, (*c*) no values.

Section 1.1 Follow-Up Exercises

Solve the following first-degree equations.

1 $x - 5 = 2x - 8$ 2 $10 - 2x = 8 - 3x$
3 $2x + 4 = 16 - x$ 4 $-5x + 2 = 16 - 3x$
5 $2(x - 3) = 3(x + 4)$ 6 $5(3 - x) = 3(5 + x)$
7 $6 - 2t = 4t + 12$ 8 $3y - 10 = 6y + 20$
9 $3 - 5t = 3t - 5$ 10 $10y - 20 = 6y + 4$
11 $3t + 10 = 4t - 6$ 12 $3(2t - 8) = 4(7 + t)$

13 $(x+6)-(5-2x)+2=0$

14 $\dfrac{x}{6}-5=\dfrac{x}{2}-7$

15 $\dfrac{t-3}{2}+\dfrac{t+3}{4}=\dfrac{8-t}{3}+2$

16 $3-\dfrac{x}{2}=\dfrac{x}{3}-2$

17 $\dfrac{v}{2}-3=5+\dfrac{v}{2}$

18 $4+x=3+\dfrac{x}{2}$

19 $3(x-2)=(x+3)/2$

20 $(t-3)/2=(4-3t)/4$

21 $3(12-x)-16=2$

22 $2(y+1)-3(y-1)=5-y$

23 $3x+1=2-(x-4)+3x$

24 $3(x-2)+4(2-x)=x+2(x+1)$

1.2 SOLVING SECOND-DEGREE EQUATIONS IN ONE VARIABLE

A second-degree equation involving the variable x has the generalized form

$$ax^2+bx+c=0$$

where a, b, and c are constants with the added provision that $a \neq 0$. Second-degree equations are usually called **quadratic equations.** If a equals zero, the x^2 term disappears and the equation is no longer of degree 2. Examples of second-degree equations are

$$6x^2-2x+1=0$$

$$3x^2=12$$

$$2x^2-1=5x+9$$

Solving Quadratic Equations

A quadratic equation (excluding an identity) can have **no real roots, one real root,** or **two real roots.** A number of different procedures can be used to determine the roots of a quadratic equation. We will discuss two of the procedures. The first step, in either case, is to rewrite the equation in the form $ax^2+bx+c=0$.

Factoring Method If the left side of the quadratic equation can be factored, the roots can be identified very easily. Consider the quadratic equation

$$x^2-4x=0$$

The left side of the equation can be factored, resulting in

$$x(x-4)=0$$

The factored form of the equation suggests that the product of the two terms equals 0. The product will equal 0 if either of the two factors equals 0. For this equation the first factor is 0 when $x=0$, and the second factor is 0 when $x=4$. Thus, the two roots are 0 and 4.

EXAMPLE 4 Determine the roots of the equation

$$x^2 + 6x + 9 = 0$$

SOLUTION

The left side of the equation can be factored such that

$$(x + 3)(x + 3) = 0$$

Setting each factor equal to 0, we find that there is one root to the equation, and it occurs when $x = -3$.

❑

Quadratic Formula When the quadratic expression cannot be factored, or if you are unable to identify the factors, you can apply the *quadratic formula*. The quadratic formula will allow you to identify all roots of an equation of the form

$$ax^2 + bx + c = 0$$

The quadratic formula is

$$x = \frac{-b \pm \sqrt{b^2 - 4ac}}{2a} \tag{1.1}$$

The following examples illustrate the use of the formula.

EXAMPLE 5 Given the quadratic equation $x^2 - 2x - 48 = 0$, the coefficients are $a = 1$, $b = -2$, and $c = -48$. Substituting these into the quadratic formula, the roots of the equation are computed as

$$x = \frac{-(-2) \pm \sqrt{(-2)^2 - 4(1)(-48)}}{2(1)}$$

$$= \frac{2 \pm \sqrt{4 + 192}}{2} = \frac{2 \pm \sqrt{196}}{2} = \frac{2 \pm 14}{2}$$

Using the plus sign, we get

$$x = \tfrac{16}{2} = 8$$

Using the minus sign, we obtain

$$x = -\tfrac{12}{2} = -6$$

Thus, these are the only real values of x which satisfy the quadratic equation.

EXAMPLE 6 To find the roots of the equation $x^2 - 2x + 1 = 0$, the coefficients are $a = 1$, $b = -2$, and $c = 1$. Substituting into the quadratic formula yields

$$x = \frac{-(-2) \pm \sqrt{(-2)^2 - 4(1)(1)}}{2(1)}$$

$$= \frac{2 \pm \sqrt{4 - 4}}{2}$$

$$= \frac{2 \pm 0}{2}$$

$$= 1$$

Since the radicand equals zero, the use of the $\pm$ sign results in the same root, 1.

EXAMPLE 7 To find the roots of the equation $x^2 - x + 10$, the coefficients are $a = 1$, $b = -1$, and $c = 10$. Substituting into the quadratic formula gives

$$x = \frac{-(-1) \pm \sqrt{(-1)^2 - 4(1)(10)}}{2(1)}$$

$$= \frac{1 \pm \sqrt{1 - 40}}{2}$$

$$= \frac{1 \pm \sqrt{-39}}{2}$$

Because there is no real square root of -39, we conclude that there are no values of x which satisfy the quadratic equation.

❑

PRACTICE EXERCISE
Solve the following second-degree equations.
(a) $x^2 + 3x + 2 = 0$
(b) $3x^2 - 2x + 5 = 0$
(c) $x^2 + 10x + 25 = 0$
Answer: (a) $x = -1, -2$, (b) no values, (c) $x = -5$.

The expression under the radical of the quadratic formula, $b^2 - 4ac$, is called the ***discriminant.*** The value of the discriminant helps us determine the number of roots of a quadratic equation.

INTERPRETATIONS OF THE DISCRIMINANT
For a quadratic equation of the form $ax^2 + bx + c = 0$:

  I   *If $b^2 - 4ac > 0$, there are two real roots.*
  II   *If $b^2 - 4ac = 0$, there is one real root.*
  III *If $b^2 - 4ac < 0$, there are no real roots.*

Section 1.2 Follow-Up Exercises

Solve the following quadratic equations using factoring.

1 $x^2 + x - 6 = 0$
2 $x^2 - 25 = 0$
3 $x^2 + 2x + 1 = 0$
4 $x^2 + 3x - 4 = 0$
5 $x^2 - 3x - 10 = 0$
6 $t^2 - 2t - 8 = 0$
7 $2t^2 + 9t + 4 = 0$
8 $5r^2 + 2r - 3 = 0$
9 $6y^2 - 9y - 6 = 0$
10 $x^2 + 10x + 25 = 0$
11 $r^2 - 16 = 0$
12 $3t^2 + 9t + 6 = 0$
13 $x^2 - 2x + 15 = 0$
14 $2x^2 - 17x - 1 = 0$
15 $4y^2 + 18y - 10 = 0$
16 $x^2 + 4x - 21 = 0$

Solve the following quadratic equations using the quadratic formula.

17 $x^2 + 8x + 12 = 0$
18 $x^2 + 12x + 36 = 0$
19 $r^2 + 2r + 1 = 0$
20 $t^2 - 2t + 1 = 0$
21 $x^2 + x + 10 = 0$
22 $x^2 + 3x + 5 = 0$
23 $x^2 + 3x - 4 = 0$
24 $9x^2 - 3x = 2$
25 $x^2 + 1 = x$
26 $3r^2 = 14r - 8$
27 $x^2 = 2x - 2$
28 $4t^2 + 3t = 1$
29 $y^2 + 2 = 2y$
30 $x^2 + 4x + 5 = 0$
31 $x^2 - 2x = -5$
32 $4x^2 - 64 = 0$

1.3 INEQUALITIES AND THEIR SOLUTION

In this section we will discuss *inequalities, interval notation,* and *solving inequalities.*

Inequalities

Inequalities express the condition that two quantities are not equal. One way of expressing this condition is by using the ***inequality symbols*** $<$ and $>$. The following illustrate the use and interpretation of these symbols:

Inequality	Interpretation
(a) $3 < 5$	"3 is less than 5"
(b) $x > 100$	"the value of x is greater than 100"
(c) $0 < y < 10$	"the value of y is greater than zero and less than 10"

These inequalities are ***strict inequalities*** since the items being compared can never equal one another. Case (*a*) illustrates an ***absolute inequality,*** which is always true. A ***conditional inequality*** is true under certain conditions. The inequality in case (*b*) holds when the variable x has a value greater than 100. If $x = 150$, the inequality is true; if $x = -25$, the inequality is not true. Case (*c*) illustrates what is termed a ***double inequality.***

One use of inequalities is to facilitate the comparison of numbers. Figure 1.1 illustrates a *real number line.* **Given two real numbers a and b, if $a < b$, a lies**

Figure 1.1

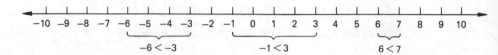

to the left of *b* on a real number line. Sample inequalities are illustrated in Fig. 1.1.

The following statements are equivalent to one another:

> *a* is less than *b*.
> *b* is greater than *a*.
> $a < b$
> $b > a$
> $b - a > 0$
> $a - b < 0$
> *a* lies to the left of *b* on a real number line.
> *b* lies to the right of *a* on a real number line.

Another type of inequality relationship is expressed by the symbols $\geq$ and $\leq$. Such inequality relationships allow for the possibility that two quantities are equal. The following illustrate these types of inequalities.

Inequality	Interpretation
(a) $x + 3 \geq 15$	"the quantity $(x + 3)$ is greater than *or* equal to 15"
(b) $y \leq x$	"the value of *y* is less than *or* equal to the value of *x*"

Interval Notation

An **interval** is the set of real numbers that lies between two numbers *a* and *b*. This can be specified using the following notation:

$$(a, b) = \{x \mid a < x < b\}$$

The notation **(*a*, *b*)** represents the **open interval** with endpoints *a* and *b*. The notation $\{x \mid a < x < b\}$ indicates that the open interval with endpoints *a* and *b* "consists of real numbers *x* *such that* (|) *x* is greater than *a* and *x* is less than *b*." By "open" interval, we mean that the endpoint values are *not included* in the interval.

A **closed interval** is one which includes the endpoint values. The notation **[*a*, *b*]** represents the closed interval which includes endpoints *a* and *b*. More precisely, this closed interval can be denoted as

$$[a, b] = \{x \mid a \leq x \leq b\}$$

Half-open intervals include one endpoint but not the other. The notation **(*a*, *b*]** represents the half-open interval which contains the endpoint *b* but not *a*. The notation **[*a*, *b*)** represents the half-open interval which includes *a* but not *b*. Figure 1.2 illustrates the graphical representation of several intervals. Notice that two

Figure 1.2

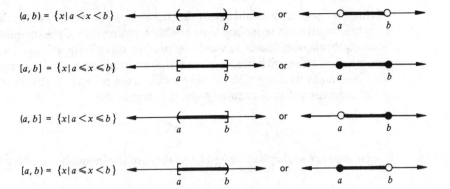

$(a, b) = \{x \mid a < x < b\}$

$[a, b] = \{x \mid a \leq x \leq b\}$

$(a, b] = \{x \mid a < x \leq b\}$

$[a, b) = \{x \mid a \leq x < b\}$

number line representations are illustrated, one using parentheses and brackets and the other using open (O) and solid (●) circles. The open-circle notation indicates that the endpoint is not included in the interval. The solid circle indicates that the endpoint is included.

EXAMPLE 8 Sketch the following intervals: (a) $(-2, 1)$, (b) $[1, 3]$, (c) $[-3, 0)$.

SOLUTION
Figure 1.3 presents the graphic representation of the intervals.

Figure 1.3

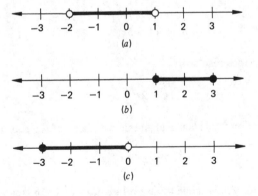

(a)

(b)

(c)

Solving Inequalities

In this section we will look at ways of solving inequalities. Solving inequalities is very similar to solving equations. What we are attempting to determine is the set of values which satisfy the inequality. Given a first-degree inequality in one variable, such as

$$2x + 3 \geq -5$$

solving the inequality would determine the values of x which satisfy the inequality. To solve, we attempt to isolate the variable x on one side of the inequality using the same algebraic operations as would be used in solving equations. The only difference when working with inequalities is that multiplication or division of both sides of an inequality by a negative number requires the reversal of the direction or *sense* of the inequality. To illustrate, given the inequality

$$-2 < 3$$

if both sides are multiplied by (-1), the sense of the inequality is reversed, resulting in

$$2 > -3$$

Similarly, to solve the inequality

$$-2x < 6$$

both sides of the inequality are divided by -2 to isolate x. The sense of the inequality must be reversed, resulting in

$$x > -3$$

EXAMPLE 9   To determine the values of x which satisfy the inequality $3x + 10 \leq 5x - 4$, 4 may be added to both sides to form

$$3x + 14 \leq 5x$$

Subtracting $3x$ from both sides results in

$$14 \leq 2x$$

Finally, dividing both sides by 2 yields the algebraic definition of the solution set

$$7 \leq x$$

That is, the original inequality is satisfied by any values of x which are greater than or equal to 7. Figure 1.4 illustrates the solution graphically.

Figure 1.4   Solution for the inequality $3x + 10 \leq 5x - 4$.

EXAMPLE 10   To determine the values of x which satisfy the inequality $6x - 10 \geq 6x + 4$, the addition of 10 to both sides yields

$$6x \geq 6x + 14$$

Subtracting $6x$ from both sides results in

$$0 \geq 14$$

which is a false statement. Hence, there are no values for x which satisfy the inequality.

EXAMPLE 11 To determine the values of x which satisfy the inequality $4x + 6 \geq 4x - 3$, 6 is subtracted from both sides to yield

$$4x \geq 4x - 9$$

and subtracting $4x$ from both sides gives us

$$0 \geq -9$$

The variable x has disappeared, and we are left with an inequality which is true all the time. This indicates that *the original inequality is true for any and all (real) values of x.*

☐

PRACTICE EXERCISE
Solve the inequality $2x - 5 \geq 3x + 2$. *Answer: $x \leq 7$.*

EXAMPLE 12 To determine the values of x which satisfy the double inequality $-2x + 1 \leq x \leq 6 - x$, we first find the solution set for each inequality.

The values of x satisfying the left inequality are determined as

$$-2x + 1 \leq x$$

$$1 \leq 3x$$

or

$$\tfrac{1}{3} \leq x$$

Those values satisfying the right inequality are

$$x \leq 6 - x$$

$$2x \leq 6$$

or

$$x \leq 3$$

The values which satisfy the double inequality consist of those values of x which satisfy both inequalities, or $\tfrac{1}{3} \leq x \leq 3$. Figure 1.5 illustrates the solution.

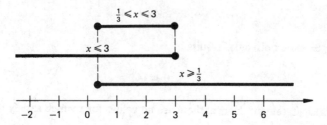

Figure 1.5   Solution for $-2x + 1 \le x \le 6 - x$.

| EXAMPLE 13 | To determine the solution to the double inequality $2x - 4 \le x \le 2x - 10$, we first solve the left inequality |

$$2x - 4 \le x$$

$$x \le 4$$

The values of x satisfying the right inequality are

$$x \le 2x - 10$$

or

$$10 \le x$$

From Fig. 1.6, we see that there are no values common to the solutions of the two inequalities. Therefore, there are no values which satisfy the double inequality.

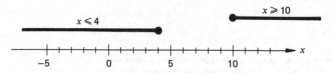

Figure 1.6   No solution for $2x - 4 \le x \le 2x - 10$.

□

PRACTICE EXERCISE
Solve the inequality $10 \le x + 5 \le 30$. *Answer:* $5 \le x \le 25$.

Second-Degree Inequalities

If an inequality involves a higher-order algebraic expression, we can often solve it if the algebraic expression can be rewritten in factored form. The following examples illustrate solutions for second-degree inequalities.

| EXAMPLE 14 | To solve the inequality |

$$x^2 - 5x + 6 \le 0$$

we first factor the left side, yielding

$$(x - 3)(x - 2) \leq 0$$

The following attributes of the two factors on the left side will result in the inequality being satisfied.

	Factor		
	(x − 3)	**(x − 2)**	**Product**
Condition 1	= 0	Any value	0
Condition 2	Any value	= 0	0
Condition 3	> 0	< 0	< 0
Condition 4	< 0	> 0	< 0

Condition 1:

$$x - 3 = 0 \quad \text{when} \quad x = 3$$

Condition 2:

$$x - 2 = 0 \quad \text{when} \quad x = 2$$

Condition 3:

$$x - 3 > 0 \quad and \quad x - 2 < 0$$

when
$$x > 3 \quad and \quad x < 2$$

Condition 4:

$$x - 3 < 0 \quad and \quad x - 2 > 0$$

when
$$x < 3 \quad and \quad x > 2$$

Figure 1.7 summarizes the results of these four conditions. Conditions 1 and 2 result in the product equaling zero when x equals 3 and 2, respectively. There are no values of x which

Figure 1.7

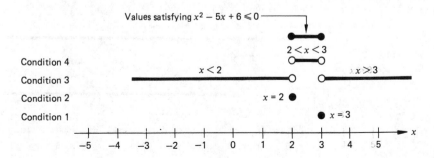

result in the sign attributes of condition 3. Finally, the values of x which satisfy condition 4 are $2 < x < 3$. Combining the values from conditions 1, 2, and 4, the inequality is satisfied if $2 \le x \le 3$.

EXAMPLE 15 To solve the inequality

$$x^2 - 2x - 15 > 0$$

we first factor the left side, resulting in

$$(x - 5)(x + 3) > 0$$

Compared with Example 14, this is a strict inequality. The left side of the inequality will be positive if the two factors have the same sign.

	Factor		
	$(x - 5)$	**$(x + 3)$**	**Product**
Condition 1	> 0	> 0	> 0
Condition 2	< 0	< 0	> 0

Condition 1:

$$x - 5 > 0 \quad and \quad x + 3 > \; 0$$

when

$$x > 5 \quad and \quad x > -3$$

Condition 2:

$$x - 5 < 0 \quad and \quad x + 3 < \; 0$$

when

$$x < 5 \quad and \quad x < -3$$

  Figure 1.8 summarizes the results of these two conditions. The values of x which satisfy condition 1 are $x > 5$. Those satisfying condition 2 are $x < -3$. Combining the results from the two conditions, the solution to the original inequality is $x < -3$ and $x > 5$.

Figure 1.8

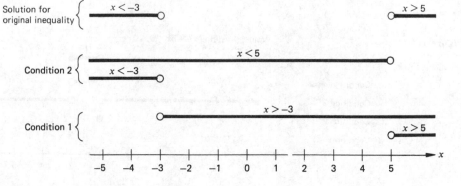

> **PRACTICE EXERCISE**
> Solve the inequality $x^2 + x - 12 \leq 0$. *Answer:* $-4 \leq x \leq 3$.

Section 1.3 Follow-up Exercises

Sketch the following intervals.

1 $(-3, 0)$

3 $(-5, -2)$

5 $(0, 3)$

7 $[-4, -3]$

9 $[-0.5, 0.5]$

11 $[1, 2]$

13 $(0, 4]$

15 $[-5, -1)$

2 $(3, 5)$

4 $(-2, 1)$

6 $(5, 10)$

8 $[-2, 3]$

10 $[2, 5]$

12 $[-5, -3]$

14 $[2, 6)$

16 $(-3, 4]$

Solve the following inequalities.

17 $3x - 2 \leq 4x + 8$

19 $x \geq x + 5$

21 $-4x + 10 \geq -10 + x$

23 $15x + 6 \geq 10x - 24$

25 $12 \geq x + 16 \geq -20$

27 $50 \leq 4x - 6 \leq 25$

29 $-10 \leq x + 8 \leq 15$

31 $0 \geq 20 - x \geq -20$

18 $x + 6 \geq 10 - x$

20 $2x \leq 2x - 10$

22 $3x + 6 \leq 3x - 5$

24 $-4x + 10 \leq x \leq 2x + 6$

26 $35 \leq 2x + 5 \leq 80$

28 $6x - 9 \leq 12x + 9 \leq 6x + 81$

30 $25 \leq 5 - x \leq 10$

32 $10 + x \leq 2x - 5 \leq 25$

Solve the following second-degree inequalities.

33 $x^2 - 16 \leq 0$

35 $x^2 + 3x - 18 \leq 0$

37 $x^2 - 2x - 3 \geq 0$

39 $2x^2 - 3x - 2 < 0$

41 $x^2 + 2x - 15 > 0$

43 $4x^2 - 100 < 0$

34 $x^2 - 9 \geq 0$

36 $x^2 + 2x - 8 \geq 0$

38 $x^2 + 4x - 12 \leq 0$

40 $2x^2 - x - 10 > 0$

42 $2x^2 + 5x + 3 < 0$

44 $6x^2 + x - 12 > 0$

1.4 ABSOLUTE VALUE RELATIONSHIPS

The **absolute value** of a number is its distance, which must be greater than or equal to zero, from zero on a real number line. The absolute value of a number a is denoted by $|a|$. Using this definition, we can confirm in Fig. 1.9 that $|3| = 3$ and $|-3| = 3$. A more formal definition follows.

Figure 1.9

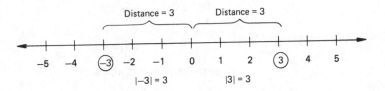

> **DEFINITION: ABSOLUTE VALUE**
> For any real number a,
>
> $$|a| = \begin{cases} a & \text{if } a \geq 0 \\ -a & \text{if } a < 0 \end{cases}$$

Another way of thinking of absolute value is that the absolute value of a number is the magnitude or size of the number, disregarding the sign.

Some Properties of Absolute Values

The following are selected properties of absolute values.

Property 1

$$|a| \geq 0$$

EXAMPLE 16

$$|-5| = 5 \geq 0$$

$$|10| = 10 \geq 0$$

$$|0| = 0 \geq 0$$

❑

Property 2

$$|-a| = |a|$$

EXAMPLE 17

$$|-4| = |4| = 4$$

❑

Property 3

$$|x - y| = |y - x|$$

EXAMPLE 18

$$|12 - 5| = |7| = 7$$

$$|5 - 12| = |-7| = 7$$

❑

Property 4

$$|ab| = |a||b|$$

EXAMPLE 19

$$|3(-5)| = |-15| = 15$$

Using Property 4,

$$|3(-5)| = |3||-5| = (3)(5) = 15$$

❑

Property 5

$$\left|\frac{a}{b}\right| = \frac{|a|}{|b|}$$

EXAMPLE 20

$$\left|\frac{-25}{10}\right| = \frac{|-25|}{|10|} = \frac{25}{10} = 2.5$$

❑

Solving Equations and Inequalities Involving Absolute Values

Suppose that we want to solve the equation

$$|x| = 4$$

Given the definition of absolute value, x must equal either -4 or 4.

EXAMPLE 21

To solve the equation

$$|x - 5| = 3$$

we know that $x - 5 = \pm 3$. That is, either

$$x - 5 = 3 \quad \text{or} \quad x - 5 = -3$$

Solving both equations, we find

$$x = 8 \quad \text{or} \quad x = 2$$

To check this result, substitution of the two values into the original equation yields

$$|8 - 5| = 3 \quad \text{and} \quad |2 - 5| = 3$$
$$|3| = 3 \quad \text{and} \quad |-3| = 3$$
$$3 = 3 \quad \text{and} \quad 3 = 3$$

PRACTICE EXERCISE
Solve the equation $|5 - 2x| = 9$. *Answer:* $x = -2, 7$.

EXAMPLE 22

Solve the following equation:

$$|10 - 2x| = |x + 5|$$

SOLUTION

Because $(10 - 2x)$ and $(x + 5)$ have the same absolute value, they are either equal or of opposite sign. Thus, the solution to the given equation requires that

$$10 - 2x = \pm(x + 5)$$

Solving for x under the two conditions gives

$$10 - 2x = (x + 5) \quad \text{or} \quad 10 - 2x = -(x + 5)$$
$$5 = 3x \quad \text{or} \quad 10 - 2x = -x - 5$$
$$\tfrac{5}{3} = x \quad \text{or} \quad 15 = x$$

Therefore, the equation is satisfied when $x = \tfrac{5}{3}$ or 15.

◻

PRACTICE EXERCISE
Solve the equation $|x + 3| = |5 - x|$. *Answer: $x = 1$*.

EXAMPLE 23   Solve the inequality $|x| < 4$.

SOLUTION

Because $|x|$ represents the distance of x from 0 on a real number line, the solution to this inequality consists of all real numbers with distance from 0 on a real number line less than 4. Figure 1.10 indicates that the values satisfying the inequality are $-4 < x < 4$.

Figure 1.10

EXAMPLE 24   Solve the inequality $|x| \geq 2$.

SOLUTION

The values of x which satisfy this inequality consist of all real numbers located 2 or more units from zero on a real number line. Figure 1.11 indicates that the values satisfying the inequality are $x \leq -2$ and $x \geq 2$.

Figure 1.11

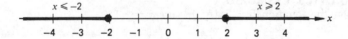

EXAMPLE 25   Solve the inequality $|x - 4| > 6$.

SOLUTION

Viewing this example in a manner similar to Example 24, we are seeking values of x which result in the number $(x - 4)$ being more than 6 units from zero on a real number line. That is, we are seeking values of x such that

$$x - 4 < -6 \quad \text{or} \quad x - 4 > 6$$

or
$$x < -2 \quad \text{or} \quad x > 10$$

Figure 1.12 illustrates the solution to the inequality.

Figure 1.12

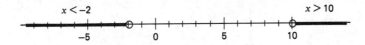

PRACTICE EXERCISE
Solve the inequality $|2x + 3| < 5$. *Answer:* $-4 < x < 1$.

Section 1.4 Follow-up Exercises

Solve the following equations.

1 $\|x\| = 8$	**2** $\|x\| = 10$
3 $\|x\| = -5$	**4** $\|x\| = -6$
5 $\|x - 6\| = 3$	**6** $\|x - 2\| = 4$
7 $\|x + 3\| = 7$	**8** $\|2x - 7\| = 1$
9 $\|x - 4\| = \|-3x + 8\|$	**10** $\|x + 7\| = \|x - 5\|$
11 $\|2x + 5\| = \|x - 4\|$	**12** $\|3x - 10\| = \|2x - 7\|$
13 $\|5 - 3x\| = \|-2x + 7\|$	**14** $\|x\| = \|-x + 5\|$

Solve the following inequalities.

15 $\|x\| < 1$	**16** $\|x\| > 8$
17 $\|x\| > -4$	**18** $\|x\| < -2$
19 $\|x\| \le 2.5$	**20** $\|2x\| \ge 12$
21 $\|x - 5\| < 10$	**22** $\|4 - 2x\| < 2$
23 $\|2x - 3\| > 5$	**24** $\|3x - 8\| > 4$
25 $\|y + 1\| \le -9$	**26** $\|6t - 15\| \le -6$
27 $\|t/2\| \ge 12$	**28** $\|y - 5\| \ge 3$
29 $\|x^2 - 2\| \ge 2$	**30** $\|x^2 - 8\| \le 8$

1.5 RECTANGULAR COORDINATE SYSTEMS

Throughout this book the visual model will be used as often as possible to reinforce understanding of different mathematical concepts. The visual model will most frequently take the form of graphical representation. To prepare for graphical representation, we will now discuss rectangular coordinate systems.

The Cartesian Plane

Consider a plane onto which are drawn a horizontal line and a vertical line, as in Fig. 1.13. The two lines are real number lines, which intersect at their respective

Figure 1.13
Cartesian plane.

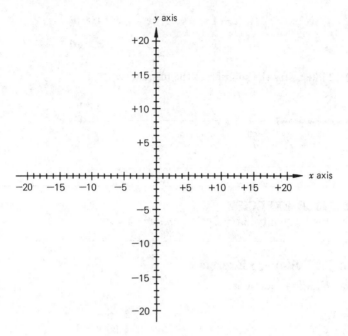

zero points. The horizontal line is called the ***horizontal axis.*** As labeled in Fig. 1.13, it is more commonly referred to as the ***x axis.*** The vertical line is the ***vertical axis*** and in this figure would be called the ***y axis.*** The two axes together are referred to as ***coordinate axes.*** Note that the horizontal axis is scaled with positive values to the right of the vertical axis and negative values to the left. Similarly, the vertical axis is scaled with positive values above the horizontal axis and negative values below.

The plane containing the coordinate axes is often called a ***coordinate plane*** or ***cartesian plane.*** The coordinate plane can be thought of as consisting of an infinite number of points, each point specified by its location relative to the two axes. The location of any point p is specified by the ***ordered pair*** of values (x, y). The first member of the ordered pair is known as the ***abscissa,*** or more commonly as the ***x coordinate.*** As indicated in Fig. 1.14, the abscissa is the ***directed distance*** along a horizontal line drawn from the vertical axis to P. The second member of the ordered pair is the ***ordinate,*** or ***y coordinate.*** The ordinate represents the directed distance along a vertical line drawn from the horizontal axis to P. Together, the ***coordinates*** (x, y) specify the location or address of a point P in a coordinate plane. The system of coordinates used in a coordinate plane is called a ***cartesian,*** or ***rectangular, coordinate system.***

To locate a point P having coordinates (a, b), first draw an imaginary vertical line through the horizontal axis at a. Then draw an imaginary horizontal line through the vertical axis at b. Point P occurs at the intersection of these two lines, as shown in Fig. 1.15. Notice for P that $a < 0$ and $b < 0$.

Figure 1.16 is a graphical representation of several sample points. The point of intersection of the two axes has coordinates $(0, 0)$ and is known as the ***origin.*** Also convince yourself that *the x coordinate of any point on the y axis is 0 and the y coordinate of any point on the x axis is 0.* Finally, note that the axes divide the

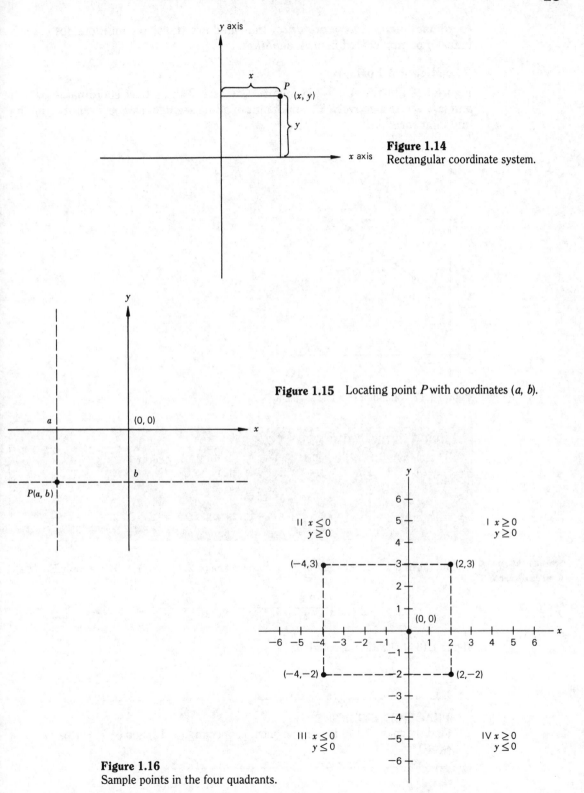

Figure 1.14
Rectangular coordinate system.

Figure 1.15   Locating point *P* with coordinates (*a, b*).

Figure 1.16
Sample points in the four quadrants.

coordinate plane into *quadrants.* Indicated are the sign conditions for coordinates of points located in each quadrant.

The Midpoint Formula

Figure 1.17 illustrates a line segment PQ, where P and Q have coordinates (x_1, y_1) and (x_2, y_2), respectively. The midpoint of a line segment can be located using the midpoint formula.

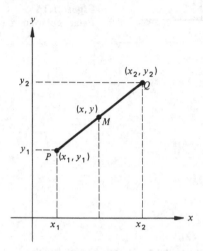

Figure 1.17   Midpoint of a line segment.

DEFINITION: MIDPOINT FORMULA
The midpoint M of the line segment connecting two points having coordinates (x_1, y_1) and (x_2, y_2), respectively, has coordinates

$$\left(\frac{x_1 + x_2}{2}, \frac{y_1 + y_2}{2} \right) \tag{1.2}$$

EXAMPLE 26   To find the midpoint of the line segment connecting $(-2, 6)$ and $(1, -9)$, we apply Eq. (1.2).

$$\left(\frac{-2 + 1}{2}, \frac{6 + (-9)}{2} \right) = \left(\frac{-1}{2}, \frac{-3}{2} \right)$$

Figure 1.18 illustrates the solution.

❑

PRACTICE EXERCISE
Find the midpoint of the line segment connecting $(4, 12)$ and $(-2, -18)$. *Answer:* $(1, -3)$.

Figure 1.18

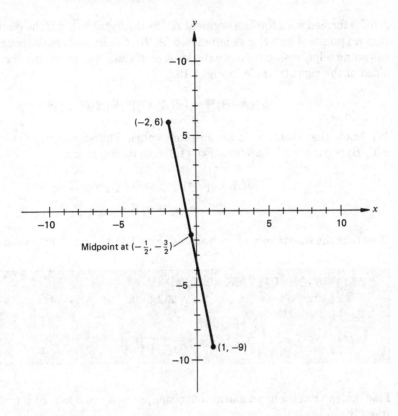

The Distance Formula

Given two points in a cartesian plane, the distance separating the two points can be determined based upon the **Pythagorean theorem.** In Fig. 1.19, suppose that we are interested in finding the distance separating points A and B. The *right triangle*

Figure 1.19
Pythagorean
theorem.

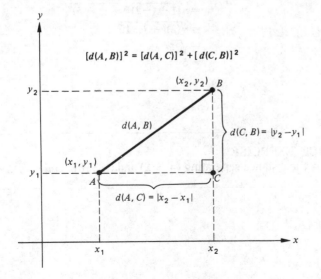

ABC is formed with the line segment *AB* as the *hypotenuse*. If the distance separating two points *A* and *B* is denoted as $d(A, B)$, the Pythagorean theorem states the following relationship between the lengths of the hypotenuse and the two opposing sides of the right triangle in Fig. 1.19.

$$[d(A, B)]^2 = [d(A, C)]^2 + [d(C, B)]^2 \qquad (1.3)$$

We know that distance is an absolute value. Therefore, $d(A, C) = |x_2 - x_1|$ and $d(C, B) = |y_2 - y_1|$. Therefore, Eq. (1.3) can be restated as

$$[d(A, B)]^2 = |x_2 - x_1|^2 + |y_2 - y_1|^2$$
$$= (x_2 - x_1)^2 + (y_2 - y_1)^2 \qquad (1.4)$$

If we take the square root of both sides of Eq. (1.4), the result is the *distance formula*.

DEFINITION: DISTANCE FORMULA
The distance between two points *A* and *B*, having coordinates (x_1, y_1) and (x_2, y_2), respectively, is

$$d(A, B) = \sqrt{(x_2 - x_1)^2 + (y_2 - y_1)^2} \qquad (1.5)$$

EXAMPLE 27 Find the length of the line segment connecting points *A* and *B* located at $(-2, 5)$ and $(1, 1)$, respectively.

SOLUTION
Applying the distance formula gives

$$d(A, B) = \sqrt{(x_2 - x_1)^2 + (y_2 - y_1)^2}$$
$$= \sqrt{[1 - (-2)]^2 + (1 - 5)^2}$$
$$= \sqrt{(3)^2 + (-4)^2}$$
$$= \sqrt{25}$$
$$= 5$$

□

PRACTICE EXERCISE
Determine the distance separating $(4, -2)$ and $(-3, 6)$. *Answer:* $\sqrt{113} = 10.63$.

Section 1.5 Follow-up Exercises

Find the midpoint of the line segment connecting the following points.

1 $(-1, 3)$ and $(4, 4)$ **2** $(5, 2)$ and $(3, 6)$

3 $(10, 4)$ and $(5, -6)$ **4** $(-1, -3)$ and $(2, 15)$

5 $(20, 40)$ and $(-5, -10)$ **6** $(-5, 24)$ and $(-1, -8)$

7 $(0, 6)$ and $(-4, 24)$ **8** $(4, 2)$ and $(-6, -16)$

9 $(5, 0)$ and $(7, -16)$ **10** $(3, -2)$ and $(-1, 12)$

11 $(6, 3)$ and $(9, -9)$ **12** $(0, 4)$ and $(4, 0)$

13 $(-2, -4)$ and $(2, 4)$ **14** $(5, 5)$ and $(-2, -2)$

Find the distance separating the following points.

15 $(4, 6)$ and $(0, 0)$ **16** $(2, 6)$ and $(4, 8)$

17 $(0, 0)$ and $(-3, -4)$ **18** $(-1, -3)$ and $(4, 3)$

19 $(3, -2)$ and $(-3, 5)$ **20** $(10, 5)$ and $(20, -10)$

21 $(-4, -2)$ and $(6, 10)$ **22** $(3, 12)$ and $(-1, 8)$

23 $(10, 0)$ and $(0, -4)$ **24** $(5, 1)$ and $(1, -4)$

25 $(-2, 4)$ and $(1, 0)$ **26** $(2, 2)$ and $(10, 8)$

27 $(5, 2)$ and $(0, 6)$ **28** $(4, 4)$ and $(-5, -8)$

29 $(7, 2)$ and $(-1, 4)$ **30** $(3, 6)$ and $(-2, 4)$

❏ KEY TERMS AND CONCEPTS

abscissa (x coordinate)   24

absolute inequality   11

absolute value   19

cartesian (rectangular) coordinate
   system   23

closed interval   12

conditional equation   5

conditional inequality   11

coordinate or cartesian plane   24

degree of equation   6

degree of polynomial   5

discriminant   10

distance formula   27

double inequality   11

equation   4

equivalent equations   5

extraneous roots   5

false statement (contradiction)   5

half-open interval   12

identity   4

inequalities   11

midpoint formula   26

open interval   12

ordinate (y coordinate)   24

origin   24

Pythagorean theorem   27

quadratic equation   8

quadratic formula   9

roots   4

strict inequality   11

❏ ADDITIONAL EXERCISES

Section 1.1

Solve the following first-degree equations.

1 $6x - 4 = 5x + 2$ **2** $-10 + 4x = 3x - 8$

  3 $4x = 3x + 6$ **4** $-2x + 8 = 2x - 4$
  5 $5y = 10y - 30$ **6** $4(y - 3) = y + 9$
  7 $6x + 20 = 40 + 8x$ **8** $15x - 4(2x + 14) = 0$
  9 $-3y - 5(y + 4) = 4$ **10** $3(x - 4) + 2(2x + 1) = 11$
11 $30x + 50(x - 6) = -20$ **12** $4(5 - x) + 2x - 10 = -2x + 10$

Section 1.2

Solve the following second-degree equations.

13 $x^2 - 36 = 0$ **14** $x^2 + 14x + 49 = 0$
15 $x^2 - 5x + 4 = 0$ **16** $4x^2 + 2x - 30 = 0$
17 $7x^2 - 70 = 21x$ **18** $2x^2 + 3x - 10 = x^2 + 6x + 30$
19 $-6x^2 + 4x - 10 = 0$ **20** $-5x^2 + 10x - 20 = 0$
21 $5x^2 - 17.5x - 10 = 0$ **22** $x^2 + 64 = 0$
23 $8x^2 + 2x - 15 = 0$ **24** $-x^2 - 2x + 35 = 0$
25 $2a^2 + 2a - 12 = 0$ **26** $5a^2 - 2a + 16 = 0$
27 $3a^2 - 3a - 18 = 0$ **28** $x^2 + 2x - 48 = 0$
29 $x^2 - 2x + 10 = 0$ **30** $5x^2 - 20x + 15 = 0$

Section 1.3

Solve the following inequalities.

31 $x - 8 \le 2x + 4$ **32** $3x + 4 \le 4x - 2$
33 $4x + 5 \ge 2x - 3$ **34** $9x - 5 \ge 6x + 4$
35 $-2x + 10 \ge x - 8$ **36** $5x - 4 \ge 3x + 8$
37 $-4 \le 2x + 2 \le 10$ **38** $4 \le -x + 3 \le 12$
39 $x + 5 \le x + 1 \le 6$ **40** $-x + 3 \le 2x + 3 \le 9$

Solve the following second-degree inequalities.

41 $x^2 - 25 \ge 0$ **42** $x^2 - 16 \ge 0$
43 $x^2 - 5x + 4 \le 0$ **44** $x^2 - x - 20 \le 0$
45 $2x^2 - 5x - 12 \ge 0$ **46** $5x^2 - 13x - 6 \ge 0$
47 $12x^2 - 5x - 2 \le 0$ **48** $3x^2 + x - 10 \le 0$

Section 1.4

Solve the following equations.

49 $|x - 5| = 4$ **50** $|10 - 2x| = 14$
51 $|x + 8| = 2$ **52** $|x - 5| = -4$
53 $|x - 4| = |8 - 2x|$ **54** $|3x - 6| = |x + 6|$
55 $|x| = |9 - x|$ **56** $|2x + 5| = |-x|$

Solve the following inequalities.

57 $|x| \le 10$ **58** $|-x| \ge 12$
59 $|x + 5| \le 8$ **60** $|x - 15| \le 10$
61 $|3x - 5| \ge 3$ **62** $|2x + 9| \ge 5$
63 $|3x - 6| \le -4$ **64** $|5x - 3| \le -9$

Section 1.5

Find the midpoint of the line segment connecting the following points.

65 $(-3, 10)$ and $(2, -15)$

66 $(-1, 3)$ and $(1, -9)$

67 $(4, 4)$ and $(-2, 2)$

68 $(0, 4)$ and $(2, 0)$

69 $(4, -8)$ and $(2, 4)$

70 (a, a) and (b, b)

71 (a, b) and $(3a, 3b)$

72 (a, b) and $(-a, -b)$

Find the distance separating the following points.

73 $(2, 4)$ and $(-4, 6)$

74 $(-2, 2)$ and $(3, -3)$

75 $(6, 2)$ and $(3, 6)$

76 $(-1, -2)$ and $(-4, -6)$

77 $(10, 5)$ and $(-10, 5)$

78 $(5, -10)$ and $(20, 10)$

79 (a, b) and $(a, 3b)$

80 $(5a, 2b)$ and $(0, 2b)$

❏ CHAPTER TEST

1 Solve the equation $5x = 5x + 10$.

2 Solve the equation $x^2 - 2x + 5 = 0$.

3 Solve the equation $x^2 - 7x + 12 = 0$.

4 Solve the following inequality:

$$-2 \le x - 6 \le x + 1$$

5 Solve the following inequality:

$$x^2 + 3x + 2 \le 0$$

6 Solve the equation $|x - 12| = |4 - x|$.

7 Solve the following inequality:

$$|x + 12| \le 8$$

8 Given the points $(-4, 8)$ and $(6, -12)$:

(*a*) Determine the midpoint of the line segment connecting the points.

(*b*) Determine the distance separating the two points.

CHAPTER 2

LINEAR EQUATIONS

CHAPTER OBJECTIVES

❏ Provide a thorough understanding of the algebraic and graphical characteristics of linear equations

❏ Provide the tools which will allow one to determine the equation which represents a linear relationship

❏ Illustrate a variety of applications of linear equations

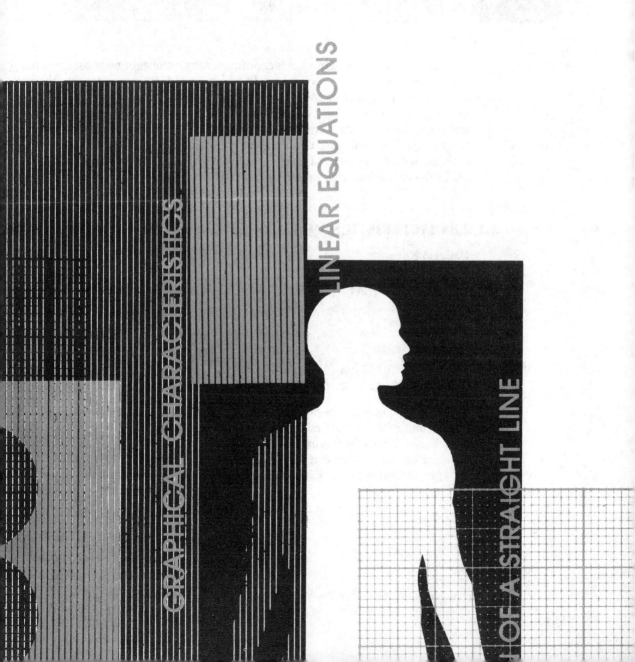

The study of linear mathematics is significant for a number of reasons. First, many real-world phenomena, which we might be interested in representing mathematically, either are linear or can be approximated reasonably well using linear relationships. As a result, linear mathematics is widely applied. Second, the analysis of linear relationships is generally easier than that of nonlinear relationships. Lastly, the methods of analyzing nonlinear relationships are sometimes similar to, or extensions of, those used in linear mathematics. Consequently, a good understanding of linear mathematics is prerequisite to the study of nonlinear mathematics.

2.1 CHARACTERISTICS OF LINEAR EQUATIONS

General Form

LINEAR EQUATION WITH TWO VARIABLES
A linear equation involving two variables x and y has the standard form

$$ax + by = c \qquad (2.1)$$

where a, b, and c are constants and a and b cannot *both* equal zero.

Linear equations are *first-degree* equations. Each variable in the equation is raised (implicitly) to the first power. The presence of terms having exponents other than 1 (e.g., x^2) or of terms involving a product of variables (e.g., $2xy$) would exclude an equation from being considered linear.

The following are all examples of linear equations involving two variables:

---

* Most chapters begin with a **Motivating Scenario.** Their purpose is to provide an example of the type of application a student should be able to solve *upon completion* of the chapter.

	Eq. (2.1) Parameters		
	a	b	c
$2x + 5y = -5$	2	5	-5
$-x + \frac{1}{2}y = 0$	-1	$\frac{1}{2}$	0
$x/3 = 25$	$\frac{1}{3}$	0	25
$\sqrt{2}u - 0.05v = 3.76$	$\sqrt{2}$	-0.05	3.76
$2s - 4t = -\frac{1}{2}$	2	-4	$-\frac{1}{2}$

(*Note:* The names of the variables may be different from x and y.)

The following are examples of equations which are *not* linear. Can you explain why?

$$2x + 3xy - 4y = 10$$

$$x + y^2 = 6$$

$$\sqrt{u} + \sqrt{v} = -10$$

$$ax + \frac{b}{y} = c$$

The form of an equation may not always be obvious. Initially, the equation

$$2x = \frac{5x - 2y}{4} + 10$$

might not appear to be linear. However, multiplying both sides of the equation by 4 and moving all variables to the left side yields $3x + 2y = 40$, which is in the form of Eq. (2.1).

Representation Using Linear Equations

Given a linear equation having the form $ax + by = c$, the **solution set** for the equation is the set of all ordered pairs (x, y) which satisfy the equation. Using **set notation** the solution set S can be specified as

$$S = \{(x, y) \mid ax + by = c\} \qquad (2.2)$$

Verbally, this set notation states that the **solution set** S consists of **elements** (x, y) *such that* (the vertical line) the equation $ax + by = c$ is satisfied. Stated differently, Eq. (2.2) expresses that S consists of all **ordered pairs** (x, y) such that $ax + by = c$. For any linear equation, S consists of an infinite number of elements; that is, **there is an infinite number of pairs of values (x, y) which satisfy a linear equation having the form $ax + by = c$.**

To determine any pair of values which satisfy a linear equation, assume a value for one of the variables, substitute this value into the equation, and solve for the corresponding value of the other variable. This method assumes that both variables are included in the equation (i.e., $a \neq 0$ and $b \neq 0$).

EXAMPLE 1   Given the equation

$$2x + 4y = 16$$

(a) Determine the pair of values which satisfies the equation when $x = -2$.
(b) Determine the pair of values which satisfies the equation when $y = 0$.

SOLUTION

(a) Substituting $x = -2$ into the equation, we have

$$2(-2) + 4y = 16$$
$$4y = 20$$
$$y = 5$$

When $x = -2$, the pair of values satisfying the equation is $x = -2$ and $y = 5$, or $(-2, 5)$.
(b) Substituting $y = 0$ into the equation gives

$$2x + 4(0) = 16$$
$$2x = 16$$
$$x = 8$$

When $y = 0$, the pair of values satisfying the equation is $(8, 0)$.

EXAMPLE 2   **(Production Possibilities)** A company manufactures two different products. For the coming week 120 hours of labor are available for manufacturing the two products. Work-hours can be allocated for production of either product. In addition, since both products generate a good profit, management is interested in using all 120 hours during the week. Each unit produced of product A requires 3 hours of labor and each unit of product B requires 2.5 hours.
(a) Define an equation which states that total work-hours used for producing x units of product A and y units of product B equal 120.
(b) How many units of product A can be produced if 30 units of product B are produced?
(c) If management decides to produce one product only, what is the maximum quantity which can be produced of product A? The maximum of product B?

SOLUTION

(a) We can define our variables as follows:

> $x =$ number of units produced of product A
> $y =$ number of units produced of product B

The desired equation has the following structure.

> Total hours used in producing products A and $B = 120$ **(2.3)**

More specifically,

$$\boxed{\begin{array}{c} \text{Total hours used} \\ \text{in producing} \\ \text{product } A \end{array} + \begin{array}{c} \text{total hours used} \\ \text{in producing} \\ \text{product } B \end{array} = 120}$$ **(2.4)**

Since the total hours used in producing a product equals hours required per unit produced times number of units produced, Eq. (2.4) reduces to

$$\boxed{3x + 2.5y = 120}$$ **(2.5)**

(*b*) If 30 units of product *B* are produced, then $y = 30$. Therefore

$$3x + 2.5(30) = 120$$

$$3x = 45$$

$$x = 15$$

Thus, one pair of values satisfying Eq. (2.5) is (15, 30). In other words, *one combination* of the two products which will fully utilize the 120 hours is 15 units of product *A* and 30 units of product *B*.

(*c*) If management decides to manufacture product *A* only, no units of product *B* are produced, or $y = 0$. If $y = 0$,

$$3x + 2.5(0) = 120$$

$$3x = 120$$

$$x = 40$$

Therefore 40 is the maximum number of units of product *A* which can be produced using the 120 hours.

If management decides to manufacture product *B* only, $x = 0$ and

$$3(0) + 2.5y = 120$$

or $$y = 48 \text{ units}$$

EXAMPLE 3

We stated earlier that there is an infinite number of pairs of values (x, y) which satisfy any linear equation. In Example 2, are there any members of the solution set which might not be realistic in terms of what the equation represents?

SOLUTION

In Example 2, x and y represent the number of units produced of the two products. Since *negative* production is not possible, negative values of x and y are not meaningful. There are negative values which satisfy Eq. (2.5). For instance, if $y = 60$, then

$$3x + 2.5(60) = 120$$

$$3x = -30$$

$$x = -10$$

In addition to negative values, it is possible to have decimal or fractional values for x and y. For example, if $y = 40$,

$$3x + 2.5(40) = 120$$

$$3x = 20$$

$$x = 6\tfrac{2}{3}$$

Depending upon the nature of the products and how they are sold, fractional values may or may not be acceptable.

❑

POINT FOR THOUGHT & DISCUSSION

Give examples of types of products where only integer values would be reasonable. Give an example of a product for which noninteger values are reasonable.

Linear Equations with n Variables

LINEAR EQUATION WITH n VARIABLES

A linear equation involving n variables $x_1, x_2, x_3, \ldots, x_n$ has the general form

$$a_1 x_1 + a_2 x_2 + a_3 x_3 + \cdots + a_n x_n = b \qquad (2.6)$$

where $a_1, a_2, a_3, \ldots, a_n$ and b are constants and *not all* $a_1, a_2, a_3, \ldots, a_n$ equal zero.

Each of the following is an example of a linear equation:

$$3x_1 - 2x_2 + 5x_3 = 0$$

$$-x_1 + 3x_2 - 4x_3 + 5x_4 - x_5 + 2x_6 = -80$$

$$5x_1 - x_2 + 4x_3 + x_4 - 3x_5 + x_6 - 3x_7 + 10x_8 - 12x_9 + x_{10} = 1{,}250$$

Given a linear equation involving n variables, as defined by Eq. (2.6), the solution set S can be specified as

$$S = \{(x_1, x_2, x_3, \ldots, x_n) \mid a_1 x_1 + a_2 x_2 + a_3 x_3 + \cdots + a_n x_n = b\} \qquad (2.7)$$

As with the two-variable case, there are infinitely many elements in the solution set. An element in S is represented by a collection of values $(x_1, x_2, x_3, \ldots, x_n)$, one for each of the n variables in the equation. One way of identifying specific elements in S is to assume values for $n-1$ of the variables, substitute these into the equation, and solve for the value of the remaining variable.

EXAMPLE 4   Given the equation

$$2x_1 + 3x_2 - x_3 + x_4 = 16$$

(a) What values satisfy the equation when $x_1 = 2$, $x_2 = -1$, and $x_3 = 0$?
(b) Determine all members of the solution set which have values of 0 for three of the four variables.

SOLUTION

(a) Substituting the given values for x_1, x_2, and x_3 into the equation yields

$$2(2) + 3(-1) - (0) + x_4 = 16$$

or

$$x_4 = 15$$

The corresponding element of the solution set is $(2, -1, 0, 15)$.
(b) If $x_1 = x_2 = x_3 = 0$, then

$$2(0) + 3(0) - (0) + x_4 = 16$$

or

$$x_4 = 16$$

If $x_1 = x_2 = x_4 = 0$,

$$2(0) + 3(0) - x_3 + (0) = 16$$

or

$$x_3 = -16$$

If $x_1 = x_3 = x_4 = 0$, then

$$2(0) + 3x_2 - (0) + (0) = 16$$

or

$$3x_2 = 16$$

and

$$x_2 = \tfrac{16}{3}$$

If $x_2 = x_3 = x_4 = 0$,

$$2x_1 + 3(0) - (0) + (0) = 16$$

or

$$2x_1 = 16$$

and

$$x_1 = 8$$

Therefore, the elements of the solution set which have three of the four variables equaling 0 are $(0, 0, 0, 16)$, $(0, 0, -16, 0)$, $(0, \tfrac{16}{3}, 0, 0)$, and $(8, 0, 0, 0)$.

◻

PRACTICE EXERCISE
In Example 2 (Production Possibilities), assume that a third product (product C) is also to be produced. Because of the additional product, management has authorized an additional 30 labor hours. If each unit of product C requires 3.75 labor hours, (*a*) determine the equation which requires that all 150 labor hours be used in producing the three products and (*b*) determine the maximum number of units which could be produced of each product.
Answer: (*a*) If $z =$ number of units produced of product C, $3x + 2.5y + 3.75z = 150$, (*b*) 50 units of A, 60 units of B, and 40 units of C.

Section 2.1 Follow-up Exercises

Determine which of the following equations are linear.

1 $-3y = 0$

2 $\sqrt{2}x + 6y = -25$

3 $-5x + 24y = 200$

4 $-x^2 + 3y = 40$

5 $2x - 3xy + 5y = 10$

6 $\sqrt{4x} - 3y = -45$

7 $u - 3v = 20$

8 $r/2 + s/5 = \frac{3}{7}$

9 $m/2 + (2m - 3n)/5 = 0$

10 $(x + 2y)/3 - 3x/4 = 2x - 5y$

11 $40 - 3y = \sqrt{24}$

12 $0.0003x - 2.3245y = x + y - 3.2543$

13 $2x_1 - 3x_2 + x_3 = 0$

14 $(x_1 - 3x_2 + 5x_3 - 2x_4 + x_5)/25 = 300$

15 $(x_1 + x_2 - x_3x_1) = 5$

16 $3x_2 - 4x_1 = 5x_3 + 2x_2 - x_4 + 36$

17 $\sqrt{x^2 + 2xy + y^2} = 25$

18 $(2x_1 - 3x_2 + x_3)/4 = (x_2 - 2x_4)/5 + 90$

19 Consider the equation $8x = 120$ as a two-variable equation having the form of Eq. (2.1).
   (*a*) Define a, b, and c.
   (*b*) What pair of values satisfies the equation wnen $y = 10$?
   (*c*) What pair of values satisfies the equation when $x = 20$?
   (*d*) Verbalize the somewhat unique nature of the solution set for this equation.

20 Rework Example 2 if product A requires 2 hours per unit and product B requires 4 hours per unit.

21 Given the equation $4x_1 - 2x_2 + 6x_3 = 0$:
   (*a*) What values satisfy the equation when $x_1 = 2$ and $x_3 = 1$?
   (*b*) Define all elements of the solution set in which the values of two variables equal 0.

22 Given the equation $x_1 - 3x_2 + 4x_3 - 2x_4 = -60$:
   (*a*) What values satisfy the equation when $x_1 = 20$, $x_2 = 6$, and $x_3 = -4$?
   (*b*) Determine all elements of the solution set for which the values of three variables equal zero.

23 The equation $x_3 = 20$ is one of a set of related equations involving four variables x_1, x_2, x_3, and x_4.
   (*a*) What values satisfy the equation when $x_1 = 4$, $x_2 = 2$, and $x_4 = 15$?
   (*b*) What values satisfy the equation when $x_3 = 10$?
   (*c*) Determine all elements of the solution set for which the values of three variables equal zero.

24 **Product Mix** A company manufactures two products, A and B. Each unit of A requires 3 labor hours and each unit of B requires 5 labor hours. Daily manufacturing capacity is 150 labor hours.
   (*a*) If x units of product A and y units of product B are manufactured each day and all labor hours are to be used, determine the linear equation that requires the use of 150 labor hours per day.

(b) How many units of A can be made each day if 25 units of B are manufactured each day?

(c) How many units of A can be made each *week* if 12 units of B are manufactured each day? (Assume a 5-day work week.)

25 Nutrition Planning A dietitian is considering three food types to be served at a meal. She is particularly concerned with the amount of one vitamin provided at this meal. One ounce of food 1 provides 8 milligrams of the vitamin; an ounce of food 2 provides 24 milligrams; and, an ounce of food 3 provides 16 milligrams. The *minimum daily requirement* (MDR) for the vitamin is 120 milligrams.

(a) If x_j equals the number of ounces of food type j served at the meal, determine the equation which ensures that the meal satisfies the MDR exactly.

(b) If only one of the three food types is to be included in the meal, how much would have to be served (in each of the three possible cases) to satisfy the MDR?

26 Emergency Airlift The Red Cross wants to airlift supplies into a South American country which has experienced an earthquake. Four types of supplies, each of which would be shipped in containers, are being considered. One container of a particular item weighs 120, 300, 250, and 500 pounds, respectively, for the four items. If the airplane to be used has a weight capacity of 60,000 pounds and x_j equals the number of containers shipped of item j:

(a) Determine the equation which ensures that the plane will be loaded to its weight capacity.

(b) If it is decided to devote this plane to one supply item only, how many containers could be shipped of each item?

27 Airlift Revisited In Exercise 26, each container of a supply item requires a specific volume of space. Suppose containers of the four items require 30, 60, 50, and 80 cubic feet, respectively. If the volume capacity of the plane is 15,000 cubic feet:

(a) Determine the equation which ensures that the volume capacity of the plane is filled exactly.

(b) If it is decided to devote this plane to one supply item only, how many containers could be shipped of each item if volume capacity is the only consideration?

(c) Using the information from Exercise 26, what is the maximum number of containers which could be shipped of each item if both weight and volume are considered? Indicate in each case whether weight or volume capacity is the constraining factor.

28 Personnel Hiring A software consulting firm has received a large contract to develop a new airline reservation system for a major airline. In order to fulfill the contract, new hiring of programmer analysts, senior programmer analysts, and software engineers will be required. Each programmer analyst position will cost $40,000 in salary and benefits. Each senior programmer analyst position will cost $50,000 and each software engineer position $60,000. The airline has budgeted $1.2 million per year for the new hirings. If x_j equals the number of persons hired for job category j (where $j = 1$ corresponds to programmer analysts, etc.):

(a) Determine the equation which ensures that total new hires will exactly consume the budget.

(b) If it were desired to spend the entire budget on one type of position, how many persons of each type could be hired?

(c) If exactly 10 programmer analysts are needed for the contract, what is the maximum number of senior programmer analysts that could be hired? Maximum number of software engineers?

29 Public Transportation New York City has received a federal grant of $100 million for improving public transportation. The funds are to be used only for the purchase of new buses, the purchase of new subway cars, or the repaving of city streets. Costs are estimated at $150,000 per bus, $180,000 per subway car, and $250,000 per mile for repaving. City officials want to determine different ways of spending the grant money.

(a) Define the decision variables and write the equation which ensures complete expenditure of the federal grant.

(b) If it has been determined that 100 buses and 200 new subway cars will be purchased, how many miles of city streets can be repaved?

(c) If officials wish to spend all of the grant money on one type of improvement, what are the different possibilities?

30 **Political Campaign** A candidate for the position of governor of a midwestern state has an advertising budget of $1.5 million. The candidate's advisors have identified four advertising options: newspaper advertisements, radio commercials, television commercials, and billboard advertisements. The costs for these media options average $1,500, $2,500, $10,000, and $7,500, respectively. If x_j equals the number of units purchased of media option j:

(a) Write an equation which requires total advertising expenditures of $1.5 million.

(b) If it has been determined that 100 newspaper ads, 300 radio ads, and 50 billboard ads will be used, how many television ads can they purchase?

(c) If 50 billboard ads are to be purchased, what is the maximum number of newspaper ads that can be purchased? Maximum number of radio ads? TV ads?

2.2 GRAPHICAL CHARACTERISTICS

Graphing Two-Variable Equations

A linear equation involving two variables graphs as a straight line in two dimensions. To graph this type of linear equation, *(1) identify and plot the coordinates of any two points which lie on the line, (2) connect the two points with a straight line, and (3) extend the straight line in both directions as far as necessary or desirable for your purposes.* The coordinates of the two points are found by identifying any two members of the solution set. Each element in the solution set graphs as a point (x, y) in 2-space, where x and y are the respective values of the two variables. For example, if the values of $x = 1$ and $y = 3$ satisfy an equation, the graphical representation of this member of the solution set is the point $(1, 3)$.

| EXAMPLE 5 | The graph of the equation |

$$2x + 4y = 16$$

is found by first identifying any two pairs of values for x and y which satisfy the equation.

> **NOTE**   Aside from the case where the right side of the equation equals 0, the easiest points to identify (algebraically) are those found by setting one variable equal to 0 and solving for the value of the other variable. That is, let $x = 0$ and solve for the value of y; then let $y = 0$ and solve for the value of x. Observe that the resulting ordered pairs, $(0, y)$ and $(x, 0)$, are points on the y and x axes, respectively.

Letting $x = 0$, the corresponding value for y is 4, and letting $y = 0$ results in $x = 8$. Thus $(0, 4)$ and $(8, 0)$ are two members of the solution set, and their graphical representation is

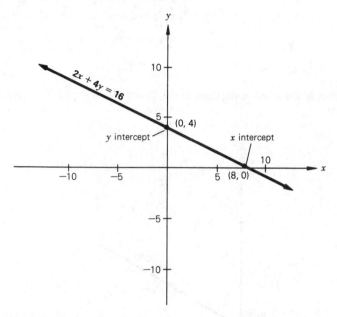

Figure 2.1   Graph of the linear equation $2x + 4y = 16$.

indicated by the two points in Fig. 2.1. The two points have been connected by a straight line, and the line has been extended in both directions.

Just as $(0, 4)$ and $(8, 0)$ are members of the solution set for the equation $2x + 4y = 16$, the coordinates of every point lying on the line represent other members of the solution set. How many unique points are there on the line? There are infinitely many, which is entirely consistent with our earlier statement that there are an infinite number of pairs of values for x and y which satisfy any linear equation. In summary, all pairs of values (x, y) that belong to the solution set of a linear equation are represented graphically by the points lying on the line representing the equation. In Fig. 2.1, the coordinates of any point *not* lying on the line are not members of the solution set for $2x + 4y = 16$.

EXAMPLE 6   Graph the linear equation $4x - 7y = 0$.

SOLUTION

This equation is an example where two different points will not be identified by setting each variable equal to 0 and solving for the remaining variable. Watch what happens! If $x = 0$,

$$4(0) - 7y = 0 \quad \text{or} \quad y = 0$$

If $y = 0$,

$$4x - 7(0) = 0 \quad \text{or} \quad x = 0$$

Both cases have yielded the same point, $(0, 0)$. Therefore, to identify a second point, a value other than zero must be assumed for one of the variables. If we let $x = 7$,

$$4(7) - 7y = 0$$

$$-7y = -28$$

$$y = 4$$

Two members of the solution set, then, are (0, 0) and (7, 4). Figure 2.2 illustrates the graph of the equation.

◻

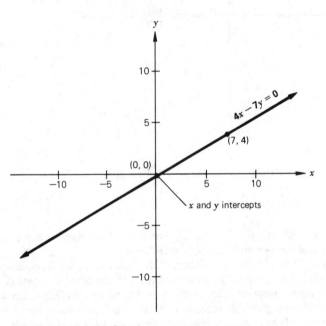

Figure 2.2 Graph of the linear equation $4x - 7y = 0$.

NOTE   Any two-variable linear equation having the form $ax + by = 0$ graphs as a straight line which *passes through the origin.* The unique property of this equation is that the right side, c, equals zero.

Intercepts

In describing the graphical appearance of a straight line, two significant attributes are the *x intercept* and *y intercept.* These can be described both graphically and algebraically.

DEFINITION: x INTERCEPT
The x intercept of an equation is the point where the graph of the equation crosses the x axis. The x intercept represents the ordered pairs found by setting $y = 0$.

> **DEFINITION: y INTERCEPT**
> The y intercept of an equation is the point where the graph of the equation crosses the y axis. The y intercept represents the ordered pairs found by setting $x = 0$.

For a two-variable linear equation there exist (except for two special cases) one x intercept and one y intercept. In Fig. 2.1, the x intercept is $(8, 0)$, and the y intercept is $(0, 4)$. In Fig. 2.2, the x and y intercepts both occur at the same point, the origin. The x intercept is $(0, 0)$, and the y intercept is $(0, 0)$. Examine both figures and verify that the x intercept represents a point having a y value of 0 and that the y intercept represents a point with x value of 0.

The Equation $x = k$

A linear equation of the form $ax = c$ is a special case of Eq. (2.1) where $b = 0$. For this equation there is no y term. Dividing both sides of the equation by a yields the simplified form

$$x = c/a$$

Since c and a are constants, we can let $c/a = k$ and write the equation in the equivalent form

$$\boxed{x = k} \tag{2.8}$$

where k is a real number. This linear equation is special in the sense that x equals k regardless of the value of y. Perhaps this is understood more easily if Eq. (2.8) is rewritten as

$$x + 0y = k$$

The variable y may assume any value as long as $x = k$. That is the only condition required by the equation. As a result, *any equation of this form graphs as a vertical line crossing the x axis at $x = k$.* Figure 2.3 illustrates two equations of this type. *Note that for these equations there is an x intercept $(k, 0)$ but no y intercept* (unless $k = 0$). What happens when $k = 0$?

The Equation $y = k$

A linear equation of the form $by = c$ is also a special case of Eq. (2.1) where $a = 0$; i.e., there is no x term. After both sides of the equation are divided by b, the general reduced form of this case is

$$\boxed{y = k} \tag{2.9}$$

where k is a real number. This equation indicates that y equals k for any value of x. Again, we can see this more readily if Eq. (2.9) is rewritten as

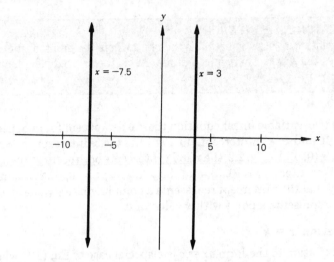

Figure 2.3   Graphs of sample equations of the form $x = k$.

$$0x + y = k$$

The variable x may assume any value as long as $y = k$. *Any equation of this form graphs as a horizontal line crossing the y axis at $y = k$.* Figure 2.4 illustrates two such equations. *Note that equations of this form have no x intercepts* (unless $k = 0$). What happens when $k = 0$?

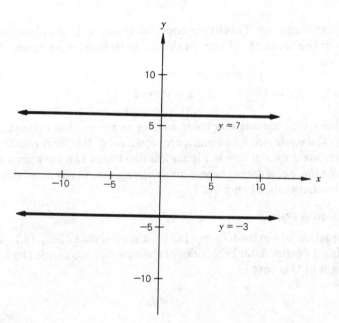

Figure 2.4   Graphs of sample equations of the form $y = k$.

Slope

Any straight line, with the exception of vertical lines, can be characterized by its *slope*. By "slope" we are referring to the *inclination of a line* — whether it rises or falls as you move from left to right along the x axis — and *the rate at which the line rises or falls* (in other words, how steep the line is).

The slope of a line may be **positive, negative, zero, or undefined. A line with a positive slope rises from left to right, or runs uphill.** For such a line the value of y increases as x increases (or conversely, y decreases as x decreases). **A line having a negative slope falls from left to right, or runs downhill.** For such a line the value of y decreases as x increases (or conversely y increases as x decreases). This means that x and y behave in an **inverse** manner; as one increases, the other decreases and vice versa. **A line having a zero slope is horizontal.** As x increases or decreases, y stays constant (the special case: $y = k$). **Vertical lines (of the form $x = k$) have a slope which is undefined.** Since x is constant, we cannot observe the behavior of y as x changes. These slope relationships are illustrated in Fig. 2.5.

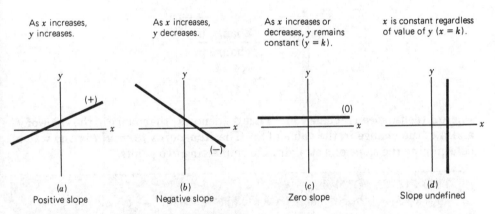

As x increases, y increases.

As x increases, y decreases.

As x increases or decreases, y remains constant ($y = k$).

x is constant regardless of value of y ($x = k$).

(a) Positive slope (b) Negative slope (c) Zero slope (d) Slope undefined

Figure 2.5 Slope conditions for straight lines.

The slope of a line is quantified by a real number. The sign of the slope (number) indicates whether the line is rising or falling. The magnitude (absolute value) of the slope indicates the relative steepness of the line. *The slope tells us the rate at which the value of y changes* relative to *changes in the value of x*. The larger the absolute value of the slope, the steeper the angle at which the line rises or falls. In Fig. 2.6a lines AB and CD both have positive slopes, but the slope of CD is larger than that for AB. Similarly, in Fig. 2.6b lines MN and OP both have negative slopes, but OP has the larger slope in an absolute value sense; it is more steeply sloped.

Given any two points which lie on a (nonvertical) straight line, the slope can be computed as a ratio of the change in the value of y while moving from one point to the other divided by the corresponding change in the value of x, or

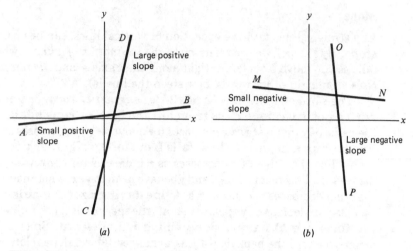

Figure 2.6 Slope: Comparing relative steepness.

$$\text{Slope} = \frac{\text{change in } y}{\text{change in } x}$$
$$= \frac{\Delta y}{\Delta x}$$

where Δ (delta) means "change in." Thus Δy denotes "the change in the value of y" and Δx "the change in the value of x." The **two-point formula** is one way of determining the slope of a straight line connecting two points.

> **TWO-POINT FORMULA**
> The slope m of the straight line connecting two points (x_1, y_1) and (x_2, y_2) is
> $$m = \frac{\Delta y}{\Delta x} = \frac{y_2 - y_1}{x_2 - x_1} \qquad (2.10)$$
> where $x_1 \neq x_2$.

Figure 2.7 illustrates the computation of Δx and Δy for the line segment PQ.

EXAMPLE 7

To compute the slope of the line connecting $(2, 4)$ and $(5, 12)$, arbitrarily identify one point as (x_1, y_1) and the other as (x_2, y_2). Given the location of the two points in Fig. 2.8, let's label $(2, 4)$ as (x_1, y_1) and $(5, 12)$ as (x_2, y_2).

In moving from $(2, 4)$ to $(5, 12)$, the change in the value of y is

$$\Delta y = y_2 - y_1 = 12 - 4 = 8$$

Figure 2.7
Measuring Δx
and Δy.

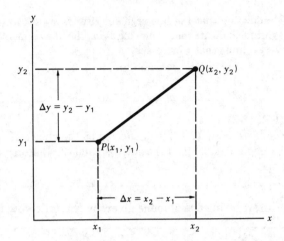

Figure 2.8

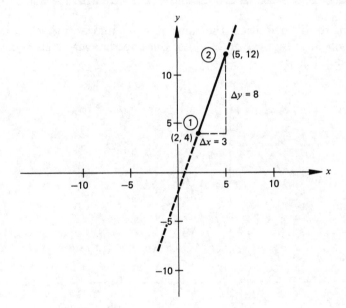

Similarly, the corresponding change in x is

$$\Delta x = x_2 - x_1 = 5 - 2 = 3$$

Hence,

$$m = \frac{\Delta y}{\Delta x} = \frac{8}{3}$$

The slope is positive, indicating that the line segment rises from left to right. The sign combined with the magnitude indicate that in moving along the line segment, y increases at a rate of 8 units for every 3 units that x increases.

◻

PRACTICE EXERCISE
Verify that the result from Eq. (2.10) is unaffected by the choice of (x_1, y_1) and (x_2, y_2). In Example 7, label (5, 12) as (x_1, y_1) and (2, 4) as (x_2, y_2) and recompute the slope.

Another way of interpreting the slope is given by the following definition.

DEFINITION: SLOPE
The *slope* is the change in the value of y if x increases by 1 unit.

According to this definition, the value of $m = \frac{8}{3}$ indicates that if x increases by 1 unit, y will *increase* by $\frac{8}{3}$ or $2\frac{2}{3}$ units. Observe this with the sequence of points identified in Fig. 2.9.

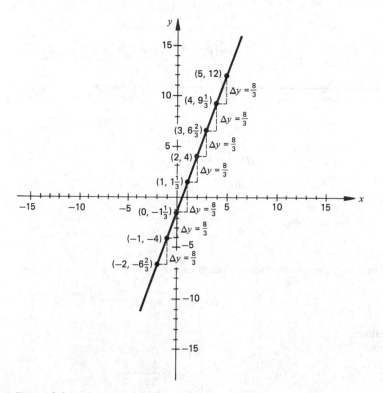

Figure 2.9 y increases by $\frac{8}{3}$ for each unit x increases.

> **NOTE** Along any straight line the slope is constant. That is, if a line is said to have a slope of -2, the slope of the line segment connecting any two points on the line will always equal -2.

EXAMPLE 8

(**Slope Undefined**) We have already seen that the slope of a linear equation having the form $y = k$ is 0. For a horizontal line, the value of y is always the same, and the numerator of the two-point formula, $y_2 - y_1$, always equals 0. We also examined the other special case of a linear equation, $x = k$. We verified that any linear equation having this form graphs as a vertical line crossing the x axis at $x = k$. *The slope of any vertical line is undefined.* This can be verified by attempting to use the two-point formula to determine the slope of the line described by $x = 5$. If we choose the two points $(x_1, y_1) = (5, 0)$ and $(x_2, y_2) = (5, -1)$, substitution into Eq. (2.10) gives

$$m = \frac{-1 - 0}{5 - 5}$$

$$= \frac{-1}{0}$$

which is not defined.

$\square$

Section 2.2 Follow-up Exercises

In Exercises 1–20, identify the x and y intercepts for the given linear equation.

1 $3x - 4y = 24$

2 $-2x + 5y = -20$

3 $-x + 3y = 9$

4 $4x + 2y = 36$

5 $-4x = 12$

6 $-10x + 300 = 0$

7 $x - 2y = 0$

8 $5x - 3y = 0$

9 $-8x + 5y = -20$

10 $(x + y)/2 = 3x - 2y + 16$

11 $2x - 3y = -18 + x$

12 $-3x + 4y - 10 = 7x - 2y + 50$

13 $15y - 90 = 0$

14 $(x - 2y)/3 - 12 = (2x + 4y)/3$

15 $ax + by = t$

16 $cx - dy = e$

17 $px = q$

18 $dx - ey + f = gx - hy$

19 $-ry = s$

20 $-e + fx - gy = h$

For Exercises 21–36, graph the given linear equation.

21 $2x - 3y = -12$

22 $-3x + 6y = -30$

23 $x - 2y = -8$

24 $-8x + 3y = 24$

25 $-x - 4y = 10$

26 $4x + 3y = -36$

27 $3x + 8y = 0$

28 $10x - 5y = 0$

29 $-5x + 2y = 0$

30 $8x - 4y = 0$

31 $-4x = 24$

32 $-2y = -9$

33 $-5y = -17.5$

34 $8x = 20$

35 $-nx = t, n > 0, t > 0$

36 $my = q, m > 0, q < 0$

37 What is the equation of the x axis? The y axis?

In Exercises 38–59, compute the slope of the line segment connecting the two points. Interpret the meaning of the slope in each case.

38 (2, 8) and (−2, −8)   **39** (−3, 10) and (2, −5)
40 (3, 5) and (−1, 15)   **41** (10, −3) and (12, 4)
42 (−2, 3) and (1, −9)   **43** (5, 8) and (−3, 28)
44 (4, −3) and (10, −12)   **45** (8, −24) and (5, −15)
46 (−2, 8) and (3, 8)   **47** (−5, −4) and (−5, 6)
48 (−4, 20) and (−4, 30)   **49** (5, 0) and (−25, 0)
50 (0, 30) and (0, −15)   **51** (5, 0) and (0, −10)
52 (a, b) and (−a, b)   **53** (0, 0) and (a, b)
54 (d, −c) and (0, 0)   **55** (−5, −5) and (5, 5)
56 (3, b) and (−10, b)   **57** (−a, −b) and (a, −b)
58 ($a + b$, c) and (a, c)   **59** ($c + d$, −$c − d$) and ($a + b$, −$a − b$)

2.3 SLOPE-INTERCEPT FORM

From a Different Vantage Point

In this section we discuss another form of expressing linear equations. In Sec. 2.1 we stated the general form of a two-variable linear equation as

$$ax + by = c \qquad (2.1)$$

Solving Eq. (2.1) for the variable y, we get

$$by = c - ax$$

or

$$y = \frac{c}{b} - \frac{ax}{b} \qquad (2.11)$$

For any linear equation the terms c/b and $-a/b$ on the right side of Eq. (2.11) have special significance, provided that $b \neq 0$. If $x = 0$, $y = c/b$. Thus, the term c/b in Eq. (2.11) is the ordinate of the y intercept. Similarly, what happens to the value of y if x increases by one unit in Eq. (2.11)? The value of y changes by $-a/b$. Thus, $-a/b$ is the slope for the equation.

This information is obtained from any linear equation of the form of Eq. (2.1) *if it can be solved for y.* Equation (2.11) is called the **slope-intercept form** of a linear equation. Equation (2.11) can be generalized in a simpler form:

$$y = mx + k \qquad (2.12)$$

where *m represents the slope* of the *line* and *k is the y coordinate of the y intercept.*

To illustrate this form, the equation

$$5x + y = 10$$

can be rewritten in the slope-intercept form as

$$y = -5x + 10$$

Hence, the slope is -5 and the y intercept equals $(0, 10)$.

NOTE   Why did the author use the letter k instead of b in Eq. (2.12)? So as not to confuse it with the b in Eq. (2.1)! Students have often seen the slope-intercept form stated as $y = mx + b$.

PRACTICE EXERCISE
Choose two points which satisfy the equation $5x + y = 10$ and verify that the slope equals -5 using Eq. (2.10).

EXAMPLE 9   The equation $y = 2x/3$ can be rewritten in slope-intercept form as

$$y = (\tfrac{2}{3})x + 0$$

The absence of the isolated constant on the right side implicitly suggests that $k = 0$. The graph of this equation is a line having a slope of $\tfrac{2}{3}$ and a y intercept $(0, 0)$.

EXAMPLE 10   The special case of a linear equation $y = k$ is in the slope-intercept form. To realize this, you must recognize that this equation can be written in the form $y = 0x + k$. The absence of the x term on the right side implicitly suggests that $m = 0$; i.e., the slope of the line having this form equals zero. We confirmed this in Sec. 2.2 when we discussed the graphical characteristics of this case. The y intercept is $(0, k)$ for such equations.

EXAMPLE 11   For the special case $x = k$, it is impossible to solve for the slope-intercept form of the linear equation. The variable y is not a part of the equation. Our conclusion is that it is impossible to determine the slope and y intercept for equations having this form. Look back at Fig. 2.3 to see whether this conclusion is consistent with our earlier findings.

❏

Interpreting the Slope and y Intercept

In many applications of linear equations, the slope and y intercept have meaningful interpretations. Take, for example, the salary equation

$$y = 3x + 25$$

where $y = $ *weekly salary, dollars*
and $x = $ *number of units sold during 1 week*
The salary equation is linear and is written in slope-intercept form. Graphically, the equation is represented by the line in Fig. 2.10, which has a slope of $+3$ and y

Figure 2.10
Salary function.

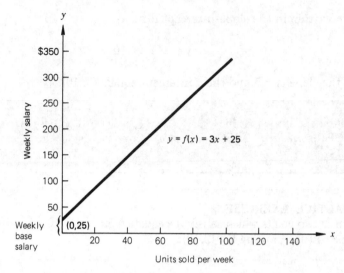

intercept equal to (0, 25). Notice that this equation has been graphed only for nonnegative values of x and y. Can you suggest why this would be appropriate?

Think back to the definition of *slope*. Since the slope represents the change in y associated with a unit increase in x, the slope of $+3$ means that weekly salary y increases by \$3 for each additional unit sold. The y coordinate of the y intercept represents the value of y when $x = 0$, or the salary which would be earned if no units were sold. This is the base salary per week.

EXAMPLE 12   A police department estimates that the total cost C of owning and operating a patrol car can be estimated by the linear equation

$$C = 0.40x + 18,000$$

where $C = $ *total cost, dollars*
and $x = $ *number of miles driven*
This equation is in slope-intercept form with a slope of 0.40 and C intercept (which is equivalent to the y intercept) of (0, 18,000). The slope suggests that total cost increases at a rate of \$0.40 for each additional mile driven. The C intercept indicates a cost of \$18,000 if the car is driven zero miles.

$\square$

Section 2.3 Follow-up Exercises

For Exercises 1–24, rewrite each equation in slope-intercept form and determine the slope and y intercept.

1 $3x - 2y = 15$ **2** $-x + 5y = 27.5$
3 $4x - 3y = 18$ **4** $2x - 7y = -21$
5 $-x + y = 8$ **6** $2x - y = -5$
7 $(x + 2y)/2 = -6$ **8** $(-2x + y)/3 = 2$
9 $(3x - 5y)/4 = -5$ **10** $(-x + 2y)/4 = 3x - y$

11 $2x = (5x - 2y)/4$

12 $(-x + 3y)/2 = 10 - 2x$

13 $4x - 3y = 0$

14 $8x + 3y = 24$

15 $3x - 6y + 10 = x$

16 $3y - 5x + 20 = 4x - 2y + 5$

17 $2x + 3y = 4x + 3y$

18 $-5x + y - 12 = 2y - 5x$

19 $8y - 24 = 0$

20 $3x + 6 = 0$

21 $mx + ny = p$

22 $mx - n = 0$

23 $c - dy = 0$

24 $dx = cy - f$

25 Women in the Labor Force The number of women in the labor force is expected to increase during the 1990s, but not as dramatically as occurred during the 1970s. One forecasting consultant uses the linear equation $n = 29.6 + 1.20t$ to predict the number of women between the ages of 35 and 44 who will be in the labor force. In this equation, n equals the number of women (aged 35 to 44) in the labor force (measured in millions) and t equals time measured in years *since* 1981 ($t = 0$ corresponds to 1981). If n is plotted on the vertical axis:

(*a*) Graph the equation.

(*b*) Identify the slope and y intercept (n intercept, here).

(*c*) Interpret the meaning of the slope and n intercept in this application.

(*d*) Predict the number of women in this age group who will be in the labor force in 1995. In the year 2000.

26 The chamber of commerce for a summer resort is trying to determine how many tourists will be visiting each season over the coming years. A marketing research firm has estimated that the number of tourists can be predicted by the equation $p = 275,000 + 7,500t$, where $p =$ number of tourists per year and $t =$ years (measured *from* this current season). Thus $t = 0$ identifies the current season, $t = 1$ is the next season, etc. If p is plotted on the vertical axis:

(*a*) Graph the equation.

(*b*) Identify the slope and y intercept (p intercept, here).

(*c*) Interpret the meaning of the slope and p intercept in this application.

27 Think Metric! $C = \frac{5}{9}F - \frac{160}{9}$ is an equation relating temperature in Celsius units to temperature measured on the Fahrenheit scale. Let $C =$ degrees Celsius and $F =$ degrees Fahrenheit; assume the equation is graphed with C measured on the vertical axis.

(*a*) Identify the slope and C intercept.

(*b*) Interpret the meaning of the slope and C intercept for purposes of converting from Fahrenheit to Celsius temperatures.

(*c*) Solve the equation for F and rework parts a and b if F is plotted on the vertical axis.

28 A police department believes that the number of serious crimes which occur each month can be estimated by the equation

$$c = 1,200 - 12.5p$$

where c equals the number of serious crimes expected per month and p equals the number of officers assigned to preventive patrol. If c is graphed on the vertical axis:

(*a*) Identify the slope and interpret its meaning.

(*b*) Identify the c intercept and interpret its meaning.

(*c*) Identify the p intercept and interpret its meaning.

29 The book value of a machine is expressed by the equation

$$V = 60,000 - 7,500t$$

where V equals the book value in dollars and t equals the age of the machine expressed in years.

(*a*) Identify the *t* and *V* intercepts.
(*b*) Interpret the meaning of the intercepts.
(*c*) Interpret the meaning of the slope.
(*d*) Sketch the function.

30 SAT Scores One small college has observed a downward trend in the average SAT score of applicants to the college. Analysis has resulted in the equation

$$s = 620 - 4.5t$$

where *s* equals the average SAT score for a given year and *t* equals time measured in years since 1985 ($t = 0$).
(*a*) Identify the *t* and *s* intercepts.
(*b*) Interpret the meaning of the intercepts. (Does your interpretation of the *t* intercept make sense?)
(*c*) Interpret the meaning of the slope.
(*d*) Sketch the equation.

31 Product Mix A company produces two products. Weekly labor availability equals 150 labor-hours. Each unit of product 1 requires 3 labor-hours and each unit of product 2 requires 4.5 labor-hours. If management wishes to use all labor-hours, the equation

$$3x + 4.5y = 150$$

is a statement of this requirement, where *x* equals the number of units produced of product 1 and *y* equals the number of units produced of product 2. Rewrite the equation in slope-intercept form and interpret the meaning of the slope and *y* intercept. Solve for the *x* intercept and interpret its meaning.

32 Portfolio Management A portfolio manager is concerned that two stocks generate an annual income of $15,000 for a client. The two stocks earn annual dividends of $2.40 and $3.50 per share, respectively. If *x* equals the number of shares of stock 1 and *y* equals the number of shares of stock 2, the equation

$$2.4x + 3.5y = 15,000$$

states that total annual dividend income from the two stocks should total $15,000. Rewrite the equation in slope-intercept form and interpret the meaning of the slope and *y* intercept in this application. Solve for the *x* intercept and interpret its meaning.

2.4 DETERMINING THE EQUATION OF A STRAIGHT LINE

In this section we show how to determine the equation for a linear relationship. The way in which you determine the equation depends upon the information available. The following sections discuss different possibilities. In each instance we will be seeking the slope-intercept form. Thus, we will need to identify the slope-intercept *parameters m* and *k*.

Slope and Intercept

The easiest situation is one in which you know the slope *m* and *y* intercept (0, *k*) of the line representing an equation. To determine the linear equation in this almost trivial case, simply substitute *m* and *k* into the slope-intercept form, Eq. (2.12). I you are interested in stating the equation in the standard form of Eq. (2.1), simply rearrange the terms of the slope-intercept equation.

EXAMPLE 13 Determine the equation of the straight line which has a slope of -5 and a y intercept of $(0, 15)$.

SOLUTION

Substituting values of $m = -5$ and $k = 15$ into Eq. (2.12) gives

$$y = -5x + 15$$

Restated in the form of Eq. (2.1), an equivalent form of this equation is

$$5x + y = 15$$

❑

Slope and One Point

If given the slope and one point which lies on a line, we can substitute the known slope m and coordinates of the given point into Eq. (2.12) and solve for k.

EXAMPLE 14 Given that the slope of a straight line is -2 and one point lying on the line is $(2, 8)$ we can substitute these values into Eq. (2.12), yielding

$$8 = (-2)(2) + k$$

or

$$12 = k$$

Since $m = -2$ and $k = 12$, the slope-intercept equation is

$$y = -2x + 12$$

And, as before, we can rewrite this equation in the equivalent form

$$2x + y = 12$$

❑

NOTE You may be wondering which form of the linear equation—Eq. (2.1) or Eq. (2.12)—is the correct or preferred form. Both are correct! The preferred form depends on what you intend to do with the equation. Depending on the type of analysis to be conducted, one of these forms may be more appropriate than the other.

EXAMPLE 15 If the slope of a straight line is zero and one point lying on the line is $(5, -30)$, the equation of the line can be found by first substituting the zero slope and coordinates $(5, -30)$ into Eq. (2.12).

$$-30 = (0)(5) + k$$

or

$$-30 = k$$

Since $m = 0$ and $k = -30$, the slope-intercept equation is

$$y = 0x + (-30)$$

or

$$y = -30$$

EXAMPLE 16 **(Point-Slope Formula)** Given a nonvertical straight line with slope m and containing the point (x_1, y_1), the slope of the line connecting (x_1, y_1) with any other point (x, y) on the line would be expressed as

$$m = \frac{y - y_1}{x - x_1}$$

Rearranging this equation results in

$$y - y_1 = m(x - x_1) \tag{2.13}$$

which is the ***point-slope formula*** for a straight line. This formula can be used to determine the equation of a nonvertical straight line given the slope and one point lying on the line. Suppose that a line has a slope of 5 and contains the point $(-4, 10)$. Substituting into Eq. (2.13) and solving for y,

$$y - 10 = 5[x - (-4)]$$

$$y - 10 = 5x + 20$$

$$y = 5x + 30$$

which is the slope-intercept form of the equation.

EXAMPLE 17 Considering the linear equation $3x - 6y = 24$:
(*a*) What is the slope of the line represented by the given equation?
(*b*) What is the slope of any line parallel to the given line?
(*c*) What is the slope of any line perpendicular to the given line?
(*d*) How many different lines are perpendicular to this line?
(*e*) Find the equation of the line which is perpendicular to the given line *and* which passes through the point $(2, 5)$.

SOLUTION
(*a*) The given equation can be restated in slope-intercept form as

$$-6y = 24 - 3x$$

or

$$y = -4 + \tfrac{1}{2}x$$

From this equation the slope equals $+\tfrac{1}{2}$, and the y intercept occurs at $(0, -4)$.

(b)

> **PARALLEL LINES**
> Two lines are parallel if they have the same slope.

Since the slope of the given line equals $+\frac{1}{2}$, any parallel lines will have a slope of $+\frac{1}{2}$.

(c)

> **PERPENDICULAR LINES**
> If a line has a slope m_1 $(m_1 \neq 0)$, the slope of any line perpendicular to the given line has a slope equal to the negative reciprocal of the given line, or $m_2 = -1/m_1$.

Since $m_1 = \frac{1}{2}$, the slope of any line perpendicular to the line $3x - 6y = 24$ is

$$m_2 = -\frac{1}{\frac{1}{2}}$$
$$= -2$$

(d) Because there is an infinite set of lines with $m = -2$, an infinite number of lines are perpendicular to this line.

(e) The line we are interested in has a slope equal to -2 and one point on the line is (2, 5). Substituting these values into the point-slope formula, Eq. (2.13), gives

$$y - 5 = (-2)(x - 2)$$
$$y - 5 = -2x + 4$$
$$y = -2x + 9$$

or, alternatively,

$$2x + y = 9$$

Figure 2.11 illustrates the lines in this example.

☐

Two Points

A more likely situation is that some data points which lie on a line have been gathered and we wish to determine the equation of the line. Assume that we are given the coordinates of two points which lie on a straight line. We can determine the slope of the line by using the two-point formula [Eq. (2.10)]. As soon as we know the slope, the y intercept can be determined by using *either* of the two data points and proceeding as we did in the previous section.

Figure 2.11

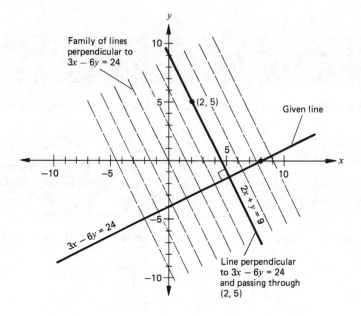

EXAMPLE 18   To determine the equation of the straight line which passes through $(-4, 2)$ and the origin, we substitute the coordinates into the two-point formula, resulting in

$$m = \frac{0-2}{0-(-4)}$$

$$= \frac{-2}{4} = -\frac{1}{2}$$

Substituting $m = -\frac{1}{2}$ and the coordinates $(-4, 2)$ into Eq. (2.13) yields

$$y - 2 = (-\tfrac{1}{2})[x - (-4)]$$

$$y - 2 = -\tfrac{1}{2}x - 2$$

$$y = -\tfrac{1}{2}x$$

Thus, the slope-intercept form of the equation is

$$y = -\tfrac{1}{2}x$$

NOTE   In this last example you might have realized that the origin is the y intercept. How would this have simplified the analysis?

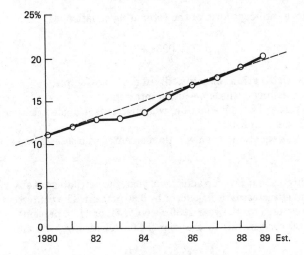

Figure 2.12 Percentage of total electricity produced in the United States attributable to nuclear sources. (Sources: Chicago Tribune, North American Electricity Council.)

EXAMPLE 19 **(Nuclear Power Utilization; Motivating Scenario)** Figure 2.12 illustrates sample data gathered by the North American Electricity Council. The graph illustrates the percentage of total electricity produced in the United States generated by nuclear power sources. The percentage appears to be increasing approximately at a linear rate over time. The council wants to determine a linear equation which approximates the relationship between the percentage of electricity generated by nuclear sources (p) and time (t) measured in years. An analyst has chosen to fit a line through the data points for 1981 and 1986. The values for p were 0.124 and 0.172 for the years 1981 and 1986, respectively.

In determining the estimating equation, let t equal years measured since 1980 (i.e., $t = 0$ corresponds to 1980, $t = 1$ corresponds to 1981, etc.). Using this definition, the two data points have coordinates (1, 0.124) and (6, 0.172). The slope-intercept relationship for this example will have the form

$$p = mt + k \qquad (2.14)$$

Given the two data points, the slope is

$$m = \frac{p_2 - p_1}{t_2 - t_1}$$
$$= \frac{0.172 - 0.124}{6 - 1}$$
$$= \frac{0.048}{5} = 0.0096$$

Using a slightly different, but equivalent, procedure to the point-slope formula, we substitute $m = 0.0096$ and the coordinates (1, 0.124) into Eq. (2.14):

$$0.124 = (0.0096)(1) + k$$
$$0.1144 = k$$

Therefore, the slope-intercept form of the estimating equation is

$$p = 0.0096t + 0.1144$$

EXAMPLE 20   **(Nuclear Power Utilization, continued)** In the last example:
(a) Interpret the meaning of the slope and p intercept.
(b) According to this estimating equation, what percentage of electricity is expected from nuclear sources in the year 2000?
(c) According to this equation, when will the percentage surpass 25 percent?

SOLUTION

(a) The slope indicates that for each additional year, the percentage of electricity attributable to nuclear sources increases by 0.0096, or by 0.96 percent. The p intercept indicates that the estimated percentage for the year 1980 was 0.1144, or 11.44 percent.
(b) A t value of 20 corresponds to the year 2000. Substituting this into the estimating equation gives

$$p = 0.0096(20) + 0.1144$$
$$= 0.192 + 0.1144$$
$$= 0.3064$$

This estimating equation predicts that 30.64 percent will be attributable to nuclear sources in the year 2000.
(c) Letting $p = 0.25$ yields

$$0.25 = 0.0096t + 0.1144$$
$$0.1356 = 0.0096t$$
$$\frac{0.1356}{0.0096} = t$$
$$14.125 = t$$

Therefore, when $t = 14.125$, the percentage will equal 0.25. Therefore, the percentage will surpass 25 percent sometime during the fifteenth year, or during 1995.

NOTE   Estimation plays a very important part in applying mathematical analysis to the world around us. Although the tools of mathematical analysis are most often very precise, the relationships which we analyze are not always exact. There are many applications in which the mathematical relationships are determined precisely. However, we often must estimate the relationships which exist between variables that are of interest. There are scientific procedures which can be used to develop our estimates. The use of such procedures enhances the likelihood that our estimates are reasonable. However, as a person actually conducting mathematical analysis or as a person who is the recipient of the results of such analyses, one should be aware that estimated relationships are usually accompanied by some measure of error. An attempt should be made to understand the magnitude of potential error which is associated with estimates and to consider the effects of such error in drawing conclusions from the mathematical analysis.

As we move through the text, you will sometimes develop mathematical relationships and sometimes be given relationships. Try to retain a questioning attitude about the source of each relationship. Be curious about their origins. Anticipate the implications of errors in the relationships. The author will try to reinforce this perspective.

Section 2.4 Follow-up Exercises

In Exercises 1–36, determine the slope-intercept form of the linear equation, given the listed attributes.

1 Slope $= -2$, y intercept $= (0, 10)$

2 Slope $= 4$, y intercept $= (0, -5)$

3 Slope $= \frac{1}{2}$, y intercept $= (0, \frac{3}{4})$

4 Slope $= -\frac{5}{2}$, y intercept $= (0, -20)$

5 Slope $= -r$, y intercept $= (0, -t/2)$

6 Slope undefined, infinite number of y intercepts

7 Slope $= -3$, $(4, -2)$ lies on line

8 Slope $= 5$, $(-3, 12)$ lies on line

9 Slope $= \frac{3}{2}$, $(-5, -8)$ lies on line

10 Slope $= -\frac{1}{2}$, $(-4, 0)$ lies on line

11 Slope $= 2.5$, $(-2, 5)$ lies on line

12 Slope $= -3.25$, $(1.5, -7.5)$ lies on line

13 Slope $= 5.6$, $(2.4, -4.8)$ lies on line

14 Slope $= -8.2$, $(-0.75, 16.3)$ lies on line

15 Slope $= w$, (p, q) lies on line

16 Slope $= -a$, $(4, -4)$ lies on line

17 Slope undefined, $(-3, -5)$ lies on line

18 Slope $= 0$, $(20, -10)$ lies on line

19 Slope $= 0$, (u, v) lies on line

20 Slope undefined, $(-t, v)$ lies on line

21 $(-4, 5)$ and $(-2, -3)$ lie on line

22 $(3, -2)$ and $(-12, 1)$ lie on line

23 $(20, 240)$ and $(15, 450)$ lie on line

24 $(-12, 760)$ and $(8, -1,320)$ lie on line

25 $(0.234, 20.75)$ and $(2.642, 18.24)$ lie on line

26 $(5.76, -2.48)$ and $(3.74, 8.76)$ lie on line

27 (a, b) and (c, d) lie on line

28 $(a, -3)$ and $(a, 15)$ lie on line

29 $(-d, b)$ and (e, b) lie on line

30 (p, r) and $(-p, r)$ lie on line

31 Passes through $(2, -4)$ and is parallel to the line $3x - 4y = 20$

32 Passes through $(-2, 10)$ and is parallel to the line $5x - y = 0$

33 Passes through $(7, 2)$ and is parallel to the line (a) $x = 7$ and (b) $y = 6$

34 Passes through $(20, -30)$ and is perpendicular to the line $4x + 2y = -18$

35 Passes through $(-8, -4)$ and is perpendicular to the line $8x - 2y = 0$

36 Passes through $(7, 2)$ and is perpendicular to the line (a) $x = 7$ and (b) $y = 6$

37 Depreciation The value of a machine is expected to decrease at a linear rate over time. Two data points indicate that the value of the machine at $t = 0$ (time of purchase) is $18,000 and its value in 1 year will equal $14,500.

(a) Determine the slope-intercept equation ($V = mt + k$) which relates the value V of the machine to its age t.

(b) Interpret the meaning of the slope and V intercept.

(c) Solve for the t intercept and interpret its meaning.

38 Depreciation The value of a machine is expected to decrease at a linear rate over time. Two data points indicate that the value of the machine 1 year after the date of purchase will be $84,000 and its value after 5 years is expected to be $36,000.

(a) Determine the slope-intercept equation ($V = mt + k$) which relates the value V of the machine to its age t, stated in years.

(b) Interpret the meaning of the slope and V intercept.

(c) Determine the t intercept and interpret its meaning.

39 If C equals degrees Celsius and F equals degrees Fahrenheit, assume that the relationship between the two temperature scales is linear and is being graphed with F on the vertical axis. Two data points on the line relating C and F are (5, 41) and (25, 77). Using these points, determine the slope-intercept equation which allows transformation from Celsius temperature to equivalent Fahrenheit temperature. Identify and interpret the meaning of the slope, C intercept, and F intercept.

40 College Retirement The largest retirement program for college professors is the Teachers Insurance and Annuity Association/College Retirement Equities Fund (TIAA/CREF). One of the investment options in this program is the CREF Money Market Account, which was initiated in 1988. Figure 2.13 illustrates the performance of this investment during the first 10 quarters of its existence. Note that V is the value of a share (unit) in this fund and that the data points reflect the end-of-month values. It appears that the value of this money market has been increasing at an approximately linear rate. If the data points (6, 11.32) and (9, 12.04) are chosen to estimate the relationship between the value of a share V and time t, measured in quarters since the inception of the CREF Money Market Fund ($t = 0$ corresponds to March 31, 1988):

Figure 2.13   CREF money market account per share (unit) quarter-end values.

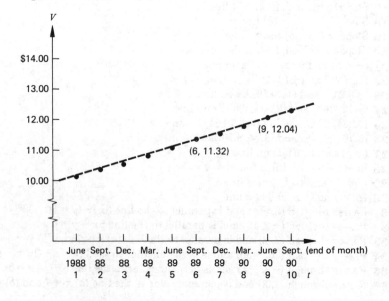

(a) Determine the slope-intercept form of the estimating equation.

(b) Identify and interpret the meaning of the slope.

(c) Forecast the value per share on June 30, 1991, and March 31, 1992.

41 College Retirement (continued) The CREF Money Market Fund was established on March 31, 1988. The initial value per share was set at $10.00.

(a) Using the equation found in part *a* of the previous exercise, estimate the value per share on March 31, 1988. How much error is there in the estimate?

(b) Similarly, determine the actual values per share on June 30, 1991, and March 31, 1992,* and compare with the forecasts in part *c* of the previous exercise. How much error was there?

(c) While you have access to the data in part *b*, test the accuracy of the estimating equation for other quarterly data points.

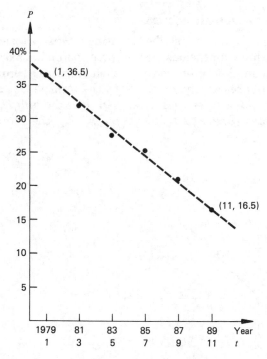

Figure 2.14 Percentage of high school students who have used marijuana in the past 30 days.

42 Marijuana Usage among High School Students Figure 2.14 illustrates some survey data regarding the usage of marijuana among high school students. A sample of high school students was taken every 2 years between 1979 and 1989. The data in Figure 2.14 reflects the percentage of students surveyed who indicated they had used marijuana during the previous 30 days. The data points suggest that the percentage of students having used marijuana is decreasing at an approximately linear rate over time. If the

---

* Contact the Teachers Insurance and Annuity Association/College Retirement Equities Fund, 730 Third Avenue, New York, New York 10017 (try calling 1-800-842-2733).

data points for 1979 (1, 36.5) and 1989 (11, 16.5) are used to estimate the linear equation which relates the percentage of students P to time t ($t = 1$ corresponding to 1979):

(a) Determine the slope-intercept form of the estimating equation.

(b) Forecast the expected percentage for 1991 and 1995.

(c) Interpret the meaning of the slope and P intercept.

2.5 LINEAR EQUATIONS INVOLVING MORE THAN TWO VARIABLES

When linear equations involve more than two variables, the algebraic properties remain basically the same but the visual or graphical characteristics change considerably or are lost altogether.

Three-Dimensional Coordinate Systems

Three-dimensional space can be described by using a ***three-dimensional coordinate system.*** In three dimensions we use three coordinate axes which are all perpendicular to one another, intersecting at their respective zero points. Figure 2.15 illustrates a set of axes which are labeled by the variables x_1, x_2, and x_3. The point of intersection of the three axes is referred to as the ***origin.*** Using three-component coordinates ***(ordered triples),*** (x_1, x_2, x_3), the coordinates of the origin are $(0, 0, 0)$.

Figure 2.15
Coordinate axis
system in three
dimensions.

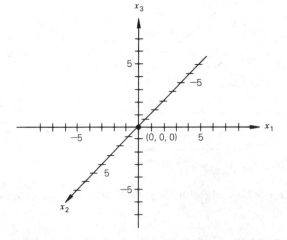

Observe that graphing three dimensions on paper (two dimensions) requires a certain perspective that may be difficult to see at first. We might have drawn Fig. 2.15 such that we were looking right "down the barrel" of the x_2 axis. In that case we would have no sense of depth or location relative to the x_2 axis. Therefore, we rotate the coordinate axes by turning the x_3 axis clockwise. This allows us to have a sense of depth when the x_2 axis is drawn at an angle.

Just as the coordinate axes in two dimensions divide 2-space into quadrants, the axes in three dimensions divide 3-space into ***octants.*** This is illustrated in Fig.

Figure 2.16
Octants of
3-space.

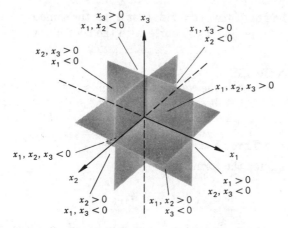

2.16. Note the sign characteristics in each octant. The three-component coordinates allow for specifying the location or address of any point in three dimensions.

As with two-dimensional coordinates, each component of (x_1, x_2, x_3) specifies the location of a point relative to each axis. Carefully examine Fig. 2.17. In order to assist in understanding this figure, a ***rectangular polyhedron*** has been sketched. Along with several other points, we are interested in the locations of the corner points of this polyhedron. Obviously, G is located at the origin, having coordinates $(0, 0, 0)$. Point F lies directly on the x_2 axis, 4 units out. Its coordinates are $(0, 4, 0)$. Point A forms the upper left corner of one end ($ABGH$) of the polyhedron. Since H lies on the x_1 axis and A is vertically above H, we can conclude that the x_1 coordinate of A is -5 *and* the x_2 coordinate of A is 0. Finally, points $A, B, C,$ and D all seem to be at the same *height* (relative to the x_3 axis). Because point B lies

Figure 2.17
Sample points
in 3-space.

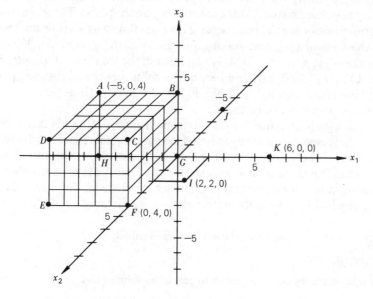

on the x_3 axis at a height of 4, we conclude that A has the same x_3 coordinate. Thus, A is located at $(-5, 0, 4)$. See if you agree with the coordinates of I and k.

PRACTICE EXERCISE
Test your skills and define the coordinates of points B, C, D, E, and J. *Answer:* $B\,(0, 0, 4)$, $C\,(0, 4, 4)$, $D\,(-5, 4, 4)$, $E\,(-5, 4, 0)$, $j\,(0, -4, 0)$.

Equations Involving Three Variables

Linear equations having the form

$$a_1 x_1 + a_2 x_2 + a_3 x_3 = b$$

graph as **planes** in three dimensions. *The number of variables in an equation determines the number of dimensions required to graphically represent the equation.* Three variables require three dimensions. It is not so important that you actually be able to graph in three dimensions. It is more important that (1) you are able to *recognize* a linear equation involving three variables, (2) you are aware that linear equations involving three variables graph as planes in three dimensions, (3) you know what a *plane* is, and (4) you have some feeling for how planes can be represented graphically. A plane, of course, is a flat surface like the ceiling, walls, and floor of the room in which you are currently sitting or lying. Instead of the two points needed to graph a line, three points are necessary to define a plane. The three points must not be **collinear;** i.e., they must not lie on the same line. Take, for example, the equation

$$2x_1 + 4x_2 + 3x_3 = 12 \qquad\qquad (2.15)$$

If we can identify three members of the solution set for this equation, they will specify the coordinates of three points lying on the plane. Three members which are identified easily are the intercepts. These are found by setting any two of the three variables equal to 0 and solving for the remaining variable. Verify that when $x_1 = x_2 = 0, x_3 = 4$, or $(0, 0, 4)$ is a member of the solution set. Similarly, verify that $(6, 0, 0)$ and $(0, 3, 0)$ are members of the solution set and thus are points lying on the plane representing Eq. (2.15). Figure 2.18 shows these points and a portion of the plane which contains them.

When graphing equations involving two variables, we identified two points and connected them with a straight line. However, we saw that in order to represent *all* members of the solution set, the line must extend an infinite distance in each direction. The same is true with the solution set for three-variable equations. To represent all members of the solution set for the equation $2x_1 + 4x_2 + 3x_3 = 12$, the plane in Fig. 2.18 must extend an infinite distance in all directions.

EXAMPLE 21   Graph the linear equation $x_1 = 0$ in three dimensions.

SOLUTION

In this problem we are being asked to graph the solution set

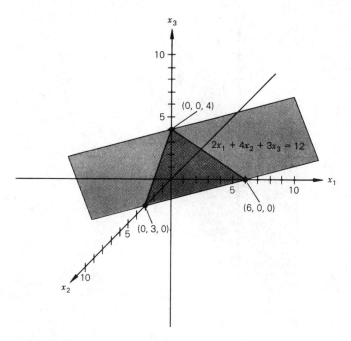

Figure 2.18 Graph of plane representing the linear equation $2x_1 + 4x_2 + 3x_3 = 12$.

$$S = \{(x_1, x_2\ x_3)|x_1 = 0\}$$

In order to graph the equation, we again need to identify three noncollinear points which satisfy the equation. We see that as long as $x_1 = 0$, x_2 and x_3 can equal *any* values. For example, $(0, 0, 0)$, $(0, 2, 0)$, and $(0, 0, 4)$ all satisfy the equation. Figure 2.19 illustrates the graph of the equation. The equation $x_1 = 0$ graphs as a plane perpendicular to the x_1 axis and passing through $x_1 = 0$. This is the $x_2 x_3$ *plane* (the plane which includes among its points all points lying on the x_2 axis *and* the x_3 axis).

◻

> Any equation of the form $x_1 = k$ graphs in 3-space as a plane perpendicular to the x_1 axis, intersecting it at $x_1 = k$.
>
> Any equation of the form $x_j = k$, where $j = 1, 2, or\ 3$, will graph as a plane which is perpendicular to the x_j axis at $x_j = k$. Figures 2.20 to 2.22 illustrate this property.

Equations Involving More than Three Variables

When more than three variables exist ($n > 3$), graphing requires more than three dimensions. Even though we cannot envision the graphical representation of such equations, the term **hyperplane** is used to describe the ("would-be") geometric

Figure 2.19
The plane $x_1 = 0$.

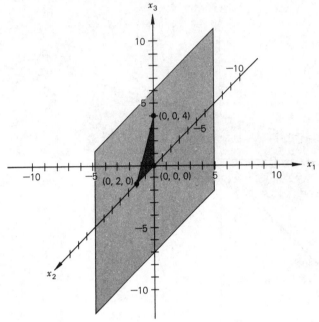

Figure 2.20
Planes of the
form $x_1 = k$.

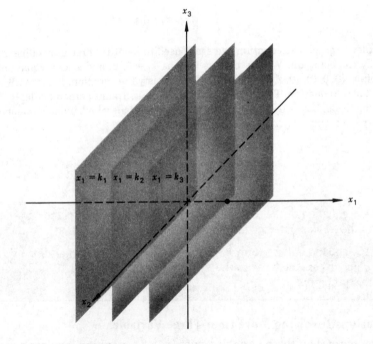

Figure 2.21
Planes of the
form $x_2 = k$.

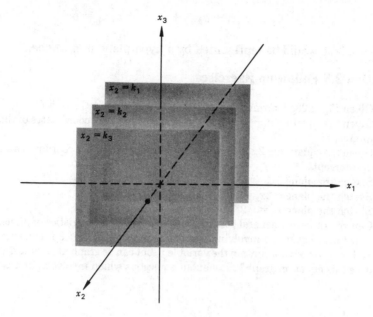

Figure 2.22
Planes of the
form $x_3 = k$.

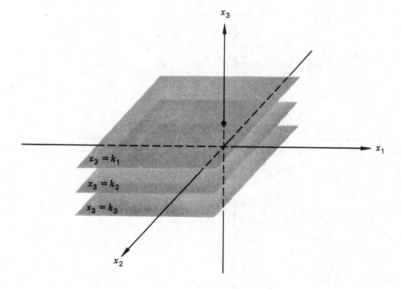

representation of the equation. Mathematicians would, for instance, say that the equation

$$x_1 + x_2 + x_3 + x_4 = 10$$

is represented by a hyperplane in 4-space or four dimensions. Or, in general, an equation of the form

$$a_1 x_1 + a_2 x_2 + \cdots + a_n x_n = b$$

where $n > 3$, would be represented by a hyperplane in **n-space.**

Section 2.5 Follow-up Exercises

1 Given Fig. 2.23, determine the coordinates of points A through I.
2 Given the equation $x_1 - 2x_2 + 4x_3 = 10$, determine the coordinates of the x_1, x_2, and x_3 intercepts.
3 Given the equation $-2x_1 + 3x_2 - x_3 = -15$, determine the coordinates of the x_1, x_2, and x_3 intercepts.
4 Sketch the plane $3x_1 = 9$.
5 Sketch the plane $-2x_2 = -8$.
6 Sketch the plane $x_3 = -2$.
*7 Can you draw any general conclusions about the characteristics of planes which represent linear equations involving two of the three variables? For example, the equation $x_1 + x_2 = 5$ does not contain the variable x_3 but can be graphed in three dimensions. How does this equation graph? How about equations which involve x_1 and x_3? x_2 and x_3?

Figure 2.23

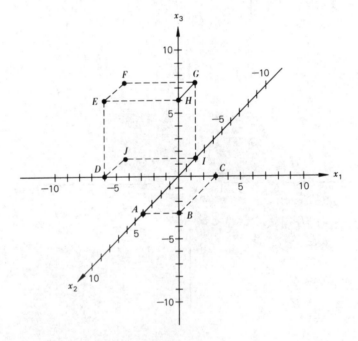

2.6 ADDITIONAL APPLICATIONS

The more exposure you have to word problems, the more skilled you will become in formulating them. The following examples illustrate the formulation of linear equations for different types of applications. Study these carefully and try as many of these types of problems as you can, both at the end of this section and at the end of the chapter.

EXAMPLE 22 **Emergency Airlift** The International Red Cross is making plans to airlift emergency food and medical supplies into a large South American city which has recently suffered from extensive flooding. Four items will be airlifted in containers to aid in the recovery from flooding. The four items and their respective volumes per container are shown in the table. The first plane to be sent into the area has a volume capacity of 6,000 cubic feet. Determine an equation whose solution set contains all possible combinations of the four items which will fill the plane to its volume capacity.

Item	Volume/Container, ft^3
Blood	20
Medical supply kits	30
Food	8
Water	6

SOLUTION

The first step in almost every word problem is to define the unknowns or variables which are to be used. It is useful to ask yourself what decisions need to be made in the problem. If these decisions can be identified, they hold the key to defining the variables.

In this example, the decision facing the Red Cross personnel deals with how many containers of each item should be sent on the first plane. Since the Red Cross wishes to ship as many supplies as possible on this first plane, it is interested in identifying the different combinations which will fill the plane to capacity (volumewise).

Verbally, the equation we are seeking should have the form

$$\boxed{\text{Volume of supplies shipped} = 6{,}000 \text{ cubic feet}}$$

We can be more specific by rewriting the equation as

$$\boxed{\begin{array}{l}\text{Volume of blood} + \text{volume of medical supply kits} \\ \qquad\qquad + \text{volume of food} + \text{volume of water} = 6{,}000\end{array}}$$

If we let

$x_1 = $ *number of containers of blood*

$x_2 = $ *number of containers of medical supply kits*

$x_3 = $ *number of containers of food*

$x_4 = $ *number of containers of water*

the equation can be stated in its correct mathematical form as

$$20x_1 + 30x_2 + 8x_3 + 6x_4 = 6{,}000$$

Verify that *each term* on the left side of the equation is formed by using the relationship

$$\boxed{\begin{array}{l}\text{Total volume} \\ \text{of item } j\end{array} = \left(\begin{array}{c}\text{volume per container} \\ \text{of item } j\end{array}\right)\left(\begin{array}{c}\text{number of containers} \\ \text{of item } j\end{array}\right)}$$

---

74 CHAPTER 2 LINEAR EQUATIONS

EXAMPLE 23 **Investment Portfolio** A local university has $5 million to invest in stocks. The board of trustees has approved six different types of stocks in which the university may invest. The current prices per share for each type of stock are indicated in the following table. Determine the equation for which the solution set includes all the different combinations of the six stocks which could be purchased for exactly $5 million.

Stock	Price per Share
1	$ 35
2	60
3	125
4	100
5	500
6	250

SOLUTION

The general form of the equation should be total dollars spent on the six stocks equals $5 million, or more specifically,

Total dollars spent on stock 1 + total dollars spent on stock 2
+ · · · + total dollars spent on stock 6 = $5 million

The basic decision to be made concerns the number of shares of each security to be purchased so as to expend the full $5 million. Therefore, let's generalize our variables as

$$x_j = number\ of\ shares\ purchased\ of\ stock\ j$$

where $j = 1, 2, 3, 4, 5,$ or 6.

Using these variables, we state the equation as

$$35x_1 + 60x_2 + 125x_3 + 100x_4 + 500x_5 + 250x_6 = 5,000,000$$

Note that each term on the left side of the equation has the form

Total dollars spent on stock j = (price per share)(number of shares purchased)

EXAMPLE 24 **Court Scheduling** A metropolitan district court sorts its cases into three categories. Court records have enabled the court clerk to provide estimates of the average number of hours required to process each type of case. Type 1 cases average 16 hours, type 2 average 8 hours, and type 3 average 4.5 hours. For the coming month 850 hours are available in the six different courtrooms in the building. Determine an equation whose solution set includes all the different combinations of the three types of cases which would schedule the courts to their capacity.

SOLUTION

The general form of the equation should be

Total court hours scheduled = 850

Letting x_1, x_2, and x_3 equal the number of cases scheduled of types 1, 2, and 3, respectively, the equation is

$$16x_1 + 8x_2 + 4.5x_3 = 850$$

EXAMPLE 25 **Nutrition Planning** A dietitian at a local school is planning luncheon menus. He has eight choices of items which may be served at any one meal. One concern of the dietitian is meeting various nutritional requirements. Our dietitian is interested in determining the various quantities of each of the eight foods which would provide exactly 45 milligrams of a required vitamin. The vitamin content per serving of each of the eight food items is shown in the table. Determine the equation whose solution set satisfies this requirement.

Food Type	1	2	3	4	5	6	7	8
mg/serving	5	7.5	3	4.5	9	10	2.5	6

SOLUTION

Letting x_j = number of servings of food j, where j = 1, 2, 3, 4, 5, 6, 7, or 8, the equation is

$$5x_1 + 7.5x_2 + 3x_3 + 4.5x_4 + 9x_5 + 10x_6 + 2.5x_7 + 6x_8 = 45$$

❑

Section 2.6 Follow-up Exercises

1 In a realistic sense, what are the maximum and minimum possible quantities of each item in Example 22?
2 Assume that the plane in Example 22 can only carry 40,000 pounds of cargo and that the items weigh 150, 100, 60, and 70 pounds per container, respectively. State the equation whose solution set contains all combinations of the four items which will add up to equal the weight capacity of the plane.
3 In a realistic sense, what are the maximum and minimum allowable values for each variable in the equation developed in Example 23?
4 The expected annual dividends per share of each of the preceding stocks are shown in the following table. Assume that the board of trustees desires to earn annual dividends of $1,000,000 from its investments. Using the same variables as in the example, develop the equation whose solution set includes all possible combinations of the six stocks which will generate annual dividends equal to $1,000,000.

Stock	1	2	3	4	5	6
Expected annual dividend	$5	$8	$4	$7.50	$30	$40

5 In which of the last four examples (Examples 22–25) should the variables be restricted to integer values?
6 A student is taking five courses and is facing the crunch of final exams. She estimates that she has 40 hours available to study. If x_j = the number of hours allocated to studying for course j, state the equation whose solution set specifies all possible allocations of time among the five courses which will exhaust the 40 hours available.

7 **Product Mix** A firm produces three products. Product A requires 5 hours of produc tion time, product B requires 3.5 hours, and product C requires 7.5 hours for each unit produced. If 240 hours are available during the coming week, determine the equation whose solution set specifies all possible quantities of the three products which can be produced using the 240 hours. What are the maximum quantities which could be produced of *each* product if only one product is produced?

8 **Transportation** A manufacturer distributes its product to four different wholesalers. The monthly capacity is 40,000 units of the product. Decisions need to be made about how many units should be shipped to each of the wholesalers. Determine the equation whose solution set specifies the different quantities which might be shipped if all 40,000 units are to be distributed.

9 **Advertising** A national firm is beginning an advertising campaign using television, radio, and newspapers. The goal is to have 10 million people see the advertisements. Past experience indicates that for every $1,000 allocated to TV, radio, and newspaper advertising, 25,000, 18,000, and 15,000 people, respectively, will see the advertisement. The decisions that need to be made are how much money should be allocated to each form of advertising in order to reach 10 million people. Determine the equation whose solution set specifies all the different advertising allocations which will result in the achievement of this goal. If only one medium is to be used, how much money would have to be spent for each medium to reach 10 million people?

10 **Agricultural Planning** An agricultural company has a goal of harvesting 500,000 bushels of soybeans during the coming year. The company has three farms available to meet this goal. Because of climate differences and other factors, the yields per acre in the different locations are 45, 30, and 36 bushels, respectively, for farms 1, 2, and 3. The decision which needs to be made concerns how many acres should be planted in soybeans at each farm in order to meet the company's goal. State the equation which allows for specifying the different possibilities for meeting the 500,000-bushel goal. If the entire soybean goal is to be achieved using just one farm, how many acres would be required at each farm?

❑ **KEY TERMS AND CONCEPTS**

❑ **IMPORTANT FORMULAS**

$$ax + by = c \qquad \text{Linear equation: two variables} \qquad (2.1)$$

$$a_1 x_1 + a_2 x_2 + \cdots + a_n x_n = b \qquad \text{Linear equation: } n \text{ variables} \qquad (2.2)$$

$$m = \frac{y_2 - y_1}{x_2 - x_1} \quad \text{Two-point formula} \tag{2.10}$$

$$y = mx + k \quad \text{Slope-intercept form of linear equation} \tag{2.12}$$

$$y - y_1 = m(x - x_1) \quad \text{Point-slope formula} \tag{2.13}$$

❏ ADDITIONAL EXERCISES

Section 2.1

In Exercises 1–12, determine whether the equation is linear.

1 $x/3 - y/4 = 2x - y + 12$

2 $(x + 4y)/8 = y$

3 $2/x - 3/y = 24$

4 $0.2x - 0.5y = 10 - 4/x$

5 $x_1 - x_2/3 + 5x_3 = x_4 - 2x_5$

6 $2/(x - 3y) = 10 + x/3$

7 $(x - y + 13)/3 + 5y = -3(x + 12)$

8 $x_1 - 4x_2 + 3x_1 x_3 = 5x_3 - 100$

9 $\sqrt{10} + 10x - 4y = -4$

10 $(x_1 - 6x_2 + 5x_3)/20 = 2/(x_1 - 3x_2)$

11 $\sqrt{x^2 - 2x + 1} + y/2 = 20 - x + 8y$

12 $\sqrt{x^2 - 4x + 4} = \sqrt{y^2 + 6y + 9}$

13 A company manufactures two different products, A and B. Each unit of product A costs $6 to produce, and each unit of product B costs $4. The company insists that total costs for the two products be $500.
 (a) Define the cost equation which states that the total cost of producing x units of product A and y units of product B equals $500.
 (b) Assuming the company has agreed to fill an order for 50 units of product A, how many units of product B should be produced if total costs are to be kept at $500?

14 A local travel agent has been authorized to sell three new vacation packages for a major airline. The three packages are priced at $800, $950, and $1,200, respectively. The airline has promised a sizable bonus commission if total sales by the travel agent equal $100,000 or more. If x_1, x_2, and x_3 equal the number of packages sold of types 1, 2, and 3, respectively:
 (a) Define the equation which states that total sales equal $100,000.
 (b) If the airline specifies that the agent must sell 20 of the $1,200 packages and 10 of the $950 packages in order to qualify for the bonus, how many of the $800 packages will be necessary to qualify?
 (c) One strategy being considered by the agent is to sponsor a charter flight in which all persons would select the same package. Given that three different charters could be planned, how many persons would have to sign up for each in order to qualify for the bonus?

15 Fund-Raising A local theater company is attempting to raise $1 million for expansion of seating capacity. They have undertaken a fund-raising drive to obtain the funds. Their campaign is soliciting donations in three different categories. The "Friend" category requires a donation of $1,000; the "Patron" category requires a donation of $5,000; the "Sustaining Member" category requires a donation of $10,000. If x_j equals the number of donors in category j ($j = 1$ for "Friend"):

(a) Determine the equation which ensures that donations from the three categories equal $1 million.

(b) If the goal is to be realized with donations from only one category of giving, how many donors would be required in each category to provide the entire $1,000,000?

Section 2.2

In Exercises 16–28, identify the x and y intercepts if they exist and graph the equation.

16 $-3x = y/2$

17 $x/3 = -4$

18 $(y - 4)/2 = 4x + 3$

19 $3x - 6y = 0$

20 $4x - 2y = -10$

21 $2x - 3y + 20 = -5x + 2y - 8$

22 $5 - 3x + 6y = -x + 5 - 2y$

23 $5y = 2y + 24$

24 $-6x + 24 = -12 + 3x$

25 $-2x + 3y = -36$

26 $(x - 6y)/2 = -3y + 10$

27 $x + y - 20 = 0$

28 $(2x - 4y)/2 = 10 + (-x + 3y)/3$

In Exercises 29–40, compute the slope of the line segment connecting the two points. Interpret the meaning of the slope.

29 $(5, 2)$ and $(-10, 5)$

30 $(-3, 8)$ and $(1, -14)$

31 $(-b, a)$ and $(-b, 3a)$

32 $(2a, 3b)$ and $(-3a, 3b)$

33 $(4, -5)$ and $(-2, 25)$

34 $(-2, 40)$ and $(3, 75)$

35 $(4.38, 2.54)$ and $(-1.24, 6.32)$

36 $(-15.2, 4.5)$ and $(8.62, -1.6)$

37 (m, n) and $(-m, -n)$

38 $(-2a, 4b)$ and $(4b, -2a)$

39 $(0, t)$ and $(-t, 0)$

40 $(-4, c)$ and $(-4, b)$

Section 2.3

In Exercises 41–52, rewrite each equation in slope-intercept form and determine the slope and y intercept.

41 $2x - 5y + 10 = -4y + 2x - 5$

42 $3x - 8y = 24 + x - 3y$

43 $(x - 4y)/3 = (5x - 2y)/2$

44 $3x - 6y = 36 + x$

45 $8x - 4y = 60 - 3x + y$

46 $x/2 = 20 - y/3$

47 $mx - ny = p$

48 $ax + by = c + dx + ey$

49 $30x - 4y + 24 = 8y + 30x - 12$

50 $-cx + cy = c$

51 $y/2 + 3x - 10 = (x + y)/2$

52 $x - 3y = 3y + 5x - 40$

53 A local dairy association enlists the help of a marketing research firm to predict the demand for milk. The research firm finds that the local demand for milk can be predicted by the equation $q = -4,000p + 10,000$, where p represents the price per quart (in dollars) and q represents the number of quarts purchased per week.
(a) Graph the equation.
(b) Identify the slope and q intercept.
(c) Interpret the meaning of the slope and q intercept in this application.

54 A manufacturing firm has 120 hours per week available in one of its departments. Two products are processed through this department. Product A requires 4 hours per unit and product B 6 hours per unit in this department. If x equals the number of units of product A produced per week and y equals the number of units of product B produced per week:
(a) Determine the equation which states that total time expended for producing these two products equals 120 hours per week.
(b) Rewrite this equation in slope-intercept form and identify the slope and y intercept.
(c) Interpret the meaning of the slope and y intercept in this application.

55 Starting Salaries Average starting salaries for students majoring in business have been increasing. The equation which predicts average starting salary is

$$s = 20,250 + 1,050t$$

where s equals average starting salary and t is time measured in years since 1990 ($t = 0$).
(a) Identify the s and t intercepts for this equation.,
(b) Interpret these values where they are meaningful.

56 A company produces two products. Each product requires a certain amount of a raw material. Product A requires 3 pounds of the raw material and product B 4 pounds. For any given week, the availability of the raw material is 2,400 pounds. If x equals the number of units produced of product A and y the number of units of product B:

(a) Determine the equation which states that total raw material used each week equals 2,400 pounds.

(b) Rewrite the equation in slope-intercept form and identify the slope and y intercept.

(c) Interpret the values of the slope and y intercept.

57 Capital Investments A large car rental agency is preparing to purchase new cars for the coming year. The capital budget for these purchases is $20 million. Two types of cars are to be purchased, one costing $12,000 and the other $14,500. If x equals the number of type 1 cars purchased and y the number of type 2 cars:

(a) Determine the equation which states that the total amount spent on new purchases equals $20 million.

(b) Rewrite the equation in slope-intercept form.

(c) Identify the slope, y intercept, and x intercept and interpret their meaning.

Section 2.4

In Exercises 58–73, use the given information to determine the slope-intercept form of the linear equation.

58 Slope undefined and line passes through $(-3, 5)$

59 Slope undefined and line passes through the origin

60 Slope equals $\frac{1}{2}$, y intercept at $(0, -20)$

61 Slope equals zero, y intercept at $(0, 5)$

62 x intercept at $(4, 0)$ and $(-2, 8)$ lies on line

63 x intercept at $(-3, 0)$ and $(8, -4)$ lies on line

64 $(-3, 6)$ and $(-1, 2)$ lie on line

65 $(-2, -18)$ and $(5, 24)$ lie on line

66 $(-4, 2c)$ and $(10, 2c)$ lie on line

67 $(3a, -5)$ and $(3a, 10)$ lie on line

68 $(-2.38, 10.52)$ and $(1.52, 6.54)$ lie on line

69 $(24.5, -100.6)$ and $(16.2, 36.5)$ lie on line

70 Passes through $(-6, 4)$ and is perpendicular to $3x - 2y = 0$

71 Passes through $(3, 10)$ and is perpendicular to $4x - 2y = -12$

72 Passes through $(-2, 8)$ and is parallel to $-4x + 8y = 20$

73 Passes through $(-4, -1)$ and is parallel to $8x - 2y = 0$

74 An economist believes there is a linear relationship between the market price of a particular commodity and the number of units suppliers of the commodity are willing to bring to the marketplace. Two sample observations indicate that when the price equals $15 per unit, the weekly supply equals 30,000 units; and when the price equals $20 per unit, the weekly supply equals 48,000 units.

(a) If price per unit, p, is plotted on the horizontal axis and the quantity supplied q is plotted on the vertical axis, determine the slope-intercept form of the equation of the line which passes through these two points.

(b) Interpret the slope of the equation in this application.

(c) Predict the weekly supply if the market price equals $25 per unit.

75 High-Tech Athletic Shoes A large sporting goods retailer with multiple stores is attempting to predict the demand for the latest in a seemingly never-ending stream of high-tech basketball shoes, the Nike-Bok Turbo Air-Pump. It is estimated that 300 pairs will be sold per day in the retailer's stores if the new shoe is priced at $200. At a price of $175, 375 pairs are expected to be sold.

(a) If price is plotted on the horizontal axis, determine the slope-intercept form of the equation for demand.

(b) Predict the expected demand at a price of $225. At a price of $160.

(c) Identify the p intercept and interpret its meaning.

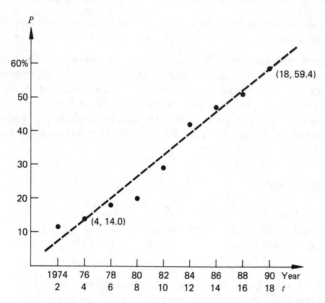

Figure 2.24 Percentage of TV households with cable television.

76 Cable Television Figure 2.24 illustrates some data gathered by Nielsen Media Research regarding the growth of cable television. Observation of the data points indicates that the growth in the percentage of TV households having cable has been approximately linear. Using the data points for 1976 and 1990 to estimate the linear relationship between the percentage P and time t ($t = 0$ corresponds to 1972):

(a) Determine the slope-intercept form of the linear estimating equation.

(b) Interpret the meaning of the slope and P intercept.

(c) Forecast the expected percentage for 1995. For the year 2000.

(d) When is the percentage expected to exceed 80?

Section 2.6

77 A retail store sells four products. Let x_1, x_2, x_3, and x_4 represent the number of units sold, respectively, of the four products. The profits earned from each unit sold of the four products are \$12, \$5, \$8, and \$10, respectively. Target profits for the firm are \$60,000.
   (a) Using x_1, x_2, x_3, and x_4, define an equation which states that total profit from selling the four products equals \$60,000.
   (b) Give the range of values (maximum and minimum) possible for each variable in the equation developed in part a.

78 A woman who has recently inherited \$200,000 decides to invest her inheritance in stocks. She is considering eight stocks, the prices of which are given in the following table.

Stock	1	2	3	4	5	6	7	8
Price per share	\$25	\$50	\$42.50	\$35	\$80	\$17.50	\$120	\$100

Determine the equation whose solution set contains all possible combinations of the eight stocks which can be purchased for \$200,000. (Be sure to define your variables.)

79 **Personnel Management** The head of personnel has been given a budget allotment of \$500,000 to staff an engineering department. Four types of employees are needed: senior engineers at an annual salary of \$60,000 each, junior engineers at an annual salary of \$32,500 each, drafters at an annual salary of \$20,000 each, and secretaries at a salary of \$15,000 each. Write an equation whose solution set contains the possible combinations of employees which could be hired for \$500,000. (Be sure to define your variables.)

❏ CHAPTER TEST

1 Given the equation $8x - 2y = -48$:
   (a) Determine the x and y intercepts.
   (b) Graph the equation.

2 Given the equation $(x + y)/3 = 24 - x$:
   (a) Rewrite the equation in slope-intercept form.
   (b) Identify the slope and y intercept.
   (c) Interpret the meaning of the slope.

3 Given two points $(3, 18)$ and $(5, -14)$:
   (a) Determine the equation of the straight line which passes through the two points.
   (b) Identify the slope, y intercept, and x intercept.

4 The equation $P = 240,000 - 7,500t$ expresses the relationship between the estimated worldwide population P of an exotic bird which has been declared an endangered species and time t measured in years since 1990 ($t = 0$ corresponds to 1990). Identify and interpret the meaning of the slope, P intercept, and t intercept.

5 Determine the equation of the straight line which is perpendicular to the line $3x - 2y = -28$ and which passes through the point $(-5, 20)$.

6 A manufacturer has a monthly supply of 750,000 pounds of a raw material used in making four products. The number of pounds required to manufacture a unit of each product equals 10, 15, 7.5, and 18, respectively. If x_1, x_2, x_3, and x_4 equal the number of units produced of each product:

    (*a*) Define the equation whose solution set includes the possible combinations of the four products which would exhaust the monthly supply of the raw material.

    (*b*) What is the maximum amount which could be made of each product if only one product is produced and the supply of raw material is the only limiting constraint?

CHAPTER 3

SYSTEMS OF LINEAR EQUATIONS

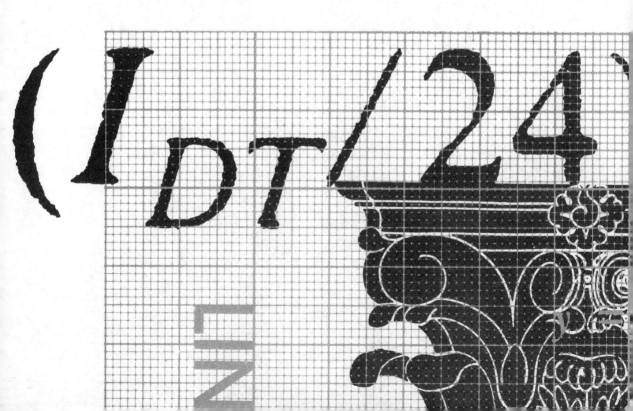

CHAPTER OBJECTIVES

❏ Provide an understanding of the nature of systems of equations and their graphical representation (where appropriate)

❏ Provide an understanding of the different solution set possibilities for systems of equations

❏ Provide an appreciation of the graphical interpretation of solution sets

❏ Present procedures for determining solution sets for systems of equations

❏ Illustrate some applications of systems of linear equations

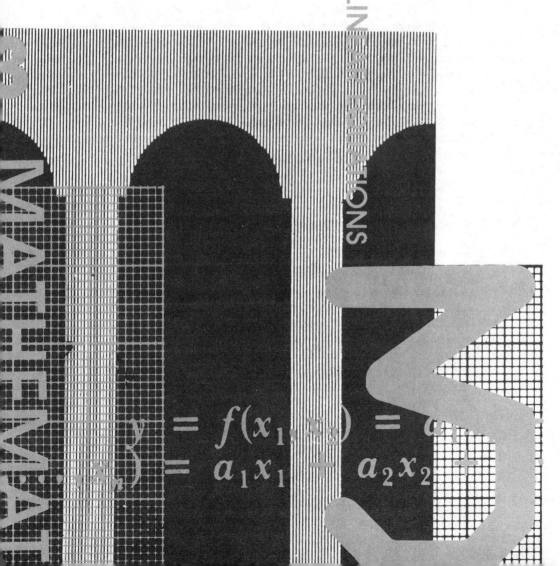

MOTIVATING SCENARIO: Emergency Airlift (continued)	Example 22 in Chap. 2 (page 73) was concerned with airlifting emergency supplies into a South American city. From Example 22 we know that the volume capacity of the plane equals 6,000 cubic feet. Another consideration is that the weight capacity of the plane is 40,000 pounds. In addition, the amount of money available for the purchase of supplies totals $150,000. Initial reports indicate that the most needed item is water. To respond to this need, Red Cross officials have specified that the number of containers of water shipped should be twice the combined number shipped of blood and medical supply kits. *Red Cross officials want to determine if there is some combination of the four items which will fill the plane to its weight and volume capacities, expend the full budget of $150,000, and satisfy the requirement regarding the shipment of water.* [Example 16]

In business, economics, or social science applications, we sometimes are interested in determining whether there are values of variables which satisfy several attributes. It may be that each attribute can be represented by using an equation, expressed in terms of the different variables. Together, the set of equations represents all of the attributes of interest. In this chapter, we will be concerned with the process used to determine whether there are values of variables which jointly satisfy a set of equations. For example, in the Motivating Scenario, we will see whether there are quantities of the four items which satisfy the attributes of weight capacity, volume capacity, budget, and water requirements.

3.1 TWO-VARIABLE SYSTEMS OF EQUATIONS

Systems of Equations

A *system of equations* is a set consisting of more than one equation. One way to characterize a system of equations is by its *dimensions.* If a system of equations consists of m equations and n variables, we say that this system is an *"m by n" system,* or that it has dimensions $m \times n$. A system of equations involving 2 equations and 2 variables is described as having dimensions 2×2. A system consisting of 15 equations and 10 variables is said to be a (15×10) system.

In solving systems of equations, we are interested in identifying values of the variables that satisfy all equations in the system simultaneously. For example, given the two equations

$$5x + 10y = 20$$

$$3x + 4y = 10$$

we may wish to identify any values of x and y which satisfy both equations at the same time. Using set notation, we would want to identify the *solution set S,* where

$$S = \{(x, y)|5x + 10y = 20 \quad \text{and} \quad 3x + 4y = 10\}$$

As you will see in this chapter, the solution set S for a system of linear equations may be a **null set,** a **finite set,** or an **infinite set.***

There are quite a few solution procedures which may be used in solving systems of equations. In this chapter, we concentrate on two different procedures. Other procedures will be presented in Chap. 9.

We begin our discussion in this chapter with the simplest systems, two equations and two variables. Our discussions will emphasize both the graphical and algebraic aspects of each situation. These procedures will be extended later in the chapter to acquaint us with how larger systems of equations are handled. We will also discuss a variety of applications of systems of equations.

Graphical Analysis

From Chap. 2 we know that a linear equation involving two variables graphs as a straight line. Thus a (2×2) system of linear equations is represented by two straight lines in two dimensions. In solving for the values of the two variables which satisfy *both* equations, we are graphically trying to determine whether the two lines have any points in common.

For (2×2) systems of equations three different types of solution sets might exist. Figure 3.1 illustrates the three possibilities. In Fig. 3.1*a*, the two lines intersect. The *coordinates* of the point of intersection (x_1, y_1) represent the solution for the system of equations, i.e., the pair of values for x and y which satisfy *both* equations. When there is just one pair of values for the variables which satisfy the system of equations, the system is said to have a **unique solution.**

In Fig. 3.1*b*, the two lines are parallel to each other. You should remember from Chap. 2 that parallel lines have the same slope; and provided that they have different y intercepts, the lines have no points in common. If a (2×2) system of

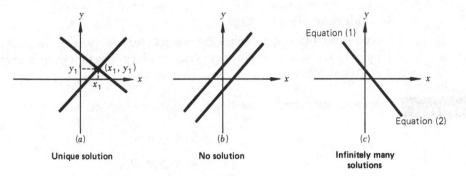

Figure 3.1 Solution set possibilities for a (2×2) system of equations.

---

* A null set contains no elements (it is empty), a finite set consists of a limited number of elements, and an infinite set consists of an infinite number of elements.

equations has these characteristics, the system is said to have **no solution.** That is, there are no values for the variables which satisfy both equations. The equations in such a system are said to be **inconsistent.**

The final possibility for a (2×2) system is illustrated in Fig. 3.1c. In this case both equations graph as the same line, and they are considered to be **equivalent equations.** An infinite number of points are common to the two lines, and the system is said to have **infinitely many solutions.** Being represented by the same line implies that both lines have the same slope *and* the same y intercept. Two equations *can* look very different from each other and still be equivalent to one another. For example, the two equations

$$-6x + 12y = -24$$

and

$$1.5x - 3y = 6$$

are equivalent. Verify that the slope and the y intercept are the same for both.

Another way of summarizing the three cases illustrated in Fig. 3.1 is as follows.

SLOPE-INTERCEPT RELATIONSHIPS

Given a (2×2) system of linear equations (in slope-intercept form),

$$y = m_1 x + k_1 \tag{3.1}$$

$$y = m_2 x + k_2 \tag{3.2}$$

where m_1 and m_2 represent the respective slopes of the two lines and k_1 and k_2 represent the respective y intercepts.

   I   *There is a* unique solution *to the system if $m_1 \neq m_2$.*
  II *There is* no solution *to the system if $m_1 = m_2$ but $k_1 \neq k_2$.*
 III *There are* infinitely many solutions *if $m_1 = m_2$ and $k_1 = k_2$.*

Graphical Solutions

Graphical solution approaches are possible for two-variable systems of equations. However, you must be accurate in your graphics. The following example illustrates a graphical solution.

EXAMPLE 1   Graphically determine the solution to the system of equations

$$2x + 4y = 20 \tag{3.3}$$

$$3x + y = 10 \tag{3.4}$$

The x and y intercepts are, respectively, $(10, 0)$ and $(0, 5)$ for Eq. (3.3). Similarly, the intercepts for Eq. (3.4) are $(\frac{10}{3}, 0)$ and $(0, 10)$. When these are plotted in Fig. 3.2 and connected, the two lines appear to cross at $(2, 4)$.

A problem with graphical solutions is that it may be difficult to read the precise coordinates of the points of intersection. This is especially true when the coordinates are not

Figure 3.2

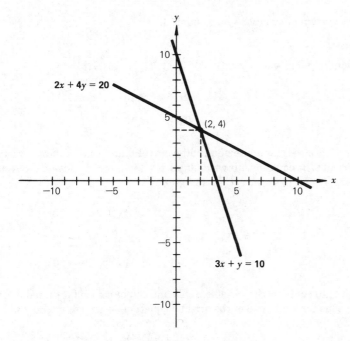

integers. This is why algebraic solution procedures are preferred from the standpoint of identifying *exact* solutions. However, whether you use graphical or algebraic procedures, there is always a check on your answer: Substitute your answer into the original equations to see whether they are satisfied by the values. Substituting $x = 2$ and $y = 4$ into Eqs. (3.3) and (3.4), we get

$$2(2) + 4(4) = 20$$

or

$$20 = 20$$

and

$$3(2) + (4) = 10$$

or

$$10 = 10$$

Therefore our solution is correct.

❑

The Elimination Procedure

One popular solution method for two- and three-variable systems is the ***elimination procedure.*** Given a (2×2) system of equations, the two equations, or multiples of the two equations, are added so as to *eliminate* one of the two variables. The resultant equation is stated in terms of the remaining variable. This equation can be solved for the remaining variable, the value of which can be substituted back into one of the original equations to solve for the value of the eliminated variable. The solution process is demonstrated in the following example, after which the procedure will be formalized.

EXAMPLE 2　　Solve the system of equations in Example 1.

SOLUTION

The original system was

$$2x + 4y = 20 \tag{3.3}$$

$$3x + \ y = 10 \tag{3.4}$$

The objective of the elimination procedure is to eliminate one of the two variables by adding (multiples of) the equations. If we *multiply* Eq. (3.4) by -4 and *add* the resulting equation [Eq. (3.4a)] to Eq. (3.4), we get Eq. (3.5):

$$2x + 4y = \ \ 20 \tag{3.3}$$

$$[-4 \cdot \text{Eq. (3.4)}] \rightarrow \quad \underline{-12x - 4y = -40} \tag{3.4a}$$

$$-10x \qquad = -20 \tag{3.5}$$

Equation (3.5) contains the variable x only and can be solved for the value $x = 2$. Substituting this value for x into one of the original equations—let's select Eq. (3.3)—we find that

$$2(2) + 4y = 20$$

$$4y = 16$$

or

$$y = \ 4$$

Therefore the unique solution to the system, as we determined graphically, is $x = 2$ and $y = 4$.

❏

PRACTICE EXERCISE

Verify that the solution is exactly the same if x is selected for elimination. To eliminate x, multiply Eqs. (3.3) and (3.4) by -3 and 2, respectively.

The elimination procedure can be generalized as follows for a (2×2) system of equations.

ELIMINATION PROCEDURE FOR (2×2) SYSTEMS

I　*Select a variable to eliminate.*

II　*Multiply (if necessary) the equations by constants so that the coefficients on the selected variable are the negatives of one another in the two equations, and add the two resulting equations.*

III　(A)　*If adding the equations results in a new equation having one variable, there is a* **unique solution** *to the system. Solve for the* *value of the remaining variable, and substitute this value back into one of the original equations to determine the value of the variable that was originally eliminated.*

(B) *If adding the equations results in the* **identity** *$0 = 0$, the two original equations are* **equivalent** *to each other and there are* **infinitely many solutions** *to the system.*

(C) *If adding the equations results in a* **false statement,** *say, $0 = 5$, the equations are* **inconsistent** *and there is* **no solution.** *See Fig. 3.3.*

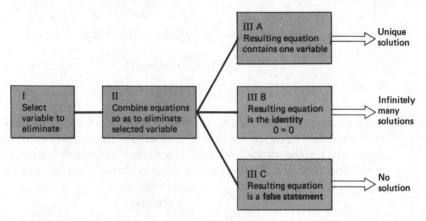

Figure 3.3 Elimination procedure for (2×2) systems.

EXAMPLE 3

(Infinitely Many Solutions) Solve the following system of equations by the elimination procedure.

$$3x - 2y = 6 \tag{3.6}$$

$$-15x + 10y = -30 \tag{3.7}$$

SOLUTION

Choosing x as a variable to eliminate, Eq. (3.6) is multiplied by 5 and added to Eq. (3.7).

$$[5 \cdot \text{Eq. (3.7)}] \rightarrow \qquad 15x - 10y = 30 \tag{3.6a}$$

$$\underline{-15x + 10y = -30} \tag{3.7}$$

$$0 = 0$$

When Eqs. (3.6a) and (3.7) are added, both variables are eliminated on the left side of the equation and we are left with the identity $0 = 0$. From step **IIB** of the solution procedure we conclude that the two equations are equivalent and there are infinitely many solutions.

◻

In order to specify sample members of the solution set, we could assume an arbitrary value for either x or y and substitute this value into one of the original equations, solving for the corresponding value of the other variable. For example, verify that if we let $y = 3$, substitution of this value into either Eq. (3.6) or (3.7) will result in the corresponding value $x = 4$. Thus, one member of the solution set is (4, 3). A more general way of specifying the solution set is to solve either of the original equations for one of the variables. The result is an equation which states the value of one variable in terms of the value of the second variable. To illustrate, if Eq. (3.6) is solved for x, the result is

$$3x = 2y + 6$$

or
$$x = \tfrac{2}{3}y + 2$$

Therefore, one way of generalizing the solution set is

$$\boxed{\begin{array}{l} y \text{ arbitrary} \\[4pt] x = \tfrac{2}{3}y + 2 \end{array}}$$

Very simply, this notation states that y may be assigned *any* real value and the corresponding value for x is obtained by substitution into the equation $x = \tfrac{2}{3}y + 2$. Alternatively, the solution set might be generalized by solving either of the original equations for y. Verify that the resulting generalization would have the form

$$\boxed{\begin{array}{l} x \text{ arbitrary} \\[4pt] y = \tfrac{3}{2}x - 3 \end{array}}$$

EXAMPLE 4   **(No Solution Set)** Solve the following system of equations by the elimination procedure.

$$6x - 12y = 24 \qquad (3.8)$$

$$-1.5x + 3y = 9 \qquad (3.9)$$

SOLUTION

Multiplying Eq. (3.9) by 4 and adding this multiple to Eq. (3.8) yields

$$6x - 12y = 24 \qquad (3.8)$$

$$[4 \cdot \text{Eq. (3.9)}] \rightarrow \quad \underline{-6x + 12y = 36} \qquad (3.9a)$$

$$0x + 0y = 60$$

or
$$0 = 60$$

Since $0 = 60$ is a false statement, there is no solution to the system of equations.

❑

PRACTICE EXERCISE
Rewrite Eqs. (3.8) and (3.9) in slope-intercept form and confirm that they
have the same slope but different y intercepts.

$(m \times 2)$ Systems, $m > 2$

When there are more than two equations ($m > 2$) involving two variables, each
equation still graphs as a line in two dimensions. For example, Fig. 3.4 illustrates
two (3×2) systems. In Fig. 3.4a the three lines all intersect at the same point, and
there is a unique solution. In Fig. 3.4b there are points which are common to
different pairs of lines, but there is no point common to all three, which means that
there is no solution. A possible, but unlikely, situation is that the m equations are
all equivalent to one another and all graph as the same line.

The solution procedure is relatively simple for these systems.

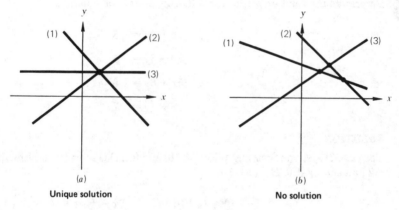

(a) Unique solution

(b) No solution

Figure 3.4 Solution possibilities for (3×2) systems.

SOLUTION PROCEDURE FOR $(m \times 2)$ SYSTEMS, $m > 2$

   I   *Select any two of the m equations and solve simultaneously.*
 II   **(A)** *If in step I there is a unique solution, substitute the values found
         into the remaining equations in the system. If all remaining equa-
         tions are satisfied by these values, they represent a unique solu-
         tion. If the values fail to satisfy any of the remaining equations,
         there is no solution to the system.*
     (B) *If in step I there is no solution, there is no solution for the system.*
     (C) *If in step I there are infinitely many solutions, two different equa-
         tions should be selected and step I should be repeated. See Fig. 3.5.*

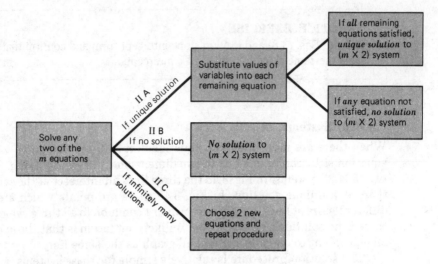

Figure 3.5   Elimination procedure for ($m \times 2$) systems, $m > 2$.

| EXAMPLE 5 |

Determine the solution set for the following system of equations:

$$x + 2y = 8 \tag{3.10}$$

$$2x - 3y = -5 \tag{3.11}$$

$$-5x + 6y = 8 \tag{3.12}$$

$$x + y = 7 \tag{3.13}$$

SOLUTION

The (2×2) system consisting of Eqs. (3.10) and (3.11) is solved by multiplying Eq. (3.10) by -2 and adding it to Eq. (3.11), or

$$[-2 \cdot \text{Eq. (3.10)}] \rightarrow \quad -2x - 4y = -16$$

$$\underline{2x - 3y = -5}$$

$$-7y = -21$$

$$y = 3$$

Substituting back into Eq. (3.10) yields

$$x + 2(3) = 8$$

or

$$x = 2$$

The solution (2, 3) is tested by substituting into Eq. (3.12). Since

$$-5(2) + 6(3) = 8$$

or

$$8 = 8$$

the point (2, 3) satisfies the first three equations. Substituting into Eq. (3.13) gives

$$2 + 3 \neq 7$$

or $$5 \neq 7$$

Since (2, 3) does not satisfy Eq. (3.13), there is no unique solution to the system of equations. Figure 3.6 illustrates the situation. Note that the lines representing Eqs. (3.10)—(3.12) intersect at the point (2, 3); however (2, 3) does not lie on the line representing Eq. (3.13).

❑

Figure 3.6
No solution for
(4 × 2) system.

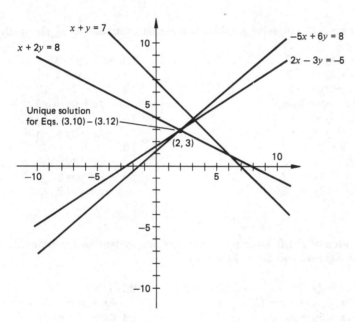

POINT FOR THOUGHT & DISCUSSION

In the solution procedure for ($m \times 2$) systems, $m > 2$, justify from a graphical standpoint why (*a*) the signal of no solution for the selected equations in step IIB would lead us to conclude that there is no solution to the entire system and (*b*) why the signal of an infinite number of solutions for the selected equations in step IIC is inconclusive regarding the solution set, requiring selection of a different pair of equations in step I.

Section 3.1 Follow-up Exercises

In Exercises 1–10, determine the nature of the solution set—unique, infinite, or no solution—by comparing the slope and y coordinates of the y intercepts for the lines representing the two equations.

1 $5x + 5y = 0$
  $x \quad\quad = -y$

2 $2x - 9y = 108$
  $8x + 6y = \quad 48$

3 $4x - 2y = \quad 8$
  $x + 2y = 12$

4 $3x - 9y = 24$
  $-x + 3y = \quad 0$

5 $x - \quad 3y = \quad\quad 8$
  $-4x + 12y = -24$

6 $4x - 2y = 36$
  $-2x + \quad y = 20$

7 $-3x \quad\quad = y + 2$
  $9x + 3y = -6$

8 $x + y = 20$
  $2x - y = 12$

9 $-x \quad\quad = -y$
  $3x + 3y = 0$

10 $4x - \quad y = 10$
  $2x + 3y = 18$

In Exercises 11–20, solve graphically and check your answer algebraically.

11 $2x - 3y = -13$
  $4x + 2y = \quad -2$

12 $3x + 2y = 8$
  $x - \quad y = 1$

13 $-x + 2y = -2$
  $3x \quad\quad = 6y + 6$

14 $x - 2y = 0$
  $-3x + 6y = 5$

15 $3x + 4y = \quad 5$
  $4x + \quad y = -2$

16 $x - 2y = \quad 4$
  $-4x + 8y = -10$

17 $4x - 2y = \quad 10$
  $-2x + \quad y = -5$

18 $x + \quad y = \quad 0$
  $-2x + 3y = 10$

19 $-x + \quad 3y = \quad 2$
  $4x - 12y = -8$

20 $-x + y = 0$
  $2x + y = 9$

Solve each of the following systems. For any system having infinitely many solutions, specify a generalized form of the solution.

21 $4x - 2y = 20$
  $-2x \quad\quad = -y + 15$

22 $4x - \quad y = 17$
  $5x + 3y = \quad 0$

23 $-2x + 5y = 20$
  $4x + \quad y = \quad 4$

24 $6x - \quad 8y = 4$
  $6 \; + 12y = 9x$

25 $2x - \quad y = \quad 9$
  $x + 3y = -6$

26 $2x + 4y = -8$
  $-3x + 2y = \quad 4$

27 $12x - 4y = 18$
  $-4x + \quad y = \quad 6$

28 $2x - \quad y = \quad 4$
  $-6x + 3y = -12$

29 $x - \quad y = 2$
  $2x + \quad y = 1$
  $7x - 5y = 6$

30 $x - 2y = -7$
  $3x + \quad y = \quad 0$
  $2x + 3y = \quad 7$

31 $x + \quad y = \quad 3$
  $2x - \quad y = 12$
  $x - 4y = 13$
  $-2x + 5y = \quad 0$

32 $x + \quad y = \quad 4$
  $2x - 3y = \quad 3$
  $4x - 2y = 10$
  $-x + 3y = \quad 0$

33 $x - \quad y = \quad 1$
  $x + 2y = -8$
  $3x - 2y = \quad 0$
  $2x - 5y = \quad 11$
  $-4x + 3y = -1$

34 $x - \quad y = \quad 8$
  $2x + \quad y = \quad 4$
  $3x + 2y = \quad 4$
  $x + 2y = -4$
  $5x - 2y = \quad 20$

3.2 GAUSSIAN ELIMINATION METHOD

In this section we discuss the gaussian elimination procedure. Although it may appear to be a bit tedious compared with the elimination procedure, it is generalizable to solve systems of any size. In addition, the computational aspects of this procedure are somewhat standardized, making for easy programming and computer implementation.

The General Idea

The *gaussian elimination method* begins with the original system of equations and transforms it, using ***row operations,*** into an equivalent system from which the solution may be read directly. Remember that an *equivalent* system is one which has the same solution set as the original system. Figure 3.7 shows the transformation — i.e., change in form — which is desired in solving a (2 × 2) system. In contrast to the elimination procedure discussed in the last section, the transformed system still has dimensions of 2 × 2. The row operations, however, have transformed the coefficients on the variables so that only one variable remains in each equation; and the value of that variable (v_1 or v_2 in Fig. 3.7) is given by the right side of the equation. Note the coefficients on each variable in the "transformed system."

Figure 3.7
Gaussian elimination
transformation for
(2 × 2) systems.

$$\begin{bmatrix} a_1 x_1 + b_1 x_2 = c_1 \\ a_2 x_1 + b_2 x_2 = c_2 \end{bmatrix} \quad \text{Original system}$$

Gaussian transformation

$$\begin{bmatrix} \textcircled{1} x_1 + \textcircled{0} x_2 = v_1 \\ \textcircled{0} x_1 + \textcircled{1} x_2 = v_2 \end{bmatrix} \quad \text{Transformed system}$$

or

$$\begin{bmatrix} x_1 \qquad = v_1 \\ \qquad x_2 = v_2 \end{bmatrix} \quad \{(v_1, v_2)\} \text{ is the solution set}$$

The following row operations are all that are needed in the gaussian elimination procedure. Given an original system of equations, the application of these operations results in an equivalent system of equations.

BASIC ROW OPERATIONS

   I   *Both sides of an equation may be multiplied by a nonzero constant.*
  II *Nonzero multiples of one equation may be added to another equation.*
 III *The order of equations may be interchanged.*

Let's work a simple example and then generalize and streamline the procedure.

EXAMPLE 6   Solve the following system of equations by the gaussian elimination method:

$$2x - 3y = -7 \tag{3.14}$$

$$x + y = 4 \tag{3.15}$$

SOLUTION

According to Fig. 3.7, we want to transform the given system to have the form

$$1x + 0y = v_1 \qquad\qquad x \quad = v_1$$
$$\qquad\qquad\text{or}$$
$$0x + 1y = v_2 \qquad\qquad y = v_2$$

On the left-hand side of the transformed equations the nonzero variable coefficients appear in a diagonal pattern from the upper left to the lower right. The objective of the gaussian elimination method is to transform the original system into this *diagonal form*. Using row operation I, we can multiply Eq. (3.14) by $\frac{1}{2}$ and the coefficient on the variable x becomes 1. The resulting equivalent system of equations is

$$[\tfrac{1}{2} \cdot \text{Eq. (3.14)}] \rightarrow \qquad 1 x - \tfrac{3}{2}y = -\tfrac{7}{2} \tag{3.14a}$$

$$x + y = 4 \tag{3.15}$$

The coefficient of x can be transformed to zero in Eq. (3.15) by applying row operation II. If Eq. (3.14a) is multiplied by -1 and added to Eq. (3.15), the resulting equivalent system is

$$1 \ x - \tfrac{3}{2}y = -\tfrac{7}{2} \tag{3.14a}$$

$$[-1 \cdot \text{Eq. (3.14a)} + \text{Eq. (3.15)}] \rightarrow \qquad 0 x + \tfrac{5}{2}y = \tfrac{15}{2} \tag{3.15a}$$

Using row operation I, we can multiply Eq. (3.15a) by $+\frac{2}{5}$. The coefficient on y becomes 1 in this equation:

$$1x - \tfrac{3}{2}y = -\tfrac{7}{2} \tag{3.14a}$$

$$[\tfrac{2}{5} \cdot \text{Eq. (3.15a)}] \rightarrow \qquad 0x + 1 y = 3 \tag{3.15b}$$

Finally, the coefficient of y can be transformed to zero in Eq. (3.14a) by applying row operation II. If Eq. (3.15b) is multiplied by $\frac{3}{2}$ and added to Eq. (3.14a), the transformed system is

$$[\tfrac{3}{2} \cdot \text{Eq. (3.15b)} + \text{Eq. (3.14a)}] \rightarrow \qquad 1x + 0 y = 1 \tag{3.14b}$$

$$0x + 1y = 3 \tag{3.15b}$$

There is a reason for carrying these zero coefficients through the transformation. You will see why very shortly. However, when these zero terms are dropped from Eqs. (3.14b) and (3.15b), the final system has the diagonal form

$$\begin{aligned} x \quad &= 1 \\ y &= 3 \end{aligned}$$

which gives the solution to the system.

□

NOTE To create a coefficient of 1 on x in Eq. (3.14), we could have begun the solution process by interchanging Eqs. (3.14) and (3.15) [rule III]. Remember that changing the order of equations has no impact on the solution set.

The Method

The general idea of the gaussian elimination method is to transform an original system of equations into diagonal form by repeatedly applying the three basic row operations. This procedure can be streamlined if we use a type of shorthand notation to represent the system of equations. The approach eliminates the variables and represents a system of equations by using the variable coefficients and right-side constants only. For example, the system of equations

$$\begin{aligned} 2x + 5y &= 10 \\ 3x - 4y &= -5 \end{aligned}$$

would be written as

$$\left.\begin{array}{rr} 2 & 5 \\ 3 & -4 \end{array}\right| \begin{array}{r} 10 \\ -5 \end{array}$$

The vertical line is used to separate the left and right sides of the equations. Each column to the left of the vertical line contains all the coefficients for *one* of the variables in the system.

For the general (2 × 2) system portrayed in Fig. 3.7, the gaussian transformation would appear as in Fig. 3.8. The primary objective is to change the array of coefficients $\begin{pmatrix} a_1 & b_1 \\ a_2 & b_2 \end{pmatrix}$ into the form $\begin{pmatrix} 1 & 0 \\ 0 & 1 \end{pmatrix}$. Although there are variations on the gaussian elimination method, and for any given problem on which you may be tempted to try a shortcut, the following procedure will always work.

$$\left.\begin{array}{cc} a_1 & b_1 \\ a_2 & b_2 \end{array}\right| \begin{array}{c} c_1 \\ c_2 \end{array}$$ Original system

↓ Gaussian transformation

$$\left.\begin{array}{cc} 1 & 0 \\ 0 & 1 \end{array}\right| \begin{array}{c} v_1 \\ v_2 \end{array}$$ Transformed system

Figure 3.8 Coefficient transformation for (2 × 2) system.

A GAUSSIAN ELIMINATION PROCEDURE FOR (2 × 2) SYSTEMS

I   *Given the (2 × 2) system of equations, create an array which contains the variable coefficients and the right-side constants as shown below:*

$$\begin{pmatrix} a_1 & b_1 & \bigm| & c_1 \\ a_2 & b_2 & \bigm| & c_2 \end{pmatrix}$$

II   *Transform the coefficients into diagonal form one column at a time beginning with column 1. First $\begin{pmatrix} a_1 \\ a_2 \end{pmatrix}$ should be transformed into $\begin{pmatrix} 1 \\ 0 \end{pmatrix}$ and then $\begin{pmatrix} b_1 \\ b_2 \end{pmatrix}$ should be transformed into $\begin{pmatrix} 0 \\ 1 \end{pmatrix}$. The process of transforming a column into the desired form is sometimes referred to as pivoting.*

(A)   *In any column transformation, first create the element which equals 1. This is accomplished by multiplying the row (equation) in which the 1 is desired by the reciprocal of the coefficient currently in that position.* If the original element in this position equals zero, first apply row operation III and interchange rows to create a nonzero element in this position. Then, multiply the row by the reciprocal of the coefficient.

(B)   *Create the zero in the column by multiplying the row found in step IIA by the negative of the value currently in the position where the 0 is desired. Add this multiple to the row in which the 0 is desired.*

Let's illustrate step **IIB**, since it tends to be confusing. Assume in the following system

$$\begin{array}{rr|r} 1 & 6 & 10 \\ 5 & 3 & 12 \end{array}$$ (1)
                                                        (2)

we desire a zero where the 5 appears in column 1. We can create the zero by multiplying row 1 by the negative of 5 , or -5, and adding this multiple of row 1 to row 2, or

$$\left.\begin{array}{rr|r} -5 & -30 & -50 \\ 5 & 3 & 12 \\ \hline 0 & -27 & -38 \end{array} \quad \begin{array}{r} -5 \cdot \text{row 1} \\ \text{row 2} \\ \hline \text{new row 2 or 2a} \end{array}\right\} \quad \textbf{Step IIB}$$

The revised system would be

$$\begin{array}{rr|r} 1 & 6 & 10 \\ 0 & -27 & -38 \end{array}$$ (1)
                                                            (2a)

The following examples illustrate the entire procedure.

EXAMPLE 7 Solve the following system by the gaussian elimination method.

$$5x + 20y = \;\;\;25$$
$$4x - \;\;7y = -26$$

SOLUTION

Let's rewrite the system without the variables.

$$\left.\begin{array}{rr|r} 5 & 20 & 25 \\ 4 & -7 & -26 \end{array}\right.\begin{array}{l} R_1 \\ R_2 \end{array}$$

Note the labeling of rows 1 and 2 by R_1 and R_2. This will be convenient for summarizing the row operations used in the transformation process.

A $\textcircled{1}$ is created in column 1 by multiplying row 1 by $\frac{1}{5}$. The new (equivalent) system is

Equivalent system 1

$$\left.\begin{array}{rr|r} \textcircled{1} & 4 & 5 \\ 4 & -7 & -26 \end{array}\right.\begin{array}{l} R_{1a} = \frac{1}{5}R_1 \\ R_2 \end{array} \qquad \text{(Step IIA)}$$

A $\textcircled{0}$ is created in row 2 of column 1 by multiplying row 1 of the new system (R_{1a}) by -4 and adding this row multiple to row 2. The new system is

Equivalent system 2

$$\left.\begin{array}{rr|r} 1 & 4 & 5 \\ \textcircled{0} & -23 & -46 \end{array}\right.\begin{array}{l} R_{1a} \\ R_{2a} = -4R_{1a} + R_2 \end{array} \qquad \text{(Step IIB)}$$

Moving to the second column, a $\textcircled{1}$ is created in row 2 by multiplying that row by $-\frac{1}{23}$. The resulting system is

Equivalent system 3

$$\left.\begin{array}{rr|r} 1 & 4 & 5 \\ 0 & \textcircled{1} & 2 \end{array}\right.\begin{array}{l} R_{1a} \\ R_{2b} = -\frac{1}{23}R_{2a} \end{array} \qquad \text{(Step IIA)}$$

Finally, a $\textcircled{0}$ is created in the second column of row 1 by multiplying the latest (R_{2b}) by -4 and adding this to row 1, or

Equivalent system 4

$$\left.\begin{array}{rr|r} 1 & \textcircled{0} & -3 \\ 0 & 1 & 2 \end{array}\right.\begin{array}{l} R_{1b} = -4R_{2b} + R_{1a} \\ R_{2b} \end{array} \qquad \text{(Step IIB)}$$

The original system has been rewritten in the equivalent diagonal form

$$x = -3$$
$$y = 2$$

which is the solution to the original system.

> **NOTE**   Remember that at any stage in the gaussian elimination process we have a system of equations which is equivalent to the original system. By this, we mean that the modified system of equations has the same solution set as the original system. This is illustrated graphically in Fig. 3.9.

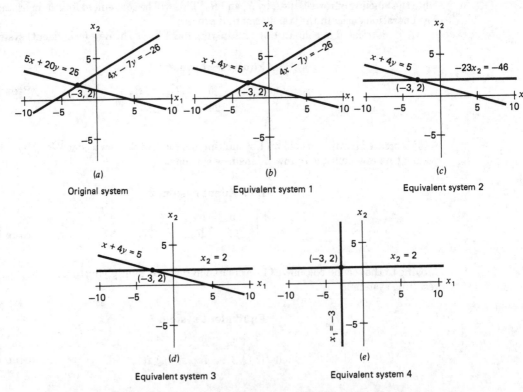

Figure 3.9   Equivalent systems for equations using gaussian elimination method.

EXAMPLE 8

(**Infinitely Many Solutions**) In Example 3 we found that the system

$$3x - 2y = 6$$
$$-15x + 10y = -30$$

had infinitely many solutions. Let's see how this is discovered when using the gaussian elimination method.

First, we rewrite the system as an array

$$\begin{array}{rr|rl} 3 & -2 & 6 & R_1 \\ -15 & 10 & -30 & R_2 \end{array}$$

A $①$ is created in column 1 by multiplying row 1 by $\frac{1}{3}$.

$$\begin{array}{rr|rll} ① & -\frac{2}{3} & 2 & R_{1a} = \frac{1}{3}R_1 & \textbf{(Step IIA)} \\ -15 & 10 & -30 & R_2 & \end{array}$$

A $⓪$ is created in row 2 of column 1 by multiplying the new row 1 by 15 and adding this multiple to row 2. The resulting system is

$$\begin{array}{rr|rll} 1 & -\frac{2}{3} & 2 & R_{1a} & \\ ⓪ & 0 & 0 & R_{2a} = 15R_{1a} + R_2 & \textbf{(Step IIB)} \end{array}$$

Moving to column 2, we wish to create a $①$ in row 2. Note that the current entry is zero. If we interchange rows 1 and 2 to create a nonzero entry in this position, we will undo what we achieved in column 1. We cannot achieve complete diagonalization. If we rewrite this equivalent system with the variables included, we get

$$x - \tfrac{2}{3}y = 2$$

$$0x + 0y = 0$$

Since any ordered pair (x, y) of real numbers satisfies the second equation, the first equation represents the only restriction on the solution set. The original two-equation system has been reduced to an equivalent system containing one equation. The fact that the second equation has been transformed into the identity $0 = 0$ is the signal that there are infinitely many solutions to the original system. As in Example 3, we can solve for x in the first equation and specify the solution set as

$$\boxed{\begin{array}{l} y \text{ arbitrary} \\ x = \tfrac{2}{3}y + 2 \end{array}}$$

EXAMPLE 9

(No Solution) In Example 4 we found that the system

$$6x - 12y = 24$$

$$-1.5x + 3y = 9$$

had no solution. Using the gaussian elimination method, the system is written

$$\begin{array}{rr|rl} 6 & -12 & 24 & R_1 \\ -1.5 & 3 & 9 & R_2 \end{array}$$

A $\textcircled{1}$ is created in column 1 by multiplying row 1 by $\frac{1}{6}$.

$$\begin{array}{rr|r} \textcircled{1} & -2 & 4 \\ -1.5 & 3 & 9 \end{array} \quad \begin{array}{l} R_{1a} = \frac{1}{6}R_1 \\ R_2 \end{array} \qquad \text{(Step IIA)}$$

A $\textcircled{0}$ is created in row 2 of column 1 by multiplying the new row 1 by 1.5 and adding this multiple to row 2.

$$\begin{array}{rr|r} 1 & -2 & 4 \\ \textcircled{0} & 0 & 15 \end{array} \quad \begin{array}{l} R_{1a} \\ R_{2a} = 1.5R_{1a} + R_2 \end{array} \qquad \text{(Step IIB)}$$

As with Example 8, we can go no further in diagonalizing this system. In fact, row 2 in this equivalent system represents a contradiction:

$$0x + 0y = 15$$

or

$$0 = 15$$

This is the same signal as with the elimination procedure discussed earlier. This system is inconsistent, having no solution.

$\square$

Section 3.2 Follow-up Exercises

Determine the solution sets for each of the following systems of equations using the gaussian elimination method.

1 $3x - 2y = 7$
   $2x + 4y = 10$

2 $2x + 4y = -16$
   $x - 2y = 16$

3 $-2x + 5y = 40$
   $3x - 2y = -5$

4 $5x - 2y = -12$
   $-3x + y = 7$

5 $-x + 2y = 4$
   $5x - 10y = -20$

6 $6x - 8y = 14$
   $-3x + 4y = -7$

7 $24x - 15y = 30$
   $-8x + 5y = -20$

8 $-x + 2y = -1$
   $5x - 10y = 6$

9 $5x - 3y = 17$
   $-2x + 5y = -22$

10 $4x - y = 11$
    $3x + 5y = -9$

11 $-x + 2y = -8$
    $3x - 6y = 24$

12 $8x - 6y = 24$
    $-4x + 3y = 10$

13 $8x - 3y = 6$
    $3x + 5y = -10$

14 $5x - 2y = 19$
    $x + 3y = -3$

15 $12x - 6y = 21$
    $-4x + 2y = -7$

16 $2x - 4y = 8$
    $-x + 2y = 10$

17 $x - y = 0$
    $3x + 4y = -21$

18 $x - 5y = 8$
    $3x + y = -8$

19 $3x - 5y = 9$
    $x + 2y = -4$

20 $12x + 20y = -8$
    $-3x - 5y = 2$

3.3 *n*-VARIABLE SYSTEMS, $n \geq 3$

Graphical Analysis for Three-Variable Systems

With three variables each linear equation graphs as a *plane* in three dimensions. In solving a system of three-variable equations, we are looking for any points common to the associated planes. Let's first consider (2 × 3) systems, or those represented by two planes. For (2 × 3) systems there cannot be a unique solution. There is no way in which two planes can intersect at only one point. Think about it! *The solution sets for (2 × 3) systems contain either no elements (no solution) or infinitely many solutions.* Figure 3.10 illustrates different possibilities for these types of systems.

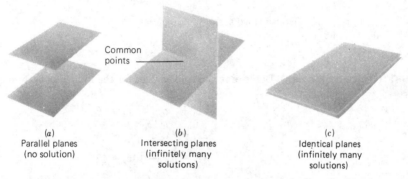

(a)	(b)	(c)
Parallel planes (no solution)	Intersecting planes (infinitely many solutions)	Identical planes (infinitely many solutions)

Figure 3.10   Possible solution sets for (2 × 3) systems.

For (m × 3) systems, where m ≥ 3, it is possible to have a unique solution, no solution, or infinitely many solutions. Figure 3.11 illustrates different solution possibilities for (3 × 3) systems.

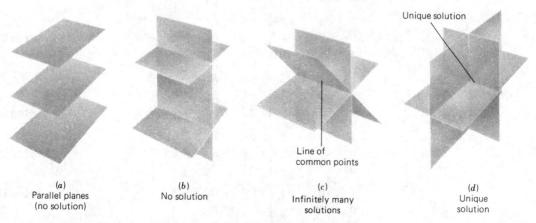

(a)	(b)	(c)	(d)
Parallel planes (no solution)	No solution	Infinitely many solutions	Unique solution

Figure 3.11   Possible solution sets for (3 × 3) systems.

Gaussian Elimination Procedure for (3×3) Systems

The gaussian elimination method for (3×3) systems of equations attempts to transform the system into diagonal form as shown in Fig. 3.12. The transformation should occur column by column moving from left to right. The following example illustrates the procedure.

$$\begin{array}{ccc|c} a_1 & b_1 & c_1 & d_1 \\ a_2 & b_2 & c_2 & d_2 \\ a_3 & b_3 & c_3 & d_3 \end{array} \quad \text{Original system}$$

Figure 3.12 Coefficient transformation for (3×3) system.

Gaussian transformation

$$\begin{array}{ccc|c} 1 & 0 & 0 & v_1 \\ 0 & 1 & 0 & v_2 \\ 0 & 0 & 1 & v_3 \end{array} \quad \text{Transformed system}$$

EXAMPLE 10 **(Unique Solution)** Determine the solution set for the following system of equations

$$\begin{aligned} x_1 + x_2 + x_3 &= 6 \\ 2x_1 - x_2 + 3x_3 &= 4 \\ 4x_1 + 5x_2 - 10x_3 &= 13 \end{aligned}$$

SOLUTION

For this example, the successive transformations will simply be listed with the corresponding row operations indicated to the right of each row.

$$\begin{array}{ccc|c} ① & 1 & 1 & 6 \\ 2 & -1 & 3 & 4 \\ 4 & 5 & -10 & 13 \end{array} \quad \begin{array}{l} R_1 \\ R_2 \\ R_3 \end{array} \qquad \textbf{(Step IIA unnecessary)}$$

$$\begin{array}{ccc|c} 1 & 1 & 1 & 6 \\ ⓪ & -3 & 1 & -8 \\ 4 & 5 & -10 & 13 \end{array} \quad \begin{array}{l} R_1 \\ R_{2a} = -2R_1 + R_2 \\ R_3 \end{array} \qquad \textbf{(Step IIB)}$$

$$\begin{array}{ccc|c} 1 & 1 & 1 & 6 \\ 0 & -3 & 1 & -8 \\ ⓪ & 1 & -14 & -11 \end{array} \quad \begin{array}{l} R_1 \\ R_{2a} \\ R_{3a} = -4R_1 + R_3 \end{array} \qquad \textbf{(Step IIB)}$$

$$\begin{array}{ccc|c} 1 & 1 & 1 & 6 \\ 0 & ① & -\frac{1}{3} & \frac{8}{3} \\ 0 & 1 & -14 & -11 \end{array} \quad \begin{array}{l} R_1 \\ R_{2b} = -\frac{1}{3}R_{2a} \\ R_{3a} \end{array} \qquad \textbf{(Step IIA)}$$

$$\begin{array}{ccc|c} 1 & ⓪ & \frac{4}{3} & \frac{10}{3} \\ 0 & 1 & -\frac{1}{3} & \frac{8}{3} \\ 0 & 1 & -14 & -11 \end{array} \quad \begin{array}{l} R_{1a} = -R_{2b} + R_1 \\ R_{2b} \\ R_{3a} \end{array} \qquad \textbf{(Step IIB)}$$

$$
\begin{array}{ccc|c}
1 & 0 & \frac{4}{3} & \frac{10}{3} \\
0 & 1 & -\frac{1}{3} & \frac{8}{3} \\
0 & \circledcirc & -\frac{41}{3} & -\frac{41}{3}
\end{array}
\quad
\begin{array}{l}
R_{1a} \\
R_{2b} \\
R_{3b} = -R_{2b} + R_{3a}
\end{array}
\qquad \textbf{(Step IIB)}
$$

$$
\begin{array}{ccc|c}
1 & 0 & \frac{4}{3} & \frac{10}{3} \\
0 & 1 & -\frac{1}{3} & \frac{8}{3} \\
0 & 0 & \textcircled{1} & 1
\end{array}
\quad
\begin{array}{l}
R_{1a} \\
R_{2b} \\
R_{3c} = -\frac{3}{41} R_{3b}
\end{array}
\qquad \textbf{(Step IIA)}
$$

$$
\begin{array}{ccc|c}
1 & 0 & \frac{4}{3} & \frac{10}{3} \\
0 & 1 & \circledcirc & 3 \\
0 & 0 & 1 & 1
\end{array}
\quad
\begin{array}{l}
R_{1a} \\
R_{2c} = \frac{1}{3} R_{3c} + R_{2b} \\
R_{3c}
\end{array}
\qquad \textbf{(Step IIB)}
$$

$$
\begin{array}{ccc|c}
1 & 0 & \circledcirc & 2 \\
0 & 1 & 0 & 3 \\
0 & 0 & 1 & 1
\end{array}
\quad
\begin{array}{l}
R_{1b} = -\frac{4}{3} R_{3c} + R_{1a} \\
R_{2c} \\
R_{3c}
\end{array}
\qquad \textbf{(Step IIB)}
$$

The system has a unique solution when $x_1 = 2$, $x_2 = 3$, and $x_3 = 1$.

EXAMPLE 11 **(No Solution)** Determine the solution set for the following system of equations.

$$
\begin{aligned}
-2x_1 + x_2 + 3x_3 &= 12 \\
x_1 + 2x_2 + 5x_3 &= 10 \\
6x_1 - 3x_2 - 9x_3 &= 24
\end{aligned}
$$

SOLUTION

As with the last example, the successive transformations are listed with the corresponding row operations indicated to the right of each row.

$$
\begin{array}{ccc|c}
-2 & 1 & 3 & 12 \\
1 & 2 & 5 & 10 \\
6 & -3 & -9 & 24
\end{array}
\quad
\begin{array}{l}
R_1 \\
R_2 \\
R_3
\end{array}
$$

$$
\begin{array}{ccc|c}
\textcircled{1} & -\frac{1}{2} & -\frac{3}{2} & -6 \\
1 & 2 & 5 & 10 \\
6 & -3 & -9 & 24
\end{array}
\quad
\begin{array}{l}
R_{1a} = -\frac{1}{2} R_1 \\
R_2 \\
R_3
\end{array}
\qquad \textbf{(Step IIA)}
$$

$$
\begin{array}{ccc|c}
1 & -\frac{1}{2} & -\frac{3}{2} & -6 \\
0 & \frac{5}{2} & \frac{13}{2} & 16 \\
6 & -3 & -9 & 24
\end{array}
\quad
\begin{array}{l}
R_{1a} \\
R_{2a} = R_2 - R_{1a} \\
R_3
\end{array}
\qquad \textbf{(Step IIB)}
$$

$$
\begin{array}{ccc|c}
1 & -\frac{1}{2} & -\frac{3}{2} & -6 \\
0 & \frac{5}{2} & \frac{13}{2} & 16 \\
\circledcirc & 0 & 0 & 60
\end{array}
\quad
\begin{array}{l}
R_{1a} \\
R_{2a} \\
R_{3a} = R_3 - 6R_{1a}
\end{array}
\qquad \textbf{(Step IIB)}
$$

At this stage, row 3 in the transformed system has the form

$$
0x_1 + 0x_2 + 0x_3 = 60
$$

No ordered triple of real numbers will satisfy this equation. This *false statement,* or contradiction, signals that there is no solution set for the original system of equations.

❑

PRACTICE EXERCISE

In this example, we did not transform the system as far as we could have. We stopped because of the false statement observed in row 3. Continue the transformation process as far as you can and verify that the system cannot be diagonalized completely.

EXAMPLE 12 (**Infinitely Many Solutions**) Determine the solution set for the system of equations

$$x_1 + x_2 + x_3 = 20$$
$$2x_1 - 3x_2 + x_3 = -5$$
$$6x_1 - 4x_2 + 4x_3 = 30$$

SOLUTION

Notice with this example that we are combining some of the row operations in the transformation process in order to conserve on space. We figure that you are starting to get the knack of it by now.

①	1	1	20	R_1	**(Step IIA unnecessary)**
2	−3	1	−5	R_2	
6	−4	4	30	R_3	
1	1	1	20	R_1	
⓪	−5	−1	−45	$R_{2a} = R_2 - 2R_1$	**(Step IIB)**
⓪	−10	−2	−90	$R_{3a} = R_3 - 6R_1$	**(Step IIB)**
1	1	1	20	R_1	
0	①	$\frac{1}{5}$	9	$R_{2b} = -\frac{1}{5}R_{2a}$	**(Step IIA)**
0	−10	−2	−90	R_{3a}	
1	⓪	$\frac{4}{5}$	11	$R_{1a} = R_1 - R_{2b}$	**(Step IIB)**
0	1	$\frac{1}{5}$	9	R_{2b}	
0	⓪	0	0	$R_{3b} = R_{3a} + 10R_{2b}$	**(Step IIB)**

At this stage, it becomes impossible to continue the diagonalization process. We cannot create a 1 in column 3 without changing the first two columns. The transformed row 3 (R_{3b}) represents the equation

$$0x_1 + 0x_2 + 0x_3 = 0$$

which is satisfied by all ordered triples (x_1, x_2, x_3). The only restrictions placed on the solution are those represented within the first two rows (R_{1a} and R_{2b}). The fact that the third row has been transformed into the identity $0 = 0$ signals that there are infinitely many solutions to the original system.

SPECIFYING SOLUTIONS WITH INCOMPLETE DIAGONALIZATION
Gaussian elimination methods should be applied from left to right to put as many columns as possible in proper form. When complete diagonalization cannot be achieved (and there is no signal indicating no solution), the solution set can be specified as follows:

1 *For any columns not in proper form, the corresponding variables may assume any (arbitrary) values.*
2 *For those columns in proper form, the values of the corresponding variables can be expressed in terms of the variables in step 1.*

In this example, column 3 cannot be transformed into proper form. Therefore, x_3 can be assigned any arbitrary value and the values of x_1 and x_2 stated in terms of that value. The system of equations corresponding to the final gaussian transformation is

$$x_1 \qquad + \tfrac{4}{5}x_3 = 11$$
$$x_2 + \tfrac{1}{5}x_3 = \ 9$$

If we solve these two equations for x_1 and x_2, respectively, we get

$$x_1 = 11 - \tfrac{4}{5}x_3$$
$$x_2 = \ \ 9 - \tfrac{1}{5}x_3$$

Since the values of x_1 and x_2 depend on the value of x_3, a generalized way of specifying the solution to the original system of equations is

$$\boxed{\begin{array}{l} x_3 \text{ arbitrary} \\ x_1 = 11 - \tfrac{4}{5}x_3 \\ x_2 = \ \ 9 - \tfrac{1}{5}x_3 \end{array}}$$

For example, one solution is $(-5, 5, 20)$. By letting $x_3 = 20$,

$$x_1 = 11 - \tfrac{4}{5}(20)$$
$$= 11 - 16$$
$$= -5$$

and

$$x_2 = 9 - \tfrac{1}{5}(20)$$
$$= 9 - 4$$
$$= 5$$

Fewer than Three Equations

In the section on graphical analysis we concluded that a (2×3) system has either no solution or infinitely many solutions. The following examples illustrate solution identification using the gaussian elimination method.

EXAMPLE 13 (No Solution) Determine the solution set for the system of equations

$$-4x_1 + 6x_2 + 2x_3 = 8$$
$$2x_1 - 3x_2 - x_3 = -14$$

SOLUTION

$$
\begin{array}{rrr|rl}
-4 & 6 & 2 & 8 & R_1 \\
2 & -3 & -1 & -14 & R_2 \\
① & -\frac{3}{2} & -\frac{1}{2} & -2 & R_{1a} = -\frac{1}{4}R_1 \\
2 & -3 & -1 & -14 & R_2 \\
1 & -\frac{3}{2} & -\frac{1}{2} & -2 & R_{1a} \\
⓪ & 0 & 0 & -10 & R_{2a} = R_2 - 2R_{1a}
\end{array}
$$

Row 2 in the transformed system is a false statement. No ordered triple will satisfy the equation

$$0x_1 + 0x_2 + 0x_3 = -10$$

This signals that there is no solution to the original system of equations.

EXAMPLE 14 (Infinitely Many Solutions) Determine the solution set for the system

$$2x_1 - 4x_2 - 2x_3 = 6$$
$$-x_1 + 2x_2 + 3x_3 = 9$$

SOLUTION

$$
\begin{array}{rrr|rl}
2 & -4 & -2 & 6 & R_1 \\
-1 & 2 & 3 & 9 & R_2 \\
① & -2 & -1 & 3 & R_{1a} = \frac{1}{2}R_1 \\
-1 & 2 & 3 & 9 & R_2 \\
1 & -2 & -1 & 3 & R_{1a} \\
⓪ & 0 & 2 & 12 & R_{2a} = R_2 + R_{1a}
\end{array}
$$

With column 1 in appropriate diagonal form, it is impossible to transform column 2 to proper form without disturbing column 1. Therefore, we move to column 3 and attempt to transform to proper diagonal form.

$$\begin{array}{ccc|c c}
1 & -2 & -1 & 3 & R_{1a} \\
0 & 0 & \textcircled{1} & 6 & R_{2b} = \frac{1}{2}R_{2a} \\
\\
1 & -2 & \textcircled{0} & 9 & R_{1b} = R_{1a} + R_{2b} \\
0 & 0 & 1 & 6 & R_{2b}
\end{array}$$

This is as far as we can go in diagonalizing. With incomplete diagonalization (and no signal of no solution for the system), we conclude that there are infinitely many solutions. Since we were unable to transform column 2 to the proper diagonal form, the solution set can be generalized from R_{1b} and R_{2b} as

$$\boxed{\begin{aligned}
& x_2 \text{ arbitrary} \\
& x_1 = 9 + 2x_2 \\
& x_3 = 6
\end{aligned}}$$

EXAMPLE 15   **(Infinitely Many Solutions)** Determine the solution set for the system

$$-10x_1 + 25x_2 - 15x_3 = 35$$
$$2x_1 - 5x_2 + 3x_3 = -7$$

SOLUTION

$$\begin{array}{ccc|c c}
-10 & 25 & -15 & 35 & R_1 \\
2 & -5 & 3 & -7 & R_2 \\
\\
\textcircled{1} & -\frac{5}{2} & \frac{3}{2} & -3.5 & R_{1a} = -\frac{1}{10}R_1 \\
2 & -5 & 3 & -7 & R_2 \\
\\
1 & -\frac{5}{2} & \frac{3}{2} & -3.5 & R_{1a} \\
\textcircled{0} & 0 & 0 & 0 & R_{2a} = R_2 - 2R_{1a}
\end{array}$$

This is as far as we can go. Without a signal of no solution, we conclude that there are infinitely many solutions. Because columns 2 and 3 cannot be put into proper form, x_2 and x_3 can be assigned arbitrary values and the value of x_1 expressed in terms of these two. The solution set can be generalized as

$$\boxed{\begin{aligned}
& x_2 \text{ arbitrary} \\
& x_3 \text{ arbitrary} \\
& x_1 = \tfrac{5}{2}x_2 - \tfrac{3}{2}x_3 - 3.5
\end{aligned}}$$

❑

n-Variable Systems, $n > 3$

With more than three variables ($n > 3$), the graphical frame of reference disappears. However, the gaussian elimination procedure is a valid solution method for these systems. For these cases, the possible solution sets are similar to the cases studied for three variables. For example if $m = n$ (the number of variables and equations are equal), it is possible to have a unique solution, infinitely many solutions, or no solution. The indications of each of these cases are exactly the same as with (3×3) systems.

When the number of equations is less than the number of variables $(m < n)$, there will be either no solution or infinitely many solutions. And when the number of equations is greater than the number of variables $(m > n)$, there may be no solution, infinitely many, or a unique solution. The objectives, aspects of interpretation, and general nature of the gaussian elimination procedure are the same for each of these situations. Beyond three-variable systems manual computation procedures are tedious. Computerized solution procedures are readily available to solve larger systems.

Section 3.3 Follow-up Exercises

Determine the solution set for each of the following systems of equations. For any system having infinitely many solutions, specify a generalized form of the solution.

1 $-2x_1 + x_2 + 3x_3 = 10$
$10x_1 - 5x_2 - 15x_3 = 30$
$x_1 + x_2 - 3x_3 = 25$

2 $x_1 + x_2 + x_3 = 2$
$x_1 - 3x_2 + 2x_3 = 7$
$4x_1 - 2x_2 - x_3 = 9$

3 $x_1 - x_2 + x_3 = -5$
$3x_1 + x_2 - x_3 = 25$
$2x_1 + x_2 + 3x_3 = 20$

4 $-4x_1 - 12x_2 + 4x_3 = -40$
$x_1 + x_2 - 6x_3 = 10$
$x_1 + 3x_2 - x_3 = 10$

5 $x_1 - 3x_2 + x_3 = 2$
$2x_1 - 4x_2 + 3x_3 = 7$
$-3x_1 + x_2 + 2x_3 = 9$

6 $x_1 + x_2 + x_3 = 0$
$3x_1 - x_2 + 2x_3 = -1$
$x_1 + 2x_2 + 3x_3 = -5$

7 $x_1 + x_2 + x_3 = 3$
$2x_1 - x_2 + 3x_3 = 13$
$3x_1 - 2x_2 + x_3 = 17$

8 $2x_1 + 4x_2 - 2x_3 = 10$
$3x_1 - x_2 + 4x_3 = 12$
$-x_1 - 2x_2 + x_3 = 0$

9 $5x_1 - 4x_2 + 6x_3 = 24$
$3x_1 - 3x_2 + x_3 = 54$
$-2x_1 + x_2 - 5x_3 = 30$

10 $-x_1 + 3x_2 + x_3 = 7$
$3x_1 - 9x_2 - 3x_3 = 14$
$4x_1 + 2x_2 - 2x_3 = 24$

11 $2x_1 + x_2 - 2x_3 = 3$
$3x_1 - x_2 - 2x_3 = 4$
$x_1 + x_2 + x_3 = 6$

12 $4x_1 + 2x_2 - 5x_3 = 13$
$x_1 + x_2 + x_3 = 2$
$2x_1 - x_2 - 3x_3 = 3$

13 $10x_1 + 5x_2 - 15x_3 = 60$
$6x_1 + 4x_2 + x_3 = 48$
$-4x_1 - 2x_2 + 6x_3 = -36$

14 $x_1 - x_2 + x_3 = 10$
$-3x_1 + x_2 - 2x_3 = 17$
$-4x_1 + 2x_2 - 3x_3 = 7$

15 $3x_1 - 6x_2 + 3x_3 = -30$
$-5x_1 + 10x_2 - 5x_3 = 50$

16 $-2x_1 + 4x_2 - 2x_3 = 20$
$x_1 - 2x_2 + x_3 = 30$

17 $8x_1 - 4x_2 + 16x_3 = 50$
$-2x_1 + x_2 - 4x_3 = 20$

18 $x_1 + x_2 + x_3 = 25$
$-x_1 + 3x_2 + x_3 = 15$

19 $3x_1 - x_2 + 2x_3 = -3$
$-15x_1 + 5x_2 - 10x_3 = 15$

20 $-x_1 + 2x_2 + x_3 = -4$
$4x_1 - 8x_2 - 4x_3 = 10$

21 What solution set possibilities exist for (a) a (5×3) system of equations, (b) a (4×8) system, (c) a (25×25) system, (d) a (100×75) system, and (e) a $(4,000 \times 1,000)$ system?

22 What solution set possibilities exist for (a) a (30×40) system of equations, (b) a $(2,500 \times 1,000)$ system, (c) a (600×30) system, (d) a $(450 \times 1,200)$ system, and (e) a (75×75) system?

3.4 SELECTED APPLICATIONS

EXAMPLE 16

(Emergency Airlift; Motivating Scenario) The motivating scenario at the beginning of this chapter discussed the emergency airlift of supplies to a South American city which had experienced an earthquake. Table 3.1 indicates the four items being considered for the first airplane to be sent to the city as well as the volume, weight, and cost per container of each item.

TABLE 3.1 Item	Volume per Container, ft³	Weight per Container, lb	Cost per Container, $
Blood	20	150	1,000
Medical supply kits	30	100	300
Food	8	60	400
Water	6	70	200

Recall from Example 22 (Chap. 2) that the volume capacity of the plane is 6,000 cubic feet. The weight capacity is 40,000 pounds. In addition, the amount of money available for the purchase of supplies totals $150,000. Initial reports indicate that the most needed item is water. To respond to this need, Red Cross officials have specified that the number of containers of water shipped should be twice the combined number shipped of blood and medical supply kits. *Red Cross officials want to determine if there is some combination of the four items which will fill the plane to its weight and volume capacities, expend the full budget of $150,000, and satisfy the requirement regarding the shipment of water.*

SOLUTION
If

$$x_1 = number\ of\ containers\ of\ blood$$

$$x_2 = number\ of\ containers\ of\ medical\ supply\ kits$$

$$x_3 = number\ of\ containers\ of\ food$$

$$x_4 = number\ of\ containers\ of\ water$$

the system of equations which represents the requirements in this problem are

$$20x_1 + 30x_2 + 8x_3 + 6x_4 = 6,000 \quad \text{(volume)}$$

$$150x_1 + 100x_2 + 60x_3 + 70x_4 = 40,000 \quad \text{(weight)}$$

$$1,000x_1 + 300x_2 + 400x_3 + 200x_4 = 150,000 \quad \text{(funds)}$$

$$x_4 = 2(x_1 + x_2) \quad \text{(water)}$$

Before solving this (4×4) system of equations, we are going to make the following changes:

1. The water requirement equation is rewritten with x_1 and x_2 brought to the left side of the equation.
2. The rearranged water equation will be positioned as the first of the four equations.

The resulting system of equations, written in array form, is

$$
\begin{array}{rrrr|r}
-2 & -2 & 0 & 1 & 0 \\
20 & 30 & 8 & 6 & 6{,}000 \\
150 & 100 & 60 & 70 & 40{,}000 \\
1{,}000 & 300 & 400 & 200 & 150{,}000
\end{array}
$$

To reduce the magnitude of some of the numbers, the third and fourth equations are divided by 10 and 100, respectively, to yield

$$
\begin{array}{rrrr|r}
-2 & -2 & 0 & 1 & 0 \\
20 & 30 & 8 & 6 & 6{,}000 \\
15 & 10 & 6 & 7 & 4{,}000 \\
10 & 3 & 4 & 2 & 1{,}500
\end{array}
\quad
\begin{array}{l}
R_1 \\
R_2 \\
R_3 \\
R_4
\end{array}
$$

$$
\begin{array}{rrrr|r}
1 & 1 & 0 & -0.5 & 0 \\
0 & 10 & 8 & 16 & 6{,}000 \\
0 & -5 & 6 & 14.5 & 4{,}000 \\
0 & -7 & 4 & 7 & 1{,}500
\end{array}
\quad
\begin{array}{l}
R_{1a} = -\frac{1}{2}R_1 \\
R_{2a} = R_2 - 20R_{1a} \\
R_{3a} = R_3 - 15R_{1a} \\
R_{4a} = R_4 - 10R_{1a}
\end{array}
$$

$$
\begin{array}{rrrr|r}
1 & 1 & 0 & -0.5 & 0 \\
0 & 1 & 0.8 & 1.6 & 600 \\
0 & -5 & 6 & 14.5 & 4{,}000 \\
0 & -7 & 4 & 7 & 1{,}500
\end{array}
\quad
\begin{array}{l}
R_{1a} \\
R_{2b} = \frac{1}{10}R_{2a} \\
R_{3a} \\
R_{4a}
\end{array}
$$

$$
\begin{array}{rrrr|r}
1 & 0 & -0.8 & -2.1 & -600 \\
0 & 1 & 0.8 & 1.6 & 600 \\
0 & 0 & 10 & 22.5 & 7{,}000 \\
0 & 0 & 9.6 & 18.2 & 5{,}700
\end{array}
\quad
\begin{array}{l}
R_{1b} = R_{1a} - R_{2b} \\
R_{2b} \\
R_{3b} = R_{3a} + 5R_{2b} \\
R_{4b} = R_{4a} + 7R_{2b}
\end{array}
$$

$$
\begin{array}{rrrr|r}
1 & 0 & -0.8 & -2.1 & -600 \\
0 & 1 & 0.8 & 1.6 & 600 \\
0 & 0 & 1 & 2.25 & 700 \\
0 & 0 & 9.6 & 18.2 & 5{,}700
\end{array}
\quad
\begin{array}{l}
R_{1b} \\
R_{2b} \\
R_{3c} = \frac{1}{10}R_{3b} \\
R_{4b}
\end{array}
$$

$$
\begin{array}{rrrr|r}
1 & 0 & 0 & -0.3 & -40 \\
0 & 1 & 0 & -0.2 & 40 \\
0 & 0 & 1 & 2.25 & 700 \\
0 & 0 & 0 & -3.4 & -1{,}020
\end{array}
\quad
\begin{array}{l}
R_{1c} = R_{1b} + 0.8R_{3c} \\
R_{2c} = R_{2b} - 0.8R_{3c} \\
R_{3c} \\
R_{4c} = R_{4b} - 9.6R_{3c}
\end{array}
$$

$$
\begin{array}{rrrr|r}
1 & 0 & 0 & -0.3 & -40 \\
0 & 1 & 0 & -0.2 & 40 \\
0 & 0 & 1 & 2.25 & 700 \\
0 & 0 & 0 & 1 & 300
\end{array}
\quad
\begin{array}{l}
R_{1c} \\
R_{2c} \\
R_{3c} \\
R_{4d} = -\dfrac{1}{3.4}R_{4c}
\end{array}
$$

$$\begin{array}{cccc|c} 1 & 0 & 0 & 0 & 50 \\ 0 & 1 & 0 & 0 & 100 \\ 0 & 0 & 1 & 0 & 25 \\ 0 & 0 & 0 & 1 & 300 \end{array} \quad \begin{array}{l} R_{1d} = R_{1c} + 0.3R_{4d} \\ R_{2d} = R_{2c} + 0.2R_{4d} \\ R_{3d} = R_{3c} - 2.25R_{4d} \\ R_{4d} \end{array}$$

The solution to the system of equations is $x_1 = 50$, $x_2 = 100$, $x_3 = 25$, and $x_4 = 300$. The mathematical recommendation is that Red Cross officials place 50 containers of blood, 100 containers of medical supply kits, 25 containers of food, and 300 containers of water on the first plane.

❑

Product Mix Problem

A variety of applications are concerned with determining the quantities of different products which satisfy specific requirements. In the following example we are interested in determining the quantities of three products which will fully utilize available production capacity.

EXAMPLE 17 A company produces three products, each of which must be processed through three different departments. Table 3.2 summarizes the hours required per unit of each product in each department. In addition, the weekly capacities are stated for each department in terms of work-hours available. What is desired is to determine whether there are any combinations of the three products which would exhaust the weekly capacities of the three departments.

TABLE 3.2

Department	Product 1	Product 2	Product 3	Hours Available per Week
A	2	3.5	3	1,200
B	3	2.5	2	1,150
C	4	3	2	1,400

SOLUTION

If we let x_j = *number of units produced per week of product j*, the conditions to be satisfied are expressed by the following system of equations.

$$2x_1 + 3.5x_2 + 3x_3 = 1{,}200 \quad \text{(department } A)$$

$$3x_1 + 2.5x_2 + 2x_3 = 1{,}150 \quad \text{(department } B)$$

$$4x_1 + 3x_2 + 2x_3 = 1{,}400 \quad \text{(department } C)$$

❑

PRACTICE EXERCISE

Verify that by solving these equations simultaneously, the solution set consists of one solution, which is $x_1 = 200$, $x_2 = 100$, and $x_3 = 150$, or (200, 100, 150). Interpret the solution set for the production supervisor of this company.

Blending Model

Some applications involve the mixing of ingredients or components to form a final blend having specific characteristics. Examples include the blending of gasoline and other petroleum products, the blending of coffee beans, and the blending of whiskeys. Very often the blending requirements and relationships are defined by linear equations or linear inequalities. The following example illustrates a simple application.

EXAMPLE 18   A coffee manufacturer is interested in blending three different types of coffee beans into a final coffee blend. The three component beans cost the manufacturer $1.20, $1.60, and $1.40 per pound, respectively. The manufacturer wants to blend a batch of 40,000 pounds of coffee and has a coffee-purchasing budget of $57,600. In blending the coffee, one restriction is that the amount used of component 2 should be twice that of component 1 (the brewmaster believes this to be critical in avoiding a bitter flavor).

The objective is to determine whether there is a combination of the three components which will lead to a final blend (1) consisting of 40,000 pounds, (2) costing $57,600, and (3) satisfying the blending restriction on components 1 and 2.

If x_j equals the number of pounds of component j used in the final blend, Eq. (3.16) specifies that the total blend should weigh 40,000 pounds:

$$x_1 + x_2 + x_3 = 40{,}000 \tag{3.16}$$

Equation (3.17) specifies that the total cost of the three components should equal $57,600:

$$1.20x_1 + 1.60x_2 + 1.40x_3 = 57{,}600 \tag{3.17}$$

The recipe restriction is stated as

$$x_2 = 2x_1$$

or alternatively,
$$-2x_1 + x_2 = 0 \tag{3.18}$$

❑

PRACTICE EXERCISE
Verify that when Eqs. (3.16)–(3.18) are solved simultaneously, the solution is $x_1 = 8{,}000$, $x_2 = 16{,}000$, and $x_3 = 16{,}000$. Interpret this solution for the brewmaster.

Portfolio Model

A *portfolio* of stocks is the set of stocks owned by an investor. In selecting the portfolio for a particular investor, consideration is often given to such factors as the amount of money to be invested, the attitude the investor has about risk (is he or she a risk taker?), and whether the investor is interested in long-term growth or short-run return. This type of problem is similar to the product mix example. The products are the stocks or securities available for investment.

EXAMPLE 19

When people invest money, there are professionals, such as stockbrokers, who may be consulted for advice about the portfolio which best meets an investor's needs. Suppose that an investor has consulted with a local investment expert. After talking with the client, the investment expert determines that the client is interested in a portfolio which will have the following attributes: (1) total value of the portfolio at the time of purchase is $50,000, (2) expected annual growth in market value equals 12 percent, and (3) average risk factor is 10 percent. Three investment alternatives have been identified with relative growth and risk rates as shown in Table 3.3.

TABLE 3.3

Investment	Expected Annual Growth in Market Value	Expected Risk
1	16%	12%
2	8	9
3	12	8

To determine the portfolio, *let's define x_j as the number of dollars invested in investment j.* The first attribute can be stated in equation form as

$$x_1 + x_2 + x_3 = 50,000 \tag{3.19}$$

Attribute (2) is a little more difficult to formulate. Let's precede the formulation by looking at a simple example. Suppose that you put $100 in a bank and it earns interest of 6 percent per year. Also suppose you put $200 in a certificate of deposit which earns interest at a rate of 8 percent per year. To determine the *average* percent return on your $300 investment, we must compute total interest and divide by the original investment, or

$$\text{Average percent return} = \frac{\text{dollars of interest earned}}{\text{total dollars invested}}$$

For this example, the average annual percent return is computed as

$$\frac{0.06(100) + 0.08(200)}{100 + 200} = \frac{6 + 16}{300} = \frac{22}{300} = 0.0733, \quad \text{or} \quad 7.33\%$$

To compute the average percent growth in our example, we must determine the annual interest (in dollars) for each investment, sum these, and divide by the total dollars invested, or

$$\text{Average percent growth} = \frac{0.16x_1 + 0.08x_2 + 0.12x_3}{x_1 + x_2 + x_3}$$

Since Eq. (3.19) specifies that $x_1 + x_2 + x_3 = 50,000$ and since the investor desires an average percent growth of 12 percent, we can rewrite the equation as

$$\frac{0.16x_1 + 0.08x_2 + 0.12x_3}{50,000} = 0.12$$

or, multiplying both sides of the equation by 50,000, we get

$$0.16x_1 + 0.08x_2 + 0.12x_3 = 6,000 \tag{3.20}$$

This equation states that the total annual increase in market value for the three investments must equal \$6,000 (or 12 percent of \$50,000).

The weighted risk condition is determined in exactly the same manner. To calculate average risk per dollar invested, each dollar must be multiplied by the risk factor associated with the investment of that dollar. These must be summed for all different investments and divided by the total investment. This relationship is generalized by the equation.

$$\text{Average risk} = \frac{\substack{\text{sum of weighted risks} \\ \text{for all investments}}}{\text{total dollars invested}}$$

This equation can be stated in our example as

$$\frac{0.12x_1 + 0.09x_2 + 0.08x_3}{50,000} = 0.10$$

or

$$0.12x_1 + 0.09x_2 + 0.08x_3 = 5,000 \tag{3.21}$$

❑

PRACTICE EXERCISE
Verify that when Eqs. (3.19)–(3.21) are solved simultaneously, $x_1 = 20,000$, $x_2 = 20,000$, and $x_3 = 10,000$. Interpret this solution for the investor.

Section 3.4 Follow-Up Exercises

1 A company produces three products, each of which must be processed through three departments. Table 3.4 summarizes the labor-hours required per unit of each product in each department. The monthly labor-hour capacities for the three departments are 1,800, 1,450, and 1,900 hours, respectively. Determine whether there is a combination of the three products which could be produced monthly so as to consume the full monthly labor availabilities of all departments.

TABLE 3.4	Department	Product 1	Product 2	Product 3
	A	3	2	5
	B	4	1	3
	C	2	4	1

2 A company produces three products, each of which must be processed through three different departments. Table 3.5 summarizes the hours required per unit of each product in each department. Monthly labor-hour capacities for the three departments are 1,600, 800, and 1,800 hours, respectively. Determine whether there is a combination of the three products which could be produced monthly to consume all of the labor-hours in each department.

TABLE 3.5	Department	Product 1	Product 2	Product 3
	A	4	5	2
	B	3	2	3
	C	1	4	2

3 A company produces three products, each of which must be processed through one department. Table 3.6 summarizes the labor-hour and raw material requirements per unit of each product. Each month there are 1,500 labor-hours and 3,800 pounds of the raw material available. If combined monthly production for the three products should equal 500 units, determine whether there are any combinations of the three products which would exhaust the monthly availabilities of labor and raw material and meet the production goal of 500 units.

TABLE 3.6		Product		
		1	2	3
	Labor-hours/unit	3	2	4
	Pounds of raw material/unit	10	8	6

4 A company produces three products, each of which must be processed through one department. Table 3.7 summarizes the labor-hour and raw material requirements per unit of each product. Each month there are 1,300 labor-hours and 4,700 pounds of raw material available. If combined monthly production for the three products should equal 400 units, determine whether there are any combinations of the three products which would exhaust the monthly availabilities of labor and raw material and meet the production goal of 400 units.

TABLE 3.7		Product		
		1	2	3
	Labor-hours/unit	5	2	4
	Pounds of raw material/unit	15	10	12

5 A blending process is to combine three components in such a way as to create a final blend of 60,000 gallons. The three components cost $2.00, $1.50, and $1.25 per gallon, respectively. Total cost of the components should equal $90,000. Another requirement in the blending is that the number of gallons used of component 1 should be twice the amount used of component 3. Determine whether there is a combination of the three components which will lead to a final blend of 60,000 gallons costing $90,000 and satisfying the blending restrictions.

6 An investor has $500,000 to spend. Three investments are being considered, each having an expected annual interest rate. The interest rates are 15, 10, and 18 percent, respectively. The investor's goal is an average return of 15 percent in the three investments. Because of the high return on investment alternative 3, the investor wants the amount in this alternative to equal 40 percent of the total investment. Determine whether there is a meaningful investment strategy which will satisfy these requirements.

7 Diet Mix Problems A dietitian is planning the menu for the evening meal at a university dining hall. Three main items, each having different nutritional content, will be served. The goal is that the nutritional content of the meal meet the minimum daily levels for three different vitamins. Table 3.8 summarizes the vitamin content *per ounce* of each food. In addition, the minimum daily requirements (MDRs) of the three vitamins are indicated. Determine the number of ounces of each food to be included in the meal such that minimum daily requirements levels are met for the three vitamins.

TABLE 3.8			mg/oz		
Vitamin	MDR	Food 1	Food 2	Food 3	
1	29	5	3	2	
2	20	2	1	3	
3	21	1	5	2	

8 Bacteria Culture A bacteria culture contains three types of bacteria. Each type requires certain amounts of carbon, phosphate, and nitrogen to survive. The daily requirements are shown in Table 3.9. Each day the culture is supplied with 100,000 units of a carbon source, 135,000 units of a phosphate source, and 230,000 units of a nitrogen source. Determine how many units of each type of bacteria can be supported in the culture.

TABLE 3.9	Bacteria Type	Carbon, Units/Day	Phosphate, Units/Day	Nitrogen, Units/Day
	A	2	4	3
	B	3	1	5
	C	6	2	8

3.5 FINAL NOTES

As we conclude our discussion of systems of linear equations, a few observations should be offered:

❏ *Systems of equations are very specific in their requirements.* Indeed, there are many applications in which the relationships of interest are strict equalities. However, we will see in later chapters that many applications involve relationships which are less restrictive, being represented mathematically by inequalities. For example, we presented examples and exercises in this chapter which stated conditions requiring that all labor-hours in a set of departments or all raw material resources be consumed in producing a set of products. Similarly, we looked at applications which stated that the total budget in a program be expended. In many applications, these relationships would be stated as inequalities. For resources in a production process, the requirement might be stated as

Quantity of resource used ≤ quantity of resource available

Similarly, a budget requirement might be stated as the inequality

Amount spent ≤ *amount available*

❏ *In an actual application, there may be no solution capable of implementation.* The requirements of a system of equations may be too specific to be satisfied. This will be indicated by either (1) a no-solution signal by the solution method or (2) a solution which contains values which are not feasible in the application (e.g., $x_1 = -500$, where x_1 equals the number of units produced of a product).

❏ *Fractional or decimal values for variables may be a problem in an application.* Although the solutions for many of the examples and exercises in this chapter conveniently involved integer values for the variables in a system of equations, *a more likely outcome* is the occurrence of fractional or decimal values. This could be a problem for the implementation of the mathematical result. If the decision variable represents something which is easily divisible, decimal results may not be a problem. For example, if a mathematical result of $x_2 = 10.23$ represents the number of gallons or pounds of some component to use in a blending process, the mathematical result may be easily implemented. On the other hand, if x_2 represents the recommended number of Boeing 747s that an airline should purchase, the mathematical result cannot be implemented. The mathematical result must be examined in light of the realities of the situation and a recommendation made which is capable of implementation.

❏ KEY TERMS AND CONCEPTS

basic row operations 97	infinitely many solutions 88
dimensions 86	no solution 88
elimination procedure 89	solution set 86
equivalent equations 97	system of equations 86
gaussian elimination method 97	unique solution 87
inconsistent equations 88	

❏ ADDITIONAL EXERCISES

Section 3.1

In Exercises 1–10, determine the nature of the solution set (unique, infinite, or no solution) by comparing the slopes and y intercepts of the corresponding lines.

1 $3x - 4y = 20$
   $-9x + 12y = -40$

2 $-x + 3y = -4$
   $5x - 15y = -20$

3 $4x - 2y = 18$
   $2x + y = 10$

4 $2x + 3y = -24$
   $5x - 4y = 36$

5 $x - 2y = 0$
  $3x + 4y = 0$

6 $16x - 4y = 24$
  $-4x + y = 10$

7 $x - 2y = 10$
  $-4x + 8y = -6$

8 $3x - 4y = 0$
  $-6x + 8y = 0$

9 $-12x + 2y = -48$
  $6x - y = 24$

10 $8x - 3y = 60$
  $20x + 8y = -100$

Solve the following systems of equations. For any systems having infinitely many solutions, specify the generalized form of the solution.

11 $4x - 2y = -40$
  $3x + 4y = 25$

12 $x + y = 6$
  $3x - 2y = 3$

13 $6x + 3y = 6$
  $2x + 4y = 14$

14 $5x - 2y = 18$
  $3x + y = 2$

15 $2x - 3y = -1$
  $-10x + 15y = 5$

16 $x - 2y = -8$
  $-4x + 8y = 10$

17 $x + y = 1$
  $3x - 2y = 18$
  $-x + 3y = -13$
  $5x - y = 23$
  $-x - 4y = 8$

18 $x + y = 0$
  $2x - 3y = -10$
  $x + 2y = 2$
  $-5x + y = 12$
  $3x - 2y = 10$

19 $3x + y = 10$
  $x - y = -2$
  $2x + 3y = 16$
  $x + y = 4$

20 $x + y = -1$
  $2x - 3y = 8$
  $x + 2y = -3$
  $4x - 3y = 10$

Section 3.2

Solve the following systems using the gaussian elimination method.

21 $-3x + 2y = 22$
  $5x + 4y = 0$

22 $5x - 8y = 1$
  $4x + 2y = 26$

23 $5x - 3y = -2$
  $-25x + 15y = 10$

24 $x - 2y = 4$
  $-5x + 10y = 10$

25 $3x - 4y = 8$
  $-2x + 2y = -6$

26 $4x - 8y = -32$
  $-x + 2y = 10$

27 $15x - 3y = 24$
  $-5x + y = -8$

28 $x - y = 3$
  $5x + 2y = -20$

29 $6x - y = 26$
  $2x + 3y = 2$

30 $4x + 5y = -5$
  $6x - 3y = 45$

31 $-x + 2y = -4$
  $5x - 10y = 20$

32 $-2x + y = -3$
  $12x - 6y = 14$

33 $7x - 4y = 1$
  $4x + 2y = -8$

34 $8x - 3y = 49$
  $2x + 3y = 1$

Section 3.3

Solve the following systems using gaussian elimination.

35 $2x_1 - x_2 + x_3 = -2$
  $x_1 + 4x_2 - x_3 = 5$
  $x_1 + x_2 + x_3 = 6$

36 $x_1 + x_2 + x_3 = 1$
  $3x_1 - 2x_2 + x_3 = -1$
  $x_1 + 3x_2 - x_3 = 11$

37 $\begin{aligned} 3x_1 - x_2 + 2x_3 &= 6 \\ x_1 + x_2 + x_3 &= 10 \\ -9x_1 + 3x_2 - 6x_3 &= -18 \end{aligned}$

38 $\begin{aligned} -x_1 + 2x_2 + x_3 &= 10 \\ x_1 - 4x_2 + x_3 &= 6 \\ 3x_1 - 6x_2 - 3x_3 &= 25 \end{aligned}$

39 $\begin{aligned} 2x_1 - x_2 + x_3 &= -5 \\ x_1 - 3x_2 + 2x_3 &= -12 \\ 3x_1 + 2x_2 + x_3 &= 1 \end{aligned}$

40 $\begin{aligned} 2x_1 + x_2 - x_3 &= 0 \\ x_1 + 2x_2 - x_3 &= -2 \\ x_1 + 4x_2 + 2x_3 &= -1 \end{aligned}$

41 $\begin{aligned} x_1 - 2x_3 + x_3 &= -4 \\ 2x_1 + 6x_2 - x_3 &= 12 \\ -3x_1 + 6x_2 - 3x_3 &= 10 \end{aligned}$

42 $\begin{aligned} 3x_1 - 2x_2 + x_3 &= 6 \\ x_1 + x_2 + x_3 &= 10 \\ -6x_1 + 4x_2 - 2x_3 &= -10 \end{aligned}$

43 $\begin{aligned} 3x_1 + 2x_2 + x_3 &= 5 \\ x_1 + x_2 + x_3 &= 3 \\ 2x_1 - x_2 + x_3 &= 9 \end{aligned}$

44 $\begin{aligned} x_1 - x_2 + x_3 &= 0 \\ 2x_1 + x_2 - 2x_3 &= 10 \\ x_1 + 2x_2 + 3x_3 &= 40 \end{aligned}$

45 $\begin{aligned} 3x_1 + 5x_2 + 2x_3 &= 5 \\ x_1 + x_2 + x_3 &= 2 \\ 2x_1 + 3x_2 &= 2 \\ 3x_1 - 2x_3 &= 6 \end{aligned}$

46 $\begin{aligned} x_1 + x_2 + x_3 &= 6 \\ 2x_1 + x_2 - 2x_3 &= -2 \\ x_1 - x_2 - x_3 &= -4 \\ 5x_1 - 2x_2 + 2x_3 &= 7 \end{aligned}$

47 $\begin{aligned} 4x_1 - x_2 + 3x_3 &= 15 \\ x_1 - 2x_2 + x_3 &= 8 \\ x_1 + x_2 + x_3 &= 2 \\ 6x_1 - x_2 + 2x_3 &= 16 \end{aligned}$

48 $\begin{aligned} x_1 + x_2 + x_3 &= 1 \\ x_1 - x_2 - x_3 &= 7 \\ 2x_1 + x_2 + x_3 &= 5 \\ 4x_1 - 2x_2 + 3x_3 &= 7 \end{aligned}$

*49 Suppose that a (3×3) system of equations is represented by three planes which intersect along a common line. How many variables would be specified as arbitrary in the generalized solution?

*50 Suppose in Exercise 49 the three planes are identical. How many variables would be specified as arbitrary?

*51 Suppose that a (2×4) system of equations can be thought of as being represented by identical hyperplanes in 4-space. How many variables would be specified as arbitrary in the generalized solution?

*52 Suppose that a $(m \times n)$ system of equations can be thought of as being represented by m identical hyperplanes in n-space. How many variables would be specified as arbitrary in the generalized solution?

Section 3.4

53 A coffee manufacturer is interested in blending three types of coffee beans into 10,000 pounds of a final coffee blend. The three component beans cost $2.40, $2.60, and $2.00 per pound, respectively. The manufacturer wants to blend the 10,000 pounds at a total cost of $21,000. In blending the coffee, one restriction is that the amounts used of component beans 1 and 2 be the same. Determine whether there is a combination of the three types of beans which will lead to a final blend of 10,000 pounds costing $21,000 and satisfying the blending restriction.

54 **Diet Mix** A dietitian is planning a meal which will consist of three food types. In planning the meal, the dietitian wants to satisfy the minimum daily requirements (MDRs) for three vitamins. Table 3.10 summarizes the vitamin content per ounce of each food type, stated in milligrams (mg). Determine whether there are any combinations of the three foods which will satisfy the MDR for the three vitamins exactly.

55 A distiller wants to blend three component bourbons into a premium whiskey. Assuming that there is no loss in the blending process, it is desired to blend 50,000 liters of the

TABLE 3.10		*Vitamin Content/Ounce, mg*	
Food Type	Vitamin 1	Vitamin 2	Vitamin 3
1	4	2	1
2	6	8	6
3	3	4	2
MDR	52	56	34

whiskey. The only blending requirement is that the amount used of bourbon 1 should be twice that used of bourbon 3. In addition, $130,000 has been allocated to purchase the component bourbons. The three bourbons cost $2.50, $2.00, and $3.00 per liter, respectively. Determine whether there is a combination of the three bourbons which will produce the desired 50,000 liters. If so, what quantities should be used?

56 Blending: Lawn Care A manufacturer of lawn fertilizers is going to blend three stock fertilizers into a custom blend. Each of the stock fertilizers is characterized by its plant food content and weed killer content. The percentages are (by weight)

> *Stock fertilizer 1: 50 percent plant food and 20 percent weed killer*
> *Stock fertilizer 2: 60 percent plant food and 10 percent weed killer*
> *Stock fertilizer 3: 40 percent plant food and 30 percent weed killer*

A batch of 10,000 pounds of the custom blend is to be produced which has a plant food content of 48 percent and a weed killer content of 22 percent. Determine the quantities of the stock fertilizers which should be blended to satisfy these requirements.

57 Trust Fund Management A trust fund has $200,000 to invest. Three alternative investments have been identified, earning 10 percent, 7 percent, and 8 percent, respectively. A goal has been set to earn an annual income of $16,000 on the total investment. One condition set by the trust is that the combined investment in alternatives 2 and 3 should be triple the amount invested in alternative 1. Determine the amount of money which should be invested in each option to satisfy the requirements of the trust fund.

❑ CHAPTER TEST

1 Solve the following system of equations graphically.

$$x + 5y = -4$$
$$-3x + 2y = -5$$

2 Solve the following system of equations.

$$5x - 2y = 25$$
$$4x + y = 7$$
$$2x - 5y = 31$$
$$x + y = -2$$

3 (*a*) What are the solution set possibilities for a (20×15) system of equations?
(*b*) For a (15×20) system?

4 Solve the following system of equations using the gaussian elimination method.

$$x_1 - 2x_2 + x_3 = 10$$
$$3x_1 - 2x_2 + 4x_3 = 20$$
$$-3x_1 + 6x_2 - 3x_3 = -30$$

5 The following are results of the gaussian elimination method. Interpret their meaning.

$$(a)\ \begin{array}{ccc|c} 1 & 6 & 4 & 10 \\ 0 & 1 & -2 & -5 \\ 0 & 0 & 0 & 16 \end{array} \qquad (b)\ \begin{array}{cccc|c} 1 & 0 & 0 & 0 & 4 \\ 0 & 1 & 0 & 0 & -2 \\ 0 & 0 & 1 & 0 & 1 \\ 0 & 0 & 0 & 1 & 3 \end{array}$$

6 A company produces three products, each of which must be processed through three different departments. Table 3.11 summarizes the hours required per unit of each product in each department as well as the weekly capacities in each department. Formulate the system of equations and solve for any combinations of the three products that would consume the weekly labor availability in all departments. Interpret your results.

TABLE 3.11		Product			Hours Available
Department	*A*	*B*	*C*		per Week
1	6	2	2		70
2	7	4	1		60
3	5	5	3		55

Computer-Based Exercises

Solve the following systems of equations using an appropriate software package.

1
$$x_1 + x_2 + x_3 + x_4 + x_5 = 5$$
$$2x_1 - x_3 + 3x_4 = 12$$
$$2x_2 - 5x_3 + x_4 - x_5 = 12$$
$$3x_1 + 2x_2 - x_3 + 5x_4 - 3x_5 = 23$$
$$5x_1 - 4x_2 + 3x_3 + 4x_5 = -6$$

2
$$x_1 + x_2 + x_3 + x_4 + x_5 = 30$$
$$x_1 + 2x_2 + 3x_3 + 4x_4 + 5x_5 = 130$$
$$3x_1 - 2x_3 + 4x_5 = 105$$
$$5x_2 - 3x_3 + 2x_4 - 2x_5 = -5$$
$$4x_1 - 10x_2 + x_3 - 5x_4 - x_5 = -55$$

3
$$x_1 + x_2 + x_3 + x_4 + x_5 + x_6 = 0$$
$$3x_1 - 2x_3 + 4x_4 - 3x_6 = 0$$
$$5x_1 - 2x_2 + 3x_3 - 4x_4 + x_6 = 26$$

$$2x_1 + x_2 - x_3 + 5x_4 + 3x_5 - x_6 = -2$$
$$3x_2 - 4x_3 + 2x_4 = -18$$
$$x_1 - 2x_2 + x_3 - 3x_4 + 5x_5 = 24$$

4
$$x_1 + x_2 + x_3 + x_4 + x_5 + x_6 = 12$$
$$2x_1 - x_2 + 3x_3 - 2x_5 = 5$$
$$x_1 + x_3 - 2x_4 + x_5 - 4x_6 = -8$$
$$x_1 + 2x_2 + 3x_3 + 4x_4 + 5x_5 + 6x_6 = 46$$
$$4x_1 - 2x_3 + x_4 - x_5 + 3x_6 = 6$$
$$x_1 - x_2 + x_3 - x_4 + x_5 - x_6 = 0$$

5
$$2x_1 + x_2 + 2x_3 + x_4 + 2x_5 + x_6 + 2x_7 + x_8 = 30$$
$$x_1 + x_2 + x_3 + x_4 + x_5 + x_6 + x_7 + x_8 = 20$$
$$3x_1 - 2x_3 + x_4 - 2x_5 + x_7 = 10$$
$$5x_3 - 2x_4 + x_5 - 2x_7 = 3$$
$$x_1 + x_2 = 7$$
$$x_5 + x_6 + x_7 = 6$$
$$5x_1 - 3x_2 + 6x_3 - 2x_4 + x_5 - 5x_6 - x_7 + 2x_8 = 17$$
$$x_1 - x_2 + x_3 - x_4 + x_5 - x_6 + x_7 - x_8 = 0$$

6
$$2x_1 - 2x_2 + x_3 + 5x_4 - x_5 + x_6 - 3x_7 + x_8 = 100$$
$$x_1 + x_2 + x_3 + x_4 + x_5 + x_6 + x_7 + x_8 = 50$$
$$2x_1 + 3x_2 - 2x_5 + 3x_6 = 50$$
$$x_1 + x_2 + x_3 = 15$$
$$x_6 + x_7 + x_8 = 25$$
$$5x_1 - 3x_2 + 2x_3 + x_4 - 2x_5 - x_6 + 5x_7 + 6x_8 = 165$$
$$4x_1 - 3x_2 + 5x_3 - 3x_5 + 2x_7 = 35$$
$$x_1 - x_2 + x_3 - x_4 + x_5 - x_6 + x_7 - x_8 = -50$$

7 The XYZ Manufacturing Company manufactures five different products. Each of the products must be processed through five different departments, A through E. Table 3.12 indicates the number of hours required to produce a unit of each product in each department. Also indicated is the number of production hours available each week in each of the departments. The company wants to determine whether there are any quantities of the five products which can be produced each week that will result in total utilization of the hours available in all departments.

(a) Formulate the appropriate system of linear equations.

(b) Determine the combinations of the five products which will utilize the five departments to capacity. How will each department's weekly capacity be allocated among the five products?

TABLE 3.12	Product					Hours Available
Department	1	2	3	4	5	per Week
A	2	1	4	3	2	330
B	4	2	3	2	1	330
C	5	4	2	4	3	440
D	3	2	2	2	3	320
E	1	1	1	1	1	130

8 A company manufactures five different products. Each of the products must be processed through five different departments, *A* through *E*. Table 3.13 indicates the number of hours required to produce a unit of each product in each department. Also indicated is the number of production hours available each month in each of the departments. The company wants to determine the combination of the five products which can be produced each month such that there will be total utilization of all hours available in the departments.

(*a*) Formulate the appropriate system of linear equations.

(*b*) Determine the combinations of the five products which satisfy the system of equations. How will each department's capacity be allocated among the five products?

TABLE 3.13

Department	Product 1	2	3	4	5	Hours Available per Month
A	3	4	2	1	4	1,150
B	5	3	4	2	1	1,050
C	2	5	3	6	4	2,200
D	4	4	5	4	3	1,700
E	2	5	5	5	4	2,000

9 A dietitian is planning the menu for the noon meal at a high school. Six main items are being considered for inclusion in the meal, each characterized by different nutritional content. The goal is that the nutritional content of the meal meet the minimum daily requirement (MDR) levels for six different vitamins. Table 3.14 summarizes the vitamin content per ounce of each food, stated in milligrams (mg). In addition, the minimum daily requirement for the six vitamins is indicated, again stated in milligrams. The problem is to determine the quantities of each food which should be included in the meal to satisfy the six vitamin requirements.

(*a*) Formulate the appropriate system of equations for this problem.

(*b*) What quantities of each food should be included?

TABLE 3.14

Vitamin	MDR	Food 1	2	3	4	5	6
1	23	4	3	0	2	4	1
2	34	5	3	4	0	0	2
3	32	0	2	6	4	3	4
4	16	0	0	2	3	5	2
5	39	5	6	2	0	3	5
6	26	2	3	2	4	2	2

10 A coffee manufacturer is interested in blending five types of coffee beans into a final blend of 120,000 pounds of coffee. The five component beans cost $2, $3, $4, $2, and $2 per pound, respectively. The budget for purchasing the five components is $300,000. In blending the coffee, three restrictions have been stated: (1) the combination of components 1 and 2 should constitute exactly half of the final blend; (2) components 1 and 5 together should constitute exactly 25 percent of the final blend; and (3) the amount of component 4 used in the blend should be exactly three times the amount used of component 3.

(a) Formulate the system of equations which states all requirements in this blending problem.

(b) How many pounds of each component should be used in the final blend?

APPENDIX: ELIMINATION PROCEDURE FOR (3 × 3) SYSTEMS

The elimination procedure for (3×3) systems is similar to that for (2×2) systems. The aim is to start with the (3×3) system and to reduce this to an equivalent system having two variables and two equations. With one of the three variables eliminated, the same procedure as used for (2×2) systems is employed to eliminate a second variable, resulting in a (1×1) system. After you solve for the remaining variable, its value is substituted sequentially back through the (2×2) system and finally the (3×3) system to determine the values of the other two variables. Figure 3.13 illustrates the process schematically. In Fig. 3.13, x_1 is eliminated first, followed by x_2. This order is not required; it is simply illustrative.

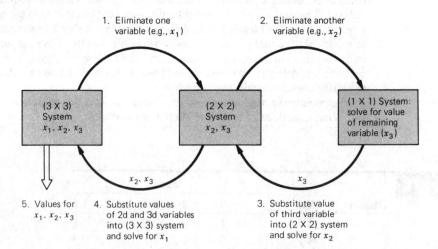

Figure 3.13   Elimination procedure for (3×3) systems.

The elimination procedure for a (3 × 3) system is as follows.

ELIMINATION PROCEDURE FOR 3 × 3 SYSTEMS

I   *Add multiples of any two of the three equations in order to eliminate one of the three variables. The result should be an equation involving the other two variables.*

II *Repeat step I with* another *pair of the original equations, eliminating the same variable as in step I. This second pair of equations will include one of the two equations used in step I and the equation not used in step I.*

III *The results of steps I and II should be a (2 × 2) system. Use the procedure for (2 × 2) systems (page 90) to determine the values for the remaining two variables.*

IV *Substitute the values of these two variables into one of the original equations. Solve for the value of the third variable.*

If during any phase of the elimination procedure an identity results [see step IIIB of the (2 × 2) procedure], then the solution set contains an infinite number of elements. An exception to this is the case in which step I results in an identity and step II a false statement, or contradiction. What is the graphical implication of these two results? *If at any stage a false statement results [step III C of the (2 × 2) procedure], then there is no solution to the original system of equations.*

EXAMPLE 20

Unique Solution Determine the solution set for the following system of equations.

$$x_1 + x_2 + x_3 = 6 \tag{3.22}$$

$$2x_1 - x_2 + 3x_3 = 4 \tag{3.23}$$

$$4x_1 + 5x_2 - 10x_3 = 13 \tag{3.24}$$

SOLUTION

Although it makes no difference which variable is eliminated first, let's eliminate x_2. If Eqs. (3.22) and (3.23) are added, the resultant Eq. (3.25) is stated in terms of x_1 and x_3:

$$
\begin{aligned}
x_1 + x_2 + x_3 &= 6 \\
\underline{2x_1 - x_2 + 3x_3} &= \underline{4} \\
3x_1 \qquad + 4x_3 &= 10
\end{aligned}
\tag{3.25}
$$

Multiplying Eq. (3.23) by $+5$ and adding it to Eq. (3.24) yields the new Eq. (3.26), as follows:

$$[5 \cdot \text{Eq. (3.23)}] \rightarrow \quad 10x_1 - 5x_2 + 15x_3 = 20$$

$$\underline{\qquad 4x_1 + 5x_2 - 10x_3 = 13}$$

$$14x_1 \qquad + \ 5x_3 = 33 \qquad \textbf{(3.26)}$$

Since x_2 has been eliminated, the system has been reduced to the (2×2) system

$$3x_1 + 4x_3 = 10 \qquad \textbf{(3.25)}$$

$$14x_1 + 5x_3 = 33 \qquad \textbf{(3.26)}$$

By proceeding as we did in Sec. 3.2, x_3 can be eliminated if we multiply Eq. (3.25) by $+5$ and Eq. (3.26) by -4. When the two equations are added, x_3 is eliminated and Eq. (3.27) is formed:

$$[5 \cdot \text{Eq. (3.25)}] \rightarrow \quad 15x_1 + 20x_3 = \quad 50$$

$$[-4 \cdot \text{Eq. (3.26)}] \rightarrow \quad \underline{-56x_1 - 20x_3 = -132}$$

$$-41x_1 \qquad = -82 \qquad \textbf{(3.27)}$$

Solving Eq. (3.27) for x_1, we get $x_1 = 2$. If this value is substituted into Eq. (3.25), the value of x_3 is determined in the following manner:

$$3(2) + 4x_3 = 10$$

$$4x_3 = \ 4$$

$$x_3 = \ 1$$

Substituting the values of $x_1 = 2$ and $x_3 = 1$ into Eq. (3.22) yields

$$2 + x_2 + 1 = 6$$

or
$$x_2 = 3$$

Verify that the solution set consists of $x_1 = 2$, $x_2 = 3$, and $x_3 = 1$ by substituting these values into Eqs. (3.23) and (3.24).

EXAMPLE 21 **(No Solution)** Determine the solution set for the following system of equations:

$$-2x_1 + x_2 + 3x_3 = 12 \tag{3.28}$$

$$x_1 + 2x_2 + 5x_3 = 10 \tag{3.29}$$

$$6x_1 - 3x_2 - 9x_3 = 24 \tag{3.30}$$

SOLUTION

Variable x_1 can be eliminated by multiplying Eq. (3.29) by $+2$ and adding it to Eq. (3.28), as follows:

$$-2x_1 + x_2 + 3x_3 = 12$$

$$[2 \cdot \text{Eq. (3.29)}] \rightarrow \quad \underline{2x_1 + 4x_2 + 10x_3 = 20}$$

$$5x_2 + 13x_3 = 32 \tag{3.31}$$

Similarly, x_1 can be eliminated by multiplying Eq. (3.29) by -6 and adding the resulting equation to Eq. (3.30), or

$$[-6 \cdot \text{Eq. (3.29)}] \rightarrow \quad -6x_1 - 12x_2 - 30x_3 = -60$$

$$\underline{6x_1 - 3x_2 - 9x_3 = 24}$$

$$-15x_2 - 39x_3 = -36 \tag{3.32}$$

Eliminating x_1 leaves the (2 × 2) system

$$5x_2 + 13x_3 = 32 \tag{3.31}$$

$$-15x_2 - 39x_3 = -36 \tag{3.32}$$

To eliminate x_2, Eq. (3.31) is multiplied by $+3$ and added to Eq. (3.32), or

$$[3 \cdot \text{Eq. (3.31)}] \rightarrow \quad 15x_2 + 39x_3 = 96$$

$$\underline{-15x_2 - 39x_3 = -36}$$

$$0 = 60 \tag{3.33}$$

Note that Eq. (3.33) is a contradiction, meaning that there is no solution to the original system of equations.

EXAMPLE 22   **(Infinitely Many Solutions)** Determine the solution set for the system of equations

$$x_1 + x_2 + x_3 = 20 \qquad\qquad \textbf{(3.34)}$$

$$2x_1 - 3x_2 + x_3 = -5 \qquad\qquad \textbf{(3.35)}$$

$$6x_1 - 4x_2 + 4x_3 = 30 \qquad\qquad \textbf{(3.36)}$$

SOLUTION

Verify that x_3 can be eliminated and Eq. (3.37) can be found by multiplying Eq. (3.34) by -1 and adding this new equation to Eq. (3.35):

$$x_1 - 4x_2 = -25 \qquad\qquad \textbf{(3.37)}$$

Also verify that Eq. (3.38) is formed by multiplying Eq. (3.34) by -4 and adding this to Eq. (3.36):

$$2x_1 - 8x_2 = -50 \qquad\qquad \textbf{(3.38)}$$

To eliminate x_1 from Eqs. (3.37) and (3.38), Eq. (3.37) may be multiplied by -2 and added to Eq. (3.38). When these operations are performed, Eq. (3.39) is an identity:

$$-2x_1 + 8x_2 = 50$$

$$\underline{2x_1 - 8x_2 = -50}$$

$$0 = 0 \qquad\qquad \textbf{(3.39)}$$

This is the signal that there are infinitely many solutions to the original system.

    To determine particular members of the solution set, return to one of the last meaningful equations generated during the elimination procedure [Eqs. (3.37) and (3.38)]. Then, solve for one of the variables in terms of the other. To illustrate, let's solve Eq. (3.37) for x_1.

$$x_1 = 4x_2 - 25 \qquad\qquad \textbf{(3.40)}$$

Now, we substitute the right side of this equation into one of the original equations wherever x_1 appears. If we substitute into Eq. (3.34), we get

$$(4x_2 - 25) + x_2 + x_3 = 20$$

$$5x_2 + x_3 = 45$$

Solving for x_3 gives

$$x_3 = 45 - 5x_2 \qquad\qquad \textbf{(3.41)}$$

Equations (3.40) and (3.41) state x_1 and x_2 in terms of x_3. Thus, one way in which we can specify the solution set is

x_2 arbitrary

$x_1 = 4x_2 - 25$

$x_3 = 45 - 5x_2$

Using this specification, verify that *one* solution to the original system is $x_1 = -5$, $x_2 = 5$, and $x_3 = 20$.

❑

CHAPTER 4

MATHEMATICAL FUNCTIONS

CHAPTER OBJECTIVES

❏ Enable the reader to understand the nature and notation of mathematical functions

❏ Provide illustrations of the application of mathematical functions

❏ Provide a brief overview of important types of functions and their characteristics

❏ Discuss the graphical representation of functions

<table>
<tr><td>

MOTIVATING SCENARIO:
Military Buildup

</td><td>

In the beginning of the Persian Gulf crisis during 1990, the United States deployed hundreds of thousands of troops to Saudi Arabia. Because of the potential for chemical warfare, there was an urgent need for gas masks for use by the troops. The Defense Department negotiated a contract with one manufacturer to supply two types of gas masks. The costs of the two types were $175 and $225, respectively. Because of the urgency of the need, the contract specified that if the combined number of masks supplied each week exceeded 5,000, the government would pay the manufacturer a bonus of $50,000 plus an additional $25 for each unit greater than 5,000. *What is desired is a formula which states the mathematical relationship between the weekly dollar sales to the government and the number of units supplied of the two types of gas masks. [Example 10]*

</td></tr>
</table>

The application of mathematics rests upon the ability to identify a relevant mathematical representation of a real-world phenomenon. This representation is often called a ***mathematical model.*** A model is relevant if it successfully captures those attributes of the phenomenon which are significant to the model builder. Just as a model airplane portrays the physical likeness of an actual airplane, a mathematical model of a demand function represents the significant interrelationships between, say, the price of a commodity and the quantity demanded.

It is important to repeat that mathematical models can mirror a reality exactly; often, however, they *approximate* the reality. If a model is a good approximation, it can be very useful in studying the reality and making decisions related to it. If a model does not approximate well, it is important that you understand this. Whether you are conducting the mathematical analysis of a problem yourself or being provided with the results of a mathematical analysis, it is important to understand the assumptions, strengths, and limitations of the models being used. *Ask questions!* Be an informed analyst/decision maker.

4.1 FUNCTIONS

In mathematical models, the significant relationships typically are represented by ***mathematical functions,*** or, more simply, ***functions.*** Functions form a cornerstone of much of what follows in this book. It is the purpose of this chapter to introduce this important topic.

Functions Defined

A function can be viewed as an input-output device. An input (or set of inputs) is provided to a mathematical rule which transforms (manipulates) the input(s) into a specific output. (See Fig. 4.1.) Consider the equation $y = x^2 - 2x + 1$. If selected values of x are *input,* the equation yields corresponding values of y as outputs. To illustrate:

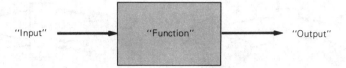

Figure 4.1 Input-output representation of function.

Input	Corresponding output
If $x = 1$	$y = (1)^2 - 2(1) + 1 = 0$
If $x = -5$	$y = (-5)^2 - 2(-5) + 1 = 36$
If $x = 10$	$y = (10)^2 - 2(10) + 1 = 81$

The equation provides the *rule* which allows one to transform a value of x into a corresponding value of y. The rule for *this* equation might be verbalized as "take the input value and square it, subtract two times the input value, and add 1." Note that for any input value, a unique output value is determined.

> **DEFINITION: FUNCTION**
> A *function* is a mathematical rule that assigns to each input value *one and only one* output value.
>
> **DEFINITION: DOMAIN/RANGE**
> The *domain* of a function is the set consisting of all possible input values.
> The *range* of a function is the set of all possible output values.

The assigning of output values to corresponding input values is often referred to as a ***mapping***. The notation

$$f: x \rightarrow y$$

represents the mapping of the set of input values x into the set of output values y, using the mapping rule f.

Figure 4.2 illustrates some important points regarding functions. The mapping indicated in Fig. 4.2a complies with the definition of a function. For each indicated value in the domain there corresponds a unique value in the range of the function. Similarly, the mapping in Fig. 4.2b complies with the definition. The fact that two different values in the domain "transform" into the same value in the range does not violate the definition. However, the mapping in Fig. 4.2c does not represent a function since one value in the domain results in two different values in the range.

The Nature and Notation of Functions

Functions, as we will treat them, suggest that the value of something depends upon the value of *one or more* other things. There are uncountable numbers of functional

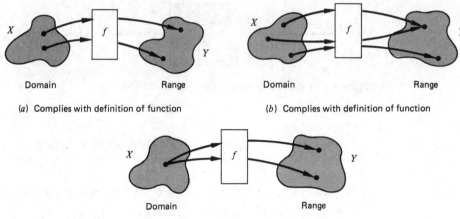

(a) Complies with definition of function

(b) Complies with definition of function

(c) Does not comply with definition of function

Figure 4.2
Sample mappings.

relationships in the world about us. The size of the crowds at a beach may depend upon the temperature and the day of the week, quantities sold of a product may depend upon the price charged for the product and the prices of competing brands, grades may depend upon the amount of time that a student studies, city tax rates may depend upon the level of municipal spending, and the dollars paid by a state in unemployment benefits depends upon the number of persons unemployed.

The language of mathematics provides a succinct way of describing how variables are functionally related. The equation

$$y = f(x)$$

denotes a functional relationship between the variables x and y. A verbal translation of this equation is "*y equals f of x*" or "*y is a function of x.*" *This equation is not to be interpreted as "y equals f times x.*" When we say that y is *a function of x*, we mean that the value of the variable y *depends upon* and is uniquely determined by the value of the variable x; x is the input variable and y the output variable. The respective roles of the two variables result in the variable x being called the ***independent variable*** and the variable y being called the ***dependent variable.*** Alternatively, y is often referred to as the ***value*** of the function. "f" is the ***name*** of the function or mapping rule.

Although y usually represents the dependent variable, x the independent variable, and f the name of the function, *any* letter may be used to represent the dependent and independent variables and the function name. The equation

$$u = g(v)$$

is a way of stating that the value of a dependent variable u is determined by the value of the independent variable v. And the name of the function or rule relating the two variables is g.

EXAMPLE 1 Imagine that you have taken a job as a salesperson. Your employer has stated that your salary will depend upon the number of units you sell each week. If we let

$$y = weekly\ salary\ in\ dollars$$

$$x = number\ of\ units\ sold\ each\ week$$

the dependency stated by your employer can be represented by the equation

$$y = f(x)$$

where f is the name of the salary function.

❑

Suppose your employer has given you the following equation for determining your weekly salary:

$$y = f(x) = 3x + 25 \qquad\qquad (4.1)$$

Given any value for x, substitution of this value into f will result in the corresponding value of y. For instance, if we want to compute your weekly salary when 100 units are sold, substitution of $x = 100$ into Eq. (4.1) yields

$$y = 3(100) + 25$$
$$= \$325$$

> For the function $y = f(x)$, the value of y which corresponds to the input value $x = b$ is denoted by $f(b)$.

In Eq. (4.1), the salary associated with selling 75 units can be denoted by $f(75)$. To evaluate $f(75)$, simply substitute the value 75 into Eq. (4.1) wherever the letter x appears, or

$$f(75) = 3(75) + 25$$
$$= \$250$$

Similarly, the value of y corresponding to $x = 0$ is denoted as $f(0)$ and is computed as $f(0) = 3(0) + 25 = \$25$.

Figure 4.3 is a schematic diagram of the salary function illustrating the input-output nature.

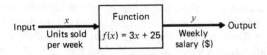

Figure 4.3 Weekly salary function.

EXAMPLE 2   Given the functional relationship

$$z = h(t)$$
$$= t^2 + t - 10$$

(a) $h(0) = (0)^2 + (0) - 10 = -10$
(b) $h(-5) = (-5)^2 + (-5) - 10 = 25 - 5 - 10 = 10$
(c) $h(u + v) = (u + v)^2 + (u + v) - 10$
$$= u^2 + 2uv + v^2 + u + v - 10 = u^2 + u + 2uv + v + v^2 - 10$$

Note in part c that the input value for t is the sum $u + v$. To evaluate $h(u + v)$, the procedure is exactly the same as for parts a and b. Wherever t appears in the function, we substitute the quantity $u + v$.

PRACTICE EXERCISE
For $t = u(v) = 2v^2 - 5v$, determine (a) $u(-5)$ and (b) $u(x - y)$.
Answer: (a) 75, (b) $2x^2 - 5x - 4xy + 5y + 2y^2$.

EXAMPLE 3   A small city police department is contemplating the purchase of an additional patrol car. Police analysts estimate the purchase cost of a fully equipped car (subcompact, but high-powered) to be $18,000. They also have estimated an average operating cost of $0.40 per mile.
(a) Determine the mathematical function which represents the total cost C of owning and operating the car in terms of the number of miles x it is driven.
(b) What are projected total costs if the car is driven 50,000 miles during its lifetime?
(c) If it is driven 100,000 miles?

SOLUTION
(a) In this example, we are asked to determine the function which relates total cost C to miles driven x. For now, we will exclude any consideration of salvage (or resale) value. The first question is: Which variable depends upon the other? A rereading of the problem and some thought about the two variables should lead to the conclusion that total cost is dependent upon the number of miles driven, or

$$C = f(x)$$

At this stage you may be able to write the cost function as

$$C = f(x) = 0.40x + 18,000$$

If you cannot write the cost function immediately, assume some sample values for mileage (independent variable) and determine the associated cost (dependent variable). Examine the respective values of the variables and see whether a pattern begins to emerge. If it does, then *articulate your mental model* (or more simply, write out the function).
Let's try this approach. What would total cost equal if the car were driven 0 miles (assuming it was purchased)? Your mental model should respond "$18,000." What would the total cost equal if the car were driven 10,000 miles? $22,000. What if it were driven 20,000

miles? $26,000. If you are having no difficulty arriving at these answers, indeed you have some mental cost model. Now is the time to express that model mathematically. The total cost of owning and operating the police car is the sum of two component costs — purchase cost and operating cost. And the type of computation you should have been making when responding to each question was to multiply the number of miles by $0.40 and add this result to the $18,000 purchase cost. Or

$$C = f(x)$$
$$= \text{total operating cost} + \text{purchase cost}$$
$$= (\text{operating cost per mile}) \, (\text{number of miles}) + \text{purchase cost}$$

or $\qquad C = 0.40x + 18,000$

(b) If the car is driven 50,000 miles, total costs are estimated to equal

$$C = f(50,000)$$
$$= 0.40(50,000) + 18,000$$
$$= \$38,000$$

(c) Similarly, at 100,000 miles

$$C = f(100,000)$$
$$= 0.40(100,000) + 18,000$$
$$= \$58,000$$

❑

Domain and Range Considerations

Earlier, the domain of a function was defined as the set of all possible input values. Because we will focus upon *real-valued* functions, the domain consists of all real values of the independent variable for which the dependent variable is defined and real. To determine the domain, it is sometimes easier to identify those values *not* included in the domain (i.e., find the exceptions). Given the domain, the range of a function is the corresponding set of values for the dependent variable. Identifying the range can be more difficult than defining the domain. We will have less concern with this process now. We will discuss the range in more detail when we examine the graphical representation of functions later in this chapter.

EXAMPLE 4

Given the function

$$y = f(x)$$
$$= x^2 - 2x + 1$$

any real value may be substituted for x with a corresponding and unique value of y resulting. If D is defined as the domain of f,

$$D = \{x \,|\, x \text{ is real}\}$$

EXAMPLE 5   The function

$$u = f(v)$$
$$= \frac{1}{v^2 - 4}$$

has the form of a quotient. Any values of v which result in the denominator equaling 0 would be excluded from the domain because division by 0 is undefined. The denominator equals 0 when $v^2 - 4 = 0$ or when v assumes a value of either $+2$ or -2. The domain of the function includes all real numbers *except* $+2$ and -2, or $D = \{v|v$ is real and $v \neq \pm 2\}$.

EXAMPLE 6   For the function

$$y = f(x)$$
$$= \sqrt{x - 5}$$

x can assume any value for which the expression under the square root sign is positive or zero. (Why is this?) To determine these values,

$$x - 5 \geq 0$$

when

$$x \geq 5$$

Thus the domain of the function includes all real numbers which are greater than or equal to 5, or $D = \{x|x$ is real and $x \geq 5\}$.

EXAMPLE 7   The function

$$y = f(x) = \sqrt{x^2 + x - 12}$$

is defined for all values of x which result in $x^2 + x - 12 \geq 0$. Equivalently, the values are those for which

$$(x + 4)(x - 3) \geq 0$$

The product of the two factors will equal zero when either of the factors equals zero. Thus, two members of the domain are $x = -4$ and $x = 3$. The product will be positive under two circumstances: Both factors are positive or both factors are negative. That is,

$$(x + 4) \quad (x - 3) \quad > 0$$

when

$$+ \qquad +$$

or

$$- \qquad -$$

The two factors are positive, respectively, when

$$x + 4 > 0 \quad \text{and} \quad x - 3 > 0$$

or

$$x > -4 \quad \text{and} \quad x > 3$$

Figure 4.4

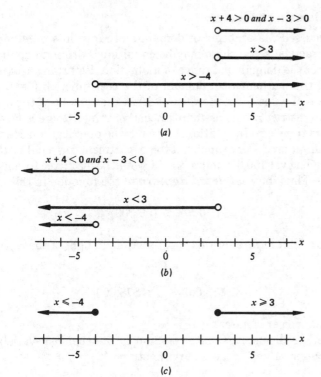

Domain for $f(x) = \sqrt{x^2 + x - 12}$

Using a number line in Fig. 4.4a to portray these results, we see that the values of x resulting in both factors being positive are $x > 3$. Similarly, the two factors are negative when

$$x + 4 < 0 \qquad \text{and} \qquad x - 3 < 0$$

or

$$x < -4 \qquad \text{and} \qquad x < 3$$

Figure 4.4b illustrates that both factors are negative when $x < -4$.

Figure 4.4c merges the results of our analysis (including values of x causing the radicand to equal zero) to illustrate that the domain of $f(x)$ is

$$D = \{x | x \leq -4 \text{ or } x \geq 3\}$$

❑

PRACTICE EXERCISE

Determine the domain for the function

$$y = f(x) = \sqrt{10 - x}$$

Answer: $x \leq 10$.

Restricted Domain and Range

We have discussed the concepts of domain and range in a purely mathematical sense. In a practical sense there may be conditions within an application which further restrict the domain and range of a function. Returning again to the patrol car example, the mathematical domain of the cost function $C = 0.40x + 18,000$ includes any real value for x. However, within the context of the application we would have to restrict x from assuming negative values (there is no such thing as negative miles traveled). In addition, there may be practical considerations which set an upper limit on x. For example, if the department has a policy that no patrol car will be driven over 150,000 miles, then x would be restricted to values no greater than 150,000. Thus the ***restricted domain*** of the function in this application is

$$0 \le x \le 150,000$$

The ***restricted range*** for this cost function, in light of the restrictions on x, would be

$$\$18,000 \le C \le \$78,000$$

assuming that the car is purchased.

In applied problems, it is quite common for independent variables to be restricted to *integer values*. In the salary function

$$\begin{aligned} y &= f(x) \\ &= 3x + 25 \end{aligned}$$

presented earlier, it is likely that the number of units sold each week, x, would be restricted to integer values. Thus, the domain of $f(x)$ might be defined as

$$D = \{x | x \text{ is an integer} \quad and \quad 0 \le x \le u\}$$

The lower limit on x is zero, excluding the possibility of negative sales, and u is an upper limit on sales which might reflect such considerations as maximum sales potential within the salesperson's district.

Note that for this function the integer restriction on the domain of f results in a range which is restricted to integer values. The range has a lower limit of 25 and an upper limit equal to $3u + 25$, or

$$R = \{25, 28, 31, 34, \dots, 3u + 25\}$$

Also, the maximum salary is a function of the maximum sales potential.

EXAMPLE 8   **(Wage Incentive Plan)** A manufacturer offers a wage incentive to persons who work on one particular product. The standard time to complete one unit of the product is 15 hours. Laborers are paid at the rate of $8 per hour up to a maximum of 15 hours for each unit of the product. If a unit of the product requires more than 15 hours, the laborer is only paid for the 15 hours the unit should have required. The manufacturer has built in a wage incentive for completion of a unit in less than 15 hours. For each hour under the 15-hour standard, the hourly wage for a worker increases by $1.50. Assume that the $1.50-per-hour incentive applies to any incremental savings including fractions of an hour [e.g., if a unit is completed in 14.5 hours, the hourly wage rate would equal $8 + 0.5($1.50) = $8.75]. Determine the function $w = f(n)$, where w equals the hourly wage rate in dollars and n equals the number of hours required to complete one unit of the product.

SOLUTION

The wage rate function has a restricted domain of $n \geq 0$, since negative production times are meaningless. In addition, the function will be described in two parts. The wage incentive applies only when production time is less than 15 hours. Thus, if $n \geq 15$, $w = 8$. If production time is less than 15 hours, the wage rate is determined as

$$w = 8 + 1.5(\text{number of hours under 15-hour standard})$$

or

$$w = 8 + 1.5(15 - n) = 30.5 - 1.5n$$

Let's test this part of the function. If a unit is produced in 13 hours, the standard has been beaten by 2 hours and the laborer should earn $11 per hour. Substituting $n = 13$ into the function gives

$$w = 30.5 - 1.5(13)$$
$$= 30.5 - 19.5$$
$$= 11.0$$

Thus, the formal statement of the wage function is

$$w = f(n) = \begin{cases} 30.5 - 1.5n & 0 \leq n < 15 \\ 8 & n \geq 15 \end{cases}$$

❑

Multivariate Functions

For many mathematical functions the value of a dependent variable depends upon more than one independent variable. Functions which contain more than one independent variable are called **multivariate functions.** In most real-world applications, multivariate functions are the most appropriate to use. For example, stating that profit is dependent only upon the number of units sold probably oversimplifies the situation. Many variables usually interact with one another in order to determine the profit for a firm.

One class of multivariate functions is that of **bivariate functions.** Bivariate functions (as compared with **univariate functions**) have two independent variables. The notation

$$z = f(x, y)$$

suggests that the dependent variable z depends upon the values of the two independent variables x and y. An example of a bivariate function is

$$z = f(x, y) = x^2 - 2xy + y^2 - 5$$

The notation for evaluating multivariate functions is very similar to that of functions of one independent variable. If we wish to evaluate $f(x, y)$ when $x = 0$ and $y = 0$, this is denoted by $f(0, 0)$. For the previous function

$$f(0, 0) = (0)^2 - 2(0)(0) + (0)^2 - 5$$
$$= -5$$

$$f(-10, 5) = (-10)^2 - 2(-10)(5) + 5^2 - 5$$
$$= 100 + 100 + 25 - 5$$
$$= 220$$

$$f(u, v) = u^2 - 2uv + v^2 - 5$$

Figure 4.5 illustrates the input-output nature of bivariate functions.

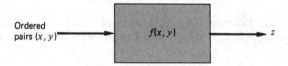

Figure 4.5   Input-output nature of bivariate functions.

PRACTICE EXERCISE
Given $z = f(x, y) = x^3 - x^2 y + 5y$, determine $f(5, -2)$.
Answer: 165.

When specifying the domain for bivariate functions, we are seeking the combinations of ordered pairs for which the function is defined, as illustrated in Fig. 4.6. For example, consider the function

$$z = f(x, y) = \frac{x^2 - 2y^2 + 4xy}{3x - y} \tag{4.2}$$

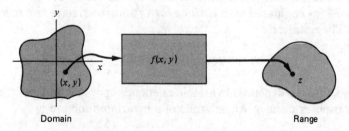

Figure 4.6   Representation of domain for $f(x, y)$.

The numerator of this function is defined for any combinations of real values for x and y. Similarly, the denominator is defined for any real values for x and y. However, the function f is not defined for values where $3x - y = 0$.

The domain of f could be specified as

$$D = \{(x, y) | 3x - y \neq 0\}$$

Graphically, the domain is represented by Fig. 4.7.

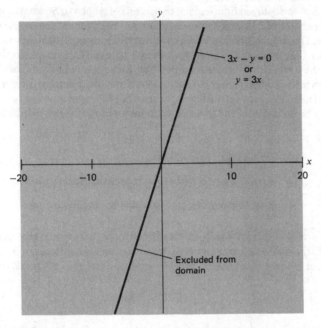

Figure 4.7 Domain for $f(x, y) = \dfrac{x^2 - 2y^2 + 4xy}{3x - y}$.

As the number of independent variables increases, the convention of using a different letter to represent each independent variable can become cumbersome. Consequently, a convenient way of representing multivariate functions is the use of *subscripted variables*. A general way of denoting a function where the value of a dependent variable y depends on the value of n independent variables is

$$y = f(x_1, x_2, x_3, \ldots, x_n)$$

The *subscript* is the positive integer index located to the right of and below each x. The index simply numbers the independent variables and enables you to distinguish one from another. We will frequently make use of subscripted notation in this book.

EXAMPLE 9   Given the function

$$y = f(x_1, x_2, x_3, x_4)$$
$$= x_1^2 - 2x_1 x_2 + x_3^2 x_4 - 25$$

$$f(-2, 0, 1, 4) = (-2)^2 - 2(-2)(0) + (1)^2(4) - 25$$
$$= 4 - 0 + 4 - 25$$
$$= -17$$

EXAMPLE 10   **(Military Buildup; Motivating Scenario)** In the beginning of the Persian Gulf crisis, the United States deployed hundreds of thousands of troops to Saudi Arabia. Because of the potential for chemical warfare, there was an urgent need for gas masks for use by the troops. The Defense Department negotiated a contract with one manufacturer to supply two types of gas masks. The costs of the two types were $175 and $225, respectively. Because of the urgency of the need, the contract specified that if the combined number of masks supplied each week exceeded 5,000, the government would pay the manufacturer a bonus of $50,000 plus an additional $25 for each unit greater than 5,000. Determine the function which states weekly sales in dollars as a function of the number of masks provided of each type.

SOLUTION

If
$$S = weekly\ sales,\ dollars$$

$$x_1 = number\ of\ type\ 1\ gas\ masks\ supplied\ per\ week$$

$$x_2 = number\ of\ type\ 2\ gas\ masks\ supplied\ per\ week$$

the function $S = f(x_1, x_2)$ will be defined in two parts, as was the case in Example 8. Payment to the manufacturer depends upon the combined weekly output. If combined *weekly* output is less than or equal to 5,000 units, payment is at the regular rate, or

$$S = 175x_1 + 225x_2$$

If combined weekly output exceeds 5,000 units, the manufacturer receives a lump sum bonus of $50,000 plus an additional $25 per unit in excess of 5,000. Mathematically, this is stated as

$$S = 175x_1 + 225x_2 + 50,000 + 25(x_1 + x_2 - 5,000)$$
$$= 175x_1 + 225x_2 + 50,000 + 25x_1 + 25x_2 - 125,000$$
$$= 200x_1 + 250x_2 - 75,000$$

The complete sales function is

$$S = f(x_1, x_2) = \begin{cases} 175x_1 + 225x_2 & x_1 + x_2 \leq 5,000 \\ 200x_1 + 250x_2 - 75,000 & x_1 + x_2 > 5,000 \end{cases}$$

Suppose that during a given week the manufacturer supplies 1,500 type 1 masks and 3,000 type 2 masks. Since $x_1 + x_2 = 4,500 < 5,000$, the manufacturer would receive compensation equal to

$$f(1,500, 3,000) = 175(1,500) + 225(3,000)$$
$$= 262,500 + 675,000$$
$$= \$937,500$$

If 3,000 masks of each type were supplied during a given week,

$$f(3,000, 3,000) = 200(3,000) + 250(3,000) - 75,000$$
$$= 600,000 + 750,000 - 75,000$$
$$= \$1,275,000$$

❑

POINT FOR THOUGHT & DISCUSSION	Examine the sales function for the condition $x_1 + x_2 > 5,000$. Although the \$25 bonus applies only to units in excess of 5,000, it appears that all units receive the \$25 bonus. Also, where is the \$50,000 bonus in the function? What does the $-75,000$ represent? The rearrangement and simplification of this function appear to distort the logic of the relationships. Clarify the logic for us!

For the remainder of this chapter, the functions discussed will contain one independent variable. Later in the book we will return to functions involving more than one independent variable.

Section 4.1 Follow-up Exercises

In Exercises 1–16, determine $f(0)$, $f(-2)$, and $f(a + b)$.

1 $f(x) = 5x - 10$ **2** $f(x) = 3x + 5$
3 $f(x) = -x + 4$ **4** $f(x) = -x/2$
5 $f(x) = mx + b$ **6** $f(x) = mx$
7 $f(x) = x^2 - 9$ **8** $f(x) = -x^2 + 2x$
9 $f(t) = t^2 + t - 5$ **10** $f(r) = tr^2 - ur + v$
11 $f(u) = u^3 - 10$ **12** $f(u) = -2u^3 + 5u$
13 $f(n) = n^4$ **14** $f(t) = 100$
15 $f(x) = x^3 - 2x + 4$ **16** $f(x) = 25 - x^2/2$

In Exercises 17–40, determine the domain of the function.

17 $f(x) = -10$ **18** $f(x) = 25$
19 $f(x) = 5x - 10$ **20** $f(x) = -x + 3$
21 $f(x) = mx + b$ **22** $f(x) = -ax$
23 $f(x) = 25 - x^2$ **24** $f(x) = x^2 - 4$
25 $f(x) = \sqrt{x + 4}$ **26** $f(x) = \sqrt{-2x + 25}$
27 $f(t) = \sqrt{-t - 8}$ **28** $f(t) = \sqrt{9 - t^2}$
29 $f(r) = \sqrt{r^2 + 9}$ **30** $f(r) = \sqrt{25 - r^2}$
31 $f(x) = 10/(4 - x)$ **32** $f(x) = (x - 4)/(x^2 - 6x - 16)$
33 $f(u) = (3u - 5)/(-u^2 + 2u + 5)$ **34** $f(t) = \sqrt{-t - 10}/(-3t^3 + 5t^2 + 10t)$
35 $f(x) = \sqrt{2.5x - 20}/(x^3 + 2x^2 - 15x)$ **36** $h(v) = \sqrt{10 - v/3}/(v^5 - 81v)$
37 $g(h) = \sqrt{h^2 - 4}/(h^3 + h^2 - 6h)$ **38** $f(x) = \sqrt{x^2 - x - 6}$
39 $f(x) = \sqrt{x^2 + 8x + 15}$ **40** $h(r) = \sqrt{r^2 - 16}$

41 The function $C(x) = 15x + 80,000$ expresses the total cost $C(x)$ (in dollars) of manufacturing x units of a product. If the maximum number of units which can be produced equals 50,000, state the restricted domain and range for this cost function.

42 Demand Function The function $q = f(p) = 280,000 - 35p$ is a *demand function* which expresses the quantity demanded of a product q as a function of the price charged for the product p, stated in dollars. Determine the restricted domain and range for this function.

43 Demand Function The function $q = f(p) = 180,000 - 30p$ is a demand function which expresses the quantity demanded of a product q as a function of the price charged for the product p, stated in dollars. Determine the restricted domain and range for this function.

44 Insurance Premiums An insurance company has a simplified method for determining the annual premium for a term life insurance policy. A flat annual fee of $150 is charged for all policies plus $2.50 for each thousand dollars of the amount of the policy. For example, a $20,000 policy would cost $150 for the fixed fee plus $50, which corresponds to the face value of the policy. If p equals the annual premium in dollars and x equals the face value of the policy (stated in thousands of dollars), determine the function which can be used to compute annual premiums.

45 In Exercise 44, assume that the smallest policy which will be issued is a $10,000 policy and the largest is a $500,000 policy. Determine the restricted domain and range for the function found in Exercise 44.

46 The local electric company uses the following method for computing monthly electric bills for one class of customers. A monthly service charge of $5 is assessed for each customer. In addition, the company charges $0.095 per kilowatt hour. If c equals the monthly charge stated in dollars and k equals the number of kilowatt hours used during a month:

    (*a*) Determine the function which expresses a customer's monthly charge as a function of the number of kilowatt hours.

    (*b*) Use this function to compute the monthly charge for a customer who uses 850 kilowatt hours.

47 Referring to Exercise 46, assume that the method for computing customer bills applies for customers who use between 200 and 1,500 kilowatt hours per month. Determine the restricted domain and range for the function in that exercise.

48 Auto Leasing A car rental agency leases automobiles at a rate of $15 per day plus $0.08 per mile driven. If y equals the cost in dollars of renting a car for one day and x equals the number of miles driven in one day:

    (*a*) Determine the function $y = f(x)$ which expresses the daily cost of renting a car.

    (*b*) What is $f(300)$? What does $f(300)$ represent?

    (*c*) Comment on the restricted domain of this function.

49 In manufacturing a product, a firm incurs costs of two types. Fixed annual costs of $250,000 are incurred regardless of the number of units produced. In addition, each unit produced costs the firm $6. If C equals total annual cost in dollars and x equals the number of units produced during a year:

    (*a*) Determine the function $C = f(x)$ which expresses annual cost.

    (*b*) What is $f(200,000)$? What does $f(200,000)$ represent?

    (*c*) State the restricted domain and restricted range of the function if maximum production capacity is 300,000 units per year.

50 Wage Incentive Plan A producer of a perishable product offers a wage incentive to drivers of its trucks. A standard delivery takes an average of 20 hours. Drivers are paid at the rate of $10 per hour up to a maximum of 20 hours. There is an incentive for drivers to make the trip in less (but not too much less!) than 20 hours. For each hour under 20, the hourly wage increases by $2.50. (The $2.50-per-hour increase in wages applies for fractions of hours. That is, if a trip takes 19.5 hours, the hourly wage increases by $0.5 \times$ $2.50, or $1.25.) Determine the function $w = f(n)$, where w equals the hourly wage (in dollars) and n equals the number of hours to complete the trip.

51 Membership Drive A small health club is trying to stimulate new memberships. For a limited time, the normal annual fee of $300 per year will be reduced to $200. As an additional incentive, for each new member in excess of 60, the annual charge for each new member will be further reduced by $2. Determine the function $p = f(n)$, where p equals the membership fee for new members and n equals the number of new members.

52 Given $f(x, y) = x^2 - 6xy + 2y^2$, determine (a) $f(0, 0)$, (b) $f(-1, 2)$, and (c) $f(5, 10)$.

53 Given $g(u, v) = 2u^2 + 5uv + v^3$, determine (a) $g(0, 0)$, (b) $g(-5, 2)$, (c) $g(5, 10)$, and (d) $g(x, y)$.

54 Given $v(h, g) = h^2/2 - 5hg + g^2 + 10$, determine (a) $v(0, 0)$, (b) $v(4, 2)$, and (c) $v(-2, -5)$.

55 Given $f(x_1, x_2, x_3) = (x_1 - x_2 + 2x_3)^2$, determine (a) $f(1, 1, 1)$, (b) $f(2, 3, -1)$, and (c) $f(2, 0, -4)$.

56 Given $f(x_1, x_2, x_3) = x_1^3 + 2x_1^2 x_2 - 3x_2 x_3 - 10$, determine (a) $f(0, 2, -3)$, (b) $f(-2, 1, 5)$, and (c) $f(3, 0, -5)$.

57 Given $f(x_1, x_2, x_3, x_4) = 2x_1 x_2 - 5x_2 x_4 + x_1 x_3 x_4$, determine (a) $f(0, 1, 0, 1)$ and (b) $f(2, 1, 2, -3)$.

58 Given $f(a, b, c, d) = 4ab - a^2 bd + 2c^2 d$, determine (a) $f(1, 2, 3, 4)$ and (b) $f(2, 0, 1, 5)$.

59 Given $f(x_1, x_2, x_3, x_4) = x_1 x_2 - 5x_3 x_4$, determine (a) $f(1, 10, 4, -5)$, (b) $f(2, 2, 2, 2)$, and (c) $f(a, b, c, d)$.

60 A company estimates that the number of units it sells each year is a function of the expenditures on radio and TV advertising. The specific function is

$$z = f(x, y) = 20{,}000x + 40{,}000y - 20x^2 - 30y^2 - 10xy$$

where z equals the number of units sold annually, x equals the amount spent for TV advertising, and y equals the amount spent for radio advertising (both in thousands of dollars).

(a) Determine the expected annual sales if $50,000 is spent on TV advertising and $20,000 is spent on radio advertising.

(b) What are expected sales if $80,000 and $100,000 are spent, respectively?

61 Pricing Model A manufacturer sells two related products, the demand for which is characterized by the following two demand functions:

$$q_1 = f_1(p_1, p_2) = 250 - 4p_1 - p_2$$

$$q_2 = f_2(p_1, p_2) = 200 - p_1 - 3p_2$$

where p_j equals the price (in dollars) of product j and q_j equals the demand (in thousands of units) for product j.

(a) How many units are expected to be demanded of each product if $20/unit is charged for product 1 and $40/unit is charged for product 2?

(b) How many units are expected if the unit prices are $40 and $30, respectively?

62 Family Shelter A women's resource center which provides housing for women and children who come from abusive homes is undertaking a grass roots fund-raising effort within the community. One component of the campaign is the sale of two types of candy bars. The profit from the candy is $0.50 and $0.75 per bar, respectively, for the two types. The supplier of the candy has offered an incentive if the total number of candy bars sold exceeds 2,000. For each bar over 2,000, an additional \$0.25 will be earned by the center. Determine the function $P = f(x_1, x_2)$, where P equals the total profit in dollars and x_j equals the number of bars sold of type j. If 750 and 900 bars, respectively, are sold, what is the profit expected to equal? If 1,500 and 2,250, respectively, are sold?

63 A salesperson is paid a base weekly salary and earns a commission on each unit sold of three different products. The base salary is \$60 and the commissions per unit sold are \$2.50, \$4.00, and \$3.00, respectively. If S equals the salesperson's weekly salary and x_j equals the number of units sold of product j during a given week, determine the salary function $S = f(x_1, x_2, x_3)$. What weekly salary would be earned if the salesperson sells 20, 35, and 15 units, respectively, of the three products?

64 In the previous exercise, assume that the salesperson can earn a bonus if combined sales for the three products exceeds 50 units for the week. The bonus equals \$25 plus \$1.25 additional commission for all units sold in excess of 50. Determine the weekly salary function $S = f(x_1, x_2, x_3)$. What salary would be earned for the 20, 35, and 15 units sold in the previous exercise?

4.2 TYPES OF FUNCTIONS

Functions can be classified according to their structural characteristics. A discussion of some of the more common functions follows. A more thorough treatment of these functions is provided in Chaps. 5 and 6. *Exponential* and *logarithmic* functions will be discussed in Chap. 7.

Constant Functions

A *constant function* has the general form

$$\boxed{y = f(x) = a_0}$$

(4.3)

where a_0 is real. For example, the function

$$y = f(x) = 20$$

is a constant function. Regardless of the value of x, the range consists of the single value 20. That is,

$$f(-10) = 20$$
$$f(1,000) = 20$$
$$f(a + b) = 20$$

As shown in Fig. 4.8, every value in the domain maps into the same value in the range for constant functions.

EXAMPLE 11   (**Marginal Revenue**) An important concept in economics is that of *marginal revenue*. Marginal revenue is the *additional revenue derived from selling one more unit of a product or service*. If each unit of a product sells at the same price, the marginal revenue is always equal to the price. For example, if a product is sold for \$7.50 per unit, the marginal revenue function can be stated as the *constant function*

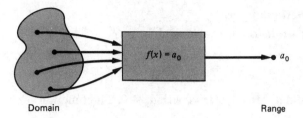

Figure 4.8 Mapping for constant function.

$$MR = f(x) = 7.5$$

where MR equals marginal revenue and x equals number of units sold of the product.

❏

Linear Functions

A *linear function* has the general (slope-intercept) form

$$y = f(x) = a_1 x + a_0 \qquad (4.4)$$

where a_0 and a_1 are constants. You should recognize this form from Chap. 2. Graphically, this function is represented by a straight line having slope a_1 and y intercept $(0, a_0)$. The function

$$y = f(x) = -2x + 15$$

is a linear function which graphs as a line having slope of -2 and y intercept at $(0, 15)$.

EXAMPLE 12 **(Total Cost)** Accountants and economists often define *total cost* (dollars flowing out of an organization) in terms of two components: *total variable cost* and *total fixed cost.* These two components must be added to determine *total cost.* The cost of owning and operating the patrol car in Example 3 is an example of a *linear* total cost function. The cost function

$$C(x) = 0.40x + 18{,}000$$

represented total variable costs by the term *0.40x* and fixed cost by the term *18,000.* Compare the structure of the cost function with Eq. (4.4) to confirm that it is in fact an example of a linear function.

❏

We will focus upon this important class of functions in the next chapter.

Quadratic Functions

A *quadratic function* has the general form

$$y = f(x) = a_2 x^2 + a_1 x + a_0$$ (4.5)

where a_2, a_1, and a_0 are constants with $a_2 \neq 0$. The function

$$y = f(x) = 3x^2 - 20x + 100$$

is a quadratic function with $a_2 = 3$, $a_1 = -20$, and $a_0 = 100$. The function

$$y = f(x) = -\frac{x^2}{2}$$

is a quadratic function with $a_2 = -\frac{1}{2}$ and $a_1 = a_0 = 0$.

EXAMPLE 13   (Demand Function) A *demand function* is a mathematical relationship expressing the way in which the quantity demanded of an item varies with the price charged for it. The demand function for a particular product is

$$q_d = f(p)$$

or $$q_d = p^2 - 70p + 1,225$$

where q_d equals the number of units demanded and p equals the price stated in dollars. Note that this particular demand function is *quadratic*. Relative to Eq. (4.5), $a_2 = 1$, $a_1 = -70$, and $a_0 = 1,225$. According to this function, the quantity demanded at a price of $10 is expected to equal

$$f(10) = (10)^2 - 70(10) + 1,225$$
$$= 100 - 700 + 1,225$$
$$= 625 \text{ units}$$

At a price of $30,

$$f(30) = (30)^2 - 70(30) + 1,225$$
$$= 900 - 2,100 + 1,225$$
$$= 25 \text{ units}$$

❑

PRACTICE EXERCISE
What price would have to be charged to eliminate any demand for the product? Answer: $35.

Quadratic functions will be examined in detail in Chap. 6.

Cubic Functions

A *cubic function* has the general form

$$y = f(x) = a_3 x^3 + a_2 x^2 + a_1 x + a_0 \qquad (4.6)$$

where a_3, a_2, a_1, and a_0 are constants with $a_3 \neq 0$. The function

$$y = f(x) = x^3 - 40x^2 + 25x - 1{,}000$$

is a cubic function with $a_3 = 1$, $a_2 = -40$, $a_1 = 25$, and $a_0 = -1{,}000$.

EXAMPLE 14   (**Epidemic Control**) An epidemic is spreading through a large western state. Health officials estimate that the number of persons who will be afflicted by the disease is a function of time since the disease was first detected. Specifically, the function is

$$n = f(t) = 0.05t^3 + 1.4$$

where n equals the number of persons estimated to be afflicted (stated in hundreds) and t equals the number of days since the disease was first detected. This approximating function is assumed to be reasonably accurate for $0 \leq t \leq 30$. After 30 days, the disease has historically run its course. How many persons are expected to have caught the disease after 10 days?

SOLUTION

The number expected to have caught the disease after 10 days is

$$
\begin{aligned}
f(10) &= 0.05(10)^3 + 1.4 \\
&= 0.05(1{,}000) + 1.4 \\
&= 50 + 1.4 \\
&= 51.4 \text{ (hundred persons)}
\end{aligned}
$$

◻

PRACTICE EXERCISE

How many persons are expected to have caught the disease after 30 days? What interpretation can be given to the constant 1.4 in the function?

Answer: 1,351.4, or 135,140 persons. Approximately 140 persons will have caught the disease by the time it is detected.

Polynomial Functions

Each of the previous functions is an example of a polynomial function. A *polynomial function of degree n* has the general form

$$y = f(x) = a_n x^n + a_{n-1} x^{n-1} + \cdots + a_1 x + a_0 \qquad (4.7)$$

where $a_n, a_{n-1}, \ldots, a_1, a_0$ are constants with $a_n \neq 0$. *The exponent on each x must be a nonnegative integer* and the degree of the polynomial is the highest power (exponent) in the function. The function

$$y = f(x) = x^5$$

is a polynomial function of degree 5 with $a_0 = a_1 = a_2 = a_3 = a_4 = 0$ and $a_5 = 1$.

Observe that constant, linear, quadratic, and cubic functions are polynomial functions of degree 0, 1, 2, and 3, respectively.

Rational Functions

A *rational function* has the general form

$$y = f(x) = \frac{g(x)}{h(x)} \qquad \text{(4.8)}$$

where g and h are both polynomial functions. Rational functions are so named because of their ratio structure. The function

$$y = f(x) = \frac{2x}{5x^3 - 2x + 10}$$

is an illustration of a rational function where $g(x) = 2x$ and $h(x) = 5x^3 - 2x + 10$.

EXAMPLE 15   **(Disability Rehabilitation)** Physical therapists often find that the rehabilitation process is characterized by a diminishing-returns effect. That is, regained functionality usually increases with the length of a therapy program but eventually in decreased amounts relative to additional program efforts. For one particular disability, therapists have developed a mathematical function which describes the cost C of a therapy program as a function of the percentage of functionality recovered, x. The function is a *rational* function having the form

$$C = f(x)$$
$$= \frac{5x}{120 - x} \qquad 0 \le x \le 100$$

where C is measured in thousands of dollars. For example, the therapy cost to gain a 30 percent recovery is estimated to equal

$$f(30) = \frac{5(30)}{120 - 30}$$
$$= \frac{150}{90}$$
$$= 1.667 \text{ (thousand dollars)}$$

PRACTICE EXERCISE
Determine the cost of therapy to gain 10 percent of functionality. 60 percent.
100 percent. *Answer:* $454; $5,000; $25,000.

Combinations of Functions

Aside from the functional forms mentioned thus far, functions may be combined
algebraically to form a resultant function. If

$$f(x) = 3x - 5 \qquad g(x) = x^2 - 2x + 1 \qquad h(x) = x^3 \qquad \text{and} \qquad j(x) = 1/2x^4$$

these functions can be combined in certain ways to form new functions. The
following are examples of *sum, difference, product,* and *quotient functions.*

1 $p(x) = f(x) + g(x) = (3x - 5) + (x^2 - 2x + 1) = x^2 + x - 4$
(sum function)
2 $q(x) = h(x) - j(x) = x^3 - 1/2x^4$ (difference function)
3 $r(x) = f(x)h(x) = (3x - 5)x^3 = 3x^4 - 5x^3$ (product function)
4 $s(x) = h(x)/j(x) = x^3/(1/2x^4) = x^3(2x^4/1) = 2x^7$ (quotient function)

The domain for a sum, difference, or product function consists of the set of
values for the independent variable for which *both* functions are defined. For
quotient functions having a general form $u(x)/v(x)$, the domain similarly consists
of the values of x for which both u and v are defined, *except* for values resulting in
$v(x) = 0$.

Composite Functions

In addition to combining functions algebraically to form new functions, component
functions can be related in another way. A *composite function* exists when one
function can be viewed as a function of the values of another function.

If $y = g(u)$ and $u = h(x)$, the composite function $y = f(x) = g(h(x))$ is created
by substituting $h(x)$ into the function $g(u)$ wherever u appears. *And for $f(x) =$
$g(h(x))$ to be defined, x must be in the domain of h and $h(x)$ must be in the domain of g.*
That is, the input value x must allow for a unique and definable output value u, and
the resulting u when input to $g(u)$ must yield a unique and definable output y.
Figure 4.9 illustrates schematically the nature of composite functions.

To illustrate these functions, assume that the function

$$y = g(x) = 2x + 50$$

indicates that a salesperson's weekly salary y is determined by the number of units
x sold each week. Suppose that an analysis has revealed that the quantity sold each
week by a salesperson is dependent upon the price charged for the product. This
function h is given by the rule

$$x = h(p) = 150 - 2.5p$$

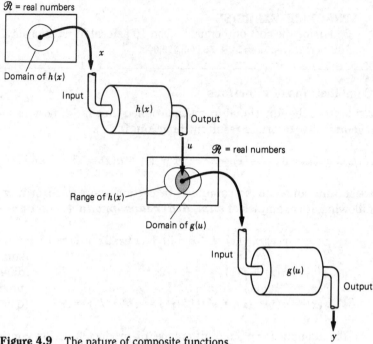

Figure 4.9   The nature of composite functions.

where p equals the price, stated in dollars.

Thus, to compute a salesperson's salary, the initial *input* is the selling price for the given week. This will determine the *output* for $h(p)$, the number of units expected to be sold. This output becomes an input to $g(x)$ to determine the weekly salary. These relationships are illustrated in Fig. 4.10a. For example, suppose that the price during a given week is \$30. The number of units expected to be sold during the week is

$$x = h(30)$$
$$= 150 - 2.5(30) = 150 - 75 = 75 \text{ units}$$

Since the number of units expected to be sold is known, the weekly salary is computed as

$$y = g(75)$$
$$= 2(75) + 50 = \$200$$

Since weekly salary depends upon the number of units sold each week and the number of units sold depends upon the price per unit, weekly salary actually can be stated directly as a function of the price per unit. Or

$$y = f(p) = g(h(p))$$

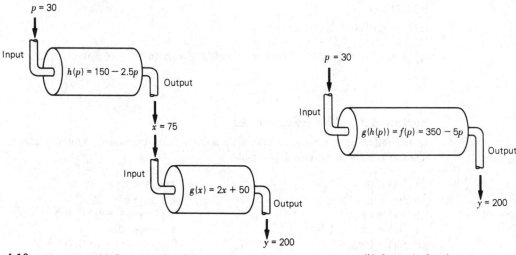

Figure 4.10 (a) Component functions (b) Composite function

To define this function, we substitute $h(p)$ into $g(x)$ wherever x appears. That is,

$$y = g(150 - 2.5p)$$
$$= 2(150 - 2.5p) + 50$$
$$= 300 - 5p + 50$$

or $\qquad y = f(p) = 350 - 5p$

The function $f(p)$ is a composite function, having been formed by combining $g(x)$ and $h(p)$. We can compute the expected weekly salary directly from $f(p)$ if we know the selling price for a given week, as shown in Fig. 4.10b. At a price of \$30,

$$y = f(30)$$
$$= 350 - 5(30) = 350 - 150 = \$200$$

which is the same value as determined before.

EXAMPLE 16 If $y = g(u) = u^2 - 2u + 10$ and $u = h(x) = x + 1$, the composite function $y = f(x) = g(h(x))$ is found by substituting $h(x)$ into $g(u)$ wherever u appears.

$$y = f(x) = g(h(x))$$
$$= g(x + 1)$$
$$= (x + 1)^2 - 2(x + 1) + 10$$
$$= x^2 + 2x + 1 - 2x - 2 + 10$$
$$= x^2 + 9$$

EXAMPLE 17 If $y = g(u) = 2u^3$ and $u = h(x) = x^2 - 2x + 5$, determine (a) $g(h(x))$, (b) $g(h(2))$, and (c) $g(h(-3))$.

SOLUTION

(a) $y = g(h(x)) = 2(x^2 - 2x + 5)^3$

(b) $g(h(2)) = 2[(2)^2 - 2(2) + 5]^3 = 2(5)^3 = 2(125) = 250$

(c) $g(h(-3)) = 2[(-3)^2 - 2(-3) + 5]^3 = 2(20)^3 = 2(8,000) = 16,000$

❑

Section 4.2 Follow-up Exercises

In the following exercises, classify (if possible) each function by type (constant, linear, quadratic, cubic, polynomial, rational).

1 $f(x) = 2^x$

2 $f(x) = -24$

3 $f(x) = (x - 5)/2$

4 $f(x) = x^2 - 25$

5 $f(x) = 2x^0$

6 $f(x) = x^5 + 2x^3 - 100$

7 $f(x) = 10 - x/4$

8 $f(x) = 10/x$

9 $f(x) = \log_{10} x$

10 $f(x) = (x^4 - 5x^2)/(x^6 + 5)$

11 $g(h) = -25/h^5$

12 $h(s) = 3 - 4s + s^2 - s^3/4$

13 $v(t) = x^2/\sqrt{x^3}$

14 $f(u) = (5u - 3)^0/4$

15 $f(n) = 50/(4)^3$

16 $g(h) = \sqrt{100}/(5)^2$

17 $f(x) = 10^x$

18 $f(x) = x^{16}/\sqrt{x}$

19 $f(t) = t^5/(36 - t^8)$

20 $f(x) = 3^{2x}$

21 $f(x) = \log_{10}(x + 5)$

22 $v(h) = \log_e h$

23 $f(x) = [(x - 9)^0]^3$

24 $f(x) = [(x + 4)^5]^0$

25 Given the general form of a constant function stated by Eq. (4.3), determine the domain for these functions.

26 Given the general form of a polynomial function stated by Eq. (4.7), determine the domain for such functions.

27 Given the general form of a rational function stated by Eq. (4.8), determine the domain for these functions.

28 Total profit from planting x_j acres at farm j is expressed by the function

$$P(x_1, x_2, x_3) = 500x_1 + 650x_2 + 450x_3 - 300,000$$

(a) What is total profit if 200 acres are planted at farm 1, 250 acres at farm 2, and 150 acres at farm 3?

(b) What is total profit if 500, 300, and 700 acres are planted, respectively, at the three farms?

(c) Identify one combination of plantings which would result in profit equaling zero.

29 The value of a truck is estimated by the function

$$V = f(t) = 20,000 - 3,000t$$

where V equals the value stated in dollars and t equals the age of the truck expressed in years.

(a) What class of function is this?

(b) What is the value after 3 years?

(c) When will the value equal 0?

30 A police department has determined that the number of serious crimes which occur per week can be estimated as a function of the number of police officers assigned to preventive patrol. Specifically, the mathematical function is

$$c = f(p) = 900 - 3.5p$$

where c equals the number of crimes expected per week and p equals the number of officers assigned to preventive patrol.

(a) What class of function is this?

(b) What is the expected number of crimes if 150 officers are assigned to preventive patrol?

(c) How many officers would have to be assigned if it is desired to reduce weekly crime levels to 500?

(d) How many officers would have to be assigned to reduce weekly crime levels to 0?

31 Total revenue from selling a particular product depends upon the price charged per unit. Specifically, the revenue function is

$$R = f(p) = 1{,}500p - 50p^2$$

where R equals total revenue in dollars and p equals price, also stated in dollars.

(a) What class of function is this?

(b) What is total revenue expected to equal if the price equals $10?

(c) What price(s) would result in total revenue equaling zero?

32 **Supply Functions** A supply function indicates the number of units of a commodity that suppliers are willing to bring to the marketplace as a function of the price consumers are willing to pay. The following function is a supply function

$$q_s = 0.5p^2 - 200$$

where q_s equals number of units supplied (stated in thousands) and p equals the selling price.

(a) What class of function is this?

(b) What quantity would be supplied if the market price is $30? $50?

(c) What price would result in 0 units being brought to the marketplace?

33 The profit function for a firm is

$$P(q) = -10q^2 + 36{,}000q - 45{,}000$$

where q equals the number of units sold and P equals annual profit in dollars.

(a) What class of function is this?

(b) What is profit expected to equal if 1,500 units are sold?

34 **Salvage Value** A major airline purchases a particular type of plane for $75 million. The company estimates that the salvage (resale) value of the plane is estimated well by the function

$$S = f(x) = 72 - 0.0006x$$

where S equals the salvage value (in millions of dollars) and x equals the number of hours of flight time for the plane.

(a) What type of function is this?

(b) What is the salvage value expected to equal after 10,000 hours of flight time?

(c) How many hours would the plane have to be flown for the salvage value to equal zero?

(d) What interpretation would you give to the y intercept? Why do you think this does not equal 75?

35 The demand function for a product is

$$q_d = p^2 - 90p + 2{,}025 \qquad 0 \le p \le 45$$

where q_d equals the number of units demanded and p equals the price per unit, stated in dollars.

(a) What type of function is this?

(b) How many units will be demanded at a price of $30?

(c) What price(s) would result in zero demand for the product?

36 An epidemic is spreading through a herd of beef cattle. The number of cattle expected to be afflicted by the disease is estimated by the function

$$n = f(t) = 0.08t^3 + 5$$

where n equals the number of cattle afflicted and t equals the number of days since the disease was first detected. How many cattle are expected to be afflicted after 10 days? After 20 days?

37 Given $f(x) = x^2 - 3$ and $g(x) = 10 - 2x$, determine (a) $f(x) + g(x)$, (b) $f(x) \cdot g(x)$, and (c) $f(x)/g(x)$.

38 Given $f(x) = \sqrt{x}$ and $g(x) = 3/(x - 1)$, determine (a) $f(x) - g(x)$, (b) $f(x) \cdot g(x)$, and (c) $f(x)/g(x)$.

39 If $y = g(u) = u^2 - 4u + 10$ and $u = h(x) = x - 4$, determine (a) $g(h(x))$, (b) $g(h(-2))$, and (c) $g(h(1))$.

40 Given $y = g(u) = 3u^2 + 4u$ and $u = h(x) = x + 8$, determine (a) $g(h(x))$, (b) $g(h(-2))$, and (c) $g(h(1))$.

41 If $y = g(u) = u^2 + 2u$ and $u = h(x) = x^3$, determine (a) $g(h(x))$, (b) $g(h(0))$, and (c) $g(h(2))$.

42 Given $c = h(s) = s^2 - 8s + 5$ and $s = f(t) = 10$, determine (a) $h(f(t))$, (b) $h(f(3))$, and (c) $h(f(-2))$.

43 Given $y = g(u) = (2)^u$ and $u = h(x) = x + 2$, determine (a) $g(h(x))$, (b) $g(h(3))$, and (c) $g(h(-2))$.

44 Given $y = g(u) = (u - 5)^2$ and $u = h(x) = x^2 + 1$, determine (a) $g(h(x))$, (b) $g(h(5))$, and (c) $g(h(-3))$.

4.3 GRAPHICAL REPRESENTATION OF FUNCTIONS

Throughout this book the visual model is used as often as possible to reinforce your understanding of different mathematical concepts. The visual model will most frequently take the form of a graphical representation. In this section we discuss the graphical representation of functions involving two variables.

Graphing Functions in Two Dimensions

Functions of one or two independent variables can be represented graphically. This graphical portrayal brings an added dimension to the understanding of mathematical functions. You will come to appreciate the increased understanding and insight that graphs provide.

Graphical representation requires a dimension for each independent variable contained in a function and one for the functional value, or dependent variable. Thus, functions of one independent variable are graphed in two dimensions, or **2-space.** Functions of two independent variables can be graphed in three dimensions, or **3-space.** When a function contains more than three variables, the graphical representation is lost.

Functions which contain two variables are graphed on a set of rectangular coordinate axes. Typically, the vertical axis is selected to represent the dependent variable in the function; the horizontal axis is usually selected to represent the independent variable.

To graph a mathematical function, one can simply *assign* different values from the domain for the independent variable and *compute* the corresponding value for the dependent variable. The resulting ordered pairs of values for the two variables represents values which satisfy the function. They also specify the coordinates of points which lie on the graph of the function. To *sketch* the function, determine an *adequate number of ordered pairs* of values which satisfy the function; locate their coordinates relative to a pair of axes. *Connect these points by a smooth curve to determine a sketch of the graph of the function.* (Although this approach will suffice for now, later we learn more efficient ways of sketching functions.)

EXAMPLE 18 We already have knowledge about how to graph linear relationships. To sketch the linear function

$$y = f(x) = 2x - 4$$

we simply need the coordinates of two points. You should recognize this as the slope-intercept form of a straight line with slope of 2 and y intercept at $(0, -4)$. Setting y equal to zero, we identify the x intercept as $(2, 0)$. Figure 4.11 is the graphical representation of the function.

EXAMPLE 19 To sketch the **quadratic function**

$$u = f(v)$$
$$= 10v^2 + 20v - 100$$

sample pairs of values for u and v are computed as shown in Table 4.1. These points are plotted in Fig. 4.12 and have been connected to provide a sketch of the function. Note that the horizontal axis is labeled with the independent variable v and the vertical axis with the dependent variable u. As opposed to the linear function in the previous example, this quadratic function obviously cannot be represented by a straight line.

TABLE 4.1	v	-6	-5	-4	-3	-2	-1	0	1	2	3	4
	u	140	50	-20	-70	-100	-110	-100	-70	-20	50	140

❏

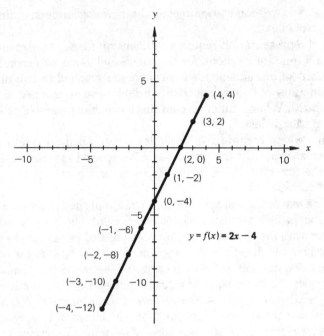

Figure 4.11 Sketch of $f(x) = 2x - 4$.

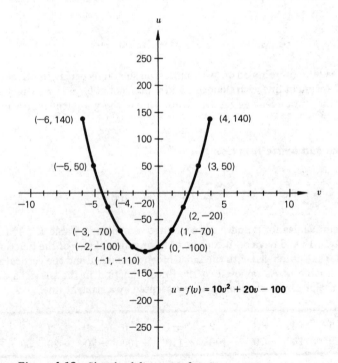

Figure 4.12 Sketch of $f(v) = 10v^2 + 20v - 100$.

NOTE A few points regarding the graphing of functions should be made. **First,** it is always useful to determine the set of sample points you want to graph *prior to scaling the axes.* By doing this, you determine the range of values which you wish to graph for the two variables. Once you have determined these ranges, you can determine the appropriate scale to use on each axis. **Secondly,** the two axes need not be scaled the same. The units on one axis may represent millions and those on the other axis single units. Similarly, the intervals used to scale each axis need not be the same width (examine the scaling in Fig. 4.12). If you overlook this possibility, your graph may run beyond the boundaries of your paper. (The author has had situations during class in which graphs required points below the floor or above the ceiling.) **Finally,** the unit of measurement for one variable does not have to be the same as that of the other variable. The cost function in the patrol car example would show cost *in dollars* on one axis and miles driven on the other axis.

POINT FOR THOUGHT & DISCUSSION What would be the effect on the shape of the graph in the last example if the vertical axis is scaled the same as the horizontal axis? What if the horizontal axis is scaled the same as the vertical axis? Given Fig. 4.12, determine the range of $f(v)$.

EXAMPLE 20 To sketch the *cubic function*

$$y = f(x) = x^3$$

sample points are computed as shown in Table 4.2. These points are plotted, resulting in the sketch of f in Fig. 4.13.

TABLE 4.2

x	0	1	2	3	-1	-2	-3
y	0	1	8	27	-1	-8	-27

EXAMPLE 21 As we have seen in this chapter, the functional relationship existing between variables is sometimes described by more than one equation. To illustrate, assume that y equals a salesperson's weekly salary in dollars and x equals the number of units of a product sold during the week. Given that the weekly salary depends upon the number of units sold, assume that the following function applies.

$$y = f(x) = \begin{cases} 2x + 50 & \text{where } 0 \le x < 40 \\ 2.25x + 75 & \text{where } x \ge 40 \end{cases}$$

If the number of units sold during a week is less than 40, the salesperson receives a base salary of $50 and a commission of $2 per unit sold. If the number of units sold during a week

Figure 4.13
Sketch of $f(x) = x^3$.

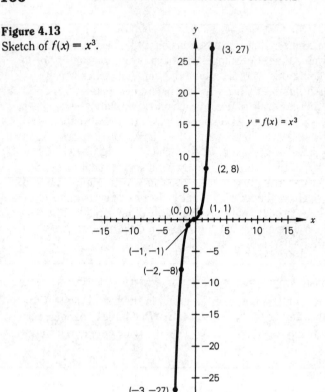

is 40 or more, a bonus of $25 raises the base portion of the salary to $75. In addition, the commission on *all* units increases to $2.25 per unit.

Figure 4.14 illustrates the graph of the function. Note that the graph uses quadrant I only, where x and y are both nonnegative. The sketch of the function is in two linear "pieces." Each piece of the graph is valid for a certain portion of the domain of the function. The open circle (○) at the end of the first segment is used to indicate that that point is *not* part of the graph. It corresponds to the break in the function at $x = 40$. The point corresponding to $x = 40$ is the first point on the second segment of the function, and this is denoted by the solid circle (●). Sketch this function yourself to verify its shape.

◻

Vertical Line Test

By the definition of a function, to each element in the domain there should correspond one and only one element in that range. This property allows for a simple graphical check to determine whether a graph represents a mathematical function. If a vertical line is drawn through any value in the domain, it will intersect the graph of the function at one point only. In contrast, if a vertical line intersects a curve at more than one point, the curve is not the graph of a function. The curve in Fig. 4.15 does not represent a function since the dashed vertical line intersects the curve at two points. Two values in the range, y_1 and y_2, are associated with one value in the domain, x_0. The graph depicts a *relation* but not a function.

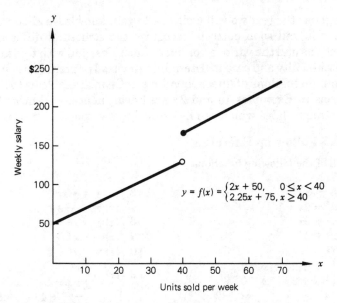

Figure 4.14 Piecewise linear function.

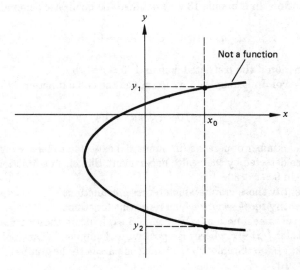

Figure 4.15 Vertical line test for function.

In this section we have been introduced to the graphical representation of mathematical functions. The procedure which has been presented must be called a "brute force" method in that it is necessary to determine an "adequate" number of points in order to get a reasonable idea of the shape of the graph of a function. However, it does work! The question of how many points are adequate will be answered with experience.

Throughout the text we will continue to gain knowledge about mathematical functions. You will soon come to recognize the structural differences between linear functions and the various nonlinear functions, and with this knowledge will come greater facility and ease in determining a visual or graphical counterpart. For example, our discussions of linear equations in Chap. 2 enable us to recognize that the functions in Examples 18 and 21 are linear; hence, we know that they will sketch as straight lines which can be defined by two points.

Section 4.3 Follow-up Exercises

Sketch each of the following functions.

1 $f(x) = 8 - 3x$

2 $f(x) = 4 - x/2$

3 $f(x) = x^2 - 2x + 1$

4 $f(x) = x^2 - 9$

5 $f(x) = x^2 + 5x$

6 $f(x) = -x^2 + 4$

7 $f(x) = x^3 + 2$

8 $f(x) = -x^3 - 1$

9 $f(x) = x^4$

10 $f(x) = -x^4 + 2$

11 $f(x) = \begin{cases} x + 2 & x \geq 0 \\ -x + 2 & x < 0 \end{cases}$

12 $f(x) = \begin{cases} x^2 & x \geq 0 \\ -x - 2 & x < 0 \end{cases}$

13 $f(x) = \begin{cases} 4 & x \leq -2 \\ |x| & -2 < x < 2 \\ -4 & x \geq 2 \end{cases}$

14 $f(x) = \begin{cases} -x^2 + 4 & -2 < x < 2 \\ -x - 3 & x \leq -2 \\ x - 3 & x \geq 2 \end{cases}$

15 Demand Function In Example 13 we examined the quadratic demand function

$$q_d = f(p) = p^2 - 70p + 1,225$$

Sketch this function if the restricted domain is $0 \leq p \leq 20$.

16 Epidemic Control In Example 14 we examined the cubic function

$$n = f(t) = 0.05t^3 + 1.4$$

where n was the number of persons (in hundreds) expected to have caught a disease t days after it was detected by the health department. Sketch this function, assuming a restricted domain $0 \leq t \leq 30$.

17 In Fig. 4.16, identify those graphs which represent functions.

18 In Fig. 4.17, identify those graphs which represent functions.

19 Given the graph of some function $f(x)$ in Fig. 4.18, explain how the graph would change if we wanted to sketch $f(x) + c$, where c is a positive real number. What would the graph of $f(x) - c$ look like? [*Hint:* Graph $f(x) = x^2$ and compare with the graphs of $g(x) = x^2 + 1$ and $h(x) = x^2 - 1$.]

20 Given the graph of some function $f(x)$ in Fig. 4.19, explain how the graph would change if we wanted to sketch $-f(x)$. [*Hint:* Graph $f(x) = x^2$ and compare with the graph of $g(x) = -x^2$.]

Figure 4.16

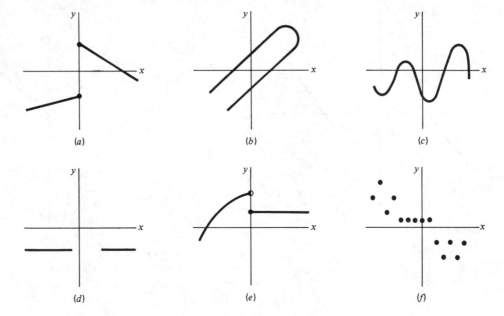

(a) (b) (c)

(d) (e) (f)

Figure 4.17

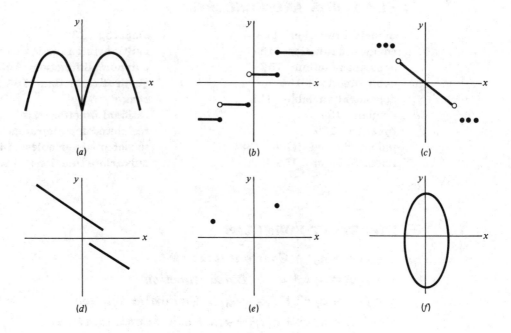

(a) (b) (c)

(d) (e) (f)

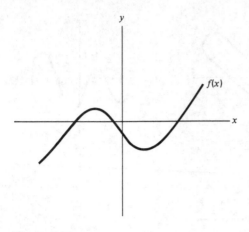

Figure 4.18

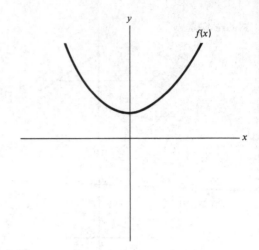

Figure 4.19

❏ KEY TERMS AND CONCEPTS

bivariate function   145
composite function   157
constant function   152
cubic function   155
dependent variable   138
domain   137
function   137
independent variable   138
linear function   153

mapping   137
multivariate function   145
polynomial function   155
quadratic function   154
range   137
rational function   156
restricted domain/range   144
subscripted variables   147
univariate functions   145

❏ IMPORTANT FORMULAS

$y = f(x) = a_0$ *Constant function*   (4.3)

$y = f(x) = a_1 x + a_0$ *Linear function*   (4.4)

$y = f(x) = a_2 x^2 + a_1 x + a_0$ *Quadratic function*   (4.5)

$y = f(x) = a_3 x^3 + a_2 x^2 + a_1 x + a_0$ *Cubic function*   (4.6)

$y = f(x) = a_n x^n + a_{n-1} x^{n-1} + \cdots + a_1 x + a_0$ *Polynomial function*

(4.7)

$y = f(x) = \dfrac{g(x)}{h(x)}$ *For g, h polynomials*   *Rational function*   (4.8)

❏ ADDITIONAL EXERCISES

Section 4.1

In Exercises 1–12, determine (a) $f(1)$, (b) $f(-2)$, and (c) $f(a - b)$.

1 $f(x) := -5x + 2$ **2** $f(x) = -x^2 + 3x + 10$

3 $f(t) = 10 - t + t^3$ **4** $f(u) = \sqrt{8}$

5 $f(x) = x/(-2 - x)$ **6** $f(r) = r^3 - 5r^2 + 3$

7 $f(x) = \sqrt{x^2 - 4}$ **8** $f(v) = (v - 1)/(-4 + v^2)$

9 $f(z) = z^4/\sqrt{64}$ **10** $f(x) = 2^x$

11 $f(x) = (x - 2)^{2x}$ **12** $f(t) = (3t^2 + 5)/(1 - t^2 + t^{10})$

In Exercises 13–24, determine the domain of the function.

13 $f(x) = 100^x$ **14** $f(x) = 4^x/(x^9 - 256x)$

15 $f(x) = \sqrt{x/(8 - x)}$ **16** $f(x) = \sqrt{x^2 - 3x + 2}/(4 - x^2)$

17 $f(x) = \sqrt{(x - 2)/(x^2 - 16)}$ **18** $f(x) = \sqrt{x^3 - 8}$

19 $f(x) = (2)^{2x}$ **20** $f(x) = e^{x^2}$, where $e = 2.718 \ldots$

21 $g(u) = \sqrt{10 - \dfrac{u}{4}} \Big/ (u^2 - 64)$ **22** $h(t) = (2)^t/\sqrt{t^2 - 7t - 12}$

23 $f(v) = \sqrt{v^2 - 9}/(v^3 - 4v)$ **24** $r(s) = \sqrt{25 - s}/(3)^s$

25 Given $f(a, b) = 3a^2 - 2ab + 5b^2$, determine (a) $f(-2, 3)$ and (b) $f(x + y, x - y)$.

26 Given $f(a, b, c, d) = a^3 - 2abc + cd^2 - 5$, determine (a) $f(-1, 2, 0, 1)$ and (b) $f(0, 0, 0, 0)$.

27 Given $f(x_1, x_2, x_3) = x_1^2 - 2x_1 x_2 + x_2 x_3 - 10$, determine (a) $f(0, 1, -3)$ and (b) $f(10, -10, 10)$.

28 Given $h(x, y, z) = x^2 y z^3$, determine (a) $h(2, -3, 1)$ and (b) $h(a + b, a, b)$.

29 A local radio station has been given the exclusive right to promote a concert in the city's civic arena, which seats 30,000 persons. The commission for the radio station is $5,000 plus $2.50 for each ticket sold to the concert.
(a) Determine the function $C = f(n)$, where C equals the commission paid to the radio station, stated in dollars, and n equals the number of tickets sold.
(b) Determine the restricted domain and range for this function.

30 A salesperson has been hired to sell three products. The salesperson is paid on a commission basis, earning $2.50, $3.00, and $2.00 per unit, respectively, for products 1, 2, and 3. In addition, the salesperson receives a base salary of $40 per week. x_j equals the number of units sold per week of product j for $j = 1, 2, 3$, and s equals the weekly salary in dollars.
(a) Determine the salary function $s = f(x_1, x_2, x_3)$.
(b) If maximum weekly sales for the three products are estimated at 20, 35, and 25 units, determine the restricted domain and range for the salary function.

Section 4.2

In Exercises 31–44, classify each function by type (constant, linear, quadratic, etc.), if possible.

31 $g(u) = (u^2 - 10u)/3$

32 $f(x) = (24 - x + 3x^2 + x^{10})/(x - 50)$

33 $g(h) = \log_{10}(0.01)$

34 $f(t) = (5t^3 - 3t^2)^0$

35 $f(s) = (0.25)^{s^2 + 3s}$

36 $f(x) = \log_5(x + 1)$

37 $f(x) = 3x/(25 - x^5)$

38 $f(n) = (5)^n/25$

39 $f(x) = (x^2 - 4)^4$

40 $f(x) = \sqrt[3]{x^{12}}$

41 $f(x) = 7.5(x^5)^0$

42 $g(h) = (3h/2)/15$

43 $f(x) = \sqrt{x}/x^3$

44 $f(t) = (t^2 - 3t + 12)/\sqrt{(t^4 + t^2)^0}$

45 University Enrollments A university projects that enrollments are going to decline as the pool of college-aged applicants begins to shrink. They have estimated the number of applications for coming years to behave according to the function

$$a = f(t) = 6,500 - 250t$$

where a equals the number of applications for admission to the university and t equals time in years measured from this current year ($t = 0$).
(a) This function is an example of what class of functions?
(b) What is the expected number of applications 5 years from today? 10 years?
(c) Do you think this function is accurate as a predictor indefinitely into the future? What kinds of factors would influence the restricted domain on t?

46 Gun Control With crime rates on the rise, it is estimated that the number of handguns in circulation is increasing. The FBI uses the function

$$n = f(t) = 25.5 + 0.025t$$

where n equals the estimated number of handguns in circulation (stated in millions) and t represents time measured in years from this current year ($t = 0$).
(a) This function is an example of what class of functions?
(b) What is the estimated number of handguns 20 years from now?
(c) How long will it take for the number of handguns to equal 26.5 million?

47 The total cost of producing x units of a product is estimated by the cost function

$$C = f(x) = 60x + 0.2x^2 + 25,000$$

where C equals total cost measured in dollars.
(a) This function is an example of what class of functions?
(b) What is the cost associated with producing 25,000 units?
(c) What is the cost associated with producing 0 units? What term might be used to describe this cost?

48 A retailer has determined that the annual cost C of purchasing, owning, and maintaining one of its products behaves according to the function

$$C = f(q) = \frac{20,000}{q} + 0.5q + 50,000$$

where q is the size (in units) of each order purchased from suppliers.
(a) What is the annual cost if the order size equals 1,000 units? 2,000 units?
(b) See whether you can determine the restricted domain for this function. (*Hint:* You will not be able to come up with an actual number for the upper limit; rather, discuss factors which will influence that value.)

49 If $y = g(u) = u^3 - 5u$ and $u = h(x) = x^2 - 4$, determine (a) $g(h(x))$ and (b) $g(h(2))$.

50 If $y = g(u) = u - 5$ and $u = h(x) = x^2 - 3x + 6$, determine (a) $g(h(x))$ and (b) $g(h(-2))$.

51 If $y = f(r) = r^2$, $r = g(s) = s^2 - 4$, and $s = h(x) = x - 3$, determine the composite functions (a) $g(h(x))$ and (b) $f(g(h(x)))$.

Section 4.3

Sketch each of the following functions.

52 $f(x) = -3$

54 $f(x) = -x^5$

56 $f(x) = \begin{cases} -x & x < 0 \\ 2 & 0 \le x < 3 \\ x & x \ge 3 \end{cases}$

53 $f(x) = -x^2 + 9$

55 $f(x) = 2^x$

57 $f(x) = \begin{cases} x^2 & x < 0 \\ -x^2 & x \ge 0 \end{cases}$

58 A salesperson receives a base monthly salary of $250 plus a commission of $4 for each unit sold. If monthly sales exceed 500 units, the salesperson receives a bonus of $150 plus an additional $2.50 for all units sold in excess of 500 units.

(a) Formulate the monthly compensation function $S = f(x)$, where S equals the monthly compensation in dollars and x equals the number of units sold per month.

(b) Sketch the compensation function.

❑ CHAPTER TEST

1 Determine the domain of the function

$$f(x) = \sqrt{x^2 + x - 6}$$

2 The function $V = f(t) = 36,000 - 4,500t$ states that the value of a piece of equipment is a function of its age. V equals the value (in dollars) and t equals the age of the equipment (in years). Determine the restricted domain and range for this function.

3 A travel agent is planning a week-long ski trip for students at a local university. The package includes airfare, transfers, lodging, breakfast and dinner each day, and lift tickets. The price is $450 per person unless the number of persons signing up exceeds 150. In this case, the package price for all persons is reduced by $2.50 for each person in excess of 150. If p equals the price of the package in dollars and n equals the number of persons signing up for the tour, determine the function $p = f(n)$.

4 If $y = g(u) = u^2/(1 - 2u)$ and $u = h(x) = x + 5$, determine the composite functions (a) $g(h(x))$ and (b) $g(h(-5))$.

5 Classify each of the following functions (constant, linear, etc.), if possible.

(a) $f(x) = x^6/\sqrt{x}$

(b) $f(x) = x^3/(1 - 3x + x^2)$

(c) $f(x) = 1/(10 - x + x^2)^{-1}$

(d) $f(x) = (120 - 5x)/24$

6 Sketch the function

$$f(x) = \begin{cases} x^2 & x \ge 0 \\ -x & x < 0 \end{cases}$$

CHAPTER 5

LINEAR FUNCTIONS: APPLICATIONS

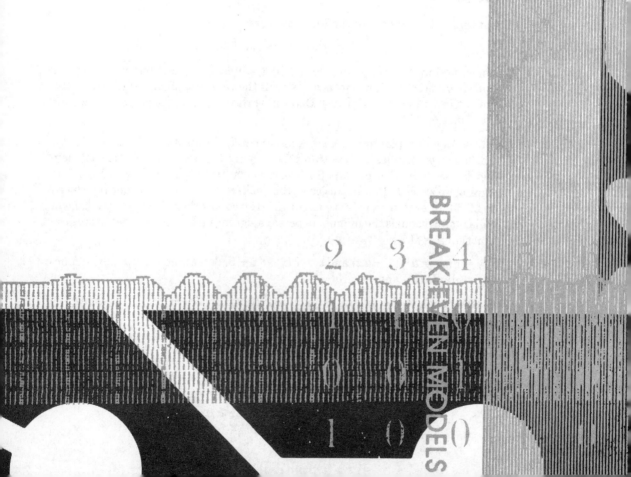

CHAPTER OBJECTIVES

❏ Present a discussion of the characteristics of linear functions

❏ Present a wide variety of applications of linear functions

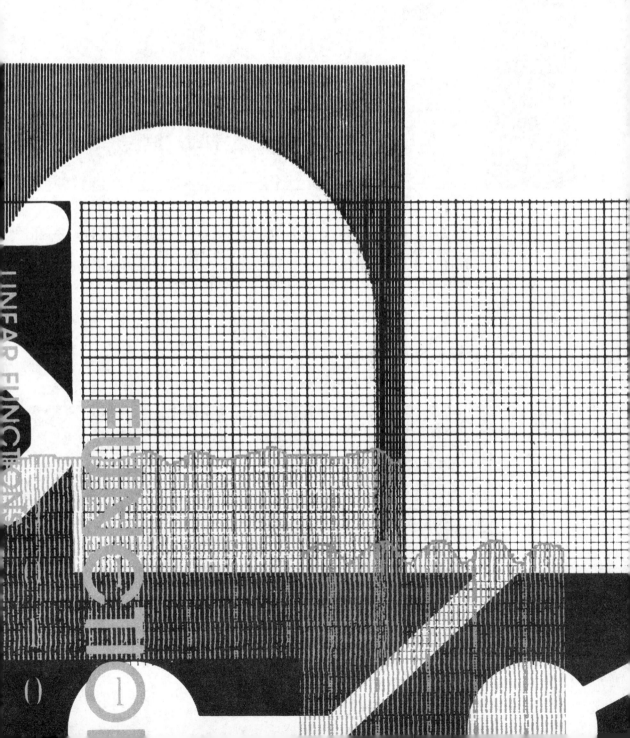

MOTIVATING SCENARIO: Federal Income Taxes	In 1990, the federal tax rates for a married couple filing jointly were as shown in the table.

Taxable Income		
Over	But not Over	Tax Rate
$ 0	$ 32,450	15%
32,450	78,400	28
78,400	162,770	33
162,770		28

What is desired is a formula, or set of formulas, which would enable a married couple to calculate their federal taxes once they know their taxable income. [Example 10]

In this chapter we extend the material presented in Chaps. 2 and 4 by presenting a discussion of linear functions. After reviewing the form and assumptions underlying these functions, we will see examples which illustrate the applications of these models in business, economics, and other areas.

5.1 LINEAR FUNCTIONS

General Form and Assumptions

DEFINITION: LINEAR FUNCTION INVOLVING ONE INDEPENDENT VARIABLE

A *linear function f* involving one independent variable x and a dependent variable y has the general form

$$y = f(x) = a_1 x + a_0 \tag{5.1}$$

where a_1 and a_0 are constants, $a_1 \neq 0$.

Equation (5.1) should be familiar from the previous chapter. In addition, you should recognize this as the slope-intercept form of a linear equation with slope a_1 and y intercept occurring at $(0, a_0)$. *For a linear function having the form of Eq. (5.1), a change in the value of y is directly proportional to a change in the value of x.* This rate of change is constant and represented by the slope a_1.

Example 1 in Chap. 4 presented the linear salary function

$$y = f(x) = 3x + 25$$

Figure 5.1
Linear salary
function.

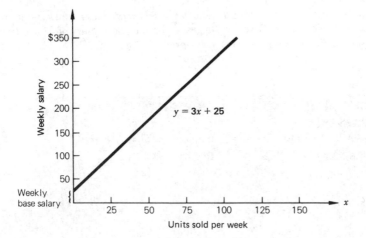

where y is defined as weekly salary in dollars and x represents the number of units sold per week. In this salary function, the salesperson is paid a base salary of $25 per week and a commission of $3 per unit sold. The change in the person's weekly salary is directly proportional to the change in the number of units sold. That is, the slope of 3 indicates the increase in weekly salary associated with each additional unit sold. The graph of the salary function appears in Fig. 5.1. Note that the graph is in the first quadrant, restricting x and y to nonnegative values. Does this make sense?

DEFINITION: LINEAR FUNCTION INVOLVING TWO INDEPENDENT VARIABLES
A linear function f involving two independent variables x_1 and x_2 and a dependent variable y has the general form

$$y = f(x_1, x_2) = a_1 x_1 + a_2 x_2 + a_0 \qquad (5.2)$$

where a_1 and a_2 are (nonzero) constants and a_0 is a constant.

For a linear function of the form of Eq. (5.2) the variable y depends jointly on the values of x_1 and x_2. The value of y varies in direct proportion to changes in the values of x_1 and x_2. Specifically, if x_1 increases by 1 unit, y will change by a_1 units. And if x_2 increases by 1 unit, y will change by a_2 units.

EXAMPLE 1 Assume that a salesperson's salary depends on the number of units sold of each of two products. More specifically, assume that the salary function

$$y = f(x_1, x_2)$$

is

$$y = 5x_1 + 3x_2 + 25$$

where $y = $ *weekly salary*, $x_1 = $ *number of units sold of product 1*, and $x_2 = $ *number of units sold of product 2*. This salary function suggests a base weekly salary of $25 and commissions per unit sold of $5 and $3, respectively, for products 1 and 2.

❑

> **DEFINITION: LINEAR FUNCTION INVOLVING n INDEPENDENT VARIABLES**
> A linear function f involving n independent variables $x_1, x_2, \ldots, x_n$ and a dependent variable y has the general form
>
> $$y = f(x_1, x_2, \ldots, x_n)$$
>
> or $$y = a_1 x_1 + a_2 x_2 + \cdots + a_n x_n + a_0 \qquad (5.3)$$
>
> where $a_1, a_2, \ldots, a_n$ are (nonzero) constants and a_0 is a constant.

Linear Cost Functions

Organizations are concerned with *costs* because they reflect dollars flowing out of the organization. These outflows usually pay for salaries, raw materials, supplies, rent, heat, utilities, and so forth. As mentioned earlier, accountants and economists often define total cost in terms of two components: **total variable cost** and **total fixed cost.** These two components must be added to determine total cost. The cost function for owning and operating the patrol car in Example 3 of Chap. 4 is an example of a linear cost function. The cost function,

$$C(x) = 0.40x + 18,000$$

had variable costs which varied with the number of miles driven and fixed costs of $18,000.

Total variable costs vary with the level of output and are computed as the product of **variable cost per unit of output** and the level of output. In a production setting, variable cost per unit is usually composed of raw material and labor costs. In the example of the patrol car, variable cost per mile consisted of operating costs per mile such as gasoline, oil, maintenance costs, and depreciation.

Linear cost functions are very often realistic, although they ignore the possibility of **economies** or **diseconomies of scale.** That is, linear cost functions imply **constant returns to scale.** Constant returns to scale imply that regardless of the number of units produced, the variable cost for each unit is the same. This assumption ignores the possibility that the elements of the production process (laborers or machines) may become more efficient as the number of units produced increases or that buying raw materials in large quantities may result in quantity discounts which in turn may lower the variable cost per unit produced (this is an example of economies of scale). The cost function for the patrol car assumes that operating costs per mile will be $0.40 regardless of the number of miles driven. We might expect that over the life of a piece of equipment, such as the patrol car, it will become less efficient and will require greater maintenance. This should translate into a higher variable cost per unit. Some cost models recognize these potential "nonlinearities" by using some measure of *average variable cost per unit*. In other situations a set of linear cost functions might be developed, each appropriate in certain cases depending on the level of output selected.

The following example illustrates the formulation of a linear cost function.

EXAMPLE 2 A firm which produces a single product is interested in determining the function that expresses annual total cost y as a function of the number of units produced x. Accountants indicate that fixed expenditures each year are $50,000. They also have estimated that raw material costs for each unit produced are $5.50, and labor costs per unit are $1.50 in the assembly department, $0.75 in the finishing room, and $1.25 in the packaging and shipping department.

The total cost function will have the form

$$y = C(x)$$
$$= \text{total variable cost} + \text{total fixed cost}$$

Total variable costs consist of two components: raw material costs and labor costs. Labor costs are determined by summing the respective labor costs for the three departments. Total cost is defined by the function

$$y = \text{total raw material cost} + \text{total labor cost} + \text{total fixed cost}$$

$$= \begin{matrix} \text{total raw} \\ \text{material cost} \end{matrix} + \begin{matrix} \text{labor cost} \\ \text{(assembly dept.)} \end{matrix} + \begin{matrix} \text{labor cost} \\ \text{(finishing room)} \end{matrix}$$
$$+ \begin{matrix} \text{labor cost} \\ \text{(shipping dept.)} \end{matrix} + \begin{matrix} \text{total fixed} \\ \text{cost} \end{matrix}$$

or $\qquad y = 5.50x + (1.50x + 0.75x + 1.25x) + 50,000$

which simplifies to

$$y = f(x) = 9x + 50,000$$

The 9 represents the combined variable cost per unit of $9.00. That is, for each additional unit produced, total cost will increase by $9.

❑

Linear Revenue Functions

The money which flows into an organization from either selling products or providing services is often referred to as ***revenue.*** The most fundamental way of computing total revenue from selling a product (or service) is

$$\text{Total revenue} = (\text{price})(\text{quantity sold})$$

An assumption in this relationship is that the selling price is the same for all units sold.

Suppose a firm sells n products. If x_i equals the number of units sold of product i and p_j equals the price of product j, the function which allows you to compute total revenue from the sale of the n products is

$$R = p_1 x_1 + p_2 x_2 + p_3 x_3 + \cdots + p_n x_n \tag{5.4}$$

This revenue function can be stated more concisely using *summation notation* as

$$R = \sum_{j=1}^{n} p_j x_j \tag{5.5}$$

Those of you seeing summation notation for the first time may want to refer to Appendix B for an introduction to this concept.

EXAMPLE 3   A local car rental agency, Hurts Renta-Lemon, is trying to compete with some of the larger national firms. Management realizes that many travelers are not concerned about frills such as windows, hubcaps, radios, and heaters. I. T. Hurts, owner and president of Hurts, has been recycling used cars to become part of the fleet. Hurts has also simplified the rental rate structure by charging a flat $9.95 per day for the use of a car. Total revenue for the year is a linear function of the number of car-days rented out by the agency, or *if R = annual revenue in dollars and d = number of car-days rented during the year,*

$$R = f(d) = 9.95d$$

❑

Linear Profit Functions

Profit for an organization is the difference between total revenue and total cost. Stated in equation form,

$$\boxed{\text{Profit} = \text{total revenue} - \text{total cost}} \tag{5.6}$$

If $\text{Total revenue} = R(x)$

and $\text{Total cost} = C(x)$

where x equals the quantity produced and sold, then profit is defined as

$$P(x) = R(x) - C(x) \tag{5.7}$$

When total revenue exceeds total cost, profit is positive. In such cases the profit may be referred to as a **net gain,** or **net profit.** When total cost exceeds total revenue, profit is negative. In such cases, the negative profit may be referred to as a **net loss,** or **deficit.** When the revenue and cost are linear functions of the same variable(s), the profit function is a linear function of the same variable(s).

EXAMPLE 4

A firm sells a single product for $65 per unit. Variable costs per unit are $20 for materials and $27.50 for labor. Annual fixed costs are $100,000. Construct the profit function stated in terms of x, the number of units produced and sold. What profit is earned if annual sales are 20,000 units?

SOLUTION

If the product sells for $65 per unit, total revenue is computed by using the linear function

$$R(x) = 65x$$

Similarly, total annual cost is made up of material costs, labor costs, and fixed costs:

$$C(x) = 20x + 27.50x + 100,000$$

which reduces to the linear cost function

$$C(x) = 47.50x + 100,000$$

Thus the profit function is computed as

$$P(x) = R(x) - C(x)$$
$$= 65x - (47.50x + 100,000)$$
$$= 17.50x - 100,000$$

Notice that $P(x)$ is a linear function. The slope of 17.50 indicates that for each additional unit produced and sold, total profit increases by $17.50. In business and economics, this is referred to as the **marginal profit** (the addition to total profit from selling the next unit). If the firm sells 20,000 units during the year,

$$P(20,000) = 17.50(20,000) - 100,000$$
$$= 350,000 - 100,000$$
$$= 250,000$$

EXAMPLE 5

(Agricultural Planning) A corporate agricultural organization has three separate farms which are to be used during the coming year. Each farm has unique characteristics which make it most suitable for raising one crop only. Table 5.1 indicates the crop selected for each farm, the annual cost of planting 1 acre of the crop, the expected revenue to be derived from each acre, and the fixed costs associated with operating each farm. In addition to the fixed costs associated with operating each farm, there are annual fixed costs of $75,000 for the corporation as a whole. Determine the profit function for the three-farm operation if $x_j = $ the number of acres planted at farm j, $r_j = $ revenue per acre at farm j, $c_j = $ cost per acre at farm j, and $F_j = $ fixed cost at farm j.

TABLE 5.1

Farm	Crop	Cost/Acre (c_j)	Revenue/Acre (r_j)	Fixed Cost (F_j)
1	Soybeans	$ 900	$1,300	$150,000
2	Corn	1,100	1,650	175,000
3	Potatoes	750	1,200	125,000

SOLUTION

Total revenue comes from the sale of crops planted at each of the three farms, or

$$R(x_1, x_2, x_3) = r_1 x_1 + r_2 x_2 + r_3 x_3$$
$$= 1{,}300x_1 + 1{,}650x_2 + 1{,}200x_3$$

Total costs are the sum of those at the three farms plus the corporate fixed costs, or

$$C(x_1, x_2, x_3) = c_1 x_1 + F_1 + c_2 x_2 + F_2 + c_3 x_3 + F_3 + 75{,}000$$
$$= 900x_1 + 150{,}000 + 1{,}100x_2 + 175{,}000 + 750x_3 + 125{,}000 + 75{,}000$$
$$= 900x_1 + 1{,}100x_2 + 750x_3 + 525{,}000$$

Total profit is a linear function computed as

$$P(x_1, x_2, x_3) = R(x_1, x_2, x_3) - C(x_1, x_2, x_3)$$
$$= 1{,}300x_1 + 1{,}650x_2 + 1{,}200x_3 - (900x_1 + 1{,}100x_2 + 750x_3 + 525{,}000)$$
$$= 400x_1 + 550x_2 + 450x_3 - 525{,}000$$

$\square$

Section 5.1 Follow-up Exercises

1 Write the general form of a linear function involving five independent variables.

2 Assume that the salesperson in Example 1 (page 177) has a salary goal of $800 per week. If product B is not available one week, how many units of product A must be sold to meet the salary goal? If product A is unavailable, how many units must be sold of product B?

3 Assume in Example 1 (page 177) that the salesperson receives a bonus when combined sales from the two products exceed 80 units. The bonus is $2.50 per unit for each unit over 80. With this incentive program, the salary function must be described by two different linear functions. What are they, and when are they valid?

4 For Example 4 (page 181), how many units must be produced and sold in order to (a) earn a profit of $1.5 million, and (b) earn zero profit (break even)?

5 A manufacturer of microcomputers produces three different models. The following table summarizes wholesale prices, material cost per unit, and labor cost per unit. Annual fixed costs are $25 million.

	Microcomputer		
	Model 1	Model 2	Model 3
Wholesale price/unit	$500	$1,000	$1,500
Material cost/unit	175	400	750
Labor cost/unit	100	150	225

(a) Determine a joint total revenue function for sales of the three different microcomputer models.

(b) Determine an annual total cost function for manufacturing the three models.

(c) Determine the profit function for sales of the three models.

(d) What is annual profit if the firm sells 20,000, 40,000 and 10,000 units, respectively, of the three models?

6 For Example 5 (page 181), the board of directors has voted on the following planting program for the coming year: 1,000 acres will be planted at farm 1, 1,600 at farm 2, and 1,550 at farm 3.

(a) What are the expected profits for the program?

(b) A summer drought has resulted in the revenue yields per acre being reduced by 20, 30, and 10 percent, respectively, at the three farms. What is the profit expected from the previously mentioned planting program?

7 Automobile Leasing A car-leasing agency purchases new cars each year for use in the agency. The cars cost $15,000 new. They are used for 3 years, after which they are sold for $4,500. The owner of the agency estimates that the variable costs of operating the cars, exclusive of gasoline, are $0.18 per mile. Cars are leased for a flat fee of $0.33 per mile (gasoline not included).

(a) Formulate the total revenue function associated with renting one of the cars a total of x miles over a 3-year period.

(b) Formulate the total cost function associated with renting a car for a total of x miles over 3 years.

(c) Formulate the profit function.

(d) What is profit if a car is leased for 60,000 miles over a 3-year period?

(e) What mileage is required in order to earn zero profit for 3 years?

8 A company produces a product which it sells for $55 per unit. Each unit costs the firm $23 in variable expenses, and fixed costs on an annual basis are $400,000. If x equals the number of units produced and sold during the year:

(a) Formulate the linear total cost function.

(b) Formulate the linear total revenue function.

(c) Formulate the linear profit function.

(d) What does annual profit equal if 10,000 units are produced and sold during the year?

(e) What level of output is required in order to earn zero profit?

9 A gas station sells unleaded regular gasoline and unleaded premium. The price per gallon charged by the station is $1.299 for unleaded regular and $1.379 for unleaded premium. The cost per gallon from the supplier is $1.219 for unleaded regular and $1.289 for premium. If x_1 equals the number of gallons sold of regular and x_2 the number of gallons sold of premium:

(a) Formulate the revenue function from selling x_1 and x_2 gallons, respectively, of the two grades of gasoline.

(b) Formulate the total cost function from purchasing x_1 and x_2 gallons, respectively, of the two grades.

(c) Formulate the total profit function.

(d) What is total profit expected to equal if the station sells 100,000 gallons of unleaded regular and 40,000 of unleaded premium?

5.2 OTHER EXAMPLES OF LINEAR FUNCTIONS

In this section we will see, by example, other applications of linear functions.

EXAMPLE 6 (Straight-Line Depreciation) When organizations purchase equipment, vehicles, buildings, and other types of "capital assets," accountants usually allocate the cost of the item over the period the item is used. For a truck costing $20,000 and having a useful life of 5 years, accountants might allocate $4,000 a year as a cost of owning the truck. The cost allocated to any given period is called ***depreciation***.

Accountants also keep records of each major asset and its current, or "book," value. For instance, the value of the truck may appear on accounting statements as $20,000 at the time of purchase, $20,000 − $4,000 = $16,000 1 year from the date of purchase, and so forth. Depreciation can also be thought of as the amount by which the book value of an asset has decreased.

Although there are a variety of depreciation methods, one of the simplest is ***straight-line depreciation.*** Under this method the rate of depreciation is constant. This implies that the book value declines as a linear function over time. *If V equals the book value (in dollars) of an asset* and *t equals time (in years) measured from the purchase date* for the previously mentioned truck,

$$V = f(t)$$
$$= \text{purchase cost} - \text{depreciation}$$
or
$$= 20{,}000 - 4{,}000t$$

The graph of this function appears in Fig. 5.2.

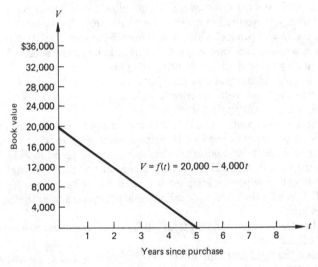

Figure 5.2 Book value function based upon straight-line depreciation.

PRACTICE EXERCISE
Define the restricted domain and range for this function. *Answer:* Domain = $\{t \mid 0 \le t \le 5\}$; range = $\{V \mid 0 \le V \le 20{,}000\}$.

EXAMPLE 7

(Linear Demand Functions) As discussed in Example 13 in Chap. 4, a ***demand function*** is a mathematical relationship expressing the way in which the quantity demanded of an item varies with the price charged for it. The relationship between these two variables — *quantity demanded* and *price per unit* — is usually *inverse;* i.e., a *decrease* in price results in

Figure 5.3
Linear demand
function.

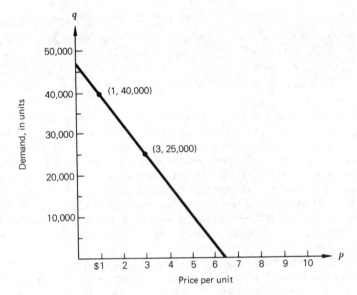

an *increase* in demand. The purpose of special sales is almost always to stimulate demand. If supermarkets reduced the price of filet mignon to $0.75 per pound, there would likely be a significant increase in the demand for that item. On the other hand, *increases* in the price of a product usually result in a *decrease* in the demand. The phrase *pricing people out of the market* refers to the customers lost as a result of price increases. If filet mignon were suddenly to triple in price with all other factors such as income levels held constant, many people currently capable of purchasing it would be priced out of the market.

There are exceptions to this behavior, of course. The demand for products or services which are considered *necessities* is likely to fluctuate less with moderate changes in price. Items such as prescription medical drugs, medical services, and certain food items are examples of this class of products.

Although most demand functions are nonlinear, there are situations in which the demand relationship either is, or can be approximated reasonably well by, a linear function. Figure 5.3 illustrates a linear demand function with two sample data points. Although most economics books measure price on the vertical axis and quantity demanded on the horizontal axis, we will reverse the labeling of the axes, as illustrated in Fig. 5.3. The reason for this is that most consumers view the demand relationship as having the form

$$\text{Quantity demanded} = f(\text{price per unit})$$

That is, consumers respond to price. Thus, quantity demanded, the dependent variable, is plotted on the vertical axis.

Verify, using the methods of Chap. 2, that the demand function in Fig. 5.3 has the form

$$q = f(p) = 47{,}500 - 7{,}500p$$

◻

EXAMPLE 8   **(Linear Supply Functions)** A *supply function* relates market price to the quantities
that suppliers are willing to produce and sell. The implication of supply functions is that
what is brought to the market depends upon the price people are willing to pay. As opposed
to the inverse nature of price and quantity demanded, the quantity which suppliers are
willing to supply usually varies directly with the market price. *All other factors being equal,
the higher the market price, the more a supplier would like to produce and sell; and the lower
the price people are willing to pay, the less the incentive to produce and sell.* Assume that you
own a lobster boat. All other factors considered equal, how much incentive is there to take
your boat and crew out if lobster is wholesaling at $0.25 per pound? How much incentive is
there if it is wholesaling at $10 per pound?

As with demand functions, supply functions can be approximated sometimes using
linear functions. Figure 5.4 illustrates a sample supply function. Note that by labeling the
vertical axis q, it is suggested that

$$\text{Quantity supplied} = f(\text{market price})$$

Figure 5.4   Linear supply function.

EXAMPLE 9   **(Market Equilibrium: Two Competing Products)** Given supply and demand functions for a product, **market equilibrium** exists if there is a price at which the quantity
demanded equals the quantity supplied. This example demonstrates the market equilibrium
for two competing products. Suppose that the following demand and supply functions have
been estimated for two competing products.

Figure 5.5 Linear supply function.

$$q_{d1} = f_1(p_1, p_2) = 100 - 2p_1 + 3p_2 \quad \text{(demand, product 1)}$$
$$q_{s1} = h_1(p_1) = 2p_1 - 4 \quad \text{(supply, product 1)}$$

$$q_{d2} = f_2(p_1, p_2) = 150 + 4p_1 - p_2 \quad \text{(demand, product 2)}$$
$$q_{s2} = h_2(p_2) = 3p_2 - 6 \quad \text{(supply, product 2)}$$

where q_{d1} = *quantity demanded of product 1*
$\quad q_{s1}$ = *quantity supplied of product 1*
$\quad q_{d2}$ = *quantity demanded of product 2*
$\quad q_{s2}$ = *quantity supplied of product 2*
$\quad p_1$ = *price of product 1, dollars*
$\quad p_2$ = *price of product 2, dollars*

Notice that the demand and supply functions are linear. Also note that the quantity demanded of a given product depends not only on the price of the product but also on the price of the competing product. The quantity supplied of a product depends only upon the price of that product.

Market equilibrium would exist in this two-product marketplace if prices existed (and were offered) such that

$$q_{d1} = q_{s1}$$

and

$$q_{d2} = q_{s2}$$

Supply and demand are equal for product 1 when

$$100 - 2p_1 + 3p_2 = 2p_1 - 4$$

or
$$4p_1 - 3p_2 = 104 \tag{5.8}$$

Supply and demand are equal for product 2 when

$$150 + 4p_1 - p_2 = 3p_2 - 6$$

or
$$-4p_1 + 4p_2 = 156 \tag{5.9}$$

If Eqs. (5.8) and (5.9) are solved simultaneously, equilibrium prices are identified as $p_1 = 221$ and $p_2 = 260$. This result suggests that if the products are priced accordingly, the quantities demanded and supplied will be equal for *each* product.

☐

PRACTICE EXERCISE
Given the prices identified above, calculate q_{d1}, q_{s1}, q_{d2}, and q_{s2}. Are equilibrium conditions satisfied? *Answer:* $q_{d1} = q_{s1} = 438$, $q_{d2} = q_{s2} = 774$; yes.

POINT FOR THOUGHT & DISCUSSION   Explain the logic underlying the assumptions in the demand and supply functions. That is, why is the demand for one product affected by the price of the product as well as the price of the competing product? Explain the logic of the plus sign for the "competing" product in each demand function. What is assumed by the inclusion of only one price variable in the supply functions? Under what circumstances would it be appropriate for both prices to be included in these functions?

EXAMPLE 10   (Federal Income Taxes; Motivating Scenario) In 1990, the federal tax rates for a married couple filing jointly were (repeating the table in the Motivating Scenario) as given in Table 5.2. What is desired is a mathematical function which allows a married couple to calculate their tax liability, given their taxable income.

TABLE 5.2   **1990 Federal Tax Rates (Married Filing Jointly)**

Taxable Income		Tax Rate
Over	But not Over	
$ 0	$ 32,450	15%
32,450	78,400	28
78,400	162,770	33
162,770		28

SOLUTION

Let
$$x = \textit{taxable income, dollars}$$
$$T = \textit{federal income tax liability, dollars}$$

We want to identify the function

$$T = f(x)$$

First, we must understand the information in Table 5.2. If a couple's taxable income is $0–$32,450, they must pay federal income tax equal to 15 percent of the taxable income. If their taxable income is greater than \$32,450 but not greater than \$78,400, they must pay 15 percent on the first \$32,450 and 28 percent on all income over \$32,450. If their taxable income is greater than \$78,400 but not greater than \$162,770, they must pay 15 percent on the first \$32,450, 28 percent on the next \$45,950 (\$78,400 − \$32,450), and 33 percent on all income over \$78,400. Thus, the tax rate applies only to income which falls within the corresponding range.

The tax function will be stated by four component functions, one for each of the taxable income ranges stated in Table 5.2. For example, if taxable income is \$0–\$32,450,

$$T = 0.15x$$

If taxable income is greater than \$32,450 but not more than \$78,400,

$$\begin{aligned} T &= 0.15(32,450) + 0.28(x - 32,450) \\ &= 4,867.5 + 0.28x - 9,086 \\ &= 0.28x - 4,218.5 \end{aligned}$$

If taxable income is greater than \$78,400 but not more than \$162,770,

$$\begin{aligned} T &= 0.15(32,450) + 0.28(45,950) + 0.33(x - 78,400) \\ &= 4,867.5 + 12,866 + 0.33x - 25,872 \\ &= 0.33x - 8,138.5 \end{aligned}$$

If taxable income is greater than \$162,770,

$$\begin{aligned} T &= 0.15(32,450) + 0.28(45,950) + 0.33(84,370) + 0.28(x - 162,770) \\ &= 4,867.5 + 12,866 + 27,842.1 + 0.28x - 45,575.6 \\ &= 0.28x \end{aligned}$$

The complete tax liability function is

$$T = f(x)$$

$$= \begin{cases} 0.15x & 0 < x \le 32,450 \\ 0.28x - 4,218.5 & 32,450 < x \le 78,400 \\ 0.33x - 8,138.5 & 78,400 < x \le 162,770 \\ 0.28x & 162,770 < x \end{cases}$$

Figure 5.6 presents a graph of this tax liability function for persons who filed in the category of "Married, filing jointly" for 1990.

EXAMPLE 11 (**Social Security Taxes**) Figure 5.7 is a graph of Social Security taxes collected in the years 1980–1989. The amount collected annually appeared to be increasing, approximately, at a linear rate. In 1980, the Social Security taxes collected were \$150 billion and in 1989, \$352 billion. Using these two data points, develop the linear function which estimates the Social Security taxes collected as a function of time since 1980.

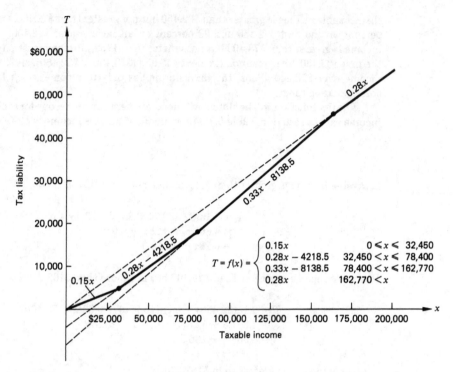

Figure 5.6   1990 tax liability: Married filing jointly.

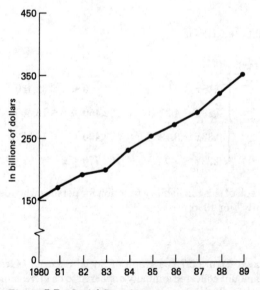

Figure 5.7   Social Security taxes.
(Data: Office of Management & Budget, DRI/McGraw-Hill)

SOLUTION

If we define

$$S = Social\ Security\ taxes\ collected,\ billions\ of\ dollars$$

$$t = time\ measured\ in\ years\ since\ 1980$$

we want to determine the linear function having the form

$$S = f(t) = a_1 t + a_0$$

The two data points (t, S) are (0, 150) and (9, 352). By observation, the value of a_0 equals 150. Substituting the data point for 1989 into the slope-intercept form gives

$$352 = a_1(9) + 150$$

$$202 = 9a_1$$

$$22.44 = a_1$$

Thus, the linear approximating function is

$$S = f(t) = 22.44t + 150$$

❑

Section 5.2 Follow-up Exercises

1 A piece of machinery is purchased for $80,000. Accountants have decided to use a straight-line depreciation method with the machine being fully depreciated after 6 years. Letting V equal the book value of the machine and t the age of the machine, determine the function $V = f(t)$. (Assume no salvage value.)

2 **Straight-Line Depreciation with Salvage Value** Many assets have a **resale,** or **salvage, value** even after they have served the purposes for which they were originally purchased. In such cases, the allocated cost over the life of the asset is the difference between the purchase cost and the salvage value. The cost allocated each time period is the allocated cost divided by the useful life. In Example 6, suppose that it is estimated that the truck (which cost $20,000) can be resold for $2,500 at the end of 5 years. The total cost to be allocated over the 5-year period is the purchase cost less the resale value, or $20,000 − $2,500 = $17,500. Using straight-line depreciation, the annual depreciation will be

$$\frac{Purchase\ cost - salvage\ value}{Useful\ life\ (in\ years)} = \frac{20,000 - 2,500}{5}$$

$$= \frac{17,500}{5}$$

$$= 3,500$$

The function expressing the book value V as a function of time t is

$$V = f(t) = 20{,}000 - 3{,}500t \qquad 0 \le t \le 5$$

In Exercise 1 assume that the machine will have a salvage value of $7,500 at the end of 6 years. Determine the function $V = f(t)$ for this situation.

3 A piece of machinery is purchased for $300,000. Accountants have decided to use a straight-line depreciation method with the machine being fully depreciated after 8 years. Letting V equal the book value of the machine and t the age of the machine, determine the function $V = f(t)$. Assume there is no salvage value.

4 Assume in Exercise 3 that the machine can be resold after 8 years for $28,000. Determine the function $V = f(t)$.

5 A company purchases cars for use by its executives. The purchase cost this year is $25,000. The cars are kept 3 years, after which they are expected to have a resale value of $5,600. If accountants use straight-line depreciation, determine the function which describes the book value V as a function of the age of the car t.

6 A police department believes that arrest rates R are a function of the number of plainclothes officers n assigned. The *arrest rate* is defined as the percentage of cases in which arrests have been made. It is believed that the relationship is linear and that each additional officer assigned to the plainclothes detail results in an increase in the arrest rate of 1.20 percent. If the current plainclothes force consists of 16 officers and the arrest rate is 36 percent:
(a) Define the function $R = f(n)$.
(b) Interpret the meaning of the R intercept.
(c) Determine the restricted domain and range for the function.
(d) Sketch the function.

7 Two points on a linear demand function are ($20, 80,000) and ($30, 62,500).
(a) Determine the demand function $q = f(p)$.
(b) Determine what price would result in demand of 50,000 units.
(c) Interpret the slope of the function.
(d) Define the restricted domain and range for the function.
(e) Sketch $f(p)$.

8 Two points (p, q) on a linear demand function are ($24, 60,000) and ($32, 44,400).
(a) Determine the demand function $q = f(p)$.
(b) What price would result in demand of 80,000 units?
(c) Interpret the slope of the function.
(d) Determine the restricted domain and range.
(e) Sketch $f(p)$.

9 Two points on a linear supply function are ($4.00, 28,000) and ($6.50, 55,000).
(a) Determine the supply function $q = f(p)$.
(b) What price would result in suppliers offering 45,000 units?
(c) Determine and interpret the p intercept.

10 Two points (p, q) on a linear supply function are ($3.50, 116,000) and ($5.00, 180,000).
(a) Determine the supply function $q = f(p)$.
(b) What price would result in suppliers offering 135,000 units for sale?
(c) Interpret the slope of the function.
(d) Determine and interpret the p intercept.
(e) Sketch $f(p)$.

11 **Alimony/Child Support** Recent surveys indicate that payment of alimony or child support tends to decline with time elapsed after the divorce decree. One survey uses the estimating function

$$p = f(t) = 90 - 12.5t$$

where p equals the percentage of cases in which payments are made and t equals time measured in years after the divorce decree.
(a) Interpret the p intercept.
(b) Interpret the slope.
(c) In what percentage of cases is alimony/child support paid after 5 years?
(d) Sketch $f(t)$.

12 Sports Injuries A survey of high school and college football players suggests that the number of career-ending injuries in this sport is increasing. In 1980 the number of such injuries was 925; in 1988 the number was 1,235. If it is assumed that the injuries are increasing at a linear rate:
(a) Determine the function $n = f(t)$, where n equals the estimated number of injuries per year and t equals time measured in years since 1980.
(b) Interpret the meaning of the slope of this function.
(c) When is it expected that the number of such injuries will go over the 1,500 mark?

13 Marriage Prospects Data released by the Census Bureau in 1986 indicated the likelihood that never-married women would eventually marry. The data indicated that the older the woman, the less the likelihood of marriage. Specifically, two statistics indicated that women who were 45 and never-married had an 18 percent chance of marriage and women 25 years old had a 78 percent chance of marriage. Assume that a linear fit to these two data points provides a reasonable approximation for the function $p = f(a)$, where p equals the probability of marriage and a equals the age of a never-married woman.
(a) Determine the linear function $p = f(a)$.
(b) Interpret the slope and p intercept.
(c) Do the values in part b seem reasonable?
(d) If the restricted domain on this function is $20 \le a \le 50$, determine $f(20)$, $f(30)$, $f(40)$, and $f(50)$.

14 Two-Income Families Figure 5.8 illustrates the results of a survey regarding two-income families. The data reflect the percentage of married couples with wives who work for four different years. The percentage appears to be increasing approximately at a linear rate. Using the data points for 1960 and 1988:
(a) Determine the linear function $P = f(t)$, where P equals the estimated percentage of married couples with wives who work and t equals time measured in years since 1950 ($t = 0$ corresponds to 1950).

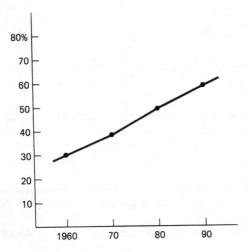

Figure 5.8 Percentage of married couples with wives who work.

(b) Interpret the meaning of the slope and the P intercept.

(c) When is it expected that the percentage will exceed 75 percent?

15 **Education Expenditures** Figure 5.9 illustrates the expenditures per student in U.S. public schools over a three-decade period. The expenditures are stated in "constant dollars" which filter out the effects of inflation. The increase in expenditures per student appears to have occurred approximately at a linear rate. In 1958, the expenditures per student were $1,750; in 1984, the expenditures were $3,812.50. Using these two data points:

(a) Determine the linear approximating function $E = f(t)$, where E equals the estimated expenditure per student in dollars and t equals time measured in years *since* 1955 ($t = 0$ corresponds to 1955).

(b) Interpret the slope and E intercept.

(c) According to this function, what are expenditures per student expected to equal in the year 2000?

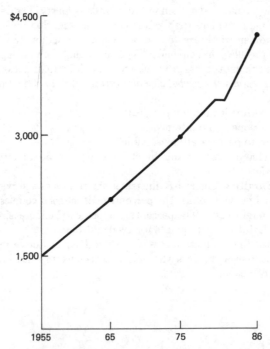

Figure 5.9   Expenditures per student in U.S. public schools (constant dollars). (Source: U.S. Dept. of Education, National Center for Education Statistics)

16 **Walt Disney Company** Figure 5.10 portrays the annual operating profits for Walt Disney Company between 1986 and 1990 (estimated). During this period, annual operating profits appear to have increased approximately in a linear manner. In 1987, annual operating profits were $0.762 billion; in 1989, they were $1.220 billion. Using these two data points:

(a) Determine the linear function $P = f(t)$, where P equals the estimated annual operating profits and t equals time measured in years *since* 1986.

(b) Interpret the slope and P intercept.

(c) Using this function, estimate annual operating profits for Walt Disney Company in the year 2000.

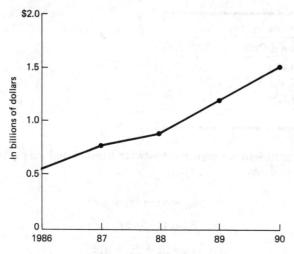

Figure 5.10   Annual operation profits, Walt Disney Company.
(Data: Company Reports, Wertheim Schroder & Co.)

17   Economic Downturn A general trend of economic decline within New York City is
reflected by Fig. 5.11. This figure indicates the vacancy rate for offices in Manhattan
during the period 1985–1990. The increase in the vacancy rate appears to be approximately linear. The vacancy rate in 1986 was 9.4 percent; in 1989, the rate was 13.2
percent. Using these two data points:
(a) Determine the linear function $V = f(t)$, where V equals the estimated vacancy rate
    (in percent) and t equals time measured in years since 1985.
(b) Interpret the slope and V intercept.
(c) Using the function, estimate the vacancy rate in 1995.

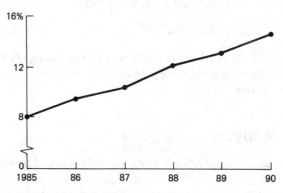

Figure 5.11   Percentage of Manhattan offices vacant.
(Business Week, June 18, 1990)

18   Federal Income Taxes Table 5.3 contains the 1990 federal tax rates for a single
person. Determine the function $T = f(x)$, where T equals the tax liability (in dollars)
for a single person and x equals taxable income (in dollars).

TABLE 5.3	Taxable Income		
	Over	**But not Over**	**Tax Rate**
	$ 0	$19,450	15%
	19,450	47,050	28
	47,050	97,620	33
	97,620		28

19 **Market Equilibrium** Given the following demand and supply functions for two competing products,

$$q_{d1} = 82 - 3p_1 + p_2$$

$$q_{s1} = 15p_1 - 5$$

$$q_{d2} = 92 + 2p_1 - 4p_2$$

$$q_{s2} = 32p_2 - 6$$

determine whether there are prices which bring the supply and demand levels into equilibrium for the two products. If so, what are the equilibrium quantities?

*20 **Market Equilibrium: Three Competing Products** The following are the demand and supply functions for three competing products.

$$q_{d1} = 46 - 10p_1 + 2p_2 + 2p_3$$

$$q_{s1} = 12p_1 - 16$$

$$q_{d2} = 30 + 2p_1 - 6p_2 + 4p_3$$

$$q_{s2} = 6p_2 - 22$$

$$q_{d3} = 38 + 2p_1 + 4p_2 - 8p_3$$

$$q_{s3} = 6p_3 - 10$$

Determine whether there are prices which would bring the supply and demand levels into equilibrium for each of the three products. If so, what are the equilibrium demand and supply quantities?

5.3 BREAK-EVEN MODELS

In this section we will discuss **break-even models,** a set of planning tools which can be, and has been, very useful in managing organizations. One significant indication of the performance of a company is reflected by the so-called bottom line of the income statement for the firm; that is, how much profit is earned! Break-even analysis focuses upon the profitability of a firm. Of specific concern in break-even analysis is identifying the level of operation or level of output that would result in a **zero profit.** This level of operations or output is called the **break-even point.** The break-even point is a useful reference point in the sense that it represents the level of operation at which **total revenue equals total cost.** Any changes from this level of operation will result in either a profit or a loss.

Break-even analysis is valuable particularly as a planning tool when firms are contemplating expansions such as offering new products or services. Similarly, it is useful in evaluating the pros and cons of beginning a new business venture. In each instance the analysis allows for a projection of profitability.

Assumptions

In this discussion we will focus upon situations in which both the total cost function and the total revenue function are linear. The use of a linear total cost function implies that variable costs per unit either are constant or can be assumed to be constant. The linear cost function assumes that total variable costs depend upon the level of operation or output. It is also assumed that the fixed-cost portion of the cost function is constant over the level of operation or output being considered.

The linear total revenue function assumes that the selling price per unit is constant. Where the selling price is not constant, average price is sometimes chosen for purposes of conducting the analysis.

Another assumption is that price per unit is greater than variable cost per unit. *Think about that for a moment.* If price per unit is less than variable cost per unit, a firm will lose money on every unit produced and sold. A break-even condition could never exist.

Break-Even Analysis

In break-even analysis the primary objective is to determine the break-even point. The break-even point may be expressed in terms of (1) *volume of output* (or level of activity), (2) *total dollar sales,* or possibly (3) *percentage of production capacity.* For example, it might be stated that a firm will break even at 100,000 units of output, when total sales equal $2.5 million or when the firm is operating at 60 percent of its plant capacity. We will focus primarily on the first of these three ways.

The methods of performing break-even analysis are rather straightforward, and there are alternative ways of determining the break-even point. The usual approach is as follows:

1 *Formulate total cost as a function of x, the level of output.*
2 *Formulate total revenue as a function of x.*
3 *Since break-even conditions exist when total revenue equals total cost, set C(x) equal to R(x) and solve for x. The resulting value of x is the break-even level of output and might be denoted by x_{BE}.*

An alternative to step 3 is to construct the profit function $P(x) = R(x) - C(x)$, set $P(x)$ equal to zero, and solve for x_{BE}.

The following example illustrates both approaches.

EXAMPLE 12 A group of engineers is interested in forming a company to produce smoke detectors. They have developed a design and estimate that variable costs per unit, including materials, labor, and marketing costs, are $22.50. Fixed costs associated with the formation, operation, and management of the company and the purchase of equipment and machinery total $250,000.

They estimate that the selling price will be $30 per detector.

(a) Determine the number of smoke detectors which must be sold in order for the firm to break even on the venture.

(b) Preliminary marketing data indicate that the firm can expect to sell approximately 30,000 smoke detectors over the life of the project if the detectors are sold for $30 per unit. Determine expected profits at this level of output.

SOLUTION

(a) If x equals the number of smoke detectors produced and sold, the total revenue function is represented by the equation

$$R(x) = 30x$$

The total cost function is represented by the equation

$$C(x) = 22.50x + 250,000$$

The break-even condition occurs when total revenue equals total cost, or when

$$\boxed{R(x) = C(x)} \tag{5.10}$$

For this problem the break-even point is computed as

$$30x = 22.50x + 250,000$$

or

$$7.50x = 250,000$$

and

$$x_{BE} = 33,333.33 \text{ units}$$

The alternative approach is first to write the profit function and set it equal to 0, as follows:

$$\begin{aligned} P(x) &= R(x) - C(x) \\ &= 30x - (22.50x + 250,000) \\ &= 7.50x - 250,000 \end{aligned}$$

Setting the profit function P equal to 0, we have

$$7.50x - 250,000 = 0$$

$$7.50x = 250,000$$

or

$$x_{BE} = 33,333.33 \text{ units}$$

This is the same result, and our conclusion is that *given the assumed cost and price parameters (values),* the firm must sell 33,333.33 units in order to break even.

PRACTICE EXERCISE
Verify that total revenue and total costs both equal $1,000,000 (taking rounding into account) at the break-even point.

(b) With sales projected at 30,000 smoke detectors,

$$P(30,000) = 7.5(30,000) - 250,000$$
$$= 225,000 - 250,000 = -25,000$$

This suggests that if all estimates — price, cost, and demand — hold true, the firm can expect to lose $25,000 on the venture.

EXAMPLE 13   **(Graphical Approach)** The essence of break-even analysis is illustrated quite effectively by graphical analysis. Figure 5.12a illustrates the total revenue function, Fig. 5.12b, the total cost function, and Fig. 5.12c, a composite graph showing both functions, for Example 12. Note in Fig. 5.12b that the fixed-cost component is distinguished from the variable-cost component. At *any* level of output x, the vertical distance within the darker shaded area indicates the fixed cost of $250,000. To this is added the total variable cost, which is represented by the vertical distance at x within the lighter area. The sum of these two vertical distances represents the total cost $C(x)$.

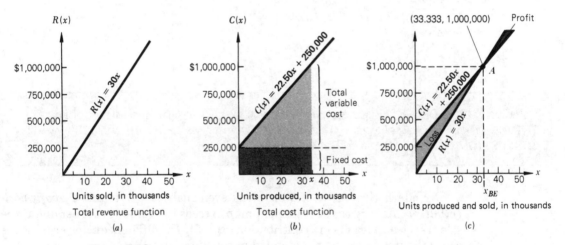

Figure 5.12

In Fig. 5.12c the two functions are graphed on the same set of axes. The point where the two functions intersect represents the one level of output where total revenue and total cost are equal. This is the break-even point. For all points to the left of the break-even point the cost function C has a value greater than the revenue function R. In this region the vertical distance separating the two functions represents the loss which would occur at a given level of output. To the right of $x = 33.333$, $R(x)$ is higher than $C(x)$, or $R(x) > C(x)$. For levels of output greater than $x = 33.333$ the vertical distance separating $R(x)$ and $C(x)$ represents the profit at a given level of output.

Figure 5.13 illustrates the profit function P for this example. The break-even point is identified by the x coordinate of the x intercept. Note that to the left of the break-even point the profit function is below the x axis, indicating a negative profit, or loss. To the right, $P(x)$ is above the x axis, indicating a positive profit.

◻

Figure 5.13
Profit function.

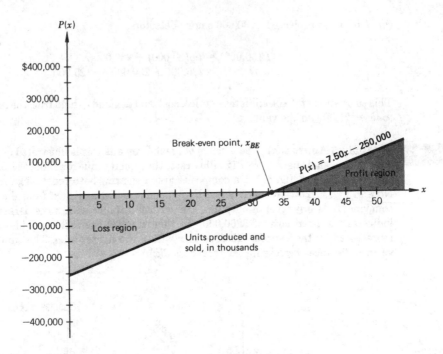

POINT FOR THOUGHT & DISCUSSION

Discuss any changes in Fig. 5.13 and the break-even point if (*a*) the price per unit increases (decreases), (*b*) the fixed cost increases (decreases), and (*c*) the variable cost per unit increases (decreases).

An alternative way of viewing break-even analysis is in terms of **profit contribution.** As long as the price per unit p exceeds the variable cost per unit v, the sale of each unit results in a contribution to profit. The difference between p and v is called the **profit margin.** Or stated in equation form,

$$\text{Profit margin} = p - v \qquad p > v \qquad\qquad (5.11)$$

The profit margin generated from the sale of units must first be allocated to recover any fixed costs which exist. At lower levels of output, the **total profit contribution** (profit margin for all units sold) is typically less than fixed costs, meaning that total profit is negative (see Fig. 5.13). Only when *total* profit contribution exceeds fixed cost will a positive profit exist. Because of this orientation — that the profit margin per unit contributes first to recovering fixed costs, after which it contributes to profit — profit margin is often called the **contribution to fixed cost and profit.**

With this perspective in mind, the computation of the break-even point can be thought of as determining the number of units to produce and sell in order to recover the fixed costs. The calculation of the break-even point is thus

$$\text{Break-even level} \atop \text{of output} = \frac{\text{fixed cost}}{\text{contribution to} \atop \text{fixed cost and profit}}$$

or
$$x_{BE} = \frac{FC}{p - v}$$
(5.12)

In fact, if you solve break-even problems by setting $R(x) = C(x)$ or by setting profit $P(x)$ equal to zero, the eventual calculation performed is that of Eq. (5.12). Applying Eq. (5.12) to Example 12 gives

$$x_{BE} = \frac{250,000}{30.000 - 22.50}$$

$$= \frac{250,000}{7.5}$$

$$= 33,333.33$$

If you look back at Example 12, the final computation reduced to the one above, regardless of the approach taken.

EXAMPLE 14 (**Convention Planning**) A professional organization is planning its annual convention to be held in San Francisco. Arrangements are being made with a large hotel in which the convention will be held. Registrants for the 3-day convention will be charged a flat fee of $500 per person, which includes registration fee, room, all meals, and tips. The hotel charges the organization $20,000 for the use of the facilities such as meeting rooms, ballroom, and recreational facilities. In addition, the hotel charges $295 per person for room, meals, and tips. The professional organization appropriates $125 of the $500 fee as annual dues to be deposited in the treasury of the national office. Determine the number of registrants necessary for the organization to recover the fixed cost of $20,000.

SOLUTION

The contribution to fixed cost and profit is the registration fee (price) per person less the cost per person charged by the hotel less the national organization's share per registrant, or

$$\text{Contribution per registrant} = \text{registration fee} - {\text{hotel charge} \atop \text{per person}} - \text{annual dues}$$
$$= 500 - 295 - 125 = \$80$$

Therefore, according to Eq. (5.12), the number of registrants required to recover the fixed cost is

$$x_{BE} = \frac{20,000}{80} = 250 \text{ persons}$$

EXAMPLE 15 (**The Movie "Dick Tracy"**) The movie "Dick Tracy," starring Warren Beatty and Madonna, was released in the summer of 1990. It was estimated that the movie cost the Walt Disney Company $45 million to produce and market. It was estimated that the movie would have to gross $100 million at the box office to "break even." *What percentage of the box office gross was Disney expecting to earn on this movie?*

SOLUTION

This example requires a slightly different type of analysis related to the break-even concept. If *x equals the percentage of the box office gross,* the information we are given is that Disney would break even if the box office gross equalled $100 million. The costs they need to recover are the $45 million production and marketing costs. Therefore, to break even

$$(\text{Box office gross})(\text{Disney \% of gross}) = 45 \text{ million}$$

or,
$$100x = 45$$

$$x = 45/100 = 0.45$$

Thus, Disney's share of the box office gross must equal 45%.

EXAMPLE 16   **(In-House Computer vs. Service Bureau Decision)** A large medical group practice has 30 full-time physicians. Currently, all billing of patients is done manually by clerks. Because of the heavy volume of billing the business manager believes it is time to convert from manual to computerized patient billing. Two options are being considered: (1) the group practice can lease its own computer and software and do the billing itself or (2) the group can contract with a computer service bureau which will do the patient billing.

The costs of each alternative are a function of the number of patient bills. The lowest bid submitted by a service bureau would result in an annual flat fee of $3,000 plus $0.95 per bill processed. With the help of a computer consultant, the business manager has estimated that the group can lease a small business computer system and the required software at a cost of $15,000 per year. Variable costs of doing the billing in this manner are estimated at $0.65 per bill.

If *x equals the number of patient bills per year,* the annual billing cost using a service bureau is represented by the function

$$S(x) = 3,000 + 0.95x$$

The annual cost of leasing a computer system and doing the billing in-house is expressed by the function

$$L(x) = 15,000 + 0.65x$$

These two alternatives are equally costly when

$$S(x) = L(x)$$

or
$$3,000 + 0.95x = 15,000 + 0.65x$$

$$0.30x = 12,000$$

$$x = 40,000$$

Thus, if the expected number of patient bills per year exceeds 40,000, the lease option is the

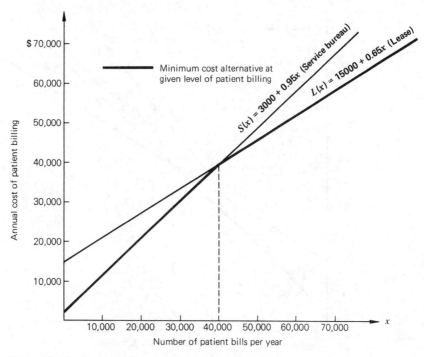

Figure 5.14 Patient-billing cost functions: two options.

less costly. If the number of patient bills is expected to be less than 40,000, the service bureau option is the less costly. Figure 5.14 illustrates the two cost functions.

◻

POINT FOR THOUGHT & DISCUSSION	Suppose the patient-billing volume is expected to be 35,000 per year. What reasons favoring the *lease* option could you present to the business manager? Discuss potential advantages and disadvantages which are not quantifiable for the *lease* and *service bureau* options.

EXAMPLE 17 **(Patient-Billing Revisited: Three Alternatives)** Suppose in the previous example that the business manager is not convinced that computer processing is the most cost-effective means of handling patient billing. She estimates that processing bills manually costs the group practice $1.25 per bill, or

$$M(x) = 1.25x$$

If this method is considered as a third option, let's examine the implications. The three cost functions are graphed together in Fig. 5.15. If you study this figure carefully, you should reach the following conclusions:

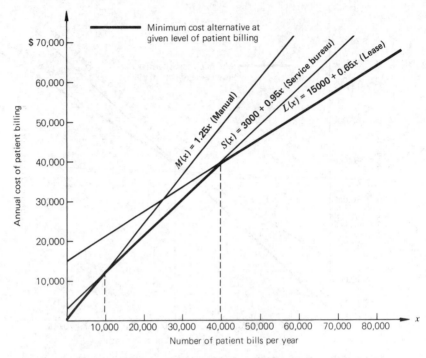

Figure 5.15   Patient-billing cost functions: three options.

1 *The least costly option at any level of patient bills is highlighted by the heavy line segments.*
2 *If the number of patient bills per year is expected to be less than 10,000, the manual system is the least costly.*
3 *If the number of patient bills is expected to be between 10,000 and 40,000, the service bureau arrangement is the least costly.*
4 *If the number is expected to exceed 40,000, the lease arrangement is the least costly.*

❑

NOTE   When performing break-even analysis for more than two alternatives (as in this example), it is strongly advised that you first sketch the relevant functions. Sometimes, the interaction which occurs between such functions is not always what might be expected. A sketch will quickly show the interaction.

EXAMPLE 18   **(Multiproduct Analysis)** Our discussion in this section has been limited to single-product/service situations. For multiproduct situations, break-even analysis can be performed when a ***product mix*** is known. The product mix expresses the ratio of output levels for the different products. For example, a firm having three products might produce 3 units of product *A* and 2 units of product *B* for each unit of product *C*. In this situation we might define *1* ***unit of product mix*** as consisting of 3 units of product *A*, 2 units of *B*, and 1 unit of

TABLE 5.4		Product		
		A	B	C
Price/unit		$40	$30	$55
Variable cost/unit		30	21	43
Profit margin		$10	$ 9	$12

C. If a product mix unit can be defined, we can conduct break-even analysis using this as the measure of output.

Suppose that these three products have the price and cost attributes shown in Table 5.4. Combined fixed cost for the three products is $240,000. Since *1 unit of the product mix* consists of 3 units of *A*, 2 units of *B*, and 1 unit of *C*, the *profit contribution per unit of product mix equals*

$$3(\$10) + 2(\$9) + 1(\$12) = \$60$$

If we let *x equal the number of units of product mix,* the profit function for the three products is

$$P(x) = 60x - 240,000$$

The break-even point occurs when $P(x) = 0$, or

$$60x - 240,000 = 0$$

$$60x = 240,000$$

$$x = 4,000$$

The firm will break even when it produces 4,000 product mix units, or 12,000 units of *A*, 8,000 units of *B*, and 4,000 units of *C*.

❏

The analysis presented in Example 18 presumes that a product mix is known. If the product mix is not known exactly but can be approximated, this analysis can still be of value as a planning tool.

Section 5.3 Follow-up Exercises

1 A firm sells a product for $45 per unit. Variable costs per unit are $33 and fixed costs equal $450,000. How many units must be sold in order to break even?

2 An enterprising college student has decided to purchase a local car wash business. The purchase cost is $150,000. Car washes will be priced at $5.50, and variable cost per car (soap, water, labor, etc.) is expected to equal $1.50. How many cars must be washed in order to recover the $150,000 purchase price?

3 A charitable organization is planning a raffle to raise $10,000. Five hundred chances will be sold on a new car. The car will cost the organization $15,000. How much should each ticket cost if the organization wishes to net a profit of $10,000?

4 A publisher has a fixed cost of $250,000 associated with the production of a college mathematics book. The contribution to profit and fixed cost from the sale of each book is $6.25.

(a) Determine the number of books which must be sold in order to break even.

(b) What is the expected profit if 50,000 books are sold?

5 A local university football team has added a national power to next year's schedule. The other team has agreed to play the game for a guaranteed fee of $100,000 plus 25 percent of the gate receipts. Assume the ticket price is $12.

(a) Determine the number of tickets which must be sold to recover the $100,000 guarantee.

(b) If college officials hope to net a profit of $240,000 from the game, how many tickets must be sold?

(c) If a sellout of 50,000 fans is assured, what ticket price would allow the university to earn the desired profit of $240,000?

(d) Again assuming a sellout, what would total profit equal if the $12 price is charged?

6 **Make or Buy Decision** Assume that a manufacturer can purchase a needed component from a supplier at a cost of $9.50 per unit, or it can invest $60,000 in equipment and produce the item at a cost of $7.00 per unit.

(a) Determine the quantity for which total costs are equal for the *make* and *buy* alternatives.

(b) What is the minimum cost alternative if 15,000 units are required? What is the minimum cost?

(c) If the number of units required of the component is close to the break-even quantity, what factors might influence the final decision to make or buy?

7 A local civic arena is negotiating a contract with a touring ice-skating show, Icey Blades. Icey Blades charges a flat fee of $60,000 per night plus 40 percent of the gate receipts. The civic arena plans to charge one price for all seats, $12.50 per ticket.

(a) Determine the number of tickets which must be sold each night in order to break even.

(b) If the civic arena has a goal of clearing $15,000 each night, how many tickets must be sold?

(c) What would nightly profit equal if average attendance is 7,500 per night?

8 In the previous exercise, assume that past experience with this show indicates that average attendance should equal 7,500 persons.

(a) What ticket price would allow the civic arena to break even?

(b) What ticket price would allow them to earn a profit of $15,000?

9 **Equipment Selection** A firm has two equipment alternatives it can choose from in producing a new product. One automated piece of equipment costs $200,000 and produces items at a cost of $4 per unit. Another semiautomated piece of equipment costs $125,000 and produces items at a cost of $5.25 per unit.

(a) What volume of output makes the two pieces of equipment equally costly?

(b) If 80,000 units are to be produced, which piece of equipment is less costly? What is the minimum cost?

10 **Robotics** A manufacturer is interested in introducing the robotics technology into one of its production processes. The process is one which would provide a "hostile environment" for humans. To be more specific, the process involves exposure to extremely high temperatures as well as to potentially toxic fumes. Two robots which appear to have the capabilities for executing the functions of the production process have been identified. There appear to be no significant differences in the speeds at which the two models work. One robot costs $180,000 and has estimated maintenance

costs of $100 per hour of operation. The second type of robot costs $250,000 with maintenance costs estimated at $80 per hour of operation.

(a) At what level of operation (total production hours) are the two robots equally costly? What is the associated cost?

(b) Define the levels of operation for which each robot would be the less costly.

11 **Computer Software Development** A firm has a computer which it uses for a variety of purposes. One of the major costs associated with the computer is software development (writing computer programs). The vice president for information systems wants to evaluate whether it is less costly to have his own programming staff or to have programs developed by a software development firm. The costs of both options are a function of the number of lines of code (program statements). The vice president estimates that in-house development costs $1.50 per line of code. In addition, annual overhead costs for supporting the programmers equal $30,000. Software developed outside the firm costs, on average, $2.25 per line of code.

(a) How many lines of code per year make costs of the two options equal?

(b) If programming needs are estimated at 30,000 lines per year, what are the costs of the two options?

(c) In part b what would the in-house cost per line of code have to equal for the two options to be equally costly?

12 **Sensitivity Analysis** Because the parameters (constants) used in mathematical models are frequently estimates, actual results may differ from those projected by the mathematical analysis. To account for some of the uncertainties which may exist in a problem, analysts often conduct *sensitivity analysis*. The objective is to assess how much a solution might change if there are changes in model parameters.

Assume in the previous exercise that software development costs by outside firms might actually fluctuate by ±20 percent from the $2.25 per-line estimate.

(a) Recompute the break-even point if costs are 20 percent higher or lower and compare your result with the original answer.

(b) Along with the uncertainty in part a, in-house variable costs might increase by as much as 30 percent because of a new union contract. Determine the combined effects of these uncertainties.

13 **Video Games** A leading manufacturer of video games is about to introduce four new games. The accompanying table summarizes price and cost data. Combined fixed costs equal $500,000. A marketing research study predicts that for each unit sold of Black Hole, 1.5 units of Haley's Comet, 3 units of Astervoids, and 4 units of PacPerson will be sold.

(a) How many product mix units must be sold to break even?

(b) How does this translate into sales of individual games?

	Game			
	PacPerson	Astervoids	Haley's Comet	Black Hole
Selling price	$50	$45	$30	$20
Variable cost/unit	20	15	10	10

14 A company produces three products which sell in a ratio of 4 units of product 2 and 2 units of product 3 for each unit sold of product 1. The following table summarizes price and cost data for the three products. If fixed costs are estimated at $2.8 million, determine the number of units of each product needed to break even.

	Product		
	1	2	3
Selling price	$40	$32	$55
Variable cost/unit	20	24	46

*15 A company is considering the purchase of a piece of equipment to be used to manufacture a new product. Four machines are being considered. The following table summarizes the purchase cost of each machine and the associated variable cost of production if the machine is used to produce the new product. Determine the ranges of output over which each machine would be the least costly alternative. Sketch the four cost functions.

	Purchase Cost	Variable Cost/Unit
Machine 1	$ 80,000	$10.00
Machine 2	120,000	9.00
Machine 3	200,000	7.50
Machine 4	300,000	5.50

❏ KEY TERMS AND CONCEPTS

break-even models 196
break-even point 196
contribution to fixed cost and
   profit 200
cost 178
demand function 184
depreciation (straight-line) 183
economies of scale 178

fixed cost 178
linear function 176
profit 180
profit margin (contribution) 200
revenue 179
salvage value 191
supply function 186
variable cost 178

❏ IMPORTANT FORMULAS

$$y = f(x) = a_1 x + a_0 \tag{5.1}$$

$$y = f(x_1, x_2) = a_1 x_1 + a_2 x_2 + a_0 \tag{5.2}$$

$$y = f(x_1, x_2, \ldots, x_n) = a_1 x_1 + a_2 x_2 + \cdots + a_n x_n + a_0 \tag{5.3}$$

$$P(x) = R(x) - C(x) \tag{5.7}$$

$$\text{Profit margin} = p - v \quad p > v \tag{5.11}$$

$$\left.\begin{array}{l} R(x) = C(x) \\[2ex] x_{BE} = \dfrac{FC}{p - v} \end{array}\right\} \quad \text{Break-even conditions}$$

$$\tag{5.10}$$

$$\tag{5.12}$$

❏ ADDITIONAL EXERCISES

Section 5.1

1 A firm sells a product for $80 per unit. Raw material costs are $12.50 per unit, labor costs are $27.50 per unit, and annual fixed costs are $360,000.
   (a) Determine the profit function $P(x)$, where x equals the number of units sold.
   (b) How many units would have to be sold to earn an annual profit of $250,000?

2 A firm produces three products which sell, respectively, for $25, $35, and $50. Labor requirements for each product are, respectively, 3.0, 4.0, and 3.5 hours per unit. Assume labor costs are $5 per hour and annual fixed costs are $75,000.
   (a) Construct a joint total revenue function for the sales of the three products.
   (b) Determine an annual total cost function for production of the three products.
   (c) Determine the profit function for the three products. Is there anything unusual about this function?
   (d) What is annual profit if 20,000, 10,000, and 30,000 units are sold, respectively, of the three products?

Section 5.2

3 A city has purchased a new asphalt paving machine for $300,000. The city comptroller states that the machine will be depreciated using a straight-line method. At the end of 8 years, the machine will be sold with an expected salvage value of $60,000.
   (a) Determine the function $V = f(t)$ which expresses the book value of the machine V as a function of its age t.
   (b) What is the book value expected to be when the machine is 6 years old?

4 The birthrate in a particular country has been declining in recent years. In 1985 the birthrate was 36.4 births per 1,000 people. In 1990 the birthrate was 34.6 births per 1,000 people. Assume R equals the birthrate per 1,000 and t equals time measured in years since 1985 ($t = 0$ for 1985).
   (a) Determine the linear estimating function $R = f(t)$.
   (b) Interpret the meaning of the slope.
   (c) If the linear pattern continues, what is the expected birthrate in the year 2000?
   (d) What is the restricted domain for this function?
   (e) Sketch the function.

5 **Gun Control** With crime rates on the rise, the number of handguns in circulation appears to be increasing. A 10-year survey of citizens within one U.S. city indicated a surprisingly linear increase in the number of handguns over time. In 1980 the estimated number of handguns was 450,000; in 1990 the estimated number was 580,000. Let n equal the number of handguns possessed by the residents of the city and let t represent time measured in years since 1980 ($t = 0$ for 1980).
   (a) Using the two data points, determine the linear estimating function $n = f(t)$.
   (b) Interpret the meaning of the slope.
   (c) If the number of guns continues to increase at the same rate, when will the number of guns surpass 750,000?

6 **Drug Addiction** The Department of Health in one state estimates that the number of cocaine users in the state is increasing approximately at a linear rate. The estimated number of users in 1980 was 950,000; the estimated number in 1985 was 1,025,000.
   (a) Determine the linear estimating function $n = f(t)$, where n equals the number of users and t equals time measured in years ($t = 0$ for 1980), using the two data points.
   (b) Interpret the meaning of the slope.

(c) If the number of users continues to increase according to this function, when will the number equal 1,250,000?

7 Alcoholism Since 1980 there has been a seemingly linear increase in the percentage of the population of one European city who are alcoholics. In 1980 the percentage was 10.5 percent. In 1990 the percentage had risen to 12.9 percent. If p equals the percentage of the population who are alcoholics and t represents time in years since 1980 ($t = 0$ for 1980):

(a) Determine the linear estimating function $p = f(t)$.

(b) Interpret the meaning of the slope.

(c) If the pattern of growth continues, forecast the percentage of alcoholics expected in 1996. What is the forecast for 2000?

8 HMO Popularity A health maintenance organization (HMO) provides health care to individuals and families on a prepaid basis. Typically, the subscriber pays an insurance premium for which most health care services are provided. These organizations typically emphasize preventive health care, and subscribers usually do not pay for office visits. A survey indicates that this type of insurance plan is being selected by more individuals. In 1980 there were 24 million individuals covered by these types of plans. In 1985 the number was 28.4 million. If the growth is assumed to be occurring at a linear rate:

(a) Determine the estimating function $n = f(t)$, where n equals the number of individuals covered by HMO plans and t equals time measured in years ($t = 0$ for 1980).

(b) What is the number of individuals covered by HMOs expected to be in the year 2000?

9 Producer Price Index The producer price index (PPI) measures wholesale prices for goods that will not undergo further processing and are ready for sale to final users. The PPI measures the cost of a selection of goods that would have cost $100 in 1982. Figure 5.16 is a graph of monthly data for the PPI between May 1, 1988, and October 1, 1990. During this period, the increase in the PPI appeared to be somewhat linear. The PPI was 111.4 on February 1, 1989, and 122.3 on October 1, 1990. Using these two data points:

(a) Determine the linear function $I = f(t)$, where I equals the producer price index and t equals time measured in *months* since January 1, 1988 ($t = 0$ corresponds to January 1, 1988).

(b) Interpret the meaning of the slope and I intercept.

(c) According to this estimating function, what was the PPI expected to equal on January 1, 1992?

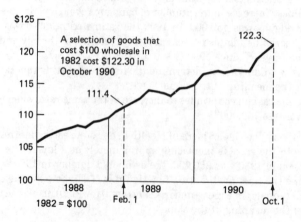

Figure 5.16 Producer price index. (Source: Bureau of Labor Statistics)

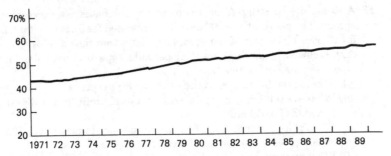

Figure 5.17 Percentage of women (age 20–24 years) in the work force. (Source: Bureau of Labor Statistics)

10 Women in the Work Force The numbers of women in the U.S. work force has been increasing steadily for several decades. Figure 5.17 illustrates data regarding the percentage of women, age 20–24, in the work force between 1971 and 1989. By observation, the increase in this age bracket was approximately linear. In 1978, 50 percent of women in this age group were in the work force; in 1985, 55.4 percent were. Using these two data points:

(a) Determine the linear estimating function $P = f(t)$, where P equals the percentage of women, age 20–24, in the work force and t equals time measured in years since 1971.

(b) Interpret the slope and P intercept.

(c) Using this estimating function, predict the percentage in the year 2000.

11 Federal Income Taxes The table below shows the 1991 federal tax rates for persons who submit as "Married filing jointly."

Taxable Income		*Tax Rate*
Over	**But not Over**	
$ 0	$34,000	15%
34,000	82,150	28
82,150		31

Determine the function $T = f(x)$, where T equals the tax liability (in dollars) for persons who are married, filing jointly, and x equals taxable income (in dollars).

Section 5.3

12 Advertising Campaign A firm is developing a TV advertising campaign. Development costs (fixed costs) are $150,000, and the firm must pay $15,000 per minute for television spots. The firm estimates that for each minute of advertising additional sales of $70,000 result. Of this $70,000, $47,500 is absorbed to cover the variable cost of producing the items and $15,000 must be used to pay for the minute of advertising. Any remainder is the contribution to fixed cost and profit.

(a) How many minutes of advertising are necessary to recover the development costs of the advertising campaign?

(b) If the firm uses 15 one-minute spots, determine total revenue, total costs (production and advertising), and total profit (or loss) resulting from the campaign.

13 **Automobile Leasing** A car-leasing agency purchases new cars each year for use in the agency. The cars cost $15,000 new. They are used for 3 years, after which they are sold for $3,600. The owner of the agency estimates that the variable costs of operating the cars, exclusive of gasoline, are $0.16 per mile. Cars are leased at a flat rate of $0.33 per mile (gasoline not included).

  (a) What is the break-even mileage for the 3-year period?

  (b) What are total revenue, total cost, and total profit for the 3-year period if a car is leased for 50,000 miles?

  (c) What price per mile must be charged in order to break even if a car is leased for 50,000 miles over a period of 3 years?

  (d) What price per mile must be charged in order to earn a profit of $5,000 per car over its 3-year lifetime if it is leased for a total of 50,000 miles?

14 A high-technology electronics firm needs a special microprocessor for use in a micro-computer it manufactures. Three alternatives have been identified for satisfying its needs. It can purchase the microprocessors from a supplier at a cost of $10 each. The firm also can purchase one of two pieces of automated equipment and manufacture the microprocessors. One piece of equipment costs $80,000 and would have variable costs per microprocessor of $8. A more highly automated piece of equipment costs $120,000 and would result in variable costs of $5 per unit. Determine the minimum-cost alternatives for different ranges of output (as determined in Example 17).

15 A new entrant to the "designer jeans" market is Françoise Strauss, a French cousin of Levi's great-grandnephew. Françoise plans on marketing three styles of jeans. The following table summarizes price and cost data. Combined fixed costs equal $7.5 million. A market research study projects a product mix such that for each pair of style A, two pairs of B and four pairs of C will be sold. How many pairs of each style must be sold in order to break even?

	Style		
	A	**B**	**C**
Selling price	$45	$36	$28
Variable cost/pair	19	17	14

16 A company is planning on producing and selling three products. The following table summarizes price and cost data for the three products. Company officials estimate that the three products will sell in a mix such that 3 units of product 2 and 5 of product 3 will be sold for each 2 sold of product 1. If fixed costs are estimated at $3.7 million, determine the number of units of each product needed to break even.

	Product		
	1	**2**	**3**
Selling price	$85	$80	$95
Variable cost/pair	50	40	67

❏ CHAPTER TEST

1 A company sells a product for $150 per unit. Raw material costs are $40 per unit, labor costs are $55 per unit, shipping costs are $15 per unit, and annual fixed costs are $200,000.

   (a) Determine the profit function $P = f(x)$, where x equals the number of units sold.

   (b) How many units must be sold in order to earn an annual profit of $750,000?

2 Ridership on a small regional airline has been declining, approximately, at a linear rate. In 1981 the number of passengers was 245,000; in 1986 the number was 215,000. If n equals the number of passengers using the airline per year and t equals time measured in years ($t = 0$ for 1981):

   (a) Determine the linear estimating function $n = f(t)$.

   (b) Interpret the meaning of the slope.

   (c) What is the number of riders expected to equal in the year 2000?

   (d) It is estimated that the airline will go out of business if ridership falls below 180,000. According to your function in part a, when will this happen?

3 **Increasing Inmate Population** Figure 5.18 reflects an increase in the average number of inmates within the jails of Baltimore. Beginning in 1984, there was a significant increase in the average number. The increase between 1984 and 1988 could be approximated by a linear function. If the average number of inmates was 1,720 in 1984 and 2,590 in 1988:

   (a) Determine the estimating function $N = f(t)$, where N equals the average number of persons in jail and t equals time measured in years since 1984.

   (b) Interpret the slope and N intercept.

   (c) According to this function, what was the average number of inmates expected to equal in 1990?

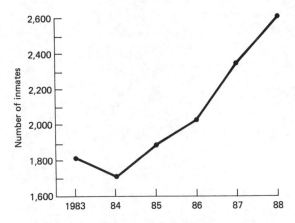

Figure 5.18   Average city jail population.
(Source: Baltimore City Jail)

4 A company is considering the purchase of a piece of equipment to be used to produce a new product. Three machines are being considered. Machine 1 costs $100,000, and the associated variable cost of production for the new product is estimated at $25. Machine 2 costs $150,000 with a variable cost of production of $22.50. Machine 3 costs $180,000 with a variable cost of $21. Determine the ranges of output over which each machine would be the least cost alternative. Sketch the three cost functions on one graph.

5 Assume that a revenue function $R(x)$ and a total cost function $C(x)$ have been used in performing break-even analysis. What is the expected effect on the break-even level of output x_{BE} if (all other things held constant):
(a) The price per unit decreases?
(b) The variable cost per unit decreases?
(c) The fixed cost decreases?

MINICASE

AUTOMOBILE REPLACEMENT DECISION

Given the high cost of gasoline, automobile owners are constantly searching for ways of economizing on driving expenses. One costly alternative for many persons is to purchase a car which gets significantly better gasoline mileage than does their present car. Chamberlain developed a mathematical formula to calculate how many years a person would have to drive a new car to make the gasoline savings offset the cost of trading in the old car and buying a new one.* The variables to be used in this analysis are

y = number of years to justify the purchase of a new car
m = gasoline mileage of present car, miles per gallon
n = gasoline mileage of new car, miles per gallon
c = net cost of new car (purchase cost less proceeds from sale of present car)
d = average number of miles driven per year
p = gasoline price per gallon

Chamberlain determines this "break-even" period using the general relationship

$$\begin{matrix} \text{Cost of gasoline} \\ \text{for present car} \\ \text{during break-even period} \end{matrix} = \begin{matrix} \text{cost of gasoline} \\ \text{for new car during} \\ \text{break-even period} \end{matrix} + \begin{matrix} \text{net cost} \\ \text{of new car} \end{matrix}$$

or

$$f_{old}(y) = f_{new}(y)$$

1 Using the variables defined previously, determine the expressions for $f_{old}(y)$ and $f_{new}(y)$.

2 Set $f_{old}(y)$ equal to $f_{new}(y)$ and derive the general formula for the break-even period y.

3 Determine the break-even period if the present car has a value of $8,000 and gets 14 miles per gallon. The new car has a purchase cost of $16,000 and gets an estimated 46 miles per gallon. Gasoline prices are currently $1.50 per gallon. Assume annual mileage of 24,000 miles.

4 Examine the sensitivity of the break-even period to changes in the price of gasoline. To test this, assume that the actual price may vary by ±25 percent from the $1.50 figure. Is there much of a change in the break-even period?

5 Rework parts 1–3 if the unknown is x, the total miles driven. That is, suppose we wish to develop a formula that allows one to determine the number of miles (rather than number of years) a new car would have to be driven. Average mileage per year d is not part of this model.

6 List assumptions of this model. What assumptions are unrealistic? Why? What cost factors have not been considered?

---

* "Should the Gas Guzzler Go?" (letter), *Science*, vol. 207, p. 1028, March 1980.

CHAPTER 6

QUADRATIC AND POLYNOMIAL FUNCTIONS

CHAPTER OBJECTIVES

❑ Generally, introduce the reader to nonlinear functions

❑ More specifically, provide an understanding of the algebraic and graphical characteristics of quadratic and polynomial functions

❑ Illustrate a variety of applications of these types of functions

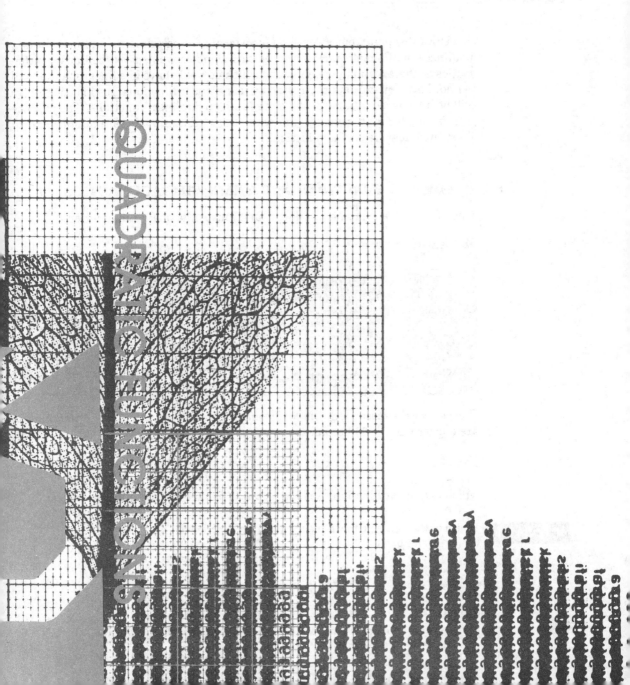

Until this point, our attentions have been focused primarily on linear (versus nonlinear) mathematics. However, as useful and convenient as the linear mathematics is, there are many phenomena which do not behave in a linear manner and cannot be adequately approximated by using linear functions. In this chapter we will be introduced to some of the more common nonlinear functions. The purpose of this chapter is to acquaint you with the attributes of these functions and to illustrate a few areas of application.

6.1 QUADRATIC FUNCTIONS AND THEIR CHARACTERISTICS

One of the more common nonlinear functions is the quadratic function.

Mathematical Form

> **DEFINITION: QUADRATIC FUNCTION**
> A *quadratic function* involving the independent variable x and the dependent variable y has the general form
>
> $$y = f(x) = ax^2 + bx + c \qquad (6.1)$$
>
> where a, b, and c are constants, $a \neq 0$.

Can you see the reason why the coefficient of x^2 cannot equal 0? If $a = 0$ and b and c are nonzero, Eq. (6.1) becomes

$$y = bx + c$$

which is a linear function. As long as $a \neq 0$, b and c can assume any values.

EXAMPLE 1 Which of the following functions are quadratic functions?
(a) $y = f(x) = 5x^2$
(b) $u = f(v) = -10v^2 - 6$
(c) $y = f(x) = 7x - 2$
(d) $g = f(h) = -6$
(e) $y = f(x) = x^2 + 10x$
(f) $y = f(x) = 6x^2 - 40x + 15$
(g) $y = f(x) = 2x^3 + 4x^2 - 2x + 5$

SOLUTION
(a) $y = 5x^2$ is a quadratic function. According to the general form of Eq. (6.1), $a = 5$, $b = 0$, and $c = 0$.

(b) $u = -10v^2 - 6$ is a quadratic function where, according to Eq. (6.1), $a = -10$, $b = 0$, and $c = -6$.

(c) $y = 7x - 2$ is *not* a quadratic function since there is no x^2 term, or $a = 0$.

(d) $g = -6$ is not a quadratic function for the same reason as in part c.

(e) $y = x^2 + 10x$ is a quadratic function where $a = 1$, $b = 10$, and $c = 0$.

(f) $y = 6x^2 - 40x + 15$ is a quadratic function where $a = 6$, $b = -40$, and $c = 15$.

(g) $y = 2x^3 + 4x^2 - 2x + 5$ is not a quadratic function because it contains the third-degree term $2x^3$.

NOTE   In Chap. 4 we stated the general form of a quadratic function as

$$y = f(x) = a_2 x^2 + a_1 x + a_0 \qquad a_2 \neq 0 \qquad (4.4)$$

You should observe that Eqs. (4.4) and (6.1) are equivalent. Only the names of the coefficients have been changed.

PRACTICE EXERCISE
Which of the following are quadratic functions?
(a) $y = f(x) = (50 - 0.001x)x$
(b) $y = f(x) = 3x(1 - x^2)$
(c) $s = g(t) = [2t(t^2 - 1)/t]\,(10)$
Answer: (a) and (c).

Graphical Representation

All quadratic functions having the form of Eq. (6.1) graph as **parabolas**. Figure 6.1 illustrates two parabolas which have different orientations. A parabola which "opens" upward, such as that in Fig. 6.1a, is said to be **concave up.** A parabola which "opens" downward, such as that in Fig. 6.1b, is said to be **concave down.*** The point at which a parabola either "bottoms out" when it is concave up or "peaks out" when it is concave down is called the **vertex** of the parabola. Points A and B are the respective **vertices** for the two parabolas in Fig. 6.1.

Figure 6.1
Parabolas.

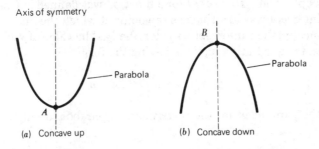

(a) Concave up   (b) Concave down

---

* A more detailed discussion of the *concavity* of functions is found in Chap. 16.

Given a quadratic function of the form of Eq. (6.1), the coordinates of the vertex of the corresponding parabola are

$$\left(\frac{-b}{2a}, \frac{4ac - b^2}{4a}\right) \qquad\qquad (6.2)$$

where a, b, and c are the parameters or constants in Eq. (6.1).

A parabola is a curve having a particular symmetry. In Fig. 6.1, the dashed vertical line which passes through the vertex is called the **axis of symmetry.** This line separates the parabola into two symmetrical halves. If you were able to fold one side of the parabola using the axis of symmetry as a hinge, you would find that the two halves coincide (i.e., they are mirror images of each other).

Sketching the graph of a quadratic function can be accomplished by using the "brute force" methods discussed in Chap. 4. However, there are certain key attributes of quadratic functions which enable us to sketch the corresponding parabola with relative ease. If the *concavity* of the parabola, *y intercept, x intercept(s),* and *vertex* can be determined, a reasonable sketch of the parabola can be drawn.

Concavity   Once a function has been recognized as having the general quadratic form of Eq. (6.1), the concavity of the parabola can be determined by the sign of the coefficient on the x^2 term. If $a > 0$, the function will graph as a parabola which is *concave up.* If $a < 0$, the parabola is *concave down.*

EXAMPLE 2

In Example 1, the following functions were determined to be quadratic.
(a) $y = f(x) = 5x^2$ (b) $u = f(v) = -10v^2 - 6$
(c) $y = f(x) = x^2 + 10x$ (d) $y = f(x) = 6x^2 - 40x + 15$
What is the concavity of the parabola representing each quadratic function?

SOLUTION

(a) The graph of $y = 5x^2$ is concave *up* since $a = +5$.
(b) The graph of $u = -10v^2 - 6$ is concave *down* since $a = -10$.
(c) The graph of $y = x^2 + 10x$ is concave *up* since $a = +1$.
(d) The graph of $y = 6x^2 - 40x + 15$ is concave *up* since $a = +6$.

$\square$

---

y **intercept**   The *y intercept* for a function was defined in Chap. 2. Graphically, the y intercept was identified as the point at which the line intersects the y axis. Algebraically, the y intercept was identified as the value of y when x equals 0, or $f(0)$. Thus, for quadratic functions having the form

$$y = f(x) = ax^2 + bx + c$$

$f(0) = c$, or the y intercept for the corresponding parabola occurs at $(0, c)$.

EXAMPLE 3

What are the y intercepts for the quadratic functions in Example 2?

SOLUTION

(*a*) For $y = 5x^2$, $f(0) = 0$, or the y intercept occurs at $(0, 0)$.

(*b*) For $u = -10v^2 - 6$, $f(0) = -6$, or the u intercept occurs at $(0, -6)$.

(*c*) For $y = x^2 + 10x$, $f(0) = 0$, or the y intercept occurs at $(0, 0)$.

(*d*) For $y = 6x^2 - 40x + 15$, $f(0) = 15$, or the y intercept occurs at $(0, 15)$.

❑

x **intercept(s)** The *x intercept* for a function was also defined in Chap. 2 as the point(s) at which the graph of a line crosses the x axis. Equivalently, the x intercept represents the value(s) of x when y equals 0. For quadratic functions, there may be one x intercept, two x intercepts, or *no x* intercept. These possibilities are shown in Fig. 6.2.

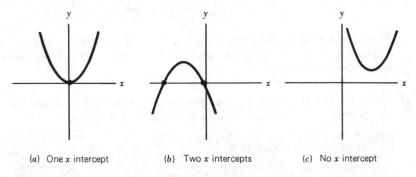

(*a*) One *x* intercept   (*b*) Two *x* intercepts   (*c*) No *x* intercept

Figure 6.2   *x* intercept possibilities for quadratic functions.

There are a number of ways of determining the x intercepts for a quadratic function if any exist. The x intercepts are found by determining the roots of the equation

$$ax^2 + bx + c = 0 \qquad\qquad (6.3)$$

Two methods of determining the roots to Eq. (6.3) were discussed in Chap. 1. These methods are illustrated in the following example.

EXAMPLE 4

(Determining Roots by Factoring) Some quadratic functions can be factored (Sec. A.3, Appendix A) into either two binomials or a binomial and a monomial. If a quadratic function can be factored, it is an easy matter to determine the roots of Eq. (6.3). For example, the values of x which satisfy the quadratic equation

$$6x^2 - 2x = 0$$

can be determined by first factoring $2x$ from the expression on the left-hand side of the equation, yielding

$$(2x)(3x - 1) = 0$$

By setting each factor equal to 0, the roots of the equation are identified as

$$2x = 0 \quad \text{or} \quad x = 0$$

and

$$3x - 1 = 0 \quad \text{or} \quad x = \tfrac{1}{3}$$

Thus, the quadratic function $y = f(x) = 6x^2 - 2x$ graphs as a parabola having x intercepts at $(0, 0)$ and $(\tfrac{1}{3}, 0)$.

Similarly, the values of x satisfying the equation

$$x^2 + 6x + 9 = 0$$

are found by factoring the left side of the equation, yielding

$$(x + 3)(x + 3) = 0$$

When the factors are set equal to 0, it is found that the only value which satisfies the equation is $x = -3$. The quadratic function $y = f(x) = x^2 + 6x + 9$ graphs as a parabola with one x intercept at $(-3, 0)$.

EXAMPLE 5   (**Determining Roots Using the Quadratic Formula**) The **quadratic formula** will *always* identify the real roots of a quadratic equation *if any exist*.

ALGEBRA FLASHBACK

The **quadratic formula**, which is used to identify roots of equations of the form of Eq. (6.3), is

$$x = \frac{-b \pm \sqrt{b^2 - 4ac}}{2a}$$

Using the quadratic formula, (1) if $b^2 - 4ac > 0$, there will be two real roots; (2) if $b^2 - 4ac = 0$, there will be one real root; and (3) if $b^2 - 4ac < 0$, there will be no real roots (Sec. 1.2).

Given the quadratic function $f(x) = x^2 - x - 3.75$, the x intercepts occur when $x^2 - x - 3.75 = 0$. Referring to Eq. (6.3), $a = 1$, $b = -1$, $c = -3.75$. Substituting these values into the quadratic formula, the roots of the equation are computed as

$$x = \frac{-(-1) \pm \sqrt{(-1)^2 - 4(1)(-3.75)}}{2(1)}$$

$$= \frac{1 \pm \sqrt{1 + 15}}{2} = \frac{1 \pm \sqrt{16}}{2} = \frac{1 \pm 4}{2}$$

or, using the plus sign, we have

$$x = \tfrac{5}{2} = 2.5$$

Using the minus sign gives

$$x = -\tfrac{3}{2} = -1.5$$

Thus, there are two values of x which satisfy the quadratic equation. The parabola representing the quadratic function will cross the x axis at $(2.5, 0)$ and $(-1.5, 0)$.

Vertex The location of the vertex of a parabola can be determined by formula using Eq. (6.2) or perhaps by observation of the x intercepts (later in the book, we will demonstrate a very easy procedure for locating the vertex). The prospect of memorizing Eq. (6.2) will not inspire you and should probably be done only as a last resort. If pushed to that last resort, the reality is that you only need to remember the formula for the x coordinate of the vertex, $-b/2a$. One way of remembering this is that it is the same as the front half of the quadratic formula as shown below.

$$\begin{array}{c} x \text{ coordinate} \\ \text{of vertex} \end{array} \longrightarrow \dfrac{-b \pm \sqrt{b^2 - 4ac}}{2a}$$

Once you have the x coordinate of the vertex, the y coordinate can be found by substituting the x coordinate into the quadratic function [i.e., $f(-b/2a)$].

Because of the symmetry of parabolas, whenever a parabola has two x intercepts, *the x coordinate of the vertex lies midway between the two x intercepts*. When a parabola has one x intercept, *the vertex occurs at the x intercept*. Figure 6.3 illustrates these relationships.

Figure 6.3
x intercept/vertex relationships.

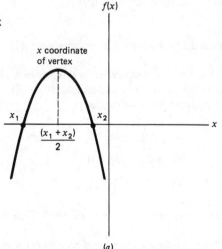

(a)

Two x intercepts: x coordinate of vertex midway between

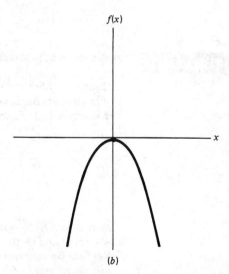

(b)

One x intercept: x intercept is vertex

EXAMPLE 6   Let's locate the vertex for the parabola representing the function

$$f(x) = 2x^2 + 3x - 20$$

First by formula, $a = 2$, $b = 3$, and $c = -20$. Thus, the x coordinate of the vertex will be

$$x = \frac{-b}{2a} = \frac{-3}{2(2)} = \frac{-3}{4} = -0.75$$

The y coordinate of the vertex follows as

$$
\begin{aligned}
y &= f(-0.75) \\
&= 2(-0.75)^2 + 3(-0.75) - 20 \\
&= 1.125 - 2.25 - 20 \\
&= -21.125
\end{aligned}
$$

Thus, the vertex is located at $(-0.75, -21.125)$.

Let's see if the x intercepts for this function are helpful in locating the vertex. Factor the left side of the quadratic function:

$$2x^2 + 3x - 20 = 0$$

when

$$(2x - 5)(x + 4) = 0$$

Setting the two factors to zero results in roots of $x = 2.5$ and $x = -4$. The x coordinate of the vertex lies midway between these two values. On a number line, the midpoint on the interval $[a_1, a_2]$ is $(a_1 + a_2)/2$. Thus, using the two x coordinates, the x coordinate of the vertex is

$$\frac{2.5 + (-4)}{2} = \frac{-1.5}{2} = -0.75$$

EXAMPLE 7   Suppose we want to sketch the quadratic function $f(x) = 3x^2 + 6x - 45$.

Concavity: *Since $a = 3 > 0$, the parabola representing this function will be concave up.*
y intercept: *Because $f(0) = -45$, the y intercept occurs at $(0, -45)$.*
x intercept(s): *Factoring the quadratic function gives*

$$3x^2 + 6x - 45 = 0$$

$$3(x^2 + 2x - 15) = 0$$

$$3(x - 3)(x + 5) = 0$$

Setting each factor equal to zero yields roots of $x = -5$ and $x = 3$. Thus, x intercepts occur at $(-5, 0)$ and $(3, 0)$.
Vertex: *Because there are two x intercepts, the x coordinate will lie midway between the two. Given that $(-5 + 3)/2 = -2/2 = -1$, the x coordinate of the vertex is $x = -1$.*

The y coordinate of the vertex is

$$f(-1) = 3(-1)^2 + 6(-1) - 45$$
$$= 3 - 6 - 45$$
$$= -48$$

Combining the information about these key attributes yields the sketch of the function shown in Fig. 6.4.

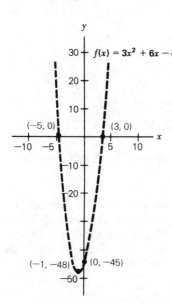

Figure 6.4   Sketch of $f(x) = 3x^2 + 6x - 45$.

PRACTICE EXERCISE
Given $f(x) = 2x^2 - x - 15$, determine: (*a*) concavity, (*b*) *y* intercept, (*c*) *x* intercept(s), and (*d*) location of vertex. *Answer:* (*a*) up, (*b*) (0, −15), (*c*) (−2.5, 0) and (3, 0), (*d*) (0.25, −15.125).

EXAMPLE 8

(Escalating NBA Salaries; Motivating Scenario) As discussed in the beginning of this chapter, NBA players have been benefiting salarywise from the prosperity of the National Basketball Association. Figure 6.5 illustrates the increase in the average player's salary. It appears that the salary data could be approximated reasonably well by using a quadratic function. For the years 1981, 1985, and 1988, the average player salaries were $190,000, $310,000, and $600,000, respectively. Using these data points, we want to determine the quadratic function which can be used to estimate average player salaries over time. Stated from a different perspective, we want to determine if there is a parabola which passes through the three data points. We are seeking to determine the quadratic function

$$s = f(t) = at^2 + bt + c \tag{6.4}$$

where $s = $ *average player salary, thousands of dollars*
and $t = $ *time measured in years since 1981*

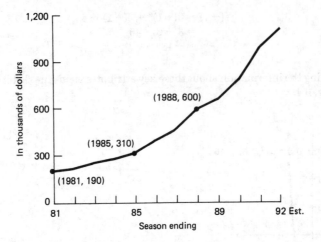

Figure 6.5   Average salary per NBA player. (Data: NBA Players' Association)

Notice that we are redefining the measurement of time from actual calendar years to years *since* 1981 (i.e., 1981 → $t = 0$, 1982 → $t = 1$, etc.). To specify this function, we need values for the three parameters a, b, and c. To determine these parameters, we need at least three data points. The salary data for the three years of interest is translated into the three data points $(0, 190)$, $(4, 310)$, and $(7, 600)$. To determine the parameters, we substitute each of the data points into Eq. (6.4).

$$190 = a(0)^2 + b(0) + c$$

$$310 = a(4)^2 + b(4) + c$$

$$600 = a(7)^2 + b(7) + c$$

Simplifying these equations gives

$$c = 190$$

$$16a + 4b + c = 310$$

$$49a + 7b + c = 600$$

Since the value of c is determined immediately (one of our data points was the s intercept), we can substitute $c = 190$ into the other two equations and solve for a and b. Verify that the resulting values are $a = 9.524$ and $b = -8.096$.

Our conclusion is that there is a quadratic function which is satisfied by the three data points, or equivalently, there is a parabola which passes through the three data points. The quadratic estimating function is

$$s = f(t) = 9.524t^2 - 8.096t + 190$$

PRACTICE EXERCISE
Using this function, estimate the average salary in 1995. *Answer:* $1,943.36 in thousands of dollars, or $1,943,360.

POINT FOR THOUGHT & DISCUSSION Suppose that three points (x_1, y_1), (x_2, y_2), and (x_3, y_3) are given. Upon substituting these into Eq. (6.1) to solve for a, b, and c, what conclusion would you reach if there is no solution to the system of three equations?

Section 6.1 Follow-Up Exercises

In Exercises 1–16, determine which functions are quadratic and identify values for the parameters a, b, and c.

1 $f(x) = 4x^2 - 20$

2 $f(x) = \dfrac{x^2}{4} + 3x$

3 $g(h) = 3h - 450$

4 $h(u) = 24 - u + u^2/3$

5 $g(s) = (s - 4)^2$

6 $f(x) = (3 - 2x + x^2)^2$

7 $f(x) = (x^2 - 4x + 5)/3$

8 $v(s) = s^2 + (s + 5)/3$

9 $h(x) = (x^2 - 2x)^2$

10 $f(x) = 25 - x^3$

11 $f(x) = \sqrt{x^4 - 4x^2 + 4}$

12 $g(h) = 1{,}000 - h^2$

13 $h(v) = 4v - 5 + 20v^2$

14 $f(x) = x^2 - 2x + x^3$

15 $g(h) = \sqrt[3]{(h - 4)^6}$

16 $f(x) = \sqrt[4]{(2 - x)^8}$

In Exercises 17–32, determine the concavity of the parabola representing the quadratic function, its y intercept, its x intercepts if any exist, and the coordinates of the vertex. Sketch the parabola.

17 $f(x) = -x^2$

18 $f(x) = -x^2 + 6x - 9$

19 $f(x) = x^2 - 3x + 2$

20 $f(x) = x^2 - 4$

21 $f(x) = -x^2 + 5$

22 $f(x) = x^2 - 4x + 4$

23 $f(x) = \dfrac{x^2}{2} - 10x$

24 $f(x) = \dfrac{-x^2}{2} + 24x$

25 $f(x) = -x^2 - 9$

26 $f(x) = 3x^2 + 7x - 20$

27 $f(x) = 2x^2 + 4$

28 $f(x) = -3 - x^2$

29 $f(x) = 6x^2 + x - 12$

30 $f(x) = 4x^2 + 5x - 6$

31 $f(x) = -x^2 - 5x$

32 $f(x) = x^2 - 7x + 10$

33 Sketch the following quadratic functions, making note of the values of $|a|$ for each and the relative steepness of the three parabolas.

$$f(x) = x^2 \qquad f(x) = 0.01x^2 \qquad f(x) = 100x^2$$

34 Determine the equation of the quadratic function which passes through the points $(0, 10)$, $(1, 6)$, and $(-2, 24)$.

35 Determine the equation of the quadratic function which passes through the points $(1, -1)$, $(-3, -33)$, and $(2, -8)$.

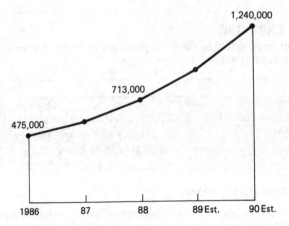

Figure 6.6   Number of people working at home 35 or more hours per week. (Sources: National Work-At-Home Survey by Link Resources Inc.)

36 Working at Home The number of persons working out of their homes has been increasing rapidly in recent years. Figure 6.6 illustrates data gathered in a study regarding the number of persons working at home 35 or more hours per week. The data appears to be almost quadratic in appearance. Using the data for 1986 and 1988 and the projected value for 1990, determine the quadratic estimating function $n = f(t)$, where n equals the number of persons working 35 or more hours per week at home (stated in thousands) and t equals time measured in years since 1986. According to this function, what is the number of persons working at home expected to equal in 1995? In 2000?

37 Computers in Public Schools A survey taken during 1990 indicated the increased availability of computers in U.S. public school classrooms. Figure 6.7 displays results of

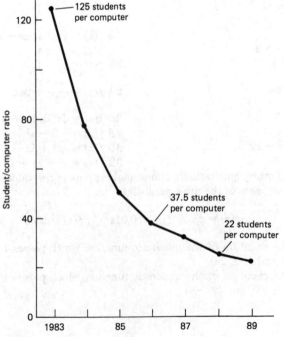

Figure 6.7   Student-computer ratio. (Source: Quality Education Data)

this survey. During the 1983–1984 academic year, the number of students per computer was 125. For the 1986–1987 academic year, the number had dropped to 37.5. For the 1989–1990 academic year, the number was 22 students per computer. Using these three data points, determine the quadratic estimating function $n = f(t)$, where n equals the number of students per computer and t equals time measured in years since the 1983–1984 academic year (i.e., $t = 0$ corresponds to 1983–1984). Using this function, estimate the number of students per computer during the 1990–1991 academic year. Based on this result, what conclusion can you reach?

38 U.S. Cellular Phone Industry The cellular telephone industry has grown rapidly since the late 1980s. Figure 6.8 presents data regarding the number of subscribers between 1985 and 1990. The pattern of growth in the number of subscribers looks as if it could be approximated reasonably well by using a quadratic function. Using the data points for 1985, 1987, and 1989 (200,000, 950,000, and 2,600,000 subscribers, respectively), determine the quadratic estimating function $n = f(t)$, where n equals the number of subscribers *(stated in millions)* and t equals time measured in years *since* 1985. According to this estimating function, how many subscribers are expected in the year 1995?

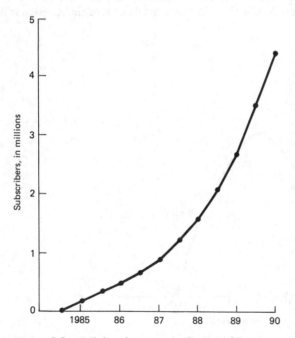

Figure 6.8 Cellular phone use in the United States.
(Source: Cellular Telecommunications Industry Association)

6.2 QUADRATIC FUNCTIONS: APPLICATIONS

In this section examples will be presented which illustrate a few areas of application of quadratic functions.

EXAMPLE 9   **(Quadratic Revenue Function)** Suppose that the demand function for a product is

$$q = f(p)$$

or

$$q = 1,500 - 50p$$

where *q equals the quantity demanded in thousands of units* and *p equals the price in dollars.* Total revenue *R* from selling *q* units is stated as the product of *p* and *q*, or

$$R = pq$$

Because the demand function states *q* as a function of *p*, total revenue can be stated as a function of price, or

$$
\begin{aligned}
R &= h(p) \\
&= p \cdot f(p) \\
&= p(1,500 - 50p) \\
&= 1,500p - 50p^2
\end{aligned}
$$

You should recognize this as a quadratic function. The total revenue function is sketched in Fig. 6.9. Notice that the restricted domain of the function is $0 \le p \le 30$. Does this make sense?

❑

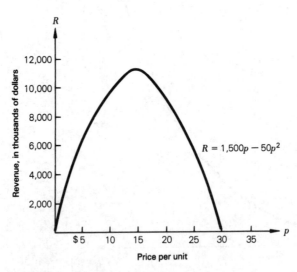

Figure 6.9   Quadratic revenue function.

PRACTICE EXERCISE
Given the *p* intercepts in Fig. 6.9, what value of *p* maximizes *R*? How many units would be demanded at this price? What is the maximum value of *R*?
Answer: $15; 750 (thousand) units; $11.25 million.

EXAMPLE 10

(Quadratic Supply Functions) Market surveys of suppliers of a particular product have resulted in the conclusion that the supply function is approximately quadratic in form. Suppliers were asked what quantities they would be willing to supply at different market prices. Results of the survey indicated that at market prices of $25, $30, and $40 the quantities which suppliers would be willing to offer to the market were 112.5, 250.0, and 600.0 (thousand) units, respectively.

We can determine the equation of the quadratic supply function by substituting the three price-quantity combinations into the general equation

$$q_s = f(p)$$

or

$$q_s = ap^2 + bp + c$$

The resulting system of equations is

$$625a + 25b + c = 112.5$$

$$900a + 30b + c = 250$$

$$1{,}600a + 40b + c = 600$$

which, when solved, yields values of $a = 0.5$, $b = 0$, and $c = -200$. Thus the quadratic supply function, shown in Fig. 6.10, is represented by

$$q_s = f(p) = 0.5p^2 - 200$$

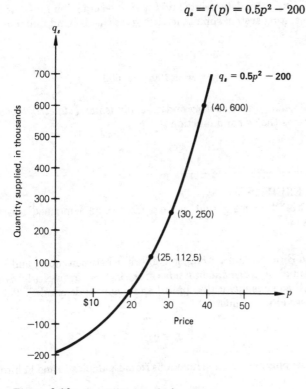

Figure 6.10   Quadratic supply function.

The quantity supplied at any market price can be estimated by substituting the price into the supply function. For example, the quantity supplied at a price of $50 is estimated to be

$$f(50) = 0.5(50)^2 - 200$$
$$= 0.5(2,500) - 200 = 1,250 - 200 = 1,050 \text{ (thousand) units}$$

POINTS FOR THOUGHT & DISCUSSION   What is the restricted domain of the supply function? Is it the same as indicated in Fig. 6.10? Interpret the meaning of the p intercept. Does this interpretation seem reasonable? Interpret the meaning of the q_s intercept. Does this interpretation make sense?

EXAMPLE 11   **(Quadratic Demand Functions)** Related to the previous example, a consumer survey was conducted to determine the demand function for the same product. Researchers asked consumers if they would purchase the product at various prices and from their responses constructed estimates of market demand at various market prices. After sample data points were plotted, it was concluded that the demand relationship was estimated best by a quadratic function. Researchers concluded that the quadratic representation was valid for prices between $5 and $45.

Three data points chosen for "fitting" the curve were (5, 2,025) (10, 1,600), and (20, 900). Substituting these data points into the general equation for a quadratic function and solving the resulting system simultaneously gives the demand function

$$q_d = g(p)$$

or

$$q_d = p^2 - 100p + 2,500$$

where *p equals the selling price in dollars* and q_d *equals demand stated in thousands of units*. Figure 6.11 illustrates the demand function.

PRACTICE EXERCISE
At a price of $30, how many units are expected to be demanded? *Answer:* 400 (thousand) units.

EXAMPLE 12   **(Supply-Demand Equilibrium)** Market equilibrium between supply and demand can be estimated for the supply and demand functions in the last two examples by determining the market price which equates quantity supplied and quantity demanded. This equilibrium condition is expressed by the equation

$$\boxed{q_s = q_d} \qquad \text{(6.5)}$$

If we substitute the supply and demand functions from Examples 10 and 11 into Eq. (6.5), we get

$$0.5p^2 - 200 = p^2 - 100p + 2,500$$

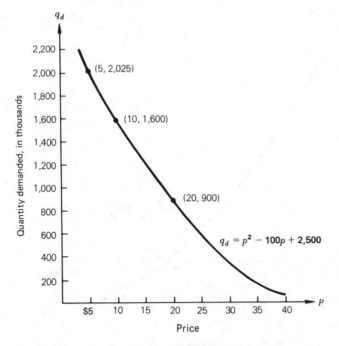

Figure 6.11 Quadratic demand function.

The equation can be rearranged so that

$$0.5p^2 - 100p + 2{,}700 = 0 \qquad \textbf{(6.6)}$$

The quadratic formula can be used to determine the roots to Eq. (6.6) as follows:

$$p = \frac{-(-100) \pm \sqrt{(-100)^2 - 4(0.5)(2{,}700)}}{2(0.5)}$$

$$= \frac{100 \pm \sqrt{4{,}600}}{1} = 100 \pm 67.82$$

The two resulting values are $p = \$32.18$ and $p = \$167.82$. The second root is outside the relevant domain of the demand function ($5 \leq p \leq 45$) and is therefore meaningless. However, $q_s = q_d$ when the selling price is \$32.18. Substitution of $p = 32.18$ into the supply and demand functions results in values of $q_s = 317.77$ and $q_d = 317.55$. (Rounding is the reason for the difference between these two values.) Thus market equilibrium occurs when the market price equals \$32.18 and the quantity supplied and demanded equals 317,770 units. Figure 6.12 illustrates the two functions.

❑

POINT FOR THOUGHT & DISCUSSION   What is it about the graphical behavior of $f(p)$ and $g(p)$ that results in two roots to Eq. (6.6)?

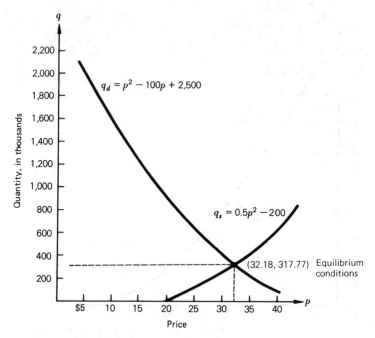

Figure 6.12   Supply-demand equilibrium.

EXAMPLE 13   **(Emergency Response: Location Model)** Figure 6.13 illustrates the relative locations of three cities along a coastal highway. The three cities are popular resorts, and their populations swell during the summer months. The three cities believe their emergency rescue and health treatment capabilities are inadequate during the vacation season. They have decided to support jointly an emergency response facility which dispatches rescue trucks and trained paramedics. A key question concerns the location of the facility.

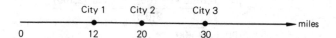

Figure 6.13   Relative location of cities.

In choosing the location, it has been agreed that the distance from the facility to each city should be kept as short as possible to ensure quick response times. Another consideration is the size of the summer population of each city, since this is one measure of the potential need for emergency response services. The larger the summer population of a city, the greater the desire to locate the facility close to the city. Analysts have decided that the criterion for selecting the location is to minimize the sum of the products of the summer populations of each town and the square of the distance between the town and the facility. This can be stated more succinctly as

$$\text{Minimize } S = \sum_{j=1}^{3} p_j d_j^2 = p_1 d_1^2 + p_2 d_2^2 + p_3 d_3^2$$

where p_j equals the summer population for city j, stated in thousands, and d_j is the distance between city j and the rescue facility.

If the summer populations are, respectively, 150,000, 100,000, and 200,000 for the three cities, compute the general expression for S. (*Hint: Let x equal the location of the facility relative to the zero point of the scale in Fig. 6.13, and let x_j equal the location of city j. The distance between the facility and city j is calculated by the equation $d_j = x - x_j$.*)

SOLUTION

With x defined as the unknown location of the proposed facility, S can be stated as a function of x. The function is defined as

$$S = f(x)$$

$$= \sum_{j=1}^{3} p_j(x - x_j)^2 = p_1(x - x_1)^2 + p_2(x - x_2)^2 + p_3(x - x_3)^2$$

$$= 150(x - 12)^2 + 100(x - 20)^2 + 200(x - 30)^2$$

$$= 150x^2 - 3,600x + 21,600 + 100x^2 - 4,000x$$
$$+ 40,000 + 200x^2 - 12,000x + 180,000$$

or $\qquad S = 450x^2 - 19,600x + 241,600$

Note that this function is quadratic and it will graph as a parabola which is concave up. S will be minimized at the vertex of the parabola, or where

$$x = \frac{-b}{2a} = \frac{-(-19,600)}{2(450)}$$

$$= \frac{19,600}{900} = 21.77$$

Figure 6.14 presents a sketch of f. According to Fig. 6.15, the emergency response facility will be located 21.77 miles to the right of the zero point, or 1.77 miles to the right of city 2.

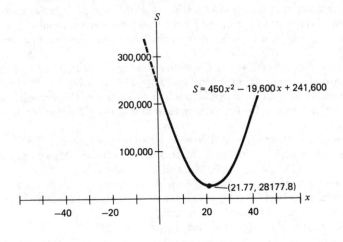

Figure 6.14 Criterion function: Emergency response location model.

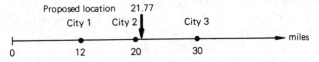

Figure 6.15 Optimal location of emergency response facility.

We will reconsider this example in Chap. 17 and solve it by another method.

◻

Section 6.2 Follow-up Exercises

1 The demand function for a particular product is

$$q = f(p) = 600,000 - 2,500p$$

where q is stated in units and p is stated in dollars. Determine the quadratic total revenue function, where R is a function of p, or $R = g(p)$. What is the concavity of the function? What is the q intercept? What does total revenue equal at a price of $50? How many units will be demanded at this price? At what price will total revenue be maximized? (*Hint:* Does the vertex correspond to maximum R?)

2 The weekly demand function for a particular product is

$$q = f(p) = 2,400 - 15p$$

where q is stated in units and p is stated in dollars. Determine the quadratic total revenue function, where R is a function of p, or $R = g(p)$. What is the concavity of the function? What is the q intercept? What does total revenue equal at a price of $50? How many units will be demanded at this price? At what price will total revenue be maximized? (*Hint:* Does the vertex correspond to maximum R?)

3 The monthly demand function for a particular product is

$$q = f(p) = 30,000 - 25p$$

where q is stated in units and p is stated in dollars. Determine the quadratic total revenue function, where R is a function of p, or $R = g(p)$. What is the concavity of the function? What is the q intercept? What does total revenue equal at a price of $60? How many units will be demanded at this price? At what price will total revenue be maximized?

4 Total revenue in Exercise 1 can be stated in terms of either price p or demand q. Restate total revenue as a function of q rather than p. That is, determine the function $R = h(q)$. (*Hint:* Solve for p in the demand function and multiply this expression by q.)

5 In Exercise 2, restate the function for total revenue as a function of q. (See Exercise 4 for a hint.)

6 In Exercise 3, restate the total revenue function as a function of q.

7 The supply function $q_s = f(p)$ for a product is quadratic. Three points which lie on the function are (30, 1,500), (40, 3,600), and (50, 6,300).
   (*a*) Determine the equation for the supply function.
   (*b*) Make any observations you can about the restricted domain of the function.
   (*c*) Compute and interpret the p intercept.
   (*d*) What quantity will be supplied at a price of $60?

8 The supply function $q_s = f(p)$ for a product is quadratic. Three points which lie on the supply function are (60, 2,750), (70, 6,000), and (80, 9,750).
   (*a*) Determine the equation for the function.
   (*b*) Make any observations you can about the restricted domain of the function.
   (*c*) Compute and interpret the p intercept.
   (*d*) What quantity will be supplied at a price of $75?

9 The supply function $q_s = f(p)$ for a product is a quadratic. Three points which lie on the supply function are (40, 600), (50, 3,300), and (80, 15,000).
   (a) Determine the equation for the function.
   (b) Make any observations you can about the restricted domain of the function.
   (c) Compute and interpret the p intercept.
   (d) What quantity will be supplied at a price of $100?

10 The demand function $q_d = f(p)$ for a product is quadratic. Three points which lie on the function are (5, 1,600), (10, 900), and (20, 100). Determine the equation for the demand function. What quantity will be demanded at a market price of $25?

11 The demand function $q_d = f(p)$ for a product is quadratic. Three points which lie on the function are (10, 2,700), (20, 1,200), and (30, 300). Determine the equation for the demand function. What quantity will be demanded at a market price of $5?

12 The demand function $q_d = f(p)$ for a product is quadratic. Three points which lie on the function are (10, 3,800), (30, 1,000), and (15, 2,800). Determine the equation for the demand function. What quantity will be demanded at a market price of $20?

13 The supply and demand functions for a product are $q_s = p^2 - 400$ and $q_d = p^2 - 40p + 2,600$. Determine the market equilibrium price and quantity.

14 The supply and demand functions for a product are $q_s = 4p^2 - 500$ and $q_d = 3p^2 - 20p + 1,000$. Determine the market equilibrium price and quantity.

15 In Example 13, assume that the criterion is to minimize the sum of the squares of the distances separating the emergency response facility and the three cities; that is,

$$S = \sum_{j=1}^{3} d_j^2$$

   (a) Determine the distance function $S = f(x)$.
   (b) Determine the location which minimizes S.

Figure 6.16
HMO location
model.

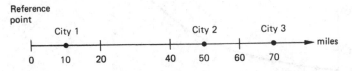

16 Health Maintenance Organization Figure 6.16 illustrates the relative locations of three cities. A large health maintenance organization (HMO) wishes to build a satellite clinic to service the three cities. The location of the clinic x should be such that the sum of the squares of the distances between the clinic and each city is minimized. This criterion can be stated as

$$\text{Minimize } S = \sum_{j=1}^{3} (x_j - x)^2$$

where x_j is the location of the city j and x is the location of the clinic.
   (a) Determine the distance function $S = f(x)$.
   (b) Determine the location which minimizes S.

17 In Exercise 16, assume that the location is to be chosen so as to minimize the sum of the products of the number of HMO members in each city and the square of the distance separating the city and the HMO facility, or

$$\text{Minimize } S = \sum_{j=1}^{3} p_j d_j^2$$

where p_j equals the number of HMO members in city j and d_j equals the distance separating city j and the HMO facility. If the number of members in each city are 10,000, 6,000, and 18,000, respectively:

(a) Determine the distance function $S = f(x)$.

(b) Determine the location which minimizes S.

18 **National Basketball Association** Example 8 in this chapter examined the rapidly increasing salaries of players in the National Basketball Association (NBA). The reason for these increases is largely attributable to the increased earnings by teams within the NBA. Figure 6.17 indicates the gross revenues of the NBA between 1981 and 1990. From the graph, it appears that the revenue function could be approximated by a quadratic function. For the season ending in 1981, league revenues were \$110 million. For the years ending 1986 and 1989 revenues were \$210 million and \$400 million, respectively. Using these data points, determine the estimating function $R = f(t)$, where R equals league revenues (in millions of dollars) and t equals years measured since the season ending in 1981 ($t = 0$ corresponds to the 1980–1981 season). Using this function, project league earnings in 1995.

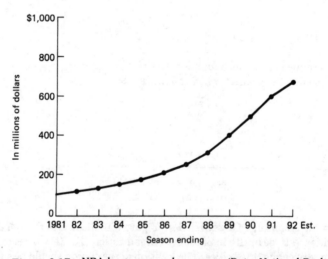

Figure 6.17   NBA league annual revenues. (Data: National Basketball Association)

19 **Employment in Cellular Telephone Industry** Figure 6.18 shows the increase in employment within the cellular telephone industry. For the years 1985, 1987, and 1989, the number of persons employed equaled 2,000, 5,000, and 13,500, respectively. Using these data points, determine the quadratic estimating function $n = f(t)$, where n equals the number of persons employed and t equals time measured in years since 1985. According to this estimating function, predict the number of employees expected in 1996.

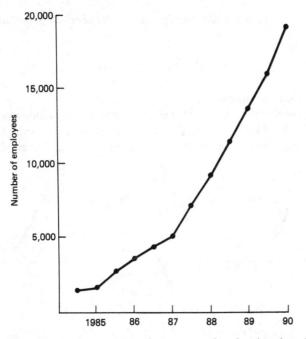

Figure 6.18 Number of persons employed within the cellular telephone industry.

6.3 POLYNOMIAL AND RATIONAL FUNCTIONS

Polynomial Functions

Linear and quadratic functions are examples of the general set of functions called *polynomial functions.*

> **DEFINITION: POLYNOMIAL FUNCTION**
> A *polynomial function* of degree n involving the independent variable x and the dependent variable y has the general form
>
> $$y = f(x)$$
>
> where
>
> $$f(x) = a_n x^n + a_{n-1} x^{n-1} + \cdots + a_1 x + a_0 \qquad (6.7)$$
>
> a_j equals a constant for each j, n is a positive integer, and $a_n \neq 0$.

The degree of a polynomial is the exponent of the highest-powered term in the expression. A linear function is a *first-degree polynomial function,* whereas a quadratic function is a *second-degree function*. A *third-degree function* such as

$$y = x^3 - 2x^2 + 5x + 10$$

is referred to as a *cubic function*.

Cubic functions of the form $y = f(x)$ tend to exhibit a behavior similar to those portrayed in Fig. 6.19. Although we will learn more about the graphical characteristics of functions in the coming chapters, let's explore one attribute of polynomial functions.

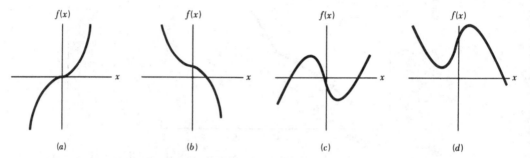

Figure 6.19   Graphical characteristics of cubic functions.

ULTIMATE DIRECTION ATTRIBUTE

The *ultimate direction* of a function *f* refers to the behavior of $f(x)$ as x assumes larger and larger positive values and as x assumes larger and larger negative values. For polynomial functions the ultimate behavior of $f(x)$ is determined by the behavior of the highest-powered term in the function. This is based on the observation that as x becomes more positive (or negative), eventually the highest-powered term will contribute more to the value of $f(x)$ than all other terms in the function.

For polynomial functions of the form

$$f(x) = a_n x^n + a_{n-1} x^{n-1} + \cdots + a_1 x + a_0$$

the ultimate behavior depends on the term $a_n x^n$. Figure 6.20 illustrates the different possibilities. The sign of a_n as well as whether n is odd or even are the significant factors in determining the ultimate direction. When *n is even*, $x^n > 0$ for positive or negative x; for these cases the sign of a_n determines the sign on $a_n x^n$. When n is odd, $x^n > 0$ if $x > 0$ and $x^n < 0$ if $x < 0$. Coupled with the sign of a_n, the ultimate direction of the graph of $f(x)$ (n odd) will be different for $x > 0$ and $x < 0$.

In Fig. 6.19, cases *(a)* and *(c)* represent cubic functions where $a_3 > 0$, and cases *(b)* and *(d)* represent those where $a_3 < 0$.

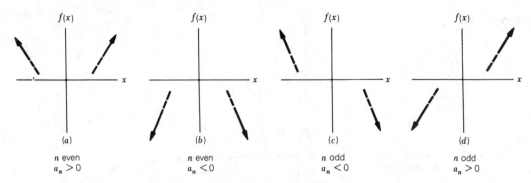

Figure 6.20 Ultimate direction attributes for polynomial functions of the form $f(x) = a_n x^n + a_{n-1} x^{n-1} + \cdots + a_1 x + a_0$.

EXAMPLE 14

For the following polynomial functions, determine the ultimate direction of $f(x)$ and sketch f.

(a) $f(x) = \dfrac{x^5}{5} + \dfrac{x^4}{8} - 2.5x^3$

(b) $g(x) = \dfrac{x^4}{4} - 8x^2 + 10$

SOLUTION

(a) The ultimate direction for $f(x)$ is determined by the term $x^5/5$. As x assumes larger and larger (positive and negative) values, this term will eventually become dominant in determining the value of $f(x)$. Because the exponent on this term is odd and the coefficient $(\frac{1}{5})$ is positive, the ultimate direction will correspond to the situation in Fig. 6.20d. Positive values of x result in positive values for $f(x)$; negative values of x result in negative values for $f(x)$.

If sufficient numbers of values of x are substituted into f and the resulting ordered pairs plotted, the sketch of f should appear as shown in Fig. 6.21.

Figure 6.21

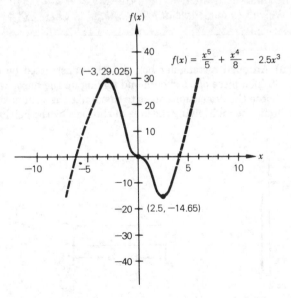

$f(x) = \dfrac{x^5}{5} + \dfrac{x^4}{8} - 2.5x^3$

Figure 6.22

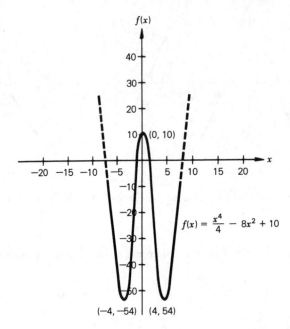

$$f(x) = \frac{x^4}{4} - 8x^2 + 10$$

$(0, 10)$

$(-4, -54)$ $(4, 54)$

(*b*) The ultimate direction of $g(x)$ is determined by the term $x^4/4$. Since the degree of this term is even and the coefficient ($\frac{1}{4}$) is positive, the ultimate direction will correspond to the situation in Fig. 6.20a. Both positive *and* negative values of x result in positive values for $f(x)$. If enough ordered pairs of values satisfying g are identified and plotted, the graph of g should appear as shown in Fig. 6.22.

NOTE   The behavior of polynomial functions between the "tails" of ultimate direction can be determined more easily than by the "brute force" method of plotting many ordered pairs. We will examine this in greater detail in Chap. 16.

EXAMPLE 15   (**Container Design**) An open rectangular box is to be constructed by cutting square corners from a 60 × 60-inch piece of cardboard and folding up the flaps, as shown in Fig. 6.23. The goal is to choose the dimensions so as to maximize the volume of the box. The volume of the box is found by multiplying the area of the base by the height of the box, or

Figure 6.23

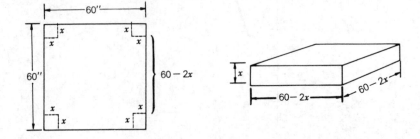

Figure 6.24
Volume-of-container
function.

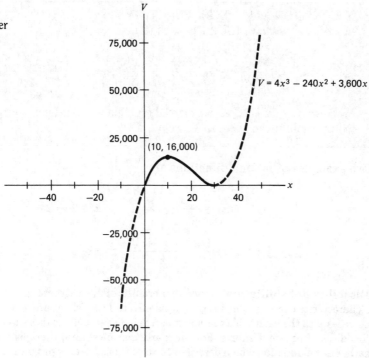

$$V = f(x) = (60 - 2x)(60 - 2x)(x)$$
$$= (3,600 - 240x + 4x^2)(x)$$
$$= 3,600x - 240x^2 + 4x^3$$

This cubic function is sketched in Fig. 6.24. It can be seen that the volume will be maximized at 16,000 cubic inches when $x = 10$. In Chap. 17, we will show how this type of problem can be solved by using differential calculus.

◻

PRACTICE EXERCISE
What are the dimensions of the box having the greatest volume? What is the restricted domain for the volume function? The restricted range? *Answer:* 40 × 40 × 10 inches; $0 \leq x \leq 30$; $0 \leq V \leq 16,000$.

Rational Functions

As mentioned in Chap. 4, **rational functions** are functions expressed as the **ratio** or quotient of two polynomials.

> **DEFINITION: RATIONAL FUNCTION**
> A *rational function* has the general form
>
> $$f(x) = \frac{g(x)}{h(x)} = \frac{a_n x^n + a_{n-1} x^{n-1} + \cdots + a_1 x + a_0}{b_m x^m + b_{m-1} x^{m-1} + \cdots + b_1 x + b_0} \qquad (6.8)$$
>
> where g is an nth-degree polynomial function and h is a nonzero mth-degree polynomial function.

Two examples of rational functions are

$$f(x) = \frac{x}{x^2 - 4} \qquad x \neq 2, -2$$

$$g(x) = \frac{x^3 - 5x + 10}{x} \qquad x \neq 0$$

EXAMPLE 16   **(Disability Rehabilitation)** Physical therapists often find that the rehabilitation process is characterized by a diminishing-returns effect. That is, regained functionality usually increases with the length of a therapy program but eventually in decreased amounts relative to additional program efforts. For one particular disability, therapists have developed a mathematical function which describes the cost C of a therapy program as a function of the percentage of functionality recovered, x. The function is a rational function having the form

$$C = f(x)$$

or
$$C = \frac{5x}{120 - x} \qquad 0 \leq x \leq 100$$

where C is measured in thousands of dollars. For example, the therapy cost to gain a 10 percent recovery is estimated to equal

$$f(10) = \frac{5(10)}{120 - 10}$$

$$= \frac{50}{110} = 0.454 \text{ (thousand dollars)}$$

The cost to gain a 60 percent recovery is estimated to equal

$$f(60) = \frac{5(60)}{120 - 60}$$

$$= \frac{300}{60} = 5.0 \text{ (thousand dollars)}$$

A sketch of this cost function appears in Fig. 6.25.

Figure 6.25
Cost of
rehabilitation.

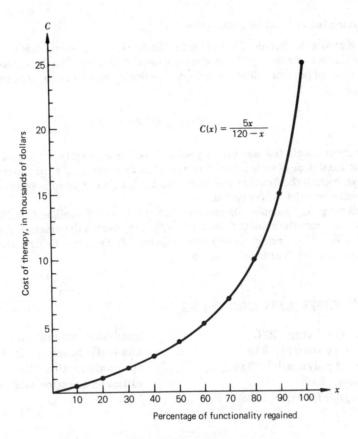

$$C(x) = \frac{5x}{120 - x}$$

Cost of therapy, in thousands of dollars

Percentage of functionality regained

Section 6.3 Follow-up Exercises

In Exercises 1–16, (*a*) determine the degree of the function and (*b*) determine the ultimate direction for the function.

1 $f(x) = -x^3/4$

2 $f(x) = -x^4/2$

3 $f(x) = x^8$

4 $f(x) = -x^4$

5 $f(x) = x^7 - 7x^3 + 8x^2 - 5x$

6 $f(x) = x^6 - 2x^5 + x^3 - 10x$

7 $f(x) = x^9 - x^7 + x^2 - 1{,}000$

8 $f(x) = x^{10} - 5x^4 - 100$

9 $f(x) = -(x^5 - 3x^2 + 5x - 4)/7$

10 $f(x) = x^{12} - \dfrac{5}{2}x^6 + 1$

11 $f(x) = -5x^8 + 2x^3$

12 $f(x) = -3x^{10} + 12x^9 - 1$

13 $f(x) = -(x + 1)^5$

14 $f(x) = -(x^3 - 2)^4$

15 $f(x) = (2x^3 - 4x)^2$

16 $f(x) = -(x - 2x^4)^3$

17 Sketch the function $f(x) = -x^3$.

18 Sketch the function $f(x) = -x^5$.

19 Sketch the function $f(x) = x^6$.

20 Sketch the function $f(x) = -x^4$.

21 Sketch the function $f(x) = 5x^5 - 10$.

22 Sketch the function $f(x) = -x^4 + 8$.

23 Sketch the function $f(x) = 4x^3 + 5$.

24 Sketch the function $f(x) = -x^6 + 6$.

25 Sketch the rational function $f(x) = 3x/(100 - x)$.

26 Sketch the rational function $f(x) = 1/(x-1)$.

27 Influenza Outbreak The Center for Disease Control reports that one strain of flu is making its way through the eastern part of the country. The Center believes that the number of persons catching the flu during this outbreak can be estimated by the function

$$n = f(t) = 0.04t^3 + 2.5$$

where n equals the number of persons who have caught the flu and t equals time measured in days since initial detection. The flu is expected to run its course within 30 days. Sketch the function and determine the number of persons expected to catch the disease over the 30-day period.

28 Referring to Example 15, assume that the piece of cardboard is 30×60 inches. (a) Construct the volume function $V = f(x)$. (b) Sketch the function. (c) Estimate the value of x which results in maximum volume. (d) Estimate the maximum volume and the associated dimensions of the box.

❏ **KEY TERMS AND CONCEPTS**

axis of symmetry   220
concave up (down)   219
degree of polynomial   239
parabola   219
polynomial function   239

quadratic formula   222
quadratic function   218
rational function   243
ultimate direction attribute   240

❏ **IMPORTANT FORMULAS**

$$y = f(x) = ax^2 + bx + c \quad a \neq 0 \quad \text{Quadratic function} \tag{6.1}$$

$$\left(\frac{-b}{2a}, \frac{4ac - b^2}{4a}\right) \quad \text{Vertex of parabola} \tag{6.2}$$

$$x = \frac{-b \pm \sqrt{b^2 - 4ac}}{2a} \quad \text{Quadratic formula}$$

$$f(x) = a_n x^n + a_{n-1}x^{n-1} + \cdots + a_1 x + a_0 \quad \text{Polynomial function} \tag{6.7}$$

$$f(x) = \frac{g(x)}{h(x)} = \frac{a_n x^n + a_{n-1}x^{n-1} + \cdots + a_1 x + a_0}{b_m x^m + b_{m-1}x^{m-1} + \cdots + b_1 x + b_0}$$

$$\text{Rational function} \tag{6.8}$$

❏ **ADDITIONAL EXERCISES**

Section 6.1

In Exercises 1–12, determine the concavity of the corresponding parabola, its y intercept and x intercepts, and the coordinates of the vertex.

1 $f(x) = 5x^2 - 3x + 100$

2 $f(x) = \dfrac{5x^2}{100} + 75$

3 $f(x) = -6x^2$

4 $f(x) = 2x^2 + 4x + 7$

5 $f(x) = 3x^2 - 4x + 5$

6 $f(x) = -x^2 + 16$

7 $f(x) = \dfrac{x^2}{2} - 10x$

8 $f(x) = -10 + 5x + 16x^2$

9 $f(x) = -x^2 - 5x$

10 $f(x) = 25x^2 + 8x - 12$

11 $f(x) = ex^2 + fx + g, \quad e, f, g > 0$

12 $f(x) = -x^2/d, \; d > 0$

13 Determine the equation of the quadratic function which passes through the points $(0, -20)$, $(5, -120)$, and $(-3, -56)$.

*14 Verify that the coordinates of the vertex of a parabola are $\left(\dfrac{-b}{2a}, \dfrac{4ac - b^2}{4a}\right)$, where

$f(x) = ax^2 + bx + c$.

*15 Given the quadratic equation $ax^2 + bx + c = 0$, show (prove) that the roots (if any exist) can be determined using the quadratic formula

$$x = \frac{-b \pm \sqrt{b^2 - 4ac}}{2a}$$

Section 6.2

16 A ball is thrown straight up into the air. The height of the ball can be described as a function of time according to the function $h(t) = -16t^2 + 128t$, where $h(t)$ is height measured in feet and t is time measured in seconds.
(a) What is the height 2 seconds after the ball is thrown?
(b) When will the ball attain its greatest height?
(c) When will the ball hit the ground ($h = 0$)?

17 An object is dropped from a bridge which is 400 feet high. The height of the object can be determined as a function of time (since being dropped) according to the function $h(t) = 400 - 16t^2$, where $h(t)$ is height measured in feet and t is time measured in seconds.
(a) What is the height of the ball after 4 seconds?
(b) How long does it take for the ball to hit the water?

18 The demand function for a particular product is

$$q = f(p) = 480,000 - 3,000p$$

where q is stated in units and p is stated in dollars. Determine the quadratic total revenue function $R = g(p)$. What does total revenue equal when $p = \$100$?

19 The demand function for a particular product is

$$q = f(p) = 1,800 - 7.5p$$

where q is stated in units and p is stated in dollars. Determine the quadratic total revenue function $R = h(q)$. (Note that q is the independent variable.)

20 The supply function $q_s = f(p)$ for a product is quadratic. Three points which lie on the graph of the supply function are (20, 150), (30, 400), and (40, 750). Determine the equation of the supply function.

21 The demand function for a product is

$$q_d = p^2 - 70p + 1{,}225$$

(a) How many units will be demanded if a price of \$20 is charged?
(b) Determine the q_d intercept and interpret its meaning.
(c) Determine the p intercept(s) and interpret.
(d) Estimate the restricted domain.

22 The demand function for a product is

$$q_d = p^2 - 90p + 2{,}025$$

(a) How many units will be demanded if a price of \$30 is charged?
(b) Determine the q_d intercept and interpret its meaning.
(c) Determine the p intercept(s) and interpret.
(d) Estimate the restricted domain.

23 The supply and demand functions for a product are $q_s = p^2 - 100$ and $q_d = p^2 - 40p + 400$. Determine the market equilibrium price and quantity.

24 The supply and demand functions for a product are $q_s = p^2 - 525$ and $q_d = p^2 - 70p + 1{,}225$. Determine the market equilibrium price and quantity.

25 A local travel agent is organizing a charter flight to a well-known resort. The agent has quoted a price of \$300 per person if 100 or fewer sign up for the flight. For every person over the 100, the price for *all* will decrease by \$2.50. For instance, if 101 people sign up, each will pay \$297.50. Let x equal the number of persons above 100.
(a) Determine the function which states price per person p as a function of x, or $p = f(x)$.
(b) In part a, is there any restriction on the domain?
(c) Formulate the function $R = h(x)$, which states total ticket revenue R as a function of x.
(d) What value of x results in the maximum value of R?
(e) What is the maximum value of R?
(f) What price per ticket results in maximum R?

26 Wage Incentive Plan A producer of a perishable product offers a wage incentive to the drivers of its trucks. A standard delivery route takes an average of 20 hours. Drivers are paid at the rate of \$10 per hour up to a *maximum* of 20 hours (if the trip requires 30 hours, the drivers receive payment for only 20 hours). There is an incentive for drivers to make the trip in less than 20 hours. For each hour under 20, the hourly wage increases by \$1. Assume x equals the number of hours required to complete the trip.
(a) Determine the function $w = f(x)$, where w equals the hourly wage in dollars.
(b) Determine the function which states the driver's salary for the trip as a function of x.
(c) What trip time x will maximize the driver's salary for the trip?
(d) What hourly wage is associated with this trip time?
(e) What is the maximum salary?

27 Nonlinear Break-Even Analysis The total cost function for producing a product is

$$C = f(x) = 100x^2 + 1,300x + 1,000$$

where x equals the number of units produced (in thousands) and C equals total cost (in thousands of dollars). Each unit of the product sells for $2,000. Using x as defined above, formulate the total revenue function (stated in thousands of dollars) and determine

(a) The level(s) of production required to break even
(b) The level of output that would result in maximum profit
(c) The expected maximum profit

28 Peak Summer Electrical Needs in the United States Figure 6.26 is a graph of the peak summer demands for electricity in the United States as compiled by the North American Electric Reliability Council. It appears that the peak demand is increasing approximately in a quadratic manner. If the peak summer demands in 1981, 1985, and 1988 were 427, 450, and 522 (thousands of megawatts), respectively:

(a) Use the three data points to determine the quadratic estimating function $D = f(t)$, where D equals the peak U.S. summer demand (in thousands of megawatts) and t equals time measured in years since 1981.
(b) Using the function from part a, estimate peak summer demand in 1995 and in 2000.

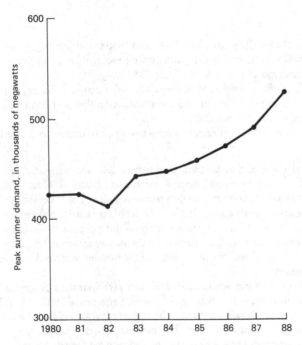

Figure 6.26 Peak summer electricity demand in the United States.

29 Credit Card Usage The use of consumer credit cards has been increasing steadily. Figure 6.27 illustrates the combined annual receivables (charges) for MasterCard and Visa for the years 1983–1990.

Figure 6.27

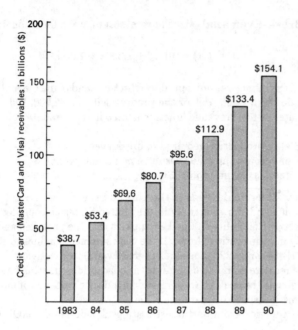

(a) Use the data for the years 1984, 1986, and 1989 to determine a *quadratic* estimating function $C = f(t)$, where C equals annual receivables in billions of dollars and t equals time measured in years since 1983.

(b) Use the data for the years 1986 and 1988 to determine a *linear* estimating function $C = g(t)$, where C equals annual receivables in billions of dollars and t equals time measured in years since 1983.

(c) Use the functions from parts a and b to estimate annual receivables for the year 1995.

30   Credit Card Usage, Continued In Exercise 29, two estimating functions (linear and quadratic) were developed for the combined annual receivables for MasterCard and Visa. An issue of interest is to determine which of the functions provides the best estimate of annual receivables. One way of deciding this is to measure the error associated with each function in estimating the 8 data points in Figure 6.27. There are different measures of error. One measure is the *sum of squares of deviations* between the actual annual receivables for the 8 years and the annual receivables predicted using the estimating functions.

(a) Given the quadratic estimating function $f(t)$ found in Exercise 29, determine estimates of annual receivables for each of the years 1983–1990. For each year, determine the difference between the actual receivables and the estimated receivables and square the difference. The measure of error associated with $f(t)$ is determined by summing the squares of the differences for the 8 years.

(b) Using the linear estimating function from Exercise 29, determine estimates of annual receivables for each of the years 1983–1990. As in part a, determine the sum of squares of the differences between the actual receivables and those estimated using $g(t)$.

(c) Based upon the results of parts a and b, which estimating function has the least error?

Section 6.3

In Exercises 31–38, determine (a) the degree of the function and (b) the ultimate direction of the function.

31 $f(x) = 8x^6 - 4x^3$

32 $f(x) = -x^5/25 - 3x^4 + 65$

33 $f(x) = 4x^5 + 5x^3 - 2x$

34 $f(x) = x^9 - 9x^7 + 5x^3 - 500$

35 $f(x) = -x^8 + 40{,}000x^5 - 25x$

36 $f(x) = -x^7 + 4x^5 + 2x^3 - x + 5$

37 $f(x) = (x^7 - 5x^6 + 3x^5 - 5x^4)/100$

38 $f(x) = -x^6/3 + 2x^3 + 4x$

39 Sketch the function $f(x) = -x^3/2 + 10$.

40 Sketch the function $f(x) = -x^5/4 + 5$.

41 Sketch the function $f(x) = x^8$.

42 Sketch the function $f(x) = -x^7$.

43 Sketch the rational function $f(x) = 5x/(200 - x)$.

44 Sketch the rational function $f(x) = 3/(x - 3)$.

❏ CHAPTER TEST

1 Given the quadratic function

$$f(x) = 4x^2 + 5x - 20$$

determine (a) the concavity, (b) y intercept, (c) x intercept(s), and (d) coordinates of the vertex of the associated parabola. (e) Sketch the parabola.

2 The demand function for a product is

$$q = f(p) = 360{,}000 - 45p$$

where q equals the quantity demanded and p equals price in dollars.

(a) Determine the quadratic revenue function $R = g(p)$.

(b) What price should be charged to maximize total revenue?

(c) Sketch the total revenue function.

3 Japanese Prosperity In recent years, the prosperity within Japan has resulted in heavy investment by the Japanese in other countries around the world. Figure 6.28 is a sketch showing the amount of money invested in Europe during the 1980s. In 1980, 1984, and 1987 the amounts invested were \$0.6, \$2.1, and \$6.25 billion, respectively. Using these three data points, determine the quadratic estimating function $I = f(t)$, where I equals the Japanese investment (in billions of dollars) and t equals time measured in years since 1980. Using this function, estimate the expected investment in the year 1995.

4 Determine the degree and ultimate direction for the following functions.

(a) $f(x) = -\dfrac{(x^3 - 4)^5}{-3}$

(b) $f(x) = -(x^3)^4 + 15x^7 - 20x$

5 An open rectangular box is to be constructed by cutting square corners from a 24×16-inch piece of sheet metal and folding up the flaps as shown in Fig. 6.29. The desire is to choose the dimensions so as to maximize the volume of the box. Formulate the function $V = f(x)$, where V equals the volume of the box and x equals the width of the square corner.

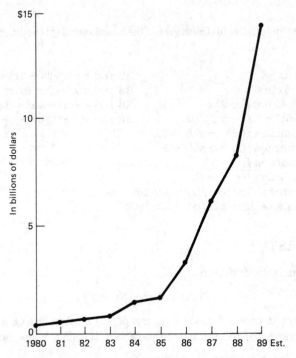

Figure 6.28 Japan's direct investment in Europe.
(Source: Ministry of Finance, Japan)

Figure 6.29

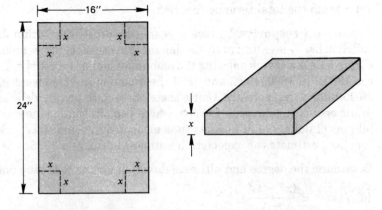

MINICASE

RETAIL MERCHANDISING WARS

Sears, K Mart, and Wal-Mart are the three leading retail merchandisers in the United States. Sears, long the leader in sales dollars, began losing its edge in the middle 1980s. Figure 6.30 is a sketch of merchandise sales revenues (in billions of dollars) for the three retailers between 1983 and 1988. Although K Mart seemed to be closing the gap on Sears, Wal-Mart was making significant advances on both Sears and K Mart. One analyst believes that sales for Sears and K Mart were increasing at an approximately linear rate during this time period and that the sales for Wal-Mart were increasing in a quadratic fashion.

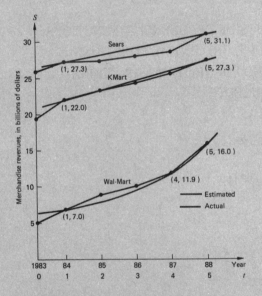

Figure 6.30 Annual merchandise revenues for Sears, K Mart, and Wal-Mart.

1. Using the data points for 1984 and 1988, determine the linear estimating function $S_1 = f(t)$, where S_1 equals annual sales for Sears (in billions of dollars) and t equals time in years measured since 1983.

2. Using the data points for 1984 and 1988, determine the corresponding linear estimating function $S_2 = g(t)$ for K Mart.

3. Using the data points for 1984, 1987, and 1988, determine the quadratic estimating function $S_3 = h(t)$ for Wal-Mart.

4. Using the estimating functions for S_1, S_2, and S_3, project annual sales for the years 1989–1995.

5. Using the functions $f(t)$ and $g(t)$, estimate the point in time when annual sales for K Mart equals that for Sears.

6. Using the functions $f(t)$, $g(t)$, and $h(t)$, estimate the points in time when annual sales for Wal-Mart equals those for K Mart and Sears.

7. Using an appropriate reference, look up annual sales data for Sears, K Mart, and Wal-Mart and determine the error in the estimates associated with part 4 for 1989 and 1990.

CHAPTER 7

EXPONENTIAL AND LOGARITHMIC FUNCTIONS

CHAPTER OBJECTIVES

❏ Discuss the nature of exponential functions: their structural characteristics and graphical behavior

❏ Present a variety of applications of exponential functions

❏ Discuss the nature of logarithms and the equivalence between exponential and logarithmic forms

❏ Discuss the characteristics of logarithmic functions

❏ Present a variety of applications of logarithmic functions

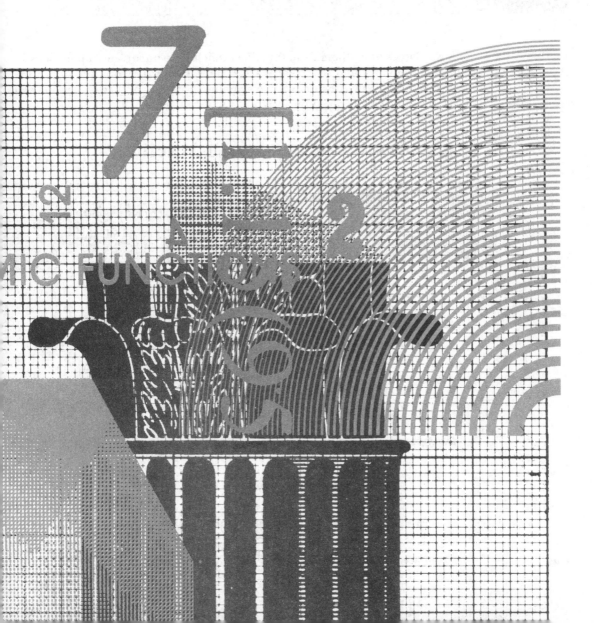

The Super Bowl is an annual extravaganza which becomes the focal point for hundreds of millions of sports fans each January. Coming into Super Bowl XXV in 1991, the cost of a 30-second advertising spot was $800,000. Why would any company spend so much money for a 30-second spot? Because the Super Bowl attracts an incredibly large worldwide TV audience. Prior to Super Bowl XXV, Super Bowls accounted for 5 of the 10 largest audiences in television history and 17 of the top 50. In Example 16, actual data will be presented reflecting the increasing cost of advertising for this event. What is desired is to determine a function which can be used to estimate the cost of Super Bowl ads in future years.

Two classes of mathematical functions which have important applications in business, economics, and the sciences are *exponential functions* and *logarithmic functions.* In this chapter we will examine the nature of these functions and illustrations of their application.

7.1 CHARACTERISTICS OF EXPONENTIAL FUNCTIONS

ALGEBRA FLASHBACK
Some important properties of exponents and radicals are repeated here for your review. Assume that a and b are positive numbers and m and n are real-valued.

$$\text{Property 1:} \quad b^m \cdot b^n = b^{m+n}$$

Examples: $\quad 2^2 2^3 = 2^{2+3} = 2^5 = 32$

$x^5 x^{-3} = x^{5-3} = x^2$

$$\text{Property 2:} \quad \frac{b^m}{b^n} = b^{m-n} \qquad b \neq 0$$

Examples: $\quad \dfrac{3^6}{3^3} = (3)^{6-3} = 3^3 = 27$

$\dfrac{x^4}{x} = x^{4-1} = x^3$

$$\text{Property 3:} \quad (b^m)^n = b^{mn}$$

Examples: $\quad (10^3)^2 = 10^{(3)(2)} = 10^6 = 1,000,000$

$(x^{-2})^5 = x^{(-2)(5)} = x^{-10}$

$$\boxed{\text{Property 4:} \quad a^m b^m = (ab)^m}$$

Examples: $\quad 3^4 2^4 = [(3)(2)]^4 = 6^4 = 1{,}296$

$\qquad\qquad x^2 y^2 = (xy)^2$

$$\boxed{\text{Property 5:} \quad b^{m/n} = \sqrt[n]{b^m}}$$

Examples: $\quad 8^{2/3} = \sqrt[3]{8^2} = \sqrt[3]{64} = 4$

$\qquad\qquad x^{1/4} = \sqrt[4]{x}$

$$\boxed{\text{Property 6:} \quad \sqrt[n]{b^m} = (\sqrt[n]{b})^m}$$

Example: $\quad \sqrt[3]{27^2} = (\sqrt[3]{27})^2 = (3)^2 = 9$

$$\boxed{\text{Property 7:} \quad b^0 = 1 \qquad b \neq 0}$$

Examples: $\quad 5{,}000^0 = 1$

$\qquad\qquad (xy)^0 = 1 \qquad$ provided $xy \neq 0$

$$\boxed{\text{Property 8:} \quad b^{-m} = \frac{1}{b^m} \qquad b \neq 0}$$

Examples: $\quad (2)^{-3} = \dfrac{1}{2^3} = \dfrac{1}{8}$

$\qquad\qquad \dfrac{1}{x^{-2}} = \dfrac{1}{1/x^2} = 1\,\dfrac{x^2}{1} = x^2$

If you are a little rusty on exponents, radicals, and their properties, you are urged to review Sec. A.5 in Appendix A.

Characteristics of Exponential Functions

DEFINITION: EXPONENTIAL FUNCTION
A function of the form

$$f(x) = b^x$$

where $b > 0$, $b \neq 1$, and x is any real number, is called an **exponential function** to the base b.

Examples of exponential functions include

$$f(x) = 10^x$$
$$g(x) = (0.5)^x$$

EXAMPLE 1 **(Medication Excretion)** In many natural processes, the rate at which something grows or decays depends upon its current value. The manner in which the body discharges medications is an example of this type of process. For one particular type of medication, suppose that half of the medication in the body is excreted from the bloodstream by the kidneys every 3 hours. Thus, for an initial dose of 100 milligrams, the content in the body after 3 hours would be

$$(100)(\tfrac{1}{2}) = 50 \text{ milligrams}$$

After 6 hours, the amount remaining in the system would be

$$(100)(\tfrac{1}{2})(\tfrac{1}{2}) = 25 \text{ milligrams}$$

After 9 hours, the amount remaining in the system would be

$$(100)(\tfrac{1}{2})(\tfrac{1}{2})(\tfrac{1}{2}) = 12.5 \text{ milligrams}$$

After *n 3-hour periods,* the amount of medication remaining in the system would be described by the function

$$A = f(n) = 100(\tfrac{1}{2})^n$$

where *A equals the number of milligrams of medication remaining in the system* and *n equals the number of (3-hour) periods since the initial dosage was administered.* It should be noted that even though we calculated values for *A* for 3-hour increments of time, the excretion of the medication happens *continuously.* Thus, our function is valid for integer as well as noninteger values for *n.* Thus, the amount remaining in the system after 10.5 hours is

$$
\begin{aligned}
A = f(3.5) &= (100)(0.5)^{3.5} \\
&= (100)(0.5)^3(0.5)^{0.5} \\
&= (100)(0.125)(\sqrt{0.5}) \\
&= (100)(0.125)(0.707) \\
&= 8.8375 \text{ milligrams}
\end{aligned}
$$

❑

PRACTICE EXERCISE
Estimate the amount of medication in the system after 15 hours. After 22.5 hours. *Answer:* 3.125 milligrams; 0.5523 milligram.

There are different classes of exponential functions. One important class is that having the form

$$y = f(x) = ab^{mx} \tag{7.1}$$

where *a, b,* and *m* are real-valued constants. One restriction is that $b > 0$ but $b \neq 1$.

In order to get a feeling for the behavior of exponential functions, let's examine some of the form $y = b^x$ [assuming $a = m = 1$ in Eq. (7.1)].

EXAMPLE 2 The exponential function $f(x) = 2^x$ can be sketched by determining a set of ordered pairs which satisfy the function. Table 7.1 indicates a sample of values assumed for x and the corresponding values for $f(x)$. Note that Property 8 must be used to evaluate 2^x when $x < 0$.

Note from Fig. 7.1 that f is an *increasing* function. That is, any increase in the value of x results in an increase in the value of $f(x)$. In addition, the graph of f is **asymptotic** to the negative x axis. As x approaches negative infinity (denoted $x \rightarrow -\infty$), $f(x)$ approaches but never quite reaches a value of 0.

TABLE 7.1

x	0	1	2	3	-1	-2	-3
$f(x) = 2^x$	1	2	4	8	0.5	0.25	0.125

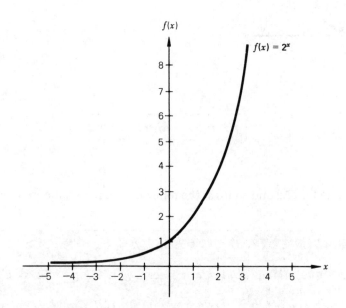

Figure 7.1

EXAMPLE 3 $f(x) = b^x$ **where** $b > 1$ Figure 7.2 illustrates graphs of the three exponential functions $f, g,$ and h.

$$f(x) = 2^x$$
$$g(x) = 2.5^x$$
$$h(x) = 3^x$$

Notice that each function has a positive base b with the only difference between them being the magnitude of b. These are graphed together to illustrate some characteristics of the set of functions.

Figure 7.2
$f(x) = b^x$, $b > 1$.

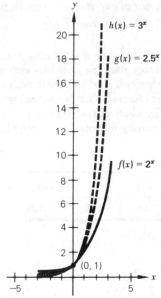

Examine Fig. 7.2 and confirm the following characteristics of this set of functions.

$$f(x) = b^x \quad where \quad b > 1 \qquad (7.2)$$

☐

CHARACTERISTICS OF FUNCTIONS $f(x) = b^x$ where $b > 1$

I   *Each function is defined for all values of x. The domain of f is the set of real numbers.*

II   *The graph of f lies entirely above the x axis (the range is the set of positive real numbers).*

III   *The graph of f is asymptotic to the x axis. That is, the value of y approaches but never reaches a value of 0 as x approaches negative infinity.*

IV   *The y intercept occurs at (0, 1).*

V   *y is an increasing function of x; that is, over the domain of the function any increase in x is accompanied by an increase in y. More precisely, this property suggests that for $x_1 < x_2$, $f(x_1) < f(x_2)$.*

VI   *The larger the magnitude of the base b, the greater the rate of increase in f(x) as x increases in value.*

The class of functions is particularly useful in modeling ***growth processes.***
We will see examples of these types of applications in the next section.

EXAMPLE 4 $f(x) = b^x$ **where** $0 < b < 1$ Figure 7.3 illustrates the graphs of the three exponential functions

$$f(x) = (0.2)^x$$

$$g(x) = (0.6)^x$$

$$h(x) = (0.9)^x$$

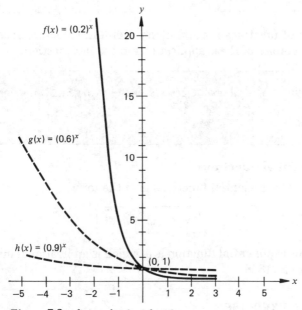

Figure 7.3 $f(x) = b^x, 0 < b < 1.$

These functions are representative of the set of exponential functions

$$f(x) = b^x \qquad 0 < b < 1$$ (7.3)

The three functions illustrated differ only in the magnitude of b.

❑

Examine Fig. 7.3 and confirm the following characteristics for this set of functions.

> **CHARACTERISTICS OF FUNCTIONS $f(x) = b^x$, $0 < b < 1$**
>
> I   *Each function is defined for all values of x (the domain is the set of real numbers).*
> II *The graph of f is entirely above the x axis (the range is the set of positive real numbers).*

Continued

III   *The graph of f is asymptotic to the x axis. That is, the value of y approaches but never reaches a value of 0 as x approaches positive infinity.*
IV   *The y intercept occurs at (0, 1).*
V   *y is a decreasing function of x; that is, any increase in the value of x is accompanied by a decrease in the value of y. More precisely, this property suggests that for $x_1 < x_2$, $f(x_1) > f(x_2)$.*
VI   *The smaller the magnitude of the base b, the greater the rate of decrease in f(x) as x increases in value.*

This class of functions is particularly useful in modeling ***decay processes.*** We will see examples of these applications in the next section.

POINT FOR THOUGHT & DISCUSSION   What are the graphical characteristics of the exponential function $f(x) = b^x$, where $b = 1$?

Base-*e* Exponential Functions

A special class of exponential functions is of the form

$$y = f(x) = ae^{mx}$$ (7.4)

The base of this exponential function is *e*, which is an irrational number approximately equal to 2.71828.

PRACTICE EXERCISE
The number *e* is the value of $(1 + 1/n)^n$ as *n* approaches ∞. To understand this behavior, perform calculations so as to complete the table below.

n	$\left(1 + \dfrac{1}{n}\right)^n$
1	2
2	____
3	____
4	____
5	____
50	____
500	____
5,000	____

50,000	_____
500,000	_____

Answer: 2.25, 2.37037, 2.441406, 2.488320, 2.691588, 2.715569, 2.718010, 2.718255, 2.718279.

Base-*e* exponential functions are particularly appropriate in modeling continuous growth and decay processes — such as bacteria growth, population growth, radioactive decay, and endangered species population decline — and the continuous compounding of interest in financial applications.

Although we will not dwell on the origin of this constant, you will gain insights as we go along as to why such an unusual constant is used as the base for such a popular class of exponential functions. And, indeed, base-*e* exponential functions are more widely applied than any other class of exponential functions.

Figure 7.4

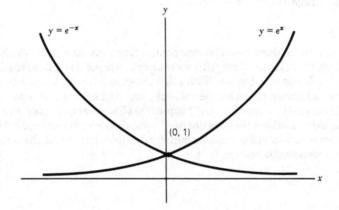

Two special exponential functions in this class are $y = e^x$ and $y = e^{-x}$. Figure 7.4 illustrates the graphs of these two functions. In order to sketch these functions, values for $(2.71828 \ldots)^x$ or $(2.71828 \ldots)^{-x}$ must be computed. This could be a very tedious process. However, since these calculations are performed frequently, values of e^x and e^{-x} are readily available. Most hand calculators have e^x or e^{-x} functions. In the event that you do not have access to an appropriate calculator, values are also available from tables such as Table 1 inside the front book cover. You should become comfortable calculating values for e^x and e^{-x} either from a calculator or table.

EXAMPLE 5   **(Modified Exponential Functions)** Certain applications of exponential functions involve functions of the form

$$y = f(x) = 1 - e^{-mx} \qquad (7.5)$$

TABLE 7.2

x	0	1	2	3	4
e^{-x}	1	0.3679	0.1353	0.0498	0.0183
$1 - e^{-x}$	0	0.6321	0.8647	0.9502	0.9817

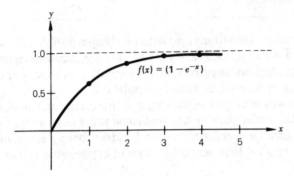

Figure 7.5   Modified exponential function.

In order to illustrate these modified exponential functions, let's graph the function $f(x) = 1 - e^{-x}$ where $x \geq 0$. Table 7.2 contains some sample data points for this function. The graph of the function is shown in Fig. 7.5. Notice that the graph of the function is asymptotic to the line $y = 1$. As x increases in value, the value of y approaches but never quite reaches a value of 1. This is because the second term e^{-x} approaches but never quite reaches a value of 0 as x increases in value. Perhaps the behavior of e^{-x} can be understood better if e^{-x} is rewritten as $1/e^{x}$. As x increases in value the denominator becomes larger and the quotient $1/e^{x}$ approaches but never quite reaches 0.

❑

Conversion to Base-*e* Functions

There are instances where base-*e* exponential functions are preferred to those having another base *b*. Exponential functions having a base other than *e* can be transformed into equivalent base-*e* functions. This is because any positive number *b* can be expressed equivalently as some power of the base *e*; that is, we can find an exponent *n* such that $e^{n} = b$, where $b > 0$.

To illustrate, suppose we have an exponential function

$$f(x) = 3^{x}$$

where the base equals 3. To convert *f* to an equivalent base-*e* function we must express the base in terms of *e*. We want to determine the exponent *n* which results in

$$e^{n} = 3$$

From Table 1 (inside the book cover) we find that

$$e^{1.1} = 3.0042$$

or

$$e^{1.1} \doteq 3$$

Therefore, the original function can be expressed as

$$f(x) = 3^x \doteq (e^{1.1})^x$$

or

$$f(x) \doteq e^{1.1x}$$

To test the equivalence of these functions, let's compute $f(2)$ using the base-3 and base-e forms.

Base 3: $\qquad\qquad\qquad f(2) = 3^2 = 9$

Base e: $\qquad\qquad\qquad f(2) = e^{1.1(2)} = e^{2.2}$

From Table 1: $\qquad\qquad\qquad e^{2.2} = 9.0250$

The difference $(9.0250 - 9 = 0.0250)$ can be attributed to the fact that we were unable to find the precise value of n resulting in $e^n = 3$ from Table 1. Our value of $n = 1.1$ is close, but approximate. More detailed tables, a hand calculator with an e^x function, or the use of logarithms (which we will illustrate later) could have assisted in making our approximation a better one.

Section 7.1 Follow-up Exercises

1 Which of the following functions can be considered to be exponential functions? For those which are not, indicate why.
(a) $y = f(x) = (\pi)^{x^2}$, where $\pi = 3.14 \ldots$
(b) $y = h(x) = \sqrt[3]{0.50}$
(c) $y = v(t) = (4)^{t^2 - 2t + 1}$
(d) $u = v(t) = \sqrt[4]{t^3}$
(e) $g = h(x) = 1/e^{2x-1}$
(f) $y = f(x) = \sqrt{2}x^5$
(g) $y = f(x) = (0.5)^{x-6}$
(h) $y = h(z) = 10^{\sqrt{z}}$
2 (a) Sketch the functions

$$y = f(x) = 2^x$$

$$y = g(x) = 2^{1.5x}$$

$$y = h(x) = 2^{2x}$$

(b) If these functions are compared with Eq. (7.1), the differences are in the values of the parameter m. Examine the sketches from part a. What conclusions can be drawn regarding the behavior of exponential functions and the value of m?

3 (*a*) Sketch the functions

$$y = f(x) = 2^x$$

$$y = g(x) = 0.5(2)^x$$

$$y = h(x) = 2(2)^x$$

(*b*) If these functions are compared with Eq. (7.1), the differences are in the values of the parameter *a*. From the sketches in part *a*, what conclusions can be drawn regarding the behavior of exponential functions and the value of *a*?

4 Refer to Eq. (7.2) and the characteristics of such functions. Let's describe the changes in such functions if a constant is added. That is, given the exponential function

$$f(x) = b^x + c \qquad \text{where} \qquad b > 1$$

(*a*) Describe characteristics I–IV for these functions when $c > 0$.
(*b*) Describe characteristics I–IV when $c < 0$.

5 Refer to Eq. (7.3) and the characteristics of such functions. Given the exponential function

$$f(x) = b^x + c \qquad \text{where} \qquad 0 < b < 1$$

(*a*) Describe characteristics I–IV for these functions when $c > 0$.
(*b*) Describe characteristics I–IV when $c < 0$.

For each of the following exponential functions, compute $f(0)$, $f(-3)$, and $f(1)$.

6 $f(x) = 3^{x^2}$ **7** $f(x) = 2^{x/2}$
8 $f(x) = e^x$ **9** $f(x) = e^{-x/2}$
10 $f(x) = e^{x-2}$ **11** $f(x) = e^{x^2/2}$
12 $f(x) = 1 - e^{0.5x}$ **13** $f(x) = 10(1 - e^{2x})$
14 $f(x) = e^{-x}/2$ **15** $f(x) = 5e^{-x}/2$
16 $f(x) = (2)^{x^2-2x+1}$ **17** $f(x) = (3)^{4-x^2}$
18 $f(x) = 4(1 - e^x)$ **19** $f(x) = 3(4 - e^{2x})$
20 $f(x) = 10 - x - e^x$ **21** $f(x) = x^2 + 3x - 4e^x$

Sketch the following functions.

22 $f(x) = e^{x/2}$ **23** $f(x) = e^x/2$
24 $f(x) = 0.5e^{x^2}$ **25** $f(x) = -2e^x$
26 $f(x) = 5(1 - e^x)$ **27** $f(x) = -2(1 - e^x)$
28 $f(x) = e^{-x/2}$ **29** $f(x) = 1 - e^{0.5x}$
30 $f(x) = 2(1 - e^{-x})$ **31** $f(x) = 4(1 - e^x)$

Convert each of the following exponential functions into equivalent base-*e* exponential functions.

32 $f(x) = (1.6)^x$ **33** $f(x) = (2)^{x^2}$
34 $f(x) = (0.6)^x$ **35** $f(x) = (2.25)^{x/2}$
36 $f(t) = 5(1.6)^{t^2}$ **37** $f(t) = 10(0.3)^t$

38 $f(t) = 2.5(20)^t$

39 $f(t) = -2(90)^t$

40 $f(u) = 3(0.5)^u$

41 $f(u) = 5(0.6)^u(1.6)^{2u}$

42 $f(z) = 2(16)^{z^2}$

43 $f(z) = (10)^z(5)^z$

44 $f(x) = (0.4)^x(0.8)^{x/2}$

45 $f(x) = (0.5)^{x/2}(3)^{x^2}$

46 Sketch the function developed in Example 1.

47 Prescription Drug Excretion For a particular prescription drug, half of the amount of the drug in the bloodstream is excreted by the kidneys every 4 hours. Given an initial dosage of 300 milligrams:

(a) Determine the function $A = f(t)$, where A equals the amount of the drug in the bloodstream (in milligrams) and t equals time since the dosage was administered, measured in increments of 4 hours.

(b) What amount is in the system after 8 hours? After 10 hours? After 24 hours?

(c) Sketch the function.

7.2 APPLICATIONS OF EXPONENTIAL FUNCTIONS

As mentioned earlier, exponential functions have particular application to *growth processes* and *decay processes*. Examples of growth processes include population growth, appreciation in the value of assets, inflation, growth in the rate at which particular resources are used (such as energy), and growth in the gross national product (GNP). Examples of decay processes include the declining value of certain assets such as machinery, the decline in the rate of incidence of certain diseases as medical research and technology improve, the decline in the purchasing power of a dollar, and the decline in the efficiency of a machine as it ages.

When a growth process is characterized by a constant percent increase in value, it is referred to as an ***exponential growth process.*** When a decay process is characterized by a constant percent decline in value, it is referred to as an ***exponential decay process.*** If the population of a country is growing constantly at a rate of 8 percent, the growth process can be described by an *exponential growth function.* If the incidence of infant mortality is declining continually at a rate of 5 percent, the decay process can be described by an *exponential decay function.*

Although exponential growth and decay functions are usually stated as a function of time, the independent variable may represent something other than time. Regardless of the nature of the independent variable, the effect is that *equal increases in the independent variable result in constant percentage changes (increases or decreases) in the value of the dependent variable.*

The following examples illustrate some of the areas of application of exponential functions.

EXAMPLE 6 (Compound Interest) The equation

$$S = P(1 + i)^n \tag{7.6}$$

can be used to determine the amount S that an investment of P dollars will grow to if it receives interest of i percent per compounding period for n compounding periods, assuming reinvestment of any accrued interest. S is referred to as the *compound amount* and P as the *principal.* If S is considered to be a function of n, Eq. (7.6) can be viewed as having the form of Eq. (7.1). That is,

$$S = f(n)$$

or
$$S = ab^{mn}$$

where $a = P$, $b = 1 + i$, and $m = 1$.

Assume that $P = \$1,000$ and $i = 0.08$ per year. Equation (7.6) becomes

$$S = f(n) = (1,000)(1.08)^n$$

To determine the value of S given any value of n, it is necessary to evaluate the exponential term $(1.08)^n$. If we want to know to what sum $\$1,000$ will grow after 25 years, we must evaluate $(1.08)^{25}$. Because this type of calculation is so common, values for the expression $(1 + i)^n$ can be determined by special function keys on many hand calculators or by tables. Table I on page T0 can be used to evaluate $(1 + 0.08)^{25}$. From Table I,

$$(1 + 0.08)^{25} = 6.8485$$

and
$$f(25) = 1,000(1.08)^{25} = 1,000(6.8485) = \$6,848.50$$

Figure 7.6 is a sketch of this function.

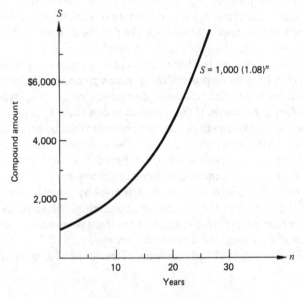

Figure 7.6 Compound amount: $\$1000$ invested at 8 percent per year, compounded annually.

EXAMPLE 7 (**Compound Interest: Continuous Compounding**) When compounding of interest occurs more than once a year, we can restate Eq. (7.6) as

$$S = P\left(1 + \frac{i}{m}\right)^{mt} \tag{7.7}$$

where *i equals the annual interest rate, m equals the number of compounding periods per year,* and *t equals the number of years.* The product *mt* equals the number of compounding periods over *t* years.

Banks often advertise **continuous compounding** for savings accounts as a way of promoting business. Continuous compounding means that compounding is occurring all the time. Another way of thinking of continuous compounding is that there are an infinite number of compounding periods each year. In Eq. (7.7), continuous compounding would suggest that we determine the value of S as m approaches $+\infty$.

It can be shown that for continuous compounding, Eq. (7.7) simplifies to

$$S = f(t) = Pe^{it} \tag{7.8}$$

In Example 6, we computed the amount that a $1,000 investment would grow to if invested at 8 percent per year for 25 years compounded annually. If the $1,000 earns 8 percent per year compounded continuously, it will grow to a sum

$$S = \$1,000e^{0.08(25)}$$
$$= 1,000e^{2.0}$$

From Table 1 (inside the cover of the book) we have

$$e^{2.0} = 7.3891$$

and
$$S = \$1,000(7.3891)$$
$$= \$7,389.10$$

Comparing this value with that found in Example 6, continuous compounding results in additional interest of $7,389.10 − $6,848.50 = $540.60 over the 25-year period. The annual and continuous compounding functions are shown together in Fig. 7.7.

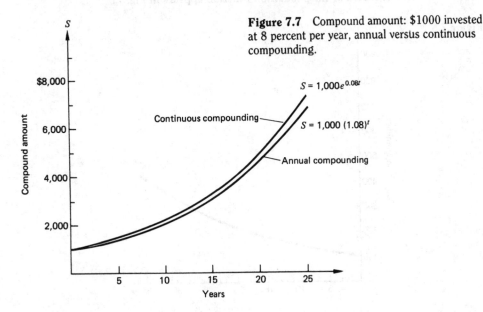

Figure 7.7 Compound amount: $1000 invested at 8 percent per year, annual versus continuous compounding.

EXAMPLE 8

(Exponential Growth Process: Population) As mentioned at the beginning of this section, *exponential growth processes* are characterized by a constant percentage increase in value over time. Such processes may be described by the general function

$$V = f(t)$$

or

$$\boxed{V = V_0 e^{kt}} \tag{7.9}$$

where *V equals the value of the function at time t, V_0 equals the value of the function at t = 0, k is the percentage rate of growth,* and *t is time measured in the appropriate units* (hours, days, weeks, years, etc.).

The population of a country was 100 million in 1970. It has been growing since that time exponentially at a constant rate of 4 percent per year. The function which estimates the size of the population P (in millions) is

$$P = f(t)$$
$$= 100e^{0.04t}$$

where 100 (million) is the population at $t = 0$ (1970) and 0.04 is the percentage rate of exponential growth.

The projected population for 1995 (assuming continued annual growth at the same rate) is found by evaluating $f(25)$, where $t = 25$ corresponds to 1995. The projected population for the country is

$$P = f(25)$$
$$= 100e^{0.04(25)}$$
$$= 100e$$
$$= 271.83 \text{ (million)}$$

A sketch of the population function appears in Fig. 7.8.

Figure 7.8

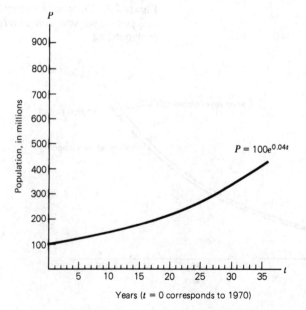

Years ($t = 0$ corresponds to 1970)

PRACTICE EXERCISE
To confirm the nature of exponential growth functions (i.e., *equal increases* in the independent variable result in *constant percentage increases* in the dependent variable), compute $f(1)$, $f(2)$, and $f(3)$ which reflect equal increases of 1 in the independent variable. Then calculate the percent increase in $f(t)$ between $t = 1$ and $t = 2$ and between $t = 2$ and $t = 3$. Are these the same?

EXAMPLE 9

(Exponential Growth Processes, continued) A question of interest in growth functions is, How long will it take for the value of the function to increase by some multiple? In Example 8 there may be a question concerning how long it will take for the population to double. In Eq. (7.9), the value V_0 will double when

$$\frac{V}{V_0} = 2$$

Dividing both sides of Eq. (7.9) by V_0, we get

$$\frac{V}{V_0} = e^{kt}$$

Therefore the value will double when

$$e^{kt} = 2$$

To illustrate this, the population in Example 8 will double when $V = 200$, or

$$200 = 100e^{0.04t}$$

Dividing both sides by 100 yields

$$2 = e^{0.04t}$$

From Table 1 we see that

$$e^{0.69} \doteq 2$$

In order to determine how long it takes the population of 100 million to double, we must find the value of t which makes

$$e^{0.04t} = 2$$

or

$$e^{0.04t} \doteq e^{0.69}$$

These expressions will be equal when their exponents are equal, or when

$$0.04t = 0.69$$

or $t = 17.25$ years*

❑

PRACTICE EXERCISE
What relationship would exist between V and V_0 if the value tripled? Quadrupled? Increased by 50 percent? Determine how long it will take for these to occur in the previous example. *Answer: $V/V_0 = 3$; $V/V_0 = 4$; $V/V_0 = 1.50$; (using Table 1) 27.5 years, 35 years, 10.25 years.*

EXAMPLE 10   (**Exponential Decay Functions**) An *exponential decay process* is one characterized by a constant percentage decrease in value over time. Such processes may be described by the general function

$$V = f(t)$$

or $V = V_0 e^{-kt}$ **(7.10)**

where V equals the value of the function at time t, V_0 equals the value of the function at $t = 0$, and k is the percentage rate of decay (sometimes called the *decay constant*). Compare Eq. (7.10) with Eq. (7.9) and note the differences.

The resale value V (stated in dollars) of a certain type of industrial equipment has been found to behave according to the function $V = f(t) = 100,000e^{-0.1t}$, where t = years since original purchase.
(*a*) What was the original value of a piece of the equipment?
(*b*) What is the expected resale value after 5 years? After 10 years?

SOLUTION

(*a*) The original value is the value of V when $t = 0$. At $t = 0$

$$V = 100,000e^{-0.1(0)}$$
$$= 100,000e^0$$
$$= 100,000$$

Thus, the original value $V_0 = \$100,000$.
(*b*) $f(5) = 100,000e^{-0.1(5)}$
$$= 100,000e^{-0.5}$$
$$= 100,000(0.6065) \quad \text{(from Table 1)}$$
$$= \$60,650$$
  $f(10) = 100,000e^{-0.1(10)}$
$$= 100,000e^{-1}$$
$$= 100,000(0.3679) \quad \text{(from Table 1)}$$
$$= \$36,790$$

---

* Solving for t can be easier if you understand logarithms. An alternative and equivalent approach would be to find the natural logarithms (Sec. 7.3) of both sides of the equation $e^{0.04t} = 2$ and equate these, solving for t. We will study this soon.

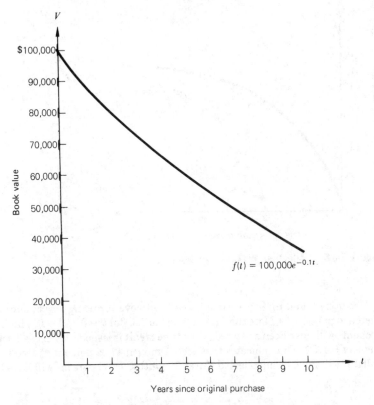

Figure 7.9 Depreciation function.

Figure 7.9 presents a graph of this decay function.

EXAMPLE 11

(**Bill Collection**) A major financial institution offers a credit card which can be used internationally. The question arose among executives as to how long it takes to collect the accounts receivable for credit issued in any given month. Data gathered over a number of years have indicated that the collection percentage for credit issued in any month is an exponential function of the time since the credit was issued. Specifically, the function which approximates this relationship is

$$P = f(t)$$

or

$$P = 0.95(1 - e^{-0.7t}) \qquad t \geq 0$$

where *P equals the percentage of accounts receivable (in dollars) collected t months after the credit is granted.* Sample data points for this function are shown in Table 7.3. The values for

TABLE 7.3	t	0	1	2	3	4
	$e^{-0.7t}$	1	0.4996	0.2466	0.1225	0.0608
	$1 - e^{-0.7t}$	0	0.5034	0.7534	0.8775	0.9392
	$0.95(1 - e^{-0.7t})$	0	0.4782	0.7156	0.8336	0.8922

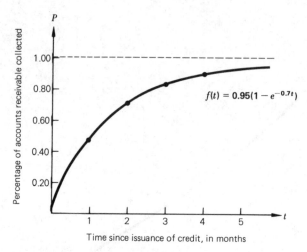

Figure 7.10 Collection response function.

$e^{-0.7t}$ can be found in Table 1 inside the front cover or using a calculator. The function is sketched in Fig. 7.10. Note the figures in Table 7.3. For $t = 0$, $f(t) = 0$, which suggests that no accounts will have been collected at the time credit is issued. When $t = 1$, the function has a value of 0.4782. This indicates that after 1 month 47.82 percent of accounts receivable (in dollars) will have been collected. After 2 months, 71.56 percent will have been collected.

☐

POINT FOR THOUGHT & DISCUSSION What value does P approach as t increases without limit? Why will the value of P never equal 1? Do you think that a restricted domain would apply in this type of application?

Section 7.2 Follow-up Exercises

1 An investment of $200,000 is made which earns interest at the rate of 8 percent per year. If interest is compounded continuously:
   (a) Determine the exponential function which states the compound amount as a function of years of investment t.
   (b) What will the $200,000 grow to if it is invested for 5 years? 10 years?

2 An investment of $500,000 is made which earns interest at the rate of 7.5 percent per year. If interest is compounded continuously:
   (a) Determine the exponential function which states the compound amount S as a function of years of investment t.
   (b) What will the $500,000 grow to if it is invested for 10 years? 20 years?

3 In Exercise 1, determine the length of time required for the investment to double in value. To quadruple.

4 In Exercise 2, determine the length of time required for the investment to double in value. To triple.

5 An investment of $1 million which earns interest at the rate of 8.5 percent per year is made. If interest is compounded continuously, (a) what will the investment grow to if it

is invested for 10 years? (*b*) 25 years? (*c*) How long will it take for the investment to increase by 50 percent?

6 An investment of $250,000 which earns interest at the rate of 10 percent per year is made. If the interest is compounded continuously, (*a*) what will the investment grow to if it is invested for 10 years? (*b*) 20 years? (*c*) How long is required for the investment to increase by 150 percent?

7 **Population Growth** The population *P* of a South American country has been growing exponentially at a constant rate of 2.5 percent per year. The population on January 1, 1985, was 40 million.
  (*a*) Write the general exponential growth function $P = f(t)$ for the population of the country, where *t* equals time measured in years since January 1, 1985.
  (*b*) If the rate and pattern of growth continue, what is the population expected to equal at the beginning of 1995? In the beginning of the year 2010?

8 In Exercise 7, determine the year in which the population can be expected to double. In what year will the population have increased by 50 percent?

9 **Solid Waste** Within a major U.S. city, the annual tonnage of solid waste (garbage) has been increasing at an exponential rate of 8 percent per year. Assume the current daily tonnage is 2,500 tons and the rate and pattern of growth continue.
  (*a*) What daily tonnage will be expected 10 years from now?
  (*b*) Current capacity for handling solid waste is 4,000 tons per day. When will this capacity no longer be sufficient?

10 **Salvage Value** The resale value *V* of a piece of industrial equipment has been found to behave according to the function

$$V = 250,000e^{-0.06t}$$

where *t* = years since original purchase.
  (*a*) What was the original value of the piece of equipment?
  (*b*) What is the expected resale value after 5 years?

11 In Exercise 10, how long does it take for the resale value of the asset to reach 25 percent of its original value?

12 **Endangered Species** The Department of the Interior for the United States estimated that the number of deer of a particular species was 60,000 at the beginning of 1980. Scientists estimate that the population of the species is decreasing exponentially at a rate of 4 percent per year.
  (*a*) Write the decay function $P = f(t)$, where *P* equals the number of deer and *t* equals time (in years) measured from 1980.
  (*b*) What is the population expected to equal in the year 2000 if the decay rate remains constant?

13 In Exercise 12, when is it expected that the population will equal 30,000?

14 For the past 3 years, real estate prices in one area of the country have been increasing at an exponential rate of 4 percent per year. A home was purchased 3 years ago for $120,000.
  (*a*) What is its estimated value today?
  (*b*) Assuming appreciation continues at the same rate, what will its value be 5 years from today?

15 **Fund-Raising** A national charity is planning a fund-raising campaign. Past experience indicates that total contributions raised are a function of the length of time that a campaign is conducted. Within one city a response function has been determined which indicates the percentage of the population *R* who will make a donation as a function of the number of days *t* of the campaign. The function is

$$R = 0.5(1 - e^{-0.05t})$$

(a) What percentage of the population will make a donation after 10 days? After 20 days?

(b) What is the upper bound on the value of R?

16 Credit Card Collections A major bank offers a credit card which can be used domestically and internationally. Data gathered over time indicate that the collection percentage for credit issued in any month is an exponential function of the time since the credit was issued. Specifically, the function approximating this relationship is

$$P = f(t) = 0.92(1 - e^{-0.10t}) \qquad t \geq 0$$

where P equals the percentage of accounts receivable (in dollars) collected t months after the credit is granted.

(a) What percentage is expected to be collected after 1 month?

(b) What percentage is expected after 3 months?

(c) What value does P approach as t increases without limit?

17 Advertising Response A large recording company sells tapes and compact discs (CDs) by direct mail only. Advertising is done through network television. Much experience with this type of sales approach has allowed analysts to determine the expected response to an advertising program. Specifically, the response function for classical music CDs and tapes is $R = f(t) = 1 - e^{-0.05t}$, where R is the percentage of customers in the target market actually purchasing the CD or tape and t is the number of times an advertisement is run on national TV.

(a) What percentage of the target market is expected to buy a classical music offering if advertisements are run one time on TV? 5 times? 10 times? 20 times?

(b) Sketch the response function $R = f(t)$.

18 Demand Function The demand function for a particular commodity is

$$q = f(p) = 10,000 e^{-0.1p}$$

(a) What is demand expected to equal at a price of $5?

(b) What is demand expected to equal at a price of $20?

19 Exponential Revenue Function Refer to the previous exercise. Using the demand function, construct the total revenue function $R = f(p)$. What is revenue expected to equal at a price of $10? What is the demand expected to equal at this price?

*20 Gompertz Function A model sometimes used to represent restricted growth is the Gompertz function. This restricted growth model has a general form

$$y = f(t) = pe^{-ce^{-kt}}$$

where p, c, and k are constants. If $p = 500$, $c = 0.2$, and $k = 0.1$, determine (a) $f(0)$ and (b) $f(10)$.

*21 Given the general Gompertz function in Exercise 20, (a) determine $f(0)$ and (b) determine the value that y approaches as t gets larger and larger.

7.3 LOGARITHMS AND LOGARITHMIC FUNCTIONS

In this section we will discuss logarithms, their properties, their use in solving exponential equations, logarithmic functions, and selected applications.

Logarithms

A **logarithm** is the *power* to which a *base* must be raised in order to yield a given number (i.e., a logarithm is an exponent). Consider the equation

$$2^3 = 8$$

The exponent 3 can be considered as the logarithm, to the base 2, of the number 8. That is, *3 is the power to which 2 must be raised in order to generate the number 8.* We can state this logarithm property as

$$3 = \log_2 8$$

In general,

$$\boxed{y = b^x \Longleftrightarrow x = \log_b y \quad \text{for} \quad b > 0}$$

We will concern ourselves with situations where the base b is restricted to positive values other than 1.

EXAMPLE 12　Following are statements of *equivalent* pairs of exponential and logarithmic equations.

Logarithmic Equation	Exponential Equation
$4 = \log_2 16$	$\Longleftrightarrow 2^4 = 16$
$2 = \log_{10} 100$	$\Longleftrightarrow 10^2 = 100$
$3 = \log_3 27$	$\Longleftrightarrow 3^3 = 27$
$-1 = \log_{10} 0.1$	$\Longleftrightarrow 10^{-1} = 0.1$

Exponential Equation	Logarithmic Equation
$10^4 = 10{,}000$	$\Longleftrightarrow 4 = \log_{10} 10{,}000$
$4^3 = 64$	$\Longleftrightarrow 3 = \log_4 64$
$5^2 = 25$	$\Longleftrightarrow 2 = \log_5 25$
$10^{-2} = 0.01$	$\Longleftrightarrow -2 = \log_{10} 0.01$

The two most commonly used bases for logarithms are base 10 and base *e*. Most of us have probably had some experience with *base-10,* or **common logarithms.*** Logarithms which use $e = 2.718 \ldots$ as the base are called **natural logarithms.**† Logarithms of this form arise from the use of exponential functions which employ *e* as the base.

Common logarithms can be denoted by

$$x = \log_{10} y$$

---

* These are sometimes referred to as *briggsian* logarithms (after H. Briggs, who first used them).
† Natural logs are named after John Napier, the Scot, as *napierian* logarithms.

However, because most logarithm computations (other than base-e) involve the base 10, a more common way of expressing such logarithms is

$$x = \log y$$

where the base, though not indicated, is implicitly 10. Base-e or natural logarithms can be denoted by

$$x = \log_e y$$

but are more commonly denoted by

$$x = \ln y$$

A logarithm having a base b other than 10 or e would be expressed as

$$x = \log_b y$$

Procedures for determining values of common logarithms will not be presented. The examples in this book will deal solely with natural logarithms. Table 2 (inside the back cover) contains values of natural logarithms. As an alternative to tables, most hand calculators have a natural logarithm function key for determining these values.

$\square$

Properties of Logarithms

The use of logarithms can result in certain efficiencies when computations of very large or very small numbers are required. These efficiencies can be attributed partly to certain properties of logarithms. Some of the more important properties follow.

$$\boxed{\text{Property 1:} \quad \log_b uv = \log_b u + \log_b v}$$

Examples:

$$\log_{10}[(100)(1{,}000)] = \log_{10} 100 + \log_{10} 1{,}000$$
$$= 2 + 3 = 5$$

$$\ln 8{,}000 = \ln[(40)(200)]$$
$$= \ln 40 + \ln 200$$
$$= 3.6889 + 5.2983 \quad \text{(from Table 2)}$$
$$= 8.9872$$

$$\boxed{\text{Property 2:} \quad \log_b \frac{u}{v} = \log_b u - \log_b v}$$

Examples:

$$\log_{10} \frac{10{,}000}{100} = \log_{10} 10{,}000 - \log_{10} 100$$

$$= 4 - 2 = 2$$

$$\ln 37.5 = \ln \frac{75}{2}$$
$$= \ln 75 - \ln 2$$
$$= 4.3175 - 0.6931 \qquad \text{(from Table 2)}$$
$$= 3.6244$$

Property 3: $\log_b u^n = n \log_b u$

Examples: $\log_{10} 100^2 = 2 \log_{10} 100$
$$= 2(2) = 4$$

$\ln 10,000 = \ln(100)^2$
$$= 2 \ln 100$$
$$= 2(4.6052) \qquad \text{(from Table 2)}$$
$$= 9.2104$$

Property 4: $\log_b b = 1$

Examples: $\log_{10} 10 = 1 \qquad (10^1 = 10)$
$$\ln e = 1 \qquad (e^1 = e)$$

Property 5: $\log_b 1 = 0$

Examples: $\log_{10} 1 = 0 \qquad (10^0 = 1)$
$$\ln 1 = 0 \qquad (e^0 = 1)$$

Property 6: $b^{\log_b x} = x$

Example: $10^{\log_{10} 100} = 10^2 = 100$

Property 7: $\log_b b^x = x$

Examples: $\log_2 2^5 = 5$
$$\ln e^3 = 3$$

Solving Logarithmic and Exponential Equations

Throughout the book we have had to solve for the roots of equations. Usually, these equations have been of the polynomial form (most frequently linear, quadratic, or cubic). The following examples illustrate the solution of logarithmic and exponential equations.

EXAMPLE 13 To solve the logarithmic equation

$$\ln x^2 + \ln x = 9$$

Property 3 is applied, resulting in

$$2 \ln x + \ln x = 9$$

$$3 \ln x = 9$$

$$\ln x = 3$$

From Table 2, $\ln 20 = 2.9957$. Therefore we can state that $x \doteq 20$ is the root of the given equation.

A more precise solution could be obtained by stating the exponential equation which is equivalent to the equation $\ln x = 3$. The equivalent equation is

$$e^3 = x$$

From Table 1 or by calculator, the more precise solution is $x = 20.086$.

EXAMPLE 14   To solve the logarithmic equation

$$\ln(x^2 + 2) - \ln x^2 = 2$$

Property 2 is applied to the left side of the equation, resulting in

$$\ln \frac{x^2 + 2}{x^2} = 2$$

Our understanding of logarithmic relationships allows us to rewrite this equation in the equivalent exponential form

$$e^2 = \frac{x^2 + 2}{x^2}$$

or from Table 1,

$$7.3891 = \frac{x^2 + 2}{x^2}$$

$$7.3891 \, x^2 = x^2 + 2$$

$$6.3891 \, x^2 = 2$$

$$x^2 = \frac{2}{6.3891}$$

$$x^2 = 0.3130$$

and

$$x = \pm 0.5595$$

EXAMPLE 15   To solve the exponential equation

$$e^{2x} = 5$$

the natural logarithm is taken of both sides of the equation, yielding

$$\ln e^{2x} = \ln 5$$

or
$$2x = \ln 5$$

From Table 2 or by calculator, ln 5 equals 1.6094 and

$$2x = 1.6094$$

$$x = 0.8047$$

EXAMPLE 16

(The Super Bowl: The Incredible Cost of Participation; Motivating Scenario) The Super Bowl is an annual extravaganza which becomes the focal point for hundreds of millions of sports fans each January. Coming into Super Bowl XXV held in 1991, the cost of a 30-second advertising spot was $800,000. Figure 7.11 is a graph showing the cost of 30-second spots for each of the first 25 Super Bowls. Why would anyone spend so much money for a 30-second spot? Because the Super Bowl attracts incredibly large TV audiences. Prior to Super Bowl XXV, Super Bowls accounted for 5 of the 10 largest audiences in television history and 17 of the top 50.

From the graph, it appears that the cost per 30-second spot is increasing approximately at an exponential rate over time. What is desired is to determine an exponential function which can be used to approximate the cost of advertising over time.

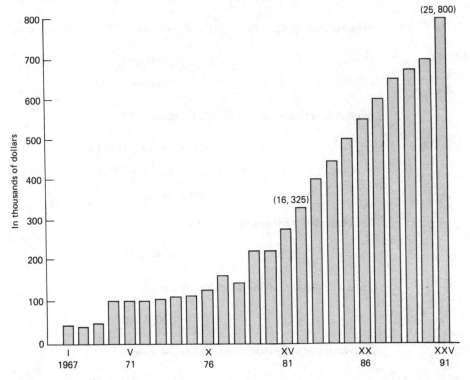

Figure 7.11 The cost of 30 seconds of advertising during the Super Bowl. (Source: Nielsen Media Research)

SOLUTION

Let's determine an estimating function which assumes exponential growth in the cost of 30-second advertising spots. That is, we will determine a function having the form

$$C = f(t) = C_0 e^{it} \tag{7.11}$$

where *C equals the cost per 30-second spot (in thousands of dollars)* and *t equals the Super Bowl number.* This exponential function has two parameters which need to be determined, C_0 and *i*. We will need at least two data points to determine these parameters of the estimating function. Let's choose the data points associated with Super Bowls XVI and XXV. For Super Bowl XVI the advertising cost was $325,000, and for Super Bowl XXV the cost was $800,000. Thus, our two data points are (16, 325) and (25, 800).

Substituting these data points into Eq. (7.11), we get

$$325 = C_0 e^{16i}$$

$$800 = C_0 e^{25i}$$

Taking the natural logarithm of both sides of each equation gives

$$\ln(325) = \ln(C_0 e^{16i}) = \ln(C_0) + \ln(e^{16i}) \quad \text{or} \quad 5.7838 = \ln(C_0) + 16i \tag{7.12}$$

$$\ln(800) = \ln(C_0 e^{25i}) = \ln(C_0) + \ln(e^{25i}) \quad \text{or} \quad 6.6846 = \ln(C_0) + 25i \tag{7.13}$$

We need to solve Eqs. (7.12) and (7.13) for C_0 and *i*. If Eq. (7.12) is subtracted from Eq. (7.13),

$$0.9008 = 9i$$

$$0.1001 = i$$

Substituting this value into Eq. (7.13) gives

$$6.6846 = \ln(C_0) + 25(0.1001)$$

$$6.6846 = \ln(C_0) + 2.5025$$

$$4.1821 = \ln(C_0)$$

The equivalent exponential equation is

$$e^{4.1821} = C_0$$

Using a calculator having an e^x function yields

$$65.5033 = C_0$$

Therefore, our estimating function has the form

$$C = f(t) = 65.5033 e^{0.1001t}$$

❑

PRACTICE EXERCISE
Using our answer, estimate advertising costs for Super Bowls XXVI and XXX.
Answer: $884,215; $1,319,622.

EXAMPLE 17

(Bacterial Growth) Many types of bacteria are believed to grow exponentially according to functions of the form

$$P = f(t) = P_0 e^{kt} \qquad (7.14)$$

where *P equals the population at time t, P_0 gives the population at t = 0, and k is the growth constant (percentage rate of growth).* Exponential growth functions were discussed in Sec. 7.2. Determine the period of time required for an initial population to double in size.

SOLUTION
If an initial population doubles,

$$\frac{P}{P_0} = 2$$

If we divide both sides of Eq. (7.14) by P_0,

$$\frac{P}{P_0} = e^{kt}$$

The population will double when

$$e^{kt} = 2$$

Finding the natural logarithm of both sides of this equation gives

$$kt = \ln 2$$

and the time required for doubling is

$$t = \frac{\ln 2}{k} \qquad (7.15)$$

If for a given bacteria the growth constant equals 0.4 and *t* is stated in hours, the time required for the population to double is

$$t = \frac{\ln 2}{0.4}$$

$$= \frac{0.6932}{0.4} = 1.733 \text{ hours}$$

□

> **POINT FOR THOUGHT & DISCUSSION**   Is the doubling time determined by Eq. (7.15) appropriate only for the initial population to double in size, or is this generalized, given the population at *any* time?

EXAMPLE 18   **(Half-Life)** An exponential decay function has the general form

$$V = V_0 e^{-kt}$$ (7.16)

where *V equals the value of the function at time t, V_0 equals the value of the function at t = 0, and k is the decay constant (percentage rate of decay).* Many natural processes are characterized by exponential decay behavior. One such process is the decay of certain radioactive substances. One measure frequently cited when examining a radioactive substance is its **half-life.** This is the time required for the amount of a substance to be reduced by a factor of $\frac{1}{2}$. For exponential decay functions of the form of Eq. (7.16), the half-life is a function of the decay constant.

Suppose the amount of a radioactive substance is determined by Eq. (7.16). The amount of the substance will halve itself when

$$\frac{V}{V_0} = 0.5$$

or when

$$e^{-kt} = 0.5$$

Taking the natural logarithm of both sides of this equation gives

$$-kt = \ln 0.5$$

and

$$t = \frac{\ln 0.5}{-k}$$ (7.17)

The decay constant for strontium 90 is $k = 0.0244$, where *t* is measured in years. An amount of strontium 90 will decay to one-half its size when

$$t = \frac{\ln 0.5}{-0.0244}$$

$$= \frac{-0.6932}{-0.0244} \doteq 28.40 \text{ years}$$

$\square$

Logarithmic Functions

When a dependent variable is expressed as a function of the logarithm of another variable, the function is referred to as a **logarithmic function.**

> A *logarithmic function* with base b has the form
>
> $$y = f(x) = \log_b u(x) \qquad (7.18)$$
>
> where $u(x) > 0$, $b > 0$, but $b \neq 1$.

The following are examples of logarithmic functions:

$$f(x) = \log x$$
$$f(x) = \ln x$$
$$f(x) = \log(x - 1)$$
$$f(x) = \ln(x^2 - 2x + 1)$$

EXAMPLE 19 Suppose we want to graph the function $y = \ln x$ where $x > 0$. The function $y = \ln x$ can be graphed using two procedures. If values of $\ln x$ are available (from tables or a hand calculator), the function can be graphed directly. Using Table 2 at the end of the book, sample values of $\ln x$ are shown in Table 7.4. The general shape of this function is indicated in Fig. 7.12.

TABLE 7.4

x	0.1	0.5	1	10	100	200	300
$\ln x$	-2.3026	-0.6932	0	2.3026	4.6052	5.2983	5.7038

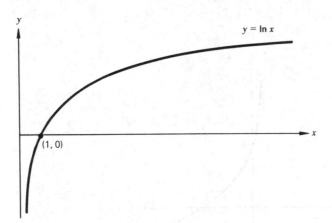

Figure 7.12

An alternative procedure for graphing a logarithmic function is to rewrite the function in its equivalent exponential form. The equivalent exponential form of $y = \ln x$ is

$$e^y = x$$

TABLE 7.5	y	-1	-0.5	0	1	2	3	4
	$x = e^y$	0.3679	0.6065	1.000	2.7183	7.3891	20.086	54.598

By *assuming values for y* and computing the corresponding values of x, a set of data points may be generated. Using Table 1 or a calculator, sample data points have been identified and are shown in Table 7.5. If these points are graphed (remembering that x is the independent variable in the function of interest, $y = \ln x$), the sketch will be identical to that of Fig. 7.12.

An examination of Fig. 7.12 indicates that the natural logarithm function $y = \ln x$ is an increasing function; also, $y > 0$ when $x > 1$, $y = 0$ when $x = 1$, and $y < 0$ when $0 < x < 1$.

EXAMPLE 20   To sketch the logarithmic function

$$y = 5 - 3 \ln(x + 1) \qquad x > -1$$

ordered pairs of values (x, y) are determined. Table 7.6 presents sample values. Figure 7.13 presents a sketch of the function. Note that the function is a decreasing function and that the curve has a vertical asymptote of $x = -1$.

TABLE 7.6	x	-0.5	0	1	2	5	10
	$y = 5 - 3 \ln(x + 1)$	7.0796	5	2.9204	1.7042	-0.3754	-2.1937

Figure 7.13

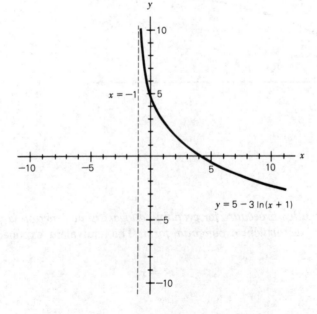

$y = 5 - 3 \ln(x + 1)$

EXAMPLE 21 (**Welfare Management**) A newly created state welfare agency is attempting to determine the number of analysts to hire to process welfare applications. Efficiency experts estimate that the average cost C of processing an application is a function of the number of analysts x. Specifically, the cost function is

$$C = f(x) = 0.001x^2 - 5 \ln x + 60$$

Given this logarithmic function:
(a) Determine the average cost per application if 20 analysts are used.
(b) Determine the average cost if 50 analysts are used.
(c) Sketch the function.

SOLUTION

(a)
$$f(20) = 0.001(20)^2 - 5 \ln(20) + 60$$
$$= 0.40 - 5(2.9957) + 60$$
$$= \$45.42$$

(b)
$$f(50) = 0.001(50)^2 - 5 \ln(50) + 60$$
$$= 2.50 - 5(3.9120) + 60$$
$$= \$42.94$$

(c) Figure 7.14 presents a sketch of this function. We will revisit this application in Chap. 17.

Figure 7.14

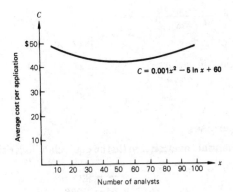

$$C = 0.001x^2 - 5 \ln x + 60$$

Average cost per application

Number of analysts

EXAMPLE 22 (**Memory Retention**) An experiment was conducted to determine the effects of elapsed time on a person's memory. Subjects were asked to look at a picture which contained many different objects. At different time intervals following this, they would be asked to recall as many objects as they could. Based on the experiment, the following function was developed,

$$R = f(t) = 84 - 25 \ln t \qquad t \geq 1$$

where R represents the average percentage recall and t equals time since studying the picture (in hours).
(a) What is the average percentage recall 1 hour after studying the picture?
(b) What is the percentage after 10 hours?
(c) Sketch the function.

SOLUTION

(*a*) $f(1) = 84 - 25 \ln(1) = 84 - 25(0) = 84$ percent

(*b*) $f(10) = 84 - 25 \ln(10) = 84 - 25(2.3026) = 26.435$ percent

(*c*) Figure 7.15 presents a sketch of the function.

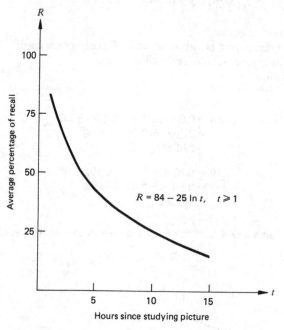

Figure 7.15   Hours since studying the picture.

Section 7.3 Follow-Up Exercises

For each of the following exponential equations, write the equivalent logarithmic equation.

1 $5^2 = 25$

3 $4^3 = 64$

5 $7^3 = 343$

7 $8^4 = 4{,}096$

9 $4^6 = 4{,}096$

11 $5^{-2} = 0.04$

13 $(0.2)^{-4} = 625$

15 $(0.4)^{-3} = 15.625$

17 $2^{-4} = 0.0625$

19 $(0.2)^{-3} = 125$

2 $2^5 = 32$

4 $6^4 = 1{,}296$

6 $3^5 = 243$

8 $5^4 = 625$

10 $10^3 = 1{,}000$

12 $2^{-3} = \frac{1}{8}$

14 $(0.5)^{-3} = 8$

16 $(0.1)^{-4} = 10{,}000$

18 $4^{-3} = 0.015625$

20 $(0.25)^{-2} = 16$

For each of the following logarithmic equations, write the equivalent exponential equation.

21 $\log_2 128 = 7$

23 $\log_4 64 = 3$

22 $\log_2 64 = 6$

24 $\log_4 256 = 4$

25 $\log_3 729 = 6$
27 $\log_2 0.0625 = -4$
29 $\log_5 3125 = 5$
31 $\log_{0.1} 10,000 = -4$
33 $\log_{0.2} 25 = -2$
35 $\log_{0.25} 64 = -3$
37 $\ln 5 = 1.6094$
39 $\ln 100 = 4.6052$

26 $\log_3 81 = 4$
28 $\log_2 0.25 = -2$
30 $\log_5 625 = 4$
32 $\log_{0.1} 1,000 = -3$
34 $\log_{0.2} 625 = -4$
36 $\log_{0.25} 16 = -2$
38 $\ln 3 = 1.0986$
40 $\ln 20 = 2.9957$

Using Table 2 or a calculator, determine the following.

41 $\ln 600$
43 $\ln 80$
45 $\ln 0.1$
47 $\ln 0.75$
49 $\ln 0.01$
51 $\ln 10,000$
53 $\ln 750$
55 $\ln 2,400$
57 $\ln 1,000,000$
59 $\ln 675$

42 $\ln 40$
44 $\ln 200$
46 $\ln 17.5$
48 $\ln 0.5$
50 $\ln 160$
52 $\ln 425$
54 $\ln 150$
56 $\ln 1,600$
58 $\ln 25,000$
60 $\ln 1,050$

Solve the following equations.

61 $3 \ln 2x - 4 = 2 \ln 2x$
63 $x^2 \ln x - 4 \ln x = 0$
65 $\ln(x^2 + 3) - \ln x^2 = 1$
67 $e^{-2x} = 40$
69 $e^{-0.25x} = 16$
71 $e^{2.5x} = 40$
73 $3e^{-2x} = 75$
75 $x^2 \ln x - 9 \ln x = 0$
77 $\ln(x - 3) - \ln x = 1.5$
79 $3e^{-0.5x} = 10$

62 $\ln x^3 - \ln x = 2$
64 $x \ln x - \ln x = 0$
66 $\ln(x + 1) - \ln x = 0.5$
68 $e^{3x} = 20$
70 $5e^{x^2} = 400$
72 $3e^{2x} = 60$
74 $5e^{-1.5x} = 125$
76 $2 \ln x^4 - \ln x^2 = 12$
78 $x^4 \ln x - \ln x = 0$
80 $10e^{5x} = 25$

Sketch the following logarithmic functions.

81 $f(x) = \ln(x/4)$
83 $f(x) = -2 \ln(x + 8)$
85 $f(x) = \ln(x^2 + 10)$
87 $f(x) = \ln(x^2 - 5) + 10$
89 $f(x) = \ln(x + 3)/2$

82 $f(x) = \ln x^3$
84 $f(x) = 10 - 3 \ln x$
86 $f(x) = \ln(3x - 5)$
88 $f(x) = \ln(10 - x) - 5$
90 $f(x) = 10 - \ln x$

91 **Super Bowl Revisited** Given the result in Example 16, in what Super Bowl is it expected that advertising costs will go over $1 million per 30-second spot?

92 **Defense Downscaling** Figure 7.16 illustrates actual and projected expenditures for the purchase, development, testing, and maintenance of weapons. The data for fiscal year (FY) 1989 and FY 1990 are actual, while the data for fiscal years after 1990 are projected by the Electronic Industries Association. The downscaling in expenditures could be approximated by an exponential decay function. Using data for 1991 and 1996, determine the exponential estimating function $V = f(t) = V_0 e^{-it}$, where V equals the annual expenditures (in billions) and t equals time measured since FY 1989.

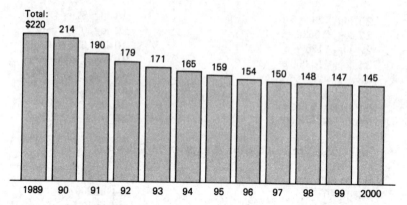

Figure 7.16 Total Pentagon procurement expenditures (in billions of dollars). (Source: Electronic Industries Association)

93 Birthrates Figure 7.17 illustrates data on the number of live births for unmarried women (age 20–24) per 1,000 live births to women of all ages between the mid-1960s and mid-1970s. The increase in birthrates can be approximated by using an exponential growth function. Given that the birthrates to these women per thousand total births were 71 and 92 during 1966 and 1971, respectively, determine the exponential growth function

$$R = f(t) = R_0 e^{it}$$

where R equals the estimated birthrate and t equals time measured in years since 1966.

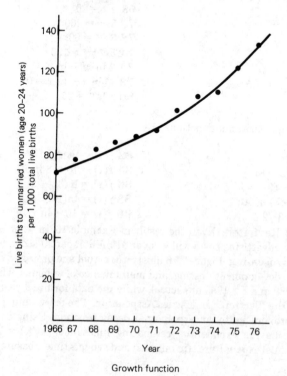

Figure 7.17 Live births for unmarried women (age 20–24 years) per 1,000 births to all women. (Source: Division of Vital Statistics, National Center for Health Statistics)

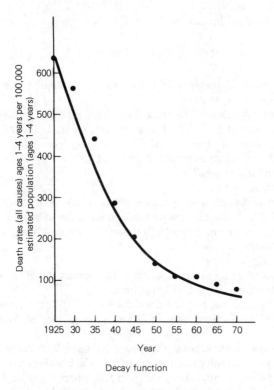

Figure 7.18 Infant mortality rates. (Source: Division of Vital Statistics, National Center for Health Statistics)

Decay function

94 Infant Mortality Rates Figure 7.18 presents data on death rates in the United States for children age 1–4. The death rates are deaths per 100,000 children in this age group. During this century, there has been a steady decline in this rate, which can be estimated by an exponential decay function. Given that the death rates per 100,000 were 202 and 94, respectively, in the years 1945 and 1970, determine the estimating function

$$R = f(t) = R_0 e^{-it}$$

where R equals the death rate per 100,000 and t equals time measured in years since 1940.

*95 Given the general logarithmic function

$$y = a \ln(x + b) - c$$

where a, b, and c are constants, determine the expression for (a) the x intercept and (b) the y intercept.

*96 Given the general exponential function

$$y = -e^{kx} + c$$

where k and c are constants, determine the expression for (a) the x intercept and (b) the y intercept.

97 A company is hiring persons to work in its plant. For the job that persons will perform, efficiency experts estimate that the average cost C of performing the task is a function of the number of persons hired x. Specifically,

$$C = f(x) = 0.005x^2 - 0.49 \ln x + 5$$

(a) What is the expected average cost if 10 persons are hired? 20 persons?
(b) Sketch this average cost function.

98 Bacterial Growth A culture of bacteria *E. coli* is being grown in a medium consisting of inorganic salts and glucose. The bacterium has an initial population of 10^6 per milliliter and it grows at an exponential rate with growth constant $k = 0.7$.
(a) Determine the exponential growth function $f(t)$, where t is in hours.
(b) What is the doubling time?
(c) What is the tripling time?

99 Bacterial Growth A particular bacterium grows at an exponential rate with growth constant $k = 0.6$. The bacterium has an initial population of 10^5 per milliliter.
(a) Determine the exponential growth function $f(t)$, where t is in hours.
(b) What is the doubling time?
(c) What is the tripling time?

100 A yeast culture grows at an exponential rate. The population of the culture doubles after 5 hours. Determine the growth constant k.

101 A radioactive substance has a decay constant $k = 0.350$. If t is measured in hours, determine the half-life for the substance. What is the quarter-life (time to reduce the amount by $\frac{1}{4}$)?

102 A radioactive isotope used to check the thyroid gland has a decay constant $k = 0.150$. If a tracer of 4 units of the isotope is introduced into the bloodstream:
(a) Determine the exponential decay function $f(t)$, where t is in days.
(b) What amount of radioactivity is expected to be in the blood after 8 days?
(c) What is the half-life for the isotope?

103 A radioactive substance has a half-life of 20,000 years. Determine the decay constant k.

104 The amount of a particular drug contained in the bloodstream can be described by an exponential decay function, where t is in hours. If the half-life of the drug is 4 hours, what is the decay constant k?

105 Memory Retention An experiment similar to that discussed in Example 22 resulted in an estimating function

$$R = f(t) = 90 - 20 \ln t \qquad t \geq 1$$

where R equals the average percentage recall and t equals time measured in hours since studying the picture.
(a) What is the average percentage recall after 1 hour? After 5 hours? After 10 hours?
(b) Sketch this function.

❏ KEY TERMS AND CONCEPTS

base-*e* exponential functions   262
common logarithm   277
continuous compounding   269
conversion to base-*e* functions   264
exponential decay process   272
exponential function   257

exponential growth process   270
logarithm   277
modified exponential function   263
natural logarithm   277
solving logarithmic and exponential
      equations   279

❏ IMPORTANT FORMULAS

$y = f(x) = 1 - e^{-mx}$	Modified exponential function	(7.5)
$S = P(1 + i)^n$	Compound amount	(7.6)
$S = Pe^{it}$	Compound amount (continuous compounding)	(7.8)
$V = V_0 e^{kt}$	Exponential growth process	(7.9)
$V = V_0 e^{-kt}$	Exponential decay process	(7.10)
$t = \dfrac{\ln 2}{k}$	Doubling time (exponential growth)	(7.15)
$t = \dfrac{\ln 0.5}{-k}$	Half-life (exponential decay)	(7.17)
$y = f(x) = \log_b a(x)$	Logarithmic function	(7.18)

❏ ADDITIONAL EXERCISES

Section 7.1

In Exercises 1–10, determine $f(-3)$, $f(0)$, and $f(2)$.

1 $f(x) = e^{x/3}$

2 $f(x) = e^{x^2/4}$

3 $f(x) = 5(1 - e^x)$

4 $f(x) = 4(1 - e^{2x})$

5 $f(x) = e^{-2x}$

6 $f(x) = e^{-0.5x}$

7 $f(x) = x^2 e^x$

8 $f(x) = x^3 e^{-x}$

9 $f(x) = 4 - 3^x$

10 $f(x = x/e^x$

Convert each of the following into base-e exponential functions.

11 $f(x) = (4.8)^x$

12 $f(x) = 30(2.56)^{x^2}$

13 $f(x) = 5(0.4)^{x/2}$

14 $f(x) = -6(85)^{x/3}$

15 $f(x) = 10^x$

16 $f(x) = 5(200)^{x^2}$

17 $f(x) = 40/(5)^x$

18 $f(x) = 15/(40)^x$

19 $f(x) = 5/(1.31)^{2x}$

20 $f(x) = (31.5)^x$

Section 7.2

21 Present Value: Continuous Compounding Assuming continuous compounding, the present value P of S dollars t years in the future can be expressed by the function

$$P = f(t) = Se^{-it}$$

The question implied by this function is, "What amount of money P must be invested today so as to grow to an amount S in t years?" It is assumed that the money will receive interest at the rate of i percent per year, compounded continuously. Assuming interest of 10 percent per year compounded continuously, what sum should be deposited today if $100,000 is desired 12 years from now?

22 Present Value: Continuous Compounding Refer to Exercise 21 for a description of the present value concept. Assuming interest of 8 percent per year compounded continuously, how much money must be invested today in order to accumulate $50,000 in 6 years?

23 The population of a particular species of fish has been estimated at 500 million. Scientists estimate that the population is growing exponentially at a rate of 6 percent per year.
(a) Determine the exponential growth function $P = f(t)$, where P equals the fish population (in millions) and t equals time measured in years since today.
(b) If the rate and pattern of growth continue, what is the fish population expected to equal in 25 years?

24 Public Utilities The number of new telephones installed each day in a particular city is currently 250. Telephone company officials believe that the number of new installations will increase exponentially at a rate of 5.5 percent per year.
(a) Determine the exponential estimating function $N = f(t)$, where N equals the number of installations per day and t equals time measured in years.
(b) If the pattern continues, what is the daily rate of installation expected to be 5 years from today? 20 years from today?

25 Endangered Species The population of a particular class of endangered wildlife species has been decreasing exponentially at a rate of 4 percent per year.
(a) If the current population is estimated at 180,000, determine the exponential decay function $P = f(t)$, where P is the estimated population of the species and t equals time measured in years.
(b) What is the population expected to equal in 4 years? In 10 years?

26 Exponential Demand/Revenue Functions The demand function for a particular product is

$$q = f(p) = 200{,}000e^{-0.15p}$$

where q equals demand (in units) and p equals price (in dollars).
(a) What is demand expected to equal at a price of $20?
(b) Construct the total revenue function $R = f(p)$.
(c) What is total revenue expected to equal at a price of $25? What is demand at this price?

27 Police Patrol Allocation A police department has determined that the average daily crime rate depends on the number of officers assigned to each shift. The function describing this relationship is

$$N = f(x)$$
$$= 300 - 8xe^{-0.03x}$$

where N equals the average daily crime rate and x equals the number of officers assigned to each shift. What is the average daily crime rate if 20 officers are assigned? If 40 officers are assigned?

28 Product Reliability A manufacturer of batteries used in portable radios, toys, flashlights, etc., estimates that the percentage P of manufactured batteries having a useful life of at least t hours is described by the function

$$P = f(t) = e^{-0.25t}$$

What percentage of batteries is expected to last at least 5 hours? At least 10 hours?

29 Learning Curves Exponential functions can be used to describe the learning process. One such exponential function has the form

$$y = f(x) = a - be^{-kx}$$

where a, b, and k are positive. For this general *learning curve* function, y represents some measure of the degree of learning and x the number of learning reinforcements.

  Industrial engineers have studied a particular job on an assembly line. The function

$$y = f(x) = 120 - 80e^{-0.30x}$$

is the learning curve function which describes the number of units completed per hour y for a typical employee as a function of the number of hours experience x the employee has with the job. (*a*) What is the hourly rate after 5 hours of experience? (*b*) After 10 hours?

30 Given the learning curve function in the previous exercise, sketch $f(x)$. Is there an upper limit on the value of y?

31 A marketing research organization believes that if a company spends x million dollars on TV advertising, total profit can be estimated by the function

$$P = f(x) = 50x^2e^{-0.5x}$$

What is estimated profit if $5 million is spent for TV advertising? If $10 million is spent?

Section 7.3

Solve the following equations.

32 $\ln x^4 - \ln x^2 = 8$

33 $x^2 \ln x - 64 \ln x = 0$

34 $x \ln x - 6 \ln x = 0$

35 $e^{x^2} = 400$

36 $e^{-5x} = 80$

37 $5e^{-0.2x} = 20$

38 $x \ln x + 5 \ln x = 0$

39 $e^{3x^2} = 150$

40 Physician Availability Figure 7.19 illustrates the relative growth in the number of

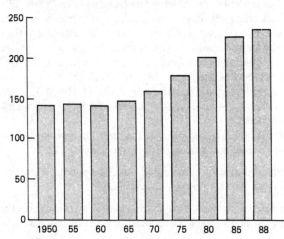

Figure 7.19 Number of physicians per 100,000 population. (Source: American Medical Association)

physicians in the United States per 100,000 population. The number of physicians per 100,000 population can be estimated by an exponential growth function. Using the data from 1955 (144 physicians) and 1970 (162 physicians):

(a) Determine the exponential estimating function $n = f(t)$, where n equals the number of physicians per 100,000 population and t equals years since 1950.

(b) According to this function, estimate the number of physicians in the year 2000.

41 Drug Enforcement Crackdown Figure 7.20 illustrates an increasing trend in the number of drug arrests each year within the United States. The number of arrests can be estimated as growing at an exponential rate. If the number of arrests was 800,000 in 1985 and 1,340,000 in 1989:

(a) Determine the exponential estimating function $A = f(t)$, where A equals the number of arrests per year (in thousands) and t equals time measured in years since 1985.

(b) According to the function in part a, estimate the number of arrests for 1995. Estimate the number for the year 2000.

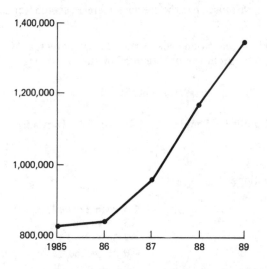

Figure 7.20 Number of arrests for drug abuse violations. (Sources: FBI Uniform Crime Reports, DEA)

42 Savings and Loan Problems During the late 1980s and early 1990s, the financial institutions suffered through difficulties associated with bad loans. Figure 7.21 indicates the annual amounts (in billions of dollars) of real estate repossessed by savings and loan associations in the United States. As can be seen, the amounts repossessed increased at a rapid rate. The increase can be approximated as an exponential growth process. Using the data from 1982 ($2.6 billion) and 1989 ($33.0 billion):

(a) Determine the exponential estimating function $R = f(t)$, where R equals the amount of real estate repossessed (in billions of dollars) and t equals time measured in years since 1982.

(b) Using the function in part a, estimate the amounts repossessed for 1984 and 1986. How do these values compare with the data in Fig. 7.21?

43 A bacterial culture grows at an exponential rate. The population of the culture doubles after 6 hours. Determine the growth constant k.

44 A radioactive isotope used to check the thyroid gland has a decay constant $k = 0.250$. If a tracer of 30 units of the isotope is introduced into the bloodstream:

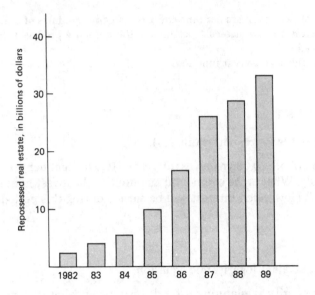

Figure 7.21 Repossessed real estate by savings and loans. (Sources: FDIC and Office of Thrift Supervision)

(*a*) Determine the exponential decay function $f(t)$, where t is stated in days.
(*b*) What amount of the isotope is expected to be in the blood after 10 days?
(*c*) What is the half-life of the isotope?

45 A radioactive substance has a half-life of 36,000 years. Determine the decay constant k.

46 Memory Retention An experiment was conducted to determine the effects of elapsed time on a person's memory. Subjects were asked to look at a picture which contained many different objects. At different time intervals following this, they would be asked to recall as many objects as they could. Based on the experiment, the following function was developed.

$$R = f(t) = 94 - 22 \ln t \qquad t \geq 1$$

For this function R represents the average percentage recall as a function of time since studying the picture (measured in hours). (*a*) What is the average percentage recall after 1 hour? (*b*) After 10 hours? (*c*) Sketch f.

47 A new state welfare agency wants to determine how many analysts to hire for processing of welfare applications. It is estimated that the average cost C of processing an application is a function of the number of analysts x. Specifically, the cost function is

$$C = 0.005x^2 - 16 \ln x + 70$$

(*a*) What is the average cost if 20 analysts are hired? 30 analysts?
(*b*) Sketch the average cost function.

48 A firm has estimated that the average production cost per unit $\overline{C}$ fluctuates with the number of units produced x. The average cost function is

$$\overline{C} = 0.002x^2 - 1,000 \ln x + 7,500$$

where $\overline{C}$ is stated in dollars per unit and x is stated in hundreds of units.
(*a*) Determine the average cost per unit if 100 units are produced. If 500 units are produced.
(*b*) Sketch the average cost function.

❏ CHAPTER TEST

1 Given $f(x) = 1,800e^{-2x}$, determine $f(5)$.

2 An investment of $20,000 receives interest of 10.5 percent per year compounded continuously. What is the compound amount S if the investment is for a period of 25 years? How much interest will be earned during this period?

3 Given

$$3^8 = 6,561$$

write the equivalent logarithmic equation.

4 A national charity is planning a fund-raising campaign in a major city. The population of the city is 2.5 million. The percentage of the population who will make a donation is described by the function

$$R = 1 - e^{-0.075x}$$

where R equals the percentage of the population and x equals the number of days the campaign is conducted. Past experience indicates that the average contribution per donor is $2. Costs of the campaign are estimated at $6,000 per day. Formulate the function $N = f(x)$ which expresses net proceeds N (total contributions minus total costs) as a function of x.

5 Solve the equation

$$\ln x^4 + \ln x = 24$$

6 The value of a piece of equipment is declining exponentially according to a function of the form $V = f(t) = V_0e^{-it}$, where V equals the value of the equipment (in dollars) and t equals the age of the equipment in years. When the equipment was 2 years old, its value was $200,000. When it was 5 years old, its value was $120,000.
(*a*) Determine the function $f(t)$.
(*b*) When is it expected that the value will equal $50,000?

TIME OF DEATH?

If an object is placed in a cooler environment, the temperature of the object decreases toward the temperature of the surrounding environment. The mathematical model which describes this process is

$$T = f(t) = ae^{-kt} + C \qquad (7.18)$$

where T equals the temperature of the object t time units after it is placed in the cooler environment, which is of temperature C. Figure 7.22 is a sketch of the function.

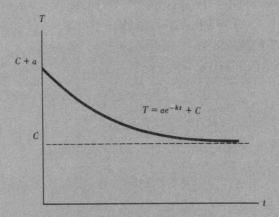

Figure 7.22

The laws of physics underlying this process have many important applications. One important application involves determining the time of death of a person, a function typically performed by a coroner. Suppose that a person's body is discovered in an apartment. The coroner arrives at 3:00 p.m. and finds that the temperature of the body is 84.6°F and the temperature of the apartment is 68°F. The coroner waits 1 hour and then retakes the temperature of the body, finding it to be 83.8°F. The requirement is to determine the time of death. [*Hints:* First, assume that the body temperature at the time of death was 98.6°F and that in Eq. (7.18) $t = 0$ corresponds to the time of death. This information, along with the room temperature, allows for determining a and C in Eq. (7.18). Second, we know that t hours after death the body temperature was 84.6°F and $t + 1$ hours after death it was 83.8°F.]

MATHEMATICS OF FINANCE

CHAPTER OBJECTIVES

❏ Provide an understanding of the time value of money

❏ Provide an understanding of the mathematics of interest computations for single-payment and annuity cash flow structures

❏ Understand the nature of and computations related to mortgage loans

❏ Introduce cost-benefit analysis and related considerations

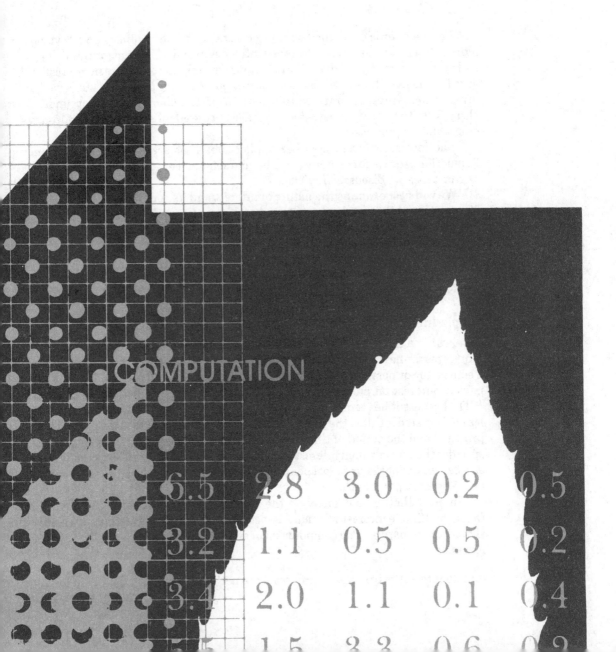

COMPUTATION

6.5 2.8 3.0 0.2 0.5

3.2 1.1 0.5 0.5 0.2

3.4 2.0 1.1 0.1 0.4

This chapter is concerned with interest rates and their effects on the value of money. Interest rates have widespread influence over decisions made by businesses and by us in our personal lives. Corporations pay millions of dollars in interest each year for the use of money they have borrowed. We earn money on sums we have invested in savings accounts, certificates of deposit, and money market funds. We also pay for the use of money which we have borrowed for school loans, mortgages, or credit card purchases.

The interest concept also has applications that are not related to money. Population growth, for example, may be characterized by an "interest rate," or rate of growth, as we discussed in Chap. 7.

We will first examine the nature of interest and its computation. Then we will discuss several different investment situations and computations related to each. Next, a special section will discuss computations related to mortgages. Finally, the last section will discuss cost-benefit analysis.

8.1 INTEREST AND ITS COMPUTATION

Simple Interest

Interest is a fee which is paid for having the use of money. We pay interest on mortgages for having the use of the bank's money. We use the bank's money to pay a contractor or person from whom we are purchasing a home. Similarly, the bank pays us interest on money invested in savings accounts or certificates of deposit (CDs) because it has temporary access to our money. The amount of money that is lent or invested is called the *principal.* Interest is usually paid in proportion to the principal and the period of time over which the money is used. The *interest rate* specifies the rate at which interest accumulates. The interest rate is typically stated as a *percentage of the principal* per period of time, for example, 18 percent per year or 1.5 percent per month.

Interest that is paid solely on the amount of the principal is called *simple interest.* Simple interest is usually associated with loans or investments which are short-term in nature. The computation of simple interest is based on the following formula:

Simple interest = principal × interest rate per time period
× number of time periods

or
$$I = Pin \qquad (8.1)$$

where I = *simple interest, dollars*
P = *principal, dollars*
i = interest rate per time period
n = *number of time periods of loan*
It is essential that the time periods for i and n be consistent with each other. That is, if i is expressed as a percentage per year, n should be expressed in number of years. Similarly, if i is expressed as a percentage per month, n must be stated in number of months.

EXAMPLE 1 A credit union has issued a 3-year loan of $5,000. Simple interest is charged at a rate of 10 percent per year. The principal plus interest is to be repaid at the end of the third year. Compute the interest for the 3-year period. What amount will be repaid at the end of the third year?

SOLUTION

Using the variable definitions from Eq. (8.1), we get P = $5,000, i = 0.10 per year, and n = 3 years. Therefore

$$I = (\$5,000)(0.10)(3)$$
$$= \$1,500$$

The amount to be repaid is the principal *plus* the accumulated interest, or

$$A = P + I$$
$$= \$5,000 + \$1,500 = \$6,500$$

EXAMPLE 2 A person "lends" $10,000 to a corporation by purchasing a bond from the corporation. Simple interest is computed quarterly at a rate of 3 percent per quarter, and a check for the interest is mailed each quarter to all bondholders. The bonds expire at the end of 5 years, and the final check includes the original principal plus interest earned during the last quarter. Compute the interest earned each quarter and the total interest which will be earned over the 5-year life of the bonds.

SOLUTION

In this problem P = $10,000, i = 0.03 per quarter, and the period of the loan is 5 years. Since the time period for i is a quarter (of a year), we must consider 5 years as 20 quarters. And since we are interested in the amount of interest earned over one quarter, we must let n = 1. Therefore, quarterly interest equals

$$I = (\$10,000)(0.03)(1)$$
$$= \$300$$

To compute total interest over the 5-year period, we multiply the per-quarter interest of $300 by the number of quarters, 20, to obtain

$$\text{Total interest} = \$300 \times 20 = \$6,000$$

□

Compound Interest

A common procedure for computing interest is by *compounding interest*. Under this procedure, the interest is reinvested. The interest earned each period is added to the principal for purposes of computing interest for the next period. The amount of interest computed using this procedure is called the **compound interest.**

A simple example will illustrate this procedure. Assume that we have deposited $8,000 in a credit union which pays interest of 8 percent per year *compounded* quarterly. Assume that we want to determine the amount of money we will have on deposit at the end of 1 year if all interest is left in the savings account. At the end of the first quarter, interest is computed as

$$I_1 = (\$8,000)(0.08)(0.25)$$
$$= \$160$$

Note that n was defined as 0.25 *year*. With the interest left in the account, the principal on which interest is earned in the second quarter is the original principal plus the $160 in interest earned during the first quarter, or

$$P_2 = P_1 + I_1 = \$8,160$$

Interest earned during the second quarter is

$$I_2 = (\$8,160)(0.08)(0.25)$$
$$= \$163.20$$

Table 8.1 summarizes the computations for the four quarters. Note that for each quarter the accumulated principal plus interest is referred to as the **compound amount.** Notice that total interest earned during the four quarters equals $659.46.

TABLE 8.1			
Quarter	**(P)** Principal	**(I)** Interest	**(S = P + I)** Compound Amount
1	$8,000.00	$160.00	$8,000.00 + $160.00 = $8,160.00
2	8,160.00	163.20	8,160.00 + 163.20 = 8,323.20
3	8,323.20	166.46	8,323.20 + 166.46 = 8,489.66
4	8,489.66	169.79	8,489.66 + 169.79 = 8,659.46

In this example, simple interest for the year would have been equal to

$$I = (\$8,000)(0.08)(1)$$
$$= \$640$$

The difference between the simple interest and the compound interest is $659.46 − $640.00 = $19.46. Compound interest exceeds simple interest in this example by almost $20 over the 1-year period.

The Power of Compound Growth

As you move through this chapter, the power of compound growth will become apparent. As observed in the last example, *compound interest is greater than simple interest*. Figure 8.1 illustrates the growth in a $10,000 investment which earns interest at 10 percent per year over a period of 10 years. Notice the significant increase in the value of the investment under compounding versus simple interest.

The *frequency* of compounding of interest has an effect on the value of an investment. Frequency refers to how often interest is computed and earned. The frequency of compounding usually ranges from "annual" compounding, in which interest is computed and added to the investment base once per year, to "continuous" compounding (first mentioned in Chap. 7). Intuitively, continuous compounding can be thought of as occurring an infinite number of times during a year. Figure 8.1 illustrates that **the value of an investment increases with increased frequency of compounding.** However, Fig. 8.1 also illustrates that the effects of continuous compounding are not as great as one might expect, compared with other compounding frequencies.

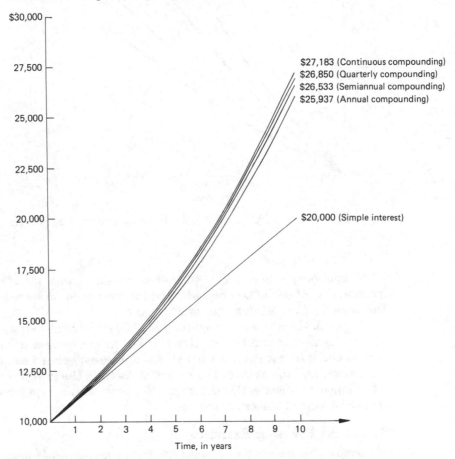

Figure 8.1 Value of $10,000 investment: Simple interest vs. compounding of interest, 10 percent.

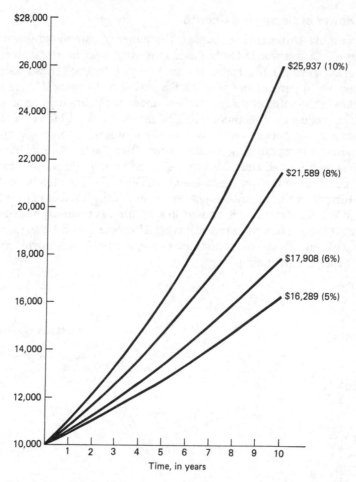

Figure 8.2   Value of $10,000 investment with annual compounding of interest.

A final observation in Fig. 8.1 is that, compared with the effects of simple interest, **the effects of compounding of interest become more significant as the period of investment becomes longer.**

Figure 8.2 illustrates the value of a $10,000 investment over a 10-year period with compounding at different interest rates. As can be seen in this figure, **the higher the interest rate, the greater the interest earned each compounding period and the greater the rate of growth in the investment.** As in Fig. 8.1, this figure illustrates that the longer the period of investment, the more significant the effects of higher interest rates.

Section 8.1 Follow-up Exercises

1 A company has issued a 5-year loan of $90,000 to a new vice president to finance a home improvement project. The terms of the loan are that it is to be paid back in full at the end of 5 years with simple interest computed at the rate of 8 percent per year. Determine the interest which must be paid on the loan for the 5-year period.

2 A student has received a $30,000 loan from a wealthy aunt in order to finance his 4-year college program. The terms are that the student repay his aunt in full at the end of 8 years with simple interest computed at a rate of 4 percent per year. Determine the interest which must be paid on the 8-year loan.

3 A woman has purchased $150,000 worth of corporate bonds. The bonds expire in 20 years, and simple interest is computed semiannually at a rate of 7 percent per 6-month period. Interest checks are mailed to bondholders every 6 months. Determine the interest the woman can expect to earn every 6 months. How much interest can she expect over the 20-year period?

4 A major airline is planning to purchase new airplanes. It wants to borrow $800 million by issuing bonds. The bonds are for a 10-year period with simple interest computed quarterly at a rate of 2 percent per quarter. Interest is to be paid each quarter to bondholders. How much will the airline have to pay in quarterly interest? How much interest will it pay over the 10-year period?

5 A $10,000 certificate of deposit earns interest of 8 percent per year, compounded semiannually. Complete the following table with regard to semiannual compounding. What is total interest over the 2-year period?

Semiannual Period	(P) Principal	(I) Interest	(S = P + I) Compound Amount
1	$10,000	$400	$10,400.00
2			
3			
4			

6 The sum of $500,000 has been placed in an investment which earns interest at the rate of 12 percent per year, compounded quarterly. Complete the following table with regard to quarterly compounding. What is total interest for the year?

Quarter	(P) Principal	(I) Interest	(S = P + I) Compound Amount
1	$500,000	$15,000	$515,000
2			
3			
4			

7 Refer to Exercise 5.
  (a) Determine the compound amount after 2 years if interest is compounded quarterly instead of semiannually.
  (b) Under which compounding plan, semiannually or quarterly, is total interest higher? By how much?

8 Refer to Exercise 6.
  (a) Determine the compound amount after 1 year if interest is compounded semiannually instead of quarterly.
  (b) Under which compounding plan is total interest higher? By how much?

8.2 SINGLE-PAYMENT COMPUTATIONS

This section discusses the relationship between a sum of money at the present time and its value at some time in the future. The assumption in this and the remaining sections is that any interest is computed on a compounding basis.

Compound Amount

Assume that a sum of money is invested and that it earns interest which is compounded. One question related to such an investment is, *What will the value of the investment be at some point in the future?* The value of the investment is the original investment (principal) plus any earned interest. In our example illustrating calculations of compound interest in Sec. 8.1, we called this the **compound amount.** Given any principal invested at the beginning of a time period, the compound amount at the end of the period was calculated as

$$\boxed{S = P + iP} \tag{8.2}$$

Let's redefine our variables and then develop a generalized formula which can be used to calculate the compound amount. Let

$P = principal, dollars$

$i \;= interest\ rate\ per\ compounding\ period$

$n = number\ of\ compounding\ periods\ (number\ of\ periods\\ \quad in\ which\ the\ principal\ has\ earned\ interest)$

$S = compound\ amount$

A **period,** for purposes of these definitions, may be any unit of time. If interest is compounded annually, a year is the appropriate period. If it is compounded monthly, a month is the appropriate period. It is again important to emphasize that *the definition of a period must be the same for both i and n.*

Suppose there has been an investment of P dollars which will earn interest at the rate of i percent per compounding period. From Eq. (8.2) we determined that the compound amount after one period is

$$S = P + iP$$

Factoring P from the terms on the right side of the equation, we can restate the compound amount as

$$S = P(1 + i) \tag{8.3}$$

If we are interested in determining the compound amount after two periods, it may be computed using the equation

$$\frac{\text{Compound amount}}{\text{after two periods}} = \frac{\text{compound amount}}{\text{after one period}}$$

$$+ \quad \text{interest earned} \atop \text{during the second period}$$

or $$S = P(1 + i) + i[P(1 + i)]$$

Factoring P and $1 + i$ from both terms on the right side of the equation gives us

$$S = P(1 + i)[1 + i]$$
or $$S = P(1 + i)^2 \tag{8.4}$$

Similarly, if we wish to determine the compound amount after three periods, it may be computed using the equation

$$\text{Compound amount} \atop \text{after three periods} = \text{compound amount} \atop \text{after two periods}$$

$$+ \quad \text{interest earned} \atop \text{during the third period}$$

or $$S = P(1 + i)^2 + i[P(1 + i)^2]$$

Factoring P and $(1 + i)^2$ from the terms on the right side of the equation, we have

$$S = P(1 + i)^2[1 + i]$$
or $$S = P(1 + i)^3 \tag{8.5}$$

The compound amount formulas developed so far are summarized below.

COMPOUND AMOUNT FORMULAS
Compound amount after one period $= P(1 + i)$.
Compound amount after two periods $= P(1 + i)^2$.
Compound amount after three periods $= P(1 + i)^3$.

And the pattern continues so that the following definition is possible.

DEFINITION: COMPOUND AMOUNT
If an amount of money P earns interest compounded at a rate of i percent per period, it will grow after n periods to the compound amount S, where

$$S = P(1 + i)^n \tag{8.6}$$

Equation (8.6) is often referred to as the ***compound amount formula.*** This relationship may be portrayed graphically as in Fig. 8.3. Notice in Eq. (8.6) that given the principal P and interest rate per compounding period i, the compound amount S is an exponential function of the number of compounding periods n, or

$$S = f(n)$$

Figure 8.3

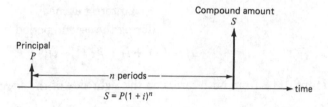

Suppose that $1,000 is invested in a savings bank which earns interest at a rate of 8 percent per year compounded annually. If all interest is left in the account, what will the account balance be after 10 years?

EXAMPLE 3

SOLUTION

In each of these examples we will assume that the investment is made at the very beginning of a compounding period. By using Eq. (8.6), the compound amount after 10 years (periods) is

$$S = \$1,000(1 + 0.08)^{10}$$

The question now becomes, How do we evaluate $(1 + 0.08)^{10}$? Among possible alternatives are these:

1 *Sit close to a pencil sharpener, load up on paper, and use a brute-force hand calculation approach.*
2 *Use an electronic calculator.*
3 *Rewrite the equation by finding the logarithm of both sides and solving for the log of S.*
4 *Use an electronic calculator with financial functions.*

These kinds of computations are fairly common, especially for banking and financial institutions. For those of you not having an electronic calculator with financial functions, sets of tables are available which provide values of $(1 + i)^n$ for given values of i and n. Table I on page T0 provides values of $(1 + i)^n$ values. The expression $(1 + i)^n$ is called the **compound amount factor**.

For our problem we simply find the column associated with an interest rate per compounding period of 8 percent and the row corresponding to 10 compounding periods. Figure 8.4 is an excerpt from these tables. The value of $(1 + 0.08)^{10}$ is 2.15892. Therefore

Figure 8.4
Excerpt from
Table II.

n	i .08
1	1.08000
2	1.16640
3	1.25971
4	1.36049
5	1.46933
6	1.58687
7	1.71382
8	1.85093
9	1.99900
10	2.15892
11	2.33164
12	2.51817
13	2.71962
14	2.93719

$$S = (\$1,000)(2.15892)$$
$$= \$2,158.92$$

The $1,000 investment will grow to $2,158.92, meaning that interest of $1,158.92 will be earned.

EXAMPLE 4   A long-term investment of $250,000 has been made by a small company. The interest rate is 12 percent per year, and interest is compounded quarterly. If all interest is reinvested at the same *rate* of interest, what will the value of the investment be after 8 years?

> **NOTE**   In almost all cases, the interest rate in a problem will be stated as an annual interest rate or as the interest rate per compounding period. When the former case occurs, the interest rate per compounding period is computed by the formula
>
> $$i = \frac{\text{interest rate per year}}{\text{number of compounding periods per year}}$$

SOLUTION

In the compound amount formula, Eq. (8.6), i is defined as the interest rate per compounding period and n as the number of compounding periods. In this problem compounding occurs every quarter of a year. The interest rate per quarter equals the annual interest rate divided by the number of compounding periods per year, or

$$i = \frac{0.12}{4} = 0.03$$

The number of compounding periods over the 8-year period is $8 \times 4 = 32$. Applying Eq. (8.6) gives

$$S = \$250,000(1 + 0.03)^{32}$$

From Table I, $(1 + 0.03)^{32} = 2.57508$, and

$$S = \$250,000(2.57508) = \$643,770$$

Over the 8-year period, interest of $643,770 − $250,000, or $393,770, is expected to be earned.

❑

> **PRACTICE EXERCISE**
> An investment of $100,000 earns interest of 6 percent per year compounded semiannually. If all interest is reinvested, what will the value of the investment be after 5 years? *Answer:* $134,392.

Present Value

The compound amount formula

$$S = P(1 + i)^n$$

is an equation involving four variables — S, P, i, and n. Given the values of any three of these four, the equation can be solved for the remaining variable. To illustrate this point, suppose that a person can invest money in a savings account at the rate of 10 percent per year compounded quarterly. Assume that the person wishes to deposit a lump sum at the beginning of the year and have that sum grow to $20,000 over the next 10 years. The question becomes, How much money should be deposited? Since we are given values for S, i, and n, we need to solve the equation for P. Doing this, we have

$$P = \frac{S}{(1 + i)^n} \qquad (8.7)$$

For the situation mentioned, $S = \$20,000$, $n = 40$ (10×4 compounding periods over the 10 years), and $i = 0.10/4 = 0.025$. The problem is illustrated in Fig. 8.5. From Table I, we have $(1 + 0.025)^{40} = 2.68506$, and

$$P = \frac{\$20,000}{2.68506}$$

$$= \$7,448.62$$

Figure 8.5
Present value
problem.

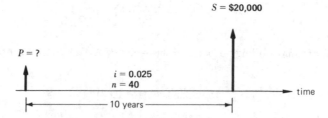

In order to accumulate $20,000 after 10 years, $7,448.62 will have to be deposited.

Although Eq. (8.7) can be applied using the compound amount factor, it can be rewritten in the form of Eq. (8.7a).

$$P = S\left[\frac{1}{(1 + i)^n}\right] \qquad (8.7a)$$

The factor in brackets, $1/(1 + i)^n$, or $(1 + i)^{-n}$, is called the ***present value factor.*** Table II (on page T4) presents selected values for this factor. The values in this table are simply the reciprocal values for those in Table I.

To solve the same problem using Table II, the appropriate value is found for $i = 0.025$ and $n = 40$. From Table II, $(1 + 0.025)^{-40} = 0.37243$. Thus

$$P = (\$20,000)(0.37243)$$
$$= \$7,448.60$$

Note that this answer is not *exactly* the same as computed using Eq. (8.7) and Table I—they are different by $0.02. This is due to rounding differences for the values in the two tables.

EXAMPLE 5 A young man has recently received an inheritance of $200,000. He wants to take a portion of his inheritance and invest it for his later years. His goal is to accumulate $300,000 in 15 years. How much of the inheritance should be invested if the money will earn 12 percent per year compounded semiannually? How much interest will be earned over the 15 years?

SOLUTION

For this problem $S = \$300,000$, $n = 30$, and $i = 0.12/2 = 0.06$. Using Eq. (8.7a) and the appropriate value from Table II, we get

$$P = (\$300,000)(0.17411)$$
$$= \$52,233$$

Over the 15-year period, interest of $300,000 − $52,233, or $247,767, will be earned.

❑

In these applications P can be considered to be the *present value* of S. That is, P and S can be considered as *equivalent* if you consider the interest which can be earned on P during n compounding periods. In Example 5, $52,233 at the time of the inheritance is considered to be the present value of $300,000 15 years hence. It is considered the present value because if the $52,233 is invested at that point in time and it earns 12 percent per year compounded semiannually, it will grow to a value of $300,000 in 15 years.

The present value concept implies that a dollar in hand today is not equivalent to a dollar in hand at some point in the future. We can understand this in an *investment sense* from the preceding discussions. As consumers, we can also appreciate this idea by observing the effects of inflation in prices over time. Businesses often have to evaluate proposed projects which will generate cash flows in different periods. Today, $20,000 of revenue is not equivalent to $20,000 in revenue 10 years from now. Thus, businesses very often use the present value concept to translate all cash flows associated with a project into equivalent dollars at one common point in time. We will examine this in greater detail in Sec. 8.5.

PRACTICE EXERCISE
What sum of money should be invested today at 8 percent per year compounded quarterly if the goal is to have a compound amount of $50,000 after 5 years? *Answer:* $33,648.50.

Other Applications of the Compound Amount Formula

The following examples illustrate other applications of the compound amount formula. They illustrate problems in which the parameters i and n are unknown.

EXAMPLE 6   When a sum of money is invested, there may be a desire to know how long it will take for the principal to grow by a given percentage. Suppose we want to know how long it will take for an investment P to double itself, given that it receives compound interest of i percent per compounding period. If an investment doubles, the ratio of the compound amount S to the principal P is 2, or

$$\frac{S}{P} = 2$$

Given the compound amount formula

$$S = P(1 + i)^n$$

if both sides are divided by P, we get

$$\frac{S}{P} = (1 + i)^n$$

Because the ratio of S/P equals the compound amount factor, an investment will double when

$$(1 + i)^n = 2$$

Given the interest rate per compounding period for the investment, the number of compounding periods required would be found by selecting the appropriate column of Table I and determining the value of n for which $(1 + i)^n = 2$.

EXAMPLE 7   A lump sum of money is invested at a rate of 10 percent per year compounded quarterly. (*a*) How long will it take the investment to double? (*b*) To triple? (*c*) To *increase by* 50 percent?

SOLUTION

(*a*) For this investment the interest rate per compounding period is $0.10/4 = 0.025$. Referring to the column in Table I which corresponds to $i = 2.5$ percent, we read down, looking for a compound amount factor equal to 2. There is no value of n for which $(1 + 0.025)^n$ equals exactly 2. However, for $n = 28$ the compound amount factor equals 1.99650, and for $n = 29$ the compound amount factor equals 2.04641. This suggests that after 28 quarters (or 7 years) the sum will almost have doubled in value. After 29 quarters ($7\frac{1}{4}$ years) the initial sum will have grown to slightly more than double its original value.

(*b*) The sum will triple when $S/P = 3$ or when $(1 + 0.025)^n = 3$. Examining the same column in Table I, we find that the investment will be slightly less than triple in value after 11 years [for $n = 44$, $(1 + 0.025)^{44} = 2.96381$] and slightly more than triple after 11.25 years [for $n = 45$, $(1 + 0.025)^{45} = 3.03790$].

(*c*) For an investment to *increase by* 50 percent

$$S = P + 0.5P$$

or
$$S = 1.5P$$

and
$$\frac{S}{P} = 1.5$$

Refer again to Table I. An investment will increase by slightly less than 50 percent after 4 years $[(1 + 0.025)^{16} = 1.48451]$ and by slightly more than 50 percent after $4\frac{1}{4}$ years $[(1 + 0.025)^{17} = 1.52162]$.

PRACTICE EXERCISE
A sum of money is invested at a rate of 7 percent per year compounded annually. How long will it take for the investment to double? *Answer:* between 10 and 11 years.

EXAMPLE 8

(College Enrollments) The board of regents of a southern state is planning the future college-level needs for the state. They have observed that the number of students attending state-operated schools — junior colleges, 4-year schools, and the state university — has been increasing at the rate of 7 percent per year. There are currently 80,000 students enrolled in the various schools. Assuming continued growth at the same rate, how long will it take for enrollments to reach 200,000 students?

SOLUTION
Defining current enrollments as $P = 80,000$ and future enrollments as $S = 200,000$ we have

$$\frac{S}{P} = \frac{200,000}{80,000}$$
$$= 2.5$$

Therefore, from Table I with $i = 7$ percent,

$$(1 + 0.07)^{13} = 2.40985 \quad \text{at} \quad n = 13$$
$$(1 + 0.07)^{14} = 2.57853 \quad \text{at} \quad n = 14$$

Enrollments will have increased beyond the 200,000 level after 14 years.

EXAMPLE 9

A person wishes to invest $10,000 and wants the investment to grow to $20,000 over the next 10 years. At what annual interest rate would the $10,000 have to be invested for this growth to occur, assuming annual compounding?

SOLUTION
In this problem S, P, and n are specified, and the unknown is the interest rate i. Substituting the known parameters into the compound amount formula gives

$$20,000 = 10,000(1 + i)^{10}$$

or
$$\frac{20,000}{10,000} = (1 + i)^{10}$$

$$2 = (1 + i)^{10}$$

To determine the interest rate, return to Table I and focus upon the row of values associated with $n = 10$. Read across the row until a value of 2 is found in the table. There is no compound amount factor which equals 2 exactly; however,

$$(1 + 0.07)^{10} = 1.96715 \quad \text{when} \quad i = 7 \text{ percent}$$

$$(1 + 0.08)^{10} = 2.15892 \quad \text{when} \quad i = 8 \text{ percent}$$

The original investment will grow to $20,000 over the 10 years if it is invested at an interest rate between 7 and 8 percent. A process called *interpolation* can be used to approximate the exact interest rate required. Although we will not devote time to this topic, it is discussed in the Minicase at the end of the chapter.

❑

Effective Interest Rates

Interest rates are typically stated as annual percentages. The stated annual rate is usually referred to as the **nominal rate.** We have seen that when interest is compounded semiannually, quarterly, and monthly, the interest earned during a year is greater than if compounded annually. When compounding is done more frequently than annually, an **effective annual interest rate** can be determined. This is the interest rate compounded annually which is equivalent to a nominal rate compounded more frequently than annually. The two rates would be considered equivalent if both result in the same compound amount.

Let r equal the effective annual interest rate, i the nominal annual interest rate, and m the number of compounding periods per year. The equivalence between the two rates suggests that if a principal P is invested for n years, the two compound amounts would be the same, or

$$P(1 + r)^n = P\left(1 + \frac{i}{m}\right)^{nm}$$

Dividing both sides of the equation by P results in

$$(1 + r)^n = \left(1 + \frac{i}{m}\right)^{nm}$$

Taking the nth root of both sides results in

$$1 + r = \left(1 + \frac{i}{m}\right)^{m}$$

and by rearranging, the effective annual interest rate can be computed as

$$\boxed{r = \left(1 + \frac{i}{m}\right)^{m} - 1} \tag{8.8}$$

In Example 4 the investment was made with a nominal interest rate of 12 percent per year compounded quarterly. For this investment $i = 0.12$ and $m = 4$. The effective annual interest rate is

$$r = \left(1 + \frac{0.12}{4}\right)^4 - 1$$
$$= (1 + 0.03)^4 - 1$$

From Table I we can determine that $(1 + 0.03)^4 = 1.12551$. Thus,

$$r = 1.12551 - 1$$
$$= 0.12551$$

The effective annual rate is 12.551 percent.

PRACTICE EXERCISE
The nominal interest rate on an investment is 7 percent per year. What is the effective annual interest rate if interest is compounded semiannually? *Answer:* 7.122 percent.

Section 8.2 Follow-up Exercises

1 A sum of $8,000 is invested in a savings account which pays interest at a rate of 9 percent per year compounded annually. If the amount is kept on deposit for 6 years, what will the compound amount equal? How much interest will be earned during the 6 years?

2 A sum of $20,000 is invested in a savings account which pays interest at a rate of 8 percent per year compounded annually. If the amount is kept on deposit for 10 years, what will the compound amount equal? How much interest will be earned during the 10 years?

3 A sum of $25,000 is invested in a savings account which pays interest at a rate of 8 percent per year compounded annually. If the amount is kept on deposit for 15 years, what will the compound amount equal? How much interest will be earned during the 15 years?

4 A sum of $50,000 is invested in a credit union which pays interest at a rate of 10 percent per year compounded annually. If the amount is kept on deposit for 5 years, what will the compound amount equal? How much interest will be earned during the 5 years?

5 A company invests $500,000 in a money market fund which is expected to yield interest at a rate of 10 percent per year compounded quarterly. If the interest rate projections are valid, to what amount should the $500,000 grow over the next 10 years? How much interest will be earned during this period?

6 A university endowment fund has invested $4 million in United States government certificates of deposit. Interest of 8 percent per year, compounded semiannually, will be earned for 10 years. To what amount will the investment grow during this period? How much interest will be earned?

7 An individual invests $25,000 in a money market fund which is expected to yield interest at a rate of 12 percent per year compounded quarterly. If the interest remains stable, to what amount should the $25,000 grow over the next 5 years? How much interest will be earned during this period?

8 An organization invests $500,000 in an investment which is expected to yield interest at a rate of 9 percent per year compounded semiannually. If the money is invested for 10 years, to what amount should it grow? How much interest will be earned during this period?

9 The compound amount factor $(1 + i)^n$ is the amount to which $1 would grow after n periods if it earns compound interest of i percent per period. Determine the compound amount and the interest earned if $1 is invested for 8 years at 12 percent per year (*a*) compounded semiannually, (*b*) compounded quarterly, and (*c*) compounded bimonthly.

10 Compute the compound amount and interest if $1 million is invested under the different conditions mentioned in Exercise 9.

11 The number of students at a local university is currently 15,000. Enrollments have been growing at a rate of 3.5 percent per year. If enrollments continue at the same rate, what is the student population expected to be 10 years from now?

12 A sales representative for the college division of a large publisher had sales of 20,000 books this past year. Her sales have been increasing at the rate of 10 percent per year. If her sales continue to grow at this rate, how many books should she expect to sell 5 years from now?

13 Consumer prices have been increasing at an average rate of 6 percent per year compounded quarterly. The base price on a particular model Chevrolet is $14,500. If prices on this model increase at the same rate as other consumer prices, what will the expected base price of this same model be 5 years from now?

14 If consumer prices are increasing at the rate of 6 percent per year compounded semiannually, an item which costs $25 today will cost what amount in 10 years?

15 If a savings account awards interest of 6 percent per year compounded quarterly, what amount must be deposited today in order to accumulate $20,000 after 5 years? How much interest will be earned during these 5 years?

16 If a credit union awards interest of 7 percent per year compounded semiannually, what amount must be deposited today in order to accumulate $40,000 after 10 years? How much interest will be earned during these 10 years?

17 What sum must be deposited today at 10 percent per year compounded quarterly if the goal is to have a compound amount of $50,000 6 years from today? How much interest will be earned during this period?

18 What sum must be deposited today at 8 percent per year compounded annually if the goal is to have a compound amount of $200,000 12 years from today? How much interest will be earned during this period?

19 What sum must be deposited today at 18 percent per year compounded monthly if the goal 5 years from today is to have a compound amount of $200,000? How much interest will be earned during this period?

20 What sum must be deposited today at 9 percent per year compounded semiannually if the goal 8 years from today is to have a compound amount of $100,000? How much interest will be earned during this period?

21 A sum of $80,000 earns interest at a rate of 8 percent per year compounded semiannually. How long will it take for the investment to grow to $150,000?

22 A sum of $25,000 earns interest at a rate of 7 percent per year compounded annually. How long will it take for the investment to grow to $60,000?

23 A sum of $40,000 earns interest at a rate of 10 percent per year compounded quarterly. How long will it take for the investment to grow to $100,000?

24 A sum of $250,000 earns interest at a rate of 12 percent per year compounded quarterly. How long will it take for the investment to grow to $400,000?

25 The nominal interest rate on an investment is 16 percent per year. Determine the effective annual interest rate if (*a*) interest is compounded semiannually and (*b*) interest is compounded quarterly.

26 The nominal interest rate on an investment is 14 percent per year. Determine the effective annual interest rate if (*a*) interest is compounded semiannually and (*b*) interest is compounded quarterly.

27 The nominal interest rate on an investment is 6 percent per year. Determine the effective annual interest rate if (*a*) interest is compounded semiannually and (*b*) interest is compounded quarterly.

28 If $400,000 is to grow to $750,000 over a 10-year period, at what annual rate of interest must it be invested, given that interest is compounded semiannually?

29 If $2,000 is to grow to $5,000 over a 12-year period, at what annual rate of interest must it be invested, given that interest is compounded annually?

30 If $500,000 is to grow to $700,000 over a 5-year period, at what annual rate of interest must it be invested, given that interest is compounded quarterly?

31 If $60,000 is to grow to $180,000 over a 10-year period, at what annual interest rate must it be invested, given that interest is compounded semiannually?

32 If $250,000 is to grow to $700,000 over an 8-year period, at what annual interest rate must it be invested, given that interest is compounded quarterly?

8.3 ANNUITIES AND THEIR FUTURE VALUE

An *annuity* is a series of periodic payments. Examples of annuities include regular deposits to a savings account, monthly car, mortgage, or insurance payments, and periodic payments to a person from a retirement fund. Although an annuity may vary in dollar amount, *we will assume that an annuity involves a series of equal payments. We will also assume that the payments are all made at the end of a compounding period.* One may certainly argue that the end of one period coincides with the beginning of the next period. The important point is that the payment does not qualify for interest in the previous period but will earn full interest during the next period. Figure 8.6 illustrates a series of payments R, each of which equals $1,000. These might represent year-end deposits in a savings account or quarterly tax payments by a self-employed person to the IRS.

Figure 8.6
Annuity.

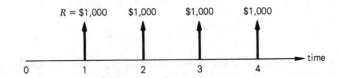

The Sum of an Annuity

Just as we had an interest in determining the future value of a lump-sum investment in Sec. 8.2, there is often some benefit in determining the future value or sum of an annuity. Example 10 illustrates a problem of this type.

EXAMPLE 10 A person plans to deposit $1,000 in a tax-exempt savings plan at the end of this year and an equal sum at the end of each following year. If interest is expected to be earned at the rate of 6 percent per year compounded annually, to what sum will the investment grow at the time of the fourth deposit?

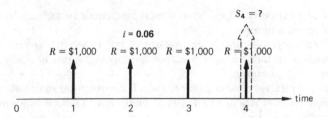

Figure 8.7   Annuity and its future value.

SOLUTION

Figure 8.7 illustrates the annuity and the timing of the deposits. Let S_n equal the sum to which the deposits will have grown *at the time of the nth deposit*. We can determine the value of S_n by applying the compound amount formula to *each* deposit, determining its value at the time of the nth deposit. These compound amounts may be summed for the four deposits to determine S_4. Figure 8.8 summarizes these calculations.

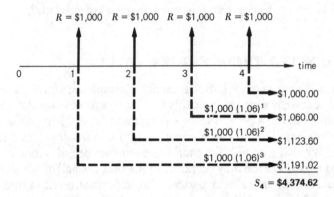

Figure 8.8   Calculating the future value of each payment R.

Note that the first deposit earns interest for 3 years, although the fourth deposit earns no interest. Of the $4,374.62, the interest which has been earned on the first three deposits is $374.62.

☐

The procedure used to determine S in Example 10 is manageable but impractical when the number of payments becomes large. Let's see whether we can develop a simpler approach to determine S_n. Note in Example 10 that S_4 was determined by the sum

$$S_4 = 1,000 + 1,000(1 + 0.06) + 1,000(1 + 0.06)^2 + 1,000(1 + 0.06)^3 \tag{8.9}$$

Let

$R = $ *amount of an annuity*

$i = $ *interest rate per period*

n = number of annuity payments (also the
number of compounding periods)

S_n = sum (future value) of the annuity after n periods (payments)

If we wish to determine the sum S_n that a series of deposits R (made at the end of each period) will grow to after n periods, first examine Eq. (8.9) for the four-period case. The comparable expression for the n-period case is

$$S_n = R + R(1 + i) + R(1 + i)^2 + \cdots + R(1 + i)^{n-1}$$

Factoring R from the terms on the right side gives

$$S_n = R[1 + (1 + i) + (1 + i)^2 + \cdots + (1 + i)^{n-1}] \qquad \textbf{(8.10)}$$

Multiplying both sides of the equation by $(1 + i)$ yields

$$(1 + i)S_n = (1 + i)R[1 + (1 + i) + (1 + i)^2 + \cdots + (1 + i)^{n-1}]$$

which simplifies to

$$S_n + iS_n = R[(1 + i) + (1 + i)^2 + (1 + i)^3 + \cdots + (1 + i)^n] \qquad \textbf{(8.11)}$$

Subtracting Eq. (8.10) from (8.11) results in

$$iS_n = R(1 + i)^n - R$$

or

$$iS_n = R[(1 + i)^n - 1]$$

Solving for S_n, we get

$$S_n = R\left[\frac{(1 + i)^n - 1}{i}\right] \qquad \textbf{(8.12)}$$

The special symbol $s_{\overline{n}|i}$, which is verbalized as "s sub n angle i," is frequently used to abbreviate the **series compound amount factor.** Thus, Eq. (8.12) can be rewritten more simply as

$$\boxed{S_n = Rs_{\overline{n}|i}} \qquad \textbf{(8.13)}$$

Table III (on page T8) contains selected values for this factor. Values for $s_{\overline{n}|i}$ are commonly available on calculators having financial functions.

EXAMPLE 11   Re-solve Example 10 using Eq. (8.13).

SOLUTION

Since $i = 0.06$ and $n = 4$, the appropriate entry in Table III is 4.37462. Substituting this value and $R = \$1,000$ into Eq. (8.13) yields

$$S_4 = (\$1,000)s_{\overline{4}|0.06}$$
$$= (\$1,000)(4.37462)$$
$$= \$4,374.62$$

which is the same answer as before.

$\square$

NOTE An assumption in this section is that interest is computed at the time of each payment. Annual payments earn interest compounded annually, quarterly payments earn interest compounded quarterly, and so forth. Differences between the timing of payments and interest computation (e.g., annual deposits to an account which earns interest compounded quarterly) can be handled by means other than those discussed in this chapter.

EXAMPLE 12 A teenager plans to deposit $50 in a savings account at the end of each quarter for the next 6 years. Interest is earned at a rate of 8 percent per year compounded quarterly. What should her account balance be 6 years from now? How much interest will she earn?

SOLUTION

In this problem $R = \$50$, $i = 0.08/4 = 0.02$, and $n = $ (6 years)(4 quarters per year) = 24 compounding periods. Using Table III,

$$S_4 = (\$50)s_{\overline{24}|0.02}$$
$$= \$50(30.42186)$$
$$= \$1,521.09$$

Over the 6-year period she will make 24 deposits of $50 for a total of $1,200. Interest for the period will be $1,521.09 − $1,200.00 = $321.09.

$\square$

PRACTICE EXERCISE
An investment earns interest of 7 percent per year, compounded annually. If $5,000 is invested at the end of each year, to what sum will the investment have grown at the time of the tenth deposit? *Answer:* $69,082.25.

Determining the Size of an Annuity

As with the compound amount formula, Eq. (8.13) can be solved for any of the four parameters, given values for the other three. For example, we might have a goal of accumulating a particular sum of money by some future time. If the rate of interest which can be earned is known, the question becomes, *What amount should be deposited each period in order to reach the goal?*

To solve such a problem, Eq. (8.13) can be solved for R, or

$$R = \frac{S_n}{s_{\overline{n}|i}}$$

This can be rewritten as

$$R = S_n \left[\frac{1}{s_{\overline{n}|i}} \right] \tag{8.14}$$

where the expression in brackets is the reciprocal of the series compound amount factor. This factor is often called the **sinking fund factor.** This is because the series of deposits used to accumulate some future sum of money is often called a **sinking fund.** Values for the sinking fund factor $[1/s_{\overline{n}|i}]$ are found in Table IV (on page T12).

EXAMPLE 13 A corporation wants to establish a sinking fund beginning at the end of this year. Annual deposits will be made at the end of this year and for the following 9 years. If deposits earn interest at the rate of 8 percent per year compounded annually, how much money must be deposited each year in order to have $12 million at the time of the 10th deposit? How much interest will be earned?

SOLUTION

Figure 8.9 indicates the cash flows for this problem. In this problem $S_{10} = \$12$ million, $i = 0.08$, and $n = 10$. Using Eq. (8.14) and Table IV,

$$R = \$12,000,000[1/s_{\overline{10}|0.08}]$$
$$= \$12,000,000(0.06903)$$
$$= \$828,360$$

Since 10 deposits of $828,360 will be made during this period, total deposits will equal $8,283,600. Because these deposits plus accumulated interest will equal $12 million, interest of $12,000,000 − $8,283,600 = $3,716,400 will be earned.

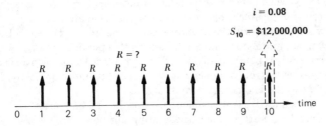

Figure 8.9 Determining the size of an annuity.

EXAMPLE 14 Assume in the last example that the corporation is going to make quarterly deposits and that interest is earned at the rate of 8 percent per year compounded quarterly. How much money should be deposited each quarter? How much less will the company have to deposit over the 10-year period as compared with annual deposits and annual compounding?

SOLUTION

For this problem $S_{40} = \$12$ million, $i = 0.08/4 = 0.02$, and $n = 40$. Using Table IV and substituting into Eq. (8.14) yields

$$R = \$12,000,000(0.01656)$$
$$= \$198,720$$

Since there will be 40 deposits of $198,720, total deposits over the 10-year period will equal $7,948,800. Compared with annual deposits and annual compounding in Example 13, total deposits required to accumulate the $12 million will be $8,283,600 − $7,948,800 = $334,800 less under the quarterly plan.

❑

> **PRACTICE EXERCISE**
> How much money must be deposited at the end of each quarter to accumulate $25,000 after 4 years? Assume interest of 8 percent per year compounded quarterly. *Answer:* $1341.25.

Section 8.3 Follow-up Exercises

1 A person wishes to deposit $5,000 per year in a savings account which earns interest of 8 percent per year compounded annually. Assume the first deposit is made at the end of this current year and additional deposits at the end of each following year.
   (a) To what sum will the investment grow at the time of the 10th deposit?
   (b) How much interest will be earned?

2 A company wants to deposit $500,000 per year in an investment which earns interest of 10 percent per year compounded annually. Assume the first deposit is made at the end of the current year and additional deposits at the end of each following year.
   (a) To what sum will the investment grow at the time of the 10th deposit?
   (b) How much interest will be earned?

3 A mother wishes to set up a savings account for her son's education. She plans on investing $750 when her son is 6 months old and every 6 months thereafter. The account earns interest of 8 percent per year, compounded semiannually.
   (a) To what amount will the account grow by the time of her son's 18th birthday?
   (b) How much interest will be earned during this period?

4 A local university is planning to invest $500,000 every 3 months in an investment which earns interest at the rate of 12 percent per year compounded quarterly. The first investment will be at the end of this current quarter.
   (a) To what sum will the investment grow at the end of 5 years?
   (b) How much interest will be earned during this period?

5 A person wants to deposit $10,000 per year for 6 years. If interest is earned at the rate of 10 percent per year, compute the amount to which the deposits will grow by the end of the 6 years if:
   (a) Deposits of $10,000 are made at the end of each year with interest compounded annually.
   (b) Deposits of $5,000 are made at the end of each 6-month period with interest compounded semiannually.
   (c) Deposits of $2,500 are made at the end of every quarter with interest compounded quarterly.

6 A corporation wants to invest $10 million per year for 5 years. If interest is earned at the rate of 14 percent per year, compute the amount to which the deposits will grow if:

   (*a*) Deposits of $10 million are made at the end of each year with interest compounded annually.

   (*b*) Deposits of $5 million are made at the end of each 6-month period with interest compounded semiannually.

   (*c*) Deposits of $2.5 million are made at the end of each quarter with interest compounded quarterly.

7 How much money must be deposited at the end of each year if the objective is to accumulate $100,000 by the time of the fifth deposit? Assume interest is earned at the rate of 15 percent per year compounded annually. How much interest will be earned on the deposits?

8 How much money must be deposited at the end of each year if the objective is to accumulate $250,000 after 10 years? Assume interest is earned at the rate of 10 percent per year compounded annually. How much interest will be earned?

9 How much money must be invested at the end of each quarter if the objective is to accumulate $1,500,000 after 10 years? Assume that the investment earns interest at the rate of 8 percent per year compounded quarterly. How much interest will be earned?

10 How much money must be deposited at the end of each quarter if the objective is to accumulate $600,000 after 8 years? Assume interest is earned at the rate of 10 percent per year compounded quarterly. How much interest will be earned?

11 A family wants to begin saving for a trip to Europe. The trip is planned for 3 years from now, and the family wants to accumulate $10,000 for the trip. If 12 deposits are made quarterly to an account which earns interest at the rate of 8 percent per year compounded quarterly, how much should each deposit equal? How much interest will be earned on their deposits?

12 A major city wants to establish a sinking fund to pay off debts of $75 million which come due in 8 years. The city can earn interest at the rate of 10 percent per year compounded semiannually. If the first deposit is made 6 months from now, what semiannual deposit will be required to accumulate the $75 million? How much interest will be earned on these deposits?

8.4 ANNUITIES AND THEIR PRESENT VALUE

Just as there are problems relating annuities to their equivalent future value, there are applications which relate an annuity to its present value equivalent. For example, we may be interested in determining the size of a deposit which will generate a series of payments (an annuity) for college, retirement years, and so forth. Or given that a loan has been made, we may be interested in determining the series of payments (annuity) necessary to repay the loan with interest. This section discusses problems of these types.

The Present Value of an Annuity

The *present value of an annuity* is an amount of money today which is equivalent to a series of equal payments in the future. Assume you have won a lottery and lottery officials give you the choice of having a lump-sum payment today or a series of payments at the end of each of the next 5 years. The two alternatives would be considered equivalent (in a monetary sense) if by investing the lump sum today you could generate (with accumulated interest) annual withdrawals equal to the five installments offered by the lottery officials. *An assumption is that the final withdrawal would deplete the investment completely.* Consider the following example.

EXAMPLE 15   **(Lottery)** A person recently won a state lottery. The terms of the lottery are that the winner will receive annual payments of $20,000 at the end of this year and each of the following 3 years. If the winner could invest money today at the rate of 8 percent per year compounded annually, what is the present value of the four payments?

SOLUTION

Figure 8.10 illustrates the situation. *If A is defined as the present value of the annuity,* we might determine the value of A by computing the present value of each $20,000 payment. Applying Eq. (8.7a) and using values from Table II, we find the sum of the four present values is $66,242.60. We can conclude that a deposit today of $66,242.60 which earns interest at the rate of 8 percent per year compounded annually could generate a series of four withdrawals of $20,000 at the end of each.of the next four years.

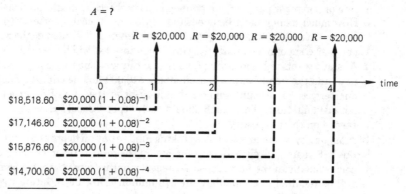

Figure 8.10   Calculating the present value of each payment R.

As with the future value of an annuity, the approach of summing the present values of each payment is possible but impractical. A more general and efficient method of determining the present value of an annuity follows.

$$R = amount\ of\ an\ annuity$$

$$i = interest\ rate\ per\ compounding\ period$$

$$n = number\ of\ annuity\ payments\ (also,\ the\ number\ of\ compounding\ periods)$$

$$A = present\ value\ of\ the\ annuity$$

Equation (8.12), which determines the future value or sum of an annuity, is restated in the following:

$$S_n = R\,\frac{(1+i)^n - 1}{i}$$ (8.15)

Figure 8.11

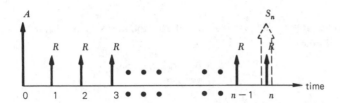

Looking at Fig. 8.11, we can think of S_n as being the equivalent future value of the annuity. If we know the value of S_n, the present value A of the annuity should simply be the present value of S_n, or

$$A = S_n(1+i)^{-n}$$

Substituting the expression for S_n from Eq. (8.15) yields

$$A = R\,\frac{(1+i)^n - 1}{i}(1+i)^{-n}$$ (8.16)

or

$$A = R\left[\frac{(1+i)^n - 1}{i(1+i)^n}\right]$$ (8.17)

Equation (8.17) can be used to compute the present value A of an annuity consisting of n equal payments R, each made at the end of n periods. The expression in brackets is referred to as the **series present worth factor**. The special symbol $a_{\overline{n}|i}$ (verbalized as "a sub n angle i") is frequently used to abbreviate the series present worth factor. Thus, Eq. (8.17) can be rewritten more simply as

$$A = Ra_{\overline{n}|i}$$ (8.18)

Selected values for $a_{\overline{n}|i}$ are found in Table V (on page T16).

EXAMPLE 16 Rework Example 15 using Eq. (8.18).

SOLUTION

Since $i = 0.08$ and $n = 4$, the present value of the four payments is found by applying Eq. (8.18) and using Table V.

$$A = \$20,000(3.31213)$$
$$= \$66,242.60$$

EXAMPLE 17 Parents of a teenage girl want to deposit a sum of money which will earn interest at the rate of 9 percent per year compounded semiannually. The deposit will be used to generate a series of eight semiannual payments of $2,500 beginning 6 months after the deposit. These payments will be used to help finance their daughter's college education. What amount must be deposited to achieve their goal? How much interest will be earned on this deposit?

SOLUTION

For this problem, $R = \$2,500$, $i = 0.09/2 = 0.045$, and $n = 8$. Using Table V and substituting into Eq. (8.18) gives

$$A = (\$2,500)(6.59589)$$
$$= \$16,489.73$$

Since the $16,489.73 will generate eight payments totaling $20,000, interest of $20,000 − $16,489.73 or $3,510.27 will be earned.

$\square$

PRACTICE EXERCISE
Determine the present value of a series of 8 annual payments of $30,000 each, the first of which begins 1 year from today. Assume interest of 6 percent per year compounded annually. *Answer:* $186,293.70.

Determining the Size of the Annuity

There are problems in which we may be given the present value of an annuity and need to determine the size of the corresponding annuity. For example, given a loan of $10,000 which is received today, what quarterly payments must be made to repay the loan in 5 years if interest is charged at the rate of 10 percent per year, compounded quarterly? The process of repaying a loan by installment payments is referred to as *amortizing a loan.*

For the loan example, the quarterly payments can be calculated by solving for R in Eq. (8.18). Solving for R, we get

$$R = \frac{A}{a_{\overline{n}|i}}$$

or

$$R = A\left[\frac{1}{a_{\overline{n}|i}}\right] \qquad (8.19)$$

The expression in brackets is sometimes called the *capital-recovery factor.* Table VI (on page T20) contains selected values for this factor.

EXAMPLE 18   Determine the quarterly payment necessary to repay the previously mentioned $10,000 loan. How much interest will be paid on the loan?

SOLUTION

For this problem $A = \$10,000$, $i = 0.10/4 = 0.025$, and $n = 20$. Using Table VI and substituting into Eq. (8.19) gives

$$R = \$10,000(0.06415)$$
$$= \$641.50$$

There will be 20 payments totaling $12,830; thus interest will equal $2,830 on the loan.

EXAMPLE 19 (**Retirement Planning**) An employee has contributed with her employer to a retirement plan. At the date of her retirement, the total retirement benefits are $250,000. The retirement program provides for investment of this sum at an interest rate of 12 percent per year compounded semiannually. Semiannual disbursements will be made for 30 years to the employee or, in the event of her death, to her heirs. What semiannual payment should be generated? How much interest will be earned on the $250,000 over the 30 years?

SOLUTION

For this problem $A = \$250,000$, $i = 0.12/2 = 0.06$, and $n = 60$. Using Table VI and substituting into Eq. (8.19), we get

$$R = \$250,000(0.06188)$$
$$= \$15,470$$

Total payments over the 30 years will equal $60 \times \$15,470$, or $928,200. Thus, interest of $928,200 − $250,000, or $678,200, will be earned over the 30 years.

$\square$

PRACTICE EXERCISE
Given $1 million today, determine the equivalent series of 10 annual payments beginning in 1 year. Assume interest of 6 percent per year, compounded annually. *Answer:* $135,870.

Mortgages

Sooner or later, many of us succumb to the "American dream" of owning a home. However, interest rates, as well as high real estate prices in some areas, cause this to be an expensive dream. Aside from the numerous pleasures of home ownership, there is *at least* one time during each month when we cringe from the effects of owning a home. That time is when we sign a check for the monthly mortgage payment. And whether we realize it or not, we spend an incredible amount of money to realize our dream.

Given a mortgage loan, many homeowners do not realize how the amount of their mortgage payment is calculated. It is calculated in the same way as were the loan payments in the last section. That is, they are calculated by using Eq. (8.19). Interest is typically compounded monthly, and the interest rate per compounding period can equal unusual fractions or decimal answers. If the annual interest rate is 8.5 percent, the value of i is $0.085/12 = \frac{17}{24}$ of a percent, or 0.0070833. Obviously Table VI cannot be used for these interest rates.

Table VII (on page T24) is an extension of Table VI designed specifically for determining mortgage payments. Note that the interest rates are stated as annual percentages.

EXAMPLE 20   A person pays $100,000 for a new house. A down payment of $30,000 leaves a mortgage of $70,000 with interest computed at 10.5 percent per year compounded monthly. Determine the monthly mortgage payment if the loan is to be repaid over (a) 20 years, (b) 25 years, and (c) 30 years. (d) Compute total interest under the three different loan periods.

SOLUTION

(a) From Table VII, the *monthly payment per dollar of mortgage* is 0.00998380 (corresponding to $n = 20 \times 12 = 240$ payments). Therefore,

$$R = \$70,000(0.00998380)$$
$$= \$698.87$$

(b) For 25 years (or 300 monthly payments),

$$R = \$70,000(0.00944182)$$
$$= \$659.27$$

(c) For 30 years (or 360 monthly payments)

$$R = \$70,000(0.00914739)$$
$$= \$640.32$$

(d) Total payments are

$$(240)(\$698.87) = \$167,728.80 \quad \text{for 20 years}$$

$$(300)(\$659.27) = \$197,781.00 \quad \text{for 25 years}$$

$$(360)(\$640.32) = \$230,515.20 \quad \text{for 30 years}$$

Because these payments are all repaying a $70,000 loan, interest on the loan is

$$\$167,728.80 - \$70,000 = \$97,728.80 \quad \text{for 20 years}$$

$$\$197,781.00 - \$70,000 = \$127,781.00 \quad \text{for 25 years}$$

$$\$230,515.20 - \$70,000 = \$160,515.20 \quad \text{for 30 years}$$

EXAMPLE 21   In the previous problem, determine the effects of a decrease in the interest rate to 10 percent (a) on monthly payments for the 25-year mortgage and (b) on total interest for the 25-year mortgage.

SOLUTION

(a) For $i = 10.00$ and 25 years,

$$R = (70,000)(0.00908700)$$
$$= \$636.09$$

Therefore, monthly payments are less by an amount of

$$\$659.27 - \$636.09 = \$23.18$$

(b) Total payments over the 25 years will equal

$$300(636.09) = \$190,827.00$$

The total interest is $120,827, which is $6,954 less than with the 10.5 percent mortgage.

EXAMPLE 22 (**Maximum Affordable Loan**) A couple estimates that they can afford a monthly mortgage of $750. Current mortgage interest rates are 10.25 percent. If a 30-year mortgage is obtainable, what is the maximum mortgage loan this couple can afford?

SOLUTION

The formula for computing the monthly mortgage payment is

$$\text{Monthly payment} = \left(\begin{array}{c}\text{dollar amount of}\\\text{mortgage loan}\end{array}\right)\left(\begin{array}{c}\text{monthly payment per}\\\text{dollar of mortgage loan,}\\\text{Table VII}\end{array}\right)$$

or

$$R = A\left(\begin{array}{c}\text{Table VII}\\\text{factor}\end{array}\right) \tag{8.20}$$

In this problem, A is the unknown. If Eq. (8.20) is rearranged,

$$A = \frac{R}{\text{Table VII factor}}$$

$$= \frac{750}{0.00896101} = \$83,695.92$$

The Advantage of the Biweekly Mortgage Payment

Variations on the way in which mortgages are paid off can lead to some interesting and, quite frankly, remarkable results. We have already alluded to the effects of shortening the life of the mortgage from the traditional 25- or 30-year mortgages. Consider a $100,000 mortgage with interest of 10 percent. A 30-year mortgage would require a monthly payment of $877.57, total payments of $315,925.92, and total interest payments of $215,925.92. A 15-year mortgage would require monthly payments of $1,074.61, total payments of $193,428.90, and total interest of $93,428.90. Thus, by adding $197.04 to the monthly payments of a 30-year loan, the loan can be repaid in half the time. The additional $197.04 per month adds up to $35,467.20 over the 15 years. However, the result is a reduction in total payments (and in total interest) of $315,925.92 − $193,428.90, or $122,497.02.

An equally intriguing concept is that of the biweekly mortgage payment. By paying half of the monthly mortgage payment every 2 weeks, a homeowner would make 26 payments during a calendar year. One effect of this is the payment of the equivalent of one extra month in mortgage payments during the calendar year. The other remarkable effect is that the combination of smaller, more frequent payments and the equivalent of an additional monthly payment results in a dramatic

TABLE 8.2	Assumption: $100,000 Mortgage, (30 yr, 10%)	
Years into Mortgage	Remaining Principal (Monthly Mortgage)	Remaining Principal (Biweekly Mortgage)
3	$98,152	$95,035
5	96,575	90,770
10	90,939	75,533
15	81,665	50,223
20	66,407	8,847

acceleration in the payoff of the loan. Table 8.2 presents data pertaining to a $100,000 mortgage issued at an interest rate of 10 percent for 30 years. Illustrated are sample figures showing the reduction in the principal of the loan over time.

Section 8.4 Follow-up Exercises

1 Determine the present value of a series of 10 annual payments of $25,000 each which begins 1 year from today. Assume interest of 9 percent per year compounded annually.

2 Determine the present value of a series of 20 annual payments of $8,000 each which begins 1 year from today. Assume interest of 7 percent per year compounded annually.

3 Determine the present value of a series of 25 semiannual payments of $10,000 each which begins in 6 months. Assume interest of 10 percent per year compounded semiannually.

4 Determine the present value of a series of 15 payments of $5,000 each which begins 6 months from today. Assume interest of 9 percent per year compounded semiannually.

5 Determine the present value of a series of 30 quarterly payments of $500 which begins 3 months from today. Assume interest of 10 percent per year compounded quarterly.

6 Determine the present value of a series of 20 quarterly payments of $2,500 each which begins 3 months from today. Assume interest of 8 percent per year compounded quarterly.

7 Determine the present value of a series of 60 monthly payments of $2,500 each which begins 1 month from today. Assume interest of 12 percent per year compounded monthly.

8 Determine the present value of a series of 36 monthly payments of $5,000 each which begins 1 month from today. Assume interest of 18 percent per year compounded monthly.

9 A person wants to buy a life insurance policy which would yield a large enough sum of money to provide for 20 annual payments of $50,000 to surviving members of the family. The payments would begin 1 year from the time of death. It is assumed that interest could be earned on the sum received from the policy at a rate of 8 percent per year compounded annually.
   (a) What amount of insurance should be taken out so as to ensure the desired annuity?
   (b) How much interest will be earned on the policy benefits over the 20-year period?

10 Assume in Exercise 9 that semiannual payments of $25,000 are desired over the 20-year period, and interest is compounded semiannually.
   (a) What amount of insurance should be taken out?
   (b) How does this amount compare with that for Exercise 9?
   (c) How much interest will be earned on the policy benefits?
   (d) How does this compare with that for Exercise 9?

11 Given $250,000 today, determine the equivalent series of 10 annual payments which could be generated beginning in 1 year. Assume interest is 12 percent compounded annually.

12 Given $5 million today, determine the equivalent series of 20 annual payments which could be generated beginning in 1 year. Assume interest of 10 percent compounded annually.

13 Given $7,500,000 today, determine the equivalent series of 20 quarterly payments which could be generated beginning in 3 months. Assume interest of 10 percent per year compounded quarterly.

14 Given $20 million today, determine the equivalent series of 20 semiannual payments which could be generated beginning in 6 months. Assume interest of 9 percent per year compounded semiannually.

15 Given $4 million today, determine the equivalent series of 24 quarterly payments which could be generated beginning in 3 months. Assume interest of 12 percent per year compounded quarterly.

16 Given $250,000 today, determine the equivalent series of 24 semiannual payments which could be generated beginning 6 months from today. Assume interest of 9 percent per year compounded semiannually.

17 (a) Determine the annual payment necessary to repay a $350,000 loan if interest is computed at 9 percent per year compounded annually. Assume the period of the loan is 6 years.
  (b) How much interest will be paid over the 6-year period?

18 (a) Determine the monthly car payment necessary to repay a $12,000 automobile loan if interest is computed at 12 percent per year compounded monthly. Assume the period of the loan is 4 years.
  (b) How much interest will be paid over the 4-year period?

19 (a) Determine the monthly car payment necessary to repay a $15,000 automobile loan if interest is computed at 12 percent per year compounded monthly. Assume the period of the loan is 3 years.
  (b) How much interest will be paid over the 3-year period?

20 (a) Determine the quarterly payment necessary to repay a $25,000 loan if interest is computed at the rate of 14 percent per year compounded quarterly. Assume the loan is to be repaid in 10 years.
  (b) How much interest will be paid over the 10-year period?

For Exercises 21–28 compute the monthly mortgage payment, total payments, and total interest.

21 Mortgage loan of $80,000 at 10 percent per year for 20 years
22 Mortgage loan of $100,000 at 11.25 percent per year for 30 years
23 Mortgage loan of $90,000 at 9.5 percent per year for 25 years
24 Mortgage loan of $200,000 at 10.75 percent per year for 30 years
25 Mortgage loan of $90,000 at 10 percent per year for 25 years
26 Mortgage loan of $150,000 at 9.75 percent per year for 30 years
27 Mortgage loan of $120,000 at 10.5 percent per year for 20 years
28 Mortgage loan of $160,000 at 12 percent per year for 25 years
29 to 36 Rework Exercises 21–28, computing the difference between the amount of the monthly mortgage payment and the *total* interest paid if the interest rate increases by 1 percent.
37 A couple estimates that they can afford a mortgage payment of $750 per month. They

can obtain a 25-year mortgage at an interest rate of 10.5 percent. What is the largest mortgage loan they can afford?

38 A couple estimates that they can afford a mortgage payment of $1,000 per month. They can obtain a 30-year mortgage at an interest rate of 10.25 percent. What is the largest mortgage loan they can afford?

39 A person estimates that she can afford a mortgage payment of $1,500 per month. She can obtain a 25-year mortgage at an interest rate of 11.5 percent. What is the largest mortgage loan she can afford?

40 A person estimates that he can afford a mortgage payment of $1,200 per month. He can obtain a 30-year mortgage at an interest rate of 11.25 percent. What is the largest mortgage loan he can afford?

8.5 COST-BENEFIT ANALYSIS

When organizations evaluate the financial feasibility of investment decisions, the time value of money is an essential consideration. This is particularly true when a project involves cash flow patterns which extend over a number of years. This section will discuss one way in which such multiperiod investments can be evaluated.

Discounted Cash Flow

Consider an investment decision characterized by the cash flow pattern shown in Fig. 8.12. An initial investment of $50,000 is expected to generate a net (after expenses) return of $15,000 at the end of 1 year and an equal return at the end of the following 3 years. Thus, a $50,000 investment is expected to return $60,000 over a 4-year period. Because the cash inflows occur over a 4-year period, the dollars during the different periods cannot be considered equivalent. To evaluate this project properly, the time value of the different cash flows must be accounted for.

One approach to evaluating a project like this is to translate all cash flows into equivalent dollars at a common base period. This is called a *discounted cash flow method.* For example, this project might be evaluated by restating all cash flows in terms of their equivalent values at $t = 0$, the time of the investment. The original $50,000 is stated in terms of dollars at $t = 0$. However, each of the $15,000 cash inflows must be restated in terms of its equivalent value at $t = 0$.

In order to discount all cash flows to a common base period, an interest rate must be assumed for the intervening period. Frequently this interest rate is an assumed minimum desired rate of return on investments. For example, management might state that a minimum desired rate of return on all investments is 10 percent per year. How this figure is obtained by management is another issue in

Figure 8.12
Net cash flows.

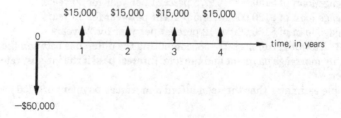

itself. Sometimes it is a reflection of the known rate of return which can be earned on alternative investments (e.g., bonds or money market funds).

Let's assume that the minimum desired rate of return for the project in Fig. 8.12 is 8 percent per year. Our discounted cash flow analysis will compute the **net present value** (NPV) of all cash flows associated with a project. The net present value is the algebraic sum of the present value of all cash flows associated with a project; cash *inflows* are treated as positive cash flows and cash *outflows* as negative cash flows. *If the net present value of all cash flows is **positive** at the assumed minimum rate of return, the actual rate of return from the project exceeds the minimum desired rate of return. If the net present value for all cash flows is **negative,** the actual rate of return from the project is less than the minimum desired rate of return.*

In our example, we discount the four $15,000 figures at 8 percent. By computing the present value of these figures, we are, in effect, determining the amount of money we would have to invest today ($t = 0$) at 8 percent in order to generate those four cash flows. Given that the net cash return values are equal, we can treat this as the computation of the present value of an annuity. Using Table V and Eq. (8.17), with $n = 4$ and $i = 0.08$,

$$A = 15,000(3.31213)$$
$$= \$49,681.95$$

This value suggests that an investment of $49,681.95 would generate an annual payment of 15,000 at the end of each of the following 4 years. In this example, an investment of $50,000 is required.

The net present value for this project combines the present values of all cash flows at $t = 0$, or

$$NPV = \frac{\text{present value}}{\text{of cash inflows}} - \frac{\text{present value}}{\text{of cash outflows}} \qquad (8.21)$$

Thus

$$NPV = \$49,681.80 - \$50,000$$
$$= -\$318.20$$

This negative value indicates that the project will result in a rate of return less than the minimum desired return of 8 percent per year, compounded annually.

EXAMPLE 23 (**Nonuniform Cash Flow Patterns**) The previous example resulted in net cash inflows which were equal over 4 years. The cash flow patterns for most investments tend to be irregular, both with regard to amount of money and timing of cash flows. Consider the cash flow pattern illustrated in Fig. 8.13. For this investment project, a $1 million investment results in no cash flow during the first year. However, at the end of each of the following 5 years the investment generates a stream of positive net returns. These returns are not equal to one another, increasing to a maximum of $450,000 at the end of the fourth year and decreasing finally to $100,000 at the end of the sixth year.

Suppose that the minimum desired return on investments is 12 percent. In order to

Figure 8.13
Net cash flows.

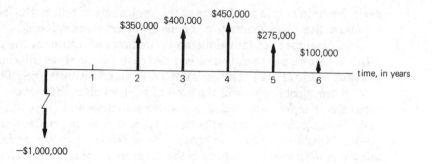

evaluate the desirability of this project, we must discount all cash flows to their equivalent values at $t = 0$. In contrast with the previous example, each net return figure must be discounted separately. The present value of each is computed as shown in Table 8.3.

TABLE 8.3	n	**Net Return**	**Present Value Factor** $(1 + 0.12)^{-n}$	**Present Value**
	2	$350,000	0.79719	$ 279,016.50
	3	400,000	0.71178	284,712.00
	4	450,000	0.63552	285,984.00
	5	275,000	0.56743	156,043.25
	6	100,000	0.50663	50,663.00
				$1,056,418.75

The net present value of all cash flows is

$$\text{NPV} = \$1,056,418.75 - \$1,000,000 = \$56,418.75$$

Because the net present value is positive, this project will result in a rate of return which exceeds the minimum desired rate of return of 12 percent per year, compounded annually. Another way of viewing this is that $1,056,418.75, invested at 12 percent per year, will generate the indicated net returns; this investment only requires $1,000,000.

❑

Extensions of Discounted Cash Flow Analysis

The net present value approach is one of a variety of methods available to evaluate long-range investment decisions. Although this analysis allows one to determine whether a project satisfies the minimum-desired-rate-of-return criterion, it does not provide a measure of the exact rate of return. Knowing the actual rate of return is desirable, especially if there is a set of competing investment opportunities which differ with respect to amount of investment and investment time horizon. Methods for computing the actual rate of return are simple extensions of the net present value technique. The actual rate of return from a project is the one which results in a net present value of 0. This can be found using a trial-and-error approach. The net present value of a project is computed using different interest rates until the NPV equals (approximately) 0.

Another consideration in evaluating such projects is the impact of taxes. Although some organizations evaluate projects on a *before-tax* basis, most find that the best analysis is on an *after-tax* basis. Considering investment credits as well as a variety of possible depreciation methods, an after-tax analysis usually is most appropriate.

Section 8.5 Follow-up Exercises

In Exercises 1–6, determine whether the investment project depicted by the cash flow diagram satisfies the minimum desired rate of return criterion. What is the NPV at the indicated interest rate?

1 Cash flow depicted in Fig. 8.14, 10 percent per year minimum rate of return
2 Cash flow depicted in Fig. 8.15, 8 percent per year minimum rate of return
3 Cash flow depicted in Fig. 8.16, 12 percent per year minimum rate of return
4 Cash flow depicted in Fig. 8.17, 9 percent per year minimum rate of return
5 Cash flow depicted in Fig. 8.18, 10 percent per year minimum rate of return
6 Cash flow depicted in Fig. 8.19, 14 percent per year minimum rate of return
*7 Estimate the actual rate of return generated by the project depicted in Fig. 8.14.
*8 Estimate the actual rate of return generated by the project depicted in Fig. 8.15.

Figure 8.14

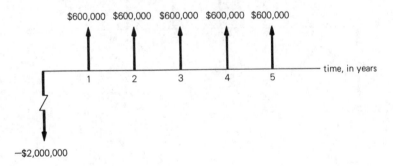

Figure 8.15

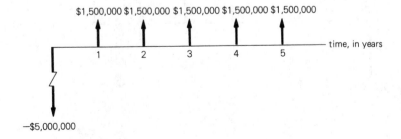

Figure 8.16

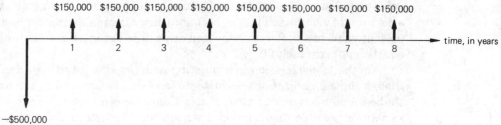

Figure 8.17

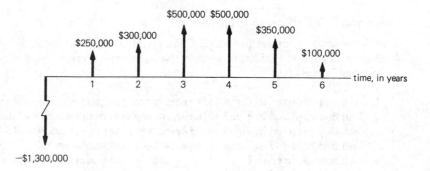

Figure 8.18

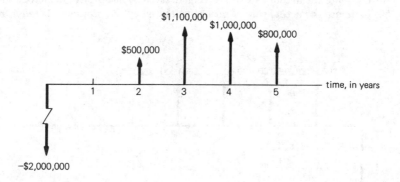

Figure 8.19

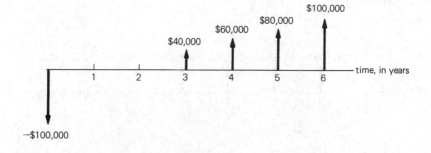

❑ **KEY TERMS AND CONCEPTS**

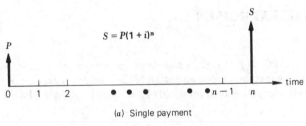

(a) Single payment

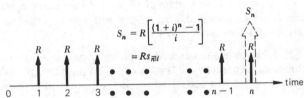

(b) Annuity and its future value

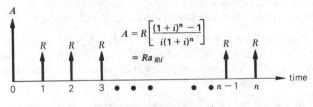

(c) Annuity and its present value

Figure 8.20 Summary of single-payment and annuity cash flow situations.

❑ **IMPORTANT FORMULAS**

$$I = Pin \quad \textbf{Simple interest} \tag{8.1}$$

$$S = P(1 + i)^n \quad \textbf{Compound amount} \tag{8.6}$$

$$P = S\left[\frac{1}{(1+i)^n}\right] \quad \text{Present value} \tag{8.7a}$$

$$r = \left(1 + \frac{i}{m}\right)^m - 1 \quad \text{Effective annual interest rate} \tag{8.8}$$

$$S_n = Rs_{\overline{n}|i} \quad \text{Sum of an annuity} \tag{8.13}$$

$$R = S_n\left[\frac{1}{s_{\overline{n}|i}}\right] \quad \text{Annuity to generate a desired sum} \tag{8.14}$$

$$A = Ra_{\overline{n}|i} \quad \text{Present value of an annuity} \tag{8.18}$$

$$R = A\left[\frac{1}{a_{\overline{n}|i}}\right] \quad \text{Annuity equivalent to a present sum} \tag{8.19}$$

❏ ADDITIONAL EXERCISES

Section 8.2

1 A sum of $25,000 is invested in a savings account which pays interest at a rate of 7 percent per year compounded annually. If the amount is kept on deposit for 15 years, what will the compound amount equal? How much interest will be earned during the 15 years?

2 A sum of $40,000 is invested at a rate of 12 percent per year compounded annually. If the investment is for a period of 5 years, what will the compound amount equal? How much interest will be earned during the 5 years?

3 A sum of $20,000 is invested in a credit union which pays interest at a rate of 6 percent per year compounded quarterly. If the amount is invested for a period of 10 years, what will the compound amount equal? How much interest will be earned during the 10 years?

4 **Personal Computers** Sales of personal computers costing less than $1,000 were estimated to equal $500 million in 1990. One analyst estimates that sales will grow at a rate of 20 percent per year over the next 5 years. Predict annual sales for the year 1995.

5 **Fire Protection** The number of fires reported each year in a major United States city has been increasing at a rate of 7 percent per year. The number of fires reported for the year 1990 was 5,000. If the number of fires continues to increase at the same rate, how many will be expected in 1996?

6 A sum of $2 million has been invested at an interest rate of 12 percent per year. If the investment is made for a period of 10 years, determine the compound amount if interest is compounded (*a*) annually, (*b*) semiannually, (*c*) quarterly, and (*d*) bimonthly.

7 Prices for a particular commodity have been increasing at an annual rate of 6 percent compounded annually. The current price of the commodity is $75. What was the price of the same item 5 years ago?

8 A sum of money will be deposited today at 9 percent per year. The goal is to have this sum grow to $75,000 in 5 years. What sum must be deposited if interest is compounded (*a*) annually and (*b*) semiannually?

9 **Real Estate** Real estate prices within one state have been increasing at an average rate of 8 percent per year. How long will it take for current prices to increase by 100 percent if the prices continue to increase at the same rate?

10 **Alcoholism** A state health agency has gathered data on the number of known alcoholics in the state. The number is currently 150,000. Data indicate that this number has been increasing at a rate of 4.5 percent per year and is expected to increase at the same rate in the future. How long will it take for the number of alcoholics in the state to reach a level of 240,000?

11 If a savings account awards interest of 6 percent per year compounded quarterly, what amount must be deposited today in order to accumulate $25,000 after 8 years? How much interest will be earned?

12 If a credit union awards interest of 7 percent per year compounded semiannually, what amount must be deposited today in order to accumulate $10,000 after 5 years? How much interest will be earned?

13 What sum must be deposited today at 10 percent per year compounded quarterly if the goal is to have a compound amount of $50,000 6 years from today? How much interest will be earned during this period?

14 What sum must be deposited today at 12 percent per year compounded quarterly if the goal is to have a compound amount of $3 million 10 years from today? How much interest will be earned during this period?

15 **Public Utilities** A major water utility estimates that the average daily consumption of water within a certain city is 30 million gallons. It has projected that the average daily consumption will equal 40 million gallons in 5 years. What annual rate of growth has the utility used in making its estimate of future consumption?

16 If compounding is done annually, at what interest rate must a sum be invested if it is to double in value over the next 6 years?

17 The nominal interest rate on an investment is 14 percent per year. Determine the effective annual interest rate if interest is compounded (*a*) semiannually and (*b*) quarterly.

18 The nominal interest rate on an investment is 10 percent per year. Determine the effective annual interest rate if interest is compounded (*a*) semiannually and (*b*) quarterly.

Section 8.3

19 Quarterly deposits of $3,500 are to be made in an account which earns interest at the rate of 8 percent per year compounded quarterly. To what sum will the investment grow by the time of the 20th deposit? How much interest will be earned during this period?

20 A person wishes to deposit $3,000 per year for 10 years. If interest is earned at the rate of 12 percent per year, compute the amount that the deposits will grow to by the end of 10 years if:
   (*a*) Deposits of $3,000 are made at the end of each year with interest compounded annually.
   (*b*) Deposits of $1,500 are made at the end of each 6-month period with interest compounded semiannually.

(c) Deposits of $750 are made at the end of every quarter with interest compounded quarterly.

21 A person wishes to deposit $7,500 per year in a savings account which earns interest of 8 percent per year compounded annually. Assume the first deposit is made at the end of this current year and additional deposits at the end of each following year.
(a) To what sum will the investment grow at the time of the 10th deposit?
(b) How much interest will be earned?

22 A corporation is planning to invest $2 million every 3 months in an investment which earns interest at the rate of 10 percent per year compounded quarterly. The first investment will be at the end of this current quarter.
(a) To what sum will the investment grow at the end of 5 years?
(b) How much interest will be earned during this period?

23 How much money must be deposited at the end of each 6-month period if the objective is to accumulate $80,000 by the time of the 10th deposit? Assume that interest is earned at the rate of 9 percent per year compounded semiannually. How much interest will be earned on these deposits?

24 A small community wants to establish a sinking fund to pay off debts of $20 million associated with the construction of a sewage treatment plant. The community can earn interest at the rate of 8 percent per year compounded quarterly. The debt comes due in 7 years. If the first deposit is made 3 months from now, what quarterly deposit will be required to accumulate the $20 million? How much interest will be earned on these deposits?

25 Interest can be earned on a savings account at the rate of 7 percent per year compounded annually. A person wishes to make deposits of $2,000 at the end of each year. How long will it take for the deposits and accumulated interest to grow to a sum which will exceed $20,000?

26 Interest can be earned on an investment at the rate of 12 percent per year compounded quarterly. If deposits of $10,000 are made at the end of each quarter, how long will it take for the deposits and accumulated interest to grow to a sum which exceeds $300,000?

27 How much money must be invested at the end of each quarter if the objective is to accumulate $250,000 after 8 years? Assume interest is earned at the rate of 10 percent per year compounded quarterly. How much interest will be earned?

28 How much money must be invested at the end of each 6-month period if the objective is to accumulate $1.2 million after 8 years? Assume interest is earned at the rate of 18 percent per year compounded semiannually. How much interest will be earned?

Section 8.4

29 Determine the present value of a series of 20 annual payments of $50,000 each which begins 1 year from today. Assume interest of 9 percent per year compounded annually.

30 Determine the present value of a series of 10 annual payments of $2,500 each which begins 1 year from today. Assume interest of 8 percent per year compounded annually.

31 Determine the present value of a series of 24 quarterly payments of $2,500 each which begins in 3 months. Assume interest is 10 percent per year compounded quarterly.

32 A person recently won a state lottery. The terms of the lottery are that the winner will receive annual payments of $50,000 at the end of this year and each of the following 19

years. If money can be invested today at the rate of 11 percent per year compounded annually, what is the present value of the 20 lottery payments?

33 Given $400,000 today, determine the equivalent series of 24 semiannual payments which could be generated beginning in 6 months. Assume interest can be earned at a rate of 7 percent per year compounded semiannually.

34 (a) Determine the monthly car payment necessary to repay a $16,000 automobile loan if interest is computed at 12 percent per year compounded monthly. Assume the period of the loan is 4 years.
   (b) How much interest will be paid over the 4-year period?

35 A lump sum of $400,000 is invested at the rate of 11 percent per year compounded annually. How many annual withdrawals of $50,000 can be made? (Assume that the first withdrawal occurs in 1 year.)

36 A family has inherited $300,000. If they choose to invest the $300,000 at 12 percent per year compounded quarterly, how many quarterly withdrawals of $25,000 can be made? (Assume that the first withdrawal is 3 months after the investment is made.)

In Exercises 37–44, determine the (a) monthly mortgage payment, (b) total payments, and (c) total interest over the life of the loan.

37 Loan of $140,000, 10.5 percent interest, 25 years

38 Loan of $80,000, 9.5 percent interest, 20 years

39 Loan of $160,000, 8.5 percent interest, 15 years

40 Loan of $150,000, 10.5 percent interest, 30 years

41 Loan of $95,000, 10.75 percent interest, 25 years

42 Loan of $100,000, 12.5 percent interest, 20 years

43 Loan of $100,000, 11.75 percent interest, 30 years

44 Loan of $125,000, 12.0 percent interest, 20 years

In Exercises 45–50, determine the maximum affordable loan.

45 Mortgage payment of $800, 10 percent interest, 25 years

46 Mortgage payment of $1,200, 9.75 percent interest, 25 years

47 Mortgage payment of $1,250, 9.75 percent interest, 20 years

48 Mortgage payment of $1,500, 10.5 percent interest, 25 years

49 Mortgage payment of $1,350, 11.5 percent interest, 20 years

50 Mortgage payment of $2,000, 10.5 percent interest, 25 years

Section 8.5

51 Determine whether the investment project depicted in Fig. 8.21 has a rate of return greater than or equal to 12 percent per year. What is the NPV at this interest rate?

52 Determine whether the investment project depicted in Fig. 8.22 has a rate of return greater than or equal to 14 percent per year. What is the NPV at this interest rate?

Figure 8.21

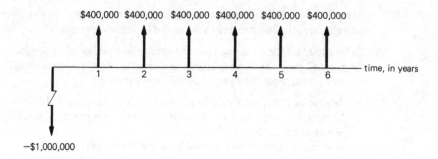

Figure 8.22

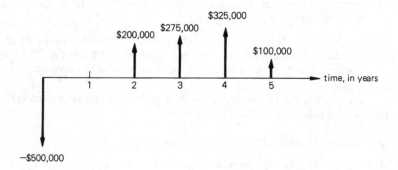

*53 Estimate the actual rate of return generated by the project depicted in Fig. 8.23.

*54 Estimate the actual rate of return generated by the project depicted in Fig. 8.24.

Figure 8.23

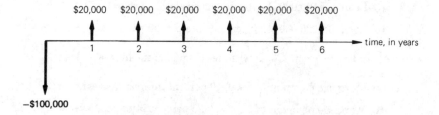

Figure 8.24

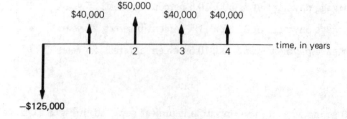

❑ **CHAPTER TEST**

1 A principal of $50,000 has been invested at an interest rate of 9 percent per year compounded semiannually. If the principal is invested for 8 years, determine the compound amount at the end of this period.

2 Real estate prices in one locality have been increasing at a rate of 7 percent per year compounded annually. A house which sells for $100,000 today would have sold for what price 3 years ago?

3 Quarterly deposits of $5,000 are to be made in an account which earns interest at the rate of 12 percent per year compounded quarterly.
(*a*) To what sum will the investment grow by the time of the 20th deposit?
(*b*) How much interest will be earned during this period?

4 A sinking fund is to be established to repay debts totaling $80,000. The debts come due in 5 years. If interest can be earned at the rate of 10 percent per year compounded semiannually, what semiannual deposit will be required to accumulate the $80,000? (Assume the first deposit is made in 6 months.) How much interest will be earned on these deposits?

5 Given $125,000 today, determine the equivalent series of 10 annual payments which could be generated beginning in 1 year. Assume interest of 11 percent per year compounded annually.

6 The nominal interest rate on an investment is 12 percent per year. Determine the effective annual interest rate if interest is compounded quarterly.

7 A mortgage loan of $150,000 is available at an annual interest rate of 10.5 percent. What is the difference between the monthly mortgage payments if the loan is for 20 years versus 30 years?

8 Does the investment project depicted in Fig. 8.25 have a rate of return greater than or equal to 10 percent? What is the NPV at this interest rate?

Figure 8.25

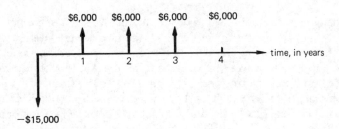

MINICASE

XYZ CORPORATION

The XYZ Corporation is considering three alternative investments characterized by the data shown in the accompanying table. Notice that the three investments have equal initial cash outlays, equal lifetimes, and equal dollar returns. Note that the patterns of dollar returns are different for the three investments.

	Alternative		
	1	2	3
Initial investment	$380,000	$380,000	$380,000
Cash inflows*			
Year 1	$180,000	$220,000	$140,000
Year 2	180,000	180,000	180,000
Year 3	180,000	140,000	220,000
Total cash inflows	$540,000	$540,000	$540,000

* Assume that cash inflows occur at the end of each year.

Required:

a *XYZ Corporation has a minimum desired rate of return on investments of 15 percent. Determine the NPV of each of these investments and determine which meet the rate of return criterion.*

b *Use* linear interpolation *to estimate the actual rates of return for the three investment alternatives. Linear interpolation is a trial-and-error method of estimating actual rates of return when such rates are different from those available from tables (or calculators). To illustrate, we used the discounted cash flow method to conclude that the investment portrayed in Fig. 8.12 resulted in a rate of return less than 8 percent per year. Our basis for this conclusion was that the NPV for the investment at 8 percent per year was −$318.20.*

Remember, the actual rate of return on an investment is one which results in an NPV of 0. For the investment in Fig. 8.12, the next lowest interest rate available from our tables is 7 percent. If the NPV is computed at a 7 percent rate, you will find that it equals +$808.15. We can conclude from this that the actual rate of return is between 7 and 8 percent per year. Linear interpolation assumes that interest rates are proportionate to NPV dollars. For this example, the actual interest rate lies somewhere between 7 and 8 percent, in proportion to the location of an NPV of 0 (located between the NPV of $808.15 at 7 percent and the NPV of −$318.20 at 8 percent).

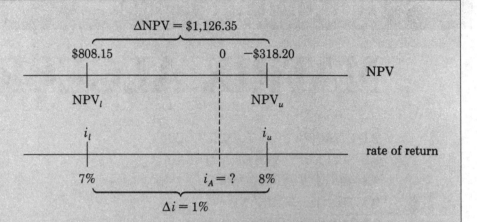

Given the lower interest rate i_l and upper interest rate i_u which bound the actual rate of return, and their respective net present values NPV_l and NPV_u, one method of estimating the actual rate of return i_A is

$$i_A = i_l + \frac{NPV_l}{\Delta NPV}\,\Delta i$$

For this example, the estimated actual rate of return is

$$i_A = 0.07 + \frac{808.15}{1126.35}\,0.01 = 0.07 + 0.717(0.01)$$

$$= 0.07 + 0.00717$$
$$= 0.07717$$

The actual interest rate is approximately 7.717 percent.

CHAPTER 9

MATRIX ALGEBRA

CHAPTER OBJECTIVES

❏ Provide an understanding of the nature of a matrix and matrix representation of data

❏ Provide an understanding of the algebra of matrices

❏ Present a variety of applications of matrices and matrix algebra

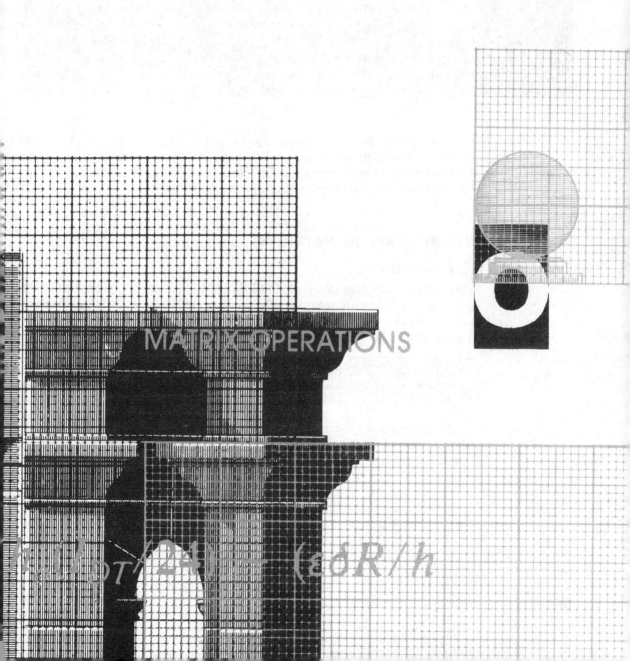

MATRIX OPERATIONS

MOTIVATING SCENARIO:
Brand-Switching Analysis

Brand-switching analysis concerns itself with the purchasing behavior of consumers who make repeated purchases of a product or service. Examples of such products or services are gasoline, detergents, soft drinks, and fast-food meals. Brand-switching analysis focuses upon brand loyalty and the degree to which consumers are willing to switch to competing products. Firms often try to project the effects that promotion campaigns, such as rebates or advertising programs, will have on the sales of their products. *If information is available concerning the rates of gains from and losses to all competitors, a firm can (a) predict its market share at some time in the future, (b) predict the rate at which the firm will increase or decrease its market share in the future, and (c) determine whether market shares will ever reach equilibrium levels in which each firm or brand retains a constant share of the market.*

This chapter discusses matrix algebra and its applications. The nature of matrices is presented, followed by a discussion of different types of matrices, the algebra of matrices, and some specialized matrix concepts. The last section in the chapter presents several applications of matrix algebra.

9.1 INTRODUCTION TO MATRICES

What Is a Matrix?

Whenever one is dealing with data, there should be concern for organizing them in such a way that they are meaningful and can be readily identified. Summarizing data in a tabular form can serve this function. A *matrix* is a common device for summarizing and displaying numbers or data.

> **DEFINITION: MATRIX**
> A *matrix* is a rectangular array of elements.

The elements of a matrix are usually, but not always, real numbers. Consider the test scores for five students on three examinations. These could be displayed in the following matrix.

$$
\begin{array}{c}
 \\
\text{Student}
\end{array}
\begin{array}{c}
 \\
\begin{array}{c} 1 \\ 2 \\ 3 \\ 4 \\ 5 \end{array}
\end{array}
\overset{\displaystyle \text{Test}}{
\begin{array}{ccc}
1 & 2 & 3 \\
\left(\begin{array}{ccc}
75 & 82 & 86 \\
91 & 95 & 100 \\
65 & 70 & 68 \\
59 & 80 & 99 \\
75 & 76 & 74
\end{array}\right)
\end{array}}
$$

The matrix contains the set of test scores enclosed by the large parentheses. The array is rectangularly shaped, having five rows (one for each student) and three columns (one for each test). Each row contains the three test scores for a particular student. Each column contains the five scores on a particular test.

GENERALIZED FORM OF A MATRIX

A matrix A containing elements a_{ij} has the general form

$$\mathbf{A} = \begin{pmatrix} a_{11} & a_{12} & \cdots & a_{1n} \\ a_{21} & a_{22} & \cdots & a_{2n} \\ \cdot & \cdot & \cdots & \cdot \\ a_{m1} & a_{m2} & \cdots & a_{mn} \end{pmatrix}$$

This generalized matrix is represented as having m rows and n columns. The subscripts on an element a_{ij} indicate the location of the element within the matrix. Element a_{ij} is located at the intersection of row i and column j of the matrix. For example, a_{21} is located at the intersection of row 2 and column 1. The element a_{35} would be located in row 3 and column 5 of the matrix.

PRACTICE EXERCISE

If the student test score matrix is named **S** and the elements are denoted by s_{ij}, what are the elements s_{12}, s_{32}, s_{43}, and s_{16}? *Answer:* 82, 70, 99, no s_{16} element.

Matrix **names** are usually represented by uppercase letters and the elements of a matrix by lowercase, subscripted letters. A matrix is characterized further by its **dimension.** The dimension or **order** indicates the number of rows and the number of columns contained within the matrix. If a matrix has m rows and n columns, it is said to have dimension **$m \times n$,** which is read **"m by n."** The student test score matrix has dimension (5 $\times$ 3), or it is referred to as a "5 by 3" matrix.

Purpose of Studying Matrix Algebra

Matrices provide a convenient medium for the storage, display, and manipulation of data. The test score data are stored conveniently in the previous matrix and the matrix provides a clear and compact method for displaying these data. Most data stored in computers are stored in a matrix format. In the FORTRAN language you reserve storage space for arrays in the memory of the computer by use of the DIMENSION statement. The statement "DIMENSION $A(20, 30)$" reserves space for a matrix **A** which has dimension of (20 $\times$ 30). In the language BASIC, the statement "DIM A (20, 30)" accomplishes the same objective.

When data are stored within matrices, there is often a need to display them. If the data are stored within a matrix in some logical pattern, the retrieval of individual items or groups of items can be relatively easy. Frequently there is a need to manipulate data which are stored in a matrix. For instance, an instructor may want to determine a class average on a given test or a student average for the three tests using the student test score data in the previously defined matrix. Matrix algebra allows for manipulating data and for performing computations while keeping the data in a matrix form. This is convenient, especially in computerized applications.

EXAMPLE 1 (U.S. Energy Consumption) The matrix **E** that follows displays average daily energy consumption by energy source for four different regions of the country during 1987. The figures are in millions of barrels per day and represent the amount of oil that would yield the equivalent energy. They have been rounded to the nearest 100,000 barrels. **E** is a (4×5) matrix.

$$
\mathbf{E} = \begin{array}{c c c c c}
\begin{array}{c} \text{American} \\ \text{oil and} \\ \text{gas} \end{array} & \text{Coal} & \begin{array}{c} \text{Imported} \\ \text{oil and} \\ \text{gas} \end{array} & \begin{array}{c} \text{Hydroelectric,} \\ \text{solar, geother-} \\ \text{mal, and syn-} \\ \text{thetic fuels} \end{array} & \text{Nuclear} \\
\end{array}
$$

$$
\mathbf{E} = \begin{pmatrix}
6.5 & 2.8 & 3.0 & 0.2 & 0.5 \\
3.2 & 1.1 & 0.5 & 0.5 & 0.2 \\
3.4 & 2.0 & 1.1 & 0.1 & 0.4 \\
5.5 & 1.5 & 3.3 & 0.6 & 0.2
\end{pmatrix}
\begin{array}{l}
\text{Northeast} \\
\text{South} \\
\text{Midwest} \\
\text{West}
\end{array}
$$

☐

The following sections discuss different types of matrices and their manipulation.

9.2 SPECIAL TYPES OF MATRICES

Vectors

One special class of matrices is called a *vector*. A vector is a matrix having only one row or one column.

> **DEFINITION: ROW VECTOR**
> A *row vector* is a matrix having only one row. A row vector **R** having n elements r_{ij} has dimension $(1 \times n)$ and the general form
>
> $$\mathbf{R} = (r_{11} \quad r_{12} \quad r_{13} \quad \cdots \quad r_{1n})$$

Notice that the generalized elements of a $(1 \times n)$ row vector can be expressed by r_{1j}, where $j = 1, \ldots n$.

The three test scores for student 1 might be saved in the (1×3) row vector **A** as

$$\mathbf{A} = (75 \quad 82 \quad 86)$$

The following (1×5) row vector is a submatrix from Example 1 summarizing the equivalent average daily energy consumption for the Northeast during 1987.

$$\mathbf{B} = (6.5 \quad 2.8 \quad 3.0 \quad 0.2 \quad 0.5)$$

DEFINITION: COLUMN VECTOR

A *column vector* is a matrix having only one column. A column vector $\mathbf{C}$ having m elements c_{ij} has dimension $m \times 1$ and the general form

$$\mathbf{C} = \begin{pmatrix} c_{11} \\ c_{21} \\ \vdots \\ c_{m1} \end{pmatrix}$$

For the previous test score matrix the scores of the five students on the first examination might be represented by the (5×1) column vector

$$\mathbf{T} = \begin{pmatrix} 75 \\ 91 \\ 65 \\ 59 \\ 75 \end{pmatrix}$$

Square Matrices

DEFINITION: SQUARE MATRIX

A *square matrix* is a matrix having the same number of rows and columns.

If the dimension of a matrix is $(m \times n)$, a square matrix is such that $m = n$. The following matrices are square.

$$\mathbf{A} = (3) \qquad \mathbf{B} = \begin{pmatrix} 1 & 3 \\ -5 & 4 \end{pmatrix} \qquad \mathbf{C} = \begin{pmatrix} 2 & 0 & -3 \\ 1 & -4 & 5 \\ 0 & 2 & 6 \end{pmatrix}$$

If a matrix $\mathbf{A}$ is square, we sometimes concern ourselves with a subset of elements a_{ij} which lie along the ***primary diagonal*** of the matrix. These elements are located in positions where $i = j$, for example, $a_{11}, a_{22}, a_{33}, a_{44}, \ldots, a_{nn}$. The elements on the primary diagonal of matrix $\mathbf{B}$ are $b_{11} = 1$ and $b_{22} = 4$. The elements on the primary diagonal of matrix $\mathbf{C}$ are $c_{11} = 2$, $c_{22} = -4$, and $c_{33} = 6$.

DEFINITION: IDENTITY MATRIX

An *identity matrix* $\mathbf{I}$, sometimes called a *unit matrix*, is a square matrix for which the elements along the primary diagonal all equal 1 and all other elements equal 0.

If e_{ij} denotes a generalized element within an identity matrix, then

$$e_{ij} = \begin{cases} 1 & \text{if } i = j \\ 0 & \text{if } i \neq j \end{cases}$$

The matrices

$$\mathbf{I} = \begin{pmatrix} 1 & 0 \\ 0 & 1 \end{pmatrix} \quad \text{and} \quad \mathbf{I} = \begin{pmatrix} 1 & 0 & 0 \\ 0 & 1 & 0 \\ 0 & 0 & 1 \end{pmatrix}$$

are (2×2) and (3×3) identity matrices.

Although we will see different applications of the identity matrix, one important property involves the multiplication of an identity matrix and another matrix. Multiplication of matrices is a legitimate algebraic operation under certain circumstances. Given a matrix $\mathbf{A}$ and an identity matrix $\mathbf{I}$, *if the product $\mathbf{AI}$ is defined, $\mathbf{AI} = \mathbf{A}$*. Similarly, *if the product $\mathbf{IA}$ is defined, then $\mathbf{IA} = \mathbf{A}$*. The identity matrix $\mathbf{I}$ is to matrix multiplication as the number 1 is to multiplication in the real number system; that is, $(a)(1) = (1)(a) = a$.

Transpose of a Matrix

There are times when the data elements in a matrix need to be rearranged. The rearrangement may be simply to see the array of numbers from a different perspective or to manipulate the data in a later stage. One rearrangement is to form the *transpose* of a matrix.

DEFINITION: TRANSPOSE
Given the $(m \times n)$ matrix $\mathbf{A}$ with elements a_{ij}, the *transpose* of $\mathbf{A}$, denoted by $\mathbf{A}^T$, is an $(n \times m)$ matrix which contains elements a_{ij}^t where $a_{ij}^t = a_{ji}$.

EXAMPLE 2   To find the transpose of the matrix

$$\mathbf{A} = \begin{pmatrix} 3 & 2 \\ 4 & 0 \\ 1 & -2 \end{pmatrix}$$

we first determine the dimension of $\mathbf{A}^T$. Since $\mathbf{A}$ is a (3×2) matrix, $\mathbf{A}^T$ will be a (2×3) matrix having the form

$$\mathbf{A}^T = \begin{pmatrix} a_{11}^t & a_{12}^t & a_{13}^t \\ a_{21}^t & a_{22}^t & a_{23}^t \end{pmatrix}$$

Using the previous definition, we get

$$\begin{aligned}
a_{11}^t &= a_{11} = 3 & a_{21}^t &= a_{12} = 2 \\
a_{12}^t &= a_{21} = 4 & a_{22}^t &= a_{22} = 0 \\
a_{13}^t &= a_{31} = 1 & a_{23}^t &= a_{32} = -2
\end{aligned}$$

or
$$A^T = \begin{pmatrix} 3 & 4 & 1 \\ 2 & 0 & -2 \end{pmatrix}$$

☐

Study the matrices A and A^T in Example 2. Do you notice any pattern? What you should observe is that *the rows of A become the columns of A^T*. Equivalently, *the columns of A become the rows of A^T*. These relationships will be true for any matrix and its transpose, and they provide an easy method for determining the transpose.

EXAMPLE 3 Let's apply this logic in finding the transpose of

$$B = \begin{pmatrix} 3 & 0 & 6 \\ 5 & 1 & 3 \\ 2 & -1 & 4 \end{pmatrix}$$

To form the transpose of B, rows 1, 2, and 3 become columns 1, 2, and 3 of B^T, or

$$B^T = \begin{pmatrix} 3 & 5 & 2 \\ 0 & 1 & -1 \\ 6 & 3 & 4 \end{pmatrix}$$

Equivalently, columns 1, 2, and 3 of B can be thought of as becoming rows 1, 2, and 3 of B^T. Both perspectives are valid.

☐

Section 9.2 Follow-up Exercises

Determine the dimension of each of the following matrices and find the transpose.

1 $(8 \quad -8 \quad 5 \quad 3)$

2 $\begin{pmatrix} 2 & 6 \\ -3 & 8 \end{pmatrix}$

3 $\begin{pmatrix} 0 & 1 \\ 5 & 2 \\ -6 & 8 \\ -2 & 4 \end{pmatrix}$

4 $\begin{pmatrix} 2 & 10 & -1 \\ -3 & -5 & 0 \\ 4 & -8 & 2 \end{pmatrix}$

5 $\begin{pmatrix} 1 & 0 & 0 \\ 0 & 1 & 0 \\ 0 & 0 & 1 \end{pmatrix}$

6 $\begin{pmatrix} -6 & 3 & 2 & 4 \\ 2 & 3 & 3 & 4 \\ 2 & -1 & 5 & 8 \end{pmatrix}$

7 $\begin{pmatrix} 1 \\ 2 \\ 3 \\ 4 \end{pmatrix}$

8 $\begin{pmatrix} 1 & 3 & 5 & 7 & 9 \\ 2 & 4 & 6 & 8 & 10 \end{pmatrix}$

9 $\begin{pmatrix} 1 & 3 & 5 \\ 6 & 4 & 2 \\ 0 & 1 & 2 \\ 4 & 6 & 3 \\ 5 & 1 & 2 \end{pmatrix}$

10 $\begin{pmatrix} 6 & 1 & 2 & 3 & 5 \\ 2 & 0 & 4 & 6 & 1 \\ 3 & 1 & -2 & 3 & 5 \\ 4 & 3 & 2 & 1 & 0 \end{pmatrix}$

11 Find a (2×4) matrix $\mathbf{A}$ for which

$$a_{ij} = \begin{cases} i+j & \text{if } i=j \\ 0 & \text{if } i \neq j \end{cases}$$

12 Find a (5×3) matrix $\mathbf{B}$ for which

$$b_{ij} = \begin{cases} i-j & \text{if } i=j \\ 2i+j & \text{if } i \neq j \end{cases}$$

9.3 MATRIX OPERATIONS

In this section we will discuss some of the operations of matrix algebra.

Matrix Addition and Subtraction

PROPERTY OF MATRIX ADDITION (SUBTRACTION)
Two matrices may be added or subtracted if and only if they have the same dimension.

If $\mathbf{A}$ and $\mathbf{B}$ are $(m \times n)$ matrices added to form a new matrix $\mathbf{C}$, $\mathbf{C}$ will have the same dimension as $\mathbf{A}$ and $\mathbf{B}$. The elements of $\mathbf{C}$ are found by adding the corresponding elements of $\mathbf{A}$ and $\mathbf{B}$. That is,

$$c_{ij} = a_{ij} + b_{ij} \quad \text{for all } i \text{ and } j$$

If a matrix $\mathbf{B}$ is subtracted from a matrix $\mathbf{A}$ to form a new matrix $\mathbf{C}$, the elements of $\mathbf{C}$ are found by subtracting corresponding elements of $\mathbf{B}$ from $\mathbf{A}$, or

$$c_{ij} = a_{ij} - b_{ij} \quad \text{for all } i \text{ and } j$$

EXAMPLE 4

Given

$$\mathbf{A} = \begin{pmatrix} 1 & 3 \\ 4 & -2 \end{pmatrix} \quad \text{and} \quad \mathbf{B} = \begin{pmatrix} -3 & 2 \\ 0 & 4 \end{pmatrix}$$

$$\mathbf{A} + \mathbf{B} = \begin{pmatrix} 1 & 3 \\ 4 & -2 \end{pmatrix} + \begin{pmatrix} -3 & 2 \\ 0 & 4 \end{pmatrix}$$

$$= \begin{pmatrix} 1+(-3) & 3+2 \\ 4+0 & -2+4 \end{pmatrix} = \begin{pmatrix} -2 & 5 \\ 4 & 2 \end{pmatrix}$$

EXAMPLE 5 Using the same matrices,

$$\mathbf{B} - \mathbf{A} = \begin{pmatrix} -3 & 2 \\ 0 & 4 \end{pmatrix} - \begin{pmatrix} 1 & 3 \\ 4 & -2 \end{pmatrix}$$

$$= \begin{pmatrix} -3 - (1) & 2 - (3) \\ 0 - (4) & 4 - (-2) \end{pmatrix}$$

$$= \begin{pmatrix} -4 & -1 \\ -4 & 6 \end{pmatrix}$$

EXAMPLE 6 The Department of Energy has projected energy consumption figures for the year 2000. The matrix **P** displays average daily energy consumption by energy source for the same regions of the United States as indicated in Example 1. As before, these figures are in millions of barrels of oil per day that would yield the equivalent energy.

	American oil and gas	Coal	Imported oil and gas	Hydroelectric, solar, geothermal, and synthetic fuels	Nuclear	
P =	5.9	4.8	2.0	0.7	1.2	Northeast
	2.9	1.9	0.2	0.9	0.5	South
	2.3	2.4	0.5	0.5	0.9	Midwest
	6.0	1.9	2.9	1.0	0.6	West

The matrix computation **P − E** reflects the estimated change in average daily energy consumption by energy source between 1987 and 2000.

$$\mathbf{P} - \mathbf{E} = \begin{pmatrix} 5.9 & 4.8 & 2.0 & 0.7 & 1.2 \\ 2.9 & 1.9 & 0.2 & 0.9 & 0.5 \\ 2.3 & 2.4 & 0.5 & 0.5 & 0.9 \\ 6.0 & 1.9 & 2.9 & 1.0 & 0.6 \end{pmatrix} - \begin{pmatrix} 6.5 & 2.8 & 3.0 & 0.2 & 0.5 \\ 3.2 & 1.1 & 0.5 & 0.5 & 0.2 \\ 3.4 & 2.0 & 1.1 & 0.1 & 0.4 \\ 5.5 & 1.5 & 3.3 & 0.6 & 0.2 \end{pmatrix}$$

$$= \begin{pmatrix} -0.6 & 2.0 & -1.0 & 0.5 & 0.7 \\ -0.3 & 0.8 & -0.3 & 0.4 & 0.3 \\ -1.1 & 0.4 & -0.6 & 0.4 & 0.5 \\ 0.5 & 0.4 & -0.4 & 0.4 & 0.4 \end{pmatrix}$$

□

PRACTICE EXERCISE
Interpret the meaning of the values in the difference matrix **P − E**.

Scalar Multiplication

A *scalar* is a real number. *Scalar multiplication* of a matrix is the multiplication of a matrix by a scalar. The product is found by multiplying each element in the matrix by the scalar. For example, if k is a scalar and **A** the following (3 × 2) matrix, then

$$k\mathbf{A} = k \cdot \begin{pmatrix} 5 & 3 \\ -2 & 1 \\ 0 & 4 \end{pmatrix} = \begin{pmatrix} 5k & 3k \\ -2k & k \\ 0 & 4k \end{pmatrix}$$

EXAMPLE 7   (Energy Forecasts) One private policy research foundation projects that energy consumption will increase by 20 percent in each region and for each energy source between 1987 and 1992. If consumption increases by 20 percent in each region and for each energy source, the consumption in 1992 will equal 120 percent of that for 1987. Thus, the projected consumption in 1992 can be determined by the scalar multiplication 1.2E, or

$$\mathbf{R} = 1.2 \begin{pmatrix} 6.5 & 2.8 & 3.0 & 0.2 & 0.5 \\ 3.2 & 1.1 & 0.5 & 0.5 & 0.2 \\ 3.4 & 2.0 & 1.1 & 0.1 & 0.4 \\ 5.5 & 1.5 & 3.3 & 0.6 & 0.2 \end{pmatrix} = \begin{pmatrix} 7.80 & 3.36 & 3.60 & 0.24 & 0.60 \\ 3.84 & 1.32 & 0.60 & 0.60 & 0.24 \\ 4.08 & 2.40 & 1.32 & 0.12 & 0.48 \\ 6.50 & 1.80 & 3.96 & 0.72 & 0.24 \end{pmatrix}$$

□

PRACTICE EXERCISE
In Example 7, what scalar multiplication would forecast a 10 percent reduction in energy consumption across the board? Answer: $\mathbf{R} = 0.9\mathbf{E}$.

The Inner Product

DEFINITION: INNER PRODUCT

Let $A = (a_{11}, a_{12}, \ldots, a_{1n})$ and $B = \begin{pmatrix} b_{11} \\ b_{21} \\ \vdots \\ b_{n1} \end{pmatrix}$; then the *inner product*,

written $A \cdot B$, is

$$A \cdot B = a_{11}b_{11} + a_{12}b_{21} + \cdots + a_{1n}b_{n1}$$

From this definition, three points should be noted:

1   *The inner product is defined only if the row and column vectors contain the same number of elements.*
2   *The inner product results when a row vector is multiplied by a column vector and the resulting product is a scalar quantity.*
3   *The inner product is computed by multiplying corresponding elements in the two vectors and algebraically summing.*

Consider the multiplication of the following vectors:

$$\mathbf{AB} = (5 - 2) \begin{pmatrix} 4 \\ 6 \end{pmatrix}$$

To find the inner product, the first element in the row vector is multiplied by the first element in the column vector; the resulting product is added to the product of element 2 in the row vector and element 2 in the column vector. For the vectors indicated, the inner product is computed as $a_{11}b_{11} + a_{12}b_{21}$, or

$$
(5 \quad -2) \begin{pmatrix} 4 \\ 6 \end{pmatrix} = (5)(4) + (-2)(6) = 8
$$

EXAMPLE 8 Given the row and column vectors

$$
\mathbf{M} = (5 \quad -2 \quad 0 \quad 1 \quad 3) \qquad \text{and} \qquad \mathbf{N} = \begin{pmatrix} -2 \\ -4 \\ 10 \\ 20 \\ 6 \end{pmatrix}
$$

the inner product is computed as

$$
\mathbf{M} \cdot \mathbf{N} = (5 \quad -2 \quad 0 \quad 1 \quad 3) \begin{pmatrix} -2 \\ -4 \\ 10 \\ 20 \\ 6 \end{pmatrix}
$$

$$
= (5)(-2) + (-2)(-4) + (0)(10) + (1)(20) + (3)(6) = 36
$$

❑

PRACTICE EXERCISE

Given $\mathbf{S} = (-5 \quad 3 \quad 0 \quad 2)$ and $\mathbf{V} = (3 \quad -1 \quad 4 \quad 2)$, find the inner product $\mathbf{SV}^{\mathrm{T}}$. *Answer:* -14.

Matrix Multiplication

Assume that a matrix $\mathbf{A}$ having dimension $m_A \times n_A$ is to be multiplied by a matrix $\mathbf{B}$ having dimension $m_B \times n_B$.

PROPERTIES OF MATRIX MULTIPLICATION

I *The matrix product AB is defined if and only if the number of columns of A equals the number of rows of B, or if $n_A = m_B$.*

II *If the multiplication can be performed (that is, $n_A = m_B$), the resulting product will be a matrix having dimension $m_A \times n_B$.*

The first multiplication property states the *necessary and sufficient condition* for matrix multiplication. If $n_A \neq m_B$, the matrices cannot be multiplied.

$$
\begin{array}{ccc}
\mathbf{A} & \cdot & \mathbf{B} \\
(m_A \times n_A) & \underset{\underset{n_A = m_B}{\boxed{}}}{?} & (m_B \times n_B)
\end{array}
\qquad
\begin{array}{l}
\text{Test for necessary} \\
\text{and sufficient condition}
\end{array}
$$

Property II defines the dimension of a product matrix.

$$
\begin{array}{cccc}
\mathbf{A} & \cdot & \mathbf{B} & = \quad \mathbf{C} \\
(m_A \times n_A) & \underset{\underset{n_A = m_B}{\boxed{}}}{\checkmark} & (m_B \times n_B) & (m_A \times n_B)
\end{array}
\qquad
\begin{array}{l}
\text{Product matrix} \\
\text{dimension}
\end{array}
$$

To determine elements of the product matrix, the following computational rule can be used.

COMPUTATIONAL RULE
If $\mathbf{AB} = \mathbf{C}$, an element c_{ij} of the product matrix is equal to the *inner product* of row i of matrix $\mathbf{A}$ and column j of matrix $\mathbf{B}$. (See Fig. 9.1)

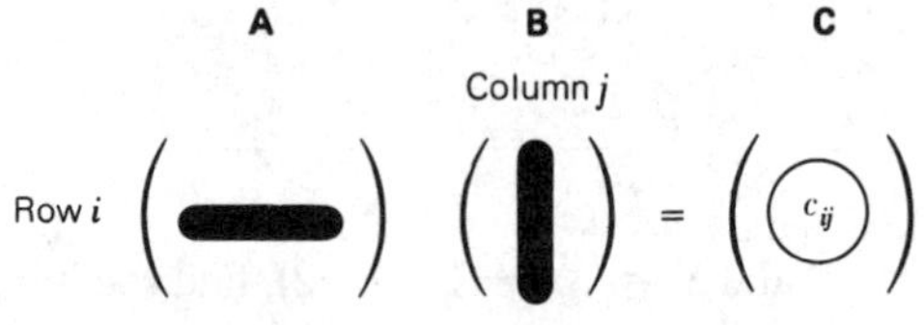

Figure 9.1 Matrix multiplication: computation of c_{ij} using the inner product.

EXAMPLE 9

To find the matrix product $\mathbf{AB}$, where

$$
\mathbf{A} = \begin{pmatrix} 2 & 4 \\ 3 & 1 \end{pmatrix} \quad \text{and} \quad \mathbf{B} = \begin{pmatrix} -4 \\ 2 \end{pmatrix}
$$

we first check to determine whether the multiplication is possible. $\mathbf{A}$ is a (2×2) matrix and $\mathbf{B}$ is a (2×1) matrix.

$$
\begin{array}{ccc}
\mathbf{A} & \mathbf{B} & \mathbf{C} \\
(2 \times 2) & (2 \times 1) & = (2 \times 1)
\end{array}
$$

The product is defined because the number of columns of **A** equals the number of rows of **B**. The resulting product matrix will be of dimension (2×1) and will have the general form

$$\mathbf{C} = \begin{pmatrix} c_{11} \\ c_{21} \end{pmatrix}$$

To find c_{11}, the inner product of row 1 of **A** and column 1 of **B** is computed, or

$$\begin{pmatrix} 2 & 4 \\ 3 & 1 \end{pmatrix} \begin{pmatrix} -4 \\ 2 \end{pmatrix} = \begin{pmatrix} 0 \end{pmatrix}$$

Similarly, c_{21} is found by computing the inner product between row 2 of **A** and column 1 of **B**, or

$$\begin{pmatrix} 2 & 4 \\ 3 & 1 \end{pmatrix} \begin{pmatrix} -4 \\ 2 \end{pmatrix} = \begin{pmatrix} 0 \\ -10 \end{pmatrix}$$

❑

NOTE   As you first attempt matrix multiplication problems, you may find it helpful to write out the general form of the product matrix. We did this by first stating the general form of **C** as $\begin{pmatrix} c_{11} \\ c_{21} \end{pmatrix}$. With the elements identified in this manner, the subscripts of each element indicate how each element can be computed.

EXAMPLE 10   Determine the matrix product **BA** for the matrices in Example 9.

SOLUTION

The product **BA** involves multiplying a (2×1) matrix by a (2×2) matrix, or

$$\begin{array}{ccc} \mathbf{B} & . & \mathbf{A} \\ (2 \times 1) & & (2 \times 2) \\ | & & | \\ 1 & \neq & 2 \end{array}$$

Since the number of columns of **B** does not equal the number of rows of **A**, the product **BA** is not defined.

❑

NOTE   This example illustrates that the commutative property which holds for the multiplication of real numbers *does not necessarily* hold for matrix multiplication. We *cannot* state automatically that $\mathbf{AB} = \mathbf{BA}$ for any two matrices **A** and **B**.

EXAMPLE 11   Find, if possible, the product $\mathbf{PI} = \mathbf{T}$, where

$$\mathbf{P} = \begin{pmatrix} 1 & 0 & -1 \\ 2 & 6 & -2 \\ 0 & 10 & 1 \\ 3 & 4 & 5 \end{pmatrix} \quad \text{and} \quad \mathbf{I} = \begin{pmatrix} 1 & 0 & 0 \\ 0 & 1 & 0 \\ 0 & 0 & 1 \end{pmatrix}$$

SOLUTION

$\mathbf{P}$ is a (4×3) matrix, and $\mathbf{I}$ is a (3×3) identity matrix. Since the number of columns of $\mathbf{P}$ equals the number of rows of $\mathbf{I}$, the multiplication can be performed and the product matrix $\mathbf{T}$ will have dimension 4×3. Thus,

$$\begin{array}{ccccc} \mathbf{P} & \cdot & \mathbf{I} & = & \mathbf{T} \\ (4 \times 3) & & (3 \times 3) & & (4 \times 3) \end{array}$$

$\mathbf{T}$ will have the general form

$$\mathbf{T} = \begin{pmatrix} t_{11} & t_{12} & t_{13} \\ t_{21} & t_{22} & t_{23} \\ t_{31} & t_{32} & t_{33} \\ t_{41} & t_{42} & t_{43} \end{pmatrix}$$

Some sample elements are computed in the following:

$$t_{11} = (1 \quad 0 \quad -1) \begin{pmatrix} 1 \\ 0 \\ 0 \end{pmatrix} = (1)(1) + (0)(0) + (-1)(0) = 1$$

$$t_{12} = (1 \quad 0 \quad -1) \begin{pmatrix} 0 \\ 1 \\ 0 \end{pmatrix} = (1)(0) + (0)(1) + (-1)(0) = 0$$

$$t_{13} = (1 \quad 0 \quad -1) \begin{pmatrix} 0 \\ 0 \\ 1 \end{pmatrix} = (1)(0) + (0)(0) + (-1)(1) = -1$$

Verify that the product matrix $\mathbf{T}$ is

$$\mathbf{T} = \begin{pmatrix} 1 & 0 & -1 \\ 2 & 6 & -2 \\ 0 & 10 & 1 \\ 3 & 4 & 5 \end{pmatrix}$$

> **NOTE** This example illustrates the property mentioned earlier concerning identity matrices. That is, if an identity matrix is multiplied by another matrix, the product will be the other matrix. In this example $PI = T$. But $P = T$; thus $PI = P$.

PRACTICE EXERCISE

Given $A = \begin{pmatrix} 2 & -3 \\ 4 & 5 \end{pmatrix}$ and $B = \begin{pmatrix} 1 & 3 \\ 7 & 4 \end{pmatrix}$, find the product AB. *Answer:*
$\begin{pmatrix} -19 & -6 \\ 39 & 32 \end{pmatrix}$.

EXAMPLE 12

(Course Average Computations) The instructor who gave the three tests to five students is preparing course averages. She has decided to weight the first two tests at 30 percent each and the third at 40 percent. The instructor wishes to compute the final averages for the five students using matrix multiplication. The matrix of grades is

$$G = \begin{pmatrix} 75 & 82 & 86 \\ 91 & 95 & 100 \\ 65 & 70 & 68 \\ 59 & 80 & 99 \\ 75 & 76 & 74 \end{pmatrix}$$

and the examination weights are placed in the row vector

$$W = (0.30 \quad 0.30 \quad 0.40)$$

The instructor needs to multiply these matrices in such a way that the first examination score for *each* student is multiplied by 0.30, the second examination score by 0.30, and the last score by 0.40. Verify for yourself that the products GW and WG are not defined. If, however, W had been stated as a column vector, the matrix product GW would lead to the desired result.

We can transform W into a column vector by simply finding its transpose. The product GW^T is defined, it leads to a (5×1) product matrix, and most important, it performs the desired computations.

$$
\begin{array}{ccccc}
G & \cdot & W^T & = & A \\
(5 \times 3) & & (3 \times 1) & & (5 \times 1)
\end{array}
$$

The final averages are computed as

$$\begin{pmatrix} 75 & 82 & 86 \\ 91 & 95 & 100 \\ 65 & 70 & 68 \\ 59 & 80 & 99 \\ 75 & 76 & 74 \end{pmatrix} \begin{pmatrix} 0.30 \\ 0.30 \\ 0.40 \end{pmatrix} = \begin{pmatrix} 75(0.3) + 82(0.3) + 86(0.4) \\ 91(0.3) + 95(0.3) + 100(0.4) \\ 65(0.3) + 70(0.3) + 68(0.4) \\ 59(0.3) + 80(0.3) + 99(0.4) \\ 75(0.3) + 76(0.3) + 74(0.4) \end{pmatrix} = \begin{pmatrix} 81.5 \\ 95.8 \\ 67.7 \\ 81.3 \\ 74.9 \end{pmatrix}$$

The averages are 81.5, 95.8, 67.7, 81.3, and 74.9, respectively, for the five students.

☐

PRACTICE EXERCISE
Compute the product $\mathbf{WG}^T$. Doesn't this yield the same result as $\mathbf{GW}^T$?

Representation of an Equation

An equation may be represented using the inner product. The expression

$$3x_1 + 5x_2 - 4x_3$$

can be represented by the inner product

$$(3 \quad 5 \quad -4) \begin{pmatrix} x_1 \\ x_2 \\ x_3 \end{pmatrix}$$

where the row vector contains the coefficients for each variable in the expression and the column vector contains the variables. Multiply the two vectors to verify that the inner product does result in the original expression.

To represent the *equation*

$$3x_1 + 5x_2 - 4x_3 = 25$$

we can equate the inner product with a (1×1) matrix containing the right-side constant, or

$$(3 \quad 5 \quad -4) \begin{pmatrix} x_1 \\ x_2 \\ x_3 \end{pmatrix} = (25)$$

Remember that for two matrices to be equal, they must have the same dimension. The inner product always results in a (1×1) matrix, which in this case contains one element—the expression $3x_1 + 5x_2 - 4x_3$.

A linear equation of the form $a_1 x_1 + a_2 x_2 + a_3 x_3 + \cdots + a_n x_n = b$ can be represented in matrix form as

$$(a_1 \quad a_2 \quad a_3 \quad \cdots \quad a_n) \begin{pmatrix} x_1 \\ x_2 \\ x_3 \\ \vdots \\ x_n \end{pmatrix} = b \qquad (9.1)$$

Representation of Systems of Equations

Whereas single equations may be represented by using the inner product, a system of equations can be represented by using matrix multiplication. The system

$$5x_1 + 3x_2 = 15$$
$$4x_1 - 2x_2 = 12$$

can be represented as

$$\begin{pmatrix} 5 & 3 \\ 4 & -2 \end{pmatrix} \begin{pmatrix} x_1 \\ x_2 \end{pmatrix} = \begin{pmatrix} 15 \\ 12 \end{pmatrix}$$

If we perform the matrix multiplication on the left side of the matrix equation, the result is

$$\begin{pmatrix} 5x_1 + 3x_2 \\ 4x_1 - 2x_2 \end{pmatrix} = \begin{pmatrix} 15 \\ 12 \end{pmatrix}$$

For these two (2×1) matrices to be equal, the corresponding elements must be equal (i.e., $5x_1 + 3x_2 = 15$ and $4x_1 - 2x_2 = 12$, *the information communicated by the original pair of equations*).

An $(m \times n)$ system of equations having the form

$$a_{11}x_1 + a_{12}x_2 + \cdots + a_{1n}x_n = b_1$$
$$a_{21}x_1 + a_{22}x_2 + \cdots + a_{2n}x_n = b_2$$
$$\cdots \cdots \cdots \cdots \cdots \cdots \cdots \cdots \cdots$$
$$a_{m1}x_1 + a_{m2}x_2 + \cdots + a_{mn}x_n = b_m$$

can be represented by the matrix equation

$$AX = B$$

where **A** is an $(m \times n)$ matrix containing the variable coefficients on the left side of the set of equations, **X** is an n-component column vector containing the n variables, and **B** is an m-component column vector containing the right-side constants for the m equations. This representation appears as

$$\begin{pmatrix} a_{11} & a_{12} & \cdots & a_{1n} \\ a_{21} & a_{22} & \cdots & a_{2n} \\ \cdots & \cdots & \cdots & \cdots \\ a_{m1} & a_{m2} & \cdots & a_{mn} \end{pmatrix} \begin{pmatrix} x_1 \\ x_2 \\ \vdots \\ x_n \end{pmatrix} = \begin{pmatrix} b_1 \\ b_2 \\ \vdots \\ b_m \end{pmatrix} \qquad (9.2)$$

EXAMPLE 13 The system of equations

$$
\begin{aligned}
x_1 - 2x_2 \qquad\quad + 3x_4 + x_5 &= 100 \\
x_1 \qquad\quad - 3x_3 + x_4 \qquad &= 60 \\
4x_2 - x_3 + 2x_4 + x_5 &= 125
\end{aligned}
$$

can be represented in the matrix form $\mathbf{AX} = \mathbf{B}$ as

$$
\begin{pmatrix} 1 & -2 & 0 & 3 & 1 \\ 1 & 0 & -3 & 1 & 0 \\ 0 & 4 & -1 & 2 & 1 \end{pmatrix}
\begin{pmatrix} x_1 \\ x_2 \\ x_3 \\ x_4 \\ x_5 \end{pmatrix}
=
\begin{pmatrix} 100 \\ 60 \\ 125 \end{pmatrix}
$$

Verify that this representation is valid and that the 0s must be included in the $\mathbf{A}$ matrix when a variable does not appear in a particular equation.

EXAMPLE 14 Matrix representation of equations is not restricted to linear equations. The quadratic equation

$$
10x^2 - 4x + 50 = 0
$$

can be represented by the equivalent matrix equation

$$
(10 \quad -4 \quad 50)
\begin{pmatrix} x^2 \\ x \\ 1 \end{pmatrix}
= (0)
$$

$\square$

Section 9.3 Follow-up Exercises

Perform the following matrix operations wherever possible.

1 $-\begin{pmatrix} 4 & -2 \\ 5 & 8 \end{pmatrix} - \begin{pmatrix} 3 & 12 \\ 8 & 4 \end{pmatrix}$

2 $\begin{pmatrix} 5 & -8 \\ 2 & 14 \end{pmatrix} + \begin{pmatrix} -6 & -2 \\ 10 & -4 \end{pmatrix} - \begin{pmatrix} -10 & 5 \\ 21 & -8 \end{pmatrix}$

3 $-3\begin{pmatrix} 4 & -3 \\ -1 & -4 \end{pmatrix} + 8\begin{pmatrix} 12 & 10 \\ -2 & -4 \end{pmatrix}$

4 $3k\begin{pmatrix} -a & b \\ -b & 2a \end{pmatrix} - 2k\begin{pmatrix} a & b \\ b & 2a \end{pmatrix}$

5 $5\begin{pmatrix} -2 & 10 \\ 8 & 15 \end{pmatrix} - 3\begin{pmatrix} 20 & -25 \\ -10 & 15 \end{pmatrix}$

6 $\begin{pmatrix} 7 & 5 \\ 8 & 4 \end{pmatrix} - 8\begin{pmatrix} 4 & 3 \\ 2 & 4 \end{pmatrix} + 6\begin{pmatrix} 1 & -8 \\ 2 & -4 \end{pmatrix}$

7 $(7 \quad -3)\begin{pmatrix} 4 \\ 8 \end{pmatrix}$

8 $(1 \quad -2 \quad -3)\begin{pmatrix} 8 \\ 5 \\ 4 \end{pmatrix}$

9 $(3 \quad -2)\begin{pmatrix} 4 \\ -4 \\ -3 \end{pmatrix}$

10 $(18 \quad -4 \quad -6)\begin{pmatrix} 2 \\ 3 \end{pmatrix}$

11 $(a \quad b)\begin{pmatrix} x \\ y \end{pmatrix}$

12 $(a_1 \quad a_2 \quad a_3)\begin{pmatrix} x_1 \\ x_2 \\ x_3 \end{pmatrix}$

13 $(-4 \quad 2 \quad -8 \quad 4)\begin{pmatrix} 0 \\ 1 \\ 2 \\ 3 \end{pmatrix}$

14 $(1 \quad -8 \quad 6 \quad -5 \quad -2)\begin{pmatrix} 6 \\ -2 \\ 4 \\ 8 \\ 4 \end{pmatrix}$

15 $(a \quad b \quad c \quad d)\begin{pmatrix} e \\ f \\ g \\ h \end{pmatrix}$

16 $(1 \quad 0 \quad -5 \quad 0 \quad 8)\begin{pmatrix} 0 \\ -2 \\ 0 \\ 6 \\ 0 \end{pmatrix}$

17 $\begin{pmatrix} 4 & 0 \\ -2 & 9 \end{pmatrix}\begin{pmatrix} 2 & 6 \\ -7 & 8 \end{pmatrix}$

18 $\begin{pmatrix} 8 & -3 \\ -2 & 0 \end{pmatrix}\begin{pmatrix} 8 \\ -8 \end{pmatrix}$

19 $(20 \quad -8)\begin{pmatrix} 1 & 0 \\ 0 & 1 \end{pmatrix}$

20 $\begin{pmatrix} 12 & 10 \\ -1 & -8 \end{pmatrix}\begin{pmatrix} 1 & 0 \\ 0 & 1 \end{pmatrix}$

21 $\begin{pmatrix} 10 & -5 \\ 0 & 13 \end{pmatrix}\begin{pmatrix} 1 & 0 & 1 \\ 0 & 1 & 0 \end{pmatrix}$

22 $\begin{pmatrix} 12 & 0 \\ 4 & 0 \\ -2 & 15 \end{pmatrix}\begin{pmatrix} 7 & 12 \\ 8 & -4 \end{pmatrix}$

23 $\begin{pmatrix} 4 & 4 \\ -2 & -3 \end{pmatrix}\begin{pmatrix} 6 & -2 \\ 4 & 20 \\ 8 & 4 \end{pmatrix}$

24 $(1 \quad 8 \quad -2)\begin{pmatrix} 0 & -2 & 7 \\ 3 & -4 & 10 \\ 1 & 2 & -3 \end{pmatrix}$

25 $\begin{pmatrix} 2 & -1 & 6 \\ 1 & 0 & -4 \\ 3 & -2 & -1 \end{pmatrix}\begin{pmatrix} 1 & 0 & 10 \\ 0 & -3 & 0 \\ 1 & 4 & 1 \end{pmatrix}$

26 $\begin{pmatrix} 1 & 0 & 0 \\ 0 & 1 & 0 \\ 0 & 0 & 1 \end{pmatrix}\begin{pmatrix} 1 & -2 & 8 \\ 3 & 12 & 4 \end{pmatrix}$

27 $\begin{pmatrix} a_{11} & a_{12} \\ a_{21} & a_{22} \end{pmatrix}\begin{pmatrix} x_1 \\ x_2 \end{pmatrix}$

28 $\begin{pmatrix} a_{11} & a_{12} & a_{13} \\ a_{21} & a_{22} & a_{23} \end{pmatrix}\begin{pmatrix} x_1 \\ x_2 \\ x_3 \end{pmatrix}$

29 $\begin{pmatrix} 2 & -7 & 3 \\ 0 & 2 & 5 \end{pmatrix}\begin{pmatrix} 6 & -1 \\ 2 & 0 \\ 4 & 8 \\ 2 & 6 \end{pmatrix}$

30 $\begin{pmatrix} 5 & -3 \\ 2 & 4 \end{pmatrix}\begin{pmatrix} 1 & -2 \\ 3 & 4 \\ 5 & -6 \end{pmatrix}$

31 $\begin{pmatrix} -2 & 2 \\ 3 & -1 \end{pmatrix}\begin{pmatrix} 1 & -4 & 0 \\ 5 & -2 & 3 \end{pmatrix}$

32 $\begin{pmatrix} 3 \\ 2 \\ 1 \end{pmatrix}(1 \quad -2 \quad 3)$

33 $\begin{pmatrix} 2 & -1 \\ 3 & 2 \\ 0 & -4 \\ 3 & -2 \end{pmatrix}\begin{pmatrix} 2 & -1 & 0 & 5 & 8 \\ 6 & 2 & 0 & -2 & 0 \end{pmatrix}$

34 $\begin{pmatrix} a & b \\ c & d \end{pmatrix}\begin{pmatrix} e & f \\ g & h \\ i & j \end{pmatrix}$

35 $\begin{pmatrix} 2 & 5 & -7 \\ 1 & 0 & -2 \\ 4 & 8 & 2 \end{pmatrix}\begin{pmatrix} -3 & 2 & 5 & 0 \\ 1 & -2 & 3 & -4 \\ -3 & 4 & -2 & 1 \end{pmatrix}$

36 $\begin{pmatrix} 2 & 8 & -1 \\ 0 & 4 & 0 \end{pmatrix}\begin{pmatrix} -1 & 0 & 3 \\ 4 & 0 & -1 \\ -1 & 0 & 1 \end{pmatrix}$

Rewrite the following systems of equations in matrix form.

37 $x - 3y = 15$
  $2x + 3y = -10$

39 $5x_1 - 2x_2 + 3x_3 = 12$
  $3x_1 - x_2 - 2x_3 = 15$

41 $ax_1 + bx_2 = c$
  $dx_1 + ex_2 = f$
  $gx_1 + hx_2 = i$

43 $a_1 x^2 + a_2 x + a_3 = b_1$
  $a_4 x^2 + a_5 x + a_6 = b_2$

45 $5x^3 - 2x^2 + x = 100$
  $3x^3 \qquad = -18$
  $\qquad 5x^2 \quad = 125$

38 $2x = 4$
  $3x + 4y = 15$

40 $5x_1 - 8x_2 = 48$
  $2x_1 - 4x_3 = 25$

42 $ax_1 + bx_2 + cx_3 + dx_4 + ex_5 = f$
  $gx_1 \qquad - hx_3 \qquad + ix_5 = j$

44 $a_{11} x^2 + a_{12} x + a_{13} = b_1$
  $a_{21} x^2 + a_{22} x + a_{23} = b_2$
  $a_{31} x^2 + a_{32} x + a_{33} = b_3$

46 $a_{11} x_1 + a_{12} x_2 + a_{13} x_3 + a_{14} x_4 = b_1$
  $a_{21} x_1 + a_{22} x_2 + a_{23} x_3 + a_{24} x_4 = b_2$
  $a_{31} x_1 + a_{32} x_2 + a_{33} x_3 + a_{34} x_4 = b_3$
  $a_{41} x_1 + a_{42} x_2 + a_{43} x_3 + a_{44} x_4 = b_4$

47 If $A = \begin{pmatrix} 2 & 1 \\ 3 & 4 \end{pmatrix}$, $B = \begin{pmatrix} 4 & 0 \\ 1 & 2 \end{pmatrix}$, and $C = \begin{pmatrix} 1 & 1 \\ 1 & 3 \end{pmatrix}$, verify that (a) $A(BC) = (AB)C$ and (b) $A(B + C) = AB + AC$.

9.4 THE DETERMINANT

An important concept in matrix algebra is that of the **determinant**. *If a matrix is square, the elements of the matrix may be combined to compute a real-valued number called the determinant.* The determinant concept is of particular interest in solving simultaneous equations.

The determinant of the matrix

$$A = \begin{pmatrix} 2 & 5 \\ 3 & -2 \end{pmatrix}$$

can be denoted either by the symbol Δ, or by placing vertical lines around the elements of the matrix. The determinant of A can be denoted by either

$$\Delta \quad \text{or} \quad \begin{vmatrix} 2 & 5 \\ 3 & -2 \end{vmatrix}$$

Equivalently, the determinant can be represented by placing vertical lines around the matrix name. Thus, the determinant of A could be represented as

$$|A| = \begin{vmatrix} 2 & 5 \\ 3 & -2 \end{vmatrix}$$

There are different ways of finding the value of a determinant. First let's discuss specific techniques for handling (1×1), (2×2), and (3×3) matrices and follow with the more generalized **cofactor procedure.**

The Determinant of a (1 × 1) Matrix

The determinant of a (1 × 1) matrix is simply the value of the one element contained in the matrix. If $\mathbf{A} = (5)$, $\Delta = 5$. If $\mathbf{M} = (-10)$, $\Delta = -10$.

The Determinant of a (2 × 2) Matrix

Given a (2 × 2) matrix having the form

$$\mathbf{A} = \begin{pmatrix} a_{11} & a_{12} \\ a_{21} & a_{22} \end{pmatrix}$$

$$\boxed{\Delta = a_{11}a_{22} - a_{21}a_{12}} \qquad (9.3)$$

The computation involves a cross multiplication of elements on the two diagonals, as indicated:

$$\Delta = \begin{pmatrix} a_{11} \\ \qquad \searrow \times \\ \qquad\qquad a_{22} \end{pmatrix} - \begin{pmatrix} \qquad\qquad a_{12} \\ \times \nearrow \\ a_{21} \end{pmatrix}$$

EXAMPLE 15   If

$$\mathbf{A} = \begin{pmatrix} 1 & -2 \\ 3 & 4 \end{pmatrix}$$

then

$$\Delta = (1)(4) - (3)(-2)$$
$$= 4 + 6 = 10$$

❑

PRACTICE EXERCISE

Find the determinant of $\mathbf{A} = \begin{pmatrix} 2 & 4 \\ 4 & 8 \end{pmatrix}$. *Answer:* $\Delta = 0$.

The Determinant of a (3 × 3) Matrix

Given the (3 × 3) matrix

$$\mathbf{A} = \begin{pmatrix} a_{11} & a_{12} & a_{13} \\ a_{21} & a_{22} & a_{23} \\ a_{31} & a_{32} & a_{33} \end{pmatrix}$$

the determinant may be found by the following process:

1   *Rewrite the first two columns of the matrix to the right of the original matrix.*
2   *Locate the elements on the three primary diagonals (P_1, P_2, P_3) and those on the three secondary diagonals (S_1, S_2, S_3).*

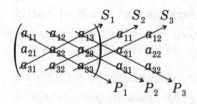

3   *Multiply the elements on each primary and each secondary diagonal.*
4   *The determinant equals the sum of the products for the three primary diagonals minus the sum of the products for the three secondary diagonals.*

Algebraically the determinant is computed as

$$\Delta = a_{11}a_{22}a_{33} + a_{12}a_{23}a_{31} + a_{13}a_{21}a_{32} - a_{31}a_{22}a_{13} \\ - a_{32}a_{23}a_{11} - a_{33}a_{21}a_{12}$$

(9.4)

EXAMPLE 16 To find the determinant of the matrix

$$\mathbf{A} = \begin{pmatrix} 3 & 1 & 2 \\ -1 & 2 & 4 \\ 3 & -2 & 1 \end{pmatrix}$$

the first two columns are rewritten to the right of the original (3 × 3) matrix:

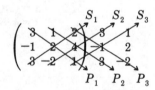

The three primary and secondary diagonals are identified and the determinant is computed as

$$\Delta = [(3)(2)(1) + (1)(4)(3) + (2)(-1)(-2)] - [(3)(2)(2) + (-2)(4)(3) \\ + (1)(-1)(1)] \\ = (6 + 12 + 4) - (12 - 24 - 1) = 22 - (-13) = 35$$

❑

PRACTICE EXERCISE

Find the determinant of $\mathbf{B} = \begin{pmatrix} 4 & 1 & -2 \\ 2 & 5 & -1 \\ -6 & 7 & 3 \end{pmatrix}$. *Answer:* $\Delta = 0$.

> **NOTE** The methods examined for (1×1), (2×2), and (3×3) matrices apply only for matrices of those dimensions. We cannot extend the (3×3) procedure to deal with (4×4), (5×5), or square matrices of a higher order. The following optional section discusses a more generalized procedure.

The Method of Cofactors

This section discusses an alternative, more generalized computational procedure which can be applied for all square matrices of size (2×2) or higher. For any square matrix **A** there can be found a ***matrix of cofactors*** which we will denote as $\mathbf{A}_c$. The matrix of cofactors will have the same dimension as **A** and will consist of elements a'_{ij} which are called ***cofactors***. For each element a_{ij} contained in **A** there will be a corresponding cofactor a'_{ij}.

The cofactor associated with an element a_{ij} is determined as follows.

PROCEDURE FOR FINDING COFACTOR ASSOCIATED WITH ELEMENT a_{ij}

1 *Either mentally or with a pencil, cross off row i and column j in the original matrix. Focus upon the remaining elements in the matrix. The remaining elements form a **submatrix** of the original matrix.*
2 *Find the determinant of the remaining submatrix. This determinant is called the **minor** of the element a_{ij}.*
3 *The cofactor a'_{ij} is found by multiplying the minor by either $+1$ or -1 depending on the position of the element a_{ij}. A formula for computing the cofactor is*

$$a'_{ij} = (-1)^{i+j} (the\ minor)$$

(The essence of this formula is that if $i + j$ is an even number, the minor is multiplied by $+1$, retaining its sign; if $i + j$ is odd, the minor is multiplied by -1, changing its sign.)

EXAMPLE 17 To find the matrix of cofactors for the (2×2) matrix

$$\mathbf{A} = \begin{pmatrix} 5 & -4 \\ 2 & -2 \end{pmatrix}$$

let's begin with the cofactor corresponding to element a_{11}. Crossing off row 1 and column 1 leaves the (1×1) submatrix (-2). The determinant of this submatrix

$$\begin{pmatrix} 5 & -4 \\ 2 & -2 \end{pmatrix} \rightarrow \underset{\text{Submatrix}}{(-2)} \rightarrow \underset{\text{Minor}}{\Delta = -2}$$

equals -2 and is therefore the *minor*. The cofactor is computed by formula as

$$a'_{11} = (-1)^{1+1}(-2) = (-1)^2(-2)$$
$$= (1)(-2) = -2$$

Equivalently (and more simply), we could reason that the cofactor a'_{11} equals the value of the minor, either retaining the sign or changing the sign. Again, the address of the cofactor is the key. *If the sum of the subscripts is even, retain the sign; if odd, change the sign.* For the cofactor a'_{11}, the sum of the subscripts is $1 + 1 = 2$, which is even. Therefore, retain the sign on the minor: $a'_{11} = -2$.

For the remaining elements:

To Find	Submatrix	By Formula
a'_{12}	$\begin{pmatrix} 5 & -4 \\ 2 & -2 \end{pmatrix}$	$a'_{12} = (-1)^{1+2}(2)$ $= (-1)(2) \quad = -2$
a'_{21}	$\begin{pmatrix} 5 & -4 \\ 2 & -2 \end{pmatrix}$	$a'_{21} = (-1)^{2+1}(-4)$ $= (-1)(-4) \quad = 4$
a'_{22}	$\begin{pmatrix} 5 & -4 \\ 2 & -2 \end{pmatrix}$	$a'_{22} = (-1)^{2+2}(5)$ $= (1)(5) \quad = 5$

or

To Find	Minor	Position	Sign on Minor	Cofactor
a'_{12}	2	odd	change	-2
a'_{21}	-4	odd	change	4
a'_{22}	5	even	retain	5

The matrix of cofactors is

$$\mathbf{A}_c = \begin{pmatrix} -2 & -2 \\ 4 & 5 \end{pmatrix}$$

EXAMPLE 18

To find the matrix of cofactors for the (3×3) matrix in Example 16, let's begin with the element a_{11}. Crossing off row 1 and column 1, we are left with a (2×2) submatrix:

$$\mathbf{A} = \begin{pmatrix} 3 & 1 & 2 \\ 1 & 2 & 4 \\ 3 & -2 & 1 \end{pmatrix} \rightarrow \begin{pmatrix} 2 & 4 \\ -2 & 1 \end{pmatrix}$$
$$\text{Submatrix}$$

By formula, the cofactor is computed as

$$a'_{11} = (-1)^{1+1} \begin{vmatrix} 2 & 4 \\ -2 & 1 \end{vmatrix}$$

$$= (-1)^2[(2)(1) - (-2)(4)] = 1(10) = 10$$

Equivalently,

Minor: $\Delta = (2)(1) - (-2)(4) = 10$

Position: $(1, 1) \rightarrow$ even $\rightarrow$ retain sign

Cofactor: $a'_{11} = 10$

For element a_{12}, row 1 and column 2 are crossed off, resulting in

$$\begin{pmatrix} 3 & 1 & 2 \\ -1 & 2 & 4 \\ 3 & -2 & 1 \end{pmatrix} \rightarrow \begin{pmatrix} -1 & 4 \\ 3 & 1 \end{pmatrix}$$
Submatrix

By formula, the cofactor a'_{12} is computed as

$$a'_{12} = (-1)^{1+2} \begin{vmatrix} -1 & 4 \\ 3 & 1 \end{vmatrix}$$

$$= (-1)^3[(-1)(1) - (3)(4)] = -1(-13) = 13$$

Equivalently,

Minor: $\Delta = (-1)(1) - (3)(4) = -13$

Position: $(1, 2) \rightarrow$ odd $\rightarrow$ change sign

Cofactor: $a'_{12} = 13$

Now it is your turn! Verify that the matrix of cofactors is

$$\mathbf{A}_c = \begin{pmatrix} 10 & 13 & -4 \\ -5 & -3 & 9 \\ 0 & -14 & 7 \end{pmatrix}$$

❏

The Determinant and Cofactors   We began this section with the objective of finding a generalized approach for computing the determinant. The *method of cofactor expansion* allows you to compute the determinant of a matrix as follows.

METHOD OF COFACTOR EXPANSION

1   *Select* any *row or column of the matrix.*
2   *Multiply each element in the row (column) by its corresponding cofactor and sum these products to yield the determinant.*

For the $(m \times m)$ matrix **A**, the determinant can be found by expanding along any *row i* according to the equation

$$\Delta = a_{i1} a'_{i1} + a_{i2} a'_{i2} + \cdots + a_{im} a'_{im}$$ (9.5)

Similarly, the determinant can be found by expanding in any column *j* according to the equation

$$\Delta = a_{1j} a'_{1j} + a_{2j} a'_{2j} + \cdots + a_{mj} a'_{mj}$$ (9.6)

The value of the determinant is the same regardless of the row or column selected for cofactor expansion!

NOTE   If your objective is to find the determinant, it is not necessary to compute the entire matrix of cofactors! You need to determine only the cofactors for the row or column selected for expansion.

EXAMPLE 19   Matrix **A** and its matrix of cofactors $\mathbf{A}_c$ from Example 17 follow.

$$\mathbf{A} = \begin{pmatrix} 5 & -4 \\ 2 & -2 \end{pmatrix} \qquad \mathbf{A}_c = \begin{pmatrix} -2 & -2 \\ 4 & 5 \end{pmatrix}$$

The determinant of **A** can be found by expanding along row 1 as

$$\begin{aligned}\Delta &= a_{11} a'_{11} + a_{12} a'_{12} \\ &= (5)(-2) + (-4)(-2) = -2\end{aligned}$$

Equivalently, the determinant can be found by expanding down column 2, or

$$\begin{aligned}\Delta &= a_{12} a'_{12} + a_{22} a'_{22} \\ &= (-4)(-2) + (-2)(5) = -2\end{aligned}$$

EXAMPLE 20   The matrices **A** and $\mathbf{A}_c$ from Example 18 are

$$\mathbf{A} = \begin{pmatrix} 3 & 1 & 2 \\ -1 & 2 & 4 \\ 3 & -2 & 1 \end{pmatrix} \qquad \mathbf{A}_c = \begin{pmatrix} 10 & 13 & -4 \\ -5 & -3 & 9 \\ 0 & -14 & 7 \end{pmatrix}$$

The determinant of **A** is computed by expanding down column 3 as

$$\Delta = (2)(-4) + (4)(9) + (1)(7)$$
$$= -8 + 36 + 7 = 35$$

☐

PRACTICE EXERCISE
Verify that the value of the determinant is the same if you expand down the other two columns or across any of the rows.

EXAMPLE 21 (**Row or Column Selection: Exploit Structure!**) The selection of a row or column for expanding by cofactors shouldn't always be arbitrary. Often you can take advantage of the content or form of a matrix. For example, let's find the determinant of the (4 × 4) matrix

$$\mathbf{A} = \begin{pmatrix} 3 & 0 & 1 & 2 \\ 6 & -2 & -5 & 4 \\ -1 & 0 & 2 & 4 \\ 3 & 0 & -2 & 1 \end{pmatrix}$$

This might appear to be an overwhelming task. However, bear with me! If we select column 2 for expansion, we will have to find only one cofactor—that corresponding to a_{22}. That is, expanding down column 2

$$\Delta = (0)a'_{12} + (-2)a'_{22} + (0)a'_{32} + (0)(a'_{42})$$
$$= (-2)a'_{22}$$

Do you understand why it makes no difference what a'_{12}, a'_{32}, and a'_{42} equal? Regardless of their values, we will multiply them by zero.

To find a'_{22}, we cross off row 2 and column 2 and are left with the (3 × 3) submatrix

$$\begin{pmatrix} 3 & 1 & 2 \\ -1 & 2 & 4 \\ 3 & -2 & 1 \end{pmatrix}$$

Compare this matrix with the one in the last example. Since probably you have just about reached your limit, this problem has been "doctored." We have already computed the determinant of this submatrix as 35 in Example 20. Thus, since the "minor" equals 35 and a'_{22} is considered to be in an "even" position, the cofactor

$$a'_{22} = 35$$

and

$$\Delta = (-2)a'_{22}$$
$$= (-2)(35) = -70$$

☐

Properties of Determinants

Certain properties hold for matrices and their determinants. *Given the square matrix A:*

Property 1: If all elements of any row or column equal zero, $\Delta = 0$.

> **PRACTICE EXERCISE**
> Verify for the matrix $A = \begin{pmatrix} 5 & 0 \\ 10 & 0 \end{pmatrix}$ that $\Delta = 0$.

Property 2: If any two rows (or columns) are interchanged, the sign of the determinant changes.

> **PRACTICE EXERCISE**
> Given the matrix $A = \begin{pmatrix} 1 & 5 \\ -6 & 15 \end{pmatrix}$, interchange columns 1 and 2 to form the matrix B. Compute the determinants of A and B and compare.

Property 3: Given that the determinant of A equals Δ, if all elements in any row or column are multiplied by a constant k, the determinant of the resulting matrix equals $k\Delta$.

> **PRACTICE EXERCISE**
> Given the matrix $A = \begin{pmatrix} 3 & -6 \\ 5 & 12 \end{pmatrix}$, multiply each element in column 2 by -5 to form the matrix B. Compute Δ_A and Δ_B and compare.

Property 4: If any multiple of one row (column) is added to another row (column), the value of the determinant is unchanged.

> **PRACTICE EXERCISE**
> Given the matrix $A = \begin{pmatrix} 1 & 2 \\ 3 & 4 \end{pmatrix}$, multiply row 2 by -3 and add the result to row 1, forming a new matrix B. Compute Δ_A and Δ_B and compare.

Property 5: If any row (column) is a multiple of another row (column), the determinant equals zero.*

---

* A special case of this property occurs when two rows (columns) are equal to one another.

PRACTICE EXERCISE

In the matrix $\mathbf{A} = \begin{pmatrix} 3 & -2 & 4 \\ 6 & 8 & 1 \\ -9 & 6 & -12 \end{pmatrix}$, note that row 3 equals (-3) times row 1. Compute Δ.

These properties can be useful in computing the value of the determinant. For example, the magnitude of the numbers being manipulated can be reduced if all elements in a row or column have a common factor. This can be achieved also by adding (subtracting) multiples of one row (column) to another. Significant efficiencies can be introduced if, prior to using the method of cofactors, multiples of rows (columns) are combined to create a row (column) containing mostly zeros (as occurred in Example 21). To illustrate this, consider the following example.

EXAMPLE 22   Suppose that we wish to find the determinant of the matrix

$$\mathbf{A} = \begin{pmatrix} 2 & 6 & 1 \\ 3 & 5 & 1 \\ 6 & -5 & 3 \end{pmatrix}$$

If we wish to use the method of cofactors, expansion along any row or column requires the evaluation of three cofactors. However, using Property 4, we can multiply column 3 by -2 and add this multiple to column 1, resulting in the matrix

$$\mathbf{B} = \begin{pmatrix} 0 & 6 & 1 \\ 1 & 5 & 1 \\ 0 & -5 & 3 \end{pmatrix}$$

According to Property 4, the determinant of this matrix is the same as that for the original matrix $\mathbf{A}$. Expanding by cofactors down column 1, we only need to determine the cofactor for element (2, 1). Thus, the determinant of $\mathbf{A}$ equals the determinant of $\mathbf{B}$, or

$$\begin{aligned} \Delta &= (0)b'_{11} + (1)b'_{21} + (0)b'_{31} \\ &= (1)(-23) \\ &= -23 \end{aligned}$$

☐

PRACTICE EXERCISE
Find the determinant of $\mathbf{A}$ by using the original matrix and confirm that the answers are the same.

Cramer's Rule

Given a system of linear equations of the form

$$\mathbf{AX} = \mathbf{B}$$

where $\mathbf{A}$ is an $(n \times n)$ square matrix of coefficients, Cramer's rule provides a method of solving the system by using determinants. To solve for the value of the jth variable, form the matrix $\mathbf{A}_j$ by replacing the jth column of $\mathbf{A}$ with the column vector $\mathbf{B}$. If we denote the determinant of $\mathbf{A}_j$ by Δ_j, the value of the jth variable is determined as

$$x_j = \frac{\Delta_j}{\Delta} \tag{9.7}$$

If $\Delta \neq 0$, the given system of equations has a unique solution. If $\Delta = 0$, the computation of Eq. (9.7) is undefined. If $\Delta = 0$ and $\Delta_1 = \Delta_2 = \cdots = \Delta_n = 0$, the system has infinitely many solutions. If $\Delta = 0$ and any $\Delta_j \neq 0$, then the system has no solution.

EXAMPLE 23   The following system of equations

$$3x_1 + 2x_2 = 80$$

$$2x_1 + 4x_2 = 80$$

can be rewritten in the matrix form $\mathbf{AX} = \mathbf{B}$ as

$$\begin{pmatrix} 3 & 2 \\ 2 & 4 \end{pmatrix} \begin{pmatrix} x_1 \\ x_2 \end{pmatrix} = \begin{pmatrix} 80 \\ 80 \end{pmatrix}$$

Applying Eq. (9.7) gives

$$x_1 = \frac{\Delta_1}{\Delta} = \frac{\begin{vmatrix} 80 & 2 \\ 80 & 4 \end{vmatrix}}{\begin{vmatrix} 3 & 2 \\ 2 & 4 \end{vmatrix}} = \frac{(80)(4) - (80)(2)}{(3)(4) - (2)(2)} = \frac{160}{8} = 20$$

$$x_2 = \frac{\Delta_2}{\Delta} = \frac{\begin{vmatrix} 3 & 80 \\ 2 & 80 \end{vmatrix}}{\begin{vmatrix} 3 & 2 \\ 2 & 4 \end{vmatrix}} = \frac{(3)(80) - (2)(80)}{(3)(4) - (2)(2)} = \frac{80}{8} = 10$$

EXAMPLE 24   Given the system of equations

$$x_1 + 2x_2 \qquad\quad = \quad 5$$

$$x_1 \qquad - x_3 = -15$$

$$-x_1 + 3x_2 + 2x_3 = \quad 40$$

the value of x_1 can be determined by using Cramer's rule as

$$x_1 = \frac{\Delta_1}{\Delta} = \frac{\begin{vmatrix} 5 & 2 & 0 \\ -15 & 0 & -1 \\ 40 & 3 & 2 \end{vmatrix}}{\begin{vmatrix} 1 & 2 & 0 \\ 1 & 0 & -1 \\ -1 & 3 & 2 \end{vmatrix}}$$

Find the determinant of the two matrices to verify that

$$x_1 = \frac{-5}{1} = -5$$

☐

PRACTICE EXERCISE
Apply Cramer's rule to verify that $x_2 = 5$ and $x_3 = 10$ for this system of equations.

Section 9.4 Follow-up Exercises

Find the determinant of each of the following matrices.

1 $A = (-5)$

2 $A = (b)$

3 $T = \begin{pmatrix} 8 & 3 \\ -2 & -4 \end{pmatrix}$

4 $S = \begin{pmatrix} -7 & 12 \\ -4 & 8 \end{pmatrix}$

5 $N = (28)$

6 $T = (-a)$

7 $B = \begin{pmatrix} 1 & 0 \\ 0 & 1 \end{pmatrix}$

8 $A = \begin{pmatrix} a & a \\ a & a \end{pmatrix}$

9 $C = \begin{pmatrix} 2 & -6 & 10 \\ 4 & 0 & -2 \\ 3 & -2 & 8 \end{pmatrix}$

10 $B = \begin{pmatrix} 4 & -3 & 10 \\ -2 & 0 & -1 \\ 7 & 12 & 8 \end{pmatrix}$

11 $D = \begin{pmatrix} 1 & -2 & 8 \\ -2 & 10 & -5 \\ 4 & -8 & 12 \end{pmatrix}$

12 $A = \begin{pmatrix} 3 & 10 & 3 \\ 2 & -6 & -6 \\ 1 & -3 & 8 \end{pmatrix}$

13 $A = \begin{pmatrix} 2 & 0 & 8 \\ 4 & -1 & -2 \\ 0 & 5 & -4 \end{pmatrix}$

14 $B = \begin{pmatrix} 2 & 4 & 7 \\ -1 & 3 & 2 \\ 4 & -2 & 0 \end{pmatrix}$

15 $C = \begin{pmatrix} -1 & -2 & -3 \\ 3 & -4 & 6 \\ 0 & 0 & 8 \end{pmatrix}$

16 $D = \begin{pmatrix} 2 & 6 & -5 \\ 5 & 0 & 10 \\ 3 & 2 & -3 \end{pmatrix}$

Find the matrix of cofactors for each of the following matrices.

17 $\begin{pmatrix} 8 & -2 \\ 10 & -4 \end{pmatrix}$

18 $\begin{pmatrix} -1 & -2 \\ -4 & 10 \end{pmatrix}$

19 $\begin{pmatrix} 1 & 0 \\ 0 & 1 \end{pmatrix}$

20 $\begin{pmatrix} a & b \\ c & d \end{pmatrix}$

21 $\begin{pmatrix} 2 & -4 & -2 \\ -2 & 0 & 4 \\ 4 & 3 & -3 \end{pmatrix}$

22 $\begin{pmatrix} 1 & 0 & 0 \\ 0 & 1 & 0 \\ 0 & 0 & 1 \end{pmatrix}$

23 $\begin{pmatrix} 1 & 0 & -1 \\ 0 & -1 & 0 \\ -1 & 0 & 1 \end{pmatrix}$

24 $\begin{pmatrix} 2 & 10 & -4 \\ 0 & 8 & 10 \\ 4 & 2 & -2 \end{pmatrix}$

25 $\begin{pmatrix} 4 & -8 \\ 2 & -6 \end{pmatrix}$

26 $\begin{pmatrix} 5 & -2 \\ -8 & 4 \end{pmatrix}$

27 $\begin{pmatrix} 10 & 8 \\ -5 & 3 \end{pmatrix}$

28 $\begin{pmatrix} -5 & 10 \\ -4 & -8 \end{pmatrix}$

29 $\begin{pmatrix} 4 & 10 & -2 \\ 8 & -3 & -5 \\ -6 & -3 & 3 \end{pmatrix}$

30 $\begin{pmatrix} 3 & -2 & 1 \\ 0 & 6 & 4 \\ 4 & 2 & 1 \end{pmatrix}$

31 $\begin{pmatrix} 5 & 0 & 20 \\ -5 & 8 & 0 \\ 10 & -5 & -10 \end{pmatrix}$

32 $\begin{pmatrix} 10 & -4 & 2 \\ 0 & 7 & -3 \\ -6 & -2 & -3 \end{pmatrix}$

33 $\begin{pmatrix} 10 & 3 & -2 \\ 3 & -2 & 7 \\ 1 & 2 & -3 \end{pmatrix}$

34 $\begin{pmatrix} -2 & 5 & 3 \\ 3 & -1 & 4 \\ -4 & 0 & 4 \end{pmatrix}$

35 to 52 Using the matrix of cofactors found, respectively, in Exercises 17–34, find the determinant of the original matrix.

53 Find the determinant of

$$A = \begin{pmatrix} 2 & 7 & -2 & 0 \\ 1 & -2 & -3 & 0 \\ 3 & 3 & 6 & 9 \\ 6 & -3 & -2 & 0 \end{pmatrix}$$

54 Find the determinant of

$$B = \begin{pmatrix} 7 & -4 & 2 & -3 \\ 0 & 0 & -3 & 0 \\ 4 & -1 & 2 & 4 \\ 6 & 2 & 8 & 4 \end{pmatrix}$$

55 Find the determinant of

$$A = \begin{pmatrix} 4 & 2 & 0 & -1 & -5 \\ 3 & 4 & -3 & 2 & -2 \\ -1 & 2 & 0 & 3 & 4 \\ 0 & 0 & 0 & -4 & 0 \\ 2 & 3 & 0 & 5 & 7 \end{pmatrix}$$

56 Find the determinant of

$$A = \begin{pmatrix} 0 & 0 & 8 & 0 & 0 \\ 3 & 0 & 2 & 0 & 9 \\ 2 & 6 & -2 & 0 & -6 \\ 3 & -3 & 4 & 0 & 3 \\ 2 & 3 & 12 & 1 & -4 \end{pmatrix}$$

In the following exercises, solve the system of equations by using Cramer's rule.

57 $3x_1 - 2x_2 = -13$
$4x_1 + 6x_2 = 0$

58 $5x_1 - 4x_2 = -8$
$3x_1 + 5x_2 = 47$

59 $x_1 - 5x_2 = -85$
$2x_1 + 4x_2 = 40$

60 $4x_1 - 8x_2 = -4$
$5x_1 + 3x_2 = 34$

61 $-x_1 + 2x_2 = -4$
$4x_1 - 8x_2 = 18$

62 $3x_1 - 2x_2 = 5$
$-9x_1 + 6x_2 = 15$

63 $x_1 + 3x_2 - 2x_3 = 17$
$2x_1 - 4x_2 + x_3 = -16$
$5x_1 + 2x_2 - 4x_3 = 21$

64 $x_1 + x_2 + x_3 = -1$
$3x_1 - 4x_2 + 2x_3 = -5$
$3x_1 - 2x_2 + x_3 = 2$

65 $3x_1 + 5x_3 = 14$
$x_1 + 4x_2 - 2x_3 = -10$
$x_1 + x_2 + x_3 = 2$

66 $3x_1 + 2x_2 + x_3 = 11$
$5x_1 - 3x_3 = 19$
$4x_2 + 6x_3 = 0$

67 $-x_1 + 2x_2 - 3x_3 = 24$
$2x_1 - 4x_2 + x_3 = 12$
$3x_1 - 6x_2 + 9x_3 = 16$

68 $x_1 + 4x_2 - 3x_3 = 16$
$8x_1 - 12x_2 + 4x_3 = -10$
$-2x_1 + 3x_2 - x_3 = 14$

9.5 THE INVERSE OF A MATRIX

For *some* matrices there can be identified another matrix called the ***multiplicative inverse matrix,*** or more simply, the ***inverse.*** The relationship between a matrix **A** and its inverse (denoted by A^{-1}) is that the product of **A** and A^{-1}, in either order, results in the identity matrix, or

$$AA^{-1} = A^{-1}A = I \qquad (9.8)$$

The inverse is similar to the *reciprocal* in the algebra of real numbers. Multiplying a quantity b by its reciprocal $1/b$ results in a product of 1. In matrix algebra multiplying a matrix by its inverse results in the identity matrix.

> **IMPORTANT OBSERVATIONS REGARDING THE INVERSE**
>
> I *For a matrix A to have an inverse, it must be square.*
> II *The inverse of A will also be square and of the same dimension as A.*
> III *Not every square matrix has an inverse.*

A square matrix will have an inverse provided that all rows or columns are ***linearly independent;*** that is, no row (or column) is a linear combination (multiple) of the remaining rows (or columns). If any rows (or columns) are linearly dependent [are linear combinations (multiples) of other rows (columns)], the matrix will not have an inverse. If a matrix has an inverse, it is said to be a ***nonsingular matrix.*** If a matrix does not have an inverse, it is said to be a ***singular matrix.***

EXAMPLE 25   We can verify that matrix **B**, which follows, is the inverse of matrix **A** by finding the products **AB** and **BA**.

$$\mathbf{A} = \begin{pmatrix} 3 & 7 \\ 2 & 5 \end{pmatrix} \qquad \mathbf{B} = \begin{pmatrix} 5 & -7 \\ -2 & 3 \end{pmatrix}$$

$$\mathbf{AB} = \begin{pmatrix} 3 & 7 \\ 2 & 5 \end{pmatrix}\begin{pmatrix} 5 & -7 \\ -2 & 3 \end{pmatrix} = \begin{pmatrix} 1 & 0 \\ 0 & 1 \end{pmatrix}$$

$$\mathbf{BA} = \begin{pmatrix} 5 & -7 \\ -2 & 3 \end{pmatrix}\begin{pmatrix} 3 & 7 \\ 2 & 5 \end{pmatrix} = \begin{pmatrix} 1 & 0 \\ 0 & 1 \end{pmatrix}$$

Because both products result in a (2 × 2) identity matrix, we can state that *matrix **B** is the inverse of **A**,* or

$$\mathbf{B} = \mathbf{A}^{-1}$$

Similarly, we can make the equivalent statement that ***A** is the inverse of **B**,* or

$$\mathbf{A} = \mathbf{B}^{-1}$$

❑

Determining the Inverse

There are several methods for determining the inverse of a matrix. One method is based on the gaussian elimination procedure discussed in Sec. 3.3. Let's develop the general procedure by using an example. If you are fuzzy on the gaussian procedure, a rereading of Sec. 3.3 is advised.

EXAMPLE 26   Let's return to the matrix **A** in the last example. If there is another matrix **B** which is the inverse of **A**, it will have dimension (2 × 2). Let's label the elements of **B** as follows.

$$\mathbf{B} = \begin{pmatrix} b_{11} & b_{12} \\ b_{21} & b_{22} \end{pmatrix}$$

If $\mathbf{A}^{-1} = \mathbf{B}$,

$$\mathbf{AB} = \mathbf{I} \qquad \text{or} \qquad \begin{pmatrix} 3 & 7 \\ 2 & 5 \end{pmatrix}\begin{pmatrix} b_{11} & b_{12} \\ b_{21} & b_{22} \end{pmatrix} = \begin{pmatrix} 1 & 0 \\ 0 & 1 \end{pmatrix}$$

If we multiply on the left side of the equation, the result is

$$\begin{pmatrix} 3b_{11} + 7b_{21} & 3b_{12} + 7b_{22} \\ 2b_{11} + 5b_{21} & 2b_{12} + 5b_{22} \end{pmatrix} = \begin{pmatrix} 1 & 0 \\ 0 & 1 \end{pmatrix}$$

For these two matrices to be equal, their respective elements must equal one another; that is,

$$3b_{11} + 7b_{21} = 1 \qquad\qquad\qquad\qquad\textbf{(9.9)}$$

$$2b_{11} + 5b_{21} = 0 \qquad\qquad\qquad\qquad\textbf{(9.10)}$$

$$3b_{12} + 7b_{22} = 0 \qquad\qquad\qquad\qquad\textbf{(9.11)}$$

$$2b_{12} + 5b_{22} = 1 \qquad\qquad\qquad\qquad\textbf{(9.12)}$$

To determine the values of b_{11} and b_{21}, we need to solve Eqs. (9.9) and (9.10) simultaneously. To determine b_{12} and b_{22}, we must solve Eqs. (9.11) and (9.12).

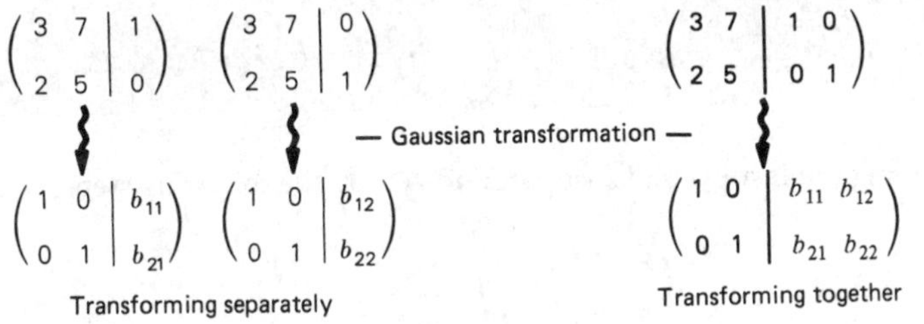

$$\begin{pmatrix} 3 & 7 & | & 1 \\ 2 & 5 & | & 0 \end{pmatrix} \quad \begin{pmatrix} 3 & 7 & | & 0 \\ 2 & 5 & | & 1 \end{pmatrix} \qquad\qquad \begin{pmatrix} 3 & 7 & | & 1 & 0 \\ 2 & 5 & | & 0 & 1 \end{pmatrix}$$

— Gaussian transformation —

$$\begin{pmatrix} 1 & 0 & | & b_{11} \\ 0 & 1 & | & b_{21} \end{pmatrix} \quad \begin{pmatrix} 1 & 0 & | & b_{12} \\ 0 & 1 & | & b_{22} \end{pmatrix} \qquad\qquad \begin{pmatrix} 1 & 0 & | & b_{11}\; b_{12} \\ 0 & 1 & | & b_{21}\; b_{22} \end{pmatrix}$$

Transforming separately   Transforming together

Figure 9.2 Gaussian transformation.

If we were to solve these two systems individually by the gaussian elimination method, the transformations would proceed as shown in Fig. 9.2. For each system we would perform row operations to transform the array of coefficients $\begin{pmatrix} 3 & 7 \\ 2 & 5 \end{pmatrix}$ into a (2×2) identity matrix. Since *both* systems have the same matrix of coefficients on the left side, *the same row operations will be used to solve both systems.* The process can be streamlined by augmenting the right-side constants for the first system with those for the second system of equations, as shown in the following.

$$\begin{pmatrix} 3 & 7 & | & 1 & 0 \\ 2 & 5 & | & 0 & 1 \end{pmatrix}$$

By doing this, the row operations need to be performed one time only.

Upon transforming the matrix of coefficients on the left side into a (2×2) identity matrix, the first column of values on the right side would contain the solution for the first system of equations (b_{11} and b_{21}) and the second column the solution for the second system of equations (b_{12} and b_{22}). The transformed matrices would appear as

$$\begin{pmatrix} 1 & 0 & | & b_{11} & b_{12} \\ 0 & 1 & | & b_{21} & b_{22} \end{pmatrix}$$

and the (2×2) matrix to the right of the vertical line is the matrix **B**, or the inverse of **A**.

GAUSSIAN REDUCTION PROCEDURE
To determine the inverse of an $(m \times m)$ matrix **A**:

I   *Augment the matrix A with an $(m \times m)$ identity matrix, resulting in*

$$(A|I)$$

II   *Perform row operations on the entire augmented matrix so as to transform A into an $(m \times m)$ identity matrix. The resulting matrix will have the form*

$$(I|A^{-1})$$

where A^{-1} can be read to the right of the vertical line.

EXAMPLE 27   Continuing the last example, we can find $\mathbf{A}^{-1}$ by the following steps:

$$\begin{pmatrix} 3 & 7 & 1 & 0 \\ 2 & 5 & 0 & 1 \end{pmatrix}$$

$$\begin{pmatrix} ① & \frac{7}{3} & \frac{1}{3} & 0 \\ 2 & 5 & 0 & 1 \end{pmatrix} \qquad \text{(multiply row 1 by } \tfrac{1}{3})$$

$$\begin{pmatrix} 1 & \frac{7}{3} & \frac{1}{3} & 0 \\ ⓪ & \frac{1}{3} & -\frac{2}{3} & 1 \end{pmatrix} \qquad \begin{array}{l}\text{(multiply row 1 by } -2 \\ \text{and add to row 2)}\end{array}$$

$$\begin{pmatrix} 1 & \frac{7}{3} & \frac{1}{3} & 0 \\ 0 & ① & -2 & 3 \end{pmatrix} \qquad \text{(multiply row 2 by 3)}$$

$$\begin{pmatrix} 1 & ⓪ & 5 & -7 \\ 0 & 1 & -2 & 3 \end{pmatrix} \qquad \begin{array}{l}\text{(multiply row 2 by } -\tfrac{7}{3} \\ \text{and add to row 1)}\end{array}$$

The inverse of **A** is

$$\mathbf{A}^{-1} = \begin{pmatrix} 5 & -7 \\ -2 & 3 \end{pmatrix}$$

as was indicated in Example 25.

NOTE   If the matrix is singular (does not have an inverse), it will not be possible to transform **A** into an identity matrix.

EXAMPLE 28   Consider the matrix

$$\mathbf{B} = \begin{pmatrix} 2 & -4 & 6 \\ 6 & 1 & 5 \\ 1 & -2 & 3 \end{pmatrix}$$

Note the linear dependence between rows 1 and 3. Row 1 is a multiple (2) of row 3. From our previous discussion we can anticipate that **B** will not have an inverse. Let's test this hypothesis by applying the gaussian procedure.

$$\left(\begin{array}{rrr|rrr} 2 & -4 & 6 & 1 & 0 & 0 \\ 6 & 1 & 5 & 0 & 1 & 0 \\ 1 & -2 & 3 & 0 & 0 & 1 \end{array}\right)$$

$$\left(\begin{array}{rrr|rrr} ① & -2 & 3 & \frac{1}{2} & 0 & 0 \\ 6 & 1 & 5 & 0 & 1 & 0 \\ 1 & -2 & 3 & 0 & 0 & 1 \end{array}\right) \quad \text{(multiply row 1 by } \tfrac{1}{2})$$

$$\left(\begin{array}{rrr|rrr} 1 & -2 & 3 & \frac{1}{2} & 0 & 0 \\ ⓪ & 13 & -13 & -3 & 1 & 0 \\ 1 & -2 & 3 & 0 & 0 & 1 \end{array}\right) \quad \begin{array}{l}\text{(multiply row 1 by } -6) \\ \text{and add to row 2)}\end{array}$$

$$\left(\begin{array}{rrr|rrr} 1 & -2 & 3 & \frac{1}{2} & 0 & 0 \\ 0 & 13 & -13 & -3 & 1 & 0 \\ ⓪ & 0 & 0 & -\frac{1}{2} & 0 & 1 \end{array}\right) \quad \begin{array}{l}\text{(multiply row 1 by } -1) \\ \text{and add to row 3)}\end{array}$$

$$\left(\begin{array}{rrr|rrr} 1 & -2 & 3 & \frac{1}{2} & 0 & 0 \\ 0 & ① & -1 & -\frac{3}{13} & \frac{1}{13} & 0 \\ 0 & 0 & 0 & -\frac{1}{2} & 0 & 1 \end{array}\right) \quad \text{(multiply row 2 by } \tfrac{1}{13})$$

$$\left(\begin{array}{rrr|rrr} 1 & ⓪ & 1 & \frac{1}{26} & \frac{2}{13} & 0 \\ 0 & 1 & -1 & -\frac{3}{13} & \frac{1}{13} & 0 \\ 0 & 0 & 0 & -\frac{1}{2} & 0 & 1 \end{array}\right) \quad \begin{array}{l}\text{(multiply row 2 by 2 and} \\ \text{add to row 1)}\end{array}$$

At this point it becomes impossible to generate a 1 in the third column of row 3. A 1 could be placed in this position by adding a multiple of row 1 or 2 to row 3. However, this would result in nonzero values in columns 1 or 2 of row 3. Try it if you need convincing. Our conclusion is that **B** has no inverse.

❑

Finding the Inverse Using Cofactors (Optional)

Another method for determining the inverse of a matrix utilizes the matrix of cofactors.

> **THE COFACTOR METHOD**
> The cofactor procedure for finding the inverse of a square matrix **A** is as follows.
>
>   I   *Determine the matrix of cofactors A_c for the matrix A.*
>   II *Determine the **adjoint matrix** A_j which is the transpose of A_c:*
>
> $$A_j = A_c^T$$
>
>   III *The inverse of A is found by multiplying the adjoint matrix by the reciprocal of the determinant of A, or*
>
> $$A^{-1} = \frac{1}{\Delta} A_j \qquad\qquad (9.13)$$

Notice from Eq. (9.13) that when the determinant of **A**, Δ, equals zero, the computation of the inverse is not defined. Thus, if $\Delta = 0$, **the matrix has no inverse.**

EXAMPLE 29   Let's determine the inverse of the matrix

$$\mathbf{A} = \begin{pmatrix} 4 & 3 \\ -2 & -1 \end{pmatrix}$$

The cofactor matrix $\mathbf{A}_c$ is

$$\mathbf{A}_c = \begin{pmatrix} -1 & 2 \\ -3 & 4 \end{pmatrix}$$

The corresponding adjoint matrix is

$$\mathbf{A}_j = \begin{pmatrix} -1 & -3 \\ 2 & 4 \end{pmatrix}$$

The determinant of **A** is

$$\Delta = (4)(-1) - (-2)(3)$$
$$= 2$$

Therefore,

$$\mathbf{A}^{-1} = \frac{1}{2} \begin{pmatrix} -1 & -3 \\ 2 & 4 \end{pmatrix}$$
$$= \begin{pmatrix} -\frac{1}{2} & -\frac{3}{2} \\ 1 & 2 \end{pmatrix}$$

EXAMPLE 30 To determine the inverse of the (3×3) matrix

$$\mathbf{B} = \begin{pmatrix} 1 & 2 & 0 \\ 1 & 0 & -1 \\ -1 & 3 & 2 \end{pmatrix}$$

the matrix of cofactors $\mathbf{B}_c$ is

$$\mathbf{B}_c = \begin{pmatrix} 3 & -1 & 3 \\ -4 & 2 & -5 \\ -2 & 1 & -2 \end{pmatrix}$$

The adjoint matrix $\mathbf{B}_j$ is the transpose of $\mathbf{B}_c$, or

$$\mathbf{B}_j = \begin{pmatrix} 3 & -4 & -2 \\ -1 & 2 & 1 \\ 3 & -5 & -2 \end{pmatrix}$$

Verify that the determinant of $\mathbf{B}$ equals 1.
  Therefore,

$$\mathbf{B}^{-1} = \frac{1}{1} \begin{pmatrix} 3 & -4 & -2 \\ -1 & 2 & 1 \\ 3 & -5 & -2 \end{pmatrix}$$

$$= \begin{pmatrix} 3 & -4 & -2 \\ -1 & 2 & 1 \\ 3 & -5 & -2 \end{pmatrix}$$

❑

NOTE **(A Potential Timesaver!)**
In step I of the cofactor method, pause after identifying one row or column of cofactors and compute Δ. If $\Delta = 0$, you're finished! The inverse does not exist. If $\Delta \neq 0$, proceed to find the remaining cofactors.

The Inverse and Systems of Equations

In Sec. 9.3 we discussed the matrix representation of systems of equations. The matrix inverse can be used to determine the solution set for a system of equations. Given a system of equations of the form $\mathbf{AX} = \mathbf{B}$, where $\mathbf{A}$ is a *square* matrix of coefficients, both sides of the matrix equation may be multiplied by $\mathbf{A}^{-1}$, yielding

$$\mathbf{A}^{-1}\mathbf{AX} = \mathbf{A}^{-1}\mathbf{B} \tag{9.14}$$

Because $\mathbf{A}^{-1}\mathbf{A} = \mathbf{I}$, Eq. (9.14) can be written as

$$\mathbf{IX} = \mathbf{A}^{-1}\mathbf{B}$$

or

$$\boxed{\mathbf{X} = \mathbf{A}^{-1}\mathbf{B}} \tag{9.15}$$

*That is, the solution vector for the system of equations can be found by multiplying the inverse of the matrix of coefficients $\mathbf{A}$ by the vector of right-side constants $\mathbf{B}$. If $\mathbf{A}^{-1}$ does not exist, the equations (more specifically, the matrix of coefficients) are linearly dependent and there is **either no solution or infinitely many solutions.***

EXAMPLE 31   Consider the system of equations

$$4x_1 + 3x_2 = 4$$
$$-2x_1 - x_2 = 0$$

This system of equations may be written in matrix form as

$$\mathbf{AX} = \mathbf{B}$$

or

$$\begin{pmatrix} 4 & 3 \\ -2 & -1 \end{pmatrix}\begin{pmatrix} x_1 \\ x_2 \end{pmatrix} = \begin{pmatrix} 4 \\ 0 \end{pmatrix}$$

To solve this system of equations by the inverse method, we must determine $\mathbf{A}^{-1}$. Conveniently, $\mathbf{A}$ is the matrix examined in Example 29 and $\mathbf{A}^{-1}$ has been computed. Therefore, the solution vector $\mathbf{X}$ is calculated as

$$\mathbf{X} = \mathbf{A}^{-1}\mathbf{B}$$
$$= \begin{pmatrix} -\frac{1}{2} & -\frac{3}{2} \\ 1 & 2 \end{pmatrix}\begin{pmatrix} 4 \\ 0 \end{pmatrix}$$
$$= \begin{pmatrix} -2 \\ 4 \end{pmatrix}$$

The solution to the system of equations is $x_1 = -2$ and $x_2 = 4$.

EXAMPLE 32   Consider the system of equations

$$x_1 + 2x_2 = 5$$
$$x_1 - x_3 = -15$$
$$-x_1 + 3x_2 + 2x_3 = 40$$

The matrix of coefficients is

$$\mathbf{A} = \begin{pmatrix} 1 & 2 & 0 \\ 1 & 0 & -1 \\ -1 & 3 & 2 \end{pmatrix}$$

and again the example has been contrived (see Example 30). The solution vector is calculated as

$$X = \begin{pmatrix} 3 & -4 & -2 \\ -1 & 2 & 1 \\ 3 & -5 & -2 \end{pmatrix} \begin{pmatrix} 5 \\ -15 \\ 40 \end{pmatrix}$$

$$= \begin{pmatrix} -5 \\ 5 \\ 10 \end{pmatrix}$$

or $x_1 = -5$, $x_2 = 5$, and $x_3 = 10$.

❑

The inverse procedure for solving systems of equations, when compared with the methods discussed in Chap. 3 (elimination-substitution method and gaussian elimination method), is less discriminating. The Chap. 3 procedures give clear and distinct signals for each type of solution set (unique, infinite, no solution). The inverse procedure does not distinguish between no solution set and an infinite solution set. If the determinant of **A** does not equal zero, a unique solution exists. If $\Delta = 0$, we can only state that there exists *either* no solution or an infinite number of solutions.

Section 9.5 Follow-up Exercises

Determine the inverse, if it exists, for the following matrices, using the gaussian procedure.

1 $\begin{pmatrix} 1 & -1 \\ 2 & -3 \end{pmatrix}$
2 $\begin{pmatrix} 2 & 3 \\ 4 & 7 \end{pmatrix}$

3 $\begin{pmatrix} 4 & 2 \\ -2 & -1 \end{pmatrix}$
4 $\begin{pmatrix} 40 & 8 \\ 30 & 6 \end{pmatrix}$

5 $\begin{pmatrix} 1 & -1 \\ 1 & 1 \end{pmatrix}$
6 $\begin{pmatrix} -1 & 3 \\ 2 & -4 \end{pmatrix}$

7 $\begin{pmatrix} 0 & 3 & 1 \\ 1 & 1 & 0 \\ 2 & 3 & 3 \end{pmatrix}$
8 $\begin{pmatrix} 1 & 0 & -1 \\ -1 & 1 & -1 \\ -1 & 0 & 2 \end{pmatrix}$

9 $\begin{pmatrix} 4 & 3 \\ 5 & 4 \end{pmatrix}$
10 $\begin{pmatrix} -2 & 3 \\ 6 & -9 \end{pmatrix}$

11 $\begin{pmatrix} 1 & 1 & 1 \\ 2 & -1 & 1 \\ 2 & 3 & 4 \end{pmatrix}$
12 $\begin{pmatrix} 10 & -2 & 6 \\ 1 & -5 & 3 \\ -5 & 1 & -3 \end{pmatrix}$

Determine the inverse of the following matrices by using the matrix-of-cofactors approach.

13 $\begin{pmatrix} 3 & 7 \\ 2 & 5 \end{pmatrix}$
14 $\begin{pmatrix} 3 & -15 \\ 5 & 25 \end{pmatrix}$

15 $\begin{pmatrix} 3 & 5 & 2 \\ 4 & 1 & 0 \\ -9 & -15 & -6 \end{pmatrix}$ **16** $\begin{pmatrix} -5 & 6 & -7 \\ 10 & -11 & 13 \\ -1 & 1 & -1 \end{pmatrix}$

17 $\begin{pmatrix} 3 & -5 \\ 4 & 2 \end{pmatrix}$ **18** $\begin{pmatrix} 5 & 2 \\ 3 & 1 \end{pmatrix}$

19 $\begin{pmatrix} 1 & -1 \\ -4 & 4 \end{pmatrix}$ **20** $\begin{pmatrix} -3 & 1 \\ 15 & -5 \end{pmatrix}$

21 $\begin{pmatrix} 1 & 1 & 1 \\ 3 & 0 & -4 \\ 1 & 2 & 5 \end{pmatrix}$ **22** $\begin{pmatrix} 1 & 1 & 1 \\ -2 & 3 & -1 \\ 5 & -4 & 2 \end{pmatrix}$

Using the results of Exercises 1 to 22, determine the solution to the systems of equations in Exercises 23–44, respectively (if one exists).

23 $\begin{aligned} x_1 - x_2 &= -1 \\ 2x_1 - 3x_2 &= -5 \end{aligned}$

24 $\begin{aligned} 2x_1 + 3x_2 &= 1 \\ 4x_1 + 7x_2 &= 3 \end{aligned}$

25 $\begin{aligned} 4x_1 + 2x_2 &= 24 \\ -2x_1 - x_2 &= 10 \end{aligned}$

26 $\begin{aligned} 40x_1 + 8x_2 &= 80 \\ 30x_1 + 6x_2 &= 60 \end{aligned}$

27 $\begin{aligned} x_1 - x_2 &= -11 \\ x_1 + x_2 &= 59 \end{aligned}$

28 $\begin{aligned} -x_1 + 3x_2 &= 5 \\ 2x_1 - 4x_2 &= 0 \end{aligned}$

29 $\begin{aligned} 3x_2 + x_3 &= 1 \\ x_1 + x_2 &= 2 \\ 2x_1 + 3x_2 + 3x_3 &= 7 \end{aligned}$

30 $\begin{aligned} x_1 - x_3 &= -10 \\ -x_1 + x_2 - x_3 &= -40 \\ -x_1 + 2x_3 &= 40 \end{aligned}$

31 $\begin{aligned} 4x_1 + 3x_2 &= 17 \\ 5x_1 + 4x_2 &= 22 \end{aligned}$

32 $\begin{aligned} -2x_1 + 3x_2 &= 10 \\ 6x_1 - 9x_2 &= 20 \end{aligned}$

33 $\begin{aligned} x_1 + x_2 + x_3 &= 2 \\ 2x_1 - x_2 + x_3 &= 9 \\ 2x_1 + 3x_2 + 4x_3 &= 4 \end{aligned}$

34 $\begin{aligned} 10x_1 - 2x_2 + 6x_3 &= 10 \\ x_1 - 5x_2 + 3x_3 &= 20 \\ -5x_1 + x_2 - 3x_3 &= 15 \end{aligned}$

35 $\begin{aligned} 3x_1 + 7x_2 &= -3 \\ 2x_1 + 5x_2 &= -3 \end{aligned}$

36 $\begin{aligned} 3x_1 - 15x_2 &= 25 \\ 5x_1 + 25x_2 &= 40 \end{aligned}$

37 $\begin{aligned} 3x_1 + 5x_2 + 2x_3 &= 20 \\ 4x_1 + x_2 &= 40 \\ -9x_1 - 15x_2 - 6x_3 &= 30 \end{aligned}$

38 $\begin{aligned} -5x_1 + 6x_2 - 7x_3 &= 25 \\ 10x_1 - 11x_2 + 13x_3 &= -45 \\ -x_1 + x_2 - x_3 &= 4 \end{aligned}$

39 $\begin{aligned} 3x_1 - 5x_2 &= 22 \\ 4x_1 + 2x_2 &= 12 \end{aligned}$

40 $\begin{aligned} 5x_1 + 2x_2 &= -14 \\ 3x_1 + x_2 &= -9 \end{aligned}$

41 $\begin{aligned} x_1 - x_2 &= 10 \\ -4x_1 + 4x_2 &= 12 \end{aligned}$

42 $\begin{aligned} -3x_1 + x_2 &= 20 \\ 15x_1 - 5x_2 &= 12 \end{aligned}$

43 $\begin{aligned} x_1 + x_2 + x_3 &= -2 \\ 3x_1 - 4x_3 &= 22 \\ x_1 + 2x_2 + 5x_3 &= -18 \end{aligned}$

44 $\begin{aligned} x_1 + x_2 + x_3 &= 1 \\ -2x_1 + 3x_2 - x_3 &= -6 \\ 5x_1 - 4x_2 + 2x_3 &= 11 \end{aligned}$

45 The solution to a system of equations having the form $\mathbf{AX} = \mathbf{B}$ can be found by the matrix multiplication

$$\mathbf{X} = \begin{pmatrix} 2 & -3 \\ -1 & 2 \end{pmatrix} \begin{pmatrix} 17 \\ 10 \end{pmatrix}$$

What was the original system of equations? What is the solution set?

46 The solution to a system of equations having the form $\mathbf{AX} = \mathbf{B}$ can be found by the matrix multiplication

$$\mathbf{X} = \begin{pmatrix} -0.5 & -1.5 \\ 1 & 2 \end{pmatrix} \begin{pmatrix} 8 \\ -6 \end{pmatrix}$$

What was the original system of equations? What is the solution set?

47 The solution to a system of equations having the form $\mathbf{AX} = \mathbf{B}$ can be found by the matrix multiplication

$$\mathbf{X} = \begin{pmatrix} 0 & -1 & 1 \\ -1 & 1 & 2 \\ 1 & 0 & -2 \end{pmatrix} \begin{pmatrix} 3 \\ 2 \\ 4 \end{pmatrix}$$

What was the original system of equations? What is the solution set?

48 The solution to a system of equations having the form $\mathbf{AX} = \mathbf{B}$ can be found by the matrix multiplication

$$\mathbf{X} = \begin{pmatrix} 3 & -4 & -2 \\ -1 & 2 & 1 \\ 3 & -5 & -2 \end{pmatrix} \begin{pmatrix} -2 \\ 5 \\ -18 \end{pmatrix}$$

What was the original system of equations? What is the solution set?

9.6 SELECTED APPLICATIONS

This section provides some illustrations of applications of matrix algebra. Unlike many other applications of mathematics, there are no set formulas or approaches to solving all matrix applications. Each application is somewhat unique. You will find that a certain level of trial and error may be required in working through the logic underlying an application. Although many approaches can be used, the author recommends that you consider the following suggestions when wrestling with an application.

Suggestions for Solving Matrix Applications

1 *Determine the desired output information that you wish to generate using matrix computations.*

2 *Examine the matrix data which are available to you and assess whether they contain the component information necessary to generate the desired outputs. You may have to extract relevant data from the matrices you are given and place them in newly defined matrices.*

3 *Perhaps, do some computations of the output information you desire* in a nonmatrix manner. *This may help you to understand the logic underlying the computations.*

4 *If you do not see, immediately, how the matrices can be combined to yield the desired information, test different matrix combinations for* computational and output compatibility *(e.g., if you believe that component matrices need to be multiplied, try to identify different matrix products which are defined. Examine rearrangements of the matrices, such as the matrix transpose. For those products which are defined, examine the dimension of the product* *matrix. Does the product matrix contain the number of pieces of output information that you desire? If so, next examine the actual computation to see if the matrix operation processes the data in the logical manner needed to yield the desired output information.)*

EXAMPLE 33   **(Election Projection)** A political pollster is watching a closely contested mayoral race in a particular city. Recent surveys indicate voters' preferences in the city's six voting districts. The matrix **P** displays these preferences.

$$
\begin{array}{c}
\text{District} \\
\begin{array}{cccccc}
1 & 2 & 3 & 4 & 5 & 6
\end{array} \\
\mathbf{P} = \begin{pmatrix}
0.40 & 0.35 & 0.30 & 0.50 & 0.30 & 0.36 \\
0.42 & 0.40 & 0.25 & 0.30 & 0.30 & 0.32 \\
0.18 & 0.25 & 0.45 & 0.20 & 0.40 & 0.32
\end{pmatrix}
\begin{array}{l}
\text{Democrat} \\
\text{Republican} \\
\text{Independent}
\end{array}
\end{array}
$$

Each column indicates the percentages of voters in each district who are expected to vote for the different mayoral candidates. For example, column 3 indicates that in district 3, 30 percent of voters are expected to vote for the Democratic candidate, 25 percent for the Republican candidate, and 45 percent for the Independent candidate.

Given these voter preferences, it is possible to project the election outcome if the number of citizens expecting to vote in each district is known. The vector **V** contains current estimates of these numbers.

$$
\mathbf{V} = \begin{pmatrix}
30,000 \\
60,000 \\
70,000 \\
45,000 \\
55,000 \\
40,000
\end{pmatrix}
$$

The election outcome can be projected by the matrix multiplication **PV**, or

$$
\begin{pmatrix}
0.40 & 0.35 & 0.30 & 0.50 & 0.30 & 0.36 \\
0.42 & 0.40 & 0.25 & 0.30 & 0.30 & 0.32 \\
0.18 & 0.25 & 0.45 & 0.20 & 0.40 & 0.32
\end{pmatrix}
\begin{pmatrix}
30,000 \\
60,000 \\
70,000 \\
45,000 \\
55,000 \\
40,000
\end{pmatrix}
=
\begin{pmatrix}
107,400 \\
96,900 \\
95,700
\end{pmatrix}
$$

These results suggest that the Democratic candidate has an edge of more than 10,000 votes over the other two candidates.

☐

PRACTICE EXERCISE

The pollster believes that the relative voting preferences within each district will not change significantly by the time of the election. Thus, changes in the projected outcome will be influenced primarily by voter turnout in each district. The campaign manager for the independent candidate believes that voter turnout in districts 3 and 5 can be increased significantly with an intensified campaign which emphasizes the importance of voting. Since districts 3 and 5 have a decided preference for the independent candidate, the hope is that the election can be swung. A survey organization estimates that the proposed "voter awareness" campaign is likely to increase turnout to levels of 35,000, 66,000, 82,000, 48,000, 70,000, and 45,000 voters, respectively, in each district. Project the election outcome under these circumstances. *Answer:* Democrat (122, 900); Republican (111, 400); Independent (117, 700).

EXAMPLE 34 (**Production Planning**) A company manufactures five products. The company has divided its sales force into three sales districts. The matrix **S** below summarizes expected sales for each of the five products in each sales region for the coming month.

$$
\begin{array}{c}
\text{Region} \\
\begin{array}{ccc}
1 & 2 & 3
\end{array} \\
\mathbf{S} = \begin{pmatrix}
500 & 200 & 350 \\
400 & 300 & 100 \\
250 & 425 & 50 \\
100 & 150 & 350 \\
200 & 175 & 225
\end{pmatrix}
\begin{array}{l}
1 \\
2 \\
3 \\
4 \\
5
\end{array} \text{Product}
\end{array}
$$

Each product is manufactured by using combinations of four standard components. The matrix **R** indicates the number of units of each component used in producing each product.

$$
\begin{array}{c}
\text{Component} \\
\begin{array}{cccc}
1 & 2 & 3 & 4
\end{array} \\
\mathbf{R} = \begin{pmatrix}
1 & 0 & 2 & 0 \\
1 & 1 & 1 & 0 \\
2 & 1 & 0 & 3 \\
0 & 2 & 1 & 1 \\
1 & 2 & 3 & 1
\end{pmatrix}
\begin{array}{l}
1 \\
2 \\
3 \\
4 \\
5
\end{array} \text{Product}
\end{array}
$$

The manufacture of each component requires the consumption of certain resources. The matrix **P** indicates the quantities of each of three standard parts and the number of production labor-hours and assembly labor-hours used to produce one unit of each component.

$$
\begin{array}{c}
\text{Resource} \\
\begin{array}{ccccc}
\text{Part} & \text{Part} & \text{Part} & \text{Prod.} & \text{Assem.} \\
1 & 2 & 3 & \text{Labor} & \text{Labor}
\end{array} \\
\mathbf{P} = \begin{pmatrix}
2 & 0 & 1 & 2 & 3 \\
1 & 3 & 2 & 5 & 1 \\
0 & 2 & 1 & 4 & 2 \\
0 & 4 & 1 & 1 & 6
\end{pmatrix}
\begin{array}{l}
1 \\
2 \\
3 \\
4
\end{array} \text{Component}
\end{array}
$$

The matrix **C** contains the costs of the five resources in matrix **P**. Part 1 costs $25; part 2, $15; part 3, $30; each labor-hour used in the production department costs $10; and each labor-hour in the assembly department costs $8.

$$\mathbf{C} = (\$25 \quad \$15 \quad \$30 \quad \$10 \quad \$8)$$

Management of the company wants to manipulate the data in these matrices to calculate (a) the total expected demand for each final product, (b) the quantities needed of each of the four components, (c) the resource requirements to produce the four components, and (d) the total cost of producing the quantities of the five products needed for the month. Figure 9.3 provides a schematic diagram of this production planning process.

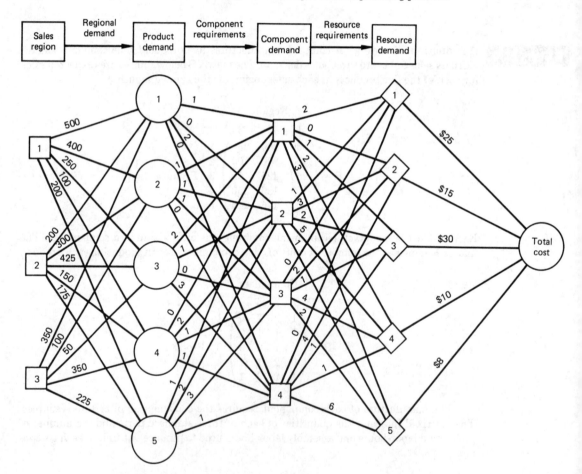

Figure 9.3   Production planning process.

SOLUTION

(*a*) Even though we could determine the expected demand for each product by adding the elements in each row of **S**, the purpose is to generate this information by using matrix operations. Multiplying **S** by a (3 × 1) column vector in which all elements equal 1 will yield the expected demands for the five products.

$$D = S \begin{pmatrix} 1 \\ 1 \\ 1 \end{pmatrix}$$

$$= \begin{pmatrix} 500 & 200 & 350 \\ 400 & 300 & 100 \\ 250 & 425 & 50 \\ 100 & 150 & 350 \\ 200 & 175 & 225 \end{pmatrix} \begin{pmatrix} 1 \\ 1 \\ 1 \end{pmatrix}$$

$$= \begin{pmatrix} 1{,}050 \\ 800 \\ 725 \\ 600 \\ 600 \end{pmatrix}$$

(b) Given that there are 4 components used in the production processes, we need to generate 4 data items which represent the quantities needed of each component. The component requirements matrix C_r can be found by multiplying D^T by the matrix R, or

$$C_r = D^T R$$

$$= (1{,}050 \quad 800 \quad 725 \quad 600 \quad 600) \begin{pmatrix} 1 & 0 & 2 & 0 \\ 1 & 1 & 1 & 0 \\ 2 & 1 & 0 & 3 \\ 0 & 2 & 1 & 1 \\ 1 & 2 & 3 & 1 \end{pmatrix}$$

$$= (3{,}900 \quad 3{,}925 \quad 5{,}300 \quad 3{,}375)$$

which indicates that 3,900 units of component 1 will be required, 3,925 of component 2, 5,300 of component 3, and 3,375 of component 4.

(c) In calculating the total resource requirements, we are seeking total needs for the three parts used in the production of the four components as well as the production and assembly labor-hours required. These five items can be calculated by multiplying the component requirements matrix C_r by the matrix P, or

$$R_r = C_r P$$

$$= (3{,}900 \quad 3{,}925 \quad 5{,}300 \quad 3{,}375) \begin{pmatrix} 2 & 0 & 1 & 2 & 3 \\ 1 & 3 & 2 & 5 & 1 \\ 0 & 2 & 1 & 4 & 2 \\ 0 & 4 & 1 & 1 & 6 \end{pmatrix}$$

$$= (11{,}725 \quad 35{,}875 \quad 20{,}425 \quad 52{,}000 \quad 46{,}475)$$

This calculation indicates that 11,725 units of part 1 will be required, 35,875 units of part 2, 20,425 units of part 3, 52,000 production labor-hours, and 46,475 assembly labor-hours.

(d) The total production cost can be computed by multiplying R_r by the transpose of the cost matrix C, or

$$T = R_r C^T$$

$$= (11{,}725 \quad 35{,}875 \quad 20{,}425 \quad 52{,}000 \quad 46{,}475) \begin{pmatrix} 25 \\ 15 \\ 30 \\ 10 \\ 8 \end{pmatrix}$$

$$= \$2{,}335{,}800$$

EXAMPLE 35

(Brand-Switching Analysis; Motivating Scenario) Brand-switching analysis concerns itself with the purchasing behavior of consumers who make repeated purchases of a product or service. Examples of such products or services are gasoline, detergents, soft drinks, and fast-food meals. Brand-switching analysis focuses upon brand loyalty and the degree to which consumers are willing to switch to competing products. Firms often try to project the effects that promotion campaigns, such as rebates or advertising programs, will have on the sales of their products. If information is available concerning the rates of gains from and losses to all competitors, a firm can (a) predict its market share at some time in the future, (b) predict the rate at which the firm will increase or decrease its market share in the future, and (c) determine whether market shares will ever reach equilibrium levels in which each firm or brand retains a constant share of the market.

Using consumer surveys, it may be possible to determine a *matrix of transition probabilities* (or *transition matrix*) which reflects the chance that a company will retain its customers, the chance that a company will gain customers from other companies, and the chance that it will lose customers to competing companies. Consider the following matrix of transition probabilities for two competing brands:

$$\mathbf{T} = \begin{pmatrix} p_{11} & p_{12} \\ p_{21} & p_{22} \end{pmatrix}$$

Let p_{ij} equal the percentage of brand i consumers who will purchase brand j during the next period. This definition implies that a consumer purchases brand i during one period and then purchases brand j during the next period. A period may be defined as any appropriate time interval, such as a week or month. (In fact, the transition matrix may reflect consumer choices during the next purchase cycle.) When $i = j$, p_{ij} represents the percentage of brand i consumers who remain loyal to brand i and purchase it again. Thus, p_{11} and p_{22} represent the percentage of original customers retained in the next period by brands 1 and 2, respectively, p_{12} represents the percentage of customers purchasing brand 1 in the previous period who purchase brand 2 in the next period, and p_{21} represents the percentage of customers purchasing brand 2 in the last period who purchase brand 1 in the next period.

To illustrate, the following transition matrix **T** indicates that brand 1 retains 80 percent of its customers but loses 20 percent to brand 2. Brand 2 retains 90 percent of its customers and loses 10 percent of its customers to brand 1.

$$\mathbf{T} = \begin{pmatrix} 0.80 & 0.20 \\ 0.10 & 0.90 \end{pmatrix}$$

If market shares are known for the two brands, it is possible to use the transition matrix to project market shares in the next period. Suppose that these are the only two brands on the market and that in the last period brand 1 had 40 percent of the market and brand 2 had 60 percent of the market. If these market shares are represented in the (1×2) share vector **S**, the expected market shares in the next period can be computed by the product **ST**, or

$$(0.40 \quad 0.60)\begin{pmatrix} 0.80 & 0.20 \\ 0.10 & 0.90 \end{pmatrix} = [0.40(0.80) + 0.60(0.10) \quad 0.40(0.20) + 0.60(0.90)]$$
$$= (0.38 \quad 0.62)$$

Carefully note how the new market shares are calculated. The 38 percent for brand 1 results from brand 1 retaining 80 percent of its previous share and from gaining 10 percent of the previous share of brand 2 customers.

If the switching behavior is constant for a number of periods, the matrix of transition probabilities will remain the same. Under these conditions the vector of market shares after n periods, S_n, can be computed as

$$S_n = S\overbrace{TTT \cdots T}^{n}$$
$$= S \cdot T^n$$

or

Given the most recent market shares S and the matrix of transition probabilities T for three competing brands, we can determine the market shares at the end of each of the next two periods as follows:

$$S = (0.30 \quad 0.40 \quad 0.30) \qquad T = \begin{pmatrix} 0.90 & 0.05 & 0.05 \\ 0.05 & 0.85 & 0.10 \\ 0.05 & 0.15 & 0.80 \end{pmatrix}$$

For the next period,

$$S_1 = (0.30 \quad 0.40 \quad 0.30)\begin{pmatrix} 0.90 & 0.05 & 0.05 \\ 0.05 & 0.85 & 0.10 \\ 0.05 & 0.15 & 0.80 \end{pmatrix}$$
$$= [0.30(0.90) + 0.40(0.05) + 0.30(0.05)$$
$$0.30(0.05) + 0.40(0.85) + 0.30(0.15)$$
$$0.30(0.05) + 0.40(0.10) + 0.30(0.80)]$$
$$= (0.305 \quad 0.400 \quad 0.295)$$

For the second period,

$$S_2 = (0.305 \quad 0.400 \quad 0.295)\begin{pmatrix} 0.90 & 0.05 & 0.05 \\ 0.05 & 0.85 & 0.10 \\ 0.05 & 0.15 & 0.80 \end{pmatrix}$$
$$= (0.30925 \quad 0.39950 \quad 0.29125)$$

After two periods, brands 2 and 3 will have experienced slight decreases in their market shares, while brand 1 will have increased its share.

EXAMPLE 36 (**Population Migration — Equilibrium Conditions**) Another application of matrix algebra deals with population migration, where the population may consist of persons, wildlife, etc. The migration patterns might be represented by a ***transition matrix*** similar to that characterizing brand-switching behavior. Given such a transition matrix along with a

Figure 9.4
Population shifts.

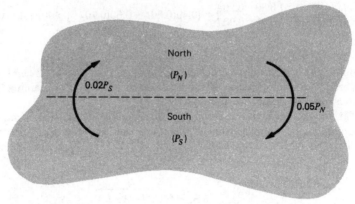

population vector describing the population totals for each relevant region, it becomes possible to project the dynamics of population shifts over time. If the migration patterns are stable over time (the transition matrix does not change), the equilibrium condition can ultimately occur where the population of each region becomes stable. At equilibrium the population increases in each region are offset by the decreases during each time period. The following simplified example illustrates this condition.

Because of the increasing cost of energy, the population within one European country seems to be shifting from the north to the south, as shown in Fig. 9.4. The transition matrix **S** describes the migration behavior observed between the two regions.

$$\begin{array}{cc} \text{To north} & \text{To south} \end{array}$$
$$\mathbf{S} = \begin{pmatrix} 0.95 & 0.05 \\ 0.02 & 0.98 \end{pmatrix} \begin{array}{l} \text{From north} \\ \text{From south} \end{array}$$

The value of 0.95 in **S** indicates that 95 percent of those living in the north during one year will still be living in the north the following year. The 0.05 represents the remaining 5 percent who would move from the north to the south. The 0.98 indicates that 98 percent of those living in the south during one year will still be living in the south the following year. The 0.02 indicates annual migration to the north of 2 percent of the population living in the south.

NOTE   To simplify the analysis, we will assume that the population of the country is constant or that the 0.95 and 0.98 parameters reflect net effects which account for births, deaths, immigration, and emigration during the year.

If P_N represents the population of the northern region of the country and P_S the population of the southern region in any given year, the projected population for each region in the following year is found by the matrix multiplication

$$\mathbf{PS} = \mathbf{P}' \tag{9.16}$$

or $\qquad (P_N \quad P_S)\begin{pmatrix} 0.95 & 0.05 \\ 0.02 & 0.98 \end{pmatrix} = (P'_N \quad P'_S) \tag{9.17}$

Equilibrium occurs when $P_N = P'_N$ and $P_S = P'_S$. If we expand Eq. (9.17), equilibrium will occur when

$$0.95P_N + 0.02P_S = P_N \qquad \text{(9.18)}$$

and $$0.05P_N + 0.98P_S = P_S \qquad \text{(9.19)}$$

We have yet to specify any population figures for this country. In order to determine the equilibrium condition, we only need the total population. Let's assume that the population of the country is 70 million persons, or

$$P_N + P_S = 70 \qquad \text{(9.20)}$$

Thus, Eq. (9.20) must be included with Eqs. (9.18) and (9.19). To solve this (3×2) system, it can be shown that only two of the three equations are needed — Eqs. (9.20) and either (9.18) or (9.19). Thus, the solution to the system

$$0.95P_N + 0.02P_S = P_N \qquad \text{(9.18)}$$

$$P_N + \quad P_S = 70 \qquad \text{(9.20)}$$

will yield the equilibrium populations. Solve (by matrix or nonmatrix methods) the system and verify that $P_N = 20$ and $P_S = 50$.

To prove that equilibrium exists at these values, we can project the population for the next year using Eq. (9.16), or

$$(20 \quad 50)\begin{pmatrix} 0.95 & 0.05 \\ 0.02 & 0.98 \end{pmatrix} = (19 + 1 \quad 1 + 49) = (20 \quad 50)$$

☐

POINTS FOR THOUGHT & DISCUSSION

Recall that we did not specify initial population distributions for the country. It turns out that equilibrium *values* are independent of these initial conditions. Given the initial figures, however, an interesting question is how much time is required to reach equilibrium. This is an issue we will not address. However, we would speculate that the nearer the initial population distribution to the equilibrium distribution, the shorter the *time to equilibrium.*

Discuss the assumptions of this model. Which assumptions do you have reservations about? Does there seem to be any value in using a model such as this?

EXAMPLE 37

(Input-Output Analysis) A Nobel Prize recipient, Wassily Leontief, is most noted for his input-output model of an economy. An assumption of the model is that whatever is produced will be consumed. Demand for an industry's output can come from two sources: (1) demand from different industries and (2) demand from sources other than industries. To illustrate this, consider the energy sector. Power companies generate energy which is (1) needed to operate their own plants, (2) needed to supply other industries with their electrical needs, and (3) needed for other consumers like us.

The first two of these are examples of ***interindustry demand*** and the last is ***nonindustry demand.*** The objective of input-output analysis is typically to determine how much an industry should produce so that both types of demand are satisfied exactly. That is, how much should be produced in order to bring supply and demand into balance?

(Read this paragraph very carefully.) The interindustry demand is usually summarized in a **technological** or **input-output matrix.** An example is the following (3 × 3) matrix **A**.

$$
\begin{array}{cc}
 & \text{User} \\
 & \begin{array}{ccc} 1 & 2 & 3 \end{array} \\
\text{Supplier} \begin{array}{c} 1 \\ 2 \\ 3 \end{array} & \left(\begin{array}{ccc} 0.3 & 0.3 & 0.2 \\ 0.1 & 0.2 & 0.3 \\ 0.2 & 0.1 & 0.4 \end{array} \right) = \mathbf{A}
\end{array}
$$

Assume that the output of an industry is measured in dollars. If a_{ij} is the general element in the input-output matrix, a_{ij} *represents the amount of industry i's output required in producing one dollar of output in industry j.* This matrix represents a three-industry situation. The element $a_{11} = 0.3$ suggests that of every dollar of output *from* industry 1, 30 percent of this value is contributed by industry 1. The element a_{12} suggests that for every dollar of output *from* industry 2, 30 percent is contributed by industry 1. The element a_{13} indicates that every dollar of output *from* industry 3 requires 20 cents of output from industry 1. The element $a_{21} = 0.1$ indicates that for every dollar of output *from* industry 1, 10 percent is contributed by industry 2. The element $a_{31} = 0.2$ indicates that for every dollar of output *from* industry 1, 20 percent is provided by industry 3. Try to interpret the remaining elements.

Let x_j equal the output from industry j (in dollars) and let d_j equal the nonindustry demand (in dollars) for the output of industry j. A set of simultaneous equations may be formulated which when solved would determine the levels of output x_j at which total supply and demand would be in equilibrium. The equations in this system would have the general form

> Industry output = interindustry demand + nonindustry demand

For the three-industry example the system would be

$$
\begin{array}{ccc}
 & \overbrace{\text{Interindustry}}^{} & \text{Nonindustry} \\
 & \text{demand} & \text{demand} \\
x_1 = & \overbrace{0.3x_1 + 0.3x_2 + 0.2x_3} + & \overbrace{d_1} \\
x_2 = & 0.1x_1 + 0.2x_2 + 0.3x_3 + & d_2 \\
x_3 = & 0.2x_1 + 0.1x_2 + 0.4x_3 + & d_3
\end{array}
\qquad (9.21)
$$

Rearranging these equations gives

$$
0.7x_1 - 0.3x_2 - 0.2x_3 = d_1
$$
$$
-0.1x_1 + 0.8x_2 - 0.3x_3 = d_2
$$
$$
-0.2x_1 - 0.1x_2 + 0.6x_3 = d_3
$$

Given a set of nonindustry demand values d_j, these equations may be solved to determine the equilibrium levels of output.

Note for a moment the structure of Eq. (9.21). If $\mathbf{X}$ is a column vector containing elements x_1, x_2, and x_3, and $\mathbf{D}$ is a column vector containing elements d_1, d_2, and d_3, Eq. (9.21) has the form

$$\mathbf{X} = \mathbf{AX} + \mathbf{D}$$

This matrix equation can be simplified as follows:

$$\mathbf{X} - \mathbf{AX} = \mathbf{D}$$

$$\mathbf{IX} - \mathbf{AX} = \mathbf{D}$$

$$(\mathbf{I} - \mathbf{A})\mathbf{X} = \mathbf{D}$$

$$\boxed{\mathbf{X} = (\mathbf{I} - \mathbf{A})^{-1}\mathbf{D}} \qquad (9.22)$$

That is, assuming a square input-output matrix $\mathbf{A}$, the equilibrium levels of output may be found by (1) forming the matrix $(\mathbf{I} - \mathbf{A})$, (2) finding $(\mathbf{I} - \mathbf{A})^{-1}$ if it exists, and (3) multiplying $(\mathbf{I} - \mathbf{A})^{-1}$ times the nonindustry demand vector $\mathbf{D}$.

You should verify that for the three-industry example, the determinant of $(\mathbf{I} - \mathbf{A})$ equals 0.245 and

$$(\mathbf{I} - \mathbf{A})^{-1} = \begin{pmatrix} 1.837 & 0.816 & 1.020 \\ 0.490 & 1.551 & 0.939 \\ 0.694 & 0.531 & 2.163 \end{pmatrix}$$

Given the input-output matrix for the three-industry example, suppose that the levels for nonindustry demands are

$$d_1 = \$50,000,000$$

$$d_2 = \$30,000,000$$

$$d_3 = \$60,000,000$$

The equilibrium levels can be determined as

$$\mathbf{X} = \begin{pmatrix} 1.837 & 0.816 & 1.020 \\ 0.490 & 1.551 & 0.939 \\ 0.694 & 0.531 & 2.163 \end{pmatrix} \begin{pmatrix} 50,000,000 \\ 30,000,000 \\ 60,000,000 \end{pmatrix}$$

$$= \begin{pmatrix} 177,530,000 \\ 127,370,000 \\ 180,410,000 \end{pmatrix}$$

Industry 1 should produce \$177,530,000 worth of output; industry 2, \$127,370,000; and industry 3, \$180,410,000.

Using the coefficients in the original input-output matrix, the interindustry demand can be calculated (in millions of dollars) as

		User		
		1	2	3
	1	53.259	38.211	36.082
Supplier	2	17.753	25.474	54.123
	3	35.506	12.737	72.164

If you add the interindustry demand plus the nonindustry demand, you will find the totals slightly different from the computed equilibrium values. These differences may be attributed to rounding errors in calculating $(\mathbf{I} - \mathbf{A})^{-1}$.

EXAMPLE 38   (Network Applications) A *network* consists of a set of *nodes* and a set of *arcs* which connect nodes. The nodes can represent cities, highway intersections, computers, water reservoirs, or less tangible items such as project milestones. Nodes typically represent points where some type of flow originates, is relayed, or terminates. Arcs in a network can represent roads, air routes, power lines, pipelines, and so forth. Figure 9.5 illustrates different representations of nodes and arcs. In 9.5a no specific flow orientation is indicated. The arc in this case is called an *undirected arc.* In 9.5b the arc is called a *directed arc* because flow is in one direction. In 9.5c the arc is *bidirected* because flows can be in both directions.

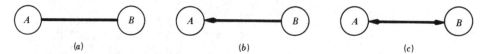

Figure 9.5   Node-arc representation.

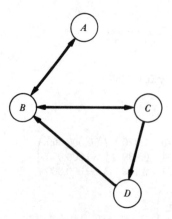

Figure 9.6   Commuter airline routes.

Figure 9.6 is a network diagram which illustrates the route structure for a small regional commuter airline servicing four cities. The nodes represent the different cities, and the arcs represent the routes servicing the cities. The bidirected arc connecting nodes A and B indicates that the airline flies from A to B and from B to A.

The essence of these node-arc relationships can be summarized in what is called an *adjacency matrix.* The adjacency matrix has a row and a column for each node. The elements of the matrix consist of 0s or 1s, depending on whether there is a directed arc from one node to another. In this example, an element in position (i, j) is assigned a value of 1 if there is service from city i to city j; otherwise, a value of 0 is assigned. Compare the adjacency matrix with Fig. 9.6. The adjacency matrix summarizes all *nonstop service* between cities on the airline's routes.

$$
\text{From} \quad
\begin{array}{c}
\\ A \\ B \\ C \\ D
\end{array}
\begin{array}{c}
\overset{\text{To}}{\begin{array}{cccc} A & B & C & D \end{array}} \\
\begin{pmatrix}
0 & 1 & 0 & 0 \\
1 & 0 & 1 & 0 \\
0 & 1 & 0 & 1 \\
0 & 1 & 0 & 0
\end{pmatrix}
\end{array}
\quad \text{Adjacency matrix}
$$

Now, if the adjacency matrix is multiplied times itself, an interesting result occurs.

$$
\begin{pmatrix}
0 & 1 & 0 & 0 \\
1 & 0 & 1 & 0 \\
0 & 1 & 0 & 1 \\
0 & 1 & 0 & 0
\end{pmatrix}
\begin{pmatrix}
0 & 1 & 0 & 0 \\
1 & 0 & 1 & 0 \\
0 & 1 & 0 & 1 \\
0 & 1 & 0 & 0
\end{pmatrix}
= \text{From}
\begin{array}{c}
\\ A \\ B \\ C \\ D
\end{array}
\begin{array}{c}
\overset{\text{To}}{\begin{array}{cccc} A & B & C & D \end{array}} \\
\begin{pmatrix}
1 & 0 & 1 & 0 \\
0 & 2 & 0 & 1 \\
1 & 1 & 1 & 0 \\
1 & 0 & 1 & 0
\end{pmatrix}
\end{array}
$$

The product matrix summarizes the *one-stop service* between all cities. For example, the product matrix indicates that there is a one-stop route from city C to city A. A check of Fig. 9.6 confirms that service is available from C to A with a stop at city B.

Let's see if this makes sense. If we examine the inner product resulting in element $(3, 1)$ of the product matrix, row 3 of the first matrix indicates the presence (absence) of direct flights *from* city C to other cities. Column 1 of the second matrix indicates the presence (absence) of flights *to* city A from other cities. The multiplication seeks matches of pairs of flights.

$$
\begin{array}{c}
A \\ B \\ C \\ D
\end{array}
\left(
\begin{array}{cccc}
& & & \\
& & & \\
\boxed{\begin{array}{cccc} 0 & 1 & 0 & 1 \end{array}} & & & \\
& & &
\end{array}
\right)
\begin{pmatrix}
0 \\ 1 \\ 0 \\ 0
\end{pmatrix}
=
\begin{array}{c}
A \\ B \\ C \\ D
\end{array}
\left(
\begin{array}{c}
\\ \\ ① \\
\end{array}
\right)
$$

In this case we look for matches between flights *from* city C to another city with flights *from* that destination city *to* city A. Perhaps we can illustrate this by expanding the inner product computation as shown in Fig. 9.7. Note in this figure that the only pairs of flights which match are C to B and B to A. Study the multiplication procedure until you understand how (and why) each element is calculated.

Although not particularly meaningful, the matrix indicates two one-stop routes from city B to city B. These reflect the round-trip routes to cities A and C.

Cubing the matrix will result in a product matrix which summarizes the number of

Inner product components	(0)	(0)	+	(1)	(1)	+	(0)	(0)	+	(1)	(0)	$\cdot = 1$
Corresponding flight pairs	$C \to A$	$A \to A$		$C \to B$	$B \to A$		$C \to C$	$C \to A$		$C \to D$	$D \to A$	

Figure 9.7   Inner product computation for row 3 and column 1.

"two-stop" routes between all cities. For example, our results thus far have indicated no nonstop or one-stop service from A to D. This product matrix suggests that there is a two-stop route $(A \to B \to C \to D)$.

$$\begin{pmatrix} 0 & 1 & 0 & 0 \\ 1 & 0 & 1 & 0 \\ 0 & 1 & 0 & 1 \\ 0 & 1 & 0 & 0 \end{pmatrix} \begin{pmatrix} 1 & 0 & 1 & 0 \\ 0 & 2 & 0 & 1 \\ 1 & 1 & 1 & 0 \\ 1 & 0 & 1 & 0 \end{pmatrix} = \begin{array}{c} \\ A \\ B \\ C \\ D \end{array} \begin{array}{cccc} & \text{To} & & \\ A & B & C & D \\ \end{array} \begin{pmatrix} 0 & 2 & 0 & 1 \\ 2 & 1 & 2 & 0 \\ 1 & 2 & 1 & 1 \\ 0 & 2 & 0 & 1 \end{pmatrix}$$

In this particular example, two-stop routes are not always meaningful. To illustrate, the 2 two-stop routes indicated from city A to city B are $A \to B \to A \to B$ and $A \to B \to C \to B$. Another example is that a two-stop route can occur when flying from city C to city A. The route is from $C \to D \to B \to A$. The more direct route, though, is $C \to B \to A$.

This example is very simple, not really justifying matrix methods. However, it is simplified for purposes of illustration. These methods, especially when executed on a computer, bring considerable efficiencies for larger-scale problems, such as might occur for major airlines like United, American, Northwest, and Delta.

❑

Section 9.6 Follow-up Exercises

1 Regarding the election discussed in Example 33, the Independent candidate made a particularly strong showing in a recent televised debate among the three candidates. The political pollster has observed a shift in voter preferences as indicated in the following matrix.

$$\mathbf{P} = \begin{pmatrix} & & \text{District} & & & \\ 1 & 2 & 3 & 4 & 5 & 6 \\ 0.35 & 0.33 & 0.30 & 0.44 & 0.25 & 0.30 \\ 0.40 & 0.38 & 0.27 & 0.35 & 0.28 & 0.35 \\ 0.25 & 0.29 & 0.43 & 0.21 & 0.47 & 0.35 \end{pmatrix} \begin{array}{l} \text{Democrat} \\ \text{Republican} \\ \text{Independent} \end{array}$$

Assuming the original estimates of voter turnout in the six districts, project the outcome of the election. Does it seem that the debate has had any effect on the outcome?

2 The following matrix is a matrix of transition probabilities related to a market dominated by two firms.

$$\mathbf{T} = \begin{pmatrix} 0.70 & 0.30 \\ 0.25 & 0.75 \end{pmatrix}$$

Assume brand 1 currently has 70 percent of the market and brand 2 has the remaining 30 percent.

  (*a*) Predict market shares in the next period.

  (*b*) Predict market shares after four periods.

  (c) Assuming that the transition matrix remains stable, will a market equilibrium be reached? If so, what are the expected equilibrium shares? (*Hint:* If p_1 and p_2 represent the market shares for brands 1 and 2, $p_1 + p_2 = 1$.)

3 Examine the following matrices of transition for two different market situations. For each matrix, it is assumed that there are three brands dominating the market. By observation, see if you can predict (without formal computation) what equilibrium conditions will be

$$T_1 = \begin{pmatrix} 0.80 & 0.15 & 0.05 \\ 0.20 & 0.70 & 0.10 \\ 0 & 0 & 1.00 \end{pmatrix} \qquad T_2 = \begin{pmatrix} 0.80 & 0.10 & 0.10 \\ 0 & 0.50 & 0.50 \\ 0 & 0.50 & 0.50 \end{pmatrix}$$

4 The following matrix illustrates the transition probabilities associated with a market dominated by three brands.

$$T = \begin{pmatrix} 0.2 & 0.6 & 0.2 \\ 0.1 & 0.5 & 0.4 \\ 0.2 & 0.3 & 0.5 \end{pmatrix}$$

Assume brand 1 currently has 40 percent of the market, brand 2 has 40 percent, and brand 3 has 20 percent.

  (*a*) Predict market shares after the next period.

  (b) Assuming the transition matrix remains stable, will a market equilibrium be reached? If so, what are the expected equilibrium shares?

5 In Example 36 assume that the transition matrix describing migration behavior is

$$\begin{array}{cc} & \text{To north} \quad \text{To south} \\ S = \left(\begin{array}{cc} 0.90 & 0.10 \\ 0.05 & 0.95 \end{array} \right) & \begin{array}{l} \text{From north} \\ \text{From south} \end{array} \end{array}$$

Determine whether the populations will attain an equilibrium condition, and if so, the populations of the two regions.

6 Referring to Example 37, assume that the nonindustry demands are $100 million, $60 million, and $80 million, respectively. Determine the equilibrium levels of output for the three industries. Also, determine the interindustry demands for the three industries.

7 Figure 9.8 is a network diagram which illustrates the route structure for a small regional commuter airline which services four cities. Using this figure, construct an adjacency matrix. Square the adjacency matrix and verbally summarize the one-stop service which exists between all cities.

8 Figure 9.9 is a network diagram which illustrates the route structure for a small commuter airline which services five cities. Using this figure, construct an adjacency matrix. Square the adjacency matrix and verbally summarize the one-stop service which exists between all cities.

9 The input-output matrix for a three-industry economy is

Figure 9.8

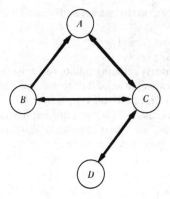

Figure 9.9

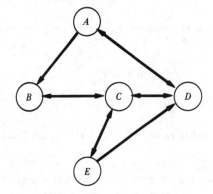

$$
\text{Supplier} \begin{array}{c} \\ 1 \\ 2 \\ 3 \end{array}
\overset{\displaystyle \text{User}}{
\overset{\displaystyle \begin{array}{ccc} 1 & 2 & 3 \end{array}}{
\begin{pmatrix} 0.25 & 0.30 & 0.20 \\ 0.20 & 0.30 & 0.20 \\ 0.40 & 0.10 & 0.25 \end{pmatrix}}} = \mathbf{A}
$$

If nonindustry demands are respectively $100,000,000, $60,000,000, and $150,000,000, (a) determine the equilibrium levels of output for the three industries and (b) determine the interindustry demands for the three industries.

10 The following transition matrix indicates the annual shifts in population within three regions of a country.

$$
\text{From} \begin{array}{c} \\ 1 \\ 2 \\ 3 \end{array}
\overset{\displaystyle \text{To}}{
\overset{\displaystyle \begin{array}{ccc} 1 & 2 & 3 \end{array}}{
\begin{pmatrix} 0.90 & 0.06 & 0.04 \\ 0.08 & 0.86 & 0.06 \\ 0.03 & 0.02 & 0.95 \end{pmatrix}}}
$$

Determine whether the populations will attain an equilibrium condition and the relative population shares of the three regions.

11 Suppose that the national office of a car rental corporation is planning its maintenance program for the next year. Executives are interested in determining the company's needs for certain repair parts and expected costs for these categories of parts. The company rents midsized, compact, and subcompact cars. The matrix **N** indicates the number of each size of car available for renting in four regions of the country.

$$
N = \begin{matrix}
 & \text{Midsized} & \text{Compact} & \text{Subcompact} & \\
 & \begin{pmatrix} 16{,}000 & 40{,}000 & 50{,}000 \\ 15{,}000 & 30{,}000 & 20{,}000 \\ 10{,}000 & 10{,}000 & 15{,}000 \\ 12{,}000 & 40{,}000 & 30{,}000 \end{pmatrix} & & \begin{matrix} \text{East} \\ \text{Midwest} \\ \text{South} \\ \text{West} \end{matrix}
\end{matrix}
$$

Four repair parts of particular interest because of their cost and frequency of replacement are fan belts, spark plugs, batteries, and tires. On the basis of studies of maintenance records in different parts of the country, analysts have determined the average number of repair parts needed per car during a year. These are summarized in the matrix **R**:

$$
R = \begin{matrix}
 & \text{Midsized} & \text{Compact} & \text{Subcompact} & \\
 & \begin{pmatrix} 1.7 & 1.6 & 1.5 \\ 12.0 & 8.0 & 5.0 \\ 0.9 & 0.75 & 0.5 \\ 4.0 & 6.5 & 6.0 \end{pmatrix} & & \begin{matrix} \text{Fan belts} \\ \text{Plugs} \\ \text{Batteries} \\ \text{Tires} \end{matrix}
\end{matrix}
$$

(*a*) Perform a matrix computation which determines the total demand for each size of car.

(*b*) Perform a matrix computation which calculates the total number of each repair part required for the fleet.

(*c*) If the matrix **C** contains the cost per unit for fan belts, plugs, batteries, and tires, perform a matrix computation which computes total combined costs for all repair parts.

$$ C = (\$1.25 \quad \$0.80 \quad \$30.00 \quad \$35.00) $$

(*d*) Perform a matrix computation which calculates total costs for each category of repair part. (Hint: This will require the formulation of a new matrix which contains the results from part *b*.)

❑ KEY TERMS AND CONCEPTS

❏ **ADDITIONAL EXERCISES**

Section 9.3

1 The matrices S_1 and S_2 represent annual sales for a firm's three products by region, stated in millions of dollars. S_1 represents sales for the firm's first year of operation and S_2 the sales for the second year of operation.

$$
S_1 = \begin{pmatrix} 2.6 & 4.8 & 1.8 & 0.9 \\ 3.2 & 4.4 & 2.5 & 2.8 \\ 2.4 & 3.6 & 3.8 & 2.5 \end{pmatrix}
\qquad
S_2 = \begin{pmatrix} 3.6 & 2.5 & 3.0 & 2.5 \\ 4.5 & 5.0 & 3.5 & 3.8 \\ 2.9 & 3.0 & 4.6 & 4.0 \end{pmatrix}
$$

with columns labeled Region 1 2 3 4.

(a) Compute $S_2 - S_1$ and interpret the meaning of the resulting matrix.
(b) Compute $S_1 + S_2$ and interpret the meaning of the resulting matrix.
(c) Management had projected a 30 percent increase in sales for all products in all regions for the second year of operation. Using matrix operations, compute the difference between the projected sales levels and the actual levels for the second year and interpret the results. Identify the regions and products which were below management's expectations.

Given the following matrices,

$$
A = \begin{pmatrix} 2 & -8 \\ 4 & 7 \end{pmatrix}
\quad
B = \begin{pmatrix} 1 & 0 & 0 \\ 0 & 1 & 0 \\ 0 & 0 & 1 \end{pmatrix}
\quad
C = \begin{pmatrix} 2 & -1 & 8 \\ 3 & 10 & -2 \end{pmatrix}
\quad
D = \begin{pmatrix} 3 & -2 & -1 \\ 0 & 1 & 0 \\ 2 & 8 & 11 \end{pmatrix}
$$

$$
E = \begin{pmatrix} -1 & 7 \\ 4 & -1 \end{pmatrix}
\quad
F = \begin{pmatrix} 0 & 0 & 1 \\ 0 & 1 & 0 \\ 1 & 0 & 0 \end{pmatrix}
\quad
G = \begin{pmatrix} 2 & -8 \\ 7 & 10 \\ 3 & -1 \end{pmatrix}
\quad
H = \begin{pmatrix} 1 & -2 & 3 \\ 3 & 2 & -1 \\ 7 & 8 & 1 \end{pmatrix}
$$

perform the following matrix computations (if possible).

2 $2A + E$	3 $4E - A$
4 $B + D - F + H$	5 $3B - 2F + D$
6 AB	7 BA
8 CD	9 $DC^T B$
10 BF	11 DH
12 CH	13 GC
14 CG	15 $C^T G$
16 DF	17 $ACBD$
18 $DBCA$	19 $C^T BD$
20 GE	21 EG

Given the following matrices,

$$A = \begin{pmatrix} 5 & -4 \\ -2 & 10 \end{pmatrix} \quad B = \begin{pmatrix} 3 & 6 & 2 \\ -2 & 5 & -1 \\ 0 & -2 & 4 \end{pmatrix} \quad C = \begin{pmatrix} 3 & -2 \\ 4 & 1 \\ -3 & 5 \\ 2 & -4 \end{pmatrix}$$

$$D = \begin{pmatrix} 5 & -5 & 10 & -4 \\ 2 & -4 & 3 & 5 \end{pmatrix} \quad E = \begin{pmatrix} 4 \\ -2 \end{pmatrix} \quad F = \begin{pmatrix} 3 & 1 & 2 \\ -2 & 0 & 4 \end{pmatrix}$$

perform the following matrix computations (if possible).

22 $5C^T - 2D$	23 $3C - 4D^T$
24 AE	25 EA
26 CA	27 AC^T
28 DC	29 CD
30 FA	31 $E^T F$
32 BF^T	33 $E^T AF$
34 BC	35 BF
36 FB	37 CAD
38 $F^T AE$	39 $D^T C$
40 $C^T D^T AE$	41 $BF^T AE$

42 State the following matrix equation in algebraic form.

$$\begin{pmatrix} 3 & -2 & 4 & 0 \\ -1 & 2 & 5 & 3 \\ 4 & -3 & 2 & 1 \end{pmatrix} \begin{pmatrix} x_1 \\ x_2 \\ x_3 \\ x_4 \end{pmatrix} = \begin{pmatrix} 10 \\ -5 \\ 24 \end{pmatrix}$$

43 State the following matrix equation in algebraic form.

$$\begin{pmatrix} 1 & 8 & -2 & 0 & -4 \\ 2 & 10 & 1 & -1 & 10 \end{pmatrix} \begin{pmatrix} x_1 \\ x_2 \\ x_3 \\ x_4 \\ x_5 \end{pmatrix} = \begin{pmatrix} 300 \\ 175 \end{pmatrix}$$

44 State the following matrix equation in algebraic form.

$$\begin{pmatrix} 3 & -2 & 1 \\ 4 & 10 & -2 \\ 8 & -3 & 0 \\ -3 & 1 & -2 \\ 0 & 4 & 12 \\ 7 & -1 & 1 \end{pmatrix} \begin{pmatrix} x_1 \\ x_2 \\ x_3 \end{pmatrix} = \begin{pmatrix} 25 \\ 100 \\ 55 \\ 75 \\ 250 \\ 15 \end{pmatrix}$$

Section 9.4

Find the matrix of cofactors for each of the following matrices.

45 $\begin{pmatrix} 25 & -10 \\ -5 & 20 \end{pmatrix}$

46 $\begin{pmatrix} 8 & -4 \\ -3 & 12 \end{pmatrix}$

47 $\begin{pmatrix} 8 & -2 & 9 \\ 2 & 18 & 2 \\ 3 & -4 & 8 \end{pmatrix}$

48 $\begin{pmatrix} 4 & 12 & -7 \\ 6 & 10 & 0 \\ 3 & -7 & -8 \end{pmatrix}$

49 $\begin{pmatrix} 16 & -8 \\ -2 & 20 \end{pmatrix}$

50 $\begin{pmatrix} 40 & -80 \\ -20 & 26 \end{pmatrix}$

51 $\begin{pmatrix} 7 & 4 & -2 \\ 0 & -8 & 3 \\ 2 & -6 & -4 \end{pmatrix}$

52 $\begin{pmatrix} 20 & 15 & -10 \\ 0 & -20 & 0 \\ -5 & 30 & -6 \end{pmatrix}$

Find the determinant for each of the following matrices.

53 $\begin{pmatrix} -6 & 25 \\ -10 & -20 \end{pmatrix}$

54 $\begin{pmatrix} -2 & -3 \\ -6 & -8 \end{pmatrix}$

55 $\begin{pmatrix} -10 & 16 \\ 5 & -8 \end{pmatrix}$

56 $\begin{pmatrix} x+1 & x+2 \\ x-3 & x-1 \end{pmatrix}$

57 $\begin{pmatrix} 1 & -1 & 1 \\ -1 & 1 & -1 \\ 1 & -1 & 1 \end{pmatrix}$

58 $\begin{pmatrix} 2 & 0 & 1 \\ 1 & -1 & -2 \\ 0 & 2 & 1 \end{pmatrix}$

59 $\begin{pmatrix} 2 & 0 & -1 \\ 5 & 2 & 3 \\ -10 & 0 & 5 \end{pmatrix}$

60 $\begin{pmatrix} -3 & 4 & 1 \\ 2 & 0 & -2 \\ 5 & 3 & 1 \end{pmatrix}$

61 $\begin{pmatrix} 3 & -1 & 2 & 0 \\ 4 & 6 & 0 & 1 \\ 7 & -1 & 0 & 5 \\ 0 & 1 & 0 & 1 \end{pmatrix}$

62 $\begin{pmatrix} 1 & 0 & 0 & 0 \\ 0 & 1 & 0 & 0 \\ 0 & 0 & 1 & 0 \\ 0 & 0 & 0 & 1 \end{pmatrix}$

Solve the following systems of equations, using Cramer's rule.

63 $\begin{aligned} x_1 + x_2 &= -1 \\ 2x_1 - x_2 &= 7 \end{aligned}$

64 $\begin{aligned} -x_1 + 2x_2 &= 24 \\ 3x_1 - 6x_2 &= 10 \end{aligned}$

65 $\begin{aligned} 2x_1 - 3x_2 + x_3 &= 1 \\ x_1 + x_2 + x_3 &= 2 \\ 3x_1 \qquad - 4x_3 &= 17 \end{aligned}$

66 $\begin{aligned} x_1 + x_2 + x_3 &= 10 \\ 5x_1 - 2x_2 + x_3 &= 3 \\ 3x_1 + x_2 - 4x_3 &= -1 \end{aligned}$

Section 9.5

For the following matrices, find the inverse (if it exists).

67 $\begin{pmatrix} 0 & 1 \\ 8 & 3 \end{pmatrix}$

68 $\begin{pmatrix} 4 & 8 \\ 7 & 2 \end{pmatrix}$

69 $\begin{pmatrix} 1 & 8 & 7 \\ 2 & 3 & 5 \\ 7 & 4 & -1 \end{pmatrix}$

70 $\begin{pmatrix} 4 & -4 & 3 \\ -3 & 2 & -2 \\ 1 & -1 & 0 \end{pmatrix}$

71 $\begin{pmatrix} 8 & 10 \\ 11 & 13 \end{pmatrix}$

72 $\begin{pmatrix} 4 & 2 & 6 \\ 9 & 3 & 7 \\ -1 & 0 & 5 \end{pmatrix}$

73 $\begin{pmatrix} 4 & 1 & 2 \\ 2 & 5 & 0 \\ 1 & 2 & 3 \end{pmatrix}$

74 $\begin{pmatrix} 2 & -1 & 2 \\ 4 & 3 & 1 \\ 10 & -5 & 10 \end{pmatrix}$

75 $\begin{pmatrix} -12 & 3 & -6 \\ 3 & 2 & 1 \\ 4 & -1 & 2 \end{pmatrix}$

76 $\begin{pmatrix} 10 & 4 & 2 \\ -5 & 6 & 4 \\ 2 & 1 & 2 \end{pmatrix}$

Using the results from Exercises 67–76, determine the solution of the following systems of equations.

77 $\quad x_2 = 4$
$\quad 8x_1 + 3x_2 = 12$

78 $4x_1 + 8x_2 = 12$
$\quad 7x_1 + 2x_2 = 9$

79 $\quad x_1 + 8x_2 + 7x_3 = 0$
$\quad 2x_1 + 3x_2 + 5x_3 = 4$
$\quad 7x_1 + 4x_2 - x_3 = 2$

80 $\quad 4x_1 - 4x_2 + 3x_3 = 7$
$\quad -3x_1 + 2x_2 - 2x_3 = -7$
$\quad x_1 - x_2 = 1$

81 $\quad 8x_1 + 10x_2 = 4$
$\quad 11x_1 + 13x_2 = 4$

82 $\quad 4x_1 + 2x_2 + 6x_3 = 2$
$\quad 9x_1 + 3x_2 + 7x_3 = 20$
$\quad -x_1 + 5x_3 = -10$

83 $4x_1 + x_2 + 2x_3 = 5$
$\quad 2x_1 + 5x_2 = -13$
$\quad x_1 + 2x_2 + 3x_3 = 1$

84 $\quad 2x_1 - x_2 + 2x_3 = 20$
$\quad 4x_1 + 3x_2 + x_3 = 15$
$\quad 10x_1 - 5x_2 + 10x_3 = 10$

85 $-12x_1 + 3x_2 - 6x_3 = 18$
$\quad 3x_1 + 2x_2 + x_3 = 10$
$\quad 4x_1 - x_2 + 2x_3 = 14$

86 $10x_1 + 4x_2 + 2x_3 = 46$
$\quad -5x_1 + 6x_2 + 4x_3 = 9$
$\quad 2x_1 + x_2 + 2x_3 = 10$

Section 9.6

87 **College Admissions** The admissions office for a large university plans on admitting 9,000 students next year. The column vector **M** indicates the expected breakdown of the new students into the categories of in-state males (ISM), in-state females (ISF), out-of-state males (OSM), and out-of-state females (OSF):

$$\mathbf{M} = \begin{pmatrix} 3{,}600 \\ 3{,}150 \\ 1{,}000 \\ 1{,}250 \end{pmatrix} \begin{matrix} \text{ISM} \\ \text{ISF} \\ \text{OSM} \\ \text{OSF} \end{matrix}$$

Admissions personnel expect the students to choose their majors within the colleges of business (B), engineering (E), and arts and sciences (A&S) according to the percentages given in the matrix **P**:

$$
\mathbf{P} =
\begin{array}{c}
\quad\ \ \text{ISM}\ \ \ \text{ISF}\ \ \ \text{OSM}\ \ \ \text{OSF} \\
\left(
\begin{array}{cccc}
0.30 & 0.30 & 0.30 & 0.24 \\
0.20 & 0.10 & 0.30 & 0.06 \\
0.50 & 0.60 & 0.40 & 0.70
\end{array}
\right)
\begin{array}{l}
\text{B} \\
\text{E} \\
\text{A\&S}
\end{array}
\end{array}
$$

Using matrix operations, compute the number of students expected to enter each college.

88 Refer to Exercise 87: The housing office estimates that students will select housing alternatives according to the percentages in **H**:

$$
\mathbf{H} =
\begin{array}{c}
\qquad\qquad\ \ \text{Fraternity} \\
\text{Dorm}\ \ \ \text{or sorority}\ \ \ \text{Off campus} \\
\left(
\begin{array}{ccc}
0.40 & 0.20 & 0.40 \\
0.70 & 0.20 & 0.10
\end{array}
\right)
\begin{array}{l}
\text{IS} \\
\text{OS}
\end{array}
\end{array}
$$

Perform a matrix multiplication which will compute the number of new students expected to choose the different housing options.

89 A company manufactures three products, each of which requires certain amounts of three raw materials as well as labor. The matrix **R** summarizes the requirements per unit of each product.

$$
\mathbf{R} =
\begin{array}{c}
\qquad \text{Raw material} \\
\ 1\ \ \ 2\ \ \ 3\ \ \ \text{Labor} \\
\left(
\begin{array}{cccc}
2 & 3 & 2 & 6 \\
3 & 2 & 8 & 8 \\
4 & 2 & 5 & 4
\end{array}
\right)
\begin{array}{l}
\text{Product } A \\
\text{Product } B \\
\text{Product } C
\end{array}
\end{array}
$$

Raw material requirements are stated in pounds per unit and labor requirements in hours per unit. The three raw materials cost $2, $8, and $2.50 per pound, respectively. Labor costs are $8 per hour. Assume 800, 2,000, and 600 units of products A, B, and C are to be produced.

(a) Perform a matrix multiplication which computes total quantities of the four resources required to produce the desired quantities of products A, B, C.

(b) Using your answer from part a, perform a matrix multiplication which calculates the combined total cost of production.

90 **Hospital Administration** A local hospital has gathered data regarding people admitted for in-patient services. The vector **P** indicates the percentages of all patients admitted to different hospital units. The vector **S** indicates the average length of patient stay (in days) for each hospital unit.

$$
\mathbf{P} =
\left(
\begin{array}{c}
0.18 \\
0.10 \\
0.24 \\
0.48
\end{array}
\right)
\begin{array}{l}
\text{Obstetrical} \\
\text{Cardiac} \\
\text{Pediatric} \\
\text{Other}
\end{array}
\qquad
\mathbf{S} = (3\quad 16\quad 2\quad 4)
$$

The vector **C** summarizes current daily patient cost for the different hospital units:

$$\mathbf{C} = (\$680 \quad \$1,400 \quad \$540 \quad \$360)$$

If 300 new patients are admitted, perform a matrix multiplication to compute:

(a) The numbers of patients admitted to each hospital unit.

(b) The total number of patient-days expected.

(c) Total cost per day for the 300 patients.

91 Social Interaction A group of executives has been surveyed regarding the persons who have direct influence over their decision making. The following matrix summarizes their responses.

	Person whose opinion is sought					
	A	B	C	D	E	F
A	0	1	0	1	0	0
B	1	0	0	0	1	0
C	1	0	0	0	0	1
D	0	1	1	0	0	1
E	0	0	0	1	0	0
F	1	0	0	0	1	0

An entry of 1 indicates that the person represented by the corresponding column has some direct influence over the decision making of the person represented by the row. An entry of 0 indicates no direct influence. As with Example 38, this matrix is similar to the adjacency matrix. The square of the matrix would indicate indirect influences on decision making involving an intermediary. Square the matrix and verbally summarize the indirect influences.

92 The technological matrix for a three-industry input-output model is

$$\mathbf{A} = \begin{pmatrix} 0.5 & 0 & 0.2 \\ 0.2 & 0.8 & 0.12 \\ 1 & 0.4 & 0 \end{pmatrix}$$

If the nonindustry demand for the output of these industries is $d_1 = \$5$ million, $d_2 = \$3$ million, and $d_3 = \$4$ million, determine the equilibrium output levels for the three industries.

93 Wildlife Migration Scientists have been studying the migration habits of a particular species of wildlife. An annual census is conducted in three different regions inhabited by the species. A stable pattern of changes has been observed in their movements. This is reflected in the following transition matrix.

		To region		
		1	2	3
From region	1	0.90	0.05	0.05
	2	0.10	0.80	0.10
	3	0.05	0.10	0.85

Assume the populations of the three regions were 40,000, 20,000, and 30,000 during the last census. Predict the populations of each region at the time of the next census and 2 years from now.

94 Figure 9.10 is a network diagram which illustrates the route structure for a commercial bus company which services eight cities. Using this figure, construct an adjacency matrix. Square the adjacency matrix and verbally summarize the one-stop service which exists between all cities.

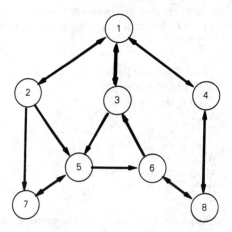

Figure 9.10   Bus routes.

❏ CHAPTER TEST

1 Find the transpose of **A** if

$$A = \begin{pmatrix} 4 & 1 & -6 & 8 \\ 0 & -3 & 10 & 5 \\ 6 & -2 & -4 & 19 \end{pmatrix}$$

2 Find the inner product:

$$(a \quad b \quad c \quad d) \begin{pmatrix} e \\ f \\ g \\ h \end{pmatrix}$$

3 Given the matrices

$$A = \begin{pmatrix} 2 & -3 \\ 3 & 14 \end{pmatrix} \quad B = \begin{pmatrix} 2 & -6 \\ 1 & 10 \\ 9 & -8 \end{pmatrix} \quad C = \begin{pmatrix} -1 & 0 & 0 \\ 0 & 1 & 0 \\ 0 & 0 & -1 \end{pmatrix}$$

determine, if possible, (*a*) **AB**, (*b*) **BA**, (*c*) **BC**, and (*d*) **CA**.

4 Write the following system of equations as a matrix product:

$$x_1 \qquad\qquad - x_4 = 20$$
$$x_2 + x_3 \qquad = 15$$
$$x_3 + x_4 = 18$$
$$x_4 = \ 9$$

5 Find the determinant for the matrix

$$A = \begin{pmatrix} 0 & 10 & -2 \\ 6 & -2 & 1 \\ 0 & 8 & -10 \end{pmatrix}$$

6 Find A^{-1} if

$$A = \begin{pmatrix} 20 & -8 \\ -5 & 2 \end{pmatrix}$$

7 The solution to a system of equations having the form $AX = B$ can be found by the matrix multiplication

$$X = \begin{pmatrix} 5 & -7 \\ -2 & 3 \end{pmatrix}\begin{pmatrix} 15 \\ 11 \end{pmatrix}$$

What was the original system of equations?

Computer-Based Exercises

Using an appropriate software package, solve the following exercises.

1 Given the following matrices

$$A = \begin{pmatrix} 2 & -3 & 2 & 1 & 4 \\ 1 & 2 & 3 & 4 & 5 \\ 2 & 0 & -1 & 5 & 4 \\ 4 & 2 & 3 & 1 & 0 \\ 5 & 4 & 3 & 2 & 1 \end{pmatrix} \qquad B = \begin{pmatrix} 3 & 0 & 1 \\ 2 & 2 & 5 \\ 1 & 4 & 3 \\ 6 & 5 & 0 \\ 1 & 2 & 3 \end{pmatrix}$$

$$C = \begin{pmatrix} 1 & -5 & 10 & 0 & 2 & 4 & -5 & 3 \\ 2 & 0 & -2 & 1 & 3 & 1 & 2 & -2 \\ 3 & -1 & 5 & 1 & -2 & 0 & -2 & 4 \\ 4 & 3 & 2 & 2 & 0 & 3 & 3 & 0 \\ 5 & 2 & 4 & 0 & 6 & 5 & 1 & 5 \end{pmatrix}$$

compute:
(a) **AB**
(b) **AC**
(c) $B^T C$

(d) $C^T A^T$
(e) A^2
(f) A^3

2 Given the following matrices

$$E = \begin{pmatrix} 4 & -3 & 5 & 8 & -5 & 4 \\ 0 & 10 & -4 & -3 & 6 & -5 \\ -15 & 8 & -8 & 20 & 25 & 3 \\ 6 & 14 & 15 & -9 & 10 & 9 \\ 0 & 24 & -8 & -5 & 30 & 18 \\ 28 & -7 & 16 & 0 & 10 & 12 \end{pmatrix} \qquad F = \begin{pmatrix} -8 & 35 & 25 \\ 15 & -15 & -20 \\ 0 & 12 & 24 \\ 18 & -6 & 10 \\ 24 & 16 & -9 \\ 15 & 20 & -5 \end{pmatrix}$$

$$G = \begin{pmatrix} 10 & 20 & 30 & 35 & 20 & -15 & -18 & -12 & 0 & 0 \\ -8 & 10 & -8 & 10 & 12 & 28 & 0 & 16 & 14 & 18 \\ 25 & 10 & 35 & 26 & -6 & -14 & -15 & 20 & 12 & 10 \\ 0 & 15 & 25 & 10 & 5 & 16 & 8 & -5 & -8 & -2 \\ 4 & 3 & 5 & -8 & 10 & 12 & -5 & 20 & 15 & 0 \\ 10 & 12 & -6 & 20 & 0 & 6 & 2 & 3 & 8 & 14 \end{pmatrix}$$

compute:

(a) **EF**

(b) **EG**

(c) **G**^{T}**F**

(d) **EE**^{T}

(e) **E**^{2}

(f) **E**^{3}

(g) **G**^{T}**E**

3 Given the following system of equations having the matrix form $\mathbf{AX = B}$,

$$
\begin{array}{rcl}
x_1 + x_2 + x_3 + x_4 + x_5 + x_6 + x_7 + x_8 + x_9 + x_{10} &=& 8 \\
2x_1 - x_2 + x_3 \quad\quad - 4x_5 + 2x_6 \quad\quad - x_8 &=& -8 \\
3x_2 \quad - x_4 + 5x_5 \quad\quad - 2x_7 \quad + 3x_9 &=& 6 \\
x_1 \quad\quad + x_5 \quad\quad\quad - 3x_8 \quad + 4x_{10} &=& -9 \\
5x_1 - 2x_2 + 4x_3 &=& 0 \\
3x_6 - 5x_7 + 2x_8 - x_9 + x_{10} &=& 5 \\
5x_3 - 2x_4 + 6x_5 - 4x_6 &=& 1 \\
x_1 - x_2 + 3x_3 + 2x_4 - 5x_5 &=& -9 \\
3x_1 + 5x_2 - x_3 + 2x_4 + x_5 - x_6 + 2x_7 - 5x_8 - x_9 + x_{10} &=& 7 \\
x_2 \quad - x_4 \quad + x_6 \quad\quad - 2x_8 \quad + x_{10} &=& -5
\end{array}
$$

(a) Find $\mathbf{A}^{-1}$.

(b) Perform the computation $\mathbf{A}^{-1}\mathbf{B}$ to determine the solution to the system.

4 Given the following system of equations having the matrix form $\mathbf{AX = B}$,

$$
\begin{array}{rcl}
x_1 + x_2 + x_3 + x_4 + x_5 + x_6 + x_7 + x_8 &=& 11 \\
2x_1 - x_2 + x_3 + 4x_4 &=& 11 \\
x_5 + x_6 + x_7 + x_8 &=& 5 \\
x_1 \quad + x_3 \quad + x_5 \quad + x_7 &=& 8 \\
x_2 \quad + x_4 \quad + x_6 \quad + x_8 &=& 3 \\
3x_1 + 2x_2 + x_3 + x_4 + 2x_5 + x_6 &=& 10 \\
x_3 + x_4 + x_5 + x_6 &=& 5 \\
5x_2 + 2x_3 \quad - x_5 \quad + 6x_7 - 5x_8 &=& 31
\end{array}
$$

(a) Find $\mathbf{A}^{-1}$.

(b) Perform the computation $\mathbf{A}^{-1}\mathbf{B}$ to determine the solution to the system.

5 The following adjacency matrix indicates the direct connections one airline has among 10 different cities.

$$
\begin{array}{c}
\text{To}\\
\begin{array}{cccccccccc}
1 & 2 & 3 & 4 & 5 & 6 & 7 & 8 & 9 & 10
\end{array}
\end{array}
$$

From	1	2	3	4	5	6	7	8	9	10
1	0	1	1	0	0	0	0	1	0	1
2	1	0	0	1	1	0	0	1	1	0
3	0	1	0	0	1	0	1	0	0	1
4	0	0	1	0	1	1	0	0	1	0
5	0	0	0	0	0	0	1	1	1	0
6	1	0	0	1	0	0	1	0	0	0
7	0	0	0	0	1	0	0	1	1	1
8	0	0	0	0	0	1	0	0	1	0
9	0	0	0	0	0	1	0	1	0	0
10	1	0	0	0	0	0	1	0	0	0

(a) Square the matrix and verbally summarize *one-stop* service between all cities.

(b) Cube the matrix and verbally summarize *two-stop* service between all cities.

6 A company is trying to decide how many of each of 10 products to produce during the coming quarter (3 months). It is not necessary that all 10 products be produced. However, the following constraints exist. First, total production of the 10 products should equal 10,000 units during the quarter. The number of units produced of product 1 should be twice that of product 2. Combined production of products 4, 5, and 8 should equal 4,000 units, and the number of units produced of product 10 should equal the combined production of products 1 and 2. In addition, it is desired that the company's six departments be utilized to capacity during the quarter. The following table summarizes the number of hours required to produce the different products in the six departments as well as the number of hours available in each department.

| | Product | | | | | | | | | | Hours |
	1	2	3	4	5	6	7	8	9	10	Available
Department 1	4	2	3	1	0	0	0	0	0	0	10,000
Department 2	1	1	1	1	2	1	0	0	0	1	9,000
Department 3	0	0	4	2	2	0	1	2	0	1	11,000
Department 4	0	1	3	2	2	5	0	1	1	0.5	8,000
Department 5	0.5	2	2	1	2	1	3	0	0	0	6,000
Department 6	0	0	4	2	1	1	2	1	0	0	4,000

(a) Formulate the system of equations which, when solved, will determine the number of units of the 10 products to produce.

(b) Use the matrix inverse method to solve the system.

(c) Verbally summarize the results.

7 The input-output matrix for a six-industry economy is shown in the following.

$$
\begin{array}{c}
 & & \text{User} \\
 & & \begin{array}{cccccc} 1 & 2 & 3 & 4 & 5 & 6 \end{array} \\
\text{Supplier}\; \begin{array}{c} 1 \\ 2 \\ 3 \\ 4 \\ 5 \\ 6 \end{array} &
\left(\begin{array}{cccccc}
0.180 & 0.005 & 0.000 & 0.003 & 0.000 & 0.010 \\
0.005 & 0.290 & 0.020 & 0.002 & 0.004 & 0.015 \\
0.030 & 0.170 & 0.450 & 0.008 & 0.010 & 0.006 \\
0.035 & 0.040 & 0.020 & 0.040 & 0.010 & 0.050 \\
0.010 & 0.001 & 0.040 & 0.250 & 0.360 & 0.250 \\
0.120 & 0.080 & 0.100 & 0.120 & 0.180 & 0.230
\end{array}\right)
\end{array}
$$

If nonindustry demands are, respectively, $20 billion, $10 billion, $30 billion, $5 billion, $12 billion, and $30 billion, (*a*) determine the equilibrium levels of output for the six industries and (*b*) determine the interindustry demands for the six industries.

HUMAN RESOURCE PLANNING

A national discount retailing company has gathered data regarding the movement of its employees throughout its organization. The following table reflects annual movement patterns (transitions) for a subset of company positions. In the table, columns indicate the position held during one year and rows indicate the positions held during the immediately following year. The elements within the table reflect the transition probabilities from one year to the next. For example, the first element in the table indicates that 95 percent of all persons who are store managers in one year will be in store manager positions in the following year. The other element in the first column indicates that of all persons who are store managers in one year, 5 percent will leave the company in the following year.

In 1992 there were 500 store managers, 850 assistant store managers, 3,600 department managers, 14,500 salespersons, 8,600 cashiers, 1,600 buyers, 3,000 assistant buyers, and 6,000 stock persons.

1 Discuss the meaning of the elements which equal zero (empty cells) in the table.

2 Interpret the meaning of each of the elements in the first row of the table.

3 Of the persons who are part of the 1992 work force, predict the numbers who will be in each job category in 1993, 1994, and 1995.

4 Given the projections for 1995, what job categories reflect increased supplies from within the organization, compared with 1992? Which categories reflect decreases?

5 If the projected needs for 1995 are 550 store managers, 920 assistant store managers, 3,900 department managers, 15,200 salespersons, 9,200 cashiers, 1,700 buyers, 3,200 assistant buyers, and 6,800 stock persons, what level of external hiring is anticipated to fulfill 1995 needs?

Job Held Next Year	Job Held Current Year							
	Store Mgr.	Asst. Store Mgr.	Dept. Mgr.	Sales-person	Cashier	Buyer	Asst. Buyer	Stock Person
Store mgr.	0.95	0.20				0.05		
Asst. store mgr.		0.70	0.20			0.10	0.10	
Dept. mgr.			0.65	0.10				
Sales-person				0.80	0.30			
Cashier					0.45			0.20
Buyer						0.75	0.20	
Asst. buyer			0.10		0.10		0.65	
Stock person								0.70
Leave firm	0.05	0.10	0.05	0.10	0.15	0.10	0.05	0.10

LINEAR PROGRAMMING: AN INTRODUCTION

CHAPTER OBJECTIVES

❑ Provide an understanding of the structure and assumptions underlying linear programming models

❑ Illustrate the graphical representation of linear inequalities

❑ Provide an understanding of graphical solution procedures for linear programming problems

❑ Illustrate the nature and significance of special phenomena which can arise with linear programming models

❑ Give examples of applications of linear programming models

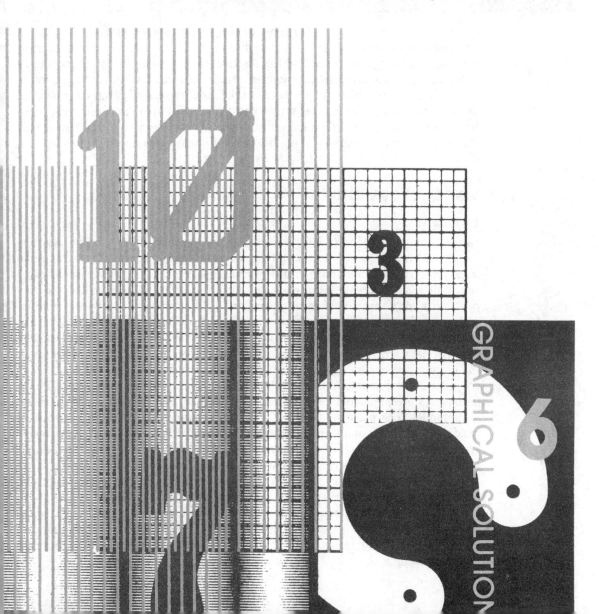

This chapter introduces the topic of *linear programming,* a topic which integrates much of the material we have discussed in the previous chapters. Linear programming is a powerful and widely applied mathematical modeling technique. Also, the technique provides an important framework which is relevant to the more generalized area of modeling techniques called *mathematical programming.* In this chapter we will discuss the nature and structure of linear programming problems. We will illustrate graphical solution procedures for solving small problems. Finally, we will see a variety of applications of linear programming models.

10.1 LINEAR PROGRAMMING

Introduction

Linear programming (LP) is a mathematical optimization technique. By "optimization" technique we mean a method which attempts to maximize or minimize some objective, e.g., maximize profits or minimize costs. Linear programming is a subset of a larger area of mathematical optimization procedures called *mathematical programming.* Linear programming is a powerful and widely applied technique. There have been extensive applications of linear programming within the military and the petroleum industry. Although these sectors have been perhaps the heaviest users of linear programming, the services sector and public sector of the economy have also applied the methods extensively.

In any linear programming problem certain decisions need to be made. These decisions are represented by *decision variables x_j* used in the linear programming model. The basic structure of a linear programming problem is either to maximize or to minimize an *objective function* while satisfying a set of *constraining conditions,* or *constraints.* The objective function is a mathematical representation of the overall goal stated as a function of the decision variables x_j. The objective function may represent goals such as profit level, total revenue, total cost, pollution levels, market share, and percent return on investment.

The set of constraints, also stated in terms of x_j, represents conditions which must be satisfied when determining levels for the decision variables. For example, in attempting to maximize profits from the production and sale of a group of products, sample constraints might reflect limited labor resources, limited raw materials, and limited demand for the products. Other conditions needing to be satisfied take the form of **requirements.** For example, in determining the quantities of different products to produce, minimum production quantities may be specified. The constraints of an LP problem can be represented by equations or by inequalities ($\leq$ and/or $\geq$ types).

These problems are called *linear* programming problems because the objective function and constraints are all linear. A simple linear programming problem is stated below:

$$
\begin{aligned}
\text{Maximize} \quad & z = 4x_1 + 2x_2 \\
\text{subject to} \quad & x_1 + 2x_2 \leq 24 \\
& 4x_1 + 3x_2 \geq 30
\end{aligned}
$$

The objective is to maximize z, which is stated as a linear function of the two decision variables x_1 and x_2. In choosing values for x_1 and x_2, however, two constraints must be satisfied. The constraints are represented by the two linear inequalities.

A Scenario

Product mix problems represent an important group of applications of mathematical modeling. We discussed product mix examples in earlier chapters. Let's illustrate the LP treatment of this type of problem in a simplified example. A firm manufactures two products, each of which must be processed through departments 1 and 2. Table 10.1 summarizes labor-hour requirements per unit for each product in each department. Also presented are weekly labor-hour capacities in each department and respective profit margins for the two products. The problem is to determine the number of units to produce of each product so as to maximize total contribution to fixed cost and profit.

TABLE 10.1	Product *A*	Product *B*	Weekly Labor Capacity
Department 1	3 h per unit	2 h per unit	120 h
Department 2	4 h per unit	6 h per unit	260 h
Profit margin	$5 per unit	$6 per unit	

If we let x_1 and x_2 equal the number of units produced and sold, respectively, of products A and B, then total profit contribution can be found by adding the contributions from both products. The contribution from each product is computed by multiplying the profit margin per unit times the number of units produced and sold. *If z is defined as the total contribution to fixed cost and profit,* we have

$$
z = 5x_1 + 6x_2
$$

From the information given in the statement of the problem, the only restrictions in deciding the number of units to produce are the weekly labor capacities in the two departments. From our discussions in earlier chapters, you should be able to verify that these restrictions can be represented by the inequalities

$$3x_1 + 2x_2 \leq 120 \qquad \text{(department 1)}$$

$$4x_1 + 6x_2 \leq 260 \qquad \text{(department 2)}$$

Although there is no formal statement of such a restriction, implicitly we know that x_1 and x_2 cannot be negative. We must account for this type of restriction in formulating the model.

By combining the objective function and constraints, the LP model which represents the problem is stated as follows:

Maximize	$z = 5x_1 + 6x_2$	
subject to	$3x_1 + 2x_2 \leq 120$	**(10.1)**
	$4x_1 + 6x_2 \leq 260$	**(10.2)**
	$x_1 \geq 0$	**(10.3)**
	$x_2 \geq 0$	**(10.4)**

Structural Constraints and Nonnegativity Constraints

The linear programming model is concerned with maximizing or minimizing a linear objective function subject to two types of constraints: (1) *structural constraints* and (2) *nonnegativity constraints,* one for each decision variable. Structural constraints reflect such factors as resource limitations and other conditions imposed by the problem setting. Inequalities (10.1) and (10.2) in the previous formulation are structural constraints. Nonnegativity constraints guarantee that each decision variable will not be negative. Constraints (10.3) and (10.4) are nonnegativity constraints. In almost all problems the nonnegativity restriction makes sense intuitively. Techniques are available to handle those unusual cases where a variable is allowed to assume negative values.

10.2 GRAPHICAL SOLUTIONS

When a linear programming model is stated in terms of two decision variables, it can be solved by graphical procedures. The graphical approach provides an effective visual frame of reference, and it is extremely helpful in understanding the kinds of phenomena which can occur in solving linear programming problems. In this section we will develop the graphical solution approach. Before discussing the graphical solution method, we will discuss the graphics of linear inequalities.

The Graphics of Linear Inequalities

When a linear inequality involves two variables, the solution set can be described graphically. For example, the inequality

$$-4x + 3y \leq -24$$

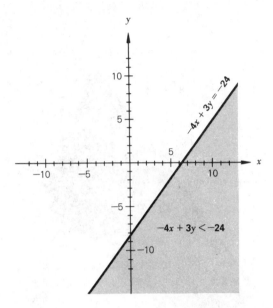

Figure 10.1   Closed half-space representing $-4x + 3y \le -24$.

has a solution set represented by the shaded, ***closed half-space*** (half-plane) in Fig. 10.1. The solution set can be divided into two subsets. One subset consists of all pairs of values (x, y) which satisfy the *equality part* or the equation $-4x + 3y \, \textcircled{=} \, -24$. This subset is represented by the straight line in Fig. 10.1. The other subset consists of all pairs of values (x, y) which satisfy the *inequality part*, or the inequality $-4x + 3y \, \textcircled{<} \, -24$. This subset is represented by the shaded area below and to the right of the straight line in Fig. 10.1.

Our conclusion is as follows: *(1) Linear inequalities which involve two variables can be represented graphically in two dimensions by a closed half-space of the cartesian plane, and (2) the half-space consists of the boundary line representing the equality part of the inequality and all points on one side of the boundary line (representing the strict inequality).*

The procedure for determining the appropriate half-space is as follows:

1   *Graph the boundary line which represents the equation.*
2   *Determine the side of the line satisfying the strict inequality.* To *make this determination, a point can be selected arbitrarily on either side of the line and its coordinates substituted into the inequality (the* origin *is a convenient choice if it does not lie on the line). If the coordinates* satisfy *the inequality, that side of the line is included in the* **permissible half-space.** *If the coordinates do not satisfy the inequality, the permissible half-space lies on the other side of the line.*

To illustrate step 2, had we chosen the origin $(0, 0)$ as a test point, the coordinates do not satisfy the inequality $[-4(0) + 3(0) \not\le -24]$. Since our test point fails to satisfy the inequality, it does not lie within the permissible half-space. Thus, in Fig. 10.1 the permissible half-space is to the right of and below the line.

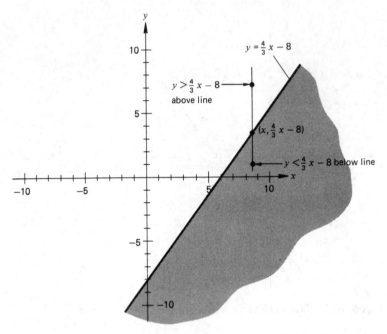

Figure 10.2   Slope-intercept determination of closed half-space for $-4x + 3y \le -24$.

Another way of determining which side of the line satisfies the strict inequality is to solve for the slope-intercept form of the constraint. Given the constraint

$$-4x + 3y \le -24$$

the slope-intercept form is

$$y \le \tfrac{4}{3}x - 8$$

As can be seen in Fig. 10.2, ordered pairs (x, y) which satisfy the slope-intercept form of the *constraint equation*

$$y = \tfrac{4}{3}x - 8$$

are represented by the boundary line. Ordered pairs (x, y) which satisfy the strict inequality

$$y < \tfrac{4}{3}x - 8$$

lie below the line.

   To generalize this approach:

Given a linear inequality of the form $ax + by$ ($\leq$ or $\geq$) c, solve for the slope-intercept form of the inequality. If the slope-intercept inequality has the form

I $\quad y \leq \dfrac{-a}{b}x + \dfrac{c}{b}$, *the corresponding half-space lies below the boundary line.*

II $\quad y \geq \dfrac{-a}{b}x + \dfrac{c}{b}$, *the corresponding half-space lies above the boundary line.*

EXAMPLE 1

A firm manufactures two products. The products must be processed through one department. Product A requires 4 hours per unit, and product B requires 2 hours per unit. Total production time available for the coming week is 60 hours. A restriction in planning the production schedule, therefore, is that total hours used in producing the two products cannot exceed 60; or, *if x_1 equals the number of units produced of product A and x_2 equals the number of units produced of product B,* the restriction is represented by the inequality

$$4x_1 + 2x_2 \leq 60$$

There are two other restrictions implied by the variable definitions. Since each variable represents a production quantity, neither variable can be negative. These restrictions are represented by the inequalities $x_1 \geq 0$ and $x_2 \geq 0$.

The solution set of the original inequality represents the different combinations of the two products which can be manufactured while not exceeding the 60 hours. Figure 10.3 illustrates the solution set graphically. Check to see whether the permissible half-space has

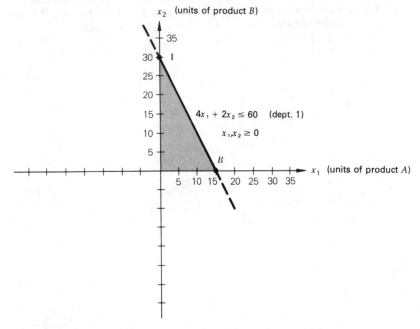

Figure 10.3 Production possibilities: Department 1.

been identified correctly. The points satisfying the inequality $4x_1 + 2x_2 \leq 60$ would be the half-space including all points on and to the left of the line. However, the restriction that both variables not be negative confines us to the portion of the half-space in the first quadrant. Thus, the shaded area represents the combinations of products A and B which can be produced. A further distinction can be made in Fig. 10.3. All combinations of the two products represented by points on $\overline{AB}$ would use all 60 hours. Any points in the interior of the shaded area represent combinations of the two items which will require fewer than 60 hours. Is the origin a possible decision?

□

Systems of Linear Inequalities

In linear programming problems, we will be dealing with *systems* of linear inequalities. Our interest will be in determining the solution set which satisfies all the inequalities in the system of constraints. To illustrate the graphical representation of systems of linear inequalities, consider the following examples.

EXAMPLE 2   Assume that the products in the previous example also need to be processed through another department in addition to the original department. In this second department, assume that product A requires 3 hours per unit and that product B requires 5 hours per unit. If the second department has 75 hours available each week, the inequality describing production possibilities in this department is

$$3x_1 + 5x_2 \leq 75 \qquad \text{(department 2)}$$

The solution set for this inequality is illustrated in Fig. 10.4. As with Fig. 10.3, the shaded area represents all combinations of products A and B which can be manufactured in the second department while not exceeding the 75 hours available.

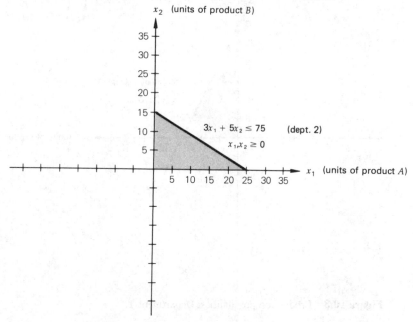

Figure 10.4   Production possibilities: Department 2.

If our objective is to determine the combinations of the two products which can be processed through *both* departments, we are looking for the solution set for the system of linear inequalities

$$4x_1 + 2x_2 \leq 60$$

$$3x_1 + 5x_2 \leq 75$$

$$x_1 \qquad \geq 0$$

$$x_2 \geq 0$$

Figure 10.5 illustrates the composite of the two solution sets in Figs. 10.3 and 10.4. The solution set for the system contains the set of points which are common to the solution sets in these figures. And in Fig. 10.5 the solution set for the system is the shaded area *ABCD*.

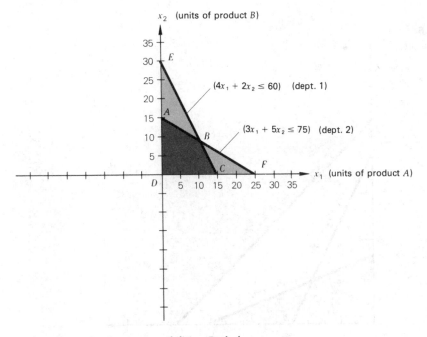

Figure 10.5 Production possibilities: Both departments.

Why are combinations of the two products within area *AEB* not possible? Why are combinations within *BFC* not possible? Is there any unique production characteristic associated with the combination of products represented by point *B*? How about the combinations along $\overline{AB}$? Those along $\overline{BC}$?

EXAMPLE 3 Graphically determine the solution set for the following system.

$$2x_1 + 5x_2 \leq 20$$
$$2x_1 + 2x_2 \geq 24$$
$$2x_1 + x_2 = 10$$
$$x_1 \geq 0$$
$$x_2 \geq 0$$

SOLUTION

Figure 10.6 illustrates the system. Note that the third member of this system is an equation whose solution set is represented by a line. Also, there are no points common to the first two inequalities. Therefore, the solution set contains no elements. There are no points (x_1, x_2) which satisfy all the relationships in the system.

Figure 10.6
No solution set.

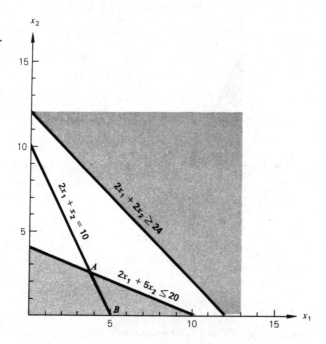

Region of Feasible Solutions

In Sec. 10.1 we formulated a two-variable, product mix LP problem. The formulation is rewritten as follows:

$$\text{Maximize} \quad z = 5x_1 + 6x_2$$
$$\text{subject to} \quad 3x_1 + 2x_2 \leq 120 \quad \text{(department 1)} \quad \textbf{(10.5)}$$

$$4x_1 + 6x_2 \leq 260 \qquad \text{(department 2)} \qquad \textbf{(10.6)}$$

$$x_1, x_2 \geq 0 \qquad\qquad\qquad\qquad \textbf{(10.7)}$$

where x_1 and x_2 represent the number of units produced of products A and B. Since the problem involves two decision variables, we can determine the optimal solution graphically. *The first step in the graphical procedure is to identify the solution set for the system of constraints.* This solution set is often called the **region of feasible solutions.** It includes all combinations of the decision variables which satisfy the structural and nonnegativity constraints. These combinations can be thought of as *candidates* for the optimal solution. The solution set for inequalities (10.5)–(10.7) is indicated in Fig. 10.7. This is the region of feasible solutions for the linear programming problem.

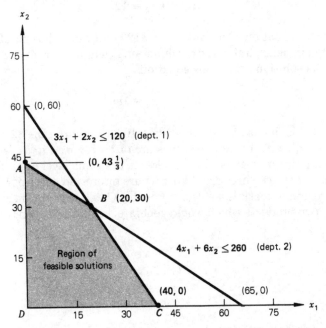

Figure 10.7 Region of feasible solutions (product mix problem).

NOTE   The coordinates of points A and C are identified as the intercept values. Because of inaccuracies in the graph, it may be difficult to read the exact coordinates of some points such as point B. To determine the exact coordinates of such points, the *equations* of the lines intersecting at the point must be solved simultaneously. To determine the coordinates (20, 30), the equality portions of (10.5) and (10.6) are solved simultaneously.

Each point within the region of feasible solutions in Fig. 10.7 represents a combination of the two products which can be produced. The problem is to determine the combination(s) which maximize the value of the objective function.

Incorporating the Objective Function

The LP solution procedure involves a search of the region of feasible solutions for the optimal solution. Before we present the search procedure, let's first examine some characteristics of objective functions. In the product mix problem, let's identify combinations of the two products which would generate some predetermined profit level. For instance, if we wanted to determine the different combinations of the two products which would generate a profit of $120, we would set the objective function equal to 120:

$$5x_1 + 6x_2 = 120$$

If we graph this equation, the result is the $120 profit line shown in Fig. 10.8. If we are interested in determining the combinations yielding a profit of $180, we would determine the solution set for the equation

$$5x_1 + 6x_2 = 180$$

The graph of this line is also shown in Fig. 10.8. Similarly, the $240 profit line is indicated in Fig. 10.8. These three lines are often referred to as *isoprofit lines* because each point on a given line represents the same level of profit.

Note that for these three profit lines we are interested in the portions which lie within the region of feasible solutions. For the $240 line there are some combinations of the two products which would generate a combined profit of $240 but are

Figure 10.8
Isoprofit lines.

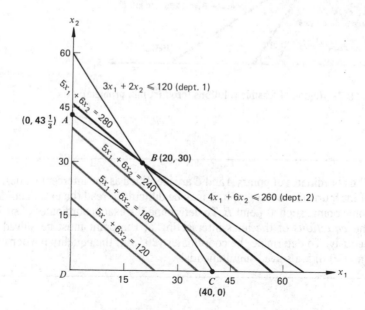

not within the region of feasible solutions. An example of such a combination is 48 units of product A and no units of product B. Even though this combination would generate a profit of $240, there are insufficient hours (in department 1) to produce these quantities.

Observe from the three profit lines that profit levels increase as the lines move outward from the origin. It also appears that the three profit lines are parallel to one another. This can be verified quickly by rewriting the profit function

$$z = 5x_1 + 6x_2 \qquad \textbf{(10.8)}$$

in the slope-intercept form. In Fig. 10.8 x_2 is equivalent to y (if we had named our variables x and y). If we solve Eq. (10.8) for x_2, we get

$$x_2 = \frac{-5}{6}x_1 + \frac{z}{6}$$

The slope for the objective function is $-\frac{5}{6}$, and it is not influenced by the value of z. It is determined solely by the coefficients of the two variables in the objective function.

The x_2 (or y) intercept is defined by $(0, z/6)$. From this, we can see that as z changes in value, so does the x_2 intercept. If z increases in value, so does the x_2 intercept, meaning that the isoprofit line moves up and to the right. If we are interested in maximizing profit, we want to move the profit line as far outward as possible while still touching a point within the region of feasible solutions. In sliding outward from the $240 line, the last point to be touched is B, with coordinates $(20, 30)$. This point lies on the $280 profit line. *Our conclusion: Profit is maximized at a value of $280 when 20 units and 30 units are manufactured, respectively, of products A and B.*

EXAMPLE 4 (**Minimization Problem**) Determine the optimal solution to the linear programming problem

$$\begin{aligned} \text{Minimize} \qquad & z = 3x_1 + 6x_2 \\ \text{subject to} \qquad & 4x_1 + x_2 \geq 20 \\ & x_1 + x_2 \leq 20 \\ & x_1 + x_2 \geq 10 \\ & x_1, x_2 \geq 0 \end{aligned}$$

SOLUTION

Figure 10.9a illustrates the region of feasible solutions for the set of constraints. In an effort to determine the optimal solution, let's determine the orientation of the objective function. Let's assume an arbitrary value for z, say 60. The equation

$$3x_1 + 6x_2 = 60$$

Figure 10.9

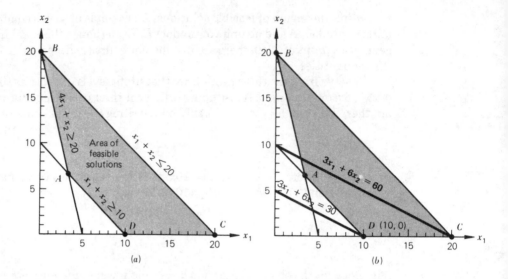

is graphed in Fig. 10.9*b*. To determine the direction of movement of the objective function, we can choose a point on either side of the line and determine the corresponding value of *z*. If we select the origin, we find the value of the objective function at (0, 0) is

$$z = 3(0) + 6(0)$$
$$= 0$$

The value of *z* at the origin is less than 60, and our conclusion is that *movement of the objective function toward the origin results in lower values of z.* Since we want to minimize *z*, we will want to move the objective function, parallel to itself, as close to the origin as possible while still having it touch a point in the region of feasible solutions. The last point touched before the function moves entirely out of the region of feasible solutions is *D*, or (10, 0). Given that the minimum value for *z* occurs at (10, 0), the minimum value is computed as

$$z = 3(10) + 6(0)$$
$$= 30$$

❑

Corner-Point Solutions

The search procedure can be simplified if we take advantage of the joint character-istics of the region of feasible solutions and the objective function. A ***convex set*** is a set of points such that if any two arbitrarily selected points within the set are connected by a straight line, all elements on the line segment are also members of the set. Figure 10.10 illustrates the difference between a convex set and a noncon-vex set. The set of points in Fig. 10.10*b* represents a convex set. If any two points within the set are connected by a line segment, each point on the line segment will also be a member of the set. In contrast to this, Fig. 10.10*a* illustrates a nonconvex set. For this set there are many pairs of points like *A* and *B* for which the connecting line segment contains points that are *not* members of the set.

Figure 10.10

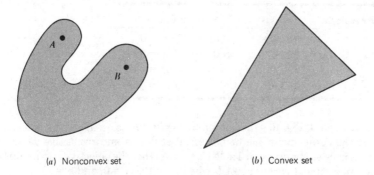

(a) Nonconvex set   (b) Convex set

This leads us to the following statements which are of fundamental importance in linear programming.

1 *The solution set for a group of linear inequalities is a convex set. Therefore, the region of feasible solutions (if one exists) for a linear programming problem is a convex set.*

2 *Given a linear objective function in a linear programming problem, the optimal solution will always include a corner point on the region of feasible solutions. This is true regardless of the slope of the objective function and for either maximization or minimization problems.*

The second statement simply implies that when a linear objective function is shifted through a convex region of feasible solutions, the last point touched before it moves entirely outside the area will include at least one corner point.

Therefore, the ***corner-point method*** for solving linear programming problems is as follows:

CORNER-POINT METHOD

I *Graphically identify the region of feasible solutions.*

II *Determine the coordinates of each corner point on the region of feasible solutions.*

III *Substitute the coordinates of the corner points into the objective function to determine the corresponding value of z.*

IV *An optimal solution occurs in a maximization problem at the corner point yielding the highest value of z and in a minimization problem at the corner point yielding the lowest value of z.*

EXAMPLE 5

In the product mix example in this section, the objective function to be maximized is $z = 5x_1 + 6x_2$. Corner points on the region of feasible solutions were $(0, 0)$, $(0, 43\frac{1}{3})$, $(20, 30)$, and $(40, 0)$. Substituting these into the objective function, we arrive at the figures in Table 10.2. Note that an optimal solution occurs at $x_1 = 20$ and $x_2 = 30$, resulting in a maximum value for z of 280.

TABLE 10.2	Corner Point	(x_1, x_2)	$z = 5x_1 + 6x_2$
	A	(0, 0)	$5(0) + 6(0) = 0$
	B	$(0, 43\frac{1}{3})$	$5(0) + 6(43\frac{1}{3}) = 260$
	C	(20, 30)	$5(20) + 6(30) = 280^*$
	D	(40, 0)	$5(40) + 7(0) = 200$

EXAMPLE 6   For Example 4, Fig. 10.9 indicates four corner points on the region of feasible solutions. By using the corner-point method, the corner points and respective values of the objective function are summarized in Table 10.3. Given that the objective is to minimize z, the optimal solution occurs at corner point D when $x_1 = 10$, $x_2 = 0$, and $z = 30$.

Had the objective been to maximize z in this problem, a maximum value of 120 would have resulted at corner point B when $x_1 = 0$ and $x_2 = 20$.

TABLE 10.3	Corner Point	(x_1, x_2)	$z = 3x_1 + 6x_2$
	A	$(3\frac{1}{3}, 6\frac{2}{3})$	$3(3\frac{1}{3}) + 6(6\frac{2}{3}) = 50$
	B	(0, 20)	$3(0) + 6(20) = 120$
	C	(20, 0)	$3(20) + 6(0) = 60$
	D	(10, 0)	$3(10) + 6(0) = 30^*$

☐

Alternative Optimal Solutions

In the corner-point method it was stated that an optimal solution will always occur at a corner point on the region of feasible solutions. There is the possibility of more than one optimal solution in a linear programming problem. Figure 10.11 illustrates a case where the objective function has the same slope as constraint (2). If the objective function is improved by moving out, away from the origin, the last points touched before the objective function moves outside the region of feasible solutions are all points on $\overline{AB}$. In this situation there would exist an infinite number of points, each resulting in the same maximum value of z. For situations such as this, we say that there are *alternative optimal solutions* to the problem.

Figure 10.11
Alternative
optimal solutions.

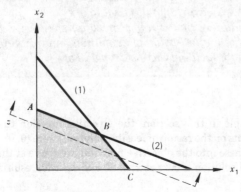

Two conditions need to be satisfied in order for alternative optimal solutions to exist: *(1) the objective function must be parallel to a constraint which forms an edge or boundary on the region of feasible solutions; (2) the constraint must form a boundary on the region of feasible solutions in the direction of optimal movement of the objective function; that is, the constraint must be a binding constraint preventing further improvement in the value of the objective function.* This second condition would be violated in Fig. 10.11 if the problem were one of minimization, i.e., if we desired to shift the objective function in the other direction. Even though the objective function is parallel to constraint (2), the constraint does not prevent us from moving toward the origin.

When using the corner-point method, alternative optimal solutions are signaled when a *tie* occurs for the optimal value of the objective function. The alternative optimal solutions occur at the "tying" corner points, as well as along the entire line segment connecting the two points.

| **EXAMPLE 7** | Solve the following linear programming problem by the corner-point method. |

$$\text{Maximize} \quad z = 20x_1 + 15x_2$$

$$\text{subject to} \quad 3x_1 + 4x_2 \leq 60 \tag{1}$$

$$4x_1 + 3x_2 \leq 60 \tag{2}$$

$$x_1 \leq 10 \tag{3}$$

$$x_2 \leq 12 \tag{4}$$

$$x_1, x_2 \geq 0 \tag{5}$$

SOLUTION

The region of feasible solutions is shown in Fig. 10.12. The corner points and their respective values for z are summarized in Table 10.4. Note that there is a tie for the highest value of z between points D and E. The slope of the objective function is the same as for constraint (2). In Fig. 10.12 there are infinitely many alternative optimal solutions along $\overline{DE}$.

| **POINTS FOR THOUGHT & DISCUSSION** | There are several implications of alternative optimal solutions. One issue is the set of criteria which might be chosen to select the solution to implement. These criteria can include tangible as well as intangible factors. To serve as a basis for discussion, return to the product mix problem on page 430 and re-solve, using the new objective function $z = 4x_1 + 6x_2$. You should verify that alternative optimal solutions exist along $\overline{AB}$ in Fig. 10.7. Discuss the implications of selecting point A versus point B. Consider such issues as the number of production hours consumed in each department and the *mix* of products to be offered to consumers. |

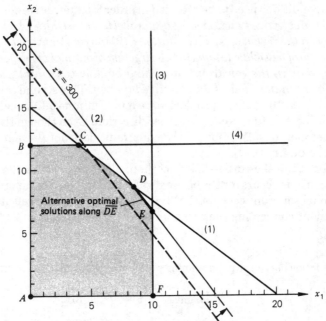

Figure 10.12   Alternative optimal solutions.

TABLE 10.4	Corner Point	(x_1, x_2)	$z = 20x_1 + 15x_2$
	A	$(0, 0)$	$20(0) + 15(0) = 0$
	B	$(0, 12)$	$20(0) + 15(12) = 180$
	C	$(4, 12)$	$20(4) + 15(12) = 260$
	D	$(\frac{60}{7}, \frac{60}{7})$	$20(\frac{60}{7}) + 15(\frac{60}{7}) = 300^*$
	E	$(10, \frac{20}{3})$	$20(10) + 15(\frac{20}{3}) = 300^*$
	F	$(10, 0)$	$20(10) + 15(0) = 200$

No Feasible Solution

The system of constraints in a linear programming problem may have no points which satisfy all constraints. In such cases, there are no points in the solution set, and the linear programming problem is said to have *no feasible solution.* Figure 10.13 illustrates a problem having no feasible solution. Constraint (1) is a "less than or equal to" type, and constraint (2) is a "greater than or equal to" type. A problem can certainly have both types of constraints. In this case the set of points satisfying one constraint includes none of the points satisfying the other.

Unbounded Solutions

Figure 10.14 illustrates what is termed an *unbounded solution space.* The two constraints appear to be ($\geq$) types with the resulting solution space extending *outward* an infinite distance—having no bound. Given an unbounded solution space, the optimal value of the objective function may be bounded or unbounded. If in Fig. 10.14 the *direction of improvement* in the objective function is toward the

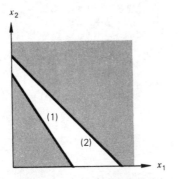

Figure 10.13 No feasible solution.

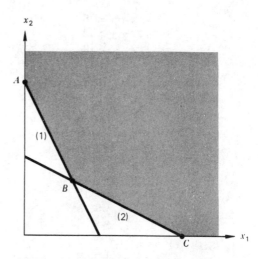

Figure 10.14 Unbounded solution space.

origin—typically a minimization objective—there would be a bound on the value of z and it would be realized at corner point A, B, or C. However, if the direction of improvement is outward, away from the origin—typically maximization—the objective function can be shifted out an infinite distance. Thus, there is no bound on the value of the objective function and the problem is said to have an **unbounded solution.** To repeat for emphasis, *an unbounded solution* occurs when there is no bound or limit on the value of the objective function.

> **NOTE**   An unbounded solution space is a *necessary condition,* but not a sufficient condition, for the occurrence of an unbounded solution.

Section 10.2 Follow-up Exercises

In Exercises 1–10, graphically determine the permissible half-space which satisfies the inequality.

1 $2x + 3y \le 24$

2 $-4x + 4y \ge 36$

3 $0.5x - y \ge 6$

4 $2x + 2.5y \le 40$

5 $1.5x + 4y \le -18$

6 $-x + 2y \ge 7$

7 $-x + 2y \ge -8$

8 $8x - 2y \ge -40$

9 $-2x + 6y \le -24$

10 $8x - 4y \le 40$

In Exercises 11–20, graphically determine the solution space (if one exists).

11 $2x - 4y \le 20$
   $3x + 2y \le 18$

12 $4x + 2y \le 28$
   $3x + 4y \le 48$

13 $5x + 2y \le 20$
   $3x + 4y \le 32$

14 $3x - 2y \le 12$
   $x + 2y \ge 4$

15 $x + y \geq 8$
  $2x + y \geq 12$
  $x \quad \leq 10$
  $x \quad \geq 2$
  $y \leq 10$

16 $x + y \geq 2$
  $x + y \leq 7$
  $x + 2y \geq 14$
  $y \leq 5$
  $x \quad \geq 0$
  $y \geq 0$

17 $4x + 3y \leq 24$
  $x + y \geq 4$
  $x \quad \leq 6$
  $y \leq 6$
  $x \quad \geq 2$
  $y \geq 1$

18 $6x + 3y \leq 24$
  $x \quad \geq 1$
  $x \quad \leq 4$
  $y \geq 1$
  $y \leq 5$

19 $4x - 2y \geq 12$
  $x + y \leq 8$
  $y \leq 6$
  $x \quad \geq 0$
  $y \geq 0$

20 $2x + 2y \leq 16$
  $3x - y \geq 18$
  $x + 2y = 10$
  $x \quad \geq 0$
  $y \geq 0$

For the following LP problems, graph the region of feasible solutions (if one exists) and solve by the corner-point method.

21 Maximize $z = 4x_1 + 8x_2$
subject to $x_1 + x_2 \leq 20$
  $2x_1 + x_2 \leq 32$
  $x_1, x_2 \geq 0$

22 Minimize $z = 5x_1 + 3x_2$
subject to $3x_1 + 2x_2 \geq 60$
  $4x_1 + 5x_2 \geq 90$
  $x_1, x_2 \geq 0$

23 Maximize $z = 30x_1 + 20x_2$
subject to $3x_1 + x_2 \leq 18$
  $x_1 + x_2 \leq 12$
  $x_1 \quad \geq 2$
  $x_2 \geq 5$
  $x_1, x_2 \geq 0$

24 Maximize $z = 10x_1 + 16x_2$
subject to $x_1 \quad \leq 400$
  $x_2 \geq 200$
  $x_1 + x_2 = 500$
  $x_1, x_2 \geq 0$

25 Minimize $z = 20x_1 + 8x_2$
subject to $x_1 + x_2 \geq 20$
  $2x_1 + x_2 \leq 48$
  $x_1 \quad \leq 20$
  $x_1 + x_2 \leq 30$
  $x_1, x_2 \geq 0$

26 Minimize $z = 2x_1 + 5x_2$
subject to $x_1 + x_2 \leq 16$
  $x_1 \quad \leq 12$
  $x_1 \quad \geq 8$
  $x_2 \leq 10$
  $x_2 \geq 4$
  $x_1, x_2 \geq 0$

27 Maximize $z = 4x_1 + 8x_2$
subject to $2x_1 + x_2 \leq 30$
  $x_1 + 2x_2 \leq 24$
  $x_1, x_2 \geq 0$

28 Maximize $z = 16x_1 + 25x_2$
subject to $2x_1 + 4x_2 \leq 40$
  $x_1 + 2x_2 \leq 30$
  $1.5x_1 + x_2 \geq 50$
  $x_1, x_2 \geq 0$

29 Maximize $z = 8x_1 + 4x_2$
subject to $20x_1 + 10x_2 \leq 60$
  $40x_1 + 32x_2 \leq 160$
  $x_1 \quad \leq 2.5$
  $x_2 \leq 4$
  $x_1, x_2 \geq 0$

30 Minimize $z = 10x_1 + 10x_2$
subject to $x_1 + x_2 \geq 12$
  $4x_1 + x_2 \geq 24$
  $x_1 \quad \geq 3$
  $x_2 \leq 18$
  $5x_1 + 4x_2 \leq 120$
  $x_1, x_2 \geq 0$

31 Maximize $z = 18x_1 + 30x_2$
subject to
$$x_1 + x_2 \geq 48$$
$$6x_1 + 9x_2 \leq 216$$
$$15x_1 + 10x_2 \leq 360$$
$$x_1, x_2 \geq 0$$

32 Maximize $z = 6x_1 + 3x_2$
subject to
$$4x_1 + 6x_2 \leq 48$$
$$x_1 + x_2 \geq 15$$
$$x_1, x_2 \geq 0$$

33 A firm manufactures two products. Each product must be processed through two departments. Product A requires 2 hours per unit in department 1 and 4 hours per unit in department 2. Product B requires 3 hours per unit in department 1 and 2 hours per unit in department 2. Departments 1 and 2 have, respectively, 60 and 80 hours available each week. Profit margins for the two products are, respectively, $3 and $4 per unit. If x_j equals the number of units produced of product j, (a) formulate the linear programming model for determining the product mix which maximizes total profit and (b) solve using the corner-point method. (c) Fully interpret the results indicating the recommended product mix. What percentage of daily capacity will be utilized in each department?

34 In Exercise 33, assume that an additional requirement is that the number of units produced of product B must be at least as great as the number produced of product A. What is the mathematical constraint representing this condition? Add this constraint to the original set and re-solve Exercise 33.

35 The dietitian at a local penal institution is preparing the menu for tonight's *light* meal. Two food items will be served at the meal. The dietitian is concerned about achieving the minimum daily requirement of two vitamins. Table 10.5 summarizes vitamin content per ounce of each food, the minimum daily requirements of each, and cost per ounce of each food. If x_j equals the number of ounces of food j:

   (a) Formulate the linear programming model for determining the quantities of the two foods which will minimize the cost of the meal while ensuring that at least minimum levels of both vitamins will be satisfied.

   (b) Solve using the corner-point method, indicating what the minimum-cost meal will consist of and its cost. What percentages of the minimum daily requirements for each vitamin will be realized?

36 In Exercise 35, assume that the amount included of food 1 must be at least 50 percent greater than that of food 2. What is the mathematical constraint representing this condition? Add this constraint to the original set and re-solve Exercise 35.

TABLE 10.5	Food 1	Food 2	Minimum Daily Requirement
Vitamin 1	2 mg/oz	3 mg/oz	18 mg
Vitamin 2	4 mg/oz	2 mg/oz	22 mg
Cost per oz	$0.12	$0.15	

10.3 APPLICATIONS OF LINEAR PROGRAMMING

In this section, we will present several areas of application of linear programming. We will revisit some of these scenarios in Chap. 11 when we discuss computer solution methods.

Diet-Mix Models

The classic diet mix problem involves determining the items which should be included in a meal so as to (1) minimize the cost of the meal while (2) satisfying certain nutritional requirements. The nutritional requirements usually take the form of daily vitamin requirements, restrictions encouraging variety in the meal (e.g., do not serve each person 10 pounds of boiled potatoes), and restrictions which consider taste and logical companion foods. The following example illustrates a simple diet mix problem.

EXAMPLE 8   **(Diet Mix Model)** A dietitian is planning the menu for the evening meal at a university dining hall. Three main items will be served, all having different nutritional content. The dietitian is interested in providing at least the minimum daily requirement of each of three vitamins in this one meal. Table 10.6 summarizes the *vitamin content per ounce of each type of food,* the cost per ounce of each food, and minimum daily requirements (MDR) for the three vitamins. Any combination of the three foods may be selected as long as the total serving size is at least 9 ounces.

TABLE 10.6

Food	Vitamin			Cost per Oz, $
	1	**2**	**3**	
1	50 mg	20mg	10 mg	0.10
2	30 mg	10 mg	50 mg	0.15
3	20 mg	30 mg	20 mg	0.12
Minimum daily requirement (MDR)	290 mg	200 mg	210 mg	

The problem is to determine the number of ounces of each food to be included in the meal. The objective is to minimize the cost of each meal subject to satisfying minimum daily requirements of the three vitamins as well as the restriction on minimum serving size.

To formulate the linear programming model for this problem, let x_j equal the number of ounces included of food j. The objective function should represent the total cost of the meal. Stated in dollars, the total cost equals the sum of the costs of the three items, or

$$z = 0.10x_1 + 0.15x_2 + 0.12x_3$$

Since we are interested in providing *at least* the minimum daily requirement for each of the three vitamins, there will be three "greater than or equal to" constraints. The constraint for each vitamin will have the form

> Milligrams of vitamin intake $\geq$ MDR

or

Milligrams from food 1 + milligrams from food 2 + milligrams from food 3 $\geq$ MDR

The constraints are, respectively,

$$50x_1 + 30x_2 + 20x_3 \geq 290 \quad \text{(vitamin 1)}$$

$$20x_1 + 10x_2 + 30x_3 \geq 200 \quad \text{(vitamin 2)}$$

$$10x_1 + 50x_2 + 20x_3 \geq 210 \quad \text{(vitamin 3)}$$

The restriction that the serving size be at least 9 ounces is stated as

$$x_1 + x_2 + x_3 \geq 9 \quad \text{(minimum serving size)}$$

The complete formulation of the problem is as follows:

Minimize	$z = 0.10x_1 + 0.15x_2 + 0.12x_3$
subject to	$50x_1 + 30x_2 + 20x_3 \geq 290$
	$20x_1 + 10x_2 + 30x_3 \geq 200$
	$10x_1 + 50x_2 + 20x_3 \geq 210$
	$x_1 + x_2 + x_3 \geq 9$
	$x_1, x_2, x_3 \geq 0$

Note that the nonnegativity constraint has been included in the formulation. This ensures that negative quantities of any of the foods will not be recommended.

This is a very simplified problem involving the planning of one meal, the use of just three food types, and consideration of only three vitamins. In actual practice, models have been formulated which consider (1) menu planning over longer periods of time (daily, weekly, etc.), (2) the interrelationships among all meals served during a given day, (3) the interrelationships among meals served over the entire planning period, (4) many food items, and (5) many nutritional requirements. The number of variables and number of constraints for such models can become extremely large.

❑

Transportation Models

Transportation models are possibly the most widely used linear programming models. Oil companies commit tremendous resources to the implementation of such models. The classic example of a transportation problem involves the shipment of some *homogeneous commodity* from *m sources of supply*, or **origins**, to *n points of demand*, or **destinations**. By *homogeneous* we mean that there are no significant differences in the quality of the item provided by the different sources of supply. The item characteristics are essentially the same.

In the classic problem each origin can supply any of the destinations. Also, the demand at each destination may be supplied jointly from a combination of the origins or totally from one origin. Each origin usually has a specific capacity which represents the maximum number of units it can supply. Each destination has a specified demand which represents the number of units needed.

Given that each origin can supply units to each destination, some measure of the cost or effort of shipping a unit is specified for each *origin-destination combination*. This may take the form of a dollar cost, distance between the two points, or time required to move from one point to another. A typical problem is concerned with determining the number of units which should be supplied from each origin to each destination. The objective is to minimize the total transportation or delivery costs while ensuring that (1) the number of units shipped from any origin does not exceed the number of units available at that origin and (2) the demand at each destination is satisfied. Example 9 illustrates a simple transportation model.

(Highway Maintenance) A medium-size city has two locations in the city at which salt and sand stockpiles are maintained for use during winter icing and snowstorms. During a storm, salt and sand are distributed from these two locations to four different city zones. Occasionally additional salt and sand are needed. However, it is usually impossible to get additional supplies during a storm since they are stockpiled at a central location some distance outside the city. City officials hope that there will not be back-to-back storms.

The director of public works is interested in determining the minimum cost of allocating salt and sand supplies during a storm. Table 10.7 summarizes the cost of supplying 1 ton of salt or sand from each stockpile to each city zone. In addition, stockpile capacities and normal levels of demand for each zone are indicated (in tons).

TABLE 10.7

| | Zone | | | | |
	1	2	3	4	Maximum Supply, Tons
Stockpile 1	$2.00	$3.00	$1.50	$2.50	900
Stockpile 2	4.00	3.50	2.50	3.00	750
Demand, tons	300	450	500	350	

In formulating the linear programming model for this problem, there are eight decisions to make — how many tons should be shipped from each stockpile to each zone. In some cases the best decision may be to ship no units from a particular stockpile to a given zone. Let's define our decision variables a little differently. *Let x_{ij} equal the number of tons supplied from stockpile i to zone j.* For example, x_{11} equals the number of tons supplied by stockpile 1 to zone 1. Similarly, x_{23} equals the number of tons supplied by stockpile 2 to zone 3. This *double-subscripted variable* conveys more information than the variables $x_1, x_2, \ldots, x_8$. Once you understand that the first subscript corresponds to the stockpile number and the second subscript to the city zone, the meaning of a variable such as x_{24} is more obvious.

Given this definition of the decision variables, the total cost of distributing salt and sand has the form

$$\text{Total cost} = 2x_{11} + 3x_{12} + 1.5x_{13} + 2.5x_{14} + 4x_{21} + 3.5x_{22} + 2.5x_{23} + 3x_{24}$$

This is the objective function that we wish to minimize.

One class of constraints deals with the capacities of the different stockpiles. For each stockpile a constraint should be formulated specifying that total shipments not exceed available supply. For stockpile 1, the sum of the shipments to all zones cannot exceed 900 tons, or

$$x_{11} + x_{12} + x_{13} + x_{14} \leq 900 \quad \text{(stockpile 1)}$$

The same constraint for stockpile 2 is

$$x_{21} + x_{22} + x_{23} + x_{24} \leq 750 \quad \text{(stockpile 2)}$$

The final class of constraints should guarantee that each zone receives the quantity demanded. For zone 1, the sum of the shipments from stockpiles 1 and 2 should equal 300 tons, or

$$x_{11} + x_{21} = 300 \qquad \text{(zone 1)}$$

The same constraints for the other three zones are, respectively,

$$x_{12} + x_{22} = 450 \qquad \text{(zone 2)}$$

$$x_{13} + x_{23} = 500 \qquad \text{(zone 3)}$$

$$x_{14} + x_{24} = 350 \qquad \text{(zone 4)}$$

The complete formulation of the linear programming model is as follows:

$$
\begin{aligned}
\text{Minimize} \quad & z = 2x_{11} + 3x_{12} + 1.5x_{13} + 2.5x_{14} + 4x_{21} + 3.5x_{22} + 2.5x_{23} + 3x_{24} \\
\text{subject to} \quad & x_{11} + x_{12} + x_{13} + x_{14} \le 900 \\
& x_{21} + x_{22} + x_{23} + x_{24} \le 750 \\
& x_{11} + x_{21} = 300 \\
& x_{12} + x_{22} = 450 \\
& x_{13} + x_{23} = 500 \\
& x_{14} + x_{24} = 350 \\
& x_{11}, x_{12}, x_{13}, x_{14}, x_{21}, x_{22}, x_{23}, x_{24} \ge 0
\end{aligned}
$$

❑

Capital Budgeting Models

Capital budgeting (rationing) decisions involve the allocation of limited investment funds among a set of competing investment alternatives. Usually, the alternatives available in any given time period are each characterized by an investment cost and some estimated benefit. The determination of investment costs is relatively easy. Estimating benefits can be more difficult, especially when projects are characterized by less tangible returns (e.g., programs having social benefits). Typically, the problem is to select the set of alternatives which will maximize overall benefits subject to budgetary constraints and other constraints which may affect the choice of projects.

EXAMPLE 10 (**Grant Awards**) A federal agency has a budget of $1 billion to award in the form of grants for innovative research in the area of energy alternatives. A management review team consisting of scientists and economists has made a preliminary review of 200 applications, narrowing the field to six finalists. Each of the six finalists' projects has been evaluated and scored in relation to potential benefits expected over the next 10 years. These estimated benefits are shown in Table 10.8. They represent the *net benefit per dollar invested* in each alternative. For example, the value of 4.4 associated with alternative 1 suggests that each

	TABLE 10.8			
Project	**Project Classification**	**Net Benefit per \$ Invested**	**Requested Level of Funding, \$ Millions**	
1	Solar	4.4	220	
2	Solar	3.8	180	
3	Synthetic fuels	4.1	250	
4	Coal	3.5	150	
5	Nuclear	5.1	400	
6	Geothermal	3.2	120	

dollar invested in that alternative will return a net (after subtracting the dollar investment) benefit of \$4.40 over the next 10 years.

Table 10.8 also shows the requested level of funding (in millions of dollars). These figures represent the *maximum* amount which can be awarded to any project. The agency can award any amount up to the indicated maximum for a given project. Similarly, the president has mandated that the nuclear project should be funded to at least 50 percent of the requested amount. The agency's administrator has a strong interest in solar projects and has requested that the combined amount awarded to the two solar projects be at least \$300 million.

The problem is to determine the amounts of money to award to each project in order to maximize total *net benefits,* measured in dollars. If we *let x_j equal the number of dollars (in millions) awarded to project j,* the objective function is

$$\text{Maximize} \quad z = 4.4x_1 + 3.8x_2 + 4.1x_3 + 3.5x_4 + 5.1x_5 + 3.2x_6$$

Notice that z equals total net benefits, in *millions of dollars.*

Structural constraints include the following types. First, the budget equals \$1 billion and the total sum awarded cannot exceed this amount. Stated mathematically,

$$x_1 + x_2 + x_3 + x_4 + x_5 + x_6 \leq 1,000 \qquad \text{(total budget)} \qquad \textbf{(10.9)}$$

There must be a constraint for *each* project which reflects the maximum possible award. For project 1 the constraint is

$$x_1 \leq 220 \qquad \text{(maximum award, project 1)}$$

Five additional constraints must be included for the remaining projects. To assure the president's concern about the nuclear project, we must include the constraint

$$x_5 \geq 0.5(400)$$

or $\qquad\qquad\qquad x_5 \geq 200 \qquad \text{(minimum award, nuclear)}$

Finally, the administrator's interest in the solar projects is assured by the constraint

$$x_1 + x_2 \geq 300 \qquad \text{(minimum award, solar projects)}$$

The complete formulation for this problem is

$$
\begin{aligned}
\text{Maximize} \quad & z = 4.4x_1 + 3.8x_2 + 4.1x_3 + 3.5x_4 + 5.1x_5 + 3.2x_6 \\
\text{subject to} \quad & x_1 + x_2 + x_3 + x_4 + x_5 + x_6 \le 1{,}000 \\
& x_1 \le 200 \\
& x_2 \le 180 \\
& x_3 \le 250 \\
& x_4 \le 150 \\
& x_5 \le 400 \\
& x_6 \le 120 \\
& x_5 \ge 200 \\
& x_1 + x_2 \ge 300 \\
& x_1, \quad x_2, \quad x_3, \quad x_4, \quad x_5, \quad x_6 \ge 0
\end{aligned}
$$

□

POINTS FOR THOUGHT & DISCUSSION Is it reasonable to state the total budget constraint [Eq. (10.9)] as a strict equality? Why or why not?

Blending Models

Linear programming has found wide application in an area referred to as **_blending models._** Blending models are formulated to determine an optimal combination of component ingredients to be blended into a final product. Blending models have been used in the blending of petroleum products, feed mixes for agricultural use, fertilizers and grass seeds, spirits, teas and coffees, and so forth. The objective with such models often is to minimize the cost of the blend. Typical constraints include lot size requirements for each blend, technological (or recipe) requirements, and limited availability of component ingredients. The following example illustrates a blending model.

EXAMPLE 11 (Petroleum Blending; Motivating Scenario) A small refinery is about to blend four petroleum products into three final blends of gasoline. Although the blending formulas are not precise, there are some restrictions which must be adhered to in the blending process. These blending restrictions are as follows:

1 *Component 2 should constitute no more than 40 percent of the volume of blend 1.*
2 *Component 3 should constitute at least 25 percent of the volume of blend 2.*
3 *Component 1 should be exactly 30 percent of blend 3.*
4 *Components 2 and 4, together, should constitute at least 60 percent of the volume of blend 1.*

There is a limited availability of components 2 and 3, 1,500,000 liters and 1,000,000 liters, respectively. The production manager wants to blend a total of 5,000,000 liters. Of this total at least 2,000,000 liters of final blend 1 should be produced. The wholesale price per liter from the sale of each final blend equals $0.26, $0.22, and $0.20, respectively. The input components cost $0.15, $0.18, $0.12, and $0.14 per liter, respectively. The problem is to determine the number of liters of each component to be used in each final blend so as to maximize the total profit contribution from the production run.

In formulating this problem, we will use the double-subscripted variable x_{ij} *to represent the number of liters of component i used in final blend j.* A major assumption in this model is that there is no volume loss in the blending process. That is, if 3 liters of component products are combined, the result is a final blend of exactly 3 liters. Refer to Fig. 10.15 to assist your understanding of the blending relationships.

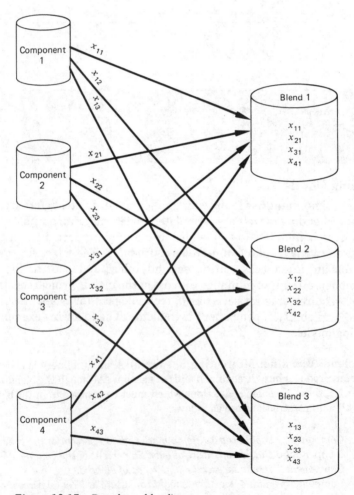

Figure 10.15 Petroleum-blending process.

The objective function has the form

Total profit contribution

= total revenue from the 3 blends	− total cost of the 4 components			
= total revenue from blend 1	+ total revenue from blend 2	+ total revenue from blend 3	− cost of component 1	− cost of component 2
			− cost of component 3	− cost of component 4
= $0.26 (no. of liters of blend 1)	+ $0.22 (no. of liters of blend 2)	+ $0.20 (no. of liters of blend 3)	− $0.15 (no. of liters of component 1)	− $0.18 (no. of liters of component 2)
			− $0.12 (no. of liters of component 3)	− $0.14 (no. of liters of component 4)

Examine the expressions in parentheses in the following equations and verify that the objective function is

$$z = 0.26(x_{11} + x_{21} + x_{31} + x_{41}) + 0.22(x_{12} + x_{22} + x_{32} + x_{42})$$
$$+ 0.20(x_{13} + x_{23} + x_{33} + x_{43}) - 0.15(x_{11} + x_{12} + x_{13})$$
$$- 0.18(x_{21} + x_{22} + x_{23}) - 0.12(x_{31} + x_{32} + x_{33})$$
$$- 0.14(x_{41} + x_{42} + x_{43})$$

which can be simplified by combining like terms to yield

$$z = 0.11x_{11} + 0.07x_{12} + 0.05x_{13} + 0.08x_{21} + 0.04x_{22} + 0.02x_{23} + 0.14x_{31}$$
$$+ 0.10x_{32} + 0.08x_{33} + 0.12x_{41} + 0.08x_{42} + 0.06x_{43} \quad \text{(total profit)}$$

Regarding structural constraints, the total production run must equal 5,000,000 liters, or

$$x_{11} + x_{12} + x_{13} + x_{21} + x_{22} + x_{23} + x_{31} + x_{32} + x_{33} + x_{41} + x_{42} + x_{43} = 5,000,000 \quad \textbf{(10.10)}$$

Recipe restriction (1) is represented by the inequality

$$\text{Amount of component 2 used in blend 1} \leq \text{40\% of amount of final blend 1}$$

or

$$x_{21} \leq 0.40(x_{11} + x_{21} + x_{31} + x_{41})$$

This constraint simplifies to the form

$$-0.4x_{11} + 0.6x_{21} - 0.4x_{31} - 0.4x_{41} \leq 0 \quad \textbf{(10.11)}$$

Recipe restriction (2) is stated as

$$\begin{array}{l}\text{Amount of component 3} \\ \text{used in blend 2}\end{array} \geq \begin{array}{l}\text{25\% of amount} \\ \text{of final blend 2}\end{array}$$

or

$$x_{32} \geq 0.25(x_{12} + x_{22} + x_{32} + x_{42})$$

or

$$-0.25x_{12} - 0.25x_{22} + 0.75x_{32} - 0.25x_{42} \geq 0 \qquad \textbf{(10.12)}$$

Restrictions (3) and (4) are represented similarly by Eq. (10.13) and inequality (10.14):

$$x_{13} = 0.3(x_{13} + x_{23} + x_{33} + x_{43})$$

or

$$0.7x_{13} - 0.3x_{23} - 0.3x_{33} - 0.3x_{43} = 0 \qquad \textbf{(10.13)}$$

and

$$x_{21} + x_{41} \geq 0.6(x_{11} + x_{21} + x_{31} + x_{41})$$

or

$$-0.6x_{11} + 0.4x_{21} - 0.6x_{31} + 0.4x_{41} \geq 0 \qquad \textbf{(10.14)}$$

The limited availabilities of components 2 and 3 are represented by the inequalities (10.15) and (10.16):

$$x_{21} + x_{22} + x_{23} \leq 1{,}500{,}000 \qquad \text{(component 1 availability)} \qquad \textbf{(10.15)}$$

$$x_{31} + x_{32} + x_{33} \leq 1{,}000{,}000 \qquad \text{(component 2 availability)} \qquad \textbf{(10.16)}$$

Finally, the minimum production requirement for final blend 1 is represented by

$$x_{11} + x_{21} + x_{31} + x_{41} \geq 2{,}000{,}000 \qquad \textbf{(10.17)}$$

The complete formulation of this blending model is

$$\begin{aligned}
\text{Maximize} \quad z = \ & 0.11x_{11} + 0.07x_{12} + 0.05x_{13} + 0.08x_{21} + 0.04x_{22} + 0.02_{23} \\
& + 0.14x_{31} + 0.10x_{32} + 0.08x_{33} + 0.12x_{41} + 0.08x_{42} + 0.06x_{43}
\end{aligned}$$

subject to

$$x_{11} + x_{12} + x_{13} + x_{21} + x_{22} + x_{23} + x_{31} + x_{32} + x_{33} + x_{41} + x_{42} + x_{43} = 5{,}000{,}000$$

$$-0.4x_{11} \qquad +0.6x_{21} \qquad\qquad -0.4x_{31} \qquad -0.4x_{41} \qquad \leq 0$$

$$-0.25x_{12} \qquad -0.25x_{22} \qquad +0.75x_{32} \qquad -0.25x_{42} \qquad \geq 0$$

$$0.7x_{13} \qquad -0.3x_{23} \qquad -0.3x_{33} \qquad -0.3x_{43} = 0$$

$$-0.6x_{11} \qquad +0.4x_{21} \qquad -0.6x_{31} \qquad +0.4x_{41} \qquad \geq 0$$

$$x_{21} + x_{22} + x_{23} \qquad \leq 1{,}500{,}000$$

$$x_{31} + x_{32} + x_{33} \qquad \leq 1{,}000{,}000$$

$$x_{11} \qquad +x_{21} \qquad +x_{31} \qquad +x_{41} \qquad \geq 2{,}000{,}000$$

$$x_{11}, x_{12}, x_{13}, x_{21}, x_{22}, x_{23}, x_{31}, x_{32}, x_{33}, x_{41}, x_{42}, x_{43} \geq 0$$

As you begin to test your skills in formulating linear programming models, the following guidelines are suggested.

TIPS ON FORMULATING LINEAR PROGRAMMING MODELS

1 *Read the statement of the problem carefully.*
2 *Identify the decision variables. These are the decisions that need to be made. Once these decisions are identified, list them providing a written mathematical definition (e.g., $x_1 =$ number of units produced and sold per week of product 1, $x_2 =$ number of units produced and sold per week of product 2).*
3 *Identify the objective. What is to be maximized or minimized (e.g., maximize total weekly profit from producing products 1 and 2)?*
4 *Identify the structural constraints. What conditions must be satisfied when we assign values to the decision variables? You may want to write a verbal description of the restriction before writing the mathematical representation (e.g., total production of product 1 $\geq$ 100 units; then $x_1 \geq 100$). Also, be comfortable with the fact that the structural constraints for a given LP model can express a wide variety of units. That is, given the set of decision variables x_j, structural constraints can be formulated which express conditions measured in dollars, hours, units produced, etc. You must simply make sure that the dimension for any given constraint is consistent on both sides of the constraint.*
5 *Write out the mathematical model. Depending on the problem, you may start by defining the objective function or the structural constraints. Don't forget to include the nonnegativity constraint.*

Section 10.3 Follow-up Exercises

1 For Example 10, modify the formulation if the following are true.
   (a) Each project should receive at least 20 percent of the requested level of funding.
   (b) The amount awarded for the synthetic-fuels project should be at least as much as that awarded for the coal project.
   (c) Combined funding for the geothermal project and the synthetic project should be at least $30 million.
   (d) Funding for the nuclear project should be at least 40 percent greater than funding for the geothermal project.
   (e) Funding for project 2 should be no more than 80 percent of the funding for project 1.
2 For Example 11, modify the formulation if the following additional conditions must be satisfied.
   (a) No more than 2 million liters should be made of final blend 1.
   (b) Components 1 and 4, together, should constitute at least 40 percent of final blend 3.
   (c) Components 2 and 3 should constitute no more than 60 percent of final blend 2.
   (d) Total revenue from blend 1 should exceed $250,000.

3 A dietitian is planning the menu for the noon meal at an elementary school. He plans to serve three main items, all having different nutritional content. The dietitian is interested in providing at least the minimum daily requirement of each of three vitamins in this one meal. Table 10.9 summarizes the vitamin content per ounce of each type of food, the cost per ounce, and the minimum daily requirement for each vitamin. Any combination of the three foods may be selected as long as the total serving size is at least 6.0 ounces.

Formulate the linear programming problem which when solved would determine the number of ounces of each food to serve. The objective is to minimize the cost of the meal while satisfying minimum daily requirement levels of the three vitamins as well as the restriction on the minimum serving size.

TABLE 10.9

Food	Vitamin 1	Vitamin 2	Vitamin 3	Cost per Ounce, $
1	20 mg	10 mg	20 mg	0.15
2	40 mg	25 mg	30 mg	0.18
3	30 mg	15 mg	25 mg	0.22
MDR	240 mg	120 mg	180 mg	

4 A leading processor of sugar has two plants which supply four warehouses. Table 10.10 summarizes weekly capacities at each plant, weekly requirements at each warehouse, and shipping cost per ton (in dollars) between any plant and any warehouse. If x_{ij} equals the number of tons shipped from plant i to depot j, formulate the linear programming model which allows for determining the distribution schedule which results in minimum shipping cost. Weekly plant capacities are not to be violated, and warehouse requirements are to be satisfied.

TABLE 10.10

	Warehouse 1	Warehouse 2	Warehouse 3	Warehouse 4	Weekly Supply, Tons
Plant 1	20	15	10	25	2,800
Plant 2	30	25	20	15	3,500
Weekly demand, tons	1,400	1,600	1,000	1,500	

5 A chemical company manufactures liquid oxygen at three different locations in the South. It must supply four storage depots in the same region. Table 10.11 summarizes shipping cost per 1,000 gallons between any plant and any depot as well as monthly capacity at each plant and monthly demand at each depot. If x_{ij} equals the number of gallons (in thousands) shipped from plant i to depot j, formulate the linear programming model which allows for determining the minimum-cost allocation schedule. Plant capacities are not to be violated, and depot demands are to be satisfied by the schedule.

6 A firm manufactures three products which must be processed through some or all of four departments. Table 10.12 indicates the number of hours a unit of each product requires in the different departments and the number of pounds of raw material required. Also listed

TABLE 10.11

	Depot				
	1	2	3	4	Supply, 1,000 Gal
Plant 1	50	40	35	20	1,000
Plant 2	30	45	40	60	1,400
Plant 3	60	25	50	30	1,800
Demand, 1,000 gal	800	750	650	900	

TABLE 10.12

	Product			
	A	B	C	Weekly Availability
Department 1	2.5	4	2	120 h
Department 2		2	2	160 h
Department 3	3	1		100 h
Department 4	2	3	2.5	150 h
Pounds of raw material per unit	5.5	4.0	3.5	500 lbs
Selling price	$60	$50	$75	
Labor cost per unit	20	27	36	
Material cost per unit	21	8	7	

are labor and material costs per unit, selling price, and weekly capacities of both labor-hours and raw materials. If the objective is to maximize total weekly profit, formulate the linear programming model for this exercise.

7 Referring to Exercise 6, write the constraints associated with each of the following conditions.

(a) Combined weekly production must be at least 50 units.

(b) The number of units of product A must be no more than twice the quantity of product B.

(c) Since products B and C are usually sold together, production levels of both should be the same.

(d) The number of units of product B should be no more than half of the total weekly production.

8 A regional truck rental agency is planning for a heavy demand during the summer months. The agency has taken truck counts at different cities and has compared these with projected needs for each city (all trucks are the same size). Three metropolitan areas are expected to have more trucks than will be needed during the summer, although four cities are expected to have fewer trucks than will be demanded. To prepare for these months, trucks can be relocated from surplus areas to shortage areas by hiring drivers. Drivers are paid a flat fee which depends on the distance between the two cities. In addition, they receive per diem (daily) expenses. Table 10.13 summarizes costs of having a truck delivered between two cities. Also shown are the projected surpluses for each city which has an oversupply and projected shortages for each city needing additional trucks. (Note that total surplus exceeds total shortage.)

If the objective is to minimize the cost of reallocating the trucks, formulate the linear programming model which would allow for solving the problem. (*Hint:* Let x_{ij} equal the number of trucks delivered from surplus area i to shortage area j.)

9 A coffee manufacturer blends four component coffee beans into three final blends of coffee. The four component beans cost the manufacturer $0.65, $0.80, $0.90, and $0.75 per

TABLE 10.13

	Shortage Area				Surplus of Trucks
	1	2	3	4	
Surplus city 1	$100	$250	$200	$150	120
Surplus city 2	200	175	100	200	125
Surplus city 3	300	180	50	400	100
Shortage of trucks	60	80	75	40	

pound, respectively. The weekly availabilities of the four components are 80,000, 40,000, 30,000, and 50,000 pounds, respectively. The manufacturer sells the three blends at wholesale prices of $1.25, $1.40, and $1.80 per pound, respectively. Weekly output should include at least 50,000 pounds of final blend 3.

The following are blending restrictions which must be followed by the brewmaster.
(a) Component 2 should constitute at least 30 percent of final blend 3 and no more than 20 percent of final blend 1.
(b) Component 3 should constitute exactly 25 percent of final blend 3.
(c) Component 4 should constitute at least 40 percent of final blend 1 and no more than 18 percent of final blend 2.

The objective is to determine the number of pounds of each component which should be used in each final blend so as to maximize weekly profit. Formulate this as an LP model, carefully defining your decision variables.

BancOhio National Bank

AN APPLICATION OF LINEAR PROGRAMMING IN THE BANKING INDUSTRY*

Most large banks have centralized check-processing centers. Checks are delivered from different branch offices to the check-processing center using a variety of modes of transportation. Upon arrival at the check-processing center the checks are (1) encoded with the dollar amount in magnetic ink on the bottom of the check, (2) microfilmed, and (3) processed by a sorting machine which reads significant information off the check pertaining to bookkeeping and the routing of the check. Checks which are drawn against other banks are then forwarded to the Federal Reserve or a clearinghouse.

The timely processing of checks is important to minimize what is referred to as "float." Float refers to the amount of money represented by outstanding checks which are in the process of collection. There is an opportunity cost for the banks when checks are in the process of collection. If the banks had these funds, they could be investing them and earning interest. These lost investments can cost large banks considerable amounts of money each year. The efficient processing of checks depends upon the scheduling of encoder operators. The scheduling is complicated by (1) high variability of check volumes on a daily and hourly basis and (2) a fixed number of available encoding machines.

The BancOhio National Bank in Columbus, with assets approaching $6 billion and with more than 200 branches statewide, successfully applied linear programming in the development of a shift-scheduling system for its encoders. The model developed for the bank determines the number of full-time and part-time encoder clerks to assign to each of a set of predetermined shifts such that weekly wages (regular time and overtime) and float costs are minimized. Bank officials estimated first-year savings from the modified shift schedules to be **$1 million.**

---

* L. J. Krajewski and L. P. Ritzman, "Shift Scheduling in Banking Operations: A Case Application," *Interfaces*, vol. 10, no. 2 (April 1980), pp 1–8.

❏ KEY TERMS AND CONCEPTS

alternative optimal solutions 436

blending models 447

capital budgeting models 445

(closed) half-space 425

convex set 434

corner-point method 435

decision variables 422

diet mix models 441

isoprofit lines 432

linear programming (LP) 422

mathematical programming 422

no feasible solution 438

nonnegativity constraints 424

objective function 422

permissible half-space 425

region of feasible solutions 431

structural constraints 424

transportation models 443

unbounded solution 439

unbounded solution space 438

❏ ADDITIONAL EXERCISES

Section 10.2

For the following LP problems, graph the region of feasible solutions (if one exists) and solve by the corner-point method.

1 Maximize $z = 8x_1 + 3x_2$
subject to
$$x_1 = x_2$$
$$2x_1 + 5x_2 \le 40$$
$$x_1, x_2 \ge 0$$

2 Minimize $z = 15x_1 + 6x_2$
subject to
$$x_1 + x_2 \ge 24$$
$$x_2 \ge 4$$
$$3x_1 + 2x_2 \le 30$$
$$x_1, x_2 \ge 0$$

3 Maximize $z = 2x_1 + 4x_2$
subject to
$$2x_1 + 2x_2 \le 10$$
$$-x_1 + x_2 \ge 8$$
$$x_1, x_2 \ge 0$$

4 Maximize $z = 3x_1 + 3x_2$
subject to
$$4x_1 + 3x_2 \ge 12$$
$$2x_1 + 3x_2 \ge -6$$
$$x_1, x_2 \ge 0$$

5 Minimize $z = 4x_1 + 4x_2$
subject to
$$x_1 + 3x_2 \le 24$$
$$3x_1 + x_2 \ge 26$$
$$x_1 - x_2 = 6$$
$$x_1, x_2 \ge 0$$

6 Maximize $z = 3x_1 + 2x_2$
subject to
$$x_1 + 2x_2 \ge 6$$
$$9x_1 + 6x_2 \le 108$$
$$x_1 \ge 8$$
$$x_2 \ge 4$$
$$x_1, x_2 \ge 0$$

7 Maximize $z = 4x_1 + 3x_2$
subject to
$$x_1 + x_2 \geq 4$$
$$8x_1 + 6x_2 \leq 48$$
$$x_1 \geq 3$$
$$x_2 \leq 6$$
$$x_1, x_2 \geq 0$$

8 Minimize $z = 3x_1 + 2x_2$
subject to
$$x_1 + x_2 \leq 10$$
$$9x_1 + 6x_2 \geq 18$$
$$x_1 \geq 2$$
$$x_2 \geq 4$$
$$x_1 \leq 6$$
$$x_2 \leq 8$$
$$x_1, x_2 \geq 0$$

9 Minimize $z = 5x_1 + 8x_2$
subject to
$$x_1 + x_2 \geq 6$$
$$3x_1 + 2x_2 \leq 30$$
$$2x_1 + x_2 \leq 5$$
$$x_1, x_2 \geq 0$$

10 Maximize $z = 8x_1 + 4x_2$
subject to
$$3x_1 + 4x_2 \leq 24$$
$$2x_1 + 3x_2 \geq 12$$
$$x_1 \leq 6$$
$$4x_1 + 3x_2 \geq 36$$
$$x_1, x_2 \geq 0$$

Section 10.3

11 A producer of machinery wishes to maximize the profit from producing two products, product A and product B. The three major inputs for each product are steel, electricity, and labor-hours. Table 10.14 summarizes the requirements per unit of each product, available resources, and profit margin per unit. The number of units of product A should be no more than 80 percent of the number of product B. Formulate the linear programming model for this situation.

TABLE 10.14

	Product		Monthly Total Available
	A	**B**	
Energy	100 kWh	200 kWh	20,000 kWh
Steel	60 lb	80 lb	10,000 lb
Labor	2.5 h	2 h	400 h
Profit per unit	$30	$40	

12 In a certain area there are two warehouses which supply food to five grocery stores. Table 10.15 summarizes the delivery cost per truckload from each warehouse to each store, the required number of truckloads per store per week, and the maximum number of truckloads available per week per warehouse. Formulate a linear programming model that would determine the number of deliveries from each warehouse to each store which would minimize total delivery cost.

TABLE 10.15

	Store					Maximum Number of Truckloads
	1	**2**	**3**	**4**	**5**	
Warehouse A	$40	$30	$45	$25	$50	100
Warehouse B	$50	$35	$40	$20	$40	250
Required number of truckloads	80	50	75	45	80	

13 Capital Expansion A company is considering the purchase of some additional machinery as part of a capital expansion program. Four types of machines are being considered. Table 10.16 indicates relevant attributes of the four machines.

TABLE 10.16		Machine		
	A	**B**	**C**	**D**
Cost	$50,000	$35,000	$60,000	$80,000
Square footage required	200	150	250	280
Daily output, units	10,000	8,000	25,000	18,000

The total budget for this program is $750,000. The maximum available floor space is 16,000 square feet. The company wants to maximize the output (total number of units produced) resulting from the purchase of the new machines. Define your decision variables carefully and formulate the LP model for this problem.

14 A national car rental firm is planning for the summer season. An analysis of current inventories of subcompact cars in seven cities along with projections of demands during the summer in these same cities indicate that three of these areas will be short of their needs while four of the cities will have surplus numbers of subcompact automobiles. In order to prepare for the summer season, company officials have decided to relocate cars from those cities expected to have surpluses to those expected to have shortages. The cars can be relocated by contracting with an auto transport firm. Bids have been received from the trucking firm which indicate the cost of relocating a car from a given surplus city to a given shortage city. Table 10.17 summarizes these costs along with the surpluses and shortages for the mentioned cities.

Let x_{ij} equal the number of cars relocated from surplus area i to shortage area j. If the objective is to minimize the cost of relocating these cars such that each shortage area will have its needs satisfied, formulate the LP model for this problem.

TABLE 10.17		Shortage Area				
	1	**2**	**3**	**4**	**Suplus of Cars**
Surplus city 1	$50	$40	$25	$30	300
Surplus city 2	$40	$25	$35	$45	150
Surplus city 3	$35	$50	$40	$25	250
Shortage of cars	250	150	125	175	

15 Financial Portfolio A person is interested in investing $500,000 in a mix of investments. Table 10.18 indicates the investment choices and estimated rates of return for each. The investor wants at least 35 percent of her investment to be in government bonds. Because of the higher perceived risk of the two stocks, she has specified that the combined investment in these not exceed $80,000. The investor also has a hunch that interest rates are going to remain high and has specified that at least 20 percent of the

TABLE 10.18	Investment	Projected Rate of Return
	Mutual fund A	0.12
	Mutual fund B	0.09
	Money market fund	0.08
	Government bonds	0.085
	Stock A	0.16
	Stock B	0.18

investment should be in the money market fund. Her final investment condition is that the amount invested in mutual fund A should be no more than the amount invested in mutual fund B. The problem is to decide the amount of money to invest in each alternative so as to maximize total annual return (in dollars). Carefully define your variables and formulate the LP model for this problem.

16 Assignment Model A company is interested in assigning five sales representatives to five different sales districts. Management has estimated the total sales each representative would generate if assigned to the different districts for a 1-year-period. Table 10.19 summarizes these sales estimates (in $1,000 units). The desire is to assign each representative to one sales district in such a way that annual sales are maximized. Let $x_{ij} = 1$ if representative i is assigned to district j and let $x_{ij} = 0$ if representative i is not assigned to district j. Formulate the LP model for this problem (remembering that each representative must be assigned and each district must be assigned a representative).

TABLE 10.19

Sales Rep.	District 1	2	3	4	5
1	200	150	100	90	160
2	120	180	160	120	100
3	100	200	120	100	140
4	150	100	220	80	180
5	80	120	100	150	200

17 Cargo Loading The owner of a cargo ship is considering the nature of the next shipment. Four different commodities are being offered for shipment. Table 10.20 summarizes their weight, volume, and revenue-generating characteristics. The cargo ship has three cargo holds, each characterized by weight and volume capacities. The forward hold has a weight capacity of 100 tons and volume capacity of 6,000 cubic feet. The center hold has a weight capacity of 140 tons and volume capacity of 8,000 cubic feet. The aft hold has a weight capacity of 80 tons and volume capacity of 5,000 cubic feet. The problem is to decide how much of each commodity should be accepted for shipment if the objective is to maximize total revenue. Specifically, it must be decided how many tons of each commodity should be placed in each hold while not exceeding the weight and volume capacities. Let x_{ij} equal the number of tons of commodity i placed in hold j and formulate the LP model for this problem.

TABLE 10.20

Commodity	Weight Offered, Tons	Volume, Ft³/Ton	Revenue, $ per Ton
1	200	70	$1,250
2	100	50	900
3	80	60	1,000
4	150	75	1,200

❑ CHAPTER TEST

1 Graphically determine the solution space, if one exists, for this system of inequalities:

$$x_1 - x_2 \geq 5$$

$$x_1 + x_2 \leq 15$$

$$x_1 \leq 20$$

$$x_2 \geq 5$$

2 A company manufactures and sells five products. Costs per unit, selling price, and hourly labor requirements per unit produced are given in Table 10.21. If the objective is to maximize total profit, formulate a linear programming model having the following constraints: at least 20 units of product A and at least 10 units of product B must be produced; sufficient raw materials are not available for total production in excess of 75 units; the number of units produced of products C and E must be equal; combined production of A and B should be no more than 50 percent of combined production of C, D, and E; the amount produced of C must be at least that of A; and labor availability in departments 1 and 2 equal 120 and 150 hours, respectively.

TABLE 10.21			Product		
	A	**B**	**C**	**D**	**E**
Cost per unit	$50	$80	$300	$25	$10
Selling price	$70	$90	$350	$50	$12
Dept. 1 labor hours/unit	2	1	0.5	1.6	0.75
Dept. 2 labor hours/unit	1.5	0.8	1.5	1.2	2.25

3 Solve the following problem using the corner-point method.

$$\text{Maximize} \quad z = 1.5x_1 + 3x_2$$

$$\text{subject to} \quad x_1 + x_2 \geq 10$$

$$5x_1 + 10x_2 \leq 120$$

$$x_1 \leq 8$$

$$x_2 \leq 6$$

$$x_1, x_2 \geq 0$$

4 For the following LP phenomena, discuss their meaning and how they appear graphically.
(a) Alternative optimal solutions
(b) No feasible solution
(c) Unbounded solution

AIR TRAFFIC CONTROLLER SCHEDULING

Officials of a major metropolitan airport are reviewing their staffing needs for air traffic controllers. The most recent labor agreement has specified hiring for eight different shifts consisting of eight hours per shift. FAA regulations specify that air controllers work 2-hour intervals with an hour break between. Airport officials specify that each controller work the first 2 hours of a shift. Thus a controller is active for 6 of the 8 hours.

Figure 10.16 illustrates the hours of each shift and the periods during which a controller is on duty. Based on an analysis of daily traffic volume and FAA guidelines, airport officials have determined the minimum number of controllers on duty for each hour of the day. These are also indicated in Fig. 10.16.

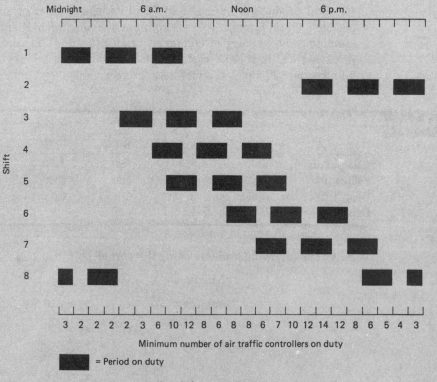

Figure 10.16 Air traffic controller shift requirements.

The base pay per controller is $80 per shift with differentials for certain shifts. Any shifts *beginning* between 4 p.m. and 11 p.m. (inclusive) are paid at a 10 percent premium for the entire shift; any *beginning* between midnight and 6 a.m. (inclusive) receive a 20 percent premium.

Airport officials want to determine the number of controllers to hire for each shift so as to meet the hourly requirements at a minimum cost per day.

a Formulate an appropriate model.

b Suppose that the FAA will allow overtime shifts of 3 hours provided the first hour of the 3 hours is off duty. Those working overtime receive credit for an additional half day of work and are paid time and a half (based on the pay/shift determined in part *a*) for the overtime period. The decisions under these assumptions are how many controllers should begin a regular shift and how many an overtime shift (of 11 hours) for each of the eight shifts. Formulate this problem.

CHAPTER 11

THE SIMPLEX AND COMPUTER SOLUTION METHODS

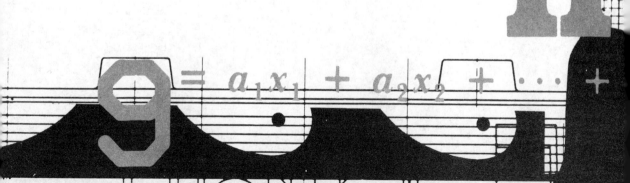

CHAPTER OBJECTIVES

❏ Provide an understanding of the simplex method of solving LP problems

❏ Illustrate the ways in which special LP phenomena evidence themselves when using the simplex method

❏ Discuss computer solution methods for LP problems, with particular emphasis on the interpretation of computer results

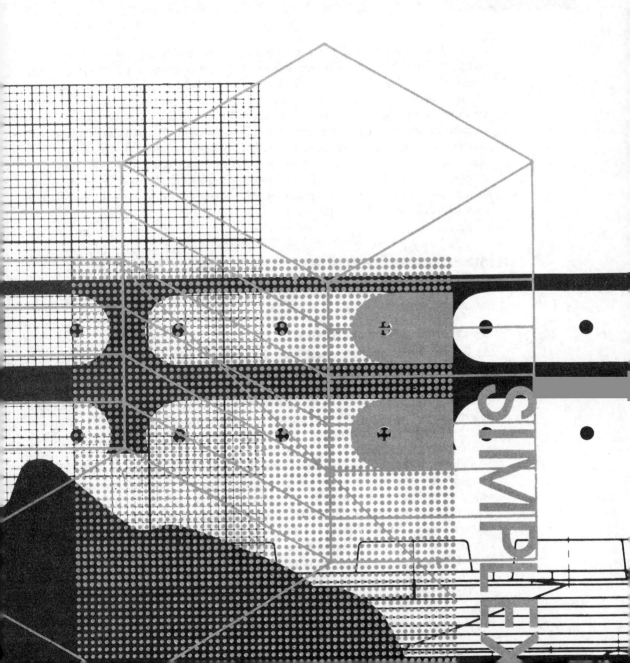

Example 10 in Chap. 10 presented an application in which a federal agency wished to award $1 billion in grants for innovative research in the area of energy alternatives. A management review team narrowed a set of 200 competing applications to a field of 6 finalists. Given a set of considerations and requirements associated with granting awards, we formulated the LP model for this application. In this chapter we wish to determine the optimal solution to the problem (Example 14).

In this chapter we will examine the simplex method of solving linear programming problems. First, we will overview the nature of and requirements for using the simplex method. Following this, the technique will be presented and illustrated for solving maximization problems which contain all "less than or equal to" constraints. Next, the procedure for solving minimization problems and problems containing other types of constraints will be presented. This will be followed by a discussion of the way in which special LP phenomena, such as alternative optimal solutions, are identified through the simplex method. We will discuss computer solution methods for LP models, with particular emphasis placed on the interpretation of the output from computer-based solutions. Finally, we will discuss the *dual* LP problem and its significance.

11.1 SIMPLEX PRELIMINARIES

Our purpose in this section is to overview the simplex method and discuss requirements necessary to use it.

Overview of the Simplex Method

As indicated in Chap. 10, graphical solution procedures are applicable only for linear programming problems involving two variables. We can discuss the geometry of three-variable problems; however, most of us are not skilled at three-dimensional graphics. And beyond three variables, there is no geometric frame of reference. Since most realistic applications of linear programming involve far more than two variables, there is a need for a solution procedure other than the graphical method.

The most popular nongraphical procedure is called the *simplex method.* The simplex method is an algebraic procedure for solving systems of equations where an objective function is to be optimized. It is an *iterative* process, which identifies a feasible starting solution. The procedure then searches to see whether there exists a better solution. "Better" is measured by whether the value of the objective function can be improved. If a better solution is signaled, the search resumes. The generation of each successive solution requires solving a system of linear equations. The search continues until no further improvement is possible in the objective function.

Graphically, you may envision the procedure as searching different corner points on the region of feasible solutions. The solutions found at each iteration of the simplex method represent such corner points. Not all corner points are exam-

Figure 11.1

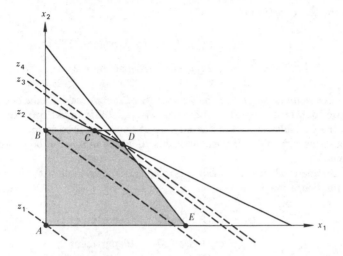

ined, however. The search chooses only a subset of these corner points, selecting a new one if and only if the objective function is at least as good as the current corner point. This idea is illustrated by Fig. 11.1. If we assume an objective of maximization, the simplex method might move from an initial solution at corner point A to points B, C, and finally point D. Note the *isoprofit lines* z_1, z_2, z_3, and z_4. The isoprofit line moves outward, away from the origin, with each successive corner point. This illustrates a situation in which the value of z is increasing with each successive solution. In a minimization problem successive solutions would have objective function values which are typically decreasing.

Requirements of the Simplex Method

There are three requirements for solving a linear programming problem by the simplex method:

REQUIREMENTS OF SIMPLEX METHOD

   I   *All constraints must be stated as equations.*
  II *The right side of a constraint cannot be negative.*
 III *All variables are restricted to nonnegative values.*

Regarding the *first requirement,* the simplex method is a special routine for solving systems of linear equations. Most linear programming problems contain constraints which are inequalities. Before we solve by the simplex method, these inequalities must be restated as equations. The transformation from inequalities to equations varies, depending on the nature of the inequality.

REQUIREMENT I ($\leq$ CONSTRAINTS)
For each "less than or equal to" ($\leq$) constraint a nonnegative variable, called a *slack variable,* is added to the left side of the constraint. This variable serves the function of balancing the two sides of the equation.

EXAMPLE 1

Consider the two constraints

$$2x_1 + 3x_2 \leq 50 \qquad \text{(department 1)}$$

$$4x_1 + 2x_2 \leq 60 \qquad \text{(department 2)}$$

where x_1 and x_2 equal, respectively, the number of units produced of products A and B. Assume that the two constraints represent limited labor availability in two departments; the coefficients on the variables represent the number of hours required to produce a unit of each product, and the right sides of the constraints equal the number of hours available in each department.

The treatment of these constraints is to add a slack variable to the left side of each. Or the constraints are rewritten as

$$2x_1 + 3x_2 + S_1 = 50 \qquad \text{(department 1)}$$

$$4x_1 + 2x_2 + S_2 = 60 \qquad \text{(department 2)}$$

The slack variables S_1 and S_2 keep the two sides of their respective equations in balance. They also have a meaning which is easy to understand. They represent, in this problem, the number of unused hours in each department. For example, $x_1 = 5$ and $x_2 = 10$ suggest producing 5 units of product A and 10 units of product B. If these values are substituted into the two constraints, we have

$$\left. \begin{array}{l} 2(5) + 3(10) + S_1 = 50 \\ 4(5) + 2(10) + S_2 = 60 \end{array} \right\} \qquad \begin{array}{l} \text{(department 1)} \\ \text{(department 2)} \end{array}$$

or

$$\left. \begin{array}{l} 40 + S_1 = 50 \\ 40 + S_2 = 60 \end{array} \right\} \qquad \begin{array}{l} \text{(department 1)} \\ \text{(department 2)} \end{array}$$

or

$$\left. \begin{array}{l} S_1 = 10 \\ S_2 = 20 \end{array} \right\} \qquad \begin{array}{l} \text{(department 1)} \\ \text{(department 2)} \end{array}$$

In other words, 40 hours would be used for production in each department. The slack variables would have to assume respective values of $S_1 = 10$ and $S_2 = 20$ to balance the equations. The interpretation of these values is that producing 5 units of product A and 10 units of product B will result in 10 hours being left over in department 1 and 20 hours being left over in department 2.

❏

PRACTICE EXERCISE
In Example 1, suppose that $x_1 = 7$ and $x_2 = 12$. What values must S_1 and S_2 assume? Discuss the interpretation of this result. *Answer:* $S_1 = 0$ and $S_2 = 8$; if 7 units of product A and 12 units of product B are produced, all hours will be used in department 1 while all but 8 will be used in department 2.

Note that slack variables become additional variables in the problem and must be treated like any other variables. This means that they are subject to requirement III; that is, they cannot assume negative values.

> **REQUIREMENT I (≥ CONSTRAINTS)**
> For each "greater than or equal to" (≥) constraint a nonnegative variable E, called a *surplus variable*, is subtracted from the left side of the constraint. This variable serves the same function as a slack variable: it keeps the two sides of the equation in balance. In addition to subtracting a surplus variable, a nonnegative variable A, called an **artificial variable**, is added to the left side of the constraint.

The artificial variable has no real meaning in the problem; its only function is to provide a convenient starting point (initial solution) for the simplex.

EXAMPLE 2 Assume in Example 1 that combined production of the two products must be at least 25 units. The constraint representing this third condition is

$$x_1 + x_2 \geq 25$$

Before we solve by the simplex method, the inequality must be transformed into the equivalent equation

$$x_1 + x_2 - E_3 + A_3 = 25$$

The subscripts on E_3 and A_3 indicate the constraint number. If $x_1 = 20$ and $x_2 = 35$, the surplus variable E_3 must equal 30 in order to balance the two sides of the equation. The interpretation of the surplus variable E_3 is that combined production of 20 units of product A and 35 units of product B exceeds the minimum requirement by 30 units. As with slack variables, surplus variables often have a meaningful interpretation in an application. The associated value of A_3 is zero. We will come to a better understanding of artificial variables as we move through this chapter.

❑

> **REQUIREMENT I (= CONSTRAINTS)**
> For each "equal to" (=) constraint, an **artificial** variable is added to the left side of the constraint.

EXAMPLE 3 Transform the following constraint set into the standard form required by the simplex method:

$$x_1 + x_2 \leq 100$$

$$2x_1 + 3x_2 \geq 40$$

$$x_1 - 2x_2 = 25$$

$$x_1, x_2 \geq 0$$

SOLUTION

The transformed constraint set is

$$x_1 + x_2 + S_1 \qquad\qquad = 100 \qquad\qquad \textbf{(1)}$$

$$2x_1 + 3x_2 \qquad - E_2 + A_2 \qquad = 40 \qquad\qquad \textbf{(2)}$$

$$x_1 - 2x_2 \qquad\qquad + A_3 = 25 \qquad\qquad \textbf{(3)}$$

$$x_1, x_2, S_1, E_2, A_2, A_3 \geq \; 0$$

Note that each supplemental variable (slack, surplus, artificial) is assigned a subscript which corresponds to the constraint number. Also, the nonnegativity restriction (requirement III) applies to all supplemental variables.

❑

Requirement II of the simplex method states that the right side of any constraint equation not be negative. If a constraint has a negative right side, the constraint can be multiplied by -1 to make the right side positive.

| EXAMPLE 4 | For the following constraints, make the right side positive. |

(a) $2x_1 - 5x_2 \leq -10$ (b) $x_1 + 6x_2 \geq -100$ (c) $5x_1 - 2x_2 = -28$

SOLUTION

(a) Multiplying the constraint by -1 results in

$$-2x_1 + 5x_2 \geq 10$$

(b) Multiplying the constraint by -1 results in

$$-x_1 - 6x_2 \leq 100$$

(c) Multiplying the constraint by -1 results in

$$-5x_1 + 2x_2 = 28$$

Notice for the inequality constraints, the sense of the inequality reverses when multiplied by a negative number.

❑

Requirement III of the simplex method states that all variables be restricted to nonnegative values. There are specialized techniques for dealing with variables which *can* assume negative values; however, we will not examine these methods. The only point which should be repeated is that slack, surplus, and artificial variables are also restricted to being nonnegative.

---

* The labeling of supplemental variables may vary slightly with different textbooks.

EXAMPLE 5

An LP problem has 5 decision variables, 10 ($\leq$) constraints, 8 ($\geq$) constraints, and 2 ($=$) constraints. When this problem is restated to comply with requirement I of the simplex method, how many variables will there be and of what types?

SOLUTION

There will be 33 variables: 5 decision variables, 10 slack variables associated with the 10 ($\leq$) constraints, 8 surplus variables associated with the 8 ($\geq$) constraints, and 10 artificial variables associated with the ($\geq$) and ($=$) constraints.

☐

PRACTICE EXERCISE
An LP problem has 8 decision variables, 20 ($\leq$) constraints, 10 ($\geq$) constraints, and 5 ($=$) constraints. When restated to comply with requirement I, how many variables will there be? *Answer:* 53

Basic Feasible Solutions

When LP problems have been converted to the *standard form* where all constraints are restated as equalities and supplemental variables have been added, the resulting system of constraint equations has more variables than equations.

EXAMPLE 6

Consider, again, the two-variable product-mix problem used to illustrate the graphical solution procedure in Sec. 10.2.

$$\text{Maximize} \quad z = 5x_1 + 6x_2$$

$$\text{subject to} \quad 3x_1 + 2x_2 \leq 120 \quad \text{(department 1)} \tag{1}$$

$$4x_1 + 6x_2 \leq 260 \quad \text{(department 2)} \tag{2}$$

$$x_1, x_2 \geq 0 \tag{3}$$

Let's focus on the set of structural constraints for a moment. When transformed into standard form, these will be rewritten as

$$3x_1 + 2x_2 + S_1 \quad\ = 120 \tag{4}$$

$$4x_1 + 6x_2 \quad + S_2 = 260 \tag{5}$$

The rewritten set of constraints is a (2×4) system of equations, with more variables than equations.

☐

Let's illustrate the concepts of (*a*) a **feasible solution,** (*b*) a **basic solution,** and (*c*) a **basic feasible solution.** If we refer to Eqs. (4) and (5) in Example 6, *a feasible solution is any set of values for the four variables which satisfies the two equations and the nonnegativity constraints.* For example, if x_1 and x_2 are both set equal to 5, these values can be substituted into (4) and (5) to solve for the corresponding values of S_1 and S_2.

$$3(5) + 2(5) + S_1 = 120$$
$$4(5) + 6(5) + S_2 = 260$$
$$25 + S_1 = 120$$
$$50 + S_2 = 260$$

or

$$S_1 = 95$$
$$S_2 = 210$$

Thus, one feasible solution for the system is $x_1 = 5$, $x_2 = 5$, $S_1 = 95$, and $S_2 = 210$.

A **basic solution** *is found by assigning values of zero to a subset of variables and solving for the corresponding values of the remaining variables.* The number of variables assigned values of zero is such that the number of remaining variables equals the number of equations (i.e., the resulting system is square). For Eqs. (4) and (5), a basic solution would be found by setting any two of the four variables equal to zero and solving for the remaining two. For example, if S_2 and x_2 are set equal to zero, substitution yields

$$3x_1 + 2(0) + S_1 = 120$$
$$4x_1 + 6(0) + (0) = 260$$

Solving for x_1 in the second equation and substituting into the first,

$$x_1 = 65$$
$$3(65) + S_1 = 120$$
$$S_1 = -75$$

Thus, one basic solution for the system is $x_1 = 65$, $x_2 = 0$, $S_1 = -75$, and $S_2 = 0$. The variables set equal to zero (x_2 and S_2) are considered the **nonbasic variables** for this solution. Those solved for (x_1 and S_1) are the **basic variables.** Notice that this particular basic solution is not feasible because it does not satisfy the nonnegativity constraint.

A **basic feasible solution** *is any basic solution which satisfies the nonnegativity constraint.* For example, if we set x_1 and x_2 equal to zero, substitution into the two equations yields

$$3(0) + 2(0) + S_1 = 120$$
$$4(0) + 6(0) + S_2 = 260$$

or

$$S_1 = 120$$
$$S_2 = 260$$

Thus, another basic solution for the system is $x_1 = 0$, $x_2 = 0$, $S_1 = 120$, and $S_2 = 260$. The variables x_1 and x_2 are the nonbasic variables, and S_1 and S_2 are the basic variables. Since all four variables satisfy the nonnegativity restriction, this is a basic feasible solution.

Let's state some definitions which are significant to our coming discussions. Assume the standard form of an LP problem which has m structural constraints and a total of n' decision variables and supplemental variables.

DEFINITION: FEASIBLE SOLUTION

A *feasible solution* is any set of values for the n' variables which satisfies both the structural and nonnegativity constraints.

DEFINITION: BASIC SOLUTION

A *basic solution* is any solution obtained by setting $(n' - m)$ variables equal to 0 and solving the system of equations for the values of the remaining m variables. The m variables solved for are called *basic variables.* These variables are said to constitute a *basis. The remaining (n' − m)* variables, or those which have been assigned values of 0, are called *nonbasic variables.*

DEFINITION: BASIC FEASIBLE SOLUTION

A *basic feasible solution* is a basic solution which also satisfies the nonnegativity constraints.

It can be shown that the optimal solution to a linear programming problem is included in the set of basic feasible solutions. Thus, the optimal solution can be found by performing a search of the set of basic feasible solutions. This is what the simplex method accomplishes. It begins with a basic feasible solution consisting of two pools of variables — m basic variables and $(n' - m)$ nonbasic variables. The simplex method determines whether the objective function can be improved by exchanging a basic variable and a nonbasic variable. If an exchange will result in an improvement, an existing basic variable is set equal to 0 (becoming a nonbasic variable), an existing nonbasic variable is included in the pool of basic variables, and the system of equations is re-solved with the new set of basic variables to form a new basic feasible solution. A determination is made again regarding whether a better solution exists. If so, another exchange takes place and the process repeats itself. The simplex method is termed an *iterative process* because a specified set of solution steps are repeated until a conclusion is reached regarding the solution for the problem. The exchange of variables which takes place at each iteration is summarized in Fig. 11.2.

EXAMPLE 7 Figure 11.3 illustrates the region of feasible solutions for an LP problem.
(a) Identify the nature (type) of each constraint.
(b) What supplemental variables will be added when the problem is restated to comply with requirement I of the simplex method?
(c) How many basic variables and nonbasic variables will there be in any basic feasible solution?
(d) What are the basic and nonbasic variables associated with corner point B? With corner point F?

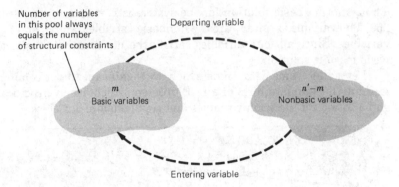

Figure 11.2   Simplex exchange of variables.

Figure 11.3

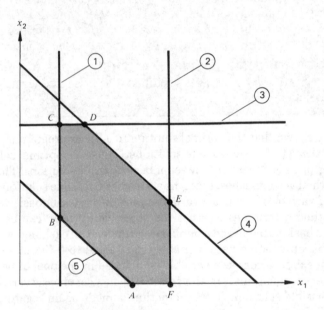

SOLUTION

(a) From Fig. 11.3, constraints (1) and (5) appear to be ($\geq$) types, while constraints (2), (3), and (4) are ($\leq$) types.

(b) Using the convention that the subscripts on supplemental variables indicate the constraint number, the following variables would be introduced when satisfying requirement I of the simplex.

Constraint	Supplemental Variables
(1)	E_1 and A_1
(2)	S_2
(3)	S_3
(4)	S_4
(5)	E_5 and A_5

(c) Because there are five structural constraints, there will be five basic variables. The total number of variables (decision variables plus supplementals) equals nine: $x_1, x_2, E_1, A_1,$ $S_2, S_3, S_4, E_5,$ and A_5. If there are five basic variables, the remaining four variables must be nonbasic.

(d) The best approach to identifying the basic variables is to first determine which of the decision variables is basic. By observation of corner point B, the x_1 and x_2 coordinates appear to both be positive. Therefore, these two variables are basic. Now we turn to the supplemental variables. We are looking for three more variables which have positive values at corner point B. Look at the location of the corner point relative to the locations of the constraints. *If the corner point does not lie on a constraint line, that constraint has either slack or surplus, depending on the type of constraint. If the corner point lies on a constraint line, there is no slack or surplus in the constraint.* For point B:

Constraint	Relationship to Corner Point	Conclusion
(1)	B lies on it	No surplus ($E_1 = 0$)
(2)	B lies to left	Slack exists ($S_2 > 0$)
(3)	B lies under	Slack exists ($S_3 > 0$)
(4)	B lies under	Slack exists ($S_4 > 0$)
(5)	B lies on it	No surplus ($E_5 = 0$)

Therefore, the basic variables are $x_1, x_2, S_2, S_3,$ and S_4. The nonbasic variables are $E_1,$ $E_5, A_1,$ and A_5. **For any corner point on a region of feasible solutions, any artificial variables will equal zero, meaning that they are nonbasic.**

Observation of corner point F should lead you to conclude that x_1 is positive and x_2 equals zero. Looking at corner point F relative to the five constraints gives the following table.

Constraint	Relationship to Corner Point	Conclusion
(1)	F lies to right	Surplus exists ($E_1 > 0$)
(2)	F lies on line	No slack ($S_2 = 0$)
(3)	F lies below	Slack exists ($S_3 > 0$)
(4)	F lies below	Slack exists ($S_4 > 0$)
(5)	F lies to right	Surplus exists ($E_5 > 0$)

Therefore, the basic variables are $x_1, E_1, S_3, S_4,$ and E_5. The nonbasic variables are $x_2,$ $S_2, A_1,$ and A_5.

☐

PRACTICE EXERCISE
What are the basic and nonbasic variables associated with corner point D? *Answer:* basic — x_1, x_2, E_1, S_2, E_5; nonbasic — $S_3, S_4, A_1,$ and A_5.

Section 11.1 Follow-up Exercises

1 Given the following LP problem, rewrite the constraint set in standard form incorporating all supplemental variables.

$$\text{Maximize} \quad z = 5x_1 - 3x_2 + 7x_3$$

$$\text{subject to} \quad x_1 + x_2 + x_3 \geq 15$$

$$3x_1 \quad\quad - 2x_3 \leq -5$$

$$4x_1 - 2x_2 + x_3 \leq 24$$

$$x_1, x_2, x_3 \geq 0$$

2 Given the following LP problem, rewrite the constraint set in standard form incorporating all supplemental variables.

$$\text{Minimize} \quad z = 3x_1 + 5x_2$$

$$\text{subject to} \quad x_1 + x_2 \geq 10$$

$$2x_1 - 3x_2 \leq 6$$

$$2x_1 + 5x_2 = 18$$

$$3x_1 + 2x_2 \leq -3$$

$$x_1, x_2 \geq 0$$

3 Given the following LP problem, rewrite the constraint set in standard form incorporating all supplemental variables.

$$\text{Minimize} \quad z = 3x_1 - 5x_2 + 2x_3 + 7x_4$$

$$\text{subject to} \quad x_1 + x_2 + x_3 + x_4 \geq 25$$

$$-x_1 + 3x_2 \quad\quad - 2x_4 \leq -20$$

$$3x_1 \quad\quad\quad - 4x_4 = 10$$

$$5x_1 - x_2 + 3x_3 + 8x_4 \leq 125$$

$$x_1 \quad\quad\quad\quad \geq 5$$

$$x_3 \quad\quad \leq 30$$

$$x_1, x_2, x_3, x_4 \geq 0$$

4 Given the following LP problem, rewrite the constraint set in standard form incorporating all supplemental variables.

$$\text{Maximize} \quad z = 6x_1 + 4x_2 + 9x_3$$

$$\text{subject to} \quad 3x_1 - x_2 + x_3 \geq 16$$

$$x_1 + x_2 + x_3 \leq 20$$

$$4x_1 \quad\quad - 2x_3 \geq 6$$

$$4x_2 + 2x_3 \geq 10$$

$$2x_1 + 5x_2 - x_3 \leq 40$$

$$x_1 \quad\quad\quad\quad \leq 4$$

$$x_3 \geq 2$$

$$x_1, x_2, x_3 \geq 0$$

5 Given the formulation of the diet mix application (Example 8) on page 443, rewrite the constraint set in standard form incorporating all supplemental variables.

6 Given the formulation of the highway maintenance application (Example 9) on page 445, rewrite the constraint set in standard form incorporating all supplemental variables.

7 Given the formulation of the grant awards application (Example 10) on page 447, rewrite the constraint set in standard form incorporating all supplemental variables.

8 Given the formulation of the petroleum-blending application (Example 11) on page 450, rewrite the constraint set in standard form incorporating all supplemental variables.

9 An LP problem has 15 decision variables, 20 ($\leq$) constraints, 12 ($\geq$) constraints, and 8 ($=$) constraints. When rewritten in standard form, how many variables will be included? How many supplemental variables of each type?

10 An LP problem has 8 decision variables, 16 ($\leq$) constraints, 10 ($\geq$) constraints, and 3 ($=$) constraints. When rewritten in standard form, how many variables will be included? How many supplemental variables of each type?

11 Given the region of feasible solutions in Fig. 11.4:
 (a) Identify the nature (type) of each constraint.
 (b) What supplemental variables will be added when the problem is restated to comply with requirement I?
 (c) How many basic variables and nonbasic variables will there be in any basic feasible solution?
 (d) What are the basic and nonbasic variables associated with corner points A, B, C, and D?

Figure 11.4

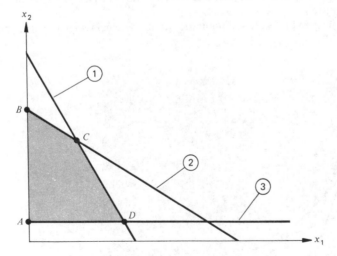

12 Given the region of feasible solution in Fig. 11.5:
 (a) Identify the nature (type) of each constraint.
 (b) What supplemental variables will be added when the problem is restated to comply with requirement I?

Figure 11.5

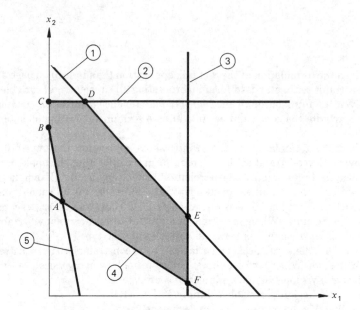

(c) How many basic variables and nonbasic variables will there be in any basic feasible solution?

(d) What are the basic and nonbasic variables associated with corner points B and D?

13 Given the region of feasible solutions in Fig. 11.6:

(a) Identify the nature (type) of each constraint.

(b) What supplemental variables will be added when the problem is restated to comply with requirement I?

(c) How many basic variables and nonbasic variables will there be in any basic feasible solution?

(d) What are the basic and nonbasic variables associated with corner points A and C?

Figure 11.6

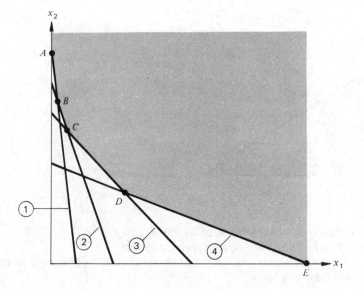

14 Given the region of feasible solutions in Fig. 11.7:
   (a) Identify the nature (type) of each constraint.
   (b) What supplemental variables will be added when the problem is restated to comply with requirement I?
   (c) How many basic variables and nonbasic variables will there be in any basic feasible solution?
   (d) What are the basic and nonbasic variables associated with each of the corner points?

Figure 11.7

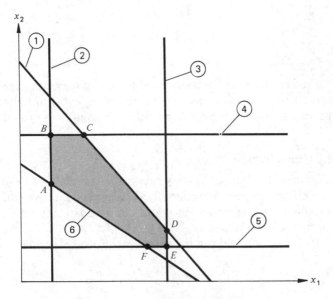

11.2 THE SIMPLEX METHOD

Before we begin our discussions of the simplex method, let's provide a generalized statement of an LP model. Given the definitions

$x_j = j$th decision variable

$c_j =$ coefficient on jth decision variable in the objective function

$a_{ij} =$ coefficient in the ith constraint for the jth variable

$b_i =$ right-hand-side constant for the ith constraint

the generalized LP model can be stated as follows:

Optimize (maximize or minimize)
$$z = c_1 x_1 + c_2 x_2 + \cdots + c_n x_n$$
subject to

$$a_{11}x_1 + a_{12}x_2 + \cdots + a_{1n}x_n (\leq, \geq, =) b_1 \quad (1)$$
$$a_{21}x_1 + a_{22}x_2 + \cdots + a_{2n}x_n (\leq, \geq, =) b_2 \quad (2)$$
$$\vdots \qquad \vdots \qquad\qquad \vdots \qquad \vdots \quad \vdots$$
$$a_{m1}x_1 + a_{m2}x_2 + \cdots + a_{mn}x_n (\leq, \geq, =) b_m \quad (m)$$
$$x_1 \geq 0$$
$$x_2 \geq 0$$
$$\vdots \quad \vdots \quad \vdots$$
$$x_n \geq 0$$

This generalized model has n decision variables and m structural constraints. Note that each structural constraint has only one of the $(\leq, \geq, =)$ conditions assigned to it. As we discuss the simplex method, we will occasionally use the notation of this model.

Solution by Enumeration

Consider a problem having m ($\leq$) constraints and n variables. Prior to solving by the simplex method, the m constraints would be changed into equations by adding m slack variables. This restatement results in a constraint set consisting of m equations and $m + n$ variables.

In Sec. 10.2 we examined the following linear programming problem:

$$\text{Maximize} \quad z = 5x_1 + 6x_2$$
$$\text{subject to} \quad 3x_1 + 2x_2 \leq 120$$
$$4x_1 + 6x_2 \leq 260$$
$$x_1, x_2 \geq 0$$

Before we solve this problem by the simplex method, the constraint set must be transformed into the equivalent set

$$3x_1 + 2x_2 + S_1 = 120$$
$$4x_1 + 6x_2 + S_2 = 260$$
$$x_1, x_2, S_1, S_2 \geq 0$$

The constraint set involves two equations and four variables. Note that the slack variables, in addition to the decision variables, are restricted to being nonnegative.

Of all the possible solutions to the constraint set, it can be proved that an optimal solution occurs when two of the four variables in this problem are set equal to zero and the system is solved for the other two variables. The question is, Which two variables should be set equal to 0 (should be *nonbasic variables*)? Let's enumerate the different possibilities. If S_1 and S_2 are set equal to 0, the constraint equations become

$$3x_1 + 2x_2 = 120$$

$$4x_1 + 6x_2 = 260$$

Solving for the corresponding *basic variables* x_1 and x_2 results in $x_1 = 20$ and $x_2 = 30$.

If S_1 and x_1 are set equal to 0, the system becomes

$$2x_2 = 120$$

$$6x_2 + S_2 = 260$$

Solving for the corresponding basic variables x_2 and S_2 results in $x_2 = 60$ and $S_2 = -100$.

Table 11.1 summarizes the *basic solutions,* that is, all the solution possibilities given that two of the four variables are assigned values of 0. Notice that solutions 2 and 5 are not feasible. They each contain a variable which has a negative value, violating the nonnegativity restriction. However, solutions 1, 3, 4, and 6 are *basic feasible solutions* to the linear programming problem and are candidates for the optimal solution.

TABLE 11.1	Solution	Nonbasic Variables	Basic Variables
	1	S_1, S_2	$x_1 = 20, x_2 = 30$
	*2	x_1, S_1	$x_2 = 60, S_2 = -100$
	3	x_1, S_2	$x_2 = 43\frac{1}{3}, S_1 = 33\frac{1}{3}$
	4	x_2, S_1	$x_1 = 40, S_2 = 100$
	*5	x_2, S_2	$x_1 = 65, S_1 = -75$
	6	x_1, x_2	$S_1 = 120, S_2 = 260$

Figure 11.8 is the graphical representation of the set of constraints. In this figure the points of intersection (A, B, C, D, E, F) between the structural constraint lines and the nonnegativity constraints (x_1 and x_2 axes) represent the set of basic solutions. Solutions 1, 3, 4, and 6 in Table 11.1 are the basic feasible solutions and correspond to the four corner points on the region of feasible solutions in Fig. 11.8. Specifically, solution 1 corresponds to corner point C, solution 3 corresponds to corner point B, solution 4 corresponds to corner point D, and solution 6 corresponds to corner point A. Solutions 2 and 5, which are not feasible, correspond to points E and F in Fig. 11.8.

The important thing to note is that by setting all combinations of two different variables equal to 0 and solving for the remaining variables, a set of potential solutions (basic solutions) was identified for the LP problem. A subset of these solutions was automatically disqualified because it contained infeasible solutions (2 and 5). However, the remaining basic feasible solutions corresponded to the corner points on the region of feasible solutions. Since we know that an optimal solution will occur at at least one of these corner points, further examination of these will reveal an optimal solution.

For a maximization problem having m ($\leq$) constraints and n decision variables, the addition of m slack variables results in m constraint equations containing

Figure 11.8

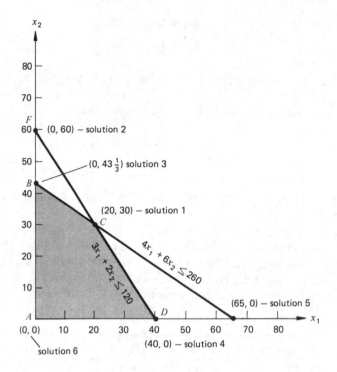

$m + n$ variables. *An optimal solution to this problem can be found by setting n of the variables equal to 0 and solving for the remaining m variables.* In the process of selecting different combinations of n variables to be set equal to 0, the simplex (1) *will never select a combination which will result in an infeasible solution* and (2) *will ensure that each new combination selected will result in a solution which has an objective function value at least as good as the current solution.*

The Algebra of the Simplex Method

Before presenting the simplex method formally, let's discuss the algebra upon which the method is based. The simplex arithmetic is based upon the gaussian elimination method discussed in Sec. 3.2. A rereading of this section is suggested if you are rusty with this method.

Let's return to the following maximization problem:

$$\text{Maximize} \quad z = 5x_1 + 6x_2 + 0S_1 + 0S_2$$
$$\text{subject to} \quad 3x_1 + 2x_2 + S_1 \qquad\quad = 120$$
$$4x_1 + 6x_2 \qquad + S_2 = 260$$
$$x_1, x_2, S_1, S_2 \geq \quad 0$$

Note that the slack variables have been assigned objective function coefficients of 0. *Although there can be exceptions, slack variables are usually assigned coefficients of 0 in the objective function.* The reason is that these variables typically contribute nothing to the value of the objective function.

Let's illustrate how the gaussian elimination procedure can be used to identify the set of basic solutions. If we represent the system of constraint equations by showing only the variable coefficients and right-hand-side constants, we have

$$
\begin{array}{cccc}
\underline{x_1} & \underline{x_2} & \underline{S_1} & \underline{S_2} \\
3 & 2 & 1 & 0 \quad | \quad 120 \qquad R_1 \\
4 & 6 & 0 & 1 \quad | \quad 260 \qquad R_2
\end{array}
$$

Remember that the gaussian elimination procedure uses *row operations* to transform the original system of equations into an equivalent system. Upon completion of the gaussian procedure (assuming a unique solution), the equivalent system has the properties that only one variable remains in each equation and that the right side of the equation equals the value of that variable. In looking at the variable coefficients for this system of constraint equations (R_1 and R_2), a convenient starting point would be to declare x_1 and x_2 as nonbasic variables, setting them equal to 0. The variable coefficients for S_1 and S_2 are already in the desired form. If x_1 and x_2 both equal 0, the corresponding values of the basic variables S_1 and S_2 are $S_1 = 120$ and $S_2 = 260$.

$$
\begin{array}{cccc}
\underline{x_1} & \underline{x_2} & \underline{S_1} & \underline{S_2} \\
3 & 2 & 1 & 0 \quad | \quad 120 \\
4 & 6 & 0 & 1 \quad | \quad 260
\end{array}
$$

Assume that we wish to set x_1 and S_1 equal to 0 and solve for x_2 and S_2. Compared with the original solution, we wish to replace S_1 with x_2 as a basic variable. And we would like the coefficients for x_2 and S_2 to have the form

$$
\begin{array}{cccc}
\underline{x_1} & \underline{x_2} & \underline{S_1} & \underline{S_2} \\
 & ① & & 0 \quad | \\
 & ⓪ & & 1 \quad |
\end{array}
$$

Since the coefficients for S_2 are already in the desired form, we need only to change those for x_2 from $\binom{2}{6}$ to $\binom{1}{0}$. To create the ①, we multiply row 1 by $\frac{1}{2}$, resulting in

$$
\begin{array}{cccc}
\underline{x_1} & \underline{x_2} & \underline{S_1} & \underline{S_2} \\
\frac{3}{2} & 1 & \frac{1}{2} & 0 \quad | \quad 60 \qquad R'_1 = \frac{1}{2}R_1 \\
4 & 6 & 0 & 1 \quad | \quad 260 \qquad R_2
\end{array}
$$

The 6 is transformed into ⓪ by multiplying the new row 1 (R'_1) by -6 and adding this multiple to row 2, or

$$
\begin{array}{cccc}
\underline{x_1} & \underline{x_2} & \underline{S_1} & \underline{S_2} \\
\frac{3}{2} & 1 & \frac{1}{2} & 0 \quad | \quad 60 \qquad R'_1 \\
-5 & 0 & -3 & 1 \quad | \quad -100 \qquad R'_2 = R_2 - 6R'_1
\end{array}
$$

Because x_1 and S_1 were set equal to 0, the values of x_2 and S_2 can be read directly as $x_2 = 60$ and $S_2 = -100$. If you refer to Table 11.1, you will see that this solution is the same as solution 2.

x_1	x_2	S_1	S_2	
$\frac{3}{2}$	1	$\frac{1}{2}$	0	60
-5	0	-3	1	-100

If we set S_1 and S_2 equal to 0, x_1 will replace S_2 as a basic variable. For the new basis we would like the coefficients on x_1 and x_2 to have the form

x_1	x_2	S_1	S_2
⓪	1	█	█
①	0	█	█

Starting with the last basic solution, the coefficients on x_2 are in the desired form, and we need to change those for x_1 from $\begin{pmatrix} \frac{3}{2} \\ -5 \end{pmatrix}$ to $\begin{pmatrix} 0 \\ 1 \end{pmatrix}$. The ① is created by multiplying R_2' by $-\frac{1}{5}$, resulting in

x_1	x_2	S_1	S_2		
$\frac{3}{2}$	1	$\frac{1}{2}$	0	60	R_1'
1	0	$\frac{3}{5}$	$-\frac{1}{5}$	20	$R_2'' = -\frac{1}{5}R_2'$

The ⓪ is created by multiplying R_2'' by $-\frac{3}{2}$ and adding this multiple to R_1', or

x_1	x_2	S_1	S_2		
0	1	$-\frac{2}{5}$	$\frac{3}{10}$	30	$R_1'' = R_1' - \frac{3}{2}R_2''$
1	0	$\frac{3}{5}$	$-\frac{1}{5}$	20	R_2''

With S_1 and S_2 set equal to 0, the values of x_1 and x_2 are read directly as $x_1 = 20$ and $x_2 = 30$, which corresponds to solution 1 in Table 11.1.

Incorporating the Objective Function

In solving by the simplex method, the objective function and constraints are combined to form a system of equations. The objective function is one of the equations, and z becomes an additional variable in the system. In rearranging the variables in the objective function so that they are all on the left side of the equation, the problem is represented by the system of equations

$$z - 5x_1 - 6x_2 - 0S_1 - 0S_2 = \quad 0 \qquad \textbf{(0)}$$

$$3x_1 + 2x_2 + \ S_1 \qquad = 120 \qquad \textbf{(1)}$$

$$4x_1 + 6x_2 \qquad + \ S_2 = 260 \qquad \textbf{(2)}$$

Note that the objective function is labeled as Eq. (0).

The objective is to solve this (3×5) system of equations so as to maximize the value of z. Since we are particularly concerned about the value of z and will want to know its value for any solution, z will always be a basic variable. The standard practice, however, is not to refer to z as a basic variable. The terms *basic variable* and *nonbasic variable* are usually reserved for the other variables in the problem.

The simplex operations are performed in a tabular format. The initial table, or *tableau,* for our problem is shown in Table 11.2. Note that there is one row for each equation and the table contains the coefficients of each variable in the equations. The b_i *column* contains the right-hand-side constants of the equations, where b_i is the right-hand-side constant for equation i or row i.

TABLE 11.2

Basic Variables	z	x_1	x_2	S_1	S_2	b_i	Row Number
	1	-5	-6	0	0	0	(0)
S_1	0	3	2	1	0	120	(1)
S_2	0	4	6	0	1	260	(2)

In a maximization problem having all $(\leq)$ constraints, the starting solution will have a set of basic variables consisting of the slack variables in the problem. By setting x_1 and x_2 equal to 0 in our problem, the initial solution is $S_1 = 120$, $S_2 = 260$, and $z = 0$. The basic variables and the rows in which their values are read are noted in the first column of the tableau.

Given any intermediate solution, the simplex method compares the nonbasic variables with the set of basic variables. The purpose is to determine whether any nonbasic variable should replace a basic variable. *A nonbasic variable will replace a basic variable only if (1) it appears that the objective function will be improved and (2) the new solution is feasible.*

*In any simplex tableau, the row (0) coefficients for all variables (excluding z) represent the change in the current value of the objective function if the variable (in that column) increases in value by one unit. The sign on the row (0) coefficients is the opposite of the actual direction of change. That is, a negative coefficient suggests an **increase** in the value of z; a positive coefficient suggests a **decrease**.*

RULE 1: OPTIMALITY CHECK IN MAXIMIZATION PROBLEM
In a maximization problem, the optimal solution has been found if all row (0) coefficients for the variables are greater than or equal to 0. If any row (0) coefficients are negative for nonbasic variables, a better solution can be found by assigning a positive quantity to these variables.

Since the row (0) coefficients for x_1 and x_2 are -5 and -6, respectively, the optimal solution has not been found in Table 11.2.

> **RULE 2: NEW BASIC VARIABLE IN MAXIMIZATION PROBLEM**
> In a maximization problem the nonbasic variable which will replace a basic variable is the one having the *most negative* row (0) coefficient. "Ties" may be broken arbitrarily.

In selecting a nonbasic variable to become a basic variable, the simplex method chooses the one which will result in the largest *marginal* (per-unit) improvement in z. Since an improvement of 6 units is better than an improvement of 5 units, the simplex would choose x_2 to become a basic variable in the next solution. *In the simplex tableau the column representing the new basic variable will be called the key column.*

If z will increase by 6 units for *each* unit of x_2, we would like x_2 to become as large as possible. The simplex method will allow x_2 to increase in value until one of the current basic variables is driven to a value of 0 (becoming a nonbasic variable). If Eqs. (1) and (2) are rewritten as a function of the nonbasic variables, we can observe the effects that changes in x_2 will have on the values of the current basic variables:

$$S_1 = 120 - 3x_1 - 2x_2 \tag{1a}$$

$$S_2 = 260 - 4x_1 - 6x_2 \tag{2a}$$

Equation (1a) indicates that S_1 currently equals 120 but will decrease in value by 2 units for each unit that x_2 increases. If x_2 is allowed to grow to a value of 120/2, or 60 units, S_1 will be driven to a value of 0. Equation (2a) indicates that S_2 currently equals 260 but will decrease in value by 6 units for each unit that x_2 increases. S_2 will be driven to a value of 0 if x_2 is allowed to grow to a value of 260/6, or $43\frac{1}{3}$ units. The question is, Which basic variable will be driven to a value of 0 (will become nonbasic) first as x_2 is allowed to increase? The answer is S_2, when $x_2 = 43\frac{1}{3}$. If x_2 were allowed to grow to a value of 60, substitution into Eq. (2a) would result in

$$S_2 = 260 - 6(60)$$
$$= -100$$

Since S_2 would be negative, this solution is not feasible. Our conclusion is that x_2 should replace S_2 as a basic variable in the next solution.

Using the tableau structure, the decision of which basic variable to replace is made by focusing upon the key column and the b_i column. The partial tableau in Table 11.3 illustrates a column which is supposed to represent the *key column* for an

TABLE 11.3			$\downarrow$ —— *Key Column*		
$\cdots$	x_k	$\cdots$	b_i	Row Number	
$\cdots$	a_{0k}	$\cdots$	b_0	(0)	
$\cdots$	a_{1k}	$\cdots$	b_1	(1)	
$\cdots$	a_{mk}	$\cdots$	b_m	(m)	

intermediate solution. The key column element a_{ik} represents the constant appearing in row i of the key (k) column. Similarly, b_i values represent the right-side constants for row i.

For any column, the a_{ij} values ($i = 1$ to m) are called marginal rates of substitution. *These values indicate the changes required in each of the current basic variables if the variable (in the key column) increases by one unit. As with the row (0) coefficients, the sign on these marginal rates of substitution is the opposite of the actual direction of change. A positive a_{ij} value indicates a* decrease *in the ith basic variable; a negative a_{ij} value indicates an* increase *in the ith basic variable.*

TABLE 11.4

Basic Variables	z	x_1	x_2	S_1	S_2	b_i	Row Number	b_i/a_{ik}
	1	−5	−6	0	0	0	(0)	
S_1	0	3	②	1	0	⑫⓪	(1)	120/2 = 60
S_2	0	4	⑥	0	1	②⑥⓪	(2)	260/6 = $43\frac{1}{3}$*

(Key Column — x_2)
(Departing Variable) S_2

If we focus on the key column in Table 11.4, the coefficient in row (0) suggests that the current value of z will *increase* by 6 units if x_2 increases in value by one unit. The a_{ik} values in rows (1) and (2) are 2 and 6, respectively. These indicate that if x_2 increases in value by one unit, (1) the value of S_1 will *decrease* by 2 units (from its current value of 120) and (2) the value of S_2 will *decrease* by 6 units (from its current value of 260).

RULE 3: DEPARTING BASIC VARIABLE
The basic variable to be replaced is found by determining the row i associated with

$$\min \frac{b_i}{a_{ik}} \qquad i = 1, \ldots, m$$

where $a_{ik} > 0$. In addition to identifying the departing basic variable, the minimum b_i/a_{ik} value is the maximum number of units which can be introduced of the incoming basic variable.

Rule 3 suggests that the ratio b_i/a_{ik} should be determined for rows (1) to (m) *where $a_{ik} > 0$*. We only consider positive a_{ik} values because these are associated with basic variables which *decrease* in value as additional units of the incoming variable are introduced. The minimum ratio should be identified and the corresponding row i, noted. The departing basic variable is the one whose value is currently read from this row. In Table 11.4 the ratios b_i/a_{ik} are computed for rows (1) and (2). The minimum ratio is $43\frac{1}{3}$, associated with row (2). Since the value of S_2 is currently read from row (2), S_2 is the departing basic variable.

POINTS FOR THOUGHT & DISCUSSION   Why do you think that rule 3 suggests identifying the *minimum* b_i/a_{ik} ratio? In Table 11.4 select the departing variable by identifying the maximum ratio and see what happens.

In the next solution we will want to read the value of the new basic variable x_2 from row (2). Thus, we need to apply the gaussian elimination procedures to transform the column of coefficients under x_2 from $\begin{pmatrix} -6 \\ 2 \\ 6 \end{pmatrix}$ to $\begin{pmatrix} 0 \\ 0 \\ 1 \end{pmatrix}$. The ① is created by multiplying row (2) by $\frac{1}{6}$. The two ⓪s are created by multiplying the *new* row (2) by $+6$ and -2 and adding these multiples, respectively, to rows (0) and (1). The next solution appears in Table 11.5. Note in Table 11.5 that a shorthand notation to the right of each row indicates how the elements of that row were computed. The notation R_j is used to represent row j.

Note also in Table 11.5 that x_2 has replaced S_2 in the column of basic variables. We can read the values of z, S_1, and x_2 from the b_i column as $z = 260$, $S_1 = 33\frac{1}{3}$, and $x_2 = 43\frac{1}{3}$. Since x_1 and S_2 are nonbasic variables, their values are zero for this solution.

TABLE 11.5

Basic Variables	z	x_1	x_2	S_1	S_2	b_i	Row Number	
	1	-1	0	0	1	260	(0)	$R_0' = R_0 + 6R_2'$
S_1	0	$\frac{10}{6}$	0	1	$-\frac{1}{3}$	$33\frac{1}{3}$	(1)	$R_1' = R_1 - 2R_2'$
x_2	0	$\frac{4}{6}$	1	0	$\frac{1}{6}$	$43\frac{1}{3}$	(2)	$R_2' = \frac{1}{6}R_2$

PRACTICE EXERCISE
Substitute the values $x_1 = 0$ and $x_2 = 43\frac{1}{3}$ into the original formulation and verify (*a*) that the first constraint has slack equal to $33\frac{1}{3}$ units, (*b*) that the second constraint is satisfied as an equality ($S_2 = 0$), and (*c*) that the value of z equals 260.

With this new solution, the first thing to check is whether it is optimal. Applying rule 1, we conclude that the solution is not optimal because of the -1 coefficient for x_1 in row (0). Since x_1 has the only negative coefficient in row (0), it will become the new basic variable. To determine the departing basic variable, we focus on the elements in the x_1 column. As shown in Table 11.6, the minimum b_i/a_{ik} ratio is 20, and this minimum ratio is associated with row (1). Since the value of S_1 is currently read from row (1), S_1 is identified as the departing variable.

In the next solution we will want to read the value of the new basic variable x_1 from row (1). As such, the column of coefficients under x_1 should be transformed from $\begin{pmatrix} -1 \\ \frac{10}{6} \\ \frac{4}{6} \end{pmatrix}$ to $\begin{pmatrix} 0 \\ 1 \\ 0 \end{pmatrix}$.

TABLE 11.6

		↓ ── Key Column						
Basic Variables	**z**	x_1	x_2	S_1	S_2	b_i	**Row Number**	b_i/a_{ik}
	1	-1	0	0	1	260	(0)	
S_1	0	$\frac{10}{6}$	0	1	$-\frac{1}{3}$	$33\frac{1}{3}$	(1)	$33\frac{1}{3} \div \frac{10}{6} = 20^*$
x_2	0	$\frac{4}{6}$	1	0	$\frac{1}{6}$	$43\frac{1}{3}$	(2)	$43\frac{1}{3} \div \frac{4}{6} = 65$

The ① is created by multiplying row (1) by $\frac{6}{10}$. The two ⓪s are created by multiplying the new row (1) by $+1$ and $-\frac{4}{6}$ and adding these multiples, respectively, to rows (0) and (2). (See equations to the right of Table 11.7.) The new solution appears in Table 11.7.

TABLE 11.7

Basic Variables	**z**	x_1	x_2	S_1	S_2	b_i	**Row Number**	
	1	0	0	$\frac{6}{10}$	$\frac{24}{30}$	280	(0)	$R_0'' = R_0' + R_1''$
x_1	0	1	0	$\frac{6}{10}$	$-\frac{6}{30}$	20	(1)	$R_1'' = \frac{6}{10}R_1'$
x_2	0	0	1	$-\frac{24}{60}$	$\frac{3}{10}$	30	(2)	$R_2'' = R_2' - \frac{4}{6}R_1''$

Note in Table 11.7 that x_1 has replaced S_1 in the column of basic variables. With S_1 and S_2 being nonbasic the values of z, x_1, and x_2 are read from the b_i column as $z = 280$, $x_1 = 20$, and $x_2 = 30$. Our next step is to check for optimality. Applying rule 1, we conclude that this solution is optimal because *all row (0) coefficients are greater than or equal to 0*. This answer agrees with the one we found when solving graphically in Sec. 10.2. The objective function is maximized at a value of 280 when $x_1 = 20$, $x_2 = 30$, $S_1 = 0$, and $S_2 = 0$. Figure 11.9 illustrates the simplex iterative progression to the optimal solution.

Summary of Simplex Procedure

Let's generalize the ***simplex procedure for maximization problems having all ($\leq$) constraints.*** First, add slack variables to each constraint and the objective function and place the variable coefficients and right-hand-side constants in a simplex tableau. Then:

1 *Identify the initial solution by declaring each of the slack variables as basic variables. All other variables are nonbasic in the initial solution.*
2 *Determine whether the current solution is optimal by applying rule 1 [**Are all row (0) coefficients ≥ 0?**]. If it is optimal, stop! If it is not optimal, proceed to step 3.*
3 *Determine the nonbasic variable which should become a basic variable in the next solution by applying rule 2 [**most negative row (0) coefficient**].*
4 *Determine the basic variable which should be replaced in the next solution by applying rule 3 (**min b_i/a_{ik} ratio where $a_{ik} > 0$**).*
5 *Apply the gaussian elimination operations to generate the new solution (or new tableau). Go to step 2.*

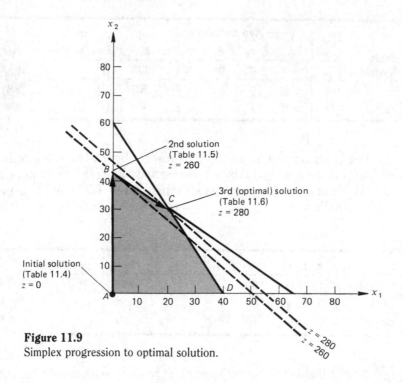

Figure 11.9
Simplex progression to optimal solution.

EXAMPLE 8   Let's solve the following linear programming problem using the simplex method.

$$\text{Maximize} \quad z = 2x_1 + 12x_2 + 8x_3$$

$$\text{subject to} \quad 2x_1 + 2x_2 + x_3 \leq 100$$

$$x_1 - 2x_2 + 5x_3 \leq 80$$

$$10x_1 + 5x_2 + 4x_3 \leq 300$$

$$x_1, x_2, x_3 \geq 0$$

Rewriting the problem in standard form with slack variables added in, we have the following.

$$\text{Maximize} \quad z = 2x_1 + 12x_2 + 8x_3 + 0S_1 + 0S_2 + 0S_3$$

$$\text{subject to} \quad 2x_1 + 2x_2 + x_3 + S_1 = 100$$

$$x_1 - 2x_2 + 5x_3 + S_2 = 80$$

$$10x_1 + 5x_2 + 4x_3 + S_3 = 300$$

$$x_1, x_2, x_3, S_1, S_2, S_3 \geq 0$$

The objective function should be restated by moving all variables to the left side of the equation. The initial simplex tableau is shown in Table 11.8.

TABLE 11.8

Basic Variables	z	x_1	x_2	x_3	S_1	S_2	S_3	b_i	Row Number	b_i/a_{ik}
	1	-2	-12	-8	0	0	0	0	(0)	
S_1	0	2	2	1	1	0	0	100	(1)	$100/2 = 50*$
S_2	0	1	-2	5	0	1	0	80	(2)	
S_3	0	10	5	4	0	0	1	300	(3)	$300/5 = 60$

Key Column points to the x_2 column.

❏ **Step 1** In the initial solution x_1, x_2, and x_3 are nonbasic variables having values of 0. The basic variables are the slack variables with $S_1 = 100$, $S_2 = 80$, $S_3 = 300$, and $z = 0$.

❏ **Step 2** Since all row (0) coefficients are *not* greater than or equal to 0, the initial solution is not optimal.

❏ **Step 3** The most negative coefficient in row (0) is -12, and it is associated with x_2. Thus, x_2 will become a basic variable in the next solution and its column of coefficients becomes the key column.

❏ **Step 4** In computing the b_i/a_{ik} ratios, the minimum ratio is 50, and it corresponds to row (1). Thus S_1 will become a nonbasic variable in the next solution. Note that no ratio was computed for row (2) because the a_{ik} value was negative.

❏ **Step 5** The new solution is found by transforming the coefficients in the x_2 column from $\begin{pmatrix} -12 \\ 2 \\ -2 \\ 5 \end{pmatrix}$ to $\begin{pmatrix} 0 \\ 1 \\ 0 \\ 0 \end{pmatrix}$.

Table 11.9 indicates the next solution. The shorthand notation to the right of each row indicates how the elements of that row were computed. In this solution the basic variables and their values are $x_2 = 50$, $S_2 = 180$, $S_3 = 50$, and $z = 600$. Continuing the simplex procedure, we next return to step 2.

TABLE 11.9

Basic Variables	z	x_1	x_2	x_3	S_1	S_2	S_3	b_i	Row Number		b_i/a_{ik}
	1	10	0	-2	6	0	0	600	(0)	$R_0' = R_0 + 12R_1'$	
x_2	0	1	1	$\frac{1}{2}$	$\frac{1}{2}$	0	0	50	(1)	$R_1' = \frac{1}{2}R_1$	$50 \div \frac{1}{2} = 100$
S_2	0	3	0	6	1	1	0	180	(2)	$R_2' = R_2 + 2R_1'$	$180 \div 6 = 30*$
S_3	0	5	0	$\frac{3}{2}$	$-\frac{5}{2}$	0	1	50	(3)	$R_3' = R_3 - 5R_1'$	$50 \div \frac{3}{2} = 33\frac{1}{3}$

Key Column points to the x_3 column.

❏ **Step 2** Since the row (0) coefficient for x_3 is negative, this solution is not optimal.

❏ **Step 3** The variable x_3 will become a basic variable in the next solution since it has the only negative coefficient in row (0). The x_3 column becomes the new key column.

❏ **Step 4** In computing the b_i/a_{ik} ratios, the minimum ratio is 30, and it corresponds to row (2). Thus, S_2 will become a nonbasic variable in the next solution.

❑ **Step 5** The new solution is found by transforming the coefficients in the x_3 column from $\begin{pmatrix} -2 \\ \frac{1}{2} \\ 6 \\ \frac{3}{2} \end{pmatrix}$ to $\begin{pmatrix} 0 \\ 0 \\ 1 \\ 0 \end{pmatrix}$.

Table 11.10 indicates the next solution. The basic variables and their values in this solution are $x_2 = 35$, $x_3 = 30$, and $S_3 = 5$. The value of z is 660. We return to step 2.

TABLE 11.10

Basic Variables	z	x_1	x_2	x_3	S_1	S_2	S_3	b_i	Row Number	
	1	11	0	0	$\frac{38}{6}$	$\frac{2}{6}$	0	660	(0)	$R_0'' = R_0' + 2R_2''$
x_2	0	$\frac{3}{4}$	1	0	$\frac{5}{12}$	$-\frac{1}{12}$	0	35	(1)	$R_1'' = R_1' - \frac{1}{2}R_2''$
x_3	0	$\frac{1}{2}$	0	1	$\frac{1}{6}$	$\frac{1}{6}$	0	30	(2)	$R_2'' = \frac{1}{6}R_2'$
S_3	0	$\frac{17}{4}$	0	0	$-\frac{11}{4}$	$-\frac{1}{4}$	1	5	(3)	$R_3'' = R_3' - \frac{3}{2}R_2''$

❑ **Step 2** Since all row (0) coefficients are greater than or equal to 0 in Table 11.10, the current solution is the optimal solution.

The objective function is maximized at a value of 660 when $x_1 = 0$, $x_2 = 35$, $x_3 = 30$, $S_1 = 0$, $S_2 = 0$, and $S_3 = 5$.

❑

Maximization Problems with Mixed Constraints

We chose the simplest problem structure to illustrate the simplex method: a maximization problem with all ($\leq$) constraints. For a maximization problem having a mix of ($\leq$, $\geq$, and $=$) constraints the simplex method itself does not change. The only change is in transforming constraints to the standard equation form with appropriate supplemental variables. *Recall that for each ($\geq$) constraint, a surplus variable is subtracted and an artificial variable is added to the left side of the constraint. For each ($=$) constraint, an artificial variable is added to the left side.* An additional column is added to the simplex tableau for each supplemental variable. Also, surplus and artificial variables must be assigned appropriate objective function coefficients (c_j values). Surplus variables usually are assigned an objective function coefficient of 0: For maximization problems, artificial variables are assigned a large negative coefficient, which we will denote by $-M$, where $|M|$ is assumed to be much larger than any other coefficient. This assignment is to make the artificial variables extremely unattractive, given the maximization objective.

Computationally, the simplex method proceeds exactly as discussed. The only difference which will be noted is the identification of the initial basis.

INITIAL BASIS IN SIMPLEX

In any linear programming problem, the initial set of basic variables will consist of all the slack variables and all the artificial variables which appear in the problem.

EXAMPLE 9

Let's solve the following maximization problem which has a mix of a ($\leq$) constraint and a ($\geq$) constraint.

$$\text{Maximize} \quad z = 8x_1 + 6x_2$$
$$\text{subject to} \quad 2x_1 + x_2 \geq 10$$
$$3x_1 + 8x_2 \leq 96$$
$$x_1, x_2 \geq 0$$

Rewriting the problem with constraints expressed as equations,

$$\text{Maximize} \quad z = 8x_1 + 6x_2 + 0E_1 - MA_1 + 0S_2$$
$$\text{subject to} \quad 2x_1 + x_2 - E_1 + A_1 \qquad = 10$$
$$3x_1 + 8x_2 \qquad\qquad + S_2 = 96$$
$$x_1, x_2, E_1, A_1, S_2 \geq 0$$

If all variables in the objective function are moved to the left side of the equation, the initial tableau for this problem appears as in Table 11.11. Note that the artificial variable is one of the basic variables in the initial solution. However, *for any problem containing artificial variables, the row (0) coefficients for the artificial variables will not equal zero in the initial tableau.* Consequently, we do not have the desired columns of an identity matrix in the columns associated with the basic variables. In Table 11.11, the $+M$ coefficient must be changed to 0 using row operations. If row (1) is multiplied by $-M$ and added to row (0), the desired result is achieved. Table 11.12 shows the result of this row operation. For this initial solution A_1 and S_2 are basic variables; $x_1, x_2,$ and E_1 are nonbasic variables; and the value of z is $-10M$.

TABLE 11.11

Need to change to zero

Basic Variables	z	x_1	x_2	E_1	A_1	S_2	b_i	Row Number
	1	−8	−6	0	ⓜ	0	0	(0)
A_1	0	2	1	−1	1	0	10	(1)
S_2	0	3	8	0	0	1	96	(2)

TABLE 11.12

Key Column   Transformed to zero

Basic Variables	z	x_1	x_2	E_1	A_1	S_2	b_i	Row Number		b_i/a_{ik}
	1	−8 −2M	−6 −M	M	⓪	0	−10M	(0)	$R_0' = R_0 - MR_1$	
A_1	0	2	1	−1	1	0	10	(1)	R_1	10/2 = 5*
S_2	0	3	8	0	0	1	96	(2)	R_2	96/3 = 32

Applying rule 1, we see that the initial solution is not optimal. The row (0) coefficients for both x_1 and x_2 are negative. Applying rule 2, the row (0) coefficient for x_1 is the most negative and x_1 is identified to become the new basic variable. The x_1 column becomes the

TABLE 11.13

Key Column ↓

Basic Variables	z	x_1	x_2	E_1	A_1	S_2	b_i	Row Number		b_i/a_{ik}
	1	0	-2	-4	$4+M$	0	40	(0)	$R_0'' = R_0' + (8+2M)R_1'$	
x_1	0	1	$\frac{1}{2}$	$-\frac{1}{2}$	$\frac{1}{2}$	0	5	(1)	$R_1' = \frac{1}{2}R_1$	—
S_2	0	0	$\frac{13}{2}$	$\frac{3}{2}$	$-\frac{3}{2}$	1	81	(2)	$R_2' = R_2 - 3R_1'$	$81/\frac{3}{2} = 54$

new key column. When b_i/a_{ik} ratios are calculated, A_1 is identified as the departing basic variable. Conducting the simplex row operations, Table 11.13 indicates the next solution. For this solution, z equals 40 and the basic variables are $x_1 = 5$ and $S_2 = 81$.

Applying rule 1, the negative row (0) coefficients for both x_2 and E_1 indicate that the current solution is not optimal. Applying rule 2, E_1 is identified to become the new basic variable. In determining the departing basic variable, the only relevant b_i/a_{ik} ratio is associated with row (2). Thus, S_2 is identified as the departing basic variable. Table 11.14 displays the results of performing the simplex row operations. Since all row (0) coefficients are greater than or equal to zero, the solution shown in this table is optimal. The objective function is maximized at a value of 256 when $x_1 = 32$ and $E_1 = 54$.

TABLE 11.14

Basic Variables	z	x_1	x_2	E_1	A_1	S_2	b_i	Row Number	
	1	0	$\frac{46}{3}$	0	M	$\frac{8}{3}$	256	(0)	$R_0''' = R_0'' + 4R_2''$
x_1	0	1	$\frac{8}{3}$	0	0	$\frac{1}{3}$	32	(1)	$R_1'' = R_1' + \frac{1}{2}R_2''$
E_1	0	0	$\frac{13}{3}$	1	-1	$\frac{2}{3}$	54	(2)	$R_2'' = \frac{2}{3}R_2'$

❑

PRACTICE EXERCISE
Solve the problem in this example graphically. Determine the locations in your graph which correspond to the solutions displayed in Tables 11.12–11.14.

Minimization Problems

The simplex procedure changes only slightly when minimization problems are solved. Aside from assigning artificial variables objective function coefficients of $+M$, the only difference relates to the interpretation of row (0) coefficients. The following two rules are modifications of rule 1 and rule 2. These apply for minimization problems.

RULE 1A: OPTIMALITY CHECK IN MINIMIZATION PROBLEM
In a minimization problem, the optimal solution has been found if all row (0) coefficients for the variables are less than or equal to 0. If any row (0) coefficients are positive for nonbasic variables, a better solution can be found by assigning a positive quantity to these variables.

> ### RULE 2A: NEW BASIC VARIABLE IN MINIMIZATION PROBLEM
> In a minimization problem, the nonbasic variable which will replace a current basic variable is the one having the **largest positive** row (0) coefficient. Ties may be broken arbitrarily.

EXAMPLE 10 Solve the following linear programming problem using the simplex method.

$$\text{Minimize} \quad z = 5x_1 + 6x_2$$
$$\text{subject to} \quad x_1 + x_2 \geq 10$$
$$2x_1 + 4x_2 \geq 24$$
$$x_1, x_2 \geq 0$$

SOLUTION

This problem is first rewritten with the constraints expressed as equations as follows:

$$\text{Minimize} \quad z = 5x_1 + 6x_2 + 0E_1 + 0E_2 + MA_1 + MA_2$$
$$\text{subject to} \quad x_1 + x_2 - E_1 \qquad + A_1 \qquad = 10$$
$$2x_1 + 4x_2 \qquad - E_2 \qquad + A_2 = 24$$
$$x_1, x_2, E_1, E_2, A_1, A_2 \geq 0$$

TABLE 11.15

Need to Be Transformed to Zero

Basic Variables	z	x_1	x_2	E_1	E_2	A_1	A_2	b_i	Row Number
	1	−5	−6	0	0	$-M$	$-M$	0	(0)
A_1	0	1	1	−1	0	1	0	10	(1)
A_2	0	2	4	0	−1	0	1	24	(2)

If all variables in the objective function are moved to the left side of the equation, the initial tableau for this problem appears as in Table 11.15. Note that the artificial variables are the basic variables in this initial solution. As in Example 9, we do not have the desired columns of an identity matrix in the basic variable columns. These $-M$ coefficients in row (0) must be changed to 0 using row operations if the value of z is to be read from row (0). In Table 11.15 we can accomplish this by multiplying rows (1) and (2) by $+M$ and adding these multiples to row (0). Table 11.16 shows the resulting tableau. In this initial solution the nonbasic variables are x_1, x_2, E_1, and E_2. The basic variables are the two artificial variables with $A_1 = 10$, $A_2 = 24$, and $z = 34M$.

Applying rule 1A, we conclude that this solution is not optimal. The row (0) coefficients for x_1 and x_2 are both positive. (Remember that M is an extremely large number.) In applying rule 2A, x_2 is identified as the new basic variable. The x_2 column becomes the new key column. The minimum b_i/a_{ik} value equals 6 and is associated with row (2). Thus, A_2 will be

TABLE 11.16

Key Column → x_2; Transformed to zero → A_1, A_2

Basic Variables	z	x_1	x_2	E_1	E_2	A_1	A_2	b_i	Row Number		b_i/a_{ik}
	1	$-5+3M$	$-6+5M$	$-M$	$-M$	⓪	⓪	$34M$	(0)	$R_0' = R_0 + MR_1 + MR_2$	
A_1	0	1	1	-1	0	1	0	10	(1)	R_1	$10/1 = 10$
A_2	0	2	4	0	-1	0	1	24	(2)	R_2	$24/4 = 6*$

the departing basic variable. Table 11.17 indicates the next solution. In Table 11.17 the nonbasic variables are $x_1, E_1, E_2,$ and A_2. For this solution $A_1 = 4$, $x_2 = 6$, and $z = 36 + 4M$.

TABLE 11.17

Key Column → x_1

Basic Variables	z	x_1	x_2	E_1	E_2	A_1	A_2	b_i	Row Number		b_i/a_{ik}
	1	$-2 + \dfrac{M}{2}$	0	$-M$	$-\dfrac{3}{2} + \dfrac{M}{4}$	0	$\dfrac{3}{2} - \dfrac{5M}{4}$	$36 + 4M$	(0)	$R_0'' = R_0' + (6 - 5M)R_2'$	
A_1	0	$\frac{1}{2}$	0	-1	$\frac{1}{4}$	1	$-\frac{1}{4}$	4	(1)	$R_1' = R_1 - R_2'$	$4/\frac{1}{2} =$
x_2	0	$\frac{1}{2}$	1	0	$-\frac{1}{4}$	0	$\frac{1}{4}$	6	(2)	$R_2' = \frac{1}{4}R_2$	$6/\frac{1}{2} =$

Applying rule 1A, we see that this solution is not optimal. The row (0) coefficients for x_1 and E_2 are both positive. In applying rule 2A, x_1 is identified as the new basic variable. The x_1 column becomes the new key column. The b_i/a_{ik} ratios are $4 \div \frac{1}{2} = 8$ and $6 \div \frac{1}{2} = 12$ for rows (1) and (2). Since the minimum ratio is associated with row (1), A_1 is identified as the departing basic variable. Table 11.18 indicates the new solution. The solution in Table 11.18 has $x_1 = 8$, $x_2 = 2$, and $z = 52$.

Applying rule 1A, we conclude that this solution is optimal. All row (0) coefficients are less than or equal to 0 for the basic and nonbasic variables. Therefore, z is minimized at a value of 52 when $x_1 = 8$ and $x_2 = 2$.

TABLE 11.18

Basic Variables	z	x_1	x_2	E_1	E_2	A_1	A_2	b_i	Row Number	
	1	0	0	-4	$-\frac{1}{2}$	$4 - M$	$\frac{1}{2} - M$	52	(0)	$R_0''' = R_0'' + \left(2 - \dfrac{M}{2}\right)R_1''$
x_1	0	1	0	-2	$\frac{1}{2}$	2	$-\frac{1}{2}$	8	(1)	$R_1'' = 2R_1'$
x_2	0	0	1	1	$-\frac{1}{2}$	-1	$\frac{1}{2}$	2	(2)	$R_2'' = R_2' - \frac{1}{2}R_1''$

□

Section 11.2 Follow-up Exercises

1 Given the linear programming problem:

$$\text{Maximize} \quad z = 14x_1 + 10x_2$$

$$\text{subject to} \qquad 5x_1 + 4x_2 \leq 48$$

$$2x_1 + 5x_2 \leq 26$$

$$x_1, x_2 \geq 0$$

(a) Transform the ($\leq$) constraints into equations.
(b) Enumerate all solutions for which two variables have been set equal to 0.
(c) From part b identify the basic feasible solutions.
(d) Graph the original constraint set and confirm that the basic feasible solutions are corner points on the area of feasible solutions.
(e) What is the optimal solution?

2 Given the linear programming problem:

$$\text{Maximize} \qquad z = 6x_1 + 4x_2$$

$$\text{subject to} \qquad 6x_1 + 10x_2 \leq 90$$

$$12x_1 + 8x_2 \leq 96$$

$$x_1, x_2 \geq 0$$

(a) Transform the ($\leq$) constraints into equations.
(b) Enumerate all solutions for which two variables have been set equal to 0.
(c) From part b identify the basic feasible solutions.
(d) Graph the original constraint set and confirm that the basic feasible solutions are corner points on the area of feasible solutions.
(e) What is the optimal solution?

For Exercises 3–8 solve by the simplex method.

3 Maximize $\quad z = 4x_1 + 2x_2$
subject to $\quad x_1 + x_2 \leq 50$
$\qquad\qquad 6x_1 \qquad\; \leq 240$
$\qquad\qquad x_1, x_2 \geq \quad 0$

4 Maximize $\quad z = 4x_1 + 4x_2$
subject to $\quad 4x_1 + 8x_2 \leq 24$
$\qquad\qquad 24x_1 + 16x_2 \leq 96$
$\qquad\qquad x_1, x_2 \geq 0$

5 Maximize $\quad z = 10x_1 + 12x_2$
subject to $\quad x_1 + x_2 \leq 150$
$\qquad\qquad 3x_1 + 6x_2 \leq 300$
$\qquad\qquad 4x_1 + 2x_2 \leq 160$
$\qquad\qquad x_1, x_2 \geq \quad 0$

6 Maximize $\quad z = 6x_1 + 8x_2 + 10x_3$
subject to $\quad x_1 + 2.5x_2 \qquad\quad \leq 1{,}200$
$\qquad\qquad 2x_1 + \quad 3x_2 + 4x_3 \leq 2{,}600$
$\qquad\qquad x_1, x_2, x_3 \geq \qquad 0$

7 Maximize $\quad z = 10x_1 + 3x_2 + 4x_3$
subject to $\quad 8x_1 + 2x_2 + 3x_3 \leq 400$
$\qquad\qquad 4x_1 + 3x_2 \qquad\quad \leq 200$
$\qquad\qquad\qquad\qquad\quad x_3 \leq \quad 40$
$\qquad\qquad x_1, x_2, x_3 \geq \quad 0$

8 Maximize $z = 4x_1 - 2x_2 + x_3$
    subject to $6x_1 + 2x_2 + 2x_3 \leq 240$
          $2x_1 - 2x_2 + 4x_3 \leq 40$
          $2x_1 + 2x_2 - 2x_3 \leq 80$
              $x_1, x_2, x_3 \geq 0$

9 (a) Solve the following linear programming problem using the simplex method.

$$\text{Minimize} \qquad z = 3x_1 + 6x_2$$

$$\text{subject to} \qquad 4x_1 + x_2 \geq 20$$

$$x_1 + x_2 \leq 20$$

$$x_1 + x_2 \geq 10$$

$$x_1, x_2 \geq 0$$

(b) Verify the solution in part a by solving graphically.

10 (a) Solve the following linear programming problem using the simplex method.

$$\text{Minimize} \qquad z = 6x_1 + 10x_2$$

$$\text{subject to} \qquad x_1 \qquad \leq 12$$

$$2x_2 = 36$$

$$3x_1 + 2x_2 \geq 54$$

$$x_1, x_2 \geq 0$$

(b) Verify the solution in part a by solving graphically.

11.3 SPECIAL PHENOMENA

In Sec. 10.2 certain phenomena which can arise when solving LP problems were discussed. Specifically, the phenomena of *alternative optimal solutions, no feasible solution,* and *unbounded solutions* were presented. In this section we discuss the manner in which these phenomena occur when solving by the simplex method.

Alternative Optimal Solutions

In Sec. 10.2 we described circumstances where there is more than one optimal solution to an LP problem. This situation, termed *alternative optimal solutions,* results when the objective function is parallel to a constraint which binds in the direction of optimization. In two-variable problems we are made aware of alternative optimal solutions with the *corner-point method* when a "tie" occurs for the optimal corner point.

When using the simplex method, alternative optimal solutions are indicated when

1 *an optimal solution has been identified, and*
2 *the row (0) coefficient for a nonbasic variable equals zero.*

The first condition confirms that there is no better solution than the present solution. The presence of a 0 in row (0) for a nonbasic variable indicates that the nonbasic variable can become a basic variable (can become positive) and the current value of the objective function (known to be optimal) will not change.

EXAMPLE 11 Consider the LP problem:

$$\text{Maximize} \quad z = 6x_1 + 4x_2$$

$$\text{subject to} \quad x_1 + x_2 \leq 5$$

$$3x_1 + 2x_2 \leq 12$$

$$x_1, x_2 \geq 0$$

TABLE 11.19

		↓— **Key Column**					Row	
Basic Variables	z	x_1	x_2	S_1	S_2	b_i	Number	b_i/a_{ik}
	1	**−6**	**−4**	**0**	**0**	**0**	(0)	
S_1	0	1	1	1	0	5	(1)	$5/1 = 5$
S_2	0	3	2	0	1	12	(2)	$12/3 = 4*$

Table 11.19 presents the initial simplex solution with $S_1 = 5$ and $S_2 = 12$ being the basic variables; x_1 and x_2 are nonbasic variables. By rule 1 we see that a better solution exists. Using rule 2, x_1 is selected as the new basic variable and the x_1 column becomes the key column. The b_i/a_{ik} ratios are computed with S_2 identified as the departing variable. The elements in the x_1 column are transformed from $\begin{pmatrix} -6 \\ 1 \\ 3 \end{pmatrix}$ to $\begin{pmatrix} 0 \\ 0 \\ 1 \end{pmatrix}$ using row operations, with the resulting simplex tableau shown in Table 11.20. In this new solution, x_1 has replaced S_2 in the basis; the basic variables are $S_1 = 1$ and $x_1 = 4$ with a resulting objective function value of $z = 24$.

TABLE 11.20   Alternative optimal solution indication

Basic Variables	z	x_1	x_2	S_1	S_2	b_i	Row Number		b_i/a_{ik}
	1	0	⓪	0	2	24	(0)	$R_0' = R_0 + 6R_2'$	
S_1	0	0	$\frac{1}{3}$	1	$-\frac{1}{3}$	1	(1)	$R_1' = R_1 - R_2'$	$1\frac{1}{3} = 3*$
x_1	0	1	$\frac{2}{3}$	0	$\frac{1}{3}$	4	(2)	$R_2' = \frac{1}{3}R_2$	$4\frac{2}{3} = 6$

Applying rule 1, we see that this is the optimal solution since all row (0) coefficients are greater than or equal to 0. However, for the nonbasic variable x_2, the row (0) coefficient equals 0. This suggests that x_2 can assume positive values (become a basic variable) and the current (optimal) value of z will not change.

> When alternative optimal solutions are indicated by the simplex method, the other optimal corner-point alternatives can be generated by treating the non-basic variable with the zero coefficient as if it were a new basic variable.

If in Table 11.20 we treat the x_2 column as a key column associated with the entry of the new basic variable x_2, b_i/a_{ik} ratios are computed as usual and S_1 is identified as the departing variable. Table 11.21 indicates the new solution. The optimality check (rule 1) indicates that the solution is optimal with $x_1 = 2$, $x_2 = 3$, and $z = 24$. And for the nonbasic variable S_1, the coefficient of 0 in row (0) indicates that there is an alternative optimal solution in which S_1 would be a basic variable. If we were to treat the S_1 column as the key column and iterate to a new solution, we would return to the first optimal solution found in Table 11.20.

TABLE 11.21

Basic Variables	z	x_1	x_2	S_1	S_2	b_i	Row Number	
	1	0	0	0	2	24	(0)	$R_0'' = R_0'$
x_2	0	0	1	3	−1	3	(1)	$R_1'' = 3R_1'$
x_1	0	1	0	−2	1	2	(2)	$R_2'' = R_2' - \frac{2}{3}R_1''$

□

PRACTICE EXERCISE

Verify the two corner-point alternative optimal solutions found in this example by solving the problem graphically.

Multiple alternative optimal (corner-point) solutions may exist in a problem. This situation may be indicated by (1) more than one nonbasic variable having a 0 coefficient in row (0) of an optimal tableau, or (2) during the course of generating successive alternative optimal solutions using the simplex method, 0 coefficients appear in row (0) for nonbasic variables which have not appeared previously in an optimal solution.

No Feasible Solution

In Sec. 10.2 we indicated that a problem has *no feasible solution* if there are no values for the variables which satisfy all the constraints. Although such a condition may be obvious by inspection in small problems, it is considerably more difficult to identify in large-scale problems.

The condition of no feasible solution is signaled in the simplex method when an artificial variable appears in an optimal basis at a positive level (value). The following example illustrates this condition.

EXAMPLE 12	Let's solve the following LP problem, which, by inspection, has no feasible solution.

$$\text{Maximize} \quad z = 10x_1 + 20x_2$$

$$\text{subject to} \quad x + x_2 \le 5$$

$$x_1 + x_2 \ge 20$$

$$x_1, x_2 \ge 0$$

Tables 11.22–11.24 present the simplex iterations in solving this problem. In Table 11.24, all row (0) coefficients are greater than or equal to 0, indicating an optimal solution. However, one of the basic variables is A_2 and it has a value of 15; that is, the optimal solution is $x_2 = 5$, $A_2 = 15$, and $z = -15M + 100$. Artificial variables have no meaning in an LP problem, and the assignment of $A_2 = 15$ signals no feasible solution to the problem.

TABLE 11.22

Needs to be set equal to zero

Basic Variables	z	x_1	x_2	S_1	E_2	A_2	b_i	Row Number	
	1	−10	−20	0	0	$\textcircled{M}$	0	(0)	R_0
S_1	0	1	1	1	0	0	5	(1)	R_1
A_2	0	1	1	0	−1	1	20	(2)	R_2

TABLE 11.23

Basic Variables	z	x_1	Key Column x_2	S_1	E_2	A_2	b_i	Row Number		b_i/a_{ik}
	1	$-10 - M$	$-20 - M$	0	M	0	$-20M$	(0)	$R_0' = R_0 + MR_2$	
S_1	0	1	1	1	0	0	5	(1)	R_1	$5/1 = 5^*$
A_2	0	1	1	0	−1	1	20	(2)	R_2	$20/1 = 20$

TABLE 11.24

Basic Variables	z	x_1	x_2	S_1	E_2	A_2	b_i	Row Number	
	1	10	0	$M + 20$	M	0	$-15M + 100$	(0)	$R_0'' = R_0' + (M + 20)R_1'$
x_2	0	1	1	1	0	0	5	(1)	$R_1' = R_1$
A_2	0	0	0	−1	−1	1	15	(2)	$R_2' = R_2 - R_1'$

◻

Unbounded Solutions

Unbounded solutions exist when (1) there is an *unbounded solution space* and (2) improvement in the objective function occurs with movement in the direction of the unbounded portion of the solution space.

If at any iteration of the simplex method the a_{ik} values are all 0 or negative for the variable selected to become the new basic variable, there is an unbounded solution for the LP problem. Do you remember our discussion about a_{ik} values in Sec. 11.2? The a_{ik} values indicate the marginal changes in the values of current basic variables for each unit introduced of the new basic variable. Positive a_{ik} values indicate *marginal decreases* in the values of the corresponding basic variables, negative a_{ik} values indicate *marginal increases,* and a_{ik} values of 0 indicate *no change.* Once a new basic variable is identified (rule 2), the departing basic variable is found by computing the minimum b_i/a_{ik} value, where $a_{ik} > 0$ (rule 3). When applying rule 3, we are focusing upon those basic variables which will decrease in value ($a_{ik} > 0$) as the new basic variable is introduced. We wish to determine the maximum number of units to introduce before an existing basic variable is driven to 0. If the a_{ik} values are all 0 or negative, none of the current basic variables will decrease in value and there is no limit on the number of units of the new variable which can be entered. Since the new variable was selected on the basis of a promised improvement in z and there is no limit on the number of units which can be entered, there is no limit on how much the objective function can be improved; hence, there is an unbounded solution.

<table>
<tr><td>**EXAMPLE 13**</td><td>Consider the following problem:</td></tr>
</table>

$$\text{Maximize} \quad z = -2x_1 + 3x_2$$

$$\text{subject to} \quad x_1 \qquad\quad \leq 10$$

$$2x_1 - x_2 \leq 30$$

$$x_1, x_2 \geq 0$$

Table 11.25 presents the initial simplex tableau. Row (0) indicates that the nonbasic variable x_2 will result in an improved value of z. However, the a_{ik} values are 0 and -1. These suggest that for *each unit* introduced of x_2, S_1 will not change and S_2 will *increase* by 1 unit. Neither of these basic variables will be driven to zero. This signals an unbounded solution.

TABLE 11.25				*Key Column*		
	z	x_1	x_2	S_1	S_2	b_i
Basic Variables	1	2	-3	0	0	0
S_1	0	1	0	1	0	10
S_2	0	2	-1	0	1	30

□

PRACTICE EXERCISE
Verify that this problem has an unbounded solution by solving graphically.

Condensed Tableaus

The simplex method can become a bit tedious when solving problems by hand. The problems we have looked at in this section have been small and easily manageable. There is a way of reducing the computational burden when using the simplex method. It involves condensing the size of the tableau. The tableau can be reduced in size because of our focus on nonbasic variables in each tableau. Given any intermediate solution to a linear programming problem, our concern is with the effects associated with introducing nonbasic variables to the basis. Therefore, why carry the identity matrix associated with the basic variables in the tableau? We can treat the identity matrix columns as "phantom" columns, remembering that they should appear in the tableau; however, it is easier (and more efficient) to not recompute these column values at each iteration.

The condensed tableau which will be illustrated here includes columns for each nonbasic variable and an identity matrix column for the departing basic variable. To illustrate this approach, let's rework the problem solved in Example 1. The solution to this problem is shown in Tables 11.26–11.28. If you compare these tableaus with the original tableaus (Tables 11.8–11.10), you will find them very

TABLE 11.26

Basic Variables	x_1	x_2 (Key Column)	x_3	S_1	b_i (Departing Basic Variable)	Row Number	b_i/a_{ik}
	-2	-12	-8	0	0	(0)	
S_1	2	2	1	1	100	(1)	$100/2 = 50*$
S_2	1	-2	5	0	80	(2)	
S_3	10	5	4	0	300	(3)	$300/5 = 60$

TABLE 11.27

Basic Variables	x_1	x_3	S_1	S_2 (Departing Basic Variable)	b_i	Row Number	b_i/a_{ik}
	10	-2	6	0	600	(0)	
x_2	1	$\frac{1}{2}$	$\frac{1}{2}$	0	50	(1)	$50 \div \frac{1}{2} = 100$
S_2	3	6	1	1	180	(2)	$180 \div 6 = 30*$
S_3	5	$\frac{3}{2}$	$-\frac{5}{2}$	0	50	(3)	$50 \div \frac{3}{2} = 33\frac{1}{3}$

TABLE 11.28

Basic Variables	x_1	S_1	S_2		b_i	Row Number
	11	$\frac{38}{6}$	$\frac{2}{6}$		660	(0)
x_2	$\frac{3}{4}$	$\frac{5}{12}$	$-\frac{1}{12}$		35	(1)
x_3	$\frac{1}{2}$	$\frac{1}{6}$	$\frac{1}{6}$		30	(2)
S_3	$\frac{17}{4}$	$-\frac{11}{4}$	$-\frac{1}{4}$		5	(3)

similar. A few comments are necessary, though, for clarification. In each tableau the column representing the departing basic variable is filled in only after the minimum b_i/a_{ik} ratio has been identified. In Table 11.26 the minimum ratio of 50 identified S_1 as the departing basic variable. Since the column elements associated with S_1 will change when we generate the next tableau (it will no longer be an identity matrix column), we insert this "phantom" column into the tableau so that we can perform the simplex arithmetic on it. Notice that in Table 11.27 x_2 has replaced S_1 as a basic variable and the new S_1 column has replaced the x_2 column as one of the nonbasic variables.

The simplex arithmetic is exactly the same as with the full-scale tableau in that basic row operations are performed to transform the column elements for the entering basic variable into the appropriate identity matrix column.

Section 11.3 Follow-up Exercises

In the following exercises, solve by the simplex method.

1 Maximize $z = 4x_1 + 2x_2$
   subject to $x_1 + x_2 \leq 15$
                      $2x_1 + x_2 \leq 20$
                      $x_1, x_2 \geq 0$

2 Minimize $z = 4x_1 + 6x_2$
   subject to $3x_1 + x_2 \geq 15$
                      $2x_1 + 3x_2 \geq 17$
                      $x_1, x_2 \geq 0$

3 Maximize $z = 6x_1 + 3x_2$
   subject to $x_1 + 2x_2 \leq 20$
                      $4x_1 + 2x_2 \leq 32$
                      $x_1 \quad\quad \leq 8$
                      $x_1, x_2 \geq 0$

4 Maximize $z = 5x_1 + 3x_2$
   subject to $4x_1 + 3x_2 \leq 24$
                      $3x_1 + x_2 \geq 20$
                      $x_1, x_2 \geq 0$

5 Minimize $z = 4x_1 - 3x_2$
   subject to $2x_1 - 4x_2 \geq 20$
                      $4x_1 + 3x_2 \leq 12$
                      $x_1, x_2 \geq 0$

6 Maximize $z = 5x_1 + 3x_2$
   subject to $-x_1 + 2x_2 \leq 10$
                      $x_2 \leq 5$
                      $x_1, x_2 \geq 0$

Solve the following problems by the simplex method using the condensed tableau.

7 Maximize $z = 25x_1 + 50x_2$
   subject to $2x_1 + 2x_2 \leq 1000$
                      $3x_1 \quad\quad \leq 600$
                      $x_1 + 3x_2 \leq 600$
                      $x_1, x_2 \geq 0$

8 Minimize $z = 6x_1 + 8x_2 + 16x_3$
   subject to $2x_1 + x_2 \quad\quad \geq 5$
                      $x_2 + 2x_3 \geq 4$
                      $x_1, x_2, x_3 \geq 0$

11.4 COMPUTER SOLUTION METHODS

In actual applications, LP problems are solved by computer methods. There are many efficient computer codes available today through computer manufacturers, software "houses," and universities. As a user of LP models, one need not always be concerned about the internal nuts and bolts of the solution method.* Rather,

---

* However, there is no question that one *can* be a better user of these packages if one *does* understand the nuts and bolts.

effective use of these models can be made if a person (1) fully understands linear programming and its assumptions, (2) is skillful in recognizing an LP problem, (3) is skillful in problem formulation, (4) can arrange for solution by a computer package, and (5) is capable of interpreting the output from such packages.

An Illustration of an LP Package

As mentioned previously, many LP computer packages are available. You should check with your computer center to see which ones are available on your system. This section illustrates one interactive LP package available for use on several microcomputer systems.*

EXAMPLE 14 (**Grant Awards; Motivating Scenario**) Example 10 in Chap. 10 presented an application in which a federal agency wanted to award $1 billion in grants for innovative research in the area of energy alternatives. To assist in following this example, the formulation of the problem is repeated in Fig. 11.10.

Maximize

$$z = 4.4x_1 + 3.8x_2 + 4.1x_3 + 3.5x_4 + 5.1x_5 + 3.2x_6$$

subject to

$$
\begin{array}{lll}
x_1 + x_2 + x_3 + x_4 + x_5 + x_6 & \leq 1000 & (1) \\
x_1 & \leq 220 & (2) \\
x_2 & \leq 180 & (3) \\
x_3 & \leq 250 & (4) \\
x_4 & \leq 150 & (5) \\
x_5 & \leq 400 & (6) \\
x_6 & \leq 120 & (7) \\
x_5 & \geq 200 & (8) \\
x_1 + x_2 & \geq 300 & (9) \\
x_1, x_2, x_3, x_4. x_5, x_6 & \geq 0 &
\end{array}
$$

Figure 11.10 Formulation of grant awards model.

Figure 11.11 indicates the results of the analysis output by the computer package. Note that the problem structure is summarized along with the actual formulation of the model. The results show the values of each decision variable followed by a summary of each constraint, the value of each slack or surplus variable, and the maximum value of the objective function. The total net benefit of the grant award program is maximized at a value of $4,527 million.

---

* The program "LINP1" is one of several programs included in *Computer Models for Management Science*, 2nd ed., by Warren Erikson and Owen P. Hall, Jr. (Addison-Wesley, Reading, Mass., 1986).

Optimal Decisions

- ❏ $x_1 = 220$ [award \$220 million to project 1 (solar)].

- ❏ $x_2 = 130$ [award \$130 million to project 2 (solar)].

- ❏ $x_3 = 250$ [award \$250 million to project 3 (synthetic fuels)].

- ❏ $x_4 = 0$ [no award to project 4 (coal)].

- ❏ $x_5 = 400$ [award \$400 million to project 5 (nuclear)].

- ❏ $x_6 = 0$ [no award to project 6 (geothermal)].

```
***************************

LINEAR PROGRAMMING ANALYSIS

***************************


   **   INFORMATION ENTERED   **

NUMBER OF CONSTRAINTS        9
NUMBER OF VARIABLES          6
NUMBER OF <= CONSTRAINTS     7
NUMBER OF  = CONSTRAINTS     0
NUMBER OF >= CONSTRAINTS     2

MAXIMIZATION PROBLEM

   4.4 X 1   + 3.8 X 2   + 4.1 X 3   + 3.5 X 4   + 5.1 X 5   +
3.2 X 6


SUBJECT TO

   1 X 1    + 1 X 2   + 1 X 3     + 1 X 4   + 1 X 5   + 1 X 6
                     <= 1000

   1 X 1    + 0 X 2   + 0 X 3     + 0 X 4   + 0 X 5   + 0 X 6
                     <= 220

   0 X 1    + 1 X 2   + 0 X 3     + 0 X 4   + 0 X 5   + 0 X 6
                     <= 180

   0 X 1    + 0 X 2   + 1 X 3     + 0 X 4   + 0 X 5   + 0 X 6
                     <= 250

   0 X 1    + 0 X 2   + 0 X 3     + 1 X 4   + 0 X 5   + 0 X 6
                     <= 150

   0 X 1    + 0 X 2   + 0 X 3     + 0 X 4   + 1 X 5   + 0 X 6
                     <= 400

   0 X 1    + 0 X 2   + 0 X 3     + 0 X 4   + 0 X 5   + 1 X 6
                     <= 120

   0 X 1    + 0 X 2   + 0 X 3     + 0 X 4   + 1 X 5   + 0 X 6
                     >= 200

   1 X 1    + 1 X 2   + 0 X 3     + 0 X 4   + 0 X 5   + 0 X 6
                     >= 300
```

Figure 11.11 Results of grant awards problem.

VARIABLE	VARIABLE VALUE	ORIGINAL COEFF
X 1	220	4.4
X 2	130	3.8
X 3	250	4.1
X 4	0	3.5
X 5	400	5.1
X 6	0	3.2

CONSTRAINT NUMBER	ORIGINAL RHS	SLACK OR SURPLUS	SHADOW PRICE
1	1000	0	3.8
2	220	0	.61
3	180	50	0
4	250	0	.3
5	150	150	0
6	400	0	1.3
7	120	120	0
8	200	200	0
9	300	50	0

OBJECTIVE FUNCTION VALUE : 4527

Figure 11.11 Continued.

Slack/Surplus Analysis

❏ There is zero slack in constraint (1), which suggests that the total budget of 1,000 ($ millions) is allocated.

❏ There is zero slack in constraint (2); the maximum possible award of $220 million is recommended for project 1 (solar).

❏ There is slack of 50 in constraint (3); the $130 million award to project 2 (solar) is $50 million less than the maximum which could have been awarded.

❏ There is zero slack in constraint (4); the maximum possible award of $250 million is recommended for project 3 (synthetic fuels).

❏ There is slack of 150 in constraint (5); with no award recommended for project 4 (coal), the agency falls $150 million short of the maximum amount permitted.

❏ There is zero slack in constraint (6); the maximum possible award of $400 million is recommended for project 5 (nuclear).

❏ There is slack of 120 in constraint (7); with no award recommended for project 6 (geothermal), the agency falls $120 million short of the maximum amount permitted.

❏ There is surplus of 200 in constraint (8); the $400 million award to project 5 (nuclear) exceeds the minimum required by $200 million.

❏ There is surplus of 50 in constraint (9); the combined award of $350 million to projects 1 and 2 (solar) exceeds the minimum required by $50 million.

❏

Shadow Prices

The solution to an LP problem is based upon certain assumptions and estimates. Once a solution is obtained, it should be analyzed carefully in light of these assumptions and estimates. This phase of the solution process is called *postoptimality analysis.* An important type of postoptimality analysis is the examination of shadow prices.

> **DEFINITION: SHADOW PRICE**
> A *shadow price* is the amount the optimal value of the objective function would change if the right-hand side of a constraint were increased by one unit.

Since many ($\leq$) constraints represent limited resources, shadow prices are often thought of as representing the economic value of having an additional unit of a resource. Figure 11.11 illustrates the shadow prices for the grant awards example.

If you examine the original formulation in Fig. 11.10, it will help in understanding the shadow price analysis. The shadow price of 3.8 for constraint (1) suggests that if the total amount of money available for grants was increased from \$1,000 million to \$1,001 million, total net benefits would be increased by \$3.8 million, increasing from the original maximum of \$4,527 million to \$4,530.8 million.

The shadow price of 0.61 for constraint (2) suggests that total net benefits would be increased by \$0.6100 million if the maximum investment allowed in project 1 was increased from \$220 to \$221 million. Remember in the optimal solution that the maximum of \$220 million was awarded to project 1.

The shadow price of 0 associated with constraint (3) indicates that the value of the objective function would not change if the maximum investment allowed in project 2 was increased from \$180 to \$181 million. This makes sense since the optimal solution currently recommends an award of \$130 million, \$50 million less than the original maximum.

Interpret the remaining shadow prices for yourself. You should also make the following observations. *The shadow prices are positive for any constraint which has been satisfied as an equality in the optimal solution (no slack or surplus exists).* The optimal solution has pushed these constraints to their limits, suggesting potential value from being able to increase these limits. *The shadow prices equal 0 for any constraint which is not satisfied as an equality (slack or surplus exists).* The optimal solution has not pushed these constraints to their respective limits. Increasing these limits farther results in no additional improvement to the objective function.

Another point to be made is that shadow prices represent *marginal returns.* They indicate the change in the optimal value of the objective function, given a unit increase in the right-hand-side constant of the corresponding constraint. The shadow price is valid for a particular range of changes in the right-hand-side constant. The shadow price of 3.8 does not suggest that an increase of \$100 million in the right-hand-side constant of constraint (1) will result in an increase of 3.8(100) = \$380 million in total net benefits. The valid range for each shadow price can be determined by another type of postoptimality analysis discussed in the next section.

Sensitivity Analysis

The parameters (constants) used in a linear programming model frequently are best estimates of their actual values. For example, the profit contribution assumed for a product may be based upon best estimates of selling price and variable cost per unit. Variable cost per unit must assume certain wage rates, expected processing times, and material costs. As another example, estimates of labor availability in different departments may not reflect uncertainties associated with absenteeism and personnel shifts. The point is that the parameters used to derive an optimal solution frequently cannot be determined with certainty.

Therefore, once a solution has been derived using these "assumed" values, it should be examined to determine the effects if the parameters take on values other than those used in the original formulation. This postsolution analysis is called *sensitivity analysis.* If the analysis reveals that the optimal *basis* and/or *objective function value* are affected only slightly by significant changes in the values of parameters, we say that the solution is *insensitive.* If, however, the basis and/or objective function do vary significantly with relatively minor changes in parameters, the solution is judged to be *sensitive* and probably deserves additional scrutiny.

Many LP packages provide sensitivity analysis. Figure 11.12 illustrates these results for the grant awards example. This particular package conducts sensitivity

```
SENSITIVITY ANALYSIS

OBJECTIVE FUNCTION COEFICIENTS

          LOWER   ORIGINAL     UPPER
VARIABLE  LIMIT   COEFFICIENT  LIMIT

X 1        3.8       4.4       NO LIMIT
X 2        0         3.8        5.1
X 3        3.8       4.1       NO LIMIT
X 4       NO LIMIT   3.5        3.8
X 5        3.8       5.1       NO LIMIT
X 6       NO LIMIT   3.2        3.8


     RIGHT HAND SIDE

CONSTRAINT  LOWER    ORIGINAL   UPPER
NUMBER      LIMIT    VALUE      LIMIT

   1         950      1000      1050
   2         170       220       350
   3         130       180      NO LIMIT
   4         200       250       300
   5         0         150      NO LIMIT
   6         350       400       450
   7         0         120      NO LIMIT
   8        NO LIMIT   200       400
   9        NO LIMIT   300       350


   ** END OF ANALYSIS **
```

Figure 11.12 Sensitivity analysis for grant awards problem.

analysis for objective function coefficients of all decision variables and right-hand-side constants for the structural constraints. Let's focus first on the analysis for the objective function coefficients.

> For objective function coefficients, sensitivity analysis determines by how much each coefficient can change and have the current basis remain the same; that is, the same pool of basic variables remains optimal and their values are unchanged. Another way of thinking about this is that we are concerned with how much these coefficients can change and still have the same corner point on the region of feasible solutions remain optimal.

For each decision variable, the original objective function coefficient is shown along with the *lower limit* and the *upper limit* for the coefficient. For example, the coefficient for x_1 is 4.4 in the original formulation. This parameter can *decrease* to a lower limit of 3.8 and the current pool of basic variables with their current values will remain optimal. Regarding increases in the parameter, the upper limit indicates that it can increase (without limit) to any value and the current set of basic variables will remain optimal. If we think about this intuitively, the original objective function coefficient (net benefit per dollar invested) of 4.4 was sufficiently high to result in an award of $220 million, the maximum possible for that project. The sensitivity analysis suggests that a decrease in c_1 might eventually make other competing projects more attractive, resulting in either no award to project 1 or a reduced award. However, if c_1 increases, the attractiveness of project 1 is greater and the decision to award the maximum of $220 million is further reinforced. We can say that the current optimal solution is *somewhat sensitive* to decreases in c_1 but *insensitive* to increases.

For x_4 this section indicates that the original objective function coefficient of 3.5 can increase to a value of 3.8 or decrease without limit. A way of interpreting this is as follows: x_4 is a nonbasic variable, suggesting that with its original net return coefficient of 3.5, project 4 was not sufficiently attractive to receive an award. This analysis implies that if the net return coefficient is increased beyond a value of 3.8, project 4 will become more attractive and will likely be awarded a grant. Similarly, any decrease in the net benefit figure will reinforce that project 4 is not worthy of investment.

PRACTICE EXERCISE
Interpret the results of the objective function sensitivity analysis for the remaining variables.

The final section of sensitivity analysis in Fig. 11.12 pertains to the right-hand-side constants of the structural constraints.

> For right-hand-side constants, sensitivity analysis determines by how much each value can change and have the optimal basis remain *feasible*.

For constraint (1), the sensitivity analysis indicates that the right-hand-side constant of 1,000 can *decrease* by as much as 50 to a lower limit of 950 or increase by as much as 50 to an upper limit of 1,050 and the optimal basis will remain feasible. If it falls below 950 or exceeds 1,050, the optimal basis will no longer be feasible. Stated differently, the optimal basis will remain feasible as long as the total amount to be awarded is between $950 million and $1,050 million. For this parameter, the optimal solution appears to be relatively sensitive to both decreases and increases.

This type of sensitivity analysis also links to shadow prices. The range of permissible variation for the right-hand-side constant also indicates the range over which the corresponding shadow price is valid. For example, Fig. 11.12 suggests that the shadow price of 3.8 assigned to constraint (1) is valid for decreases of as much as 50 and increases up to 50. In other words, the constant can increase to 1,050 and the optimal value of z will equal $4,527 + 50(3.8) = 4,717$. By the same token, if this constant decreases to 950, the optimal value of z will equal $4,527 - 50(3.8) = 4,337$.

> **PRACTICE EXERCISE**
> Interpret the sensitivity analysis for the right-hand-side constants of constraints (3) and (7).

Section 11.4 Follow-up Exercises

The following exercises are contingent upon the availability of a computerized LP package.

1 In Sec. 10.2 the LP problem on page 430 was solved using the corner-point method. Verify this result by solving with an LP package.

2 In Sec. 10.2 the LP problem in Example 4 (page 433) was solved. Verify the result by solving with an LP package.

3 Example 7 on page 437 solved an LP problem and found alternative optimal solutions. Solve using an LP package. Does your package explicitly signal alternative optimal solutions?

4 The LP problem

$$\text{Maximize} \quad z = 6x_1 + 4x_2$$

$$\text{subject to} \quad x_1 + x_2 \geq 10$$

$$3x_1 + 2x_2 \geq 15$$

$$x_1, x_2 \geq 0$$

has an unbounded solution. Solve using an LP package and determine whether your package explicitly signals this result.

5 Solve the LP problem

$$\text{Maximize} \quad z = 2x_1 + 12x_2 + 8x_3$$

$$\text{subject to} \quad 2x_1 + 2x_2 + x_3 \leq 100$$

$$x_1 - 2x_2 + 5x_3 \leq 80$$

$$10x_1 + 5x_2 + 4x_3 \leq 300$$

$$x_1, x_2, x_3 \geq 0$$

If the LP package has the capabilities, determine the shadow prices for the three constraints and conduct sensitivity analysis on the objective function coefficients and the right-hand-side constants.

6 Given the formulation of the diet mix application (Example 8) on page 443, determine the optimal solution and interpret the results.

7 Given the formulation of the highway maintenance application (Example 9) on page 445, determine the optimal solution and interpret the results.

8 Given the formulation of the petroleum-blending application (Example 11) on page 450, determine the optimal solution and interpret the results.

11.5 THE DUAL PROBLEM

Every LP problem has a related problem called the **dual problem** or, simply, the **dual.** Given an original LP problem, referred to as the **primal problem,** or **primal,** the dual can be formulated from information contained in the primal. The dual problem is significant for numerous theoretical reasons and also for practical purposes. One property of the dual is that when it is solved, it provides all essential information about the solution to the primal problem. Similarly, the solution to the primal provides all essential information about the solution to the dual problem. Given an LP problem, its solution can be determined by solving *either* the original problem or its dual. The structural properties of the two problems may result in a decided preference regarding which problem should be solved. Even with computer-based solution methods, computational efficiencies can arise from solving one form of the problem.

Formulation of the Dual

The parameters and structure of the primal provide all the information necessary to formulate the dual. Figure 11.13 illustrates the formulation of a maximization problem and the dual of the problem. Let's make some observations regarding the relationships between these primal and dual problems.

1 *The primal is a maximization problem and the dual is a minimization problem.* **The sense of optimization is always opposite for corresponding primal and dual problems.**

2 *The primal consists of two variables and three constraints and the dual consists of three variables and two constraints.* **The number of variables in the primal always equals the number of constraints in the dual. The number of constraints in the primal always equals the number of variables in the dual.**

3 *The objective function coefficients for x_1 and x_2 in the primal equal the right-hand-side constants for constraints (1) and (2) in the dual.* **The objective function coefficient for the jth primal variable equals the right-hand-side constant for the jth dual constraint.**

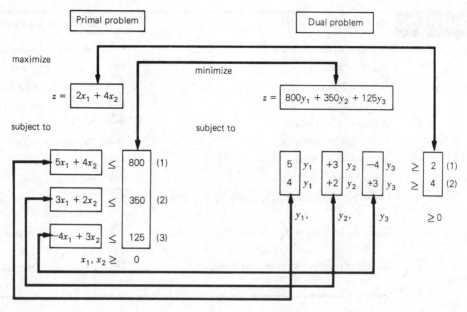

Figure 11.13 Some primal-dual relationships.

4 *The right-hand-side constants for constraints (1) – (3) in the primal equal the objective function coefficients for the dual variables y_1, y_2, and y_3.* **The right-hand-side constant for the ith primal constraint equals the objective function coefficient for the ith dual variable.**

5 *The variable coefficients for constraint (1) of the primal equal the column coefficients for the dual variable y_1. The variable coefficients for constraints (2) and (3) of the primal equal the column coefficients of the dual variables y_2 and y_3.* **The coefficients a_{ij} in the primal are the transpose of those in the dual. That is, the row coefficients in the primal become column coefficients in the dual, and vice versa.**

Even though this problem has a primal which is a maximization type, the primal (original problem) may be a minimization problem. The rules of transformation actually should be stated in terms of how to transform from a maximization problem to the corresponding minimization problem, or vice versa. Table 11.29 summarizes the symmetry of the two types of problems and their relationships.

Relationships 4 and 8 indicate that an equality constraint in one problem corresponds to an **unrestricted variable** in the other problem. An unrestricted variable can assume a value which is positive, negative, or 0. Similarly, relationships 3 and 7 indicate that a problem may have **nonpositive variables** (for example, $x_j \leq 0$). Unrestricted and nonpositive variables appear to violate the third requirement of the simplex method, the nonnegativity restriction. Although this is true, for problems containing any of these special types of variables there are methods which allow us to adjust the formulation to satisfy the third requirement.

TABLE 11.29	**Maximization Problem**		**Minimization Problem**
	Number of constraints	$\underset{(1)}{\Longleftrightarrow}$	Number of variables
	($\leq$) constraint	$\underset{(2)}{\Longleftrightarrow}$	Nonnegative variable
	($\geq$) constraint	$\underset{(3)}{\Longleftrightarrow}$	Nonpositive variable
	($=$) constraint	$\underset{(4)}{\Longleftrightarrow}$	Unrestricted variable
	Number of variables	$\underset{(5)}{\Longleftrightarrow}$	Number of constraints
	Nonnegative variable	$\underset{(6)}{\Longleftrightarrow}$	($\geq$) constraint
	Nonpositive variable	$\underset{(7)}{\Longleftrightarrow}$	($\leq$) constraint
	Unrestricted variable	$\underset{(8)}{\Longleftrightarrow}$	($=$) constraint
	Objective function coefficient for jth variable	$\underset{(9)}{\Longleftrightarrow}$	Right-hand-side constant for jth constraint
	Right-hand-side constant for ith constraint	$\underset{(10)}{\Longleftrightarrow}$	Objective function coefficient for ith variable
	Coefficient in constraint i for variable j	$\underset{(11)}{\Longleftrightarrow}$	Coefficient in constraint j for variable i

EXAMPLE 15   Given the primal problem

$$\text{Minimize} \quad z = 10x_1 + 20x_2 + 15x_3 + 12x_4$$

$$\text{subject to} \quad x_1 + x_2 + x_3 + x_4 \geq 100 \tag{1}$$

$$2x_1 \qquad - x_3 + 3x_4 \leq 140 \tag{2}$$

$$x_1 + 4x_2 \qquad - 2x_4 = 50 \tag{3}$$

$$x_1, x_3, x_4 \geq 0$$

$$x_2 \quad \text{unrestricted}$$

verify that the corresponding dual is

$$\text{Maximize} \quad z = 100y_1 + 140y_2 + 50y_3$$

$$\text{subject to} \quad y_1 + 2y_2 + y_3 \leq 10$$

$$y_1 \qquad + 4y_3 = 20$$

$$y_1 - y_2 \qquad \leq 15$$

$$y_1 + 3y_2 - 2y_3 \leq 12$$

$$y_1 \geq 0$$

$$y_2 \leq 0$$

$$y_3 \quad \text{unrestricted}$$

$\square$

Primal-Dual Solutions

It was indicated earlier that the solution to the primal problem can be obtained from the solution to the dual problem and vice versa. Let's illustrate this by example. Consider the primal problem:

$$\text{Maximize} \quad z = 5x_1 + 6x_2$$

$$\text{subject to} \quad 3x_1 + 2x_2 \le 120 \quad \text{(1)}$$

$$4x_1 + 6x_2 \le 260 \quad \text{(2)}$$

$$x_1, x_2 \ge 0$$

The corresponding dual is

$$\text{Minimize} \quad z = 120y_1 + 260y_2$$

$$\text{subject to} \quad 3y_1 + 4y_2 \ge 5$$

$$2y_1 + 6y_2 \ge 6$$

$$y_1, y_2 \ge 0$$

Table 11.30 presents the final (optimal) tableau for the dual problem. Note from this tableau that z is minimized at a value of 280 when $y_1 = \frac{3}{5}$ and $y_2 = \frac{4}{5}$. The primal problem was solved earlier in the chapter. Table 11.7, which summarizes the optimal solution, is repeated for convenience. Let's illustrate how the solution to each problem can be read from the optimal tableau of the corresponding dual problem.

PRIMAL-DUAL PROPERTY 1

If feasible solutions exist for both the primal and dual problems, then both problems have an optimal solution for which the objective function values are equal. A peripheral relationship is that if one problem has an unbounded solution, its dual has no feasible solution.

TABLE 11.30

Basic Variables	z	y_1	y_2	E_1	A_1	E_2	A_2	b_i	Row Number
	1	0	0	-20	$(20) - M$	-30	$(30) - M$	280	(0)
y_1	0	1	0	$-\frac{3}{5}$	$\frac{3}{5}$	$\frac{2}{5}$	$-\frac{2}{5}$	$\frac{3}{5}$	(1)
y_2	0	0	1	$\frac{1}{5}$	$-\frac{1}{5}$	$-\frac{3}{10}$	$\frac{3}{10}$	$\frac{4}{5}$	(2)

Final dual tableau

TABLE 11.7

Basic Variables	z	x_1	x_2	S_1	S_2	b_i	Row Number
	1	0	0	$\frac{6}{10}$	$\frac{24}{30}$	280	(0)
x_1	0	1	0	$\frac{6}{10}$	$-\frac{6}{30}$	20	(1)
x_2	0	0	1	$-\frac{24}{60}$	$\frac{3}{10}$	30	(2)

Final primal tableau

For this primal-dual pair of problems, note that the optimal values for their respective objective functions both equal 280.

> **PRIMAL-DUAL PROPERTY 2**
> The optimal values for decision variables in one problem are read from row (0) of the optimal tableau for the other problem.

The optimal values $y_1 = \frac{2}{5}$ and $y_2 = \frac{4}{5}$ are read from Table 11.7 as the row (0) coefficients for the slack variables S_1 and S_2. The optimal values $x_1 = 20$ and $x_2 = 30$ are read from Table 11.30 as the negatives of the row (0) coefficients for the surplus variables E_1 and E_2. These values can be read, alternatively, under the respective artificial variables, as the portion (term) of the row (0) coefficient *not* involving M.

Epilogue

The dual is a slippery, but significant, topic in linear programming. The purpose has been to acquaint you with the topic and to overview some important properties of the dual. If you choose to take a course in linear or mathematical programming, you will probably receive a more detailed treatment of the dual and its implications.

Section 11.5 Follow-up Exercises

For the following primal problems, formulate the corresponding dual problem.

1 Maximize $z = 3x_1 + 4x_2 + 2x_3$
   subject to
   $$x_1 + x_2 + x_3 \le 45$$
   $$4x_1 + 5x_2 - 3x_3 \le 30$$
   $$-x_1 + 3x_2 - 4x_3 \le 50$$
   $$x_1, x_2, x_3 \ge 0$$

2 Minimize $z = 4x_1 + 3x_2 + 5x_3$
   subject to
   $$2x_1 + x_2 - 5x_3 \ge 300$$
   $$x_1 + x_2 + x_3 \ge 75$$
   $$x_1, x_2, x_3 \ge 0$$

3 Maximize $z = 20x_1 + 15x_2 + 18x_3 + 10x_4$
   subject to
   $$5x_1 - 3x_2 + 10x_3 + 4x_4 \le 60$$
   $$x_1 + x_2 + x_3 = 25$$
   $$-x_2 + 4x_3 + 7x_4 \ge 35$$
   $$x_1, x_2, x_3 \ge 0$$
   $$x_4 \quad \text{unrestricted}$$

4 Minimize $z = 6x_1 + 4x_2$
   subject to
   $$x_1 + x_2 \le 45$$
   $$5x_1 - 4x_2 = 10$$
   $$-3x_1 + 5x_2 \ge 75$$
   $$3x_1 + 6x_2 \ge 30$$
   $$x_1 \quad \text{unrestricted}$$
   $$x_2 \ge 0$$

5 Minimize $z = 4x_1 + 5x_2 + 2x_3 + 3x_4 + x_5$
   subject to
   $$x_1 + x_2 + x_3 + x_4 + x_5 = 45$$
   $$3x_1 + 5x_2 - 2x_4 \le 24$$
   $$7x_3 - 5x_4 + 3x_5 \ge 20$$
   $$x_1, x_2, x_4 \ge 0$$
   $$x_3, x_5 \quad \text{unrestricted}$$

6 Maximize $\quad z = x_1 + 4x_2 + 6x_3 + 2x_4 + 3x_5 + 2x_6$

subject to
$$
\begin{aligned}
x_1 + \quad\quad x_3 + \quad\quad x_5 \quad\quad &= 40 \\
3x_1 + 5x_2 + 2x_3 - x_4 + 3x_5 - 3x_6 &= 70 \\
4x_2 - 5x_3 \quad\quad + 2x_5 - 4x_6 &= 35 \\
x_1, x_3, x_4, x_6 &\geq 0 \\
x_2, x_5 \quad \text{unrestricted}
\end{aligned}
$$

7 Given the following primal problem:

$$\text{Maximize} \quad z = 5x_1 + 3x_2$$

$$\text{subject to} \quad 2x_1 + 4x_2 \leq 32$$

$$3x_1 + 2x_2 \leq 24$$

$$x_1, x_2 \geq 0$$

(a) Formulate the corresponding dual problem.

(b) Solve the primal problem using the simplex method.

(c) Determine the optimal solution to the dual problem from the optimal tableau of the primal problem.

(d) Solve the dual problem using the simplex method to verify the result obtained in part c. Read the optimal solution to the primal problem from the optimal tableau of the dual.

8 Given the following primal problem:

$$\text{Minimize} \quad z = 4x_1 + 3x_2$$

$$\text{subject to} \quad 4x_1 + 2x_2 \geq 80$$

$$3x_1 + x_2 \geq 50$$

$$x_1, x_2 \geq 0$$

(a) Formulate the corresponding dual problem.

(b) Solve the primal problem using the simplex method.

(c) Determine the optimal solution to the dual problem from the optimal tableau of the primal problem.

(d) Solve the dual problem using the simplex method to verify the result obtained in part c. Read the optimal solution to the primal problem from the optimal tableau of the dual.

❏ KEY TERMS AND CONCEPTS

alternative optimal solutions 496

artificial variable 467

basic feasible solution 471

basic solution 471

basic variable 471

basis 471

dual (problem) 510

gaussian elimination method 480

key column 484

no feasible solution 498

nonbasic variable 471

nonpositive variables 511

postoptimality analysis 506

primal (problem) 510

shadow price 506

simplex method 464

slack variable 465

solution by enumeration 478

surplus variable 467

tableau 483

unbounded solution 499

unrestricted variable 511

❑ ADDITIONAL EXERCISES

Section 11.1

1 Given the following LP problem, rewrite the constraint set in standard form incorporating all supplemental variables.

$$\text{Maximize} \quad z = 6x_1 - 2x_2 + 8x_3$$

$$\text{subject to} \quad 2x_1 + x_2 + x_3 \geq 50$$

$$2x_1 - 3x_2 \leq -15$$

$$6x_1 + 4x_2 + 5x_3 \geq 40$$

$$x_1 + 2x_2 \geq 15$$

$$-2x_2 \leq -4$$

$$x_1, x_2, x_3 \geq 0$$

2 Given the following LP problem, rewrite the constraint set in standard form incorporating all supplemental variables.

$$\text{Minimize} \quad z = 5x_1 + 2x_2 + 3x_3$$

$$\text{subject to} \quad 5x_1 - 3x_2 + 6x_3 \leq 30$$

$$x_1 + x_2 + x_3 \geq 14$$

$$3x_1 - 4x_3 \geq -15$$

$$x_2 \leq 10$$

$$x_1 = 2x_3$$

$$x_1, x_2, x_3 \geq 0$$

3 Given the following LP problem, rewrite the constraint set in standard form incorporating all supplemental variables.

$$\text{Maximize} \quad z = 10x_1 + 12x_2 + 8x_3$$

$$\text{subject to} \quad 3x_1 - 2x_2 + x_3 \geq 12$$

$$x_1 - 3x_2 \leq 20$$

$$x_1 + x_2 + x_3 \geq 8$$

$$4x_1 - 5x_2 + 3x_3 \leq 25$$

$$x_1 \geq 5$$

$$x_2 \leq 6$$

$$x_3 \geq 2$$

$$x_1, x_2, x_3 \geq 0$$

4 An LP problem has 25 decision variables, 60 ($\leq$) constraints, 30 ($\geq$) constraints, and 10 ($=$) constraints. When rewritten in standard form, how many variables will be included? How many supplemental variables of each type?

5 An LP problem has 50 decision variables, 80 ($\leq$) constraints, 150 ($\geq$) constraints, and 20 ($=$) constraints. When rewritten in standard form, how many variables will be included? How many supplemental variables of each type?

6 An LP problem has 150 decision variables, 300 ($\leq$) constraints, 100 ($\geq$) constraints, and 50 ($=$) constraints. When rewritten in standard form, how many variables will be included? How many supplemental variables of each type?

7 Given the region of feasible solutions in Fig. 11.14:
(a) Identify the nature (type) of each constraint.
(b) What supplemental variables will be added when solved by the simplex method?
(c) How many basic and nonbasic variables will there be?
(d) Identify the basic and nonbasic variables associated with all corner points.

Figure 11.14

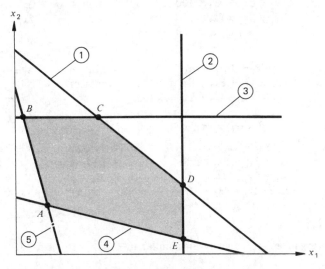

8 Given the region of feasible solutions in Fig. 11.15:
(a) Identify the nature (type) of each constraint.
(b) What supplemental variables will be added when solved by the simplex method?

Figure 11.15

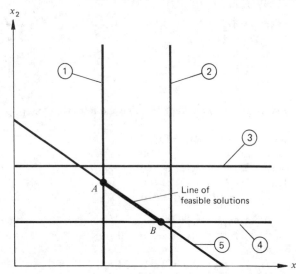

(c) How many basic and nonbasic variables will there be?

(d) Identify the basic and nonbasic variables associated with all corner points.

Section 11.2

Solve the following problems using the simplex method.

9 Maximize subject to
$$z = 5x_1 + 9x_2$$
$$4x_1 + 8x_2 \le 600$$
$$12x_1 + 8x_2 \le 960$$
$$x_1, x_2 \ge 0$$

10 Minimize subject to
$$z = 100x_1 + 75x_2$$
$$x_1 + x_2 \ge 200$$
$$x_2 \ge 100$$
$$x_1 \ge 80$$
$$x_1, x_2 \ge 0$$

11 Maximize subject to
$$z = 4x_1 + 2x_2 + 6x_3$$
$$x_1 + 2x_2 + x_3 \le 100$$
$$3x_1 + 2x_2 + 3x_3 \le 120$$
$$x_1, x_2, x_3 \ge 0$$

12 Minimize subject to
$$z = 4x_1 + 4x_2$$
$$2x_1 + 4x_2 \ge 160$$
$$2x_1 \ge 60$$
$$2x_2 \ge 40$$
$$x_1, x_2 \ge 0$$

13 Maximize subject to
$$z = 6x_1 + 12x_2 + 5x_3 + 2x_4$$
$$3x_1 + 4x_2 + 8x_3 + 6x_4 \le 1,100$$
$$8x_1 + 2x_2 + 4x_3 + 2x_4 \le 1,400$$
$$4x_1 + 6x_2 + 2x_3 + 4x_4 \le 400$$
$$x_1, x_2, x_3, x_4 \ge 0$$

14 Maximize subject to
$$z = 5x_1 + 8x_2 + x_3$$
$$x_1 + x_2 + 3x_3 \le 70$$
$$x_1 + 2x_2 + x_3 \le 100$$
$$2x_1 + x_2 + x_3 \le 80$$
$$x_1, x_2, x_3 \ge 0$$

15 Minimize subject to
$$z = 8x_1 + 4x_2 + 7x_3$$
$$4x_1 + 6x_2 + 2x_3 \ge 120$$
$$4x_1 + 2x_2 + 2x_3 \ge 80$$
$$2x_1 + 2x_2 + 4x_3 \ge 80$$
$$x_1, x_2, x_3 \ge 0$$

16 Maximize subject to
$$z = 7x_1 + 2x_2 + 5x_3$$
$$x_1 + 3x_2 + x_3 = 35$$
$$2x_1 + x_2 + x_3 \le 50$$
$$x_1 + x_2 + 2x_3 \le 40$$
$$x_1, x_2, x_3 \ge 0$$

Section 11.3

In the following exercises, solve by the simplex method.

17 Maximize subject to
$$z = 5x_1 + 10x_2$$
$$4x_1 + x_2 \le 53$$
$$x_1 + 2x_2 \le 22$$
$$x_1, x_2 \ge 0$$

18 Minimize subject to
$$z = 15x_1 + 25x_2$$
$$5x_1 + 3x_2 \ge 80$$
$$6x_1 + 10x_2 \ge 160$$
$$x_1, x_2 \ge 0$$

19 Minimize subject to
$$z = 6x_1 + 4x_2$$
$$3x_1 + 2x_2 \le 24$$
$$x_1 + 2x_2 \ge 30$$
$$x_1, x_2 \ge 0$$

20 Maximize subject to
$$z = 2x_1 + 3x_2$$
$$-x_1 + 3x_2 \ge 12$$
$$x_1 \le 5$$
$$x_1, x_2 \ge 0$$

Section 11.4

Use an appropriate software (LP) package to solve the following exercises.

21 Maximize subject to
$$z = 3x_1 + 10x_2 + 4x_3 + 6x_4$$
$$2x_1 + 2x_2 + 5x_3 + x_4 \le 50$$
$$x_1 - 2x_2 + x_3 + 5x_4 \le 40$$
$$10x_1 + 5x_2 + 2x_3 + 4x_4 \le 150$$
$$x_1, x_2, x_3, x_4 \ge 0$$

22 Minimize subject to
$$z = 5x_1 + 4x_2 + 6x_3 + 7x_4 + 4x_5$$
$$x_1 + x_2 + x_3 + x_4 + x_5 \ge 20$$
$$3x_1 + 4x_2 + 5x_3 \le 100$$
$$4x_3 + 2x_4 + 3x_5 \ge 24$$

$$x_1 \qquad\qquad\qquad \geq\ 4$$
$$x_4 \qquad\qquad \geq\ 5$$
$$x_1, x_2, x_3, x_4, x_5 \geq\ 0$$

23 Maximize $\quad z = -1x_1 + 1x_2 + 2x_3 + 2x_4 + 4x_5 + 5x_6$

subject to
$$x_1 \qquad\quad + x_4 \qquad\qquad\qquad \leq 32{,}000$$
$$x_2 \qquad\quad + x_5 \qquad\qquad \leq 20{,}000$$
$$x_3 \qquad\quad + x_6 \leq 38{,}000$$
$$-5x_1 + x_2 - 4x_3 \qquad\qquad\qquad \leq 0$$
$$28x_1 + 10x_2 - 7x_3 \qquad\qquad\qquad \geq 0$$
$$-2x_4 + 4x_5 - x_6 \leq 0$$
$$8x_4 - 10x_5 - 27x_6 \geq 0$$
$$x_1 + x_2 + x_3 \qquad\qquad\qquad \geq 30{,}000$$
$$x_1, x_2, x_3, x_4, x_5, x_6 \geq 0$$

24 Maximize $\quad z = 180{,}000x_1 + 20{,}000x_2 + 72{,}000x_3 + 80{,}000x_4$

subject to
$$30{,}000x_1 + 12{,}000x_2 + 30{,}000x_3 + 20{,}000x_4 \leq 65{,}000$$
$$40{,}000x_1 + 8{,}000x_2 + 20{,}000x_3 + 30{,}000x_4 \leq 80{,}000$$
$$40{,}000x_1 \qquad\quad + 20{,}000x_3 + 40{,}000x_4 \leq 80{,}000$$
$$30{,}000x_1 + 4{,}000x_2 + 20{,}000x_3 + 10{,}000x_4 \leq 50{,}000$$
$$x_1 \qquad\qquad\qquad\qquad\qquad \leq 1$$
$$x_2 \qquad\qquad\qquad\qquad \leq 1$$
$$x_3 \qquad\qquad\quad \leq 1$$
$$x_4 \leq 1$$
$$x_1, x_2, x_3, x_4 \geq 0$$

25 Minimize $\quad z = 464x_1 + 513x_2 + 654x_3 + 867x_4 + 352x_5 + 416x_6 + 690x_7 + 791x_8$
$$+\ 995x_9 + 682x_{10} + 388x_{11} + 685x_{12}$$

subject to
$$x_1 + x_2 + x_3 + x_4 \qquad\qquad\qquad\qquad\qquad \leq\ 75$$
$$x_5 + x_6 + x_7 + x_8 \qquad\qquad\qquad \leq 125$$
$$x_9 + x_{10} + x_{11} + x_{12} \leq 100$$
$$x_1 \qquad\qquad + x_5 \qquad\qquad + x_9 \qquad\qquad = 80$$
$$x_2 \qquad\qquad + x_6 \qquad\qquad + x_{10} \qquad = 65$$
$$x_3 \qquad\qquad + x_7 \qquad\qquad + x_{11} \qquad = 70$$
$$x_4 \qquad\qquad + x_8 \qquad\qquad + x_{12} = 85$$
$$x_1, x_2, x_3, x_4, x_5, x_6, x_7, x_8, x_9, x_{10}, x_{11}, x_{12} \geq\ 0$$

26 Solve the financial portfolio problem in Exercise 15 (page 457) and interpret the results.
27 Solve the assignment model problem stated in Exercise 16 (page 458) and interpret the results.
28 Solve the cargo-loading problem in Exercise 17 (page 458) and interpret the results.

Section 11.5

For the following problems, formulate the corresponding dual problem.

29 Minimize $\quad z = 8x_1 + 4x_2 + 5x_3 + 6x_4$

subject to
$$3x_1 - 2x_2 + 4x_3 + x_4 \leq 125$$
$$x_1 \quad + x_3 + x_4 = 75$$
$$2x_1 + x_2 - 3x_3 \qquad \leq 150$$
$$x_1, x_2, x_3 \geq\ 0$$
$$x_4 \quad \text{unrestricted}$$

30 Maximize $\quad z = x_1 + 6x_2 + 5x_3 + 3x_4 + 2x_5$

subject to $\quad 3x_1 + 4x_2 \qquad\qquad\qquad \leq\ 35$

$$5x_1 + 3x_2 + 7x_3 - 2x_4 + x_5 \geq 130$$
$$x_1 + x_2 + x_3 \qquad\qquad = 50$$
$$x_4 + x_5 \leq 20$$
$$x_1, x_2, x_3, x_4, x_5 \geq 0$$

31 Formulate the dual of the highway maintenance problem (Example 9 on page 445) presented in Chap. 10.

Formulate the dual of the grant awards problem (Example 10 on page 447) presented in Chap. 10.

33 Formulate the dual of the petroleum-blending problem (Example 11 on page 450) presented in Chap. 10.

❏ CHAPTER TEST

1 Given the following linear programming problem:

$$\text{Maximize} \quad z = 10x_1 + 8x_2 + 12x_3$$
$$\text{subject to} \quad 4x_1 - 2x_2 + x_3 \leq 25$$
$$x_1 + 3x_2 \qquad\qquad \geq -10$$
$$2x_1 \qquad + 3x_3 = -20$$
$$x_1, x_2, x_3 \geq 0$$

transform the constraint set into an equivalent system of constraint equations suitable for the simplex method.

2 Solve the following linear programming problem using the simplex method.

$$\text{Maximize} \quad z = 20x_1 + 24x_2$$
$$\text{subject to} \quad 3x_1 + 6x_2 \leq 60$$
$$4x_1 + 2x_2 \leq 32$$
$$x_1, x_2 \geq 0$$

3 You are given the linear programming problem:

$$\text{Minimize} \quad z = 5x_1 + 4x_2$$
$$\text{subject to} \quad x_1 + x_2 \geq 10$$
$$2x_1 - x_2 = 15$$
$$x_1, x_2 \geq 0$$

(a) Set up the initial simplex tableau and revise it, if necessary, so that the row (0) coefficients equal 0 for all basic variables.

(b) Which basic variable will leave first?

(c) Which nonbasic variable will enter first?

4 Describe the way in which alternative optimal solutions are indicated when using the simplex method. How is an unbounded solution indicated?

5 Given the following primal problem, formulate the corresponding dual problem.

$$\text{Minimize} \quad z = 8x_1 + 5x_2 + 6x_3$$
$$\text{subject to} \quad x_1 + x_2 + x_3 = 25$$

$$4x_1 - 5x_2 \geq 10$$

$$x_1 - x_2 + 2x_3 \leq 48$$

$$x_2 \leq 12$$

$$x_1, x_2 \geq 0$$

$$x_3 \quad \text{unrestricted}$$

6 Discuss the meaning and significance of (*a*) shadow prices and (*b*) sensitivity analysis.

MINICASE

CONTRACT AWARDS

The purchasing department for a state agency has requested bids for five different products. Three suppliers have submitted bids on the products. Table 11.31 summarizes the prices bid per unit for each product. Notice that suppliers did not necessarily submit bids for all five products. Also shown in Table 11.31 is the agency's required quantity for each item.

TABLE 11.31	Product				
Supplier	1	2	3	4	5
1	$5.00	$7.50	$3.00	—	$4.50
2	—	$7.25	$3.20	$8.75	$4.20
3	$4.80	$7.75	$3.10	$9.00	—
Agency requirement	20,000	15,000	30,000	25,000	22,000

Some suppliers have indicated maximum quantities they can provide of particular products. Supplier 1 indicates that it can provide no more than 10,000 units of product 3, supplier 2 can provide no more than 8,000 units of product 2, and supplier 3 can provide no more than 18,000 units of product 1. State purchasing regulations do not require that all units of a given product be purchased from one supplier. Similarly, they do not require that contracts be awarded to the lowest bidder.

The purchasing department wants to determine how many units of each product it should purchase from each supplier so as to satisfy agency requirements at a minimum total cost.

1 Formulate the LP model for this problem, carefully defining your variables.
2 Solve the problem using a computerized LP package and fully interpret the results. Indicate the quantities of each product purchased from and the dollar amounts awarded to each supplier. Also indicate what the minimum total costs equal.
*3 Assume that the purchasing department wishes not to award more than $300,000 in contracts to any òne supplier. Also, assume that supplier 3 stipulated that it should be awarded contracts of at least $200,000; if this requirement is not met, the supplier will withdraw all bids. Modify the original formulation and solve using a computerized LP package. Interpret your results and compare with the answer in part 2. (*Hint:* Formulate and solve two separate models — one which assumes that supplier 3 will have its requirement met, the other assuming that supplier 3 has withdrawn its bids.)

CHAPTER 12

TRANSPORTATION AND ASSIGNMENT MODELS

CHAPTER OBJECTIVES

Provide an overview of several extensions of the basic linear programming model. Included will be a discussion of the assumptions, distinguishing characteristics, methods of solution, and applications of:

- ❏ The transportation model
- ❏ The assignment model

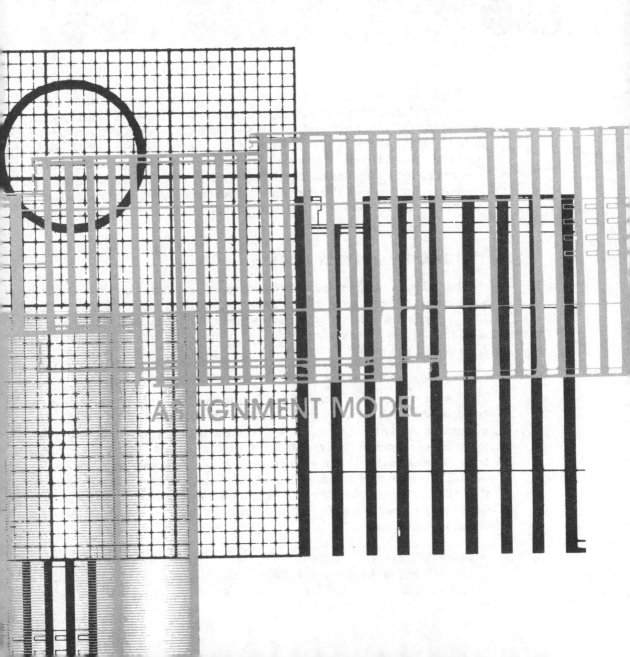

The National Collegiate Athletic Association (NCAA) Division I basketball tournament is under way. In preparing for the four regional tournaments, the Tournament Committee has selected four teams of referees who have been judged to be the most qualified for this year's tourney. Each team of officials consists of six regular referees and two alternates, who also travel to the regional tournament. Officials are paid a standard tournament fee plus travel expenses. **The committee wants to assign the four teams of officials so as to minimize total travel expenses (Example 3).**

There exists a set of mathematical programming models which are direct extensions of the standard linear programming model. This set of models is widely applied and includes, among others, the *transportation model* and the *assignment model.* These models will be overviewed in this chapter. For each model, we will discuss the general form and assumptions. One or more sample applications will be illustrated, and solution methods will be presented. The purposes of this chapter are to acquaint you with this important collection of models and to provide more experience with the area of problem formulation.

12.1 THE TRANSPORTATION MODEL

All linear programming models can be solved by the simplex method. However, some classes of linear programming problems, because of their special structure, lend themselves to solution by methods which are computationally more efficient than the simplex method. One such class of linear programming model is the *transportation model.*

General Form and Assumptions

The classic transportation model involves the shipment of some *homogeneous commodity* from a set of *origins* to a set of *destinations.* Each origin represents a source of supply for the commodity; each destination represents a point of demand for the commodity. Example 9 in Chap. 10 (page 444) is an example of a transportation model. In that example the homogeneous commodity was salt and sand to be used on roads during winter icing and snowstorms. The origins were two stockpiles, each characterized by a maximum storage capability. The destinations were the four zones of the city, each characterized by an expected need (demand) during a given storm.

> **ASSUMPTION 1**
> The standard model assumes a homogeneous commodity.

This first assumption implies that there are no significant differences in the characteristics of the commodity available at each origin. This suggests that unless other restrictions exist, each origin can supply units to any of the destinations.

The purpose in the classic transportation model is to allocate the supply available at each origin so as to satisfy the demand at each destination. Various criteria can be employed to measure the effectiveness of the allocation. Typical objectives are minimization of total transportation costs or some weighted measure of distance,* or maximization of total profit contribution from the allocation. In the example in Chap. 10 the objective was to minimize total distribution cost. Table 12.1 summarizes the cost of distributing a ton of salt or sand from each stockpile to each city zone. Also shown are stockpile capacities and normal levels of demand (modified slightly from the original example).

TABLE 12.1

Stockpile (Origin)	Zone (Destination)				Maximum Supply, Tons
	1	2	3	4	
1	$2.00	$3.00	$1.50	$2.50	900
2	4.00	3.50	2.50	3.00	750
Demand, tons	300	450	550	350	

Given that x_{ij} equals the number of tons of salt and sand distributed from stockpile i to zone j, the complete formulation of this modified problem is

Minimize $z = 2x_{11} + 3x_{12} + 1.5x_{13} + 2.5x_{14} + 4x_{21} + 3.5x_{22} + 2.5x_{23} + 3x_{24}$

subject to $x_{11} + x_{12} + x_{13} + x_{14} \qquad\qquad\qquad\qquad = 900$ (stockpile 1)

$\qquad\qquad\qquad\qquad x_{21} + x_{22} + x_{23} + x_{24} = 750$ (stockpile 2)

$x_{11} \qquad\qquad\qquad + x_{21} \qquad\qquad\qquad\quad = 300$ (zone 1)

$\qquad x_{12} \qquad\qquad\qquad\quad + x_{22} \qquad\qquad\quad = 450$ (zone 2)

$\qquad\qquad x_{13} \qquad\qquad\qquad\quad x_{23} \qquad\quad = 550$ (zone 3)

$\qquad\qquad + x_{14} \qquad\qquad\qquad\quad x_{24} = 350$ (zone 4)

$x_{11}, x_{12}, x_{13}, x_{14}, x_{21}, x_{22}, x_{23}, x_{24} \geq 0$

ASSUMPTION 2
The standard model assumes that total supply and total demand are equal.

This second assumption is required by a special solution algorithm for this type of model. Even though the example was modified to create balance between supply and demand, this assumption is rarely satisfied in an actual problem. Procedures do exist for handling imbalance when it occurs in a problem. These procedures are

---

* For example, the number of units distributed weighted (multiplied) by the distance they travel.

similar to adding slack and surplus variables to a linear programming formulation for purposes of using the simplex method.

Notice in the structural constraints that all variables have coefficients of 0 or 1. For example, the first constraint can be envisioned as having the form

$$1x_{11} + 1x_{12} + 1x_{13} + 1x_{14} + 0x_{21} + 0x_{22} + 0x_{23} + 0x_{24} = 900$$

Similarly, the last constraint has the implicit form

$$0x_{11} + 0x_{12} + 0x_{13} + 1x_{14} + 0x_{21} + 0x_{22} + 0x_{23} + 1x_{24} = 350$$

This characteristic, along with the assumed balance between supply and demand, is significant in distinguishing transportation models from other linear programming models. Solution methods have been developed which capitalize upon this structure and result in significant computational efficiencies.

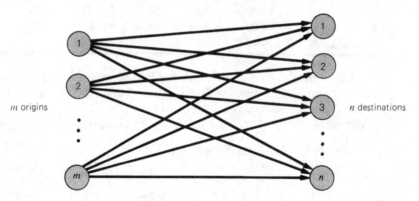

Figure 12.1   Generalized transportation model.

Let's state the generalized transportation model associated with the structure shown in Fig. 12.1. If

x_{ij} = *number of units distributed from origin i to destination j*

c_{ij} = *contribution to the objective function from distributing one unit from origin i to destination j*

s_i = *number of units available at origin i*

d_j = *number of units demanded at destination j*

m = *number of origins*

n = *number of destinations*

the generalized model can be stated as follows:

Minimize
(or maximize)

$$z = c_{11}x_{11} + c_{12}x_{12} + \cdots + c_{1n}x_{1n} + c_{21}x_{21} + c_{22}x_{22} + \cdots$$
$$+ c_{2n}x_{2n} + \cdots + c_{mn}x_{mn}$$

subject to

$$
\left.
\begin{aligned}
x_{11} + x_{12} + \cdots + x_{1n} &= s_1 \\
x_{21} + x_{22} + \cdots + x_{2n} &= s_2 \\
\vdots \qquad\qquad \vdots \quad & \vdots \\
x_{m1} + x_{m2} + \cdots + x_{mn} &= s_m
\end{aligned}
\right\} \text{supply constraints}
$$

$$
\left.
\begin{aligned}
x_{11} + x_{21} + \cdots + x_{m1} &= d_1 \\
x_{12} + x_{22} + \cdots + x_{m2} &= d_2 \\
\vdots \qquad\qquad \vdots \quad & \vdots \\
x_{1n} + x_{2n} + \cdots + x_{mn} &= d_n
\end{aligned}
\right\} \text{demand constraints}
$$

$$x_{ij} \geq 0 \qquad \text{for all } i \text{ and } j$$

Implicit in the model is the balance between supply and demand,

$$s_1 + s_2 + \cdots + s_m = d_1 + d_2 + \cdots + d_n$$

The transportation model is a very flexible model which can be applied to problems that have nothing to do with the distribution of commodities. The following example is an illustration.

EXAMPLE 1

(Job Placement Screening) A job placement agency works on a contract basis with employers. A computer manufacturer is opening a new plant and has contracted with the placement agency to process job applications for prospective employees. Because of the uneven demands in workload at the agency, it often uses part-time personnel for the purpose of processing applications. For this particular contract, five placement analysts must be hired. Each analyst has provided an estimate of the maximum number of job applications he or she can evaluate during the coming month. (The agency screens these estimates to ensure that they are reasonable.) Analysts are compensated on a piecework basis, with the rate determined by the type of application evaluated and the experience of the analyst.

Table 12.2 summarizes, by analyst, the cost of processing each type of job application. Also indicated are the maximum number of applications which can be processed by each analyst and the number of applications expected in each job category. The problem for the agency is to determine the number of applications of each job type to assign to each analyst so as to minimize the cost of processing the expected batch of job applications. Each analyst can be thought of as an origin with a maximum capacity for application screening. Each job category represents a destination having a demand for application screening.

If x_{ij} equals the number of job applications of type j assigned to analyst i, the problem can be formulated as shown in the model on the top of page 528. Notice that the total supply (the maximum number of applications which can be processed by five analysts) exceeds total demand (expected number of applications). As a result, constraints (1) to (5) cannot be stated as equalities. And according to assumption 2, total supply and demand must be brought into balance, artificially, before solving the problem.

TABLE 12.2

Placement Analyst	Type of Job Application				Maximum Number of Applications
	1 Engineer	2 Programmer/ Analyst	3 Skilled Laborer	4 Unskilled Laborer	
1	$15	$10	$8	$7	90
2	12	8	7	5	120
3	16	9	9	8	140
4	12	10	7	7	100
5	10	7	6	6	110
Expected number of applications	100	150	175	125	560 / 550

Minimize $z = 15x_{11} + 10x_{12} + 8x_{13} + 7x_{14} + \cdots + 6x_{54}$

subject to

$$x_{11} + x_{12} + x_{13} + x_{14} \qquad\qquad\qquad\qquad\qquad\qquad\qquad\qquad\qquad \le 90 \ (1)$$
$$x_{21} + x_{22} + x_{23} + x_{24} \qquad\qquad\qquad\qquad\qquad\qquad\qquad \le 120 \ (2)$$
$$x_{31} + x_{32} + x_{33} + x_{34} \qquad\qquad\qquad\qquad\qquad \le 140 \ (3)$$
$$x_{41} + x_{42} + x_{43} + x_{44} \qquad\qquad\qquad \le 100 \ (4)$$
$$x_{51} + x_{52} + x_{53} + x_{54} \le 110 \ (5)$$
$$x_{11} \qquad + x_{21} \qquad + x_{31} \qquad + x_{41} \qquad + x_{51} \qquad = 100 \ (6)$$
$$x_{12} \qquad + x_{22} \qquad + x_{32} \qquad + x_{42} \qquad + x_{52} \qquad = 150 \ (7)$$
$$x_{13} \qquad + x_{23} \qquad + x_{33} \qquad + x_{43} \qquad + x_{53} \qquad = 175 \ (8)$$
$$x_{14} \qquad + x_{24} \qquad + x_{34} \qquad + x_{44} \qquad + x_{54} = 125 \ (9)$$
$$x_{ij} \ge 0 \text{ for all}$$
$$i \text{ and } j$$

12.2 SOLUTIONS TO TRANSPORTATION MODELS

As indicated earlier, the transportation model is distinguished from other linear programming models because its structure lends itself to more efficient solution procedures than the simplex method. The simplex method *can* be used to solve transportation models. However, methods such as the *stepping stone algorithm* and a dual-based enhancement called the *MODI method** prove much more efficient.

---

* For descriptions of these methods, see Frank S. Budnick, Dennis W. McLeavey, and Richard Mojena, *Principles of Operations Research for Management,* 2d ed., Richard D. Irwin, Homewood, Ill. (1988), Chap. 7.

Initial (Starting) Solutions

The increased efficiency can occur during two different phases of solution: (1) determination of the initial solution and (2) progressing from the initial solution to the optimal solution. With the simplex method the initial solution is predetermined by the constraint structure. The initial set of basic variables will always consist of the slack and artificial variables in the problem. With transportation models the stepping stone algorithm (or the MODI method) will accept any feasible solution as a starting point. Consequently, various approaches to finding a *good* starting solution have been proposed. These include the **northwest corner method,** the **least cost method,** and **Vogel's approximation method** (see note on page 528). Particularly with the latter two methods, the hope is that some extra effort "up front" may result in a starting solution which is close to the optimal solution. The expectation is that this would reduce the time and effort required by the stepping stone algorithm to move from the initial solution to the optimal solution. None of the proposed methods has proved to be consistently more successful than the others. However, depending on the problem, they can lead to considerable efficiencies.

TABLE 12.3		Destination			
Origin	1	2	3	Supply	
1	5	10	10	55	
2	20	30	20	80	
3	10	20	30	75	
Demand	70	100	40	210	

Consider the data contained in Table 12.3 for a transportation problem involving three origins and three destinations. Assume that the elements in the body of the table represent the costs of shipping a unit from each origin to each destination. Also shown are the supply capacities of the three origins and the demands at each destination. Conveniently, the total supply and total demand are equal to one another. The problem is to determine how many units to ship from each origin to each destination so as to satisfy the demands at the three destinations while not violating the capacities of the three origins. The objective is to make these allocations in such a way as to minimize total transportation costs.

We will solve this problem using two special algorithms. In Example 2 we illustrate the northwest corner method, which can be used to determine an initial (starting) solution. In the next section, we illustrate the stepping stone algorithm, which can be used to solve these types of models. Before we begin these examples, let's discuss some requirements of the stepping stone algorithm.

REQUIREMENTS OF STEPPING STONE ALGORITHM

1 *Total supply at the origins must equal total demand at the destinations. Since this is not typically the case in actual applications, the "balance" between supply and demand often is created artificially. This is done by adding a "dummy" origin or a "dummy" destination having*

(continued)

sufficient supply (demand) to create the necessary balance. Our example has been contrived so that balance already exists.

2   Given a transportation problem with m origins and n destinations
(where m and n include any "dummy" origins or destinations added
to create balance), the number of basic variables in any given solution must equal $m + n - 1$. In our problem, any solution should contain $3 + 3 - 1 = 5$ basic variables.

EXAMPLE 2   **(Finding an Initial Solution: The Northwest Corner Method)** When solving transportation problems using the stepping stone algorithm, the work is conducted in a tabular
format, as was the case in using the simplex method. As can be seen in Table 12.4, the tabular
format for our example resembles a "bowling sheet."

TABLE 12.4

Origin	*Destination* 1		*Destination* 2		*Destination* 3		Supply
1	x_{11}	5	x_{12}	10	x_{13}	10	55
2	x_{21}	20	x_{22}	30	x_{23}	20	80
3	x_{31}	10	x_{32}	20	x_{33}	30	75
Demand	70		100		40		210

For each origin/destination combination, there is a cell which contains the value of the
corresponding decision variable x_{ij} and the objective function coefficient or unit transportation cost. As we proceed on to solve a problem, we will substitute actual values for the x_{ij}'s in
the table as we try different allocations.

As indicated earlier, there are several techniques one might employ to determine an
initial solution. The northwest corner method is illustrated in this example because it is
simple to use. Regardless of the technique used to find an initial solution, the following
conditions must exist: (1) the supply at each origin should be allocated, (2) the demand at
each destination must be satisfied, and (3) there should be exactly $m + n - 1$ basic variables
(allocations).

The *northwest corner method* is a popular (but unthinking) technique for arriving at an
initial solution. The technique starts in the upper left-hand cell (northwest corner) of a
transportation table and assigns units from origin 1 to destination 1. Assignments are
continued in such a way that the supply at origin 1 is completely allocated before moving on
to origin 2. The supply at origin 2 is completely allocated before moving to origin 3, and so on.
Similarly, a sequential allocation to the destinations assures that the demand at destination
1 is satisfied before making allocations to destination 2, and so forth. This pattern of
assignments leads to a sort of staircase arrangement of assignments in the transportation
table. Table 12.5 indicates the initial solution to our problem as derived using the northwest
corner method. Let's examine the allocations.

| TABLE 12.5 | **Initial Solution Derived Using Northwest Corner Method** | | | |

| | | *Destination* | | |
Origin	1	2	3	Supply
1	5 (55)	10	10	5̶5̶
2	20 (15)	30 (65)	20	8̶0̶ 6̶5̶
3	10	20 (35)	30 (40)	7̶5̶ 4̶0̶
Demand	7̶0̶ 1̶5̶	1̶0̶0̶ 3̶5̶	4̶0̶	210

1 *Starting in the northwest corner, the supply at origin 1 is 55 and the demand at destination 1 is 70. Thus, we allocate all available supply at origin 1 in an attempt to satisfy the demand at destination 1 ($x_{11} = 55$).*

2 *When the complete supply at origin 1 has been allocated, the next allocation will be from origin 2. The allocation in cell (1, 1) did not satisfy the demand at destination 1 completely. Fifteen additional units are demanded. Comparing the supply at origin 2 with the remaining demand at destination 1, we allocate 15 units from origin 2 to destination 1 ($x_{21} = 15$). This allocation completes the needs of destination 1, and the demand at destination 2 will be addressed next.*

3 *The last allocation left origin 2 with 65 units. The demand at destination 2 is 100 units. Thus, we allocate the remaining supply of 65 units to destination 2 ($x_{22} = 65$). The next allocation will come from origin 3.*

4 *The allocation of 65 units from origin 2 left destination 2 with unfulfilled demand of 35 units. Since origin 3 has a supply of 75 units, 35 units are allocated to complete the demand for that destination ($x_{32} = 35$). Thus, the next allocation will be to destination 3.*

5 *The allocation of 35 units from origin 3 leaves that origin with 40 remaining units. The demand at destination 3 also equals 40; thus, the final allocation of 40 units is made from origin 3 to destination 3 ($x_{33} = 40$).*

Notice in Table 12.5 that all allocations are circled in the appropriate cells. These represent the basic variables for this solution. There should be five basic variables ($m + n - 1$) to satisfy the requirement of the stepping stone algorithm. The recommended allocations and associated costs for this initial solution are summarized in Table 12.6.

❑

The Stepping Stone Algorithm

Given the initial solution generated in Table 12.6, the stepping stone algorithm performs a *marginal analysis* which examines the effects of changing the given solution. Specifically, the marginal effects of introducing a unit of a nonbasic variable are examined, as was the case with the simplex method.

❑ **Step 1: Determine the *improvement index* for each nonbasic variable (cell).** As we examine the effects of introducing one unit of a

TABLE 12.6	From Origin	To Destination	Quantity		Unit Cost	Total Cost
	1	1	55	×	$ 5.00	$ 275.00
	2	1	15	×	20.00	300.00
	2	2	65	×	30.00	1,950.00
	3	2	35	×	20.00	700.00
	3	3	40	×	30.00	1,200.00
						$4,425.00

nonbasic variable, we focus upon two marginal effects: *(1) What adjustments must be made to the values of the current basic variables* (in order to continue satisfying all supply and demand constraints)? *(2) What is the resulting change in the value of the objective function?*

TABLE 12.7	**Closed Path and Adjustments for Cell (1, 2)**

		Destination					
Origin	1		2		3		Supply
1	−1 ← 5 ← +1 (55)		10		10		55
2	+1 → 20 → −1 (15)		30 (65)		20		80
3	10		20 (35)		30 (40)		75
Demand	70		100		40		210

To illustrate, focus upon Table 12.7. Cell (1, 2) has no allocation in the initial solution and is a nonbasic cell. The question we want to ask is what trade-offs (or adjustments) with the existing basic variables would be required if 1 unit is shipped from origin 1 to destination 2? If 1 unit is allocated to cell (1, 2), a total of 56 units will be committed from origin 1, 1 more than its capacity. Thus, we must reduce shipments elsewhere from origin 1 by 1 unit. The only place we can reduce is in cell (1, 1). Thus, we adjust by decreasing the allocation in that cell by 1 unit. However, this reduction results in an undershipment to destination 1. Reducing shipments in cell (1, 1) to 54 results in a total of 69 units' being allocated to destination 1, 1 short of its demand. Thus, we compensate by increasing the allocation to cell (2, 1) by 1 unit to 16. This adjustment results in an overshipment from origin 2 (16 + 65 = 81). To adjust, we decrease shipments in cell (2, 2) by 1 unit to 64. This last adjustment returns the system to balance. The series of required adjustments is indicated in Table 12.7 by the *closed path* of directed arrows. Summarizing, to compensate for adding a unit to cell (2, 2), shipments in cell (1, 1) must be *decreased* by 1 unit, shipments in cell (2, 1) must be *increased* by 1 unit, and shipments in cell (2, 2) must be *decreased* by 1 unit.

Now that we have identified the necessary adjustments in the current basic variables, the next question is, What is the marginal effect on the value of the objective function? To determine this, we look at the closed path in Table 12.7. For each cell (i, j) receiving an increased allocation of 1 unit, costs *increase* by the corresponding cost coefficient (c_{ij}). Similarly, costs *decrease* by the value of the cost coefficient wherever allocations have been reduced by 1 unit. These effects are summarized in Table 12.8.

TABLE 12.8 | **Marginal Effects on Value of Objective Function from Introducing One Unit in Cell (1, 2)**

Cell Adjusted	Adjustment	Change in Cost
(1, 2)	+1	+$10.00
(1, 1)	−1	− 5.00
(2, 1)	+1	+ 20.00
(2, 2)	−1	− 30.00
Net change		−$ 5.00 (Improvement index)

The net (marginal) effect associated with allocating 1 unit from origin 1 to destination 2 is to reduce total cost by $5.00. This marginal change in the objective function is called the **improvement index** for cell (1, 2).

> **CALCULATION OF IMPROVEMENT INDEX**
> Trace a "closed path" which begins at the unoccupied cell of interest; moves alternately in horizontal and vertical directions, pivoting *only on occupied cells;* and terminates on the unoccupied cell. A (+1) is assigned to the unoccupied cell (indicating an increase of 1 unit) and succeeding corner points on the path are alternately assigned (−1) and (+1) values. The pluses and minuses indicate the necessary adjustments for satisfying the row (supply) and column (demand) requirements. *Note:* The direction in which the path is traced is not important. Tracing clockwise or counterclockwise results in the same path and identical adjustments.
> Once the closed path has been identified for a nonbasic cell, the improvement index for that cell is calculated by adding all objective function coefficients for cells in plus positions on the path and subtracting corresponding objective function coefficients for cells in negative positions on the path.

Table 12.9 indicates the closed path for cell (1, 3), which was nonbasic in the initial solution. Notice that adding a unit to cell (1, 3) requires the adjustments, and results in the consequences, shown in Table 12.10. Notice that adding a unit to cell (1, 3) will result in a net reduction in the value of the objective function of $15.00.

TABLE 12.9	Closed Path and Adjustments for Cell (1, 3)

Destination

Origin	1	2	3	Supply
1	−1 ← 5 ⑤⑤	10 −1	+1 ↑ 10	55
2	+1 ↓ 20 ⑮	−1 30 ⑥⑤	20	80
3	10	+1 20 ③⑤	−1 30 ④⓪	75
Demand	70	100	40	210

TABLE 12.10	Marginal Effects on Value of Objective Function from Introducing One Unit in Cell (1, 3)

Cell Adjusted	Adjustment	Change in Cost
(1, 3)	+1	+$10.00
(1, 1)	−1	− 5.00
(2, 1)	+1	+ 20.00
(2, 2)	−1	− 30.00
(3, 2)	+1	+ 20.00
(3, 3)	−1	− 30.00
Net change		−$15.00 (Improvement index)

TABLE 12.11	Closed Paths and Improvement Indices for Initial Solution

Nonbasic Cell	Closed Path	Improvement Index
(1, 2)	(1, 2) → (1, 1) → (2, 1) → (2, 2) → (1, 2)	−$ 5.00
(1, 3)	(1, 3) → (1, 1) → (2, 1) → (2, 2) → (3, 2) → (3, 3) → (1, 3)	−$15.00
(2, 3)	(2, 3) → (2, 2) → (3, 2) → (3, 3) → (2, 3)	−$20.00*
(3, 1)	(3, 1) → (3, 2) → (2, 2) → (2, 1) → (3, 1)	$ 0.00

Table 12.11 summarizes the closed paths and improvement indices for all nonbasic cells in the initial solution. Verify these paths and values for the improvement indices to make sure you understand what we have been discussing.

❏ **Step 2: If a better solution exists, determine which variable (cell) should enter the basis.** An examination of the improvement indices in Table 12.11 indicates that the introduction of three of the four nonbasic variables would lead to a reduction in total costs.

For *minimization problems,* a better solution exists if there are any negative improvement indices. An optimal solution has been found when *all* improvement indices are nonnegative. For *maximization problems,* a better solution exists if there are any positive improvement indices. An optimal solution has been found when all improvement indices are nonpositive.

As in the simplex method, we select the variable (cell) which leads to the greatest marginal improvement in the objective function.

ENTERING VARIABLE
For minimization problems, the entering variable is identified as the cell having the largest *negative* improvement index (ties may be broken arbitrarily). For maximization problems, the entering variable is the cell having the largest *positive* improvement index.

In our example, cell (2, 3), or x_{23}, is selected as the entering variable.

TABLE 12.12 | **Closed Path and Adjustments for Entering Cell (2, 3)**

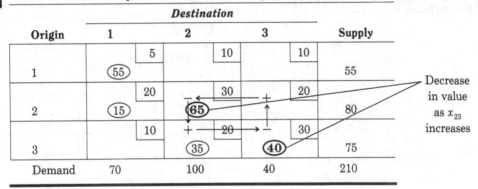

□ **Step 3: Determine the departing variable and the number of units to assign the entering variable.** This step is performed by returning to the closed path associated with the incoming cell. Table 12.12 shows the closed path for cell (2, 3). Because the stepping stone algorithm parallels the simplex exactly, we need to determine the number of units we can assign to cell (2, 3) such that the value of one of the current basic variables is driven to 0. From Table 12.12 we see that only two basic variables decrease in value as additional units are allocated to cell (2, 3): cells (2, 2) and (3, 3), both of which are in *minus* positions on the closed path. The question is which of these will go to 0 first as more units are added to cell (2, 3). See whether you can reason that when the *40th unit* is added to cell (2, 3), the value for cell (2, 2) reduces to 25, the value for cell (3, 2) increases to 75, and the value of cell (3, 3) goes to 0.

DEPARTING VARIABLE
The departing variable is identified as the smallest basic variable in a minus position on the closed path for the entering variable.

NUMBER OF UNITS TO ASSIGN ENTERING VARIABLE
The number of units equals the size of the departing variable (the smallest value in a minus position).

❏ **Step 4: Develop the new solution and return to step 1.** Again referring to the closed path for the incoming cell (2, 3), add the quantity determined in step 3 to all cells in plus positions and subtract this quantity from those in minus positions. Thus, given that the entering variable $x_{23} = 40$ from step 3, the cells on the closed path are adjusted, leading to the second solution shown in Table 12.13. When you determine a new solution, you should check the allocations along each row and column to make sure that they add to the respective supply and demand values. Also, make sure that there are $m + n - 1$ basic variables. As can be seen in Table 12.13, both of these requirements are satisfied.

TABLE 12.13	**Second Solution**

Origin	Destination 1	Destination 2	Destination 3	Supply	
1	5 (55)	10	10	55	
2	20 (15)	30 (25)	20 (40)	80	New basic variable (cell)
3	10	20 (75)	30	75	Departed basic variable (cell)
Demand	70	100	40	210	

One other piece of information which is of interest is the new value of the objective function. We have two alternatives in determining this value. We can multiply the value of each basic variable times its corresponding objective function coefficient and sum, as we did in Table 12.6. Or given that the original value of the objective function was $4,425 and that each unit introduced to cell (2, 3) decreases the value of the objective function by $20, introducing 40 units to cell (2, 3) results in a new value for total cost of

$$z = \$4,425 - (40)(\$20)$$
$$= \$4,425 - \$800$$
$$= \$3,625$$

Tables 12.14–12.18 summarize the remaining steps in solving this problem. See whether you can verify these steps and the final result.

TABLE 12.14 Closed Paths and Improvement Indices for Second Solution

Nonbasic Cell	Closed Path	Improvement Index
(1, 2)	(1, 2) → (1, 1) → (2, 1) → (2, 2) → (1, 2)	−$ 5.00*
(1, 3)	(1, 3) → (1, 1) → (2, 1) → (2, 3) → (1, 3)	+$ 5.00
(3, 1)	(3, 1) → (3, 2) → (2, 2) → (2, 1) → (3, 1)	$ 0.00
(3, 3)	(3, 3) → (2, 3) → (2, 2) → (3, 2) → (3, 3)	+$20.00

TABLE 12.15 Third Solution (z = $3,500.00)

Origin	Destination 1	Destination 2	Destination 3	Supply
1	[5] (30)	[10] (25)	[10]	55
2	[20] (40)	[30]	[20] (40)	80
3	[10]	[20] (75)	[30]	75
Demand	70	100	40	210

TABLE 12.16 Closed Paths and Improvement Indices for Third Solution

Nonbasic Cell	Closed Path	Improvement Index
(1, 3)	(1, 3) → (1, 1) → (2, 1) → (2, 3) → (1, 3)	+$ 5.00
(2, 2)	(2, 2) → (1, 2) → (1, 1) → (2, 1) → (2, 2)	+$ 5.00
(3, 1)	(3, 1) → (3, 2) → (1, 2) → (1, 1) → (3, 1)	−$ 5.00*
(3, 3)	(3, 3) → (2, 3) → (2, 1) → (1, 1) → (1, 2) → (3, 2) → (3, 3)	+$15.00

TABLE 12.17 Fourth (and Optimal) Solution (z = $3,350.00)

Origin	Destination 1	Destination 2	Destination 3	Supply
1	[5]	[10] (55)	[10]	55
2	[20] (40)	[30]	[20] (40)	80
3	[10] (30)	[20] (45)	[30]	75
Demand	70	100	40	210

TABLE 12.18	Closed Paths and Improvement Indices for Fourth Solution	
Nonbasic Cell	**Closed Path**	**Improvement Index**
(1, 1)	(1, 1) → (3, 1) → (3, 2) → (1, 2) → (1, 1)	+$ 5.00
(1, 3)	(1, 3) → (1, 2) → (3, 2) → (3, 1) → (2, 1) → (2, 3) → (1, 3)	+$10.00
(2, 2)	(2, 2) → (2, 1) → (3, 1) → (3, 2) → (2, 2)	$0.00
(3, 3)	(3, 3) → (2, 3) → (2, 1) → (3, 1) → (3, 3)	+$20.00

Because all improvement indices are nonnegative in Table 12.18, we conclude that the solution in Table 12.17 is optimal. That is, total cost will be minimized at a value of $3,350 if 55 units are shipped from origin 1 to destination 2, 40 units from origin 2 to destination 1, 40 units from origin 2 to destination 3, 30 units from origin 3 to destination 1, and 45 units from origin 3 to destination 2.

Table 12.18 also illustrates the phenomenon of *alternative optimal solutions*.

> **ALTERNATIVE OPTIMAL SOLUTIONS**
> *Given that an optimal solution has been identified for a transportation model,* alternative optimal solutions exist if any improvement indices equal 0. If the conditions for optimality exist, allocation of units to cells having improvement indices of 0 results in no change in the (optimal) value for the objective function.

In our optimal solution, Table 12.18 indicates that allocation of units to cell (2, 2) would result in no change in the total cost.

Computer Solution Methods

As we would expect, there are numerous transportation computer packages available for solving these models. Usually the input-output features are simpler than with straight linear programming packages. Figure 12.2 illustrates the input and output features of one computer package* for the transportation problem just solved. For the data input portion, the user responses are shown in color to distinguish them from the computer responses.

Section 12.2 Follow-up Exercises

1 For the problem just completed, generate an alternative optimal solution by selecting cell (2, 2) in Table 12.17 as the entering variable and introducing as many units as possible.
2 Given the data for a transportation problem in Table 12.19:
   (a) Use the northwest corner method to determine an initial solution.
   (b) Proceed on to solve for the optimal solution using the stepping stone algorithm.

---

* "TRAN1" from Warren J. Erikson and Owen P. Hall, Jr., *Computer Models for Management Science,* 2d ed., Addison-Wesley, Reading, Mass., 1986.

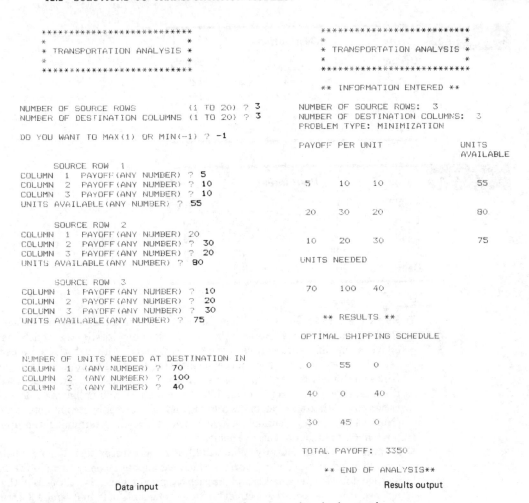

Figure 12.2 Computer-based solution for previously solved example.

	Destination			
	---	---	---	---
Origin	**1**	**2**	**3**	**Supply**
1	20	30	10	100
2	30	40	25	300
3	35	15	20	100
Demand	150	125	225	

TABLE 12.19

3 Given the data for a transportation problem in Table 12.20:
   (a) Use the northwest corner method to determine an initial solution.
   (b) Proceed on to solve for the optimal solution using the stepping stone algorithm.
4 Given the data for the transportation problem in Table 12.21:
   (a) Use the northwest corner method to determine an initial solution.
   (b) Proceed on to solve for the optimal solution using the stepping stone algorithm.

TABLE 12.20

Origin	Destination 1	2	3	Supply
1	8	6	10	125
2	4	9	8	150
3	7	6	5	95
Demand	110	85	175	

TABLE 12.21

Origin	Destination 1	2	3	Supply
1	40	20	30	500
2	60	75	45	600
3	35	50	60	400
Demand	300	500	700	

5 **Supply Exceeds Demand** When the total supply in a transportation model exceeds total demand, "balance" is created by adding a "dummy" destination which has demand equal to the difference between supply and demand. Although there are exceptions, generally the objective function coefficients assigned to the dummy column equal zero (since no shipments are going to be made to this destination). In Exercise 2, assume that the total supply at origin 1 equals 150. Add a dummy column having a demand of 50, assign costs of 0 to each cell in the column, and solve using the stepping stone algorithm. [Note: The dummy column must be included when determining the appropriate $(m + n - 1)$ number of basic variables.]

6 **Demand Exceeds Supply** When total demand exceeds total supply, "balance" is created by adding a "dummy" origin which has supply capacity equal to the difference between total supply and demand. As with the previous exercise, the cells in the dummy row are usually assigned objective function coefficients of 0. In Exercise 3, assume that the demand at destination 3 equals 225. Add a dummy row having a supply of 50 units, assign costs of 0 to each cell in the row, and solve using the stepping stone algorithm. (See the note at the end of Exercise 5.)

7 A brewery has three bottling plants which bottle generically labeled beer. The beer is distributed from the three plants to five regional warehouses. Table 12.22 summarizes distribution costs as well as weekly capacities for the plants and weekly requirements at

TABLE 12.22

Plant	Warehouse 1	2	3	4	5	Weekly Capacity, Hundreds of Cases
1	$ 20	$ 35	$ 30	$ 40	$ 42	400
2	45	30	42	36	38	350
3	38	40	36	35	50	450
Weekly requirement, hundreds of cases	150	300	200	250	175	

each warehouse, both stated in hundreds of cases. The main body of the table contains distribution costs in dollars per hundred cases. The problem is to determine the number of cases to be distributed weekly from each plant to each warehouse so as to minimize total distribution costs.

(a) Formulate the objective function and constraint set for this transportation model.

(b) By trial and error, or some other means, estimate what you believe is a good solution. (*Hint:* An optimal solution will require only seven positive variables.)

(c) Solve using the stepping stone algorithm.

(d) If a computerized transportation package (or a linear programming package) is available, solve for the optimal solution.

TABLE 12.23	Shortage City				
Surplus City	1	2	3	4	Surplus Cars
1	$30	$45	$26	$28	20
2	32	40	28	24	18
3	27	38	30	32	32
Number of cars short	10	15	12	20	

8 A car rental firm needs to relocate cars for the coming month. Three of its cities are projected to have a surplus of cars, and four cities are expected to be short. Table 12.23 indicates the cost of relocating a car from a surplus city to a shortage city. Also shown are the projected shortages and surpluses for the different cities. The problem is to determine the number of cars to relocate from each surplus city to each shortage city so as to minimize total relocation costs.

(a) Formulate the objective function and constraint set.

(b) By trial and error, or some other means, estimate what you believe is a good solution. (*Hint:* An optimal solution will require only six positive variables.)

(c) Solve using the stepping stone algorithm.

(d) If a computer package is available, solve for the optimal solution.

9 If a transportation-type problem involves 30 origins and 50 destinations, and total supply and demand are equal:

(a) How many decision variables will appear in the linear programming formulation?

(b) How many constraints?

(c) How many total variables when the problem is converted to the standard form for the simplex method?

10 The Ace Asphalt Corporation has signed a contract to supply asphalt for four road construction projects. Ace has three asphalt plants which can supply asphalt for any or all of the projects. Table 12.24 indicates the daily capacities of each plant in truckloads,

TABLE 12.24	Construction Project				Daily Capacity, Truckloads
Plant	1	2	3	4	
1	$ 80	$100	$60	$ 70	120
2	40	80	75	60	100
3	100	120	90	110	80
Daily demand, truckloads	50	40	75	60	

the daily demand for each construction project, and the profit margin per truckload shipped from each plant to each project.

Ace wishes to determine the number of trucks to allocate from each plant to each project in order to maximize total daily profit from the contract.

(a) Formulate the objective function and constraint set.
(b) By trial and error, or some other means, estimate what you believe is a good solution. (*Hint:* An optimal solution will require only six positive variables.)
(c) Solve using the stepping stone algorithm.
(d) If a computer package is available, solve for the optimal solution.

11 **Basketball Scouting** A national scouting service does scouting reports on high school basketball players. These reports can be sold to universities and colleges for recruiting purposes. The service contracts with five scouting coordinators to perform the scouting and submit written reports on players from eight different regions of the country. A fee is charged by the coordinators for each player scouted, and it varies depending on the region of the country in which the player resides. For the coming year the scouting service has identified 1,500 players who show potential to play at the college level. Table 12.25 summarizes the fees charged per scouting report, the number of players in each region, and the maximum number of players who can be assigned to each coordinator. (*Note:* Each coordinator subcontracts with free-lance scouts.) Coordinators do not necessarily scout in all regions. This is indicated in Table 12.25 by the absence of a fee.

TABLE 12.25	**Scouting Fees per Player**								
					Region				Maximum Number
Coordinator	1	2	3	4	5	6	7	8	of Players
1	$ 30	$ 40	$ 25	$ 45	$ 35				300
2	45	55		25	30	$ 40	$ 50		350
3	60		40		50	20	45	$ 30	325
4	40	40		30	50	35	40	25	250
5	50	60	30	40	80	35	25	45	400
Number of players to be scouted	150	100	250	175	225	200	180	220	

The scouting service wants to determine how many players in each region it should assign to each coordinator in order to minimize the cost of obtaining the 1,500 scouting reports.

(a) Formulate the objective function and constraint set.
(b) By trial and error, estimate what you believe is a good solution. (*Hint:* An optimal solution will require only 12 positive variables.)
(c) If a computer package is available, solve for the optimal solution. (*Hint:* If you use a transportation package, you will have to assign extremely large scouting fees, e.g., $1,000, to the empty cells in Table 12.25. This will assure that coordinators will not be assigned any players outside their regions.)

12.3 THE ASSIGNMENT MODEL AND METHODS OF SOLUTION

A special case of the transportation model is the ***assignment model.*** This model is appropriate in problems which involve the assignment of resources to tasks (e.g., assign n persons to n different tasks or jobs). Just as the special structure of the transportation model allows for solution procedures which are more efficient than the simplex method, the structure of the assignment model allows for solution methods more efficient than the transportation method.

General Form and Assumptions

The general assignment problem involves the assignment of n resources (origins) to n tasks (destinations). Typical examples of assignment problems include the assignment of salespersons to sales territories, airline crews to flights, snowplow crews to zones of a city, ambulance units to calls for service, referees and officials to sports events, and lawyers within a law firm to cases or clients. The objective in making assignments can be one of minimization or maximization (e.g., minimization of total time required to complete n tasks or maximization of total profit from assigning salespersons to sales territories). The following assumptions are significant in formulating assignment models.

> **ASSUMPTION 1**
> Each resource is assigned exclusively to one task.
>
> **ASSUMPTION 2**
> Each task is assigned exactly one resource.
>
> **ASSUMPTION 3**
> For purposes of solution, the number of resources available for assignment must equal the number of tasks to be performed.

If

$$x_{ij} = \begin{cases} 1 & \textit{if resource i is assigned to task j} \\ 0 & \textit{if resource i is not assigned to task j} \end{cases}$$

$c_{ij} = $ *objective function contribution if resource i is assigned to task j*

$n = $ *number of resources and number of tasks*

the generalized assignment model is as shown on the top of page 544.

Notice for this model that the variables are restricted to the two values of 0 (nonassignment of the resource) or 1 (assignment of the resource). This restriction on the values of the variables is quite different from the other linear programming models we have examined. Also, constraints (1) to (n) ensure that each resource is assigned to one task only. Constraints ($n + 1$) to ($n + n$) ensure that each task is assigned exactly one resource. According to assumption 3, the number of resources

$$
\begin{array}{llr}
\text{Maximize} & z = c_{11}x_{11} + c_{12}x_{12} + \cdots + c_{1n}x_{1n} + c_{21}x_{21} + \cdots + c_{nn}x_{nn} & \\
\text{(or minimize)} & & \\
\text{subject to} & x_{11} + x_{12} + \cdots + x_{1n} & = 1 \quad (1) \\
& \qquad\quad x_{21} + x_{22} + \cdots + x_{2n} & = 1 \quad (2) \\
& \qquad\qquad\qquad\qquad\qquad \ddots & \vdots \\
& \qquad\qquad\qquad\qquad x_{n1} + x_{n2} + \cdots + x_{nn} = 1 & (n) \\
& x_{11} + \qquad\qquad x_{21} + \cdots \qquad + x_{n1} & = 1 \quad (n+1) \\
& \qquad x_{12} + \qquad\qquad x_{22} + \cdots \qquad + x_{n2} & = 1 \quad (n+2) \\
& \qquad\quad \ddots \qquad\qquad\qquad \ddots \qquad\qquad \ddots & \vdots \\
& \qquad\qquad x_{1n} + \qquad\qquad x_{2n} + \cdots \qquad + x_{nn} = 1 & (n+n) \\
\end{array}
$$

$$\left.\begin{array}{}\\\\\\\\\end{array}\right\} \text{each resource assigned to one task}$$

$$\left.\begin{array}{}\\\\\\\\\end{array}\right\} \text{each task assigned one resource}$$

$$x_{ij} = 0 \text{ or } 1 \text{ for all } i \text{ and } j$$

must equal the number of tasks for purposes of solving the problem. As with the transportation model, this condition might have to be artificially imposed for a given problem. Finally, notice that all right-hand-side constants, which are equivalent to s_i and d_j values in the transportation model, equal 1. The supply of each resource is 1 unit and the demand for each task is 1 unit.

EXAMPLE 3   **(NCAA Referee Assignments; Motivating Scenario)** The National Collegiate Athletic Association (NCAA) Division I basketball tournament is under way. In preparing for the four regional tournaments, the Tournament Committee has selected four teams of referees who have been judged to be the most qualified for this year's tourney. Each team of officials consists of six regular referees and two alternates, who also travel to the regional tournament in the event that illness, injury, or other circumstances preclude the participation of any of the regular officials. The committee has selected the teams in such a way that each official is from a different athletic conference. (This is to give the appearance of unbiased officiating.)

Officials are paid a standard tournament fee plus travel expenses. The travel expenses vary, depending upon the regional tournament location they are assigned. Table 12.26 summarizes the estimated travel expenses for each team according to regional tournament assigned. The committee wants to assign the four teams of officials so as to minimize total travel expenses.

TABLE 12.26	*Regional Tournament Assignment*			
Team of Officials	**(1)** East	**(2)** Midwest	**(3)** Far West	**(4)** Southwest
1	$6,600	$7,200	$6,750	$7,050
2	6,400	6,800	7,250	7,400
3	6,950	7,000	7,400	6,950
4	7,600	6,900	7,300	7,000

This problem can be formulated as an assignment model. Let

$$x_{ij} = \begin{cases} 1 & \text{if team } i \text{ is assigned to tournament } j \\ 0 & \text{if team } i \text{ is not assigned to tournament } j \end{cases}$$

The problem formulation is as follows:

Minimize $z = 6{,}600x_{11} + 7{,}200x_{12} + 6{,}750x_{13} + 7{,}050x_{14} + 6{,}400x_{21} + \cdots + 7{,}000x_{44}$

subject to

$x_{11} + x_{12} + x_{13} + x_{14}$				$= 1$ (1)
	$x_{21} + x_{22} + x_{23} + x_{24}$			$= 1$ (2)
		$x_{31} + x_{32} + x_{33} + x_{34}$		$= 1$ (3)
			$x_{41} + x_{42} + x_{43} + x_{44} = 1$	(4)
$x_{11} +$	$x_{21} +$	$x_{31} +$	x_{41}	$= 1$ (5)
$x_{12} +$	$x_{22} +$	$x_{32} +$	x_{42}	$= 1$ (6)
$x_{13} +$	$x_{23} +$	$x_{33} +$	x_{43}	$= 1$ (7)
$x_{14} +$	$x_{24} +$	$x_{34} +$	$x_{44} = 1$	(8)

$$x_{ij} = 0 \text{ or } 1 \text{ for all } i \text{ and } j$$

Constraints (1) to (4) assure that each team of officials is assigned to one tourney site only; constraints (5) to (8) assure that each site is assigned exactly one team of officials.

◻

Solution Methods

Assignment models can be solved using various procedures. These include *total enumeration of all solutions, 0-1 programming methods,* the *simplex method, transportation methods (stepping stone),* and *special-purpose algorithms.* These techniques are listed in an order reflecting *increasing* efficiency. Assignment models have been distinguished from the standard linear programming models and transportation models because of their special structure. Again, these models lend themselves to more efficient solution procedures by special-purpose algorithms which are designed to exploit this special structure. One of the most popular methods is the ***Hungarian method,*** discussed in the next section.

Figure 12.3 illustrates a computer package for solving assignment models.* The problem solved is the referee assignment formulation in Example 3. Notice that travel expenses are minimized at a value of $27,000, when team 1 is assigned to the Far West regional tournament, team 2 to the East regional tournament, team 3 to the Southwest regional tournament, and team 4 to the Midwest regional tournament.

---

* "ASGT1" from Warren J. Erikson and Owen P. Hall, Jr., *Computer Models for Management Science,* 2d ed., Addison-Wesley, Reading, Mass., 1986.

```
C O M P U T E R   M O D E L S   F O R   M A N A G E M E N T   S C I E N C E

ASSIGNMENT MODEL                              03-29-1992 - 10:24:36


                -=*=-   INFORMATION ENTERED  -=*=-



        TOTAL NUMBER OF ROWS              :        4
        TOTAL NUMBER OF COLUMNS           :        4
        PROBLEM TYPE                      :     MINIMIZATION

                              PAYOFF VALUES

             Eas          Mid          Wst          Sws
      Tm1   6600.000    7200.000     6750.000     7050.000
      Tm2   6400.000    6800.000     7250.000     7400.000
      Tm3   6950.000    7000.000     7400.000     6950.000
      Tm4   7600.000    6900.000     7300.000     7000.000




                    -=*=-   RESULTS   -=*=-

                        ROW ASSIGNMENTS


      EasMidWstSws

  Tm1   -   -   A   -
  Tm2   A   -   -   -
  Tm3   -   -   -   A
  Tm4   -   A   -   -


                    TOTAL PAYOFF :  27000



     ----------  E N D   O F   A N A L Y S I S  ----------
```

Figure 12.3 Computer solution to referee assignment model (Example 3).

The Hungarian Method

In this section we will illustrate the Hungarian method, which is a special-purpose
algorithm used to solve assignment models. This algorithm capitalizes upon the
special structure of assignment models, providing a relatively efficient solution
procedure compared with other approaches mentioned in the previous section.

The Hungarian method is based on the concept of opportunity costs. There are
three steps in implementing the method. First, an opportunity cost table is constructed from the table of assignment costs. Second, it is determined whether an
optimal assignment can be made. If an optimal assignment cannot be made, the
third step involves a revision of the opportunity cost table. Let's illustrate the
algorithm with the following example.

EXAMPLE 4 **(Court Scheduling)** A court administrator is in the process of scheduling four court
dockets. Four judges are available to be assigned, one judge to each docket. The court
administrator has information regarding the types of cases on each of the dockets as well as
data indicating the relative efficiency of each of the judges in processing different types of

| TABLE 12.27 | **Estimated Days to Clear Docket** |

Estimated Days to Clear Docket

	Docket			
Judge	1	2	3	4
1	14	13	17	14
2	16	15	16	15
3	18	14	20	17
4	20	13	15	18

court cases. Based upon this information, the court administrator has compiled the data in Table 12.27. Table 12.27 shows estimates of the number of court-days each judge would require in order to completely process each court docket. The court administrator would like to assign the four judges so as to minimize the total number of court-days needed to process all four dockets.

SOLUTION

❏ **Step 1: Determine the opportunity cost table.** To determine the opportunity cost table, two steps are required. First, the least cost element in each row is identified and subtracted from all elements in the row. The resulting cost table is sometimes called the *row-reduced cost table*. Having accomplished this, the least cost element in each column is identified and subtracted from all other elements in the column, resulting in the *opportunity cost table*. Table 12.28 is the row-reduced cost table for our example. Table 12.29 is the opportunity cost table.

TABLE 12.28 Row-Reduced Cost Table

	Docket			
Judge	1	2	3	4
1	1	0	4	1
2	1	0	1	0
3	4	0	6	3
4	7	0	2	5

TABLE 12.29 Opportunity Cost Table

	Docket			
Judge	1	2	3	4
1	0	0	3	1
2	0	0	0	0
3	3	0	5	3
4	6	0	1	5

❏ **Step 2: Determine whether an optimal assignment can be made.** The technique for determining whether an optimal assignment is possible at this stage consists of drawing straight lines (vertically and horizontally) through the opportunity cost table in such a way as to minimize the number of lines necessary to cover all zero entries. If the number of lines equals either the number of rows or the number of columns in the table, an optimal assignment can be made. If the number of lines is less than the number of rows or columns, an optimal assignment cannot be determined, and the opportunity cost table must be revised. Table 12.30 illustrates this step as applied to Table 12.29. As shown, three lines are required to cover all zeros in the table, and an optimal assignment cannot be determined.

❏ **Step 3: Revise the opportunity cost table.** If it is not possible to determine an optimal assignment in step 2, the opportunity cost table must be modified.

TABLE 12.30

Judge	Docket 1	2	3	4
1	0	0	3	1
2	0	0	0	0
3	3	0	5	3
4	6	0	1	5

This is accomplished by identifying the smallest number in the table not covered by a straight line and subtracting this number from all numbers not covered by a straight line. Also, this same number is *added* to all numbers lying at the *intersection* of any two lines. Looking at Table 12.30, the smallest element not covered by a straight line is the 1 at the intersection of row 4 and column 3. If this number is subtracted from all elements not covered by straight lines and if it is added to the numbers found at the intersection of any two straight lines, the revised opportunity cost table is as shown in Table 12.31. Step 2 is repeated at this point.

TABLE 12.31 **Revised Opportunity Cost Table**

Judge	Docket 1	2	3	4
1	0	1	3	1
2	0	1	0	0
3	2	0	4	2
4	5	0	0	4

❏ **Step 2: Determine whether an optimal assignment can be made.** Repeating step 2, Table 12.31 shows that four lines are required to cover all of the zero elements. Thus an optimal assignment can be made. The optimal assignments may not be apparent from the table. One procedure for identifying the assignments is *to select a row or column in which there is only one zero, and make an assignment to that cell.* There is only one zero in column 4. Thus the first assignment is judge 2 to docket 4. Since no other assignments can be made in row 2 or column 4, we cross them off. With row 2 and column 4 crossed off, look for a row or column in which there is only one zero. As can be seen in Table 12.32, the re-

TABLE 12.32 **Optimal Assignments**

Judge	Docket 1	2	3	4	Final Assignments	Days
1	0	1	3	1	Judge 1 — Docket 1	14
2	0	1	0	0	Judge 2 — Docket 4	15
3	2	0	4	2	Judge 3 — Docket 2	14
4	5	0	0	4	Judge 4 — Docket 3	15
					Total days	58

maining assignments are apparent, since there are three zero elements that are the only such elements in a row or column. Thus optimal assignment of the four judges results in 58 judge-days to clear the four dockets.

❏

Summary of the Hungarian Method

❏ **Step 1:** Determine the opportunity cost table.

  a. Determine the row-reduced cost table by subtracting the least cost element in each row from all elements in the same row.

  b. Using the row-reduced cost table, identify the least cost element in each column, and subtract from all elements in that column.

❏ **Step 2:** Determine whether or not an optimal assignment can be made. Draw the minimum number of straight lines necessary to cover all zero elements in the opportunity cost table. If the number of straight lines is less than the number of rows (or columns) in the table, the optimal assignment cannot be made. Go to step 3. If the number of straight lines equals the number of rows (columns), the optimal assignments can be identified.

❏ **Step 3:** Revise the opportunity cost table. Identify the smallest element in the opportunity cost table not covered by a straight line.

  a. Subtract this element from every element not covered by a straight line.

  b. Add this element to any element(s) found at the intersection of two straight lines.

  c. Go to step 2.

The Hungarian method can be used when the objective function is to be maximized. Two alternative approaches can be used in this situation. The signs on the objective function coefficients can be changed, and the objective function can be minimized; or opportunity costs can be determined by subtracting the largest element (e.g., profit) in a row or column rather than the smallest element.

Section 12.3 Follow-up Exercises

1 Solve the NCAA referee problem (Example 3) using the Hungarian method.

2 **Court Scheduling** A district court administrator wants to assign five judges to five court dockets. The objective is to minimize the total time required to complete all of the cases scheduled on the five dockets. The administrator has made estimates of the number of days which would be required for each judge to clear each different docket. These estimates are shown in Table 12.33 and are based on the composition of case types on each docket and an analysis of the record of each judge and his or her experience in being able to complete different types of cases.

 (*a*) Formulate the objective function and constraint set.

 (*b*) Solve using the Hungarian method.

 (*c*) If a computerized assignment model package (or a transportation or linear programming package) is available, solve for the optimal solution.

3 **Charter Airplane Assignment** An airline company has five airplanes available for charter flights this weekend. Five organizations have requested the use of a plane. Based

TABLE 12.33

			Docket		
Judge	1	2	3	4	5
1	20	18	22	24	21
2	18	21	26	20	20
3	22	26	27	25	19
4	25	24	22	24	18
5	23	20	25	23	22

on an analysis of the expected revenues from each requested charter and of estimated operating expenses, the company has arrived at expected profit figures resulting from the assignment of each of the five planes to each proposed charter. These figures are displayed in Table 12.34.

If the objective is to assign the five airplanes to five of the charters:

(a) Formulate the objective function and constraints.
(b) Solve using the Hungarian method.
(c) If a computer package is available, solve for the optimal assignment.

TABLE 12.34

	Charter Request				
Airplane	1	2	3	4	5
1	$2,500	$1,000	$2,800	$3,200	$3,500
2	1,800	2,800	4,300	2,700	3,400
3	2,300	1,800	4,000	2,800	3,600
4	3,000	2,100	2,000	2,500	3,000
5	2,800	2,500	2,700	3,000	2,500

4 Sales Force Assignment A college publisher wishes to assign five sales representatives to five sales districts. Management has estimated the annual sales (in thousands of dollars) each representative would generate if assigned to the different districts. These are summarized in Table 12.35. Management wishes to assign the five representatives to the five sales districts in such a way as to maximize total annual sales.

(a) Formulate the objective function and constraints.
(b) Solve using the Hungarian method.
(c) If a computer package (linear programming, transportation, or assignment) is available, solve for the optimal assignment.

TABLE 12.35 | **Estimated Annual Sales, $ Thousands**

	Sales District				
Sales Representative	1	2	3	4	5
1	125	140	90	150	110
2	180	190	160	175	200
3	140	250	240	265	210
4	220	200	240	250	225
5	275	300	260	290	310

❏ **KEY TERMS AND CONCEPTS**

alternative optimal solution 538
assignment model 543
destination 524
Hungarian method 546
improvement index 533
northwest corner method 529

opportunity cost table 547
origin 524
row-reduced cost table 547
stepping stone algorithm 531
transportation model 524

❏ **ADDITIONAL EXERCISES**

Section 12.2

1 Given the data for a transportation problem in Table 12.36:
  (*a*) Use the northwest corner method to determine an initial solution.
  (*b*) Proceed on to solve for the optimal solution using the stepping stone algorithm.

TABLE 12.36		*Destination*			
Origin	**1**	**2**	**3**	**Supply**	
1	50	80	60	250	
2	30	70	40	350	
3	90	40	80	400	
Demand	400	250	350		

2 Given the data for a transportation problem in Table 12.37:
  (*a*) Use the northwest corner method to determine an initial solution.
  (*b*) Proceed on to solve for the optimal solution using the stepping stone algorithm.

TABLE 12.37		*Destination*			
Origin	**1**	**2**	**3**	**Supply**	
1	10	15	5	350	
2	30	10	25	500	
3	20	30	15	650	
Demand	600	400	500		

3 Maximization Problems If the objective function is to be maximized in a transportation problem, the stepping stone algorithm applies just as in minimization problems. The only difference is in the interpretation of the improvement index. In a maximization problem, there is a better solution than the current one if any improvement indices are positive. If more than one improvement index is positive, the entering cell is identified by the most positive improvement index. An optimal solution has been found when all improvement indices are nonpositive.

In Exercise 1, assume that the objective function is to be maximized. Solve for the optimal solution using the stepping stone algorithm.

4 Re-solve Exercise 2 if the objective function is to be maximized.

TABLE 12.38	Bid Price per Sheet						
Laundry Company	Nursing Home						Weekly Capacity, Sheets
	1	2	3	4	5		
1	$0.25	$0.28	$0.20	$0.30	$0.24		4,000
2	$0.24		$0.25	$0.28			3,000
3		$0.25	$0.23	$0.30	$0.26		2,800
Weekly needs, sheets	1,500	2,000	1,200	2,400	1,000		

5 **Nursing Home Management** A corporation has responsibility for managing five area nursing homes. One of the advantages of this type of arrangement is the cost-effectiveness of centralized purchasing of supplies for the five nursing homes. The corporation has requested bids for supplying clean bed linens for the five homes. Table 12.38 summarizes the bids submitted by three laundry companies. The price per sheet includes pickup and delivery costs. Two of the laundry firms have not submitted bids for all five nursing homes. The nursing homes excluded lie outside their delivery areas.

The laundry firms may receive contracts to supply some or all of the needs of any one nursing home. The problem is to determine the number of sheets per week to be supplied by each laundry for each nursing home so as to minimize total weekly cost.

(a) Formulate the objective function and constraint set.
(b) By trial and error estimate what you believe is a good solution.
(c) If a computerized transportation (or linear programming) package is available, solve for the optimal solution.

6 **Gasoline Reallocation** A major oil company is planning for the usual increase in vacation driving which occurs during the summer. In an effort to accommodate motorists' needs better, the company has surveyed its regional distributors. The survey revealed that four regions expect to have surplus supplies of gasoline, while six regions are expected to fall short of their needs. The company wants to develop a plan for reallocating supplies from surplus regions to shortage regions. Table 12.39 indicates the surplus and shortage quantities by region and the cost (stated in hundreds of dollars) per 100,000 gallons of diverting gasoline from one region to another. The problem is to determine the amount of gasoline to be reallocated from each surplus region to each shortage region so as to minimize the total cost.

TABLE 12.39	Reallocation Costs, $ Hundreds						
Surplus Region	Shortage Region						Surplus, 100,000's of Gallons
	1	2	3	4	5	6	
1	25	18	20	30	15	12	40
2	20	25	30	28	26	18	25
3	16	20	25	15	24	30	60
4	18	23	20	10	15	25	100
Shortage, 100,000's of gallons	15	20	40	30	10	25	

(a) Formulate the objective function and constraint set for this model.
(b) If a computerized transportation (or linear programming) package is available, solve for the optimal solution.

Section 12.3

7 Software Development The director of data processing for a consulting firm wants to assign four programming tasks to four of her programmers. She has estimated the total number of days each programmer would take if assigned each of the programs. Table 12.40 summarizes these estimates.

If the objective is to assign one programmer per task in such a way as to minimize the total number of days required to complete the tasks:
(a) Formulate the objective function and constraints.
(b) Solve using the Hungarian method.
(c) If a computer package is available, solve for the optimal assignment.

TABLE 12.40	**Estimated Days per Programming Task**			
	Programming Task			
Programmer	**1**	**2**	**3**	**4**
1	45	50	38	56
2	42	53	34	60
3	50	48	40	62
4	48	47	36	58

8 Detective Assignment A police chief wishes to assign five detective teams to five groups of unsolved crimes. The five teams are different with respect to number of detectives, years of experience, and methods of operation. After analyzing the types of cases in each of the five groups, the chief has estimated the percentage of cases each detective team would solve if assigned to each group. These estimates are shown in Table 12.41. The number of outstanding crimes in the five groups is 40, 30, 25, 50, and 20, respectively. If the objective is to assign the five teams in such a way as to maximize the *total number of crimes solved:*

(a) Formulate the objective function and constraints.
(b) Solve using the Hungarian method.
(c) If a computer package is available, solve for the optimal assignment.

TABLE 12.41	**Expected Percentage of Crimes Solved**				
Detective Team	**Group of Unsolved Crimes**				
	1	**2**	**3**	**4**	**5**
1	0.30	0.40	0.25	0.30	0.20
2	0.40	0.35	0.40	0.25	0.30
3	0.35	0.50	0.40	0.40	0.45
4	0.50	0.55	0.30	0.20	0.40
5	0.45	0.30	0.50	0.50	0.35

9 Table 12.42 summarizes the costs of assigning four resources to four tasks. Solve for the minimum-cost solution using the Hungarian method.

TABLE 12.42

Resource	Task 1	2	3	4
1	24	20	26	22
2	30	22	24	26
3	28	25	28	24
4	26	28	27	25

❏ CHAPTER TEST

1 Given the data for the following transportation problem, determine a solution using the northwest corner method. What are the total costs associated with the solution?

Origin	Destination 1	2	3	4	Supply
1	30	50	25	20	1200
2	40	30	35	60	1500
3	25	75	40	50	2400
4	60	15	50	30	1000
Demand	800	1900	2000	1400	

2 Solve the following transportation problem using the stepping stone algorithm.

Origin	Destination 1	2	3	4	Supply
1	20	27	35	15	300
2	18	31	23	19	500
3	12	17	14	28	450
Demand	350	400	300	200	

3 The following table summarizes the number of days required for four persons to complete four different tasks. Determine the optimal assignment of persons to tasks if the objective is to minimize the total number of days to complete all four jobs.

Person	Job 1	2	3	4
1	8	20	15	17
2	15	16	12	10
3	22	19	16	30
4	25	15	12	9

MINICASE

DISTRIBUTION WAREHOUSING

A company distributes a product from three existing warehouses to eight demand points around the country. The capacity of the three existing warehouses is almost exhausted and the company is considering adding one or more new warehouses. Table 12.43 summarizes relevant information regarding the distribution of the product. Shown in Table 12.43 are the distribution costs per unit from each existing and proposed warehouse to each demand point, monthly capacities of the warehouses, monthly overhead expenses associated with operating each warehouse, and monthly demands at the eight demand points. The company wants to determine whether to add any of the proposed warehouses and what quantities should be distributed each month from each warehouse to each demand point. The objective is to minimize the sum of monthly distribution costs and monthly overhead costs. Evaluate the four possible warehouse configurations (the existing three alone and the existing three plus one or both of the proposed warehouses). *When performing the analysis, the overhead costs will have to be added manually to the minimum distribution costs obtained from the computer for each warehouse configuration.* When summarizing the results for each warehouse configuration, include the following:

a. The recommended quantities to ship from each warehouse to each demand point.

b. Total monthly distribution cost, total monthly overhead cost, and total monthly cost.

c. Unused capacity each month for the warehouses included.

TABLE 12.43		*Demand Point*								Monthly Capacity	Monthly Overhead
		1	2	3	4	5	6	7	8		
Existing warehouse	1	$16	$12	$22	$18	$10	$8	$15	$20	18,000	$60,000
	2	20	16	19	12	13	6	20	18	24,000	$70,000
	3	18	15	16	22	9	6	24	25	32,000	$75,000
Proposed warehouse	4	12	10	16	8	5	6	14	15	30,000	$40,000
	5	14	12	14	6	7	5	16	12	36,000	$42,000
Monthly demand:		8,000	10,000	15,000	12,000	5,000	4,000	13,000	5,000		

CHAPTER 13

INTRODUCTION TO PROBABILITY THEORY

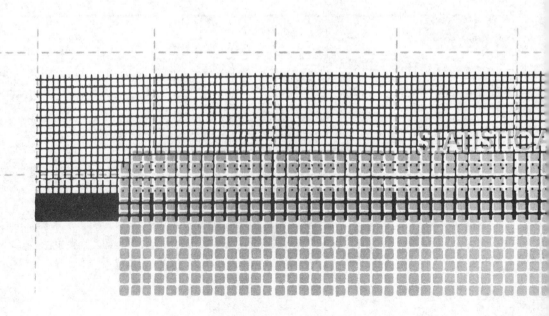

CHAPTER OBJECTIVES

❏ Review the fundamentals of set theory and set operations

❏ Provide an understanding of fundamental counting methods including permutations and combinations

❏ Introduce the notion of probability and the computation of probabilities for selected statistical environments

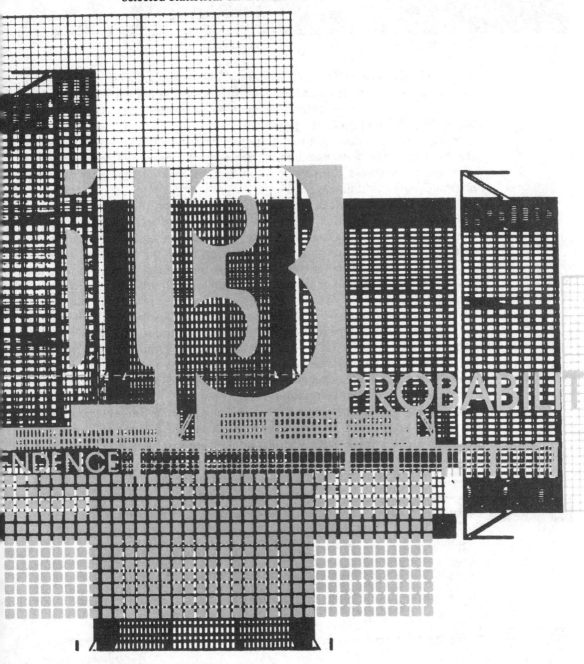

MOTIVATING SCENARIO: IRS Audit Possibilities

The Internal Revenue Service (IRS) estimates the probability of an error on personal income tax returns to be .4. Suppose that an experiment is conducted in which three returns are selected at random for purposes of an audit. The results of auditing an individual tax return are that "it contains no errors" or "it contains errors." **In a sample of three tax returns, we want to determine all possible outcomes and the probabilities of occurrence of these outcomes (Example 30).**

Much in life is characterized by uncertainty. Many phenomena in our world seem to be characterized by random behavior. Most decisions are made in an environment characterized by an absence of complete or perfect knowledge. A decision about the number of units of a product to produce is based upon *estimates* of the number of units expected to be sold. If the number to be sold was known in advance, the decision would be to manufacture precisely this quantity, incurring neither shortages nor overages. However, in actual decision-making situations exact information is rarely obtainable.

Probability concepts can be very useful in dealing with the uncertainty which characterizes most decision-making environments. Probability theory takes advantage of the fact that, for many uncertain phenomena, there are "long run" patterns. For instance, on the single flip of a fair coin it is uncertain whether a head or a tail will occur. However, with many flips of the same coin — over the long run — approximately half of the outcomes should be "heads" and approximately half "tails."

Medical research heavily relies upon long-run observations to generalize results. Table 13.1 shows the results of a study which examined the effectiveness of a new drug in reducing the recurrence of heart attacks. Persons who survived a first heart attack were given the drug for a period of 5 years. Table 13.1 summarizes the results of the study as the sample of persons increased. Notice how the *percentage*, or proportion, having no recurrence fluctuates, yet tends to stabilize, as the sample increased in size. The uncertainty of the effectiveness of the drug makes it impossible to know what the experience will be for any one person. However, the evidence after 10,000 observations suggests that approximately 63.5 percent of heart attack victims using this particular drug will not have another attack in the 5 years

TABLE 13.1	Persons Receiving the Drug, Number	Persons Having No Recurrence of Heart Attack, Number	Persons Having No Recurrence, Percentage
	100	64	64
	300	196	65.3
	500	317	63.4
	1,000	636	63.6
	5,000	3,177	63.54
	10,000	6,351	63.51

following a first heart attack. These results have greater meaning when compared with the experiences for a second sample of heart attack victims who were not given the same drug. For 10,000 persons in this second group, the percentage having no recurrence was 45.26. The comparative results are encouraging regarding the potential for the new drug.

The purpose of this chapter is to introduce the fundamentals of probability theory. Because *set theory* provides a useful vehicle for presenting and discussing probability concepts, the first section in this chapter provides a brief review of set concepts and set operations. The second section discusses special counting methods which are quite useful in probability theory. The next section introduces fundamental concepts of probability and the computation of probabilities. The final section discusses the computation of probabilities under conditions of *statistical independence* and *statistical dependence.*

13.1 INTRODUCTION TO SETS AND SET OPERATIONS

Although we used some set notation earlier when discussing *solution sets* for linear equations, we will now present a more formal review of set concepts and set operations.

Sets

A *set* is any *well-defined* collection of objects. Examples of sets include the *set* of students enrolled in History 100 at State University, the *set* of U.S. citizens who are 65 years of age or older, the *set* of NBA players who average over 20 points and over 12 rebounds per game during a given season, the *set* of cities being considered for the next World's Fair, and the *set* of real numbers. Frequently, objects belong to a set because they possess a certain attribute or attributes required for membership in the set. For a State University student to be a member of the above-mentioned set, he or she must be enrolled in History 100. For an NBA player to be a member of the previously mentioned set, he must have averaged over 20 points *and* over 12 rebounds per game during the specified season. The objects which belong to a set are called **elements** of the set.

Membership in a set is usually defined in one of two ways. The **enumeration method** simply lists all elements in a set. If we designate a set name by an uppercase letter, we might define the set of positive odd integers having a value less than 10 as

$$A = \{1, 3, 5, 7, 9\}$$

Note the use of *braces* to group the elements or members of *the set A*.

The enumeration method is convenient when the number of elements in a set is small or when it is not easy or possible to define a property which specifies the requirement(s) for membership in the set. An alternative approach to defining sets is the **descriptive property method.** With this approach, the set is defined by stating the property required for membership in the set. The set A, defined in the example above, can be redefined as "such that"

$$A = \{x | x \text{ is a positive odd integer less than } 10\}$$

Verbally, the translation of this equation is "*A is the set consisting of all elements x 'such that' (the vertical line) x is a positive odd integer having a value less than 10.*" The x to the left of the vertical line indicates the general notation for an element of the set; the expression to the right of the vertical line states the condition(s) required of an element for membership in the set.

To indicate that an object e is a member of a set S, we use the notation

$$e \in S$$

Verbally, this notation translates as "*e is a member (or element) of the set S.*" In the previously defined set A we can say:

$$9 \in A$$

The notation $e \notin S$ means that an object e is not a member of set S.

The *number of elements* contained in a set B is denoted by

$$n(B)$$

Thus for set A, $n(A) = 5$.

Special Sets

There are certain special sets which are referred to frequently when discussing the algebra of sets.

> **DEFINITION: UNIVERSAL SET**
> The *universal set* $\mathcal{U}$ is the set which contains all possible elements within a particular application under consideration.

EXAMPLE 1 If we consider an opinion survey conducted of a random sample of residents of New York City, several sets of persons might be identified. One set might consist of those persons included in the sample; another set would consist of those residents *not* included in the sample. In this application, the universal set $\mathcal{U}$ could be defined as all residents of New York City. ∎

> **DEFINITION: COMPLEMENT**
> The *complement* of a set S is the set of all elements in the universal set that are not members of set S. The complement of set S is denoted by S'.

EXAMPLE 2

In Example 1, if A represents the set of residents included in the survey, the complement of A, denoted A', is the set of all residents of New York who did not participate in the survey.

EXAMPLE 3

If the set S consists of all positive integers and the universal set is defined as all integers, then the complement S' consists of all negative integers and zero.

EXAMPLE 4

If $\mathcal{U} = \{1, 2, 3, 4, 5, 6, 7, 8, 9, 10\}$ and $A = \{1, 3, 5, 7, 9\}$, the complement of set A contains all elements which are members of $\mathcal{U}$ but not A, or $A' = \{2, 4, 6, 8, 10\}$.

❑

> **DEFINITION: NULL SET**
> The *empty*, or *null, set* $\varnothing$ is the set consisting of no elements.

EXAMPLE 5

Consider the system of equations

$$4x + 3y = 10$$
$$4x + 3y = -5$$

Since there are no values for x and y which satisfy both equations, the solution set S for the system of equations is the null set, or

$$S = \varnothing$$

Stated differently,

$$S = \{(x, y) | 4x + 3y = 10 \text{ and } 4x + 3y = -5\}$$
$$= \varnothing$$

❑

> **DEFINITION: SUBSET**
> A set A is a *subset* of a set B if and only if every element of set A is also an element of set B. This subset relationship is denoted as $A \subset B$, which may be read "A is a subset of B."

EXAMPLE 6

Given the following sets,

$$A = \{1, 2, 3, 4, 5, 6, 7, 8, 9, 10\} \qquad C = \{x | x \text{ is a real number}\}$$
$$B = \{1, 3, 5, 7, 9\} \qquad\qquad D = \{z | z - 1 = 4\}$$

we can identify the following subset relationships: $A \subset C, B \subset C, D \subset C, B \subset A, D \subset A$, and $D \subset B$. Verify these for yourself; are there any other subset relationships?

❑

> **NOTE** *By definition*, the null set is a subset of every set. Consequently, in the previous example $\emptyset \subset A$, $\emptyset \subset B$, $\emptyset \subset C$, and $\emptyset \subset D$.

Venn Diagram Representation

Venn diagrams are a convenient way of envisioning set relationships. To illustrate, Fig. 13.1 depicts a universal set $\mathcal{U}$ within which is another set A, depicted by a circular area. Figure 13.2 illustrates the complement relationship. The primary value of these figures is the information they convey about the relationships among sets. For example, if a set B is a subset of another set A, the Venn diagram representation of set B should be contained within set A. In Fig. 13.3, sets A and B are both subsets within the universal set, with set B portrayed as a subset of A.

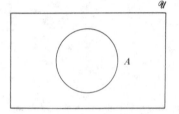

Figure 13.1 Venn diagram: Set A within universal set $\mathcal{U}$.

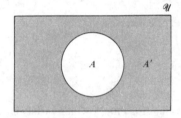

Figure 13.2 Venn diagram relationship of complement.

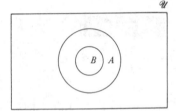

Figure 13.3 Venn diagram representation of subset relationships: $A \subset \mathcal{U}$, $B \subset \mathcal{U}$, $B \subset A$.

Set Operations

Just as there are arithmetic operations which provide the foundation for algebra, trigonometry, and other areas of study in mathematics, there is an arithmetic of set theory which allows for the development of an algebra of sets.

> **DEFINITION: SET EQUALITY**
> Two sets A and B are equal if and only if (iff) every element of A is an element of B and every element of B is an element of A. Stated symbolically,
>
> $$A = B \quad \text{iff} \quad A \subset B \ \text{ and } \ B \subset A$$

EXAMPLE 7   Given the following sets, determine whether any sets are equal.

$$A = \{1, 2\} \qquad\qquad C = \{1, 2, 3\}$$

$$B = \{x | (x - 1)(x - 2)(x - 3) = 0\} \qquad D = \{x | x^2 - 3x + 2 = 0\}$$

SOLUTION

Set B can be defined equivalently as $B = \{1, 2, 3\}$. Given this, we can make the statement that set B equals set C, or $B = C$. The roots of the quadratic equation in set D are $x = 1$ and $x = 2$. Thus, set D can be redefined as

$$D = \{1, 2\}$$

and set A equals set D, or $A = D$.

❑

DEFINITION: UNION OF SETS

The **union** of two sets A and B, denoted by $A \cup B$, is a set which consists of all elements contained in either set A or set B or both A and B.

The Venn diagram representation of $A \cup B$ is shown in Fig. 13.4.

Figure 13.4
Union of sets
A and B.

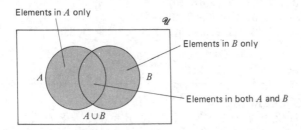

EXAMPLE 8 Given the following sets,

$$A = \{1, 2, 3, 4, 5\}$$

$$B = \{1, 3, 5, 7, 9\}$$

$$C = \{2, 4, 6, 8, 10\}$$

(a) $A \cup B = \{1, 2, 3, 4, 5\} \cup \{1, 3, 5, 7, 9\} = \{1, 2, 3, 4, 5, 7, 9\}$
(b) $A \cup C = \{1, 2, 3, 4, 5\} \cup \{2, 4, 6, 8, 10\} = \{1, 2, 3, 4, 5, 6, 8, 10\}$
(c) $B \cup C = \{1, 3, 5, 7, 9\} \cup \{2, 4, 6, 8, 10\} = \{1, 2, 3, 4, 5, 6, 7, 8, 9, 10\}$

❑

NOTE The union of any set A and its complement results in the universal set $\mathcal{U}$, or $A \cup A' = \mathcal{U}$. Also, $A \cup \varnothing = A$.

DEFINITION: INTERSECTION OF SETS

The *intersection* of two sets A and B, denoted by $A \cap B$, is the set of all elements which belong to *both* A and B.

The Venn diagram representation of the intersection of two sets A and B is the shaded area in Fig. 13.5. Note in the diagram that the intersection is the area *common to* the two sets.

Figure 13.5
Intersection of
sets A and B.

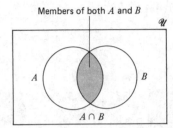

EXAMPLE 9

Given the sets A, B, and C defined in the previous example:
(a) $A \cap B = \{1, 2, 3, 4, 5\} \cap \{1, 3, 5, 7, 9\} = \{1, 3, 5\}$
(b) $A \cap C = \{1, 2, 3, 4, 5\} \cap \{2, 4, 6, 8, 10\} = \{2, 4\}$
(c) $B \cap C = \{1, 3, 5, 7, 9\} \cap \{2, 4, 6, 8, 10\} = \varnothing$
(d) $A \cap A' = \{1, 2, 3, 4, 5\} \cap \{6, 7, 8, 9, 10\} = \varnothing$
(e) $B \cap \mathcal{U} = \{1, 3, 5, 7, 9\} \cap \{1, 2, 3, 4, 5, 6, 7, 8, 9, 10\} = \{1, 3, 5, 7, 9\}$

◻

PRACTICE EXERCISE
Given $A = \{1, 3, 5, 7, -1, -3, -5, -7\}$, $B = \{-3, -2, -1, 0, 1, 2, 3\}$, and $C = \{-4, -3, -1, 2, 4\}$, determine (a) $A \cup C$, (b) $B \cap C$, (c) $A \cap B \cup C$. Answer: (a) $\{-7, -5, -3, -2, -1, 0, 1, 2, 3, 5, 7\}$, (b) $\{-3, -1, 2\}$, (c) $\{-4, -3, -1, 1, 2, 3, 4\}$.

Some other general properties of sets include the following:

1 $\boxed{A \cap A' = \varnothing}$

The intersection of any set A and its complement A' is the null set.
Graphically this is shown in Fig. 13.6. This result was illustrated in Example 9d.

Figure 13.6
$A \cap A' = \varnothing$.

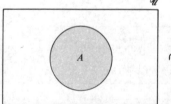 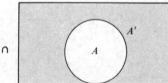

2 $\boxed{A \cap A = A}$

The intersection of any set A and itself is the same set A. *Graphically this is shown in Fig. 13.7.*

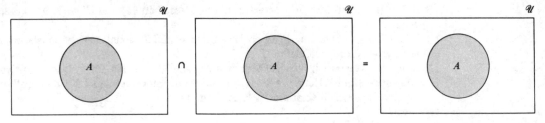

Figure 13.7 $A \cap A = A.$

3

$$A \cap \mathcal{U} = A$$

The intersection of any set A and the universal set $\mathcal{U}$ is the set A. *This is shown graphically in Fig. 13.8. This result was illustrated in Example 9e.*

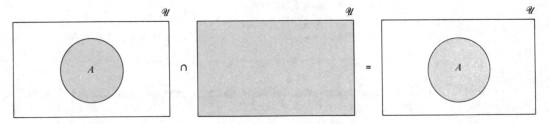

Figure 13.8 $A \cap \mathcal{U} = A.$

Section 13.1 Follow-up Exercises

In Exercises 1–5 redefine each set using the descriptive property method.

1 $A = \{1, 3, 5, 7, 9, 11, 13, 15, 17, 19\}$
2 $S = \{-3, 3, -2, 2, -1, 1, 0\}$
3 $V = \{a, e, i, o, u\}$
4 $S = \{0, -1, -4, -9, -16, -25, -36\}$
5 $C = \{-1, -8, -27, -64\}$

In Exercises 6–10 redefine each set by enumeration.

6 $A = \{a | a$ is a negative odd integer greater than $-10\}$
7 $B = \{b | b$ is a positive integer less than 8$\}$
8 $C = \{c | c$ is the name of a day of the week$\}$
9 $B = \{b | $when $a = 2, a + 3b = -7\}$
10 $M = \{m | m$ is the fourth power of a negative integer greater than $-6\}$
11 If $\mathcal{U} = \{1, 2, 3, 4, 5, 6, 7, 8, 9, 10\}$ and $B = \{b | b$ is a positive even integer less than 10$\}$, define B'.
12 If $\mathcal{U}$ equals the set of students in a mathematics class and P is the set of students who fail the course, define P'.

13 If $\mathcal{U} = \{x|x$ is an integer greater than 6 but less than 14$\}$ and $S' = \{7, 9, 10, 12, 13\}$, define S.

14 If $\mathcal{U}$ is the set consisting of all positive integers and T' equals the set consisting of all positive even integers, define T.

15 If $\mathcal{U} = \{x|x$ is a positive integer less than 20$\}$, $A = \{1, 5, 9, 19\}$, $B = \{b|b$ is a positive odd integer less than 11$\}$, and $C = \{c|c$ is a positive odd integer less than 20$\}$, define all subset relationships which exist among $\mathcal{U}$, A, B, and C.

16 Given

$$\mathcal{U} = \{2, 4, 6, 8, 10, 12, 14, 16, 18\}$$

$$A = \{4, 8, 16\}$$

$$B = \{2, 4, 6, 8, 10\}$$

draw a Venn diagram representing the sets.

17 If $\mathcal{U} = \{x|x$ is a negative integer greater than $-11\}$, $A = \{a|a$ is a negative odd integer greater than $-10\}$, and $B = \{b|b$ is a negative integer greater than $-6\}$, draw a Venn diagram representing the sets.

18 Given the following sets, (a) state which, if any, are equal and (b) define all subset relationships among A, B, C, and D.

$$A = \{3, -4\} \qquad\qquad C = \{x|x^3 + x^2 - 12x = 0\}$$

$$B = \{x|(x-3)(x+4) = 0\} \qquad D = \{0, -4, -3\}$$

19 Given the following sets, (a) state which, if any, are equal and (b) define all subset relationships among A, B, C, and D.

$$A = \{0, 1, -1\} \qquad\qquad C = \{1, 0, -1\}$$

$$B = \{b|b^3 - b = 0\} \qquad D = \{d|-d + d^3 = 0\}$$

20 Given the sets

$$A = \{1, 3, 5, 7\}$$

$$B = \{1, 2, 3, 4, 5, 6, 7, 8, 9, 10\}$$

$$C = \{2, 4, 6, 8\}$$

find:

(a) $A \cup B$ (b) $A \cup C$
(c) $B \cup C$ (d) $A \cap B$
(e) $A \cap C$ (f) $B \cap C$

21 If in Exercise 20 $\mathcal{U} = \{x|x$ is a positive integer$\}$ find:

(a) $A \cap A'$ (b) $A' \cap B'$
(c) $B' \cup C$ (d) $A \cap C'$
(e) $B' \cup A$ (f) $C' \cap A'$

22 Given any set A:

(a) $A \cup \mathcal{U} =$ (b) $A' \cap \mathcal{U} =$
(c) $A \cup \varnothing =$ (d) $A' \cup \varnothing =$
(e) $\qquad \mathcal{U}' =$

13.2 PERMUTATIONS AND COMBINATIONS

In this section, we will focus on various counting arrangements of the elements of a set. To begin this discussion, examine the following example.

EXAMPLE 10

Suppose we are playing a game in which each play involves the roll of a die followed by the flip of a coin. When the die is rolled, there are 6 possible outcomes (1, 2, 3, 4, 5, or 6). When the coin is flipped, the 2 outcomes are heads (H) or tails (T). Let's suppose that we are interested in determining all possible outcomes for this game. One way of approaching this problem is to construct a ***tree diagram*** as shown in Fig. 13.9.

The first outcome in this game is the result of rolling the die. The tree diagram represents the 6 possible outcomes with six branches emanating from left to right. Given any 1 of these outcomes, there are 2 possible outcomes in the second part of the game. These are represented in the tree diagram by pairs of branches emanating from each of the six original branches. What the tree diagram enumerates are the 12 possible combined outcomes for this game.

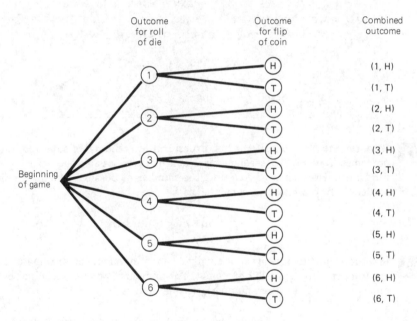

Figure 13.9 Tree diagram for die roll/coin flip example.

PRACTICE EXERCISE
Construct the tree diagram which enumerates all possible outcomes in the previous game if the coin is flipped first, followed by the roll of the die. How many combined outcomes are there?

Tree diagrams can be very effective in portraying and counting the possible outcomes of a sequence of experiments or events. They do have limitations when the number of possible outcomes increases. If we were to draw a tree diagram corresponding to the possible outcomes in the drawing of a six-digit lottery number, imagine the size and number of branches. However, we might indeed want to evaluate the number of possible outcomes to assess our chances of winning such a lottery. Fortunately, there are more efficient ways of counting these possibilities.

FUNDAMENTAL PRINCIPLE OF COUNTING
(Multiplication Rule)

I   *If two experiments are performed in order, and there are n_1 possible outcomes for the first experiment and n_2 possible outcomes for the second experiment, then there are $n_1 \cdot n_2$ possible combined outcomes. An implicit assumption is that the outcome in the first experiment in no way influences the outcome of the second experiment.*

II   *If N experiments are performed in order, with n_1, n_2, . . . , n_N possible outcomes for the N experiments, then there are*

$$n_1 \cdot n_2 \cdot \cdots \cdot n_N$$

possible combined outcomes.

EXAMPLE 11   If we refer to Example 10, the first experiment was the roll of a die and it had six possible outcomes ($n_1 = 6$). The second experiment had two possible outcomes ($n_2 = 2$). Using the fundamental principle of counting, the number of possible combined outcomes for the sequence of experiments is computed as

$$n_1 \cdot n_2 = (6)(2) = 12$$

EXAMPLE 12   If we refer to the drawing of a six-digit lottery number, the drawing of each digit is an experiment with 10 possible outcomes. The number of possible combined outcomes (or the number of possible lottery numbers) is computed as

$$n_1 \cdot n_2 \cdot n_3 \cdot n_4 \cdot n_5 \cdot n_6 = 10 \cdot 10 \cdot 10 \cdot 10 \cdot 10 \cdot 10 = 10^6$$
$$= 1,000,000$$

(What do you think about the chance of winning?)

EXAMPLE 13   **(Surgical Team)**   The chief surgeon for an upcoming transplant operation is preparing to select the supporting surgical team. Needed will be one assisting surgeon, one resident surgeon, one anesthesiologist, one surgical nurse, one assisting nurse, and one orderly. Given the date of the surgery, the chief surgeon can select from three assisting surgeons, seven resident surgeons, five anesthesiologists, six surgical nurses, ten assisting nurses, and five orderlies. How many possible supporting surgical teams can be selected?

SOLUTION

If we apply the fundamental principle of counting, the number of possible teams equals

$$3 \cdot 7 \cdot 5 \cdot 6 \cdot 10 \cdot 5 = 31,500$$

□

Having discussed the fundamental principle of counting, we will now use this principle in developing two other important counting methods.

Permutations

A *permutation* is an ordered arrangement of a set of items. Consider the three numerals 1, 2, and 3. One permutation of these numerals is 123. Another permutation is 132. All the different permutations of these three numerals are

$$123 \quad 132 \quad 213 \quad 231 \quad 312 \quad 321$$

RULE 1: PERMUTATION COUNTING
The number of permutations of n different items *taken n at a time* is denoted by $_nP_n$, where

$$_nP_n = n(n-1)(n-2) \cdots 2 \cdot 1 \qquad (13.1)$$

The logic underlying Eq. (13.1) is that in selecting the first of the n items, there are n choices. Once the first item is selected, there remain $n-1$ choices for the second item, or $n(n-1)$ possible choices for the first two items. Following selection of the second item there are $n-2$ choices for the third item, or $n(n-1)(n-2)$ possible choices for the first three items. This logic concludes that having selected the first $n-1$ items, there remains but one choice for the nth item. Equation (13.1) can be rewritten using *factorial notation* as

$$_nP_n = n! \qquad (13.2)$$

The notation $n!$ (read "n factorial") is a shorthand way of representing the product on the right side of Eq. (13.1). For example, "5 factorial" is expressed as

$$5! = 5 \cdot 4 \cdot 3 \cdot 2 \cdot 1$$

and $$10! = 10 \cdot 9 \cdot 8 \cdot 7 \cdot 6 \cdot 5 \cdot 4 \cdot 3 \cdot 2 \cdot 1$$

By definition $$0! = 1$$

The number of different permutations of the three numerals 1, 2, and 3 taken three at a time is $_3P_3 = 3! = 3 \cdot 2 \cdot 1 = 6$. Note that this equals the number of different permutations enumerated previously.

EXAMPLE 14   Six football teams compete in a particular conference. Assuming no ties, how many different end-of-season rankings are possible in the conference?

SOLUTION
The number of different rankings is

$$_6P_6 = 6!$$
$$= 6 \cdot 5 \cdot 4 \cdot 3 \cdot 2 \cdot 1 = 720$$

□

RULE 2: PERMUTATION COUNTING
The number of permutations of n different objects *taken r at a time* is denoted by $_nP_r$, where

$$_nP_r = \underbrace{n(n-1)(n-2) \cdots (n-r+1)}_{r\ factors} \qquad (13.3)$$

The logic underlying Eq. (13.3) is similar to that for Eq. (13.1). However, once $r - 1$ items have been selected, the number of different choices for the rth item equals $n - (r - 1)$, or $n - r + 1$.

An alternative statement of Eq. (13.3) is obtained by multiplying the right side of Eq. (13.3) by $(n - r)!/(n - r)!$. Thus,

$$_nP_r = n(n-1)(n-2) \cdots (n-r+1) \cdot \frac{(n-r)!}{(n-r)!}$$

$$= n(n-1)(n-2) \cdots (n-r+1) \frac{(n-r)(n-r-1) \cdots 1}{(n-r)(n-r-1) \cdots 1}$$

or

$$_nP_r = \frac{n!}{(n-r)!} \qquad (13.4)$$

EXAMPLE 15   A person wishes to place a bet which selects the first three horses to finish a race in their correct order of finish. If eight horses are in the race, how many different possibilities exist for the first three horses (assuming no ties)?

SOLUTION
According to Eq. (13.3), the number of possibilities is

$$_8P_3 = 8 \cdot 7 \cdot 6 = 336$$

or, according to Eq. (13.4),

$$_8P_3 = \frac{8!}{(8-3)!}$$

$$= \frac{8 \cdot 7 \cdot 6 \cdot 5 \cdot 4 \cdot 3 \cdot 2 \cdot 1}{5 \cdot 4 \cdot 3 \cdot 2 \cdot 1} = 8 \cdot 7 \cdot 6 = 336$$

The logic underlying this computation is that there are eight possibilities for first place. Given these eight, there remain seven possibilities for second place; and, given the selection of first and second places, there are six choices for third place.

EXAMPLE 16

A presidential candidate would like to visit seven cities prior to the next primary election date. However, it will be possible for him to visit only three of the cities. How many different itineraries can he and his staff consider?

SOLUTION

Because "order of visits" is important in planning an itinerary, the number of possible itineraries is equal to the number of permutations of seven cities taken three at a time, or

$$_7P_3 = \frac{7!}{(7-3)!}$$

$$= \frac{7 \cdot 6 \cdot 5 \cdot \cancel{4} \cdot \cancel{3} \cdot \cancel{2} \cdot \cancel{1}}{\cancel{4} \cdot \cancel{3} \cdot \cancel{2} \cdot \cancel{1}}$$

$$= 7 \cdot 6 \cdot 5$$

$$= 210$$

PRACTICE EXERCISE
In Example 16, (*a*) suppose there are eight cities rather than seven. How many different itineraries are there? (*b*) With the original seven cities assume that the candidate wishes to visit four. How many different itineraries are there? *Answer:* (*a*) 336, (*b*) 840.

Combinations

With permutations we are concerned with the number of different ways in which a set of items can be arranged. In many situations there is an interest in the number of ways in which a set of items can be selected without any particular concern for the order or arrangement of the items. For example, one may be interested in determining the number of different committees of three people which can be formed from six candidates. In this instance, the committee consisting of $\{A, B, C\}$ is the same committee as $\{B, A, C\}$, where A, B, and C represent three of the candidates. Order of selection is not significant in determining the number of different committees.

A *combination* is a set of items with no consideration given to the order or arrangement of the items. A combination of *r* items selected from a set of *n* items is a *subset* of the set of *n* items. Consider the set of four letters $\{A, B, C, D\}$. Suppose we are interested in determining the number of combinations of these four letters

ABC	ACB	BAC	BCA	CAB	CBA
ABD	ADB	BAD	BDA	DAB	DBA
ACD	ADC	CAD	CDA	DAC	DCA
BCD	BDC	DBC	DCB	CDB	CBD

Figure 13.10   Permutations of the letters $\{A, B, C, D\}$ taken three at a time.

when taken three at a time. Let's begin by enumerating all possible permutations of these four when taken three at a time. Figure 13.10 displays these permutations. According to Eq. (13.4) there should be, and are, in Fig. 13.10

$$_4P_3 = \frac{4!}{(4-3)!}$$
$$= \frac{4 \cdot 3 \cdot 2 \cdot 1}{1}$$
$$= 24$$

different permutations. Notice that each row in Fig. 13.10 represents the different permutations of three of the four letters; that is, given three different letters, there are $3! = 3 \cdot 2 \cdot 1 = 6$ different arrangements of the three letters. Since a combination of the three letters does not consider order, each row actually represents only one combination. Thus there are only four different combinations of the four letters when taken three at a time.

In general, the relationship between permutations and combinations is

$$_nP_r = {_nC_r} \cdot r!$$

or, when solving for $_nC_r$,

$$_nC_r = \frac{_nP_r}{r!} \tag{13.5}$$

Substituting the right side of Eq. (13.4) into Eq. (13.5),

$$_nC_r = \frac{n!/(n-r)!}{r!}$$

or

$$_nC_r = \frac{n!}{r!(n-r)!}$$

COMBINATIONS COUNTING RULE
The number of different combinations of r items which can be selected from n

different items is more commonly denoted by $\binom{n}{r}$, and

$$\binom{n}{r} = \frac{n!}{r!(n-r)!} \qquad (13.6)$$

To return to the committee example, the number of different combinations of six persons taken three at a time equals

$$\binom{6}{3} = \frac{6!}{3!(6-3)!}$$

$$= \frac{\cancel{6} \cdot 5 \cdot 4 \cdot \cancel{3} \cdot \cancel{2} \cdot \cancel{1}}{\cancel{3} \cdot \cancel{2} \cdot 1 \cdot \cancel{3} \cdot \cancel{2} \cdot \cancel{1}}$$

$$= 20$$

EXAMPLE 17 Super Bowl organizers are selecting game officials. From 12 officials who are eligible, 5 will be selected. How many different teams of 5 officials can be selected from the 12?

SOLUTION

The number of different teams of officials is

$$\binom{12}{5} = \frac{12!}{5!(12-5)!}$$

$$= \frac{12!}{5!7!} = \frac{12 \cdot 11 \cdot 10 \cdot 9 \cdot 8 \cdot \cancel{7} \cdot \cancel{6} \cdot \cancel{5} \cdot \cancel{4} \cdot \cancel{3} \cdot \cancel{2} \cdot \cancel{1}}{5 \cdot 4 \cdot 3 \cdot 2 \cdot 1 \cdot \cancel{7} \cdot \cancel{6} \cdot \cancel{5} \cdot \cancel{4} \cdot \cancel{3} \cdot \cancel{2} \cdot \cancel{1}}$$

$$= 792$$

> **PRACTICE EXERCISE**
> Suppose in the Super Bowl example that 15 officials are eligible and 6 will be selected. How many different teams of officials can be selected? *Answer:* 5005.

EXAMPLE 18 If we refer to the presidential candidate in Example 16, how many combinations of three cities might he and his staff consider visiting?

SOLUTION

In this problem, *order* of visits is not relevant. We are interested in determining the number of different sets of three cities which might be considered. This is a combinations problem, where

$$\binom{7}{3} = \frac{7!}{3!(7-3)!}$$

$$= \frac{7 \cdot \cancel{6} \cdot 5 \cdot \cancel{4} \cdot \cancel{3} \cdot \cancel{2} \cdot \cancel{1}}{\cancel{3} \cdot \cancel{2} \cdot 1 \cdot \cancel{4} \cdot \cancel{3} \cdot \cancel{2} \cdot \cancel{1}}$$

$$= 35$$

Some counting problems require combinations of counting methods, as illustrated by the following example.

EXAMPLE 19   (**Nutrition Planning**) Nutritionists recommend that each adult consume, on a daily basis, a minimum of (*a*) four servings from the milk group of foods, (*b*) two servings from the meat group, (*c*) four servings from the vegetable and fruit group, and (*d*) four servings from the bread and cereal group. Suppose that a chef has six items from the milk group, five items from the meat group, seven items from the vegetable and fruit group, and eight items from the bread and cereal group. If the chef is to provide no more than one serving of any given item during the day, how many different groupings of food could be considered for a given day's menu?

SOLUTION

This problem requires the use of the combinations counting rule and the fundamental principle of counting. To find the solution, we first determine the number of combinations of each food group that can be provided on any given day. For example, the number of different combinations of milk group items is

$$\binom{6}{4} = \frac{6!}{4!(6-4)!} = \frac{6 \cdot 5 \cdot \cancel{4} \cdot \cancel{3} \cdot \cancel{2} \cdot \cancel{1}}{\cancel{4} \cdot \cancel{3} \cdot \cancel{2} \cdot \cancel{1} \cdot 2 \cdot 1} = 15$$

Similarly, the numbers of different combinations for the other three food items are, respectively,

(meat) $$\binom{5}{2} = \frac{5!}{2!(5-2)!} = \frac{5 \cdot 4 \cdot \cancel{3} \cdot \cancel{2} \cdot \cancel{1}}{2 \cdot 1 \cdot \cancel{3} \cdot \cancel{2} \cdot \cancel{1}} = 10$$

(vegetable and fruit) $$\binom{7}{2} = \frac{7!}{4!(7-4)!} = \frac{7 \cdot \cancel{6} \cdot 5 \cdot \cancel{4} \cdot \cancel{3} \cdot \cancel{2} \cdot \cancel{1}}{\cancel{4} \cdot \cancel{3} \cdot \cancel{2} \cdot 1 \cdot \cancel{3} \cdot \cancel{2} \cdot \cancel{1}} = 35$$

(bread and cereal) $$\binom{8}{4} = \frac{8!}{4!(8-4)!} = \frac{8 \cdot 7 \cdot \cancel{6} \cdot 5 \cdot \cancel{4} \cdot \cancel{3} \cdot \cancel{2} \cdot \cancel{1}}{4 \cdot \cancel{3} \cdot \cancel{2} \cdot 1 \cdot \cancel{4} \cdot \cancel{3} \cdot \cancel{2} \cdot \cancel{1}} = 70$$

Given the number of different options which can be selected from each food group, we can use the fundamental principle of counting to determine the total number of different food groups.

$$\text{Total possible food groupings} = \binom{6}{4}\binom{5}{2}\binom{7}{2}\binom{8}{4}$$

$$= (15)(10)(35)(70)$$

$$= 367,500$$

Section 13.2 Follow-up Exercises

1 A game consists of flipping a coin twice. Draw a tree diagram which enumerates all possible combined outcomes for the game.

2 A game consists of flipping a coin three times in a row. Draw a tree diagram which enumerates all possible combined outcomes for the game.

3 **Emissions Control** An automobile inspection station inspects vehicles for level of air pollution emissions. Vehicles either pass (P) or fail (F) the inspection. Draw a decision tree which enumerates the possible outcomes associated with five consecutive automobile inspections.

4 **Health Profile** A cancer research project classifies persons in four categories: male or female; heavy smoker, moderate smoker, or nonsmoker; regular exercise program or no regular program; overweight or not overweight. Draw a tree diagram to enumerate all possible classifications of persons.

Use the fundamental counting principle to solve Exercises 5–8.

5 A license plate consists of two letters followed by three single-digit numbers. Determine the number of different license plate codes which are possible.

6 **College Admissions** The admissions office at a local university classifies applicants as male or female; in-state or out-of-state; preferred college within the university (Engineering, Business, Liberal Arts, Education, and Pharmacy); above-average, average, or below-average SAT scores; and request for financial aid or no request for financial aid. Determine the number of possible applicant classifications.

7 A student is planning his schedule for the fall. For the five courses he is considering there are three possible English instructors, six sociology instructors, four mathematics instructors, eight history instructors, and five political science instructors. Determine the number of different sets of instructors possible for his fall schedule.

8 Determine the number of possible seven-digit telephone numbers if none of the first three digits can equal zero and:
(a) Any digits can be used for the remaining numbers.
(b) The first digit must be odd, alternating after that between even and odd digits.
(c) All digits must be even.
(d) No digit can be repeated.

Evaluate the following factorial expressions.

9 $7!$

10 $9!$

11 $15!$

12 $(15 - 8)!$

13 $\dfrac{7!}{4!}$

14 $\dfrac{15!}{6!}$

15 $\dfrac{8! \cdot 5!}{6!}$

16 $\dfrac{15! \cdot 8!}{10!}$

17 $\dfrac{10!}{3! \cdot 6!}$

18 $\dfrac{10!}{8! \cdot 2!}$

19 $\dfrac{8!}{0! \cdot 5!}$

20 $\dfrac{9!}{3! \cdot 5!}$

In Exercises 21–28, evaluate each symbol.

21 $_6P_6$

22 $_7P_3$

23 $_8P_6$ **24** $_9P_4$

25 $\binom{5}{5}$ **26** $\binom{6}{4}$

27 $\binom{8}{4}$ **28** $\binom{7}{3}$

29 Ten horses are to be placed within a starting gate for a major sweepstakes race. How many different starting arrangements are possible?

30 A political candidate wishes to visit eight different states. In how many different orders can she visit these states?

31 The same political candidate in Exercise 30 has time and funds to visit only four states. How many different combinations of four states can she visit?

32 A credit card company issues credit cards which have a three-letter prefix as part of the card number. A sample card number is ABC1234.
   (a) If each letter of the prefix is to be different, how many prefixes are possible?
   (b) If each of the four numerals following the prefix is to be different, how many different four-digit sequences are possible?

33 Eight astronauts are being considered for the next flight team. If a flight team consists of three members, how many different combinations of astronauts could be considered?

34 A portfolio management expert is considering 30 stocks for investment. Only 15 stocks will be selected for inclusion in a portfolio. How many different combinations of stocks can be considered?

35 Four persons are to be selected for the board of directors of a local hospital. If twelve candidates have been selected, how many different groups of four could be selected for the board?

36 Given a committee of ten persons, in how many ways can we select a chairperson, vice chairperson, and recording secretary?

37 Six airline companies have submitted applications for operating over a new international route. Only two of the companies will be awarded permits to operate over the route. How many different sets of airlines could be selected?

38 **Medical Research** A major research foundation is considering funding a set of medical research projects. Fifteen applications have been submitted, but only six will receive funding. How many different sets of projects could be funded?

39 A bridge hand consists of 13 cards. How many different bridge hands can be dealt from a deck consisting of 52 cards?

*40 **Design Team** The president of a major corporation has decided to undertake the development of a major new product which will give the corporation a significant competitive edge. The president wants to appoint a special product design team which will consist of three engineers, one marketing research analyst, one financial analyst, and two production supervisors. There are eight engineers, four marketing research analysts, six financial analysts, and five production supervisors being considered for the team. How many different design teams could be created?

*41 **Education** The chairperson of a high school mathematics department wants to select eight seniors, six juniors, five sophomores, and four freshmen for the high school mathematics team. Ten seniors, eight juniors, eight sophomores, and six freshmen have applied for the team and have qualified on the basis of their mathematics grades. How many different teams could the chairperson select from this group?

13.3 BASIC PROBABILITY CONCEPTS

This section will introduce the notion of probability and some basic probability concepts.

Experiments, Outcomes, and Events

The notion of probability is associated with ***random processes,*** or ***random experiments.*** A ***random experiment*** is a process which results in one of a number of possible ***outcomes.*** The possible outcomes are known prior to the performance of the random experiment, but one cannot predict with certainty which particular outcome will result. Classic random experiments include flipping a coin, rolling a die, drawing a card from a well-shuffled deck, and selecting a ball from an urn which contains a certain number of balls. There are many random processes around us which are less obvious. Many manufacturing processes produce defective products in a random manner. The times between the arrivals of telephone calls at a telephone exchange, cars at tollbooths, and customers at supermarkets have been described as random processes. The process by which the sex of a child is determined is also random.

Each repetition of an experiment can be thought of as a ***trial.*** Each trial has an observable outcome. If we assume in the coin-flipping experiment that the coin cannot come to rest on its edge, the two possible outcomes for a trial are the occurrence of a head or the occurrence of a tail. The set of all possible outcomes for an experiment is called the ***sample space.*** The sample space for an experiment is the set of outcomes S such that any performance of the experiment (trial) results in one and only one element of S. Each element in S is referred to as a ***simple outcome.***

For the coin-flipping example, the sample space S is defined as

$$S = \{\text{head, tail}\}$$

For an experiment which measures the time between arrivals of telephone calls at a telephone exchange, the sample space might be defined as

$$S = \{t | t \text{ is measured in seconds and } t \geq 0\}$$

In the coin-flipping experiment we have a ***finite sample space*** because there are a finite number of possible outcomes. In the telephone call experiment we have an ***infinite sample space*** because there are an infinite number of possible outcomes.

Given an experiment, outcomes are frequently classified into ***events.*** An event E for an experiment is a subset of a sample space, as shown in Fig. 13.11. The way in which events are defined depends upon the set of outcomes for which probabilities are to be computed. For the coin-flipping experiment, events and outcomes are likely to be defined identically, as shown in Fig. 13.12. In Fig. 13.12 there is a one-to-one correspondence between each event and each simple outcome in the sample space.

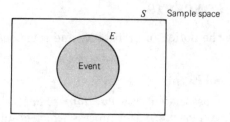

Figure 13.11   An event is a subset of the sample space.

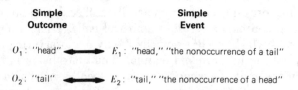

Figure 13.12   Outcome-event mapping for coin flip experiment.

If an event E consists of only one simple outcome in S, it is called a ***simple event***.

The relationship between an outcome and an event can be illustrated better when considering the experiment involving the time between arrivals of telephone calls. Events may be defined in different ways depending on the purpose of the experiment. In Fig. 13.13, the infinite sample space has been transformed into an infinite set of events using a one-for-one mapping of simple outcomes into simple events. For the same experiment, Fig. 13.14 shows a mapping of the infinite sample space into two events. Thus, it is possible that an event can be defined in such a way as to include multiple outcomes.

If an event E consists of more than one simple outcome in S, it is called a ***compound event***.

EXAMPLE 20   Suppose that an experiment consists of selecting three manufactured parts from a production process and observing whether they are *acceptable* (satisfy all production specifications) or *defective* (do not satisfy all specifications).
(*a*) Determine the sample space S.

	Simple Outcome		Simple Event	
	(Time between successive arrivals)		(Time between successive arrivals)	
O_1:	0 seconds	$\longleftrightarrow$	E_1:	0 seconds
O_2:	1 second	$\longleftrightarrow$	E_2:	1 second
O_3:	2 seconds	$\longleftrightarrow$	E_3:	2 seconds
$\vdots$	$\cdots$		$\vdots$	$\cdots$
O_{n+1}:	n seconds	$\longleftrightarrow$	E_{n+1}:	n seconds

Figure 13.13   Outcome-event mapping for telephone call experiment.

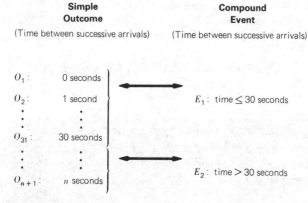

Figure 13.14 Outcome-event mapping for telephone call experiment.

(b) What outcomes are included in the event "exactly two acceptable parts"?
(c) What outcomes are included in the event "at least one defective part"?

SOLUTION

(a) Figure 13.15 is a tree diagram generating the possible outcomes for this experiment. The letter A denotes an "acceptable" part, and the letter D a "defective" part. The possible experiment outcomes are represented by the solution space

$$S = \{AAA,\ AAD,\ ADA,\ ADD,\ DAA,\ DAD,\ DDA,\ DDD\}$$

where the elements of S are the simple outcomes for the experiment.

(b) The event "exactly two acceptable parts" is the subset of S

$$E_1 = \{AAD,\ ADA,\ DAA\}$$

Figure 13.15
Tree diagram for
Example 20.

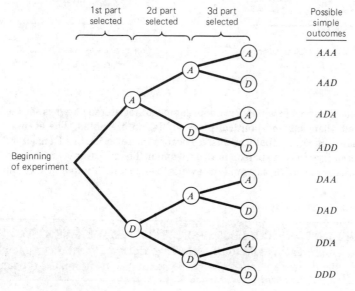

Note that this event is a compound event.

(c) The event "at least one defective part" is a compound event. This event includes all simple outcomes characterized by one, two, or three defective parts. It is the subset of S

$$E_2 = \{AAD, ADA, DAA, ADD, DAD, DDA, DDD\}$$

☐

DEFINITION: MUTUALLY EXCLUSIVE EVENTS

A set of events is said to be *mutually exclusive* if the occurrence of any one of the events precludes the occurrence of any of the other events. Specifically, the events $E_1, E_2, \ldots, E_n$ are mutually exclusive if $E_i \cap E_j = \varnothing$ for all i and j, $i \neq j$.

In flipping a coin, the two possible simple outcomes are *heads* and *tails*. Since the occurrence of a head precludes any possibility of a tail, and vice versa, the events "heads" and "tails" are mutually exclusive events. Suppose for the roll of a die, two events are defined as E_1 ("one") and E_2 ("a value less than three"). E_1 and E_2 are not mutually exclusive because the occurrence of one event does not necessarily preclude the other. For a single trial, the outcome "one" implies that both E_1 and E_2 have occurred. Figure 13.16 is a Venn diagram representation of mutually exclusive and nonmutually exclusive events in a sample space S.

Figure 13.16

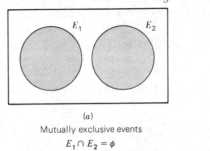

(a)
Mutually exclusive events
$E_1 \cap E_2 = \phi$

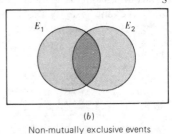

(b)
Non-mutually exclusive events
$E_1 \cap E_2 \neq \phi$

EXAMPLE 21   Consider a survey in which a random sample of registered voters is selected. For each voter selected, their sex and political party affiliation are noted. The events "Democrat" and "woman" are not mutually exclusive because the selection of a Democrat does not preclude the possibility that the person is also a woman. The two events "man" and "woman" would be mutually exclusive, as would the events "Democrat," "Republican," and "Independent."

☐

DEFINITION: COLLECTIVELY EXHAUSTIVE EVENTS

A set of events is said to be *collectively exhaustive* if their union accounts for all possible outcomes of an experiment (i.e., their union is the sample space).

The events "head" and "tail," associated with the flip of a coin, are collectively exhaustive since their union accounts for all possible outcomes. For an experiment which involves flipping a coin two times, the events $H_1 H_2$, $H_1 T_2$, and $T_1 T_2$ describe possible outcomes. This set of events is not collectively exhaustive since the union of the events does not include the outcome $T_1 H_2$.

For the same experiment, the set of events $H_1 H_2$, $H_1 T_2$, $T_1 T_2$, and $T_1 H_2$ is *both* mutually exclusive and collectively exhaustive.

TABLE 13.2

| Sex | Major | | | |
---	(B) Business	(L) Liberal Arts	(E) Preengineering	Total
Male (M)	350	300	100	750
Female (F)	250	450	50	750
Total	600	750	150	1,500

EXAMPLE 22

(College Admissions) Table 13.2 indicates some characteristics of the first-year class at a junior college. Suppose that we are going to select a student at random from this class. Events for this experiment can be defined in different ways. The sample space S consists of simple outcomes which describe the sex and major of the student. Thus,

$$S = \{MB, ML, ME, FB, FL, FE\}$$

We might also define events by sex or major. For instance, the event "male student" (M) is a compound event consisting of the simple outcomes MB, ML, and ME, or

$$M = \{MB, ML, ME\}$$

Similarly, the event "engineering student" (E) is the compound event

$$E = \{ME, FE\}$$

Determine, for each of the following sets of events, whether they are mutually exclusive and/or collectively exhaustive.
(a) $\{M, F\}$
(b) $\{B, L, E\}$
(c) $\{MB, ML, ME, B\}$

SOLUTION
(a) The two events M (male) and F (female) are mutually exclusive because the occurrence of a male student precludes the possibility of a female student, and vice versa. The two events are also collectively exhaustive since all outcomes in the sample space can be mapped into these events.
(b) The three events B (business), L (liberal arts), and E (preengineering) are mutually exclusive. The reason for this is that the occurrence of one major precludes the possibility of the other two majors. The three events are also collectively exhaustive since all outcomes in the sample space can be mapped into these events.
(c) The events MB, ML, ME, and B are not mutually exclusive because the occurrence of a

male in business (*MB*) does not preclude the occurrence of a business major (*B*). These events are not collectively exhaustive since they do not account for females majoring in liberal arts or preengineering.

☐

PRACTICE EXERCISE
Determine whether the following events are mutually exclusive and/or collectively exhaustive
(*a*) {*M, F, B, L, E*}
(*b*) {*FB, FL, FE, M*}
(*c*) {*FB, FL, FE, B, L*}
Answer: (*a*) Not mutually exclusive but collectively exhaustive; (*b*) mutually exclusive and collectively exhaustive, (*c*) neither mutually exclusive nor collectively exhaustive.

Probabilities

Although we know the possible outcomes of a random experiment, we cannot predict with certainty what outcome will occur. We can guess head or tail for the flip of a coin, but we cannot know for sure. With many random processes, however, there is long-run regularity. As mentioned earlier, the long-run expectation in flipping a coin is that approximately half of the outcomes will be heads and half tails. In the roll of a die, the long-run expectation is that each side of the die will occur approximately one-sixth of the time. These values reflect the expectation of the *relative frequency* of an event. The relative frequency of an event is the proportion of the time that the event occurs. It is computed by dividing the number of times *m* the event occurs by the number of times *n* the experiment is conducted. *The probability of an event can be thought of as the relative frequency m/n of the event over the long run.*

Given the nature of the relative frequency of an event, we can state the following probability rule.

RULE 1
The probability of an event *E*, denoted by *P*(*E*), is a number between 0 and 1 inclusive, or

$$0 \le P(E) \le 1 \qquad (13.7)$$

Two special cases of Eq. (13.7) are $P(E) = 0$ and $P(E) = 1$. *If P(E) = 0, it is certain that event E will not occur.* For example, if a coin is two-headed, $P(\text{tail}) = 0$ in a single flip of the coin. *If P(E) = 1, it is certain that event E will occur.* With the same coin, $P(\text{head}) = 1$. If $0 < P(E) < 1$, there is uncertainty about the occurrence of event *E*. For example, if $P(E) = .4$, we can state that there is a 40 percent chance (likelihood) that event *E* will occur.

EXAMPLE 23 (**College Admissions, continued**) Table 13.2 (in Example 22) indicated some characteristics of the first-year class at a junior college. Assume that a student will be selected at random from the freshman class and that each person has an equal chance of being selected. Using the relative frequency concept of probability, we can estimate the likelihood that the selected student will have certain characteristics. For example, the probability that the selected student will be a male is

$$P(M) = \frac{\text{number of males}}{\text{total number of first-year students}}$$

$$= \frac{750}{1,500} = .50$$

The probability that the selected student will be a preengineering student is

$$P(E) = \frac{\text{number of preengineering students}}{\text{total number of first-year students}}$$

$$= \frac{150}{1,500} = .10$$

The probability that the selected student will be a female majoring in business is

$$P(FB) = \frac{250}{1,500} = .167$$

Table 13.3 summarizes the probabilities of various events associated with selecting a student. Notice that the sum of the probabilities for the events M and F equals 1. Similarly, the sum of the probabilities for the events B, L, and E equals 1.

TABLE 13.3

Sex	Major			
	(B) Business	(L) Liberal Arts	(E) Preengineering	Total
Male (M)	.233	.200	.067	.500
Female (F)	.167	.300	.033	.500
Total	.400	.500	.100	1.000

POINT FOR THOUGHT & DISCUSSION Why do the probabilities for the two different sets of events $\{M, F\}$ and $\{B, L, E\}$ each total 1?

Probabilities may be classified in several ways. One classification is the distinction between **objective** and **subjective probabilities.** Objective probabilities are based upon historical experience or general knowledge. For example, probabilities assigned to the events associated with flipping a fair coin or rolling a die are based upon much historical and generally known experience. Other objective probabilities are assigned because of actual experimentation. For instance, if one has been told that a die is "loaded," one method of determining the probability of any side occurring is to roll the die many times while keeping a record of the relative frequency for each side.

Subjective probabilities are assigned in the absence of wide historical experience. They are based upon personal experiences and intuition. They are really expressions of personal judgment. A subjective probability would be the probability that you would assign to your receiving a grade of A in this course. Such an estimate would reflect your personal assessment of such factors as your aptitude in this type of course, your perception of the degree of difficulty of this course, and your assessment of the instructor and the way in which he or she will conduct the course and evaluate performance.

Some Additional Rules of Probability

> **RULE 2**
> If $P(E)$ represents the probability that an event E will occur, the probability that E will not occur, denoted by $P(E')$, is
>
> $$P(E') = 1 - P(E) \tag{13.8}$$

If the probability that a business will earn a profit during its first year of operation is estimated at .35, the probability that it will not earn a profit during the first year is $1 - .35 = .65$. Given a "fair" die, the probability of rolling a "one" equals $\frac{1}{6}$. The probability of not rolling a "one" equals $1 - \frac{1}{6} = \frac{5}{6}$.

> **RULE 3**
> If events E_1 and E_2 are mutually exclusive, the probability that either E_1 or E_2 will occur is
>
> $$P(E_1 \cup E_2) = P(E_1) + P(E_2) \tag{13.9}$$

EXAMPLE 24 (Public Works) The department of public works for a community is gearing up for winter. The department is planning its sand and salt needs for maintaining roads during and after snowstorms. An analysis of past winters has resulted in the probability estimates in Table 13.4 regarding the expected number of major snowstorms. What is the probability of three or more major snowstorms during the coming year?

TABLE 13.4

n (Number of Major Storms)	$P(n)$
0	.10
1	.25
2	.30
3	.20
More than 3	.15

SOLUTION

The simple events in Table 13.4 which correspond to the compound event "3 or more" major snowstorms are $n = 3$ and $n =$ more than 3. These two simple events are mutually exclusive. Therefore, the probability of three or more major snowstorms is

$$P(3 \text{ or more}) = P(3) + P(\text{more than } 3)$$
$$= .20 + .15 = .35$$

RULE 4
Given n mutually exclusive events $E_1, E_2, \ldots, E_n$, the probability that E_1, $E_2, \ldots$, or E_n will occur is

$$P(E_1 \cup E_2 \cup \cdots \cup E_n) = P(E_1) + P(E_2) + \cdots + P(E_n) \quad (13.10)$$

EXAMPLE 25 Given the information in Example 24, what is the probability of having fewer than three major snowstorms?

SOLUTION

The simple events corresponding to "fewer than three" storms are 0, 1, and 2 storms. Since these simple events are mutually exclusive, we can apply rule 4, resulting in

$$P(\text{less than } 3) = P(0) + P(1) + P(2) = .10 + .25 + .30 = .65$$

NOTE This problem could have been solved using the result in Example 24 and applying rule 2. That is,

$$P(\text{less than } 3) = 1 - P(3 \text{ or more})$$
$$= 1 - .35 = .65$$

When a set of events is not mutually exclusive, rule 3 must be modified to reflect the possibility that two events may occur at the same time.

RULE 5
The probability of occurrence of event E_1, event E_2, or both E_1 and E_2 is

$$P(E_1 \cup E_2) = P(E_1) + P(E_2) - P(E_1 \cap E_2) \quad (13.11)$$

Note that the intersection operator is used to denote the joint or simultaneous occurrence of events E_1 and E_2.

Rule 3 is the special case of rule 5 where $P(E_1 \cap E_2) = 0$. This is because E_1 and

Figure 13.17

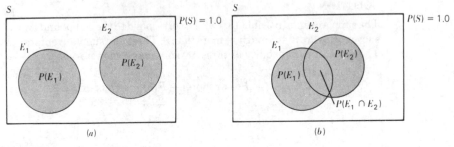

Rule 3: $P(E_1 \cup E_2) = P(E_1) + P(E_2)$
(Mutually exclusive events)

(Events not mutually exclusive)

E_2 are assumed to be mutually exclusive in rule 3, meaning that the events can never occur together. Figure 13.17 illustrates these rules using Venn diagrams. In these diagrams, areas represent probabilities, with the area of the sample space S assumed to equal 1.0.

EXAMPLE 26

In an experiment consisting of selecting one card at random from a deck of 52 cards, the events "king" and "spade" are not mutually exclusive. The probability of selecting a king, a spade, or both a king and a spade (the king of spades) is determined by applying rule 5, or

$$P(\text{king} \cup \text{spade}) = P(\text{king}) + P(\text{spade}) - P(\text{king} \cap \text{spade})$$
$$= P(\text{king}) + P(\text{spade}) - P(\text{king of spades})$$
$$= \tfrac{4}{52} + \tfrac{13}{52} - \tfrac{1}{52}$$
$$= \tfrac{16}{52} = \tfrac{4}{13}$$

Note that $P(\text{king} \cap \text{spade})$ must be subtracted to offset the double counting of the event $(\text{king} \cap \text{spade})$. When $P(\text{king})$ is computed, the king of spades is included among the four kings in the deck; and when $P(\text{spade})$ is computed, the king of spades is included among the 13 spades in the deck. Thus, double counting of the king of spades has occurred, and $P(\text{king}$ of spades) must be subtracted.

PRACTICE EXERCISE
In Example 26, what is the probability of selecting a "face" card (king, queen, or jack) or a heart? *Answer:* 22/52.

EXAMPLE 27

(Energy Conservation) A group of 2,000 people was surveyed regarding policies which might be enacted to conserve oil. Of the 2,000, 1,000 people said that gas rationing would be acceptable to them, 500 said that a federal surtax of $0.25 per gallon would be acceptable, and 275 indicated that both rationing and the surtax would be acceptable. If a person is selected at random from this group, what is the probability that the person would:
(a) Find the surtax acceptable?
(b) Find the surtax acceptable but not gas rationing?
(c) Find one or both of the alternatives acceptable?
(d) Find neither alternative acceptable?

Figure 13.18

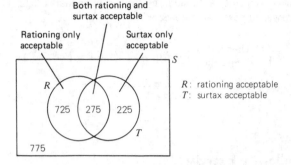

SOLUTION

A Venn diagram is useful in summarizing the survey results. Figure 13.18 is the Venn diagram for this survey. Note that the 275 finding both policies acceptable are represented by the intersection of sets R and T. The 1,000 persons in set R consist of two types of respondents: 275 persons who find both policies acceptable and 725 who find rationing acceptable, but not the surtax. Similarly, the 500 persons found in set T consist of the 275 who find both policies acceptable and 225 who approve of the surtax but not rationing. Adding together those who approve of one or both of the policies, we get a total of 1,225. The remaining persons (775) are those who approved of neither policy.

(a) Since 500 persons approved of the surtax, the probability that a person would approve is

$$P(T) = \frac{\text{number of persons approving surtax}}{\text{number of persons surveyed}} = \frac{500}{2,000} = .25$$

(b) 225 persons indicated approval of the surtax but not rationing. Thus, the probability of selecting such a person is $225/2,000 = .1125$.

(c) We can answer this part in two ways. The simplest way is to count up the number of respondents who approved of one or both policies and divide by the total, or

$$P(R \cup T) = \frac{725 + 275 + 225}{2,000} = \frac{1,225}{2,000} = .6125$$

Or we can use rule 5.

$$P(R \cup T) = P(R) + P(T) - P(R \cap T)$$

We computed $P(T)$ in part a. Verify that $P(R) = .50$ and $P(R \cap T) = .1375$. Therefore,

$$P(R \cup T) = .50 + .25 - .1375$$
$$= .6125$$

(d) Using the relative frequency approach, this probability equals $775/2{,}000 = .3875$. Or we can use our answer in part c as follows:

$$P\left(\begin{array}{c}\text{neither policy}\\ \text{acceptable}\end{array}\right) = P[(R \cup T)'] = 1 - P(R \cup T)$$

$$= 1 - .6125$$

$$= .3875$$

☐

Section 13.3 Follow-up Exercises

1 **Pollution Monitoring** A water quality inspector is conducting an experiment where he samples the water from various wells to see if it has *acceptable* (A) or *unacceptable* (U) levels of contaminants. Suppose the inspector is going to inspect three wells, one after the other, and record the quality of the water for each.
   (a) Determine the sample space S for this experiment.
   (b) Construct a tree diagram enumerating the possible outcomes.
   (c) What simple outcomes are included in the event "exactly two acceptable wells"?
   (d) What simple outcomes are included in the event "at least one acceptable well"?

2 **Recidivism** A criminal justice researcher is studying the rate of recidivism (repeat offenders) for child molestation. He is conducting an experiment where he examines the criminal record for persons convicted of child molestation. If a person has been convicted more than once, he is classified as a "recidivist" (R). If a person has not been convicted more than once for child molestation, he is classified as a "nonrecidivist" (N) in the experiment. If the researcher examines the records of three offenders, (a) determine the sample space S for the experiment, (b) construct a tree diagram enumerating the possible outcomes for the experiment, (c) determine the set of simple outcomes included in the event "two or fewer recidivists," and (d) determine the set of simple outcomes included in the event "at least one recidivist."

3 Table 13.5 indicates some characteristics of a pool of 1,000 applicants for an administrative position. Applicants are classified by sex and by highest educational degree received. Suppose that one applicant is to be selected at random in an experiment. The sample space S for this experiment consists of the simple outcomes $S = \{MC, MH, MN, FC, FH, FN\}$.
   (a) Determine the set of simple outcomes which are used to define the compound event "male applicant" (M).
   (b) Determine the set of simple outcomes which are used to define the compound event "highest degree of applicant is college degree" (C).

TABLE 13.5		*Highest Degree*		
Sex	College Degree (*C*)	High School Diploma (*H*)	No Degree (*N*)	Total
Male (*M*)	350	100	40	490
Female (*F*)	275	210	25	510
Total	625	310	65	1000

4 In Exercise 3, determine, for each of the following sets of events, whether they are mutually exclusive and/or collectively exhaustive.

(a) {M, F, H}
(b) {C, H, N, M, F}
(c) {MC, MH, MN, F}
(d) {MC, FC, C, H, N}
(e) {M, FC, FH}

5 In Exercise 3, suppose that one applicant is selected at random (each having an equal chance of being selected). What is the probability that the applicant selected will (a) be a female, (b) have a high school diploma as the highest degree, (c) be a male with no degrees, and (d) be a female with a college degree?

6 Table 13.6 indicates some characteristics of 10,000 borrowers from a major financial institution. Borrowers are classified according to the type of loan (personal or business) and level of risk. Suppose that an experiment is to be conducted where one borrower's account is selected at random. The sample space S for this experiment consists of the simple outcomes S = {PL, PA, PH, BL, BA, BH}.

(a) Determine the set of simple outcomes which are used to define the compound event "personal loan" (P).

(b) Determine the set of simple outcomes which are used to define the compound event "average-risk loan" (A).

TABLE 13.6		Credit Risk		
Type of Loan	Low Risk (L)	Average Risk (A)	High Risk (H)	Total
Personal (P)	2,400	3,600	1,600	7,600
Business (B)	650	950	800	2,400
Total	3,050	4,550	2,400	10,000

7 In Exercise 6, determine, for each of the following sets of events, whether they are mutually exclusive and/or collectively exhaustive.

(a) {P, L, A, H}
(b) {PL, BL, PA, BA, H}
(c) {P, B, H}
(d) {B, PL, PH, BL, BH}
(e) {PL, BL, PA, BH}

8 In Exercise 6, suppose that one account is selected at random (each having an equal chance of selection). What is the probability that the account selected will (a) be in the average-risk category, (b) be a personal loan, (c) be a business loan in the high-risk category, and (d) be a personal loan with a low risk?

9 **Child Care Alternatives** In 1987, the U.S. Census Bureau estimated that 9.1 million children under age five required primary child care because of employed mothers. Table 13.7 indicates the child care alternatives and the Census Bureau's estimates of the number of children using each type. If a child from this group is selected at random, what is the probability that (a) the child is cared for in his or her home, (b) the child cares for himself or herself, and (c) the child is cared for at work by the mother?

10 **Aging U.S. Population** A U.S. Census Bureau study reveals that the average age of the population is increasing. The bureau estimates that by the year 2000, there will be

TABLE 13.7	Type of Child Care	Number of Children
	In another home	3,239,600
	Day care/nursery school	2,220,400
	Child cares for self	18,200
	Mother cares for child at work	809,900
	Care in child's home	2,720,900
	Other	91,000

105.6 million households. Table 13.8 indicates projections regarding the age of the head of household. If in the year 2000 a household is selected at random, what is the probability that the head of household will be (*a*) of age 65 or older, (*b*) of age 25–34, (*c*) of age 35 or older, and (*d*) of age 45 or younger?

TABLE 13.8	Age of Head of Household	Number of Households
	15–24	4,224,000
	25–34	16,896,000
	35–44	24,288,000
	45–54	22,176,000
	55–64	14,784,000
	65 and older	23,232,000

11 Cardiac Care In order to support its request for a cardiac intensive care unit, the emergency room at a major urban hospital has gathered data on the number of heart attack victims seen. Table 13.9 indicates the probabilities of different numbers of heart attack victims being treated in the emergency room on a typical day. For a given day, what is the probability that (*a*) five or fewer victims will be seen, (*b*) five or more victims will be seen, and (*c*) no more than seven victims will be seen?

TABLE 13.9	Number of Victims Treated (n)	$P(n)$
	Fewer than 5	.08
	5	.16
	6	.30
	7	.26
	More than 7	.20

12 Fire Protection The number of fire alarms pulled each hour fluctuates in a particular city. Analysts have estimated the probability of different numbers of alarms per hour as shown in Table 13.10. In any given hour, what is the probability that (*a*) more than 8 alarms will be pulled, (*b*) between 8 and 10 alarms (inclusive) will be pulled, and (*c*) no more than 9 alarms will be pulled?

13 A card is to be drawn at random from a well-shuffled deck. What is the probability that

TABLE 13.10	Number of Alarms Pulled (*n*)	*P(n)*
	Fewer than 8	.16
	8	.20
	9	.24
	10	.28
	More than 10	.12

the card will be (*a*) a king or jack, (*b*) a face card (jack, queen, or king), (*c*) a 7 or a spade, and (*d*) a face card or a card from a red suit?

14 A survey of 2,000 consumers was conducted to determine their purchasing behavior regarding two leading soft drinks. It was found that during the past month 800 persons had purchased brand *A*, 300 had purchased brand *B*, and 100 had purchased both brand *A* and brand *B*. If a person is selected at random from this group (assuming equal chance of selection for each person), what is the probability that the person (*a*) would have purchased brand *A* during the past month, (*b*) would have purchased brand *B* but not brand *A*, (*c*) would have purchased brand *A*, brand *B*, or both, and (*d*) would not have purchased either brand?

15 Vitamin C Research In recent years there has been much controversy about the possible benefits of using supplemental doses of vitamin C. Claims have been made by proponents of vitamin C that supplemental doses will reduce the incidence of the common cold and influenza (flu). A test group of 3,000 persons received supplemental doses of vitamin C for a period of 1 year. During this period it was found that 800 such people had one or more colds, 250 people suffered from influenza, and 150 people suffered from both colds and influenza. If a person is selected at random from this test group (assuming equal likelihood of selection), what is the probability that the person (*a*) would have had one or more colds but not influenza, (*b*) would have had both colds and influenza, (*c*) would have had one or more colds but no influenza, or influenza but no colds, and (*d*) would have suffered neither colds nor influenza?

16 The sample space for an experiment consists of five simple events $E_1, E_2, E_3, E_4,$ and E_5. These events are mutually exclusive. The probabilities of occurrence of these events are $P(E_1) = .20$, $P(E_2) = .15$, $P(E_3) = .25$, $P(E_4) = .30$, and $P(E_5) = .10$. Several compound events can be defined for this experiment. They are

$$F = \{E_1, E_2, E_3\} \qquad G = \{E_1, E_3, E_5\} \qquad H = \{E_4, E_5\}$$

Determine (*a*) $P(F)$, (*b*) $P(G)$, (*c*) $P(H)$, (*d*) $P(G')$, (*e*) $P(F \cup G)$, (*f*) $P(G \cup H)$, (*g*) $P(F \cap H)$, and (*h*) $P(F \cap G)$.

17 The sample space for an experiment consists of four simple events $E_1, E_2, E_3,$ and E_4, which are mutually exclusive. The probabilities of occurrence of these events are $P(E_1) = .2$, $P(E_2) = .1$, $P(E_3) = .4$, and $P(E_4) = .3$. Several compound events can be defined for this experiment. They include

$$A = \{E_1, E_2, E_3\} \qquad B = \{E_2, E_4\} \qquad C = \{E_1, E_3, E_4\}$$

Determine (*a*) $P(A)$, (*b*) $P(A')$, (*c*) $P(B)$, (*d*) $P(C)$, (*e*) $P(C')$, (*f*) $P(A \cup B)$, (*g*) $P(B \cup C)$, and (*h*) $P(A \cap C)$.

13.4 STATES OF STATISTICAL INDEPENDENCE AND DEPENDENCE

Statistical Independence

In addition to our previous classifications, events may be classified as ***independent*** or ***dependent***.

> **DEFINITION: INDEPENDENT EVENTS**
> Two events are independent if the occurrence or nonoccurrence of one event in no way affects the likelihood (or probability) of occurrence of the other event.

The outcomes of successive flips of a fair coin are an example of independent events. The occurrence of a head or tail on any one toss of the coin has no effect on the probability of a head or tail on the next or succeeding tosses of the coin. Drawing cards from a deck or balls from an urn *with replacement* is an experiment characterized by independent events. *"With replacement"* means that the item selected is put back in the deck or the urn before the next item is selected. In the case of a deck of cards, the probability of drawing a heart on each draw is $\frac{13}{52}$ as long as the previously drawn card is replaced in the deck.

The random generation of *acceptable* or *defective* products from a production process is another example of events which may be statistically independent. For such processes, the probability of generating an acceptable (or defective) unit is assumed to remain the same for a given trial regardless of previous outcomes.

*The simple probability of an event is often called a **marginal probability**.* The marginal probability of a head is .5 when a fair coin is tossed. Assuming a random draw, the marginal probability of selecting a spade from a deck of cards equals $\frac{13}{52}$. Very often there is an interest in computing the probability of two or more events occurring at the same time or in succession. For example, we may be interested in the probability of rolling a pair of dice and having a 5 occur on one die and a 2 on the other; or we may be interested in the likelihood that five heads would occur in five successive tosses of a coin. *The probability of the joint occurrence of two or more events is called a **joint probability**.*

> **RULE 6**
> The joint probability that two independent events E_1 and E_2 will occur together or in succession equals the product of the marginal probabilities of E_1 and E_2, or
>
> $$P(E_1 \cap E_2) = P(E_1) \cdot P(E_2) \qquad (13.12)$$

EXAMPLE 28

The probability that a machine will produce a defective part equals .05. It is assumed that the production process is characterized by statistical independence. That is, the probability of any item being defective is .05, regardless of the quality of previous units. Suppose we wish to determine the probability that two consecutive parts will be defective. If the event D represents the occurrence of a defective item, then according to rule 6,

$$P(D_1 \cap D_2) = P(D_1)P(D_2)$$
$$= (.05)(.05) = .0025$$

EXAMPLE 29 (Fund-Raising) A university has started using phonathons as a major way of soliciting donations from alumni. The vice president for development estimates that the probability an alumnus, contacted by phone, will contribute is .24. Given two successive contacts, what is the probability that:

(a) The first contacted alumnus will contribute and the second not?
(b) The first will not contribute but the second will?
(c) Both will contribute?
(d) Neither will contribute?

SOLUTION

If the event C represents the occurrence of a contribution, the event C' represents the nonoccurrence of a contribution.

(a) $P(C_1 \cap C_2') = P(C_1) \cdot P(C_2')$
$= (.24)(1 - .24)$
$= (.24)(.76)$
$= .1824$

(b) $P(C_1' \cap C_2) = P(C_1') \cdot P(C_2)$
$= (.76)(.24)$
$= .1824$

(c) $P(C_1 \cap C_2) = P(C_1) \cdot P(C_2)$
$= (.24)(.24)$
$= .0576$

(d) $P(C_1' \cap C_2') = P(C_1') \cdot P(C_2')$
$= (.76)(.76)$
$= .5776$

$\square$

PRACTICE EXERCISE

Draw a tree diagram to enumerate all possible outcomes associated with making two consecutive contacts with alumni. Did we enumerate all possible outcomes in our example? Sum the four probabilities we computed. Is there any significance to your answer?

RULE 7

The joint probability that n independent events $E_1, E_2, \ldots, E_n$ will occur together or in succession is

$$P(E_1 \cap E_2 \cap \cdots \cap E_n) = P(E_1)P(E_2) \cdots P(E_n) \qquad (13.13)$$

EXAMPLE 30 (IRS Audit; Motivating Scenario) The Internal Revenue Service (IRS) estimates the probability of an error on personal income tax returns to be .4. Suppose that an experiment is conducted in which three returns are selected at random for purposes of an audit, and we wish to determine all possible outcomes and the probabilities of these outcomes.

A tool which is useful in problems such as this is a ***probability tree.*** A probability tree is a type of tree diagram which enumerates all possible outcomes in an experiment and their probabilities of occurrence. Figure 13.19 illustrates the probability tree for this experiment. Let E_i represent the outcome "error" for the ith trial and N_i the outcome "no error for the ith

Figure 13.19
Probability tree
for IRS example.

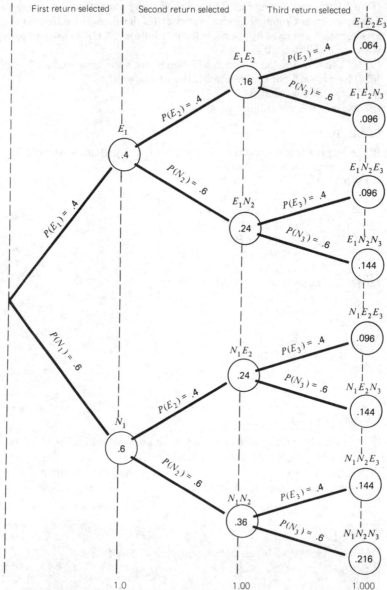

trial." For the first return selected, the two circles indicate the marginal probabilities of the two simple events which are possible.

For the second return selected, the four circles indicate the joint probabilities of the different outcomes possible when selecting two returns. If we assume that the outcomes for successive selections (trials) are independent, the probabilities of the different joint events for two selected returns can be computed using rule 6. For example,

$$P(E_1 \cap N_2) = P(E_1)P(N_2)$$
$$= (.4)(.6) = .24$$

For the third return selected, the eight circles indicate the joint probabilities of the different outcomes possible when selecting three returns. These probabilities can be computed using rule 7. For example,

$$P(E_1 \cap N_2 \cap E_3) = P(E_1)P(N_2)P(E_3)$$
$$= (.4)(.6)(.4) = .096$$

Note in Fig. 13.19 that for each trial the sum of the joint probabilities equals 1. This is because the set of events identified is collectively exhaustive *and* mutually exclusive.

☐

PRACTICE EXERCISE
Using the probability tree in Fig. 13.19, determine the probability that (*a*) two of three returns contain errors and (*b*) the three returns contain at least one error. *Answer:* (*a*) .288, and (*b*) .784.

In addition to marginal and joint probabilities, another type of probability is a **conditional probability.** *The notation* $P(E_1|E_2)$ *represents the conditional probability of event* E_1 *given that event* E_2 *has occurred.* The probability of a head on the third toss of a coin, given that the first two tosses both resulted in a head, is a conditional probability.

By definition, however, independent events have the property that the occurrence or nonoccurrence of one event has no influence on the probability of another event.

RULE 8
Given two independent events E_1 and E_2, the conditional probability of event E_1 given that event E_2 has occurred is the marginal probability of E_1, or

$$P(E_1|E_2) = P(E_1) \qquad (13.14)$$

The conditional probability of a 6 on the roll of a die given that no 6s have occurred in the last 20 rolls equals $\frac{1}{6}$. For the IRS example, the probability that the next return selected contains an error equals .4, regardless of the outcomes from previous income tax returns examined.

Statistical Dependence

In contrast to the state of statistical independence, many events are characterized as being **statistically dependent.**

DEFINITION: DEPENDENT EVENTS
Two events are **dependent** if the probability of occurrence of one event is affected by the occurrence or nonoccurrence of the other event.

Examples of experiments consisting of dependent events include drawing cards from a deck or balls from an urn *without replacement*. If a card is selected which is not a heart and the card is kept out of the deck, the probability of selecting a heart on the next draw is not the same as it was for the first draw. Given the following events related to weather conditions,

Event E_1 = it will snow

Event E_2 = the temperature will be below freezing

the probability of event E_1 is affected by the occurrence or nonoccurrence of event E_2. The following examples illustrate the computation of probabilities under conditions of statistical dependence.

TABLE 13.11

	No Exercise (NE)	Some Exercise (SE)	Regular Exercise (RE)	Total
Heart disease (HD)	700	300	100	1,100
No heart disease ($\overline{HD}$)	1,300	6,600	1,000	8,900
Total	2,000	6,900	1,100	10,000

EXAMPLE 31   A nationwide survey of 10,000 middle-aged men resulted in the data shown in Table 13.11. It is believed that the results from this survey of 10,000 men are representative of these particular attributes for the average middle-aged man in this country. We can estimate probabilities related to heart disease and exercise habits based upon relative frequencies of occurrence in the survey. For example, the probability that a middle-aged man gets no exercise is

$$P(NE) = \frac{\text{number of respondents who do no exercise}}{\text{number of men surveyed}}$$

$$= \frac{2,000}{10,000} = .20$$

The probability that a middle-aged man has heart disease is

$$P(HD) = \frac{\text{number of respondents having heart disease}}{\text{number of men surveyed}}$$

$$= \frac{1,100}{10,000} = .11$$

The joint probability that a middle-aged man exercises regularly and has heart disease is

$$P(HD \cap RE) = \frac{100}{10,000} = .01$$

Suppose we are interested in the conditional probability that a man will suffer from heart disease *given* that he does not exercise. The *given* information focuses our attentions upon the first column in Table 13.11, those respondents who do not exercise. Of the 2,000 surveyed who do not exercise, 700 suffer heart disease. Therefore, the conditional probability is computed as

$$P(HD|NE) = \frac{\text{number of respondents who do not}}{\text{number of respondents who do not exercise}}$$

$$= \frac{700}{2,000} = .35$$

The conditional probability that a man exercises regularly given that he suffers heart disease is

$$P(RE|HD) = \frac{\text{number of respondents who exercise}}{\text{number of respondents who suffer heart disease}}$$

$$= \frac{100}{1,100} = .09$$

❑

Conditional probabilities under conditions of statistical dependence can be computed using the following rule.

RULE 9

The conditional probability of event E_1 given the occurrence of event E_2 is

$$P(E_1|E_2) = \frac{P(E_1 \cap E_2)}{P(E_2)} \qquad (13.15)$$

The conditional probability is found by dividing the joint probability of events E_1 and E_2 by the marginal probability of E_2.

Indirectly, we were using Eq. (13.15) when computing the conditional probabilities in Example 31. Applying Eq. (13.15) in that example, we get

$$P(HD|NE) = \frac{P(HD \cap NE)}{P(NE)}$$

$$= \frac{700/10,000}{2,000/10,000} = \frac{.07}{.20} = .35$$

and
$$P(RE|HD) = \frac{P(RE \cap HD)}{P(HD)}$$

$$= \frac{100/10,000}{1,100/10,000} = \frac{.01}{.11} = .09$$

EXAMPLE 32   Suppose that a person selects a card at random from a deck of 52 cards and tells us that the selected card is red. The probability that the card is the king of hearts *given* that it is red can be determined using Eq. (13.15).

$$P(\text{king of hearts}|\text{red}) = \frac{P(\text{king of hearts} \cap \text{red})}{P(\text{red})}$$

$$= \frac{1/52}{26/52} = \frac{1}{26}$$

$\square$

If both sides of Eq. (13.15) are multiplied by $P(E_2)$, the resulting equation provides an expression for the joint probability $P(E_1 \cap E_2)$, or

$$\boxed{P(E_1 \cap E_2) = P(E_2)P(E_1|E_2)} \qquad (13.16)$$

EXAMPLE 33   The joint probability of selecting two aces in a row from a deck without replacement of the first card can be found using Eq. (13.16) as

$$P(A_1 \cap A_2) = P(\text{ace on first draw}) \cdot P(\text{ace on second draw } given \text{ an ace on first draw})$$
$$= P(A_1)P(A_2|A_1)$$
$$= \tfrac{4}{52} \cdot \tfrac{3}{51} = \tfrac{12}{2{,}652}$$

EXAMPLE 34   A large jar contains eight red balls, six yellow balls, and six blue balls. Two balls are to be selected at random from the jar. Assume that each ball in the jar has an equal chance of being selected and that the first ball selected will not be placed back into the jar.
(a) What is the probability that the first ball will be red and the second yellow?
(b) What is the probability that both balls will be blue?
(c) What is the probability that neither will be red?

SOLUTION

Selecting balls from a jar *without replacement* is an experiment characterized by statistical dependence.
(a) Using Eq. (13.16) gives

$$P(R_1 \cap Y_2) = P(R_1) \cdot P(Y_2|R_1)$$
$$= \tfrac{8}{20} \cdot \tfrac{6}{19} = \tfrac{48}{380} = \tfrac{12}{95}$$

(b)

$$P(B_1 \cap B_2) = P(B_1) \cdot P(B_2|B_1)$$
$$= \tfrac{6}{20} \cdot \tfrac{5}{19} = \tfrac{30}{380} = \tfrac{3}{38}$$

(c) If we let R'_j represent the event "selection of a nonred ball on the jth draw,"

$$P(R'_1 \cap R'_2) = P(R'_1) \cdot P(R'_2|R'_1)$$
$$= \tfrac{12}{20} \cdot \tfrac{11}{19} = \tfrac{132}{380} = \tfrac{34}{95}$$

EXAMPLE 35   (**Terrorist Activities**) A terrorist group operating within a foreign country has made a threat to destroy certain airplanes unless its demands are satisfied. They claim to have placed bombs on 3 of 20 planes currently on the ground at a major airport. All passengers and crews have been evacuated and a bomb squad is about to begin its search of the planes. Of interest is the probability that the first 3 planes searched will be the 3 affected planes.

If A denotes the selection of an affected plane and N the selection of a nonaffected plane, we are interested in $P(A_1 \cap A_2 \cap A_3)$. The probability that the first 2 planes selected are affected is computed as

$$P(A_1 \cap A_2) = P(A_1) \cdot P(A_2|A_1)$$
$$= \tfrac{3}{20} \cdot \tfrac{2}{19} = \tfrac{6}{380} = \tfrac{3}{190}$$

The probability that the first 3 planes selected are the affected ones is found as

$$P(A_1 \cap A_2 \cap A_3) = P(A_1 \cap A_2) \cdot P(A_3|A_1 \cap A_2)$$
$$= \tfrac{3}{190} \cdot \tfrac{1}{18} = \tfrac{3}{3,420}$$

PRACTICE EXERCISE
What is the probability that a random selection of 4 planes will result in the outcome $A_1 N_2 A_3 A_4$? *Answer:* $102/116,280 = 1/1,140$.

EXAMPLE 36

(**Space Shuttle Simulation**) While the space shuttle flights were being prepared, one contingency anticipated was replacement of defective components while in space. One scenario simulated suggested that of 10 fuel cells of a particular type used on a mission, 3 become defective while in flight. Suppose that it is impossible to know which 3 are defective. The only way of identifying the defective cells is to remove them and test them with on-board equipment. If the cells are selected at random, (*a*) construct a probability tree which summarizes the possible outcomes from selecting 3 cells and testing them. (*b*) What is the probability that none of the defective cells will be included in the sample of 3? One defective cell? Two defective cells? All 3?

SOLUTION

(*a*) If we let D_j and N_j represent defective and nondefective outcomes for the jth cell selected, Fig. 13.20 presents the probability tree summarizing the possible outcomes from selecting 3 cells. Note that the events described in this problem are statistically dependent. Also, note that the sum of the probabilities for all outcomes associated with each trial equals 1. Finally, in an experiment consisting of 3 trials (3 fuel cells selected), there are 8 different outcomes possible.

(*b*) The probability that none of the defective cells will be selected corresponds to the outcome $N_1 N_2 N_3$, which has a probability of $\tfrac{210}{720}$. The probability of 1 defective cell being selected corresponds to the outcomes $D_1 N_2 N_3$, $N_1 D_2 N_3$, and $N_1 N_2 D_3$. The sum of the probabilities for these outcomes equals $3(\tfrac{126}{720}) = \tfrac{378}{720}$. The probability of 2 defective cells being selected corresponds to the outcomes $D_1 D_2 N_3$, $D_1 N_2 D_3$, and $N_1 D_2 D_3$. The sum of the probabilities for these outcomes equals $3(\tfrac{42}{720}) = \tfrac{128}{720}$. Finally, the probability of all 3 defective cells being identified corresponds to outcome $D_1 D_2 D_3$ which has a probability of $\tfrac{6}{720}$.

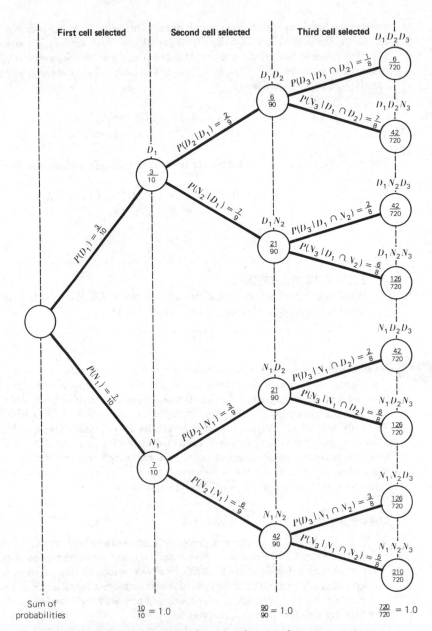

Figure 13.20   Probability tree for space shuttle example.

Section 13.4 Follow-up Exercises

1 The probability that a machine will produce a defective part equals .15. If the process is characterized by statistical independence, what is the probability that (*a*) two items in succession will not be defective, (*b*) the first three items are not defective and the fourth is defective, and (*c*) five consecutive items will not be defective?

2 **IRS Audit** An income tax return can be audited by the federal government and/or by

the state. The probability that an individual tax return will be audited by the federal government is .03. The probability that it will be audited by the state is .04. Assume that audit decisions are made independent of one another at the federal and state levels.

(*a*) What is the probability of being audited by both agencies?

(*b*) What is the probability of a state audit but not a federal audit?

(*c*) What is the probability of not being audited?

3 A coin is weighted such that $P(H) = .45$ and $P(T) = .55$. Construct a probability tree denoting all possible outcomes if the coin is tossed three times. What is the probability of two tails in three tosses? Two heads?

4 Five cards are selected at random from a deck of 52. If the drawn cards are not replaced in the deck, what is the probability of selecting an ace, king, ace, jack, and ace, in that order?

TABLE 13.12		Response			
Respondent Age	**Very Likely**	**Likely**	**Unlikely**	**Total**
20–29	850	1,700	500	3,050
30–39	700	1,100	450	2,250
40 and over	600	600	1,500	2,700
Total	2,150	3,400	2,450	8,000

5 Table 13.12 summarizes the results of a recent survey of attitudes regarding nuclear war. The question asked was, "How likely do you believe it is that a nuclear war will occur during the next 10 years?" If a respondent is selected at random from the sample of 8,000, what are the following probabilities?

(*a*) The respondent is 30 years or older.

(*b*) The respondent believes nuclear war is "likely."

(*c*) The respondent is between the ages of 30 and 39 and believes that nuclear war is "very likely."

(*d*) The respondent is between the ages of 20 and 39 and believes that nuclear war is "unlikely."

(*e*) The respondent believes that nuclear war is "unlikely," given that he or she is between the ages of 20 and 29.

(*f*) The respondent is 40 years of age or older, given that he or she believes nuclear war is "unlikely."

6 A television game show contestant has earned the opportunity to win some prizes. The contestant is shown 10 boxes, 4 of which contain prizes. If the contestant is allowed to select any 4 of the boxes, what is the probability that (*a*) four prizes will be selected, (*b*) no prizes will be selected, and (*c*) the first 3 boxes selected contain no prizes but the 4th box does?

7 For the previous exercise, draw a probability tree which summarizes the different outcomes possible when selecting 4 boxes at random and their associated probabilities. What is the probability that at least one prize will be won? Exactly one prize?

8 Refer to Example 34. Construct a probability tree which summarizes all outcomes and their probabilities for the selection of two balls.

9 Given that cards are selected at random, without replacement, from a standard 52-card deck, determine the probability that (*a*) the first 2 cards are hearts, (*b*) the first is a spade, the second a club, the third a heart, and the fourth a diamond, (*c*) 3 aces are selected in a row, and (*d*) no aces are included in the first 4 cards.

10 A graduating class consists of 52 percent women and 48 percent men. Of the men, 20 percent are engineering majors. If a graduate is selected at random from the class, what is the probability the student is a male engineering major? A male majoring in something other than engineering?

TABLE 13.13	*Current Disease*			
	Heart	Cancer	Diabetes	Total
Family history	880	440	380	1,700
No family history	920	760	620	2,300
Total	1,200	1,200	1,000	4,000

11 Table 13.13 summarizes the results of a recent health survey. Persons who were suffering from heart disease, cancer, or diabetes were asked whether there had been any known history of the disease in their family. If a person is selected at random from this sample of 4,000, what is the probability that:
(a) The person has cancer?
(b) The person had a family history of their particular disease?
(c) The person has cancer and had no family history of the disease?
(d) The person has diabetes, given that the person selected has a family history of their disease.
(e) The person has no family history of the disease, given that the person selected has heart disease.

TABLE 13.14	Product 1	Product 2	Product 3	Total
Acceptable	156	350	204	710
Unacceptable	24	40	26	90
Total	180	390	230	800

12 A sample of 800 parts has been selected from three product lines and inspected by the quality control department. Table 13.14 summarizes the results of the inspection. If a part is selected at random from this sample, what is the probability that:
(a) The part is of the product 1 type?
(b) The part is unacceptable?
(c) The part is an acceptable unit of product 3?
(d) The part is an unacceptable unit of product 1?
(e) The part is acceptable, given that the selected part is a unit of product 2?
(f) The part is product 1, given that the selected part is acceptable?
(g) The part is product 3, given that the selected part is unacceptable?
13 A pool of applicants for a welding job consists of 30 percent women and 70 percent men. Of the women, 60 percent have college degrees. Of the men, 40 percent have college degrees. What is the probability that a randomly selected applicant will be (a) a woman with a college degree and (b) a man without a college degree?
14 Suppose that E and F are events where $P(E) = .4$, $P(F) = .3$, and $P(E \cup F) = .6$. Determine (a) $P(E \cap F)$, (b) $P(E|F)$, and (c) $P(F|E)$.
15 Suppose that G and H are events where $P(G) = .25$, $P(H) = .45$, and $P(G \cup H) = .55$. Determine (a) $P(G \cap H)$, (b) $P(G|H)$, and (c) $P(H|G)$.

❏ KEY TERMS AND CONCEPTS

collectively exhaustive events 580
combination 571
complement 560
conditional probability 595
descriptive property method 559
element 559
enumeration method 559
factorial notation 569
finite sample space 577
fundamental principle of
    counting 568
infinite sample space 577
intersection of sets 563
joint probability 592
marginal probability 592
mutually exclusive events 580
null (empty) set 561
objective probability 583

outcomes 577
permutation 569
probability 582
probability tree 593
random experiment 569
relative frequency 582
sample space 577
set 559
set equality 562
statistical dependence 595
statistical independence 592
subjective probability 583
subset 561
tree diagram 567
union of sets 563
universal set 560
Venn diagram 562

❏ IMPORTANT FORMULAS

$$_nP_n = n! \tag{13.2}$$

$$_nP_r = n(n-1)(n-2)\cdots(n-r+1) \tag{13.3}$$

$$_nP_r = \frac{n!}{(n-r)!} \tag{13.4}$$

$$\binom{n}{r} = \frac{n!}{r!(n-r)!} \tag{13.6}$$

$$0 \le P(E) \le 1 \tag{13.7}$$

$$P(E') = 1 - P(E) \tag{13.8}$$

$$P(E_1 \cup E_2) = P(E_1) + P(E_2) \quad \text{(mutually exclusive)} \tag{13.9}$$

$$P(E_1 \cup E_2 \cup \cdots \cup E_n) = P(E_1) + P(E_2) + \cdots + P(E_n)$$
$$\text{(mutually exclusive)} \tag{13.10}$$

$$P(E_1 \cup E_2) = P(E_1) + P(E_2) - P(E_1 \cap E_2) \tag{13.11}$$

$$P(E_1 \cap E_2) = P(E_1)P(E_2) \quad \text{(independence)} \tag{13.12}$$

$$P(E_1 \cap E_2 \cap \cdots \cap E_n) = P(E_1)P(E_2) \cdots P(E_n)$$
$$\text{(independence)} \tag{13.13}$$

$$P(E_1|E_2) = P(E_1) \quad \text{(independence)} \tag{13.14}$$

$$P(E_1|E_2) = \frac{P(E_1 \cap E_2)}{P(E_2)} \quad \text{(dependence)} \tag{13.15}$$

$$P(E_1 \cap E_2) = P(E_2)P(E_1|E_2) \tag{13.16}$$

❏ ADDITIONAL EXERCISES

Section 13.1

1 Redefine set A using the descriptive property method if:
   (a) $A = \{\frac{1}{8}, \frac{1}{4}, \frac{1}{2}, 1, 2, 4, 8, 16, 32\}$
   (b) $A = \{\frac{1}{9}, \frac{1}{3}, 1, 3, 9, 27, 81\}$
   (c) $A = \{.001, .01, .1, 1, 10, 100, 1000\}$
   (d) $A = \{1, 4, 27, 256, 3125\}$

2 Given

$$\mathcal{U} = \{x|x \text{ is an integer greater than } -8 \text{ but less than } +9\}$$

$$A = \{a|a \text{ is an even positive integer less than } 15\}$$

$$B = \{b|b \text{ is an odd integer greater than } -5 \text{ but less than } +4\}$$

   (a) Define A'.
   (b) Define B'.

3 If $\mathcal{U}$ consists of all students enrolled in courses at a university, A consists of all male students, B consists of all students aged 35 years or over, and C consists of all engineering students, (a) define the set A', (b) define the set B', and (c) define the set C'.

4 If $\mathcal{U}$ consists of the different total scores possible on the roll of a pair of dice and B' consists of the scores of 3, 5, and 7, define B.

5 If

$$\mathcal{U} = \{1, 2, 3, 4, 5, 6, 7, 8, 9, 10\} \qquad B = \{1, 3, 5, 7, 9\}$$

$$A = \{1, 5, 9\} \qquad\qquad\qquad C = \{2, 4, 6, 8, 10\}$$

define all subset relationships which exist for these sets.

6 Draw a Venn diagram representing all the sets in Exercise 5.

7 Ten residents of a city were surveyed regarding their use of public transportation in that city. They were asked whether they had ridden the subway (S), the bus (B), or neither (N) during the past year. The responses were as follows:

Resident	1	2	3	4	5	6	7	8	9	10
Response	N	N	B	B, S	S	B, S	B, S	B, S	B	S

Draw a Venn diagram which summarizes how all residents responded to the survey.

8 Given the following sets, determine whether any sets are equal.

$$A = \{x|x^3 + 6x^2 + 9x = 0\} \qquad B = \{x|x^2 + 3x = 0\}$$

$$C = \{-3, 0\} \qquad\qquad\qquad D = \{-3, 0, 3\}$$

9 Given the sets

$$\mathcal{U} = \{x | x \text{ is a positive integer less than } 20\} \qquad A = \{5, 10, 15\}$$

$$B = \{2, 4, 6, 8, 10\} \qquad\qquad\qquad C = \{1, 5, 9, 15, 17\}$$

find:

(a) $A \cap B$

(b) $A \cup B \cup C$

(c) $A' \cap B'$

(d) $A' \cup C'$

(e) $A \cap B \cap C$

(f) $A' \cup B$

(g) $A' \cap B$

(h) $(A \cap B \cap C)'$

10 In Fig. 13.21, the numbers represent the number of elements contained in the various subsets. Determine:

(a) $n(A)$

(b) $n(A \cup B)$

(c) $n(A \cup B \cup C)$

(d) $n(\mathcal{U})$

(e) $n(A' \cup B)$

(f) $n(B' \cap C')$

(g) $n(B \cap C)$

(h) $n(A' \cap B' \cap C')$

Figure 13.21

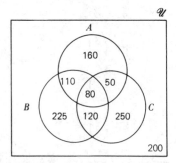

Section 13.2

11 A game consists of flipping a coin, followed by rolling a die. Draw a tree diagram which enumerates all possible outcomes for the game.

12 Stock Market Analysis A brokerage house analyzes market trends by selecting samples of stocks from different industries and noting from the previous days' trading whether there was *no change* in the price, a *decrease* in price, or an *increase* in the price of the stock. If two stocks are selected sequentially, draw a decision tree which enumerates all possible outcomes.

13 A used car wholesaler has agents who locate used cars and evaluate them for purposes of purchasing and reselling. The agents classify cars by size (full, medium, compact, and subcompact), age (0–2 years, 2–4 years, 4–6 years, and over 6 years), relative mileage for the age of the car (very high, high, average, below-average), and body condition (excellent, good, fair, and poor). Using the fundamental counting principle, determine the number of possible automobile classifications.

14 The wine steward at a gourmet restaurant is preparing to select wines for a special dinner. Four wines will be served during the evening. One wine will be served with the appetizer. Three choices are available for this wine. Four wines are being considered for serving with the salad. Five wines are being considered for the entree and three for after the dinner. How many possible combinations of wines could be considered for this meal?

Evaluate the following factorial expressions.

15 $\dfrac{10!}{3! \cdot 4!}$

16 $\dfrac{9!}{3! \cdot (6-4)!}$

17 $\dfrac{6!}{6! \cdot 0!}$

18 $\dfrac{8! \cdot 6!}{3! \cdot 4! \cdot 5!}$

19 $\dfrac{10!4!}{6!8!0!}$

20 $\dfrac{10!6!2!}{9!4!}$

21 $\dfrac{0!7!3!}{6!5!}$

22 $\dfrac{20!}{15!8!}$

23 $_8P_6 =$

24 $_6P_2 =$

25 $_7P_4 =$

26 $_8P_3 =$

27 $_6P_3 =$

28 $_8P_2 =$

29 $\dbinom{5}{2} =$

30 $\dbinom{6}{2} =$

31 $\dbinom{8}{5} =$

32 $\dbinom{7}{3} =$

33 $\dbinom{7}{2} =$

34 $\dbinom{8}{6} =$

35 A basketball coach has been frustrated in not being able to find the best 5 players to have in his starting lineup. There are 15 players on the team. If we assume that any of the 15 players can be selected for any of the five different positions in the starting lineup, how many different lineups are possible?

36 A political candidate wishes to visit six different cities. In how many different orders can the candidate visit these cities?

37 How many different telephone numbers can be dialed (or pushed) with a three-digit area code and a seven-digit regional number? (Assume 10 possible numbers for each digit.)

38 An automobile dealer has eight different car models. The dealer can display only four in the showroom. How many different combinations of cars can the dealer select for the showroom?

*39 **Olympic Tryouts** The Olympic basketball selection committee has cut the squad to 30 players. The head coach has decided that his final squad of 12 players should consist of 3 centers, 5 forwards, and 4 guards. Of the 30 remaining players, 8 are centers, 13 are forwards, and 9 are guards. How many different 12-player squads can be considered?

Section 13.3

40 Table 13.15 indicates some data gathered on a group of 3,000 victims of robbery, burglary, or both. Victims are classified as residential or business victims and by type of crime(s) committed. Suppose that a victim is selected at random from the 3,000. The sample space S for this experiment consists of the simple outcomes $S = \{R/RV, R/BV, R/RB, B/RV, B/BV, B/RB\}$.

    (a) Determine the set of simple outcomes used to define the compound event "victim of robbery, only."

    (b) Determine the set of simple outcomes used to define the compound event "victim of burglary."

TABLE 13.15	(RV) Robbery Victim	(BV) Burglary Victim	(RB) Robbery and Burglary Victim	Total
Residence (R)	250	1,200	450	1,900
Business (B)	400	250	450	1,100
Total	650	1,450	900	3,000

41 In Exercise 40, determine, for each of the following sets of events, whether they are mutually exclusive and/or collectively exhaustive.

    (a) $\{R/RV, B/RV, BV, RB\}$

    (b) $\{B/RV, B/BV, B/RB, RV, BV\}$

    (c) $\{R, B, RV, RB\}$

    (d) $\{R/RV, B/RV, R/BV, RB\}$

42 In Exercise 40, suppose that one victim is selected at random from this group. What is the probability that the victim will (a) be a residential victim, (b) be a victim of robbery only, (c) be a commercial victim of both robbery and burglary?

43 The probability that an applicant for pilot school will be admitted is .3. If three applicants are selected at random, what is the probability that (a) all three will be admitted, (b) none will be admitted, and (c) only one will be admitted?

44 A student estimates the probability of receiving an A in a course at .3 and the probability of receiving a B at .4. What is the probability that the student (a) will not receive an A, (b) will not receive a B, and (c) will receive neither an A nor a B?

45 Soviet Religion Table 13.16 reflects the religious preferences of persons living in the Soviet Union during the mid-1980s.* If a Soviet person were selected at random during the mid-1980s, what is the probability that the person (a) would be atheist or nonreligious, (b) would practice some form of religion, (c) would be a Protestant, Catholic, or Jew, and (d) would be a non-Muslim?

46 An urn contains 8 green-dotted balls, 10 green-striped balls, 12 blue-dotted balls, and 10 blue-striped balls. If a ball is selected at random from the urn, what is the probability that the ball will be (a) green or striped, (b) dotted, (c) blue or dotted?

---

* *Source: World Christian Encyclopedia.*

TABLE 13.16	Religious Preference	Number of Persons, Millions
	Orthodox	84
	Muslim	30
	Protestant	8
	Catholic	5
	Jew	3
	Other religions	1
	Atheists or nonreligious	137

47 **Credit Ratings** A credit-rating agency rates a person's credit standing as "excellent," "good," "fair," or "poor." The probability that a person will have an excellent rating is .20. The probability of a good rating is .35. What is the probability that a person (a) will not have an excellent rating, (b) will not have at least a good rating, and (c) will have no better than a good rating?

48 **Wine Cooler Preferences** A survey of 2,400 consumers was conducted to determine their purchasing behavior regarding two leading wine coolers. It was found that during the past summer 600 had purchased brand A, 400 had purchased brand B, and 100 had purchased both brands A and B. If a person is selected at random from this group, what is the probability that the person (a) would have purchased brand A, (b) would have purchased brand A but not brand B, (c) would have purchased brand A, brand B, or both, and (d) would not have purchased either brand?

49 **Home Run Production** Table 13.17 indicates some data gathered by the baseball commissioner's office. Indicated in the table is the number of home runs hit in a single baseball game and the probability of that number being hit. If a game is selected at random, what is the probability (a) that no more than three home runs will be hit, (b) that at least one home run will be hit, (c) that fewer than five home runs will be hit, and (d) that between one and three home runs (inclusive) will be hit?

TABLE 13.17	Home Runs/Game, (n)	P(n)
	0	.12
	1	.18
	2	.26
	3	.22
	4	.12
	More than 4	.10

Section 13.4

50 The probability that a customer entering a particular store will make a purchase is .40. If two customers enter the store, what is the probability that (a) they will both make a purchase, (b) neither will make a purchase, and (c) precisely one of the two will make a purchase?

51 A single die is rolled and each side has an equal chance of occurring. What is the probability of rolling four consecutive 6s?

52 In the previous exercise, what is the probability that the four rolls will result in the same outcome each time?

53 A ball is selected at random from an urn containing three red-striped balls, eight solid red balls, six yellow-striped balls, four solid yellow balls, and four blue-striped balls.
(a) What is the probability that the ball is yellow, given that it is striped?
(b) What is the probability that the ball is striped, given that it is red?
(c) What is the probability that the ball is blue, given that it is solid-colored?

54 The probability that the price of a particular stock will increase during a business day is .4. If the nature of the change in price on any day is independent of what has happened on previous days, what is the probability that the price will (a) increase 4 days in a row, (b) remain the same or decrease 3 days in a row, and (c) increase 2 days out of 3?

55 Suppose that E and F are events and $P(E) = .2$, $P(F) = .5$, and $P(E \cup F) = .60$. Determine (a) $P(E \cap F)$, (b) $P(E|F)$, and (c) $P(F|E)$.

❏ **CHAPTER TEST**

1 Given the sets $A = \{1, 2, 3, 4, 5, 6, 7, 8\}$, $B = \{-2, 0, 2, 4, 6, 8, 10\}$, and $C = \{-3, -2, -1, 0, 1, 2, 3\}$, determine the sets:
(a) $A \cap B$
(b) $A \cap B \cap C$
(c) $A \cup B \cap C$

2 What is the difference between the states of statistical independence and statistical dependence?

3 A grocer has display space for three products. He has six products that he would like to display.
(a) How many different arrangements of three products can be made? •
(b) How many different combinations of the six products could he put on display?

4 What is the probability of drawing three cards, without replacement, from a deck of cards and getting three kings?

5 An urn contains 18 red balls, 14 red-striped balls, 16 yellow balls, and 12 yellow-striped balls.
(a) Given that a ball selected from the urn is striped, what is the probability it is yellow?
(b) Given that a ball selected from the urn is not striped, what is the probability it is red?

MINICASE

THE BIRTHDAY PROBLEM

A classic application of probability theory relates to the likelihood, within a group of persons, that two or more have the same birthday. The probability that two or more persons have the same birthday obviously depends upon the size of the group of persons considered. The larger the group, the higher the probability. Let's make some assumptions related to this problem. Assume that there are 365 different birthday possibilities (which ignores leap year) and that for any given person, the 365 days are equally likely to be the person's birthday.

To determine the probability that two or more persons within a group have the same birthday, it is easier to compute the probability that no two persons have the same birthday. Think about this for a moment. The event "two or more persons have the same birthday" consists of many possibilities. These possibilities must account for "three or more," "four or more," etc. They must also account for subsets of persons having the same birthdays (e.g., two persons born on January 5 and three persons born on April 26). Thus, if one can determine the probability p that no two persons have the same birthday, the desired probability can be computed as $1 - p$.

Given a group of n persons, selected at random, the number of different birthday outcomes is

$$T = \overbrace{(365)(365)(365) \cdots (365)}^{n} = (365)^n$$

If, within this group of n persons, no two have the same birthday, there must be n different birth dates. The number of different outcomes which satisfy this event is computed as

$$O = (365)(364)(363) \cdots (365 - n + 1)$$

Thus, the probability that, within the group of n persons, *no two persons have the same birthday* equals

$$p = \frac{O}{T} = \frac{(365)(364)(363) \cdots (365 - n + 1)}{(365)^n}$$

Required:

a. Discuss the logic behind the computation of T.

b. Discuss the logic behind the computation of O.

c. Compute the probability that two or more persons will have the same birthday in a group of five randomly selected persons.

d. Compute the same probability for groups of 10, 20, 30, 40, and 50 persons.

e. What is the smallest group of persons for which the probability of two or more having the same birthday exceeds .50? Exceeds .75?

CHAPTER 14

PROBABILITY DISTRIBUTIONS

CHAPTER OBJECTIVES

❏ Introduce the notion of a random variable

❏ Provide an understanding of probability distributions and their attributes

❏ Acquaint readers with the characteristics and usage of the binomial probability distribution

❏ Acquaint readers with the characteristics and usage of the normal probability distribution

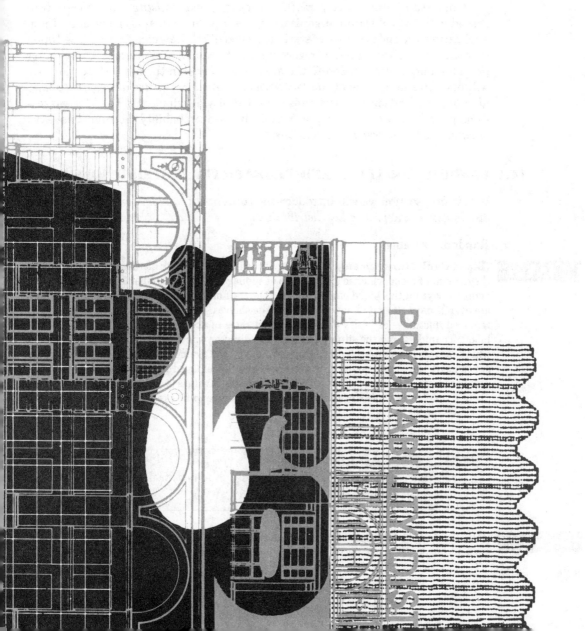

A major bank issues credit cards under the name VISACARD. It has determined that, on the average, 40 percent of all credit card accounts are paid in full following the initial billing. That is, for any given month, 40 percent of all accounts which had new charges reflected on their latest bill will not incur interest charges. If a sample of six accounts is selected at random, estimate the probability that none of the six will have incurred interest charges for their last billing (Example 18).

One area of study within mathematics is called *statistics*. Statistics is concerned with the collection, organization, description, and analysis of data. In an application, the end objective of statistics is usually to draw conclusions based upon data which have been gathered regarding the phenomenon of interest. This chapter provides a brief introduction to the area of statistics. In the first two sections we will develop the notions of *random variables, frequency distributions, probability distributions,* and special attributes of probability distributions. The last two sections introduce two of the more frequently used probability distributions—the *binomial* and the *normal distributions*.

14.1 RANDOM VARIABLES AND PROBABILITY DISTRIBUTIONS

In this first section we will introduce the concepts of *random variables, frequency distributions,* and *probability distributions*.

Random Variables

EXAMPLE 1
(Blood Bank Management) The director of a local blood bank has been concerned about the amount of blood which is lost because of spoilage. Blood is perishable, the legal lifetime being 21 days for this blood bank. Each day the staff must identify any units of blood which are over 21 days old and remove them from the shelves. The number of units which must be removed fluctuates each day depending upon rates of acquisition and rates of use of the blood. To understand the situation better, the director wants to conduct an experiment in which records are kept each day on the number of units removed from inventory. For this experiment, the simple events are the number of units removed each day. If X represents the number of units removed, the sample space consists of values 0, 1, 2, 3, ... X is an example of a *random variable* since the value of X fluctuates in no predictable manner.

□

DEFINITION: RANDOM VARIABLE
A *random variable* is a function which assigns a numerical value to each simple event in a sample space.

EXAMPLE 2
(IRS Revisited) Example 30 in Chap. 13 (page 593) discussed the likelihood that errors are made on personal income tax returns. It was estimated that the probability of an error on

TABLE 14.1	Simple Events in S	Random Variable (Number of Returns Containing Errors) X	Probability
	EEE	3	$(.4)(.4)(.4) = .064$
	EEN	2	$(.4)(.4)(.6) = .096$
	ENE	2	$(.4)(.6)(.4) = .096$
	ENN	1	$(.4)(.6)(.6) = .144$
	NEE	2	$(.6)(.4)(.4) = .096$
	NEN	1	$(.6)(.4)(.6) = .144$
	NNE	1	$(.6)(.6)(.4) = .144$
	NNN	0	$(.6)(.6)(.6) = .216$

a return selected at random is .4. The example discussed an experiment in which three returns were selected at random and examined for errors. Figure 13.19 (page 594) portrays the probability tree for this experiment.

The sample space S for this experiment consists of the set of simple events

$$S = \{EEE, EEN, ENE, ENN, NEE, NEN, NNE, NNN\}$$

where each simple event represents a possible outcome from auditing three income tax returns. Table 14.1 summarizes these simple events along with their probabilities of occurrence. For this experiment, there is probably less interest in the sequence of findings but greater interest in the number of returns found to contain errors. The number containing errors in a sample of three returns is 0, 1, 2, or 3. Thus, we might define a random variable X for this experiment, where X represents the number of returns found to contain errors. The random variable X assigns each simple event in the sample space one number, 0, 1, 2, or 3. These assignments are also shown in Table 14.1.

❏

If a random variable can assume only a countable number of distinct values, it is called a ***discrete random variable.*** The outcomes of an experiment which measures the number of units of a product demanded each day can be represented by a discrete random variable. The outcomes of an experiment which measures the number of cars passing through a tollbooth each hour can be represented by a discrete random variable.

A random variable which can assume any of the infinitely many values within some interval of real numbers is called a ***continuous random variable.*** In an experiment which selects people at random and records some attribute such as height or weight, the outcomes can be represented by a continuous random variable. The outcomes of an experiment which measures the length of time that a transistor will operate before failing can be described by a continuous random variable.

Frequency Distributions

When experiments are conducted and observations made regarding the values of selected random variables, the data can be studied to determine whether any meaningful conclusions can be reached. A common tool used in these studies is a *frequency distribution.* A frequency distribution summarizes each possible value for a random variable along with the number of occurrences, or the *frequency,* for that value.

EXAMPLE 3

(Blood Bank Management, continued) Consider the blood bank example discussed earlier. To understand the perishability situation better, the director has conducted an experiment over an 80-day period. At the end of each day *the number of units removed from inventory because of spoilage, X,* is noted. This is the random variable in the experiment. Table 14.2 is a frequency distribution summarizing the results of the experiment. This frequency distribution provides a convenient display of the data tabulated in this experiment.

TABLE 14.2 **Blood Perishability Frequency Distribution**

Units Removed from Inventory (X)	Occurrences
0	2
1	6
2	8
3	8
4	10
5	16
6	14
7	10
8	6
	80

After studying the data, the director implemented a new program of blood management, making changes in blood collection programs and blood exchange procedures with other blood banks participating in a regional blood bank cooperative. To determine whether the new policies resulted in any changes in loss rates, a similar experiment was conducted for a period of 50 days. The frequency distribution in Table 14.3 summarizes the results.

TABLE 14.3

Units Removed from Inventory (X)	Occurrences
0	3
1	5
2	8
3	12
4	9
5	5
6	4
7	3
8	1
	50

TABLE 14.4	Units Removed from Inventory (X)	Relative Frequency
	0	2/80 = 0.025
	1	6/80 = 0.075
	2	8/80 = 0.100
	3	8/80 = 0.100
	4	10/80 = 0.125
	5	16/80 = 0.200
	6	14/80 = 0.175
	7	10/80 = 0.125
	8	6/80 = 0.075
		80/80 = 1.000

TABLE 14.5	Units Removed from Inventory (X)	Relative Frequency
	0	3/50 = 0.060
	1	5/50 = 0.100
	2	8/50 = 0.160
	3	12/50 = 0.240
	4	9/50 = 0.180
	5	5/50 = 0.100
	6	4/50 = 0.080
	7	3/50 = 0.060
	8	1/50 = 0.020
		50/50 = 1.000

Although Tables 14.2 and 14.3 display data gathered on the same random variable, it is not easy to compare the results of the two experiments directly because of the differing lengths of the experiments (80 days versus 50 days). To make the data more comparable, we can compute for each experiment the *relative frequency* of each value of the random variable. This is done by dividing the number of occurrences (the frequency) by the total number of observations in the experiment. Tables 14.4 and 14.5 express the results of the experiments using relative frequencies.

We will not take time to try to reach conclusions regarding the results of these two experiments. A quick glance gives the impression that the new policies may have resulted in fewer losses (based on the 50-day experiment). Our concern in this discussion has been to illustrate the use of frequency distributions as tools for tabulating and displaying the results of studies of random processes.

❑

Probability Distributions

A *probability distribution* is a complete listing of all possible values of a random variable along with the probabilities of each value. If a discrete random variable X can assume n values $x_1, x_2, x_3, \ldots, x_n$ having respective probabilities of occur-

TABLE 14.6	**Generalized Discrete Probability Distribution**	
Value of Random Variable $(X = x_i)$	**Probability $P(x_i)$**
x_1	p_1
x_2	p_2
x_3	p_3
.	.
.	.
.	.
x_n	$\underline{p_n}$
	1.0

rence $p_1, p_2, p_3, \ldots , p_n$, the corresponding ***discrete probability distribution*** is as illustrated in Table 14.6. Since all possible values of a random variable are included in a probability distribution, the sum of the probabilities always will equal 1. The relative frequencies shown in Tables 14.4 and 14.5 can be interpreted as probabilities. Thus these tables provide examples of two discrete probability distributions.

EXAMPLE 4

In Example 2, we discussed an experiment where three personal income tax returns would be selected at random and examined to determine whether they contained any errors. Table 14.1 showed the assignment of values of the random variable to each simple event in the sample space. Given the probabilities of each simple event and the definition of the random variable, the next logical step is to construct the corresponding probability distribution.

The possible values for the random variable "number of returns containing errors" are 0, 1, 2, and 3. To construct the probability distribution, we simply identify all simple events associated with each specific value of the random variable and add the probabilities of these events to arrive at the probability of that value of the random variable. For example, the simple events associated with the value $X = 2$ are *EEN, ENE,* and *NEE*. Adding the probabilities of these three mutually exclusive events, as shown in Table 14.7, we conclude that $P(X = 2) = .288$. If a similar process is used for the other values of X, the result is the probability distribution in Table 14.8.

TABLE 14.7	**Simple Events in S**	**Random Variable (Number of Returns Containing Errors) X**	**Probability**
	EEE	3	.064
	EEN	2	.096
	ENE	2	.096
	ENN	1	.144
	NEE	2	.096
	NEN	1	.144
	NNE	1	.144
	NNN	0	.216

$P(X = 2) = .288$

TABLE 14.8	**Discrete Probability Distribution for IRS Example**	
Number of Returns Containing Errors (X)		**Probability $P(X)$**
0		.216
1		.432
2		.288
3		.064
		1.000

The following general properties should be remembered regarding discrete probability distributions.

> **PROPERTIES OF DISCRETE PROBABILITY DISTRIBUTIONS**
> Given a discrete random variable X which can assume n values x_1, x_2, x_3, . . . , x_n:
>
> 1 *Only one probability $P(X = x_i)$ is to be assigned to each value of the random variable.*
> 2 *$0 \le P(X = x_i) \le 1$ for all x_i.*
> 3 *$P(X = x_1) + P(X = x_2) + P(X = x_3) + \cdots + P(X = x_n) = 1.0$.*

Histograms

Histograms provide an effective graphical representation of frequency distributions and discrete probability distributions. A histogram is a type of bar graph consisting of a set of rectangles, one for each possible outcome or value of the random variable. The width of each rectangle equals 1, and the height equals the probability (or relative frequency) of the particular outcome. Figure 14.1 shows histograms for the two probability distributions in the blood bank example. Providing a picture of the two distributions, the histograms allow for easier comparison of the results from the two experiments. Without performing any formal analysis, the histograms seem to suggest that the newly instituted policies may have had an effect in reducing losses due to spoilage.

One further point should be made regarding histograms. When they are used to represent discrete probability distributions, the area of each rectangle equals the probability of occurrence of the corresponding value of the random variable. This is because the width of each rectangle equals 1 and the height equals the probability. Thus if you compute the area of each rectangle and add these together for all rectangles, the total area will equal 1.0.

EXAMPLE 5	**(Bank Queues)** A local bank has been concerned about the length of time its customers must wait before being serviced by a teller. A study of 500 customers has resulted in the

Figure 14.1

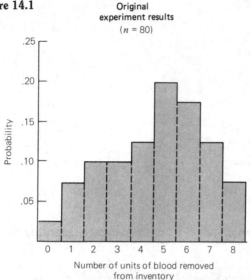

Original
experiment results
($n = 80$)

Number of units of blood removed
from inventory

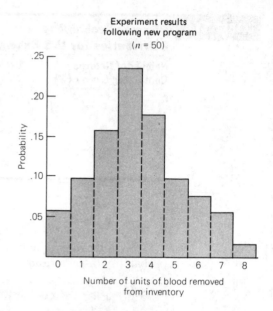

Experiment results
following new program
($n = 50$)

Number of units of blood removed
from inventory

probability distribution in Table 14.9. Waiting time (in minutes) per customer is the random variable.

(*a*) What is the probability a customer will wait for a teller?

(*b*) What is the probability that a customer will wait less than 2 minutes? More than 3 minutes?

TABLE 14.9	Waiting Time, X, Minutes	$P(X)$
	0	.32
	1	.24
	2	.18
	3	.12
	4	.09
	5	.05
		1.00

SOLUTION

(*a*) $P(\text{wait}) = P(X = 1) + P(X = 2) + P(X = 3) + P(X = 4) + P(X = 5)$
$= .24 + .18 + .12 + .09 + .05$
$= .68$

This result could have been determined as

$$P(\text{wait}) = 1 - P(\text{no wait})$$
$$= 1 - P(X = 0)$$
$$= 1 - .32$$
$$= .68$$

(b)
$$P(\text{wait} < 2 \text{ minutes}) = P(X = 0) + P(X = 1)$$
$$= .32 + .24$$
$$= .56$$

$$P(\text{wait} > 3 \text{ minutes}) = P(X = 4) + P(X = 5)$$
$$= .09 + .05$$
$$= .14$$

❑

Section 14.1 Follow-up Exercises

1 Given the following random variables for a series of experiments, which are discrete and which are continuous?
(a) The weights of students at a high school
(b) The number of cigarettes smoked on a daily basis
(c) The body temperature of a person at any given time
(d) The length of a newborn baby
(e) The number of applications for welfare received each day by a social agency
(f) The amount of water used by a community each day
(g) The number of grains of sand on a given beach each day
(h) The length of life of a size AA battery

2 **Public Works** The director of public works for a New England city has checked the city records to determine the number of major snowstorms which have occurred in each of the last 50 years. Table 14.10 presents a frequency distribution summarizing the findings.
(a) Construct the probability distribution for this study.
(b) Draw a histogram for this distribution.
(c) What is the probability that there will be more than two major storms in a given year? Three or fewer?

TABLE 14.10	Number of Storms	Frequency
	0	3
	1	5
	2	10
	3	13
	4	8
	5	16
	6	5
		60

3 **Fire Protection** The fire chief for a small volunteer fire department has compiled data on the number of false alarms called in each day for the past 360 days. Table 14.11 presents a frequency distribution summarizing the findings.
(a) Construct the probability distribution for this study.
(b) Draw a histogram for the distribution.
(c) What is the probability that fewer than four false alarms will be called in on any given day? Three or more?

TABLE 14.11	Number of False Alarms	Frequency
	0	75
	1	80
	2	77
	3	40
	4	28
	5	24
	6	20
	7	16
		360

4 **Quality Control** Production runs for a particular product are made in lot sizes of 100 units. Each unit is inspected to ensure that it is not defective in any way. The number of defective units per run seems to be random. A quality control engineer has gathered data on the number of defective units for each of the last 50 production runs. Table 14.12 presents a frequency distribution summarizing the findings.

(a) Construct the probability distribution for this study.

(b) Draw a histogram for the distribution.

(c) What is the probability that a production run will result in fewer than 10 defective units? More than 10?

TABLE 14.12	Number of Defective Units	Frequency
	5	3
	6	4
	7	4
	8	6
	9	9
	10	11
	11	8
	12	4
	13	0
	14	1
		50

5 **Drunken Driving** Local police have instituted a roadblock program for checking the sobriety of drivers. Cars are selected at random and drivers checked for signs of excessive drinking. If it is suspected that a driver has had too much to drink, standard tests of sobriety are administered. In reviewing the success of the program, a police lieutenant has compiled data on 150 roadblock efforts. The lieutenant is interested in the number of "hits," or drivers found to be legally drunk for each roadblock effort. Table 14.13 summarizes the findings.

(a) Construct the probability distribution for this study.

(b) Draw a histogram for the distribution.

TABLE 14.13	Number of Drunken Drivers	Frequency
	0	8
	1	20
	2	26
	3	28
	4	26
	5	16
	6	14
	7	10
	8	2
		150

(c) What is the probability a roadblock effort will identify any drunken drivers? Five or more?

6 Construct the discrete probability distribution which corresponds to the experiment of tossing a *fair* coin three times. Let the random variable X equal the number of heads occurring in three tosses. What is the probability of two or more heads?

7 Construct the discrete probability distribution which corresponds to the experiment of the single roll of a *pair* of dice. Assume equal likelihood of occurrence of each side of a die and let the random variable X equal the sum of the dots which appear on the pair.

8 **Unemployment** Unemployment statistics within a western state indicate that 6 percent of those eligible to work are unemployed. Suppose that an experiment is conducted where three persons are selected at random and their employment status is noted. If the random variable for this experiment is defined as the number of persons unemployed, (a) construct the probability distribution for this experiment, and determine the probability that (b) none of the three is unemployed or (c) two or more are employed.

9 Table 14.14 is a probability distribution for the random variable X.
(a) Determine the probability distribution for the random variable X^2.
(b) Determine the probability distribution for the random variable $X + 1$.

TABLE 14.14	X	P(X)
	1	.15
	2	.20
	3	.30
	4	.25
	5	.15

10 Given that a random variable can assume values of 0, 1, 2, and 3, which of the following cases satisfy the conditions for being probability distributions?
(a) $P(X = 0) = \frac{1}{8}$, $P(X = 1) = \frac{1}{3}$, $P(X = 2) = 0$, $P(X = 3) = \frac{1}{2}$
(b) $P(X = 0) = .2$, $P(X = 1) = .3$, $P(X = 2) = .2$, $P(X = 3) = .1$
(c) $P(X = 0) = .1$, $P(X = 1) = .25$, $P(X = 2) = .15$, $P(X = 3) = .2$, $P(X = 4) = .3$
(d) $P(X = 0) = .18$, $P(X = 1) = .23$, $P(X = 2) = .26$, $P(X = 3) = .33$

14.2 MEASURES OF CENTRAL TENDENCY AND VARIATION

Given a set of data gathered during an experiment, it is often of interest to describe the data using a single measure or number. The number which we choose depends upon the particular attribute we wish to describe. For some experiments, we may wish to describe the extremes of the data values. In these cases we might identify the smallest value and/or the largest value in the data set. For example, a study of multiple births in mice resulting from the use of a fertility drug might focus on the largest number of births to any one mouse. In another experiment, we may be interested in the total or sum of all values in the data set. In an experiment which records the number of points scored each game by Larry Bird or Michael Jordan in a given season, we might want to describe the data by determining the total number of points scored during the season.

The Mean

In most experiments, there is an interest in describing the center, or middle, of the set of data. There are several measures which can be used to describe this attribute. These are commonly referred to as *measures of central location* or *measures of central tendency.* Of the different measures of central location, the *arithmetic mean,* or more simply the *mean,* is the most widely used. This measure is also referred to (in everyday language) as the *average. The mean ($\bar{x}$) of a set of values $x_1, x_2, x_3, \ldots, x_n$ is the sum of the values divided by the total number of values (n) in the set,* or

$$(\bar{x}) = \frac{x_1 + x_2 + x_3 + \cdots + x_n}{n} \tag{14.1}$$

EXAMPLE 6   A student has taken five exams in a college mathematics course. The scores on the five exams were 78, 96, 82, 72, and 92. The mean score for these five exams is

$$\bar{x} = \frac{78 + 96 + 82 + 72 + 92}{5}$$

$$= \frac{420}{5} = 84.0$$

Notice, in this example, that the mean does not equal any of the actual scores. However, it provides a measure of the "average" performance of the student for the five exams.

EXAMPLE 7   The biweekly payroll for the 3,225 employees of a large manufacturing company is $3,100,000. Determine the mean biweekly salary for this firm.

SOLUTION

In this example, we are not given the individual observations (biweekly salaries) for each employee. However, we are given the total for these observations and the number of employees. Therefore, the mean biweekly salary is computed as

$$\bar{x} = \frac{\$3,100,000}{3,225}$$

$$= \$961.24 \quad \text{(rounded to the nearest cent)}$$

❑

For large data sets where different outcomes occur with different frequencies, Eq. (14.1) can be modified to reduce the computational burden. **Given a set of values** $x_1, x_2, x_3, \ldots, x_n$, **which occur with respective frequencies,** f_1, $f_2, f_3, \ldots, f_n$, **the mean is computed as**

$$\bar{x} = \frac{x_1 \cdot f_1 + x_2 \cdot f_2 + x_3 \cdot f_3 + \cdots + x_n \cdot f_n}{f_1 + f_2 + f_3 + \cdots + f_n} \qquad (14.2)$$

EXAMPLE 8 Table 14.2 (on page 616) displayed the frequency distribution for the 80-day experiment at the blood bank. We can determine the mean number of units removed from inventory per day by using Eq. (14.2).

$$\bar{x} = \frac{(0)(2) + (1)(6) + (2)(8) + (3)(8) + (4)(10) + (5)(16) + (6)(14) + (7)(10) + (8)(6)}{2 + 6 + 8 + 8 + 10 + 16 + 14 + 10 + 6}$$

$$= \frac{368}{80}$$

$$= 4.6$$

For the 80-day experiment, the mean number of units removed from inventory due to spoilage was 4.6 units per day.

❑

PRACTICE EXERCISE

Compute the mean for the following set of outcomes.

Value of x	Frequency of occurrence
20	5
30	8
40	12
50	6

Answer: 36.129.

Given a set of data, extreme values in the data set can have a severe effect on the mean. Suppose that five persons have participated in a small golf tournament and their respective scores are 72, 76, 80, 78, and 140. The mean score for this group is 89.2; however, the high score of 140 contributes a significant effect when computing the mean. The mean score for the other four players is 76.5. For data sets such as this, other kinds of measures can be used to describe central location.

The Median

One other measure of central location is the *median.* **When the total number of data items** *is odd, the median is the middle data value when the data items are arranged in increasing or decreasing order.* For the golfing example, the median score is found by first arranging the scores in increasing (or decreasing) order. As you can see, the middle item is 78, and this is the median for the data set.

$$72 \quad 76 \quad \textcircled{78} \quad 80 \quad 140$$

When the total number of data items *is even,* the median is the mean of the two middle items in the data set.

EXAMPLE 9   **(Audubon Society)** A state chapter of the Audubon Society has spent the weekend conducting a population survey for a particular species of bird. Twenty society members participated and their confirmed total sightings are listed below, arranged in increasing order of magnitude. Given an even number of data items, the two middle numbers are the ones in positions 10 and 11. For this data set, the numbers in these positions are "9" and "10." There are 9 data items less than (or equal to) the "9," and there are 9 data items greater than the "10." Therefore, the median is the mean of these two middle items, or the median equals $(9 + 10)/2 = 9.5$

$$4, 5, 5, 6, 7, 7, 9, 9, 9, \textcircled{9, 10,} 12, 13, 15, 15, 17, 20, 25, 26, 30$$

❑

The Mode

One other measure which is sometimes used to describe the central tendency of a data set is the *mode. The mode is the data value which occurs with the greatest frequency.* There are two advantages of using the mode as a measure of central location: (1) it requires no calculations and (2) it can be used to describe quantitative data and qualitative data (e.g., color, sex, and letter grade in a course).

EXAMPLE 10   In the previous example, the mode is 9. That is, the number of sightings which occurred with the greatest frequency was 9. Four Audubon members reported nine sightings. The next-highest frequency was two and this was associated with three different values in the data set: "5," "7," and "15."

EXAMPLE 11   **(Qualitative Data Sets)** A new regional airline surveyed its first 20,000 passengers in order to get feedback on the airline's performance. Each passenger was asked to evaluate the overall quality of service. Table 14.15 summarizes the responses to this question. Notice that

TABLE 14.15	Response	Respondents
	Excellent	4,532
	Very good	8,849
	Good	5,953
	Poor	531
	Terrible	135

the responses in this experiment are qualitative, as opposed to quantitative. Yet we can state that the mode, or "modal response," for this experiment is "very good." More passengers gave this response than any other.

❏

The Mean of a Discrete Probability Distribution

> **MEAN OF DISCRETE PROBABILITY DISTRIBUTION**
> If a discrete random variable X can assume n values $x_1, x_2, \ldots, x_n$ having respective probabilities of occurrence $p_1, p_2, \ldots, p_n$, the *mean value μ* (mu) of the random variable is
>
> $$\mu = x_1 p_1 + x_2 p_2 + \cdots + x_n p_n \qquad (14.3)$$

POINT FOR THOUGHT & DISCUSSION   Equation (14.3) follows directly from Eq. (14.2). Can you show this?

EXAMPLE 12   Suppose that Tables 14.4 and 14.5 (on page 617) presented the actual probability distributions for the number of units of blood spoiling on a given day in the blood bank example. Let's compute the means for these two distributions using Eq. (14.3). For Table 14.4,

$$\mu_1 = 0(.025) + 1(.075) + 2(.100) + 3(.100) + 4(.125) + 5(.200) + 6(.175) + 7(.125) \\ + 8(.075)$$
$$= 4.60 \text{ units per day}$$

For Table 14.5,

$$\mu_2 = 0(.060) + 1(.100) + 2(.160) + 3(.240) + 4(.180) + 5(.100) + 6(.080) + 7(.060) \\ + 8(.020)$$
$$= 3.42 \text{ units per day}$$

Thus, before implementation of the new program of blood management, the mean number of units of blood spoiling per day was 4.6 (the same value as computed in Example 8). After implementation of the new policies, the mean was 3.42 units per day. *On average,* there appears to have been a decrease in the rate of spoilage.

EXAMPLE 13   **(Drug Hot Line)** A city has set up a drug abuse hot line in order to offer assistance to persons who are seeking help with their drug problems. The director of the program has gathered data on the number of calls received each day. Table 14.16 contains the probability distribution for this random variable. What is the mean number of calls per day?

SOLUTION
Using Eq. (14.3), the mean number of calls per day is

$$\mu = 5(.08) + 6(.14) + 7(.18) + 8(.24) + 9(.16) + 10(.10) + 11(.08) + 12(.02)$$
$$= 7.98$$

TABLE 14.16	Calls per Day (X)	$P(X)$
	5	.08
	6	.14
	7	.18
	8	.24
	9	.16
	10	.10
	11	.08
	12	.02
		1.00

□

The Standard Deviation

For probability distributions, the mean provides a measure of the central location of the data. This is only one attribute of the data set. In order to describe the data more fully, it is useful to look at other attributes. One important attribute which the mean does not describe is the *spread* or variability in the data. Consider the two probability distributions in Table 14.17. Both of these distributions have the same mean, $\mu = 50$. However, the variation in the values of the two random variables is quite different.

TABLE 14.17	X_1	$P(X_1)$	X_2	$P(X_2)$
	49	.05	0	.05
	50	.90	50	.90
	51	.05	100	.05

In order to describe the variability that exists within a data set, some ***measure of variation*** can be used. One way of measuring variation in a data set is by noting the extremes of variation. The ***range*** for a data set is the difference between the largest and smallest values in the set. Let's refer to the five golfers and their scores of 72, 76, 78, 80, and 140. The lowest score for this group was 72 and the highest was 140. Therefore, the range for this set of data is $140 - 72 = 68$. Looking at this as a measure of variation in the scores might suggest considerable variability. The reality for this data set is that four of the scores have relatively little variation from one another. The score of 140 is atypical. This example illustrates that the range is a measure which is easy to calculate and easy to understand. However, it fails to provide any information about any data values which lie between the extremes.

An important attribute of a measure of variation is that it should be small when data values are grouped closely around the mean and should be large when data values are scattered widely about the mean. Thus one way of measuring variation might be to measure the distance each data point is from the mean. If a sample data set consists of n items $x_1, x_2, \ldots, x_n$, and the mean for this data set is $\bar{x}$, the distances of the n data points from the mean would be represented by $(x_1 - \bar{x})$, $(x_2 - \bar{x}), \ldots, (x_n - \bar{x})$. These differences are referred to as ***deviations from the mean.*** If we add these deviations together and divide by n, the result will be the ***average deviation from the mean.*** Unfortunately, it always turns out that the sum of the deviations from the mean equals zero. (See Follow-up Exercise 18.) The sum of all positive deviations from the mean always equals the sum of the negative deviations; consequently, they offset one another. Therefore, the average deviation from the mean will always equal zero. As a result, this is not a good measure of variation.

Even though the *sum* of the deviations from the mean equals zero, each *individual* deviation is a measure of variation for that particular data item. If we take the individual deviations and square them, it turns out that we can come up with a useful measure of variation. Squaring the individual deviations and dividing by n results in the *mean of the squares of the deviations*. The mean of the squares of the deviations is more commonly referred to as the ***variance. If a data set consists of n items*** $x_1, x_2, \ldots, x_n$, ***and the mean of this set is*** $\bar{x}$, ***the variance for the data set is***

$$\text{Var}(x) = \frac{(x_1 - \bar{x})^2 + (x_2 - \bar{x})^2 + \cdots + (x_n - \bar{x})^2}{n} \qquad (14.4)$$

Because we square the deviations from the mean in order to arrive at the variance, this measure of variation does not accurately reflect the actual magnitude of variation in the data set. To correct for this, another measure of variation can be defined by taking the square root of the variance. This measure of variation is called the ***standard deviation*** and it is represented by the Greek symbol σ (sigma). Thus,

$$\sigma = \sqrt{\text{Var}(x)} = \sqrt{\frac{(x_1 - \bar{x})^2 + (x_2 - \bar{x})^2 + \cdots + (x_n - \bar{x})^2}{n}} \qquad (14.5)$$

The standard deviation is the most commonly used measure of variation for a set of random numbers.

EXAMPLE 14

In Example 6, we determined the mean score for a student who had taken five exams during a college mathematics course. The mean score, given the five exam scores of 78, 96, 82, 72, and 92, was $\bar{x} = 84.0$. The variance for this set of data items is

$$
\begin{aligned}
\text{Var}(x) &= \frac{(78 - 84)^2 + (96 - 84)^2 + (82 - 84)^2 + (72 - 84)^2 + (92 - 84)^2}{5} \\
&= \frac{(-6)^2 + (12)^2 + (-2)^2 + (-12)^2 + (8)^2}{5} \\
&= \frac{36 + 144 + 4 + 144 + 64}{5} \\
&= \frac{392}{5} = 78.4
\end{aligned}
$$

The standard deviation equals the square root of the variance, or

$$\sigma = \sqrt{78.4} = 8.85$$

$\square$

There are some significant relationships involving the standard deviation which can be useful in reaching conclusions about a data set. In particular, you will see one of these relationships in Sec. 14.4. Although we could spend much more time exploring this and other measures of variation, our main focus in this chapter is on probability distributions. Thus we will conclude this section by discussing the standard deviation as a measure of variation for discrete probability distributions.

> **STANDARD DEVIATION: DISCRETE PROBABILITY DISTRIBUTION**
> Given a discrete random variable X which can assume n values $x_1, x_2, \ldots,$ x_n having respective probabilities of occurrence $p_1, p_2, \ldots, p_n$, the **standard deviation** of the random variable with mean μ is
>
> $$\sigma = \sqrt{(x_1 - \mu)^2 p_1 + (x_2 - \mu)^2 p_2 + \cdots + (x_n - \mu)^2 p_n} \qquad (14.6)$$

EXAMPLE 15 Let's compute the standard deviations for the two distributions in Table 14.17. For the first distribution,

$$\sigma = \sqrt{(49 - 50)^2(.05) + (50 - 50)^2(.90) + (51 - 50)^2(.05)}$$
$$= \sqrt{.05 + 0 + .05} = \sqrt{.10} = 0.3162$$

For the second distribution,

$$\sigma = \sqrt{(0 - 50)^2(.05) + (50 - 50)^2(.90) + (100 - 50)^2(.05)}$$
$$= \sqrt{125 + 0 + 125} = \sqrt{250} = 15.81$$

$\square$

> **PRACTICE EXERCISE**
> Compute the standard deviation for the probability distributions (Tables 14.4 and 14.5) in the blood bank example. *Answer:* $\sigma_1 = 2.107$, $\sigma_2 = 1.919$.

Section 14.2 Follow-up Exercises

For the following data sets, compute the (a) mean, (b) median, (c) mode, (d) range, and (e) standard deviation.

1 {20, 40, 60, 80, 100, 120, 140, 160, 180, 200}
2 {5, 10, 40, 20, 35, 20, 50, 0, 5, 15, 25, 30, 20, 40, 45}
3 {30, 36, 28, 18, 42, 10, 20, 52}
4 {100, 500, 800, 400, 600, 500, 700, 800, 700, 400, 800, 200}

5 {60, 63, 45, 57, 75, 105, 15, 60, 40, 80}

6 {0, −25, 40, −20, 25, −40, 20, 0, 0}

7 Determine the mean, median, and mode for the following frequency distribution.

Value of X	20	30	40	50	60
Frequency	8	12	10	16	6

8 Determine the mean, median, and mode for the following frequency distribution.

Value of X	1000	2000	3000	4000	5000
Frequency	20	15	10	25	10

9 **Forest Fire Control** Table 14.18 summarizes data from an experiment in which the Department of Environmental Management documented the number of forest fires reported each day over a 60-day period. Determine the mean, median, and mode for this data and interpret the meaning of each.

TABLE 14.18	Number of forest fires per day	0	1	2	3	4	5	6	7	8
	Frequency	10	8	6	6	5	9	7	5	4

10 **Absenteeism** An employer has been concerned about absenteeism in her firm. Union-management relations have been strained in recent months due to an inability to reach agreement on a new contract for the 50 employees in the union. Table 14.19 summarizes data the employer has gathered on daily absenteeism over the past 30 workdays. Determine the mean, median, and mode for this data and interpret the meaning of each.

TABLE 14.19	Number of absences	4	5	6	7	8
	Frequency	3	5	8	10	4

11 Table 14.20 presents a discrete probability distribution associated with the daily demand for a product.
(a) Determine the mean daily demand.
(b) What is the standard deviation of daily demand?

TABLE 14.20	Number Demanded per Day (X)	$P(X)$
	10	.08
	20	.24
	30	.28
	40	.30
	50	.10
		1.00

12 A manufactured product consists of five electrical components. Each of the components has a limited lifetime. The company has tested the product to determine the reliability of the components. Table 14.21 presents a probability distribution where the random variable X indicates the number of components which fail during the first 100 hours of operation.

(a) What is the mean number of components which fail during the first 100 hours of operation?

(b) What is the standard deviation of the random variable X?

(c) If the product will continue to operate if no more than two components fail, what percentage of the manufactured parts will continue to operate during the first 100 hours?

TABLE 14.21	Number of Components Failing (X)	$P(X)$
	0	.05
	1	.09
	2	.22
	3	.32
	4	.20
	5	.12
		1.00

13 Compute the respective means and standard deviations for the two distributions in Table 14.22.

TABLE 14.22	X_1	$P(X_1)$	X_2	$P(X_2)$
	500	1	0	.050
			100	.125
			300	.200
			500	.250
			700	.200
			900	.125
			1,000	.050
				1.000

14 For Exercise 2 in Sec. 14.1, compute (a) the mean number of major snowstorms per year and (b) the standard deviation for the probability distribution.

15 For Exercise 3 in Sec. 14.1, compute (a) the mean number of false alarms per day and (b) the standard deviation.

16 For Exercise 4 in Sec. 14.1, compute (a) the mean number of defective units per production run and (b) the standard deviation.

17 For Exercise 5 in Sec. 14.1, compute (a) the mean number of drunken drivers identified per roadblock and (b) the standard deviation.

*18 Given a data set consisting of n items $x_1, x_2, \ldots, x_n$ with mean $\bar{x}$, show that the sum of deviations from the mean equals zero.

14.3 THE BINOMIAL PROBABILITY DISTRIBUTION

This section discusses one of the more commonly used discrete probability distributions — the **binomial probability distribution.** First is a discussion of the characteristics of random processes which can be represented by the binomial distribution. This will be followed by a discussion of the binomial distribution and its applications.

Bernoulli Processes

Many random processes are characterized by trials in which there are just two mutually exclusive outcomes possible. Manufactured parts may be sampled to determine whether they are of an *acceptable* quality or *defective;* the toss of a coin results in either *head* or *tail;* patients arriving at an emergency room might be classified as *male* or *female;* survey questionnaires mailed to potential respondents may be classified as *returned* or *not returned;* and questions on a multiple-choice test may be judged as answered *correctly* or *incorrectly.*

Frequently the two possible outcomes in these situations are characterized as "successful" or "unsuccessful." The occurrence of a head in the toss of a coin may be declared a successful outcome. This assignment of labels, though, is completely arbitrary. The occurrence of a tail might just as easily, and appropriately, be termed a success (it depends on the side for which you are rooting).

Random experiments such as tossing a fair coin are examples of **Bernoulli processes.** Bernoulli processes have the following characteristics.

CHARACTERISTICS OF BERNOULLI PROCESSES

I   *There are n independent trials, each of which has two possible outcomes,* success *or* failure.
II *The probability of a* success *p remains fixed for each trial.*
III *The probability of* failure *q remains fixed for each trial and $q = 1 - p$.*
IV *The random variable X is the total number of successes in n trials.*

Head   Tail

Male   Female

Figure 14.2
Bernoulli processes are characterized by two mutually exclusive outcomes.

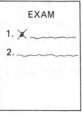

Incorrect

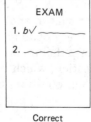

Correct

EXAMPLE 16

Assume that a single die is rolled 25 times. A successful outcome is the occurrence of a 6 in a roll of the die. This is an example of a Bernoulli process. If a fair die is assumed, the probability of a success for each trial is $\frac{1}{6}$, or $p = \frac{1}{6}$. If the probability of success equals $\frac{1}{6}$, the probability of a failure q is $q = 1 - p = 1 - \frac{1}{6} = \frac{5}{6}$. Successive rolls of a die are statistically independent. For this experiment, the random variable X is the number of times a 6 occurs in the 25 rolls.

EXAMPLE 17

One hundred parts will be selected from a manufacturing process which is judged to produce defective parts at random with a probability of .05. The process resulting in defective parts is characterized by statistical independence. If we consider the production of a nondefective part as a successful outcome, the probability of a successful outcome is $p = .95$. The probability of a failure (defective part) is $q = 1 - .95 = .05$. For this experiment, the random variable X is the number of nondefective items identified in the 100 trials.

$\square$

The Binomial Distribution

Consider a student who is taking a true-false quiz consisting of five questions. Assume that the probability of answering any one of the questions correctly is .8. Suppose we are interested in the probability that the student will answer exactly four questions correctly. This situation is a Bernoulli process where the probability of success on any trial is $p = .8$, the probability of a failure is $q = .2$, the number of trials is $n = 5$, and the random variable X equals the number of questions answered correctly.

One way of answering four questions correctly is to answer the first four correctly and the last incorrectly; or, stated in terms of success (S) or failure (F), the sequence of outcomes is $S_1 S_2 S_3 S_4 F_5$. Due to independence, the probability of this *joint event* is computed by applying rule 7 from Sec. 13.4, or

$$P(S_1 \cap S_2 \cap S_3 \cap S_4 \cap F_5) = (.8)(.8)(.8)(.8)(.2)$$
$$= .08192$$

Another way of getting four correct answers is the sequence $F_1 S_2 S_3 S_4 S_5$ and

$$P(F_1 \cap S_2 \cap S_3 \cap S_4 \cap S_5) = (.2)(.8)(.8)(.8)(.8)$$
$$= .08192$$

which is the same probability as computed for the joint event $S_1 S_2 S_3 S_4 F_5$. There are, in fact, five different ways in which four questions can be answered correctly. They are $S_1 S_2 S_3 S_4 F_5$, $S_1 S_2 S_3 F_4 S_5$, $S_1 S_2 F_3 S_4 S_5$, $S_1 F_2 S_3 S_4 S_5$, and $F_1 S_2 S_3 S_4 S_5$. Note that in applying rule 7 to compute the probability of each of these sequences, the probability of success, $p = .8$, will be a factor 4 times and the probability of a failure will be a factor once. The only difference is the order of multiplication, which has no effect on the product. Thus, the probability of answering four questions correctly is

$$p(X = 4) = \left(\begin{array}{c}\text{number of different ways}\\\text{in which four questions can}\\\text{be answered correctly}\end{array}\right)\left(\begin{array}{c}\text{probability of answering four}\\\text{questions, out of five, correctly}\\\text{in any one order}\end{array}\right)$$

$$= 5(.08192) = .4096$$

An alternative to enumerating the different ways in which four successes can occur in five trials is to recognize that this is a **combinations** question. The number of ways in which k successes can occur in n trials is found by applying Eq. (13.6), or

$$\binom{n}{k} = \frac{n!}{k!(n-k)!}$$

The number of ways in which four successes can occur in five trials is

$$\binom{5}{4} = \frac{5!}{4!(5-4)!}$$

$$= \frac{5!}{4! \, 1!} = 5$$

The *binomial probability distribution* is used to represent experiments which are Bernoulli processes. The following computational rule is fundamental to determining *binomial probabilities*.

COMPUTATION OF BINOMIAL PROBABILITIES

Given a Bernoulli process where the probability of success in any trial equals p and the probability of a failure equals q, the probability of k successes in n trials is

$$P(k,n) = \binom{n}{k} \overbrace{(p \cdot p \cdot p \cdots p)}^{k \text{ factors}} \overbrace{(q \cdot q \cdot q \cdots q)}^{n - k \text{ factors}}$$

or

$$P(k,n) = \binom{n}{k} p^k q^{n-k} \qquad\qquad \textbf{(14.7)}$$

In Eq. (14.7), the factor $\binom{n}{k}$ represents the number of ways in which k successes can occur in n trials and $p^k q^{n-k}$ represents the probability of k successes in n trials, for *any* one of these ways.

If we continue with the quiz example, the possible outcomes for the quiz are zero, one, two, three, four, or five questions correct (successes). Let's compute the probabilities for the other outcomes. The probability of zero successes (correct answers) in five trials is

$$P(0, 5) = \binom{5}{0} (.8)^0(.2)^5$$

$$= \frac{5!}{0!(5 - 0)!} (.00032) = 1(.00032) = .00032$$

The probability of one success in five trials is

$$P(1, 5) = \binom{5}{1} (.8)^1(.2)^4$$

$$= \frac{5!}{1!(5 - 1)!} (.00128) = 5(.00128) = .0064$$

The probability of two successes in five trials is

$$P(2, 5) = \binom{5}{2} (.8)^2(.2)^3$$

$$= \frac{5!}{2!(5 - 2)!} (.00512) = 10(.00512) = .0512$$

The probability of three successes in five trials is

$$P(3, 5) = \binom{5}{3} (.8)^3(.2)^2$$

$$= \frac{5!}{3!(5 - 3)!} (.02048) = 10(.02048) = .2048$$

The probability of five successes in five trials is

$$P(5, 5) = \binom{5}{5} (.8)^5(.2)^0$$

$$= \frac{5!}{0!(5 - 0)!} (.32768) = 1(.32768) = .32768$$

Table 14.23 summarizes the binomial probability distribution for this experiment. Note that the possible values for the random variable are mutually exclusive and collectively exhaustive. Also, the sum of the probabilities equals 1.0.

Figure 14.3 presents a histogram representation for this distribution. Each bar corresponds to one of the six events, and the height of the bar equals the probability of the event.

EXAMPLE 18 (**Consumer Credit; Motivating Scenario**) A major bank issues credit cards under the name VISACARD. It has determined that, on the average, 40 percent of all credit card

TABLE 14.23	**Binomial Distribution for Quiz Results**	
Number of Successes (Correct Answers on Quiz) (X)		**$P(X)$**
0		.00032
1		.00640
2		.05120
3		.20480
4		.40960
5		.32768
		1.00000

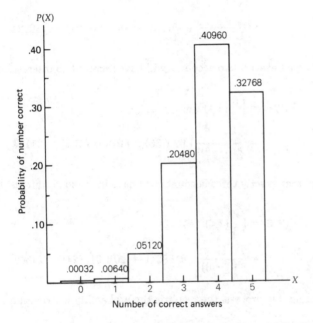

Figure 14.3 Histogram representation of a binomial distribution.

accounts are paid in full following the initial billing. That is, for any given month, 40 percent of all accounts will not incur interest charges. If a sample of six accounts is selected at random, estimate the probability that none of the six will have incurred interest charges for their last billing.

SOLUTION

This experiment can be considered to be a Bernoulli process where an account incurring no interest charges is considered a success and one incurring interest charges is considered a failure. For this experiment $p = .40$, $q = .60$, and $n = 6$.

The probability that none of the six accounts selected will have incurred no interest charges is

$$P(0, 6) = \binom{6}{0} (.4)^0 (.6)^6$$

$$= \frac{6!}{0!(6-0)!} (.046656) = (1)(.046656) = .046656$$

EXAMPLE 19 (**Consumer Credit, continued**) Construct the complete probability distribution for the experiment of selecting six consumer credit accounts.

SOLUTION

We complete the probability distribution by applying Eq. (14.7) to calculate the probabilities of 1, 2, 3, 4, 5, and 6 accounts having incurred no interest charges. The probability that exactly one account will have incurred no interest charge is

$$P(1, 6) = \binom{6}{1} (.4)^1 (.6)^5$$

$$= \frac{6!}{1!(6-1)!} (.031104) = (6)(.031104) = .186624$$

The probability that exactly two accounts will have incurred no interest charge is

$$P(2, 6) = \binom{6}{2} (.4)^2 (.6)^4$$

$$= \frac{6!}{2!(6-2)!} (.020736) = (15)(.020736) = .31104$$

The probability that exactly three accounts will have incurred no interest charges is

$$P(3, 6) = \binom{6}{3} (.4)^3 (.6)^3$$

$$= \frac{6!}{3!(6-3)!} (.013824) = (20)(.013824) = .27648$$

By continuing this process, the result is the probability distribution shown in Table 14.24. ∎

PRACTICE EXERCISE

In Example 19, what is the probability that no more than three accounts will have incurred no interest charges? More than four accounts? *Answer:* .8208, .04096.

POINT FOR THOUGHT & DISCUSSION

Is the process in Example 36 of Chap. 13 (page 599) (space shuttle simulation) a Bernoulli process? Why or why not?

TABLE 14.24	**Binomial Distribution for Consumer Credit Review**	
Number of Successes (Accounts Found to Have Incurred No Interest Charges) (X)		**$P(X)$**
0		.046656
1		.186624
2		.311040
3		.276480
4		.138240
5		.036864
6		.004096
		1.000000

Mean and Standard Deviation of the Binomial Distribution

> **MEAN OF A BINOMIAL DISTRIBUTION**
> The mean value μ for a binomial distribution is determined by the equation
>
> $$\mu = np \tag{14.8}$$

Equation (14.8) indicates that the mean number of successes in a Bernoulli process is equal to the product of the number of trials and the probability of success for each individual trial. If a binomial distribution describes the number of heads occurring in 500 flips of a fair coin, the mean for this experiment is

$$\mu = 500(.5)$$
$$= 250$$

which suggests that on the average we would expect to get 250 heads in 500 flips of a fair coin.

EXAMPLE 20   In Examples 18 and 19, the probability that an account incurred no interest charges was .4. In the selection of six accounts, $n = 6$ and $p = .4$. The mean for this binomial distribution is

$$\mu = 6(.4)$$
$$= 2.4$$

This suggests that on average (over the long run) a selection of six accounts at random should result in 2.4 having incurred no interest charges.

EXAMPLE 21   Earlier we discussed a situation where a student was taking a true-false quiz involving five questions. The probability of answering any question correctly was estimated at .8. For this Bernoulli process, the mean number of correct answers is

$$\mu = 5(.8) = 4.0$$

This result suggests that if this student took many five-question quizzes with the same probability of success, the average number of correct answers for the quizzes would equal 4.0. If the number of questions on a quiz is 36, the mean number of correct answers is

$$\mu = 36(.8) = 28.8$$

STANDARD DEVIATION OF A BINOMIAL DISTRIBUTION
The standard deviation of a binomial probability distribution is determined by the equation

$$\sigma = \sqrt{npq} \tag{14.9}$$

For Example 19, $n = 6$ and $p = .4$. The standard deviation is

$$\sigma = \sqrt{6(.4)(.6)}$$
$$= \sqrt{1.44} = 1.2$$

PRACTICE EXERCISE

Table 14.25 is an extension of Table 14.24 for the consumer credit example. Using this table and Eqs. (14.3) and (14.6), compute μ and σ. Compare your results with those found using Eqs. (14.8) and (14.9).

TABLE 14.25	Number of Successes (X)	$P(X)$	$X \cdot P(X)$	$X - \mu$	$(X - \mu)^2$	$(X - \mu)^2 \cdot P(X)$
	0	.046656				
	1	.186624				
	2	.311040				
	3	.276480				
	4	.138240				
	5	.036864				
	6	.004096				
		1.000000	$\mu = $ ___			$\sigma^2 = $ _____
						$\sigma = $ _____

EXAMPLE 22   (**Polygraph Reliability**) A manufacturer of polygraphs (lie detectors) claims that its machines can correctly distinguish truthful responses to questions from untruthful responses 85 percent of the time. If the machine is tested using a set of 50 questions, determine (*a*) the mean for this Bernoulli process and (*b*) the standard deviation.

SOLUTION

For this process, $p = .85$ and $n = 50$.

(a) The mean number of responses correctly identified is

$$\mu = 50(.85) = 42.5$$

(b) The standard deviation is

$$\sigma = \sqrt{(50)(.85)(.15)}$$
$$= \sqrt{6.375} = 2.52$$

❑

Section 14.3 Follow-up Exercises

1 Determine which of the following random variables are not variables in a Bernoulli process.
  (a) $X =$ the number of heads in the toss of a coin 20 times
  (b) $X =$ the heights of 10 students selected at random
  (c) $X =$ the number of 6s which appear in five rolls of a *pair* of dice
  (d) $X =$ scores earned by 100 different students on a standardized test
  (e) $X =$ the closing price of a stock for 10 randomly selected days
  (f) $X =$ the number of arrivals per hour at an emergency room observed for 20 randomly selected hours of operation
  (g) $X =$ the number of false alarms in a sample of 10 fire alarms where the probability that any alarm is a false alarm equals .18

2 A fair coin is to be flipped 4 times. What is the probability that exactly two heads will occur? Four heads? Two or more heads?

3 A fair die will be rolled 4 times. What is the probability that exactly two 1s will occur? Fewer than four 1s?

4 **Drunken Driving** A state has determined that of all traffic accidents in which a fatality occurs, 70 percent involve situations in which at least one driver has been drinking. If a sample of four fatal accidents is selected at random, construct the binomial distribution where the random variable X equals the number of accidents in which at least one driver was drinking.

5 **Couch Potatoes** It has been determined that 80 percent of all American households have at least one television set. If five residences are selected at random, construct the binomial distribution where the random variable X equals the number of residences having at least one television.

6 A firm which conducts consumer surveys by mail has found that 30 percent of those families receiving a questionnaire will return it. In a survey of 10 families, what is the probability that exactly five families will return the questionnaire? Exactly 6 families? Ten families?

7 A student takes a true-false examination which consists of 10 questions. The student knows nothing about the subject and chooses answers at random. Assuming independence between questions and a probability of .9 of answering any question correctly, what is the probability that the student will pass the test (assume that passing means getting seven or more correct)? If the test contains 20 questions, does the probability of passing change (14 or more correct)?

8 **Immunization** A particular influenza vaccine has been found to be 95 percent effective in providing immunity. In a random sample of five vaccinated people who have been

exposed to this strain of influenza, what is the probability that none of the five will come down with the disease?

9 An urn contains 4 red balls, 2 green balls, and 4 blue balls. If 10 balls are selected at random with replacement between each draw, what is the probability that exactly 4 red balls will be selected? What is the probability that exactly 2 green balls will be selected?

10 A manufacturing process produces defective parts randomly at a rate of 8 percent. In a sample of 10 parts, what is the probability that fewer than 2 will be defective?

11 In Exercise 10, what is the mean number of defective parts expected to equal? What is the interpretation of this value? What is the standard deviation for this distribution?

12 In a local hospital 48 percent of all babies born are males. On a particular day five babies are born. What is the probability that four or more of the babies are males? What is the mean of this distribution for $n = 5$? What is the standard deviation?

13 **Political Poll** For an upcoming U.S. senatorial election, opinion polls indicate that 50 percent of the population support the Democratic candidate, 40 percent support the Republican candidate, and 10 percent are undecided. If a sample of five persons is selected at random, what is the probability that at least four persons will be supportive of the Democratic candidate? Fewer than two persons will support the Democratic candidate?

14 **Cigarette Smoking** A local hospital has been conducting an experimental program to assist persons to stop cigarette smoking. Upon completion of the program, participants realize a 60 percent success rate. If a sample of four past participants is selected at random, what is the probability that all four will have stopped smoking? At least three will have stopped?

15 **Economic Recession** An opinion poll reveals that 80 percent of the persons in one New England state believe that the area is suffering an economic recession. If a sample of six persons is selected at random in that New England State, what is the probability that exactly half of the persons will believe that a recession exists?

14.4 THE NORMAL PROBABILITY DISTRIBUTION

With continuous probability distributions, the number of possible values for the random variable is infinite. For these distributions, probabilities are assigned only to intervals of values for the random variable. To illustrate this, think of the random variable X which equals the annual rainfall in an area, measured in inches. The number of possible amounts of rain is infinite. For example, one possible value for X is 24.000056 inches. With an infinite number of possible values for X, the likelihood of any one value is extremely small. Thus, with continuous probability distributions, statements are not made regarding the probability that the random variable will assume a specific value. Rather, statements are usually made regarding the probability that the random variable will assume a value within a defined interval. In the rainfall example, we may want to know the probability that annual rainfall will be between 24 and 25 inches.

This section discusses one of the most familiar and most widely applied continuous probability distributions — the ***normal probability distribution.***

The Normal Probability Distribution

The normal probability distribution is one of the most important distributions in modern-day probability theory. The normal probability distribution is represented

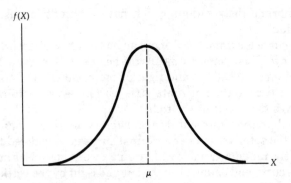

Figure 14.4   Normal curve.

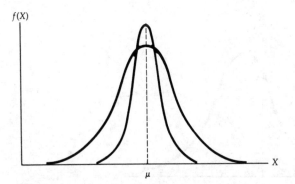

Figure 14.5   Normal distributions with equal means.

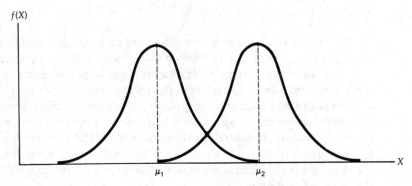

Figure 14.6   Normal distributions with equal standard deviations.

by the classic *bell-shaped curve,* or ***normal curve,*** shown in Fig. 14.4. The bell-shaped curve in Fig. 14.4 is typical of a family of bell-shaped curves which represent normal probability distributions, each different with regard to its mean and standard deviation. Figure 14.5 illustrates the graphs of two normal distributions having the same mean but different standard deviations. Figure 14.6 illustrates the graphs of two normal distributions which have different means but the same standard deviation.

The normal curve is symmetrical about an imaginary vertical line which passes through the mean μ. This symmetry means the height of the curve is the same if one moves equal distances to the left and right of the mean. The "tails" of the curve come closer and closer to the horizontal axis without ever reaching it, no matter how far one moves to the left or the right.

Areas under the curve representing a probability distribution are equivalent to probabilities. If the total area under a normal curve is considered to equal 1, the probability that the random variable will assume a value between a and b equals the area beneath the curve and bounded on the left and right by the vertical lines $X = a$ and $X = b$. This is illustrated in Fig. 14.7.

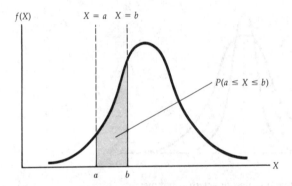

Figure 14.7   The area beneath the normal curve represents probability.

Consider a situation in which the scores on a standardized aptitude test have been found to be *normally distributed* with a mean of 70 and standard deviation of 7.5. Suppose we are interested in determining the probability that a student will score between 70 and 85 on the test. In order to determine this probability, we take advantage of a very useful property of normal probability distributions. *Any normal probability distribution with mean μ and standard deviation σ can be transformed into an equivalent **standard (unit) normal distribution** which has a mean equal to 0 and standard deviation equal to 1.* The transformation redefines each value of the random variable X in terms of its distance from the mean, *stated as a multiple of the standard deviation.*

Figure 14.8 illustrates this transformation of X in terms of another variable z. The variable z expresses X in terms of its distance from the mean, stated in multiples of the standard deviation. Note that z values to the right of the mean are positive, and those to the left of the mean are negative. A value of X which is one standard deviation to the right of the mean would be defined equivalently by a z value of 1. A point located three standard deviations to the left of the mean would be defined equivalently by a z value of -3.

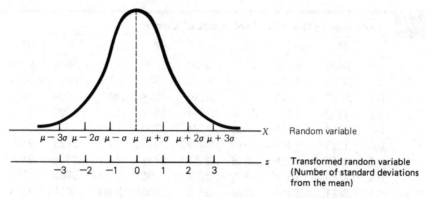

Figure 14.8 Transformation to standard (unit) normal distribution.

EXAMPLE 23 Given a random variable X with $\mu = 10$ and $\sigma = 2$, the following equivalences exist between specific values of X and corresponding z values.

X	z (Number of Standard Deviations from the Mean)
0	-5.0
5	-2.5
8	-1.0
10	0
12	1.0
15	2.5
18	4.0

Table 14.26 on page 646 provides areas under the standard normal curve. *Note that the areas given are areas under the curve between the mean and another point located z standard deviations from the mean.* The normal curve is such that 50 percent of the area is to the left of the mean and 50 percent to the right. That is, there is a 50 percent chance that the value of the random variable X will be less than the mean and a 50 percent chance it will be greater than the mean. Because of the symmetry of the normal curve, Table 14.26 can be used to determine areas between the mean and another point when the second point is to the left *or* right of the mean. A z value of 1 in Table 14.26 tells us that the area under the curve between $z = 0$ and $z = 1$ is 0.3413. *The area is the same between $z = 0$ and $z = -1$.* Figure 14.9 illustrates these areas. Note also that we can make the statement that the area under the standard normal curve *between $z = -1$ and $z = 1$* equals 0.6826.

Let's return to the original problem concerning the standardized aptitude test. Scores had been found to be normally distributed with a mean of 70 and standard deviation of 7.5. The problem was to determine the probability that a student selected at random will score between 70 and 85. To determine this probability, we

TABLE 14.26	**Area Under the Standard Normal Curve**								
	.00	.01	.02	.03	.04	.05	.06	.07	.08	.09
0.0	.0000	.0040	.0080	.0120	.0160	.0199	.0239	.0279	.0319	.0359
0.1	.0398	.0438	.0478	.0517	.0557	.0596	.0636	.0675	.0714	.0753
0.2	.0793	.0832	.0871	.0910	.0948	.0987	.1026	.1064	.1103	.1141
0.3	.1179	.1217	.1255	.1293	.1331	.1368	.1406	.1443	.1480	.1517
0.4	.1554	.1591	.1628	.1664	.1700	.1736	.1772	.1808	.1844	.1879
0.5	.1915	.1950	.1985	.2019	.2054	.2088	.2123	.2157	.2190	.2224
0.6	.2257	.2291	.2324	.2357	.2389	.2422	.2454	.2486	.2518	.2549
0.7	.2580	.2612	.2642	.2673	.2704	.2734	.2764	.2794	.2823	.2852
0.8	.2881	.2910	.2939	.2967	.2995	.3023	.3051	.3078	.3106	.3133
0.9	.3159	.3186	.3212	.3238	.3264	.3289	.3315	.3340	.3365	.3389
1.0	.3413	.3438	.3461	.3485	.3508	.3531	.3554	.3577	.3599	.3621
1.1	.3643	.3665	.3686	.3708	.3729	.3749	.3770	.3790	.3810	.3830
1.2	.3849	.3869	.3888	.3907	.3925	.3944	.3962	.3980	.3997	.4015
1.3	.4032	.4049	.4066	.4082	.4099	.4115	.4131	.4147	.4162	.4177
1.4	.4192	.4207	.4222	.4236	.4251	.4265	.4279	.4292	.4306	.4319
1.5	.4332	.4345	.4357	.4370	.4382	.4394	.4406	.4418	.4429	.4441
1.6	.4452	.4463	.4474	.4484	.4495	.4505	.4515	.4525	.4535	.4545
1.7	.4554	.4564	.4573	.4582	.4591	.4599	.4608	.4616	.4625	.4633
1.8	.4641	.4649	.4656	.4664	.4671	.4678	.4686	.4693	.4699	.4706
1.9	.4713	.4719	.4726	.4732	.4738	.4744	.4750	.4756	.4761	.4767
2.0	.4772	.4778	.4783	.4788	.4793	.4798	.4803	.4808	.4812	.4817
2.1	.4821	.4826	.4830	.4834	.4838	.4842	.4846	.4850	.4854	.4857
2.2	.4861	.4864	.4868	.4871	.4875	.4878	.4881	.4884	.4887	.4890
2.3	.4893	.4896	.4898	.4901	.4904	.4906	.4909	.4911	.4913	.4916
2.4	.4918	.4920	.4922	.4925	.4927	.4929	.4931	.4932	.4934	.4936
2.5	.4938	.4940	.4941	.4943	.4945	.4946	.4948	.4949	.4951	.4952
2.6	.4953	.4955	.4956	.4957	.4959	.4960	.4961	.4962	.4963	.4964
2.7	.4965	.4966	.4967	.4968	.4969	.4970	.4971	.4972	.4973	.4974
2.8	.4974	.4975	.4976	.4977	.4977	.4978	.4979	.4979	.4980	.4981
2.9	.4981	.4982	.4982	.4983	.4984	.4984	.4985	.4985	.4986	.4986
3.0	.49865	.4987	.4987	.4988	.4988	.4989	.4989	.4989	.4990	.4990

Figure 14.9

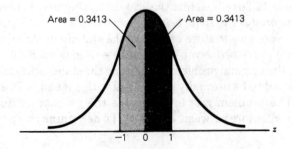

Area = 0.3413   Area = 0.3413

must transform the original distribution into the standard normal distribution. In order to do this, equivalent z values must be identified for pertinent X values. The formula enabling one to transform values of the random variable X into equivalent z values is

$$z = \frac{X - \mu}{\sigma}$$ **(14.10)**

The z value corresponding to the mean is always 0. To illustrate, the z value corresponding to a score of 70 is

$$z = \frac{70 - 70}{7.5} = 0$$

The z value corresponding to an X value of 85 is

$$z = \frac{85 - 70}{7.5}$$

$$= \frac{15}{7.5} = 2$$

This means that a score of 85 is two standard deviations above (to the right of) the mean score of 70.

Thus, the probability that a student will score between 70 and 85 is equal to the area under the standard normal curve between $z = 0$ and $z = 2$. This area, illustrated in Fig. 14.10, is read directly from Table 14.26 as 0.4772. Therefore, the probability that a student will score between 70 and 85 equals .4772. Note in Fig. 14.10 that an equivalent X scale has been drawn below the z scale to show the corresponding value of X. This is not necessary, but it helps to remind us of the pertinent values for the random variable X.

Suppose we are interested in the probability that a student will score between 0 and 85 on the examination. This probability is equal to the probability that z will be less than 2 for the standard normal distribution. The probability is the area under

Figure 14.10

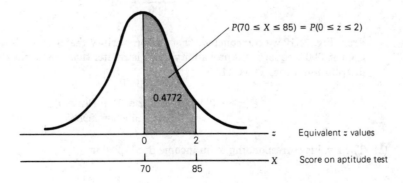

$P(70 \le X \le 85) = P(0 \le z \le 2)$

0.4772

0   2 z Equivalent z values

70   85 X Score on aptitude test

Figure 14.11

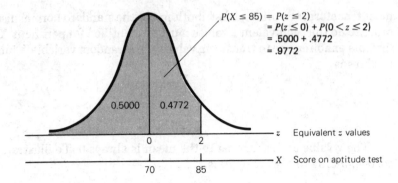

$P(X \leq 85) = P(z \leq 2)$
$= P(z \leq 0) + P(0 < z \leq 2)$
$= .5000 + .4772$
$= .9772$

0.5000   0.4772

0   2 z Equivalent z values

70   85 X Score on aptitude test

the standard normal curve, illustrated in Fig. 14.11. This area consists of the 0.5000 to the left of the mean and the 0.4772 we identified previously, or

$$P(z \leq 2) = P(z \leq 0) + P(0 < z \leq 2)$$
$$= .5000 + .4772 = .9772$$

NOTE   For problems requiring the identification of probabilities for a normally distributed variable, it is strongly advised that you make a rough sketch which identifies the equivalent area or areas under the standard normal curve.

EXAMPLE 24   A survey of per capita income indicated that the annual income for people in one state is normally distributed with a mean of $9,800 and a standard deviation of $1,600. If a person is selected at random, what is the probability that the person's annual income is (a) greater than $5,000, (b) greater than $12,200, (c) between $8,520 and $12,200, and (d) between $11,400 and $13,000?

SOLUTION
(a) The z value corresponding to an income of $5,000 is

$$z = \frac{5,000 - 9,800}{1,600}$$

$$= \frac{-4,800}{1,600} = -3$$

From Fig. 14.12 we can conclude that the probability that a person's salary is greater than $5,000 is equal to the probability that z is greater than -3 for the standard normal distribution. From Table 14.26

$$P(z > -3) = P(-3 < z \leq 0) + P(z > 0)$$
$$= .49865 + .5000 = .99865$$

(b) The z value corresponding to an income of $12,200 is

Figure 14.12

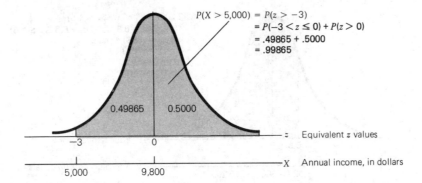

$P(X > 5{,}000) = P(z > -3)$
$= P(-3 < z \le 0) + P(z > 0)$
$= .49865 + .5000$
$= .99865$

$$z = \frac{12{,}200 - 9{,}800}{1{,}600}$$

$$= \frac{2{,}400}{1{,}600} = 1.5$$

From Fig. 14.13 we can conclude that the probability that a person's salary is greater than \$12,200 is equal to the probability that z is greater than 1.5. From Table 14.26 we can determine that $P(0 < z \le 1.5) = .4332$. Since

$$P(z > 0) = .5000$$

$$P(z > 1.5) = P(z > 0) - P(0 < z \le 1.5)$$
$$= .5000 - .4332 = .0668$$

(c) The z value corresponding to an income of \$8,520 is

$$z = \frac{8{,}520 - 9{,}800}{1{,}600}$$

$$= \frac{-1{,}280}{1{,}600} = -.8$$

Figure 14.13

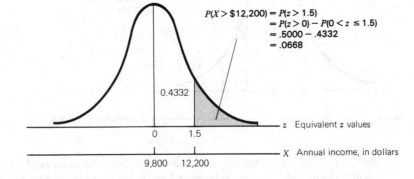

$P(X > \$12{,}200) = P(z > 1.5)$
$= P(z > 0) - P(0 < z \le 1.5)$
$= .5000 - .4332$
$= .0668$

Figure 14.14

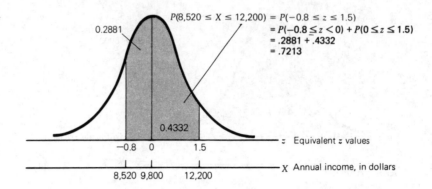

From Fig. 14.14, the probability that a person's salary is between \$8,520 and \$12,200 is equal to the probability that z is between -0.8 and 1.5, or

$$P(-0.8 \le z \le 1.5) = P(-0.8 \le z < 0) + P(0 \le z \le 1.5)$$
$$= .2881 + .4332 = .7213$$

(d) The z value corresponding to an income of \$11,400 is

$$z = \frac{11,400 - 9,800}{1,600}$$

$$= \frac{1,600}{1,600} = 1$$

The z value corresponding to an income of \$13,000 is

$$z = \frac{13,000 - 9,800}{1,600}$$

$$= \frac{3,200}{1,600} = 2$$

Figure 14.15

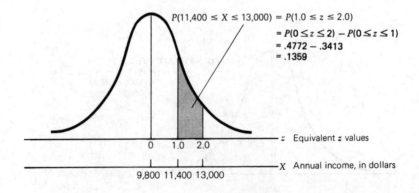

From Fig. 14.15, the probability that a person's salary is between \$11,400 and \$13,000 is equal to the probability that z is between 1 and 2. To determine this probability, we must find the area between $z = 0$ and $z = 2$ and subtract from this the area between $z = 0$ and $z = 1$. Or

$$P(1 \le z \le 2) = P(0 \le z \le 2) - P(0 \le z \le 1)$$
$$= .4772 - .3413 = .1359$$

❑

PRACTICE EXERCISE
In Example 24, what is the probability that a person's annual income is
(a) greater than \$8,200, (b) less than \$15,800, and (c) between \$9,000 and
\$13,000? *Answer:* (a) .8413, (b) .9878, and (c) .6687.

One point should be made regarding the use of Table 14.26. It was mentioned earlier that for continuous variables, the probability of occurrence of any specific value of the variable equals 0. That is, for any point a,

$$P(X = a) = 0$$

Thus, for any two constants $a < b$,

$$P(a \le X \le b) = P(a < X \le b) = P(a \le X < b) = P(a < X < b)$$

The practical implication of this is that the values in Table 14.26 represent the probabilities that z will assume values between two points z_1 and z_2, where the exact values of z_1 and z_2 may or may not be included. In Example 24a, the probabilities that a person's salary is greater than \$5,000 or greater than or equal to \$5,000 are both the same — .99865.

One final observation about normally distributed random variables is that of all the possible outcomes of the random variable X, about 68 percent are expected to occur within plus or minus one standard deviation from the mean (that is, $\mu \pm 1\sigma$), about 95 percent are expected to occur within plus or minus two standard deviations of the mean ($\mu \pm 2\sigma$), and about 99 percent within plus or minus three standard deviations of the mean ($\mu \pm 3\sigma$). This is a useful set of properties when attempting to reach generalizations about normally distributed random variables. These properties are illustrated in Fig. 14.16.

Section 14.4 Follow-up Exercises

1 Given a normal distribution where $\mu = 50$ and $\sigma = 8$, determine the z values corresponding to each of the following values of the random variable: (a) 56, (b) 42, (c) 66, (d) 36, and (e) 75.

2 Given a normal distribution where $\mu = 300$ and $\sigma = 60$, determine the z values corresponding to the following values of the random variable: (a) 320, (b) 160, (c) 365, (d) 430, and (e) 130.

3 Given a normal distribution where $\mu = 0.72$ and $\sigma = 0.08$, determine the z values corre-

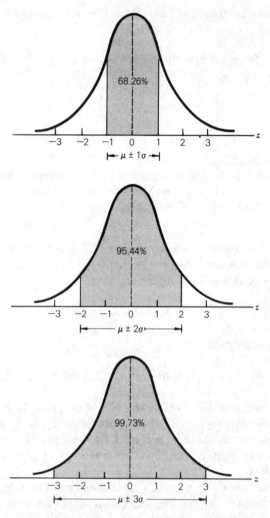

Figure 14.16 Properties of normally distributed random variables.

sponding to each of the following values of the random variable: (a) 0.84, (b) 0.62, (c) 0.50, (d) 0.90, and (e) 0.48.

4 Given a normal distribution where $\mu = 18$ and $\sigma = 4.0$, determine the z values corresponding to each of the following values of the random variable: (a) 25, (b) 12.5, (c) 22.5, (d) 17.2, and (e) 19.8.

5 For the standard normal distribution determine:
  (a) $P(z > 2.4)$ (b) $P(z < 1.2)$
  (c) $P(0.8 < z < 3.0)$ (d) $P(-2.3 \leq z \leq 2.8)$

6 For the standard normal distribution determine:
  (a) $P(z > -1.6)$ (b) $P(z < +1.3)$
  (c) $P(-1.7 < z < 0.3)$ (d) $P(-1.4 \leq z \leq 0.9)$

7 For the standard normal distribution determine:
  (a) $P(z > -0.25)$ (b) $P(z \le -0.4)$
  (c) $P(-1.5 < z < -0.6)$ (d) $P(-1.3 \le z \le 0.45)$
8 For the standard normal distribution determine:
  (a) $P(0.8 < z < 1.35)$ (b) $P(-1.35 < z < -1.25)$
  (c) $P(-0.7 \le z \le -0.25)$ (d) $P(-0.45 < z < 0.05)$
9 Given a random variable X which is normally distributed with a mean of 15 and standard deviation of 2.5, determine:
  (a) $P(X \ge 11.8)$ (b) $P(X \le 17.8)$
  (c) $P(9.6 \le X \le 16.1)$ (d) $P(8.6 \le X \le 10.9)$
10 Given a random variable X which is normally distributed with a mean of 75 and standard deviation of 5, determine:
  (a) $P(X \ge 80)$ (b) $P(X < 78.5)$
  (c) $P(66 \le X \le 72.5)$ (d) $P(80 < X < 88.5)$
11 Given a random variable X which is normally distributed with a mean of 300 and standard deviation of 20, determine:
  (a) $P(X \le 255)$ (b) $P(275 \le X \le 345)$
  (c) $P(316 \le X \le 346)$ (d) $P(270 \le X \le 295)$
12 Given a random variable X which is normally distributed with a mean of 160 and standard deviation of 8, determine:
  (a) $P(X \le 150)$ (b) $P(148 \le X \le 154)$
  (c) $P(162 \le X \le 184)$ (d) $P(154 \le X \le 172)$
13 **Birth Weights** The weights of newborn babies at a particular hospital have been observed to be normally distributed with a mean of 7.4 pounds and a standard deviation of 0.4 pound. What is the probability that a baby born in this hospital will weigh more than 8 pounds? Less than 7 pounds?
14 The annual income of workers in one state is normally distributed with a mean of $17,500 and a standard deviation of $2,000. If a worker is chosen at random, what is the probability that the worker earns more than $16,000? Less than $12,000? Between $15,000 and $20,000?
15 A manufacturer has conducted a study of the lifetime of a particular type of light bulb. The study concluded that the lifetime, measured in hours, is a random variable with a normal distribution. The mean lifetime is 650 hours with a standard deviation of 100 hours. What is the probability that a bulb selected at random would have a lifetime between 500 and 800 hours? Greater than 900 hours?
16 Grades on a national aptitude test have been found to be normally distributed with a mean of 480 and a standard deviation of 75. What is the probability that a student selected at random will score between 450 and 540? Greater than 600?
17 In a large city the number of calls for police service during a 24-hour period seems to be random. The number of calls has been found to be normally distributed with a mean of 225 and a standard deviation of 30. What is the probability that for a randomly selected day the number of calls will be fewer than 300? More than 180?
18 Annual sales (in dollars) per salesperson for a copy machine manufacturer are normally distributed with a mean of $480,000 and standard deviation of $40,000. If a salesperson is selected at random, what is the probability his or her annual sales (a) exceed $600,000, (b) are between $400,000 and $500,000, (c) are less than $450,000, or (d) are between $540,000 and $600,000?
19 **Physical Fitness** A national physical fitness test has been administered. One element of the test measured the number of push-ups a person could do. For high school seniors, the number of push-ups was normally distributed with a mean of 12.5 and a standard

deviation of 5.0. If a high school senior is selected at random, what is the probability that a senior could do (*a*) more than 16 push-ups, (*b*) more than 20 push-ups, (*c*) between 10 and 15 push-ups, and (*d*) fewer than 25 push-ups?

20 Seismology A seismologist has gathered data on the frequency of earthquakes around the world which measure 5.0 or greater on the Richter scale. The seismologist estimates that the number of earthquakes per year is normally distributed with a mean of 24 and standard deviation of 4.0. In any given year, what is the probability that there will be (*a*) more than 30 earthquakes, (*b*) fewer than 18 earthquakes, (*c*) more than 16 earthquakes, and (*d*) between 20 and 25 earthquakes?

❏ KEY TERMS AND CONCEPTS

❏ IMPORTANT FORMULAS

$$\bar{x} = \frac{x_1 + x_2 + x_3 + \cdots + x_n}{n} \tag{14.1}$$

$$\bar{x} = \frac{x_1 \cdot f_1 + x_2 \cdot f_2 + x_3 \cdot f_3 + \cdots + x_n \cdot f_n}{f_1 + f_2 + f_3 + \cdots + f_n} \tag{14.2}$$

$$\mu = x_1 p_1 + x_2 p_2 + \cdots + x_n p_n \tag{14.3}$$

$$\mathrm{Var}(x) = \frac{(x_1 - \bar{x})^2 + (x_2 - \bar{x})^2 + \cdots + (x_n - \bar{x})^2}{n} \tag{14.4}$$

$$\sigma = \sqrt{\mathrm{Var}(x)} = \sqrt{\frac{(x_1 - \bar{x})^2 + (x_2 - \bar{x})^2 + \cdots + (x_n - \bar{x})^2}{n}} \tag{14.5}$$

$$\sigma = \sqrt{(x_1 - \mu)^2 p_1 + (x_2 - \mu)^2 p_2 + \cdots + (x_n - \mu)^2 p_n} \tag{14.6}$$

$$P(k, n) = \binom{n}{k} P^k q^{n-k} \tag{14.7}$$

$$\mu = np \quad \text{(binomial)} \tag{14.8}$$

$$\sigma = \sqrt{npq} \quad \text{(binomial)} \qquad\qquad (14.9)$$

$$z = \frac{X - \mu}{\sigma} \quad \text{(normal)} \qquad\qquad (14.10)$$

❏ ADDITIONAL EXERCISES

Sections 14.1 and 14.2

1 The number of automobiles sold each day at a local dealership appears to fluctuate in a random manner. The sales manager has gathered some data over the past 150 days; a summary of the findings is given in Table 14.27.
   (a) Construct the probability distribution for this study.
   (b) Draw a histogram for this distribution.
   (c) Compute the mean, median, mode, and standard deviation.
   (d) What is the probability that at least one car will be sold on a given day? More than four? Fewer than five?

TABLE 14.27	Cars Sold per Day	Frequency
	0	24
	1	32
	2	40
	3	20
	4	12
	5	9
	6	6
	7	4
	8	3
		150

2 **Occupational Safety** A district office of OSHA receives complaints regarding potentially unsafe work conditions. The district manager has compiled some data on the number of complaints filed each day over the past year. Table 14.28 summarizes the data.
   (a) Construct the probability distribution for this study.
   (b) Draw a histogram for this distribution.
   (c) Compute the mean, median, mode, and standard deviation.
   (d) What is the probability that more than 10 complaints would be filed on a given day? Fewer than 10?

3 **Epidemiology** The Center for Disease Control in Atlanta has been gathering data on a relatively rare type of viral infection. Table 14.29 summarizes data on the number of new cases reported to the center each day for the past 400 days.
   (a) Construct the probability distribution for this study.
   (b) Draw a histogram for the probability distribution.

TABLE 14.28	Complaints Per Day	Frequency
	5	25
	6	40
	7	60
	8	48
	9	56
	10	40
	11	34
	12	20
	13	18
	14	10
	15	9

(c) Compute the mean, median, mode, and standard deviation.

(d) What is the probability that at least one new case will be reported on a given day? More than one? Fewer than five?

TABLE 14.29	New Cases Reported Per Day	Frequency
	0	35
	1	68
	2	74
	3	55
	4	48
	5	36
	6	32
	7	25
	8	17
	9	8
	10	2

4 Construct the discrete probability distribution which corresponds to the experiment of tossing a coin three times. Assume $P(H) = .7$ and let the random variable X equal the number of tails in three tosses. What is the probability of more than one head? Two or more tails?

5 Compute the mean and standard deviation for the probability distribution in Table 14.30.

TABLE 14.30	X	20	40	60	80	100
	$P(X)$	.12	.32	.26	.18	.12

6 Compute the mean and standard deviation for the probability distribution in Table 14.31.

TABLE 14.31						
X	0	5	10	15	20	25
P(X)	.04	.18	.30	.25	.17	.06

7 Compute the mean and standard deviation for the probability distribution in Table 14.32.

TABLE 14.32					
X	−10	−5	0	5	10
P(X)	.20	.30	.15	.05	.30

8 Compute the mean and standard deviation for the probability distribution in Table 14.33.

TABLE 14.33					
X	0	100	200	300	400
P(X)	.30	.10	.05	.35	.20

9 Automobile Recall An automobile manufacturer has announced a recall on 1990 vehicles because of a defect in camshafts. Since the recall announcement, the service manager for a local dealership has kept daily records of the number of responses to the recall announcement. She has constructed a probability distribution for the number of responses each day. This is shown in Table 14.34. (*a*) Compute and interpret the mean for this distribution and (*b*) compute the standard deviation.

TABLE 14.34							
Number of responses per day	0	1	2	3	4	5	6
Probability	.10	.15	.18	.22	.30	.03	.02

10 UFO Sightings A society which investigates purported sightings of unidentified flying objects (UFOs) has gathered data on the frequency of reports around the United States. Table 14.35 is a probability distribution for the number of reported sightings each day.
(*a*) Compute and interpret the mean for this distribution.
(*b*) Compute the standard deviation.

TABLE 14.35						
Number of reports per day	0	1	2	3	4	5
Probability	.42	.34	.14	.06	.03	.01

Section 14.3

11 A fair coin is to be flipped five times. What is the probability that exactly three heads will occur? No heads?

12 A turnpike authority released data which indicates that 80 percent of the vehicles which travel on the turnpike are cars. Assume that arrivals at the entrance to the turnpike occur at random with regard to the type of vehicle. In a random sample of eight vehicles which arrive at the entrance, what is the probability that all eight will be cars? That none will be cars?

13 A U.S. Customs official estimates that 30 percent of all persons returning from Europe fail to declare all purchases which are subject to duty. If five persons are randomly selected upon their return from Europe, determine the probabilities that zero, one, two, three, four, or five of these persons will fail to declare all their purchases which are subject to duty. Construct a histogram which summarizes these results.

14 A test consists of 20 true-false questions. Find the probability that a student who knows the correct answer to 10 of the questions, but guesses at the remaining questions by tossing a coin, will score 90 percent or higher on the test. (Assume the probability of a correct guess $= .6$.)

15 The IRS has determined that 60 percent of all personal income tax returns contain at least one error. If a sample of 10 returns is selected at random, what is the probability that exactly 8 will be found to contain at least one error?

16 A manufacturing process produces defective parts randomly at a rate of 12 percent. In a random sample of 10 parts, what is the probability that fewer than 2 defective parts will be found?

17 In Exercise 16, what is the mean number of defective parts expected in a sample of 10? What is the interpretation of this value? What is the standard deviation for this distribution?

18 A binomial process is characterized by $p = .7$. With a random sample of 500:
(a) Determine the mean for the experiment.
(b) Determine the standard deviation.

19 A binomial process is characterized by $p = .10$. With a random sample of 400:
(a) Determine the mean for the experiment.
(b) Determine the standard deviation.

20 A binomial process is characterized by $p = .85$. With a random sample of 150:
(a) Determine the mean for the experiment.
(b) Determine the standard deviation.

21 A binomial process is characterized by $p = .6$. With a random sample of 75:
(a) Determine the mean for the experiment.
(b) Determine the standard deviation.

22 At a small college, it has been determined that 30 percent of all students have some type of scholarship support. If a sample of 10 students is selected at random, what is the probability that more than 9 would have some type of scholarship support? What is the mean number of students having scholarships in a sample of 500 students?

Section 14.4

23 For the standard normal distribution determine:
  (a) $P(z \le 2.8)$
  (b) $P(-0.6 \le z \le 1.3)$
  (c) $P(-2.6 \le z \le -1.5)$
  (d) $P(0.24 \le z \le 2.26)$

24 For the standard normal distribution determine:
  (a) $P(z \ge -0.28)$
  (b) $P(z \le 0.86)$
  (c) $P(-1.04 \le z \le 0.76)$
  (d) $P(-0.88 \le z \le 2.16)$

25 For the standard normal distribution determine:
  (a) $P(-0.24 \le z \le -0.10)$
  (b) $P(1.20 \le z \le 1.56)$
  (c) $P(-0.7 \le z \le 0.25)$
  (d) $P(-1.80 \le z \le -1.16)$

26 Given a random variable X which is normally distributed with a mean of 120 and standard deviation of 12, determine:
  (a) $P(X \ge 105)$
  (b) $P(X \le 135)$
  (c) $P(112 \le X \le 117)$
  (d) $P(126 \le X \le 138)$

27 Given a random variable X which is normally distributed with a mean of 180 and standard deviation of 20, determine:
  (a) $P(X \ge 215)$
  (b) $P(X \ge 166)$
  (c) $P(160 \le X \le 170)$
  (d) $P(164 \le X \le 208)$

28 Given a random variable X which is normally distributed with a mean of 760 and standard deviation of 25, determine:
  (a) $P(730 \le X)$
  (b) $P(700 \le X \le 710)$
  (c) $P(767.5 \le X \le 792.5)$
  (d) $P(X \le 722.5)$

29 Blood Pressure Screening The diastolic blood pressures for a group of women aged 25 to 34 years are normally distributed with a mean of 82 mmHg and a standard deviation of 3 mmHg. If a person is selected at random from this age group of women, what is the probability that (a) her diastolic blood pressure is between 77.5 and 85 mmHg, (b) her blood pressure is less than 86.5 mmHg, and (c) her blood pressure is higher than 88 mmHg?

30 Law School Admissions The entering students for one law school class averaged 680 on the LSAT with a standard deviation of 40. The LSAT scores for this class also appear to be normally distributed.
  (a) What percentage of the class is likely to have scored above 700 on the test? Between 720 and 750? Above 650? Below 600?
  (b) One student was 1 standard deviation below the mean on the test. About what percentage of his classmates scored lower on the test than he?

31 Cigarette Smoking A recent study conducted by the Public Health Service found that males who smoke average 18 cigarettes per day. The number of cigarettes smoked per day is normally distributed with a standard deviation of 4. If a male smoker is selected at random, what is the probability that he smokes (a) more than one pack (20 cigarettes) per day, (b) less than half of a pack per day, and (c) less than $1\frac{1}{2}$ packs?

32 Personal Savings A large investment bank has conducted a study of the investment habits of individuals aged 35–44 during 1990. The study indicated that annual savings per person were normally distributed with a mean of $5,200 and a standard deviation of

$800. If an individual in this age group is selected at random, what is the probability that during 1990 the person had savings (*a*) of more than $6,000, (*b*) of less than $4,000, (*c*) between $5,000 and $7,000, or (*d*) of less than $5,000?

33 **Charitable Contributions** The Internal Revenue Service has analyzed the "charitable deduction" claims for all persons who itemized deductions on their 1991 income tax returns. The IRS found that the contributions claimed per individual wage earner were normally distributed with a mean of $840 and a standard deviation of $180. If one wage earner is selected at random, what is the probability that during 1991 this individual contributed (*a*) more than $1,020, (*b*) less than $435, (*c*) between $390 and $1,380, or (*d*) more than $300?

34 A manufacturing process produces a circular metal part. The diameters of the manufactured parts have been found to be normally distributed with a mean of 20 centimeters and a standard deviation of 0.05 centimeter. If a metal part is selected at random, what is the probability that it will have a diameter between 19.925 and 20.075 centimeters?

35 **Major League Baseball** Batting averages in the National League for one year were normally distributed with a mean of 0.275 and a standard deviation of 0.020. If a player is selected at random, what is the probability his average was (*a*) greater than 0.300, (*b*) less than 0.250, (*c*) between 0.280 and 0.320, and (*d*) less than 0.200 *or* greater than 0.350?

❏ CHAPTER TEST

1 Given the following probability distribution, (*a*) compute the mean and standard deviation and (*b*) determine the probability that the random variable will assume a value greater than 210.

X	P(X)
200	.06
210	.16
220	.24
230	.04
240	.26
250	.24

2 It has been determined that 30 percent of all American households have cable television. If four residences are selected at random, construct the binomial distribution where the random variable X equals the number of residences having cable.

3 A manufacturing process produces defective parts randomly at a rate of 8 percent. In a sample of 400 parts, what is the mean number of defective parts expected? What is the standard deviation for this distribution?

4 A fair coin is to be flipped six times. What is the probability of getting exactly four tails in the six flips?

5 A random variable X is normally distributed with a mean of 180 and a standard deviation of 40. Determine (a) $P(X \leq 162)$, (b) $P(150 \leq X \leq 160)$, and (c) $P(X \geq 204)$.

DIFFERENTIATION

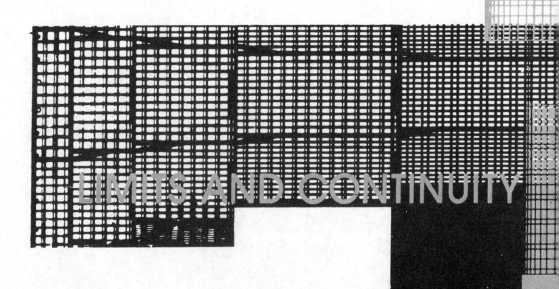

LIMITS AND CONTINUITY

❑ Introduce the concepts of limits and continuity

❑ Provide an understanding of average rate of change

❑ Provide an understanding of the derivative, including its computation and interpretation

❑ Present selected rules of differentiation and illustrate their use

❑ Introduce the nature of higher-order derivatives and their interpretation

A flu epidemic is spreading through a large midwestern state. Based upon similar epidemics which have occurred in the past, epidemiologists have formulated a mathematical function which estimates the number of persons who will be afflicted by the flu. Using this estimating function, health department officials want to predict the effects of the epidemic, including estimating the rate of affliction and the number of persons likely to be afflicted with the flu (Example 55).

This is the first of six chapters which examine the *calculus* and its application to business, economics, and other areas of problem solving. Two major areas of study within the calculus are *differential calculus* and *integral calculus.* Differential calculus focuses on *rates of change* in analyzing a situation. Graphically, differential calculus solves the following problem: *Given a function whose graph is a smooth curve and given a point in the domain of the function, what is the slope of the line tangent to the curve at this point?* You will see later that the "slope" expresses the instantaneous rate of change of the function.

Integral calculus involves summation of a special type. Graphically, the concepts of *area* in two dimensions or *volume* in three dimensions are important in integral calculus. In two dimensions, integral calculus solves the following problem: *Given a function whose graph is a smooth curve and two points in the domain of the function, what is the area of the region bounded by the curve and the x axis between these two points?*

This chapter, the following two, and Chap. 20 will discuss differential calculus and its applications. Chapter 18 will introduce integral calculus and Chap. 19 will discuss applications of integral calculus. The goal in these chapters is to provide an appreciation for what the calculus is and where it can be applied. Though it would take several semesters of intensive study to understand most of the finer points of the calculus, your coverage will enable you to understand the tools for conducting analyses at elementary levels.

This chapter is concerned with laying foundations for the remaining chapters. First, two concepts which are important in the theory of differential calculus — *limits* and *continuity* — will be presented. This discussion will be followed by an intuitive development of the concept of the *derivative.* The remainder of the chapter will provide the tools for finding derivatives as well as insights into interpreting the meaning of the derivative. Although proofs of the rules of differentiation are not presented in the main part of the chapter, selected proofs are presented in the appendix at the end of the chapter.

15.1 LIMITS

Two concepts which are important in the theory of differential and integral calculus are the *limit of a function* and *continuity.* The concept of the limit is introduced in this section. The next section expands upon this topic and introduces the concept of continuity. Since these concepts are frequently misunderstood, take care in reading these discussions.

Limits of Functions

In the calculus there is often an interest in the limiting value of a function as the independent variable approaches some specific real number. This limiting value, when it exists, is called a *limit*. The notation

$$\lim_{x \to a} f(x) = L \qquad (15.1)$$

is used to express the limiting value of a function. Equation (15.1) is read "the limit of $f(x)$, as x approaches the value a, equals L." *When investigating a limit, one is asking whether $f(x)$ approaches a specific value L as the value of x gets closer and closer to a.*

There are different procedures for determining the limit of a function. The temptation is simply to substitute the value $x = a$ into f and determine $f(a)$. This actually is a valid way of determining the limit for many but not all functions.

One approach that can be used is to substitute values of the independent variable into the function while observing the behavior of $f(x)$ as the value of x comes closer and closer to a. An important point in this procedure is that the value of the function is observed as the value of a is approached from both sides of a. The notation $\lim_{x \to a^-} f(x)$ represents the limit of $f(x)$ as x approaches a from the left (*left-hand limit*) or from below. The notation $\lim_{x \to a^+} f(x)$ represents the limit of $f(x)$ as x approaches a from the right (*right-hand limit*) or from above. *If the value of the function approaches the same number L as x approaches a from either direction, then the limit equals L.* To state it more precisely:

> **TEST FOR EXISTENCE OF A LIMIT**
> If $\lim_{x \to a^-} f(x) = L$ and $\lim_{x \to a^+} f(x) = L$, then
>
> $$\lim_{x \to a} f(x) = L$$

If the limiting values of $f(x)$ are different when x approaches a from each direction, then the function does not approach a limit as x approaches a. The following examples are illustrative.

EXAMPLE 1   In order to determine $\lim_{x \to 2} x^3$ (if it exists), let's construct a table of assumed values for x and corresponding values for $f(x)$. Table 15.1 indicates these values. Note that the value of $x = 2$ has been approached from both the left and the right. From either direction, $f(x)$ is approaching the same value, 8. Since

$$\lim_{x \to 2^-} x^3 = 8 \qquad \text{and} \qquad \lim_{x \to 2^+} x^3 = 8$$

then

$$\lim_{x \to 2} x^3 = 8$$

TABLE 15.1	**Approaching $x = 2$ from the Left**						
x	1	1.5	1.9	1.95	1.99	1.995	1.999
$f(x) = x^3$	1	3.375	6.858	7.415	7.881	7.94	7.988
	Approaching $x = 2$ from the Right						
x	3	2.5	2.1	2.05	2.01	2.005	2.001
$f(x) = x^3$	27	15.625	9.261	8.615	8.121	8.060	8.012

Note that this limit could have been determined by simply substituting $x = 2$ into f.

Figure 15.1 confirms our result. The closer x gets to the value 2, the closer the value of $f(x)$ comes to 8.

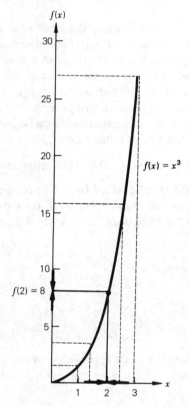

Figure 15.1 $\lim\limits_{x \to 2^-} x^3 = \lim\limits_{x \to 2^+} x^3 = 8$.

EXAMPLE 2	Given the function

$$f(x) = \begin{cases} 2x & \text{when } x \le 4 \\ 2x + 3 & \text{when } x > 4 \end{cases}$$

TABLE 15.2	**Approaching $x = 4$ from the Left**					
x	3	3.5	3.8	3.9	3.95	3.99
$f(x) = 2x$	6.0	7.0	7.6	7.8	7.9	7.98
	Approaching $x = 4$ from the Right					
x	5	4.5	4.3	4.1	4.05	4.01
$f(x) = 2x + 3$	13.0	12.0	11.6	11.2	11.1	11.02

let's determine whether $\lim\limits_{x \to 4} f(x)$ exists. A table is constructed with values for $f(x)$ determined as x approaches the value of 4 from both the left and right (Table 15.2). As x approaches the value of 4 from the left, $f(x)$ approaches a value of 8, or

$$\lim_{x \to 4^-} f(x) = 8$$

As x approaches 4 from the right, $f(x)$ approaches a value of 11, or

$$\lim_{x \to 4^+} f(x) = 11$$

Since

$$\lim_{x \to 4^-} f(x) \neq \lim_{x \to 4^+} f(x)$$

the function does not approach a limiting value as $x \to 4$, and $\lim\limits_{x \to 4} f(x)$ does not exist. Figure 15.2 shows the graph of this function. Recall from Chap. 4 that the solid circle (●) indicates that $x = 4$ is included in the domain for the lower line segment, and the open circle (O)

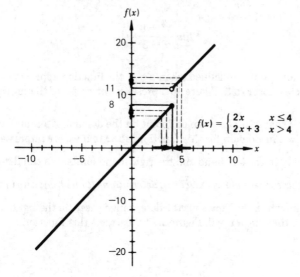

Figure 15.2 $\lim\limits_{x \to 4^-} f(x) \neq \lim\limits_{x \to 4^+} f(x)$.

indicates that $x = 4$ is not included in the domain for the upper line segment. The jump in values of the function at $x = 4$ is the reason that the limit does not exist. An important point regarding this function is that limits do exist' for $f(x)$ for values of x other than 4.

EXAMPLE 3   Let's determine whether $\lim\limits_{x \to 3} f(x)$ exists if

$$f(x) = \frac{x^2 - 9}{x - 3}$$

Since the denominator equals 0 when $x = 3$, we can conclude that the function is undefined at this point. And it would be tempting to conclude that no limit exists when $x = 3$. However, this function does approach a limit as x approaches (gets closer to) 3, even though the function is not defined at $x = 3$.

TABLE 15.3

Approaching $x = 3$ from the Left					
x	2	2.5	2.9	2.95	2.99
$f(x) = \dfrac{x^2 - 9}{x - 3}$	5.0	5.5	5.9	5.95	5.99

Approaching $x = 3$ from the Right					
x	4	3.5	3.1	3.05	3.01
$f(x) = \dfrac{x^2 - 9}{x - 3}$	7.0	6.5	6.1	6.05	6.01

Table 15.3 contains values of $f(x)$ as x approaches 3 from both the left and the right. Since

$$\lim_{x \to 3^-} f(x) = 6 \quad \text{and} \quad \lim_{x \to 3^+} f(x) = 6$$

then

$$\lim_{x \to 3} \frac{x^2 - 9}{x - 3} = 6$$

Even though the function is undefined when $x = 3$, the function approaches a value of 6 as the value of x comes closer to 3. Figure 15.3 presents the graph of the function.

EXAMPLE 4   One special case is that of limits at an endpoint on the domain of a function. Consider the function $f(x) = \sqrt{x}$. The domain for this function is $x \geq 0$. If we are interested in $\lim\limits_{x \to 0} \sqrt{x}$, we cannot determine both the left-hand and the right-hand limits. We can determine $\lim\limits_{x \to 0^+} \sqrt{x}$, but not $\lim\limits_{x \to 0^-} \sqrt{x}$. *Our concern always will be in determining limits from within the domain of a function.* In this instance, the limit must be determined based on the right-hand limit only. Since $\lim\limits_{x \to 0^+} \sqrt{x} = 0$, then $\lim\limits_{x \to 0} \sqrt{x} = 0$. Figure 15.4 illustrates this function.

$\square$

Figure 15.3

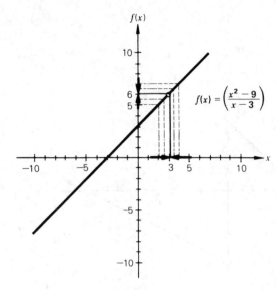

$$f(x) = \left(\frac{x^2 - 9}{x - 3}\right)$$

Figure 15.4

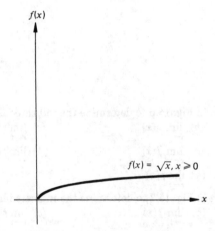

$f(x) = \sqrt{x}, x \geq 0$

A key point with the limit concept is that we are not interested in the value of $f(x)$ when $x = a$. We are interested in the behavior of $f(x)$ as x comes closer and closer to a value of a. And the notation

$$\lim_{x \to a} f(x) = L$$

means that as x gets close to a, but $x \neq a$, $f(x)$ gets close to L.

Section 15.1 Follow-up Exercises

1 Use the graph of the function in Fig. 15.5 to determine the indicated limits.

    (a) $\lim\limits_{x \to -5^-} f(x)$ (b) $\lim\limits_{x \to -5^+} f(x)$ (c) $\lim\limits_{x \to -5} f(x)$

    (d) $\lim\limits_{x \to 0^-} f(x)$ (e) $\lim\limits_{x \to 0^+} f(x)$ (f) $\lim\limits_{x \to 0} f(x)$

Figure 15.5

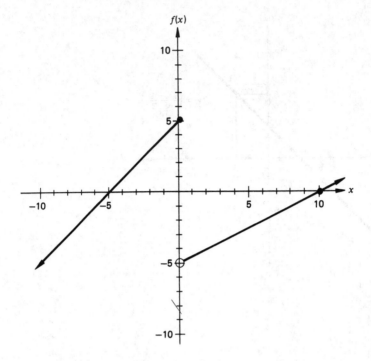

2 Use the graph of the function in Fig. 15.6 to determine the indicated limits.

(a) $\lim\limits_{x \to 0^-} f(x)$ (b) $\lim\limits_{x \to 0^+} f(x)$ (c) $\lim\limits_{x \to 0} f(x)$

(d) $\lim\limits_{x \to 3^-} f(x)$ (e) $\lim\limits_{x \to 3^+} f(x)$ (f) $\lim\limits_{x \to 3} f(x)$

(g) $\lim\limits_{x \to -5} f(x)$ (h) $\lim\limits_{x \to 1} f(x)$

3 Use the graph of the function in Fig. 15.7 to determine the indicated limits.

(a) $\lim\limits_{x \to 1^-} f(x)$ (b) $\lim\limits_{x \to 1^+} f(x)$ (c) $\lim\limits_{x \to 1} f(x)$

(d) $\lim\limits_{x \to 0^-} f(x)$ (e) $\lim\limits_{x \to 0^+} f(x)$ (f) $\lim\limits_{x \to 0} f(x)$

4 Use the graph of the function in Fig. 15.8 to determine the indicated limits.

(a) $\lim\limits_{x \to -4^-} f(x)$ (b) $\lim\limits_{x \to -4^+} f(x)$ (c) $\lim\limits_{x \to -4} f(x)$

(d) $\lim\limits_{x \to -2^-} f(x)$ (e) $\lim\limits_{x \to -2^+} f(x)$ (f) $\lim\limits_{x \to -2} f(x)$

(g) $\lim\limits_{x \to 1^-} f(x)$ (h) $\lim\limits_{x \to 1} f(x)$

For the following exercises, determine the limit (if it exists) by constructing a table of values for $f(x)$ and examining left- and right-hand limits (where appropriate).

5 $\lim\limits_{x \to 3} 2x^2$ **6** $\lim\limits_{x \to -4} (5x + 15)$

7 $\lim\limits_{x \to 5} f(x)$ where $f(x) = \begin{cases} 2x & \text{for } x < 5 \\ 20 - 2x & \text{for } x \geq 5 \end{cases}$

8 $\lim\limits_{x \to 5} f(x)$ where $f(x) = \begin{cases} 2x^2 & \text{for } x < 5 \\ 2 - x & \text{for } x \geq 5 \end{cases}$

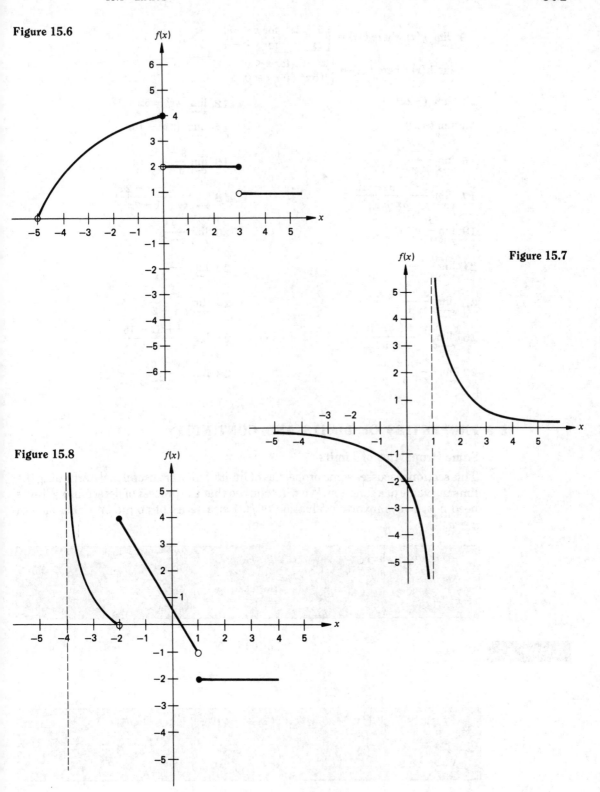

Figure 15.6

Figure 15.7

Figure 15.8

9 $\lim\limits_{x \to -1} g(x)$ where $g(x) = \begin{cases} 5 - 2x & \text{for } x < -1 \\ 2x & \text{for } x \geq -1 \end{cases}$

10 $\lim\limits_{x \to 0} h(x)$ where $h(x) = \begin{cases} x^2 & \text{for } x < 0 \\ 0.5x^2 & \text{for } x \geq 0 \end{cases}$

11 $\lim\limits_{x \to -3} (-2x^3)$ **12** $\lim\limits_{x \to 4} (4x^2 - 5x + 1)$

13 $\lim\limits_{x \to 2} (-x^3)$ **14** $\lim\limits_{x \to -2} (2x^3 - 16)$

15 $\lim\limits_{x \to 8} \dfrac{x^2 - 64}{x - 8}$ **16** $\lim\limits_{x \to 3} \dfrac{9 - x^2}{3 - x}$

17 $\lim\limits_{x \to -1} \dfrac{3x^2 - 9x - 12}{3x + 3}$ **18** $\lim\limits_{x \to -1/2} \dfrac{1 - x - 2x^2}{1 - 2x}$

19 $\lim\limits_{x \to 0} \dfrac{4}{x}$ **20** $\lim\limits_{x \to 3} \dfrac{4}{x - 3}$

21 $\lim\limits_{x \to 3} \dfrac{8}{x + 3}$ **22** $\lim\limits_{x \to 2} \dfrac{5}{4 - 2x}$

23 $\lim\limits_{x \to -7} \dfrac{x^2 - 49}{x + 7}$ **24** $\lim\limits_{x \to -2} \dfrac{x^2 - 4}{x + 2}$

25 $\lim\limits_{x \to -5} \dfrac{x^2 + 3x - 10}{x + 5}$ **26** $\lim\limits_{x \to 3} \dfrac{x^2 + 3x - 18}{x - 3}$

27 $\lim\limits_{x \to 3} \dfrac{x^2 + 3x}{x + 3}$ **28** $\lim\limits_{x \to 3} \dfrac{2x^2 + 7x - 15}{x - 3}$

15.2 PROPERTIES OF LIMITS AND CONTINUITY

Some Properties of Limits

This section discusses some properties of limits which are useful in determining the limiting value of a function. We will soon see that the process of determining limits need not always involve evaluation of $f(x)$ at a series of points on either side of $x = a$.

> **Property 1** If $f(x) = c$, where c is real,
>
> $$\lim\limits_{x \to a} (c) = c$$

EXAMPLE 5

$$\lim\limits_{x \to 3} 100 = 100$$

❑

> **Property 2** If $f(x) = x^n$, where n is a positive integer, then
>
> $$\lim\limits_{x \to a} x^n = a^n$$

EXAMPLE 6

$$\lim_{x \to -2} x^3 = (-2)^3 = -8$$

❑

Property 3 If $f(x)$ has a limit as $x \to a$ and c is real, then

$$\lim_{x \to a} c \cdot f(x) = c \cdot \lim_{x \to a} f(x)$$

EXAMPLE 7

$$\lim_{x \to 10} 5x^2 = 5 \lim_{x \to 10} x^2$$
$$= 5(10)^2 = 500$$

❑

Property 4 If $\lim_{x \to a} f(x)$ and $\lim_{x \to a} g(x)$ exist, then

$$\lim_{x \to a} [f(x) \pm g(x)] = \lim_{x \to a} f(x) \pm \lim_{x \to a} g(x)$$

EXAMPLE 8

$$\lim_{x \to -1} (x^5 - 10) = \lim_{x \to -1} x^5 - \lim_{x \to -1} 10$$
$$= (-1)^5 - 10 = -1 - 10 = -11$$

❑

Property 5 If $\lim_{x \to a} f(x)$ and $\lim_{x \to a} g(x)$ exist, then

$$\lim_{x \to a} [f(x) \cdot g(x)] = \lim_{x \to a} f(x) \cdot \lim_{x \to a} g(x)$$

EXAMPLE 9

$$\lim_{x \to 4} [(x^2 - 5)(x + 1)] = \lim_{x \to 4} (x^2 - 5) \cdot \lim_{x \to 4} (x + 1)$$
$$= [(4)^2 - 5][4 + 1]$$
$$= [16 - 5][5]$$
$$= 11(5) = 55$$

❑

Property 6 If $\lim_{x \to a} f(x)$ and $\lim_{x \to x} g(x)$ exist, then

$$\lim_{x \to a} \frac{f(x)}{g(x)} = \frac{\lim_{x \to a} f(x)}{\lim_{x \to a} g(x)} \qquad \text{provided } \lim_{x \to a} g(x) \neq 0$$

EXAMPLE 10

$$\lim_{x \to -5} \frac{x}{x^2 + 10} = \frac{\lim_{x \to -5} x}{\lim_{x \to -5} (x^2 + 10)}$$

$$= \frac{-5}{(-5)^2 + 10} = \frac{-5}{35} = \frac{-1}{7}$$

❑

As you will see in the following examples, evaluating a limit frequently requires the use of more than one of these properties.

EXAMPLE 11

$$\lim_{x \to -1} x^4 = (-1)^4 = 1 \qquad \textbf{(Property 2)}$$

EXAMPLE 12

$$\lim_{x \to 5} (x^2 - x + 10) = \lim_{x \to 5} x^2 - \lim_{x \to 5} x + \lim_{x \to 5} 10 \qquad \textbf{(Property 4)}$$

$$= 5^2 - 5 + 10 \qquad \textbf{(Properties 1 and 2)}$$

$$= 30$$

EXAMPLE 13

$$\lim_{x \to -2} 5x^3 = 5 \cdot \lim_{x \to -2} x^3 \qquad \textbf{(Property 3)}$$

$$= (5)(-2)^3 \qquad \textbf{(Property 2)}$$

$$= (5)(-8) = -40$$

EXAMPLE 14

$$\lim_{x \to 0} [(x^5 - 1)(x^3 + 4)] = \lim_{x \to 0} (x^5 - 1) \cdot \lim_{x \to 0} (x^3 + 4) \qquad \textbf{(Property 5)}$$

$$= (\lim_{x \to 0} x^5 - \lim_{x \to 0} 1)(\lim_{x \to 0} x^3 + \lim_{x \to 0} 4) \qquad \textbf{(Property 4)}$$

$$= (0 - 1)(0 + 4) \qquad \textbf{(Properties 1 and 2)}$$

$$= (-1)(4) = -4$$

EXAMPLE 15

$$\lim_{x \to 2} \frac{x^3 - 1}{x^2} = \frac{\lim_{x \to 2} (x^3 - 1)}{\lim_{x \to 2} x^2} \qquad \textbf{(Property 6)}$$

$$= \frac{\lim_{x \to 2} x^3 - \lim_{x \to 2} 1}{\lim_{x \to 2} x^2} \qquad \textbf{(Property 4)}$$

$$= \frac{2^3 - 1}{2^2} = \frac{8 - 1}{4} = \frac{7}{4} \qquad \textbf{(Properties 1 and 2)}$$

❑

The properties make the process of evaluating limits considerably easier for certain classes of functions. Limits of these types of functions may be evaluated by **substitution** to determine $f(a)$. For these classes of functions

$$\lim_{x \to a} f(x) = f(a) \qquad \textbf{(15.2)}$$

Polynomial functions are a commonly used class of functions for which Eq. (15.2) is valid. This conclusion follows from Properties 1 through 4.

EXAMPLE 16

$$\lim_{x \to -2} (3x^2 - 4x + 10) = f(-2)$$

$$= 3(-2)^2 - 4(-2) + 10 = 12 + 8 + 10 = 30$$

◻

In Example 3 we determined

$$\lim_{x \to 3} \frac{x^2 - 9}{x - 3} = 6$$

Even though the function is not defined at $x = 3$, the value of the function approaches 6 as x approaches 3. This function is an example of a family of "quotient" functions which can be simplified by factoring.

$$f(x) = \frac{x^2 - 9}{x - 3}$$

$$= \frac{(x + 3)(x - 3)}{x - 3}$$

$$= x + 3 \qquad \text{for all } x \neq 3$$

Even though $f(x) = [(x + 3)(x - 3)]/(x - 3)$ and $g(x) = x + 3$ are *not* the same function, *they are the same everywhere $f(x)$ is defined.* As illustrated in Fig. 15.9, g and f graph as identical lines except for the "hole" at $x = 3$ on the graph of f. However, since we are not interested in what happens to f at $x = 3$, we can determine the behavior of f by studying the behavior of g.

Figure 15.9

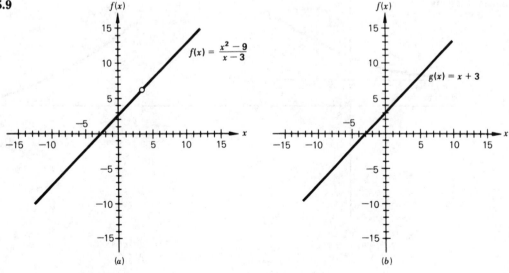

(a)   (b)

Using g as an equivalent function to f, the substitution approach to determining the limit is valid. That is,

$$\lim_{x \to 3} \frac{x^2 - 9}{x - 3} = \lim_{x \to 3} (x + 3)$$

$$= 3 + 3 = 6$$

EXAMPLE 17

$$\lim_{x \to -1} \frac{4x^2 + x - 3}{x + 1} = \lim_{x \to -1} \frac{(4x - 3)(x + 1)}{x + 1}$$

$$= \lim_{x \to -1} (4x - 3)$$

$$= 4(-1) - 3 = -7$$

❑

Limits and Infinity

Frequently there is an interest in the behavior of a function as the independent variable becomes large without limit ("approaching" either positive or negative infinity). Examine the two functions sketched in Fig. 15.10. In Fig. 15.10*a*, as x approaches negative infinity, $f(x)$ approaches but never quite reaches a value of 4. Using limit notation, we state

$$\lim_{x \to -\infty} f(x) = 4$$

We can also state that $f(x)$ has a ***horizontal asymptote*** of $y = 4$ as x approaches $-\infty$. Again, this suggests that $f(x)$ approaches but never quite reaches a value of 4 as x approaches $-\infty$.

Figure 15.10
Limits at infinity.

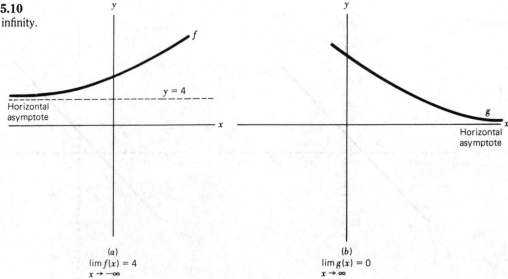

(a)
$\lim_{x \to -\infty} f(x) = 4$

(b)
$\lim_{x \to \infty} g(x) = 0$

Similarly, in Fig. 15.10*b*, *g* approaches but never quite reaches the *x* axis as *x* approaches ∞. We can state this behavior by the notation

$$\lim_{x \to \infty} g(x) = 0$$

As with *f*, *g* has a horizontal asymptote of *y* = 0 as *x* approaches ∞.
A more formal definition for a horizontal asymptote follows.

DEFINITION: HORIZONTAL ASYMPTOTE
The line *y* = *a* is a *horizontal asymptote* of the graph of *f* if and only if

$$\lim_{x \to \infty} f(x) = a$$

or

$$\lim_{x \to -\infty} f(x) = a$$

Let's examine the evaluation of limits as the independent variable approaches positive or negative infinity. Consider

$$\lim_{x \to \infty} (1/x)$$

Table 15.4 illustrates the substitution of sample values for *x*. We can see from this table that $\lim_{x \to \infty} (1/x) = 0$.

TABLE 15.4	*x*	1	10	100	1,000	10,000	100,000
	$f(x) = 1/x$	1	0.1	0.01	0.001	0.0001	0.00001

Consider the evaluation of $\lim_{x \to \infty} (1/x^2)$. Table 15.5 summarizes values of the function for different values of *x*. Again, we see that $\lim_{x \to \infty} (1/x^2) = 0$. Compared with $\lim_{x \to \infty} (1/x)$, the squaring of *x* in the denominator results in the limit being approached at a faster rate.

TABLE 15.5	*x*	1	10	100	1,000	10,000
	$f(x) = 1/x^2$	1	0.01	0.0001	0.000001	0.00000001

EXAMPLE 18 Consider the evaluation of $\lim_{x \to \infty} (3x^2 + 5x)/(4x^2 - 5)$. *One technique for evaluating this limit is to factor the monomial term of the highest degree from both the numerator and the denominator. Factoring $3x^2$ in the numerator and $4x^2$ in the denominator results in*

$$\lim_{x \to \infty} \left[\frac{3x^2}{4x^2} \cdot \frac{\left(1 + \dfrac{5x}{3x^2}\right)}{\left(1 - \dfrac{5}{4x^2}\right)} \right] = \lim_{x \to \infty} \left[\frac{3}{4} \cdot \frac{\left(1 + \dfrac{5}{3x}\right)}{\left(1 - \dfrac{5}{4x^2}\right)} \right]$$

$$= \lim_{x \to \infty} \frac{3}{4} \left[\frac{(1 + 0)}{(1 - 0)} \right]$$

$$= \frac{3}{4}$$

EXAMPLE 19   Consider the evaluation of $\lim\limits_{x \to -\infty} (5x^3 - 20x)/(3x^2 + 5)$. Using the same technique as in the previous example, we factor $5x^3$ in the numerator and $3x^2$ in the denominator. This results in

$$\lim_{x \to -\infty} \left[\frac{5x^3}{3x^2} \cdot \frac{\left(1 - \dfrac{20x}{5x^3}\right)}{\left(1 + \dfrac{5}{3x^2}\right)} \right] = \lim_{x \to -\infty} \left[\frac{5x}{3} \cdot \frac{\left(1 - \dfrac{4}{x^2}\right)}{\left(1 + \dfrac{5}{3x^2}\right)} \right]$$

$$= \lim_{x \to -\infty} \left[\frac{5x}{3} \cdot \frac{(1 - 0)}{(1 + 0)} \right]$$

$$= -\infty$$

$\square$

Although it may not be clear from the examples presented thus far, the limit of a rational function as x approaches positive or negative infinity is simply *the limit of the quotient of the monomial term of highest degree in the numerator and the monomial term of highest degree in the denominator*. This is because the highest-degree terms in both the numerator and denominator dominate all other terms as x approaches positive or negative infinity. In Example 18, we can evaluate the limit by evaluating the behavior of $3x^2/4x^2$ as $x \to \infty$. In Example 19, the limit can be found by evaluating the limit of $5x^3/3x^2$ as $x \to -\infty$.

Let's illustrate this concept specifically by evaluating

$$\lim_{x \to \infty} \frac{4x^4 - 2x^3 + 5x}{25 + 3x^2 - 7x^4}$$

Since $4x^4$ and $-7x^4$ are the monomial terms of highest degree in the numerator and denominator, respectively, the limit can be evaluated by determining

$$\lim_{x \to \infty} (4x^4/-7x^4) = \lim_{x \to \infty} (4/-7)$$

$$= \frac{-4}{7}$$

Another limit possibility is illustrated in Fig. 15.11. In this figure $f(x)$ becomes arbitrarily large as x approaches a. For this situation we state that f has a **vertical asymptote** of $x = a$ because as x approaches the value a, $f(x)$ gets large without

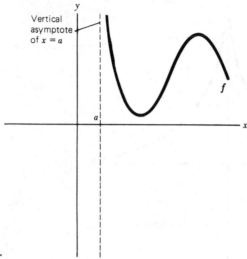

Figure 15.11
Vertical asymptote.

limit. Visually, the ordinate of the curve representing f becomes large without limit as the curve gets closer to a line at $x = a$, never quite touching it. A more formal definition of this phenomenon follows.

> **DEFINITION: VERTICAL ASYMPTOTE**
> The line $x = a$ is a *vertical asymptote* of the graph of f if and only if
>
> $$\lim_{x \to a^-} f(x) = \infty \qquad (\text{or} -\infty)$$
>
> or
>
> $$\lim_{x \to a^+} f(x) = \infty \qquad (\text{or} -\infty)$$

EXAMPLE 20 To evaluate $\lim_{x \to 0} (1/x^2)$, we must take left- and right-hand limits as x approaches 0. Table 15.6 presents selected values. We should conclude that $\lim_{x \to 0} (1/x^2) = \infty$. Graphically, f appears as in Fig. 15.12. Note that $f(x) = 1/x^2$ has a vertical asymptote of $x = 0$ and horizontal asymptotes of $y = 0$.

TABLE 15.6

Approaching $x = 0$ from the Left					
x	-1	-0.5	-0.1	-0.01	-0.001
$f(x) = \dfrac{1}{x^2}$	1	4	100	10,000	1,000,000

Approaching $x = 0$ from the Right					
x	1	0.5	0.1	0.01	0.001
$f(x) = \dfrac{1}{x^2}$	1	4	100	10,000	1,000,000

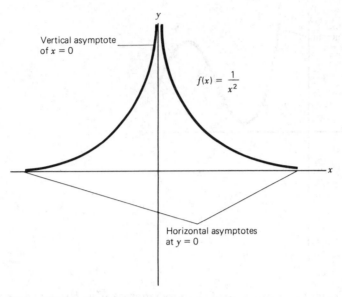

Figure 15.12   Horizontal asymptote at $y = 0$.

NOTE   A conclusion that

$$\lim_{x \to a} f(x) = \infty$$

or

$$\lim_{x \to a} f(x) = -\infty$$

acknowledges that *a limit does not exist:* that is, there is not a real-number limiting value for $f(x)$.

Continuity

In an informal sense, a function is described as **continuous** if it can be sketched without lifting your pen or pencil from the paper (i.e., it has no gaps, no jumps, and no breaks). Most of the functions that we will examine in the calculus will be continuous functions. Figure 15.13 indicates the sketches of four different functions. Those depicted in Fig. 15.13*a* and 15.13*b* are continuous since they can be drawn without lifting your pencil. Those in Fig. 15.13*c* and 15.13*d* are not continuous because of the "breaks" in the functions. A function which is not continuous is termed **discontinuous**. A more formal definition of the property of continuity follows.

Figure 15.13
Assessing
continuity.

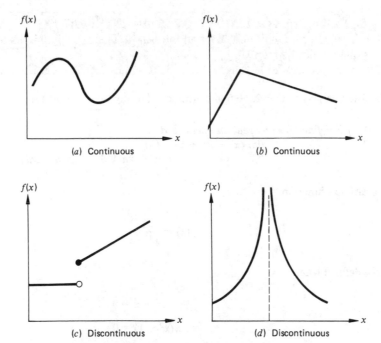

(a) Continuous

(b) Continuous

(c) Discontinuous

(d) Discontinuous

DEFINITION: CONTINUITY AT A POINT
A function f is said to be *continuous* at $x = a$ if

1 *the function is defined at $x = a$, and*
2 $\lim\limits_{x \to a} f(x) = f(a)$ (15.3)

EXAMPLE 21   In Example 1 we determined that for $f(x) = x^3$

$$\lim_{x \to 2} x^3 = 8$$

Because $f(x) = x^3$ is defined at $x = 2$ and $\lim\limits_{x \to 2} x^3 = f(2) = 8$, we can state that the function $f(x) = x^3$ is continuous at $x = 2$.

EXAMPLE 22   In Example 3 we determined that

$$\lim_{x \to 3} \frac{x^2 - 9}{x - 3} = 6$$

Because $x = 3$ is not in the domain of the function, $f(3)$ is not defined and we can state that the function $(x^2 - 9)/(x - 3)$ is *discontinuous* at $x = 3$.

□

DEFINITION: CONTINUITY OVER AN INTERVAL
A function f is continuous over an interval $[a, b]$ if it is continuous at every point within the interval.

EXAMPLE 23   The function $f(x) = x^2 - 2x + 5$ is continuous for any real x because

1. *f is defined for any real numbers, and*
2. $\lim\limits_{x \to a} (x^2 - 2x + 5) = f(a)$
   $= a^2 - 2a + 5$ *for all real a*

EXAMPLE 24   The rational function

$$f(x) = \frac{1}{x^3 - x}$$

is not defined when

$$x^3 - x = 0$$

or

$$x(x^2 - 1) = 0$$

or

$$x(x + 1)(x - 1) = 0$$

Since the left side of the equation equals 0 when $x = 0$, $x = -1$, or $x = +1$, the function is discontinuous at these three values. Figure 15.14 presents a sketch of the function. Note that this function has vertical asymptotes described by the equations $x = 1$, $x = -1$, and $x = 0$.

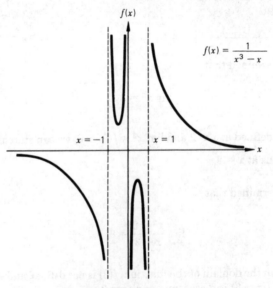

$$f(x) = \frac{1}{x^3 - x}$$

Figure 15.14   Discontinuities at $x = 0, -1, 1$.

Section 15.2 Follow-up Exercises

For the following exercises, find the indicated limit.

1 $\lim\limits_{x\to 0} (3x^2 - 5x + 3)$

2 $\lim\limits_{x\to 2} (2x^3 - 10x)$

3 $\lim\limits_{x\to 2} \left(\dfrac{x^3}{3} - 7x^2\right)$

4 $\lim\limits_{x\to 3} \dfrac{2x - 8}{x + 4}$

5 $\lim\limits_{x\to 2} \dfrac{2x^2 + 3}{x^3 + 4x - 2}$

6 $\lim\limits_{x\to -4} 250$

7 $\lim\limits_{x\to 0} 175$

8 $\lim\limits_{x\to -10} \dfrac{x^2 + 25}{x}$

9 $\lim\limits_{x\to 1} (3x^3 + 2x^2 + x)$

10 $\lim\limits_{x\to -2} (-x^4 + 8x^2)$

11 $\lim\limits_{x\to -2} (-x^3 + 5x^2 + 10)$

12 $\lim\limits_{x\to 2} (5x^3 + 10x^2)$

13 $\lim\limits_{x\to 5} \left(\dfrac{4x - 20}{3x^2 - 7x + 5}\right)$

14 $\lim\limits_{x\to -2} \left(\dfrac{10 + x^2}{5 - 8x + 2x^2}\right)$

15 $\lim\limits_{x\to -3} (6x^3 + 2x)(5x - 10)$

16 $\lim\limits_{x\to 5} \left[\left(\dfrac{x + 3}{x + 6}\right)(x^2 - 12)\right]$

17 $\lim\limits_{x\to -4} \dfrac{x^2 + 8x - 14}{2x + 7}$

18 $\lim\limits_{x\to 1} \dfrac{x^2 + 3x - 24}{x - 3}$

19 $\lim\limits_{x\to -4} \dfrac{x^2 - 16}{x + 4}$

20 $\lim\limits_{x\to -9} \dfrac{81 - x^2}{9 + x}$

21 $\lim\limits_{x\to c} (4x^3 - 5x^2 + 10)$

22 $\lim\limits_{x\to -d} (x^2 - 2x + 3)$

For the following exercises, find the indicated limit and comment on the existence of any asymptotes.

23 $\lim\limits_{x\to \infty} \dfrac{4}{x^2}$

24 $\lim\limits_{x\to -\infty} \dfrac{5x - 3}{x + 10}$

25 $\lim\limits_{x\to \infty} \dfrac{-3x}{5x + 100}$

26 $\lim\limits_{x\to -\infty} \dfrac{8x + 10}{-4x}$

27 $\lim\limits_{x\to -\infty} 2x$

28 $\lim\limits_{x\to \infty} \dfrac{x^2 + 3}{x - 4}$

29 $\lim\limits_{x\to \infty} \dfrac{-8x}{4x + 1,000}$

30 $\lim\limits_{x\to -\infty} \dfrac{5x + 10,000}{2x - 5,000}$

31 $\lim\limits_{x\to -\infty} \dfrac{100 - 3x^3}{-x^3}$

32 $\lim\limits_{x\to \infty} \dfrac{3x^3 - 500}{5,000 - x^3}$

In the following exercises, determine whether there are any discontinuities and, if so, where they occur.

33 $f(x) = 3x^2 + 2x + 1$

34 $f(x) = \dfrac{1}{x + 2}$

35 $f(x) = \dfrac{x^4}{5}$

36 $f(x) = \dfrac{3x^2}{x + 5}$

37 $f(x) = \dfrac{1}{8 - 2x}$

38 $f(x) = |x|$

39 $f(x) = \dfrac{2x + 3}{x^2 + 4x - 21}$

40 $f(x) = \dfrac{x - 2}{x^2 - 8x}$

41 $f(x) = \dfrac{4x - 3}{x^3 - x^2 - 6x}$

42 $f(x) = \dfrac{5/x}{2x^2 + 7x - 15}$

43 $f(x) = \dfrac{20}{x^2 - 3x - 10}$

44 $f(x) = \dfrac{4/x}{18 + 3x - x^2}$

45 $f(x) = \dfrac{10/(5 - x)}{4 - x^2}$

46 $f(x) = \dfrac{5/(3 - x)}{x^2 - 16}$

47 $f(x) = \dfrac{3x - 5}{x^4 - 27x}$

48 $f(x) = \dfrac{3/(x^2 - 1)}{2/(x^2 - 4)}$

15.3 AVERAGE RATE OF CHANGE

Average Rate of Change and Slope

As discussed in Chap. 2, the slope of a straight line can be determined by applying the two-point formula.

> **TWO-POINT FORMULA**
>
> $$m = \frac{\Delta y}{\Delta x} = \frac{y_2 - y_1}{x_2 - x_1}$$ (15.4)

Figure 15.15 illustrates the graph of a linear function. With linear functions the slope is constant over the domain of the function. The slope provides an *exact measure* of the rate of change in the value of y with respect to a change in the value

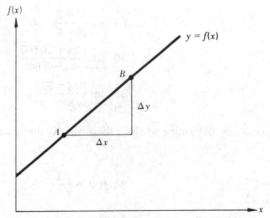

Figure 15.15   Linear function with constant slope.

of x. If the function in Fig. 15.15 represents a linear cost function and x equals the number of units produced, the slope indicates the rate at which total cost increases with respect to changes in the level of output.

With nonlinear functions the rate of change in the value of y with respect to a change in x is not constant. However, one way of partially describing nonlinear functions is by the **average rate of change** over some interval.

EXAMPLE 25   Assume that a person takes a leisurely automobile trip and that the distance traveled d can be estimated as a function of time t by the nonlinear function

$$d = f(t) = 8t^2 + 8t$$

where d is measured in miles, t is measured in hours, and $0 \le t \le 5$. During this 5-hour journey the speed of the car may change continuously (e.g., because of traffic lights, rest stops, etc.).

After 1 hour, the total distance traveled is

$$f(1) = 8(1)^2 + 8(1)$$
$$= 16 \text{ miles}$$

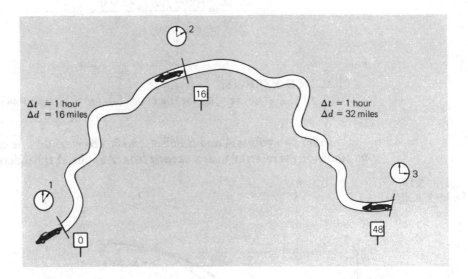

The average rate of change in the distance traveled with respect to a change in time during a time interval (better known as **average velocity**) is computed as

$$\frac{\text{Distance traveled}}{\text{Time traveled}}$$

For the first hour of this trip, the average velocity equals

$$\frac{\Delta d}{\Delta t} = \frac{f(1) - f(0)}{1 - 0} = \frac{16 - 0}{1} = 16 \text{ mph}$$

The distance traveled at the end of 2 hours is

$$f(2) = 8(2)^2 + 8(2)$$
$$= 32 + 16 = 48 \text{ miles}$$

The distance traveled *during* the second hour is

$$\Delta d = f(2) - f(1)$$
$$= 48 - 16 = 32 \text{ miles}$$

The average velocity *for the second hour* equals

$$\frac{\Delta d}{\Delta t} = \frac{32}{1} = 32 \text{ mph}$$

The average velocity for the second hour is different compared with that for the first hour.
The average velocity *during the first 2 hours* is the total distance traveled during that period divided by the time traveled, or

$$\frac{\Delta d}{\Delta t} = \frac{f(2) - f(0)}{2 - 0} = \frac{48 - 0}{2} = 24 \text{ mph}$$

❑

PRACTICE EXERCISE
What is the average velocity for the entire 5-hour trip? *Answer:* 48 mph.

Consider two points A and B in Fig. 15.16. The straight line connecting these two points on f is referred to as a **secant line.** At point A the independent variable

Figure 15.16
Secant line AB.

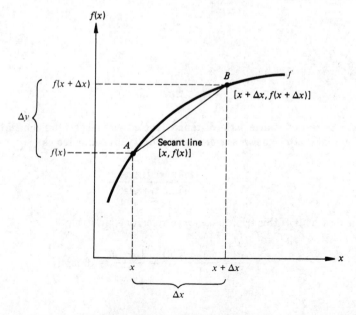

has a value of x, and the corresponding value of the dependent variable can be determined by evaluating $f(x)$. At point B the independent variable has changed in value to $x + \Delta x$, and the corresponding value of the dependent variable can be determined by evaluating $f(x + \Delta x)$. In moving from point A to point B, the change in the value of x is $(x + \Delta x) - x$, or Δx. The associated change in the value of y is $\Delta y = f(x + \Delta x) - f(x)$. The ratio of these changes is

$$\frac{\Delta y}{\Delta x} = \frac{f(x + \Delta x) - f(x)}{\Delta x}$$ (15.5)

Equation (15.5) is sometimes referred to as the ***difference quotient.***

DEFINITION: THE DIFFERENCE QUOTIENT
Given any two points on a function f having coordinates $[x, f(x)]$ and $[(x + \Delta x), f(x + \Delta x)]$, the difference quotient provides a general expression which represents

I   *the* average rate of change *in the value of y with respect to the change in x while moving from $[x, f(x)]$ to $[(x + \Delta x), f(x + \Delta x)]$;*
II   *the slope of the secant line connecting the two points.*

EXAMPLE 26

(a) Find the general expression for the difference quotient of the function $y = f(x) = x^2$.
(b) Find the slope of the line connecting $(-2, 4)$ and $(3, 9)$ using the two-point formula.
(c) Find the slope in part b using the expression for the difference quotient found in part a.

SOLUTION

(a) Given two points on the function $f(x) = x^2$ which have coordinates $(x, f(x))$ and $(x + \Delta x, f(x + \Delta x))$, we have

$$\frac{\Delta y}{\Delta x} = \frac{f(x + \Delta x) - f(x)}{\Delta x} = \frac{(x + \Delta x)^2 - x^2}{\Delta x}$$

$$= \frac{[x^2 + x(\Delta x) + x(\Delta x) + (\Delta x)^2] - x^2}{\Delta x}$$

$$= \frac{[x^2 + 2x(\Delta x) + (\Delta x)^2] - x^2}{\Delta x}$$

$$= \frac{2x(\Delta x) + (\Delta x)^2}{\Delta x}$$

Factoring Δx from each term in the numerator and canceling with Δx in the denominator, we get

$$\frac{\Delta y}{\Delta x} = \frac{f(x + \Delta x) - f(x)}{\Delta x} = \frac{\cancel{\Delta x}\,(2x + \Delta x)}{\cancel{\Delta x}}$$

$$= 2x + \Delta x$$ (15.6)

> **NOTE**   When finding the difference quotient, the evaluation of $f(x + \Delta x)$ and $f(x)$ for a specific function causes the greatest difficulty for students. If, for this function, you were asked to find $f(3)$, you would substitute the value of 3 into the function wherever the independent variable appears, or $f(3) = 3^2 = 9$. When asked to find $f(x + \Delta x)$ in this example, we substituted $x + \Delta x$ into the function where the independent variable appeared and evaluated $(x + \Delta x)^2$. Similarly, when finding $f(x)$, we substituted the value x where the independent variable appeared and evaluated x^2. *Whenever you are asked to determine the difference quotient, $f(x)$ will always be the specific function with which you are working.*

(b) Using the two-point slope formula, we get

$$\frac{\Delta y}{\Delta x} = \frac{f(3) - f(-2)}{3 - (-2)}$$

$$= \frac{(3)^2 - (-2)^2}{5}$$

$$= \frac{9 - 4}{5} = \frac{5}{5} = 1$$

Figure 15.17

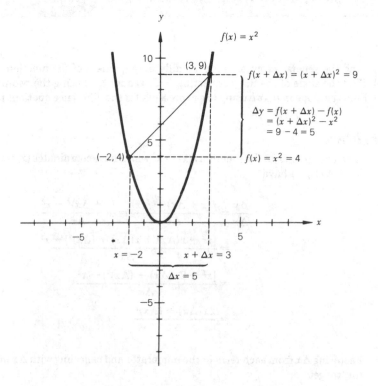

The slope of the secant line connecting $(-2, 4)$ and $(3, 9)$ on f equals 1.

(c) As shown in Fig. 15.17, let $(x, f(x))$ and $(x + \Delta x, f(x + \Delta x))$ correspond to the points

(−2, 4) and (3, 9) on the function. Assume that x corresponds to the coordinate −2 and $x + \Delta x$ corresponds to the coordinate of 3. Thus,

$$((x), f(x)) \quad ((x + \Delta x), f(x + \Delta x))$$
$$((-2), 4) \quad ((3), 9)$$

So $$x = -2$$

and $$x + \Delta x = 3$$

Therefore $$(-2) + \Delta x = 3$$

or $$\Delta x = 5$$

This value for Δx (the change in x) simply means that in moving *from* (−2, 4) to (3, 9), the value of x has *increased* by 5 units.

If we substitute the values of $x = -2$ and $\Delta x = 5$ into Eq. (15.6), the result is exactly the same as we got in part *b*.

$$\frac{\Delta y}{\Delta x} = 2x + \Delta x$$
$$= 2(-2) + 5 = 1$$

◻

PRACTICE EXERCISE
As in finding the slope of a line connecting two points, the difference quotient [Eq. (15.5)] for two specific points on a function is unaffected by which point is labeled [x, $f(x)$] and which [$(x + \Delta x)$, $f(x + \Delta x)$]. In the last example, let $x = 3$ and $(x + \Delta x) = -2$. Evaluate the difference quotient and see if you arrive at the same result as in the example.

Section 15.3 Follow-up Exercises

For each of the following functions, determine the average rate of change in the value of y in moving from $x = -1$ to $x = 2$.

1 $y = f(x) = 3x^2$

2 $y = f(x) = 5x^3$

3 $y = f(x) = x^2 - 2x + 3$

4 $y = f(x) = x^3 - 2x^2 + x + 2$

5 $y = f(x) = x^2/(x + 4)$

6 $y = f(x) = x^3/2$

7 $y = f(x) = 2x^2 + 6x + 3$

8 $y = f(x) = 2x^2 - 8x + 10$

9 $y = f(x) = 4x^2 - 2x$

10 $y = f(x) = -x^2 + 2x + 4$

11 $y = f(x) = -5x^3$

12 $y = f(x) = 3x^3 + 4x - 5$

13 $y = f(x) = x^4$

14 $y = f(x) = x^4 - 10$

15 A ball is thrown straight up into the air. The height of the ball can be described as a function of time according to the function

$$h(t) = -16t^2 + 128t$$

where $h(t)$ is height measured in feet and t is time measured in seconds.

(a) Determine the average rate of change in height between $t = 0$ and $t = 2$. Between $t = 0$ and $t = 4$. Between $t = 0$ and $t = 8$.

(b) How long does it take for the ball to hit the ground ($h = 0$)?

16 An object is dropped from a bridge which is 576 feet high. The height of the object can be determined as a function of time (since being dropped) according to the function

$$h(t) = 576 - 16t^2$$

where $h(t)$ is height measured in feet and t is time measured in seconds.

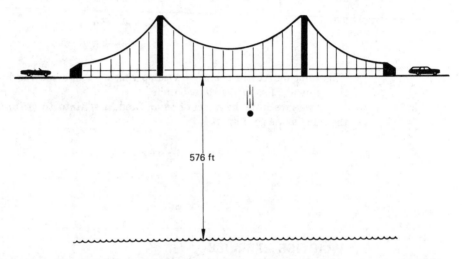

576 ft

(a) Determine the average rate of change in height between $t = 0$ and $t = 1$. Between $t = 0$ and $t = 2$. Between $t = 0$ and $t = 4$.

(b) How long does it take for the ball to hit the water ($h = 0$)?

17 Table 15.7 indicates annual sales (in dollars) for a company during a selected period. At what average rate did annual sales increase between 1988 and 1990? Between 1988 and 1989? Between 1989 and 1990? Between 1989 and 1991?

TABLE 15.7	Year	1988	1989	1990	1991
	Annual sales (millions)	$100.8	$105.4	$109.8	$116.5

18 **Baseball Attendance** Annual attendance at professional baseball games has been increasing in recent years. Figures for the years 1987 to 1991 are shown in Table 15.8. Determine the average rate of change in annual attendance between 1987 and 1991, 1987 and 1990, and 1989 and 1991.

TABLE 15.8	Year	1987	1988	1989	1990	1991
	Annual attendance (millions)	63.1	64.8	66.0	67.8	69.0

19 Minority Population Growth Hispanics are the fastest-growing minority group within the United States. If current trends continue, it is estimated that Hispanics will surpass blacks as the largest minority group somewhere around the year 2005. Table 15.9 shows estimates of the U.S. Hispanic population (in millions) in recent years. Determine the average rate of change in the Hispanic population between 1987 and 1989, 1988 and 1990, and 1987 and 1990.

TABLE 15.9	Year	1987	1988	1989	1990
	Population	19.2	19.9	21.0	22.4

20 Pay-Per-View TV Pay-per-view television (PPV) has been a growing option among TV viewers. With PPV, cable TV subscribers buy only the cable shows they wish to watch. Special events, such as championship boxing matches or live concerts, are prime PPV offerings. Figure 15.18 summarizes total revenues for the industry for four recent years. Determine the average rate of change in PPV revenues between 1987 and 1989, and between 1987 and 1990.

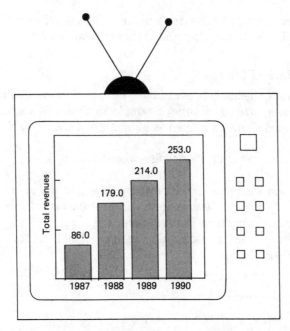

Figure 15.18 Pay-per-view revenues, $ millions.

21 A person takes an automobile trip. The distance traveled d (in miles) is described as a function of time t (in hours):

$$d = f(t) = 5t^2 + 12t \quad \text{where } 0 \le t \le 4$$

(a) What is the average speed during the first hour? Second hour?

(b) What is the average speed for the 4-hour trip?

For Exercises 22–39, (a) determine the general expression for the difference quotient, and (b) use the difference quotient to compute the slope of the secant line connecting points at $x = 1$ and $x = 3$.

22 $y = f(x) = 4x^2 + 3$

23 $y = f(x) = x^2 + 3x$

24 $y = f(x) = 10x^2 + 20x$

25 $y = f(x) = 5$

26 $y = f(x) = -3x^2 + 8x + 10$

27 $y = f(x) = 5x^2 + 20x$

28 $y = f(x) = \dfrac{x^2}{2} + 2x$

29 $y = f(x) = \dfrac{x^2}{3} + 5x$

30 $y = f(x) = ax^2 - bx$

31 $y = f(x) = mx^2 - n$

*32 $y = f(x) = x^3$

*33 $y = f(x) = -2x^3$

*34 $y = f(x) = 1/x$

*35 $y = f(x) = 5/x$

*36 $y = f(x) = 2x^3 - 10$

*37 $y = f(x) = -5x^3$

*38 $y = f(x) = -4/x$

*39 $y = f(x) = -6/x$

15.4 THE DERIVATIVE

In this section the concept of the **derivative** will be developed. This concept is fundamental to all that follows, so study this material carefully.

Instantaneous Rate of Change

A distinction needs to be made between the concepts of **average rate of change** and **instantaneous rate of change.** Example 25 discussed a situation in which the distance traveled d was estimated as a function of time t by the function

$$d = f(t) = 8t^2 + 8t \quad \text{where } 0 \le t \le 5$$

Suppose that we are interested in determining how fast the car is moving at the *instant* that $t = 1$. We might determine this instantaneous velocity by examining the average velocity during time intervals near $t = 1$.

For instance, the average velocity during the second hour (between $t = 1$ and $t = 2$) can be determined as

$$\frac{\Delta d}{\Delta t} = \frac{f(2) - f(1)}{2 - 1}$$

$$= \frac{[8(2^2) + 8(2)] - [8(1^2) + 8(1)]}{1} = \frac{48 - 16}{1} = 32 \text{ mph}$$

The average velocity between $t = 1$ and $t = 1.5$ can be determined as

$$\frac{\Delta d}{\Delta t} = \frac{f(1.5) - f(1)}{1.5 - 1}$$

$$= \frac{[(1.5)^2 + 8(1.5)] - [8(1^2) + 8(1)]}{0.5} = \frac{30 - 16}{0.5} = 28 \text{ mph}$$

The average velocity between $t = 1$ and $t = 1.1$ can be determined as

$$\frac{\Delta d}{\Delta t} = \frac{f(1.1) - f(1)}{1.1 - 1}$$

$$= \frac{[8(1.1)^2 + 8(1.1)] - [8(1^2) + 8(1)]}{0.1} = \frac{18.48 - 16}{0.1} = 24.8 \text{ mph}$$

The average velocity between $t = 1$ and $t = 1.01$ can be determined as

$$\frac{\Delta d}{\Delta t} = \frac{f(1.01) - f(1)}{1.01 - 1}$$

$$= \frac{[8(1.01)^2 + 8(1.01)] - [8(1^2) + 8(1)]}{0.01} = \frac{16.2408 - 16}{0.01} = 24.08 \text{ mph}$$

Figure 15.19

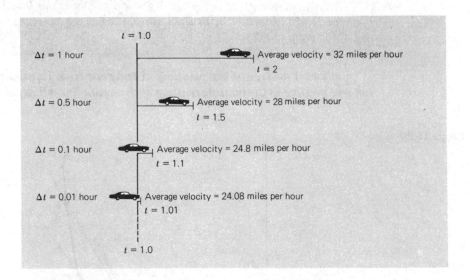

As shown in Fig. 15-19, these computations have been determining the average velocity over shorter and shorter time intervals measured *from* $t = 1$. As the time interval becomes shorter (or as the second value of t is chosen closer and closer to 1), the average velocity $\Delta d / \Delta t$ is approaching a limiting value. The *instantaneous velocity* at $t = 1$ can be defined as this limiting value. To determine this limiting value, we could compute

$$\lim_{t \to 1} \frac{f(t) - f(1)}{t - 1} = \lim_{t \to 1} \frac{(8t^2 + 8t) - [8(1)^2 + 8(1)]}{t - 1}$$

$$= \lim_{t \to 1} \frac{(8t^2 + 8t) - 16}{t - 1}$$

$$= \lim_{t \to 1} \frac{8(t^2 + t - 2)}{t - 1}$$

$$= \lim_{t \to 1} \frac{8(t+2)(t-1)}{t-1}$$

$$= \lim_{t \to 1} 8(t+2)$$

$$= 8 \lim_{t \to 1} (t+2)$$

$$= 8(1+2) = 24$$

Thus, the *instantaneous* velocity of the automobile at $t = 1$ is 24 miles per hour. Note that the average velocity is measured over a time interval and the instantaneous velocity is defined for a particular point in time. The instantaneous velocity is a "snapshot" of what is happening at a particular instant.

> **GEOMETRIC REPRESENTATION OF INSTANTANEOUS RATE OF CHANGE**
> The instantaneous rate of change of a smooth, continuous function can be represented geometrically by the slope of the line drawn tangent to the curve at the point of interest.

Let's first determine the meaning of **tangent line.** Consider Fig. 15.20. *The tangent line at A is the limiting position of the secant line AB as point B comes closer*

Figure 15.20

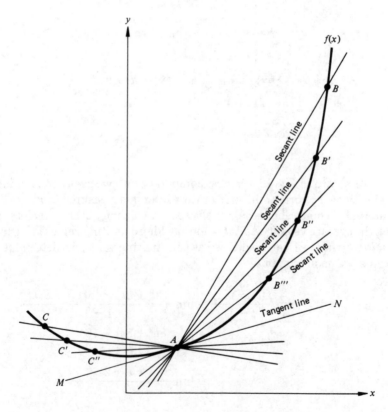

and closer to A. Note how the position of the secant line rotates in a clockwise manner as *B* is drawn closer to *A*(*AB*, *AB'*, *AB"*, and *AB'"*). The limiting position of *AB* is the line segment *MN*. This same limiting position results whether *A* is approached with secant lines from the left or right of *A*. The sequence of secant lines *AC*, *AC'*, and *AC"* has the same limiting position *MN* as *C* is drawn closer to *A*. *Since MN is the limiting position whether the approach is from the left or right, MN is the tangent line at point A.*

Not all continuous functions have unique tangent lines at each point on the function. For example, the function $y = f(x) = \sqrt{|x|}$, shown in Fig. 15.21, does not have a tangent at (0, 0). The secant lines drawn from (0, 0) to points on the left or right do not converge to the same limiting position.

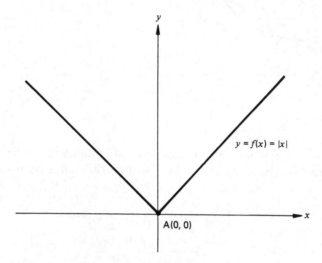

Figure 15.21 No tangent line at point *A*.

DEFINITION: SLOPE OF CURVE
The slope of a curve at $x = a$ is the slope of the tangent line at $x = a$.

Later we will have a particular interest in determining the instantaneous rate of change in functions. Since the instantaneous rate of change is represented by the slope of the tangent line at the point of interest, we will need a way of determining these slopes of tangent lines. Suppose in Fig. 15.22 that we are interested in finding the slope of the tangent line at *A*. There are several different methods we might use to determine this slope. If we were good at mechanical drawing, we might construct a tangent line at point *A* using graph paper, read from the line coordinates of any two points, and substitute these coordinates into the two-point formula.

An alternative approach would be to pick another point *B* on the curve. If we connect *A* and *B* with a secant line, the slope of the line segment *AB* can be computed and used as an "approximation" to the slope of *MN*. Obviously, the slope of *AB* is not a good approximation. However, let's continue, still referring to Fig.

Figure 15.22

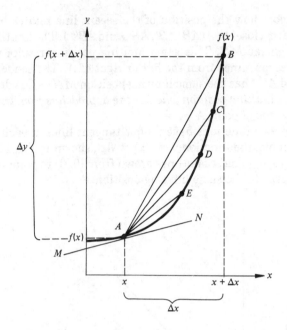

15.22. At point B the value of the independent variable is $x + \Delta x$; the distance between A and B along the x axis is Δx. Using the difference quotient, the slope of AB is

$$\frac{\Delta y}{\Delta x} = \frac{f(x + \Delta x) - f(x)}{\Delta x} \qquad (15.5)$$

Now, let's observe what happens to our approximation if a second point is chosen closer to point A. If point C is chosen as the second point, the slope of the secant line AC is still a poor approximation, but better than that of AB. And if you look at the slopes of AD and AE, you should conclude that *as the second point is chosen closer and closer to A, the approximation becomes better*. In fact, as the value of Δx gets close to 0, the slope of the tiny line segment connecting A with the second point becomes an excellent approximation. The exact tangent slope can be determined by finding the limit of the difference quotient as $\Delta x \to 0$.

DEFINITION: THE DERIVATIVE
Given a function of the form $y = f(x)$, the *derivative* of the function is

$$\frac{dy}{dx} = \lim_{\Delta x \to 0} \frac{f(x + \Delta x) - f(x)}{\Delta x} \qquad (15.7)$$

provided this limit exists.

The following points should be made regarding this definition.

> **COMMENTS ABOUT THE DERIVATIVE**
>
> I   *Equation (15.7) is the general expression for the derivative of the function f.*
> II *The derivative represents the **instantaneous rate of change** in the dependent variable given a change in the independent variable. The notation dy/dx is used to represent the instantaneous rate of change in y with respect to a change in x. This notation is distinguished from $\Delta y/\Delta x$ which represents the average rate of change.*
> III *The derivative is a **general expression for the slope** of the graph of f at any point x in the domain.*
> IV *If the limit in Eq. (15.7) does not exist, the derivative does not exist.*

The Limit Approach to Finding the Derivative

> **FINDING THE DERIVATIVE (LIMIT APPROACH)**
>
> ❏ **Step 1**   Determine the difference quotient for f using Eq. (15.5).
> ❏ **Step 2**   Find the limit of the difference quotient as $\Delta x \to 0$ using Eq. (15.7).

The following examples illustrate the ***limit approach*** for determining the derivative.

EXAMPLE 27   Find the derivative of $f(x) = -5x + 9$.

SOLUTION

The function $f(x) = -5x + 9$ is linear with a slope of -5. With the slope always -5, the derivative of f should equal -5.

❏ **Step 1** The difference quotient is

$$\frac{\Delta y}{\Delta x} = \frac{f(x + \Delta x) - f(x)}{\Delta x}$$

$$= \frac{[-5(x + \Delta x) + 9] - (-5x + 9)}{\Delta x}$$

$$= \frac{-5x - 5\,\Delta x + 9 + 5x - 9}{\Delta x}$$

$$= \frac{-5\,\Delta x}{\Delta x}$$

or

$$\frac{\Delta y}{\Delta x} = -5$$

❏ **Step 2** The derivative is the limit of the difference quotient, or

$$\frac{dy}{dx} = \lim_{\Delta x \to 0} (-5)$$
$$= -5$$

Thus, the derivative is exactly what we anticipated.

| EXAMPLE 28 | Find the derivative of $f(x) = x^2$.

SOLUTION

❏ **Step 1** In Example 26 we found that the difference quotient for $f(x) = x^2$ was

$$\frac{\Delta y}{\Delta x} = 2x + \Delta x$$

❏ **Step 2** The derivative is the limit of the difference quotient, or

$$\frac{dy}{dx} = \lim_{\Delta x \to 0} (2x + \Delta x)$$

or $\dfrac{dy}{dx} = 2x$

❏

USING AND INTERPRETING THE DERIVATIVE

To determine the instantaneous rate of change (or equivalently, the slope) at any point on the graph of a function f, substitute the value of the independent variable into the expression for dy/dx. The derivative, evaluated at $x = c$, can be denoted by $\left.\dfrac{dy}{dx}\right|_{x=c}$, which is read "the derivative of y with respect to x evaluated at $x = c$."

| EXAMPLE 29 | For the function $f(x) = x^2$:
(a) Determine the instantaneous rate of change in $f(x)$ at $x = -3$.
(b) Determine the instantaneous rate of change in $f(x)$ at $x = 0$.
(c) Determine the instantaneous rate of change in $f(x)$ at $x = +3$.

SOLUTION

In Exercise 28 we determined that $dy/dx = 2x$. Answers to parts a to c are found by substitution into this expression.

(a) $\left.\dfrac{dy}{dx}\right|_{x=-3} = 2(-3) = -6$

(b) $\left.\dfrac{dy}{dx}\right|_{x=0} = 2(0) = 0$

(c) $\left.\dfrac{dy}{dx}\right|_{x=3} = 2(3) = +6$

☐

POINTS FOR THOUGHT & DISCUSSION

From Chap. 6 we know that the function $y = x^2$ is quadratic. Sketch the function and confirm that the values which we found in Example 29 seem reasonable as representing the slope at $x = -3$, 0, and 3. The derivative expression $dy/dx = 2x$ suggests that as x becomes more negative, the slope becomes more negative; and as x becomes more positive, the slope becomes more positive. Does this seem correct in light of your sketch?

EXAMPLE 30

For the function $f(x) = -2x^2 + 3x - 10$:
(a) Determine the derivative.
(b) Determine the instantaneous rate of change in $f(x)$ at $x = 5$.
(c) Determine where on the function the slope equals 0.

SOLUTION

(a) **Step 1** The difference quotient for f is

$$\frac{\Delta y}{\Delta x} = \frac{f(x + \Delta x) - f(x)}{\Delta x}$$

$$= \frac{[-2(x + \Delta x)^2 + 3(x + \Delta x) - 10] - (-2x^2 + 3x - 10)}{\Delta x}$$

$$= \frac{[-2(x^2 + 2x\,\Delta x + \Delta x^2) + 3x + 3\,\Delta x - 10] + 2x^2 - 3x + 10}{\Delta x}$$

$$= \frac{-2x^2 - 4x\,\Delta x - 2\,\Delta x^2 + 3x + 3\,\Delta x - 10 + 2x^2 - 3x + 10}{\Delta x}$$

Simplifying the numerator gives

$$\frac{\Delta y}{\Delta x} = \frac{-4x\,\Delta x - 2(\Delta x)^2 + 3\,\Delta x}{\Delta x}$$

Factoring Δx from the numerator and simplifying yield

$$\frac{\Delta y}{\Delta x} = \frac{\cancel{\Delta x}\,(-4x - 2\,\Delta x + 3)}{\cancel{\Delta x}} = -4x - 2\,\Delta x + 3$$

Step 2 The derivative of the function is

$$\frac{dy}{dx} = \lim_{\Delta x \to 0}\ (-4x - 2\,\Delta x + 3)$$

or

$$\frac{dy}{dx} = -4x + 3$$

(b) $\left.\dfrac{dy}{dx}\right|_{x=5} = -4(5) + 3$

$\qquad\qquad = -17$

(c) The slope will equal 0 whenever $dy/dx = 0$, or for this function when $-4x + 3 = 0$. Solving for x, we have

$$-4x = -3$$

or $\qquad\qquad\qquad x = \tfrac{3}{4}$

The only point where the slope equals 0 occurs when $x = \tfrac{3}{4}$.

☐

PRACTICE EXERCISE

Verify that $x = \tfrac{3}{4}$ is the x coordinate of the vertex of the parabola representing the function $f(x) = -2x^2 + 3x - 10$ using the appropriate formula in Sec. 6.2.

If a function is not continuous at a point, it cannot have a derivative at that point. However, some functions are continuous and yet there may be points within the domain at which the derivative does not exist.

EXAMPLE 31   Consider $f(x) = |2x|$ shown in Fig. 15.23. Suppose we wish to find the derivative of f at $x = 0$. This can be determined by evaluating Eq. (15.7) at $x = 0$.

$$\frac{dy}{dx} = \lim_{\Delta x \to 0} \frac{f(0 + \Delta x) - f(0)}{\Delta x}$$

$$= \lim_{\Delta x \to 0} \frac{|2(0 + \Delta x)| - |2(0)|}{\Delta x}$$

$$= \lim_{\Delta x \to 0} \frac{|2\,\Delta x|}{\Delta x}$$

Figure 15.23

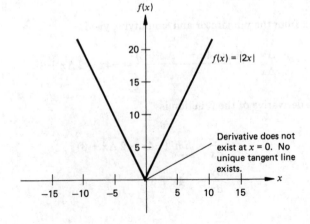

TABLE 15.10	**Approaching from the Left**			
Δx		-1	-0.5	-0.01
$\dfrac{\|2\,\Delta x\|}{\Delta x} = \dfrac{-2\,\Delta x}{\Delta x}$		-2	-2	$-2.$

	Approaching from the Right			
Δx		1	0.5	0.01
$\dfrac{\|2\,\Delta x\|}{\Delta x} = \dfrac{2\,\Delta x}{\Delta x}$		2	2	2

Evaluating both the left- and right-hand limits (Table 15.10 indicates sample values),

$$\lim_{\Delta x \to 0^-} \frac{|2\,\Delta x|}{\Delta x} = -2 \quad and \quad \lim_{\Delta x \to 0^+} \frac{|2\,\Delta x|}{\Delta x} = 2$$

Since these two limits are not the same,

$$\lim_{\Delta x \to 0} \frac{|2\,\Delta x|}{\Delta x}$$

does not exist. Therefore, for the function $f(x) = |2x|$, which is continuous over its domain, the derivative does not exist *at the point* $x = 0$. Notice that a unique tangent line cannot be drawn at $x = 0$.

❑

Section 15.4 Follow-up Exercises

In Exercises 1–24, (*a*) determine the derivative of f using the limit approach, and (*b*) determine the slope at $x = 1$ and $x = -2$.

1 $f(x) = 4x + 6$ **2** $f(x) = -20$
3 $f(x) = 8x^2$ **4** $f(x) = 5x^2$
5 $f(x) = 3x^2 - 5x$ **6** $f(x) = 10x^2 - 8x$
7 $f(x) = x^2 - 3x + 5$ **8** $f(x) = 15x^2 + 2x + 8$
9 $f(x) = -6x^2 + 3x - 1$ **10** $f(x) = -3x^2 - 10x$

11 $f(x) = 20x^2 - 10$ **12** $f(x) = \dfrac{x^2}{2} + 6x$

13 $f(x) = ax + b$ **14** $f(x) = -ax^2 + bx$
*15 $f(x) = -2/x$ ***16** $f(x) = 4/x$
*17 $f(x) = a/x$ ***18** $f(x) = 5/x^2$
*19 $f(x) = -3/x$ ***20** $f(x) = 3x^3$
*21 $f(x) = x^3/2$ ***22** $f(x) = 4/x^2$
*23 $f(x) = x^3 + 3x^2$ ***24** $f(x) = x^4$

15.5 DIFFERENTIATION

The process of finding a derivative is called ***differentiation.*** Fortunately, for us, the process does not have to be as laborious as it may have seemed when we used the limit approach. A set of rules of differentiation exists for finding the derivatives of many common functions. Although there are many functions for which the derivative does not exist, *our concern will be with functions which are differentiable.*

Rules of Differentiation

The rules of differentiation presented in this section have been developed using the limit approach. The mathematics involved in proving these rules can be fairly complicated. For our purposes it will suffice to present the rules without proof. The appendix at the end of the chapter presents proofs of selected differentiation rules for anyone interested.

The rules of differentiation apply to functions which have specific structural characteristics. A rule will state that if a function has specific characteristics, then the derivative of the function will have a resulting form. As you study these rules, remember that each function can be graphed and that the derivative is a general expression for the slope of the function. An alternative to the dy/dx notation is to let $f'(x)$ (read "***f* prime of *x***") represent the derivative of the function f at x. That is, given $f(x)$,

$$\frac{dy}{dx} = f'(x)$$

This notation will be used in presenting the rules.

> **RULE 1: CONSTANT FUNCTION**
> If $f(x) = c$, where c is any constant
>
> $$f'(x) = 0$$

EXAMPLE 32 Consider the constant function $f(x) = 5$. Applying rule 1, $f'(x) = 0$. If you consider what the function looks like graphically, this result seems reasonable. The function $f(x) = 5$ graphs as a horizontal line intersecting the y axis at $(0, 5)$. The slope at all points along such a function equals 0.

❏

> **RULE 2: POWER RULE**
> If $f(x) = x^n$, where n is a real number,
>
> $$f'(x) = nx^{n-1}$$

EXAMPLE 33 Consider $f(x) = x$. This function is the same as $f(x) = x^1$. Applying rule 2 to find the derivative, $n = 1$ and

$$f'(x) = nx^{n-1}$$
$$= 1 \cdot x^{1-1}$$
$$= x^0$$
$$= 1$$

This implies that for the function $f(x) = x$, the slope equals 1 at all points. You should recognize that $f(x) = x$ is a linear function with slope of 1.

EXAMPLE 34   Consider $f(x) = x^5$. Applying rule 2 to find the derivative, $n = 5$ and

$$f'(x) = 5x^{5-1}$$
$$= 5x^4$$

EXAMPLE 35   Consider $f(x) = 1/x^3$.

> **ALGEBRA FLASHBACK**
>
> $$\frac{1}{x^n} = x^{-n}$$

Rewriting f as $f(x) = x^{-3}$, the derivative can be found using rule 2. Since $n = -3$,

$$f'(x) = -3x^{-3-1}$$
$$= -3x^{-4}$$
$$= \frac{-3}{x^4}$$

EXAMPLE 36   Consider $f(x) = \sqrt[3]{x^2}$

> **ALGEBRA FLASHBACK**
>
> $$\sqrt[n]{x^m} = x^{m/n}$$

Rewriting f as $f(x) = x^{2/3}$, we can again apply rule 2. Since $n = \frac{2}{3}$,

$$f'(x) = \frac{2}{3}x^{2/3-1}$$
$$= \frac{2x^{-1/3}}{3}$$
$$= \frac{2}{3x^{1/3}}$$
$$= \frac{2}{3\sqrt[3]{x}}$$

PRACTICE EXERCISE
Determine $f'(x)$ if (a) $f(x) = x^6$, (b) $f(x) = 1/x^4$. *Answer:* (a) $f'(x) = 6x^5$, (b) $f'(x) = -4/x^5$.

RULE 3: CONSTANT TIMES A FUNCTION
If $f(x) = c \cdot g(x)$, where c is a constant and g is a differentiable function,

$$f'(x) = c \cdot g'(x)$$

EXAMPLE 37   Consider $f(x) = 10x^2$. Applying rule 3 to find the derivative, $c = 10$, $g(x) = x^2$, and

$$\begin{aligned}
f'(x) &= c \cdot g'(x) \\
&= 10(2x) \\
&= 20x
\end{aligned}$$

EXAMPLE 38   Consider $f(x) = -3/x$. This function can be rewritten as

$$f(x) = -3\left(\frac{1}{x}\right) = -3x^{-1}$$

Applying rules 2 and 3,

$$\begin{aligned}
f'(x) &= (-3)(-1)(x^{-1-1}) \\
&= 3x^{-2} \\
&= \frac{3}{x^2}
\end{aligned}$$

❑

RULE 4: SUM OR DIFFERENCE OF FUNCTIONS
If $f(x) = u(x) \pm v(x)$, where u and v are differentiable,

$$f'(x) = u'(x) \pm v'(x)$$

Rule 4 implies that the derivative of a function formed by the sum (difference) of two or more component functions is the sum (difference) of the derivatives of the component functions.

EXAMPLE 39   Consider the function $f(x) = x^2 - 5x$. According to rule 4, f can be expressed as $f(x) = u(x) - v(x)$, where

$$u(x) = x^2$$

and

$$v(x) = 5x$$

Therefore,

$$f'(x) = u'(x) - v'(x)$$
$$= 2x - 5$$

EXAMPLE 40 Consider $f(x) = 5x^4 - 8x^3 + 3x^2 - x + 50$. The derivative is found by extending rule 4 and differentiating each term of $f(x)$.

$$f'(x) = 5(4x^3) - 8(3x^2) + 3(2x) - 1 + 0$$
$$= 20x^3 - 24x^2 + 6x - 1$$

❑

PRACTICE EXERCISE
Find $f'(x)$ if $f(x) = 8x^3 - 4x^2 + 3x - 10$. *Answer:* $f'(x) = 24x^2 - 8x + 3$.

RULE 5: PRODUCT RULE
If $f(x) = u(x) \cdot v(x)$, where u and v are differentiable, then

$$f'(x) = u'(x) \cdot v(x) + v'(x) \cdot u(x)$$

Verbally, the derivative of a product is the first function times the derivative of the second function *plus* the second function times the derivative of the first.

EXAMPLE 41 Consider $f(x) = (x^2 - 5)(x - x^3)$. Applying rule 5, $u(x) = x^2 - 5$ and $v(x) = x - x^3$. Therefore,

$$f'(x) = u'(x)v(x) + v'(x)u(x)$$
$$= (2x)(x - x^3) + (1 - 3x^2)(x^2 - 5)$$
$$= 2x^2 - 2x^4 + x^2 - 5 - 3x^4 + 15x^2$$
$$= -5x^4 + 18x^2 - 5$$

EXAMPLE 42 In the previous example the function could have been rewritten in the equivalent form

$$f(x) = (x^2 - 5)(x - x^3)$$
$$= x^3 - x^5 - 5x + 5x^3$$
$$= -x^5 + 6x^3 - 5x$$

Finding the derivative using this form of the function does not require the use of rule 5. The derivative is

$$f'(x) = -5x^4 + 18x^2 - 5$$

which is the same result as obtained before.

❑

> **NOTE**   As with the previous two examples, many functions can be algebraically manipulated into an equivalent form. This can be useful for two reasons. ***First***, rewriting a function in an equivalent form can allow the use of derivative rules which are more efficient or easier to remember. ***Second***, finding the derivative of both the original function and an equivalent form of the function provides a check on your answer.

PRACTICE EXERCISE

Find, $f'(x)$ if $f(x) = (x^3 - 2x^5)(x^4 - 3x^2 + 10)$. *Answer:* $f'(x) = -18x^8 + 49x^6 - 115x^4 + 30x^2$.

RULE 6: QUOTIENT RULE

If $f(x) = u(x)/v(x)$, where u and v are differentiable and $v(x) \neq 0$, then

$$f'(x) = \frac{v(x) \cdot u'(x) - u(x) \cdot v'(x)}{[v(x)]^2}$$

Verbally, the derivative of a quotient is the denominator times the derivative of the numerator *minus* the numerator times the derivative of the denominator, all divided by the square of the denominator.

EXAMPLE 43   In Example 35 we used the power rule to determine that the derivative of $f(x) = 1/x^3$ is $f'(x) = -3/x^4$. Since f has the form of a quotient, we can, as an alternative approach, apply rule 6. If $u(x) = 1$ and $v(x) = x^3$,

$$\begin{aligned}
f'(x) &= \frac{v(x)u'(x) - u(x)v'(x)}{[v(x)]^2} \\
&= \frac{x^3(0) - (1)(3x^2)}{(x^3)^2} \\
&= \frac{-3x^2}{x^6} \\
&= \frac{-3}{x^4}
\end{aligned}$$

EXAMPLE 44   Consider $f(x) = (3x^2 - 5)/(1 - x^3)$. Applying rule 6 with $u(x) = 3x^2 - 5$ and $v(x) = (1 - x^3)$,

$$\begin{aligned}
f'(x) &= \frac{(1 - x^3)(6x) - (3x^2 - 5)(-3x^2)}{(1 - x^3)^2} \\
&= \frac{6x - 6x^4 + 9x^4 - 15x^2}{(1 - x^3)^2}
\end{aligned}$$

$$= \frac{3x^4 - 15x^2 + 6x}{(1 - x^3)^2}$$

❑

PRACTICE EXERCISE

Find $f'(x)$ if $f(x) = \dfrac{10 - x}{x^2 + 1}$. *Answer:* $f'(x) = (x^2 - 20x - 1)/(x^2 + 1)^2$.

Section 15.5 Follow-up Exercises

For Exercises 1–38, find $f'(x)$.

1 $f(x) = 140$

2 $f(x) = -55$

3 $f(x) = 0.55$

4 $f(x) = 4x^0$

5 $f(x) = x^3 - 4x$

6 $f(x) = -3x/4 + 9$

7 $f(x) = 2x^5$

8 $f(x) = -x/3$

9 $f(x) = \sqrt[5]{x^3}$

10 $f(x) = \sqrt[4]{x^3}$

11 $f(x) = \sqrt{x^5}$

12 $f(x) = \sqrt[3]{x^7}$

13 $f(x) = x^{10}$

14 $f(x) = x^{5/3}$

15 $f(x) = \dfrac{x^6}{3} - 2x$

16 $f(x) = \dfrac{x^4}{2} - 3x^2 + 10$

17 $f(x) = \dfrac{x^3}{2} - 100$

18 $f(x) = x^2 - \sqrt{x}$

19 $f(x) = 5/x^2$

20 $f(x) = 3/5x^3$

21 $f(x) = -10/x^4$

22 $f(x) = \sqrt{2}/x^3$

23 $f(x) = x - 1/\sqrt{x}$

24 $f(x) = 1/\sqrt[6]{x^5}$

25 $f(x) = 2/\sqrt[3]{x}$

26 $f(x) = 1/6\sqrt[3]{x}$

27 $f(x) = (x^3 - 2x)(x^5 + 6x^2)$

28 $f(x) = \left(\dfrac{x^2}{2} - 10\right)(x^3 - 2x^2 + 1)$

29 $f(x) = (x^3 - x + 3)(x^6 - 10x^4)$

30 $f(x) = (2 - x - 3x^4)(10 + x - 4x^3)$

31 $f(x) = (6x^2 - 2x + 1)(x^3/4 + 5)$

32 $f(x) = [(x + 3)/2][x^2 - 4x + 9]$

33 $f(x) = x/(1 - x^2)$

34 $f(x) = 4x/(6x^2 - 5)$

35 $f(x) = (10 - x)/(x^2 + 2)$

36 $f(x) = 3x^5/(x^2 - 2x + 1)$

37 $f(x) = 1/(4x^5 - 3x^2 + 1)$

38 $f(x) = (-x^3 + 1)/(x^5 - 20)$

For Exercises 39–48, (a) find $f'(2)$, and (b) determine values of x for which $f'(x) = 0$.

39 $f(x) = 10x - 5$

40 $f(x) = 8x^2 - 1^r + 1$

41 $f(x) = x^3/3 - 6x + 8$

42 $f(x) = 16x^4/4 - x$

43 $f(x) = a_2 x^2 + a_1 x + a_0$

44 $f(x) = -1/x$

45 $f(x) = x^2/(1 - x^2)$

46 $f(x) = x^2 - 16x + 5$

47 $f(x) = -x^3/3 + 9x$

48 $f(x) = -x/(x - 5)$

15.6 ADDITIONAL RULES OF DIFFERENTIATION

In this section we present some additional rules of differentiation which build upon the rules of Sec. 15.5.

> **RULE 7: POWER OF A FUNCTION**
> If $f(x) = [u(x)]^n$, where u is a differentiable function and n is a real number, then
>
> $$f'(x) = n \cdot [u(x)]^{n-1} \cdot u'(x)$$

This rule looks very similar to the power rule (rule 2). In fact, the power rule is the special case of this rule where $u(x)$ equals x. When $u(x) = x$, $u'(x) = 1$ and applying rule 7 results in

$$f'(x) = n(x)^{n-1}(1)$$
$$= nx^{n-1}$$

EXAMPLE 45 Consider $f(x) = \sqrt{7x^4 - 5x - 9}$. The function f can be rewritten as $f(x) = (7x^4 - 5x - 9)^{1/2}$. In this form, rule 7 applies, where $u(x) = 7x^4 - 5x - 9$. Applying rule 7,

$$f'(x) = n[u(x)]^{n-1} \cdot u'(x)$$
$$= \tfrac{1}{2}(7x^4 - 5x - 9)^{(1/2)-1}(28x^3 - 5)$$
$$= (14x^3 - \tfrac{5}{2})(7x^4 - 5x - 9)^{-1/2}$$

which can be rewritten as

$$f'(x) = \frac{14x^3 - \tfrac{5}{2}}{(7x^4 - 5x - 9)^{1/2}}$$

or

$$f'(x) = \frac{14x^3 - \tfrac{5}{2}}{\sqrt{7x^4 - 5x - 9}}$$

EXAMPLE 46 Consider the function

$$f(x) = \left(\frac{3x}{1 - x^2}\right)^5$$

This function has the form stated in rule 7, where u is the rational function $3x/(1 - x^2)$. Applying rule 7 [note that rule 6 must be used to find $u'(x)$],

$$f'(x) = 5\left(\frac{3x}{1 - x^2}\right)^4 \frac{(1 - x^2)(3) - (3x)(-2x)}{(1 - x^2)^2}$$
$$= 5\left(\frac{3x}{1 - x^2}\right)^4 \frac{3 - 3x^2 + 6x^2}{(1 - x^2)^2}$$
$$= 5\left(\frac{3x}{1 - x^2}\right)^4 \frac{3 + 3x^2}{(1 - x^2)^2}$$

❑

PRACTICE EXERCISE
Find $f'(x)$ if $f(x) = (3x^2 - 5x + 8)^{10}$. *Answer:* $f'(x) = (60x - 50)(3x^2 - 5x + 8)^9$.

RULE 8: BASE-e EXPONENTIAL FUNCTIONS
If $f(x) = e^{u(x)}$, where u is differentiable, then

$$f'(x) = u'(x)e^{u(x)}$$

EXAMPLE 47 Consider $f(x) = e^x$. Applying rule 8, the exponent $u(x) = x$ and

$$f'(x) = u'(x)e^{u(x)}$$
$$= 1e^x$$
$$= e^x$$

This result is unique in that *the function e^x and its derivative are identical.* That is, $f(x) = f'(x) = e^x$. Graphically, the interpretation is that for any value of x, the slope of the graph of $f(x) = e^x$ is exactly equal to the value of the function.

EXAMPLE 48 Consider $f(x) = e^{-x^2+2x}$. Applying rule 8,

$$f'(x) = (-2x + 2)e^{-x^2+2x}$$

❑

PRACTICE EXERCISE
Find $f'(x)$ if $f(x) = 10e^{5x^2-4x}$. *Answer:* $f'(x) = (100x - 40)e^{5x^2-4x}$.

RULE 9: NATURAL LOGARITHM FUNCTIONS
If $f(x) = \ln u(x)$, where u is differentiable, then

$$f'(x) = \frac{u'(x)}{u(x)}$$

EXAMPLE 49 Consider the function

$$f(x) = \ln x$$

If we let $u(x) = x$ and apply rule 9,

$$f'(x) = \frac{u'(x)}{u(x)}$$

$$= \frac{1}{x}$$

EXAMPLE 50	Consider $f(x) = \ln(5x^2 - 2x + 1)$. Applying rule 9,

$$u(x) = 5x^2 - 2x + 1$$

and
$$f'(x) = \frac{10x - 2}{5x^2 - 2x + 1}$$

❑

PRACTICE EXERCISE

Find $f'(x)$ if $f(x) = \ln(4x^2 - 16x)$. *Answer:* $f'(x) = (8x - 16)/(4x^2 - 16x)$.

Chain Rule

Rule 7 (power of a function) was presented expeditiously and without the attention it should have been given. Rule 7 is a special case of the more general *chain rule*.

RULE 10: CHAIN RULE
If $y = f(u)$ is a differentiable function and $u = g(x)$ is a differentiable function, then

$$\frac{dy}{dx} = \frac{dy}{du} \cdot \frac{du}{dx}$$

Recall that in Chap. 4 we examined composite functions—functions whose values depend upon other functions. The chain rule specifically applies to composite functions. Consider the two functions

$$y = f(u) = 20 - 3u$$

and
$$u = g(x) = 5x - 4$$

Note that the value of y ultimately depends upon x. We can see this by observing that if x increases by 1 unit, u increases by 5 units. And an increase in u of 5 units results in a *decrease* in y of $(3)(5) = 15$ units. To determine how the value of y responds to changes in x, we can apply the chain rule. Since

$$\frac{dy}{du} = -3$$

and
$$\frac{du}{dx} = 5$$

then
$$\frac{dy}{dx} = \frac{dy}{du} \cdot \frac{du}{dx} = (-3)(5)$$
$$= -15$$

EXAMPLE 51　Given $y = f(u) = u^2 - 2u + 1$ and $u = g(x) = x^2 - 1$,

$$\frac{dy}{dx} = \frac{dy}{du} \cdot \frac{du}{dx}$$
$$= (2u - 2)(2x)$$

We can rewrite dy/dx strictly in terms of x by substituting $u = x^2 - 1$. This results in

$$\frac{dy}{dx} = [2(x^2 - 1) - 2](2x)$$
$$= (2x^2 - 4)(2x)$$
$$= 4x^3 - 8x$$

Rule 7 would approach this problem by first restating $f(u)$ in terms of x. That is, because $u = x^2 - 1$

$$y = u^2 - 2u + 1$$
$$= (x^2 - 1)^2 - 2(x^2 - 1) + 1$$

The expression for dy/dx is found directly as

$$\frac{dy}{dx} = 2(x^2 - 1)(2x) - 4x$$
$$= 4x^3 - 4x - 4x$$
$$= 4x^3 - 8x$$

which is the same result.

EXAMPLE 52　Given $y = f(u) = u^3 - 5u$ where $u = g(x) = x^4 + 3x$,

$$\frac{dy}{dx} = \frac{dy}{du} \cdot \frac{du}{dx}$$
$$= (3u^2 - 5)(4x^3 + 3)$$

Rewriting this result as a function of x gives

$$\frac{dy}{dx} = [3(x^4 + 3x)^2 - 5](4x^3 + 3)$$

$$= [3(x^8 + 6x^5 + 9x^2) - 5](4x^3 + 3)$$
$$= (3x^8 + 18x^5 + 27x^2 - 5)(4x^3 + 3)$$
$$= 12x^{11} + 72x^8 + 108x^5 - 20x^3 + 9x^8 + 54x^5$$
$$\quad + 81x^2 - 15$$
$$= 12x^{11} + 81x^8 + 162x^5 - 20x^3 + 81x^2 - 15$$

□

PRACTICE EXERCISE
Given $y = f(u) = u^3 + 3u$ and $u = g(x) = x + 3$, find dy/dx. *Answer: $dy/dx = 3(x+3)^2 + 3 = 3x^2 + 18x + 30$.*

Section 15.6 Follow-up Exercises

In Exercises 1–38, determine $f'(x)$.

1 $f(x) = (1 - 4x^3)^5$

2 $f(x) = (7x^2 - 3x + 1)^3$

3 $f(x) = (x^3 - 2x + 5)^4$

4 $f(x) = (5x^3 + 1)^4$

5 $f(x) = \sqrt{1 - 5x^3}$

6 $f(x) = \sqrt[3]{x^2 - 2x + 5}$

7 $f(x) = 1/\sqrt{x^2 - 1}$

8 $f(x) = \sqrt{1/(x^2 + 9)}$

9 $f(x) = e^x$

10 $f(x) = e^{x^3}$

11 $f(x) = 10e^{x^2}$

12 $f(x) = -5e^{x/2}$

13 $f(x) = (5e^x)^3$

14 $f(x) = 3x^2 e^x$

15 $f(x) = 4xe^{x^2}$

16 $f(x) = (e^x - 5)^4$

17 $f(x) = e^{x^2 - 2x + 5}$

18 $f(x) = 10e^{x^3 - 2x}$

19 $f(x) = e^x/x$

20 $f(x) = 2x^2/e^x$

21 $f(x) = (e^x)^3$

22 $f(x) = \sqrt{e^{2x}}$

23 $f(x) = \ln(5x)$

24 $f(x) = \ln(x/2)$

25 $f(x) = \ln(x^2 - 3)$

26 $f(x) = \ln(x^3 - 2x^2 + 5)$

27 $f(x) = x^2 \ln x$

28 $f(x) = (x + 3) \ln x^2$

29 $f(x) = \dfrac{10x}{\ln x}$

30 $f(x) = (\ln x)^3$

31 $f(x) = \dfrac{x - 1}{\ln 3x}$

32 $f(x) = (5x - \ln x)^5$

33 $f(x) = \dfrac{\ln x}{e^{x^2}}$

34 $f(x) = \dfrac{e^{x^2 + 1}}{\ln(x + 1)}$

35 $f(x) = \sqrt{(x - 1)^5(6x - 5)}$

36 $f(x) = \sqrt[3]{\left(\dfrac{x^2}{5x - 1}\right)^2}$

37 $f(x) = (6x - 2)\sqrt{x^2 - 5x + 3}$

38 $f(x) = \sqrt[3]{\dfrac{(1 - x^3)^5 x^3}{(9 - 2x^2)}}$

In the following exercises, find dy/dx.

39 $y = f(u) = 5u + 3$ and $u = g(x) = -3x + 10$

40 $y = f(u) = u^2 - 5$ and $u = g(x) = 10x - 3$

41 $y = f(u) = u^2 - 2u + 1$ and $u = g(x) = x^2$

42 $y = f(u) = u^3$ and $u = g(x) = x^2 + 3x + 1$

43 $y = f(u) = 10 - 5u^3$ and $u = g(x) = -x + x^3$

44 $y = f(u) = u^4 - u^2 + 1$ and $u = g(x) = x^2 - 4$

45 $y = f(u) = \sqrt{u}$ and $u = g(x) = x^2/2$

46 $y = f(u) = \sqrt{u^2 - 1}$ and $u = g(x) = x^4$

47 $y = f(u) = (u - 3)^5$ and $u = g(x) = x^2 - 2x$

48 $y = f(u) = \sqrt{u^3}$ and $u = g(x) = \sqrt{x}$

49 $y = f(u) = e^u$ and $u = g(x) = 2x^2 - 5x$

50 $y = f(u) = e^{u^2}$ and $u = g(x) = x^2 - 5$

51 $y = f(u) = \ln(5u - 3)$ and $u = g(x) = 4x^3 - 3x^2$

52 $y = f(u) = 10 \ln(15 - u^3)$ and $u = g(x) = x^2 - 2x + 5$

For Exercises 53–64, (a) find $f'(2)$, and (b) determine values of x for which $f'(x) = 0$.

53 $f(x) = (5x^2 - 10)^6$

54 $f(x) = (2x - 8)^5$

55 $f(x) = \sqrt{x^2 + 21}$

56 $f(x) = \sqrt{x^2 + 5}$

57 $f(x) = e^{-x}$

58 $f(x) = e^x$

59 $f(x) = xe^{-x}$

60 $f(x) = xe^x$

61 $f(x) = \ln x - x$

62 $f(x) = \ln 30x - 3x$

63 $f(x) = \ln 2x - 2x$

64 $f(x) = \ln x - x^2/2$

15.7 INSTANTANEOUS-RATE-OF-CHANGE INTERPRETATION

We stated earlier that the derivative can be used to determine instantaneous rate of change.

EXAMPLE 53 In Example 25 the function

$$d = f(t) = 8t^2 + 8t$$

described the distance (in miles) an automobile traveled as a function of time (in hours). The instantaneous velocity of the car at any point in time is found by evaluating the derivative at that value of t. To determine the instantaneous velocity at $t = 3$, the derivative

$$f'(t) = 16t + 8$$

must be evaluated at $t = 3$, or

$$\begin{aligned} f'(3) &= 16(3) + 8 \\ &= 56 \text{ mph} \end{aligned}$$

EXAMPLE 54 An object is dropped from a cliff which is 1,296 feet above the ground. The height of the object is described as a function of time. The function is

$$h = f(t) = -16t^2 + 1,296$$

where h equals the height in feet and t equals time measured in seconds from the time the object is dropped.

(a) How far will the object drop in 2 seconds?

(b) What is the instantaneous velocity of the object at $t = 2$?

(c) What is the velocity of the object at the instant it hits the ground?

SOLUTION

(a) The change in the height is

$$\begin{aligned} \Delta h &= f(2) - f(0) \\ &= [-16(2)^2 + 1,296] - [-16(0)^2 + 1,296] \\ &= (-64 + 1,296) - 1,296 = -64 \end{aligned}$$

Thus, the object *drops* 64 feet during the first 2 seconds.

(b) Since $f'(t) = -32t$, the object will have a velocity equal to

$$f'(2) = -32(2)$$
$$= -64 \text{ feet/second}$$

at $t = 2$. The minus sign indicates the direction of velocity (down).

(c) In order to determine the velocity of the object when it hits the ground, we must know *when* it will hit the ground. The object will hit the ground when $h = 0$, or when

$$-16t^2 + 1{,}296 = 0$$

If we solve for t,

$$t^2 = \frac{1{,}296}{16} = 81$$

and $t = \pm 9$. Since a negative root is meaningless, we can conclude that the object will hit the ground after 9 seconds. The velocity at this time will be

$$f'(9) = 32(9)$$
$$= 288 \text{ feet/second}$$

EXAMPLE 55

(Tracking an Epidemic; Motivating Scenario) A flu epidemic is spreading through a large midwestern state. Based upon similar epidemics which have occurred in the past, epidemiologists have formulated a mathematical function which estimates the number of persons who will be afflicted by the flu. Specifically, the function is

$$n = f(t) = -0.3t^3 + 10t^2 + 300t + 250$$

where *n equals the number of persons afflicted, t equals time, measured in days, since initial detection by health department officials*, and the relevant (restricted) domain is $0 \le t \le 30$. Using this estimating function:

(a) What interpretation can be given to $f(0)$?

(b) How many persons are expected to contract the flu after 10 days? After 20 days?

(c) What is the average rate at which the flu is expected to spread between $t = 10$ and $t = 20$?

(d) What is the instantaneous rate at which the flu is expected to be spreading at $t = 11$? At $t = 12$?

(e) What interpretation is suggested by the results in part *d*?

SOLUTION

(a) $f(0)$ would be interpreted as the estimated number of persons having come down with the disease at the time of initial detection of the disease by health department officials. According to this function, approximately 250 persons would have been afflicted.

(b)
$$f(10) = -0.3(10)^3 + 10(10)^2 + 300(10) + 250$$
$$= -300 + 1{,}000 + 3{,}000 + 250$$
$$= 3{,}950 \text{ persons}$$

$$f(20) = -0.3(20)^3 + 10(20)^2 + 300(20) + 250$$
$$= -2{,}400 + 4{,}000 + 6{,}000 + 250$$
$$= 7{,}850 \text{ persons}$$

(c) $$\frac{\Delta n}{\Delta t} = \frac{f(20) - f(10)}{20 - 10} = \frac{7{,}850 - 3{,}950}{10} = \frac{3{,}900}{10} = 390 \text{ persons/day}$$

(d) The instantaneous rate at which the flu is spreading is estimated by $f'(t)$.

$$f'(t) = -0.9t^2 + 20t + 300$$

At $t = 11$,

$$\begin{aligned} f'(11) &= -0.9(11)^2 + 20(11) + 300 \\ &= -108.9 + 220 + 300 \\ &= 411.1 \text{ persons/day} \end{aligned}$$

At $t = 12$,

$$\begin{aligned} f'(12) &= -0.9(12)^2 + 20(12) + 300 \\ &= -129.6 + 240 + 300 \\ &= 410.4 \text{ persons/day} \end{aligned}$$

(e) The results in part d suggest that the rate at which persons are being afflicted by the disease has declined between the 11th and 12th days.

EXAMPLE 56   (**Exponential Growth Processes**) Exponential growth processes were introduced in Chap. 7. The generalized exponential growth function was presented in Eq. (7.9) as

$$V = f(t) = V_0 e^{kt}$$

We indicated that these processes are characterized by a constant percentage rate of growth. To verify this, let's find the derivative

$$f'(t) = V_0(k)e^{kt}$$

which can be written as

$$f'(t) = kV_0 e^{kt}$$

The derivative represents the instantaneous rate of change in the value V with respect to a change in t. The *percentage rate of change* would be found by the ratio

$$\frac{\text{Instantaneous rate of change}}{\text{Value of the function}} = \frac{f'(t)}{f(t)}$$

For this function,

$$\begin{aligned} \frac{f'(t)}{f(t)} &= \frac{kV_0 e^{kt}}{V_0 e^{kt}} \\ &= k \end{aligned}$$

This confirms that for an exponential growth function of the form of Eq. (7.9), k represents the percentage rate of growth. Given that k is a constant, the percentage rate of growth is the same for all values of t.

❑

Section 15.7 Follow-up Exercises

1 The function $h = f(t) = 2.5t^3$, where $0 \le t \le 30$, describes the height h (in hundreds of feet) of a rocket t seconds after it has been launched.
   (a) What is the average velocity during the time interval $0 \le t \le 10$?
   (b) What is the instantaneous velocity at $t = 10$? At $t = 20$?

2 An object is launched from ground level with an initial velocity of 256 feet per second. The function which describes the height h of the ball is

$$h = f(t) = 256t - 16t^2$$

where h is measured in feet and t is time measured in seconds since the ball was thrown.
   (a) What is the velocity of the ball at $t = 1$ second?
   (b) When will the ball return to the ground?
   (c) What is the velocity of the ball when it hits the ground?

3 A ball is dropped from the roof of a building which is 256 feet high. The height of the ball is described by the function

$$h = f(t) = -16t^2 + 256$$

where h equals the height in feet and t equals time measured in seconds from when the ball was dropped.
   (a) What is the average velocity during the time interval $1 \le t \le 2$?
   (b) What is the instantaneous velocity at $t = 3$?
   (c) What is the velocity of the ball at the instant it hits the ground?

4 **Epidemic Control** An epidemic is spreading through a large western state. Health officials estimate that the number of persons who will be afflicted by the disease is a function of time since the disease was first detected. Specifically, the function is

$$n = f(t) = 300t^3 - 20t^2$$

where n equals the number of persons and $0 \le t \le 60$, measured in days.
   (a) How many persons are expected to have caught the disease after 10 days? After 30 days?
   (b) What is the average rate at which the disease is expected to spread between $t = 10$ and $t = 30$?
   (c) What is the instantaneous rate at which the disease is expected to be spreading at $t = 20$?

5 **Population Growth** The population of a country is estimated by the function

$$P = 125e^{0.035t}$$

where P equals the population (in millions) and t equals time measured in years since 1990.

(a) What is the population expected to equal in the year 2000?

(b) Determine the expression for the instantaneous rate of change in the population.

(c) What is the instantaneous rate of change in the population expected to equal in the year 2000?

6 Investment Appreciation A rare piece of artwork has been appreciating in value over recent years. The function

$$V = 1.5e^{0.06t}$$

estimates the value V of the artwork (measured in millions of dollars) as a function of time t, measured in years since 1986.

(a) What is the value estimated to equal in the year 1996? In the year 2000?

(b) Determine the general expression for the instantaneous rate of change in the value of the artwork.

(c) At what rate is the value of the artwork expected to be increasing in the year 2000?

7 Endangered Species The population of a rare species of wildlife is declining. The function

$$P = 75,000e^{-0.025t}$$

estimates the population P of the species as a function of time, measured in years since 1980.

(a) What is the population expected to equal in the year 1996?

(b) Determine the general expression for the instantaneous rate of change in the population.

(c) At what rate is the population estimated to be declining in the year 1996?

8 Asset Depreciation The value of a particular asset is estimated by the function

$$V = 240,000e^{-0.04t}$$

where V is the value of the asset and t is the age of the asset, measured in years.

(a) What is the value of the asset expected to equal when 4 years old?

(b) Determine the general expression for the instantaneous rate of change in the value of the asset.

(c) What is the rate of change expected to equal when the asset is 10 years old?

15.8 HIGHER-ORDER DERIVATIVES

Given a function f, there are other derivatives which can be defined. This section discusses these **higher-order derivatives** and their interpretation.

The Second Derivative

The derivative f' of the function f is often referred to as the **first derivative** of the function. The adjective *first* is used to distinguish this derivative from other derivatives associated with a function. The **order** of the first derivative is 1.

The **second derivative** f'' of a function is the derivative of the first derivative. At x, it is denoted by either d^2y/dx^2 or $f''(x)$. The second derivative is found by applying the same rules of differentiation as were used in finding the first

TABLE 15.11	$f(x)$	$f'(x)$	$f''(x)$
	x^5	$5x^4$	$20x^3$
	$x^2 - 3x + 10$	$2x - 3$	2
	$mx + b$	m	0
	$x^3 - 2x^2 + 5x$	$3x^2 - 4x + 5$	$6x - 4$
	$x^{3/2}$	$\frac{3}{2}x^{1/2}$	$\frac{3}{4}x^{-1/2}$
	e^x	e^x	e^x
	$\ln x$	$1/x$	$-1/x^2$

derivative. Table 15.11 illustrates the computation of first and second derivatives for several functions.

Just as the first derivative is a measure of the instantaneous rate of change in the value of y with respect to a change in x, the *second derivative* is a measure of the instantaneous rate of change in the value of the first derivative with respect to a change in x. Described differently, the second derivative is a *measure of the instantaneous rate of change in the slope with respect to a change in x.*

Consider the function $f(x) = -x^2$. The first and second derivatives of this function are

$$f'(x) = -2x \qquad f''(x) = -2$$

Figure 15.24 illustrates the graphs of f, f', and f''.

The function f is a parabola which is concave down with the vertex at $(0, 0)$. The tangent slope is positive to the left of the vertex but becomes *less positive* as x approaches 0. To the right of the vertex the tangent slope is negative, becoming *more negative* (decreasing) as x increases. The graph of f' indicates the value of the slope at any point on f. Note that values of $f'(x)$ are positive, but becoming less positive, as x approaches 0 from the left. And $f'(x)$ becomes more and more negative as the value of x becomes more positive. Thus, the graph of f' is consistent with our observations of the sketch of f.

The second derivative is a measure of the instantaneous rate of change in the first derivative or in the slope of the graph of a function. Since $f''(x) = -2$, the rate of change in the first derivative is constant over the entire function. Specifically, $f''(x) = -2$ suggests that everywhere on the function the slope is *decreasing* at an

Figure 15.24

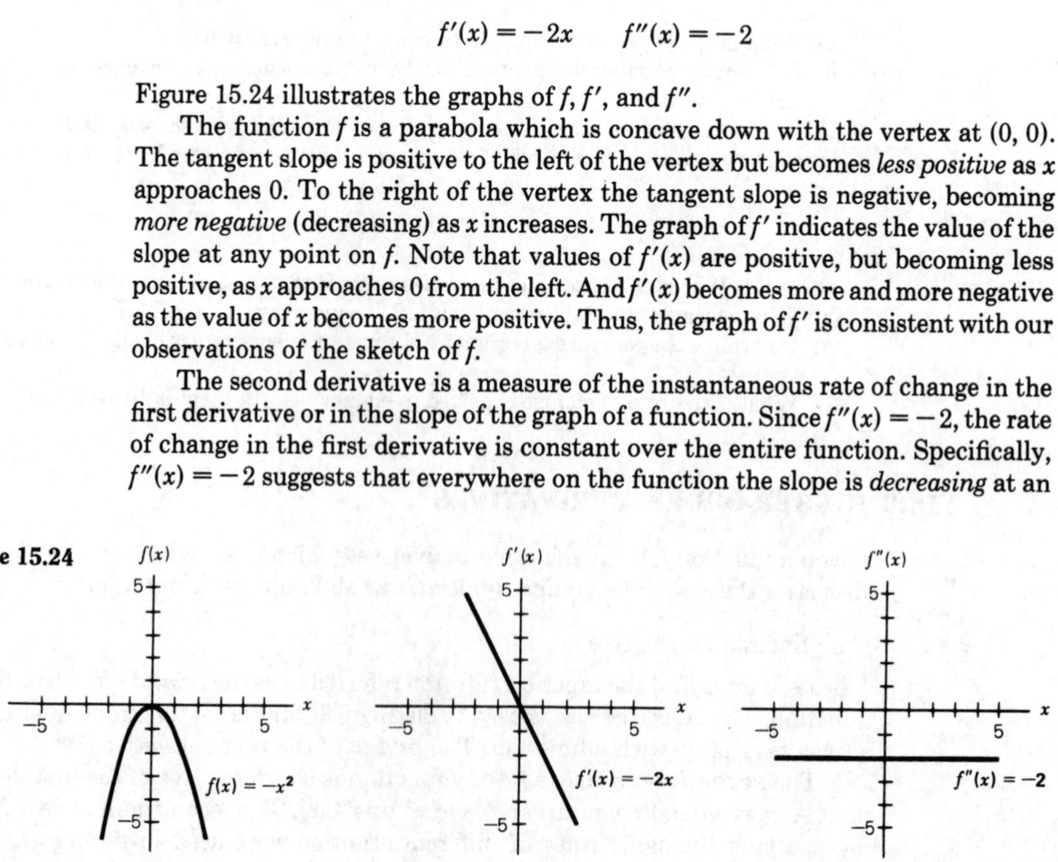

$f(x) = -x^2$

$f'(x) = -2x$

$f''(x) = -2$

instantaneous rate of 2 units for each unit that x increases. Note that the graph of f' is a linear function with slope of -2.

Some earlier examples in this chapter discussed functions having the form $d = f(t)$, where d represents the distance traveled after time t. We concluded that the instantaneous velocity at any time t was represented by the first derivative $f'(t)$. The second derivative of this type of function $f''(t)$ provides a measure of the instantaneous rate of change *in the velocity* with respect to a change in time. Measured in units of distance per unit of time squared (for example, ft/sec² and km/h²), this second derivative represents the instantaneous **acceleration** of an object. If

$$d = f(t) = t^3 - 2t^2 + 3t$$

the expression representing the **instantaneous velocity** is

$$f'(t) = 3t^2 - 4t + 3$$

and the expression representing the **instantaneous acceleration** is

$$f''(t) = 6t - 4$$

Third and Higher-Order Derivatives

Further derivatives can be determined for functions. These derivatives become less easy to understand from an intuitive standpoint. However, they will be useful to us later, and they have particular value at higher levels of mathematical analysis.

> **DEFINITION: nth-ORDER DERIVATIVE**
> The *nth-order derivative* of f, denoted by $f^{(n)}$, is found by differentiating the derivative of order $n - 1$. That is, at x,
>
> $$f^{(n)}(x) = \frac{d}{dx}[f^{(n-1)}(x)]$$

EXAMPLE 57   Given

$$f(x) = x^6 - 2x^5 + x^4 - 3x^3 + x^2 - x + 1$$

the derivatives of f are

$$f'(x) = 6x^5 - 10x^4 + 4x^3 - 9x^2 + 2x - 1$$
$$f''(x) = 30x^4 - 40x^3 + 12x^2 - 18x + 2$$
$$f'''(x) = 120x^3 - 120x^2 + 24x - 18$$
$$f^{(4)}(x) = 360x^2 - 240x + 24$$
$$f^{(5)}(x) = 720x - 240$$
$$f^{(6)}(x) = 720$$
$$f^{(7)}(x) = 0$$

All additional higher-order derivatives will also equal 0.

☐

> **NOTE**   For polynomial functions of *degree n,* the first derivative is a polynomial function of degree $n-1$. Each successive derivative is a polynomial function of degree one less than the previous derivative (until the degree of a derivative equals 0).

Section 15.8 Follow-up Exercises

For Exercises 1–30, (a) find $f'(x)$ and $f''(x)$, (b) evaluate $f'(1)$ and $f''(1)$, and (c) verbalize the meaning of $f'(1)$ and $f''(1)$.

1 $f(x) = 15$

2 $f(x) = 24 - 10x$

3 $f(x) = 4x^2 - x + 5$

4 $f(x) = x^2 - 15x + 10$

5 $f(x) = 5x^3$

6 $f(x) = 7x^3 - 2x^2 + 5x + 1$

7 $f(x) = x^4/4 - x^3/3 + 10x$

8 $f(x) = x^5/5 - x^3/3 + 100$

9 $f(x) = 1/x$

10 $f(x) = (x^2 - 2)^5$

11 $f(x) = 3x^5 - 2x^3$

12 $f(x) = 5x^4 - 10x^2$

13 $f(x) = 2/x^2$

14 $f(x) = -4/x^3$

15 $f(x) = (x - 5)^4$

16 $f(x) = (5 - 2x)^3$

17 $f(x) = \sqrt{x+1}$

18 $f(x) = \sqrt{10 - 2x}$

19 $f(x) = e^{2x}$

20 $f(x) = e^{10-2x}$

21 $f(x) = e^{x^2}$

22 $f(x) = e^{x^2/2}$

23 $f(x) = \ln 2x$

24 $f(x) = \ln 4x$

25 $f(x) = \ln(x^2 - 5)$

26 $f(x) = \ln(x^3 + 4)$

27 $f(x) = \dfrac{x}{1 - x^2}$

28 $f(x) = \dfrac{2x}{x^2 + 1}$

29 $f(x) = e^x \ln x$

30 $f(x) = e^{2x} \ln x$

31 The height of a falling object dropped from a height of 1,000 feet is described by the function

$$h = 1{,}000 - 16t^2$$

where h is measured in feet and t is measured in seconds.
(a) What is the velocity at $t = 4$?
(b) What is the acceleration at $t = 4$?

32 The height of a falling object dropped from a height of 1,200 feet is described by the function

$$h = 1{,}200 - 16t^2$$

where h is measured in feet and t is measured in seconds.
(a) What is the velocity at $t = 3$? At $t = 5$?
(b) What is the acceleration at $t = 3$? At $t = 5$?

33 A ball thrown upward from the roof of a building which is 600 feet high will be at a height of h feet after t seconds, as described by the function

$$h = f(t) = -16t^2 + 50t + 600$$

(a) What is the height of the ball after 3 seconds?

(b) What is the velocity of the ball after 3 seconds? (A negative sign implies a downward direction.)

(c) What is the acceleration of the ball at $t = 0$? At $t = 5$?

34 A ball thrown upward from the roof of a building which is 750 feet high will be at a height of h feet after t seconds, as described by the function

$$h = f(t) = -16t^2 + 50t + 750$$

(a) What is the height of the ball after 5 seconds?

(b) What is the velocity of the ball after 5 seconds? (A negative sign implies a downward direction.)

(c) What is the acceleration of the ball at $t = 0$? At $t = 5$?

In Exercises 35–50, find all higher-order derivatives.

35 $f(x) = 16x^3 - 4x^2$

36 $f(x) = 2,500$

37 $f(x) = mx + b$

38 $f(x) = -x/4 + 10$

39 $f(x) = x^5 - 5x^4 - 30x^2$

40 $f(x) = (x - 10)^3$

41 $f(x) = a_3 x^3 + a_2 x^2 + a_1 x + a_0$

42 $f(x) = (a_1 x + b_1)(a_2 x + b_2)$

43 $f(x) = 10x^4 - 2x^3 + 3$

44 $f(x) = -x^5 + 3x^2$

45 $f(x) = (ax + b)^3$

46 $f(x) = (cx - d)^4$

47 $f(x) = a_4 x^4 + a_3 x^3$

48 $f(x) = \dfrac{x^4}{4} - \dfrac{x^3}{3}$

49 $f(x) = e^x$

50 $f(x) = e^{-x}$

❏ KEY TERMS AND CONCEPTS

❏ IMPORTANT FORMULAS

$$\frac{\Delta y}{\Delta x} = \frac{f(x + \Delta x) - f(x)}{\Delta x} \qquad \text{Difference quotient} \qquad (15.5)$$

$$\frac{dy}{dx} = \lim_{\Delta x \to 0} \frac{f(x + \Delta x) - f(x)}{\Delta x} \qquad \text{Derivative} \qquad (15.7)$$

Rules of Differentiation

1 If $f(x) = c$, $f'(x) = 0$ (Constant function)

2 If $f(x) = x^n$, $f'(x) = nx^{n-1}$ (**Power rule**)

3 If $f(x) = c \cdot g(x)$, $f'(x) = c \cdot g'(x)$ (**Constant times a function**)

4 If $f(x) = u(x) \pm v(x)$, $f'(x) = u'(x) \pm v'(x)$ (**Sum or difference**)

5 If $f(x) = u(x)v(x)$, $f'(x) = u'(x)v(x) + v'(x)u(x)$ (**Product rule**)

6 If $f(x) = u(x)/v(x)$, $f'(x) = \dfrac{v(x)u'(x) - u(x)v'(x)}{[v(x)]^2}$ (**Quotient rule**)

7 If $f(x) = [u(x)]^n$, $f'(x) = n[u(x)]^{n-1}u'(x)$ (**Power of a function**)

8 If $f(x) = e^{u(x)}$, $f'(x) = u'(x)e^{u(x)}$ (**Base-e exponential functions**)

9 If $f(x) = \ln u(x)$, $f'(x) = u'(x)/u(x)$

(**Natural logarithm functions**)

10 $\dfrac{dy}{dx} = \dfrac{dy}{du} \cdot \dfrac{du}{dx}$ (**Chain rule**)

❏ ADDITIONAL EXERCISES

Section 15.1

1 Use the graph of the function in Fig. 15.25 to determine the indicated limits.

(a) $\lim\limits_{x \to -4^-} f(x)$ (b) $\lim\limits_{x \to -4^+} f(x)$ (c) $\lim\limits_{x \to -4} f(x)$

(d) $\lim\limits_{x \to -2^-} f(x)$ (e) $\lim\limits_{x \to -2^+} f(x)$ (f) $\lim\limits_{x \to -2} f(x)$

(g) $\lim\limits_{x \to 2} f(x)$ (h) $\lim\limits_{x \to 1} f(x)$

Figure 15.25

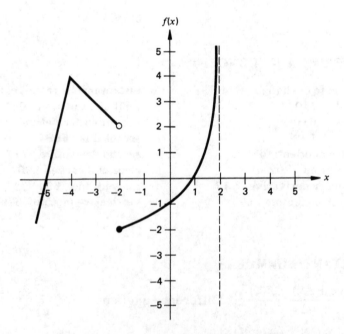

2 Use the graph of the function in Fig. 15.26 to determine the indicated limits.

(a) $\lim\limits_{x \to \infty} f(x)$ (b) $\lim\limits_{x \to -\infty} f(x)$ (c) $\lim\limits_{x \to -1^-} f(x)$

(d) $\lim\limits_{x \to -1^+} f(x)$ (e) $\lim\limits_{x \to -1} f(x)$

Figure 15.26

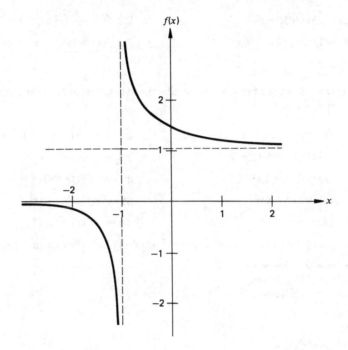

Section 15.2

For the following exercises, find the indicated limit, if it exists.

3 $\lim\limits_{x\to 3}\left[x^2\left(4x-2+\dfrac{1}{x}\right)\right]$

4 $\lim\limits_{x\to 1}\dfrac{x^5-2x+6}{x^4-6x}$

5 $\lim\limits_{x\to 5}\dfrac{2x^2+7x-15}{x+5}$

6 $\lim\limits_{x\to -1}(x^5+x^4+x^3+x^2+x+1)$

7 $\lim\limits_{x\to -2}\dfrac{x^2-3}{x+3}$

8 $\lim\limits_{x\to -3}\dfrac{2x^2+x-15}{x+3}$

9 $\lim\limits_{x\to \infty}\dfrac{1}{x-5}$

10 $\lim\limits_{x\to -\infty}\dfrac{-x^2+1}{x+1}$

11 $\lim\limits_{x\to \infty}\dfrac{5x^3+2x^2+1}{x^2-x+7}$

12 $\lim\limits_{x\to -\infty}\dfrac{2x+1}{x}$

13 $\lim\limits_{x\to -\infty}\dfrac{4x-5}{-x+3}$

14 $\lim\limits_{x\to \infty}\dfrac{1/x}{1/2x}$

In the following exercises, determine whether there are any discontinuities and, if so, where they occur.

15 $f(x)=x^7/(x^2-1)$

16 $f(x)=(x^2-9)/(16-x^4)$

17 $f(x)=x^2-3x+2$

18 $f(x)=(x^2-9)/(64-x^3)$

19 $f(x)=\begin{cases}2/x^2 & \text{if } x\neq 6\\ 10 & \text{if } x=6\end{cases}$

20 $f(x)=\begin{cases}x^2-5 & \text{if } x>5\\ -x^2 & \text{if } x\leq 5\end{cases}$

21 $f(x) = 25x^2/(24 - 8x)$ **22** $f(x) = (3 - x)/(x^2 - 3x + 2)$

23 $f(x) = 1/(x^2 + 4)$ **24** $f(x) = (x^3 - 3x^2 + 2x + 1)/(x^2 + 6)$

Section 15.3

For each of the following functions, determine the average rate of change in the value of y in moving from $x = 1$ to $x = 4$.

25 $y = f(x) = x^2 + 2x - 3$ **26** $y = f(x) = x^3 - 3x^2 + 1$

27 $y = f(x) = (x - 1)/(1 - x^2)$ **28** $y = f(x) = -x^3$

29 $y = f(x) = 4x^2 - 2x + 5$ **30** $y = f(x) = 30x^2 - 10x$

31 $y = f(x) = 4x^3 - 2x^2$ **32** $y = f(x) = -5x^3 + 2x^2 - 8$

33 $y = f(x) = -x^4$ **34** $y = f(x) = 1/x^2$

35 The population of a city has increased annually as indicated in the following table.

Year	Population, Millions
1986	2.55
1987	2.70
1988	2.80
1989	2.88
1990	2.90
1991	3.01

At what average rate did the population change between 1986 and 1991? Between 1987 and 1990?

For the following exercises, (a) determine the general expression for the difference quotient, and (b) use the difference quotient to compute the slope of the secant line connecting points at $x = 1$ and $x = 3$.

36 $f(x) = -5x^2 + 2x$ **37** $f(x) = -3/x$

38 $f(x) = 10$ **39** $f(x) = 2x + 7$

40 $f(x) = 3x^2 + x + 50$ **41** $f(x) = x^2/3$

42 $f(x) = 20x^2 + 6x + 25$ **43** $f(x) = -3x^2 + 4x + 1$

44 $f(x) = 3x^3$ **45** $f(x) = -7x^3 + 5$

*46 $f(x) = x^3 - 1$ ***47** $f(x) = 10/x$

*48 $f(x) = ax + b$ ***49** $f(x) = ax^2 + bx + c$

Section 15.4

For the following functions, (a) find the derivative using the limit approach, and (b) determine any values of x for which the slope equals 0.

50 $f(x) = -3x^2 + 6x$ **51** $f(x) = -3/x$

52 $f(x) = cx^2$ **53** $f(x) = -4x^2 - 2x$

*54 $f(x) = x^4$ ***55** $f(x) = x^4 - 4x$

56 $f(x) = (x^3/3) - 9x - 10$ **57** $f(x) = (x^3/3) - 16x + 45$

58 $f(x) = -10x^2 + 40x$ **59** $f(x) = ax^2 + bx$

Section 15.5

For the following exercises, find $f'(x)$.

60 $f(x) = \frac{11}{15}$ **61** $f(x) = 25x^2 + 15x$

62 $f(x) = 3\left(x + \dfrac{4}{x}\right)$ **63** $f(x) = 15x^2/(1 - x^2)$

64 $f(x) = 24x^4 - 15x^3 + 100$ **65** $f(x) = 5x^4/3 + 2x^3$

66 $f(x) = \sqrt{x^7}$ **67** $f(x) = x^2/(4 - x^3)$

68 $f(x) = \dfrac{5}{x^4 - 8}$ **69** $f(x) = \dfrac{x - 3x^2}{x^3 + 3x}$

70 $f(x) = \dfrac{x^2 - 10x}{x^3 + 5x^2 - x}$ **71** $f(x) = \dfrac{1}{\sqrt[3]{x}}$

72 $f(x) = x^2/\sqrt{x}$ **73** $f(x) = (x - 7)(x^3 - 3x^2 + 2x)$

74 $f(x) = (10 - x^5)(x^6 - 2x^3 + x)$ **75** $f(x) = (5 - x + 3x^2)(x^8 + 5x^3 - 10x^2)$

Section 15.6

For the following exercises, find $f'(x)$.

76 $f(x) = (x^3 + 4)^4$ **77** $f(x) = (5x^2 - 10x)^5$

78 $f(x) = (4x^3 - 3x^2 - 2x)^6$ **79** $f(x) = (8 - x^3)^{3/2}$

80 $f(x) = \sqrt{x^3 - 2x^2 + 10}$ **81** $f(x) = \sqrt[3]{x^2 + 6}$

82 $f(x) = 1/\sqrt{10 - x^3}$ **83** $f(x) = 1/\sqrt[3]{x^2 - 5x}$

84 $f(x) = e^{x^3 - 2x}/2$ **85** $f(x) = 10e^{x^2 - 2x}$

86 $f(x) = 5x^2 e^{x^2}$ **87** $f(x) = 2xe^{-x}$

88 $f(x) = 2e^{3x}/x^2$ **89** $f(x) = x^2/4e^x$

90 $f(x) = (e^{3x+1} - 5)^6$ **91** $f(x) = (1 - e^{x^3})^5$

92 $f(x) = (a - be^{-x})^c$ **93** $f(x) = x^5/e^{2x}$

94 $f(x) = \ln(ax^3 + bx^2 + cx + d)$ **95** $f(x) = \ln bx^4$

96 $f(x) = \ln(4 - 2x)$ **97** $f(x) = \ln 9x$

98 $f(x) = x^3 \ln x$ **99** $f(x) = x^2 \ln(x - 5)$

100 $f(x) = [\ln(x^2 + 1)]^4$ **101** $f(x) = (\ln x)^3$

*102 $f(x) = [(x - 5)\sqrt{x + 4}]/(1 - x)$ ***103** $f(x) = [(x^2 + 3)/(4 - x)]^3$

*104 $f(x) = [x\sqrt{x^2 + 7}]^{1/2}$ *105 $f(x) = [(x^2)^3]^4/(x^3)^2$

106 $f(x) = (x^3 - 3x^2)/\ln(x - 5)$ 107 $f(x) = (x/\ln x)^5$

108 $f(x) = x^4 \ln(2x^3 + 4)$ 109 $f(x) = \ln(5x^2 - 2x)$

For the following exercises, find dy/dx.

110 $y = f(u) = u^3 - 4$ and $u = g(x) = x^3 + 3$

111 $y = f(u) = (2u + 3)^2$ and $u = g(x) = x^2 - 3x$

112 $y = f(u) = u^{1/3}$ and $u = g(x) = x^2 - 5$

113 $y = f(u) = \dfrac{6u - 1}{2 - u^2}$ and $u = g(x) = x^2 + 2$

114 $y = f(u) = \sqrt{u^2 - 5u}$ and $u = g(x) = \dfrac{4}{x}$

Section 15.7

115 A ball thrown upward from the roof of a building which is 900 feet high will be at a height of h feet after t seconds, as described by the function

$$h = f(t) = -16t^2 + 80t + 900$$

(a) What is the height of the ball after 4 seconds?
(b) What is the velocity of the ball after 4 seconds?
(c) What is the acceleration of the ball at $t = 0$? At $t = 4$?

116 An object is launched from ground level with an initial velocity of 512 feet per second. The function which describes the height of the ball is

$$h = f(t) = 512t - 16t^2$$

where h is measured in feet and t is time measured in seconds since the ball was thrown.
(a) What is the velocity of the ball at $t = 2$ seconds?
(b) When will the ball return to the ground?
(c) What is the velocity of the ball at the instant it hits the ground?

117 **Population Growth** The population of a city is estimated by the function

$$P = f(t) = 1.2e^{0.045t}$$

where P equals the population (in millions) and t equals time measured in years since 1988.
(a) What is the population expected to equal in 1995?
(b) Determine the expression for the instantaneous rate of change in the population.
(c) At what rate is the population expected to be changing in 1995?

118 **Real Estate Devaluation** Following a rapid increase in the values of residential homes during the mid-1980s, real estate values in the Northeast began to drop in 1990. The function

$$V = f(t) = 140{,}000e^{-0.002t}$$

is a function which estimates the average value V (in \$) of a single-family residence in one particular township, where t equals time measured in months *since* January 1, 1990.

(a) What was the average value estimated to equal on July 1, 1990? On January 1, 1992?

(b) Determine the general expression for the instantaneous rate of change in the average value of a single-family residence in this town.

(c) At what rate is the value changing on January 1, 1991?

❏ CHAPTER TEST

1 Determine:

(a) $\displaystyle\lim_{x \to 3} \frac{x^2 - x - 6}{x - 3}$

(b) $\displaystyle\lim_{x \to \infty} \frac{500}{(2x - 1)/x}$

2 Determine whether there are any discontinuities on f and, if so, where they occur if

$$f(x) = \frac{x^2 - 25}{4x^2 - 2x + 5}$$

3 Determine dy/dx using the limit approach if

$$f(x) = -3x^2 - 2x$$

4 Find $f'(x)$ if:

(a) $f(x) = 6/\sqrt[5]{x^4}$

(b) $f(x) = 7x^6 - 8x^5 + 3x^3 + 90$

(c) $f(x) = (18 + x)/(x^3 + 8)^3$

(d) $f(x) = x^2 e^{4x}$

(e) $f(x) = \ln(5x^3 - x^2)$

(f) $f(x) = (e^x \ln x)^4$

5 Given $f(x) = 15x^2 - 90x - 35$, determine the locations of any points on the graph of $f(x)$ where the slope equals 0.

6 Find all higher-order derivatives of f if

$$f(x) = \frac{5x^4}{4} - \frac{3x^3}{3} + 6x^2 + 10x$$

7 Find dy/dx if

$$y = f(u) = u^4 + 3u \qquad \text{and} \qquad u = g(x) = x^2 + 10$$

APPENDIX: Proofs of Selected Rules of Differentiation

The first proof is "partial" in that it proves rule 2 when n is a positive integer. Proof of this rule for all other values of n is beyond the scope of this text.

**RULE 2
Modified**

If $f(x) = x^n$, where n is a positive integer, then $f'(x) = nx^{n-1}$.

PROOF

$$f(x) = x^n$$

$$f(x + \Delta x) = (x + \Delta x)^n$$

$$f(x + \Delta x) - f(x) = (x + \Delta x)^n - x^n$$

By the *binomial theorem,* for $n = $ a positive integer,

$$(x + \Delta x)^n = x^n + nx^{n-1}(\Delta x) + \frac{n(n-1)}{2} x^{n-2}(\Delta x)^2 + \cdots + (\Delta x)^n$$

Substituting into $f(x + \Delta x) - f(x)$ yields

$$f(x + \Delta x) - f(x) = x^n + nx^{n-1}(\Delta x) + \frac{n(n-1)}{2} x^{n-2}(\Delta x)^2$$
$$+ \cdots + (\Delta x)^n - x^n$$

Dividing by Δx and taking the limit as $\Delta x \to 0$, we get

$$\lim_{\Delta x \to 0} \frac{f(x + \Delta x) - f(x)}{\Delta x} = \lim_{\Delta x \to 0} \left[x^n + nx^{n-1} \Delta x + \frac{n(n-1)}{2} x^{n-2}(\Delta x)^2 \right.$$
$$\left. + \cdots + (\Delta x)^n - x^n \right] \Big/ \Delta x$$
$$= \lim_{\Delta x \to 0} \left[nx^{n-1} + \frac{n(n-1)}{2} x^{n-2}(\Delta x) \right.$$
$$\left. + \cdots + (\Delta x)^{n-1} \right]$$

or $\qquad\qquad\qquad f'(x) = nx^{n-1}$

RULE 3

If $f(x) = c \cdot g(x)$, where c is a constant and g is a differentiable function, then $f'(x) = c \cdot g'(x)$.

PROOF

$$f(x) = c \cdot g(x)$$

$$f(x + \Delta x) = c \cdot g(x + \Delta x)$$

$$f(x + \Delta x) - f(x) = c \cdot g(x + \Delta x) - c \cdot g(x)$$
$$= c[g(x + \Delta x) - g(x)]$$

$$\frac{f(x + \Delta x) - f(x)}{\Delta x} = \frac{c[g(x + \Delta x) - g(x)]}{\Delta x}$$

$$= c \frac{g(x + \Delta x) - g(x)}{\Delta x}$$

$$\lim_{\Delta x \to 0} \frac{f(x + \Delta x) - f(x)}{\Delta x} = \lim_{\Delta x \to 0} c \frac{g(x + \Delta x) - g(x)}{\Delta x}$$

$$= c \cdot \lim_{\Delta x \to 0} \frac{g(x + \Delta x) - g(x)}{\Delta x}$$

or

$$f'(x) = c \cdot g'(x)$$

RULE 4 If $f(x) = u(x) \pm v(x)$, where u and v are differentiable, then $f'(x) = u'(x) \pm v'(x)$. The proof presented is for the "sum" of $u(x)$ and $v(x)$.

PROOF

$$f(x) = u(x) + v(x)$$

$$f(x + \Delta x) = u(x + \Delta x) + v(x + \Delta x)$$

$$f(x + \Delta x) - f(x) = [u(x + \Delta x) + v(x + \Delta x)] - [u(x) + v(x)]$$

$$\frac{f(x + \Delta x) - f(x)}{\Delta x} = \frac{[u(x + \Delta x) + v(x + \Delta x)] - [u(x) + v(x)]}{\Delta x}$$

$$= \left[\frac{u(x + \Delta x) - u(x)}{\Delta x} + \frac{v(x + \Delta x) - v(x)}{\Delta x} \right]$$

$$\lim_{\Delta x \to 0} \frac{f(x + \Delta x) - f(x)}{\Delta x} = \lim_{\Delta x \to 0} \left[\frac{u(x + \Delta x) - u(x)}{\Delta x} + \frac{v(x + \Delta x) - v(x)}{\Delta x} \right]$$

$$= \lim_{\Delta x \to 0} \frac{u(x + \Delta x) - u(x)}{\Delta x} + \lim_{\Delta x \to 0} \frac{v(x + \Delta x) - v(x)}{\Delta x}$$

or

$$f'(x) = u'(x) + v'(x)$$

RULE 5 If $f(x) = u(x) \cdot v(x)$, where u and v are differentiable, then

$$f'(x) = u'(x) \cdot v(x) + v'(x) \cdot u(x)$$

PROOF

$$f(x) = u(x) \cdot v(x)$$

$$f(x + \Delta x) = u(x + \Delta x) \cdot v(x + \Delta x)$$

$$f(x + \Delta x) - f(x) = u(x + \Delta x) \cdot v(x + \Delta x) - u(x) \cdot v(x)$$

Adding *and* subtracting the quantity $u(x + \Delta x) \cdot v(x)$ to the right side gives

$$f(x + \Delta x) - f(x) = u(x + \Delta x) \cdot v(x + \Delta x) - u(x + \Delta x) \cdot v(x)$$
$$+ u(x + \Delta x) \cdot v(x) - u(x)v(x)$$
$$= u(x + \Delta x)[v(x + \Delta x) - v(x)]$$
$$+ v(x)[u(x + \Delta x) - u(x)]$$

$$\frac{f(x + \Delta x) - f(x)}{\Delta x} = \frac{u(x + \Delta x)[v(x + \Delta x) - v(x)] + v(x)[u(x + \Delta x) - u(x)]}{\Delta x}$$

$$= u(x + \Delta x) \frac{v(x + \Delta x) - v(x)}{\Delta x} + v(x) \frac{u(x + \Delta x) - u(x)}{\Delta x}$$

$$\lim_{\Delta x \to 0} \frac{f(x + \Delta x) - f(x)}{\Delta x} = \lim_{\Delta x \to 0} u(x + \Delta x) \frac{v(x + \Delta x) - v(x)}{\Delta x}$$

$$+ \lim_{\Delta x \to 0} v(x) \frac{u(x + \Delta x) - u(x)}{\Delta x}$$

$$= \lim_{\Delta x \to 0} u(x + \Delta x) \cdot \lim_{\Delta x \to 0} \frac{v(x + \Delta x) - v(x)}{\Delta x}$$

$$+ \lim_{\Delta x \to 0} v(x) \cdot \lim_{\Delta x \to 0} \frac{u(x + \Delta x) - u(x)}{\Delta x}$$

or

$$f'(x) = u(x) \cdot v'(x) + v(x) \cdot u'(x)$$
$$= v'(x) \cdot u(x) + u'(x) \cdot v(x)$$

RULE 6

If $f(x) = u(x)/v(x)$, where u and v are differentiable and $v(x) \neq 0$, then

$$f'(x) = \frac{v(x) \cdot u'(x) - u(x) \cdot v'(x)}{[v(x)]^2}$$

PROOF

$$f(x) = \frac{u(x)}{v(x)}$$

$$f(x + \Delta x) = \frac{u(x + \Delta x)}{v(x + \Delta x)}$$

$$f(x + \Delta x) - f(x) = \frac{u(x + \Delta x)}{v(x + \Delta x)} - \frac{u(x)}{v(x)}$$

which can be rewritten as

$$f(x + \Delta x) - f(x) = \frac{u(x + \Delta x) \cdot v(x) - u(x) \cdot v(x + \Delta x)}{v(x + \Delta x) \cdot v(x)}$$

Adding *and* subtracting the quantity $[u(x)v(x)]/[v(x + \Delta x)v(x)]$ yields

$$f(x + \Delta x) - f(x) = \frac{u(x + \Delta x) \cdot v(x) - u(x) \cdot v(x + \Delta x)}{v(x + \Delta x) \cdot v(x)}$$

$$+ \frac{u(x) \cdot v(x)}{v(x + \Delta x) \cdot v(x)} - \frac{u(x) \cdot v(x)}{v(x + \Delta x) \cdot v(x)}$$

Rearranging gives

$$f(x + \Delta x) - f(x) = \frac{u(x + \Delta x) \cdot v(x) - u(x) \cdot v(x) - v(x + \Delta x) \cdot u(x) + u(x) \cdot v(x)}{v(x + \Delta x) \cdot v(x)}$$

$$= \frac{v(x)[u(x + \Delta x) - u(x)] - u(x)[v(x + \Delta x) - v(x)]}{v(x + \Delta x) \cdot v(x)}$$

$$\frac{f(x + \Delta x) - f(x)}{\Delta x} = \frac{v(x)[u(x + \Delta x) - u(x)] - u(x)[v(x + \Delta x) - v(x)]}{v(x + \Delta x) \cdot v(x)} \bigg/ \Delta x$$

$$= \frac{v(x)\, \dfrac{u(x + \Delta x) - u(x)}{\Delta x} - u(x)\, \dfrac{v(x + \Delta x) - v(x)}{\Delta x}}{v(x + \Delta x) \cdot v(x)}$$

$$\lim_{\Delta x \to 0} \frac{f(x + \Delta x) - f(x)}{\Delta x}$$

$$= \lim_{\Delta x \to 0} \frac{v(x)\, \dfrac{u(x + \Delta x) - u(x)}{\Delta x} - u(x)\, \dfrac{v(x + \Delta x) - v(x)}{\Delta x}}{v(x + \Delta x) \cdot v(x)}$$

$$= \frac{\lim\limits_{\Delta x \to 0} v(x) \cdot \lim\limits_{\Delta x \to 0} \dfrac{u(x + \Delta x) - u(x)}{\Delta x} - \lim\limits_{\Delta x \to 0} u(x) \cdot \lim\limits_{\Delta x \to 0} \dfrac{v(x + \Delta x) - v(x)}{\Delta x}}{\lim\limits_{\Delta x \to 0} v(x + \Delta x) \cdot \lim\limits_{\Delta x \to 0} v(x)}$$

or
$$f'(x) = \frac{v(x) \cdot u'(x) - u(x) \cdot v'(x)}{v(x) \cdot v(x)}$$

$$= \frac{v(x) \cdot u'(x) - u(x) \cdot v'(x)}{[v(x)]^2}$$

$\square$

CHAPTER 16

OPTIMIZATION: METHODOLOGY

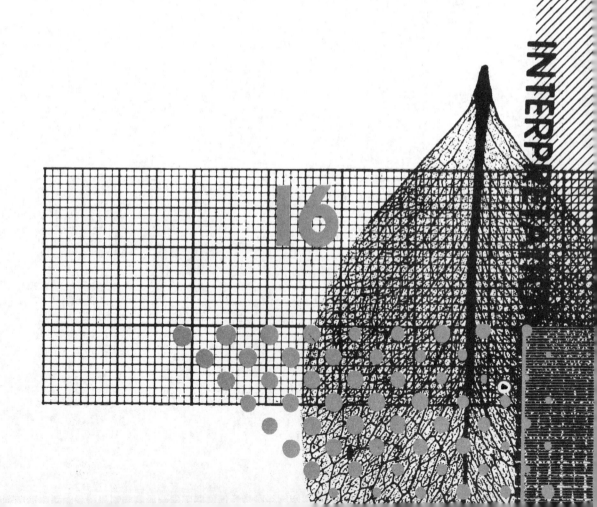

CHAPTER OBJECTIVES

❏ Enhance understanding of the meaning of first and second derivatives

❏ Reinforce understanding of the nature of concavity

❏ Provide a methodology for determining optimization conditions for mathematical functions

❏ Illustrate methods for sketching the general shape of mathematical functions

❏ Illustrate a wide variety of applications of optimization procedures

Differential calculus offers considerable insight regarding the behavior of mathematical functions. It is particularly useful in estimating the graphical representation of a function in two dimensions. This contrasts with the "brute force" methods of sketching functions, which we discussed in Chap. 4. We want to illustrate this attribute of differential calculus by sketching the function

$$f(x) = \frac{x^4}{4} - \frac{8x^3}{3} + 8x^2$$

(Example 17).

In this chapter the tools developed in Chap. 15 will be extended. We will further our understanding of the first and second derivatives. We will see how these derivatives can be useful in describing the behavior of mathematical functions. A major objective of the chapter is to develop a method for determining where a function achieves maximum or minimum values. We will show how these calculus-based **optimization** procedures facilitate the sketching of functions.

16.1 DERIVATIVES: ADDITIONAL INTERPRETATIONS

In this section we will continue to expand our understanding of derivatives.

The First Derivative

As mentioned in the previous chapter, the first derivative represents the instantaneous rate of change in $f(x)$ with respect to a change in x.

> **DEFINITION: INCREASING FUNCTION**
> The function f is said to be an *increasing function* on an interval I if for any x_1 and x_2 within the interval, $x_1 < x_2$ implies that $f(x_1) < f(x_2)$.*

Increasing functions can also be identified by slope conditions. *If the first derivative of f is positive throughout an interval, then the slope is positive and f is an increasing function on the interval.* That is, at any point within the interval, a slight increase in the value of x will be accompanied by an increase in the value of $f(x)$. The curves in Fig. 16.1a and 16.1b are the graphs of increasing functions of x because the tangent slope at any point is positive.

---

* Technically speaking, these are definitions for *strictly* increasing (decreasing) functions.

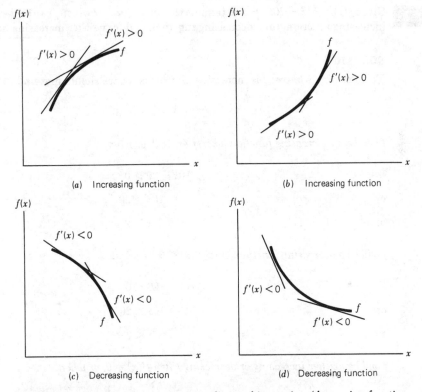

Figure 16.1 The relationship between $f'(x)$ and increasing/decreasing functions.

> **DEFINITION: DECREASING FUNCTION**
> The function f is said to be a *decreasing function* on an interval I if for any x_1 and x_2 within the interval, $x_1 < x_2$ implies that $f(x_1) > f(x_2)$.*

As with increasing functions, decreasing functions can be identified by tangent slope conditions. *If the first derivative of f is negative throughout an interval, then the slope is negative and f is a decreasing function on the interval.* That is, at any point within the interval a slight increase in the value of x will be accompanied by a decrease in the value of $f(x)$. The curves in Fig. 16.1c and d are the graphs of decreasing functions of x.

> **NOTE** If a function is increasing (decreasing) on an interval, the function is increasing (decreasing) at every point within the interval.

---

* Technically speaking, these are definitions for *strictly* increasing (decreasing) functions.

EXAMPLE 1 Given $f(x) = 5x^2 - 20x + 3$, determine the intervals over which f can be described as (a) an increasing function, (b) a decreasing function, and (c) neither increasing nor decreasing.

SOLUTION

To determine whether f is increasing or decreasing, we should first find f':

$$f'(x) = 10x - 20$$

f will be an increasing function when f'(x) > 0, or when

$$10x - 20 > 0$$
or $$10x > 20$$
or $$x > 2$$

f will be a decreasing function when f'(x) < 0, or when

$$10x - 20 < 0$$
or $$10x < 20$$
or $$x < 2$$

f will be neither increasing or decreasing when f'(x) = 0, or when

$$10x - 20 = 0$$
or $$10x = 20$$
or $$x = 2$$

Summarizing, f is a decreasing function when $x < 2$, neither increasing nor decreasing at $x = 2$, and an increasing function when $x > 2$. Sketch the graph of f to see whether these conclusions seem reasonable.

❑

The second derivative $f''(x)$ is a measure of the instantaneous rate of change in $f'(x)$ with respect to a change in x. In other words, it indicates the rate at which the slope of the function is changing with respect to a change in x—whether the slope of the function is increasing or decreasing at a particular instant.

If $f''(x)$ is negative on an interval I of f, the first derivative is decreasing on I. Graphically, the slope is decreasing in value on the interval. If $f''(x)$ is positive on an interval I of f, the first derivative is increasing on I. Graphically, the slope is increasing on the interval.

Examine Fig. 16.2. Either mentally construct tangent lines or lay a straightedge on the curve to represent the tangent line at various points. Along the curve from A to B the slope is slightly negative near A and becomes more and more negative as we get closer to B. In fact, the slope of the tangent line goes from a value

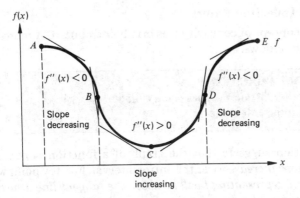

Figure 16.2 The relationship between $f''(x)$ and increasing/decreasing slope $[f'(x)]$.

of 0 at A to its most negative value at point B. Thus the slope is decreasing in value over the interval between A and B, and we would expect $f''(x)$ to be negative on this interval [i.e., $f''(x) < 0$].

Having reached its most negative value at point B, the slope continues to be negative on the interval between B and C; however, the slope becomes less and less negative, eventually equaling 0 at C. If the slope assumes values which are becoming *less negative* (e.g., $-5, -4, -3, -2, -1, 0$), the slope is *increasing* in value. As such, we would expect $f''(x)$ to be positive on this interval [i.e., $f''(x) > 0$].

Between C and D the slope becomes more and more positive, assuming its largest positive value at D. Since the slope is increasing in value on this interval, we would expect $f''(x)$ to be positive.

Between D and E the slope continues to be positive, but it is becoming less and less positive, eventually equaling 0 at E. If the slope is positive but becoming smaller (e.g., 5, 4, 3, 2, 1, 0), we would expect $f''(x)$ to be negative on the interval.

Figure 16.3 summarizes the first- and second-derivative conditions for the four regions of the function. These relationships can be difficult to understand. Take your time studying these figures and retrace the logic if necessary.

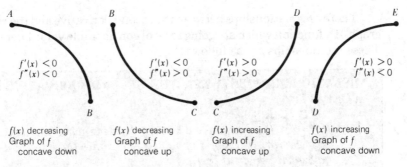

Figure 16.3 Joint characteristics for $f'(x)$ and $f''(x)$.

Concavity and Inflection Points

In Chap. 6 the concept of concavity was briefly introduced. A more formal definition of concavity follows.

DEFINITION: CONCAVITY
The graph of a function *f* is *concave up (down)* on an interval if *f′* increases (decreases) on the entire interval.

This definition suggests that the graph of a function is concave up on an interval if the slope *increases* over the entire interval. For any point within such an interval *the curve representing f will lie above the tangent line drawn at the point.* Similarly, the graph of a function is concave down on an interval if the slope *decreases* over the entire interval. For any point within such an interval *the curve representing f will lie below the tangent line drawn at the point.*

In Fig. 16.4 the graph of *f* is *concave down* between *A* and *B*, and it is *concave up* between *B* and *C*. Note that between *A* and *B* the curve lies below its tangent lines and between *B* and *C* the curve lies above its tangent lines. Point *B* is where the concavity changes from concave down to concave up. A point at which the concavity changes is called an ***inflection point.*** Thus, point *B* is an inflection point.

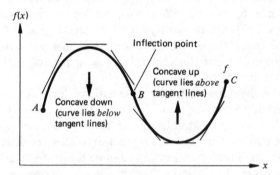

Figure 16.4   Representation of concavity conditions.

There are relationships between the second derivative and the concavity of the graph of a function which are going to be of considerable value later in this chapter. These relationships are as follows.

RELATIONSHIPS BETWEEN THE SECOND DERIVATIVE AND CONCAVITY

I   *If $f''(x) < 0$ on an interval $a \leq x \leq b$, the graph of f is concave down over that interval. For any point $x = c$ within the interval, f is said to be concave down at [c, f(c)].*

> II   If $f''(x) > 0$ on any interval $a \leq x \leq b$, the graph of f is concave up over that interval. For any point $x = c$ within the interval, f is said to be concave up at $[c, f(c)]$.
>
> III   If $f''(x) = 0$ at any point $x = c$ in the domain of f, no conclusion can be drawn about the concavity at $[c, f(c)]$.

Be very careful not to reverse the logic of these relationships! Because of relationship III we *cannot* make statements about the sign of the second derivative knowing the concavity of the graph of a function. For example, *we cannot state that if the graph of a function is concave down at $x = a$, $f''(a) < 0$.*

EXAMPLE 2   To determine the concavity of the graph of the generalized quadratic function $f(x) = ax^2 + bx + c$, let's find the first and second derivatives.

$$f'(x) = 2ax + b \quad \text{and} \quad f''(x) = 2a$$

If $a > 0$, then $f''(x) = 2a > 0$. From relationship II, the graph of f is concave up whenever $a > 0$. If $a < 0$, then $f''(x) = 2a < 0$. From relationship I, the graph of f is concave down whenever $a < 0$. This is entirely consistent with our discussions in Chap. 6 which concluded that if $a > 0$, f graphs as a parabola which is concave up. And if $a < 0$, f graphs as a parabola which is concave down.

EXAMPLE 3   For $f(x) = x^3 - 2x^2 + x - 1$, determine the concavity of the graph of f at $x = -2$ and $x = 3$.

SOLUTION

$$f'(x) = 3x^2 - 4x + 1$$

and

$$f''(x) = 6x - 4$$

Evaluating $f''(x)$ at $x = -2$ gives

$$f''(-2) = 6(-2) - 4 = -16$$

Since $f''(-2) < 0$, the graph of f is concave down at $x = -2$. To determine the concavity at $x = 3$, we find

$$f''(3) = 6(3) - 4 = 14$$

Since $f''(3) > 0$, the graph of f is concave up at $x = 3$.

EXAMPLE 4   Determine the concavity of the graph of $f(x) = x^4$ at $x = 0$.

SOLUTION

$$f'(x) = 4x^3$$

$$f''(x) = 12x^2$$

$$f''(0) = 12(0)^2 = 0$$

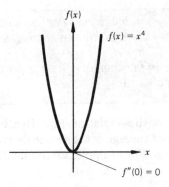

Figure 16.5   No conclusions regarding concavity.

According to relationship III, we can make no statement about the concavity at $x = 0$. However, substituting a sufficient number of values for x into f and plotting these ordered pairs, we see that f has the shape shown in Fig. 16.5. From this sketch it is obvious that the graph is concave up at $x = 0$.

❑

LOCATING INFLECTION POINTS

I   *Find all points a where $f''(a) = 0$.*
II *If $f''(x)$ changes sign when passing through $x = a$, there is an inflection point at $x = a$.*

A *necessary condition (something which must be true) for the existence of an inflection point at $x = a$ is that $f''(a) = 0$.* That is, by finding all values of x for which $f''(x) = 0$, *candidate locations* for inflection points are identified.* The condition $f''(a) = 0$ does not guarantee that an inflection point exists at $x = a$ (see Example 4). Step II confirms whether a candidate location is an inflection point.

The essence of this test is to choose points slightly to the left and right of $x = a$ and determine if the concavity is different on each side. If $f''(x)$ is positive to the left and negative to the right, or vice versa, there has been a *change* in concavity when passing through $x = a$. Thus, an inflection point exists at $x = a$.

EXAMPLE 5

In Example 4, $f''(0) = 0$ implies that $x = 0$ is a candidate location for an inflection point (**step I**).

❑ **Step II** To verify that an inflection point does *not* exist at $x = 0$ for $f(x) = x^4$, $f''(x)$ is evaluated to the left at $x = -0.1$ and to the right at $x = +0.1$.

---

* Other candidates for inflection points occur where $f''(x)$ is discontinuous. However, we will not encounter such candidates in this book.

$$f''(-0.1) = 12(-0.1)^2$$
$$= 0.12 > 0$$

and

$$f''(+0.1) = 12(+0.1)^2$$
$$= 0.12 > 0$$

Since the second derivative has the same sign to the left and right of $x = 0$, there is no inflection point at $x = 0$.

EXAMPLE 6   To determine the location(s) of all inflection points on the graph of

$$f(x) = \frac{x^4}{12} - \frac{x^3}{2} + x^2 + 10$$

we find

$$f'(x) = \frac{4x^3}{12} - \frac{3x^2}{2} + 2x$$

$$= \frac{x^3}{3} - \frac{3x^2}{2} + 2x$$

and

$$f''(x) = \frac{3x^2}{3} - \frac{6x}{2} + 2$$

$$= x^2 - 3x + 2$$

❏ **Step I** $f''(x)$ is set equal to 0 in order to find candidate locations:

$$x^2 - 3x + 2 = 0$$

or

$$(x - 1)(x - 2) = 0$$

Therefore, $f''(x) = 0$ when $x = 1$ and $x = 2$.

❏ **Step II** For $x = 1$, $f''(x)$ is evaluated to the left and right of $x = 1$ at $x = 0.9$ and $x = 1.1$.

$$f''(0.9) = (0.9)^2 - 3(0.9) + 2$$
$$= 0.81 - 2.7 + 2$$
$$= 0.11 > 0$$

$$f''(1.1) = (1.1)^2 - 3(1.1) + 2$$
$$= 1.21 - 3.3 + 2$$
$$= -0.09 < 0$$

Since the sign of $f''(x)$ changes, an inflection point exists at $x = 1$. When values of $x = 1.9$ and $x = 2.1$ are chosen for evaluating $f''(x)$ to the left and right of $x = 2$.

$$f''(1.9) = -0.09 < 0$$

$$f''(2.1) = 0.11 > 0$$

Because $f''(x)$ changes sign, we conclude that an inflection point also exists at $x = 2$.

❑

PRACTICE EXERCISE
Determine the locations of any inflection points on the graph of $f(x) = x^3/3 + x^2 - 8x$. *Answer:* Inflection point at $x = -1$.

EXAMPLE 7   **(Tracking an Epidemic: Revisited)** The Motivating Scenario in Chap. 15 discussed the spread of a flu epidemic. The function

$$n = f(t) = -0.3t^3 + 10t^2 + 300t + 250$$

was used to estimate the number of persons afflicted with the flu, n, as a function of the number of days since initial detection by health department officials, t. Determine any inflection points and interpret their meaning in this application.

SOLUTION

❑ **Step I**

$$f'(t) = -0.9t^2 + 20t + 300$$

$$f''(t) = -1.8t + 20$$

Setting $f''(t) = 0$ results in one candidate for an inflection point at $t = 11.11$.

❑ **Step II** Because $f''(11) = 0.2$ and $f''(12) = -1.6$, the sign of $f''(t)$ changes when passing through the candidate location. Thus, one inflection point exists for the function.

❑ **Interpretation** The inflection point can be interpreted as representing the point in time when the rate at which persons are being afflicted by the flu decreases. Prior to $t = 11.11$, additional persons are being afflicted *at an increasing rate*. After $t = 11.11$, additional persons are being afflicted, but *at a decreasing rate*.

❑

Concavity from a Different Perspective

We will use the terms *concave up* and *concave down* to describe the curvature attribute which we call concavity. Other terminology may be used to describe this attribute. For example, many writers distinguish between **strictly concave functions** and **strictly convex functions**.

DEFINITION: STRICTLY CONCAVE (CONVEX) FUNCTION
A function which is strictly concave (convex) has the following graphical property: Given any two points A and B which lie on the curve representing the function, if the two points are connected by a straight line, the entire line segment AB will lie below (above) the curve except at points A and B.

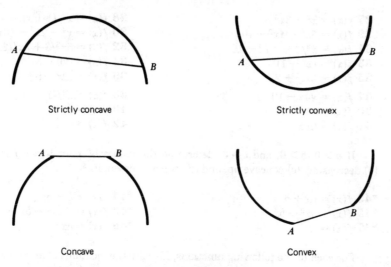

Figure 16.6 Concave and convex functions.

This definition can be loosened somewhat to define a **concave function** and a **convex function** [as opposed to *strictly* concave (convex) functions]. If the line segment *AB* is allowed to lie below (above) the curve, *or to lie on the curve*, the function is termed a concave (convex) function. Figure 16.6 illustrates these definitions.

Section 16.1 Follow-up Exercises

For each of the following functions, (*a*) determine whether *f* is increasing or decreasing at *x* = 1. Determine the values of *x* for which *f* is (*b*) an increasing function, (*c*) a decreasing function, and (*d*) neither increasing nor decreasing.

1 $f(x) = 20 - 4x$ 2 $f(x) = 15x + 16$

3 $f(x) = x^2 - 3x + 20$ 4 $f(x) = 3x^2 + 12x + 9$

5 $f(x) = x^3/3 + x^2/2$ 6 $f(x) = x^3/3 + x^2/2 - 6x$

7 $f(x) = x^4 + 2x^2$ 8 $f(x) = 3x^5$

9 $f(x) = (x + 3)^{3/2}$ 10 $f(x) = 3x^2/(x^2 - 1)$

11 $f(x) = -x^2 + 4x + 15$ 12 $f(x) = -2x^2 + 20x + 3$

13 $f(x) = 5x^2 + 40x + 50$ 14 $f(x) = 3x^2/2 - 9x + 5$

15 $f(x) = (x - 4)^{3/2}$ 16 $f(x) = (x - 5)^4$

17 $f(x) = (2x - 10)^5$ 18 $f(x) = (8x + 24)^8$

19 $f(x) = \dfrac{(4x + 20)^9}{4}$ 20 $f(x) = (2x + 18)^7$

For each of the following functions, use $f''(x)$ to determine the concavity conditions at $x = -2$ and $x = +1$.

21 $f(x) = -3x^2 + 2x - 3$ 22 $f(x) = x^3 + 12x + 1$

23 $f(x) = x^2 - 4x + 9$ 24 $f(x) = -x^2 + 5x$

25 $f(x) = \sqrt{x^2 + 10}$ 26 $f(x) = (x + 1)^3$

27 $f(x) = x^2 + 3x^3$

28 $f(x) = x^2/(1 + x)$

29 $f(x) = 5x^3 - 4x^2 + 10x$

30 $f(x) = x^4 + 2x^3 - 10x^2$

31 $f(x) = x^3/3 - x^2/2 + 10x$

32 $f(x) = 5x^3/3 + 3x^2/2 - 5x + 25$

33 $f(x) = (x^2 - 1)^3$

34 $f(x) = (20 - 3x)^5$

35 $f(x) = (3x^2 + 2)^4$

36 $f(x) = (2x - 8)^4$

37 $f(x) = \sqrt{4x - 10}$

38 $f(x) = x^3/(1 - x)$

39 $f(x) = e^x$

40 $f(x) = e^{-x}$

41 $f(x) = \ln x$

42 $f(x) = -\ln x$

If $a > 0$, $b > 0$, and $c > 0$, determine the values of x for which f is (a) increasing, (b) decreasing, (c) concave up, and (d) concave down, if:

*43 $f(x) = ax + b$

*44 $f(x) = b - ax$

*45 $f(x) = ax^2 + bx + c$

*46 $f(x) = -ax^2 - bx - c$

*47 $f(x) = ax^3$

*48 $f(x) = ax^4$

For each of the following functions, identify the locations of any inflection points.

49 $f(x) = x^3 - 9x^2$

50 $f(x) = -x^3 + 24x^2$

51 $f(x) = x^4/12 - x^3/3 - 7.5x^2$

52 $f(x) = (3 - x)^4$

53 $f(x) = (x - 5)^3$

54 $f(x) = x^5/20 - x^3/6$

55 $f(x) = -10x^4 + 100$

56 $f(x) = (x - 1)/x$

57 $f(x) = x^3 + 6x^2 - 18$

58 $f(x) = -x^3 - 30x^2$

59 $f(x) = x^4/12 + x^3/6 - 3x^2$

60 $f(x) = x^4/12 + 7x^3/6 + 5x^2$

61 $f(x) = x^4/4 - 9x^2/2 + 100$

62 $f(x) = x^6/30 - 4x^2/2$

63 $f(x) = (3x - 12)^{5/2}$

64 $f(x) = (2x - 8)^{7/2}$

65 $f(x) = (x - 5)^5$

66 $f(x) = (x + 2)^4$

67 $f(x) = e^x$

68 $f(x) = e^{-x}$

69 $f(x) = \ln x$

70 $f(x) = -\ln x$

71 Given the function shown in Fig. 16.7, indicate the values of x for which f is (a) increasing, (b) decreasing, and (c) neither increasing nor decreasing.

72 Given the function shown in Fig. 16.7, indicate the values of x for which f is (a) concave up, (b) concave down, (c) changing concavity, (d) concave, and (e) convex.

Figure 16.7

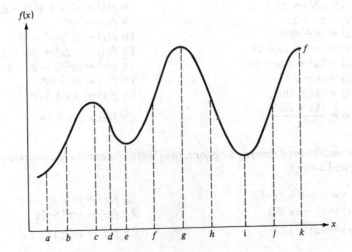

Figure 16.8

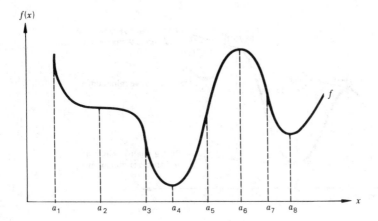

73 Given the function shown in Fig. 16.7, indicate the values of x for which f is (*a*) increasing at an increasing rate, (*b*) increasing at a decreasing rate, (*c*) decreasing at a decreasing rate, and (*d*) decreasing at an increasing rate.

74 Given the function shown in Fig. 16.8, indicate the values of x for which f is (*a*) increasing, (*b*) decreasing, and (*c*) neither increasing nor decreasing.

75 Given the function shown in Fig. 16.8, indicate the values of x for which f is (*a*) concave up, (*b*) concave down, (*c*) changing concavity, (*d*) concave, and (*e*) convex.

76 Given the function shown in Fig. 16.8, indicate the values of x for which f is (*a*) increasing at an increasing rate, (*b*) increasing at a decreasing rate, (*c*) decreasing at a decreasing rate, and (*d*) decreasing at an increasing rate.

16.2 IDENTIFICATION OF MAXIMA AND MINIMA

In this section we will examine functions with the purpose of locating maximum and minimum values.

Relative Extrema

> **DEFINITION: RELATIVE MAXIMUM**
> If f is defined on an interval (b, c) which contains $x = a$, f is said to reach a *relative (local) maximum* at $x = a$ if $f(a) \geq f(x)$ for all x within the interval (b, c).

> **DEFINITION: RELATIVE MINIMUM**
> If f is defined on an interval (b, c) which contains $x = a$, f is said to reach a *relative (local) minimum* at $x = a$ if $f(a) \leq f(x)$ for all x within the interval (b, c).

Both definitions focus upon the value of $f(x)$ within an interval. A relative maximum refers to a point where the value of $f(x)$ is greater than the values for any points which are nearby. A relative minimum refers to a point where the value of

Figure 16.9
Relative extrema.

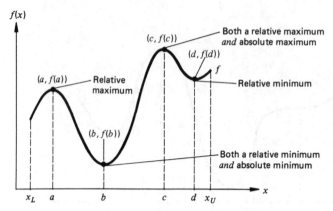

$f(x)$ is lower than the values for any points which are nearby. If we use these definitions and examine Fig. 16.9, f has **relative maxima** at $x = a$ and $x = c$. Similarly, f has **relative minima** at $x = b$ and $x = d$. Collectively, relative maxima and minima are called **relative extrema**.

DEFINITION: ABSOLUTE MAXIMUM
A function f is said to reach an **absolute maximum** at $x = a$ if $f(a) > f(x)$ for any other x in the domain of f.

DEFINITION: ABSOLUTE MINIMUM
A function f is said to reach an **absolute minimum** at $x = a$ if $f(a) < f(x)$ for any other x in the domain of f.

If we refer again to Fig. 16.9, $f(x)$ reaches an absolute maximum at $x = c$. It reaches an absolute minimum at $x = b$. It should be noted that *a point on the graph of a function can be both a relative maximum (minimum) and an absolute maximum (minimum).*

Critical Points

We will have a particular interest in relative maxima and minima. It will be important to know how to identify and distinguish between them.

NECESSARY CONDITIONS FOR RELATIVE MAXIMA (MINIMA)
Given the function f, necessary conditions for the existence of a relative maximum or minimum at $x = a$ (a contained in the domain of f) are

1. $f'(a) = 0$, *or*
2. $f'(a)$ *is undefined.*

Points which satisfy either of the conditions in this definition are *candidates* for relative maxima (minima). Such points are often referred to as **critical points**.

Figure 16.10
Critical points.

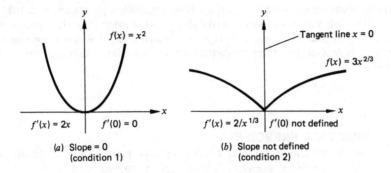

(a) Slope = 0
(condition 1)

(b) Slope not defined
(condition 2)

Points which satisfy condition 1 are those on the graph of f where the slope equals 0. Points satisfying condition 2 are exemplified by discontinuities on f or points where $f'(x)$ cannot be evaluated. Values of x in the domain of f which satisfy either condition 1 or condition 2 are called **critical values.** These are denoted with an asterisk (x^*) in order to distinguish them from other values of x. Given a critical value for f, the corresponding critical point is $[x^*, f(x^*)]$.

Figure 16.10 illustrates the graphs of two functions which have critical points at (0, 0). For the function $f(x) = x^2$, shown in Fig. 16.10a, $f'(x) = 2x$ and a critical value occurs when $x = 0$ (condition 1), where the function achieves a relative minimum.

For the function $f(x) = 3x^{2/3}$, $f'(x) = 2/x^{1/3}$. Note that condition 1 can never be satisfied since there are no points where the tangent slope equals 0. However, a critical value of $x = 0$ exists according to condition 2. The derivative is undefined (the tangent line is the vertical line $x = 0$ for which the slope is undefined). However, $f(0)$ is defined and the critical point (0, 0) is a relative minimum, as shown in Fig. 16.10b.

EXAMPLE 8 In order to determine the location(s) of any critical points on the graph of

$$f(x) = \frac{x^3}{3} - \frac{x^2}{2} - 6x + 100$$

the derivative f' is found.

$$f'(x) = \frac{3x^2}{3} - \frac{2x}{2} - 6$$

$$= x^2 - x - 6$$

$f'(x) = 0$ when

$$x^2 - x - 6 = 0$$

or

$$(x - 3)(x + 2) = 0$$

When the two factors are set equal to 0, two critical values are $x = 3$ and $x = -2$. When these values are substituted into f, the resulting critical points are $(-2, 107\frac{1}{3})$ and $(3, 86\frac{1}{2})$.

The only statements we can make about the behavior of f at these points is that the slope equals 0. Furthermore, nowhere else on the graph of f does the slope equal 0. Additional testing is necessary to determine whether there is a relative maximum or minimum at $x = 3$ and $x = -2$.

❑

> **PRACTICE EXERCISE**
> Determine the location(s) of any critical points on the graph of $f(x) = x^3/3 + x^2 - 8x$. *Answer:* Critical points at $(-4, 26\frac{2}{3})$ and $(2, -9\frac{1}{3})$.

POINT FOR THOUGHT & DISCUSSION   What comment can be made regarding the existence of critical points on constant functions [for example, $f(x) = 10$]?

Figure 16.11 illustrates the different possibilities of critical points where $f'(x) = 0$. Figures 16.11a and b illustrate relative maximum and minimum points, whereas Figs. 16.11c and d illustrate two different types of inflection points. In Fig.

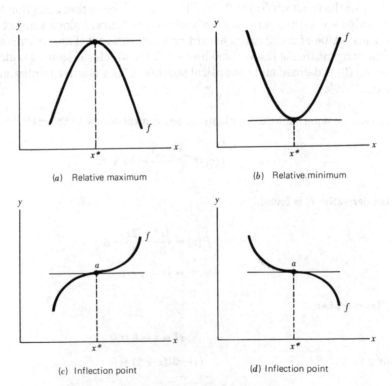

(a) Relative maximum   (b) Relative minimum

(c) Inflection point   (d) Inflection point

Figure 16.11 Critical points where $f''(x) = 0$.

16.11c the graph of the function has a slope of 0 at point *a*, and it is also changing from being concave down to concave up. In Fig. 16.11d the graph has a slope of 0 and is changing from being concave up to concave down.

Any critical point where f'(x) = 0 will be a relative maximum, a relative minimum, or an inflection point.

> **POINT FOR THOUGHT & DISCUSSION**
>
> For polynomial functions *f* of degree *n*, the largest possible number of critical points where $f'(x) = 0$ is $n - 1$. Thus, a function *f* of degree 5 can have *as many as* four points of zero slope. Why is this so?

The First-Derivative Test

In an effort to locate relative maximum or minimum points, the first step is to locate all critical points on the graph of the function. Given that a critical point may be either a relative maximum or minimum or an inflection point, some test must be devised to distinguish among these. There are a number of tests available. One test which is easy to understand intuitively is the ***first-derivative test.***

After the locations of critical points are identified, the first-derivative test requires an examination of slope conditions to the left and right of the critical point. Figure 16.12 illustrates the four critical point possibilities and their slope

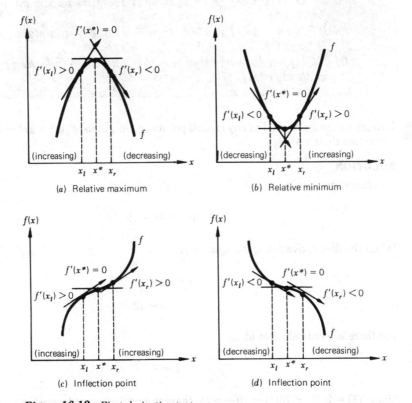

Figure 16.12 First-derivative test.

conditions to either side of x^*. *For a relative maximum, the slope is positive to the left (x_l) and negative to the right (x_r). For a relative minimum, the slope is negative to the left and positive to the right. For the inflection points, the slope has the **same sign** to the left or the right of the critical point.*

Another way of describing this test is the following:

1 *For a relative maximum, the value of the function is increasing to the left and decreasing to the right.*
2 *For a relative minimum, the value of the function is decreasing to the left and increasing to the right.*
3 *For inflection points, the value of the function is either increasing both to the left* and *right or decreasing both to the left and right.*

A summary of the first-derivative test follows.

FIRST-DERIVATIVE TEST

I   *Locate all critical values x^*.*
II *For any critical value x^*, determine the value of $f'(x)$ to the left (x_l) and right (x_r) of x^*.*
(a) *If $f'(x_l) > 0$ and $f'(x_r) < 0$, there is a **relative maximum** for f at $[x^*, f(x^*)]$.*
(b) *If $f'(x_l) < 0$ and $f'(x_r) > 0$, there is a **relative minimum** for f at $[x^*, f(x^*)]$.*
(c) *If $f'(x)$ has the same sign at both x_l and x_r, an inflection point exists at $[x^*, f(x^*)]$.*

EXAMPLE 9   Determine the location(s) of any critical points on the graph of $f(x) = 2x^2 - 12x - 10$, and determine their nature.

SOLUTION
The first derivative is

$$f'(x) = 4x - 12$$

When the first derivative is set equal to 0,

$$4x - 12 = 0$$

or
$$4x = 12$$

and there is a critical value at

$$x = 3$$

Since $f(3) = 2(3^2) - 12(3) - 10 = -28$, there is a critical point located at $(3, -28)$.

To test the critical point, let's select $x_l = 2.9$ and $x_r = 3.1$.

$$f'(2.9) = 4(2.9) - 12$$
$$= 11.6 - 12 = -0.4$$

$$f'(3.1) = 4(3.1) - 12$$
$$= 12.4 - 12 = +0.4$$

Because the first derivative is negative (-0.4) to the left of $x = 3$ and positive ($+0.4$) to the right, the point $(3, -28)$ is a relative minimum on f (note that f is a quadratic function which graphs as a parabola that is concave up).

□

NOTE   **A CAVEAT**
When selecting x_l and x_r, one must stay reasonably close to the critical value x^*. If you stray too far left or right, you may reach an erroneous result, as in Fig. 16.13, where a relative minimum could be judged to be a relative maximum. Some latitude does exist in selecting x_l and x_r. However, when more than one critical point exists, x_l and x_r should be chosen in such a way that they fall between the critical value being examined and any adjacent critical values.

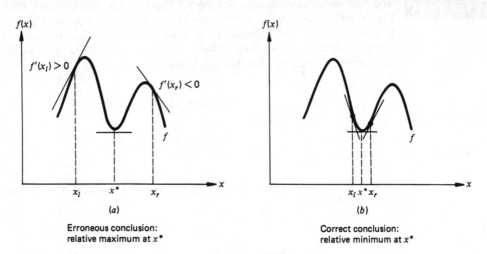

(a)

Erroneous conclusion:
relative maximum at x^*

(b)

Correct conclusion:
relative minimum at x^*

Figure 16.13   Choosing x_l and x_r for first-derivative test.

EXAMPLE 10   In Example 8 we determined that the graph of the function

$$f(x) = \frac{x^3}{3} - \frac{x^2}{2} - 6x + 100$$

has critical points at $(3, 86\frac{1}{2})$ and $(-2, 107\frac{1}{3})$. To determine the nature of these critical points, we examine the first derivative

$$f'(x) = x^2 - x - 6$$

In testing the critical point at $x = 3$, let's select $x_l = 2$ and $x_r = 4$.

$$f'(2) = (2)^2 - 2 - 6$$
$$= -4$$

$$f'(4) = (4)^2 - 4 - 6$$
$$= 6$$

Since $f'(x)$ is negative (f is decreasing) to the left of $x = 3$ and positive (f is increasing) to the right, a relative minimum occurs for f when $x = 3$.
In testing the critical point at $x = -2$, let's select $x_l = -3$ and $x_r = -1$.

$$f'(-3) = (-3)^2 - (-3) - 6$$
$$= 6$$

$$f'(-1) = (-1)^2 - (-1) - 6$$
$$= -4$$

Since $f'(x)$ is positive (f is increasing) to the left of $x = -2$ and negative (f is decreasing) to the right, a relative maximum occurs for $f(x)$ when $x = -2$.

EXAMPLE 11   The first-derivative test also is valid for those critical points where $f'(x)$ is undefined. Consider the function $f(x) = 3x^{2/3}$, the graph of which is shown in Fig. 16.10b. For this function, $f'(x) = 2/x^{1/3}$. Because $f'(x)$ is undefined when $x = 0$, a critical value exists according to condition 2. Since $f(0) = 3(0)^{2/3} = 0$, there is a critical point at $(0, 0)$. Using the first-derivative test, let's select $x_l = -1$ and $x_r = 1$.

$$f'(-1) = \frac{2}{\sqrt[3]{-1}} = \frac{2}{-1} = -2$$

$$f'(1) = \frac{2}{\sqrt[3]{1}} = \frac{2}{1} = 2$$

Since $f'(x)$ is negative to the left of $x = 0$ and positive to the right, a relative minimum occurs at the critical point $(0, 0)$.

$\square$

PRACTICE EXERCISE
In the Practice Exercise on page 748, critical points of $(-4, 26\frac{2}{3})$ and $(2, -9\frac{1}{3})$ were identified for the function $f(x) = x^3/3 + x^2 - 8x$. Determine the nature of these critical points using the first-derivative test. *Answer:* Relative maximum at $(-4, 26\frac{2}{3})$ and relative minimum at $(2, -9\frac{1}{3})$.

The Second-Derivative Test

For critical points, where $f'(x) = 0$, the most expedient test is the ***second-derivative test.*** Intuitively, the second-derivative test attempts to determine the concavity of the function at a critical point $[x^*, f(x^*)]$. We concluded in Sec. 16.1 that if $f''(x) < 0$ at a point on the graph of f, the curve is *concave down* at that point. If $f''(x) > 0$ at a point on the graph of f, the curve is *concave up* at that point. Thus, the second-derivative test suggests finding the value of $f''(x^*)$. Of greater interest, though, is the *sign* of $f''(x^*)$. If $f''(x^*) > 0$, not only is the slope equal to 0 at x^* but the function f is concave up at x^*. If we refer to the four critical point possibilities in Fig. 16.11, only one is concave up at x^*, that being the relative minimum in Fig. 16.11*b*.

If $f''(x^*) < 0$, the function is concave down at x^*. Again referring to Fig. 16.11, the only critical point accompanied by concave-down conditions is the relative maximum in Fig. 16.11*a*.

As stated in Sec. 16.1, if $f''(x^*) = 0$, no conclusion can be drawn regarding the concavity at $[x^*, f(x^*)]$. Another test such as the first-derivative test is required to determine the nature of these particular critical points. A summary of the second-derivative test follows.

SECOND-DERIVATIVE TEST

I *Find all critical values x^*, such that $f'(x^*) = 0$.*
II *For any critical value x^*, determine the value of $f''(x^*)$.*
  (a) If $f''(x^*) > 0$, the function is concave up at x^* and there is a relative minimum for f at $[x^*, f(x^*)]$.
  (b) If $f''(x^*) < 0$, the function is concave down at x^* and there is a relative maximum for f at $[x^*, f(x^*)]$.
  (c) If $f''(x^*) = 0$, no conclusions can be drawn about the concavity at x^* nor the nature of the critical point. Another test such as the first-derivative test is necessary.

EXAMPLE 12   Examine the following function for any critical points and determine their nature.

$$f(x) = -\tfrac{3}{2}x^2 + 6x - 20$$

SOLUTION

We should recognize this as a quadratic function which graphs as a parabola that is concave down. There should be one critical point which is a relative maximum. To confirm this, we find the first derivative.

$$f'(x) = -3x + 6$$

If $f'(x)$ is set equal to 0,

$$-3x + 6 = 0$$

$$-3x = -6$$

and one critical value occurs when

$$x = 2$$

The value of $f(x)$ when $x = 2$ is

$$f(2) = -\tfrac{3}{2}(2^2) + 6(2) - 20$$
$$= -6 + 12 - 20 = -14$$

The only critical point occurs at $(2, -14)$.

Using the second-derivative test yields

$$f''(x) = -3$$

and

$$f''(2) = -3 < 0$$

Since the second derivative is negative at $x = 2$, we can conclude that the graph of f is concave down at this point, and the critical point is a relative maximum. Figure 16.14 presents a sketch of the function.

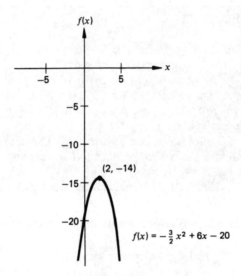

Figure 16.14   Relative maximum at $(2, -14)$.

EXAMPLE 13   Examine the following function for any critical points and determine their nature.

$$f(x) = \frac{x^4}{4} - \frac{9x^2}{2}$$

SOLUTION

If f' is identified,

$$f'(x) = \frac{4x^3}{4} - \frac{18x}{2}$$
$$= x^3 - 9x$$

$f'(x)$ equals zero when

$$x^3 - 9x = 0$$

$$x(x^2 - 9) = 0$$

or when

$$x(x + 3)(x - 3) = 0$$

If the three factors are set equal to 0, critical values are found when

$$x = 0 \qquad x = -3 \qquad x = 3$$

Substituting these critical values into f, critical points occur on the graph of f at $(0, 0)$, $(-3, -81/4)$, and $(3, -81/4)$.

The second derivative is

$$f''(x) = 3x^2 - 9$$

To test the critical point $(0, 0)$,

$$f''(0) = 3(0^2) - 9$$
$$= -9 < 0$$

Since $f''(0)$ is negative, the function is concave down at $x = 0$ and a *relative maximum* occurs at $(0, 0)$. To test the critical point $(-3, -81/4)$.

$$f''(-3) = 3(-3)^2 - 9$$
$$= 27 - 9 = 18 > 0$$

The graph of f is concave up when $x = -3$ and a *relative minimum* occurs at $(-3, -81/4)$. To test the critical point $(3, -81/4)$.

$$f''(3) = 3(3^2) - 9$$
$$= 27 - 9 = 18 > 0$$

The graph of f is concave up when $x = 3$ and a *relative minimum* occurs at $(3, -81/4)$.

To summarize, relative minima occur on f at the points $(-3, -81/4)$ and $(3, -81/4)$; and a relative maximum occurs at $(0, 0)$. Figure 16.15 is a sketch of the graph of f.

❑

PRACTICE EXERCISE
In the Practice Exercise on page 752, you were asked to determine the nature of two critical points using the first-derivative test. Use the second-derivative test to confirm these results.

EXAMPLE 14 Examine the following function for any critical points and determine their nature.

$$f(x) = -10,000e^{-0.03x} - 120x + 10,000$$

Figure 16.15

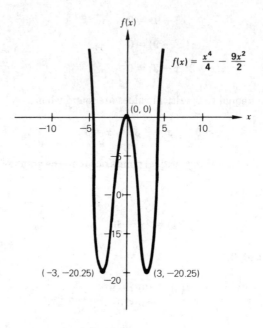

$f(x) = \dfrac{x^4}{4} - \dfrac{9x^2}{2}$

$(0, 0)$

$(-3, -20.25)$ $(3, -20.25)$

SOLUTION

If we find f' and set it equal to 0,

$$f'(x) = -10{,}000(-0.03)e^{-0.03x} - 120$$
$$= 300e^{-0.03x} - 120$$

$$300e^{-0.03x} - 120 = 0$$

when $300e^{-0.03x} = 120$

or when $e^{-0.03x} = \frac{120}{300} = 0.4$

To solve for x, take the natural logarithm of both sides of the equation.

$$-0.03x = \ln 0.4$$

Therefore, $-0.03x = -0.9163$

when $x = -0.9163/-0.03$

A critical value occurs when $x = 30.54$

The only critical point occurs when $x = 30.54$.
    Continuing with the second-derivative test, we have

$$f''(x) = 300(-0.03)e^{-0.03x}$$
$$= -9e^{-0.03x}$$

$$f''(30.54) = -9e^{-0.03(30.54)}$$
$$= -9e^{-0.9162}$$
$$= -9(0.4)$$
$$= -3.6 < 0$$

Figure 16.16

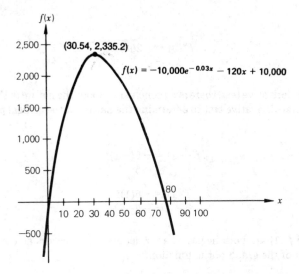

Therefore, a *relative maximum* occurs when $x = 30.54$. The corresponding value for $f(x)$ is

$$f(30.54) = -10,000e^{-0.03(30.54)} - 120(30.54) + 10,000$$
$$= -10,000(0.4) - 3,664.8 + 10,000 = 2,335.2$$

The relative maximum occurs at (30.54, 2,335.2). Figure 16.16 contains a sketch of the function.

◻

When the Second-Derivative Test Fails

If $f''(x^*) = 0$, the second derivative does not allow for any conclusion about the behavior of f at x^*. Consider the following example.

EXAMPLE 15   Examine the following function for any critical points and determine their nature.

$$f(x) = -x^5$$

SOLUTION

$$f'(x) = -5x^4$$

Setting f' equal to 0,

$$-5x^4 = 0$$

when

$$x = 0$$

Thus, a critical value exists for f when $x = 0$ and there is a critical point at (0, 0). Continuing with the second-derivative test, we get

$$f''(x) = -20x^3$$

At $x = 0$

$$f''(0) = -20(0)^3$$
$$= 0$$

Using the second-derivative test, there is no conclusion about the nature of the critical point. We can use the first-derivative test to determine the nature of the critical point. If $x_l = -1$ and $x_r = 1$, then

$$f'(-1) = -5(-1)^4$$
$$= -5$$

$$f'(1) = -5(1)^4$$
$$= -5$$

Since $f'(-1)$ and $f'(1)$ are both negative, an inflection point occurs at $x = 0$. Figure 16.17 presents a sketch of the graph of the function.

Figure 16.17

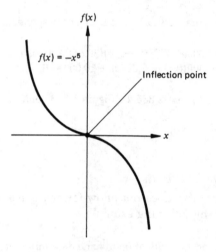

Inflection point

$f(x) = -x^5$

Higher-Order Derivative Test (Optional)

There are several ways to reach a conclusion about the nature of a critical point when the second-derivative test fails. One such method is the *higher-order derivative test*. Although not as intuitive as the first- or second-derivative tests, it is generally conclusive.

HIGHER-ORDER DERIVATIVE TEST

I   *Given a critical point [x*, f(x*)] on f, find the lowest-order derivative whose value is nonzero at the critical value x*. Denote this derivative as $f^{(n)}(x)$, where n is the order of the derivative.*

> **II** *If the order n of this derivative is even, $f(x^*)$ is a relative maximum if $f^{(n)}(x^*) < 0$ and a relative minimum if $f^{(n)}(x^*) > 0$.*
>
> **III** *If the order n of this derivative is odd, the critical point is an inflection point.*

EXAMPLE 16 Identify any critical points and determine their nature if

$$f(x) = (x - 2)^4$$

SOLUTION

First, we find f'.

$$f'(x) = 4(x - 2)^3(1)$$
$$= 4(x - 2)^3$$

If we set f' equal to 0

$$4(x - 2)^3 = 0$$

when

$$x = 2$$

At this critical value

$$f(2) = (2 - 2)^4$$
$$= (0)^4 = 0$$

Thus a critical point occurs at $(2, 0)$.

To determine the nature of the critical point, the second derivative is

$$f''(x) = 4(3)(x - 2)^2$$
$$= 12(x - 2)^2$$

Evaluating f'' at the critical value,

$$f''(2) = 12(2 - 2)^2$$
$$= 0$$

There is no conclusion based upon the second-derivative test. If we proceed using the higher-order derivative test, the third derivative is

$$f'''(x) = 24(x - 2)$$

and

$$f'''(2) = 24(2 - 2) = 0$$

Since $f'''(2) = 0$, there is no conclusion based on the third derivative.

The fourth derivative is

Figure 16.18

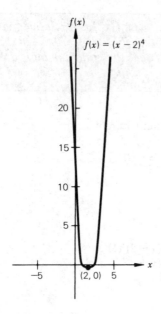

$$f^{(4)}(x) = 24$$

and
$$f^{(4)}(2) = 24$$

This is the lowest-order derivative not equaling 0 when $x = 2$. Since the order of the derivative ($n = 4$) is even, a relative maximum or minimum exists at $(2, 0)$. To determine which is the case, we look at the sign of $f^{(4)}(2)$. Because $f^{(4)}(2) > 0$, we can conclude that there is a relative minimum at $(2, 0)$. Figure 16.18 contains a sketch of the function.

❑

NOTE   The second-derivative test is actually a special case of the higher-order derivative test — the case where the lowest-order derivative not equaling 0 is the second derivative ($n = 2$).

PRACTICE EXERCISE
Apply the higher-order derivative test to determine the nature of the critical point in Example 15.

Section 16.2 Follow-up Exercises

For each of the following functions determine the location of all critical points and determine their nature.

1 $f(x) = 3x^2 - 48x + 100$
3 $f(x) = -10x^3 + 5$

2 $f(x) = x^3/3 - 5x^2 + 16x + 100$
4 $f(x) = x^2 - 8x + 4$

5 $f(x) = x^3/3 - 2.5x^2 + 4x$ **6** $f(x) = 5x^3 - 20$

7 $f(x) = 3x^4/4 - 75x^2/2$ **8** $f(x) = -x^4/4 + 9x^2/2$

9 $f(x) = -5x^5 - 10$ **10** $f(x) = x^5 - 2$

11 $f(x) = -x^2/2 + 6x + 3$ **12** $f(x) = -6x^2 - 36x + 10$

13 $f(x) = 4x^2/15 + 4$ **14** $f(x) = -x^3/10$

15 $f(x) = 2x^3/3 - x^2/2 - 10x$ **16** $f(x) = x^3/3 + 8x^2 + 60x$

17 $f(x) = 4x^5/5 - 324x$ **18** $f(x) = -2x^5/5 + 32x$

19 $f(x) = x^3/6 + 2x^2$ **20** $f(x) = x^3/3 + x^2/2 - 20x$

21 $f(x) = -2x^2 + x^4/4$ **22** $f(x) = -x^4/4 + 8x^2 + 5$

23 $f(x) = 2x^5/5 - x^4/4 - 5x^3$ **24** $f(x) = x^5/5 + 3x^4/4 - 4x^3/3$

25 $f(x) = x^5/5 - x$ **26** $f(x) = -2x^3 + 10.5x^2 + 12x$

27 $f(x) = x^6/6 - x^2/2 + 2$ **28** $f(x) = 4x^3/3 - 6x^2$

29 $f(x) = (x + 10)^4$ **30** $f(x) = (2x + 9)^3$

31 $f(x) = -(4x + 2)^3$ **32** $f(x) = (2x - 8)^3$

33 $f(x) = -(2x^2 - 8)^4$ **34** $f(x) = (x^2 - 16)^3$

35 $f(x) = e^x$ **36** $f(x) = e^{-x}$

37 $f(x) = 500e^{-0.10x} + 50x$ **38** $f(x) = -45e^{-0.2x} - 18x + 10$

39 $f(x) = e^{2x-5}$ **40** $f(x) = -100e^{-0.25x} - 50x$

41 $f(x) = xe^{-x}$ **42** $f(x) = -80e^{-0.10x} - 40x$

43 $f(x) = 40e^{-0.05x} + 6x - 10$ **44** $f(x) = 20e^{-0.05x} + 4x - 3$

45 $f(x) = 10 + \ln x$ **46** $f(x) = \ln x - 0.5x$

47 $f(x) = \ln(x^2 + 1) - x$ **48** $f(x) = \ln x - x^2/4$

49 $f(x) = 4x \ln x$ **50** $f(x) = x^2 \ln x$

51 $f(x) = \ln 5x - 10x$ **52** $f(x) = \ln 24x - x^3$

53 $f(x) = x^2 + x - \ln x$ **54** $f(x) = 0.5x^2 + 7x - 30 \ln x$

*55 $f(x) = x/(x^2 + 1)$ ***56** $f(x) = x(x + 2)^3$

*57 $f(x) = ax^2 + bx + c$, where $a > 0$, $b < 0$, $c > 0$

*58 $f(x) = ax^2 + bx + c$, where $a < 0$, $b < 0$, $c < 0$

*59 Original Equation Test It was mentioned in the last section that other techniques exist for determining the nature of critical points. One test involves comparing the value of $f(x^*)$ with the values of $f(x)$ just to the left and right of x^*. Refer to Fig. 16.11, and determine a set of rules which would allow one to distinguish among the four critical point possibilities.

*60 Compare the relative efficiencies associated with performing the original equation test and the first-derivative test of critical points.

*61 Compare the relative efficiencies associated with performing the first-derivative, second-derivative, and higher-order derivative tests of critical points.

*62 When the Second-Derivative Test Fails — An Alternative Given a critical value determined when $f'(x) = 0$ and failure of the second-derivative test to yield a conclusion, determine a set of rules which will yield a conclusion based upon checking the concavity conditions to the left and right of the critical value.

16.3 CURVE SKETCHING

Sketching functions is facilitated with the information we have acquired in this chapter. One can get a feeling for the general shape of the graph of a function without determining and plotting a large number of ordered pairs. This section discusses some of the key determinants of the shape of the graph of a function and illustrates curve-sketching procedures.

Key Data Points

In determining the general shape of the graph of a function, the following attributes are the most significant:

- ❏ Relative maxima and minima
- ❏ Inflection points
- ❏ x and y intercepts
- ❏ Ultimate direction

To illustrate, consider the function

$$f(x) = \frac{x^3}{3} - 4x^2 + 12x + 5$$

1 Relative Maxima and Minima To locate relative extrema on f, we find the first derivative

$$f'(x) = x^2 - 8x + 12$$

Setting f' equal to 0 yields

$$x^2 - 8x + 12 = 0$$

or

$$(x - 2)(x - 6) = 0$$

Critical values occur at $x = 2$ and $x = 6$. If we substitute these critical values into f,

$$f(2) = \frac{2^3}{3} - 4(2^2) + 12(2) + 5$$

$$= \tfrac{8}{3} - 16 + 24 + 5 = 15\tfrac{2}{3}$$

and

$$f(6) = \frac{6^3}{3} - 4(6^2) + 12(6) + 5$$

$$= 72 - 144 + 72 + 5 = 5$$

Thus, critical points exist at $(2, 15\tfrac{2}{3})$ and $(6, 5)$. Graphically, we know that zero slope conditions exist at these points, as shown in Fig. 16.19a.

The second derivative of f is

$$f''(x) = 2x - 8$$

To test the nature of the critical point $(2, 15\tfrac{2}{3})$,

$$f''(2) = 2(2) - 8 = -4 < 0$$

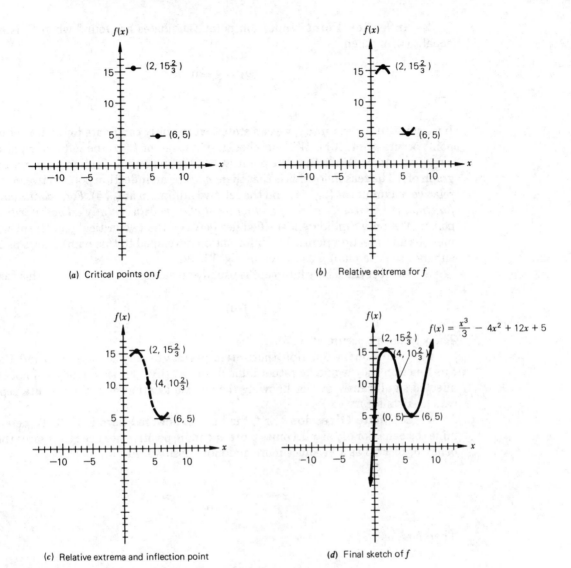

(a) Critical points on f

(b) Relative extrema for f

(c) Relative extrema and inflection point

(d) Final sketch of f

Figure 16.19   Development of sketch of $f(x) = x^3/3 - 4x^2 + 12x + 5$.

Therefore, a relative maximum occurs at $(2, 15\frac{2}{3})$. To test the nature of the critical point $(6, 5)$,

$$f''(6) = 2(6) - 8$$
$$= 4 > 0$$

Therefore a relative minimum occurs at $(6, 5)$. The information we have developed thus far allows us to develop the sketch of f to the degree shown in Fig. 16.19b.

2 Inflection Points Inflection point candidates are found when f'' is set equal to 0, or when

$$2x - 8 = 0$$

or

$$x = 4$$

If we substitute $x = 4$ into f, we can state that the only candidate for an inflection point occurs at $(4, 10\frac{1}{3})$. *Without* checking the sign of f'' to the left and right of $x = 4$, we can conclude that the point $(4, 10\frac{1}{3})$ is the only inflection point on the graph of f. The reason for this is that there *must be* an inflection point between the relative maximum at $(2, 15\frac{2}{3})$ and the relative minimum at $(6, 5)$. *For a continuous function, the concavity of the function must change between any adjacent critical points.* The only candidate identified lies between the two critical points; thus, it must be an inflection point. The information developed to this point allows us to enhance the sketch of f, as shown in Fig. 16.19c.

3 Intercepts The y intercept is usually an easy point to locate. In this case

$$f(0) = 5$$

The y intercept occurs at $(0, 5)$.

Depending on the function, the x intercepts may or may not be easy to find. For this function they would be rather difficult to identify. Our sketch of f will not be affected significantly by not knowing the precise location of the one x intercept which exists for f.

4 Ultimate Direction For f, the highest-powered term is $x^3/3$. To determine the behavior of f as x becomes more and more positive, we need to observe the behavior of $x^3/3$ as x becomes more and more positive. As

$$x \to +\infty \qquad \frac{x^3}{3} \to +\infty$$

Therefore, as

$$x \to +\infty \qquad f(x) = \frac{x^3}{3} - 4x^2 + 12x + 5 \to +\infty$$

Similarly, as

$$x \to -\infty \qquad \frac{x^3}{3} \to -\infty$$

and

$$f(x) = \frac{x^3}{3} - 4x^2 + 12x + 5 \to -\infty$$

Figure 16.19d incorporates the intercepts and the ultimate directions into our sketch.

EXAMPLE 17 (**Motivating Scenario**) Sketch the graph of the function

$$f(x) = \frac{x^4}{4} - \frac{8x^3}{3} + 8x^2$$

SOLUTION

1 Relative Maxima and Minima To locate relative extrema of f, we find the first derivative:

$$f'(x) = x^3 - 8x^2 + 16x$$

Setting f' equal to 0 yields

$$x^3 - 8x^2 + 16x = 0$$

$$x(x^2 - 8x + 16) = 0$$

or

$$x(x - 4)(x - 4) = 0$$

If the factors are set equal to 0, critical values are found at $x = 0$ and $x = 4$. The corresponding values of $f(x)$ are

$$f(0) = 0$$

and

$$f(4) = \frac{4^4}{4} - \frac{8(4^3)}{3} + 8(4^2)$$

$$= 64 - 170\tfrac{2}{3} + 128 = 21\tfrac{1}{3}$$

Therefore, critical points occur at $(0, 0)$ and $(4, 21\tfrac{1}{3})$.

Testing $x = 0$, we find

$$f''(x) = 3x^2 - 16x + 16$$

and

$$f''(0) = 16 > 0$$

A relative minimum occurs at $(0, 0)$.

Testing $x = 4$,

$$f''(4) = 3(4)^2 - 16(4) + 17$$
$$= 48 - 64 + 16 = 0$$

No conclusion can be drawn about $x = 4$ based upon the second derivative. Continuing with the higher-order derivative test,

$$f'''(x) = 6x - 16$$

and

$$f'''(4) = 6(4) - 16$$
$$= 8 > 0$$

Since the order of the derivative is odd, an inflection point occurs at $(4, 21\tfrac{1}{3})$.

    2 Inflection Points Candidates for inflection points are found by setting f'' equal to 0, or when

$$3x^2 - 16x + 16 = 0$$

$$(3x - 4)(x - 4) = 0$$

$$x = \tfrac{4}{3} \quad \text{and} \quad x = 4$$

We have already verified that an inflection point occurs at $(4, 21\tfrac{1}{3})$. Confirm for yourself that $(\tfrac{4}{3}, 8.69)$ is also an inflection point.

    3 Intercepts By computing $f(0) = (0)^4/4 - 8(0)^3/3 + 8(0)^2 = 0$, we conclude that the y intercept occurs at $(0, 0)$. To locate the x intercepts,

$$\frac{x^4}{4} - \frac{8x^3}{3} + 8x^2 = 0$$

when

$$x^2\left(\frac{x^2}{4} - \frac{8x}{3} + 8\right) = 0$$

One root to this equation is $x = 0$, suggesting that one x intercept is located at $(0, 0)$ (earlier, we should have observed that the y intercept is also an x intercept). Use the quadratic formula to verify that there are no roots for the equation

$$\frac{x^2}{4} - \frac{8x}{3} + 8 = 0$$

For f, the point $(0, 0)$ represents the only intercept.

    4 Ultimate Direction The ultimate behavior of $f(x)$ is linked to the behavior of the term $x^4/4$. As

$$x \to +\infty \qquad \frac{x^4}{4} \to +\infty \quad \text{and} \quad f(x) \to +\infty$$

As

$$x \to -\infty \qquad \frac{x^4}{4} \to +\infty \quad \text{and} \quad f(x) \to +\infty$$

Using the information gathered, we can sketch the approximate shape of f as shown in Fig. 16.20.

    ❑

Section 16.3 Follow-up Exercises

Sketch the graphs of the following functions.

1 $f(x) = x^3/3 - 5x^2 + 16x - 100$ 2 $f(x) = x^2 - 5x + 6$
3 $f(x) = x^4/4 - 25x^2/2$ 4 $f(x) = x^3/3 - 2.5x^2 + 4x$
5 $f(x) = (6x - 12)^3$ 6 $f(x) = x^6/6 - 8x^2 - 10$
7 $f(x) = -(x - 5)^3$ 8 $f(x) = (x^2 - 16)^4$

Figure 16.20

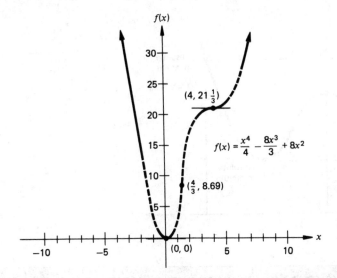

$$f(x) = \frac{x^4}{4} - \frac{8x^3}{3} + 8x^2$$

9 $f(x) = x^3/3 - 7x^2/2 - 30x$

11 $f(x) = -5x^5 + 100$

13 $f(x) = 2x^3/3 + x^2/2 - 10x$

15 $f(x) = -4x^5/5 + 324x - 250$

17 $f(x) = -e^{-x} - 10x$

19 $f(x) = \ln{(x^2 + 25)}$

10 $f(x) = -(x - 2)^4$

12 $f(x) = x^5 - 25$

14 $f(x) = x^3/3 - 8x^2 + 60x$

16 $f(x) = 2x^5/5 - 32x$

18 $f(x) = x^4/4 - 9x^2/2$

20 $f(x) = x^3/3 - 3.5x^2 - 30x$

16.4 RESTRICTED-DOMAIN CONSIDERATIONS

In this section we will examine procedures for identifying absolute maxima and minima when the domain of a function is restricted.

When the Domain Is Restricted

Very often in applied problems the domain is restricted. For example, if profit P is stated as a function of the number of units produced x, it is likely that x will be restricted to values such that $0 \le x \le x_u$. In this case x is restricted to nonnegative values (there is no production of negative quantities) which are less than or equal to some upper limit x_u. The value of x_u may reflect production capacity, as defined by limited labor, limited raw materials, or by the physical capacity of the plant itself.

In searching for the absolute maximum or absolute minimum of a function, consideration must be given not only to the relative maxima and minima of the function but also to the **endpoints** of the domain of the function. For example, examine the function graphed in Fig. 16.21. Note that the domain of the function is restricted to values between 0 and x_u and that the absolute maximum for f occurs at x_u, the right endpoint of the domain. The absolute minimum occurs at x_2, which is also a relative minimum on the function. The procedure for identifying absolute extrema follows.

Figure 16.21

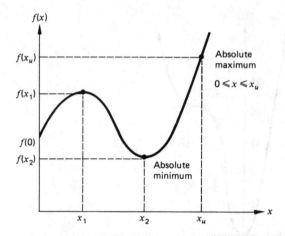

PROCEDURE FOR IDENTIFYING ABSOLUTE MAXIMUM AND MINIMUM POINTS

Given the continuous function f defined over the closed interval $[x_l,\ x_u]$:

I   *Locate all critical points $[x^*, f(x^*)]$ which lie within the domain of the function.[1] Exclude from consideration any critical values x^* which lie outside the domain.*

II *Compute the values of $f(x)$ at the two endpoints of the domain $[f(x_l)$ and $f(x_u)]$.*

III *Compare the values of $f(x^*)$ for all relevant critical points with $f(x_l)$ and $f(x_u)$. The absolute maximum is the largest of these values. The absolute minimum is the smallest of these values.*

EXAMPLE 18   Determine the locations and values of the absolute maximum and minimum for the function

$$f(x) = \frac{x^3}{3} - \frac{7x^2}{2} + 6x + 5$$

where $2 \le x \le 10$.

SOLUTION

❏ **Step I** The first derivative is

$$f'(x) = \frac{3x^2}{3} - \frac{14x}{2} + 6$$

$$= x^2 - 7x + 6$$

---

[1] Recall that critical points satisfy the condition $f'(x) = 0$ or $f'(x)$ undefined.

If f' is set equal to 0,

$$x^2 - 7x + 6 = 0$$

or

$$(x - 6)(x - 1) = 0$$

Thus,

$$x = 6 \quad \text{and} \quad x = 1$$

The only critical value *within the domain* of the function is $x = 6$.

$$f(6) = \frac{6^3}{3} - \frac{7(6^2)}{2} + 6(6) + 5$$

$$= 72 - 126 + 36 + 5 = -13$$

Thus, a critical point occurs at $(6, -13)$.
    To test $x = 6$,

$$f''(x) = 2x - 7$$

$$f''(6) = 2(6) - 7$$
$$= 5 > 0$$

Since $f''(6) > 0$, a relative minimum occurs at $(6, -13)$. Because $f'(x)$ is defined for all real x, no other critical values exist.

❏ **Step II** The values of $f(x)$ at the endpoints of the domain are

$$f(2) = \frac{2^3}{3} - \frac{7(2^2)}{2} + 6(2) + 5$$

$$= \tfrac{8}{3} - 14 + 12 + 5 = 5\tfrac{2}{3}$$

and

$$f(10) = \frac{(10)^3}{3} - \frac{7(10)^2}{2} + 6(10) + 5$$

$$= \frac{1,000}{3} - \frac{700}{2} + 65 = 48\tfrac{1}{3}$$

❏ **Step III** Comparing $f(2)$, $f(6)$, and $f(10)$, we find the absolute minimum of -13 occurs when $x = 6$ and the absolute maximum of $48\tfrac{1}{3}$ occurs when $x = 10$. Figure 16.22 presents a sketch of the function.

❏

Section 16.4 Follow-up Exercises

In the following exercises, determine the locations and values of the absolute maximum and absolute minimum for f.

1 $f(x) = 2x^2 - 4x + 5$, where $2 \le x \le 8$
2 $f(x) = -x^2 + 8x - 100$, where $-2 \le x \le 4$
3 $f(x) = x^3 - 12x^2$, where $2 \le x \le 10$

Figure 16.22

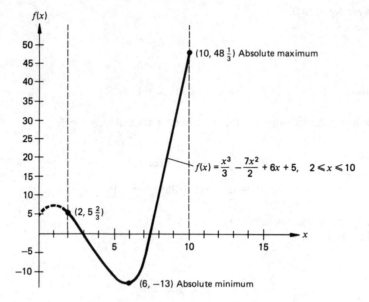

4 $f(x) = -2x^3 - 15x^2 + 10$, where $-6 \leq x \leq +2$

5 $f(x) = x^5/5 - x - 25$, where $3 \leq x \leq 8$

6 $f(x) = x^5/5 - 3x^4/4 + 2x^3/3 - 20$, where $0 \leq x \leq 5$

7 $f(x) = x^6/6 - x^5 + 2.5x^4$, where $0 \leq x \leq 4$

8 $f(x) = x^3 + 10$, where $1 \leq x \leq 5$ ·

9 $f(x) = -4x^2 + 6x - 10$, where $0 \leq x \leq 10$

10 $f(x) = x^3/3 - x^2/2 - 6x$, where $0 \leq x \leq 5$

11 $f(x) = x^4/4 - 4x^2 + 16$, where $5 \leq x \leq 10$

12 $f(x) = x^4/4 - 7x^3/3 + 5x^2$, where $0 \leq x \leq 4$

13 $f(x) = x^5/5 - 5x^4/4 - 14x^3/3 - 10$, where $0 \leq x \leq 6$

14 $f(x) = x^4/4 - 8x^2 + 25$, where $x \geq 0$

15 $f(x) = \ln(x^2 + 10)$, where $-1 \leq x \leq 4$

16 $f(x) = x^{2/3}$, where $0 \leq x \leq 4$

17 $f(x) = x^{1/2}$, where $4 \leq x \leq 16$

18 $f(x) = (x - 2)^{1/3}$, where $0 \leq x \leq 10$

❏ KEY TERMS AND CONCEPTS

absolute maximum (minimum)   746

concave functions   743

concavity   738

convex functions   743

critical points   746

critical values   747

decreasing function   735

first-derivative test   749

higher-order derivative test   758

increasing function   734

inflection point   738

relative maximum (minimum)   745

restricted-domain
     considerations   767

second-derivative test   752

ultimate direction   764

❏ ADDITIONAL EXERCISES

Section 16.1

For the following exercises, determine the intervals over which f is (a) increasing, (b) decreasing, (c) neither increasing nor decreasing, (d) concave up, and (e) concave down.

1 $f(x) = 2x^2 - 5x + 8$　　　　　　　**2** $f(x) = 20 - 4x + x^2$
3 $f(x) = x^3 - 27x$　　　　　　　　**4** $f(x) = 2x^5$
5 $f(x) = b$　　　　　　　　　　　　**6** $f(x) = 5x^2 - 20x + 100$
7 $f(x) = 4x^2 - 2x + 6$　　　　　　**8** $f(x) = x^3/3 - x^2 + x - 5$
9 $f(x) = x^3/3 + 2x^2 + 4x$　　　　**10** $f(x) = (x - 5)^4$
11 $f(x) = (x + 3)^5$　　　　　　　**12** $f(x) = (2x - 8)^3$

For the following exercises, identify the locations of any inflection points.

13 $f(x) = -x^5$　　　　　　　　　　**14** $f(x) = 3x^4/2$
15 $f(x) = 8x^3 - 2x^2 + 150$　　　**16** $f(x) = x^4/12 + x^3 + 4x^2$
17 $f(x) = x^5/20 - x^3/6$　　　　**18** $f(x) = (x + 4)^3$
19 $f(x) = (x - 1)^5$　　　　　　　**20** $f(x) = x^4/12 + x^3/6 - 3x^2 + 120$
21 $f(x) = x^3/6 - x^2 + 9$　　　　**22** $f(x) = x^6$
23 $f(x) = x^4/6 + 5x^3 + 12x^2 - 4$　　**24** $f(x) = 2x^6 - x^5$

25 $f(x) = \left(x - \dfrac{1}{2}\right)^4$　　　　　　　　**26** $f(x) = (2x - 7)^4$

Section 16.2

For the following functions, determine the location of all critical points and determine their nature.

27 $f(x) = 2x^2 - 16x + 30$　　　　**28** $f(x) = -x^2/2 + 8x + 7$
29 $f(x) = 4x^4$　　　　　　　　　　**30** $f(x) = x^4 - 25x^2/2$
31 $f(x) = x^3 - 4x^2 + 40$　　　　**32** $f(x) = -2x^3 + 3x^2/2 + 3x + 1$
33 $f(x) = x^5 - x^4 - x^3/3$　　　**34** $f(x) = (-x + 2)^6$
35 $f(x) = 3x^4 - 16x^3 + 24x^2 + 10$　　**36** $f(x) = x^4 - 20x^3 + 100x^2 + 80$
37 $f(x) = 2x + 50/x$　　　　　　　**38** $f(x) = 96\sqrt{x} - 6x$
39 $f(x) = 3x^3 - x^2/2 + 5x$　　　**40** $f(x) = 5x^3 - x^2 + 12x$
41 $f(x) = (2x - 5)^4$　　　　　　　**42** $f(x) = (x + 5)^3$
43 $f(x) = (x + 4)^5$　　　　　　　**44** $f(x) = (3x - 9)^4$
45 $f(x) = x^4/4 + x^3 + x^2 + 5$　　**46** $f(x) = -2x^3 + 7x^2 - 4x - 9$
47 $f(x) = 30x - e^x$　　　　　　　**48** $f(x) = 2.5x + e^{-0.5x}$
49 $f(x) = e^{(x - 0.2x^2)}$　　　　　**50** $f(x) = 40x - e^{0.1x} + 50$
51 $f(x) = 10x - e^{0.2x}$　　　　　**52** $f(x) = e^{(2x - 0.1x^2)}$
53 $f(x) = 40x - e^{2x}$　　　　　　**54** $f(x) = 3.5x - e^{1.5x}$
55 $f(x) = \ln 50x - 15x$　　　　　**56** $f(x) = 8x^2 \ln x$
57 $f(x) = 80x - 20 \ln x$　　　　**58** $f(x) = 0.5x^2 - 4x - 5 \ln x + 50$
59 $f(x) = 45x - 5 \ln x$　　　　　**60** $f(x) = \ln 20x - 2x$

Section 16.3

Sketch the graphs of the following functions.

61 $f(x) = x^3/3 - 7x^2/2 + 12x$

62 $f(x) = (10 - x)^3$

63 $f(x) = (x + 5)^3$

64 $f(x) = 2x^5/5 + x^4/4 - x^3 + 1$

65 $f(x) = -2x^3 + 7x^2 - 4x - 9$

66 $f(x) = x^4/4 + x^3 + x^2 + 5$

67 $f(x) = (x + 3)^3$

68 $f(x) = x^4 - 25x^2/2$

Section 16.4

For the following functions, determine the location and values of the absolute maximum and absolute minimum.

69 $f(x) = 3x^2 - 48x + 30$, where $0 \leq x \leq 10$

70 $f(x) = 2x^2 - 5x + 15$, where $-1 \leq x \leq 4$

71 $f(x) = 2x^3/3 + 3x^2 + 4x - 1$, where $-3 \leq x \leq 5$

72 $f(x) = x^3/3 - 7x^2/2 - 30x$, where $0 \leq x \leq 10$

73 $f(x) = 2x^5/5 - 27x^2$, where $-2 \leq x \leq 3$

74 $f(x) = -x^6 + x^4 + 2x^3/3$, where $-2 \leq x \leq 4$

75 $f(x) = x^4 + 5x^3 + 5.5x^2 + 6$, where $-1 \leq x \leq 1$

76 $f(x) = x^3 - 4x^2 + 5$, where $-3 \leq x \leq 5$

77 $f(x) = -2x^3 + 3x^2/2 + 3x + 1$, where $-2 \leq x \leq 4$

78 $f(x) = (-x + 2)^4$, where $0 \leq x \leq 3$

❏ CHAPTER TEST

1 Given $f(x) = x^3/3 - 3x^2 - 40x$, determine the values of x for which f is (*a*) increasing, (*b*) decreasing, and (*c*) neither increasing or decreasing.

2 Sketch a portion of a function for which (*a*) $f'(x) < 0$ and $f''(x) > 0$, and (*b*) $f'(x) > 0$ and $f''(x) > 0$.

3 For the following functions, determine the locations of all critical points and determine their nature.

(*a*) $f(x) = \dfrac{x^3}{3} - 2x^2 - 21x + 5$

(*b*) $f(x) = e^{-x^2 + 3}$

4 Given

$$f(x) = \frac{x^4}{12} - \frac{x^3}{3} - 4x^2$$

identify the locations of all inflection points.

5 Determine the locations and values of the absolute maximum and absolute minimum for

$$f(x) = -\frac{x^3}{3} - \frac{x^2}{2} + 2x - 10, \qquad -2 \le x \le 3$$

6 Sketch the function $f(x) = (x + 4)^5$.

OPTIMIZATION: APPLICATIONS

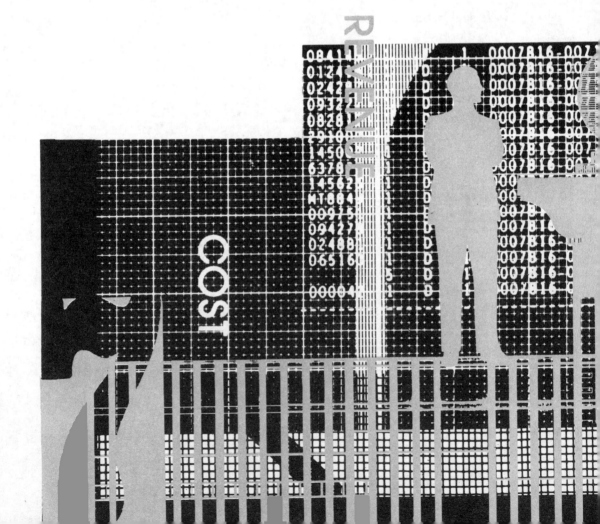

CHAPTER OBJECTIVES

❏ Illustrate a wide variety of applications of optimization procedures

❏ Reinforce skills in problem formulation

❏ Reinforce skills of interpretation of mathematical results

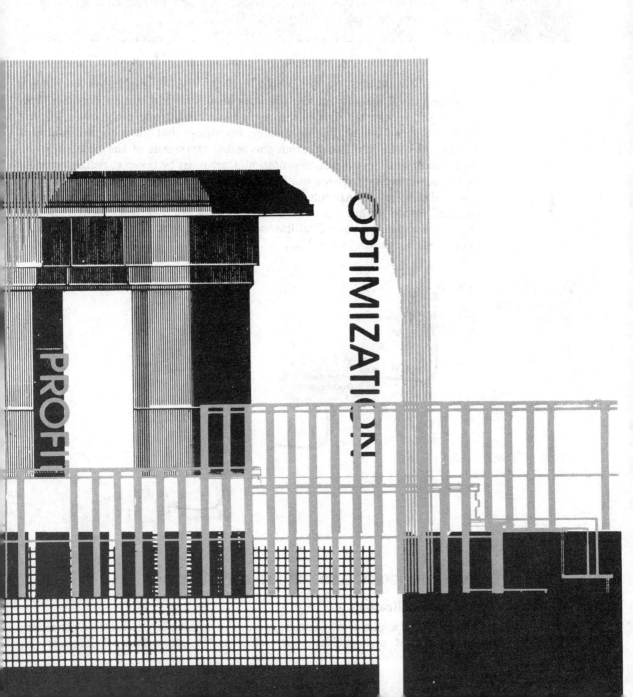

Chapter 16 provided the tools of classical optimization. That is, it gave us a method for examining functions in order to locate maximum and minimum points. This chapter will be devoted to illustrating the use of these procedures in a variety of applications. As you begin this chapter, remember that these applied problems usually require a translation from the verbal statement of the problem to an appropriate mathematical representation. Care must be taken to define variables (unknowns) precisely. Once a mathematically derived solution has been found, an essential element in the problem-solving process is the interpretation of the mathematical result, within the context of the application setting. As you proceed through this chapter, you will utilize some or all of the stages of this problem-solving process, as shown in Fig. 17.1.

Figure 17.1
Problem-solving process.

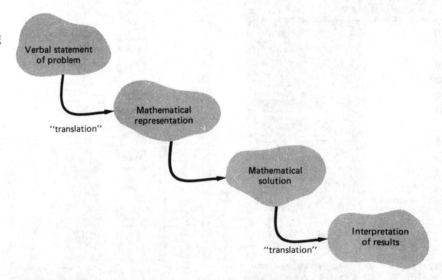

17.1 REVENUE, COST, AND PROFIT APPLICATIONS

Revenue Applications

The following applications focus on revenue maximization. Recall that the money which flows *into* an organization from either selling products or providing services is referred to as *revenue.* The most fundamental way of computing total revenue from selling a product (or service) is

> Total revenue = (price per unit)(quantity sold)

An assumption in this relationship is that the selling price is the same for all units sold.

EXAMPLE 1   The demand for the product of a firm varies with the price that the firm charges for the product. The firm estimates that annual total revenue R (stated in $1,000s) is a function of the price p (stated in dollars). Specifically,

$$R = f(p) = -50p^2 + 500p$$

(a) Determine the price which should be charged in order to maximize total revenue.
(b) What is the maximum value of annual total revenue?

SOLUTION

(a) From Chap. 6 we should recognize that the revenue function is quadratic. It will graph as a parabola which is concave down. Thus the maximum value of R will occur at the vertex. The first derivative of the revenue function is

$$f'(p) = -100p + 500$$

If we set $f'(p)$ equal to 0,

$$-100p + 500 = 0$$
$$-100p = -500$$

or a critical value occurs when

$$p = 5$$

Although we know that a relative maximum occurs when $p = 5$ (because of our knowledge of quadratic functions), let's formally verify this using the second-derivative test:

$$f''(p) = -100 \quad \text{and} \quad f''(5) = -100 < 0$$

Therefore, a relative maximum of f occurs at $p = 5$.

(b) The maximum value of R is found by substituting $p = 5$ into f, or

$$f(5) = -50(5^2) + 500(5)$$
$$= -1,250 + 2,500 = 1,250$$

Thus, annual total revenue is expected to be maximized at $1,250 (1,000s) or $1.25 million when the firm charges $5 per unit. Figure 17.2 presents a sketch of the revenue function.

❑

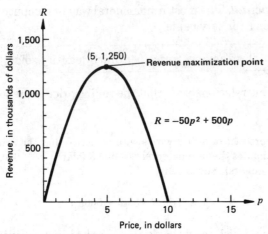

Figure 17.2   Quadratic revenue function.

NOTE   This has been mentioned earlier in the text when we have dealt with applications; however, it is worth repeating. It is quite common for students to work through a word problem, find the solution, but have no ability to interpret the results within the framework of the application. If you become caught up in the mechanics of finding a solution and temporarily lose your frame of reference regarding the original problem, reread the problem, making special note of how the variables are defined. Also review the specific questions asked in the problem. This should assist in reminding you of the objectives and the direction in which you should be heading.

EXAMPLE 2   **(Public Transportation Management)** The transit authority for a major metropolitan area has experimented with the fare structure for the city's public bus system. It has abandoned the zone fare structure in which the fare varies depending on the number of zones through which a passenger passes. The new system is a fixed-fare system in which a passenger may travel between any two points in the city for the same fare.

The transit authority has surveyed citizens to determine the number of persons who would use the bus system if the fare was fixed at different amounts. From the survey results, systems analysts have determined an approximate demand function which expresses the ridership as a function of the fare charged. Specifically, the demand function is

$$q = 10,000 - 125p$$

where q equals the average number of riders per hour and p equals the fare in cents.
(a) Determine the fare which should be charged in order to maximize hourly bus fare revenue.
(b) What is the expected maximum revenue?
(c) How many riders per hour are expected under this fare?

SOLUTION

(a) The first step is to determine a function which states hourly revenue as a function of the fare p. The reason for selecting p as the independent variable is that the question was to determine the fare which would result in maximum total revenue. Also, the fare is a *decision variable*—a variable whose value can be decided by the transit authority management.

The general expression for total revenue is, as stated before,

$$R = pq$$

But in this form, R is stated as a function of two variables—p and q. *At this time* we cannot deal with the optimization of functions involving more than one independent variable. The demand function, however, establishes a relationship between the variables p and q which allows us to transform the revenue function into one where R is stated as a function of one independent variable p. The right side of the demand function states q in terms of p. If we substitute this expression for q into the revenue function, we get

$$R = f(p)$$
$$= p(10,000 - 125p)$$

or

$$R = 10,000p - 125p^2$$

The first derivative is

$$f'(p) = 10,000 - 250p$$

If the derivative is set equal to 0,

$$10,000 - 250p = 0$$

$$10,000 = 250p$$

and a critical value occurs when

$$40 = p$$

The second derivative is found and evaluated at $p = 40$ to determine the nature of the critical point:

$$f''(p) = -250$$

$$f''(40) = -250 < 0$$

Thus, a relative maximum occurs for f when $p = 40$. Since f is everywhere concave downward, the interpretation of this result is that hourly revenue will be maximized when a fixed fare of $0.40 (40 cents) is charged.

(b)
$$f(40) = 10,000(40) - 125(40)^2$$
$$= 400,000 - 200,000 = 200,000$$

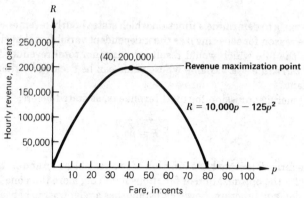

Figure 17.3   Quadratic revenue function.

Since the fare is stated in cents, the maximum expected hourly revenue is 200,000 cents, or $2,000.

(*c*) The average number of riders expected each hour with this fare is found by substituting the fare into the demand function, or

$$q = 10{,}000 - 125(40)$$
$$= 10{,}000 - 5{,}000$$
$$= 5{,}000 \text{ riders per hour}$$

Figure 17.3 presents a sketch of the hourly revenue function.

$\square$

Cost Applications

As mentioned earlier, costs represent cash *outflows* for an organization. Most organizations seek ways to minimize these outflows. This section presents applications which deal with the minimization of some measure of cost.

EXAMPLE 3   **(Inventory Management)** A common problem in organizations is determining how much of a needed item should be kept on hand. For retailers, the problem may relate to how many units of each product should be kept in stock. For producers, the problem may involve how much of each raw material should be kept available. This problem is identified with an area called *inventory control,* or *inventory management.* Concerning the question of how much "inventory" to keep on hand, there may be costs associated with having too little or too much inventory on hand.

A retailer of motorized bicycles has examined cost data and has determined a cost function which expresses the annual cost of purchasing, owning, and maintaining inventory as a function of the size (number of units) of each order it places for the bicycles. The cost function is

$$C = f(q) = \frac{4{,}860}{q} + 15q + 750{,}000$$

where C equals annual inventory cost, stated in dollars, and q equals the number of cycles ordered each time the retailer replenishes the supply.
(a) Determine the order size which minimizes annual inventory cost.
(b) What is minimum annual inventory cost expected to equal?

SOLUTION

(a) The first derivative is

$$f'(q) = -4,860q^{-2} + 15$$

If f' is set equal to 0,

$$-4,860q^{-2} + 15 = 0$$

when

$$\frac{-4,860}{q^2} = -15$$

Multiplying both sides by q^2 and dividing both sides by -15 yields

$$\frac{4,860}{15} = q^2$$

$$324 = q^2$$

Taking the square root of both sides, critical values exist at

$$\pm 18 = q$$

The value $q = -18$ is meaningless in this application (negative order quantities are not possible). The nature of the only meaningful critical point ($q = 18$) is checked by finding f'':

$$f''(q) = 9,720q^{-3}$$

$$= \frac{9,720}{q^3}$$

At the critical value,

$$f''(18) = \frac{9,720}{(18)^3}$$

$$= 1.667 > 0$$

Note that $f''(q) > 0$ for $q > 0$. Therefore the graph of f is everywhere concave upward. Thus, the minimum value for f occurs when $q = 18$. Annual inventory costs will be minimized when 18 bicycles are ordered each time the retailer replenishes the supply.

(b) Minimum annual inventory costs are determined by calculating $f(18)$, or

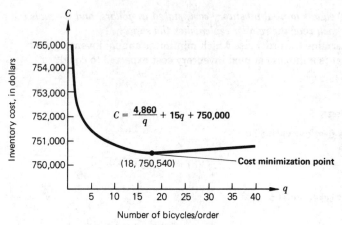

Figure 17.4   Inventory cost function.

$$f(18) = \frac{4,860}{18} + 15(18) + 750,000$$

$$= 270 + 270 + 750,000 = \$750,540$$

Figure 17.4 presents a sketch of the cost function. (The end of chapter minicase discusses assumptions which underlie the inventory cost function in this example.)

EXAMPLE 4   **(Minimizing Average Cost per Unit)** The total cost of producing q units of a certain product is described by the function

$$C = 100,000 + 1,500q + 0.2q^2$$

where C is the total cost stated in dollars. Determine the number of units of q that should be produced in order to minimize the *average cost per unit*.

SOLUTION

Average cost per unit is calculated by dividing the total cost by the number of units produced. For example, if the total cost of producing 10 units of a product equals $275, the average cost per unit is $275/10 = $27.50. Thus, the function representing average cost per unit in this example is

$$\overline{C} = f(q) = \frac{C}{q} = \frac{100,000}{q} + 1,500 + 0.2q$$

The first derivative of the average cost function is

$$f'(q) = -100,000q^{-2} + 0.2$$

If f' is set equal to 0,

$$0.2 = \frac{100,000}{q^2}$$

or

$$q^2 = \frac{100{,}000}{0.2}$$

$$= 500{,}000$$

Finding the square root of both sides, we have critical values at

$$q = \pm 707.11 \text{ (units)}$$

The value $q = -707.11$ is meaningless in this application since production level, q, must be nonnegative.

The nature of the one relevant critical point is tested with the second-derivative test:

$$f''(q) = 200{,}000q^{-3}$$

$$= \frac{200{,}000}{q^3}$$

$$f''(707.11) = \frac{200{,}000}{(707.11)^3}$$

$$= 0.00056 > 0$$

The second derivative $f''(p)$ is positive for $q > 0$, which means that the graph of f is concave upward for $q > 0$. Thus, the minimum value of f occurs when $q = 707.11$. This minimum average cost per unit is

$$f(707.11) = \frac{100{,}000}{707.11} + 1500 + 0.2(707.11)$$

$$= 141.42 + 1{,}500 + 141.42 = \$1{,}782.84$$

Figure 17.5 is a sketch of the average cost function.

Figure 17.5
Average cost
function.

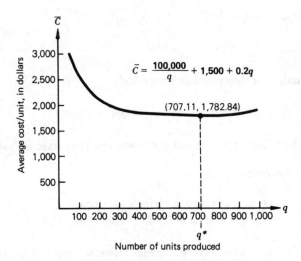

$$\bar{C} = \frac{100{,}000}{q} + 1{,}500 + 0.2q$$

(707.11, 1,782.84)

Average cost/unit, in dollars

q

q^*

Number of units produced

PRACTICE EXERCISE
For Example 4, what is the total cost of production at this level of output? What are two different ways in which this figure can be computed? *Answer:* $1,260,663.90.

Profit Applications

This section contains two examples which deal with profit maximization.

EXAMPLE 5   (**Sales Force Allocation**) Example 1 in Chap. 6 discussed the *law of diminishing returns* as an illustration of a nonlinear function. A major cosmetic and beauty supply firm, which specializes in a door-to-door sales approach, has found that the response of sales to the allocation of additional sales representatives behaves according to the law of diminishing returns. For one regional sales district, the company estimates that *annual profit P, stated in hundreds of dollars, is a function of the number of sales representatives x assigned to the district.* Specifically, the function relating these two variables is

$$P = f(x) = -12.5x^2 + 1,375x - 1,500$$

(*a*) What number of representatives will result in maximum profit for the district?
(*b*) What is the expected maximum profit?

SOLUTION
(*a*) The derivative of the profit function is

$$f'(x) = -25x + 1,375$$

If f' is set equal to 0,

$$-25x = -1,375$$

or a critical value occurs at

$$x = 55$$

Checking the nature of the critical point, we find

$$f''(x) = -25 \quad \text{and} \quad f''(55) = -25 < 0$$

Since the graph of f is everywhere concave downward, the maximum value of f occurs when $x = 55$.
(*b*) The expected maximum profit is

$$f(55) = -12.5(55)^2 + 1,375(55) - 1,500$$
$$= -37,812.5 + 75,625 - 1,500 = 36,312.5$$

We can conclude that annual profit will be maximized at a value of $36,312.5 (100s), or

$3,631,250, if 55 representatives are assigned to the district. Figure 17.6 presents a sketch of the profit function.

$\square$

EXAMPLE 6

(Solar Energy) A manufacturer has developed a new design for solar collection panels. Marketing studies have indicated that annual demand for the panels will depend on the price charged. The demand function for the panels has been estimated as

$$q = 100,000 - 200p \qquad (17.1)$$

where q equals the number of units demanded each year and p equals the price in dollars. Engineering studies indicate that the total cost of producing q panels is estimated well by the function

$$C = 150,000 + 100q + 0.003q^2 \qquad (17.2)$$

Formulate the profit function $P = f(q)$ which states the annual profit P as a function of the number of units q which are produced and sold.

SOLUTION

We have been asked to develop a function which states profit P as a function of q. As opposed to Example 5, *we must construct* the profit function. The total cost function [Eq. (17.2)] is stated in terms of q. However, we need to formulate a total revenue function stated in terms of q. The basic structure for computing total revenue is

Figure 17.6
Profit function.

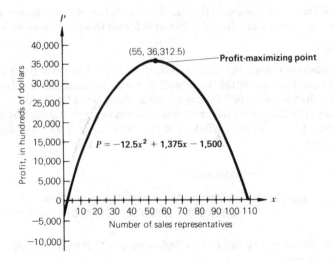

$P = -12.5x^2 + 1,375x - 1,500$

(55, 36,312.5) — Profit-maximizing point

Number of sales representatives

$$R = pq \qquad \text{(17.3)}$$

Because we want R to be stated in terms of q, we can replace p in Eq. (17.3) with an equivalent expression derived from the demand function. Solving for p in Eq. (17.1) gives

$$200p = 100{,}000 - q$$

or
$$p = 500 - 0.005q \qquad \text{(17.4)}$$

We can substitute the right side of this equation into Eq. (17.3) to yield the revenue function

$$R = (500 - 0.005q)q$$
$$= 500q - 0.005q^2$$

Now that both the revenue and cost functions have been stated in terms of q, the profit function can be defined as

$$P = f(q)$$
$$= R - C$$
$$= 500q - 0.005q^2 - (150{,}000 + 100q + 0.003q^2)$$
$$= 500q - 0.005q^2 - 150{,}000 - 100q - 0.003q^2$$

or
$$P = -0.008q^2 + 400q - 150{,}000$$

☐

PRACTICE EXERCISE
For Example 6, determine (*a*) the number of units q that should be produced to maximize annual profit; (*b*) the price that should be charged for each panel to generate a demand equal to the answer in part *a*; and (*c*) the maximum annual profit. *Answer:* (*a*) $q = 25{,}000$ units, (*b*) $p = \$375$, (*c*) $\$4{,}850{,}000$.

EXAMPLE 7   **(Restricted Domain)** Assume in the last example that the manufacturer's annual production capacity is 20,000 units. Re-solve Example 6 with this added restriction.

SOLUTION

With the added restriction, the domain of the function is defined as $0 \le q \le 20{,}000$. From Sec. 16.4, it should be recalled that we must compare the values of $f(q)$ at the endpoints of the domain with the values of $f(q^*)$ for any q^* value, where $0 \le q^* \le 20{,}000$.

The only critical point on the profit function occurs at $q = 25{,}000$, which is outside the domain. Thus, profit will be maximized at one of the endpoints. Evaluating $f(q)$ at the endpoints, we find

$$f(0) = -150{,}000$$

and
$$f(20{,}000) = -0.008(20{,}000)^2 + 400(20{,}000) - 150{,}000$$
$$= -3{,}200{,}000 + 8{,}000{,}000 - 150{,}000 = 4{,}650{,}000$$

Profit is maximized at a value of $\$4{,}650{,}000$ when $q = 20{,}000$, or when the manufacturer operates at capacity.

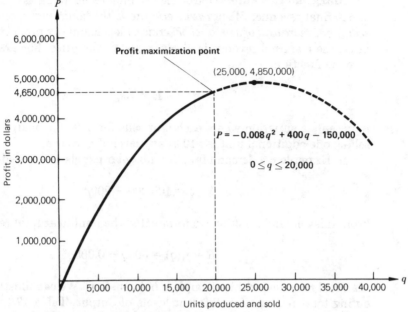

Figure 17.7 Profit function/restricted domain.

The price which should be charged is found by substituting $q = 20{,}000$ into Eq. (17.4), or

$$p = 500 - 0.005(20{,}000)$$
$$= 500 - 100 = \$400$$

Figure 17.7 presents a sketch of the profit function.

❑

Marginal Approach to Profit Maximization

An alternative approach to finding the profit maximization point involves ***marginal analysis.*** Popular among economists, marginal analysis examines *incremental effects* on profitability. Given that a firm is producing a certain number of units each year, marginal analysis would be concerned with the effect on profit if *one* additional unit is produced and sold.

To utilize the marginal approach to profit maximization, the following conditions must hold.

REQUIREMENTS FOR USING THE MARGINAL APPROACH

I *It must be possible to identify the total revenue function* and *the total cost function, separately.*

II *The revenue and cost functions must be stated in terms of the level of output or number of units produced and sold.*

Marginal Revenue One of the two important concepts in marginal analysis is marginal revenue. *Marginal revenue is the additional revenue derived from selling one more unit of a product or service.* If each unit of a product sells at the same price, the marginal revenue is always equal to the price. For example, the linear revenue function

$$R = 10q$$

represents a situation where each unit sells for $10. The marginal revenue from selling one additional unit is $10 at any level of output q.

In Example 6 a demand function for solar panels was stated as

$$q = 100,000 - 200p$$

From this demand function we formulated the nonlinear total revenue function

$$R = f_1(q) = 500q - 0.005q^2 \qquad (17.5)$$

Marginal revenue for this example is not constant. We can illustrate this by computing total revenue for different levels of output. Table 17.1 illustrates these calculations for selected values of q. The third column represents the marginal revenue associated with moving from one level of output to another. Note that although the differences are slight, the marginal revenue values are changing at each different level of output.

TABLE 17.1	**Computation of Marginal Revenue**	
Level of Output q	**Total Revenue** $f_1(q)$	**Marginal Revenue** $\Delta R = f_1(q) - f_1(q - 1)$
100	$49,950.00	
101	$50,448.995	$498.995
102	$50,947.98	$498.985
103	$51,446.955	$498.975

For a total revenue function $R(q)$, the derivative $R'(q)$ represents the instantaneous rate of change in total revenue given a change in the number of units sold. R' also represents a general expression for the slope of the graph of the total revenue function. For purposes of marginal analysis, the derivative is used to represent the marginal revenue, or

$$\boxed{MR = R'(q)} \qquad (17.6)$$

The derivative, as discussed in Chap. 15, provides an approximation to actual changes in the value of a function. As such, R' can be used to approximate the marginal revenue from selling the next unit. If we find R' for the revenue function in Eq. (17.5),

$$R'(q) = 500 - 0.010q$$

To approximate the marginal revenue from selling the 101st unit, we evaluate R' at $q = 100$, or

$$R'(100) = 500 - 0.010(100)$$
$$= 500 - 1 = 499$$

This is a very close approximation to the actual value ($498.995) for marginal revenue shown in Table 17.1.

Marginal Cost The other important concept in marginal analysis is marginal cost. *Marginal cost is the additional cost incurred as a result of producing and selling one more unit of a product or service.* Linear cost functions assume that the variable cost per unit is constant; for such functions the marginal cost is the same at any level of output. An example of this is the cost function

$$C = 150,000 + 3.5q$$

where variable cost per unit is constant at $3.50. The marginal cost for this cost function is always $3.50.

A nonlinear cost function is characterized by variable marginal costs. This can be illustrated by the cost function

$$C = f_2(q) = 150,000 + 100q + 0.003q^2 \tag{17.7}$$

which was used in Example 6. We can illustrate that the marginal costs do fluctuate at different levels of output by computing marginal cost values for selected values of q. This computation is illustrated in Table 17.2.

For a total cost function $C(q)$, the derivative $C'(q)$ represents the instantaneous rate of change in total cost given a change in the number of units produced. $C'(q)$ also represents a general expression for the slope of the graph of the total cost function. For purposes of marginal analysis, the derivative is used to represent the marginal cost, or

$$\boxed{MC = C'(q)} \tag{17.8}$$

TABLE 17.2	**Computation of Marginal Cost**		
	Level of Output q	Total Cost $f_2(q)$	Marginal Cost $\Delta C = f_2(q) - f_2(q-1)$
	100	$160,030.00	
	101	$160,130.603	$100.603
	102	$160,231.212	$100.609
	103	$160,331.827	$100.615

As with R', C' can be used to approximate the marginal cost associated with producing the next unit. The derivative of the cost function in Eq. (17.7) is

$$C'(q) = 100 + 0.006q$$

To approximate the marginal cost from producing the 101st unit, we evaluate C' at $q = 100$, or

$$C'(100) = 100 + 0.006(100)$$
$$= \$100.60$$

If we compare this value with the actual value (\$100.603) in Table 17.2, we see that the two values are very close.

Marginal Profit Analysis   As indicated earlier, marginal profit analysis is concerned with the effect on profit if one additional unit of a product is produced and sold. As long as the additional revenue brought in by the next unit exceeds the cost of producing and selling that unit, there is a net profit from producing and selling that unit and total profit increases. If, however, the additional revenue from selling the next unit is exceeded by the cost of producing and selling the additional unit, there is a net loss from that next unit and total profit decreases. A rule of thumb concerning whether or not to produce an additional unit (assuming profit is of greatest importance) is given next.

RULE OF THUMB: SHOULD AN ADDITIONAL UNIT BE PRODUCED?

I   *If MR > MC, produce the next unit.*
II *If MR < MC, do not produce the next unit.*

For many production situations, the marginal revenue exceeds the marginal cost at lower levels of output. As the level of output (quantity produced) increases, the amount by which marginal revenue exceeds marginal cost becomes smaller. Eventually, a level of output is reached at which $MR = MC$. Beyond this point $MR < MC$, and total profit begins to decrease with added output. Thus, if the point can be identified where $MR = MC$ for the last unit produced and sold, total profit will be maximized. This profit maximization level of output can be identified by the following condition.

PROFIT MAXIMIZATION CRITERION
Produce to the level of output where

$$MR = MC \qquad\qquad (17.9)$$

Stated in terms of derivatives, this criterion suggests producing to the point where

$$\boxed{R'(q) = C'(q)} \qquad \text{(17.10)}$$

This equation is a natural result of finding the point where the profit function is maximized, i.e., set the derivative of

$$P(q) = R(q) - C(q)$$

equal to 0 and solve for q.

$$P'(q) = R'(q) - C'(q)$$

and $$P'(q) = 0$$

when $$R'(q) - C'(q) = 0$$

or $$R'(q) = C'(q)$$

Let q^* be a value where $R'(q) = C'(q)$. The second derivative of P is $P''(q) = R''(q) - C''(q)$. By the second derivative test, profit will be maximized at $q = q^*$ provided

$$P''(q^*) < 0$$

or $$R''(q^*) - C''(q^*) < 0$$

or $$R''(q^*) < C''(q^*)$$

If $R''(q) < C''(q)$ for all values of $q > 0$, then profit has an absolute maximum value of $q = q^*$.

> **SUFFICIENT CONDITION FOR PROFIT MAXIMIZATION**
> Given a level of output q^* where $R'(q) = C'(q)$ (or $MR = MC$), producing q^* will result in profit maximization if
>
> $$R''(q^*) < C''(q^*) \qquad \text{(17.11)}$$

EXAMPLE 8

Re-solve Example 6 using the marginal approach.

SOLUTION

In Example 6

$$R = 500q - 0.005q^2$$

and $$C = 150{,}000 + 100q + 0.003q^2$$

Because the revenue and cost functions are distinct, and both are stated in terms of the level of output q, the two requirements for conducting marginal analysis are satisfied. We have already determined that

$$R'(q) = 500 - 0.01q$$

and
$$C'(q) = 100 + 0.006q$$

Therefore,
$$R'(q) = C'(q)$$

when
$$500 - 0.01q = 100 + 0.006q$$

$$-0.016q = -400$$

or
$$q^* = 25,000$$

Since
$$R''(q^*) = -0.01 \quad \text{and} \quad C''(q^*) = 0.006$$

$$R''(q^*) < C''(q^*)$$

or
$$-0.01 < 0.006$$

and there is a relative maximum on the profit function when $q = 25,000$. Figure 17.8 presents the graphs of $R(q)$ and $C(q)$.

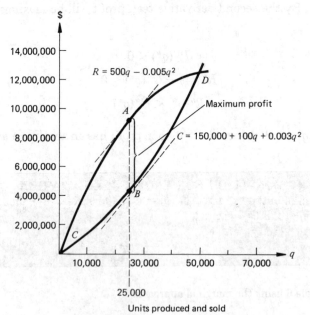

Figure 17.8 Marginal analysis: profit maximization.

Take a moment to examine Fig. 17.8. The following observations are worth noting:

1 *Points C and D represent points where the revenue and cost functions intersect. These represent break-even points.*
2 *Between points C and D the revenue function is above the cost function, indicating that total revenue is greater than total cost and a profit will be earned within this interval. For levels of output to the right of D, the cost function lies above the revenue function, indicating that total cost exceeds total revenue and a negative profit (loss) will result.*

3 *The vertical distance separating the graphs of the two functions represents the profit or loss, depending on the level of output.*

4 *In the interval $0 \leq q \leq 25{,}000$, the slope of the revenue function is positive and greater than the slope of the cost function. Stated in terms of MR and MC, MR > MC in this interval.*

5 *Also, in the interval $0 \leq q \leq 25{,}000$, the vertical distance separating the two curves becomes greater, indicating that profit is increasing on the interval.*

6 *At $q = 25{,}000$ the slopes at points A and B are the same, indicating that MR = MC. Also, at $q = 25{,}000$ the vertical distance separating the two curves is greater than at any other point in the profit region; thus, this is the point of profit maximization.*

7 *For $q > 25{,}000$ the slope of the revenue function is positive but less positive than that for the cost function. Thus, MR < MC and for each additional unit profit decreases, actually resulting in a loss beyond point D.*

EXAMPLE 9

In Example 5 we were asked to determine the number of sales representatives x which would result in maximum profit P for a cosmetic and beauty supply firm. The profit function was stated as

$$P = f(x) = -12.5x^2 + 1{,}375x - 1{,}500$$

Using the marginal approach, determine the number of representatives which will result in maximum profit for the firm.

SOLUTION

We cannot use the marginal approach in this example because we cannot identify the total revenue and total cost functions which were combined to form the profit function! Requirement 1 for using marginal analysis is not satisfied.

EXAMPLE 10

Figure 17.9 illustrates a sketch of a linear revenue function and a nonlinear cost function. To the left of q^*, the slope of the revenue function exceeds the slope of the cost function,

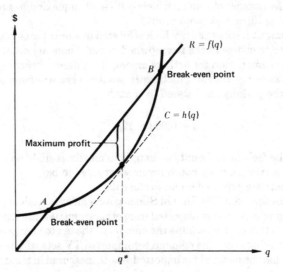

Figure 17.9 Linear revenue/quadratic cost functions.

indicating that $MR > MC$. At q^* the slopes of the two functions are the same. The vertical distance separating the two functions is greater at q^* than for any other value of q between points A and B. Points A and B are break-even points.

◻

Section 17.1 Follow-up Exercises

1 A firm has determined that total revenue is a function of the price charged for its product. Specifically, the total revenue function is

$$R = f(p) = -10p^2 + 1,750p$$

where p equals the price in dollars.
(a) Determine the price p which results in maximum total revenue.
(b) What is the maximum value for total revenue?

2 The demand function for a firm's product is

$$q = 150,000 - 75p$$

where q equals the number of units demanded and p equals the price in dollars.
(a) Determine the price which should be charged to maximize total revenue.
(b) What is the maximum value for total revenue?
(c) How many units are expected to be demanded?

3 The annual profit for a firm depends upon the number of units produced. Specifically, the function which describes the relationship between profit P (stated in dollars) and the number of units produced x is

$$P = -0.01x^2 + 5,000x - 25,000$$

(a) Determine the number of units x which will result in maximum profit.
(b) What is the expected maximum profit?

4 **Beach Management** A community which is located in a resort area is trying to decide on the parking fee to charge at the town-owned beach. There are other beaches in the area, and there is competition for bathers among the different beaches. The town has determined the following function which expresses the average number of cars per day q as a function of the parking fee p stated in cents.

$$q = 6,000 - 12p$$

(a) Determine the fee which should be charged to maximize daily beach revenues.
(b) What is the maximum daily beach revenue expected to be?
(c) How many cars are expected on an average day?

5 **Import Tax Management** The United States government is studying the import tax structure for color television sets imported from other countries into the United States. The government is trying to determine the amount of the tax to charge on each TV set. The government realizes that the demand for imported TV sets will be affected by the tax. It estimates that the demand for imported sets D, measured in hundreds of TV sets, will be related to the import tax t, measured in cents, according to the function

$$D = 80,000 - 12.5t$$

(a) Determine the import tax which will result in maximum tax revenues from importing TV sets.
(b) What is the maximum revenue?
(c) What will the demand for imported color TV sets equal with this tax?

6 A manufacturer has determined a cost function which expresses the annual cost of purchasing, owning, and maintaining its raw material inventory as a function of the size of each order. The cost function is

$$C = \frac{51,200}{q} + 80q + 750,000$$

where q equals the size of each order (in tons) and C equals the annual inventory cost.
(a) Determine the order size q which minimizes annual inventory cost.
(b) What are minimum inventory costs expected to equal?

7 In Exercise 6 assume that the maximum amount of the raw material which can be accepted in any one shipment is 20 tons.
(a) Given this restriction, determine the order size q which minimizes annual inventory cost.
(b) What are the minimum annual inventory costs?
(c) How do these results compare with those in Exercise 6?

8 A major distributor of racquetballs is thriving. One of the distributor's major problems is keeping up with the demand for racquetballs. Balls are purchased periodically from a sporting goods manufacturer. The annual cost of purchasing, owning, and maintaining the inventory of racquetballs is described by the function

$$C = \frac{280,000}{q} + 0.15q + 2,000,000$$

where q equals the order size (in dozens of racquetballs) and C equals the annual inventory cost.
(a) Determine the order size q which minimizes annual inventory cost.
(b) What are the minimum inventory costs expected to equal?

9 The distributor in Exercise 8 has storage facilities to accept up to 1,200 dozens of balls in any one shipment.
(a) Determine the order size q which minimizes annual inventory costs.
(b) What are the minimum inventory costs?
(c) How do these results compare with those in Exercise 8?

10 The total cost of producing q units of a certain product is described by the function

$$C = 5,000,000 + 250q + 0.002q^2$$

where C is the total cost stated in dollars.
(a) How many units should be produced in order to minimize the *average cost per unit*?
(b) What is the minimum average cost per unit?
(c) What is the total cost of production at this level of output?

11 The total cost of producing q units of a certain product is described by the function

$$C = 350,000 + 7,500q + 0.25q^2$$

where C is the total cost stated in dollars.

(a) Determine how many units q should be produced in order to minimize the *average cost per unit*.

(b) What is the minimum average cost per unit?

(c) What is the total cost of production at this level of output?

12 Re-solve Exercise 11 if the maximum production capacity is 1,000 units.

13 Public Utilities A cable TV antenna company has determined that its profitability depends upon the monthly fee it charges its customers. Specifically, the relationship which describes annual profit P (stated in dollars) as a function of the monthly rental fee r (stated in dollars) is

$$P = -50,000r^2 + 2,750,000r - 5,000,000$$

(a) Determine the monthly rental fee r which will lead to maximum profit.

(b) What is the expected maximum profit?

14 In Exercise 13 assume that the local public utility commission has restricted the CATV company to a monthly fee not to exceed $20.

(a) What fee leads to a maximum profit for the company?

(b) What is the effect of the utility commission's ruling on the profitability of the firm?

15 A company estimates that the demand for its product fluctuates with the price it charges. The demand function is

$$q = 280,000 - 400p$$

where q equals the number of units demanded and p equals the price in dollars. The total cost of producing q units of the product is estimated by the function

$$C = 350,000 + 300q + 0.0015q^2$$

(a) Determine how many units q should be produced in order to maximize annual profit.

(b) What price should be charged?

(c) What is the annual profit expected to equal?

16 Solve the previous exercise, using the marginal approach to profit maximization.

17 If annual capacity is 40,000 units in Exercise 15, how many units q will result in maximum profit? What is the loss in profit attributable to the restricted capacity?

18 An equivalent way of solving Example 2 is to state total revenue as a function of q, the average number of riders per hour. Formulate the function $R = g(q)$ and determine the number of riders q which will result in maximum total revenue. Verify that the maximum value of R and the price which should be charged are the same as obtained in Example 2.

19 The total cost and total revenue functions for a product are

$$C(q) = 500 + 100q + 0.5q^2$$

$$R(q) = 500q$$

(a) Using the marginal approach, determine the profit-maximizing level of output.

(b) What is the maximum profit?

20 A firm sells each unit of a product for $50. The total cost of producing x (thousand) units is described by the function

$$C(x) = 10 - 2.5x^2 + x^3$$

where $C(x)$ is measured in thousands of dollars.

(a) Use the marginal approach to determine the profit-maximizing level of output.
(b) What is total revenue at this level of output? Total cost? Total profit?

21 The profit function for a firm is

$$P(q) = -4.5q^2 + 36{,}000q - 45{,}000$$

(a) Using the marginal approach, determine the profit-maximizing level of output.
(b) What is the maximum profit?

22 The total cost and total revenue functions for a product are

$$C(q) = 5{,}000{,}000 + 250q + 0.002q^2$$

$$R(q) = 1{,}250q - 0.005q^2$$

(a) Using the marginal approach, determine the profit-maximizing level of output.
(b) What is the maximum profit?

23 The total cost and total revenue functions for a product are

$$C(q) = 40{,}000 + 25q + 0.002q^2$$

$$R(q) = 75q - 0.008q^2$$

(a) Using the marginal approach, determine the profit-maximizing level of output.
(b) What is the maximum profit?

24 Portrayed in Fig. 17.10 is a total cost function $C(q)$ and a total revenue function $R(q)$. Discuss the economic significance of the four levels of output q_1, q_2, q_3, and q_4.

Figure 17.10

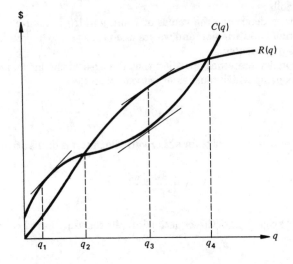

17.2 ADDITIONAL APPLICATIONS

The following examples are additional applications of optimization procedures.

EXAMPLE 11   **(Real Estate)** A large multinational conglomerate is interested in purchasing some prime boardwalk real estate at a major ocean resort. The conglomerate is interested in acquiring a rectangular lot which is located on the boardwalk. The only restriction is that the lot have an area of 100,000 square feet. Figure 17.11 presents a sketch of the layout with x equaling the boardwalk frontage for the lot and y equaling the depth of the lot (both measured in feet).

The seller of the property is pricing the lots at $5,000 per foot of frontage along the boardwalk and $2,000 per foot of depth away from the boardwalk. The conglomerate is interested in determining the dimensions of the lot which will minimize the total purchase cost.

Figure 17.11

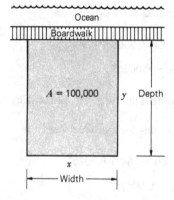

Refer to Fig. 17.11. Total purchase cost for a lot having dimensions of x feet by y feet is

$$C = 5,000x + 2,000y \qquad (17.12)$$

where C is cost in dollars.

The problem is to determine the values of x and y which minimize C. However, C is stated as a function of two variables, and we are unable, as yet, to handle functions which have two independent variables.

Because the conglomerate has specified that the area of the lot must equal 100,000 square feet, a relationship which must exist between x and y is

$$xy = 100,000 \qquad (17.13)$$

Given this relationship, we can solve for either variable in terms of the other. For instance,

$$y = \frac{100,000}{x} \qquad (17.14)$$

We can substitute the right side of this equation into the cost function wherever the variable y appears, or

$$C = f(x)$$

$$= 5{,}000x + 2{,}000 \, \frac{100{,}000}{x}$$

$$= 5{,}000x + \frac{200{,}000{,}000}{x} \tag{17.15}$$

Equation (17.15) is a restatement of Eq. (17.12) only in terms of one independent variable. We can now seek the value of x which minimizes the purchase cost C.

The first derivative is

$$C'(x) = 5{,}000 - 200{,}000{,}000x^{-2}$$

If C' is set equal to 0,

$$5{,}000 = \frac{200{,}000{,}000}{x^2}$$

$$x^2 = \frac{200{,}000{,}000}{5{,}000}$$

$$= 40{,}000$$

or critical values occur at

$$x = \pm 200$$

The critical point at $x = -200$ is meaningless. To test $x = 200$,

$$C''(x) = 400{,}000{,}000x^{-3}$$

$$= \frac{400{,}000{,}000}{x^3}$$

$$C''(200) = \frac{400{,}000{,}000}{(200)^3}$$

$$= \frac{400{,}000{,}000}{8{,}000{,}000} = 50 > 0$$

Since $C''(x) > 0$ for $x > 0$, the graph of C is concave upward for $x > 0$. Thus, the minimum value of C occurs at $x = 200$.

Total costs will be minimized when the width of the lot equals 200 feet. The depth of the lot can be found by substituting $x = 200$ into Eq. (17.14), or

$$y = \frac{100{,}000}{200}$$

$$= 500$$

If the lot is 200 feet by 500 feet, total cost will be minimized at a value of

$$C = \$5,000(200) + \$2,000(500)$$
$$= \$2,000,000$$

EXAMPLE 12 **(Emergency Response: Location Model)** Example 13 in Chap. 6 discussed a problem in which three resort cities had agreed jointly to build and support an emergency response facility which would house rescue trucks and trained paramedics. The key question dealt with the location of the facility. The criterion selected was to choose the location so as to minimize S, the sum of the products of the summer populations of each town and the square of the distance between the town and the facility. Figure 17.12 shows the relative locations of the three cities.

Figure 17.12

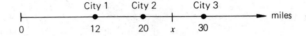

The criterion function to be minimized was determined to be

$$S = f(x) = 450x^2 - 19,600x + 241,600$$

where x is the location of the facility relative to the zero point in Fig. 17.12. (You may want to reread Example 13 on page 234.) Given the criterion function, the first derivative is

$$f'(x) = 900x - 19,600$$

If f' is set equal to 0,

$$900x = 19,600$$

and a critical value occurs at

$$x = 21.77$$

Checking the nature of the critical point, we find

$$f''(x) = 900 \text{ for } x > 0$$

In particular, $$f''(21.77) = 900 > 0$$

Thus, f is minimized when $x = 21.77$. The criterion S is minimized at $x = 21.77$, and the facility should be located as shown in Fig. 17.13.

Figure 17.13

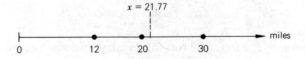

EXAMPLE 13 (**Equipment Replacement**) A decision faced by many organizations is determining the optimal point in time to replace a major piece of equipment. Major pieces of equipment are often characterized by two cost components — *capital cost* and *operating cost*. **Capital cost** is purchase cost less any salvage value. If a machine costs $10,000 and is later sold for $2,000, the capital cost is $8,000. **Operating cost** includes costs of owning and maintaining a piece of equipment. Gasoline, oil, insurance, and repair costs associated with owning and operating a vehicle would be considered operating costs.

Some organizations focus on the *average capital cost* and *average operating cost* when they determine when to replace a piece of equipment. These costs tend to trade off against one another. That is, as one cost increases, the other decreases. Average capital cost for a piece of equipment tends to decrease over time. For a new automobile which decreases in value from $12,000 to $9,000 in the first year, the average capital cost for that year is $3,000. If the automobile decreases in value to $2,000 after 5 years, the average capital cost is

$$\frac{\$12,000 - \$2,000}{5} = \frac{\$10,000}{5} = \$2,000 \text{ per year}$$

Average operating cost tends to increase over time as equipment becomes less efficient and more maintenance is required. For example, the average annual operating cost of a car tends to increase as the car ages.

A taxi company in a major city wants to determine how long it should keep its cabs. Each cab comes fully equipped at a cost of $18,000. The company estimates average capital cost and average operating cost to be a function of x, the number of miles the car is driven. The salvage value of the car, in dollars, is expressed by the function

$$S(x) = 16,000 - 0.10x$$

This means that the car decreases $2,000 in value as soon as the cab is driven, and it decreases in value at the rate of $0.10 per mile.

The average operating cost, stated in dollars per mile, is estimated by the function

$$O(x) = 0.0000003x + 0.15$$

Determine the number of miles the car should be driven prior to replacement if the objective is to minimize the *sum* of average capital cost and average operating cost.

SOLUTION

Average capital cost per mile equals the purchase cost less the salvage value, all divided by the number of miles driven, or

$$C(x) = \frac{18,000 - (16,000 - 0.10x)}{x}$$

$$= \frac{2,000 + 0.10x}{x}$$

$$= \frac{2,000}{x} + 0.10$$

The sum of average capital cost and average operating cost is

$$f(x) = O(x) + C(x)$$

$$= 0.0000003x + 0.15 + \frac{2,000}{x} + 0.10$$

$$= 0.0000003x + 0.25 + \frac{2,000}{x}$$

$$f'(x) = 0.0000003 - 2,000x^{-2}$$

If f' is set equal to 0,

$$0.0000003 = \frac{2,000}{x^2}$$

$$x^2 = \frac{2,000}{0.0000003}$$

$$= 6,666,666,666.67$$

or critical values occur when

$$x = \pm 81,649.6$$

Again, a negative value for x is meaningless. Checking the critical value $x = 81,649.6$, we have

$$f''(x) = 4,000x^{-3}$$

$$= \frac{4,000}{x^3}$$

For $x > 0$, $f''(x) > 0$ (i.e., the graph of f is concave upward for $x > 0$). Therefore f is minimized when $x = 81,649.6$,

$$f(81,649.6) = 0.0000003(81,649.6) + 0.25 + \frac{2,000}{81,649.6}$$

$$= 0.02450 + 0.25 + 0.02450 = 0.299$$

Average capital and operating costs are minimized at a value of $0.299 per mile when a taxi is driven 81,649.6 miles. Total capital and operating costs will equal (cost/mile) · (number of miles), or

$$(\$0.299)(81,649.6) = \$24,413.23$$

Figure 17.14 illustrates the two component cost functions and the total cost function. Notice that the average operating cost per mile $O(x)$ increases with increasing values of x and that average capital cost per mile $C(x)$ decreases with increasing values of x.

Figure 17.14

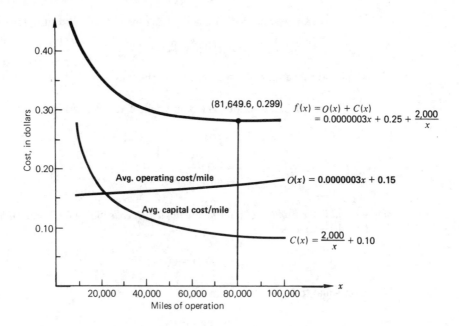

POINT FOR THOUGHT & DISCUSSION	Given what you understand about decisions of when to replace equipment, critique the assumptions used in this model. What relevant factors or considerations are not accounted for when using the results of this model?

EXAMPLE 14   (**Bill Collection**) Example 11 in Chap. 7 discussed the collection of accounts receivable for credit issued to people who use a major credit card. The financial institution determined that the percentage of accounts receivable P (in dollars) collected t months after the credit was issued is

$$P = 0.95(1 - e^{-0.7t})$$

The average credit issued in any one month is $100 million. The financial institution estimates that *for each $100 million in new credit* issued in any month, collection efforts cost $1 million per month. That is, if credit of $100 million is issued today, it costs $1 million for every month the institution attempts to collect these accounts receivable. Determine the number of months that collection efforts should be continued if the objective is to maximize the *net collections N* (dollars collected minus collection costs).

SOLUTION

Given that $100 million of credit is issued, the amount of receivables collected (in millions of dollars) equals

(Amount of credit issued)(percentage of accounts collected)

or
$$(100)(0.95)(1 - e^{-0.7t})$$

Therefore, net collections N are described by the function

Net collections = amount collected − collection costs

or
$$\begin{aligned}
N &= f(t) \\
&= (100)(0.95)(1 - e^{-0.7t}) - (1)t \\
&= 95(1 - e^{-0.7t}) - t \\
&= 95 - 95e^{-0.7t} - t
\end{aligned}$$

where t equals the number of months during which collection efforts are conducted. The first derivative is

$$f'(t) = 66.5e^{-0.7t} - 1$$

If f' is set equal to 0,

$$66.5e^{-0.7t} = 1$$

$$e^{-0.7t} = 0.01503$$

From Table 1

$$e^{-4.2} = 0.0150$$

Thus, $e^{-0.7t} \doteq 0.01503$ when

$$-0.7t = -4.2$$

and a critical value occurs at

$$t = 6$$

The only critical point on f occurs when $t = 6$. Since $f''(t) = -46.55e^{-0.7t} < 0$ for all $t > 0$, $f''(6) < 0$ and f is maximized at $t = 6$. Maximum net collections are

$$\begin{aligned}
f(6) &= 95 - 95e^{-0.7(6)} - 6 \\
&= 95 - 95(0.0150) - 6 = 95 - 1.425 - 6 = 87.575
\end{aligned}$$

or $87.575 million.

For each $100 million of credit issued, net collections will be maximized at a value of $87.575 million if collection efforts continue for 6 months.

☐

> **PRACTICE EXERCISE**
> (*a*) Verify that the critical point at $t = 6$ is a relative maximum.
> (*b*) What is the total (gross) amount collected over the 6-month period? *Answer:* (*b*) $93.575 million.

EXAMPLE 15

(**Welfare Management**) A newly created state welfare agency is attempting to determine the number of analysts to hire to process welfare applications. Efficiency experts estimate that the average cost C of processing an application is a function of the number of analysts x. Specifically, the cost function is

$$C = f(x) = 0.001x^2 - 5 \ln x + 60$$

Determine the number of analysts who should be hired in order to minimize the average cost per application.

SOLUTION
The derivative of f is

$$f'(x) = 0.002x - 5\,\frac{1}{x}$$

$$= 0.002x - \frac{5}{x}$$

If f' is set equal to 0,

$$0.002x = \frac{5}{x}$$

$$0.002x^2 = 5$$

$$x^2 = \frac{5}{0.002}$$

$$= 2{,}500$$

and a critical value occurs when

$$x = 50$$

(The root $x = -50$ is meaningless.)
    The value of $f(x)$ at the critical point is

$$f(50) = 0.001(50)^2 - 5 \ln 50 + 60$$
$$= 0.001(2{,}500) - 5(3.912) + 60 = 2.5 - 19.56 + 60 = \$42.94$$

To check the nature of the critical point,

Figure 17.15

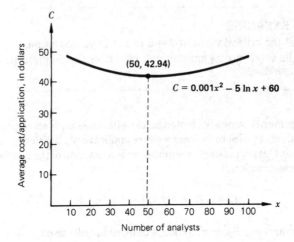

$$f''(x) = 0.002 + 5x^{-2}$$

$$= 0.002 + \frac{5}{x^2} > 0 \text{ for } x > 0$$

In particular, $f''(50) = 0.002 + \dfrac{5}{(50)^2}$

$$= 0.002 + \frac{5}{2,500} = 0.002 + 0.002 = 0.004 > 0$$

Therefore, f is minimized when $x = 50$. Average processing cost per application is minimized at a value of \$42.94 when 50 analysts are employed. Figure 17.15 presents a sketch of the average cost function.

EXAMPLE 16   **(Compensation Planning)** A producer of a perishable product offers a wage incentive to drivers of its trucks. A standard delivery takes an average of 20 hours. Drivers are paid at the rate of \$10 per hour up to a *maximum* of 20 hours. If the trip requires more than 20 hours, the drivers receive compensation for only 20 hours. There is an incentive for drivers to make the trip in less (but not too much less!) than 20 hours. For each hour under 20, the hourly wage increases by \$1.

(a) Determine the function $w = f(x)$ where w equals the hourly wage in dollars and x equals the number of hours required to complete the trip.

(b) What trip time x will maximize the driver's salary for a trip?

(c) What is the hourly wage associated with this trip time?

(d) What is the maximum salary?

(e) How does this salary compare with that received for a 20-hour trip?

SOLUTION

(a) The hourly wage function must be stated in two parts.

$$\text{Hourly wage} = \begin{cases} \$10 + \$1 \times (\text{no. of hours trip time is less than 20}) \\ \qquad\qquad (\text{when trip time is less than 20 hours}) \\ \$10 \qquad\quad (\text{when trip time is 20 hours or more}) \end{cases}$$

Given the variable definitions for x and w, this function can be restated as

$$w = f(x) = \begin{cases} 10 + 1(20 - x) & 0 \le x < 20 \\ 10 & x \ge 20 \end{cases} \qquad \begin{array}{l}\textbf{(17.16a)} \\ \textbf{(17.16b)}\end{array}$$

(b) A driver's salary S for a trip will equal \$10/hour $\times$ 20 hours = \$200 if trip time is greater than or equal to 20 hours. If the trip time is less than 20 hours,

$$\begin{aligned} S &= g(x) \\ &= wx \\ &= [10 + 1(20 - x)]x \\ &= (30 - x)x \\ &= 30x - x^2 \end{aligned} \qquad \textbf{(17.17)}$$

We need to compare the \$200 salary for $x \ge 20$ with the highest salary for a trip time of less than 20 hours.

To examine g for a relative maximum, we find the derivative

$$g'(x) = 30 - 2x$$

Setting g' equal to 0,

$$30 - 2x = 0$$
$$30 = 2x$$

and a critical value occurs when

$$15 = x$$

To check the behavior of $g(x)$ when $x = 15$,

$$g''(x) = -2 \text{ for } 0 \le x \le 20$$

and

$$g''(15) = -2 < 0$$

Therefore, a maximum value of g occurs when $x = 15$, or when a trip takes 15 hours.

(c) The hourly wage associated with a 15-hour trip is

$$\begin{aligned} w &= 10 + 1(20 - 15) \\ &= 10 + 5 = \$15 \end{aligned}$$

(d) The driver salary associated with a 15-hour trip is found by evaluating $g(15)$. If we substitute $x = 15$ into Eq. (17.17),

$$\begin{aligned} S &= 30(15) - 15^2 \\ &= 450 - 225 = \$225 \end{aligned}$$

We also could have arrived at this answer by multiplying the hourly wage of \$15 times the trip time of 15 hours.

(e) The $225 salary for a 15-hour trip is $25 more than the salary for a trip time of 20 hours or more.

EXAMPLE 17

(Pipeline Construction; Motivating Scenario) A major oil company is planning to construct a pipeline to deliver crude oil from a major well site to a point where the crude will be loaded on tankers and shipped to refineries. Figure 17.16 illustrates the relative locations of the well site A and the destination point C. Points A and C are on opposite sides of a dense forest which is approximately 25 miles wide. Point C is also 100 miles south of A. The oil company is proposing a pipeline which will run south along the east side of the forest, and at some point x will cross through the forest to point C. Construction costs are $100,000 per mile along the edge of the forest and $200,000 per mile for the section crossing through the forest. Determine the crossing point x which will result in minimum construction costs for the pipeline.

Figure 17.16

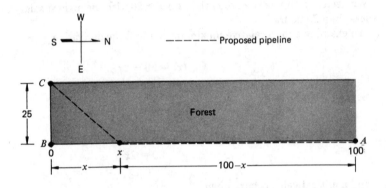

SOLUTION

Construction costs will be computed according to the formula

> Cost = $100,000/mile (miles of pipeline along the edge of the forest)
>   +$200,000/mile (miles of pipeline crossing through the forest)   **(17.18)**

The distance from point A to the crossing point x is $(100 - x)$ miles.

> **PYTHAGOREAN THEOREM**
> Given a right triangle with base a, height b, and hypotenuse c,
>
> $$c^2 = a^2 + b^2$$
>
> or
>
> $$c = \sqrt{a^2 + b^2}$$
>
> See Fig. 17.17.

Using the Pythagorean theorem, the length of the section of the pipeline from C to x is

$$\sqrt{x^2 + (25)^2}$$

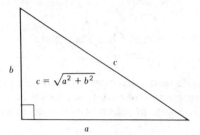

Figure 17.17 Pythagorean theorem.

Using Eq. (17.18), the total cost of construction of the pipeline, C (stated in thousands of dollars), is

$$
\begin{aligned}
C &= f(x) \\
&= 100(100 - x) + 200\sqrt{x^2 + (25)^2} \\
&= 10{,}000 - 100x + 200\sqrt{x^2 + 625} \\
&= 10{,}000 - 100x + 200(x^2 + 625)^{1/2}
\end{aligned}
\tag{17.19}
$$

To examine f for any relative minima, we find the derivative

$$
\begin{aligned}
f'(x) &= -100 + 200(\tfrac{1}{2})(x^2 + 625)^{-1/2}(2x) \\
&= -100 + 200x(x^2 + 625)^{-1/2} \\
&= -100 + \frac{200x}{(x^2 + 625)^{1/2}}
\end{aligned}
$$

Setting f' equal to zero,

$$
-100 + \frac{200x}{(x^2 + 625)^{1/2}} = 0
$$

$$
\frac{200x}{\sqrt{x^2 + 625}} = 100
$$

$$
\frac{200x}{100} = \sqrt{x^2 + 625}
$$

$$
2x = \sqrt{x^2 + 625}
\tag{17.20}
$$

If we square both sides of Eq. (17.20),

$$
4x^2 = x^2 + 625
$$

$$
3x^2 = 625
$$

$$
x^2 = \frac{625}{3} = 208.33
$$

and a (relevant) critical value occurs when $x = 14.43$. (A negative root is meaningless.)

PRACTICE EXERCISE
Use the second-derivative test to verify that f has a minimum when $x = 14.43$.

The practice exercise above should verify that a relative minimum occurs when $x = 14.43$, or the pipeline should cross through the forest after the pipeline has come south 85.57 miles. Total construction costs (in thousands of dollars) can be calculated by substituting $x = 14.43$ into Eq. (17.19), or

$$
\begin{aligned}
C &= 10,000 - 100(14.43) + 200\sqrt{(14.43)^2 + 625} \\
&= 10,000 - 1,443 + 200\sqrt{833.22} \\
&= 10,000 - 1,443 + 200(28.86) \\
&= 10,000 - 1,443 + 5,772 \\
&= 14,329 \ (\$1,000\text{s}) \\
&= \$14,329,000
\end{aligned}
$$

❑

POINT FOR THOUGHT & DISCUSSION	What alternative procedure could be used to compute the minimum total cost figure of $14,329,000?

The following example, although not an optimization application, is one of particular importance in economics.

EXAMPLE 18   (**Elasticity of Demand**) An important concept in economics and price theory is the **price elasticity of demand,** or more simply, the **elasticity of demand.** Given the demand function for a product $q = f(p)$ and a particular point (p, q) on the function, the elasticity of demand is the ratio

$$
\frac{\text{Percentage change in quantity demanded}}{\text{Percentage change in price}} \tag{17.21}
$$

This ratio is a measure of the *relative* response of demand to changes in price. Equation (17.21) can be expressed symbolically as

$$
\frac{\dfrac{\Delta q}{q}}{\dfrac{\Delta p}{p}} \tag{17.22}
$$

The point elasticity of demand is the limit of Eq. (17.22) as $\Delta p \to 0$. Using the Greek letter η (eta) to denote the *point elasticity of demand at a point* (p, q),

$$\eta = \lim_{\Delta p \to 0} \frac{\dfrac{\Delta q}{q}}{\dfrac{\Delta p}{p}}$$

$$= \lim_{\Delta p \to 0} \frac{\Delta q}{q} \cdot \frac{p}{\Delta p}$$

$$= \frac{p}{q} \lim_{\Delta p \to 0} \frac{\Delta q}{\Delta p}$$

or,

$$\eta = \frac{p}{q} \cdot f'(p) \qquad (17.23)$$

Given the demand function $q = f(p) = 500 - 25p$, let's calculate the point elasticity of demand at prices of (a) \$15, (b) \$10, and (c) \$5.

For $p = \$15$

$$\eta = \frac{p}{q} \cdot f'(p)$$

$$= \frac{15}{f(15)} \cdot (-25)$$

$$= \frac{15}{500 - 25(15)} (-25)$$

$$= \frac{-375}{125} = -3$$

The interpretation of $\eta = -3$ is that at a price of \$15, an increase in price by 1 percent would result in a *decrease* in the quantity demanded of approximately 3 percent. The percent change in demand is estimated to be three times the percent change in price.

For $p = \$10$

$$\eta = \frac{10}{f(10)} (-25)$$

$$= \frac{10}{500 - 25(10)} (-25)$$

$$= \frac{-250}{250} = -1$$

The interpretation of $\eta = -1$ is that at a price of \$10, an increase in price by 1 percent would result in a *decrease* in quantity demanded of approximately 1 percent. The percent change in demand is estimated to be the same as the percent change in price.

For p = $5

$$\eta = \frac{5}{f(5)}(-25)$$

$$= \frac{5}{500 - 25(5)}(-25)$$

$$= \frac{-125}{375} = -1/3$$

The interpretation of $\eta = -1/3$ is that at a price of \$5, an increase in price by 1 percent would result in a *decrease* in quantity demanded of approximately 0.33 percent. The percent change in demand is estimated to be less than the percent change in price.

❑

Economists classify point elasticity values into three categories.

❑ *Case 1* ($|\eta| > 1$): The percentage change in demand is greater than the percentage change in price (e.g., a 1 percent change in price results in a greater than 1 percent change in demand). In these regions of a demand function, demand is said to be *elastic.*

❑ *Case 2* ($|\eta| < 1$): The percentage change in demand is less than the percentage change in price. In these regions of the demand function, demand is said to be *inelastic.*

❑ *Case 3* ($|\eta| = 1$): The percentage change in demand equals the percentage change in price. In these regions of the demand function, demand is said to be *unit elastic.*

Section 17.2 Follow-up Exercises

1 A person wishes to fence in a rectangular garden which is to have an area of 1,500 square feet. Determine the dimensions which will create the desired area but will require the minimum length of fencing.

2 An owner of a ranch wishes to build a rectangular riding corral having an area of 5,000 square meters. If the corral appears as in Fig. 17.18, determine the dimensions x and y which will require the minimum length of fencing. (*Hint:* Set up a function for the total length of fencing required, stated in terms of x and y. Then, remembering that $xy = 5,000$, restate the length function in terms of either x or y.)

Figure 17.18

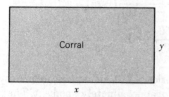

3 A small beach club has been given 300 meters of flotation barrier to enclose a swimming area. The desire is to create the largest rectangular swim area given the 300 meters of flotation barrier. Figure 17.19 illustrates the proposed layout. Note that the flotation

Figure 17.19

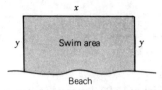

barrier is required on only three sides of the swimming area. Determine the dimensions x and y which result in the largest swim area. What is the maximum area? (*Hint:* Remember that $x + 2y = 300$.)

4 An automobile distributor wishes to create a parking area near a major U.S. port for storing new cars from Japan. The parking area is to have a total area of 1,000,000 square meters and will have dimensions as indicated in Fig. 17.20. Because of security concerns, the section of fence across the front of the lot will be more heavy-duty and taller than the fence used along the sides and rear of the lot. The cost of fence for the front is $20 per running meter and that used for the other three sides costs $12 per running meter. Determine the dimensions x and y which result in a minimum total cost of fence. What is the minimum cost? (*Hint:* $xy = 1,000,000$.)

Figure 17.20

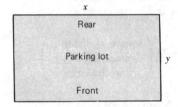

5 **Corrections Management** Figure 17.21 illustrates a recreation yard which is to be fenced within a prison. In addition to enclosing the area, a section of fence should divide the total area in half. If 3,600 feet of fence are available, determine the dimensions x and y which result in the maximum enclosed area. What is the maximum area? (*Hint:* $2x + 3y = 3,600$.)

Figure 17.21

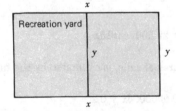

6 **Warehouse Location** A manufacturer wishes to locate a warehouse between three cities. The relative locations of the cities are shown in Fig. 17.22. The objective is to locate the warehouse so as to minimize the sum of the squares of the distances separating each city and the warehouse. How far from the reference point should the warehouse be located?

Figure 17.22

Figure 17.23

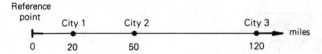

7 **Health Maintenance Organization** Figure 17.23 illustrates the relative locations of three cities. A large health maintenance organization (HMO) wishes to build a satellite clinic to service the three cities. The location of the clinic x should be such that the sum of the squares of the distances between the clinic and each city is minimized. This criterion can be stated as

$$\text{Minimize} \quad S = \sum_{j=1}^{3} (x_j - x)^2$$

where x_j is the location of city j and x is the location of the clinic. Determine the location x which minimizes S.

*8 **HMO, continued** In Exercise 7, suppose that cities 1, 2, and 3 have 10,000, 5,000, and 3,000 persons, respectively, who are members of the HMO. Assume that the HMO has established its location criterion as the minimization of

$$\sum_{j=1}^{3} \left(\begin{array}{c} \text{HMO membership} \\ \text{in city } j \end{array} \right) \left(\begin{array}{c} \text{square of distance} \\ \text{separating city } j \\ \text{and the clinic} \end{array} \right)$$

or

$$S = \sum_{j=1}^{3} n_j (x_j - x)^2$$

where n_j equals the number of members residing in city j. Determine the location x which minimizes S.

9 A police department purchases new patrol cars for $26,000. The department estimates average capital cost and average operating cost to be a function of x, the number of miles the car is driven. The salvage value of a patrol car (in dollars) is expressed by the function

$$S(x) = 22{,}500 - 0.15x$$

Average operating cost, stated in dollars per mile, is estimated by the function

$$O(x) = 0.0000006x + 0.20$$

(a) Determine how many miles the car should be driven prior to replacement if the objective is to minimize the sum of average capital cost and average operating cost per mile.

(b) What is the minimum cost per mile?

(c) What is the salvage value expected to equal?

10 **Commercial Aircraft Replacement** A major airline purchases a particular type of plane at a cost of $40,000,000. The company estimates that average capital cost and average operating cost are a function of x, the number of hours of flight time. The salvage value of a plane (in dollars) is expressed by the function

$$S(x) = 36,000,000 - 10,000x$$

Average operating cost, stated in dollars per hour of flight time, is estimated by the function

$$O(x) = 500 + 0.40x$$

(a) Determine how many hours a plane should be flown before replacement if the objective is to minimize the sum of average capital and average operating cost per hour.

(b) What is the minimum cost per hour?

(c) What is the salvage value expected to equal?

11 A university ski club is organizing a weekend trip to a ski lodge. The price for the trip is $100 if 50 or fewer persons sign up for the trip. For every traveler in excess of 50, the price for *all* will decrease by $1. For instance, if 51 persons sign up, each will pay $99. Let x equal the number of travelers in excess of 50.

(a) Determine the function which states price per person p as a function of x.

(b) In part a, is there any restriction on the domain?

(c) Formulate the function $R = h(x)$, which states total revenue R as a function of x.

(d) What value of x results in the maximum value of R?

(e) How many persons should sign up for the trip?

(f) What is the maximum value of R?

(g) What price per ticket results in the maximum revenue?

(h) Could the club generate more revenue by taking 50 or fewer persons?

12 A national charity is planning a fund-raising campaign in a major United States city having a population of 2 million. The percentage of the population who will make a donation is estimated by the function.

$$R = 1 - e^{-0.02x}$$

where R equals the percentage of the population and x equals the number of days the campaign is conducted. Past experience indicates that the average contribution in this city is $2 per donor. Costs of the campaign are estimated at $10,000 per day.

(a) How many days should the campaign be conducted if the objective is to maximize net proceeds (total contributions minus total costs) from the campaign?

(b) What are maximum net proceeds expected to equal? What percentage of the population is expected to donate?

13 A national distribution company sells CD's by mail only. Advertising is done on local TV stations. A promotion program is being planned for a major metropolitan area for a new country western recording. The target audience — those who might be interested in this type of recording — is estimated at 600,000. Past experience indicates that for this city and this type of recording the percentage of the target market R actually purchasing a CD is a function of the length of the advertising campaign t, stated in days. Specifically, this *sales response function* is

$$R = 1 - e^{-0.025t}$$

The profit margin on each CD is $1.50. Advertising costs include a fixed cost of $15,000 and a variable cost of $2,000 per day.

(a) Determine how long the campaign should be conducted if the goal is to maximize *net profit* (gross profit minus advertising costs).

(b) What is the expected maximum net profit?

(c) What percentage of the target market is expected to purchase the CD?

14 Assume in Example 14 that the average amount of credit issued each month is $50 million and monthly collection costs equal $0.5 million. Re-solve the problem.

15 A police department has determined that the average daily crime rate in the city depends upon the number of officers assigned to each shift. Specifically, the function describing this relationship is

$$N = f(x) = 500 - 10xe^{-0.025x}$$

where N equals the average daily crime rate and x equals the number of officers assigned to each shift. Determine the number of officers which will result in a minimum average daily crime rate. What is the minimum average daily crime rate?

16 A firm's annual profit is stated as a function of the number of salespersons employed. The profit function is

$$P = 20(x)e^{-0.002x}$$

where P equals profit stated in thousands of dollars and x equals the number of salespersons.

(a) Determine the number of salespersons which will maximize annual profit.

(b) What is the maximum profit expected to equal?

17 A company is hiring persons to work in its plant. For the job the persons will perform, efficiency experts estimate that the average cost C of performing the task is a function of the number of persons hired x. Specifically,

$$C = f(x) = 0.003x^2 - 0.216 \ln x + 5$$

(a) Determine the number of persons who should be hired to minimize the average cost.

(b) What is the minimum average cost?

18 A company is hiring people to work in its plant. For the job the people will perform, efficiency experts estimate that the average cost C of performing the task is a function of the number of people hired x. Specifically,

$$C = f(x) = 0.005x^2 - 0.49 \ln x + 5$$

(a) Determine the number of people who should be hired to minimize the average cost.

(b) What is the minimum average cost?

19 **Wage Incentive Plan** A manufacturer offers a wage incentive to persons who work on one particular product. The standard time to complete one unit of the product is 15 hours. Laborers are paid at the rate of $6 per hour up to a maximum of 15 hours for each unit they work on (if a laborer takes 20 hours to complete a unit, he or she is only paid for the 15 hours the unit should have taken). There is an incentive for laborers to complete a unit in less than 15 hours. For each hour under 15 the hourly wage increases by $1.50. Let x equal the number of hours required to complete a unit

(a) Determine the function $w = f(x)$ where w equals the hourly wage in dollars.

(b) What length of time x will maximize a laborer's total wages for completing one unit?

(c) What is the hourly wage associated with this time per unit x?

(*d*) What is maximum wage per unit?

(*e*) How does this salary compare with the wages earned by taking 15 or more hours per unit?

20 Pipeline Construction A major oil company is planning to construct a pipeline to deliver crude oil from a major well site to a point where the crude will be loaded on tankers and shipped to refineries. Figure 17.24 illustrates the relative locations of the well site *A* and the destination point *C*. Points *A* and *C* are on opposite sides of a lake which is 20 miles wide. Point *C* is also 200 miles south of *A* along the lake.

    The oil company is proposing a pipeline which will run south along the east side of the lake and at some point, *x*, will cross the lake to point *C*. Construction costs are $50,000 per mile along the bank of the lake and $100,000 per mile for the section crossing the lake. Determine the crossing point *x* which leads to minimum construction costs. What is the minimum construction cost?

Figure 17.24

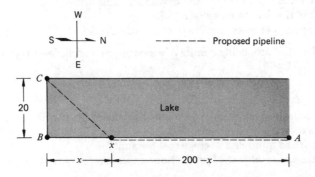

In the following exercises, (*a*) determine the general expression for the point elasticity of demand, (*b*) determine the elasticity of demand at the indicated prices, also classifying demand as elastic, inelastic, or unit elastic, and (*c*) interpret the meaning of the elasticity values found in part (*b*).

21 $q = f(p) = 1200 - 60p, p = \$5, p = \$10,$ and $p = \$15.$

22 $q = f(p) = 150 - 2.5p, p = \$15, p = \$30,$ and $p = \$45$

23 $q = f(p) = 12000 - 10p^2, p = \$10, p = \$20,$ and $p = \$30$

24 $q = f(p) = 900 - p^2, p = \$5, p = \$15,$ and $p = \$25$

25 Given the demand function $q = f(p) = 900 - 30p,$ determine the price at which the demand is unit elastic.

26 Given the demand function $q = f(p) = 80 - 1.6p,$ determine the price at which the demand is unit elastic.

❏ ADDITIONAL EXERCISES

1 A firm sells each unit of a product for $250. The cost function which describes the total cost *C* as a function of the number of units produced and sold *x* is

$$C(x) = 50x + 0.1x^2 + 150$$

(*a*) Formulate the profit function $P = f(x).$

(b) How many units should be produced and sold in order to maximize total profit?

(c) What is total revenue at this level of output?

(d) What is total cost at this level of output?

2 Re-solve Exercise 1 using the marginal approach.

3 A local travel agent is organizing a charter flight to a well-known resort. The agent has quoted a price of $300 per person if 100 or fewer sign up for the flight. For every person over the 100, the price for *all* will decrease by $2.50. For instance, if 101 people sign up, each will pay $297.50. Let x equal the number of persons above 100.

(a) Determine the function which states price per person p as a function of x, or $p = f(x)$.

(b) In part a, is there any restriction on the domain?

(c) Formulate the function $R = h(x)$, which states total ticket revenue R as a function of x.

(d) What value of x results in the maximum value of R?

(e) What is the maximum value of R?

(f) What price per ticket results in the maximum R?

4 The total cost of producing q units of a certain product is described by the function

$$C = 12,500,000 + 100q + 0.02q^2$$

(a) Determine how many units q should be produced in order to minimize the *average cost per unit*.

(b) What is the minimum average cost per unit?

(c) What is the total cost of production at this level of output?

5 A principle of economics states that the average cost per unit is minimized when the marginal cost equals the average cost. Show that this is true for the cost function in Exercise 4.

6 The quadratic total cost function for a product is

$$C = ax^2 + bx + c$$

where x equals the number of units produced and sold and C is stated in dollars. The product sells at a price of p dollars per unit.

(a) Construct the profit function stated in terms of x.

(b) What value of x results in maximum profit?

(c) What restriction assures that a relative maximum occurs at this value of x?

(d) What restrictions on a, b, c, and p assure that $x > 0$?

7 An oil field currently has 10 wells, each producing 300 barrels of oil per day. For each new well drilled, it is estimated that the yield per well will decrease by 10 barrels per day. Determine the number of new wells to drill in order to maximize total daily output for the oil field. What is the maximum output?

8 A small warehouse is to be constructed which is to have a total area of 10,000 square feet. The building is to be partitioned as shown in Fig. 17.25. Costs have been estimated based on exterior and interior wall dimensions. The costs are $200 per running foot of exterior wall plus $100 per running foot of interior wall.

(a) Determine the dimensions which will minimize the construction costs.

(b) What are the minimum costs?

Figure 17.25

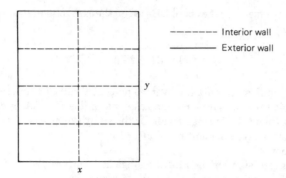

9 An open rectangular box is to be constructed by cutting square corners from a 40 × 40-inch piece of cardboard and folding up the flaps as shown in Fig. 17.26.
(*a*) Determine the value of *x* which will yield the box of maximum volume.
(*b*) What is the maximum volume?

Figure 17.26

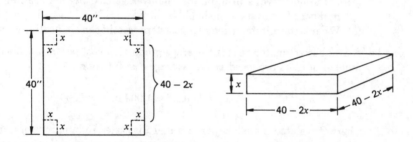

10 The demand function for a product is

$$q = f(p) = 25{,}000e^{-0.05p}$$

where *q* is quantity demanded (in units) and *p* is the price (in dollars).
(*a*) Determine the value of *p* which will result in maximum total revenue.
(*b*) What is the maximum total revenue?

11 A marketing research organization believes that if a company spends *x* million dollars on TV advertising, total profit can be estimated by the function

$$P = f(x) = 40x^2e^{-0.5x}$$

where *P* is measured in millions of dollars.
(*a*) How much should be spent on TV advertising in order to maximize total profit?
(*b*) What is the maximum profit?

12 Memory Retention An experiment was conducted to determine the effects of elapsed time on a person's memory. Subjects were asked to look at a picture which contained many different objects. After studying the picture, they were asked to recall as many of the objects as they could. At different time intervals following this, they would be asked to recall as many objects as they could. Based on the experiment, the following function was developed:

$$\overline{R} = f(t) = 84 - 25 \ln t \qquad t \geq 1$$

For this function $\overline{R}$ represents the average percent recall as a function of time since studying the picture (measured in hours). A value of $\overline{R} = 50$ would indicate that at the corresponding time t the average recall for the study group was 50 percent.

(a) What is the average percent recall after 1 hour?

(b) After 10 hours?

(c) Find the expression for the rate of change in $\overline{R}$ with respect to time.

(d) What is the maximum percent recall? Minimum?

13 A new state welfare agency wants to determine how many analysts to hire for processing of welfare applications. It is estimated that the average cost C of processing an application is a function of the number of analysts x. Specifically, the cost function is

$$C = 0.005x^2 - 16 \ln x + 70$$

(a) If the objective is to minimize the average cost per application, determine the number of analysts who should be hired.

(b) What is the minimum average cost of processing an application expected to equal?

14 A firm has estimated that the average production cost per unit $\overline{C}$ fluctuates with the number of units produced x. The average cost function is

$$\overline{C} = 0.002x^2 - 1,000 \ln x + 7,500$$

where $\overline{C}$ is stated in dollars per unit and x is stated in hundreds of units.

(a) Determine the number of units which should be produced in order to minimize the average production cost per unit?

(b) What is the minimum average cost expected to equal?

(c) What are total production costs expected to equal?

*15 Postal Service The U.S. Postal Service requires that parcel post packages conform to specified dimensions. Specifically, the length plus girth must be no greater than 84 inches.

(a) Find the dimensions (r and l) of a cylindrical package which maximize the volume of the package. (*Hint:* See Fig. 17.27 and remember that $V = \pi r^2 l$.)

(b) What is the girth of the package?

(c) What is the maximum volume?

Figure 17.27

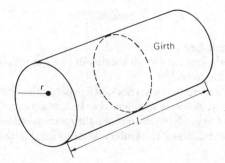

Girth

Figure 17.28

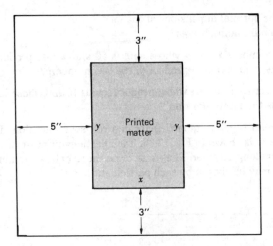

16 **Poster Problem** Figure 17.28 is a sketch of a poster which is being designed for a political campaign. The printed area should contain 1,500 square inches. A margin of 5 inches should appear on each side of the printed matter and a 3-inch margin on the top and bottom.

(a) Determine the dimensions of the printed area which minimize the *area* of the poster.

(b) What are the optimal dimensions of the poster?

(c) What is the minimum poster area?

*17 A person wants to purchase a rectangular piece of property in order to construct a warehouse. The warehouse should have an area of 10,000 square feet. Zoning ordinances specify that there should be at least 30 feet between the building and the side boundaries of the lot and at least 40 feet between the building and the front and rear boundaries of the lot. Figure 17.29 is a sketch of the proposed lot.

(a) Determine the dimensions of the warehouse which will minimize the area of the lot.

Figure 17.29

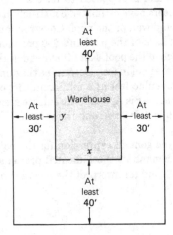

(b) What are the optimal dimensions of the lot?

(c) What is the minimum lot area?

18 Determine two numbers x and y whose sum is 50 and whose product is as large as possible. What is the maximum product of the two numbers?

19 Determine two positive numbers whose product equals 40 and whose sum is as small as possible. What is the minimum sum?

*20 Biathlon A woman is going to participate in a run-swim biathlon. The course is of variable length and is shown in Fig. 17.30. Each participant must run (or walk) from the starting point along the river. They must cross the river by swimming; however, the crossing point x may be chosen by each participant.

Figure 17.30

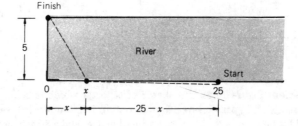

This contestant estimates that she will average 5 miles per hour for the running portion of the biathlon and 1 mile per hour for the swimming portion. She wishes to minimize her time in the event.

(a) Determine the crossing point x which will result in the minimum time.

(b) What is the minimum time expected to equal?

(c) How does this time compare with that if she chose to run the full 25 miles (crossing point is at $x = 0$)?

Hint: Time = distance ÷ speed or hours = miles ÷ miles/hour; also

$$t_{total} = t_{run} + t_{swim}$$

*21 Solid Waste Management A local city is planning to construct a solid waste treatment facility. One of the major components of the plant is a solid waste agitation pool. This pool is to be circular in shape and is supposed to have a volume capacity of 2 million cubic feet. Municipal engineers have estimated construction costs as a function of the surface area of the base and wall of the pool. Construction costs are estimated at $80 per square foot for the base of the pool and $30 per square foot of wall surface. Figure 17.31 presents a sketch of the pool. Note that r equals the radius of the pool in feet and h equals the depth of the pool in feet. Determine the dimensions r and h which provide a capacity of 2 million cubic feet at a minimum cost of construction. (*Hint:* The area A of a circle having radius r is $A = \pi r^2$, the surface area A of a circular cylinder having radius r and height h is $A = 2\pi rh$, and the volume V is $V = \pi r^2 h$.)

In the following exercises, (a) determine the general expression for the point elasticity of demand, (b) determine the elasticity of demand at the indicated prices, also classifying demand as elastic, inelastic, or unit elastic, and (c) interpret the meaning of the elasticity values found in part b.

Figure 17.31

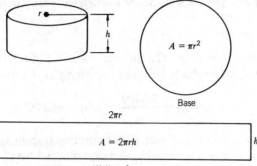

$2\pi r$

$A = 2\pi rh$ h

Wall surface

22 $q = f(p) = 2{,}500 - 80p$, $p = \$6$, $p = \$15$, $p = \$24$

23 $q = f(p) = 875 - p - 0.05p^2$, $p = \$50$, $p = \$70$, and $p = \$100$

24 Given the demand function in Exercise 22, determine the price(s) at which demand is (*a*) elastic, (*b*) inelastic, and (*c*) unit elastic.

25 Given the demand function $q = f(p) = 2400 - 40p$, determine the price(s) at which demand is (*a*) elastic, (*b*) inelastic, and (*c*) unit elastic.

❏ CHAPTER TEST

1 The demand function for a product is

$$q = f(p) = 60{,}000 - 7.5p$$

where q equals the quantity demanded and p equals the price in dollars. Formulate the total revenue function $R = g(p)$.

2 The total revenue function for a product is

$$R = f(x) = -4x^2 + 300x$$

where R is measured in hundreds of dollars and x equals the number of units sold (in 100s). The total cost of producing x (hundred) units is described by the function

$$C = g(x) = x^2 - 150x + 5{,}000$$

where C is measured in hundreds of dollars.
(*a*) Formulate the profit function $P = h(x)$.
(*b*) How many units should be produced and sold in order to maximize total profit?
(*c*) What is the maximum profit?

3 An importer wants to fence in a storage area near the local shipping docks. The area will be used for temporary storage of shipping containers. The area is to be rectangular with an area of 100,000 square feet. The fence will cost $20 per running foot.

(a) Determine the dimensions of the area which will result in fencing costs being minimized.

(b) What is the minimum cost?

4 A retailer has determined that the annual cost C of purchasing, owning, and maintaining one of its products behaves according to the function

$$C = f(q) = \frac{20,000}{q} + 0.5q + 80,000$$

where q is the size (in units) of each order purchased from suppliers.

(a) What order quantity q results in minimum annual cost?

(b) What is the minimum annual cost?

5 The demand function for a product is

$$q = f(p) = 35,000e^{-0.05p}$$

where q is quantity demanded (in units) and p is the price (in dollars).

(a) Determine the value of p which will result in maximum total revenue.

(b) What is the maximum total revenue?

6 A company is hiring persons to work in its plant. For the job the persons will perform, efficiency experts estimate that the average cost C of performing the task is a function of the number of persons hired x. Specifically,

$$C = f(x) = 0.005x^2 - 0.49 \ln x + 5$$

(a) Determine the number of persons who should be hired to minimize the average cost.

(b) What is the minimum average cost?

MINICASE

THE EOQ MODEL

The economic order quantity (EOQ) model is a classic inventory model. The purpose of the EOQ model is to find the quantity of an item to order which minimizes total inventory costs. The model assumes three different cost components; *ordering cost, carrying cost,* and *purchase cost.* Ordering costs are those associated with placing and receiving an order. These costs are largely for salaries of persons involved in requisitioning goods, processing paperwork, receiving goods, and placing goods into inventory. Ordering costs are assumed to be incurred each time an order is placed.

Carrying costs, sometimes referred to as *holding costs,* are the costs of owning and maintaining inventory. Carrying costs include such components as cost of storage space, insurance, salaries of inventory control personnel, obsolescence, and opportunity costs associated with having the investment in inventory. Carrying costs are often expressed as a percentage of the average value of inventory on hand (e.g., 25 percent per year). Purchase cost is simply the cost of the inventory items.

Although there are variations on the EOQ model, the basic model makes the following assumptions: (1) demand for items is known and is at a constant (or near constant) rate, (2) time between placing and receiving an order (*lead time*) is known with certainty, (3) order quantities are always the same size, and (4) inventory replenishment is instantaneous (i.e., the entire order is received in one batch).

If we assume a time frame of 1 year, total inventory costs are

$$TC = \text{annual ordering cost} + \text{annual carrying cost}$$
$$+ \text{annual purchase cost}$$

$$= \left(\begin{array}{c}\text{number of}\\\text{orders per year}\end{array}\right)\left(\begin{array}{c}\text{ordering}\\\text{cost per order}\end{array}\right)$$

$$+ \left(\begin{array}{c}\text{average}\\\text{inventory}\\\text{in units}\end{array}\right)\left(\begin{array}{c}\text{value}\\\text{per unit}\end{array}\right)\left(\begin{array}{c}\text{carrying}\\\text{cost in percent}\end{array}\right)$$

$$+ \left(\begin{array}{c}\text{annual}\\\text{demand}\end{array}\right)\left(\begin{array}{c}\text{purchase price}\\\text{per unit}\end{array}\right)$$

If

$$D = \textit{annual demand, units}$$

$$C_o = \textit{ordering cost per order}$$

$$C_h = \textit{carrying cost (stated as a percentage of average}$$
$$\textit{value of inventory on hand)}$$

$$p = \textit{purchase price per unit}$$

$$q = \textit{order quantity}$$

annual inventory costs can be expressed as a function of the order quantity q as follows:

$$TC = f(q) = \frac{D}{q} C_o + \frac{q}{2} pC_h + pD$$

Requirements:

1 For a given inventory item, $D = 5,000$, $C_o = \$125$, $p = \$100$, and $C_h = 0.20$. Determine the value of q which minimizes total annual inventory costs. What are minimum annual inventory costs? How many orders must be placed each year? What are annual ordering costs? Annual carrying costs?

2 The order quantity which minimizes annual inventory costs is termed the "economic order quantity" or EOQ. Using the generalized cost function, determine the general expression for the order quantity q which minimizes annual inventory cost. (*Hint:* Find the derivative with respect to q, assuming that D, C_o, p, and C_h are constant.)

3 Prove that the critical value for q does result in a relative minimum on the cost function.

4 Using the expression for q found in part 2, show that annual ordering cost equals annual carrying cost when operating at the EOQ level.

5 Annual inventory costs can be expressed in terms of the number of orders placed per year, N, recognizing that $N = D/q$. Rewrite the generalized cost function in terms of N rather than q. Determine the general expression for the value of N which minimizes annual inventory costs. Confirm that the critical value for N does result in the minimum value of the cost function.

INTEGRAL CALCULUS: AN INTRODUCTION

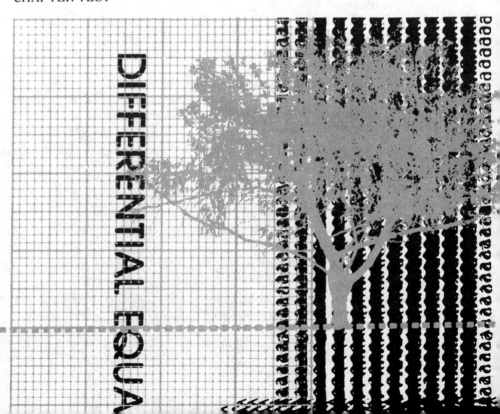

CHAPTER OBJECTIVES

❏ Introduce the nature and methods of integral calculus

❏ Present selected rules of integration and illustrate their use

❏ Illustrate other methods of integration which may be appropriate when basic rules are not

In Chap. 17 we defined *marginal cost* as the additional cost incurred as a result of producing and selling one more unit of a product or service. We also determined that given the total cost function $C = f(q)$, the marginal cost function is the derivative of the total cost function, or $MC = f'(q)$. Suppose that we are given the function

$$MC = x + 100$$

where x equals the number of units produced. It is also known that total cost equals \$40,000 when $x = 100$. *What is desired is to determine the total cost function $C = f(x)$* (Example 5).

In this chapter we will introduce a second major area of study within the calculus — *integral calculus*. As was mentioned at the beginning of Chap. 15, differential calculus is useful in considering rates of change and tangent slopes. An important concern of integral calculus is the determination of areas which occur between curves and other defined boundaries. Also, if the derivative of an unknown function is known, integral calculus may provide a way of determining the original function.

As we begin this new area of study, it will be of value to know where we are headed. First, integral calculus comprises a major area of study within the calculus. We will devote two chapters to this material. The purpose is to survey the area in such a way that you have an understanding of the concerns and methods of integral calculus, how integral calculus relates to differential calculus, and where it can be applied.

In this chapter the nature of integral calculus will be introduced first by relating it to derivatives. As there were rules for finding derivatives in differential calculus, there are rules for finding *integrals* in integral calculus. The more commonly applied rules will be presented in Secs. 18.2 and 18.3. Section 18.4 discusses procedures for finding integrals when the rules in Secs. 18.2 and 18.3 are not applicable. Finally, Sec. 18.5 focuses upon differential equations. Chapter 19 will focus upon the applications of integral calculus.

18.1 ANTIDERIVATIVES

The Antiderivative Concept

Given a function f, we are acquainted with how to find the derivative f'. There may be occasions in which we are given the derivative f' and wish to determine the original function f. Since the process of finding the original function is the *reverse* of differentiation, f is said to be an **antiderivative** of f'.

Consider the derivative

$$f'(x) = 4 \qquad \text{(18.1)}$$

By using a trial-and-error approach, it is not very difficult to conclude that the function

$$f(x) = 4x \qquad \qquad (18.2)$$

has a derivative of the form of Eq. (18.1). Another function having the same derivative is

$$f(x) = 4x + 1$$

In fact, any function having the form

$$f(x) = 4x + C \qquad \qquad (18.3)$$

where C is any constant, will have the same derivative. Thus, given the derivative in Eq. (18.1), our conclusion is that the original function was one of the *family* of functions characterized by Eq. (18.3). This family of functions is a set of linear functions whose members all have a slope of $+4$ but different y intercepts C. Figure 18.1 illustrates selected members of this family of functions.

We can also state that the function

$$f(x) = 4x + C$$

is the *antiderivative* of

$$f'(x) = 4$$

Figure 18.1

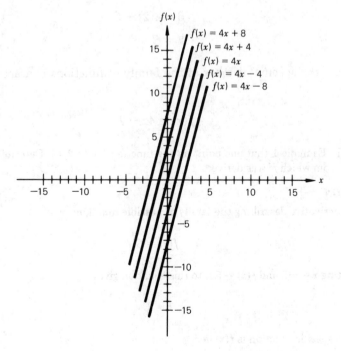

EXAMPLE 1   Find the antiderivative of $f'(x) = 0$.

SOLUTION

We know that the derivative of any constant function is 0. Therefore the antiderivative is $f(x) = C$.

EXAMPLE 2   Find the antiderivative of $f'(x) = 2x - 5$.

SOLUTION

Using a trial-and-error approach and working with each term separately, you should conclude that the antiderivative is

$$f(x) = x^2 - 5x + C$$

☐

NOTE   An easy check on your antiderivative f is to differentiate it to determine f'.

With additional information it may be possible to determine the precise function from which f' was derived. Assume for the original example we are told that $f'(x) = 4$ *and* one point on the original function is (2, 6). Since the coordinates of this point must satisfy the equation of the original function, we can solve for the specific value C by substituting $x = 2$ and $f(x) = 6$ into Eq. (18.3), or

$$f(x) = 4x + C$$
$$6 = 4(2) + C$$
$$-2 = C$$

Therefore, the specific member of the family of functions characterized by Eq. (18.3) is

$$f(x) = 4x - 2$$

EXAMPLE 3   Assume in Example 1 that one point on the function f is $(-2, 5)$. Determine the specific function from which f' was derived.

SOLUTION

The antiderivative describing the family of possible functions was

$$f(x) = C$$

Substituting $x = -2$ and $f(x) = 5$ into this equation gives

$$5 = C$$

Thus, the specific function is $f(x) = 5$.

EXAMPLE 4 Assume in Example 2 that one point on the function f is (2, 20). Determine the specific function from which f' was derived.

SOLUTION

The antiderivative describing the family of possible functions was

$$f(x) = x^2 - 5x + C$$

Substituting $x = 2$ and $f(x) = 20$ into this equation, we have

$$20 = 2^2 - 5(2) + C$$

$$20 = -6 + C$$

$$26 = C$$

Thus the original function is

$$f(x) = x^2 - 5x + 26$$

❑

Revenue and Cost Functions

In Chap. 17, we discussed the *marginal approach* for determining the profit-maximizing level of output. We stated that an expression for marginal revenue (MR) is the derivative of the total revenue function where the independent variable is the level of output. Similarly, we said that an expression for marginal cost (MC) is the derivative of the total cost function. If we have an expression for either marginal revenue or marginal cost, the respective antiderivatives will be the total revenue and total cost functions.

EXAMPLE 5 (**Marginal Cost; Motivating Scenario**) The function describing the marginal cost of producing a product is

$$MC = x + 100$$

where x equals the number of units produced. It is also known that total cost equals $40,000 when $x = 100$. Determine the total cost function.

SOLUTION

To determine the total cost function, we must first find the antiderivative of the marginal cost function, or

$$C(x) = \frac{x^2}{2} + 100x + C \tag{18.4}$$

Given that $C(100) = 40,000$, we can solve for the value of C, which happens to represent the fixed cost.

$$40,000 = \frac{(100)^2}{2} + 100(100) + C$$

$$40,000 = 5,000 + 10,000 + C$$

or
$$25,000 = C$$

The specific function representing the total cost of producing the product is

$$C(x) = \frac{x^2}{2} + 100x + 25,000$$

EXAMPLE 6   **(Marginal Revenue)** The marginal revenue function for a company's product is

$$MR = 50,000 - x$$

where x equals the number of units produced and sold. If total revenue equals 0 when no units are sold, determine the total revenue function for the product.

SOLUTION

Because the marginal revenue function is the derivative of the total revenue function, the total revenue function is the antiderivative of *MR*. Using a trial-and-error approach gives

$$R(x) = 50,000x - \frac{x^2}{2} + C \qquad \qquad \textbf{(18.5)}$$

Since we are told that $R(0) = 0$, substitution of $x = 0$ and $R = 0$ into Eq. (18.5) yields

$$0 = 50,000(0) - \frac{0^2}{2} + C$$

or
$$0 = C$$

Thus, the total revenue function for the company's product is

$$R(x) = 50,000x - \frac{x^2}{2}$$

❑

Section 18.1 Follow-up Exercises

In Exercises 1–30, find the antiderivative of the given function.

1 $f'(x) = 80$ **2** $f'(x) = -50$
3 $f'(x) = \frac{1}{5}$ **4** $f'(x) = \sqrt{70}$
5 $f'(x) = 3x$ **6** $f'(x) = -6x$
7 $f'(x) = x^2/2$ **8** $f'(x) = x^2$
9 $f'(x) = x^4$ **10** $f'(x) = x^3/3$

11 $f'(x) = x^2 - 4x$

12 $f'(x) = x^3 + x^2 + 6x$

13 $f'(x) = x^2 + 8x + 10$

14 $f'(x) = x^5$

15 $f'(x) = 9x^2 + 10x$

16 $f'(x) = 6x^2 + 2x + 20$

17 $f'(x) = 3x^2 + 18x + 12$

18 $f'(x) = 18x^2 - 10x - 100$

19 $f'(x) = 8x^3 - 6x^2$

20 $f'(x) = 4x^3 - 3x^2$

21 $f'(x) = 12x^3 - 9x^2 + 3$

22 $f'(x) = x^3 + x^2 - 4x + 1$

23 $f'(x) = 5x^4 - 9x^2 - 6$

24 $f'(x) = 20x^4 + 8x^3 - 4x$

25 $f'(x) = 30x^4 - 2x^3 + 8x - 5$

26 $f'(x) = 15x^4 - 6x^3 + 2x - 8$

27 $f'(x) = x^3/2$

28 $f'(x) = \sqrt{2}x^3$

29 $f'(x) = -18x^5 + 9x^2 - 10x$

30 $f'(x) = 36x^5 - 15x^4 + 3x^2$

In Exercises 31–50, determine f given f' and a point which satisfies f.

31 $f'(x) = 20$, $(1, 20)$

32 $f'(x) = -8x + 2$, $(2, 10)$

33 $f'(x) = 10x$, $(-2, 10)$

34 $f'(x) = x^2$, $(-4, 26)$

35 $f'(x) = 4x^3$, $(2, 15)$

36 $f'(x) = -2x^3$, $(6, 10)$

37 $f'(x) = -x^2 + 4x$, $(3, 45)$

38 $f'(x) = x^2 + 2x - 3$, $(-3, 8)$

39 $f'(x) = 6x^2 + 8x$, $(2, -20)$

40 $f'(x) = 5x^4$, $(-5, 48)$

41 $f'(x) = 9x^2 + 2x$, $(-2, 2)$

42 $f'(x) = 9x^2 - 4x + 2$, $(-8, -20)$

43 $f'(x) = 4x^3 - 3x^2 + 2x$, $(3, 80)$

44 $f'(x) = x^3 - x^2 + x + 3$, $(-1, 0)$

45 $f'(x) = 8x^3 - 3x^2$, $(2, 8)$

46 $f'(x) = 5x^4 - 6x^2$, $(5, 140)$

47 $f'(x) = -4x^3 - 12x$, $(-5, 40)$

48 $f'(x) = -8x^3 + 6x^2$, $(-6, 28)$

49 $f'(x) = 6x^5 - 3x^2$, $(1, 10)$

50 $f'(x) = 10 - x + x^2$, $(2, -7)$

51 The marginal revenue function for a company's product is

$$MR = 40{,}000 - 4x$$

where x equals the number of units sold. If total revenue equals 0 when no units are sold, determine the total revenue function for the product.

52 The function describing the marginal cost (in dollars) of producing a product is

$$MC = 8x + 800$$

where x equals the number of units produced. It is known that total cost equals \$80,000 when 40 units are produced. Determine the total cost function.

53 The function describing the *marginal profit* from producing and selling a product is

$$MP = -6x + 450$$

where x equals the number of units and MP is the marginal profit measured in dollars. When 100 units are produced and sold, *total profit* equals \$5,000. Determine the total profit function.

54 The function describing the marginal profit from producing and selling a product is

$$MP = -3x + 500$$

where x equals the number of units and MP is the marginal profit measured in dollars. When 200 units are produced and sold, total profit equals \$15,000. Determine the total profit function.

18.2 RULES OF INTEGRATION

Fortunately, we need not resort to a trial-and-error approach whenever we wish to find an antiderivative. As with differentiation, a set of rules has been developed for finding antiderivatives. If a function has a particular form, a rule may be available which allows one to determine its antiderivative very easily.

Integration

The process of finding antiderivatives is more frequently called *integration.* And the family of functions obtained through this process is called the *indefinite integral.* The notation

$$\int f(x) \, dx \tag{18.6}$$

is often used to indicate the indefinite integral of the function f. The symbol $\int$ is the *integral sign; f* is the *integrand,* or the function for which we want to find the indefinite integral; and dx, as we will deal with it, indicates the variable with respect to which the integration process is performed. Two verbal descriptions of the process indicated by Eq. (18.6) are "integrate the function f with respect to the variable x" or "find the indefinite integral of f with respect to x."

NOTE Keep in mind that finding an indefinite integral is the same as finding an antiderivative.

In Example 2 we found that the antiderivative of $2x - 5$ is $x^2 - 5x + C$. We can denote this, using integral notation, as

$$\int (2x - 5) \, dx = x^2 - 5x + C$$

A more formal definition of the indefinite integral follows.

DEFINITION: INDEFINITE INTEGRAL
Given that f is a continuous function,

$$\int f(x) \, dx = F(x) + C \tag{18.7}$$

if $F'(x) = f(x)$.

In this definition C is termed the *constant of integration.* Again, C reflects the indefinite nature of finding the antiderivative, or indefinite integral.

Rules of Integration

Following are a set of rules for finding the indefinite integral of some functions common in business and economics applications.

> **RULE 1: CONSTANT FUNCTIONS**
>
> $$\int k\, dx = kx + C$$
>
> where k is any constant.

Example 7 illustrates this rule.

EXAMPLE 7

(a) $\displaystyle\int (-2)\, dx = -2x + C$

(b) $\displaystyle\int \frac{3}{2}\, dx = \frac{3}{2}x + C$

(c) $\displaystyle\int \sqrt{2}\, dx = \sqrt{2}x + C$

(d) $\displaystyle\int 0\, dx = (0)x + C = C$

❏

> **RULE 2: POWER RULE**
>
> $$\int x^n\, dx = \frac{x^{n+1}}{n+1} + C \quad n \neq -1$$

This rule is analogous to the power rule of differentiation. *Note that this rule is not valid when $n = -1$.* We will treat this exception shortly. Verbally, the rule states that when the integrand is x raised to some real-valued power, increase the exponent of x by 1, divide by the new exponent, and add the constant of integration. Example 8 provides several illustrations of this rule.

EXAMPLE 8

(a) $\displaystyle\int x\, dx = \frac{x^2}{2} + C$

(b) $\displaystyle\int x^2\, dx = \frac{x^3}{3} + C$

(c) $\displaystyle\int \sqrt{x}\, dx = \int x^{1/2}\, dx = \frac{x^{3/2}}{\frac{3}{2}} + C = \frac{2}{3}x^{3/2} + C$

(d) $\displaystyle\int \frac{1}{x^3}\, dx = \int x^{-3}\, dx = \frac{x^{-2}}{-2} + C = -\frac{1}{2x^2} + C$

❏

> **NOTE** Do not forget the built-in checking mechanism. It takes only a few seconds and may save you from careless errors. Find the derivative of the indefinite integrals found in Example 8 and see whether they equal the respective integrands. Some algebraic manipulation may be needed to verify these results.

RULE 3

$$\int kf(x)\, dx = k \int f(x)\, dx$$

where k is any constant

Verbally, this rule states that the indefinite integral of a constant k times a function f is found by multiplying the constant by the indefinite integral of f. Another way of viewing this rule is to say that whenever a *constant* can be factored from the integrand, the constant may also be factored outside the integral. Example 9 provides some illustrations of this rule.

EXAMPLE 9   (a) $\displaystyle \int 5x\, dx = 5 \int x\, dx$

$$= 5\left(\frac{x^2}{2} + C_1\right)$$

$$= \frac{5x^2}{2} + 5C_1$$

$$= \frac{5x^2}{2} + C$$

Check   If $f(x) = \dfrac{5x^2}{2} + C$, $f'(x) = \dfrac{5}{2}(2x) = 5x$ ✓

> **NOTE** With indefinite integrals we always add the constant of integration. In using rule 3, the algebra suggests that any constant k factored out of the integral will be multiplied by the constant of integration (e.g., the $5C_1$ term in this example). This multiplication is unnecessary. We simply need a constant of integration to indicate the "indefinite nature" of the integral. Thus, the convention is to add C and not a multiple of C. In the last step the $5C_1$ term is rewritten as just C, since C can represent any constant as well as $5C_1$.

(b) $\displaystyle \int \frac{x^2}{2}\, dx = \int \frac{1}{2} x^2\, dx$

$$= \frac{1}{2} \int x^2\, dx$$

$$= \frac{1}{2} \frac{x^3}{3} + C$$

$$= \frac{x^3}{6} + C$$

Check   If $f(x) = \frac{x^3}{6} + C,$ $f'(x) = \frac{3x^2}{6} = \frac{x^2}{2}$ ✓

(c) $\displaystyle\int \frac{3}{\sqrt{x}}\, dx = \int 3x^{-1/2}\, dx$

$$= 3 \int x^{-1/2}\, dx$$

$$= 3 \frac{x^{1/2}}{\frac{1}{2}} + C$$

$$= 6x^{1/2} + C$$

Check   If $f(x) = 6x^{1/2} + C,$ $f'(x) = 6(\tfrac{1}{2})x^{-1/2}$

$$= \frac{3}{x^{1/2}}$$

$$= \frac{3}{\sqrt{x}}$$ ✓

❑

RULE 4

If $\displaystyle\int f(x)\, dx$ and $\displaystyle\int g(x)\, dx$ exist, then

$$\int [f(x) \pm g(x)]\, dx = \int f(x)\, dx \pm \int g(x)\, dx$$

The indefinite integral of the sum (difference) of two functions is the sum (difference) of their respective indefinite integrals.

EXAMPLE 10   (a) $\displaystyle\int (3x - 6)\, dx = \int 3x\, dx - \int 6\, dx$

$$= \frac{3x^2}{2} + C_1 - (6x + C_2)$$

$$= \frac{3x^2}{2} - 6x + C_1 - C_2$$

$$= \frac{3x^2}{2} - 6x + C$$

Note again that even though the two integrals technically result in separate constants of integration, these constants may be considered together as one.

Check   If $f(x) = \dfrac{3x^2}{2} - 6x + C,$ $f'(x) = 3x - 6$ ✓

(b) $\displaystyle \int (4x^2 - 7x + 6)\, dx = \int 4x^2\, dx - \int 7x\, dx + \int 6\, dx$

$$= \frac{4x^3}{3} - \frac{7x^2}{2} + 6x + C$$

Check   If $f(x) = \dfrac{4x^3}{3} - \dfrac{7x^2}{2} + 6x + C,$ $f'(x) = \dfrac{12x^2}{3} - \dfrac{14x}{2} + 6$

$$= 4x^2 - 7x + 6 \quad ✓$$

❑

Section 18.2 Follow-up Exercises

In the following exercises, find the indefinite integral (if possible).

1 $\displaystyle \int 60\, dx$

2 $\displaystyle \int -25\, dx$

3 $\displaystyle \int dx/8$

4 $\displaystyle \int dx$

5 $\displaystyle \int 8x\, dx$

6 $\displaystyle \int (x/2)\, dx$

7 $\displaystyle \int -3x\, dx$

8 $\displaystyle \int (-8x/3)\, dx$

9 $\displaystyle \int (3x + 6)\, dx$

10 $\displaystyle \int (10 - 5x)\, dx$

11 $\displaystyle \int (x/3 - 1/4)\, dx$

12 $\displaystyle \int (x/2 + 1/4)\, dx$

13 $\displaystyle \int (3x^2 - 4x + 2)\, dx$

14 $\displaystyle \int (-6x^2 + 10)\, dx$

15 $\displaystyle \int (-18x^2 + x - 5)\, dx$

16 $\displaystyle \int (10 - 6x + 15x^2)\, dx$

17 $\displaystyle \int (4x^3 + 6x^2 - 3)\, dx$

18 $\displaystyle \int (8x^3 + 6x^2 - 2x + 10)\, dx$

19 $\displaystyle \int (x^5 - 12x^3 + 3)\, dx$

20 $\displaystyle \int (8x^3 + x^2/2 + 6)\, dx$

21 $\displaystyle \int \sqrt[5]{x}\, dx$

22 $\displaystyle \int (2/\sqrt[3]{x})\, dx$

23 $\displaystyle \int dx/x^3$

24 $\displaystyle \int (20/\sqrt[3]{x^2})\, dx$

25 $\int (ax^4 + bx^2)\, dx$ 26 $\int (mx + b)\, dx$

27 $\int (a/bx^n)\, dx$ 28 $\int \sqrt[b]{x}\, dx$

29 $\int dx/x^n$ 30 $\int (a/\sqrt[b]{x})\, dx$

31 $\int 2\sqrt[3]{x}\, dx$ 32 $\int 3\sqrt{x}\, dx$

33 $\int \sqrt[5]{x^3}\, dx$ 34 $\int 8\sqrt[4]{x}\, dx$

35 $\int (4\sqrt[4]{x}/3)\, dx$ 36 $\int (3\sqrt[3]{x}/2)\, dx$

37 $\int (dx/x^4)$ 38 $\int (-8\, dx/x^3)$

39 $\int (16\, dx/x^2)$ 40 $\int (3\, dx/x^3)$

41 $\int (dx/\sqrt{x})$ 42 $\int (dx/\sqrt[3]{x})$

43 $\int (-15\, dx/\sqrt[5]{x})$ 44 $\int (-2\, dx/\sqrt{x})$

45 $\int (ax^4 + bx^3 + cx^2 + dx + e)\, dx$ 46 $\int (6\, dx/x^6)$

47 $\int (4ax^3 + 3bx^2 + 2cx)\, dx$ 48 $\int (dx/ax^2)$

49 $\int (dx/ax^n)$ 50 $\int ([1/x^2] + [2/x^3])\, dx$

18.3 ADDITIONAL RULES OF INTEGRATION

This section presents additional rules of integration and illustrates their application.

RULE 5: POWER-RULE EXCEPTION

$$\int x^{-1}\, dx = \ln x + C$$

This is the exception associated with rule 2 (the power rule) where $n = -1$ for x^n. Remember our differentiation rules? If $f(x) = \ln x$, $f'(x) = 1/x = x^{-1}$.

RULE 6

$$\int e^x\, dx = e^x + C$$

> **RULE 7**
> $$\int [f(x)]^n f'(x)\, dx = \frac{[f(x)]^{n+1}}{n+1} + C \qquad n \neq -1$$

This rule is similar to the power rule (rule 2). In fact, the power rule is the special case of this rule where $f(x) = x$. If the integrand consists of the product of a function f raised to a power n and the derivative of f, the indefinite integral is found by increasing the exponent of f by 1 and dividing by the new exponent.

EXAMPLE 11   Evaluate $\displaystyle\int (5x - 3)^3 (5)\, dx$.

SOLUTION

When an integrand is identified which contains a function raised to a power, you should immediately think of rule 7. The first step is to determine the function f. In this case, the function which is raised to the third power is

$$f(x) = 5x - 3$$

Once f has been identified, f' should be determined. In this case

$$f'(x) = 5$$

If the integrand has the form $[f(x)]^n f'(x)$, then rule 7 applies. The integrand in this example *does* have the required form, and

$$\int \overbrace{(5x-3)^3}^{f(x)}\,\overbrace{(5)}^{f'(x)}\, dx = \frac{(5x-3)^4}{4} + C$$

Check   If $f(x) = \dfrac{(5x-3)^4}{4} + C,$ $\qquad f'(x) = \dfrac{4}{4}(5x-3)^3(5)$

$$= (5x-3)^3(5) \quad \checkmark$$

EXAMPLE 12   Evaluate $\displaystyle\int \sqrt{2x^2 - 6}\,(4)\, dx$.

SOLUTION

The integrand can be rewritten as

$$\int (2x^2 - 6)^{1/2}(4)\, dx$$

Referring to rule 7, we have

$$f(x) = 2x^2 - 6 \quad \text{and} \quad f'(x) = 4x$$

For rule 7 to apply, $(2x^2 - 6)^{1/2}$ should be multiplied by f', or $4x$, in the integrand. Since the other factor in the integrand is 4 and not $4x$, we cannot evaluate the integral using rule 7.

EXAMPLE 13 Evaluate $\displaystyle\int (x^2 - 2x)^5(x - 1)\, dx$.

SOLUTION

For this integral

$$f(x) = x^2 - 2x \quad \text{and} \quad f'(x) = 2x - 2$$

Again it seems that the integrand is not in the proper form. To apply rule 7, the second factor in the integrand should be $2x - 2$, not $x - 1$. However, recalling rule 3 and using some algebraic manipulations, we get

$$\int (x^2 - 2x)^5(x - 1)\, dx = \frac{2}{2}\int (x^2 - 2x)^5(x - 1)\, dx$$

$$= \frac{1}{2}\int (x^2 - 2x)^5(2)(x - 1)\, dx$$

$$= \frac{1}{2}\int (x^2 - 2x)^5(2x - 2)\, dx \qquad (18.8)$$

What we have done is *manipulate* the integrand into the proper form. Rule 3 indicated that *constants* can be factored from inside to outside the integral sign. Similarly, we can move a constant which is a factor from outside the integral sign to inside. We multiplied the integrand by 2 and offset this by multiplying the integral by $\frac{1}{2}$. Effectively, we have simply multiplied the original integral by $\frac{2}{2}$, or 1. Thus, we have changed the appearance of the original integral but not its value.

Evaluating the integral in Eq. (18.8) gives

$$\int (x^2 - 2x)^5(x - 1)\, dx = \frac{1}{2}\int \overbrace{(x^2 - 2x)^5}^{f(x)}\overbrace{(2x - 2)}^{f'(x)}\, dx$$

$$= \frac{1}{2}\frac{(x^2 - 2x)^6}{6} + C$$

$$= \frac{(x^2 - 2x)^6}{12} + C$$

Check If $f(x) = \dfrac{(x^2 - 2x)^6}{12} + C,$ $f'(x) = \dfrac{6}{12}(x^2 - 2x)^5(2x - 2)$

$$= \frac{6(x^2 - 2x)^5(2)(x - 1)}{12}$$

$$= (x^2 - 2x)^5(x - 1) \quad \checkmark$$

EXAMPLE 14 Evaluate $\displaystyle\int (x^4 - 2x^2)^4(4x^2 - 4)\, dx$.

SOLUTION

For this integral

$$f(x) = x^4 - 2x^2 \quad \text{and} \quad f'(x) = 4x^3 - 4x$$

The integrand is not quite in the form of rule 7. There is great temptation to perform the following operations.

$$\int (x^4 - 2x^2)^4 (4x^2 - 4) \, dx = \frac{x}{x} \int (x^4 - 2x^2)^4 (4x^2 - 4) \, dx$$

$$= \frac{1}{x} \int (x^4 - 2x^2)^4 x (4x^2 - 4) \, dx$$

$$= \frac{1}{x} \int (x^4 - 2x^2)^4 (4x^3 - 4x) \, dx$$

However, we have not discussed any property which allows us to factor *variables* through an integral sign. Constants yes; variables no! Therefore, given our current rules, we cannot evaluate the integral.

❑

RULE 8

$$\int f'(x) e^{f(x)} \, dx = e^{f(x)} + C$$

This rule, as with the previous rule, requires that the integrand be in a very specific form. Rule 6 is actually the special case of this rule when $f(x) = x$.

EXAMPLE 15 Evaluate $\displaystyle\int 2x e^{x^2} \, dx$.

SOLUTION

When an integrand is identified which contains e raised to a power that is a function of x, you should immediately think of rule 8. As with rule 7, the next step is to see if the integrand has the form required to apply rule 8. For this integrand

$$f(x) = x^2 \quad \text{and} \quad f'(x) = 2x$$

Refer to rule 8: the integrand has the appropriate form, and

$$\int \overbrace{2x}^{f'(x)} \; \overbrace{e^{x^2}}^{f(x)} \; dx = e^{x^2} + C$$

Check   If $f(x) = e^{x^2} + C$, $f'(x) = (2x) e^{x^2}$ ✔

EXAMPLE 16 Evaluate $\int x^2 e^{3x^3}\, dx$.

SOLUTION
Referring to rule 8, for this integrand we have

$$f(x) = 3x^3 \quad \text{and} \quad f'(x) = 9x^2$$

The integrand is currently not in a form which is suitable for using rule 8. However,

$$\int x^2 e^{3x^3}\, dx = \frac{9}{9} \int x^2 e^{3x^3}\, dx$$

$$= \frac{1}{9} \int 9x^2 e^{3x^3}\, dx$$

which is suitable for using rule 8. Therefore,

$$\int x^2 e^{3x^3}\, dx = \tfrac{1}{9} e^{3x^3} + C$$

Check If $f(x) = \tfrac{1}{9} e^{3x^3} + C$, $f'(x) = \tfrac{1}{9} e^{3x^3}(9x^2) = x^2 e^{3x^3}$ ✔

RULE 9

$$\int \frac{f'(x)}{f(x)}\, dx = \ln f(x) + C$$

EXAMPLE 17 Evaluate $\int \dfrac{6x}{3x^2 - 10}\, dx$.

SOLUTION
Referring to rule 9, we have

$$f(x) = 3x^2 - 10 \quad \text{and} \quad f'(x) = 6x$$

Since the integrand has the form required by rule 9,

$$\int \frac{6x\, dx}{3x^2 - 10} = \ln(3x^2 - 10) + C$$

Check If $f(x) = \ln(3x^2 - 10) + C$, $f'(x) = \dfrac{6x}{3x^2 - 10}$ ✔

EXAMPLE 18 Evaluate $\int \dfrac{x - 1}{4x^2 - 8x + 10}\, dx$.

SOLUTION

Referring to rule 9, we have

$$f(x) = 4x^2 - 8x + 10 \quad \text{and} \quad f'(x) = 8x - 8$$

The integrand does not seem to comply with that required by rule 9. However, an algebraic manipulation allows us to rewrite the integrand in the required form, or

$$\int \frac{x-1}{4x^2 - 8x + 10}\, dx = \frac{8}{8} \int \frac{x-1}{4x^2 - 8x + 10}\, dx$$

$$= \frac{1}{8} \int \frac{8(x-1)}{4x^2 - 8x + 10}\, dx$$

$$= \frac{1}{8} \int \frac{8x-8}{4x^2 - 8x + 10}\, dx$$

which does comply with the form of rule 9. Therefore,

$$\int \frac{x-1}{4x^2 - 8x + 10}\, dx = \frac{1}{8} \ln(4x^2 - 8x + 10) + C$$

Check   If $f(x) = \frac{1}{8} \ln(4x^2 - 8x + 10) + C$,

$$f'(x) = \frac{1}{8}\left[\frac{8x-8}{4x^2 - 8x + 10}\right] = \frac{x-1}{4x^2 - 8x + 10} \quad \checkmark$$

❑

Section 18.3 Follow-up Exercises

In the following exercises, find the indefinite integral (if possible).

1 $\int (x+20)^5\, dx$

2 $\int \sqrt{x-30}\, dx$

3 $\int 8(8x+20)^3\, dx$

4 $\int \frac{1}{2}[10 - (x/2)]^4\, dx$

5 $\int [(x/3) + 15]^2(\frac{1}{3})\, dx$

6 $\int \sqrt{8+2x}\,(6)\, dx$

7 $\int (3x-10)^3(x)\, dx$

8 $\int x\sqrt{x+6}\, dx$

9 $\int (x^2+3)^4(2x)\, dx$

10 $\int (x^3+1)^4(3x^2)\, dx$

11 $\int (x^3+5)^3(x^2)\, dx$

12 $\int (x^2+3)^{3/2}(x)\, dx$

13 $\int (2x^2-4x)^6(x-1)\, dx$

14 $\int (x^2/4 - x/2)^5(x-1)\, dx$

15 $\int (2x/\sqrt{x^2+8})\,dx$

16 $\int 3x^2/(x^3+4)^3\,dx$

17 $\int (4x^3+8x^2)^5(x)\,dx$

18 $\int 4x^3\sqrt{x^2+1}\,dx$

19 $\int (4x^3+1)^3(12x)\,dx$

20 $\int (3x^4-5)^4(12x^2)\,dx$

21 $\int \sqrt{2x^3+3}\,(x^2)\,dx$

22 $\int \sqrt[3]{20+3x^3}\,(x^2)\,dx$

23 $\int (2x^2+8x)^3(x+2)\,dx$

24 $\int (3x-3x^3)^4(3x^2-1)\,dx$

25 $\int (x^3+3x^4)^3(3x+12x^2)\,dx$

26 $\int \sqrt{9x-3x^2}\,(3-2x)\,dx$

27 $\int e^{x^2}\,dx$

28 $\int e^{x-8}\,dx$

29 $\int e^{3x}\,dx$

30 $\int 2xe^{x^2}\,dx$

31 $\int e^{ax}\,dx$

32 $\int (x+2)e^{x^2+4x}\,dx$

33 $\int \dfrac{-x}{x^2+5}\,dx$

34 $\int \dfrac{4x}{100+x^2}\,dx$

35 $\int \dfrac{18}{6x+5}\,dx$

36 $\int \dfrac{x^3-1}{x^4-4x}\,dx$

37 $\int \dfrac{dx}{ax+b}$

38 $\int \dfrac{ax-1}{ax^2-2x}\,dx$

39 $\int \dfrac{2ax+b}{ax^2+bx}\,dx$

40 $\int \dfrac{a-bx}{2ax-bx^2}\,dx$

18.4 OTHER TECHNIQUES OF INTEGRATION (OPTIONAL)

The nine integration rules presented in Secs. 18.2 and 18.3 apply only to a subset of the functions which might be integrated. This subset includes some of the more common functions used in business and economics applications. A natural question is, What happens when our rules do not apply? This section discusses two techniques which can be employed when the other rules do not apply and when the structure of the integrand is of an appropriate form. We will also discuss the use of special tables of integration formulas.

Integration by Parts

Recall the product rule of differentiation from Chap. 15. This rule stated that if

$$f(x) = u(x)v(x)$$

then $\qquad\qquad\qquad f'(x) = v(x)u'(x) + u(x)v'(x)$

We can write this rule in a slightly different form as

$$\frac{d}{dx}[u(x)v(x)] = v(x)u'(x) + u(x)v'(x)$$

If we integrate both sides of this equation, the result is

$$u(x)v(x) = \int v(x)u'(x)\,dx + \int u(x)v'(x)\,dx$$

And rewriting this equation, we get the ***integration-by-parts formula.***

INTEGRATION-BY-PARTS FORMULA

$$\int u(x)v'(x)\,dx = u(x)v(x) - \int v(x)u'(x)\,dx \tag{18.9}$$

This equation expresses a relationship which can be used to determine integrals where the integrand has the form $u(x)v'(x)$.

The integration-by-parts procedure is a trial-and-error method which may or may not be successful for a given integrand. If the integrand is in the form of a product and the other integration rules do not apply, the following procedure can be attempted.

TRIAL-AND-ERROR INTEGRATION-BY-PARTS PROCEDURES

I   *Define two functions u and v and determine whether the integrand has the form $u(x)v'(x)$.*

II *If two functions are found such that $u(x)v'(x)$ equals the integrand, attempt to find the integral by evaluating the right side of Eq. (18.9). The key is whether you can evaluate $\int v(x)u'(x)\,dx$.*

The following examples illustrate the approach.

EXAMPLE 19   Determine $\int xe^x\,dx$.

SOLUTION

Our first observation should be that the integrand is in a product form. The temptation is to try to use rule 8, which applies for integrals of the form $\int f'(x)e^{f(x)}\,dx$. With the exponent $f(x)$ defined as x, $f'(x) = 1$ and the integrand is not in an appropriate form to apply rule 8.

Let's try to define two functions u and v such that the integrand has the form $u(x)v'(x)$.

NOTE   A hint is to examine the factors of the integrand to determine whether one of them has the form of the derivative of another function.

Let's define v' as equaling x and u as equaling e^x so that the integrand has the form

$$\int \overbrace{x}^{v'(x)} \overbrace{e^x}^{u(x)} dx$$

With these definitions we can determine v by integrating v' and u' by differentiating u, or

$$v'(x) = x \qquad \text{suggests that} \qquad v(x) = \frac{x^2}{2}$$

and $\qquad\qquad u(x) = e^x \qquad \text{suggests that} \qquad u'(x) = e^x$

With u, v, and their derivatives defined, we substitute into Eq. (18.9):

$$\int \overbrace{xe^x}^{v'(x)u(x)} dx = \overbrace{e^x \frac{x^2}{2}}^{u(x)v(x)} - \int \overbrace{\frac{x^2}{2} e^x}^{v(x)u'(x)} dx$$

An examination of $\int (x^2/2)e^x \, dx$ suggests that this integral may be as difficult to evaluate as the original integral.

So let's backtrack and start again. Let's redefine v' and u such that $v'(x) = e^x$ and $u(x) = x$:

$$\int \overbrace{x}^{u(x)} \overbrace{e^x}^{v'(x)} dx$$

Given these definitions,

$$u(x) = x \qquad \text{suggests that} \qquad u'(x) = 1$$
$$v'(x) = e^x \qquad \text{suggests that} \qquad v(x) = e^x$$

Substituting into Eq. (18.9) yields

$$\int \overbrace{xe^x}^{u(x)v'(x)} dx = \overbrace{x \ e^x}^{u(x)v(x)} - \int \overbrace{e^x(1)}^{v(x)u'(x)} dx$$

$$= xe^x - \int e^x \, dx$$

Because $\int e^x \, dx = e^x$,

$$\int xe^x \, dx = xe^x - e^x + C$$

Check   Differentiating this answer as a check, we find

$$\frac{d}{dx}[xe^x - e^x + C] = (1)e^x + e^x x - e^x$$

$$= xe^x \quad \checkmark$$

EXAMPLE 20 Determine $\int x^2 \ln x\, dx$.

SOLUTION

If we let $u(x) = \ln x$ and $v'(x) = x^2$, then

$$u'(x) = \frac{1}{x} \quad \text{and} \quad v(x) = \int x^2\, dx = \frac{x^3}{3}$$

Substituting into Eq. (18.9) gives

$$\int \overbrace{x^2 \ln x}^{v'(x)u(x)}\ dx = \overbrace{(\ln x)}^{u(x)} \overbrace{\left(\frac{x^3}{3}\right)}^{v(x)} - \int \overbrace{\frac{x^3}{3}\frac{1}{x}}^{v(x)u'(x)}\ dx$$

$$= \frac{x^3}{3}\ln x - \int \frac{x^2}{3}\, dx$$

$$= \frac{x^3}{3}\ln x - \frac{x^3}{9} + C$$

Check   Checking this answer by differentiating, we get

$$\frac{d}{dx}\left[\frac{x^3}{3}\ln x - \frac{x^3}{9} + C\right] = \frac{3x^2}{3}\ln x + \frac{1}{x}\frac{x^3}{3} - \frac{3x^2}{9}$$

$$= x^2 \ln x + \frac{x^2}{3} - \frac{x^2}{3}$$

$$= x^2 \ln x \quad \checkmark$$

EXAMPLE 21 Determine $\int \ln x\, dx$.

SOLUTION

Although the integrand is not in the form of a product, we can imagine it to have the form

$$\int \ln x (1)\, dx$$

Letting $u(x) = \ln x$ and $v'(x) = 1$, we have

$$u'(x) = \frac{1}{x}$$

and
$$v(x) = \int 1\, dx$$
$$= x$$

Substituting into Eq. (18.9) yields

$$\int \overbrace{\ln x}^{u(x)v'(x)} dx = \overbrace{(\ln x)}^{u(x)} \overbrace{(x)}^{v(x)} - \int \overbrace{x\frac{1}{x}}^{v(x)u'(x)} dx$$

$$= x \ln x - \int dx$$

$$= x \ln x - x + C$$

Check This answer can be checked by differentiating, or

$$\frac{d}{dx}[x \ln x - x + C] = (1) \ln x + \frac{1}{x}x - 1$$

$$= \ln x + 1 - 1$$

$$= \ln x \;\checkmark$$

Integration by parts is often time-consuming, given the trial-and-error approach required. You are again reminded to first examine the integrand carefully to determine whether our other rules apply before you try this procedure.

Integration by Partial Fractions

Rational functions have the form of a quotient of two polynomials. Many rational functions exist which cannot be integrated by the rules (specifically, rule 9) presented earlier. When this occurs, one possibility is that the rational function can be restated in an equivalent form consisting of more elementary functions. The following example illustrates the decomposition of a rational function into equivalent **partial fractions.**

$$f(x) = \frac{x + 3}{x^2 + 3x + 2} = \frac{2}{x + 1} - \frac{1}{x + 2}$$

Verify that the rational function

$$f(x) = \frac{x + 3}{x^2 + 3x + 2}$$

cannot be integrated using rule 9. However, the equivalent partial fractions can be integrated. Thus,

$$\int \frac{x+3}{x^2+3x+2}\,dx = \int \frac{2}{x+1}\,dx - \int \frac{dx}{x+2}$$

$$= 2\ln(x+1) - \ln(x+2) + C$$

$$= \ln\left[\frac{(x+1)^2}{(x+2)}\right] + C$$

This technique for integrating rational functions is called the ***method of partial fractions.***

Let's discuss the process of decomposing a rational function into equivalent partial fractions. To apply the method of partial fractions, the rational function must have the form of a *proper fraction.* A rational function is a proper fraction if the degree of the numerator polynomial is lower than the degree of the denominator polynomial.

ALGEBRA FLASHBACK
The **degree of a term** is the sum of the exponents on the variables contained in the term. For example, the degree of $5x^2yz^3$ is 6. The **degree of a polynomial** is the degree of the term of highest degree in the polynomial. For example, the degree of the polynomial

$$2x^3 - 4x^2 + x - 10$$

is 3.

The rational function $f(x) = x^2/(5x^3 - 2x + 1)$ is in the form of a proper fraction. In an *improper fraction,* the degree of the numerator polynomial is the same or higher than that of the denominator polynomial. The rational function $f(x) = x^3/(3x^2 - 10)$ is in the form of an improper fraction. Improper fractions can be reduced to the algebraic sum of a polynomial and a proper fraction by performing long division of the numerator and denominator functions. To illustrate, the improper fraction

$$\frac{x^3 - 2x}{x - 1}$$

can be divided as follows:

$$
\begin{array}{r}
x^2 + x - 1 \\
x - 1 \overline{)\, x^3 - 2x } \\
\underline{x^3 - x^2 } \\
x^2 - 2x \\
\underline{x^2 - x } \\
-x \\
\underline{-x + 1} \\
-1
\end{array}
$$

Thus, long division results in

$$\frac{x^3 - 2x}{x - 1} = x^2 + x - 1 + \frac{-1}{x - 1}$$

If our objective is to integrate

$$f(x) = \frac{x^3 - 2x}{x - 1}$$

we could do so as follows:

$$\int \frac{x^3 - 2x}{x - 1}\, dx = \int \left(x^2 + x - 1 + \frac{-1}{x - 1} \right) dx$$

$$= \int (x^2 + x - 1)\, dx - \int \frac{dx}{x - 1}$$

$$= \frac{x^3}{3} + \frac{x^2}{2} - x - \ln(x - 1) + C$$

Given a proper fraction, the decomposition into equivalent partial fractions requires that the denominator be factored. In general, for each factor of the denominator there is a corresponding partial fraction. The form of each factor determines the form of the equivalent partial fraction. Table 18.1 summarizes some of the possibilities.

TABLE 18.1

Form of Factor	Form of the Corresponding Partial Fraction
1 Unique linear factor $ax + b$	$\dfrac{A}{ax + b}$, A constant
2 Repeated linear factor $(ax + b)^n$	$\dfrac{A_1}{ax + b} + \dfrac{A_2}{(ax + b)^2} + \cdots + \dfrac{A_n}{(ax + b)^n}$
3 Unique quadratic factor $ax^2 + bx + c$	$\dfrac{Ax + B}{ax^2 + bx + c}$, A and B constants

EXAMPLE 22   Earlier we stated that

$$f(x) = \frac{x + 3}{x^2 + 3x + 2} = \frac{2}{x + 1} - \frac{1}{x + 2}$$

Let's derive these partial fractions. The first step is to try and factor the denominator of f. Because

$$x^2 + 3x + 2 = (x + 1)(x + 2)$$

the denominator can be factored into two *linear* factors. According to Table 18.1, the decomposition of f should result in *two* (one for each factor) partial fractions, or

$$\frac{x + 3}{(x + 1)(x + 2)} = \frac{A_1}{x + 1} + \frac{A_2}{x + 2} \qquad \textbf{(18.10)}$$

To solve for the constants A_1 and A_2, the two fractions on the right side of Eq. (18.10) are combined over the common denominator $(x + 1)(x + 2)$, yielding

$$\frac{x + 3}{(x + 1)(x + 2)} = \frac{A_1(x + 2) + A_2(x + 1)}{(x + 1)(x + 2)}$$

or
$$\frac{x + 3}{(x + 1)(x + 2)} = \frac{(A_1 + A_2)x + 2A_1 + A_2}{(x + 1)(x + 2)}$$

For the two sides of this equation to be equal, the numerators of the fractions must be equal. A_1 and A_2 must be such that

$$A_1 + A_2 = 1$$

and
$$2A_1 + A_2 = 3$$

Solving these equations, $A_1 = 2$ and $A_2 = -1$. If these values are substituted into Eq. (18.10), the result is

$$\frac{x + 3}{(x + 1)(x + 2)} = \frac{2}{x + 1} - \frac{1}{x + 2}$$

which is the result shown earlier.

EXAMPLE 23   Using the method of partial fractions, find the indefinite integral

$$\int \frac{5x + 8}{x^2 + 4x + 4}\, dx$$

SOLUTION

The first step is to verify that rules 7 and 9 do not apply. Once convinced of this, we attempt to factor the denominator. Because

$$x^2 + 4x + 4 = (x + 2)^2$$

Table 18.1 indicates that the integrand can be decomposed into the general partial fractions

$$\frac{5x + 8}{(x + 2)^2} = \frac{A_1}{x + 2} + \frac{A_2}{(x + 2)^2} \qquad \textbf{(18.11)}$$

To solve for A_1 and A_2, the two partial fractions on the right side of Eq. (18.11) are combined over the common denominator $(x+2)^2$, yielding

$$\frac{5x+8}{(x+2)^2} = \frac{A_1(x+2) + A_2}{(x+2)^2}$$

or

$$\frac{5x+8}{(x+2)^2} = \frac{A_1 x + 2A_1 + A_2}{(x+2)^2}$$

For the two sides of this equation to be equal, A_1 and A_2 must be chosen such that

$$A_1 = 5$$

and

$$2A_1 + A_2 = 8$$

Since $A_1 = 5$,

$$2(5) + A_2 = 8$$

$$A_2 = -2$$

Substituting these values into Eq. (18.11) gives

$$\frac{5x+8}{(x+2)^2} = \frac{5}{x+2} - \frac{2}{(x+2)^2}$$

Therefore,

$$\int \frac{5x+8}{(x+2)^2}\,dx = \int \frac{5}{x+2}\,dx - \int \frac{2}{(x+2)^2}\,dx$$

$$= 5\int \frac{dx}{x+2} - 2\int (x+2)^{-2}\,dx$$

$$= 5\ln(x+2) - 2\frac{(x+2)^{-1}}{-1} + C$$

$$= 5\ln(x+2) + \frac{2}{(x+2)} + C$$

EXAMPLE 24 Using the method of partial fractions, find the indefinite integral

$$\int \frac{2x^2 - 1}{x^3 + x^2}\,dx$$

SOLUTION

The first step is to verify that rule 9 does not apply for this integrand. Once convinced of this, we attempt to factor the denominator. Because

$$x^3 + x^2 = x^2(x+1)$$

we conclude that it has two factors: the quadratic factor x^2 and the linear factor $(x + 1)$. According to Table 18.1, the integrand can be decomposed into the general partial fractions

$$\frac{2x^2 - 1}{x^2(x + 1)} = \frac{A_1 x + B_1}{x^2} + \frac{A_2}{x + 1} \qquad \textbf{(18.12)}$$

To solve for the constants A_1, B_1, and A_2, the two partial fractions on the right side of Eq. (18.12) are combined over the common denominator $x^2(x + 1)$, yielding

$$\frac{2x^2 - 1}{x^2(x + 1)} = \frac{(A_1 x + B_1)(x + 1) + A_2 x^2}{x^2(x + 1)}$$

$$= \frac{A_1 x^2 + A_1 x + B_1 x + B_1 + A_2 x^2}{x^2(x + 1)}$$

or $\qquad \dfrac{2x^2 - 1}{x^2(x + 1)} = \dfrac{(A_1 + A_2)x^2 + (A_1 + B_1)x + B_1}{x^2(x + 1)}$

For the two sides of this equation to be equal, A_1, A_2, and B_1 must be such that

$$A_1 + A_2 = 2$$

$$A_1 + B_1 = 0$$

and $\qquad\qquad\qquad B_1 = -1$

Substituting the value $B_1 = -1$ into the second equation yields $A_1 = 1$. And substituting this value into the first equation results in $A_2 = 1$.

Therefore,

$$\frac{2x^2 - 1}{x^2(x + 1)} = \frac{x - 1}{x^2} + \frac{1}{x + 1}$$

and $\qquad \displaystyle\int \frac{2x^2 - 1}{x^3 + x^2}\, dx = \int \frac{x - 1}{x^2}\, dx + \int \frac{dx}{x + 1}$

Although the integrand of the first partial fraction is not in an appropriate form to apply rule 9,

$$\frac{x - 1}{x^2} = \frac{x}{x^2} - \frac{1}{x^2}$$

$$= \frac{1}{x} - \frac{1}{x^2}$$

Therefore, $\qquad \displaystyle\int \frac{(2x^2 - 1)}{x^3 + x^2}\, dx = \int \frac{dx}{x} - \int \frac{dx}{x^2} + \int \frac{dx}{x + 1}$

$$= \ln x + \frac{1}{x} + \ln(x + 1) + C$$

□

PRACTICE EXERCISE

In Example 24, the denominator was factored into $x^2(x+1)$. Equation (18.12) assumed that the x^2 factor was a unique quadratic factor, according to Table 18.1. This x^2 factor could also be viewed as a repeated linear factor, according to Table 18.1. Use the method of partial fractions to find the indefinite integral with x^2 assumed to be a repeated linear factor.

Tables of Integrals

For cases where our rules and other procedures are inadequate for determining indefinite integrals, special tables of integrals are available which may contain literally hundreds of integration formulas. Each formula applies to an integrand which has a particular functional form. To use the tables, you match the form of your integrand with the corresponding general form in the table. Once the appropriate formula has been identified, the indefinite integral follows directly from the formula.

Table 18.2 illustrates some sample integration formulas for natural logarithms and exponential forms of an integrand.

TABLE 18.2

1. $\displaystyle\int \ln x\, dx = x \ln x - x + C$

2. $\displaystyle\int x \ln x\, dx = \frac{x^2}{2} \ln x - \frac{x^2}{4} + C$

3. $\displaystyle\int x^2 \ln x\, dx = \frac{x^3}{3} \ln x - \frac{x^3}{9} + C$

4. $\displaystyle\int x^n \ln ax\, dx = \frac{x^{n+1}}{n+1} \ln ax - \frac{x^{n+1}}{(n+1)^2} + C \qquad n \neq -1$

5. $\displaystyle\int (\ln x)^2\, dx = x(\ln x)^2 - 2x \ln x + 2x + C$

6. $\displaystyle\int (\ln x)^n\, dx = x(\ln x)^n - n \int (\ln x)^{n-1}\, dx + C \qquad n \neq -1$

7. $\displaystyle\int \frac{(\ln x)^n}{x}\, dx = \frac{1}{n+1} (\ln x)^{n+1} + C \qquad n \neq -1$

8. $\displaystyle\int x^n \ln x\, dx = x^{n+1} \left[\frac{\ln x}{n+1} - \frac{1}{(n+1)^2} \right] + C \qquad n \neq -1$

9. $\displaystyle\int e^{-x}\, dx = -e^{-x} + C$

10. $\displaystyle\int e^{ax}\, dx = \frac{e^{ax}}{a} + C$

11. $\displaystyle\int xe^{ax}\, dx = \frac{e^{ax}}{a^2} (ax - 1) + C$

(Table 18.2 continued)

(Table 18.2 continued)

$$12 \int \frac{dx}{1 + e^x} = x - \ln(1 + e^x) = \ln \frac{e^x}{1 + e^x} + C$$

$$13 \int \frac{dx}{a + be^{px}} = \frac{x}{a} - \frac{1}{ap} \ln(a + be^{px}) + C$$

EXAMPLE 25   According to formula (4) in Table 18.2

$$\int x^3 \ln 5x \, dx = \frac{x^{3+1}}{3+1} \ln 5x - \frac{x^{3+1}}{(3+1)^2} + C$$

$$= \frac{x^4}{4} \ln 5x - \frac{x^4}{16} + C$$

We can check this result, as before, by differentiating.

$$\frac{d}{dx} \left(\frac{x^4}{4} \ln 5x - \frac{x^4}{16} + C \right) = \left(\frac{4x^3}{4} \ln 5x + \frac{1}{x} \frac{x^4}{4} \right) - \frac{4x^3}{16}$$

$$= x^3 \ln 5x + \frac{x^3}{4} - \frac{x^3}{4}$$

$$= x^3 \ln 5x$$

which was the original integrand.

EXAMPLE 26   According to formula (7),

$$\int \frac{(\ln x)^5}{x} \, dx = \frac{1}{5+1} (\ln x)^{5+1} + C$$

$$= \frac{(\ln x)^6}{6} + C$$

To check this result,

$$\frac{d}{dx} \left[\frac{(\ln x)^6}{6} + C \right] = \frac{1}{6} [6(\ln x)^5] \frac{1}{x}$$

$$= \frac{(\ln x)^5}{x}$$

which was the original integrand.

EXAMPLE 27   According to formula (10),

$$\int e^{-5x} \, dx = \frac{e^{-5x}}{-5} + C$$

To check this result,

$$\frac{d}{dx}\left(\frac{e^{-5x}}{-5} + C\right) = -\frac{1}{5}(-5)e^{-5x}$$
$$= e^{-5x}$$

(Note: We could have manipulated the integrand in this example so as to apply rule 8.)

☐

Section 18.4 Follow-up Exercises

In Exercises 1–14, determine the indefinite integral (if possible) using integration by parts.

1 $\displaystyle\int xe^{-x}\,dx$ **2** $\displaystyle\int 5xe^x\,dx$

3 $\displaystyle\int x\sqrt[3]{x+1}\,dx$ **4** $\displaystyle\int x\sqrt{x+1}\,dx$

5 $\displaystyle\int xe^{2x}\,dx$ **6** $\displaystyle\int xe^{-2x}\,dx$

7 $\displaystyle\int (x+4)\ln x\,dx$ **8** $\displaystyle\int x^2\ln 5x\,dx$

9 $\displaystyle\int x(x+2)^4\,dx$ **10** $\displaystyle\int x(x-4)^5\,dx$

11 $\displaystyle\int (x/\sqrt{x-3})\,dx$ **12** $\displaystyle\int (x/[x-3]^2)\,dx$

13 $\displaystyle\int (\ln x/x^2)\,dx$ **14** $\displaystyle\int (2x+5)(x+1)^{1/2}\,dx$

In Exercises 15–22, use the method of partial fractions to find the indefinite integral.

15 $\displaystyle\int \frac{5-x}{x^2+5x+6}\,dx$ **16** $\displaystyle\int \frac{5x-7}{x^2-2x-15}\,dx$

17 $\displaystyle\int \frac{7-2x}{x^2-2x+1}\,dx$ **18** $\displaystyle\int \frac{10x+25}{x^2+6x+9}\,dx$

19 $\displaystyle\int \frac{5x^2-2x+64}{x^3-16x}\,dx$ **20** $\displaystyle\int \frac{x-3}{x^3+2x^2}\,dx$

21 $\displaystyle\int \frac{4x^2-2x-6}{x^3-x}\,dx$ **22** $\displaystyle\int \frac{36-9x-5x^2}{x^3-9x}\,dx$

In Exercises 23–36, determine the indefinite integral (if possible) using Table 18.2.

23 $\displaystyle\int x^4\ln 10x\,dx$ **24** $\displaystyle\int (\ln 4x/x^2)\,dx$

25 $\displaystyle\int (\ln x/x^3)\,dx$ **26** $\displaystyle\int (\ln x)^4\,dx$

27 $\displaystyle\int (\ln x)^2 \, dx$ **28** $\displaystyle\int ([\ln x]^3/x) \, dx$

29 $\displaystyle\int x^4 \ln x \, dx$ **30** $\displaystyle\int (\ln x/x^5) \, dx$

31 $\displaystyle\int e^{2.5x} \, dx$ **32** $\displaystyle\int e^{-2x} \, dx$

33 $\displaystyle\int xe^{5x} \, dx$ **34** $\displaystyle\int (x/e^{3x}) \, dx$

35 $\displaystyle\int \frac{dx}{5 + 3e^{2x}}$ **36** $\displaystyle\int \frac{dx}{10 - 2e^x}$

18.5 DIFFERENTIAL EQUATIONS

A *differential equation* is an equation which involves derivatives and/or differentials.* In Sec. 18.1 we examined differential equations (without calling them by that name) when going from a derivative f' to the corresponding antiderivative f. We *solved* the differential equation when we found the antiderivative. In this section we will expand upon this topic using the rules of integration to assist in the solution process.

Ordinary Differential Equations

If a differential equation involves derivatives of a function of one independent variable, it is called an *ordinary differential equation.* The following equation is an example, where the independent variable is x.

$$\frac{dy}{dx} = 5x - 2 \tag{18.13}$$

Differential equations are also classified by their *order,* which is the order of the highest-order derivative appearing in the equation. Equation (18.13) is a first-order, ordinary differential equation. Another way of classifying differential equations is by their *degree.* The degree is the *power* of the highest-order derivative in the differential equation. Equation (18.13) is of *first degree.* Equation (18.14) is an ordinary differential equation of first order and second degree.

$$\left(\frac{dy}{dx}\right)^2 - 10 = y \tag{18.14}$$

EXAMPLE 28

Classify each of the following ordinary differential equations by *order* and *degree.*
(a) $dy/dx = x^2 - 2x + 1$
(b) $d^2y/dx^2 - (dy/dx) = x$
(c) $d^2y/dx^2 - (dy/dx)^3 + 2x = 0$

---

* We have regularly used notation such as dy/dx. Separately, dy and dx are called *differentials,* reflecting instantaneous changes in y and x, respectively.

SOLUTION

(a) This is an ordinary differential equation of first order and first degree.

(b) This is an ordinary differential equation of second order and first degree.

(c) This is an ordinary differential equation of second order and first degree (dy/dx is not the highest-order derivative).

$\square$

Solutions of Ordinary Differential Equations

In this section we will focus on the solutions to ordinary differential equations. A solution to a differential equation is a function not containing derivatives or differentials which satisfies the original differential equation.

Solutions to differential equations can be classified into **general solutions** and **particular solutions.** A general solution is one which contains arbitrary constants of integration. A particular solution is one which is obtained from the general solution. For particular solutions, specific values are assigned to the constants of integration based on **initial conditions** or **boundary conditions.**

Consider the differential equation

$$\frac{dy}{dx} = 3x^2 - 2x + 5$$

The *general solution* to this differential equation is found by integrating the equation, or

$$y = f(x) = \frac{3x^3}{3} - \frac{2x^2}{2} + 5x + C$$
$$= x^3 - x^2 + 5x + C$$

Given the *initial condition* that $f(0) = 15$, the *particular solution* is derived by substituting these values into the general solution and solving for C.

$$15 = 0^3 - 0^2 + 5(0) + C$$

or $$15 = C$$

The particular solution to the differential equation is

$$f(x) = x^3 - x^2 + 5x + 15$$

EXAMPLE 29 Given the differential equation

$$f''(x) = \frac{d^2y}{dx^2} = x - 5 \qquad (18.15)$$

and the *boundary conditions* $f'(2) = 4$ and $f(0) = 10$, determine the general solution and particular solution.

SOLUTION

The given equation is an ordinary differential equation of *second* order and *first* degree. If the equation is integrated, the result is

$$f'(x) = \frac{dy}{dx} = \frac{x^2}{2} - 5x + C_1 \qquad (18.16)$$

which is also a differential equation because it contains the derivative dy/dx. Thus, to rid the equation of any derivatives we must integrate Eq. (18.16).

$$f(x) = y = \frac{x^3}{6} - \frac{5x^2}{2} + C_1 x + C_2 \qquad (18.17)$$

Equation (18.17) is the *general* solution to the original differential equation. *Notice the use of subscripts on the constants of integration to distinguish them from one another.*

> **NOTE**   The general solution of an nth-order differential equation will contain n constants of integration.

To obtain the particular solution, the boundary conditions must be substituted into Eqs. (18.16) and (18.17). Starting with Eq. (18.16) and the condition that $f'(2) = 4$,

$$4 = \frac{(2)^2}{2} - 5(2) + C_1$$

$$4 = -8 + C_1$$

and
$$12 = C_1$$

If this value is substituted into Eq. (18.17), along with the other boundary condition information $f(0) = 10$,

$$10 = \frac{0^3}{6} - \frac{5(0)^2}{2} + 12(0) + C_2$$

$$10 = C_2$$

Therefore the particular solution is

$$f(x) = \frac{x^3}{6} - \frac{5x^2}{2} + 12x + 10$$

❑

In Chaps. 7 and 15 exponential growth and exponential decay functions were discussed. Exponential growth functions have the general form

$$\boxed{V = V_0 e^{kt}} \qquad (18.18)$$

where V_0 equals the value of the function when $t = 0$ and k is a positive constant. It was demonstrated in Example 56 of Chap. 15 (page 715) that these functions are characterized by a constant percentage rate of growth k, and

$$\frac{dV}{dt} = kV_0 e^{kt}$$

or

$$\boxed{\frac{dV}{dt} = kV} \qquad (18.19)$$

Equation (18.19) is a differential equation; the general solution to Eq. (18.19) is expressed by Eq. (18.18), where V_0 and k are constants.

Empirical research frequently involves observations of a process (e.g., bacteria growth, population growth or decay, and radioactivity decay) over time. The gathered data often reflect values of the function at different points in time and measures of *rates of change* in the value of the function. It is from these types of data that the actual functional relationship can be derived. Stated more simply, research frequently results in differential equations which describe, partially, the relationship among variables. The solution to these differential equations results in a complete description of the functional relationships.

EXAMPLE 30 (**Species Growth**) The population of a rare species of fish is believed to be growing exponentially. When first identified and classified, the population was estimated at 50,000. Five years later the population was estimated to equal 75,000. *If P equals the population of this species at time t, where t is measured in years,* the population growth occurs at a rate described by the differential equation

$$\frac{dP}{dt} = kP = kP_0 e^{kt}$$

Integrating this equation gives

$$\int \frac{dP}{dt} = \int kP_0 e^{kt}$$

$$P = P_0 \int ke^{kt}$$

Or the general solution is

$$P = P_0 e^{kt}$$

where P_0 and k are constants. To determine the specific value of P_0, we utilize the initial condition which states that $P = 50,000$ when $t = 0$,

$$50,000 = P_0 e^{k(0)}$$

or

$$50,000 = P_0$$

The value of k can be found by substituting the boundary condition ($P = 75,000$ when $t = 5$) along with the $P_0 = 50,000$ into the general solution.

$$75,000 = 50,000e^{k(5)}$$

$$1.5 = e^{5k}$$

If the natural logarithm of both sides of the equation is found,

$$\ln 1.5 = 5k$$

From Table 2 (inside the back cover)

$$\ln 1.5 = 0.4055$$

$$0.4055 = 5k$$

and

$$0.0811 = k$$

Thus, the particular function describing the growth of the population is

$$P = 50,000e^{0.0811t}$$

❑

An exponential decay process is characterized by a constant percentage decrease in value over time. The general function describing these decay processes is

$$V = V_0 e^{-kt} \qquad \textbf{(18.20)}$$

where V equals the value of the function at time t, V_0 equals the value of the function at $t = 0$, and k is the percentage rate of decay. The rate of change in the value of the function with respect to a change in time is

$$\frac{dV}{dt} = -kV_0 e^{-kt}$$

or

$$\frac{dV}{dt} = -kV \qquad \textbf{(18.21)}$$

Equation (18.21) is a differential equation for which the general solution is Eq. (18.20).

EXAMPLE 31   **(Drug Absorption)** A particular prescription drug was administered to a person in a dosage of 100 milligrams. The amount of the drug contained in the bloodstream diminishes over time as described by an exponential decay function. After 6 hours, a blood sample reveals that the amount in the system is 40 milligrams. If V equals the amount of the drug in the bloodstream after t hours and V_0 equals the amount in the bloodstream at $t = 0$, the decay occurs at a rate described by the function

$$\frac{dV}{dt} = -kV = -kV_0 e^{-kt}$$

The general solution to this differential equation is

$$V = V_0 e^{-kt}$$

If the initial condition ($V = 100$ at $t = 0$) is substituted into this equation, V_0 is identified as 100 and

$$V = 100e^{-kt}$$

Substituting the boundary condition ($V = 40$ when $t = 6$) yields

$$40 = 100e^{-k(6)}$$

or

$$0.4 = e^{-6k}$$

Taking the natural logarithm of both sides of the equation,

$$\ln(0.4) = -6k$$

From Table 2,

$$\ln(0.4) = -0.9163$$

and

$$-0.9163 = -6k$$

$$0.1527 = k$$

Thus, the particular function describing the drug level decay function is

$$V = 100e^{-0.1527t}$$

❑

Extension of Differential Equations

This discussion has revealed the "tip of the iceberg" regarding the topic of differential equations. We have examined only the simplest of cases. This topic frequently is the focus of an entire one-term course. The objective for us has been to introduce this topic and relate it to integral calculus.

Section 18.5 Follow-up Exercises

In Exercises 1–12, classify each differential equation by order and degree.

1 $dy/dx = x^3 - x^2 + 5x$

2 $dy/dx - d^2y/dx^2 = x + 5$

3 $(d^2y/dx^2)^3 = x^3 - dy/dx$

4 $d^2y/dx^2 = (dy/dx)^4 - x^3$

5 $dy/dx = x^4 - 5x^2 + 8x$

6 $dy/dx = -(d^2y/dx^2)^3 + 10$

7 $(dy/dx)^3 = dy/dx + x + 10$

8 $d^2y/dx^2 = 5(dy/dx)^3 - x^2 - 5$

9 $dy/dx = 28 + x - (d^2y/dx^2)^3$

10 $x^2 + x = dy/dx - (d^2y/dx^2)^3$

In Exercises 11–28, find the general solution for the differential equation.

11 $dy/dx = x^4 + 2x + 6$

12 $dy/dx = 6x^2 + 6x + 18$

13 $dy/dx = 1/x$

14 $dy/dx = (2x)(x^2 + 5)^4$

15 $dy/dx = 5x/(5x^2 + 10)$

16 $dy/dx = 6xe^{3x^2}$

17 $d^2y/dx^2 = x + 5$

18 $6x = 24x^2 + d^2y/dx^2$

19 $2(d^2y/dx^2) + x^3 = d^2y/dx^2 + 2x - 16$

20 $d^2y/dx^2 = 3x - 5$

21 $d^2y/dx^2 = e^x$

22 $d^2y/dx^2 = x^2 - x + 4$

23 $d^2y/dx^2 = e^x + 5$

24 $d^2y/dx^2 = 12x^2 - 6x + 40$

25 $3(d^2y/dx^2) - 5 = 2(d^2y/dx^2) - x$

26 $dy/dx = 2x/(x^2 - 5)$

27 $4(dy/dx) - 2x(x^2 + 15)^3 = 3\, dy/dx$

28 $3(d^2y/dx^2) + 12x^2 - 5 = 2(d^2y/dx^2) + x$

In Exercises 29–38, find the general and particular solutions for the differential equation.

29 $dy/dx = 2x, f(0) = -50$

30 $dy/dx = 6x^2 + 2x + 6, f(0) = 18$

31 $dy/dx = x^2 + 3x + 8, f(1) = 7.5$

32 $dy/dx = (6x)(3x^2 + 7), f(2) = 20$

33 $d^2y/dx^2 = 6x + 18, f'(5) = -10, f(2) = 30$

34 $d^2y/dx^2 = 15; f'(2) = 20, f(3) = -10$

35 $d^2y/dx^2 = 25e^{5x}; f'(0) = 4, f(0) = -2$

36 $3 - d^2y/dx^2 = 20x; f'(0) = -12, f(-1) = 18$

37 $d^2y/dx^2 = 6x - 9; f'(2) = 10, f(-2) = -10$

38 $dy/dx = 4x(2x^2 - 8)^3; f(2) = 20$

39 The population of a newly discovered species of rabbit appears to be growing exponentially. When first identified in a South American country, the population was estimated at 500. Two years later the population was estimated to equal 1,250. Determine the exponential growth function which describes the population P as a function of time t, measured in years since the discovery of the species.

40 The population of an endangered species of Alaskan elk appears to be declining at an exponential rate. When the decline was first suspected, the elk population was estimated to equal 2,500. Ten years later the population was estimated to equal 1,250. Determine the exponential decay function which describes the elk population as a function of time t, measured in years since the decline was first suspected.

41 The population of an endangered species of wolf appears to be declining at an exponential rate. When the decline was first suspected, the wolf population was estimated to equal 40,000. Five years later the population was estimated to equal 32,000. Determine the exponential decay function which describes the wolf population as a function of time t, measured in years since the decline was first suspected.

42 The population of a particular species of wildlife appears to be growing exponentially. When first identified, the population was estimated at 200,000. Four years later it was estimated at 320,000. Determine the exponential growth function which describes the population P as a function of time t, measured in years since the discovery of the species.

❏ KEY TERMS AND CONCEPTS

antiderivative 828
constant of integration 834
differential equation 858
indefinite integral 834
integral sign 834
integrand 834

integration 834
integration by parts 845
method of partial fractions 849
ordinary differential equations 858
solutions of ordinary differential
    equations 859

❏ IMPORTANT FORMULAS

$$\int k\,dx = kx + c \qquad k \text{ real}$$ (Rule 1)

$$\int x^n\,dx = \frac{x^{n+1}}{n+1} + C \qquad n \neq -1$$ (Rule 2)

$$\int kf(x)\,dx = k\int f(x)\,dx \qquad k \text{ real}$$ (Rule 3)

$$\int [f(x) \pm g(x)]\,dx = \int f(x)\,dx \pm \int g(x)\,dx$$ (Rule 4)

$$\int x^{-1}\,dx = \ln x + C$$ (Rule 5)

$$\int e^x\,dx = e^x + C$$ (Rule 6)

$$\int [f(x)]^n f'(x)\,dx = \frac{[f(x)]^{n+1}}{n+1} + C \qquad n \neq -1$$ (Rule 7)

$$\int f'(x)e^{f(x)}\,dx = e^{f(x)} + C$$ (Rule 8)

$$\int \frac{f'(x)}{f(x)}\,dx = \ln f(x) + C$$ (Rule 9)

❏ ADDITIONAL EXERCISES

Section 18.1

Find the antiderivative of the following functions.

1 $f'(x) = -30$

2 $f'(x) = \sqrt{2e}$

3 $f'(x) = x/3$

4 $f'(x) = x/4 + 10$

5 $f'(x) = 9x^2 + 6x + 8$

7 $f'(x) = 8x^3 + 9x^2 - 5$

9 $f'(x) = 4x(2x^2 + 10)^4$

6 $f'(x) = 8x^3 + x^2 - 2x + 20$

8 $f'(x) = 16x^3 - 9x^2 + 3x + 5$

10 $f'(x) = 6x^2(2x^3 + 4)^3$

For the following exercises, determine f given f' and a point which satisfies f.

11 $f'(x) = 4$, (2, 10)

13 $f'(x) = -4x^3/3 + 2x - 1$, (−2, 10)

15 $f'(x) = -8x^3 + 12x^2$, (−1, 25)

12 $f'(x) = x^2 + 2x$, (0, 10)

14 $f'(x) = -8x^3 + 6x^2$, (−2, 10)

16 $f'(x) = 2x(x^2 - 5)^3$, (3, 40)

Sections 18.2 and 18.3

For the following exercises, find the indefinite integral (if possible).

17 $\int 25\, dx$

18 $\int -18\, dx$

19 $\int 26x\, dx$

20 $\int (-4x + 15)\, dx$

21 $\int (-10x - 6)\, dx$

22 $\int (x/4 - 10)\, dx$

23 $\int (2x^3 + 6x^2)\, dx$

24 $\int (x^3/2 + 3x^2 + 8)\, dx$

25 $\int (x^4 - 2/x^3)\, dx$

26 $\int (x - 5)^5\, dx$

27 $\int (2x^6 + 6x^4 - x^3 + 9x^2 - 4x)\, dx$

28 $\int (ax^4 + bx^3 + cx^2 + dx + e)\, dx$

29 $\int ([1/x] - [3/x^2] + [4/x^3])\, dx$

30 $\int (1 + [1/x] - [1/x^2])\, dx$

31 $\int (2/\sqrt[3]{x})\, dx$

32 $\int (-20/\sqrt{x})\, dx$

33 $\int (dx/x^6)$

34 $\int (8/x^4)\, dx$

35 $\int (-2/x^2)\, dx$

36 $\int (-5/x^3)\, dx$

37 $\int \sqrt{x^5}\, dx$

38 $\int \sqrt[4]{x^3}\, dx$

39 $\int (2x^2 - 3)^{1/2}(4x)\, dx$

40 $\int (x^3 + 2e^x + 6)\, dx$

41 $\int (\sqrt[3]{x^2} + \sqrt[5]{x})\, dx$

42 $\int (x^2 - 10)(2x)\, dx$

43 $\int (3x^2 - 10)^3(x)\,dx$

44 $\int (2x^3 - x^2)^2(3x^2 - x)\,dx$

45 $\int (2x^2 + 2x)^5(x + 1)\,dx$

46 $\int (4x^3 - 6x^2)^3(6x - 6)\,dx$

47 $\int \dfrac{-x}{4 - x^2/2}\,dx$

48 $\int \dfrac{4 - x}{x^3}\,dx$

49 $\int \dfrac{2}{x - 5}\,dx$

50 $\int (e^2 - 3)\,dx$

51 $\int (x/\sqrt{x^2 - 8})\,dx$

52 $\int (x/[x^2 + 3]^5)\,dx$

Section 18.4

Find the indefinite integral (if possible) using integration by parts.

53 $\int x \ln x\,dx$

54 $\int xe^{ax}\,dx$

55 $\int e^x(x + 1)^2\,dx$

56 $\int xe^{-3x}\,dx$

57 $\int x^2 e^x\,dx$

Find the indefinite integral (if possible) using the method of partial fractions.

58 $\int \dfrac{10 - 2x}{x^2 - 5x - 4}\,dx$

59 $\int \dfrac{3x - 5}{x^2 + 2x - 15}\,dx$

60 $\int \dfrac{3x^2 - x + 5}{x^3 - 4x}\,dx$

61 $\int \dfrac{x - 5}{x^3 - 9x}\,dx$

Find the indefinite integral (if possible) using Table 18.2.

62 $\int 3e^{-5x}\,dx$

63 $\int (xe^{-3x}/2)\,dx$

64 $\int dx/(5 + 2e^{3x})$

65 $\int x^5 \ln x\,dx$

66 $\int ([\ln x]^4/x)\,dx$

67 $\int x^5 \ln 3x\,dx$

68 $\int ([\ln x]^2/2)\,dx$

69 $\int -2(\ln x)^3\,dx$

Section 18.5

Classify the following differential equations by order and degree.

70 $dy/dx = 8x^4 + 6x^2 - 10x + 5$

71 $(d^2y/dx^2)^3 - (d^2y/dx^2)^2 = 5x^2 - (dy/dx)$

72 $(d^2y/dx^2)^3 = (dy/dx)^4 + x + 3$

Find the general solution for each of the following differential equations.

73 $dy/dx = x^4 + 5x$

75 $d^2y/dx^2 = x^2/6 + 20x$

74 $dy/dx = x/(x^2 - 3) + 6x^2 e^{2x^3 - 4}$

76 $x^2 + 3x = d^2y/dx^2 + 30$

For each of the following differential equations, find the general and particular solutions.

77 $dy/dx = -4x + 4$, $f(2) = 18$

79 $d^2y/dx^2 = x^2/12 + 12x$
  $f'(2) = 20$
  $f(2) = 40$

78 $dy/dx = 4x^3 + 9x^2$, $f(-1) = 50$

80 $20x = 5 + (d^2y/dx^2)$
  $f'(1) = 25$
  $f(2) = 81$

❏ CHAPTER TEST

1 Given $f'(x) = 4x^3 + 2x - 20$ and the point $(5, 200)$ which satisfies f, determine f.

2 Find the following indefinite integrals

  (a) $\displaystyle\int dx/\sqrt[3]{x^5}$

  (b) $\displaystyle\int (x^4 + 10)^5 x^3 \, dx$

  (c) $\displaystyle\int e^{-10x} \, dx$

3 Find the general and particular solutions for the differential equation $6 + x = d^2y/dx^2$, where $f(2) = 24$ and $f'(3) = 5$.

4 The marginal revenue function for a company's product is

$$MR = 220{,}000 - 18x$$

where x equals the number of units sold. If total revenue equals 0 when 0 units are sold, determine the total revenue function.

5 Using integration by parts, find the indefinite integral $\int xe^{10x}\, dx$.

6 Use the method of partial fractions to find the indefinite integral

$$\int \frac{20x - 10}{x^2 - x - 6}\, dx$$

CHAPTER 19

INTEGRAL CALCULUS: APPLICATIONS

53xxxx1949330584   86.00 000042

APPROXIMATION

CHAPTER OBJECTIVES

❏ Introduce the definite integral

❏ Illustrate the application of the definite integral in measuring areas

❏ Provide a wide variety of applications of integral calculus

❏ Illustrate the relationship between integral calculus and probability theory

MOTIVATING SCENARIO: Blood Bank Management	A hospital blood bank is conducting a blood drive to replenish its inventory of blood. The hospital estimates that blood will be donated at a rate of $d(t)$ pints per day, where $$d(t) = 500e^{-0.4t}$$ where t equals the length of the blood drive in days. *If the goal of the blood drive is 1,000 units (pints), hospital administrators want to know how long it will take to reach the goal* (Example 24).

This chapter will focus on the application of integral calculus. In particular, we will discuss the *definite integral,* the use of definite integrals in calculating areas beneath and between curves, several applications utilizing integral calculus, and the application of integral calculus to probability theory.

19.1 DEFINITE INTEGRALS

In this section we will introduce the definite integral, which forms the basis for many applications of integral calculus.

The Definite Integral

The *definite integral* can be interpreted both as an area and as a limit. Consider the graph of the function $f(x) = x^2$, $x \geq 0$, shown in Fig. 19.1. Assume that we wish to determine the shaded area A under the curve between $x = 1$ and $x = 3$. One approach is to *approximate* the area by computing the areas of a set of rectangles inscribed within the shaded area. In Fig. 19.2 two rectangles have been drawn within the area of interest. The width of each rectangle equals 1, and the heights are respectively $f(1)$ and $f(2)$. Using the sum of the areas of the two rectangles to approximate the area of interest, we have

$$A^* = f(1) \cdot (1) + f(2) \cdot (1)$$
$$= (1)^2 \cdot (1) + (2)^2 \cdot (1)$$
$$= (1)(1) + (4)(1) = 5$$

where A^* is the approximate area. Note that this approximation *underestimates* the actual area. The error introduced is represented by the more lightly shaded areas.

In Fig. 19.3 four rectangles have been drawn within the area of interest. The width of each rectangle equals $\frac{1}{2}$, and the total area of the four rectangles is computed using the equation

$$A^* = f(1) \cdot (0.5) + f(1.5) \cdot (0.5) + f(2) \cdot (0.5) + f(2.5) \cdot (0.5)$$
$$= (1)^3 \cdot (0.5) + (1.5)^2 \cdot (0.5) + (2)^2 \cdot (0.5) + (2.5)^2 \cdot (0.5)$$
$$= (1)(0.5) + (2.25)(0.5) + (4)(0.5) + (6.25)(0.5)$$
$$= 0.5 + 1.125 + 2.0 + 3.125 = 6.75$$

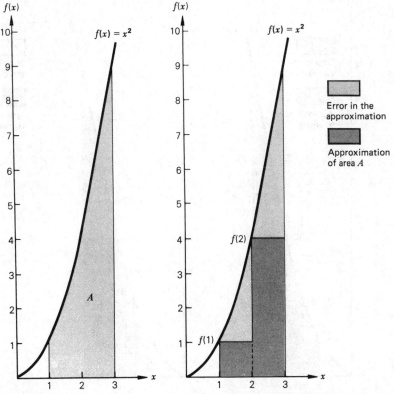

Figure 19.1

Figure 19.2 Approximation using two rectangles.

Compared with Fig. 19.2, the use of four rectangles rather than two results in a better approximation of the actual area. The more lightly shaded area (measure of error) is smaller in Fig. 19.3.

In Fig. 19.4 eight rectangles have been drawn, each having a width equal to 0.25. The area of these rectangles is computed by using the equation

$$A^* = f(1) \cdot (0.25) + f(1.25) \cdot (0.25) + \cdots + f(2.75) \cdot (0.25)$$
$$= (1)^2 \cdot (0.25) + (1.25)^2 \cdot (0.25) + \cdots + (2.75)^2 \cdot (0.25)$$
$$= (1)(0.25) + (1.5625)(0.25) + \cdots + (7.5625)(0.25) = 7.6781$$

Observe that this approximation is better than the others. In fact, if we continue to subdivide the interval between $x = 1$ and $x = 3$, making the base of each rectangle smaller and smaller, the approximation will come closer and closer to the actual area (which we will determine later to equal $8\frac{2}{3}$).

Let's now look at this process in a more general sense. Consider the function in Fig. 19.5. Suppose we are interested in determining the area beneath the curve but above the x axis between $x = a$ and $x = b$. Further suppose that the interval has been subdivided into n rectangles. Assume that the width of rectangle i is Δx_i and the height is $f(x_i)$. It is not necessary to assume that the width of each rectangle is

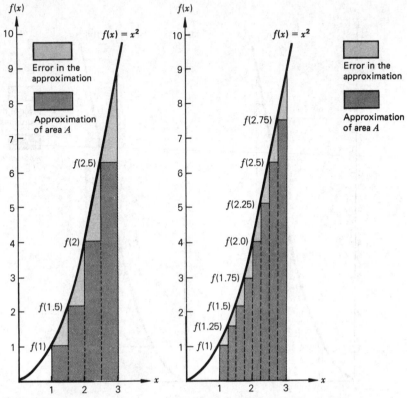

Figure 19.3 Approximation using four rectangles.

Figure 19.4 Approximation using eight rectangles.

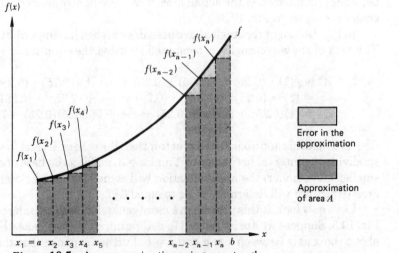

Figure 19.5 Area approximation using n rectangles.

the same. We can approximate the area of interest by summing the areas of the n rectangles, or

$$A^* = f(x_1)\,\Delta x_1 + f(x_2)\,\Delta x_2 + \cdots + f(x_n)\,\Delta x_n$$

$$= \sum_{i=1}^{n} f(x_i)\,\Delta x_i$$

As we observed for the function $f(x) = x^2$, the approximation becomes more and more accurate as the width of the rectangles becomes smaller and smaller — and hence as the number of rectangles concurrently becomes larger and larger. We can formalize this observation by stating that *when the limit exists,*

$$\lim_{n\to\infty} \sum_{i=1}^{n} f(x_i)\,\Delta x_i = A \qquad\qquad (19.1)$$

That is, the actual area under the curve, A, is the limiting value of the sum of the areas of the n inscribed rectangles as the number of rectangles approaches infinity (and the width Δx_i of each approaches 0).

Just as the summation sign Σ applies when the sum of discrete elements is desired, the definite integral implies summation for continuous functions.

DEFINITION: DEFINITE INTEGRAL

If f is a bounded function on the interval $[a, b]$, we shall define the ***definite integral*** of f as

$$\int_a^b f(x)\,dx = \lim_{n\to\infty} \sum_{i=1}^{n} f(x_i)\,\Delta x_i = A \qquad\qquad (19.2)$$

provided this limit exists, as the size of all the intervals in the subdivision approaches zero and hence the number of intervals n approaches infinity.

The left side of Eq. (19.2) presents the notation of the *definite integral.* The values a and b which appear, respectively, below and above the integral sign are called the ***limits of integration.*** The ***lower limit of integration*** is a, and the ***upper limit of integration*** is b. The notation $\int_a^b f(x)\,dx$ can be described as "*the definite integral of f between a lower limit $x = a$ and an upper limit $x = b$,*" or more simply "*the integral of f between a and b.*"

Evaluating Definite Integrals

The evaluation of definite integrals is facilitated by the following important theorem.

FUNDAMENTAL THEOREM OF INTEGRAL CALCULUS
If a function f is continuous over an interval and F is any antiderivative of f, then for any points $x = a$ and $x = b$ on the interval, where $a \leq b$,

$$\int_a^b f(x)\, dx = F(b) - F(a) \qquad (19.3)$$

According to the fundamental theorem of integral calculus, the definite integral can be evaluated by (1) determining the indefinite integral $F(x) + C$ and (2) computing $F(b) - F(a)$, sometimes denoted by $F(x)]_a^b$. As you will see in the following example, there is no need to include the constant of integration in evaluating definite integrals.

EXAMPLE 1   To evaluate $\int_0^3 x^2\, dx$, the indefinite integral is

$$F(x) = \int x^2\, dx$$

$$= \frac{x^3}{3} + C$$

Now

$$\int_0^3 x^2\, dx = \left(\frac{x^3}{3} + C \right) \Big]_0^3 = \left(\frac{3^3}{3} + C \right) - \left(\frac{0^3}{3} + C \right)$$

$$= 9 + C - C$$

$$= 9$$

❑

*When evaluating definite integrals, always subtract the value of the indefinite integral at the lower limit of integration from the value at the upper limit of integration. **The constant of integration will always drop out in this computation, as it did in this example. Thus, there is no need to include the constant in evaluating definite integrals.***

EXAMPLE 2   To evaluate $\int_1^4 (2x^2 - 4x + 5)\, dx$,

$$F(x) = \int (2x^2 - 4x + 5)\, dx$$

$$= \frac{2x^3}{3} - \frac{4x^2}{2} + 5x$$

$$= \frac{2x^3}{3} - 2x^2 + 5x$$

Therefore,

$$\int_1^4 (2x^2 - 4x + 5)\, dx = \frac{2x^3}{3} - 2x^2 + 5x \Bigg]_1^4$$

$$= \left[\frac{2(4)^3}{3} - 2(4)^2 + 5(4)\right] - \left[\frac{2(1)^3}{3} - 2(1)^2 + 5(1)\right]$$

$$= (\tfrac{128}{3} - 32 + 20) - (\tfrac{2}{3} - 2 + 5) = 30\tfrac{2}{3} - 3\tfrac{2}{3} = 27$$

□

PRACTICE EXERCISE

Evaluate $\int_1^3 (x^3 - 2x)\, dx$. *Answer:* 12.

EXAMPLE 3 To evaluate $\int_{-2}^1 e^x\, dx$,

$$F(x) = \int e^x\, dx$$

$$= e^x$$

Therefore,

$$\int_{-2}^1 e^x\, dx = e^x \Bigg]_{-2}^1$$

$$= e^1 - e^{-2}$$

or from Table 1,

$$e^1 - e^{-2} = 2.7183 - 0.1353$$

$$= 2.5830$$

EXAMPLE 4 To evaluate $\int_2^4 \dfrac{x\, dx}{x^2 - 1}$,

$$F(x) = \int \frac{x\, dx}{x^2 - 1}$$

$$= \frac{1}{2} \int \frac{2x\, dx}{x^2 - 1}$$

and by rule 9,

$$F(x) = \tfrac{1}{2} \ln(x^2 - 1)$$

Therefore,

$$\int_2^4 \frac{x\, dx}{x^2 - 1} = \frac{1}{2} \ln(x^2 - 1) \Bigg]_2^4$$

$$= \tfrac{1}{2} \ln(4^2 - 1) - \tfrac{1}{2} \ln(2^2 - 1)$$

$$= \tfrac{1}{2} \ln 15 - \tfrac{1}{2} \ln 3$$

From Table 2,

$$\int_2^4 \frac{x\, dx}{x^2 - 1} = \frac{1}{2}(2.7081) - \frac{1}{2}(1.0986)$$

$$= 1.35405 - 0.5493 = 0.80475$$

□

PRACTICE EXERCISE

Evaluate (a) $\int_{-1}^{2} e^{2x} \, dx$ and (b) $\int_{1}^{3} \frac{4x}{x^2 + 5} \, dx$. *Answer:* (a) 27.23135, (b) 1.6946.

Properties of Definite Integrals

There are several properties which can be of assistance when evaluating definite integrals. These follow, along with examples to illustrate.

PROPERTY 1
If f is defined and continuous on the interval (a, b),

$$\int_{a}^{b} f(x) \, dx = -\int_{b}^{a} f(x) \, dx \qquad (19.4)$$

EXAMPLE 5   Consider the function $f(x) = 4x^3$.

$$\int_{-2}^{1} 4x^3 \, dx = \frac{4x^4}{4} \Big]_{-2}^{1} = x^4 \Big]_{-2}^{1}$$
$$= (1)^4 - (-2)^4 = 1 - 16 = -15$$

$$\int_{1}^{-2} 4x^3 \, dx = \frac{4x^4}{4} \Big]_{1}^{-2} = x^4 \Big]_{1}^{-2}$$
$$= (-2)^4 - (1)^4 = 16 - 1 = 15$$

Thus,

$$\int_{-2}^{1} 4x^3 \, dx = -\int_{1}^{-2} 4x^3 \, dx$$

❑

PROPERTY 2

$$\int_{a}^{a} f(x) \, dx = 0 \qquad (19.5)$$

EXAMPLE 6

$$\int_{10}^{10} e^x \, dx = e^x \Big]_{10}^{10}$$
$$= e^{10} - e^{10}$$
$$= 0$$

❑

PROPERTY 3

If f is continuous on the interval (a, c) and $a < b < c$,

$$\int_a^b f(x)\,dx + \int_b^c f(x)\,dx = \int_a^c f(x)\,dx \qquad (19.6)$$

EXAMPLE 7   Show that

$$\int_0^4 3x^2\,dx = \int_0^2 3x^2\,dx + \int_2^4 3x^2\,dx$$

SOLUTION

$$\int_0^4 3x^2\,dx = \frac{3x^3}{3}\bigg]_0^4 = x^3\bigg]_0^4$$

$$= (4)^3 - (0)^3 = 64$$

$$\int_0^2 3x^2\,dx + \int_2^4 3x^2\,dx = x^3\bigg]_0^2 + x^3\bigg]_2^4$$

$$= [(2)^3 - (0)^3] + [(4)^3 - (2)^3]$$

$$= (8 - 0) + (64 - 8) = 64$$

Thus,

$$\int_0^4 3x^2\,dx = \int_0^2 3x^2\,dx + \int_2^4 3x^2\,dx$$

or

$$64 = 64$$

◻

PRACTICE EXERCISE

Show that

$$\int_0^4 3x^2\,dx = \int_0^1 3x^2\,dx + \int_1^4 3x^2\,dx$$

PROPERTY 4

$$\int_a^b cf(x)\,dx = c\int_a^b f(x)\,dx \qquad (19.7)$$

where c is constant.

EXAMPLE 8   Show that for the function in Example 7,

$$\int_0^4 3x^2 \, dx = 3 \int_0^4 x^2 \, dx$$

SOLUTION

In Example 7, $\int_0^4 3x^2 \, dx$ was determined to equal 64. We need to evaluate $3 \int_0^4 x^2 \, dx$.

$$3 \int_0^4 x^2 \, dx = 3 \left. \frac{x^3}{3} \right]_0^4$$

$$= 3 \left[\frac{(4)^3}{3} - \frac{(0)^3}{3} \right] = 3 \left(\frac{64}{3} \right) = 64$$

PROPERTY 5

If $\int_a^b f(x) \, dx$ and $\int_a^b g(x) \, dx$ exist,

$$\int_a^b [f(x) \pm g(x)] \, dx = \int_a^b f(x) \, dx \pm \int_a^b g(x) \, dx \qquad \text{(19.8)}$$

EXAMPLE 9   Suppose we wish to evaluate

$$\int_2^4 (4x - 5) \, dx + \int_2^4 (5 - 6x) \, dx$$

Because the limits of integration are the same, Eq. (19.8) indicates that the integrands can be combined algebraically using one definite integral, or

$$\int_2^4 (4x - 5) \, dx + \int_2^4 (5 - 6x) \, dx = \int_2^4 [(4x - 5) + (5 - 6x)] \, dx$$

$$= \int_2^4 (-2x) \, dx$$

$$= -x^2 \Big]_2^4$$

$$= [-(4)^2] - [-(2)^2]$$

$$= -16 + 4 = -12$$

> **PRACTICE EXERCISE**
> Evaluate
>
> $$\int_2^4 (4x-5)\, dx + \int_2^4 (5-6x)\, dx$$
>
> using the two definite integrals and verify that their sum equals -12.

Section 19.1 Follow-up Exercises

Evaluate the following definite integrals.

1 $\displaystyle\int_0^3 x\, dx$

2 $\displaystyle\int_2^4 (x+2)\, dx$

3 $\displaystyle\int_1^5 dx$

4 $\displaystyle\int_{-2}^2 8\, dx$

5 $\displaystyle\int_{-1}^3 2x\, dx$

6 $\displaystyle\int_{-1}^1 -8x\, dx$

7 $\displaystyle\int_0^2 (x^2+4x)\, dx$

8 $\displaystyle\int_0^4 (4x^3+3x^2)\, dx$

9 $\displaystyle\int_0^{49} \sqrt{x}\, dx$

10 $\displaystyle\int_{-2}^2 2e^x\, dx$

11 $\displaystyle\int_4^6 (dx/x)$

12 $\displaystyle\int_4^9 (4\, dx/x)$

13 $\displaystyle\int_{-3}^0 (2x)e^{x^2}\, dx$

14 $\displaystyle\int_{-1}^0 (3x^2)e^{x^3}\, dx$

15 $\displaystyle\int_{-1}^1 5x\, dx$

16 $\displaystyle\int_{-2}^1 2x^3\, dx$

17 $\displaystyle\int_b^b 6x^2\, dx$

18 $\displaystyle\int_c^c 4x^3\, dx$

19 $\displaystyle\int_0^4 (2x+5)\, dx$

20 $\displaystyle\int_2^4 3x^2\, dx$

21 $\displaystyle\int_6^{12} 10\, dx$

22 $\displaystyle\int_0^3 4x^3\, dx$

23 $\displaystyle\int_0^2 (x+3)^2\, dx$

24 $\displaystyle\int_9^{16} \sqrt{x}\, dx$

25 $\displaystyle\int_{-1}^2 (x^2+4x+6)\, dx$

26 $\displaystyle\int_1^4 3x^5\, dx$

27 $\displaystyle\int_2^3 -9x^2\, dx$

28 $\displaystyle\int_2^4 -3e^x\, dx$

29 $\displaystyle\int_0^4 e^x\, dx$

30 $\displaystyle\int_0^2 2xe^{x^2}\, dx$

31 $\int_2^4 (dx/2x)$

32 $\int_1^3 \dfrac{2x}{x^2-4}\,dx$

33 $\int_{-2}^2 (3x^2 + 4x + 10)\,dx$

34 $\int_{-2}^1 (4x^3 + 6x^2)\,dx$

35 $\int_0^2 5e^{5x}\,dx$

36 $\int_{-1}^2 12e^{6x}\,dx$

37 $\int_0^4 2x(x^2+4)^2\,dx$

38 $\int_0^2 -6x(3x^2+4)^3\,dx$

39 $\int_1^4 (6\,dx/x)$

40 $\int_4^{10} (-20\,dx/x)$

41 $\int_4^6 \dfrac{6x}{3x^2-5}\,dx$

42 $\int_5^{10} \dfrac{-5x\,dx}{5x^2-4}$

43 $\int_1^3 \dfrac{6x^2}{x^3+6}\,dx$

44 $\int_2^4 \dfrac{-12x^2}{2x^3+3}\,dx$

45 $\int_2^4 \dfrac{2x^2+x}{4x^3+6x^2}\,dx$

46 $\int_1^4 \dfrac{3x^2+x}{8x^3+4x^2}\,dx$

47 $\int_1^2 (mx-b)\,dx$

48 $\int_0^3 (ax^2+bx+c)\,dx$

49 $\int_2^3 \dfrac{x^2}{x^3-1}\,dx$

50 $\int_3^5 \dfrac{6x}{x^2+4}\,dx$

51 $\int_2^4 (x^2+3x+1)\,dx - \int_2^4 (2x^2-7x+1)\,dx$

52 $\int_0^3 (6x^2+4x+12)\,dx + \int_0^3 (3x^2-2x+8)\,dx$

53 $\int_2^4 (5x^3+5x^2)\,dx - \int_2^4 (8x^3+6x^2+10)\,dx$

54 $\int_{-2}^1 (5x^2+2x+3)\,dx - \int_{-2}^1 (3x^2+2x-4)\,dx$

55 $\int_2^4 (x+8)\,dx - \int_4^2 (2x-6)\,dx$

56 $\int_1^3 (x^2+8)\,dx + \int_3^1 (-2x^2+10)\,dx$

57 $\int_0^3 (3x^2+2x+4)\,dx - \int_3^0 (x^2+5x)\,dx$

58 $\int_1^4 (10x^2-8x+3)\,dx + \int_4^1 (8x^2+4x-8)\,dx$

59 $\int_0^3 5e^x\,dx - \int_3^0 2e^x\,dx$

60 $\int_1^2 (4\,dx/x) + \int_2^1 (3\,dx/x)$

19.2 DEFINITE INTEGRALS AND AREAS

One of the practical applications of integral calculus is that definite integrals can be used to compute areas. These may represent areas which are bounded by curves representing functions and/or the coordinate axes. In Sec. 19.4 we will examine situations where such areas hold particular meaning in applied problems.

Areas between a Function and the x Axis

Definite integrals can be used to compute the area between the curve representing a function and the x axis. Several different situations may occur. The treatment of each varies and is now discussed.

> **CASE 1: ($f(x) > 0$)**
> When the value of a continuous function f is positive over the interval $a \le x \le b$—that is, the graph of f lies above the x axis — the area which is bounded by f, the x axis, $x = a$, and $x = b$ is determined by
>
> $$\int_a^b f(x)\ dx$$

Figure 19.6 illustrates the situation.

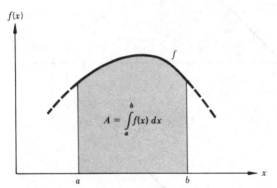

Figure 19.6 Determination of areas above x axis.

EXAMPLE 10 Determine the area beneath $f(x) = x^2$ and above the x axis between $x = 1$ and $x = 3$.

SOLUTION

This area was illustrated earlier in Fig. 19.1. The area is computed as

$$A = \int_1^3 x^2 \, dx$$

$$= \frac{x^3}{3}\Bigg]_1^3$$

$$= \frac{3^3}{3} - \frac{1^3}{3} = 9 - \tfrac{1}{3} = 8\tfrac{2}{3}$$

The (exact) area equals $8\tfrac{2}{3}$ *square* units.

EXAMPLE 11 Determine the area indicated in Fig. 19.7.

SOLUTION

Let's anticipate the answer by using familiar formulas for computing the area of a rectangle and a triangle. As shown in Fig. 19.8, the area of interest can be thought of as being composed of a rectangle of area A_2 and a triangle of area A_1. Therefore,

Figure 19.7

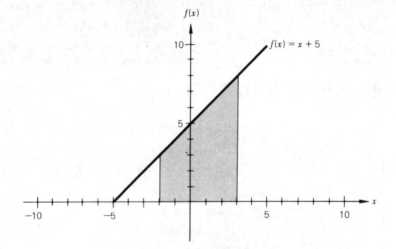

Figure 19.8

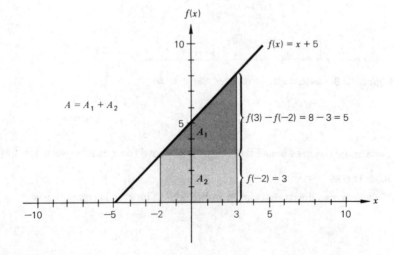

$$A = A_1 + A_2$$
$$= \tfrac{1}{2}bh + lw$$
$$= \tfrac{1}{2}(5)(5) + (5)(3) = 12.5 + 15 = 27.5 \text{ square units}$$

Using the definite integral,

$$A = \int_{-2}^{3} (x + 5)\, dx$$

$$= \frac{x^2}{2} + 5x \bigg]_{-2}^{3}$$

$$= \left[\frac{(3)^2}{2} + 5(3) \right] - \left[\frac{(-2)^2}{2} + 5(-2) \right]$$

$$= (\tfrac{9}{2} + 15) - (2 - 10) = 19.5 - (-8) = 27.5 \text{ square units}$$

$\square$

PRACTICE EXERCISE

For the function in Fig. 19.7, determine the area between the curve and the x axis which is bounded on the left by $x = 2$ and on the right by $x = 6$. *Answer:* 36.

CASE 2: ($f(x) < 0$)

When the value of a continuous function f is negative over the interval $a \le x \le b$ — that is, the graph of f lies below the x axis — the area which is bounded by f, the x axis, $x = a$, and $x = b$ is determined by

$$\int_{a}^{b} f(x)\, dx$$

However, the definite integral evaluates the area as *negative* when it lies below the x axis. Because area is absolute (or *positive*), the area will be computed as

$$-\int_{a}^{b} f(x)\, dx$$

EXAMPLE 12   Determine the area indicated in Fig. 19.9.

SOLUTION

Because f is a negative function,

$$A = -\int_{2.5}^{3.5} -\frac{x^3}{2}\, dx$$

Figure 19.9

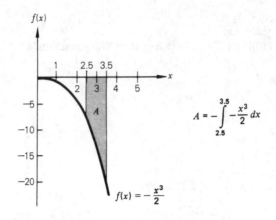

$$A = -\int_{2.5}^{3.5} -\frac{x^3}{2}\,dx$$

$$= -\left[-\frac{x^4}{8}\right]_{2.5}^{3.5}$$

$$= -\left\{\left[-\frac{(3.5)^4}{8}\right] - \left[\frac{-(2.5)^4}{8}\right]\right\}$$

$$= -[-18.7578 - (-4.8828)] = -[-13.875] = 13.875 \text{ square units}$$

❑

PRACTICE EXERCISE
In Fig. 19.9, find the area between f and the x axis bounded on the left and right by $x = 2$ and $x = 5$, respectively. *Answer:* 609/8, or 76¼.

CASE 3: ($f(x) < 0$ and $f(x) > 0$)
When the value of a continuous function f is positive over part of the interval $a \leq x \leq b$ and negative over the remainder of the interval — part of the area between f and the x axis is above the x axis and part is below the x axis —

then $\displaystyle\int_a^b f(x)\,dx$ calculates the **net area.** That is, areas above the x axis are

evaluated as positive, and those below are evaluated as negative. The two are combined algebraically to yield the net value.

EXAMPLE 13   Evaluate $\displaystyle\int_0^{15} (x - 5)\,dx$ to determine the *net* area, shown in Fig. 19.10.

SOLUTION
Again, we can anticipate the answer using the formula for the area of a triangle. Remembering that the area below the x axis (A_1) will be evaluated as negative when we integrate, we have

Figure 19.10

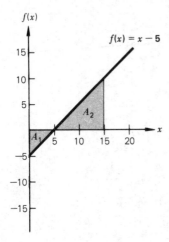

$$A = -A_1 + A_2$$
$$= -\tfrac{1}{2}(5)(5) + \tfrac{1}{2}(10)(10)$$
$$= -12.5 + 50$$
$$= 37.5 \text{ square units}$$

Evaluating the definite integral,

$$A = \int_0^{15} (x - 5)\, dx$$
$$= \frac{x^2}{2} - 5x \Bigg]_0^{15}$$
$$= \left[\frac{(15)^2}{2} - 5(15)\right] - \left[\frac{0^2}{2} - 5(0)\right]$$
$$= (112.5 - 75) - 0$$
$$= 37.5 \text{ square units}$$

❑

Finding Areas between Curves

The following examples illustrate procedures for determining areas between curves.

EXAMPLE 14   Determine the shaded area between f and g indicated in Fig. 19.11.

SOLUTION

In order to determine the area A, it is necessary to examine the composition of the area. The area cannot be determined by integrating only one of the functions. One way of determining A is shown in Fig. 19.12. If g is integrated between $x = 0$ and $x = 3$, the resulting area includes A, but it also includes an additional area which is not part of A. Having overestimated A, we need to subtract the *surplus*. The surplus area happens to be the area under f between $x = 0$ and $x = 3$. Thus, A can be determined as

Figure 19.11

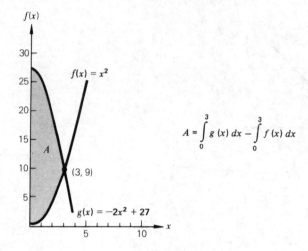

$$A = \int_0^3 g(x)\, dx - \int_0^3 f(x)\, dx$$

Figure 19.12

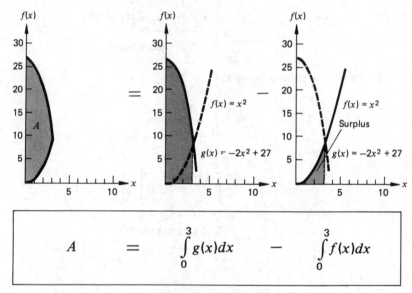

$$A \qquad = \qquad \int_0^3 g(x)\,dx \qquad - \qquad \int_0^3 f(x)\,dx$$

$$A = \int_0^3 g(x)\, dx - \int_0^3 f(x)\, dx$$

or $\qquad A = \int_0^3 (-2x^2 + 27)\, dx - \int_0^3 x^2\, dx$

Using Property 5,

$$A = \int_0^3 (-3x^2 + 27)\, dx$$

$$= -x^3 + 27x \Big]_0^3$$

$$= [-(3)^3 + 27(3)] - [-(0)^3 + 27(0)] = -27 + 81 - 0 = 54 \text{ square units}$$

Figure 19.13

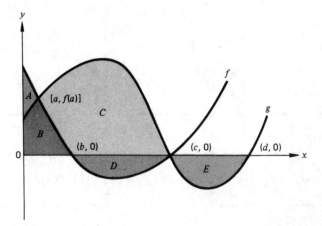

EXAMPLE 15 Referring to Fig. 19.13, determine the combination of definite integrals which would compute the size of (a) area A, (b) area B, (c) area C.

SOLUTION

This example presents no actual numbers. It really is an exercise in the logic of formulating
combinations of definite integrals to define areas.

(a) The upper boundary on A is determined by f. If f is integrated between $x = 0$ and $x = a$,
the result is area a_1, which includes A plus a surplus area a_2. The surplus area can be
determined by integrating g between $x = 0$ and $x = a$. Thus,

$$A = a_1 - a_2$$
$$= \int_0^a f(x)\, dx - \int_0^a g(x)\, dx$$

This is illustrated graphically in Fig. 19.14a.

(b) B can be viewed as consisting of two subareas, b_1 and b_2. The upper boundary on B is
determined by g up until $x = a$ and by f when $a \leq x \leq b$. If g is integrated between $x = 0$
and $x = a$, the resulting area is a portion of B. The remaining portion of B can be
determined by integrating f between $x = a$ and $x = b$. Thus,

$$B = b_1 + b_2$$
$$= \int_0^a g(x)\, dx + \int_a^b f(x)\, dx$$

This is illustrated graphically in Fig. 19.14b.

(c) The upper boundary on C is determined entirely by g. If we integrate g between $x = a$ and
$x = c$, the resulting area, c_1, includes C plus a surplus area c_2. The surplus area can be
determined by integrating f between $x = a$ and $x = b$. Thus,

$$C = c_1 - c_2$$
$$= \int_a^c g(x)\, dx - \int_a^b f(x)\, dx$$

Figure 19.14

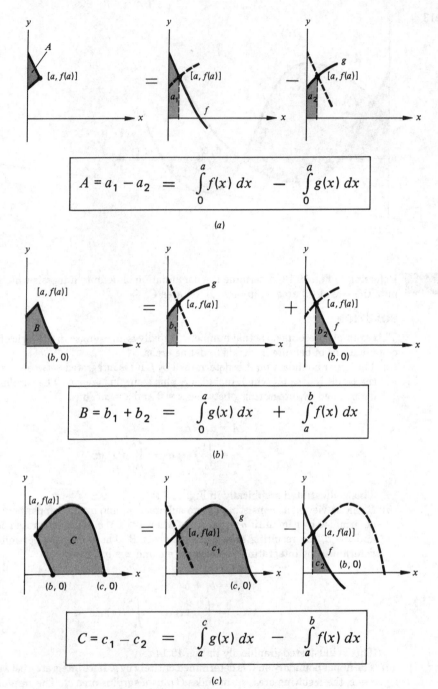

$$A = a_1 - a_2 = \int_0^a f(x)\,dx \;-\; \int_0^a g(x)\,dx$$

(a)

$$B = b_1 + b_2 = \int_0^a g(x)\,dx \;+\; \int_a^b f(x)\,dx$$

(b)

$$C = c_1 - c_2 = \int_a^c g(x)\,dx \;-\; \int_a^b f(x)\,dx$$

(c)

This is illustrated graphically in Fig. 19.14c.

PRACTICE EXERCISE
In Fig. 19.13, determine the combination of definite integrals which compute
(a) area D and (b) area E. *Answer:* $(a) - \int_b^c f(x)\,dx$, $(b) - \int_c^d g(x)\,dx$.

AREA BETWEEN TWO CURVES
If the function $y = f(x)$ lies above the function $y = g(x)$ over the interval $a \leq x \leq b$, the area between the two functions over the interval is

$$\int_a^b [f(x) - g(x)]\,dx$$

Figure 19.15 is a graphical representation of this property.

Figure 19.15
Area between two curves.

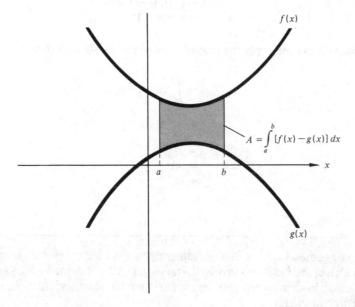

This property is really a formalization of the logic we have used earlier. For example, we utilized this property in defining area A in Example 14; similarly, the calculation of A in Example 15 utilized it. Interestingly, this property is valid for functions which lie below the x axis. To illustrate this, consider the following example.

EXAMPLE 16 Consider the functions $f(x)$ and $g(x)$ shown in Fig. 19.16. Suppose we are interested in determining the area between these two functions over the interval $0 \leq x \leq 4$. Using the formula for the area of a triangle, we can calculate the total area as the sum of the areas of the triangle above the x axis and the triangle below the x axis.

Figure 19.16

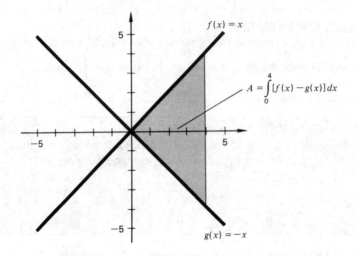

$$A = \tfrac{1}{2}(4)(4) + \tfrac{1}{2}(4)(4)$$
$$= 8 + 8 = 16$$

Alternatively, we can apply the property stated previously such that

$$A = \int_0^4 [x - (-x)]\, dx$$
$$= \int_0^4 2x\, dx$$
$$= x^2 \big]_0^4$$
$$= (4)^2 - (0)^2$$
$$= 16$$

NOTE   A suggestion in using definite integrals to compute areas is to **always draw a sketch of the functions involved.** Having a picture of the areas of interest makes it easier to identify pertinent boundaries and to understand the logic required to define the areas.

Section 19.2 Follow-up Exercises

In Exercises 1–20, (a) sketch f and (b) determine the size of the area bounded by f and the x axis over the indicated interval.

1 $f(x) = -2x + 8$, between $x = 1$ and $x = 4$
2 $f(x) = 16 - 2x$, between $x = 2$ and $x = 6$
3 $f(x) = x^2$, between $x = 2$ and $x = 8$
4 $f(x) = 10 - x^2$, between $x = -1$ and $x = 1$
5 $f(x) = 3x^3$, between $x = 2$ and $x = 4$
6 $f(x) = -2x^3$, between $x = 0$ and $x = 3$

7 $f(x) = -8x^2$, between $x = -2$ and $x = 3$
8 $f(x) = -x^2$, between $x = -2$ and $x = 2$
9 $f(x) = 10 - x^2$, between $x = -2$ and $x = 3$
10 $f(x) = 4 - x^2$, between $x = -5$ and $x = -2$
11 $f(x) = e^x$, between $x = 1$ and $x = 3$
12 $f(x) = -e^x$, between $x = 1$ and $x = 3$
13 $f(x) = -x^2 + 1$, between $x = 0$ and $x = 3$
14 $f(x) = 5x^4 - 5$, between $x = -1$ and $x = 2$
15 $f(x) = 40x - x^2$, between $x = -10$ and $x = 20$
16 $f(x) = 10x - x^2$, between $x = -5$ and $x = 5$
17 $f(x) = xe^{x^2}$, between $x = 2$ and $x = 4$
18 $f(x) = 4xe^{x^2}$, between $x = 1$ and $x = 3$
19 $f(x) = (1/x)$, between $x = 5$ and $x = 10$
20 $f(x) = (5/x)$, between $x = 2$ and $x = 5$
21 Referring to Fig. 19.17, determine the combinations of definite integrals which would compute the area of (*a*) *A*, (*b*) *B*, (*c*) *C*, (*d*) *D*.
22 Referring to Fig. 19.18, determine the combinations of definite integrals which would compute the area of (*a*) *A*, (*b*) *B*, (*c*) *C*, (*d*) *D*.

Figure 19.17

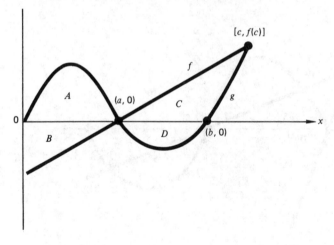

Figure 19.18

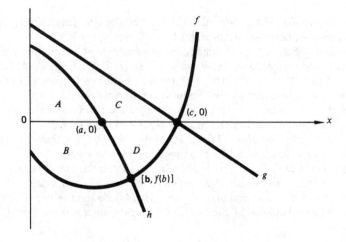

Figure 19.19

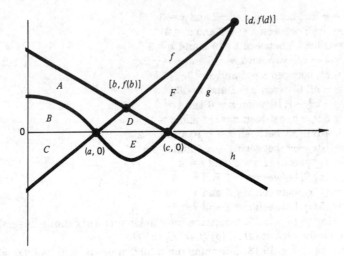

Figure 19.20

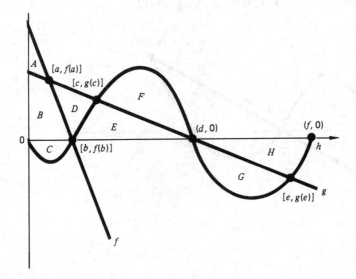

23 Referring to Fig. 19.19, determine the combinations of definite integrals which would compute the area of (*a*) *A*, (*b*) *B*, (*c*) *C*, (*d*) *D*, (*e*) *E*, (*f*) *F*.

24 Referring to Fig. 19.20, determine the combinations of definite integrals which would compute the area of (*a*) *A*, (*b*) *B*, (*c*) *C*, (*d*) *D*, (*e*) *E*, (*f*) *F*, (*g*) *G*, (*h*) *H*.

25 Given $f(x) = 2x^2$ and $g(x) = 27 - x^2$, (*a*) sketch the two functions. (*b*) For $x \geq 0$, determine the area bounded by the two functions and the *y* axis.

26 Given $f(x) = x^2$ and $g(x) = 64 - x^2$, (*a*) sketch the two functions. (*b*) For $x \geq 0$, determine the area bounded by the two functions and the *y* axis.

27 Given $f(x) = 2x$ and $g(x) = 12 - x$, (*a*) sketch the two functions. (*b*) For $x \geq 0$, determine the area bounded by the two functions and the *y* axis.

28 Given $f(x) = 3x + 2$ and $g(x) = 20 - 6x$, (*a*) sketch the two functions. (*b*) For $x \geq 0$, determine the area bounded by the two functions and the *y* axis.

29 Given $f(x) = x^2 - 10x$ and $g(x) = -x^2 + 10x$, (*a*) sketch the two functions. (*b*) Determine the area bounded by the two functions between $x = 0$ and $x = 10$.

30 Given $f(x) = x^2$ and $g(x) = 2x + 8$, for $x \geq 0$ determine the area bounded on three sides by the two functions and the y axis.

31 Find the area between $f(x) = 2x^2 - 4x + 6$ and $g(x) = -x^2 + 2x$ over the interval $0 \leq x \leq 2$.

32 Find the area between $f(x) = x + 2$ and $g(x) = 1 - 2x$ over the interval $0 \leq x \leq 5$.

33 Find the area between $f(x) = x^2 - 2x$ and $g(x) = -x - 4$ over the interval $-1 \leq x \leq 3$.

34 Find the area between $f(x) = x^2$ and $g(x) = -x^2 - 2$ over the interval $-2 \leq x \leq 2$.

35 Find the area between $f(x) = x^2 + 2$ and $g(x) = -e^x$ over the interval $-1 \leq x \leq 4$.

36 Find the area between $f(x) = 2x^2 - 2x + 10$ and $g(x) = -x^2 + x + 4$ over the interval $1 \leq x \leq 5$.

19.3 METHODS OF APPROXIMATION

There are situations in which there may not exist a rule of integration to evaluate a particular integrand, even using tables of integration rules. If one needs to evaluate a definite integral for such a function, there are methods for approximating the value of the definite integral. In this section we will examine three numerical approximation methods which might be used to evaluate

$$\int_a^b f(x)\, dx$$

Rectangle Rule

If we think of evaluating the definite integral in terms of computing the area beneath a curve, one method involves dividing the interval $a \leq x \leq b$ into n equal subintervals, each having a width $(b - a)/n$. If x_i is defined as the midpoint of interval i, we can construct a set of n rectangles each having width $(b - a)/n$ and height equal to $f(x_i)$ as shown in Fig. 19.21. To approximate the value of the definite integral, we add the areas of the n rectangles.

If we denote the width of each interval . . . $(b - a)/n$ . . . as Δx, we can express the rectangle rule as

Figure 19.21
Rectangle rule.

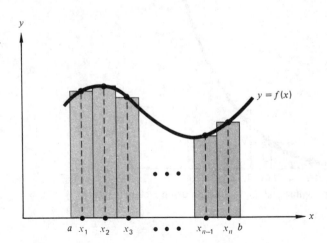

$$\int_a^b f(x)\,dx \approx f(x_1)\,\Delta x + f(x_2)\,\Delta x + \cdots + f(x_n)\,\Delta x$$

$$\approx \Delta x[\,f(x_1) + f(x_2) + \cdots + f(x_n)\,]$$

(19.9)

for n chosen sufficiently large.

EXAMPLE 17 Let's illustrate the use of this rule for a relatively simple function $f(x) = x^3$. Suppose we wish to evaluate $\int_0^4 x^3\,dx$. Let's use the rectangle rule, dividing the interval into four equal subintervals having widths equal to 1. As shown in Fig. 19.22, the midpoint values for the four subintervals are 0.5, 1.5, 2.5, and 3.5. Applying the rectangle rule, the numerical approximation of the definite integral is

$$\int_0^4 x^3\,dx \approx f(0.5)(1) + f(1.5)(1) + f(2.5)(1) + f(3.5)(1)$$

$$= (0.5)^3 + (1.5)^3 + (2.5)^3 + (3.5)^3$$
$$= 0.125 + 3.375 + 15.625 + 42.875$$
$$= 62.0$$

Confirm for yourself that the actual value of the definite integral is 64. The numerical approximation provided by the rectangle rule *underestimated* the actual value of the definite integral by 2.

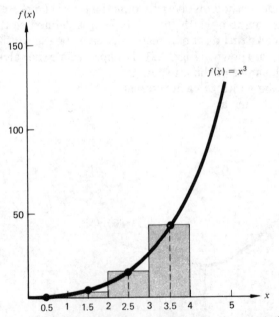

Figure 19.22 Numerical approximation using rectangle rule.

Trapezoidal Rule

Another numerical approximation method for definite integrals is the *trapezoidal rule*. As with the rectangle rule, the interval $a \leq x \leq b$ is divided into n subintervals of equal width. As shown in Fig. 19.23, the approximation consists of adding the areas of the n trapezoids defined by these subintervals. The heights of the n trapezoids are defined by the *endpoints* of the subintervals.

Figure 19.23
Trapezoidal rule.

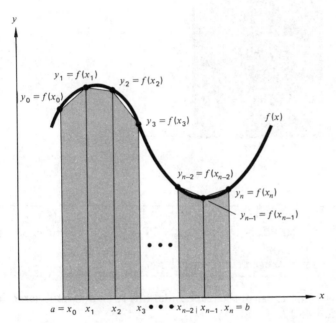

Given a trapezoid as shown in Fig. 19.24, the area of the trapezoid is defined as the product of the base of the trapezoid and the average height of the trapezoid, or

$$A = \Delta x[(y_1 + y_2)/2]^*$$

Thus, if we use the areas of the n trapezoids in Fig. 19.23 to approximate the definite integral, the definite integral is evaluated as

$$\int_a^b f(x)\, dx \approx [(b-a)/n][(y_0 + y_1)/2] + [(b-a)/n][(y_1 + y_2)/2]$$
$$+ [(b-a)/n][(y_2 + y_3)/2]$$
$$+ \cdots + [(b-a)/n][(y_{n-1} + y_n)/2]$$

or

---

* A more conventional definition states that the area of a trapezoid with parallel sides a and b and altitude h equals $\frac{1}{2}(a+b)h$.

Figure 19.24
Area of
trapezoid.

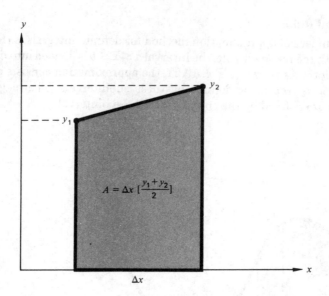

$$A = \Delta x \left[\frac{y_1 + y_2}{2}\right]$$

$$\boxed{\int_a^b f(x)\, dx \approx \left(\frac{b-a}{2n}\right)(y_0 + 2y_1 + 2y_2 + \cdots + 2y_{n-1} + y_n)}$$

$$\text{for } n \text{ chosen sufficiently large.} \qquad \textbf{(19.10)}$$

Let's illustrate the use of the trapezoidal rule to evaluate the same definite integral as in Example 17.

EXAMPLE 18   Using the same subinterval definitions as in Example 17, the endpoints are 0, 1, 2, 3, and 4. Therefore, $y_0 = f(0) = 0$, $y_1 = f(1) = 1$, $y_2 = f(2) = 8$, $y_3 = f(3) = 27$, and $y_4 = f(4) = 64$. Using Eq. (19.10),

$$\int_0^4 x^3\, dx \approx [(4-0)/2(4)][0 + 2(1) + 2(8) + 2(27) + 64]$$

$$= (1/2)(0 + 2 + 16 + 54 + 64)$$

$$= (1/2)(136)$$

$$= 68$$

Verify that the approximation provided by the trapezoidal rule *overestimates* the actual value of the definite integral by 4.

❑

Simpson's Rule

The third numerical approximation method is called *Simpson's rule*. As opposed to using rectangles or trapezoids to approximate the value of a definite integral, Simpson's rule is based upon parabolic estimation. Graphically, the interval $a \le$

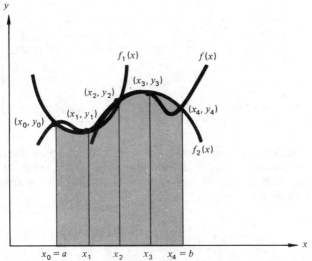

$f_1(x)$

(x_3, y_3)

$f(x)$

(x_2, y_2)

(x_1, y_1)

(x_4, y_4)

(x_0, y_0)

$f_2(x)$

$x_0 = a$ x_1 x_2 x_3 $x_4 = b$

Figure 19.25   Parabolic approximation of $f(x)$ using f_1 and f_2.

$x \leq b$ is divided into an *even number* of n subintervals which are of equal width. Then, a series of parabolas are fit to the graph, one for each *pair* of subintervals. This is illustrated in Fig. 19.25. Do you remember that three data points define a quadratic function of the form $f(x) = ax^2 + bx + c$? With Simpson's rule, a parabola is fit to the data points (x_0, y_0), (x_1, y_1), and (x_2, y_2) over the first two subintervals and another parabola is fit to the three data points (x_2, y_2), (x_3, y_3), and (x_4, y_4) over the second two subintervals in Fig. 19.25. These parabolas are used to approximate the function $f(x)$ over the interval $a \leq x \leq b$. The area under $f(x)$ over this interval is then approximated by finding the areas under the two parabolas.

Given the case where $x_1 - x_0 = x_2 - x_1$, a formula exists for determining the area under a parabola which passes through (x_0, y_0), (x_1, y_1), and (x_2, y_2). The formula does not require the actual determination of the quadratic function passing through these points. If $x_1 - x_0 = x_2 - x_1 = w$,

$$(ax^2 + bx + c)\, dx = \frac{w}{3}\, (y_0 + 4y_1 + y_2)$$

For the purpose of approximating the definite integral, $w = (b - a)/n$. Thus, if $f(x)$ is continuous over the interval $a \leq x \leq b$, and the number of subintervals n is even,

$$\int_a^b f(x)\, dx \approx \frac{b-a}{3n}\, (y_0 + 4y_1 + y_2) + \frac{b-a}{3n}\, (y_2 + 4y_3 + y_4)$$

$$+ \cdots + \frac{b-a}{3n}\, (y_{n-2} + 4y_{n-1} + y_n)$$

or

$$\int_a^b f(x)\, dx \approx \frac{b-a}{3n}\, (y_0 + 4y_1 + 2y_2 + 4y_3 + 2y_4$$
$$+ \cdots + 2y_{n-2} + 4y_{n-1} + y_n)$$

(19.11)

for n chosen sufficiently large (n even).

EXAMPLE 19   Let's use Simpson's rule to approximate the definite integral from the last two examples. Given that the subintervals are defined the same, $(x_0, y_0) = (0, 0)$, $(x_1, y_1) = (1, 1)$, $(x_2, y_2) = (2, 8)$, $(x_3, y_3) = (3, 27)$, and $(x_4, y_4) = (4, 64)$. With $a = 0$ and $b = 4$, the definite integral is estimated as

$$\int_0^4 x^3\, dx \approx \frac{4-0}{3(4)}\, [0 + 4(1) + 2(8) + 4(27) + 64]$$

$$= \frac{1}{3}\, (192)$$

$$= 64$$

which is *precisely* the value of $\int_0^4 x^3\, dx$.

Generally, there is an expectation that Simpson's rule will be more accurate than the rectangle or trapezoidal rule for a given value of n.

EXAMPLE 20   Let's illustrate the use of these approximations for a case where we have no method for finding the indefinite integral. Suppose that we wish to evaluate

$$\int_0^2 \frac{1}{x^2 + 1}\, dx$$

Let's approximate this definite integral using the three methods and subdividing into four subintervals each having width of 0.5. Figure 19.26 indicates the definition of the interval endpoints as well as the midpoints of each interval.

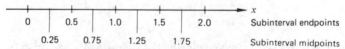

Figure 19.26   Subinterval endpoints and midpoints.

RECTANGLE RULE

Applying Eq. (19.9),

$$\int_0^2 \frac{1}{x^2 + 1}\, dx \approx (0.5)[\, f(0.25) + f(0.75) + f(1.25) + f(1.75)]$$

$$= (0.5)\left[\frac{1}{(0.25)^2+1}+\frac{1}{(0.75)^2+1}+\frac{1}{(1.25)^2+1}+\frac{1}{(1.75)^2+1}\right]$$

$$= 0.5(0.94118+0.64000+0.39024+0.24615)$$

$$= 1.10879$$

TRAPEZOIDAL RULE

Applying Eq. (19.10),

$$\int_0^2 \frac{1}{x^2+1}\,dx \approx \left[\frac{2-0}{(2)(4)}\right][f(0)+2f(0.5)+2f(1.0)+2f(1.5)+f(2.0)]$$

$$= \tfrac{1}{4}[1+2(0.8)+2(0.5)+2(0.308)+0.2]$$

$$= \tfrac{1}{4}(4.416)$$

$$= 1.104$$

SIMPSON'S RULE

Applying Eq. (19.11),

$$\int_0^2 \frac{1}{x^2+1}\,dx \approx \left[\frac{2-0}{(3)(4)}\right][f(0)+4f(0.5)+2f(1.0)+4f(1.5)+f(2.0)]$$

$$= \left(\frac{1}{6}\right)[1+4(0.8)+2(0.5)+4(0.308)+0.2]$$

$$= \frac{1}{6}(6.632)$$

$$= 1.105$$

Since we cannot evaluate this definite integral explicitly, we cannot really assess the accuracy of these approximations. Experience would indicate that the value obtained using Simpson's rule is likely to be the most accurate.

$\square$

Section 19.3 Follow-up Exercises

For each of the following exercises, (a) evaluate the definite integral using the rectangle rule (subdivide into four intervals), (b) evaluate the exact value using explicit rules of integration, and (c) calculate the error of the approximation.

1 $\displaystyle\int_0^4 x^2\,dx$

2 $\displaystyle\int_0^4 4x^2\,dx$

3 $\displaystyle\int_0^2 4x^3\,dx$

4 $\displaystyle\int_0^2 8x^3\,dx$

5 $\displaystyle\int_0^4 e^x\,dx$

6 $\displaystyle\int_1^3 3e^x\,dx$

7 $\displaystyle\int_2^4 (x^3-2x^2)\,dx$

8 $\displaystyle\int_2^4 (2x^3+3x^2)\,dx$

9 $\displaystyle\int_4^8 (2x)(x^2-5)^3\,dx$

10 $\displaystyle\int_4^8 (4x)(2x^2+10)^4\,dx$

For each of the following exercises, (a) evaluate the definite integral using the trapezoidal rule ($n = 4$), (b) evaluate the exact value using explicit rules of integration, and (c) calculate the error of the approximation.

11 $\int_1^5 (5x - 2)\, dx$

12 $\int_2^6 (20 - 4x)\, dx$

13 $\int_1^3 (4x^2 - x)\, dx$

14 $\int_3^5 (2x - 3x^2)\, dx$

15 $\int_0^4 2e^x\, dx$

16 $\int_1^5 e^{-x}\, dx$

17 $\int_0^4 (8x)(4x^2 - 5)^3\, dx$

18 $\int_0^4 (3x^2)(x^3 + 5)^3\, dx$

19 $\int_0^2 2xe^{x^2}\, dx$

20 $\int_0^2 xe^{2x^2}\, dx$

For each of the following exercises, (a) evaluate the definite integral using Simpson's rule (subdivide into four intervals), (b) evaluate the exact value using explicit rules of integration, and (c) calculate the error of the approximation.

21 $\int_0^2 (5x + 8)\, dx$

22 $\int_0^4 (4 - 2x)\, dx$

23 $\int_1^5 (4x^2 - 5x)\, dx$

24 $\int_2^6 (x^3 - 10)\, dx$

25 $\int_0^2 (5x^4 - x)\, dx$

26 $\int_2^4 (2x - x^3)\, dx$

27 $\int_0^4 5e^x\, dx$

28 $\int_4^8 -4e^{-x}\, dx$

29 $\int_4^6 6x^2(x^3 - 1)^3\, dx$

30 $\int_2^4 4x(x^2 + 3)^3\, dx$

For each of the following exercises, (a) evaluate the definite integral using explicit rules of integration; (b) approximate the definite integral using the rectangle rule, the trapezoidal rule, and Simpson's rule (subdividing into four intervals); and (c) determine the most accurate approximation method.

31 $\int_0^2 10\, dx$

32 $\int_0^4 -5\, dx$

33 $\int_2^6 (x^2 - 5)\, dx$

34 $\int_4^8 (10 - x + x^2)\, dx$

35 $\int_0^2 4x^3\, dx$

36 $\int_2^4 8x^3\, dx$

37 $\int_4^6 10e^x\, dx$

38 $\int_2^4 5e^{-x}\, dx$

39 $\int_0^4 4x^3(x^4 - 1)^3\, dx$

40 $\int_0^2 6x^2(x^3 + 6)^4\, dx$

In the following exercises, the definite integral cannot be evaluated using our rules of integration. Approximate the definite integral using each of the three approximation methods ($n = 4$).

41 $\displaystyle\int_0^2 \frac{4}{2x^2 + 1}\, dx$

42 $\displaystyle\int_0^4 \frac{10}{x^2 + 4}\, dx$

43 $\displaystyle\int_0^4 \frac{2}{x^3 + 1}\, dx$

44 $\displaystyle\int_0^2 \frac{x}{x^3 + 2}\, dx$

45 $\displaystyle\int_2^6 \frac{1}{1 + e^x}\, dx$

46 $\displaystyle\int_0^4 \frac{5}{2 + e^x}\, dx$

47 $\displaystyle\int_0^2 \sqrt{2 + x^3}\, dx$

48 $\displaystyle\int_0^4 \sqrt{10 + x^3}\, dx$

19.4 APPLICATIONS OF INTEGRAL CALCULUS

The following examples illustrate sample applications of integral calculus.

EXAMPLE 21

(Revenue) In Chap. 18 we discussed how the total revenue function can be determined by integrating the marginal revenue function. As a simple extension of this concept, assume that the price of a product is constant at a value of $10 per unit, or the marginal revenue function is

$$MR = f(x)$$
$$= 10$$

where x equals the number of units sold. Total revenue from selling x units can be determined by integrating the marginal revenue function between 0 and x. For example, the total revenue from selling 1,500 units can be computed as

$$\int_0^{1,500} 10\, dx = 10x \Big]_0^{1,500}$$
$$= 10(1,500)$$
$$= \$15,000$$

This is a rather elaborate procedure for calculating total revenue since we simply could have multiplied price by quantity sold to determine the same result. However, it does illustrate how the area beneath the marginal revenue function (Fig. 19.27) can be interpreted as total revenue or incremental revenue. The additional revenue associated with increasing sales from 1,500 to 1,800 units can be computed as

$$\int_{1,500}^{1,800} 10\, dx = 10x \Big]_{1,500}^{1,800}$$
$$= \$18,000 - \$15,000$$
$$= \$3,000$$

EXAMPLE 22

(Maintenance Expenditures) An automobile manufacturer estimates that the annual rate of expenditure $r(t)$ for maintenance on one of its models is represented by the function

Figure 19.27

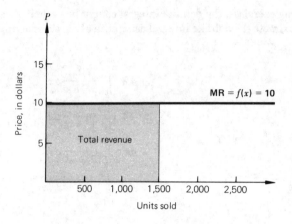

$$r(t) = 100 + 10t^2$$

where t is the age of the automobile stated in years and r(t) is measured in dollars per year. This function suggests that when the car is 1 year old, maintenance expenses are being incurred at a rate of

$$r(1) = 100 + 10(1)^2$$
$$= \$110 \text{ per year}$$

When the car is 3 years old, maintenance costs are being incurred at a rate of

$$r(3) = 100 + 10(3)^2$$
$$= \$190 \text{ per year}$$

As would be anticipated, the older the automobile, the more maintenance is required. Figure 19.28 illustrates the sketch of the rate of expenditure function.

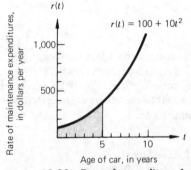

Figure 19.28   Rate of expenditure function.

The area under this curve between any two values of t is a measure of the expected maintenance cost during that time interval. The expected maintenance expenditures during the automobile's first 5 years are computed as

$$\int_0^5 (100 + 10t^2)\, dt = 100t + \frac{10t^3}{3} \Big]_0^5$$

$$= 100(5) + \frac{10(5)^3}{3} = 500 + 416.67 = \$916.67$$

Of these expenditures, those expected to be incurred *during the fifth year* are estimated as

$$\int_4^5 (100 + 10t^2)\, dt = 100t + \frac{10t^3}{3} \Big]_4^5$$

$$= \left[100(5) + \frac{10(5)^3}{3} \right] - \left[100(4) + \frac{10(4)^3}{3} \right]$$

$$= 916.67 - (400 + 213.33)$$

$$= \$303.34$$

EXAMPLE 23 **(Fund-Raising)** A state civic organization is conducting its annual fund-raising campaign for its summer camp program for the disadvantaged. Campaign expenditures will be incurred at a rate of $10,000 per day. From past experience it is known that contributions will be high during the early stages of the campaign and will tend to fall off as the campaign continues. The function describing the rate at which contributions are received is

$$c(t) = -100t^2 + 20{,}000$$

where t represents the day of the campaign, and c(t) equals the rate at which contributions are received, measured in dollars per day. The organization wishes to maximize the net proceeds from the campaign.

(a) Determine how long the campaign should be conducted in order to maximize net proceeds.

(b) What are total campaign expenditures expected to equal?

(c) What are total contributions expected to equal?

(d) What are the net proceeds (total contributions less total expenditures) expected to equal?

SOLUTION

(a) The function which describes the rate at which expenditures e(t) are incurred is

$$e(t) = 10{,}000$$

Figure 19.29 illustrates the two functions. As long as the rate at which contributions are made exceeds the rate of expenditures for the campaign, *net proceeds* are positive. Refer to Fig. 19.29. Net proceeds will be positive up until the time when the graphs of the two functions intersect. Beyond this point, the rate of expenditure exceeds the rate of contribution. That is, contributions would be coming in at a rate of less than $10,000 per day.

The graphs of the two functions intersect when

$$c(t) = e(t)$$

or when $$-100t^2 + 20{,}000 = 10{,}000$$

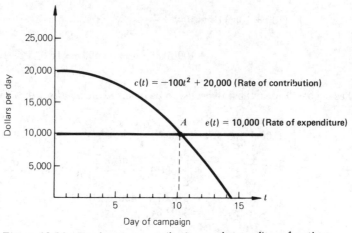

Figure 19.29 Fund-raising contributions and expenditure functions.

$$-100t^2 = -10,000$$

$$t^2 = 100$$

$$t = 10 \text{ days}$$

(Negative root meaningless.)

(b) Total campaign expenditures are represented by the area under e between $t = 0$ and $t = 10$. This could be found by integrating e between these limits or more simply by multiplying:

$$E = (\$10,000 \text{ per day})(10 \text{ days})$$
$$= \$100,000$$

(c) Total contributions during the 10 days are represented by the area under c between $t = 0$ and $t = 10$, or

$$C = \int_0^{10} (-100t^2 + 20,000) \, dt$$

$$= -100 \frac{t^3}{3} + 20,000t \Big]_0^{10}$$

$$= \frac{-100(10)^3}{3} + 20,000(10)$$

$$= -33,333.33 + 200,000 = \$166,666.67$$

(d) Net proceeds are expected to equal

$$C - E = \$166,666.67 - \$100,000$$
$$= \$66,666.67$$

EXAMPLE 24 (**Blood Bank Management; Motivating Scenario**) A hospital blood bank is conducting a blood drive to replenish its inventory of blood. The hospital estimates that blood will be donated at a rate of $d(t)$ pints per day, where

$$d(t) = 500e^{-0.4t}$$

and *t equals the length of the blood drive in days.* If the goal of the blood drive is 1,000 pints, when will the hospital reach its goal?

SOLUTION

In this problem the area between the graph of d and the t axis represents total donations of blood in pints. Unlike previous applications, the desired area is known; the unknown is the upper limit of integration, as shown in Fig. 19.30. The hospital will reach its goal when

$$\int_0^{t^*} 500e^{-0.4t}\, dt = 1,000$$

Rewriting the integrand,

$$\int_0^{t^*} -1,250(-0.4)e^{-0.4t}\, dt = 1,000$$

or

$$-1,250 \int_0^{t^*} -0.4e^{-0.4t}\, dt = 1,000$$

Evaluating the definite integral and solving for t^*,

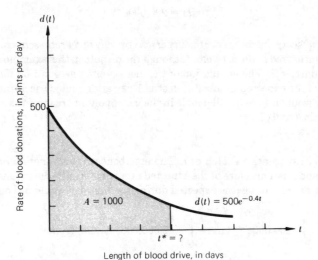

Figure 19.30 Determining upper limit of integration.

$$-1{,}250[e^{-0.4t}]_0^{t^*} = 1{,}000$$

$$-1{,}250[e^{-0.4t^*} - e^{-0.4(0)}] = 1{,}000$$

$$-1{,}250[e^{-0.4t^*} - 1] = 1{,}000$$

$$-1{,}250e^{-0.4t^*} + 1{,}250 = 1{,}000$$

$$-1{,}250e^{-0.4t^*} = -250$$

$$e^{-0.4t^*} = \frac{-250}{-1{,}250}$$

$$e^{-0.4t^*} = 0.2$$

Taking the natural logarithm of both sides of the equation (using Table 2),

$$-0.4t^* = -1.6094$$

$$t^* = \frac{-1.6094}{-0.4}$$

or $$t^* = 4.0235$$

Thus, the hospital will reach its goal in approximately 4 days.

EXAMPLE 25   (**Nuclear Power**) An electric company has proposed building a nuclear power plant on the outskirts of a major metropolitan area. As might be expected, public opinion is divided and discussions have been heated. One lobbyist group opposing the construction of the plant has presented some disputed data regarding the consequences of a catastrophic accident at the proposed plant. The lobbyist group estimates that the rate at which deaths would occur within the metropolitan area because of radioactive fallout is described by the function

$$r(t) = 200{,}000e^{-0.1t}$$

where *r(t) represents the rate of deaths in persons per day and t represents time elapsed since the accident, measured in days.* (*Note:* Although the dispute in this example is quite real, the data are all contrived!) The population of the metropolitan area is 1.5 million persons.
(*a*) Determine the expected number of deaths 1 day after a major accident.
(*b*) How long would it take for all people in the metropolitan area to succumb to the effects of the radioactivity?

SOLUTION
(*a*) Figure 19.31 presents a sketch of *r*. The area beneath this function between any two points t_1 and t_2 is a measure of the expected number of deaths during that time interval. Thus, the number of deaths expected during the first day would be computed as

$$\int_0^1 200{,}000e^{-0.1t}\, dt = \int_0^1 -2{,}000{,}000(-0.1)e^{-0.1t}\, dt$$

$$= -2{,}000{,}000 \int_0^1 (-0.1)e^{-0.1t}\, dt$$

Figure 19.31
Rate of deaths.

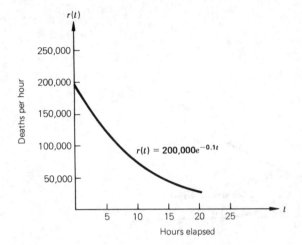

$$= -2,000,000e^{-0.1t}\Big]_0^1$$
$$= -2,000,000e^{-0.1} + 2,000,000e^0$$
$$= -2,000,000(e^{-0.1} - e^0)$$
$$= -2,000,000(0.9048 - 1)$$
$$= -2,000,000(-0.0952) = 190,400 \text{ people}$$

(b) As morbid as the thought is, the entire population would succumb after t^* days, where

$$\int_0^{t^*} 200,000e^{-0.1t}\, dt = 1,500,000$$

or when
$$-2,000,000e^{-0.1t}\Big]_0^{t^*} = 1,500,000$$

Solving for t^*, we get

$$-2,000,000e^{-0.1t^*} + 2,000,000 = 1,500,000$$

$$-2,000,000e^{-0.1t^*} = -500,000$$

$$e^{-0.1t^*} = 0.25$$

If the natural logarithm (Table 2) is found for both sides of the equation,

$$-0.1t^* = -1.3863$$

or
$$t^* = 13.863 \text{ days}$$

EXAMPLE 26 (**Consumer's Surplus**) One way of measuring the value or utility that a product holds for a consumer is the price that he or she is willing to pay for it. Economists contend that

Figure 19.32
Consumer's
surplus.

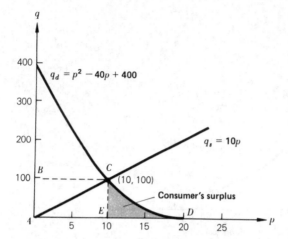

consumers actually receive bonus or surplus value from the products they purchase according to the way in which the marketplace operates.

Figure 19.32 portrays the supply and demand functions for a product. Equilibrium occurs when a price of $10 is charged and demand equals 100 units. If dollars are used to represent the value of this product to consumers, our accounting practices would suggest that the total revenue ($10 · 100 units = $1,000) is a measure of the *economic value* of this product. The area of rectangle *ABCE* represents this measure of value.

However, if you consider the nature of the demand function, there would have been a demand for the product at prices higher than $10. That is, there would have been consumers willing to pay almost $20 for the product. And additional consumers would have been drawn into the market at prices between $10 and $20. If we assume that the price these people would be willing to pay is a measure of the utility the product holds for them, they actually receive a bonus when the market price is $10. Refer again to Fig. 19.32. Economists would claim that a measure of the actual utility of the product is the area *ABCDE*. And when the market is in equilibrium, the extra utility received by consumers, referred to as the ***consumer's surplus,*** is represented by the shaded area *CDE*. This area can be found as

$$\int_{10}^{20} (p^2 - 40p + 400)\, dp = \frac{p^3}{3} - 20p^2 + 400p \Big]_{10}^{20}$$

$$= \left[\frac{(20)^3}{3} - 20(20)^2 + 400(20) \right]$$

$$- \left[\frac{(10)^3}{3} - 20(10)^2 + 400(10) \right]$$

$$= 2{,}666.67 - 2{,}333.33 = \$333.34$$

Our accounting methods would value the utility of the product at $1,000. Economists would contend that the actual utility is $1,333.34, or that the consumer's surplus equals $333.34. This measure of added, or bonus, utility applies particularly to those consumers who would have been willing to pay more than $10.

EXAMPLE 27   **(Volume of a Solid of Revolution)** Consider the function *f* in Fig. 19.33. If the half-plane bounded by *f*, the *x* axis, and the lines $x = a$ and $x = b$ is rotated about the *x* axis one full

Figure 19.33

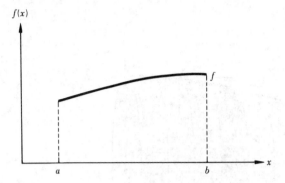

Figure 19.34
Surface of
revolution.

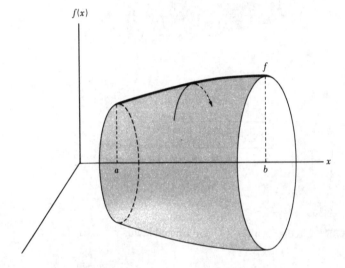

revolution, a **surface of revolution** is formed. Each point of f sweeps a circular path. The composite sweep of all points f on the interval $a \leq x \leq b$ is the surface of revolution shown in Fig. 19.34.

Corresponding to the surface of revolution is a **solid of revolution.** This is the volume which the plane sweeps out as the graph of f is rotated about the x axis. Suppose that we wish to determine the volume of the solid of revolution. We might estimate the volume by forming an approximate solid of revolution consisting of a set of right circular cylinders, as shown in Fig. 19.35. In this figure the interval $a \leq x \leq b$ has been subdivided into equal subintervals of width Δx. The height of each of the right cylinders is Δx. The radius of each right cylinder is $f(x_i)$, where x_i is the left-hand value of x for subinterval i.

As shown in Fig. 19.36, the volume of a right cylinder having radius r and height h is

$$V = \pi r^2 h$$

where π (pi) $= 3.14. \ldots$ For each right cylinder in Fig. 19.35, the volume is

$$V_i = \pi [f(x_i)]^2 \, \Delta x$$

Figure 19.35

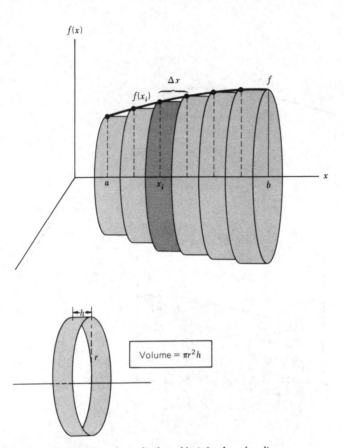

Figure 19.36   Circular cylinder of height h and radius r.

If the interval $a \le x \le b$ has been divided into n equal subintervals, the estimated volume of the solid of revolution is

$$V = \sum_{i=1}^{n} V_i = \sum_{i=1}^{n} \pi [f(x_i)]^2 \, \Delta x$$

The estimate will become more accurate as the interval $a \le x \le b$ is subdivided into a greater number of subintervals, with Δx approaching 0.

DEFINITION: VOLUME OF SOLID OF REVOLUTION
Given a function f which is continuous over the interval $a \le x \le b$, the volume of the solid of revolution generated as the function is rotated one revolution about the x axis is defined by

$$V = \lim_{n \to \infty} \sum_{i=1}^{n} \pi [f(x_i)]^2 \, \Delta x$$

Figure 19.37

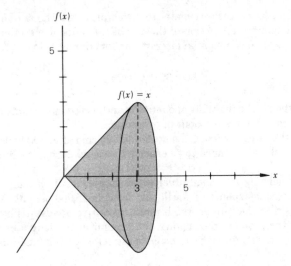

When the limit exists, it can be shown that

$$V = \int_a^b \pi[f(x)]^2 \, dx \qquad (19.12)$$

Given the function $f(x) = x$, suppose we wish to determine the volume of the solid of revolution generated as the graph of f, between $x = 0$ and $x = 3$, is rotated about the x axis. This solid of revolution is shown in Fig. 19.37. The volume is

$$V = \int_0^3 \pi(x)^2 \, dx$$

$$= \pi \frac{(x)^3}{3} \bigg]_0^3$$

$$= \pi \left[\frac{(3)^3}{3} - \frac{(0)^3}{3} \right]$$

$$= 9\pi$$

$$\doteq 28.26 \text{ cubic units}$$

□

Section 19.4 Follow-up Exercises

1 The marginal revenue function for a firm's product is

$$MR = -0.04x + 10$$

where x equals the number of units sold.

(a) Determine the total revenue from selling 200 units of the product.

(b) What is the added revenue associated with an increase in sales from 100 to 200 units?

2 A manufacturer of jet engines estimates that the rate at which maintenance costs are incurred on its engines is a function of the number of hours of operation of the engine. For one engine used on commercial aircraft, the function is

$$r(x) = 60 + 0.040x^2$$

where x equals the number of hours of operation and $r(x)$ equals the rate at which repair costs are incurred in dollars per hour of operation.
(a) Determine the rate at which costs are being incurred after 100 hours of operation.
(b) What are total maintenance costs expected to equal during the first 100 hours of operation?

3 A company specializing in a mail-order sales approach is beginning a promotional campaign. Advertising expenditures will cost the firm $5,950 per day. Marketing specialists estimate that the rate at which profit (exclusive of advertising costs) will be generated from the promotion campaign decreases over the length of the campaign. Specifically, the rate $r(t)$ for this campaign is estimated by the function

$$r(t) = -50t^2 + 10,000$$

where t represents the day of the campaign and $r(t)$ is measured in dollars per day. In order to maximize *net* profit, the firm should conduct the campaign as long as $r(t)$ exceeds the daily advertising cost.
(a) Graph the function $r(t)$ and the function $c(t) = 5,950$, which describes the rate at which advertising expenses are incurred.
(b) How long should the campaign be conducted?
(c) What are total advertising expenditures expected to equal during the campaign?
(d) What *net* profit will be expected?

4 Rework Example 23, assuming that campaign expenditures will be incurred at a rate of $5,000 per day and that

$$c(t) = -10t^2 + 9,000$$

5 Rework Example 25, assuming that $r(t) = 200,000e^{-0.05t}$.
6 Rework Example 25, assuming that the population equals 800,000 and $r(t) = 100,000e^{-0.1t}$.
7 You are given the demand function

$$q_d = p^2 - 30p + 200$$

and the supply function

$$q_s = 5p$$

where p is stated in dollars, q_d and q_s are stated in units, and $0 \leq p \leq 9$.
(a) Sketch the two functions.
(b) Determine the equilibrium price and quantity.
(c) Determine the value of the consumer's surplus if the market is in equilibrium.

8 Energy Conservation A small business is considering buying an energy-saving device which will reduce its consumption of fuel. The device will cost $32,000. Engineering estimates suggest that savings from using the device will occur at a rate of $s(t)$ dollars per year, where

$$s(t) = 20{,}000e^{-0.5t}$$

And t equals time measured in years. Determine how long it will take for the firm to recover the cost of the device (that is, when the accumulated fuel savings equal the purchase cost).

9 **Blood Bank Management** A hospital blood bank conducts an annual blood drive to replenish its inventory of blood. The hospital estimates that blood will be donated at a rate of $d(t)$ pints per day, where

$$d(t) = 300e^{-0.1t}$$

and t equals the length of the blood drive in days. If the goal for the blood drive is 2,000 pints, when will the hospital reach its goal?

10 **Forest Management** The demand for commercial forestland timber has been increasing rapidly over the past three to four decades. The function describing the rate of demand for timber is

$$d(t) = 20 + 0.003t^2$$

where $d(t)$ is stated in billions of cubic feet per year and t equals time in years ($t = 0$ corresponds to January 1, 1990).

(a) Determine the rate of demand at the beginning of 1995.
(b) Determine the rate of demand at the beginning of 2010.
(c) Determine the *total* demand for timber during the period 1990 through 2009. [*Hint:* Integrate $d(t)$ between $t = 0$ and $t = 20$.]

11 **Solid Waste Management** The rate $w(t)$ at which solid waste is being generated in a major United States city is described by the function

$$w(t) = 0.5e^{0.025t}$$

where $w(t)$ is stated in billions of tons per year and t equals time measured in years ($t = 0$ corresponds to January 1, 1990).

(a) Determine the rate at which solid waste is expected to be generated at the beginning of the year 2000.
(b) What total tonnage is expected to be generated during the 20-year period from 1990 through 2009?

12 **Epidemic Control** A health research center specializes in the study of epidemics. It estimates that for one particular type of epidemic which occurred in one region of the country, the rate at which new people were afflicted was described by the function

$$r(t) = 50e^{0.2t} - 40$$

where $r(t)$ is the rate of new afflictions, measured in people per day, and t equals time since the beginning of the epidemic, measured in days. (a) How many persons were afflicted during the first 10 days? (b) During the first 20 days?

13 **Learning Curves** People in the manufacturing industries have observed in many instances that employees assigned to a new job or task become more efficient with experience. That is, as the employee repeats the task, he or she becomes more familiar with the operations, motions, and equipment required to perform the job. Some companies have enough experience with job training that they can project how quickly an employee will learn a job. Very often a *learning curve* can be constructed which estimates

the rate at which a job is performed as a function of the number of times the job has been performed by an employee.

The *learning curve* for a particular job has been defined as

$$h(x) = \frac{20}{x} + 4 \qquad x > 0$$

where $h(x)$ equals the production rate measured in hours per unit and x equals the unit produced.

(a) Determine the production rate $h(x)$ at the time of the 10th unit ($x = 10$).

(b) Integrating the learning curve over a specified interval provides an estimate of the total number of production hours required over the corresponding range of output. Determine the total number of hours expected for producing the first 20 units by integrating $h(x)$ between $x = 1$ and $x = 20$.

(c) Sketch h.

(d) Is there any limit suggested as to how efficient an employee can become at this job?

14 Producer's Surplus Example 26 discussed the notion of consumer's surplus, which represents what economists believe to be a measure of the added utility consumers enjoy when the market is in equilibrium. Economists also suggest that producers receive a bonus or added utility when the market is in equilibrium. Figure 19.38 repeats the supply and demand functions presented in Example 26.

If you focus on the supply function q_s, it indicates that certain suppliers would be willing to supply units at prices less than the equilibrium price of $10. When the market price is $10, these suppliers earn more than they otherwise would have. If each supplier sells at the price he or she is willing to, the total revenue received would be represented by area ABC. Since total revenue at equilibrium is represented by $ABCD$, the shaded area represents a measure of the added value to suppliers. This added value is referred to as *producer's surplus*.

(a) Determine the producer's surplus in Example 26.

(b) Determine the producer's surplus for the functions described in Exercise 7.

15 Given the function $f(x) = -x^2 + 9$, find the volume of the solid of revolution between $x = 0$ and $x = 3$ if f is rotated about the x axis.

16 Given the function $f(x) = -x - 2$, find the volume of the solid of revolution between $x = -2$ and $x = 5$ if f is rotated about the x axis.

Figure 19.38

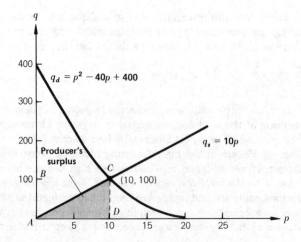

19.5 INTEGRAL CALCULUS AND PROBABILITY (OPTIONAL)

A mathematical function which determines the probability of each possible outcome of an experiment is called a ***probability density function*** (pdf). Compared with probability distributions which display each outcome and its probability, density functions can be thought of as the mathematical functions used to compute the probability. If x is a continuous random variable, its density function f must satisfy the two conditions

    1 $f(x) \geq 0$ *for all x, and*
    2   *the area under the graph of f equals 1.*

The first condition prohibits negative probabilities for any event, and the second guarantees that the events are mutually exclusive and collectively exhaustive.

    The probability that a random variable x assumes a value in the interval between a and b, where $a < b$, equals the area under the density function between $x = a$ and $x = b$. Applying the definite integral, the probability that x will assume a value between $x = a$ and $x = b$ equals $\int_a^b f(x)\, dx$, or

$$P(a \leq x \leq b) = \int_a^b f(x)\, dx$$

Technically, if we wanted to determine the probability that a normally distributed variable x, having a mean μ and standard deviation σ, will assume a value between $x = a$ and $x = b$, $a < b$, we could determine the probability by integrating the density function, or

$$P(a \leq x \leq b) = \int_a^b \frac{1}{\sqrt{2\pi}\sigma} e^{-1/2[(x-\mu)/\sigma]^2}\, dx$$

Fortunately, the equivalent conversion to the standard normal distribution and the availability of tables such as Table 14.25 eliminate any need to perform what appears to be a cumbersome integration.

EXAMPLE 28   Consider the probability density function for the random variable x

$$f(x) = \frac{2 + x}{30} \qquad 0 \leq x \leq 6$$

Determine the probability that x will assume a value between 2 and 5.

SOLUTION
The density function and the area of interest are illustrated in Fig. 19.39. The probability is computed as

$$P(2 \leq x \leq 5) = \int_2^5 \frac{2 + x}{30}\, dx$$

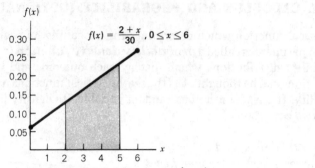

Figure 19.39 Probability density function.

$$= \frac{x}{15} + \frac{x^2}{60} \Big]_2^5$$

$$= \left(\frac{5}{15} + \frac{5^2}{60} \right) - \left(\frac{2}{15} + \frac{2^2}{60} \right)$$

$$= \frac{5}{15} + \frac{25}{60} - \frac{2}{15} - \frac{4}{60}$$

$$= \frac{20 + 25 - 8 - 4}{60} = \frac{33}{60} = .55$$

EXAMPLE 29 A 3-hour examination is given to all prospective salespeople of a national retail chain. The time x in hours required to complete the examination has been found to be random with a density function

$$f(x) = \frac{-x^2 + 10x}{36} \qquad 0 \le x \le 3$$

Determine the probability that someone will complete the test in 1 hour or less.

SOLUTION

The density function and area of interest are shown in Fig. 19.40. The probability is computed as

$$P(0 \le x \le 1) = \int_0^1 \frac{-x^2 + 10x}{36} \, dx$$

$$= \frac{-x^3}{108} + \frac{10x^2}{72} \Big]_0^1$$

$$= \frac{-(1)^3}{108} + \frac{10(1)^2}{72} = \frac{-1}{108} + \frac{10}{72} = -.009 + .139 = .13$$

❑

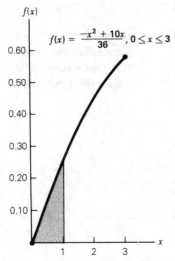

Figure 19.40 Probability density function.

Section 19.5 Follow-up Exercises

1 The probability density function for a continuous random variable x is

$$f(x) = \frac{5 - x}{4.5} \qquad 2 \le x \le 5$$

What is the probability that the random variable will assume a value greater than 3? Less than 2?

2 The probability density function for a continuous random variable x is

$$f(x) = \frac{x^2 - 10x + 25}{39} \qquad 0 \le x \le 3$$

Determine the probability that the random variable will assume a value greater than 2. Less than 1.

3 The probability density function for a continuous random variable x is

$$f(x) = \frac{1 + 2x}{36} \qquad 2 \le x \le 6$$

Determine the probability that the random variable will assume a value between 2 and 4.

❏ KEY TERMS AND CONCEPTS

definite integral 875

fundamental theorem of integral
    calculus 876

limits of integration (lower and
    upper) 875

probability density function 917

rectangle rule 895

Simpson's rule 898

trapezoidal rule 897

❏ IMPORTANT FORMULAS

$$\int_a^b f(x)\,dx = F(b) - F(a) \tag{19.3}$$

$$\int_a^b f(x)\,dx = -\int_b^a f(x)\,dx \tag{19.4}$$

$$\int_a^a f(x)\,dx = 0 \tag{19.5}$$

$$\int_a^b f(x)\,dx + \int_b^c f(x)\,dx = \int_a^c f(x)\,dx \tag{19.6}$$

$$\int_a^b cf(x)\,dx = c\int_a^b f(x)\,dx \tag{19.7}$$

$$\int_a^b [\,f(x) \pm g(x)]\,dx = \int_a^b f(x)\,dx \pm \int_a^b g(x)\,dx \tag{19.8}$$

$$\int_a^b f(x)\,dx \approx \Delta x[\,f(x_1) + f(x_2) + \cdots + f(x_n)] \quad \text{Rectangle rule} \tag{19.9}$$

$$\int_a^b f(x)\,dx \approx \left(\frac{b-a}{2n}\right)(y_0 + 2y_1 + 2y_2 + \cdots + 2y_{n-1} + y_n)$$
$$\text{Trapezoidal rule} \tag{19.10}$$

$$\int_a^b f(x)\,dx \approx \frac{b-a}{3n}(y_0 + 4y_1 + 2y_2 + 4y_3 + 2y_4 + \cdots + 2y_{n-2}$$
$$+ 4y_{n-1} + y_n), \quad n \text{ even} \quad \text{Simpson's rule} \tag{19.11}$$

❏ ADDITIONAL EXERCISES

Section 19.1

Evaluate the following integrals.

1 $\displaystyle\int_0^1 (x^2 - 3e^x - 2)\,dx$

2 $\displaystyle\int_1^3 (4x^2 + 5)\,dx$

3 $\displaystyle\int_{-2}^2 (6x + 7)\,dx$

4 $\displaystyle\int_{-1}^0 3x^2\,dx$

5 $\displaystyle\int_0^1 (-2x^5 + 3x^4)\, dx$

6 $\displaystyle\int_{-1}^1 (x-3)/x^3\, dx$

7 $\displaystyle\int_1^2 (-4x^3 + 6x^2)\, dx$

8 $\displaystyle\int_2^4 (20x^3 - 15x^2)\, dx$

9 $\displaystyle\int_0^4 (2x)(x^2 - 3)^3\, dx$

10 $\displaystyle\int_0^2 (2x)(x^2 + 5)^4\, dx$

11 $\displaystyle\int_1^2 (x^2/2 - 2x)\, dx$

12 $\displaystyle\int_{-3}^0 3x^2 e^{x^3}\, dx$

13 $\displaystyle\int_{-3}^3 (-5xe^{x^2+2}\, dx$

14 $\displaystyle\int_2^3 \frac{6x}{3x^2 - 7}\, dx$

15 $\displaystyle\int_0^1 x^{1/2}(1 + x)\, dx$

16 $\displaystyle\int_{-4}^0 e^{-1-6x}\, dx$

17 $\displaystyle\int_1^4 3x^2(x^3 - 3)^3\, dx$

18 $\displaystyle\int_0^2 12x^2(x^3 - 10)^4\, dx$

19 $\displaystyle\int_0^2 3x(3x^2 - 6)^3\, dx$

20 $\displaystyle\int_0^4 (16x)(2x^2 - 3)^4\, dx$

21 $\displaystyle\int_2^4 \frac{-4x}{x^2 + 3}\, dx$

22 $\displaystyle\int_0^2 \frac{-4x}{10 - 2x^2}\, dx$

23 $\displaystyle\int_0^2 \frac{2x + 2}{x^2 + 2x + 1}\, dx$

24 $\displaystyle\int_2^4 \frac{4x + 4}{2x^2 + 4x - 5}\, dx$

25 $\displaystyle\int_4^5 \frac{x^2 - 2x}{x^3 - 3x^2}\, dx$

26 $\displaystyle\int_4^5 \frac{2x^2 - 4x}{2x^3 - 6x^2}\, dx$

27 $\displaystyle\int_0^3 (x^2 + 7x + 4)\, dx - \int_0^3 (x^2 + 7x + 3)\, dx$

28 $\displaystyle\int_1^4 (x^2 + 6)\, dx + \int_1^4 (x^2 + x + 6)\, dx$

29 $\displaystyle\int_{-4}^3 (8x^2 + 2x)\, dx - \int_{-4}^3 (x^2 + 3x + 5)\, dx$

30 $\displaystyle\int_0^5 (x^4 - 3x^2 + 10)\, dx - \int_0^5 (4x^3 + 3x^2 - 6)\, dx + \int_0^5 (4x^4 + 6x^2 - 4)\, dx$

Section 19.2

In Exercises 31–36, (a) sketch f and (b) determine the size of the area between f and the x axis over the indicated interval.

31 $f(x) = 5 + 2x$, between $x = 3$ and $x = 5$

32 $f(x) = e^{2x}$, between $x = 0$ and $x = 2.5$

33 $f(x) = x^2$, between $x = 2$ and $x = 4$

34 $f(x) = \ln(x)$, between $x = 4$ and $x = 8$

35 $f(x) = e^{-x}$, between $x = 1$ and $x = 4$

36 $f(x) = 4x^3$, between $x = 2$ and $x = 4$

37 Given $f(x) = x^2 + 4$ and $g(x) = x + 6$, for $x \geq 0$, determine the size of the area bounded on three sides by the two functions and the y axis.

38 Given $f(x) = x^2$ and $g(x) = -x^2 + 8$, determine the size of the finite area which is bounded by the two functions (a sketch will help).

39 Given $f(x) = 10 - x^2$ and $g(x) = x^2 - 22$, for $x \geq 0$, determine the size of the area bounded on three sides by the two functions and the y axis.

40 Given $f(x) = 18 - x$ and $g(x) = 3x + 2$, determine the size of the area bounded on three sides by the two functions and the y axis.

41 Given the functions in Fig. 19.41, determine the combinations of definite integrals necessary to compute the size of (a) A, (b) B, (c) C, (d) D, (e) E, (f) F, (g) G, and (h) H.

42 Given the functions in Fig. 19.42, determine the combinations of definite integrals necessary to compute the size of (a) A, (b) B, (c) C, (d) D, (e) E, (f) F, and (g) G.

Figure 19.41

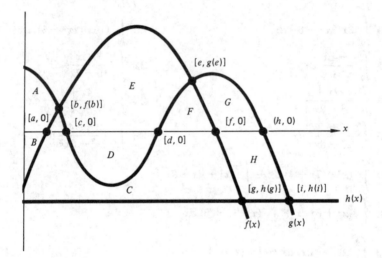

Figure 19.42

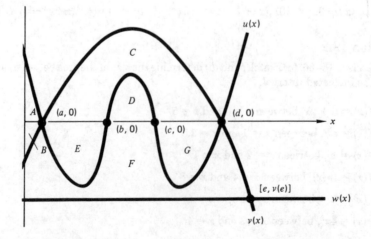

Section 19.3

For the following exercises, (a) evaluate the definite integral using explicit rules of integration; (b) approximate the definite integral using the rectangle rule, the trapezoidal rule, and Simpson's rule (subdividing into four intervals); and (c) determine the most accurate approximation method.

43 $\displaystyle\int_{2}^{6} (x^3 - 6x^2 + 4x)\,dx$

44 $\displaystyle\int_{0}^{8} (8x^2 + 6x^2)\,dx$

45 $\displaystyle\int_{0}^{2} (10 - 3x^2)\,dx$

46 $\displaystyle\int_{0}^{4} (5 + 2x - 3x^2)\,dx$

47 $\displaystyle\int_{2}^{4} 6xe^{x^2-1}\,dx$

48 $\displaystyle\int_{0}^{4} 4xe^{2x^2}\,dx$

49 $\displaystyle\int_{2}^{6} (4x - 12x^2)(x^2 - 2x^3)^4\,dx$

50 $\displaystyle\int_{2}^{10} 3x^2(x^3 - 1)^3\,dx$

51 $\displaystyle\int_{1}^{3} (8x^3 - 6x^2)\,dx$

52 $\displaystyle\int_{1}^{5} (2x - 6x^2 + 4x^3)\,dx$

For the following definite integrals, examine the effect on the error of approximation as the number of subintervals is increased from $n = 2$ to $n = 4$ to $n = 8$. Use standard rules of integration to determine the actual value of the definite integral and then approximate using the indicated method.

53 $\displaystyle\int_{0}^{8} x^2\,dx$, rectangle rule

54 $\displaystyle\int_{0}^{4} x^3\,dx$, rectangle rule

55 $\displaystyle\int_{0}^{4} (3x^2 + 4x^3)\,dx$, trapezoidal rule

56 $\displaystyle\int_{2}^{6} (10x - 4x^3)\,dx$, trapezoidal rule

57 $\displaystyle\int_{0}^{8} (8x^3 - 4x)\,dx$, Simpson's rule

58 $\displaystyle\int_{0}^{4} (-8x^3 + 6x^2)\,dx$, Simpson's rule

Section 19.4

59 The marginal revenue function for a firm's product is

$$MR = f(x) = 17.5$$

where x equals the number of units sold. Using a definite integral, determine the total revenue associated with selling 500 units.

60 The marginal revenue function for a firm's product is

$$MR = -0.04x + 10$$

where x equals the number of units sold. What is the additional revenue if sales increase from 100 units to 200 units?

61 A manufacturer of a piece of special-purpose industrial equipment estimates that the annual rate of expenditure $r(t)$ for maintenance is represented by the function

$$r(t) = 1{,}000 + 25t^2$$

where t is the age of the machine in years and $r(t)$ is measured in dollars per year.

(a) Determine the rate at which maintenance costs are being incurred when the machine is 2 years old.

(b) What are total maintenance costs expected to equal during the first 3 years?

62 Oil Consumption In 1991 the amount of oil used in a particular region of the United States was 5 billion barrels. The demand for oil was growing at an exponential rate of 10 percent per year. The function describing annual rate of consumption $c(t)$ at time t is

$$c(t) = 5e^{0.10t}$$

where t is measured in years; $t = 0$ corresponds to January 1, 1991; and $c(t)$ is measured in billions of barrels per year. If the demand for oil continues to grow at this rate, how much oil is expected to be consumed in the 20-year period January 1, 1991, to January 1, 2011?

*63 Velocity and Acceleration Given the function $s(t)$ which describes the position of a moving object as a function of time t, the velocity function is $v(t) = s'(t)$, and the acceleration function is $a(t) = v'(t) = s''(t)$.

An object in free fall experiences a constant downward acceleration of 32 feet per second per second. The acceleration function for a particular object is $a(t) = -32$. The initial velocity of the object thrown from ground level is 80 feet per second.

(a) Determine the function $v(t)$ which describes the velocity of the object at time t.

(b) Determine the velocity at $t = 2$.

(c) Determine the function $s(t)$ which describes the height of the object at time t.

(d) What is the height at $t = 1$?

64 The demand for a product has been decreasing exponentially. The annual rate of demand $d(t)$ is

$$d(t) = 250{,}000e^{-0.15t}$$

where $t = 0$ corresponds to January 1, 1992. If the demand continues to decrease at the same rate,

(a) Determine the annual rate of demand at $t = 4$.

(b) How many total units are expected to be demanded over the time interval 1992 through 2001 ($t = 0$ to $t = 10$)?

65 Unemployment Compensation A state has projected that the cost of unemployment compensation will be at a rate of $5e^{0.05t}$ million dollars per year t years from now.

(a) Compute total unemployment compensation over the next 5 years.

(b) How long will it be until total benefits paid out equal $200 million?

66 Given the function $f(t) = 2e^{0.5t}$, find the volume of the solid of revolution between $t = 0$ and $t = 5$. Sketch the solid of revolution.

67 A microcomputer manufacturer estimates that sales of its microcomputer systems will be at the rate of $\sqrt{1.2t + 10}$ thousand units per year t years from now.

(a) At what rate are sales expected to be 10 years from now?

(b) What are total sales expected to be over the next 10 years?

Section 19.5

68 The probability density function for a continuous random variable is

$$f(x) = \tfrac{3}{4}(4x - x^2 - 3) \qquad 1 \le x \le 3$$

Determine the probability that the random variable will assume a value less than 2. Greater than 1.5.

69 The probability density function for a continuous random variable x is

$$f(x) = \frac{-x^3 + 80}{256} \qquad 0 \le x \le 4$$

What is the probability that the random variable will assume a value between 2 and 4?

70 The probability density function for a continuous random variable is

$$f(x) = \frac{1 + 2x}{36} \qquad 2 \le x \le 6$$

Determine the probability that the random variable will assume a value between 3 and 6.

❑ CHAPTER TEST

1 Evaluate (*a*) $\displaystyle\int_0^4 (x^2 - 3x + 1)\, dx$ and (*b*) $\displaystyle\int_2^4 e^{x/2}\, dx.$

2 Given $f(x) = x^2$ and $g(x) = 8 + 2x$, (*a*) sketch the two functions. (*b*) For $x \ge 0$, determine the area bounded by the two functions and the y axis.

3 An automobile manufacturer estimates that the annual rate of expenditure $r(t)$ for maintenance on one of its models is represented by the function

$$r(t) = 120 + 8t^2$$

where t is the age of the automobile stated in years and $r(t)$ is measured in dollars per year.
(*a*) At what annual rate are maintenance costs being incurred when the car is 4 years old?
(*b*) What are total maintenance costs expected to equal during the first 3 years?

4 The probability density function for a continuous random variable is

$$f(x) = \frac{2 + x}{30} \qquad 0 \le x \le 6$$

Determine the probability that x will assume a value greater than 2.

5 Given the definite integral

$$\int_0^8 3x^2 \, dx$$

(a) Evaluate using explicit rules of integration.

(b) Approximate the value using the rectangle rule, trapezoidal rule, and Simpson's rule (subdividing into four intervals).

(c) Determine the most accurate approximation method.

THE SOCIAL SECURITY DILEMMA: A PROBLEM OF SOLVENCY

In the late 1970s, the U.S. government and its citizens (especially the more senior members) became greatly concerned about the Social Security program and its survival. The major problem was the growth in the population eligible for Social Security benefits. Disbursements from the Social Security trust fund reached a point where they outpaced income. The trust fund had a balance of approximately $24 billion at the beginning of 1980.

Economists confirmed that the rate of disbursement of benefits was increasing in a *linear* manner. One "think tank" projected that disbursements would be at an annual rate of $122.5 billion at the beginning of 1980. At the beginning of 1985 they estimated that disbursements would be occurring at an annual rate of $230 billion. This same group confirmed that the rate of income generation was increasing in a linear manner. They projected that income would be generated at an annual rate of $117.5 billion at the beginning of 1980 and at an annual rate of $222.5 billion at the beginning of 1985.

Required:

1 Formulate the linear function which estimates the rate of disbursements as a function of time. Formulate the linear function which estimates the rate of income generation as a function of time. Sketch the two functions.

2 Assuming the economists' assumptions were valid, determine when the trust fund would have gone broke (i.e., when the $24 billion balance would have been consumed). Assume that the $24 billion figure has been normalized to account for interest income on the trust fund balance as well as on income from Social Security contributions.

3 If the economists' assumptions were valid, what was the projected deficit expected at the beginning of 1985?

4 A number of possible remedies were proposed. One senator suggested transferring $5 billion per year from a Medicare trust fund in order to delay bankruptcy. How long would it have taken for the trust fund to go broke if such a transfer had been arranged?

5 Another proposal suggested sharp reductions in benefits for those retiring at age 62. If this plan was implemented, economists projected that disbursements at the beginning of 1985 would be reduced to an annual rate of $217.5 billion (a linear trend is still assumed). Assuming that the rate of income generation remained the same, when would the rate of disbursements equal the rate of income generation? Would the fund go broke under this proposal?

6 A final proposal suggested increasing the Social Security taxation rate, resulting in increased income for the trust fund. If increased as proposed, economists estimated that the rate of income generation would be at an annual rate of $230 billion at the beginning of 1985 (linear trend still assumed). Assuming that the rate of disbursement remained the same as originally estimated, when would the rate of disbursements equal the rate of income generation? Would the fund go broke under this proposal?

CHAPTER 20

OPTIMIZATION: FUNCTIONS OF SEVERAL VARIABLES

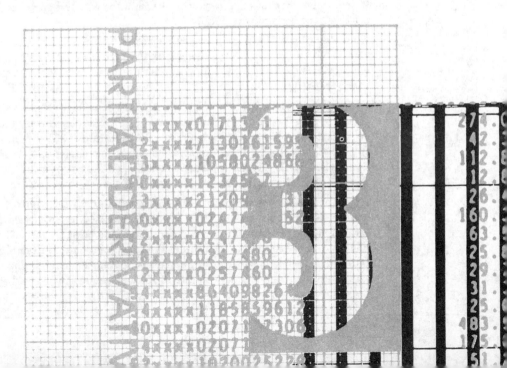

CHAPTER OBJECTIVES

❏ Provide an understanding of the calculus of functions which contain two independent variables

❏ Illustrate the graphical representation of functions in three dimensions

❏ Overview the optimization procedures for functions which contain more than two independent variables

❏ Introduce the nature of and procedures for optimization of functions subject to constraining conditions

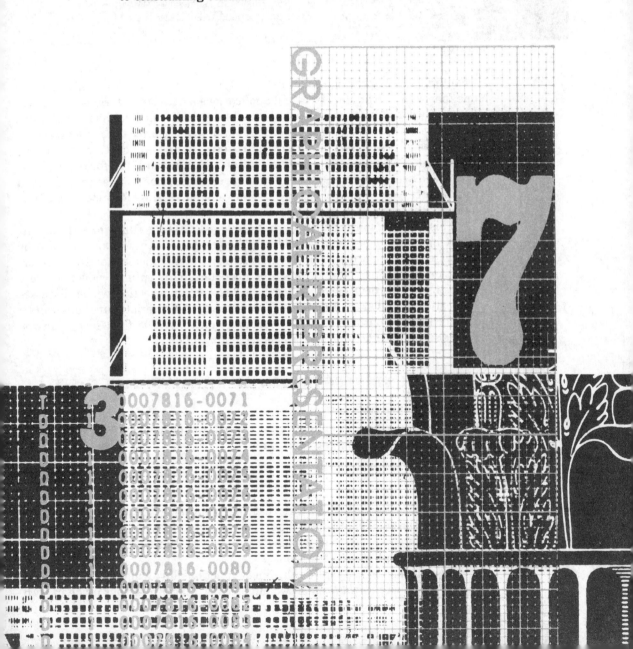

MOTIVATING SCENARIO: Method of Least Squares — Finding the Best Fit to a Set of Data Points	Throughout the text we have discussed the notion of estimating mathematical relationships. In Chaps. 2, 5, 6, and 7 we encountered actual applications in which we used sample data points to determine linear, quadratic, and exponential estimating functions. In each case, sample data points were selected for use in determining the estimating functions. There are scientific ways of developing estimating functions. Given a set of data points and an assumed functional form (e.g., linear, quadratic, etc.), the *method of least squares* is one of the most popular methods of determining the "best" fit to the data. In this chapter, you will see that the least squares model is based upon optimization procedures (Example 21).

Chapters 15 to 17 provided a methodology for examining functions which involve one independent variable. In actual applications a decision criterion or objective frequently depends on more than one variable. As mentioned in Chap. 4, functions which involve more than one independent variable are referred to as **multivariate functions,** or **functions of several variables.** Methods of differential calculus are available for examining the behavior of such functions and for determining optimal values (maxima and minima). As we examine some of these procedures in this chapter, you will see that they are very similar to those for functions of one independent variable.

This chapter will concentrate initially on **bivariate functions** (functions involving two independent variables). The graphics of these functions will be illustrated, and a discussion of the derivatives of these functions and their interpretation will follow. Next, procedures for determining optimal values of these functions will be developed. This will be followed by a section which discusses applications of bivariate functions. The discussion of bivariate functions will be extended to the optimization of n-variable functions. The final section in the chapter examines the topic of *constrained optimization.*

20.1 GRAPHICAL REPRESENTATION OF BIVARIATE FUNCTIONS

Graphical Representation

A function involving a dependent variable z and two independent variables x and y may be represented using the notation

$$z = f(x, y) \qquad (20.1)$$

We established earlier in the text that the number of variables in a function determines the number of dimensions required to graph the function. Whereas two dimensions are required to graph single-variable functions, three dimensions are required to graph bivariate functions.

We established in Chap. 4 that linear functions involving one independent variable graph as *straight lines* in two dimensions. Linear functions involving two independent variables graph as *planes* in three dimensions. Generally speaking, nonlinear functions involving one independent variable graph as *curves* in two dimensions. And nonlinear functions containing two independent variables graph as *curved surfaces* in three dimensions. Examples of nonlinear surfaces include the undulating surface of a golf green, a mogul-laden ski slope, and a billowing sail on a sailboat. An important point is that these functions are represented by *surfaces,* not solids.

Sketching Bivariate Functions

Although graphing in three dimensions is difficult, there are procedures that can be used in some instances to sketch the general shape of the graph of a bivariate function. An understanding of the graphics of these functions can assist with the material which follows.

Consider the bivariate function

$$z = f(x, y) = 25 - x^2 - y^2 \qquad (20.2)$$

where $0 \le x \le 5$ and $0 \le y \le 5$. In order to sketch this function, let's fix the value of one of the independent variables and graph the resulting function. For example, if we let $y = 0$, the function f becomes

$$z = 25 - x^2 - 0^2$$

or
$$z = 25 - x^2 \qquad (20.3)$$

By fixing the value of one of the independent variables, the function is restated in terms of the other independent variable. That is, once the value of one independent variable is specified, the value of the dependent variable varies with the value of the remaining independent variable. Given Eq. (20.3), Table 20.1 indicates selected values for x and resulting values of z.

TABLE 20.1	x	0	1	2	3	4	5
	$z = 25 - x^2$	25	24	21	16	9	0

Figure 20.1 is a partial sketch of the function with the value of y fixed at 0. If we let $y = 0$, the sketch of Eq. (20.3) must be in the xz plane. A close examination of Eq. (20.3) reveals that the relationship between z and x is quadratic. And the sketch in Fig. 20.1 is part of a parabola which is concave down.

If we let $x = 0$ in the original function, f becomes

$$z = 25 - 0^2 - y^2$$

or
$$z = 25 - y^2 \qquad (20.4)$$

Table 20.2 indicates selected values of y and the resulting values of z.

Figure 20.2 is a partial sketch of $f(x, y)$. With $x = 0$, the sketch of Eq. (20.4) is in the yz plane. Equation (20.4) indicates a quadratic relationship between y and z.

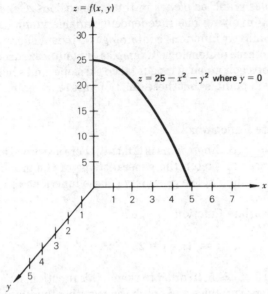

Figure 20.1   Partial sketch of $f(x, y) = 25 - x^2 - y^2$.

TABLE 20.2	y	0	1	2	3	4	5
	$z = 25 - y^2$	25	24	21	16	9	0

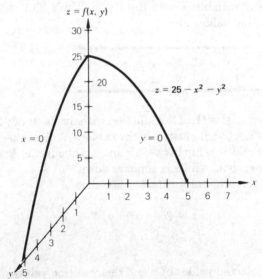

Figure 20.2   Partial sketch of $f(x, y) = 25 - x^2 - y^2$.

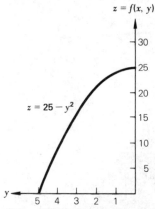

Figure 20.3 Trace with $x = 0$ viewed along the x axis.

And if you were to look at Fig. 20.2 in a direction parallel to the x axis, you would see that this equation graphs as a portion of a parabola which is concave down. Figure 20.3 indicates what you would see if sighting along the x axis.

DEFINITION: TRACE

Given $z = f(x, y)$, a *trace* is the graph of f when one variable is held constant.

Refer to Fig. 20.2. The two portions of f which are illustrated are *traces*. One is a trace where $y = 0$, and the other is a trace where $x = 0$. Each trace represents a *rib* on the surface which represents the function.

Figure 20.4 presents a sketch of the function which includes four additional traces. Letting $y = 1$, the function becomes

$$f(x, y) = 25 - x^2 - 1^2$$
$$= 24 - x^2$$

The trace representing this function is parallel to the xz plane and 1 unit out along the positive y axis. Similarly, letting $y = 3$, we have

$$f(x, y) = 25 - x^2 - 3^2$$
$$= 16 - x^2$$

The trace representing this function graphs parallel to the xz plane and 3 units out along the y axis.

Traces have also been sketched by letting $x = 1$ and $x = 3$. These six traces in combination begin to resemble a skeletal structure of the surface. And if we were to graph more traces associated with other assumed values for x and y, we would get a more accurate representation of the surface representing f, similar to the shaded portion of Fig. 20.4.

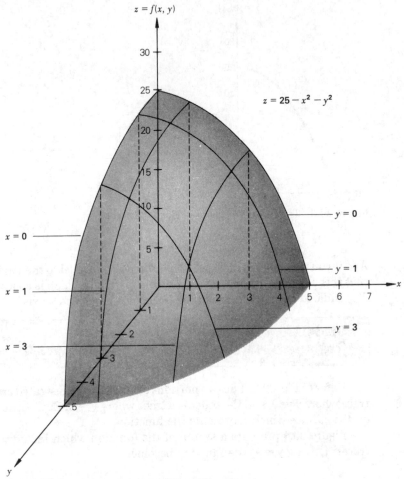

Figure 20.4   Sketch of $f(x, y) = 25 - x^2 - y^2$.

Therefore, a procedure which can *sometimes* provide a rough sketch of a function of the form $z = f(x, y)$ is to assume selected values for x and y and graph the traces which represent the resulting functions.

NOTE   An important observation regarding the graphics of a function $f(x, y)$ should be made. *Whenever x is held constant, the resulting trace graphs in a plane which is parallel to the yz plane. Whenever y is held constant, the resulting trace graphs in a plane which is parallel to the xz plane.*

Section 20.1 Follow-up Exercises

Sketch the following functions.

1 $f(x, y) = 16 - x^2 - y^2$, where $0 \le x \le 4$ and $0 \le y \le 4$
2 $f(x, y) = 9 - x^2 - y^2$, where $0 \le x \le 3$ and $0 \le y \le 3$
3 $f(x, y) = 4 - x^2 - y^2$, where $0 \le x \le 2$ and $0 \le y \le 2$
4 $f(x, y) = 25 - x^2/4 - y^2/4$, where $0 \le x \le 10$ and $0 \le y \le 10$
5 $f(x, y) = x^2 + y^2$, where $0 \le x \le 5$ and $0 \le y \le 5$

20.2 PARTIAL DERIVATIVES

Although more involved, the calculus of bivariate functions is very similar to that of single-variable functions. In this section we will discuss derivatives of bivariate functions and their interpretation.

Derivatives of Bivariate Functions

With single-variable functions, the derivative represents the instantaneous rate of change in the dependent variable with respect to a change in the independent variable. For bivariate functions, there are two *partial derivatives*. These derivatives represent the instantaneous rate of change in the dependent variable with respect to changes in the two independent variables, separately. Given a function $z = f(x, y)$, a partial derivative can be found with respect to each independent variable. The partial derivative taken with respect to x is denoted by

$$\frac{\partial z}{\partial x} \quad \text{or} \quad f_x$$

The partial derivative taken with respect to y is denoted by

$$\frac{\partial z}{\partial y} \quad \text{or} \quad f_y$$

Although both notational forms are used to denote the partial derivative, we will use the subscripted notation f_x and f_y in this chapter.

> **DEFINITION: PARTIAL DERIVATIVE**
> Given the function $z = f(x, y)$, the *partial derivative* of z with respect to x at (x, y) is
>
> $$f_x = \lim_{\Delta x \to 0} \frac{f(x + \Delta x, y) - f(x, y)}{\Delta x}$$
>
> provided the limit exists. The partial derivative of z with respect to y at (x, y) is
>
> $$f_y = \lim_{\Delta y \to 0} \frac{f(x, y + \Delta y) - f(x, y)}{\Delta y}$$
>
> provided the limit exists.

EXAMPLE 1 Consider the function

$$f(x, y) = 3x^2 + 5y^3$$

To find the partial derivative with respect to x, let's use the *limit approach* from Sec. 15.3. First, the difference quotient is formed as

$$\frac{\Delta f(x, y)}{\Delta x} = \frac{f(x + \Delta x, y) - f(x, y)}{\Delta x}$$

$$= \frac{3(x + \Delta x)^2 + 5y^3 - (3x^2 + 5y^3)}{\Delta x}$$

$$= \frac{3(x^2 + 2x\,\Delta x + \Delta x^2) + 5y^3 - 3x^2 - 5y^3}{\Delta x}$$

which simplifies to

$$\frac{\Delta f(x, y)}{\Delta x} = \frac{6x\,\Delta x + 3\,\Delta x^2}{\Delta x}$$

$$= \frac{\Delta x\,(6x + 3\,\Delta x)}{\Delta x}$$

$$= 6x + 3\,\Delta x$$

The partial derivative is

$$f_x = \lim_{\Delta x \to 0} \frac{\Delta f(x, y)}{\Delta x}$$

$$= \lim_{\Delta x \to 0} (6x + 3\,\Delta x)$$

$$= 6x$$

Note that in finding f_x, we are examining the effects of changes in x (i.e., Δx); the other independent variable y is held constant.

The partial derivative taken with respect to y can be found in a similar manner.

$$\frac{\Delta f(x, y)}{\Delta y} = \frac{\Delta f(x, y + \Delta y) - f(x, y)}{\Delta y}$$

$$= \frac{3x^2 + 5(y + \Delta y)^3 - (3x^2 + 5y^3)}{\Delta y}$$

$$= \frac{3x^2 + 5(y^3 + 3y^2\,\Delta y + 3y\,\Delta y^2 + \Delta y^3) - 3x^2 - 5y^3}{\Delta y}$$

$$= \frac{3x^2 + 5y^3 + 15y^2\,\Delta y + 15y\,\Delta y^2 + 5\,\Delta y^3 - 3x^2 - 5y^3}{\Delta y}$$

$$= \frac{\Delta y\,(15y^2 + 15y\,\Delta y + 5\,\Delta y^2)}{\Delta y}$$

$$= 15y^2 + 15y\,\Delta y + 5\,\Delta y^2$$

The partial derivative is

$$f_y = \lim_{\Delta y \to 0} \frac{\Delta f(x, y)}{\Delta y}$$

$$= \lim_{\Delta y \to 0} (15y^2 + 15y\,\Delta y + 5\,\Delta y^2)$$

$$= 15y^2$$

In determining f_y, we are examining the effects of changes in y (i.e., Δy); the other independent variable x is held constant.

EXAMPLE 2 Consider the function

$$f(x, y) = 5x^2y$$

Using the limit approach to find the partial derivatives,

$$\frac{\Delta f(x, y)}{\Delta x} = \frac{f(x + \Delta x, y) - f(x, y)}{\Delta x}$$

$$= \frac{5(x + \Delta x)^2 y - 5x^2 y}{\Delta x}$$

$$= \frac{5(x^2 + 2x\,\Delta x + \Delta x^2)y - 5x^2 y}{\Delta x}$$

$$= \frac{5x^2y + 10xy\,\Delta x + 5y\,\Delta x^2 - 5x^2y}{\Delta x}$$

$$= \frac{\Delta x\,(10xy + 5y\,\Delta x)}{\Delta x}$$

$$= 10xy + 5y\,\Delta x$$

Thus, $$f_x = \lim_{\Delta x \to 0} \frac{\Delta f(x,y)}{\Delta x}$$

$$= \lim_{\Delta x \to 0} (10xy + 5y\,\Delta x)$$

$$= 10xy$$

To find f_y

$$\frac{\Delta f(x, y)}{\Delta y} = \frac{f(x, y + \Delta y) - f(x, y)}{\Delta y}$$

$$= \frac{5x^2(y + \Delta y) - 5x^2 y}{\Delta y}$$

$$= \frac{5x^2y + 5x^2\,\Delta y - 5x^2y}{\Delta y}$$

$$= 5x^2$$

And $$f_y = \lim_{\Delta y \to 0} \frac{\Delta f(x, y)}{\Delta y}$$

$$= \lim_{\Delta y \to 0} 5x^2$$

$$= 5x^2$$ $\square$

Fortunately, partial derivatives are found more easily using the same differentiation rules we used in Chaps. 15–17. The only exception is that *when a partial derivative is found with respect to one independent variable, the other independent variable is assumed to be held constant.* For instance, in finding the partial derivative with respect to x, y is assumed to be constant. And a very important point is that *the variable which is assumed constant must be treated as a constant in applying the rules of differentiation.*

EXAMPLE 3 Find the partial derivatives f_x and f_y for the function

$$f(x, y) = 5x^2 + 6y^3$$

SOLUTION

First, to find the partial derivative with respect to x, the variable y must be assumed held constant. Differentiating term by term, we find that the derivative of $5x^2$ with respect to x is $10x$. In differentiating the second term, remember that y is assumed to be constant. Thus, this term has the general form of

$$6(\text{constant})^3$$

which is simply a constant. Since a constant does not change in value as other variables change in value—or remember from Chap. 15 that the derivative of a constant equals 0—the derivative of the second term equals 0. Thus,

$$f_x = 10x + 0$$
$$= 10x$$

In finding the partial derivative with respect to y, the variable x is assumed held constant. In differentiating term by term, $5x^2$ is viewed as being a constant since x is assumed constant and the derivative equals 0. The derivative of $6y^3$ with respect to y is $18y^2$. Thus,

$$f_y = 0 + 18y^2$$
$$= 18y^2$$

EXAMPLE 4 Find f_x and f_y for the function

$$f(x, y) = 4xy$$

SOLUTION

To find f_x, y is assumed to be constant. The term $4xy$ is in the form of a product. To differentiate such product terms, you can use two approaches. The first approach is simply to apply the product rule. In viewing $4xy$ as the product of $4x$ and y, the product rule yields

$$f_x = (4)(y) + (0)(4x)$$

or $$f_x = 4y$$

An *alternative approach* is to remember which variable is assumed constant. When y is held constant, we can rearrange $4xy$ to have the form

$$f(x, y) = \text{constant} \cdot x$$

or
$$f(x, y) = (4y)x$$

By grouping the 4 and y, this term has the general form of a constant $4y$ times x. And the derivative of a constant times x is the constant, or

$$f_x = 4y$$

To find f_y, x is assumed constant. Applying the product rule, we find

$$f_y = (0)(y) + (1)(4x)$$

or
$$f_y = 4x$$

Or by using the alternative approach, the factor $4x$ is constant (with x held constant) and f can be viewed as having the form

$$f(x, y) = \text{constant} \cdot y$$
$$= (4x)y$$

The derivative with respect to y is the constant, or

$$f_y = 4x$$

EXAMPLE 5 Find f_x and f_y if

$$f(x, y) = -10xy^3$$

SOLUTION

To find f_x, y must be assumed constant. This function can be rearranged (mentally or explicitly) to have the form

$$f(x, y) = (\text{constant})x = (-10y^3)x$$

where $-10y^3$ is constant. The derivative is

$$f_x = -10y^3$$

For f_y the function can be viewed as having the form

$$(\text{constant})y^3 \quad \text{or} \quad (-10x)y^3$$

The derivative with respect to y is

$$(\textbf{constant})(3y^2)$$

or
$$f_y = -30xy^2$$

❑

PRACTICE EXERCISE

Verify the expressions for f_x and f_y by finding them using the product rule.

EXAMPLE 6   Find f_x and f_y if

$$f(x, y) = e^{x^2 + y^2}$$

SOLUTION

Using the differentiation rules for exponential functions,

$$f_x = (2x + 0)e^{x^2 + y^2} = 2xe^{x^2 + y^2}$$

$$f_y = (0 + 2y)e^{x^2 + y^2} = 2ye^{x^2 + y^2}$$

EXAMPLE 7   Find f_x and f_y if

$$f(x, y) = (3x - 2y^2)^3$$

SOLUTION

Remembering the power of a function rule, we have

$$f_x = 3(3x - 2y^2)^2(3)$$
$$= 9(3x - 2y^2)^2$$

$$f_y = 3(3x - 2y^2)^2(-4y)$$
$$= -12y(3x - 2y^2)^2$$

❑

PRACTICE EXERCISE

Given $f(x, y) = (4x^2 - 5y^3)^4$, find f_x and f_y. *Answer:* $f_x = 32x(4x^2 - 5y^3)^3$, $f_y = -60y^2(4x^2 - 5y^3)^3$.

Interpreting Partial Derivatives

One interpretation of partial derivatives deals with the tangent slope. As with single-variable functions, the partial derivatives f_x and f_y have a tangent slope interpretation.

SLOPE INTERPRETATION OF f_x AND f_y

I f_x *is a general expression for the tangent slope of the family of traces which are parallel to the xz plane.*

> **II** f_y *is a general expression for the tangent slope of the family of traces which are parallel to the yz plane.*

The partial derivative f_x estimates the change in z given a change in x, assuming y is held constant. In Sec. 20.1, we saw that when y is held constant, the corresponding traces are graphed parallel to the xz plane. f_x represents the slope of these traces.

Similarly, f_y assumes that x is held constant. When x was held constant in Sec. 20.1, the result was a family of traces which were parallel to the yz plane. And f_y represents the slope of these traces. Figure 20.5 illustrates the slope representation.

The other interpretation of partial derivatives is the instantaneous rate of change interpretation. As with single-variable functions, partial derivatives can be used to approximate changes in the value of the dependent variable, given a change in *one* of the independent variables. For example, f_x can be used to approximate the

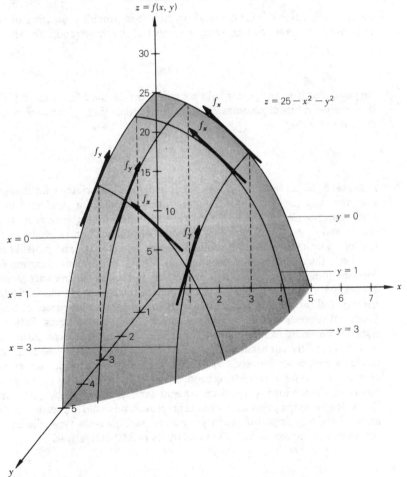

Figure 20.5 Tangent slope representation of partial derivatives.

change in $f(x, y)$, given a change in x with y assumed constant. The partial derivative f_y can be used to approximate the change in $f(x, y)$, given a change in y with x assumed constant. The following examples illustrate this interpretation.

EXAMPLE 8   **(Multiproduct Demand Interrelationships)** Until this point, we have assumed that the demand for a product only depends on the price of that product. Thus, the *demand functions* we have examined have had the form

$$q = f(p)$$

Frequently, the demand for a product or service is influenced not only by its own price but also by the prices of other products or services. Equation (20.5) is a demand function which expresses the quantity demanded of product 1 (q_1) as a function of its price (p_1) as well as the prices of two other products (p_2 and p_3), all stated in dollars.

$$\boxed{q_1 = f(p_1, p_2, p_3) = 10{,}000 - 2.5p_1 + 3p_2 + 1.5p_3} \qquad \textbf{(20.5)}$$

Partial derivatives of this demand function can provide a measure of the instantaneous response of demand to changes in the prices of the three goods. For example,

$$f_{p_1} = -2.5$$

suggests that if p_2 and p_3 are held constant, the demand for product 1 will *decrease* at an instantaneous rate of 2.5 units for every unit (dollar) that p_1 increases. Similarly, the partial derivatives

$$f_{p_2} = 3 \quad \text{and} \quad f_{p_3} = 1.5$$

indicate the instantaneous rate of change in demand associated with changes in the prices of the other two products. $f_{p_2} = 3$ suggests that demand for product 1 will *increase* at an instantaneous rate of 3 units for every unit (dollar) that p_2 increases (p_1 and p_3 held constant), and $f_{p_3} = 1.5$ indicates that demand for product 1 will *increase* at an instantaneous rate of 1.5 units for each unit (dollar) that p_3 increases (p_1 and p_2 held constant).

    Let's make a couple of observations. First, this demand function is linear and the corresponding partial derivatives are constant. That is, the instantaneous rates of change are really the same anywhere in the domain of the demand function. Second, the fact that the demand for product 1 *increases* with an increase in the prices of both product 2 and product 3 suggests an interdependence among the three products. This is the type of relationship we would expect to exist among ***competing products.*** Examples of competing products are different brands of the same product (e.g., steel-belted radial tires) or different products which can be used to satisfy a particular need (e.g., margarine vs. butter and chicken vs. beef). For competing products we would expect that as the price of one product increases, demand for it will decrease and demand for competing products will increase. Equivalently, as the price for a product decreases, we would expect demand for it to increase and demand for competing products to decrease. This is the type of behavior illustrated by the demand function in Eq. (20.5) and by its partial derivatives.

❑

> **PRACTICE EXERCISE**
> Given the demand function
>
> $$q_1 = f(p_1, p_2, p_3) = 120{,}000 - 0.5p_1^2 - 0.4p_2^2 - 0.2p_3^2$$
>
> (a) find all partial derivatives. (b) If current prices for the three products are $p_1 = 10$, $p_2 = 20$, and $p_3 = 30$, evaluate the partial derivatives and interpret their meaning. (c) What do the demand function and its partial derivatives suggest about the interdependence among the three products? *Answer: (a)* $f_{p_1} = -p_1$, $f_{p_2} = -0.8p_2$, $f_{p_3} = -0.4p_3$; *(b)* $f_{p_1}(10, 20, 30) = -10$, $f_{p_2}(10, 20, 30) = -16$, $f_{p_3}(10, 20, 30) = -12$; *(c)* they are complementary products.

EXAMPLE 9 **(Advertising Expenditures)** A national manufacturer estimates that the number of units it sells each year is a function of its expenditures on radio and TV advertising. The function specifying this relationship is

$$z = 50{,}000x + 40{,}000y - 10x^2 - 20y^2 - 10xy$$

where *z equals the number of units sold annually, x equals the amount spent for TV advertising, and y equals the amount spent for radio advertising (both in $1,000s)*.

Assume that the firm is currently spending $40,000 on TV advertising ($x = 40$) and $20,000 on radio advertising ($y = 20$). With these expenditures,

$$f(40, 20) = 50{,}000(40) + 40{,}000(20) - 10(40)^2 - 20(20)^2 - 10(40)(20)$$
$$= 2{,}000{,}000 + 800{,}000 - 16{,}000 - 8{,}000 - 8{,}000 = 2{,}768{,}000$$

or it is projected that 2,768,000 units will be sold.

Suppose we are interested in determining the effect on annual sales if $1,000 more is spent on TV advertising. The partial derivative f_x should provide us with an approximation of this effect: This partial derivative is

$$f_x = 50{,}000 - 20x - 10y$$

Because we are interested in the instantaneous rate of change, given that expenditures are currently $40,000 and $20,000, we evaluate f_x at $x = 40$ and $y = 20$:

$$f_x(40, 20) = 50{,}000 - 20(40) - 10(20)$$
$$= 50{,}000 - 800 - 200 = 49{,}000$$

Evaluating the partial derivative, we can state that an increase in TV expenditures of $1,000 should result in additional sales of *approximately* 49,000 units.

To determine how accurate this approximation is, let's evaluate $f(41, 20)$:

$$f(41, 20) = 50{,}000(41) + 40{,}000(20) - 10(41)^2 - 20(20)^2 - 10(41)(20)$$
$$= 2{,}050{,}000 + 800{,}000 - 16{,}810 - 8{,}000 - 8{,}200 = 2{,}816{,}990$$

The *actual increase* in sales is projected as

$$f(41, 20) - f(40, 20) = 2,816,990 - 2,768,000$$
$$= 48,990 \text{ units}$$

The difference between the actual increase and the increase estimated by using f_x is $48,990 - 49,000 = -10$ units. The minus sign indicates that the partial derivative *overestimated* the actual change.

Now suppose that we are interested in determining the effect if an additional $1,000 is spent on radio, rather than TV, advertising. The partial derivative taken with respect to y will approximate this change:

$$f_y = 40,000 - 40y - 10x$$

Evaluating f_y at $x = 40$ and $y = 20$, we get

$$f_y(40, 20) = 40,000 - 40(20) - 10(40)$$
$$= 40,000 - 800 - 400 = 38,800$$

Thus an increase of $1,000 in radio advertising expenditures will lead to an *approximate increase* of 38,800 units. With an increase of $1,000 in radio advertising expenses, actual sales are estimated at

$$f(40, 21) = 50,000(40) + 40,000(21) - 10(40)^2 - 20(21)^2 - 10(40)(21)$$
$$= 2,000,000 + 840,000 - 16,000 - 8,820 - 8,400 = 2,806,780 \text{ units}$$

The *actual increase* in sales is

$$f(40, 21) - f(40, 20) = 2,806,780 - 2,768,000$$
$$= 38,780 \text{ units}$$

Again, the approximate change estimated by using f_y is in error by only 20 units.

From a comparative standpoint, if $1,000 is to be allocated to either TV or radio, it appears that the greater return (in terms of increased sales) will come from TV.

❑

Second-Order Derivatives

As with single-variable functions, we can determine second-order derivatives for bivariate functions. These will be of considerable importance to us in the next section when we seek to optimize the value of a function.

For functions of the form $f(x, y)$, there are *four* different second-order derivatives. These are divided into two types: ***pure second-order partial derivatives*** and ***mixed partial derivatives.*** The two pure partial derivatives are denoted by f_{xx} and f_{yy}. The pure second-order partial derivative with respect to x, f_{xx}, is found by first finding f_x and then differentiating f_x with respect to x. Similarly, f_{yy} is found by determining the expression for f_y and then differentiating f_y with respect to y.

The two mixed partial derivatives are denoted by f_{xy} and f_{yx}. The mixed partial derivative f_{xy} is found by determining f_x and then differentiating f_x with respect to y. Similarly, f_{yx} is found by determining f_y and then differentiating f_y with respect to x. Figure 20.6 summarizes the procedures for finding second-order derivatives.

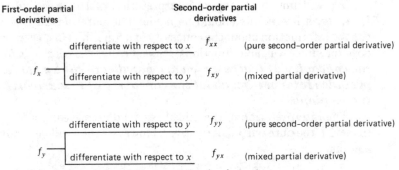

Figure 20.6 Determination of second-order derivatives.

EXAMPLE 10 Determine all first- and second-order derivatives for the function

$$f(x, y) = 8x^3 - 4x^2y + 10y^3$$

SOLUTION

We start with the first derivatives:

$$f_x = 24x^2 - 8xy$$
$$f_y = -4x^2 + 30y^2$$

The pure partial derivative f_{xx} is found by differentiating f_x with respect to x, or

$$f_x = 24x^2 - 8xy \quad \rightarrow \quad f_{xx} = 48x - 8y$$

f_{yy} is found by differentiating f_y with respect to y, or

$$f_y = -4x^2 + 30y^2 \quad \rightarrow \quad f_{yy} = 60y$$

The mixed partial derivative f_{xy} is found by differentiating f_x with respect to y, or

$$f_x = 24x^2 - 8xy \quad \rightarrow \quad f_{xy} = -8x$$

f_{yx} is found by differentiating f_y with respect to x, or

$$f_y = -4x^2 + 30y^2 \quad \rightarrow \quad f_{yx} = -8x$$

$\square$

NOTE A proposition known as **Young's theorem** states that the mixed partial derivatives f_{xy} and f_{yx} equal one another provided that f_{xy} and f_{yx} are both continuous. Notice that this was true in Example 10. This property provides a possible check on errors which may have been made in finding f_x, f_y, f_{xy}, and f_{yx}.

We will not dwell on the interpretation of these second-order derivatives. However, a few points should be made. The pure partial derivatives f_{xx} and f_{yy} convey information about the concavity of a function (just as the second derivative does for single-variable functions). Specifically, f_{xx} *offers information about the concavity of traces which are parallel to the xz plane.* Similarly, f_{yy} *provides information about the concavity of traces which are parallel to the yz plane.*

The interpretation of the mixed partial derivatives f_{xy} and f_{yx} is less intuitive than with the pure partial derivatives. However, they will be significant in the next section.

Section 20.2 Follow-up Exercises

In Exercises 1–40, determine f_x and f_y.

1 $f(x, y) = 3x^2 - 10y^3$	**2** $f(x, y) = -2x + 3y$
3 $f(x, y) = 10x^2 + 2xy - 6y^2$	**4** $f(x, y) = x^2 - 2y^3$
5 $f(x, y) = x^3y^5$	**6** $f(x, y) = 25xy^3$
7 $f(x, y) = 6x^2 - xy + 30y^2$	**8** $f(x, y) = 20x^2 + 7x^2y^3 + 5y^3$
9 $f(x, y) = -4/xy^2$	**10** $f(x, y) = -2x/3y^2$
11 $f(x, y) = (x^2 - 5y)(2x + 4y^5)$	**12** $f(x, y) = (1/x)(y^2 + 3y^3)$
13 $f(x, y) = (x - y)^4$	**14** $f(x, y) = (x^3 + 3y^2)^4$
15 $f(x, y) = \sqrt{x^2 + y^2}$	**16** $f(x, y) = -20/\sqrt[3]{x + 2y}$
17 $f(x, y) = \ln(x + y)$	**18** $f(x, y) = e^{3x-2y}$
19 $f(x, y) = e^{5x^3-2y^2}$	**20** $f(x, y) = e^{2x} \ln y$
21 $f(x, y) = 3x^4 - 8x^2y^3$	**22** $f(x, y) = 10x^2 - 25xy + 30y^3$
23 $f(x, y) = -3x^3 + 8xy^2 + y^3$	**24** $f(x, y) = 15y^3 - 5yx^3 + 25$
25 $f(x, y) = 2x^2/3y^3$	**26** $f(x, y) = -x/y^3$
27 $f(x, y) = (x - y)/x^5$	**28** $f(x, y) = (3x - y^2)/(x^2 + 1)$
29 $f(x, y) = -5x^2/(x + y)$	**30** $f(x, y) = 3y^2/(x^2 - 5y)$
31 $f(x, y) = (x - y)^3$	**32** $f(x, y) = (8x - 3y)^4$
33 $f(x, y) = \sqrt[3]{x^2 + 2y^2}$	**34** $f(x, y) = \sqrt[4]{4x^3 + 2y^2}$
35 $f(x, y) = \ln(x^2 - y^2)$	**36** $f(x, y) = \ln(x^2 - 4xy + y^2)$
37 $f(x, y) = e^{x^2y}$	**38** $f(x, y) = e^{xy^3}$
39 $f(x, y) = e^x \ln y$	**40** $f(x, y) = e^{y^2} \ln x$

In Exercises 41–60, find all second-order partial derivatives.

41 $f(x, y) = 3x^2 + 5y^3 + 10$	**42** $f(x, y) = x^2 - 3x + 4y^3 + 2y$
43 $f(x, y) = 5x^3 - 3xy + 3y^2$	**44** $f(x, y) = -20x^5 + 10xy + 6y^3$
45 $f(x, y) = x^3y^4$	**46** $f(x, y) = -10xy^3$
47 $f(x, y) = e^{x+y}$	**48** $f(x, y) = \ln(x + y)$
49 $f(x, y) = e^{xy}$	**50** $f(x, y) = e^y \ln x$
51 $f(x, y) = (x - y)^5$	**52** $f(x, y) = (y - x)^4$
53 $f(x, y) = x^4y^2$	**54** $f(x, y) = x^2y^5$
55 $f(x, y) = x/y^2$	**56** $f(x, y) = y/x^2$
57 $f(x, y) = (6x - 8y)^{3/2}$	**58** $f(x, y) = \sqrt{x + y}$
59 $f(x, y) = \ln x^3y^2$	**60** $f(x, y) = \ln 3xy^3$

61 Given $f(x, y) = 100x^2 + 200y^2 - 10xy$:

(a) Determine $f(10, 20)$.

    (b) Using partial derivatives, estimate the change expected in $f(x, y)$ if x increases by 1 unit.

    (c) Compare the actual change with the estimated change.

    (d) Repeat parts b and c, assuming a possible increase in y of 1 unit.

62 Given $f(x, y) = 20x^3 - 30y^3 + 10x^2y$:

    (a) Determine $f(20, 10)$.

    (b) Using partial derivatives, estimate the change expected in $f(x, y)$ if x increases by 1 unit.

    (c) Compare the actual change with the estimated change.

    (d) Repeat parts b and c, assuming a possible increase in y of 1 unit.

63 A firm estimates that the number of units it sells each year is a function of the advertising expenditures for TV and radio. The function expressing this relationship is

$$z = 2{,}000x + 5{,}000y - 20x^2 - 10y^2 - 50xy$$

where z equals the number of units sold, x equals the amount spent on TV advertising, and y equals the amount spent on radio advertising (the latter two variables expressed in \$1,000s). The firm is presently allocating \$50,000 to TV and \$30,000 to radio.

    (a) What are annual sales expected to equal?

    (b) Using partial derivatives, estimate the effect on annual sales if an additional \$1,000 is allocated to TV.

    (c) Using partial derivatives, estimate the effect on annual sales if an additional \$1,000 is allocated to radio.

    (d) Where does it seem that the \$1,000 is better spent?

64 Given the demand function

$$q_1 = f(p_1, p_2) = 25{,}000 - 0.1p_1^2 - 0.5p_2^2$$

    (a) Determine the partial derivatives f_{p_1} and f_{p_2}.

    (b) If $p_1 = 20$ and $p_2 = 10$, evaluate f_{p_1} and f_{p_2} and interpret their meaning.

    (c) How are the two products interrelated?

65 Given the demand function

$$q_1 = f(p_1, p_2, p_3) = 250{,}000 - 0.5p_1^2 + p_2^2 + 0.4p_3^2$$

    (a) Determine the partial derivatives f_{p_1}, f_{p_2}, and f_{p_3}.

    (b) If $p_1 = 30$, $p_2 = 10$, and $p_3 = 20$, evaluate the partial derivatives and interpret their meaning.

    (c) How are the three products interrelated?

20.3 OPTIMIZATION OF BIVARIATE FUNCTIONS

The process of finding optimum values of bivariate functions is very similar to that used for single-variable functions. This section discusses the process.

Critical Points

As with single-variable functions, we will have a particular interest in identifying relative maximum and minimum points on the surface representing a function

$f(x, y)$. Relative maximum and minimum points have the same meaning in three dimensions as in two dimensions.

> **DEFINITION: RELATIVE MAXIMUM**
> A function $z = f(x, y)$ is said to have a *relative maximum* at $x = a$ and $y = b$ if for all points (x, y) "sufficiently close" to (a, b)
>
> $$f(a, b) \geq f(x, y)$$

A relative maximum usually appears as the top or peak of a mound on the surface representing $f(x, y)$.

> **DEFINITION: RELATIVE MINIMUM**
> A function $z = f(x, y)$ is said to have a *relative minimum* at $x = a$ and $y = b$ if for all points (x, y) "sufficiently close" to (a, b)
>
> $$f(a, b) \leq f(x, y)$$

A relative minimum usually appears as the bottom of a valley on the surface representing $f(x, y)$.

Figure 20.7 illustrates both a relative maximum point and a relative minimum point. If you examine the slope conditions of a smooth surface at a relative maximum or at a relative minimum, you should conclude that a line drawn tangent at the point in any direction has a slope of 0. Given that the first partial derivatives f_x and f_y represent general expressions for the tangent slope of traces which are parallel, respectively, to the xz and yz planes, we can state the following.

Figure 20.7
Relative extrema in 3-space.

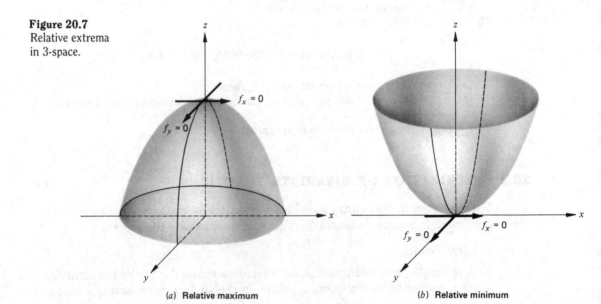

(a) Relative maximum   (b) Relative minimum

NECESSARY CONDITION FOR RELATIVE EXTREMA
A *necessary condition* for the existence of a relative maximum or a relative minimum of a function f whose partial derivatives f_x and f_y both exist is that

$$f_x = 0 \quad \text{and} \quad f_y = 0 \qquad (20.6)$$

An important part of this definition is that *both* f_x and f_y equal 0. As illustrated in Fig. 20.8 there can be an infinite number of points on a surface where f_x equals 0. In Fig. 20.8 a tangent line drawn parallel to the xz plane anywhere along the trace AB will have a slope of 0 ($f_x = 0$). However, the only point where *both* f_x and f_y equal zero is at A. At the other points along AB, a tangent line drawn parallel to the yz plane has a negative slope ($f_y < 0$).

Figure 20.8
$f_x = 0$ along trace AB.

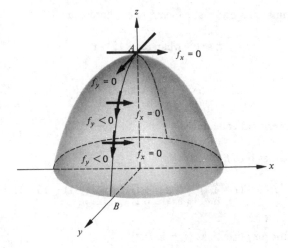

Figure 20.9
$f_y = 0$ along trace AC.

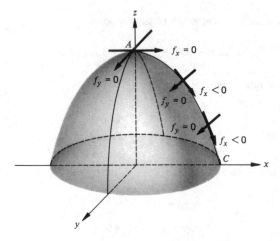

Similarly, Fig. 20.9 illustrates a trace AC along which $f_y = 0$ but $f_x < 0$ except at point A.

The values x^* and y^* at which Eq. (20.6) is satisfied are **critical values.** The corresponding point $(x^*, y^*, f(x^*, y^*))$ is a candidate for a relative maximum or minimum on f and is called a **critical point.**

EXAMPLE 11   Locate any critical points on the graph of the function

$$f(x, y) = 4x^2 - 12x + y^2 + 2y - 10$$

SOLUTION

First, find the expressions for f_x and f_y

$$f_x = 8x - 12$$
$$f_y = 2y + 2$$

To determine the values of x and y at which f_x and f_y both equal 0,

$$f_x = 0 \quad \text{when } 8x - 12 = 0$$

or a critical value for x is $x = \frac{3}{2}$

$$f_y = 0 \quad \text{when } 2y + 2 = 0$$

or the corresponding critical value for y is $y = -1$

Substituting these values into f,

$$f(\tfrac{3}{2}, -1) = 4(\tfrac{3}{2})^2 - 12(\tfrac{3}{2}) + (-1)^2 + 2(-1) - 10$$
$$= 9 - 18 + 1 - 2 - 10 = -20$$

The only critical point for f occurs at $(\tfrac{3}{2}, -1, -20)$.

EXAMPLE 12   To locate any critical points on the graph of the function

$$f(x, y) = -2x^2 - y^2 + 8x + 10y - 5xy$$

we find the first partial derivatives,

$$f_x = -4x + 8 - 5y$$
$$f_y = -2y + 10 - 5x$$

The values of x and y which make f_x and $f_y = 0$ are found by solving the following equations:

$$-4x + 8 - 5y = 0 \tag{20.7}$$
$$-2y + 10 - 5x = 0 \tag{20.8}$$

Rewriting these equations, we have

$$4x + 5y = 8 \qquad \text{(20.9)}$$

$$5x + 2y = 10 \qquad \text{(20.10)}$$

Multiplying both sides of Eq. (20.9) by -2 and both sides of Eq. (20.10) by 5 and adding the resulting equations, we get

$$-8x - 10y = -16$$
$$\underline{25x + 10y = 50}$$

$$17x = 34$$

and a critical value for x is

$$x = 2$$

If $x = 2$ is substituted into Eq. (20.10), we find

$$5(2) + 2y = 10$$

$$2y = 0$$

and the corresponding critical value for y is $\quad y = 0$

If we substitute these values into f,

$$\begin{aligned} f(2, 0) &= -2(2)^2 - (0)^2 + 8(2) + 10(0) - 5(2)(0) \\ &= -8 + 16 \\ &= 8 \end{aligned}$$

Thus, a critical point occurs at $(2, 0, 8)$.

EXAMPLE 13 To determine any critical points on the graph of the function

$$f(x, y) = 2x^2 + 4xy - x^2y - 4x$$

the first partial derivatives are identified and set equal to 0.

$$f_x = 4x + 4y - 2xy - 4 = 0 \qquad \text{(20.11)}$$

$$f_y = 4x - x^2 = 0 \qquad \text{(20.12)}$$

These two equations must be solved simultaneously. However, the equations are not linear. In Eq. (20.12), f_y will equal 0 when

$$4x - x^2 = 0$$

$$x(4 - x) = 0$$

or two critical values for x are

$$x = 0 \quad \text{and} \quad x = 4$$

To determine the values of y which correspond to these critical values of x and which make f_x equal 0, substitute these values, one at a time, into Eq. (20.11).

For $x = 0$,

$$4(0) + 4y - 2(0)y - 4 = 0$$

$$4y = 4$$

$$y = 1$$

Thus, one critical point occurs on the graph of f when $x = 0$ and $y = 1$. When $x = 0$ and $y = 1$,

$$f(0, 1) = 2(0)^2 + 4(0)(1) - (0)^2(1) - 4(0)$$
$$= 0$$

Thus one critical point occurs at $(0, 1, 0)$.

For $x = 4$,
$$4(4) + 4y - 2(4)y - 4 = 0$$

$$16 + 4y - 8y - 4 = 0$$

$$-4y = -12$$

$$y = 3$$

Since

$$f(4, 3) = 2(4)^2 + 4(4)(3) - (4)^2(3) - (4)(4)$$
$$= 32 + 48 - 48 - 16 = 16$$

another critical point occurs on f at $(4, 3, 16)$.

❑

Distinguishing among Critical Points

Once a critical point has been identified, it is necessary to determine its nature. Aside from relative maximum and minimum points, there is one other situation in

Figure 20.10
Saddle point.

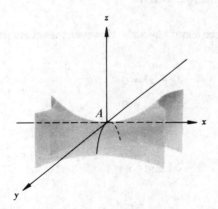

which f_x and f_y both equal 0. Figure 20.10 illustrates this situation, which is referred to as a ***saddle point***. A saddle point is a portion of a surface which has the shape of a saddle. At point A — "where you sit on the horse" — the values of f_x and f_y both equal 0. However, the function does not reach either a relative maximum or a relative minimum at A. If you slice through the surface at point A with the plane having the equation $x = 0$, the resulting edge or trace signals a relative maximum at A. However, in slicing through the surface with the plane having the equation $y = 0$, the resulting trace indicates a relative minimum at A. Figure 20.11 illustrates these observations.

The conditions which allow you to distinguish among relative maximum, relative minimum, or saddle points follow. The test of a critical point is a second-derivative test which, from an intuitive standpoint, investigates the concavity conditions at the critical point.

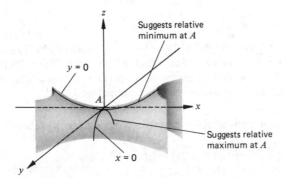

Figure 20.11   Conflicting concavity signals for saddle point.

TEST OF CRITICAL POINT
Given that a critical point of f is located at (x^*, y^*, z) where all second partial derivatives are continuous, determine the value of $D(x^*, y^*)$, where

$$D(x^*, y^*) = f_{xx}(x^*, y^*)f_{yy}(x^*, y^*) - [f_{xy}(x^*, y^*)]^2 \qquad (20.13)$$

I   If $D(x^*, y^*) > 0$, the critical point is
   (a) *a relative maximum if both $f_{xx}(x^*, y^*)$ and $f_{yy}(x^*, y^*)$ are negative,*
   (b) *a relative minimum if both $f_{xx}(x^*, y^*)$ and $f_{yy}(x^*, y^*)$ are positive.*
II   If $D(x^*, y^*) < 0$, the critical point is a saddle point.
III If $D(x^*, y^*) = 0$, other techniques (beyond the scope of this text) are required to determine the nature of the critical point.

EXAMPLE 14

In Example 11 we determined that a critical point occurs on the graph of the function

$$f(x, y) = 4x^2 - 12x + y^2 + 2y - 10$$

Determine the nature of the critical point at $(\frac{3}{2}, -1, -20)$.

SOLUTION

From Example 11,

$$f_x = 8x - 12$$
$$f_y = 2y + 2$$

The four second-order derivatives are

$$f_{xx} = 8 \qquad f_{xy} = 0$$
$$f_{yy} = 2 \qquad f_{yx} = 0$$

Evaluating $D(x^*, y^*)$, we have

$$D(\tfrac{3}{2}, -1) = (8)(2) - 0^2$$
$$= 16 > 0$$

Since $D(\tfrac{3}{2}, -1) > 0$ and $f_{xx}(\tfrac{3}{2}, -1) = 8$ and $f_{yy}(\tfrac{3}{2}, -1) = 2$, both of which are greater than 0, we can conclude that a *relative minimum* occurs at $(\tfrac{3}{2}, -1, -20)$. Figure 20.12 is a sketch of the function. This graph, as well as a number of the following, is computer-generated using the SAS Graph package and plotted on the Calcomp plotter.*

EXAMPLE 15   To determine the location and nature of any critical points for the function

$$f(x, y) = -2x^2 + 24x - y^2 + 30y$$

the first partial derivatives are

$$f_x = -4x + 24$$
$$f_y = -2y + 30$$

Setting f_x and f_y equal to zero, we have

$$f_x = -4x + 24 = 0 \qquad \text{and} \qquad f_y = -2y + 30 = 0$$

Critical values are identified at

$$x = 6 \qquad \text{and} \qquad y = 15$$

Since

$$f(6, 15) = -2(6)^2 + 24(6) - (15)^2 + 30(15)$$
$$= -72 + 144 - 225 + 450$$
$$= 297$$

---

* Statistical Analysis System, G3d subroutine.

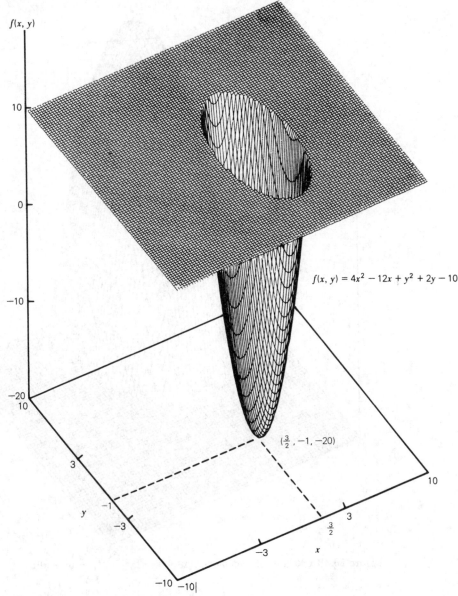

Figure 20.12 Relative minimum on $f(x, y) = 4x^2 - 12x + y^2 + 2y - 10$.

there is one critical point on the graph of f at $(6, 15, 297)$.

To determine the nature of the critical point, the second-order derivatives are

$$f_{xx} = -4 \qquad f_{xy} = 0$$
$$f_{yy} = -2 \qquad f_{yx} = 0$$

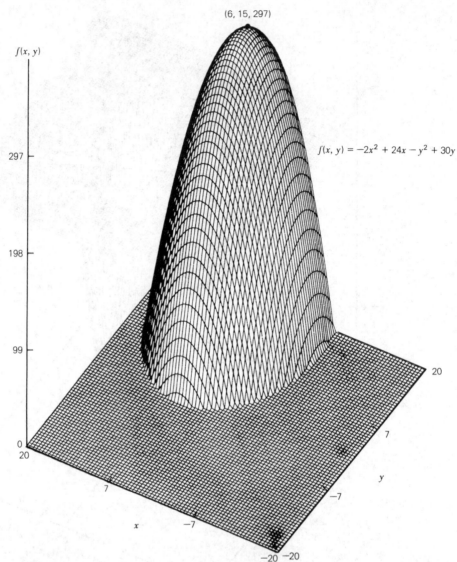

Figure 20.13 Relative maximum on $f(x, y) = -2x^2 + 24x - y^2 + 30y$.

Evaluating $D(x^*, y^*)$, we get

$$D(6, 15) = (-4)(-2) - 0^2$$
$$= 8 > 0$$

Since $D(6, 15) > 0$ and both f_{xx} and f_{yy} are negative, a relative maximum occurs at $(6, 15, 297)$. Figure 20.13 is a sketch of the function.

EXAMPLE 16 In Example 13 we determined that critical points occur on the graph of the function

$$f(x, y) = 2x^2 + 4xy - x^2y - 4x$$

at $(0, 1, 0)$ and $(4, 3, 16)$. To determine the nature of the two critical points, we must find all second derivatives. From Example 13

$$f_x = 4x + 4y - 2xy - 4$$
$$f_y = 4x - x^2$$

The second-order derivatives are

$$f_{xx} = 4 - 2y \qquad f_{xy} = 4 - 2x$$
$$f_{yy} = 0 \qquad f_{yx} = 4 - 2x$$

Evaluation of $(0, 1, 0)$:

$$\begin{aligned} D(0, 1) &= [4 - 2(1)](0) - [4 - 2(0)]^2 \\ &= (2)(0) - 4^2 \\ &= -16 < 0 \end{aligned}$$

Since $D(0, 1) < 0$, a saddle point occurs on the graph of f at $(0, 1, 0)$.

Evaluation of $(4, 3, 16)$:

$$\begin{aligned} D(4, 3) &= [4 - 2(3)](0) - [4 - 2(4)]^2 \\ &= (-2)(0) - (-4)^2 \\ &= -16 < 0 \end{aligned}$$

A second saddle point occurs on the graph of f, this one at $(4, 3, 16)$. Figure 20.14 presents a sketch of the function.

EXAMPLE 17 Given the function

$$f(x, y) = -x^2 - y^3 + 12y^2$$

determine the location and nature of all critical points.

SOLUTION
Finding the first partial derivatives and setting them equal to 0 gives us

$$f_x = -2x = 0 \quad \text{or} \quad \text{a critical value occurs at } x = 0$$
$$f_y = -3y^2 + 24y = 0$$
$$3y(-y + 8) = 0$$

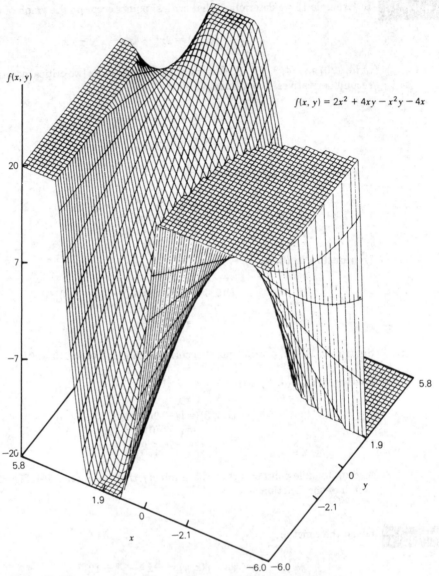

$$f(x, y) = 2x^2 + 4xy - x^2y - 4x$$

Figure 20.14   Two saddle points on $f(x, y) = 2x^2 + 4xy - x^2y - 4x$.

Or critical values occur at

$$y = 0 \quad \text{and} \quad y = 8$$

There are *two* critical points on f—one associated with the critical values $x = 0$ and $y = 0$ and the other associated with the critical values $x = 0$ and $y = 8$. Verify that the two stationary points occur at $(0, 0, 0)$ and $(0, 8, 256)$.

The second-order derivatives are

$$f_{xx} = -2 \qquad f_{xy} = 0$$
$$f_{yy} = -6y + 24 \qquad f_{yx} = 0$$

Evaluation of (0, 0, 0):

$$D(0, 0) = (-2)[-6(0) + 24] - 0^2$$
$$= (-2)(24) - 0$$
$$= -48 < 0$$

Therefore, a saddle point occurs at $(0, 0, 0)$.

Evaluation of (0, 8, 256):

$$D(0, 8) = (-2)[-6(8) + 24] - 0^2$$
$$= (-2)(-24)$$
$$= 48 > 0$$

With $D > 0$, the critical point is either a relative maximum or a relative minimum. The values of the two pure partial derivatives at the critical point are $f_{xx}(0, 8) = -2$ and $f_{yy}(0, 8) = -6(8) + 24 = -24$. Since both are negative, we conclude that a relative maximum occurs on the graph of f at $(0, 8, 256)$. Figure 20.15 is a sketch of f which provides two different views of the surface.

❑

Section 20.3 Follow-up Exercises

For the following functions, determine the location and nature of all critical points.

1 $f(x, y) = 4x^2 - y^2 + 80x + 20y - 10$
2 $f(x, y) = x^2 + xy - 5x - 2y^2 + 2y$
3 $f(x, y) = x^3/3 - 5x^2/2 + 3y^2 - 12y$
4 $f(x, y) = -8x^2 + 12xy + 44x - 12y^2 + 12y$
5 $f(x, y) = 3x^2 - 4xy + 3y^2 + 8x - 17y + 5$
6 $f(x, y) = -x^2 + 6x - 12y + y^3 + 5$
7 $f(x, y) = x^3 + y^2 - 3x + 6y + 10$
8 $f(x, y) = x^2 - 2xy + 3y^2 + 4x - 16y + 22$
9 $f(x, y) = x^3 + y^3 - 3xy$
10 $f(x, y) = x^3 + y^3 + 3xy$
11 $f(x, y) = xy + \ln x + y^2 - 10, x > 0$
12 $f(x, y) = 25x - 25xe^{-y} - 50y - x^2$
13 $f(x, y) = x^2/2 + 2y^2 - 20x - 40y + 100$
14 $f(x, y) = -3x^2 - 2y^2 + 45x - 30y - 50$
15 $f(x, y) = 3x^2 + y^2 + 3xy - 60x - 32y + 200$
16 $f(x, y) = x^2 + xy + y^2 - 3x$
17 $f(x, y) = xy - x^3 - y^2$
18 $f(x, y) = 3x^2 - 4xy + 3y^2 + 8x - 17y + 5$
19 $f(x, y) = -2x^2 + 2xy - y^2 + 4x - 6y + 10$
20 $f(x, y) = x^2 + 2xy + 5y^2 + 2x + 10y - 20$
21 $f(x, y) = x^3 - y^2 - 3x + 4y + 25$

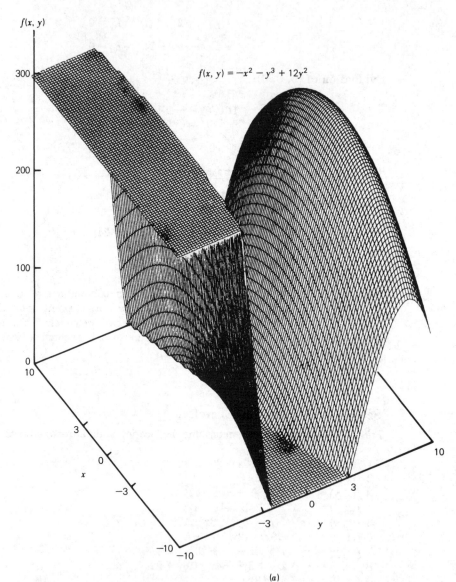

(a)

Figure 20.15 Relative maximum and saddle point on $f(x, y) = -x^2 - y^3 + 12y^2$.

22 $f(x, y) = 2x^2 + y^3 - x - 12y + 15$

23 $f(x, y) = x^2 + y^2 + xy - 6x + 6$

24 $f(x, y) = 4xy - x^3 - y^2$

25 $f(x, y) = x^3 + y^2 - 3x + 6y + 10$

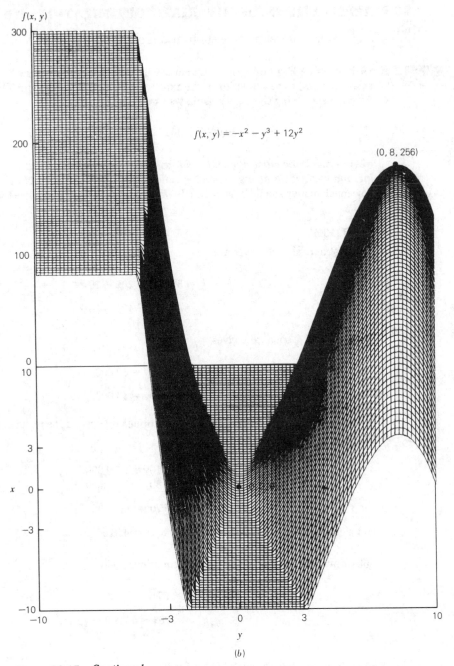

$f(x, y) = -x^2 - y^3 + 12y^2$

$(0, 8, 256)$

(b)

Figure 20.15 *Continued*

20.4 APPLICATIONS OF BIVARIATE OPTIMIZATION

This section presents some applications of the optimization of bivariate functions.

EXAMPLE 18 (**Advertising Expenditures**) Example 9 involved a manufacturer who estimated annual sales (in units) to be a function of the expenditures made for radio and TV advertising. The function specifying this relationship was stated as

$$z = 50,000x + 40,000y - 10x^2 - 20y^2 - 10xy$$

where *z equals the number of units sold each year, x equals the amount spent for TV advertising, and y equals the amount spent for radio advertising (x and y both in $1,000s)*. Determine how much money should be spent on TV and radio in order to maximize the number of units sold.

SOLUTION

The first partial derivatives are

$$f_x = 50,000 - 20x - 10y$$
$$f_y = 40,000 - 40y - 10x$$

Setting f_x and f_y equal to 0 gives

$$20x + 10y = 50,000$$
$$10x + 40y = 40,000$$

If both sides of the second equation are multiplied by -2 and the result is added to the first equation, then

$$
\begin{aligned}
20x + 10y &= 50,000 \\
-20x - 80y &= -80,000 \\
\hline
-70y &= -30,000
\end{aligned}
$$

and a critical value for *y* is $\qquad y = 428.57$

Substituting *y* into one of the original equations yields

$$20x + 10(428.57) = 50,000$$
$$20x = 50,000 - 4,285.7$$
$$= 45,714.3$$

and the corresponding critical value for *x* is

$$x = 2,285.72$$

Total sales associated with $x = 2,285.72$ and $y = 428.57$ equal

$$f(2{,}285.72, 428.57) = 50{,}000(2{,}285.72) + 40{,}000(428.57)$$
$$-10(2{,}285.72)^2 - 20(428.57)^2 - 10(2{,}285.72)(428.57)$$
$$= 65{,}714{,}296.00 \text{ units}$$

Thus a critical point occurs on the graph of f at $(2285.72, 428.57, 65{,}714{,}296)$.

To determine the nature of the critical point, the second derivatives are

$$f_{xx} = -20 \qquad f_{xy} = -10$$
$$f_{yy} = -40 \qquad f_{yx} = -10$$

Testing the critical point, we find

$$D(2{,}285.72, 428.57) = (-20)(-40) - (-10)^2$$
$$= 800 - 100 = 700 > 0$$

Since $D > 0$ and both f_{xx} and f_{yy} are negative, we can conclude that annual sales are maximized at 65,714,296 units when 2,285.72 ($1,000s) is spent for TV advertising and 428.57 ($1,000s) is spent for radio advertising. Figure 20.16 is a sketch of the sales surface.

EXAMPLE 19	**(Pricing Model)** A manufacturer sells two related products, the demands for which are estimated by the following two demand functions:

$$q_1 = 150 - 2p_1 - p_2 \qquad\qquad\qquad \textbf{(20.14)}$$

$$q_2 = 200 - p_1 - 3p_2 \qquad\qquad\qquad \textbf{(20.15)}$$

where p_j equals the price (in dollars) of product j and q_j equals the demand (in thousands of units) for product j. Examination of these demand functions indicates that the two products are related. The demand for one product depends not only on the price charged for the product itself but also on the price charged for the other product.

The firm wants to determine the price it should charge for each product in order to maximize total revenue from the sale of the two products.

SOLUTION

This revenue-maximizing problem is exactly like the single-product problems discussed in Chap. 17. The only difference is that there are two products and two pricing decisions to be made.

Total revenue from selling the two products is determined by the function

$$\boxed{R = p_1 q_1 + p_2 q_2} \qquad\qquad\qquad \textbf{(20.16)}$$

This function is stated in terms of four independent variables. As with the single-product problems, we can substitute the right side of Eqs. (20.14) and (20.15) into Eq. (20.16) to yield

$$R = f(p_1, p_2)$$
$$= p_1(150 - 2p_1 - p_2) + p_2(200 - p_1 - 3p_2)$$
$$= 150p_1 - 2p_1^2 - p_1 p_2 + 200p_2 - p_1 p_2 - 3p_2^2$$
$$= 150p_1 - 2p_1^2 - 2p_1 p_2 + 200p_2 - 3p_2^2$$

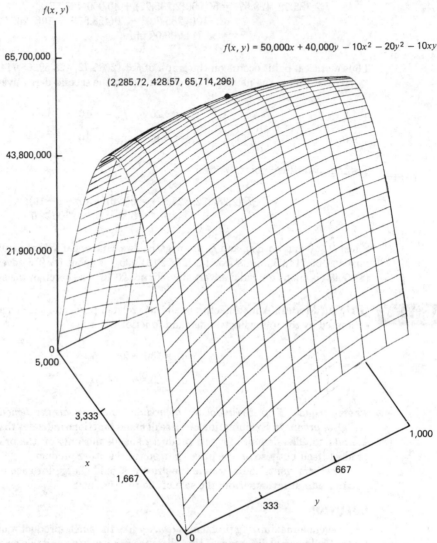

Figure 20.16   Relative maximum on $f(x, y) = 50{,}000x + 40{,}000y - 10x^2 - 20y^2 - 10xy$.

We can now proceed to examine the revenue surface for relative maximum points.
The first partial derivatives are

$$f_{p_1} = 150 - 4p_1 - 2p_2$$

$$f_{p_2} = -2p_1 + 200 - 6p_2$$

Setting f_{p_1} and f_{p_2} equal to 0, we have

$$4p_1 + 2p_2 = 150 \tag{20.17}$$

$$2p_1 + 6p_2 = 200 \tag{20.18}$$

If Eq. (20.18) is multiplied by -2 and added to Eq. (20.17), we get

$$4p_1 + 2p_2 = 150$$
$$\underline{-4p_1 - 12p_2 = -400}$$
$$-10p_2 = -250$$

or a critical value for p_2 is
$$p_2 = 25$$

Substituting $p_2 = 25$ into Eq. (20.17) yields

$$4p_1 + 2(25) = 150$$
$$4p_1 = 100$$

or a critical value for p_1 is
$$p_1 = 25$$

If these values are substituted into f,

$$f(25, 25) = 150(25) - 2(25)^2 - 2(25)(25) + 200(25) - 3(25)^2$$
$$= 4{,}375$$

A critical point occurs on f at $(25, 25, 4{,}375)$.
  The second derivatives are

$$f_{p_1 p_1} = -4 \qquad f_{p_1 p_2} = -2$$
$$f_{p_2 p_2} = -6 \qquad f_{p_2 p_1} = -2$$

And
$$D(25, 25) = (-4)(-6) - (-2)^2$$
$$= 24 - 4 = 20 > 0$$

Since $D(x^*, y^*) > 0$ and $f_{p_1 p_1}$ and $f_{p_2 p_2}$ are both negative, a relative maximum exists on the graph of f when $p_1 = 25$ and $p_2 = 25$. Revenue will be maximized at a value of \$4,375 (thousands) when each product is sold for \$25. Expected demand at these prices can be determined by substituting p_1 and p_2 into the demand equations, or

$$q_1 = 150 - 2(25) - (25) = 75 \text{ (thousand units)}$$
$$q_2 = 200 - (25) - 3(25) = 100 \text{ (thousand units)}$$

Figure 20.17 is a sketch of the revenue surface.

EXAMPLE 20 (**Satellite Clinic Location**) A large health maintenance organization (HMO) is planning to locate a satellite clinic in a location which is convenient to three suburban townships, the relative locations of which are indicated in Fig. 20.18. The HMO wants to select a preliminary site by using the following criterion: determine the location (x, y) which minimizes the *sum of the squares* of the distances from each township to the satellite clinic.

SOLUTION

The unknowns in this problem are x and y, the coordinates of the satellite clinic location. We need to determine an expression for the square of the distance separating the clinic and each

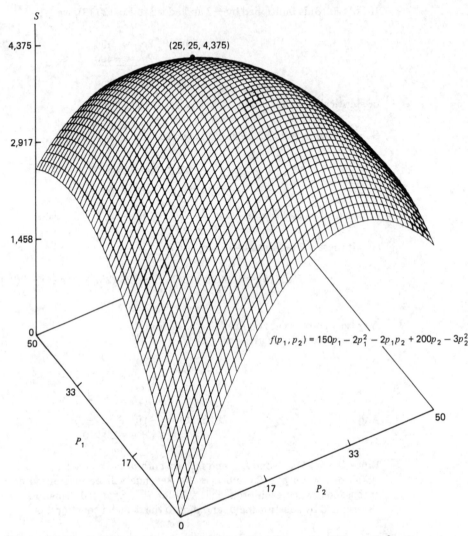

Figure 20.17 Relative maximum on revenue surface $f(p_1, p_2) = 150p_1 - 2p_1^2 - 2p_1 p_2 + 200p_2 - 3p_2^2$.

of the towns. The Pythagorean theorem* provides this for us. Given two points (x_1, y_1) and (x_2, y_2), the square of the distance d separating these two points is found using the equation

$$d^2 = (x_2 - x_1)^2 + (y_2 - y_1)^2 \qquad (20.19)$$

To illustrate, the square of the distance separating the clinic with location (x, y) and township A located at $(40, 20)$ is

$$d^2 = (x - 40)^2 + (y - 20)^2$$

---

* See Chap. 17, page 808.

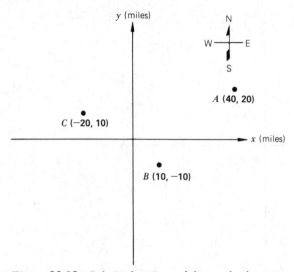

Figure 20.18 Relative locations of three suburban townships.

Finding similar expressions for the square of the distance separating townships B and C and the clinic and summing for the three townships, we get

$$s = f(x, y)$$
$$= [(x - 40)^2 + (y - 20)^2] + [(x - 10)^2 + (y + 10)^2]$$
$$+ [(x + 20)^2 + (y - 10)^2]$$

The right side of this function can be expanded or left in this form for purposes of finding derivatives. Let's leave it as it is. The first partial derivatives are

$$f_x = 2(x - 40)(1) + 2(x - 10)(1) + 2(x + 20)(1)$$
$$= 2x - 80 + 2x - 20 + 2x + 40$$
$$= 6x - 60$$

$$f_y = 2(y - 20)(1) + 2(y + 10)(1) + 2(y - 10)(1)$$
$$= 2y - 40 + 2y + 20 + 2y - 20$$
$$= 6y - 40$$

If the two partial derivatives are set equal to 0, critical values are found when $x = 10$ and $y = 6\frac{2}{3}$. The second partial derivatives are

$$f_{xx} = 6 \qquad f_{xy} = 0$$

$$f_{yy} = 6 \qquad f_{yx} = 0$$

$$D(10, 6\tfrac{2}{3}) = (6)(6) - 0^2 + 36 > 0$$

Since $D > 0$ and f_{xx} and f_{yy} are both greater than 0, we conclude that a relative minimum occurs on f when $x = 10$ and $y = 6\frac{2}{3}$, or when the satellite clinic is located as indicated in Fig. 20.19.

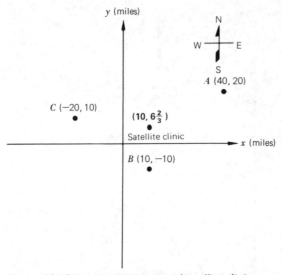

Figure 20.19   Proposed location of satellite clinic.

EXAMPLE 21

(Method of Least Squares: Finding the Best Fit to a Set of Data Points; Motivating Scenario) Organizations gather data regularly on a multitude of variables which are related to their operation. One major area of analysis deals with determining whether there are any patterns to the data—are there any apparent relationships among the variables of interest? For example, the demand functions to which we have continually referred have most likely been determined by gathering data on the demand for a product at different prices. And analysis of this data translates into an estimated demand function.

Consider the four data points (x_1, y_1), (x_2, y_2), (x_3, y_3), and (x_4, y_4) in Fig. 20.20, which have been gathered for the variables x and y. Suppose there is evidence suggesting that x and y are related and that the nature of the relationship is linear. And suppose that we would like to fit a straight line to these points, the equation of which would be used as an approximation of the actual relationship existing between x and y. The question becomes, What line best fits the data points? There are an infinite number of straight lines which can be fit to these data points, each having the general form

$$\boxed{y_p = ax + b} \tag{20.20}$$

The difference between each line would be in the slope a and/or the y coordinate of the y intercept b. Note the subscript on y_p in Eq. (20.20). This notation indicates that the line fit to the data points can be used to *predict* values of y, given a known value of x.

In Fig. 20.20, the predicted values of y, given the x coordinates of the four data points, are indicated on the line. The vertical distance separating the actual data point and the corresponding point on the line is a measure of the error introduced by using the line to predict the location of the data point. The error, indicated by the d_j values in Fig. 20.20, is called the ***deviation*** between the actual value of y and the predicted value of y for the jth data point, or

$$\boxed{d_j = y_j - y_{p_j}} \tag{20.21}$$

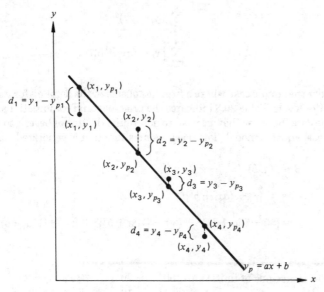

Figure 20.20 Four sample data points.

Given that we wish to find the "best" line to fit to the data points, the next question is, How do you define *best*? One of the most popular methods of finding the line of best fit is the **least squares model.** The least squares model defines *best* as the line which minimizes the sum of the squared deviations for all the sample data points.

Given a set of n data points, the method of least squares seeks the line which minimizes

$$S = d_1^2 + d_2^2 + \cdots + d_n^2$$

$$= \sum_{j=1}^{n} d_j^2$$

or
$$S = \sum_{j=1}^{n} (y_j - y_{p_j})^2 \qquad \textbf{(20.22)}$$

For any line $y_p = ax + b$ chosen to fit the data points, Eq. (20.22) can be rewritten as

$$S = f(a, b) = \sum_{j=1}^{n} [y_j - (ax_j + b)]^2 \qquad \textbf{(20.23)}$$

The method of least squares seeks the values of a and b which result in a minimum value for S. In Fig. 20.20 we would seek the line which minimizes

$$S = d_1^2 + d_2^2 + d_3^2 + d_4^2$$

$$= \sum_{j=1}^{4} d_j^2$$

$$= \sum_{j=1}^{4} (y_j - y_{p_j})^2$$

$$= \sum_{j=1}^{4} [y_j - (ax_j + b)]^2$$

Consider the simple case where a firm has observed the demand for its product at three different price levels. Table 20.3 indicates the price-demand combinations. Figure 20.21 is a graph of the data. Suppose that we wish to determine the line of best fit to these data points using the least squares model. The least squares function is generated using Eq. (20.23).

$$S = f(a, b)$$

$$= \sum_{j=1}^{3} [y_j - (ax_j + b)]^2$$

$$= [50 - (5a + b)]^2 + [30 - (10a + b)]^2 + [20 - (15a + b)]^2$$

TABLE 20.3	*y* (demand in thousands of units)	50	30	20
	x (price in dollars)	5	10	15

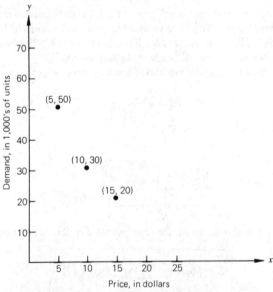

Figure 20.21 Three sample price-demand data points.

To determine the values of *a* and *b* which minimize *S*, we find the partial derivatives with respect to *a* and *b*.

$$f_a = 2[50 - (5a + b)](-5) + 2[30 - (10a + b)](-10)$$
$$+ 2[20 - (15a + b)](-15)$$

$$= -500 + 50a + 10b - 600 + 200a + 20b - 600 + 450a + 30b$$
$$= 700a + 60b - 1,700$$

$$f_b = 2[50 - (5a + b)](-1) + 2[30 - (10a + b)](-1)$$
$$\qquad + 2[20 - (15a + b)](-1)$$
$$= -100 + 10a + 2b - 60 + 20a + 2b - 40 + 30a + 2b$$
$$= 60a + 6b - 200$$

If these two derivatives are set equal to 0, the following two equations result:

$$700a + 60b = 1,700 \qquad\qquad \textbf{(20.24)}$$

$$60a + 6b = 200 \qquad\qquad \textbf{(20.25)}$$

Multiplying Eq. (20.25) by -10 and adding it to Eq. (20.24) yields

$$\begin{aligned} 700a + 60b &= 1,700 \\ -600a - 60b &= -2,000 \\ \hline 100a \quad\quad &= -300 \end{aligned}$$

$$a = -3$$

Substituting $a = -3$ into Eq. (20.25) yields

$$60(-3) + 6b = 200$$

$$6b = 380$$

$$b = 63\tfrac{1}{3}$$

To verify that the critical point results in a minimum value for S,

$$f_{aa} = 700 \qquad f_{ab} = 60$$

$$f_{bb} = 6 \qquad f_{ba} = 60$$

$$\begin{aligned} D(-3, 63\tfrac{1}{3}) &= (700)(6) - (60)^2 \\ &= 4,200 - 3,600 \\ &= 600 > 0 \end{aligned}$$

Since $D > 0$ and both f_{aa} and f_{bb} are positive, we can conclude that the sum of the squares of the deviations S is minimized when $a = -3$ and $b = 63\tfrac{1}{3}$, or when the data points are fit with a straight line having a slope of -3 and y intercept of $63\tfrac{1}{3}$. The equation of this line is

$$y_p = -3x + 63\tfrac{1}{3}$$

The minimum sum of squared deviations can be determined by substituting $a = -3$ and $b = 63\tfrac{1}{3}$ into f if that value is of interest.

❑

> **NOTE**   This example is intended to illustrate the underlying basis for this popular estimation technique. Fortunately, the actual implementation (use) of the method of least squares does not require the formulation of the sum of squares function and optimization analysis as illustrated in this example. Typically, least squares analysis is performed by inputting the sample data points into a hand calculator or any of a wide variety of statistical software packages which are available.

Section 20.4 Follow-up Exercises

1   A manufacturer estimates that annual sales (in units) are a function of the expenditures made for TV and radio advertising. The function specifying the relationship is

$$z = 40,000x + 60,000y - 5x^2 - 10y^2 - 10xy$$

where z equals the number of units sold each year, x equals the amount spent for TV advertising, and y equals the amount spent for radio advertising (both x and y in $1,000s).
   (a) Determine how much should be spent for radio and TV advertising in order to maximize the number of units sold.
   (b) What is the maximum number of units expected to equal?

2   A company sells two products. Total revenue from the two products is estimated to be a function of the numbers of units sold of the two products. Specifically, the function is

$$R = 30,000x + 15,000y - 10x^2 - 10y^2 - 10xy$$

where R equals total revenue and x and y equal the numbers of units sold of the two products.
   (a) How many units of each product should be produced in order to maximize total revenue?
   (b) What is the maximum revenue?

3   A firm sells two products. The demand functions for the two products are

$$q_1 = 110 - 4p_1 - p_2$$
$$q_2 = 90 - 2p_1 - 3p_2$$

where p_j equals the price of product j and q_j equals the demand (in thousands of units) for product j.
   (a) Determine the price which should be charged for each product in order to maximize total revenue from the two products.
   (b) How many units will be demanded of each product at these prices?
   (c) What is maximum total revenue expected to equal?

4   A company is planning to locate a warehouse which will supply three major department stores. The relative locations of the three department stores on a set of coordinate axes are (30, 10), (0, 40), and (−30, −10) where the coordinates are stated in miles. Figure 20.22 indicates the relative locations of the department stores. Determine the warehouse location (x, y) which minimizes the sum of the squares of the distances from each city to the warehouse.

Figure 20.22
Department store
locations.

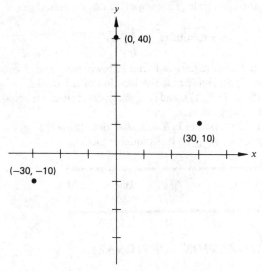

5 **Airport Location** A new airport is being planned to service four metropolitan areas. The relative locations of the metropolitan areas on a set of coordinate axes are (20, 5), (0, 30), (−30, 10), and (−5, −5), where the coordinates are stated in miles. Figure 20.23 indicates the relative locations of the four cities. Determine the airport location (x, y) which minimizes the sum of the squares of the distances from the airport to each metropolitan area.

6 In Example 20, assume that the number of HMO members living in the three townships equals 20,000, 10,000, and 30,000, respectively, for townships A, B, and C. Also assume that the HMO wishes to determine the location (x, y) which minimizes the sum of the products of the number of members in each township and the square of the distance

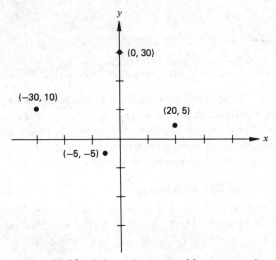

Figure 20.23 Relative locations of four metropolitan areas.

separating the towns and the clinic. This objective can be stated as

$$\text{minimize} \quad \sum_{j=1}^{3} n_j d_j^2$$

where n_j equals the number of members living in township j and d_j equals the distance from township j to the clinic. Determine the location of the clinic.

7 Given the data points $(2, 2)$, $(-3, 17)$, and $(10, -22)$, determine the equation of the line of best fit using the least squares model.

8 Given the price-demand data points in Table 20.4, determine the equation of the line of best fit to these data points using the least squares model.

TABLE 20.4			
y (demand in thousands of units)	200	160	120
x (price in dollars)	30	40	50

20.5 *n*-VARIABLE OPTIMIZATION (OPTIONAL)

When a function contains more than two independent variables, the process for identifying relative maxima and relative minima is very similar to that used for functions having two independent variables. Before discussing the process, let's define these relative extrema.

> **DEFINITION: RELATIVE MAXIMUM**
> A function $y = f(x_1, x_2, \ldots, x_n)$ is said to have a *relative maximum* at $x_1 = a_1, x_2 = a_2, \ldots, x_n = a_n$ if for all points $(x_1, x_2, \ldots, x_n)$ sufficiently close to $(a_1, a_2, \ldots, a_n)$,
>
> $$f(a_1, a_2, \ldots, a_n) \geq f(x_1, x_2, \ldots, x_n)$$
>
> **DEFINITION: RELATIVE MINIMUM**
> A function $y = f(x_1, x_2, \ldots, x_n)$ is said to have a *relative minimum* at $x_1 = a_1, x_2 = a_2, \ldots, x_n = a_n$ if for all points $(x_1, x_2, \ldots, x_n)$ sufficiently close to $(a_1, a_2, \ldots, a_n)$,
>
> $$f(a_1, a_2, \ldots, a_n) \leq f(x_1, x_2, \ldots, x_n)$$

With more than two independent variables it is not possible to graph a function. However, we may say that the function $f(x_1, x_2, \ldots, x_n)$ is represented by a *hypersurface* in $(n + 1)$ dimensions. Our interest with these functions is to identify the $(n + 1)$-dimensional equivalents to *peaks* (relative maxima) and *valleys* (relative minima) on a three-dimensional surface.

Necessary Condition for Relative Extrema

> A necessary condition for a relative maximum or a relative minimum of a function whose partial derivatives $f_{x_1}, f_{x_2}, \ldots, f_{x_n}$ all exist is that
>
> $$f_{x_1} = 0, \; f_{x_2} = 0, \; \ldots, \; f_{x_n} = 0 \qquad \text{(20.26)}$$

The necessary condition requires that all first partial derivatives of f equal 0.

EXAMPLE 22   To locate any *candidates* for relative extreme points on

$$f(x_1, x_2, x_3) = x_1^2 - 2x_1 x_2 + 2x_2^2 + 2x_1 x_3 + 4x_3^2 - 2x_3$$

the first partial derivatives are found:

$$f_{x_1} = 2x_1 - 2x_2 + 2x_3$$
$$f_{x_2} = -2x_1 + 4x_2$$
$$f_{x_3} = 2x_1 + 8x_3 - 2$$

Since all three derivatives must equal 0, the three equations

$$2x_1 - 2x_2 + 2x_3 = 0$$
$$-2x_1 + 4x_2 \quad\quad = 0$$
$$2x_1 \quad\quad + 8x_3 = 2$$

must be solved simultaneously. When the system is solved the critical values

$$x_1 = -1 \quad\quad x_2 = -\tfrac{1}{2} \quad\quad x_3 = \tfrac{1}{2}$$

are identified. Because

$$f(-1, -\tfrac{1}{2}, \tfrac{1}{2}) = -\tfrac{1}{2}$$

we can state that there is one critical point $(-1, -\tfrac{1}{2}, \tfrac{1}{2}, -\tfrac{1}{2})$ which is a candidate for a relative extreme point.

$\square$

Sufficient Conditions

As with functions containing one or two independent variables, the test of critical points requires the use of second derivatives. Specifically, the test utilizes a ***hessian matrix,*** which is a matrix of second partial derivatives having the form

$$\mathbf{H} = \begin{pmatrix} f_{x_1 x_1} & f_{x_1 x_2} & f_{x_1 x_3} & \cdots & f_{x_1 x_n} \\ f_{x_2 x_1} & f_{x_2 x_2} & f_{x_2 x_3} & \cdots & f_{x_2 x_n} \\ \cdots\cdots\cdots\cdots\cdots\cdots\cdots\cdots \\ f_{x_n x_1} & f_{x_n x_2} & f_{x_n x_3} & \cdots & f_{x_n x_n} \end{pmatrix}$$

Given the function $f(x_1, x_2, x_3, \ldots, x_n)$, the hessian matrix is square, of dimension $(n \times n)$. The principal diagonal consists of the *pure* second partial derivatives, and the nondiagonal elements are *mixed* partial derivatives. The matrix is also symmetric about the principal diagonal when the second partial derivatives are continuous. Under such circumstances the mixed partial derivatives taken with respect to the same two variables are equal. That is, $f_{x_i x_j} = f_{x_j x_i}$.

For an $(n \times n)$ hessian matrix, a set of n *submatrices* can be identified. The first of these is the (1×1) submatrix consisting of the element located in row 1 and column 1, $f_{x_1 x_1}$. Let's denote this submatrix as $\mathbf{H_1}$, where

$$\mathbf{H_1} = (f_{x_1 x_1})$$

The second submatrix is the (2×2) matrix

$$\mathbf{H_2} = \begin{pmatrix} f_{x_1 x_1} & f_{x_1 x_2} \\ f_{x_2 x_1} & f_{x_2 x_2} \end{pmatrix}$$

The third submatrix is the (3×3) matrix

$$\mathbf{H_3} = \begin{pmatrix} f_{x_1 x_1} & f_{x_1 x_2} & f_{x_1 x_3} \\ f_{x_2 x_1} & f_{x_2 x_2} & f_{x_2 x_3} \\ f_{x_3 x_1} & f_{x_3 x_2} & f_{x_3 x_3} \end{pmatrix}$$

The nth submatrix is the hessian matrix itself, or $\mathbf{H_n} = \mathbf{H}$.

The determinants of these submatrices are called ***principal minors***. The principal minor associated with the ith submatrix can be denoted as Δ_i.

SUFFICIENT CONDITION FOR RELATIVE EXTREMA
Given critical values $x_1 = a_1$, $x_2 = a_2$, $x_3 = a_3$, . . . , $x_n = a_n$ for which

$$f_{x_1} = f_{x_2} = f_{x_3} = \cdots = f_{x_n} = 0$$

and all second-order derivatives are continuous:

I   A relative maximum *exists if the principal minors (evaluated at the critical values) alternate in sign with the odd-numbered principal minors negative and the even-numbered principal minors positive. That is,*

$$\Delta_1 < 0, \qquad \Delta_2 > 0, \qquad \Delta_3 < 0, \ldots$$

II   A relative minimum *exists if the principal minors (evaluated at the critical values) are all positive. That is,*

$$\Delta_1 > 0, \qquad \Delta_2 > 0, \qquad \Delta_3 > 0, \ldots$$

III   *If neither of the first two conditions is satisfied, no conclusion can be drawn regarding the critical point. Further analysis in the neighborhood of the critical point is required to determine its nature.*

EXAMPLE 23   Continuing Example 22, the hessian matrix is

$$\mathbf{H} = \begin{pmatrix} 2 & -2 & 2 \\ -2 & 4 & 0 \\ 2 & 0 & 8 \end{pmatrix}$$

The submatrices and corresponding values of the principal minors are

$$\mathbf{H_1} = (2) \qquad\qquad \Delta_1 = 2$$

$$\mathbf{H_2} = \begin{pmatrix} 2 & -2 \\ -2 & 4 \end{pmatrix} \qquad\qquad \Delta_2 = 4$$

$$\mathbf{H_3} = \begin{pmatrix} 2 & -2 & 2 \\ -2 & 4 & 0 \\ 2 & 0 & 8 \end{pmatrix} \qquad \Delta_3 = 16$$

Since the principal minors Δ_1, Δ_2, and Δ_3 are all positive, we conclude that a relative minimum occurs at the critical point $(-1, -\frac{1}{2}, \frac{1}{2}, -\frac{1}{2})$.

EXAMPLE 24 Locate any critical points and determine their nature for the function

$$f(x_1, x_2, x_3) = -2x_1^3 + 6x_1 x_3 + 2x_2 - x_2^2 - 6x_3^2 + 5$$

SOLUTION

The first partial derivatives are found and set equal to 0, as follows:

$$f_{x_1} = -6x_1^2 + 6x_3 = 0$$
$$f_{x_2} = 2 - 2x_2 = 0$$
$$f_{x_3} = 6x_1 - 12x_3 = 0$$

If these equations are solved simultaneously, critical values occur when $x_1 = 0$, $x_2 = 1$, and $x_3 = 0$, and also when $x_1 = \frac{1}{2}$, $x_2 = 1$, and $x_3 = \frac{1}{4}$. If the corresponding values of $f(x_1, x_2, x_3)$ are computed, we can state that critical points occur at $(0, 1, 0, 6)$ and $(\frac{1}{2}, 1, \frac{1}{4}, 6\frac{1}{4})$.

To test the nature of these critical points, second partial derivatives are identified and combined into the hessian matrix

$$\mathbf{H} = \begin{pmatrix} -12x_1 & 0 & 6 \\ 0 & -2 & 0 \\ 6 & 0 & -12 \end{pmatrix}$$

Evaluation of (0, 1, 0, 6):

$$\mathbf{H_1} = (-12(0)) = (0) \qquad \text{and} \quad \Delta_1 = 0$$

$$\mathbf{H_2} = \begin{pmatrix} 0 & 0 \\ 0 & -2 \end{pmatrix} \qquad \text{and} \quad \Delta_2 = 0$$

$$\mathbf{H_3} = \begin{pmatrix} 0 & 0 & 6 \\ 0 & -2 & 0 \\ 6 & 0 & -12 \end{pmatrix} \qquad \text{and} \quad \Delta_3 = 72$$

These principal minors do not satisfy the requirements for either a relative maximum or relative minimum; thus, no conclusions can be reached regarding the critical point $(0, 1, 0, 6)$.

Evaluation of $(\frac{1}{2}, 1, \frac{1}{4}, 6\frac{1}{4})$:

$$\mathbf{H_1} = (-12(\tfrac{1}{2})) = (-6) \qquad \text{and} \quad \Delta_1 = -6$$

$$\mathbf{H_2} = \begin{pmatrix} -6 & 0 \\ 0 & -2 \end{pmatrix} \qquad \text{and} \quad \Delta_2 = 12$$

$$\mathbf{H_3} = \begin{pmatrix} -6 & 0 & 6 \\ 0 & -2 & 0 \\ 6 & 0 & -12 \end{pmatrix} \quad \text{and} \quad \Delta_3 = -72$$

Since $\Delta_1 < 0$, $\Delta_2 > 0$, and $\Delta_3 < 0$, we can conclude that a relative maximum occurs at $(\frac{1}{2}, 1, \frac{1}{4}, 6\frac{1}{4})$.

❑

Section 20.5 Follow-up Exercises

For the following functions, locate any critical points and determine their nature.

1 $f(x_1, x_2, x_3) = x_1 - 4x_1 x_2 - x_2^2 + 5x_3^2 - 2x_2 x_3$
2 $f(x_1, x_2, x_3) = 10x_1^2 + 15x_2^2 + 5x_3^2 - 60x_1 + 90x_2 - 40x_3 + 15{,}000$
3 $f(x_1, x_2, x_3) = 2x_1^2 + x_1 x_2 + 4x_2^2 + x_1 x_3 + x_3^2 + 2$
4 $f(x_1, x_2, x_3) = x_1^2 - 3x_1 x_2 + 3x_2^2 + 4x_2 x_3 + 6x_3^2$
5 $f(x_1, x_2, x_3) = 25 - x_1^2 - x_2^2 - x_3^2$
6 $f(x_1, x_2, x_3, x_4) = 8x_2^3 + 4x_1^2 + 2x_2^2 + 5x_3^2 + 3.5x_4^2 - 24x_1 + 20x_3 - 75$
7 Pricing Model A firm manufactures three competing microcomputers. Demand functions for each of the three computers are

$$q_1 = 4{,}000 - 2p_1 + p_2 + p_3$$

$$q_2 = 6{,}000 + p_1 - 3p_2 + p_3$$

$$q_3 = 5{,}000 + p_1 + p_2 - 2p_3$$

where q_i is the estimated demand (in units per year) for computer i and p_i is the price of the ith computer (in dollars per unit).
(a) Determine the prices which will result in maximum total revenue from the three computers. Verify that you have, in fact, identified a relative maximum.
(b) What quantities should be produced if these prices are charged?
(c) What is the maximum total revenue?
8 Joint Cost Model A firm manufactures three products. The joint cost function is

$$\begin{aligned} C &= f(q_1, q_2, q_3) \\ &= 10q_1^2 + 30q_2^2 + 20q_3^2 - 400q_1 - 900q_2 - 1{,}000q_3 + 750{,}000 \end{aligned}$$

where C is the total cost (in dollars) of producing q_1, q_2, and q_3 units of products 1, 2, and 3, respectively.

(a) Determine the quantities which result in a minimum total cost. Confirm that the critical point is a relative minimum.

(b) What is the expected minimum total cost?

20.6 OPTIMIZATION SUBJECT TO CONSTRAINTS (OPTIONAL)

Our discussion of calculus-based optimization methods has focused upon *unconstrained optimization.* Many applications of mathematical modeling involve the optimization of an *objective function* subject to certain constraining conditions, or more simply, *constraints.* These constraints represent restrictions that can influence the degree to which an objective function is optimized. Constraints may reflect such restrictions as limited resources (e.g., labor, materials, or capital), limited demand for products, sales goals, etc. Problems having this structure are considered to be *constrained optimization problems.* We examined a linear subset of these problems when we studied linear programming (Chaps. 10–12). In this section we will examine a method for solving certain nonlinear constrained optimization problems.

The Lagrange Multiplier Method (Equality Constraint)

Consider the constrained optimization problem.

$$
\begin{array}{ll}
\text{Maximize (or minimize)} & y = f(x_1, x_2) \\
\text{subject to} & g(x_1, x_2) = k
\end{array}
\tag{20.27}
$$

In Eq. (20.27), f is the objective function and $g(x_1, x_2) = k$ is an *equality* constraint.

One way of solving this type of problem is to combine the information in Eq. (20.27) into the composite function

$$
L(x_1, x_2, \lambda) = f(x_1, x_2) - \lambda[g(x_1, x_2) - k]
\tag{20.28}
$$

This composite function is called the *lagrangian function,* and the variable λ (lambda) is referred to as the *Lagrange multiplier.* The lagrangian function is composed of the objective function and a linear multiple of the constraint equation. Notice in the lagrangian function that λ can equal any value and the term $\lambda[g(x_1, x_2) - k]$ will equal 0, provided that (x_1, x_2) are values which satisfy the constraint. Thus, the value of the newly formed lagrangian function L will equal the value of the original objective function f.

The creation of the lagrangian function ingeniously transforms the original constrained problem into an unconstrained problem which can be solved by procedures very similar to those discussed in the previous section. That is, to solve the original problem, Eq. (20.27), partial derivatives of $L(x_1, x_2, \lambda)$ are found with respect to x_1, x_2, and λ and are then set equal to 0.

NECESSARY CONDITIONS FOR RELATIVE EXTREMA

$$L_{x_1} = 0$$

$$L_{x_2} = 0 \qquad\qquad\qquad (20.29)$$

$$L_\lambda = 0$$

EXAMPLE 25   Consider the problem

$$\text{Maximize} \qquad f(x_1, x_2) = 25 - x_1^2 - x_2^2$$
$$\text{subject to} \qquad 2x_1 + x_2 = 4$$

The Lagrange multiplier method transforms this problem into the unconstrained form

$$L(x_1, x_2, \lambda) = 25 - x_1^2 - x_2^2 - \lambda(2x_1 + x_2 - 4)$$

First partial derivatives are identified as

$$L_{x_1} = -2x_1 - 2\lambda$$
$$L_{x_2} = -2x_2 - \lambda$$
$$L_\lambda = -2x_1 - x_2 + 4$$

Critical values are found by setting the three partial derivatives equal to zero and solving simultaneously.

$$-2x_1 \qquad - 2\lambda = 0 \qquad\qquad (20.30)$$
$$-2x_2 - \lambda = 0 \qquad\qquad (20.31)$$
$$-2x_1 - x_2 + 4 = 0 \qquad\qquad (20.32)$$

Multiplying both sides of Eq. (20.31) by -2 gives

$$4x_2 + 2\lambda = 0$$

Adding this to Eq. (20.30),

$$\begin{aligned}
4x_2 + 2\lambda &= 0 \\
\underline{-2x_1 - 2\lambda} &= 0 \qquad\qquad (20.30) \\
-2x_1 + 4x_2 &= 0 \qquad\qquad (20.33)
\end{aligned}$$

Solving for x_1 in Eq. (20.33) yields

$$4x_2 = 2x_1$$
$$2x_2 = x_1 \qquad\qquad (20.34)$$

If this value for x_1 is substituted into Eq. (20.32),

$$-2(2x_2) - x_2 + 4 = 0$$

$$-5x_2 = -4$$

$$x_2 = \tfrac{4}{5} = 0.8$$

If this value is substituted into Eq. (20.34), $x_1 = 1.6$. Also, substituting $x_2 = 0.8$ into Eq. (20.31) yields $\lambda = -1.6$. Thus, $x_1 = 1.6$, $x_2 = 0.8$, and $\lambda = -1.6$ are critical values on the lagrangian function. These values for x_1 and x_2 also represent the only candidate points for a relative maximum (or minimum).

❑

Sufficient Condition

To assess the behavior of $L(x_1, x_2, \lambda)$ at any critical values, the ***bordered hessian matrix*** $\mathbf{H_B}$ must be determined, where

$$\mathbf{H_B} = \begin{pmatrix} 0 & g_{x_1} & g_{x_2} \\ g_{x_1} & L_{x_1 x_1} & L_{x_1 x_2} \\ g_{x_2} & L_{x_2 x_1} & L_{x_2 x_2} \end{pmatrix}$$

and g_{x_i} represents the partial derivative of the left side of the constraint taken with respect to x_i.

> **SUFFICIENT CONDITIONS FOR RELATIVE EXTREMA**
> Given critical values $x_1 = a_1$, $x_2 = a_2$, and $\lambda = \lambda^*$ for which $L_{x_1} = L_{x_2} = L_\lambda = 0$, the determinant of $\mathbf{H_B}$, denoted as Δ_B, is evaluated at the critical values.
>
> I   A relative maximum *exists if* $\Delta_B > 0$.
> II A relative minimum *exists if* $\Delta_B < 0$.

EXAMPLE 26   To determine the behavior of

$$L(x_1, x_2, \lambda) = 25 - x_1^2 - x_2^2 - \lambda(2x_1 + x_2 - 4)$$

at $x_1 = 1.6$, $x_2 = 0.8$, and $\lambda = -1.6$, we form the bordered hessian matrix

$$\mathbf{H_B} = \begin{pmatrix} 0 & 2 & 1 \\ 2 & -2 & 0 \\ 1 & 0 & -2 \end{pmatrix}$$

If we use methods from Chap. 9 to find the determinant, we will find that

$$\Delta_B = 10 > 0$$

which implies that $L(x_1, x_2, \lambda)$ achieves a relative maximum when $x_1 = 1.6$, $x_2 = 0.8$, and $\lambda = -1.6$. If these values are substituted into the lagrangian function,

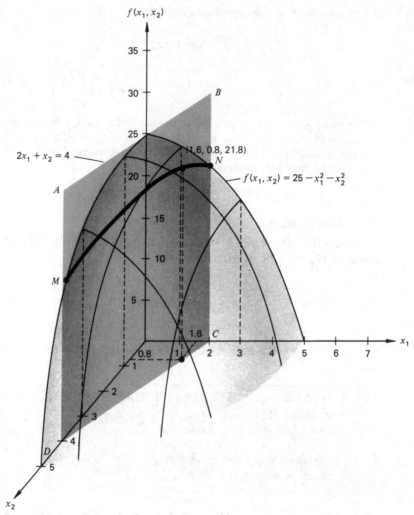

Figure 20.24   Constrained optimization problem.

$$L(1.6, 0.8, -1.6) = 25 - (1.6)^2 - (0.8)^2 - (-1.6)[2(1.6) + 0.8 - 4]$$
$$= 25 - 2.56 - 0.64 + 1.6(0) = 21.8$$

Therefore $L(x_1, x_2, \lambda)$ achieves a maximum value of 21.8. This is also the maximum value for $f(x_1, x_2)$ in the original constrained optimization problem of Example 25.

Figure 20.24 is a graphic portrayal of this problem. You may remember that the surface representing $f(x_1, x_2) = 25 - x_1^2 - x_2^2$ is the same as shown earlier in Fig. 20.4. Without the constraint, the relative maximum would occur at $(0, 0, 25)$. The constraint $2x_1 + x_2 = 4$ requires that the only values which can be considered lie on the intersection of the plane $ABCD$ and the surface representing f. Given the points of intersection (MN) between the surface $f(x_1, x_2)$ and the plane $ABCD$, the maximum value for f occurs at $(1.6, 0.8, 21.8)$.

$\square$

NOTE The structure of this problem is very much like that of Example 11 on page 798 in Chap. 17. Equation (17.12) is the objective function: Eq. (17.1) is a constraint. We solved that problem by solving for one variable in terms of the other in Eq. (17.13) and substituting into the objective function. This approach also can be used for Example 25. Why, then, the Lagrange multiplier approach? Example 25 is a relatively simple problem. The structure of the constraint(s) in a problem frequently do not allow for substitutions; hence, enter Lagrange!

n-Variable Single-Equality Constraint Case

Given a problem of the form

$$\text{Maximize (or minimize)} \quad y = f(x_1, x_2, \ldots, x_n)$$
$$\text{subject to} \quad g(x_1, x_2, \ldots, x_n) = k \tag{20.35}$$

the Lagrange multiplier method is just slightly different from the two independent variable case. The corresponding lagrangian function is

$$L(x_1, x_2, \ldots, x_n, \lambda) = f(x_1, x_2, \ldots, x_n) - \lambda[g(x_1, x_2, \ldots, x_n) - k] \tag{20.36}$$

NECESSARY CONDITION FOR RELATIVE EXTREMA

$$\begin{aligned} L_{x_1} &= 0 \\ L_{x_2} &= 0 \\ &\vdots \\ L_{x_n} &= 0 \\ L_\lambda &= 0 \end{aligned} \tag{20.37}$$

where $L_{x_1}, L_{x_2}, \ldots, L_{x_n}, L_\lambda$ all exist.

The bordered hessian matrix for the n-variable case has the form

$$\mathbf{H_B} = \begin{pmatrix} 0 & g_{x_1} & g_{x_2} & \cdots & g_{x_n} \\ g_{x_1} & L_{x_1 x_1} & L_{x_1 x_2} & \cdots & L_{x_1 x_n} \\ g_{x_2} & L_{x_2 x_1} & L_{x_2 x_2} & \cdots & L_{x_2 x_n} \\ \cdots\cdots\cdots\cdots\cdots\cdots\cdots\cdots\cdots\cdots\cdots \\ g_{x_n} & L_{x_n x_1} & L_{x_n x_2} & \cdots & L_{x_n x_n} \end{pmatrix} \tag{20.38}$$

Given the bordered hessian in Eq. (20.38), various submatrices are defined as follows:

$$\mathbf{H}_{B_2} = \begin{pmatrix} 0 & g_{x_1} & g_{x_2} \\ g_{x_1} & L_{x_1 x_1} & L_{x_1 x_2} \\ g_{x_2} & L_{x_2 x_1} & L_{x_2 x_2} \end{pmatrix}$$

$$\mathbf{H}_{B_3} = \begin{pmatrix} 0 & g_{x_1} & g_{x_2} & g_{x_3} \\ g_{x_1} & L_{x_1 x_1} & L_{x_1 x_2} & L_{x_1 x_3} \\ g_{x_2} & L_{x_2 x_1} & L_{x_2 x_2} & L_{x_2 x_3} \\ g_{x_3} & L_{x_3 x_1} & L_{x_3 x_2} & L_{x_3 x_3} \end{pmatrix}$$

$$\mathbf{H}_{B_n} = \mathbf{H}_B = \begin{pmatrix} 0 & g_{x_1} & g_{x_2} & \cdots & g_{x_n} \\ g_{x_1} & L_{x_1 x_1} & L_{x_1 x_2} & \cdots & L_{x_1 x_n} \\ g_{x_2} & L_{x_2 x_1} & L_{x_2 x_2} & \cdots & L_{x_2 x_n} \\ \cdots & \cdots & \cdots & \cdots & \cdots \\ g_{x_n} & L_{x_n x_1} & L_{x_n x_2} & \cdots & L_{x_n x_n} \end{pmatrix}$$

Principal minors for these submatrices can be denoted as $\Delta_{B_2}, \Delta_{B_3}, \ldots, \Delta_{B_n}$.

SUFFICIENT CONDITIONS FOR RELATIVE EXTREMA
Given critical values $x_1 = a_1$, $x_2 = a_2$, $\ldots$, $x_n = a_n$, and $\lambda = \lambda^*$ for which

$$L_{x_1} = L_{x_2} = \cdots = L_{x_n} = L_\lambda = 0$$

all principal minors associated with $\mathbf{H}_B$ are evaluated at the critical values.

I A relative maximum *exists if*

$$\Delta_{B_2} > 0,\ \Delta_{B_3} < 0,\ \Delta_{B_4} > 0,\ \ldots$$

II A relative minimum *exists if*

$$\Delta_{B_2} < 0,\ \Delta_{B_3} < 0,\ \ \Delta_{B_4} < 0,\ \ldots$$

EXAMPLE 27   Given the problem

$$\begin{aligned} \text{Maximize} &\quad f(x_1, x_2, x_3) = 5x_1 x_2 x_3 \\ \text{subject to} &\quad x_1 + 2x_2 + 3x_3 = 24 \end{aligned}$$

the corresponding lagrangian function is

$$L(x_1, x_2, x_3, \lambda) = 5x_1 x_2 x_3 - \lambda(x_1 + 2x_2 + 3x_3 - 24)$$

To locate any critical values, the first partial derivatives are found and set equal to 0.

$$L_{x_1} = 5x_2 x_3 - \lambda \qquad\qquad = 0$$

$$L_{x_2} = 5x_1x_3 - 2\lambda \qquad\qquad = 0$$

$$L_{x_3} = 5x_1x_2 - 3\lambda \qquad\qquad = 0$$

$$L_{x_3} = -x_1 - 2x_2 - 3x_3 + 24 = 0$$

These four equations can be rewritten as

$$5x_2x_3 = \lambda \qquad\qquad \text{(20.39)}$$

$$5x_1x_3 = 2\lambda \qquad\qquad \text{(20.40)}$$

$$5x_1x_2 = 3\lambda \qquad\qquad \text{(20.41)}$$

$$x_1 + 2x_2 + 3x_3 = 24 \qquad\qquad \text{(20.42)}$$

If both sides of Eq. (20.39) are divided by the corresponding side of Eq. (20.40),

$$\frac{5x_2x_3}{5x_1x_3} = \frac{\lambda}{2\lambda} \quad \text{or} \quad \frac{x_2}{x_1} = \frac{1}{2}$$

Thus
$$x_2 = \frac{x_1}{2} \qquad\qquad \text{(20.43)}$$

In a similar manner, both sides of Eq. (20.39) can be divided by both sides of Eq. (20.41):

$$\frac{5x_2x_3}{5x_1x_2} = \frac{\lambda}{3\lambda} \quad \text{or} \quad \frac{x_3}{x_1} = \frac{1}{3}$$

and
$$x_3 = \frac{x_1}{3} \qquad\qquad \text{(20.44)}$$

Substituting Eqs. (20.43) and (20.44) into Eq. (20.42),

$$x_1 + 2\frac{x_1}{2} + 3\frac{x_1}{3} = 24$$

$$3x_1 = 24$$

$$x_1 = 8$$

If this value is substituted into Eqs. (20.43), (20.44), and (20.38), critical values are identified for $L(x_1, x_2, x_3, \lambda)$ as $x_1 = 8$, $x_2 = 4$, $x_3 = \frac{8}{3}$, and $\lambda = \frac{160}{3}$, or $53\frac{1}{3}$.

To test the nature of this critical point, the bordered hessian is identified as

$$\mathbf{H}_B = \begin{pmatrix} 0 & 1 & 2 & 3 \\ 1 & 0 & 5x_3 & 5x_2 \\ 2 & 5x_3 & 0 & 5x_1 \\ 3 & 5x_2 & 5x_1 & 0 \end{pmatrix}$$

Evaluated at the critical values,

$$\mathbf{H}_B = \begin{pmatrix} 0 & 1 & 2 & 3 \\ 1 & 0 & \frac{40}{3} & 20 \\ 2 & \frac{40}{3} & 0 & 40 \\ 3 & 20 & 40 & 0 \end{pmatrix}$$

The bordered principal minors are

$$\Delta_{B_2} = \begin{vmatrix} 0 & 1 & 2 \\ 1 & 0 & \frac{40}{3} \\ 2 & \frac{40}{3} & 0 \end{vmatrix} = \frac{160}{3}$$

$$\Delta_{B_3} = \begin{vmatrix} 0 & 1 & 2 & 3 \\ 1 & 0 & \frac{40}{3} & 20 \\ 2 & \frac{40}{3} & 0 & 40 \\ 3 & 20 & 40 & 0 \end{vmatrix} = -4{,}800$$

Since $\Delta_{B_2} > 0$ and $\Delta_{B_3} < 0$, we can conclude that a relative maximum occurs for $L(x_1, x_2, x_3, \lambda)$ [and also for $f(x_1, x_2, x_3)$] when $x_1 = 8$, $x_2 = 4$, $x_3 = \frac{8}{3}$ and $\lambda = \frac{160}{3}$. The constrained maximum value is

$$5x_1 x_2 x_3 = 5(8)(4)(\tfrac{8}{3})$$
$$= \frac{1{,}280}{3} = 426\tfrac{2}{3}$$

□

Interpreting λ

Lambda is more than just an artificial creation allowing for the solution of constrained optimization problems. It has an interpretation which can be very useful. Given the generalized lagrangian function of Eq. (20.36),

$$\boxed{\frac{\partial L}{\partial k} = L_k = \lambda} \qquad (20.45)$$

Thus, λ can be interpreted as the instantaneous rate of change in the value of the lagrangian function with respect to a change in the right-hand-side constant k of the constraint equation. The value of $\lambda = \frac{160}{3}$ in the optimal solution to the previous example suggests that if the right-hand-side constant, 24, *increases* (decreases) by 1 unit, the optimal value for $f(x_1, x_2, x_3)$ will *increase* (decrease) by approximately $\frac{160}{3}$ units from the current maximum of $426\frac{2}{3}$.

The economic interpretation of λ can be of particular value in problems where the constraint(s) represent such things as limited resources. If there is an ability to provide additional resources, the λ values can offer guidance in allocating such resources.*

---

* For those who studied linear programming in Chaps. 10–12, λ is equivalent to a *shadow price*.

Extensions

The Lagrange approach can be extended to the case of multiple constraints and the case of constraint sets which include inequality as well as equality constraint types. This material, however, is beyond the scope of this text.

Section 20.6 Follow-up Exercises

In Exercises 1–8, examine the function for relative extrema and test for the nature of any extrema.

1. $f(x_1, x_2) = -3x_1^2 - 2x_2^2 + 20x_1 x_2$ subject to $x_1 + x_2 = 100$
2. $f(x_1, x_2) = x_1 x_2$ subject to $x_1 + x_2 = 6$
3. $f(x_1, x_2) = x_1^2 + 3x_1 x_2 - 6x_2$ subject to $x_1 + x_2 = 42$
4. $f(x_1, x_2) = 5x_1^2 + 6x_2^2 - x_1 x_2$ subject to $x_1 + 2x_2 = 24$
5. $f(x_1, x_2) = 12x_1 x_2 - 3x_2^2 - x_1^2$ subject to $x_1 + x_2 = 16$
6. $f(x_1, x_2, x_3) = x_1^2 + x_2^2 + x_3^2$ subject to $x_1 - x_2 + 2x_3 = 6$
7. $f(x_1, x_2, x_3) = x_1^2 + x_1 x_2 + 2x_2^2 + x_3^2$ subject to $x_1 - 3x_2 - 4x_3 = 16$
8. $f(x_1, x_2, x_3) = x_1 x_2 x_3$ subject to $x_1 + 2x_2 + 3x_3 = 18$
9. A company has received an order for 200 units of one of its products. The order will be supplied from the combined production of its two plants. The joint cost function for production of this particular product is

$$C = f(q_1, q_2) = 2q_1^2 + q_1 q_2 + q_2^2 + 500$$

where q_1 and q_2 equal the quantities produced at plants 1 and 2, respectively. If the objective is to minimize total costs subject to the requirement that 200 units be supplied from the two plants, what quantities should be supplied by each plant?

10. A factory manufactures two types of products. The joint cost function is

$$C = f(x_1, x_2) = x_1^2 + 2x_2^2 - x_1 x_2$$

where C is the weekly production cost stated in thousands of dollars and x_1 and x_2 equal the quantities produced of the two products each week. If combined weekly production is to equal 16 units, what quantities of each product will result in minimum total costs?

❏ KEY TERMS AND CONCEPTS

❏ IMPORTANT FORMULAS

$$D(x^*, y^*) = f_{xx}(x^*, y^*)f_{yy}(x^*, y^*) - [f_{xy}(x^*, y^*)]^2 \tag{20.13}$$

$$S = f(a, b) = \sum_{j=1}^{n} [y_j - (ax_j + b)]^2 \tag{20.23}$$

$$L(x_1, x_2, \lambda) = f(x_1, x_2) - \lambda[g(x_1, x_2) - k] \tag{20.28}$$

$$L(x_1, x_2, \ldots, x_n, \lambda) = f(x_1, x_2, \ldots, x_n) - \lambda[g(x_1, x_2, \ldots, x_n) - k] \tag{20.36}$$

❏ ADDITIONAL EXERCISES

Section 20.2

Determine f_x and f_y for the following functions.

1 $f(x, y) = 6x^2 + 8x^2y + y^3$

2 $f(x, y) = 6x^2y^4 - 8x^3y^2$

3 $f(x, y) = (5x^3 + 7y^3)^4$

4 $f(x, y) = \sqrt[3]{4xy^3 + 10x}$

5 $f(x, y) = (x^2 + y)/8xy^2$

6 $f(x, y) = 3x^2(x + y^2)^3$

7 $f(x, y) = e^{4xy}$

8 $f(x, y) = e^{x/y^2}$

9 $f(x, y) = 10x^4y^5$

10 $f(x, y) = 5x^2y^5$

11 $f(x, y) = x^3/4y^2$

12 $f(x, y) = -4y^3/x^4$

13 $f(x, y) = e^{x^2y}$

14 $f(x, y) = 10e^{4x^2y^3}$

15 $f(x, y) = \ln(x/y)$

16 $f(x, y) = \ln(x^3y^4)$

17 $f(x, y) = 20x^3/\ln y$

18 $f(x, y) = 40xy^3/\ln x$

Find all second-order partial derivatives for the following functions.

19 $f(x, y) = 10x^2y^3$

20 $f(x, y) = 5x^3 + 2xy^2 + 5y^2$

21 $f(x, y) = \ln xy$

22 $f(x, y) = e^{x^2+y^2}$

23 $f(x, y) = (3x - 2y)^4$

24 $f(x, y) = (5x - 3y)^4$

25 $f(x, y) = \sqrt{x - y}$

26 $f(x, y) = \sqrt{10x - 5y}$

27 $f(x, y) = 25x^4y$

28 $f(x, y) = x^5y^3/2$

Section 20.3

For the following functions, determine the location and nature of all critical points.

29 $f(x, y) = 2x^2 + y^2 + 80x + 40y + 10$

30 $f(x, y) = 2x^2 + xy - 9x - 2y^2 + 2y$

31 $f(x, y) = x^3/3 - 5x^2/2 + 6y^2 + 36y$

32 $f(x, y) = -4x^2 + 12xy + 44x - 12y^2 + 12y$

33 $f(x, y) = 3x^2 - 2xy + 3y^2 + 8x - 8y + 5$

34 $f(x, y) = -2x^2 + 16x - 75y + y^3 + 5$

35 $f(x, y) = x^3 + y^2 - 12x + 16y + 10$

36 $f(x, y) = x^2 - 2xy + 3y^2 + 4x - 36y + 22$

37 $f(x, y) = x^2 + 3y^2 + 4x - 9y + 10$

38 $f(x, y) = -4x^2 + 12x - 3y^2 + 36y - 5$

39 $f(x, y) = 2x^3 + y^3 - 3x^2 + 1.5y^2 - 12x - 90y$

40 $f(x, y) = xy - 1/x - 1/y$

41 $f(x, y) = 6x^2 + 30x + y^2 - 6y$

42 $f(x, y) = xy + 4\ln x + 2y^2 - 10, \qquad x > 0$

Section 20.4

43 A firm sells two products. The annual total revenue R behaves as a function of the number of units sold. Specifically,

$$R = 400x - 4x^2 + 1{,}960y - 8y^2$$

where x and y equal, respectively, the number of units sold of each product. The cost of producing the two products is

$$C = 100 + 2x^2 + 4y^2 + 2xy$$

    (*a*) Determine the number of units which should be produced and sold in order to maximize annual profit.

    (*b*) What does total revenue equal?

    (*c*) What do total costs equal?

    (*d*) What is the maximum profit?

44 A firm sells two products. The demand functions for the two products are

$$q_1 = 110 - 4p_1 - p_2$$
$$q_2 = 90 - 2p_1 - 3p_2$$

where p_j equals the price of product j in dollars and q_j equals the demand (in thousands of units) for product j.

    (*a*) Determine the prices which should be charged for each product in order to maximize total revenue from the two products.

    (*b*) How many units will be demanded of each product at these prices?

    (*c*) What is the maximum total revenue expected to equal?

45 Given the four data points $(-1, 12.5)$, $(3, 7.5)$, $(-4, 25)$, and $(10, -10)$, determine the equation of the line of best fit using the least squares model.

46 Given the data points $(10, 10)$, $(-8, 1)$, and $(2, 6)$, determine the equation of the line of best fit using the least squares model.

*47 A rectangular container is being designed which is to have a volume of 64,000 cubic inches. The objective is to minimize the amount of material used in constructing the container. Thus, the surface area is to be minimized. If x, y, and z represent the dimensions of the container (in inches), determine the dimensions which minimize the surface area. (*Hint:* $V = xyz$.)

Section 20.5

For the following functions, locate any critical points and determine their nature.

48 $f(x_1, x_2, x_3) = x_1^2 + x_2^2 + x_3^2 - 4x_1 - 8x_2 - 12x_3 + 56$

49 $f(x_1, x_2, x_3) = x_1^2 + 3x_2^2 + 3x_3^2 + 2x_1x_2 + 4x_2x_3 + 2x_1x_3$

50 $f(x_1, x_2, x_3) = x_1^2 + x_2^2 + x_3^2 + x_1x_2 + x_1x_3 + 4x_1 - 4x_2 + 8x_3$

51 $f(x_1, x_2, x_3) = 200 - x_1^2 - x_2^2 - 2x_3^2 + 20x_1 + 10x_2 + 20x_3$

Section 20.6

Examine the following functions for relative extrema and test for the nature of any extrema. What is the optimal value of λ?

52 $f(x_1, x_2) = 20x_1 + 10x_2 - x_1^2 - x_2^2$ subject to $x_1 + 2x_2 - 10 = 0$

53 $f(x_1, x_2) = -x_2^2 - 5x_1^2 + 4x_1x_2 + 16x_1 + 10x_2$ subject to $2x_1 + x_2 - 60 = 0$

54 $f(x_1, x_2) = 3x_1^2 + x_2^2 + 3x_1x_2 - 60x_1 - 32x_2 + 400$ subject to $x_1 + x_2 - 10 = 0$

55 $f(x_1, x_2) = 2x_2^2 - 6x_1^2$ subject to $2x_1 + x_2 - 4 = 0$

56 A cylindrical container is to be designed to hold 12 ounces, or 26 cubic inches, of liquid. Determine the dimensions (height and radius) which result in minimum surface area for the container. What is the minimum surface area? (Assume the container has both a top and a bottom.)

57 A firm estimates that its monthly profit is a function of the amount of money it spends on TV and radio advertising per month. The profit function is

$$P = f(x, y) = \frac{80x}{5 + x} + \frac{40y}{10 + y} - 2x - 2y$$

where P equals the monthly profit (in thousands of dollars) and x and y equal the monthly advertising expenditure for TV and radio, respectively (both in thousands of dollars). If the monthly advertising budget is $25,000, determine the amount which should be allocated to each medium in order to maximize monthly profit. What is the optimal value for λ? Interpret the meaning of this value.

*58 A warehouse which will have a volume of 850,000 cubic feet is to be constructed. The warehouse is to have a rectangular foundation of dimensions x feet by y feet and height of z feet. Construction costs are estimated based upon floor and ceiling area as well as wall area. The estimated costs are $6 per square foot of wall area, $8 per square foot of floor area, and $6 per square foot of ceiling area.
(a) Formulate the cost function for constructing the warehouse.
(b) Determine the building dimensions which result in minimum construction costs.
(c) What is the minimum cost?

❏ CHAPTER TEST

1 Give two interpretations of f_x.

2 What is a *trace*?

3 Determine f_x and f_y if

$$f(x, y) = 15x^3 - 4y^2 + 5x^2y^3$$

4 Determine all second-order partial derivatives for the function

$$f(x, y) = 8x^5 + 6x^2 + 8x^2y^3$$

5 Given the function

$$f(x, y) = 3x^2 - 4xy + 3y^2 + 8x - 17y + 5$$

(a) Locate any critical points and determine their nature.
(b) What is $f(x^*, y^*)$?

6 A researcher at a college of agriculture estimated that annual profit at a local farm can be described by the function

$$P = 1{,}600x + 2{,}400y - 2x^2 - 4y^2 - 4xy$$

where P equals annual profit in dollars, x equals the number of acres planted with soybeans, and y equals the number of acres planted with corn. Determine the number of acres of each crop which should be planted if the objective is to maximize annual profit. What is the expected maximum profit?

7 Locate any critical points and determine their nature for the function

$$f(x_1, x_2, x_3) = -5x_1^2 - 8x_2^2 - 3x_3^2 + 40x_1 - 40x_2 + 24x_3 + 100$$

8 Given the function

$$f(x_1, x_2) = -4x_1^3 + 3x_2^2 - 4x_1x_2$$

$$\text{subject to} \quad x_1 - 2x_2 - 20 = 0$$

formulate the lagrangian function.

BACK-ORDER INVENTORY MODEL

A variation of the classic EOQ model (Chap. 17 Minicase on page 824) allows for the possibility of shortages of inventory items. In this model, an inventory may be depleted while demand for the item continues to exist. This leads to shortages of the inventory item, and the model assumes that these items can be "back-ordered." When the next replenishment supply arrives, the back-ordered demand is filled first and the remaining items are placed in inventory. This type of inventory management is common for vendors such as those in the home furnishings industry. Although vendors incur additional costs by allowing shortages, the hope is to reduce carrying costs (by maintaining less inventory) and ordering costs (by ordering less frequently and in larger quantities).

Figure 20.25 illustrates a typical inventory cycle for this model. Constant demand results in a linear depletion of inventory from a maximum level of L. After t_1 time units, the inventory is depleted. Demand continues for a time t_2 before the replenishment supply of q units arrives. During t_2 the continued demand for the item results in a shortage of S units. Thus, upon arrival of the replenishment supply, S units must be allocated to cover the shortage.

If

❏ D = annual demand in units

❏ C_0 = ordering cost per order

❏ C_h = carrying cost per item per year

❏ C_s = shortage cost per item per year

❏ S = maximum shortage (in units)

❏ q = order quantity

the relevant cost function is

$$TC = \frac{\text{annual ordering}}{\text{cost}} + \frac{\text{annual carrying}}{\text{cost}} + \frac{\text{annual shortage}}{\text{cost}}$$

or
$$TC = f(q, S) = \frac{D}{q} C_0 + \frac{(q - S)^2}{2q} C_h + \frac{S^2 C_s}{2q} \tag{20.46}$$

Requirements:

a. If $D = 600{,}000$, $C_o = \$100$, $C_h = \$0.25$, and $C_s = \$2$, determine the values of q and S which minimize total annual ordering, carrying, and shortage costs. What is the minimum cost? The maximum inventory level?

b. Using Eq. (20.46), show that the general expressions for q and S which result in minimum annual inventory cost are

$$q^* = \sqrt{\frac{2DC_o}{C_h}\left(\frac{C_h + C_s}{C_s}\right)}$$

and

$$S^* = \sqrt{\frac{2C_o D C_h}{C_h C_s + C_s^2}}$$

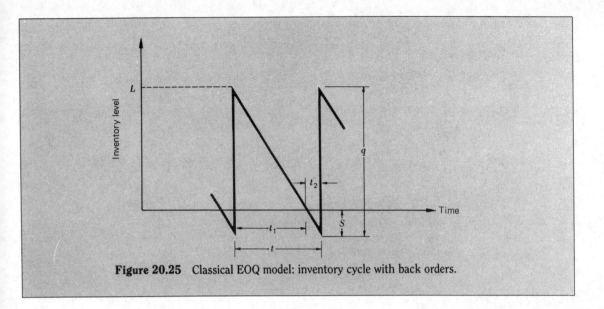

Figure 20.25 Classical EOQ model: inventory cycle with back orders.

COMPOUND INTEREST TABLES

Table I   Compound-Amount Factor $(1 + i)^n$

Table II   Present-Value Factor $(1 + i)^{-n}$

Table III   Series Compound-Amount Factor $\dfrac{(1 + i)^n - 1}{i} = s_{\overline{n}|i}$

Table IV   Sinking Fund Factor $\dfrac{i}{(1 + i)^n - 1} = \dfrac{1}{s_{\overline{n}|i}}$

Table V   Series Present-Worth Factor $\dfrac{(1 + i)^n - 1}{i(1 + i)^n} = a_{\overline{n}|i}$

Table VI   Capital-Recovery Factor $\dfrac{i(1 + i)^n}{(1 + i)^n - 1} = \dfrac{1}{a_{\overline{n}|i}}$

Table VII   Monthly Payment per Dollar of Mortgage Loan

TABLE I Compound-Amount Factor $(1 + i)^n$

n:	0.01 (1%)	0.015 (1½%)	0.02 (2%)	0.025 (2½%)	0.03 (3%)	0.035 (3½%)
1	1.01000	1.01500	1.02000	1.02500	1.03000	1.03500
2	1.02010	1.03022	1.04040	1.05062	1.06090	1.07122
3	1.03030	1.04568	1.06121	1.07689	1.09273	1.10872
4	1.04060	1.06136	1.08243	1.10381	1.12551	1.14752
5	1.05101	1.07728	1.10408	1.13141	1.15927	1.18769
6	1.06152	1.09344	1.12616	1.15969	1.19405	1.22926
7	1.07214	1.10984	1.14869	1.18869	1.22987	1.27228
8	1.08286	1.12649	1.17166	1.21840	1.26677	1.31681
9	1.09369	1.14339	1.19509	1.24886	1.30477	1.36290
10	1.10462	1.16054	1.21899	1.28008	1.34392	1.41060
11	1.11567	1.17795	1.24337	1.31209	1.38423	1.45997
12	1.12683	1.19562	1.26824	1.34489	1.42576	1.51107
13	1.13809	1.21355	1.29361	1.37851	1.46853	1.56396
14	1.14947	1.23176	1.31948	1.41297	1.51259	1.61869
15	1.16097	1.25023	1.34587	1.44830	1.55797	1.67535
16	1.17258	1.26899	1.37279	1.48451	1.60471	1.73399
17	1.18430	1.28802	1.40024	1.52162	1.65285	1.79468
18	1.19615	1.30734	1.42825	1.55966	1.70243	1.85749
19	1.20811	1.32695	1.45681	1.59865	1.75351	1.92250
20	1.22019	1.34686	1.48595	1.63862	1.80611	1.98979
21	1.23239	1.36706	1.51567	1.67958	1.86029	2.05943
22	1.24472	1.38756	1.54598	1.72157	1.91610	2.13151
23	1.25716	1.40838	1.57690	1.76461	1.97359	2.20611
24	1.26973	1.42950	1.60844	1.80873	2.03279	2.28333
25	1.28243	1.45095	1.64061	1.85394	2.09378	2.36324
26	1.29526	1.47271	1.67342	1.90029	2.15659	2.44596
27	1.30821	1.49480	1.70689	1.94780	2.22129	2.53157
28	1.32129	1.51722	1.74102	1.99650	2.28793	2.62017
29	1.33450	1.53998	1.77584	2.04641	2.35657	2.71188
30	1.34785	1.56308	1.81136	2.09757	2.42726	2.80679
31	1.36133	1.58653	1.84759	2.15001	2.50008	2.90503
32	1.37494	1.61032	1.88454	2.20376	2.57508	3.00671
33	1.38869	1.63448	1.92223	2.25885	2.65234	3.11194
34	1.40258	1.65900	1.96068	2.31532	2.73191	3.22086
35	1.41660	1.68388	1.99989	2.37321	2.81386	3.33359
36	1.43077	1.70914	2.03989	2.43254	2.89828	3.45027
37	1.44508	1.73478	2.08069	2.49335	2.98523	3.57103
38	1.45953	1.76080	2.12230	2.55568	3.07478	3.69601
39	1.47412	1.78721	2.16474	2.61957	3.16703	3.82537
40	1.48886	1.81402	2.20804	2.68506	3.26204	3.95926
41	1.50375	1.84123	2.25220	2.75219	3.35990	4.09783
42	1.51879	1.86885	2.29724	2.82100	3.46070	4.24126
43	1.53398	1.89688	2.34319	2.89152	3.56452	4.38970
44	1.54932	1.92533	2.39005	2.96381	3.67145	4.54334
45	1.56481	1.95421	2.43785	3.03790	3.78160	4.70236
46	1.58046	1.98353	2.48661	3.11385	3.89504	4.86694
47	1.59626	2.01328	2.53634	3.19170	4.01190	5.03728
48	1.61223	2.04348	2.58707	3.27149	4.13225	5.21359
49	1.62835	2.07413	2.63881	3.35328	4.25622	5.39606
50	1.64463	2.10524	2.69159	3.43711	4.38391	5.58493
51	1.66108	2.13682	2.74542	3.52304	4.51542	5.78040
52	1.67769	2.16887	2.80033	3.61111	4.65089	5.98271
53	1.69447	2.20141	2.85633	3.70139	4.79041	6.19211
54	1.71141	2.23443	2.91346	3.79392	4.93412	6.40883
55	1.72852	2.26794	2.97173	3.88877	5.08215	6.63314
56	1.74581	2.30196	3.03117	3.98599	5.23461	6.86530
57	1.76327	2.33649	3.09179	4.08564	5.39165	7.10559
58	1.78090	2.37154	3.15362	4.18778	5.55340	7.35428
59	1.79871	2.40711	3.21670	4.29248	5.72000	7.61168
60	1.81670	2.44322	3.28103	4.39979	5.89160	7.87809

TABLE I   Compound-Amount Factor $(1 + i)^n$ *(Continued)*

n:	0.04 (4%)	0.045 (4½%)	0.05 (5%)	0.06 (6%)	0.07 (7%)	0.08 (8%)
1	1.04000	1.04500	1.05000	1.06000	1.07000	1.08000
2	1.08160	1.09202	1.10250	1.12360	1.14490	1.16640
3	1.12486	1.14117	1.15762	1.19102	1.22504	1.25971
4	1.16986	1.19252	1.21551	1.26248	1.31080	1.36049
5	1.21665	1.24618	1.27628	1.33823	1.40255	1.46933
6	1.26532	1.30226	1.34010	1.41852	1.50073	1.58687
7	1.31593	1.36086	1.40710	1.50363	1.60578	1.71382
8	1.36857	1.42210	1.47746	1.59385	1.71819	1.85093
9	1.42331	1.48610	1.55133	1.68948	1.83846	1.99900
10	1.48024	1.55297	1.62889	1.79085	1.96715	2.15892
11	1.53945	1.62285	1.71034	1.89830	2.10485	2.33164
12	1.60103	1.69588	1.79586	2.01220	2.25219	2.51817
13	1.66507	1.77220	1.88565	2.13293	2.40985	2.71962
14	1.73168	1.85194	1.97993	2.26090	2.57853	2.93719
15	1.80094	1.93528	2.07893	2.39656	2.75903	3.17217
16	1.87298	2.02237	2.18287	2.54035	2.95216	3.42594
17	1.94790	2.11338	2.29202	2.69277	3.15882	3.70002
18	2.02582	2.20848	2.40662	2.85434	3.37993	3.99602
19	2.10685	2.30786	2.52695	3.02560	3.61653	4.31570
20	2.19112	2.41171	2.65330	3.20714	3.86968	4.66096
21	2.27877	2.52024	2.78596	3.39956	4.14056	5.03383
22	2.36992	2.63365	2.92526	3.60354	4.43040	5.43654
23	2.46472	2.75217	3.07152	3.81975	4.74053	5.87146
24	2.56330	2.87601	3.22510	4.04893	5.07237	6.34118
25	2.66584	3.00543	3.38635	4.29187	5.42743	6.84848
26	2.77247	3.14068	3.55567	4.54938	5.80735	7.39635
27	2.88337	3.28201	3.73346	4.82235	6.21387	7.98806
28	2.99870	3.42970	3.92013	5.11169	6.64884	8.62711
29	3.11865	3.58404	4.11614	5.41839	7.11426	9.31727
30	3.24340	3.74532	4.32194	5.74349	7.61226	10.06266
31	3.37313	3.91386	4.53804	6.08810	8.14511	10.86767
32	3.50806	4.08998	4.76494	6.45339	8.71527	11.73708
33	3.64838	4.27403	5.00319	6.84059	9.32534	12.67605
34	3.79432	4.46636	5.25335	7.25103	9.97811	13.69013
35	3.94609	4.66735	5.51602	7.68609	10.67658	14.78534
36	4.10393	4.87738	5.79182	8.14725	11.42394	15.96817
37	4.26809	5.09686	6.08141	8.63609	12.22362	17.24563
38	4.43881	5.32622	6.38548	9.15425	13.07927	18.62528
39	4.61637	5.56590	6.70475	9.70351	13.99482	20.11530
40	4.80102	5.81636	7.03999	10.28572	14.97446	21.72452
41	4.99306	6.07810	7.39199	10.90286	16.02267	23.46248
42	5.19278	6.35162	7.76159	11.55703	17.14426	25.33948
43	5.40050	6.63744	8.14967	12.25045	18.34435	27.36664
44	5.61652	6.93612	8.55715	12.98548	19.62846	29.55597
45	5.84118	7.24825	8.98501	13.76461	21.00245	31.92045
46	6.07482	7.57442	9.43426	14.59049	22.47262	34.47409
47	6.31782	7.91527	9.90597	15.46592	24.04571	37.23201
48	6.57053	8.27146	10.40127	16.39387	25.72891	40.21057
49	6.83335	8.64367	10.92133	17.37750	27.52993	43.42742
50	7.10668	9.03264	11.46740	18.42015	29.45703	46.90161
51	7.39095	9.43910	12.04077	19.52536	31.51902	50.65374
52	7.68659	9.86386	12.64281	20.69689	33.72535	54.70604
53	7.99405	10.30774	13.27495	21.93870	36.08612	59.08252
54	8.31381	10.77159	13.93870	23.25502	38.61215	63.80913
55	8.64637	11.25631	14.63563	24.65032	41.31500	68.91386
56	8.99222	11.76284	15.36741	26.12934	44.20705	74.42696
57	9.35191	12.29217	16.13578	27.69710	47.30155	80.38112
58	9.72599	12.84532	16.94257	29.35893	50.61265	86.81161
59	10.11503	13.42336	17.78970	31.12046	54.15554	93.75654
60	10.51963	14.02741	18.67919	32.98769	57.94643	101.25706

TABLE I Compound-Amount Factor $(1 + i)^n$ *(Continued)*

n:	0.09 (9%)	0.10 (10%)	0.11 (11%)	0.12 (12%)	0.13 (13%)	0.14 (14%)
1	1.09000	1.10000	1.11000	1.12000	1.13000	1.14000
2	1.18810	1.21000	1.23210	1.25440	1.27690	1.29960
3	1.29503	1.33100	1.36763	1.40493	1.44290	1.48154
4	1.41158	1.46410	1.51807	1.57352	1.63047	1.68896
5	1.53862	1.61051	1.68506	1.76234	1.84244	1.92541
6	1.67710	1.77156	1.87041	1.97382	2.08195	2.19497
7	1.82804	1.94872	2.07616	2.21068	2.35261	2.50227
8	1.99256	2.14359	2.30454	2.47596	2.65844	2.85259
9	2.17189	2.35795	2.55804	2.77308	3.00404	3.25195
10	2.36736	2.59374	2.83942	3.10585	3.39457	3.70722
11	2.58043	2.85312	3.15176	3.47855	3.83586	4.22623
12	2.81266	3.13843	3.49845	3.89598	4.33452	4.81790
13	3.06580	3.45227	3.88328	4.36349	4.89801	5.49241
14	3.34173	3.79750	4.31044	4.88711	5.53475	6.26135
15	3.64248	4.17725	4.78459	5.47357	6.25427	7.13794
16	3.97031	4.59497	5.31089	6.13039	7.06733	8.13725
17	4.32763	5.05447	5.89509	6.86604	7.98608	9.27646
18	4.71712	5.55992	6.54355	7.68997	9.02427	10.57517
19	5.14166	6.11591	7.26334	8.61276	10.19742	12.05569
20	5.60441	6.72750	8.06231	9.64629	11.52309	13.74349
21	6.10881	7.40025	8.94917	10.80385	13.02109	15.66758
22	6.65860	8.14027	9.93357	12.10031	14.71383	17.86104
23	7.25787	8.95430	11.02627	13.55235	16.62663	20.36158
24	7.91108	9.84973	12.23916	15.17863	18.78809	23.21221
25	8.62308	10.83471	13.58546	17.00006	21.23054	26.46192
26	9.39916	11.91818	15.07986	19.04007	23.99051	30.16658
27	10.24508	13.10999	16.73865	21.32488	27.10928	34.38991
28	11.16714	14.42099	18.57990	23.88387	30.63349	39.20449
29	12.17218	15.86309	20.62369	26.74993	34.61584	44.69312
30	13.26768	17.44940	22.89230	29.95992	39.11590	50.95016
31	14.46177	19.19434	25.41045	33.55511	44.20096	58.08318
32	15.76333	21.11378	28.20560	37.58173	49.94709	66.21483
33	17.18203	23.22515	31.30821	42.09153	56.44021	75.48490
34	18.72841	25.54767	34.75212	47.14252	63.77744	86.05279
35	20.41397	28.10244	38.57485	52.79962	72.06851	98.10018
36	22.25123	30.91268	42.81808	59.13557	81.43741	111.83420
37	24.25384	34.00395	47.52807	66.23184	92.02428	127.49099
38	26.43668	37.40434	52.75616	74.17966	103.98743	145.33973
39	28.81598	41.14478	58.55934	83.08122	117.50580	165.68729
40	31.40942	45.25926	65.00087	93.05097	132.78155	188.88351
41	34.23627	49.78518	72.15096	104.21709	150.04315	215.32721
42	37.31753	54.76370	80.08757	116.72314	169.54876	245.47301
43	40.67611	60.24007	88.89720	130.72991	191.59010	279.83924
44	44.33696	66.26408	98.67589	146.41750	216.49682	319.01673
45	48.32729	72.89048	109.53024	163.98760	244.64140	363.67907
46	52.67674	80.17953	121.57857	183.66612	276.44478	414.59414
47	57.41765	88.19749	134.95221	205.70605	312.38261	472.63732
48	62.58524	97.01723	149.79695	230.39078	352.99234	538.80655
49	68.21791	106.71896	166.27462	258.03767	398.88135	614.23946
50	74.35752	117.39085	184.56483	289.00219	450.73593	700.23299
51	81.04970	129.12994	204.86696	323.68245	509.33160	798.26561
52	88.34417	142.04293	227.40232	362.52435	575.54470	910.02279
53	96.29514	156.24723	252.41658	406.02727	650.36551	1037.42598
54	104.96171	171.87195	280.18240	454.75054	734.91303	1182.66562
55	114.40826	189.05914	311.00247	509.32061	830.45173	1348.23881
56	124.70501	207.96506	345.21274	570.43908	938.41045	1536.99224
57	135.92846	228.76156	383.18614	638.89177	1060.40381	1752.17115
58	148.16202	251.63772	425.33661	715.55878	1198.25630	1997.47512
59	161.49660	276.80149	472.12364	801.42583	1354.02962	2277.12163
60	176.03129	304.48164	524.05724	897.59693	1530.05347	2595.91866

TABLE I Compound-Amount Factor $(1 + i)^n$ *(Continued)*

n	0.15 (15%)	0.16 (16%)	0.17 (17%)	0.18 (18%)	0.19 (19%)	0.20 (20%)
1	1.15000	1.16000	1.17000	1.18000	1.19000	1.20000
2	1.32250	1.34560	1.36890	1.39240	1.41610	1.44000
3	1.52087	1.56090	1.60161	1.64303	1.68516	1.72800
4	1.74901	1.81064	1.87389	1.93878	2.00534	2.07360
5	2.01136	2.10034	2.19245	2.28776	2.38635	2.48832
6	2.31306	2.43640	2.56516	2.69955	2.83976	2.98598
7	2.66002	2.82622	3.00124	3.18547	3.37932	3.58318
8	3.05902	3.27841	3.51145	3.75886	4.02139	4.29982
9	3.51788	3.80296	4.10840	4.43545	4.78564	5.15978
10	4.04556	4.41144	4.80683	5.23384	5.69468	6.19174
11	4.65239	5.11726	5.62399	6.17593	6.77667	7.43008
12	5.35025	5.93603	6.58007	7.28759	8.06424	8.91610
13	6.15279	6.88579	7.69868	8.59936	9.59645	10.69932
14	7.07571	7.98752	9.00745	10.14724	11.41977	12.83918
15	8.13706	9.26552	10.53872	11.97375	13.58953	15.40702
16	9.35762	10.74800	12.33030	14.12902	16.17154	18.48843
17	10.76126	12.46768	14.42646	16.67225	19.24413	22.18611
18	12.37545	14.46251	16.87895	19.67325	22.90052	26.62333
19	14.23177	16.77652	19.74838	23.21444	27.25162	31.94800
20	16.36654	19.46076	23.10560	27.39303	32.42942	38.33760
21	18.82152	22.57448	27.03355	32.32378	38.59101	46.00512
22	21.64475	26.18640	31.62925	38.14206	45.92331	55.20614
23	24.89146	30.37622	37.00623	45.00763	54.64873	66.24737
24	28.62518	35.23642	43.29729	53.10901	65.03199	79.49685
25	32.91895	40.87424	50.65783	62.66863	77.38807	95.39622
26	37.85680	47.41412	59.26966	73.94898	92.09181	114.47546
27	43.53531	55.00038	69.34550	87.25980	109.58925	137.37055
28	50.06561	63.80044	81.13423	102.96656	130.41121	164.84466
29	57.57545	74.00851	94.92705	121.50054	155.18934	197.81359
30	66.21177	85.84988	111.06465	143.37064	184.67531	237.37631
31	76.14354	99.58586	129.94564	169.17735	219.76362	284.85158
32	87.56507	115.51959	152.03640	199.62928	261.51871	341.82189
33	100.69983	134.00273	177.88259	235.56255	311.20726	410.18627
34	115.80480	155.44317	208.12263	277.96381	370.33664	492.22352
35	133.17552	180.31407	243.50347	327.99729	440.70061	590.66823
36	153.15185	209.16432	284.89906	387.03680	524.43372	708.80187
37	176.12463	242.63062	333.33191	456.70343	624.07613	850.56225
38	202.54332	281.45151	389.99833	538.91004	742.65059	1020.67470
39	232.92482	326.48376	456.29805	635.91385	883.75421	1224.80964
40	267.86355	378.72116	533.86871	750.37834	1051.66751	1469.77157
41	308.04308	439.31654	624.62639	885.44645	1251.48433	1763.72588
42	354.24954	509.60719	730.81288	1044.82681	1489.26636	2116.47106
43	407.38697	591.14434	855.05107	1232.89563	1772.22696	2539.76527
44	468.49502	685.72744	1000.40975	1454.81685	2108.95009	3047.71832
45	538.76927	795.44383	1170.47941	1716.68388	2509.65060	3657.26199
46	619.58466	922.71484	1369.46091	2025.68698	2986.48422	4388.71439
47	712.52236	1070.34921	1602.26927	2390.31063	3553.91622	5266.45726
48	819.40071	1241.60509	1874.65504	2820.56655	4229.16030	6319.74872
49	942.31082	1440.26190	2193.34640	3328.26853	5032.70076	7583.69846
50	1083.65744	1670.70380	2566.21528	3927.35686	5988.91390	9100.43815
51	1246.20606	1938.01641	3002.47188	4634.28109	7126.80754	10920.52578
52	1433.13697	2248.09904	3512.89210	5468.45169	8480.90098	13104.63094
53	1648.10751	2607.79488	4110.08376	6452.77300	10092.27216	15725.55712
54	1895.32364	3025.04207	4808.79800	7614.27214	12009.80387	18870.66855
55	2179.62218	3509.04880	5626.29366	8984.84112	14291.66661	22644.80226
56	2506.56551	4070.49660	6582.76358	10602.11252	17007.08327	27173.76271
57	2882.55034	4721.77606	7701.83339	12510.49278	20238.42909	32608.51525
58	3314.93289	5477.26023	9011.14507	14762.38148	24083.73061	39130.21830
59	3812.17282	6353.62187	10543.03973	17419.61014	28659.63943	46956.26196
60	4383.99875	7370.20137	12335.35648	20555.13997	34104.97092	56347.51435

TABLE II   Present-Value Factor $(1 + i)^{-n}$

n	0.01 (1%)	0.015 (1½%)	0.02 (2%)	0.025 (2½%)	0.03 (3%)	0.035 (3½%)
1	0.99010	0.98522	0.98039	0.97561	0.97087	0.96618
2	0.98030	0.97066	0.96117	0.95181	0.94260	0.93351
3	0.97059	0.95632	0.94232	0.92860	0.91514	0.90194
4	0.96098	0.94218	0.92385	0.90595	0.88849	0.87144
5	0.95147	0.92826	0.90573	0.88385	0.86261	0.84197
6	0.94205	0.91454	0.88797	0.86230	0.83748	0.81350
7	0.93272	0.90103	0.87056	0.84127	0.81309	0.78599
8	0.92348	0.88771	0.85349	0.82075	0.78941	0.75941
9	0.91434	0.87459	0.83676	0.80073	0.76642	0.73373
10	0.90529	0.86167	0.82035	0.78120	0.74409	0.70892
11	0.89632	0.84893	0.80426	0.76214	0.72242	0.68495
12	0.88745	0.83639	0.78849	0.74356	0.70138	0.66178
13	0.87866	0.82403	0.77303	0.72542	0.68095	0.63940
14	0.86996	0.81185	0.75788	0.70773	0.66112	0.61778
15	0.86135	0.79985	0.74301	0.69047	0.64186	0.59689
16	0.85282	0.78803	0.72845	0.67362	0.62317	0.57671
17	0.84438	0.77639	0.71416	0.65720	0.60502	0.55720
18	0.83602	0.76491	0.70016	0.64117	0.58739	0.53836
19	0.82774	0.75361	0.68643	0.62553	0.57029	0.52016
20	0.81954	0.74247	0.67297	0.61027	0.55368	0.50257
21	0.81143	0.73150	0.65978	0.59539	0.53755	0.48557
22	0.80340	0.72069	0.64684	0.58086	0.52189	0.46915
23	0.79544	0.71004	0.63416	0.56670	0.50669	0.45329
24	0.78757	0.69954	0.62172	0.55288	0.49193	0.43796
25	0.77977	0.68921	0.60953	0.53939	0.47761	0.42315
26	0.77205	0.67902	0.59758	0.52623	0.46369	0.40884
27	0.76440	0.66899	0.58586	0.51340	0.45019	0.39501
28	0.75684	0.65910	0.57437	0.50088	0.43708	0.38165
29	0.74934	0.64936	0.56311	0.48866	0.42435	0.36875
30	0.74192	0.63976	0.55207	0.47674	0.41199	0.35628
31	0.73458	0.63031	0.54125	0.46511	0.39999	0.34423
32	0.72730	0.62099	0.53063	0.45377	0.38834	0.33259
33	0.72010	0.61182	0.52023	0.44270	0.37703	0.32134
34	0.71297	0.60277	0.51003	0.43191	0.36604	0.31048
35	0.70591	0.59387	0.50003	0.42137	0.35538	0.29998
36	0.69892	0.58509	0.49022	0.41109	0.34503	0.28983
37	0.69200	0.57644	0.48061	0.40107	0.33498	0.28003
38	0.68515	0.56792	0.47119	0.39128	0.32523	0.27056
39	0.67837	0.55953	0.46195	0.38174	0.31575	0.26141
40	0.67165	0.55126	0.45289	0.37243	0.30656	0.25257
41	0.66500	0.54312	0.44401	0.36335	0.29763	0.24403
42	0.65842	0.53509	0.43530	0.35448	0.28896	0.23578
43	0.65190	0.52718	0.42677	0.34584	0.28054	0.22781
44	0.64545	0.51939	0.41840	0.33740	0.27237	0.22010
45	0.63905	0.51171	0.41020	0.32917	0.26444	0.21266
46	0.63273	0.50415	0.40215	0.32115	0.25674	0.20547
47	0.62646	0.49670	0.39427	0.31331	0.24926	0.19852
48	0.62026	0.48936	0.38654	0.30567	0.24200	0.19181
49	0.61412	0.48213	0.37896	0.29822	0.23495	0.18532
50	0.60804	0.47500	0.37153	0.29094	0.22811	0.17905
51	0.60202	0.46798	0.36424	0.28385	0.22146	0.17300
52	0.59606	0.46107	0.35710	0.27692	0.21501	0.16715
53	0.59016	0.45426	0.35010	0.27017	0.20875	0.16150
54	0.58431	0.44754	0.34323	0.26358	0.20267	0.15603
55	0.57853	0.44093	0.33650	0.25715	0.19677	0.15076
56	0.57280	0.43441	0.32991	0.25088	0.19104	0.14566
57	0.56713	0.42799	0.32344	0.24476	0.18547	0.14073
58	0.56151	0.42167	0.31710	0.23879	0.18007	0.13598
59	0.55595	0.41544	0.31088	0.23297	0.17483	0.13138
60	0.55045	0.40930	0.30478	0.22728	0.16973	0.12693

TABLE II Present-Value Factor $(1+i)^{-n}$ *(Continued)*

n	0.04 (4%)	0.045 (4½%)	0.05 (5%)	0.06 (6%)	0.07 (7%)	0.08 (8%)
1	0.96154	0.95694	0.95238	0.94340	0.93458	0.92593
2	0.92456	0.91573	0.90703	0.89000	0.87344	0.85734
3	0.88900	0.87630	0.86384	0.83962	0.81630	0.79383
4	0.85480	0.83856	0.82270	0.79209	0.76290	0.73503
5	0.82193	0.80245	0.78353	0.74726	0.71299	0.68058
6	0.79031	0.76790	0.74622	0.70496	0.66634	0.63017
7	0.75992	0.73483	0.71068	0.66506	0.62275	0.58349
8	0.73069	0.70319	0.67684	0.62741	0.58201	0.54027
9	0.70259	0.67290	0.64461	0.59190	0.54393	0.50025
10	0.67556	0.64393	0.61391	0.55839	0.50835	0.46319
11	0.64958	0.61620	0.58468	0.52679	0.47509	0.42888
12	0.62460	0.58966	0.55684	0.49697	0.44401	0.39711
13	0.60057	0.56427	0.53032	0.46884	0.41496	0.36770
14	0.57748	0.53997	0.50507	0.44230	0.38782	0.34046
15	0.55526	0.51672	0.48102	0.41727	0.36245	0.31524
16	0.53391	0.49447	0.45811	0.39365	0.33873	0.29189
17	0.51337	0.47318	0.43630	0.37136	0.31657	0.27027
18	0.49363	0.45280	0.41552	0.35034	0.29586	0.25025
19	0.47464	0.43330	0.39573	0.33051	0.27651	0.23171
20	0.45639	0.41464	0.37689	0.31180	0.25842	0.21455
21	0.43883	0.39679	0.35894	0.29416	0.24151	0.19866
22	0.42196	0.37970	0.34185	0.27751	0.22571	0.18394
23	0.40573	0.36335	0.32557	0.26180	0.21095	0.17032
24	0.39012	0.34770	0.31007	0.24698	0.19715	0.15770
25	0.37512	0.33273	0.29530	0.23300	0.18425	0.14602
26	0.36069	0.31840	0.28124	0.21981	0.17220	0.13520
27	0.34682	0.30469	0.26785	0.20737	0.16093	0.12519
28	0.33348	0.29157	0.25509	0.19563	0.15040	0.11591
29	0.32065	0.27902	0.24295	0.18456	0.14056	0.10733
30	0.30832	0.26700	0.23138	0.17411	0.13137	0.09938
31	0.29646	0.25550	0.22036	0.16425	0.12277	0.09202
32	0.28506	0.24450	0.20987	0.15496	0.11474	0.08520
33	0.27409	0.23397	0.19987	0.14619	0.10723	0.07889
34	0.26355	0.22390	0.19035	0.13791	0.10022	0.07305
35	0.25342	0.21425	0.18129	0.13011	0.09366	0.06763
36	0.24367	0.20503	0.17266	0.12274	0.08754	0.06262
37	0.23430	0.19620	0.16444	0.11579	0.08181	0.05799
38	0.22529	0.18775	0.15661	0.10924	0.07646	0.05369
39	0.21662	0.17967	0.14915	0.10306	0.07146	0.04971
40	0.20829	0.17193	0.14205	0.09722	0.06678	0.04603
41	0.20028	0.16453	0.13528	0.09172	0.06241	0.04262
42	0.19257	0.15744	0.12884	0.08653	0.05833	0.03946
43	0.18517	0.15066	0.12270	0.08163	0.05451	0.03654
44	0.17805	0.14417	0.11686	0.07701	0.05095	0.03383
45	0.17120	0.13796	0.11130	0.07265	0.04761	0.03133
46	0.16461	0.13202	0.10600	0.06854	0.04450	0.02901
47	0.15828	0.12634	0.10095	0.06466	0.04159	0.02686
48	0.15219	0.12090	0.09614	0.06100	0.03837	0.02487
49	0.14634	0.11569	0.09156	0.05755	0.03632	0.02303
50	0.14071	0.11071	0.08720	0.05429	0.03395	0.02132
51	0.13530	0.10594	0.08305	0.05122	0.03173	0.01974
52	0.13010	0.10138	0.07910	0.04832	0.02965	0.01828
53	0.12509	0.09701	0.07533	0.04558	0.02771	0.01693
54	0.12028	0.09284	0.07174	0.04300	0.02590	0.01567
55	0.11566	0.08884	0.06833	0.04057	0.02420	0.01451
56	0.11121	0.08501	0.06507	0.03827	0.02262	0.01344
57	0.10693	0.08135	0.06197	0.03610	0.02114	0.01244
58	0.10282	0.07785	0.05902	0.03406	0.01976	0.01152
59	0.09886	0.07450	0.05621	0.03213	0.01847	0.01067
60	0.09506	0.07129	0.05354	0.03031	0.01726	0.00988

TABLE II   Present-Value Factor $(1 + i)^{-n}$ (Continued)

n	0.09 (9%)	0.10 (10%)	0.11 (11%)	0.12 (12%)	0.13 (13%)	0.14 (14%)
1	0.91743	0.90909	0.90090	0.89286	0.88496	0.87719
2	0.84168	0.82645	0.81162	0.79719	0.78315	0.76947
3	0.77218	0.75131	0.73119	0.71178	0.69305	0.67497
4	0.70843	0.68301	0.65873	0.63552	0.61332	0.59208
5	0.64993	0.62092	0.59345	0.56743	0.54276	0.51937
6	0.59627	0.56447	0.53464	0.50663	0.48032	0.45559
7	0.54703	0.51316	0.48166	0.45235	0.42506	0.39964
8	0.50187	0.46651	0.43393	0.40388	0.37616	0.35056
9	0.46043	0.42410	0.39092	0.36061	0.33288	0.30751
10	0.42241	0.38554	0.35218	0.32197	0.29459	0.26974
11	0.38753	0.35049	0.31728	0.28748	0.26070	0.23662
12	0.35553	0.31863	0.28584	0.25668	0.23071	0.20756
13	0.32618	0.28966	0.25751	0.22917	0.20416	0.18207
14	0.29925	0.26333	0.23199	0.20462	0.18068	0.15971
15	0.27454	0.23939	0.20900	0.18270	0.15989	0.14010
16	0.25187	0.21763	0.18829	0.16312	0.14150	0.12289
17	0.23107	0.19784	0.16963	0.14564	0.12522	0.10780
18	0.21199	0.17986	0.15282	0.13004	0.11081	0.09456
19	0.19449	0.16351	0.13768	0.11611	0.09806	0.08295
20	0.17843	0.14864	0.12403	0.10367	0.08678	0.07276
21	0.16370	0.13513	0.11174	0.09256	0.07680	0.06383
22	0.15018	0.12285	0.10067	0.08264	0.06796	0.05599
23	0.13778	0.11168	0.09069	0.07379	0.06014	0.04911
24	0.12640	0.10153	0.08170	0.06588	0.05323	0.04308
25	0.11597	0.09230	0.07361	0.05882	0.04710	0.03779
26	0.10639	0.08391	0.06631	0.05252	0.04168	0.03315
27	0.09761	0.07628	0.05974	0.04689	0.03689	0.02908
28	0.08955	0.06934	0.05382	0.04187	0.03264	0.02551
29	0.08215	0.06304	0.04849	0.03738	0.02889	0.02237
30	0.07537	0.05731	0.04368	0.03338	0.02557	0.01963
31	0.06915	0.05210	0.03935	0.02980	0.02262	0.01722
32	0.06344	0.04736	0.03545	0.02661	0.02002	0.01510
33	0.05820	0.04306	0.03194	0.02376	0.01772	0.01325
34	0.05339	0.03914	0.02878	0.02121	0.01568	0.01162
35	0.04899	0.03558	0.02592	0.01894	0.01388	0.01019
36	0.04494	0.03235	0.02335	0.01691	0.01228	0.00894
37	0.04123	0.02941	0.02104	0.01510	0.01087	0.00784
38	0.03783	0.02673	0.01896	0.01348	0.00962	0.00688
39	0.03470	0.02430	0.01708	0.01204	0.00851	0.00604
40	0.03184	0.02209	0.01538	0.01075	0.00753	0.00529
41	0.02921	0.02009	0.01386	0.00960	0.00666	0.00464
42	0.02680	0.01826	0.01249	0.00857	0.00590	0.00407
43	0.02458	0.01660	0.01125	0.00765	0.00522	0.00357
44	0.02255	0.01509	0.01013	0.00683	0.00462	0.00313
45	0.02069	0.01372	0.00913	0.00610	0.00409	0.00275
46	0.01898	0.01247	0.00823	0.00544	0.00362	0.00241
47	0.01742	0.01134	0.00741	0.00486	0.00320	0.00212
48	0.01598	0.01031	0.00668	0.00434	0.00283	0.00186
49	0.01466	0.00937	0.00601	0.00388	0.00251	0.00163
50	0.01345	0.00852	0.00542	0.00346	0.00222	0.00143
51	0.01234	0.00774	0.00488	0.00309	0.00196	0.00125
52	0.01132	0.00704	0.00440	0.00276	0.00174	0.00110
53	0.01038	0.00640	0.00396	0.00246	0.00154	0.00096
54	0.00953	0.00582	0.00357	0.00220	0.00136	0.00085
55	0.00874	0.00529	0.00322	0.00196	0.00120	0.00074
56	0.00802	0.00481	0.00290	0.00175	0.00107	0.00065
57	0.00736	0.00437	0.00261	0.00157	0.00094	0.00057
58	0.00675	0.00397	0.00235	0.00140	0.00083	0.00050
59	0.00619	0.00361	0.00212	0.00125	0.00074	0.00044
60	0.00568	0.00328	0.00191	0.00111	0.00065	0.00039

TABLE II Present-Value Factor $(1 + i)^{-n}$ *(Continued)*

n	0.15 (15%)	0.16 (16%)	0.17 (17%)	0.18 (18%)	0.19 (19%)	0.20 (20%)
1	0.86957	0.86207	0.85470	0.84746	0.84034	0.83333
2	0.75614	0.74316	0.73051	0.71818	0.70616	0.69444
3	0.65752	0.64066	0.62437	0.60863	0.59342	0.57870
4	0.57175	0.55229	0.53365	0.51579	0.49867	0.48225
5	0.49718	0.47611	0.45611	0.43711	0.41905	0.40188
6	0.43233	0.41044	0.38984	0.37043	0.35214	0.33490
7	0.37594	0.35383	0.33320	0.31393	0.29592	0.27908
8	0.32690	0.30503	0.28478	0.26604	0.24867	0.23257
9	0.28426	0.26295	0.24340	0.22546	0.20897	0.19381
10	0.24718	0.22668	0.20804	0.19106	0.17560	0.16151
11	0.21494	0.19542	0.17781	0.16192	0.14757	0.13459
12	0.18691	0.16846	0.15197	0.13722	0.12400	0.11216
13	0.16253	0.14523	0.12989	0.11629	0.10421	0.09346
14	0.14133	0.12520	0.11102	0.09855	0.08757	0.07789
15	0.12289	0.10793	0.09489	0.08352	0.07359	0.06491
16	0.10686	0.09304	0.08110	0.07078	0.06184	0.05409
17	0.09293	0.08021	0.06932	0.05998	0.05196	0.04507
18	0.08081	0.06914	0.05925	0.05083	0.04367	0.03756
19	0.07027	0.05961	0.05064	0.04308	0.03670	0.03130
20	0.06110	0.05139	0.04328	0.03651	0.03084	0.02608
21	0.05313	0.04430	0.03699	0.03094	0.02591	0.02174
22	0.04620	0.03819	0.03162	0.02622	0.02178	0.01811
23	0.04017	0.03292	0.02702	0.02222	0.01830	0.01509
24	0.03493	0.02838	0.02310	0.01883	0.01538	0.01258
25	0.03038	0.02447	0.01974	0.01596	0.01292	0.01048
26	0.02642	0.02109	0.01687	0.01352	0.01086	0.00874
27	0.02297	0.01818	0.01442	0.01146	0.00912	0.00728
28	0.01997	0.01567	0.01233	0.00971	0.00767	0.00607
29	0.01737	0.01351	0.01053	0.00823	0.00644	0.00506
30	0.01510	0.01165	0.00900	0.00697	0.00541	0.00421
31	0.01313	0.01004	0.00770	0.00591	0.00455	0.00351
32	0.01142	0.00866	0.00658	0.00501	0.00382	0.00293
33	0.00993	0.00746	0.00562	0.00425	0.00321	0.00244
34	0.00864	0.00643	0.00480	0.00360	0.00270	0.00203
35	0.00751	0.00555	0.00411	0.00305	0.00227	0.00169
36	0.00653	0.00478	0.00351	0.00258	0.00191	0.00141
37	0.00568	0.00412	0.00300	0.00219	0.00160	0.00118
38	0.00494	0.00355	0.00256	0.00186	0.00135	0.00098
39	0.00429	0.00306	0.00219	0.00157	0.00113	0.00082
40	0.00373	0.00264	0.00187	0.00133	0.00095	0.00068
41	0.00325	0.00228	0.00160	0.00113	0.00080	0.00057
42	0.00282	0.00196	0.00137	0.00096	0.00067	0.00047
43	0.00245	0.00169	0.00117	0.00081	0.00056	0.00039
44	0.00213	0.00146	0.00100	0.00069	0.00047	0.00033
45	0.00186	0.00126	0.00085	0.00058	0.00040	0.00027
46	0.00161	0.00108	0.00073	0.00049	0.00033	0.00023
47	0.00140	0.00093	0.00062	0.00042	0.00028	0.00019
48	0.00122	0.00081	0.00053	0.00035	0.00024	0.00016
49	0.00106	0.00069	0.00046	0.00030	0.00020	0.00013
50	0.00092	0.00060	0.00039	0.00025	0.00017	0.00011
51	0.00080	0.00052	0.00033	0.00022	0.00014	0.00009
52	0.00070	0.00044	0.00028	0.00018	0.00012	0.00008
53	0.00061	0.00038	0.00024	0.00015	0.00010	0.00006
54	0.00053	0.00033	0.00021	0.00013	0.00008	0.00005
55	0.00046	0.00028	0.00018	0.00011	0.00007	0.00004
56	0.00040	0.00025	0.00015	0.00009	0.00006	0.00004
57	0.00035	0.00021	0.00013	0.00008	0.00005	0.00003
58	0.00030	0.00018	0.00011	0.00007	0.00004	0.00003
59	0.00026	0.00016	0.00009	0.00006	0.00003	0.00002
60	0.00023	0.00014	0.00008	0.00005	0.00003	0.00002

TABLE III   Series Compound-Amount Factor $\dfrac{(1+i)^n - 1}{i} = s_{\overline{n}|i}$

			i			
n	0.01 (1%)	0.015 (1½%)	0.02 (2%)	0.025 (2½%)	0.03 (3%)	0.035 (3½%)
1	1.00000	1.00000	1.00000	1.00000	1.00000	1.00000
2	2.01000	2.01500	2.02000	2.02500	2.03000	2.03500
3	3.03010	3.04522	3.06040	3.07562	3.09090	3.10622
4	4.06040	4.09090	4.12161	4.15252	4.18363	4.21494
5	5.10101	5.15227	5.20404	5.25633	5.30914	5.36247
6	6.15202	6.22955	6.30812	6.38774	6.46841	6.55015
7	7.21354	7.32299	7.43428	7.54743	7.66246	7.77941
8	8.28567	8.43284	8.58297	8.73612	8.89234	9.05169
9	9.36853	9.55933	9.75463	9.95452	10.15911	10.36850
10	10.46221	10.70272	10.94972	11.20338	11.46388	11.73139
11	11.56683	11.86326	12.16872	12.48347	12.80780	13.14199
12	12.68250	13.04121	13.41209	13.79555	14.19203	14.60196
13	13.80933	14.23683	14.68033	15.14044	15.61779	16.11303
14	14.94742	15.45038	15.97394	16.51895	17.08632	17.67699
15	16.09690	16.68214	17.29342	17.93193	18.59891	19.29568
16	17.25786	17.93237	18.63929	19.38022	20.15688	20.97103
17	18.43044	19.20136	20.01207	20.86473	21.76159	22.70502
18	19.61475	20.48938	21.41231	22.38635	23.41444	24.49969
19	20.81090	21.79672	22.84056	23.94601	25.11687	26.35718
20	22.01900	23.12367	24.29737	25.54466	26.87037	28.27968
21	23.23919	24.47052	25.78332	27.18327	28.67649	30.26947
22	24.47159	25.83758	27.29898	28.86286	30.53678	32.32890
23	25.71630	27.22514	28.84496	30.58443	32.45288	34.46041
24	26.97346	28.63352	30.42186	32.34904	34.42647	36.66653
25	28.24320	30.06302	32.03030	34.15776	36.45926	38.94986
26	29.52563	31.51397	33.67091	36.01171	38.55304	41.31310
27	30.82089	32.98668	35.34432	37.91200	40.70963	43.75906
28	32.12910	34.48148	37.05121	39.85980	42.93092	46.29063
29	33.45039	35.99870	38.79223	41.85630	45.21885	48.91080
30	34.78489	37.53868	40.56808	43.90270	47.57542	51.62268
31	36.13274	39.10176	42.37944	46.00027	50.00268	54.42947
32	37.49407	40.68829	44.22703	48.15028	52.50276	57.33450
33	38.86901	42.29861	46.11157	50.35403	55.07784	60.34121
34	40.25770	43.93309	48.03380	52.61289	57.73018	63.45315
35	41.66028	45.59209	49.99448	54.92821	60.46208	66.67401
36	43.07688	47.27597	51.99437	57.30141	63.27594	70.00760
37	44.50765	48.98511	54.03425	59.73395	66.17422	73.45787
38	45.95272	50.71989	56.11494	62.22730	69.15945	77.02889
39	47.41225	52.48068	58.23724	64.78298	72.23423	80.72491
40	48.88637	54.26789	60.40198	67.40255	75.40126	84.55028
41	50.37524	56.08191	62.61002	70.08762	78.66330	88.50954
42	51.87899	57.92314	64.86222	72.83981	82.02320	92.60737
43	53.39778	59.79199	67.15947	75.66080	85.48389	96.84863
44	54.93176	61.68887	69.50266	78.55232	89.04841	101.23833
45	56.48107	63.61420	71.89271	81.51613	92.71986	105.78167
46	58.04589	65.56841	74.33056	84.55403	96.50146	110.48403
47	59.62634	67.55194	76.81718	87.66789	100.39650	115.35097
48	61.22261	69.56522	79.35352	90.85958	104.40840	120.38826
49	62.83483	71.60870	81.94059	94.13107	108.54065	125.60185
50	64.46318	73.68283	84.57940	97.48435	112.79687	130.99791
51	66.10781	75.78807	87.27099	100.92146	117.18077	136.58284
52	67.76889	77.92489	90.01641	104.44449	121.69620	142.36324
53	69.44658	80.09376	92.81674	108.05561	126.34708	148.34595
54	71.14105	82.29517	95.67307	111.75700	131.13749	154.53806
55	72.85246	84.52960	98.58653	115.55092	136.07162	160.94689
56	74.58098	86.79754	101.55826	119.43969	141.15377	167.58003
57	76.32679	89.09951	104.58943	123.42569	146.38838	174.44533
58	78.09006	91.43600	107.68122	127.51133	151.78003	181.55092
59	79.87096	93.80754	110.83484	131.69911	157.33343	188.90520
60	81.66967	96.21465	114.05154	135.99159	163.05344	196.51688

TABLE III   Series Compound-Amount Factor $\dfrac{(1+i)^n - 1}{i} = s_{\overline{n}|i}$ (Continued)

n	0.04 (4%)	0.045 (4½%)	0.05 (5%)	0.06 (6%)	0.07 (7%)	0.08 (8%)
1	1.00000	1.00000	1.00000	1.00000	1.00000	1.00000
2	2.04000	2.04500	2.05000	2.06000	2.07000	2.08000
3	3.12160	3.13702	3.15250	3.18360	3.21490	3.24640
4	4.24646	4.27819	4.31012	4.37462	4.43994	4.50611
5	5.41632	5.47071	5.52563	5.63709	5.75074	5.86660
6	6.63298	6.71689	6.80191	6.97532	7.15329	7.33593
7	7.89829	8.01915	8.14201	8.39384	8.65402	8.92280
8	9.21423	9.38001	9.54911	9.89747	10.25980	10.63663
9	10.58280	10.80211	11.02656	11.49132	11.97799	12.48756
10	12.00611	12.28821	12.57789	13.18079	13.81645	14.48656
11	13.48635	13.84118	14.20679	14.97164	15.78360	16.64549
12	15.02581	15.46403	15.91713	16.86994	17.88845	18.97713
13	16.62684	17.15991	17.71298	18.88214	20.14064	21.49530
14	18.29191	18.93211	19.59863	21.01507	22.55049	24.21492
15	20.02359	20.78405	21.57856	23.27597	25.12902	27.15211
16	21.82453	22.71934	23.65749	25.67253	27.88805	30.32428
17	23.69751	24.74171	25.84037	28.21288	30.84022	33.75023
18	25.64541	26.85508	28.13238	30.90565	33.99903	37.45024
19	27.67123	29.06356	30.53900	33.75999	37.37896	41.44626
20	29.77808	31.37142	33.06595	36.78559	40.99549	45.76196
21	31.96920	33.78314	35.71925	39.99273	44.86518	50.42292
22	34.24797	36.30338	38.50521	43.39229	49.00574	55.45676
23	36.61789	38.93703	41.43048	46.99583	53.43614	60.89330
24	39.08260	41.68920	44.50200	50.81558	58.17667	66.76476
25	41.64591	44.56521	47.72710	54.86451	63.24904	73.10594
26	44.31174	47.57064	51.11345	59.15638	68.67647	79.95442
27	47.08421	50.71132	54.66913	63.70577	74.48382	87.35077
28	49.96758	53.99333	58.40258	68.52811	80.69769	95.33883
29	52.96629	57.42303	62.32271	73.63980	87.34653	103.96594
30	56.08494	61.00707	66.43885	79.05819	94.46079	113.28321
31	59.32834	64.75239	70.76079	84.80168	102.07304	123.34587
32	62.70147	68.66625	75.29883	90.88978	110.21815	134.21354
33	66.20953	72.75623	80.06377	97.34316	118.93343	145.95062
34	69.85791	77.03026	85.06696	104.18375	128.25876	158.62667
35	73.65222	81.49662	90.32031	111.43478	138.23688	172.31680
36	77.59831	86.16397	95.83632	119.12087	148.91346	187.10215
37	81.70225	91.04134	101.62814	127.26812	160.33740	203.07032
38	85.97034	96.13820	107.70955	135.90421	172.56102	220.31595
39	90.40915	101.46442	114.09502	145.05846	185.64029	238.94122
40	95.02552	107.03032	120.79977	154.76197	199.63511	259.05652
41	99.82654	112.84669	127.83976	165.04768	214.60957	280.78104
42	104.81960	118.92479	135.23175	175.95054	230.63224	304.24352
43	110.01238	125.27640	142.99334	187.50758	247.77650	329.58301
44	115.41288	131.91384	151.14301	199.75803	266.12085	356.94965
45	121.02939	138.84997	159.70016	212.74351	285.74931	386.50562
46	126.87057	146.09821	168.68516	226.50812	306.75176	418.42607
47	132.94539	153.67263	178.11942	241.09861	329.22439	452.90015
48	139.26321	161.58790	188.02539	256.56453	353.27009	490.13216
49	145.83373	169.85936	198.42666	272.95840	378.99900	530.34274
50	152.66708	178.50303	209.34800	290.33590	406.52893	573.77016
51	159.77377	187.53566	220.81540	308.75606	435.98595	620.67177
52	167.16472	196.97477	232.85617	328.28142	467.50497	671.32551
53	174.85131	206.83863	245.49897	348.97831	501.23032	726.03155
54	182.84536	217.14637	258.77392	370.91701	537.31644	785.11408
55	191.15917	227.91796	272.71262	394.17203	575.92859	848.92320
56	199.80554	239.17427	287.34825	418.82235	617.24359	917.83706
57	208.79776	250.93711	302.71566	444.95169	661.45065	992.26402
58	218.14967	263.22928	318.85144	472.64879	708.75219	1072.64514
59	227.87566	276.07460	335.79402	502.00772	759.36484	1159.45676
60	237.99069	289.49795	353.58372	533.12818	813.52038	1253.21330

TABLE III   Series Compound-Amount Factor $\dfrac{(1+i)^n - 1}{i} = s_{\overline{n}|i}$ (Continued)

n	0.09 (9%)	0.10 (10%)	0.11 (11%)	0.12 (12%)	0.13 (13%)	0.14 (14%)
1	1.00000	1.00000	1.00000	1.00000	1.00000	1.00000
2	2.09000	2.10000	2.11000	2.12000	2.13000	2.14000
3	3.27810	3.31000	3.34210	3.37440	3.40690	3.43960
4	4.57313	4.64100	4.70973	4.77933	4.84980	4.92114
5	5.98471	6.10510	6.22780	6.35285	6.48027	6.61010
6	7.52333	7.71561	7.91286	8.11519	8.32271	8.53552
7	9.20043	9.48717	9.78327	10.08901	10.40466	10.73049
8	11.02847	11.43589	11.85943	12.29969	12.75726	13.23276
9	13.02104	13.57948	14.16397	14.77566	15.41571	16.08535
10	15.19293	15.93742	16.72201	17.54874	18.41975	19.33730
11	17.56029	18.53117	19.56143	20.65458	21.81432	23.04452
12	20.14072	21.38428	22.71319	24.13313	25.65018	27.27075
13	22.95338	24.52271	26.21164	28.02911	29.98470	32.08865
14	26.01919	27.97498	30.09492	32.39260	34.88271	37.58107
15	29.36092	31.77248	34.40536	37.27971	40.41746	43.84251
16	33.00340	35.94973	39.18995	42.75328	46.67173	50.98035
17	36.97370	40.54470	44.50084	48.88367	53.73906	59.11760
18	41.30134	45.59917	50.39594	55.74971	61.72514	68.39407
19	46.01846	51.15909	56.93949	63.43968	70.74941	78.96923
20	51.16012	57.27500	64.20283	72.05244	80.94683	91.02493
21	56.76453	64.00250	72.26514	81.69874	92.46992	104.76842
22	62.87334	71.40275	81.21431	92.50258	105.49101	120.43600
23	69.53194	79.54302	91.14788	104.60289	120.20484	138.29704
24	76.78981	88.49733	102.17415	118.15524	136.83147	158.65862
25	84.70090	98.34706	114.41331	133.33387	155.61956	181.87083
26	93.32398	109.18177	127.99877	150.33393	176.85010	208.33274
27	102.72313	121.09994	143.07864	169.37401	200.84061	238.49933
28	112.96822	134.20994	159.81729	190.69889	227.94989	272.88923
29	124.13536	148.63093	178.39719	214.58275	258.58338	312.09373
30	136.30754	164.49402	199.02088	241.33268	293.19922	356.78685
31	149.57522	181.94342	221.91317	271.29261	332.31511	407.73701
32	164.03699	201.13777	247.32362	304.84772	376.51608	465.82019
33	179.80032	222.25154	275.52922	342.42945	426.46317	532.03501
34	196.98234	245.47670	306.83744	384.52098	482.90338	607.51991
35	215.71075	271.02437	341.58955	431.66350	546.68082	693.57270
36	236.12472	299.12681	380.16441	484.46312	618.74933	791.67288
37	258.37595	330.03949	422.98249	543.59869	700.18674	903.50708
38	282.62978	364.04343	470.51056	609.83053	792.21101	1030.99808
39	309.06646	401.44778	523.26673	684.01020	896.19845	1176.33781
40	337.88245	442.59256	581.82607	767.09142	1013.70424	1342.02510
41	369.29187	487.85181	646.82693	860.14239	1146.48579	1530.90861
42	403.52813	537.63699	718.97790	964.35948	1296.52895	1746.23582
43	440.84566	592.40069	799.06547	1081.08262	1466.07771	1991.70883
44	481.52177	652.64076	887.96267	1211.81253	1657.66781	2271.54807
45	525.85873	718.90484	986.63856	1358.23003	1874.16463	2590.56480
46	574.18602	791.79532	1096.16880	1522.21764	2118.80603	2954.24387
47	626.86276	871.97485	1217.74737	1705.88375	2395.25082	3368.83801
48	684.28041	960.17234	1352.69958	1911.58980	2707.63342	3841.47534
49	746.86565	1057.18957	1502.49653	2141.98058	3060.62577	4380.28188
50	815.08356	1163.90853	1668.77115	2400.01825	3459.50712	4994.52135
51	889.44108	1281.29938	1853.33598	2689.02044	3910.24304	5694.75433
52	970.49077	1410.42932	2058.20294	3012.70289	4419.57464	6493.01994
53	1058.83494	1552.47225	2285.60526	3375.22724	4995.11934	7403.04273
54	1155.13009	1708.71948	2538.02184	3781.25451	5645.48485	8440.46872
55	1260.09180	1880.59142	2818.20424	4236.00505	6380.39789	9623.13434
56	1374.50006	2069.65057	3129.20671	4745.32565	7210.84961	10971.37314
57	1499.20506	2277.61562	3474.41944	5315.76473	8149.26006	12508.36538
58	1635.13352	2506.37719	3857.60558	5954.65650	9209.66387	14260.53654
59	1783.29553	2758.01490	4282.94220	6670.21528	10407.92017	16258.01165
60	1944.79213	3034.81640	4755.06584	7471.64111	11761.94979	18535.13328

TABLE III **Series Compound-Amount Factor** $\dfrac{(1+i)^n - 1}{i} = s_{\overline{n}|i}$ **(Continued)**

	i					
n	0.15 (15%)	0.16 (16%)	0.17 (17%)	0.18 (18%)	0.19 (19%)	0.20 (20%)
1	1.00000	1.00000	1.00000	1.00000	1.00000	1.00000
2	2.15000	2.16000	2.17000	2.18000	2.19000	2.20000
3	3.47250	3.50560	3.53890	3.57240	3.60610	3.64000
4	4.99337	5.06650	5.14051	5.21543	5.29126	5.36800
5	6.74238	6.87714	7.01440	7.15421	7.29660	7.44160
6	8.75374	8.97748	9.20685	9.44197	9.68295	9.92992
7	11.06680	11.41387	11.77201	12.14152	12.52271	12.91590
8	13.72682	14.24009	14.77325	15.32700	15.90203	16.49908
9	16.78584	17.51851	18.28471	19.08585	19.92341	20.79890
10	20.30372	21.32147	22.39311	23.52131	24.70886	25.95868
11	24.34928	25.73290	27.19994	28.75514	30.40355	32.15042
12	29.00167	30.85017	32.82393	34.93107	37.18022	39.58050
13	34.35192	36.78620	39.40399	42.21866	45.24446	48.49660
14	40.50471	43.67199	47.10267	50.81802	54.84091	59.19592
15	47.58041	51.65951	56.11013	60.96527	66.26068	72.03511
16	55.71747	60.92503	66.64885	72.93901	79.85021	87.44213
17	65.07509	71.67303	78.97915	87.06804	96.02175	105.93056
18	75.83636	84.14072	93.40561	103.74028	115.26588	128.11667
19	88.21181	98.60323	110.28456	123.41353	138.16640	154.74000
20	102.44358	115.37975	130.03294	146.62797	165.41802	186.68800
21	118.81012	134.84051	153.13854	174.02100	197.84744	225.02560
22	137.63164	157.41499	180.17209	206.34479	236.43846	271.03072
23	159.27638	183.60138	211.80134	244.48685	282.36176	326.23686
24	184.16784	213.97761	248.80757	289.49448	337.01050	392.48424
25	212.79302	249.21402	292.10486	342.60349	402.04249	471.98108
26	245.71197	290.08827	342.76268	405.27211	479.43056	567.37730
27	283.56877	337.50239	402.03234	479.22109	571.52237	681.85276
28	327.10408	392.50277	471.37783	566.48089	681.11162	819.22331
29	377.16969	456.30322	552.51207	669.44745	811.52283	984.06797
30	434.74515	530.31172	647.43912	790.94799	966.71217	1181.88157
31	500.95692	616.16161	758.50377	934.31863	1151.38748	1419.25788
32	577.10046	715.74746	888.44941	1103.49598	1371.15110	1704.10946
33	664.66552	831.26706	1040.48581	1303.12526	1632.66981	2045.93136
34	765.36535	965.26979	1218.36839	1538.68781	1943.87708	2456.11762
35	881.17016	1120.71295	1426.49102	1816.65161	2314.21372	2948.34115
36	1014.34568	1301.02703	1669.99450	2144.64890	2754.91433	3539.00937
37	1167.49753	1510.19135	1954.89356	2531.68570	3279.34805	4247.81125
38	1343.62216	1752.82197	2288.22547	2988.38913	3903.42418	5098.37350
39	1546.16549	2034.27348	2678.22379	3527.29918	4646.07477	6119.04820
40	1779.09031	2360.75724	3134.52184	4163.21303	5529.82898	7343.85784
41	2046.95385	2739.47840	3668.39055	4913.59137	6581.49649	8813.62941
42	2354.99693	3178.79494	4293.01695	5799.03782	7832.98082	10577.35529
43	2709.24647	3688.40213	5023.82983	6843.86463	9322.24718	12693.82635
44	3116.63344	4279.54648	5878.88090	8076.76026	11094.47414	15233.59162
45	3585.12846	4965.27391	6879.29065	9531.57711	13203.42423	18281.30994
46	4123.89773	5760.71774	8049.77006	11248.26098	15713.07483	21938.57193
47	4743.48239	6683.43257	9419.23097	13273.94796	18699.55905	26327.28631
48	5456.00475	7753.78179	11021.50024	15664.25859	22253.47527	31593.74358
49	6275.40546	8995.38687	12896.15528	18484.82514	26482.63557	37913.49229
50	7217.71628	10435.64877	15089.50167	21813.09367	31515.33633	45497.19075
51	8301.37372	12106.35258	17655.71696	25740.45053	37504.25023	54597.62890
52	9547.57978	14044.36899	20658.18884	30374.73162	44631.05777	65518.15468
53	10980.71674	16292.46803	24171.08094	35843.18331	53111.95875	78622.78562
54	12628.82425	18900.26291	28281.16470	42295.95631	63204.23091	94348.34274
55	14524.14789	21925.30498	33089.96270	49910.22844	75214.03479	113219.01129
56	16703.77008	25434.35377	38716.25636	58895.06957	89505.70140	135863.81354
57	19210.33559	29504.85038	45299.01994	69497.18209	106512.78466	163037.57625
58	22092.88593	34226.62644	53000.85333	82007.67486	126751.21375	195646.09150
59	25407.81882	39703.88667	62011.99840	96770.05634	150834.94436	234776.30980
60	29219.99164	46057.50853	72555.03813	114189.66648	179494.58379	281723.57177

TABLE IV   Sinking Fund Factor $\dfrac{i}{(1+i)^n - 1} = \dfrac{1}{s_{\overline{n}|i}}$

| | | | | i | | |
n	0.01 (1%)	0.015 (1½%)	0.02 (2%)	0.025 (2½%)	0.03 (3%)	0.035 (3½%)
1	1.00000	1.00000	1.00000	1.00000	1.00000	1.00000
2	0.49751	0.49628	0.49505	0.49383	0.49261	0.49140
3	0.33002	0.32838	0.32675	0.32514	0.32353	0.32193
4	0.24628	0.24444	0.24262	0.24082	0.23903	0.23725
5	0.19604	0.19409	0.19216	0.19025	0.18835	0.18648
6	0.16255	0.16053	0.15853	0.15655	0.15460	0.15267
7	0.13863	0.13656	0.13451	0.13250	0.13051	0.12854
8	0.12069	0.11858	0.11651	0.11447	0.11246	0.11048
9	0.10674	0.10461	0.10252	0.10046	0.09843	0.09645
10	0.09558	0.09343	0.09133	0.08926	0.08723	0.08524
11	0.08645	0.08429	0.08218	0.08011	0.07808	0.07609
12	0.07885	0.07668	0.07456	0.07249	0.07046	0.06848
13	0.07241	0.07024	0.06812	0.06605	0.06403	0.06206
14	0.06690	0.06472	0.06260	0.06054	0.05853	0.05657
15	0.06212	0.05994	0.05783	0.05577	0.05377	0.05183
16	0.05794	0.05577	0.05365	0.05160	0.04961	0.04768
17	0.05426	0.05208	0.04997	0.04793	0.04595	0.04404
18	0.05098	0.04881	0.04670	0.04467	0.04271	0.04082
19	0.04805	0.04588	0.04378	0.04176	0.03981	0.03794
20	0.04542	0.04325	0.04116	0.03915	0.03722	0.03536
21	0.04303	0.04087	0.03878	0.03679	0.03487	0.03304
22	0.04086	0.03870	0.03663	0.03465	0.03275	0.03093
23	0.03889	0.03673	0.03467	0.03270	0.03081	0.02902
24	0.03707	0.03492	0.03287	0.03091	0.02905	0.02727
25	0.03541	0.03326	0.03122	0.02928	0.02743	0.02567
26	0.03387	0.03173	0.02970	0.02777	0.02594	0.02421
27	0.03245	0.03032	0.02829	0.02638	0.02456	0.02285
28	0.03112	0.02900	0.02699	0.02509	0.02329	0.02160
29	0.02990	0.02778	0.02578	0.02389	0.02211	0.02045
30	0.02875	0.02664	0.02465	0.02278	0.02102	0.01937
31	0.02768	0.02557	0.02360	0.02174	0.02000	0.01837
32	0.02667	0.02458	0.02261	0.02077	0.01905	0.01744
33	0.02573	0.02364	0.02169	0.01986	0.01816	0.01657
34	0.02484	0.02276	0.02082	0.01901	0.01732	0.01576
35	0.02400	0.02193	0.02000	0.01821	0.01654	0.01500
36	0.02321	0.02115	0.01923	0.01745	0.01580	0.01428
37	0.02247	0.02041	0.01851	0.01674	0.01511	0.01361
38	0.02176	0.01972	0.01782	0.01607	0.01446	0.01298
39	0.02109	0.01905	0.01717	0.01544	0.01384	0.01239
40	0.02046	0.01843	0.01656	0.01484	0.01326	0.01183
41	0.01985	0.01783	0.01597	0.01427	0.01271	0.01130
42	0.01928	0.01726	0.01542	0.01373	0.01219	0.01080
43	0.01873	0.01672	0.01489	0.01322	0.01170	0.01033
44	0.01820	0.01621	0.01439	0.01273	0.01123	0.00988
45	0.01771	0.01572	0.01391	0.01227	0.01079	0.00945
46	0.01723	0.01525	0.01345	0.01183	0.01036	0.00905
47	0.01677	0.01480	0.01302	0.01141	0.00996	0.00867
48	0.01633	0.01437	0.01260	0.01101	0.00958	0.00831
49	0.01591	0.01396	0.01220	0.01062	0.00921	0.00796
50	0.01551	0.01357	0.01182	0.01026	0.00887	0.00763
51	0.01513	0.01319	0.01146	0.00991	0.00853	0.00732
52	0.01476	0.01283	0.01111	0.00957	0.00822	0.00702
53	0.01440	0.01249	0.01077	0.00925	0.00791	0.00674
54	0.01406	0.01215	0.01045	0.00895	0.00763	0.00647
55	0.01373	0.01183	0.01014	0.00865	0.00735	0.00621
56	0.01341	0.01152	0.00985	0.00837	0.00708	0.00597
57	0.01310	0.01122	0.00956	0.00810	0.00683	0.00573
58	0.01281	0.01094	0.00929	0.00784	0.00659	0.00551
59	0.01252	0.01066	0.00902	0.00759	0.00636	0.00529
60	0.01224	0.01039	0.00877	0.00735	0.00613	0.00509

TABLE IV Sinking Fund Factor $\dfrac{i}{(1+i)^n - 1} = \dfrac{1}{s_{\overline{n}|i}}$ *(Continued)*

n	0.04 (4%)	0.045 (4½%)	0.05 (5%)	0.06 (6%)	0.07 (7%)	0.08 (8%)
1	1.00000	1.00000	1.00000	1.00000	1.00000	1.00000
2	0.49020	0.48900	0.48780	0.48544	0.48309	0.48077
3	0.32035	0.31877	0.31721	0.31411	0.31105	0.30803
4	0.23549	0.23374	0.23201	0.22859	0.22523	0.22192
5	0.18463	0.18279	0.18097	0.17740	0.17389	0.17046
6	0.15076	0.14888	0.14702	0.14336	0.13980	0.13632
7	0.12661	0.12470	0.12282	0.11914	0.11555	0.11207
8	0.10853	0.10661	0.10472	0.10104	0.09747	0.09401
9	0.09449	0.09257	0.09069	0.08702	0.08349	0.08008
10	0.08329	0.08138	0.07950	0.07587	0.07238	0.06903
11	0.07415	0.07225	0.07039	0.06679	0.06336	0.06008
12	0.06655	0.06467	0.06283	0.05928	0.05590	0.05270
13	0.06014	0.05828	0.05646	0.05296	0.04965	0.04652
14	0.05467	0.05282	0.05102	0.04758	0.04434	0.04130
15	0.04994	0.04811	0.04634	0.04296	0.03979	0.03683
16	0.04582	0.04402	0.04227	0.03895	0.03586	0.03298
17	0.04220	0.04042	0.03870	0.03544	0.03243	0.02963
18	0.03899	0.03724	0.03555	0.03236	0.02941	0.02670
19	0.03614	0.03441	0.03275	0.02962	0.02675	0.02413
20	0.03358	0.03188	0.03024	0.02718	0.02439	0.02185
21	0.03128	0.02960	0.02800	0.02500	0.02229	0.01983
22	0.02920	0.02755	0.02597	0.02305	0.02041	0.01803
23	0.02731	0.02568	0.02414	0.02128	0.01871	0.01642
24	0.02559	0.02399	0.02247	0.01968	0.01719	0.01498
25	0.02401	0.02244	0.02095	0.01823	0.01581	0.01368
26	0.02257	0.02102	0.01956	0.01690	0.01456	0.01251
27	0.02124	0.01972	0.01829	0.01570	0.01343	0.01145
28	0.02001	0.01852	0.01712	0.01459	0.01239	0.01049
29	0.01888	0.01741	0.01605	0.01358	0.01145	0.00962
30	0.01783	0.01639	0.01505	0.01265	0.01059	0.00883
31	0.01686	0.01544	0.01413	0.01179	0.00980	0.00811
32	0.01595	0.01456	0.01328	0.01100	0.00907	0.00745
33	0.01510	0.01374	0.01249	0.01027	0.00841	0.00685
34	0.01431	0.01298	0.01176	0.00960	0.00780	0.00630
35	0.01358	0.01227	0.01107	0.00897	0.00723	0.00580
36	0.01289	0.01161	0.01043	0.00839	0.00672	0.00534
37	0.01224	0.01098	0.00984	0.00786	0.00624	0.00492
38	0.01163	0.01040	0.00928	0.00736	0.00580	0.00454
39	0.01106	0.00986	0.00876	0.00689	0.00539	0.00419
40	0.01052	0.00934	0.00828	0.00646	0.00501	0.00386
41	0.01002	0.00886	0.00782	0.00606	0.00466	0.00356
42	0.00954	0.00841	0.00739	0.00568	0.00434	0.00329
43	0.00909	0.00798	0.00699	0.00533	0.00404	0.00303
44	0.00866	0.00758	0.00662	0.00501	0.00376	0.00280
45	0.00826	0.00720	0.00626	0.00470	0.00350	0.00259
46	0.00788	0.00684	0.00593	0.00441	0.00326	0.00239
47	0.00752	0.00651	0.00561	0.00415	0.00304	0.00221
48	0.00718	0.00619	0.00532	0.00390	0.00283	0.00204
49	0.00686	0.00589	0.00504	0.00366	0.00264	0.00189
50	0.00655	0.00560	0.00478	0.00344	0.00246	0.00174
51	0.00626	0.00533	0.00453	0.00324	0.00229	0.00161
52	0.00598	0.00508	0.00429	0.00305	0.00214	0.00149
53	0.00572	0.00483	0.00407	0.00287	0.00200	0.00138
54	0.00547	0.00461	0.00386	0.00270	0.00186	0.00127
55	0.00523	0.00439	0.00367	0.00254	0.00174	0.00118
56	0.00500	0.00418	0.00348	0.00239	0.00162	0.00109
57	0.00479	0.00399	0.00330	0.00225	0.00151	0.00101
58	0.00458	0.00380	0.00314	0.00212	0.00141	0.00093
59	0.00439	0.00362	0.00298	0.00199	0.00132	0.00086
60	0.00420	0.00345	0.00283	0.00188	0.00123	0.00080

TABLE IV Sinking Fund Factor $\dfrac{i}{(1+i)^n-1}=\dfrac{1}{s_{\overline{n}|i}}$ (Continued)

n	0.09 (9%)	0.10 (10%)	0.11 (11%)	0.12 (12%)	0.13 (13%)	0.14 (14%)
1	1.00000	1.00000	1.00000	1.00000	1.00000	1.00000
2	0.47847	0.47619	0.47393	0.47170	0.46948	0.46729
3	0.30505	0.30211	0.29921	0.29635	0.29352	0.29073
4	0.21867	0.21547	0.21233	0.20923	0.20619	0.20320
5	0.16709	0.16380	0.16057	0.15741	0.15431	0.15128
6	0.13292	0.12961	0.12638	0.12323	0.12015	0.11716
7	0.10869	0.10541	0.10222	0.09912	0.09611	0.09319
8	0.09067	0.08744	0.08432	0.08130	0.07839	0.07557
9	0.07680	0.07364	0.07060	0.06768	0.06487	0.06217
10	0.06582	0.06275	0.05980	0.05698	0.05429	0.05171
11	0.05695	0.05396	0.05112	0.04842	0.04584	0.04339
12	0.04965	0.04676	0.04403	0.04144	0.03899	0.03667
13	0.04357	0.04078	0.03815	0.03568	0.03335	0.03116
14	0.03843	0.03575	0.03323	0.03087	0.02867	0.02661
15	0.03406	0.03147	0.02907	0.02682	0.02474	0.02281
16	0.03030	0.02782	0.02552	0.02339	0.02143	0.01962
17	0.02705	0.02466	0.02247	0.02046	0.01861	0.01692
18	0.02421	0.02193	0.01984	0.01794	0.01620	0.01462
19	0.02173	0.01955	0.01756	0.01576	0.01413	0.01266
20	0.01955	0.01746	0.01558	0.01388	0.01235	0.01099
21	0.01762	0.01562	0.01384	0.01224	0.01081	0.00954
22	0.01590	0.01401	0.01231	0.01081	0.00948	0.00830
23	0.01438	0.01257	0.01097	0.00956	0.00832	0.00723
24	0.01302	0.01130	0.00979	0.00846	0.00731	0.00630
25	0.01181	0.01017	0.00874	0.00750	0.00643	0.00550
26	0.01072	0.00916	0.00781	0.00665	0.00565	0.00480
27	0.00973	0.00826	0.00699	0.00590	0.00498	0.00419
28	0.00885	0.00745	0.00626	0.00524	0.00439	0.00366
29	0.00806	0.00673	0.00561	0.00466	0.00387	0.00320
30	0.00734	0.00608	0.00502	0.00414	0.00341	0.00280
31	0.00669	0.00550	0.00451	0.00369	0.00301	0.00245
32	0.00610	0.00497	0.00404	0.00328	0.00266	0.00215
33	0.00556	0.00450	0.00363	0.00292	0.00234	0.00188
34	0.00508	0.00407	0.00326	0.00260	0.00207	0.00165
35	0.00464	0.00369	0.00293	0.00232	0.00183	0.00144
36	0.00424	0.00334	0.00263	0.00206	0.00162	0.00126
37	0.00387	0.00303	0.00236	0.00184	0.00143	0.00111
38	0.00354	0.00275	0.00213	0.00164	0.00126	0.00097
39	0.00324	0.00249	0.00191	0.00146	0.00112	0.00085
40	0.00296	0.00226	0.00172	0.00130	0.00099	0.00075
41	0.00271	0.00205	0.00155	0.00116	0.00087	0.00065
42	0.00248	0.00186	0.00139	0.00104	0.00077	0.00057
43	0.00227	0.00169	0.00125	0.00092	0.00068	0.00050
44	0.00208	0.00153	0.00113	0.00083	0.00060	0.00044
45	0.00190	0.00139	0.00101	0.00074	0.00053	0.00039
46	0.00174	0.00126	0.00091	0.00066	0.00047	0.00034
47	0.00160	0.00115	0.00082	0.00059	0.00042	0.00030
48	0.00146	0.00104	0.00074	0.00052	0.00037	0.00026
49	0.00134	0.00095	0.00067	0.00047	0.00033	0.00023
50	0.00123	0.00086	0.00060	0.00042	0.00029	0.00020
51	0.00112	0.00078	0.00054	0.00037	0.00026	0.00018
52	0.00103	0.00071	0.00049	0.00033	0.00023	0.00015
53	0.00094	0.00064	0.00044	0.00030	0.00020	0.00014
54	0.00087	0.00059	0.00039	0.00026	0.00018	0.00012
55	0.00079	0.00053	0.00035	0.00024	0.00016	0.00010
56	0.00073	0.00048	0.00032	0.00021	0.00014	0.00009
57	0.00067	0.00044	0.00029	0.00019	0.00012	0.00008
58	0.00061	0.00040	0.00026	0.00017	0.00011	0.00007
59	0.00056	0.00036	0.00023	0.00015	0.00010	0.00006
60	0.00051	0.00033	0.00021	0.00013	0.00009	0.00005

TABLE IV   Sinking Fund Factor $\dfrac{i}{(1+i)^n-1}=\dfrac{1}{s_{\overline{n}|i}}$ *(Continued)*

			i			
n	0.15 (15%)	0.16 (16%)	0.17 (17%)	0.18 (18%)	0.19 (19%)	0.20 (20%)
1	1.00000	1.00000	1.00000	1.00000	1.00000	1.00000
2	0.46512	0.46296	0.46083	0.45872	0.45662	0.45455
3	0.28798	0.28526	0.28257	0.27992	0.27731	0.27473
4	0.20027	0.19738	0.19453	0.19174	0.18899	0.18629
5	0.14832	0.14541	0.14256	0.13978	0.13705	0.13438
6	0.11424	0.11139	0.10861	0.10591	0.10327	0.10071
7	0.09036	0.08761	0.08495	0.08236	0.07985	0.07742
8	0.07285	0.07022	0.06769	0.06524	0.06289	0.06061
9	0.05957	0.05708	0.05469	0.05239	0.05019	0.04808
10	0.04925	0.04690	0.04466	0.04251	0.04047	0.03852
11	0.04107	0.03886	0.03676	0.03478	0.03289	0.03110
12	0.03448	0.03241	0.03047	0.02863	0.02690	0.02526
13	0.02911	0.02718	0.02538	0.02369	0.02210	0.02062
14	0.02469	0.02290	0.02123	0.01968	0.01823	0.01689
15	0.02102	0.01936	0.01782	0.01640	0.01509	0.01388
16	0.01795	0.01641	0.01500	0.01371	0.01252	0.01144
17	0.01537	0.01395	0.01266	0.01149	0.01041	0.00944
18	0.01319	0.01188	0.01071	0.00964	0.00868	0.00781
19	0.01134	0.01014	0.00907	0.00810	0.00724	0.00646
20	0.00976	0.00867	0.00769	0.00682	0.00605	0.00536
21	0.00842	0.00742	0.00653	0.00575	0.00505	0.00444
22	0.00727	0.00635	0.00555	0.00485	0.00423	0.00369
23	0.00628	0.00545	0.00472	0.00409	0.00354	0.00307
24	0.00543	0.00467	0.00402	0.00345	0.00297	0.00255
25	0.00470	0.00401	0.00342	0.00292	0.00249	0.00212
26	0.00407	0.00345	0.00292	0.00247	0.00209	0.00176
27	0.00353	0.00296	0.00249	0.00209	0.00175	0.00147
28	0.00306	0.00255	0.00212	0.00177	0.00147	0.00122
29	0.00265	0.00219	0.00181	0.00149	0.00123	0.00102
30	0.00230	0.00189	0.00154	0.00126	0.00103	0.00085
31	0.00200	0.00162	0.00132	0.00107	0.00087	0.00070
32	0.00173	0.00140	0.00113	0.00091	0.00073	0.00059
33	0.00150	0.00120	0.00096	0.00077	0.00061	0.00049
34	0.00131	0.00104	0.00082	0.00065	0.00051	0.00041
35	0.00113	0.00089	0.00070	0.00055	0.00043	0.00034
36	0.00099	0.00077	0.00060	0.00047	0.00036	0.00028
37	0.00086	0.00066	0.00051	0.00039	0.00030	0.00024
38	0.00074	0.00057	0.00044	0.00033	0.00026	0.00020
39	0.00065	0.00049	0.00037	0.00028	0.00022	0.00016
40	0.00056	0.00042	0.00032	0.00024	0.00018	0.00014
41	0.00049	0.00037	0.00027	0.00020	0.00015	0.00011
42	0.00042	0.00031	0.00023	0.00017	0.00013	0.00009
43	0.00037	0.00027	0.00020	0.00015	0.00011	0.00008
44	0.00032	0.00023	0.00017	0.00012	0.00009	0.00007
45	0.00028	0.00020	0.00015	0.00010	0.00008	0.00005
46	0.00024	0.00017	0.00012	0.00009	0.00006	0.00005
47	0.00021	0.00015	0.00011	0.00008	0.00005	0.00004
48	0.00018	0.00013	0.00009	0.00006	0.00004	0.00003
49	0.00016	0.00011	0.00008	0.00005	0.00004	0.00003
50	0.00014	0.00010	0.00007	0.00005	0.00003	0.00002
51	0.00012	0.00008	0.00006	0.00004	0.00003	0.00002
52	0.00010	0.00007	0.00005	0.00003	0.00002	0.00002
53	0.00009	0.00006	0.00004	0.00003	0.00002	0.00001
54	0.00008	0.00005	0.00004	0.00002	0.00002	0.00001
55	0.00007	0.00005	0.00003	0.00002	0.00001	0.00001
56	0.00006	0.00004	0.00003	0.00002	0.00001	0.00001
57	0.00005	0.00003	0.00002	0.00001	0.00001	0.00001
58	0.00005	0.00003	0.00002	0.00001	0.00001	0.00001
59	0.00004	0.00003	0.00002	0.00001	0.00001	0.00000
60	0.00003	0.00002	0.00001	0.00001	0.00001	0.00000

TABLE V Series Present-Worth Factor $\dfrac{(1+i)^n - 1}{i(1+i)^n} = a_{\overline{n}|i}$

n	0.01 (1%)	0.015 (1½%)	0.02 (2%)	0.025 (2½%)	0.03 (3%)	0.035 (3½%)
1	0.99010	0.98522	0.98039	0.97561	0.97087	0.96618
2	1.97040	1.95588	1.94156	1.92742	1.91347	1.89969
3	2.94099	2.91220	2.88388	2.85602	2.82861	2.80164
4	3.90197	3.85438	3.80773	3.76197	3.71710	3.67308
5	4.85343	4.78264	4.71346	4.64583	4.57971	4.51505
6	5.79548	5.69719	5.60143	5.50813	5.41719	5.32855
7	6.72819	6.59821	6.47199	6.34939	6.23028	6.11454
8	7.65168	7.48593	7.32548	7.17014	7.01969	6.87396
9	8.56602	8.36052	8.16224	7.97087	7.78611	7.60769
10	9.47130	9.22218	8.98259	8.75206	8.53020	8.31661
11	10.36763	10.07112	9.78685	9.51421	9.25262	9.00155
12	11.25508	10.90751	10.57534	10.25776	9.95400	9.66333
13	12.13374	11.73153	11.34837	10.98318	10.63496	10.30274
14	13.00370	12.54338	12.10625	11.69091	11.29607	10.92052
15	13.86505	13.34323	12.84926	12.38138	11.93794	11.51741
16	14.71787	14.13126	13.57771	13.05500	12.56110	12.09412
17	15.56225	14.90765	14.29187	13.71220	13.16612	12.65132
18	16.39827	15.67256	14.99203	14.35336	13.75351	13.18968
19	17.22601	16.42617	15.67846	14.97889	14.32380	13.70984
20	18.04555	17.16864	16.35143	15.58916	14.87747	14.21240
21	18.85698	17.90014	17.01121	16.18455	15.41502	14.69797
22	19.66038	18.62082	17.65805	16.76541	15.93692	15.16712
23	20.45582	19.33086	18.29220	17.33211	16.44361	15.62041
24	21.24339	20.03041	18.91393	17.88499	16.93554	16.05837
25	22.02316	20.71961	19.52346	18.42438	17.41315	16.48151
26	22.79520	21.39863	20.12104	18.95061	17.87684	16.89035
27	23.55961	22.06762	20.70690	19.46401	18.32703	17.28536
28	24.31644	22.72672	21.28127	19.96489	18.76411	17.66702
29	25.06579	23.37608	21.84438	20.45355	19.18845	18.03577
30	25.80771	24.01584	22.39646	20.93029	19.60044	18.39205
31	26.54229	24.64615	22.93770	21.39541	20.00043	18.73628
32	27.26959	25.26714	23.46833	21.84918	20.38877	19.06887
33	27.98969	25.87895	23.98856	22.29188	20.76579	19.39021
34	28.70267	26.48173	24.49859	22.72379	21.13184	19.70068
35	29.40858	27.07559	24.99862	23.14516	21.48722	20.00066
36	30.10751	27.66068	25.48884	23.55625	21.83225	20.29049
37	30.79951	28.23713	25.96945	23.95732	22.16724	20.57053
38	31.48466	28.80505	26.44064	24.34860	22.49246	20.84109
39	32.16303	29.36458	26.90259	24.73034	22.80822	21.10250
40	32.83469	29.91585	27.35548	25.10278	23.11477	21.35507
41	33.49969	30.45896	27.79949	25.46612	23.41240	21.59910
42	34.15811	30.99405	28.23479	25.82061	23.70136	21.83486
43	34.81001	31.52123	28.66156	26.16645	23.98190	22.06269
44	35.45545	32.04062	29.07996	26.50385	24.25427	22.28279
45	36.09451	32.55234	29.49016	26.83302	24.51871	22.49545
46	36.72724	33.05649	29.89231	27.15417	24.77545	22.70092
47	37.35370	33.55319	30.28658	27.46748	25.02471	22.89944
48	37.97396	34.04255	30.67312	27.77315	25.26671	23.09124
49	38.58808	34.52468	31.05208	28.07137	25.50166	23.27656
50	39.19612	34.99969	31.42361	28.36231	25.72976	23.45562
51	39.79814	35.46767	31.78785	28.64616	25.95123	23.62862
52	40.39419	35.92874	32.14495	28.92308	26.16624	23.79576
53	40.98435	36.38300	32.49505	29.19325	26.37499	23.95726
54	41.56866	36.83054	32.83828	29.45683	26.57766	24.11330
55	42.14719	37.27147	33.17479	29.71398	26.77443	24.26405
56	42.71999	37.70588	33.50469	29.96486	26.96546	24.40971
57	43.28712	38.13387	33.82813	30.20962	27.15094	24.55045
58	43.84863	38.55554	34.14523	30.44841	27.33101	24.68642
59	44.40459	38.97097	34.45610	30.68137	27.50583	24.81780
60	44.95504	39.38027	34.76089	30.90866	27.67556	24.94473

TABLE V   Series Present-Worth Factor $\dfrac{(1+i)^n - 1}{i(1+i)^n} = a_{\overline{n}|i}$ *(Continued)*

n	0.04 (4%)	0.045 (4½%)	0.05 (5%)	0.06 (6%)	0.07 (7%)	0.08 (8%)
1	0.96154	0.95694	0.95238	0.94340	0.93458	0.92593
2	1.88609	1.87267	1.85941	1.83339	1.80802	1.78326
3	2.77509	2.74896	2.72325	2.67301	2.62432	2.57710
4	3.62990	3.58753	3.54595	3.46511	3.38721	3.31213
5	4.45182	4.38998	4.32948	4.21236	4.10020	3.99271
6	5.24214	5.15787	5.07569	4.91732	4.76654	4.62288
7	6.00205	5.89270	5.78637	5.58238	5.38929	5.20637
8	6.73274	6.59589	6.46321	6.20979	5.97130	5.74664
9	7.43533	7.26879	7.10782	6.80169	6.51523	6.24689
10	8.11090	7.91272	7.72173	7.36009	7.02358	6.71008
11	8.76048	8.52892	8.30641	7.88687	7.49867	7.13896
12	9.38507	9.11858	8.86325	8.38384	7.94269	7.53608
13	9.98565	9.68285	9.39357	8.85268	8.35765	7.90378
14	10.56312	10.22283	9.89864	9.29498	8.74547	8.24424
15	11.11839	10.73955	10.37966	9.71225	9.10791	8.55948
16	11.65230	11.23402	10.83777	10.10590	9.44665	8.85137
17	12.16567	11.70719	11.27407	10.47726	9.76322	9.12164
18	12.65930	12.15999	11.68959	10.82760	10.05909	9.37189
19	13.13394	12.59329	12.08532	11.15812	10.33560	9.60360
20	13.59033	13.00794	12.46221	11.46992	10.59401	9.81815
21	14.02916	13.40472	12.82115	11.76408	10.83553	10.01680
22	14.45112	13.78442	13.16300	12.04158	11.06124	10.20074
23	14.85684	14.14777	13.48857	12.30338	11.27219	10.37106
24	15.24696	14.49548	13.79864	12.55036	11.46933	10.52876
25	15.62208	14.82821	14.09394	12.78336	11.65358	10.67478
26	15.98277	15.14661	14.37519	13.00317	11.82578	10.80998
27	16.32959	15.45130	14.64303	13.21053	11.98671	10.93516
28	16.66306	15.74287	14.89813	13.40616	12.13711	11.05108
29	16.98371	16.02189	15.14107	13.59072	12.27767	11.15841
30	17.29203	16.28889	15.37245	13.76483	12.40904	11.25778
31	17.58849	16.54439	15.59281	13.92909	12.53181	11.34980
32	17.87355	16.78889	15.80268	14.08404	12.64656	11.43500
33	18.14765	17.02286	16.00255	14.23023	12.75379	11.51389
34	18.41120	17.24676	16.19290	14.36814	12.85401	11.58693
35	18.66461	17.46101	16.37419	14.49825	12.94767	11.65457
36	18.90828	17.66604	16.54685	14.62099	13.03521	11.71719
37	19.14258	17.86224	16.71129	14.73678	13.11702	11.77518
38	19.36786	18.04999	16.86789	14.84602	13.19347	11.82887
39	19.58448	18.22966	17.01704	14.94907	13.26493	11.87858
40	19.79277	18.40158	17.15909	15.04630	13.33171	11.92461
41	19.99305	18.56611	17.29437	15.13802	13.39412	11.96723
42	20.18563	18.72355	17.42321	15.22454	13.45245	12.00670
43	20.37079	18.87421	17.54591	15.30617	13.50696	12.04324
44	20.54884	19.01838	17.66277	15.38318	13.55791	12.07707
45	20.72004	19.15635	17.77407	15.45583	13.60552	12.10840
46	20.88465	19.28837	17.88007	15.52437	13.65002	12.13741
47	21.04294	19.41471	17.98102	15.58903	13.69161	12.16427
48	21.19513	19.53561	18.07716	15.65003	13.73047	12.18914
49	21.34147	19.65130	18.16872	15.70757	13.76680	12.21216
50	21.48218	19.76201	18.25593	15.76186	13.80075	12.23348
51	21.61749	19.86795	18.33898	15.81308	13.83247	12.25323
52	21.74758	19.96933	18.41807	15.86139	13.86212	12.27151
53	21.87267	20.06634	18.49340	15.90697	13.88984	12.28843
54	21.99296	20.15918	18.56515	15.94998	13.91573	12.30410
55	22.10861	20.24802	18.63347	15.99054	13.93994	12.31861
56	22.21982	20.33303	18.69854	16.02881	13.96256	12.33205
57	22.32675	20.41439	18.76052	16.06492	13.98370	12.34449
58	22.42957	20.49224	18.81954	16.09898	14.00346	12.35601
59	22.52843	20.56673	18.87575	16.13111	14.02192	12.36668
60	22.62349	20.63802	18.92929	16.16143	14.03918	12.37655

TABLE V   Series Present-Worth Factor $\dfrac{(1+i)^n - 1}{i(1+i)^n} = a_{\overline{n}|i}$ *(Continued)*

				i		
n	0.09 (9%)	0.10 (10%)	0.11 (11%)	0.12 (12%)	0.13 (13%)	0.14 (14%)
1	0.91743	0.90909	0.90090	0.89286	0.88496	0.87719
2	1.75911	1.73554	1.71252	1.69005	1.66810	1.64666
3	2.53129	2.48685	2.44371	2.40183	2.36115	2.32163
4	3.23972	3.16987	3.10245	3.03735	2.97447	2.91371
5	3.88965	3.79079	3.69590	3.60478	3.51723	3.43308
6	4.48592	4.35526	4.23054	4.11141	3.99755	3.88867
7	5.03295	4.86842	4.71220	4.56376	4.42261	4.28830
8	5.53482	5.33493	5.14612	4.96764	4.79877	4.63886
9	5.99525	5.75902	5.53705	5.32825	5.13166	4.94637
10	6.41766	6.14457	5.88923	5.65022	5.42624	5.21612
11	6.80519	6.49506	6.20652	5.93770	5.68694	5.45273
12	7.16073	6.81369	6.49236	6.19437	5.91765	5.66029
13	7.48690	7.10336	6.74987	6.42355	6.12181	5.84236
14	7.78615	7.36669	6.98187	6.62817	6.30249	6.00207
15	8.06069	7.60608	7.19087	6.81086	6.46238	6.14217
16	8.31256	7.82371	7.37916	6.97399	6.60388	6.26506
17	8.54363	8.02155	7.54879	7.11963	6.72909	6.37286
18	8.75563	8.20141	7.70162	7.24967	6.83991	6.46742
19	8.95011	8.36492	7.83929	7.36578	6.93797	6.55037
20	9.12855	8.51356	7.96333	7.46944	7.02475	6.62313
21	9.29224	8.64869	8.07507	7.56200	7.10155	6.68696
22	9.44243	8.77154	8.17574	7.64465	7.16951	6.74294
23	9.58021	8.88322	8.26643	7.71843	7.22966	6.79206
24	9.70661	8.98474	8.34814	7.78432	7.28288	6.83514
25	9.82258	9.07704	8.42174	7.84314	7.32998	6.87293
26	9.92897	9.16095	8.48806	7.89566	7.37167	6.90608
27	10.02658	9.23722	8.54780	7.94255	7.40856	6.93515
28	10.11613	9.30657	8.60162	7.98442	7.44120	6.96066
29	10.19828	9.36961	8.65011	8.02181	7.47009	6.98304
30	10.27365	9.42691	8.69379	8.05518	7.49565	7.00266
31	10.34280	9.47901	8.73315	8.08499	7.51828	7.01988
32	10.40624	9.52638	8.76860	8.11159	7.53830	7.03498
33	10.46444	9.56943	8.80054	8.13535	7.55602	7.04823
34	10.51784	9.60857	8.82932	8.15656	7.57170	7.05985
35	10.56682	9.64416	8.85524	8.17550	7.58557	7.07005
36	10.61176	9.67651	8.87859	8.19241	7.59785	7.07899
37	10.65299	9.70592	8.89963	8.20751	7.60872	7.08683
38	10.69082	9.73265	8.91859	8.22099	7.61833	7.09371
39	10.72552	9.75696	8.93567	8.23303	7.62684	7.09975
40	10.75736	9.77905	8.95105	8.24378	7.63438	7.10504
41	10.78657	9.79914	8.96491	8.25337	7.64104	7.10969
42	10.81337	9.81740	8.97740	8.26194	7.64694	7.11376
43	10.83795	9.83400	8.98865	8.26959	7.65216	7.11733
44	10.86051	9.84909	8.99878	8.27642	7.65678	7.12047
45	10.88120	9.86281	9.00791	8.28252	7.66086	7.12322
46	10.90018	9.87528	9.01614	8.28796	7.66448	6.12563
47	10.91760	9.88662	9.02355	8.29282	7.66768	7.12774
48	10.93358	9.89693	9.03022	8.29716	7.67052	7.12960
49	10.94823	9.90630	9.03624	8.30104	7.67302	7.13123
50	10.96168	9.91481	9.04165	8.30450	7.67524	7.13266
51	10.97402	9.92256	9.04653	8.30759	7.67720	7.13391
52	10.98534	9.92960	9.05093	8.31035	7.67894	7.13501
53	10.99573	9.93600	9.05489	8.31281	7.68048	7.13597
54	11.00525	9.94182	9.05846	9.31501	7.68184	7.13682
55	11.01399	9.94711	9.06168	8.31697	7.68304	7.13756
56	11.02201	9.95191	9.06457	8.31872	7.68411	7.13821
57	11.02937	9.95629	9.06718	8.32029	7.68505	7.13878
58	11.03612	9.96026	9.06954	8.32169	7.68589	7.13928
59	11.04231	9.96387	9.07165	8.32294	7.68663	7.13972
60	11.04799	9.96716	9.07356	8.32405	7.68728	7.14011

TABLE V Series Present-Worth Factor $\dfrac{(1+i)^n - 1}{i(1+i)^n} = a_{\overline{n}|i}$ *(Continued)*

n	0.15 (15%)	0.16 (16%)	0.17 (17%)	0.18 (18%)	0.19 (19%)	0.20 (20%)
1	0.86957	0.86207	0.85470	0.84746	0.84034	0.83333
2	1.62571	1.60523	1.58521	1.56564	1.54650	1.52778
3	2.28323	2.24589	2.20958	2.17427	2.13992	2.10648
4	2.85498	2.79818	2.74324	2.69006	2.63859	2.58873
5	3.35216	3.27429	3.19935	3.12717	3.05763	2.99061
6	3.78448	3.68474	3.58918	3.49760	3.40978	3.32551
7	4.16042	4.03857	3.92238	3.81153	3.70570	3.60459
8	4.48732	4.34359	4.20716	4.07757	3.95437	3.83716
9	4.77158	4.60654	4.45057	4.30302	4.16333	4.03097
10	5.01877	4.83323	4.65860	4.49409	4.33893	4.19247
11	5.23371	5.02864	4.83641	4.65601	4.48650	4.32706
12	5.42062	5.19711	4.98839	4.79322	4.61050	4.43922
13	5.58315	5.34233	5.11828	4.90951	4.71471	4.53268
14	5.72448	5.46753	5.22930	5.00806	4.80228	4.61057
15	5.84737	5.57546	5.32419	5.09158	4.87586	4.67547
16	5.95423	5.66850	5.40529	5.16235	4.93770	4.72956
17	6.04716	5.74870	5.47461	5.22233	4.98966	4.77463
18	6.12797	5.81785	5.53385	5.27316	5.03333	4.81219
19	6.19823	5.87746	5.58449	5.31624	5.07003	4.84350
20	6.25933	5.92884	5.62777	5.35275	5.10086	4.86958
21	6.31246	5.97314	5.66476	5.38368	5.12677	4.89132
22	6.35866	6.01133	5.69637	5.40990	5.14855	4.90943
23	6.39884	6.04425	5.72340	5.43212	5.16685	4.92453
24	6.43377	6.07263	5.74649	5.45095	5.18223	4.93710
25	6.46415	6.09709	5.76623	5.46691	5.19515	4.94759
26	6.49056	6.11818	5.78311	5.48043	5.20601	4.95632
27	6.51353	6.13636	5.79753	5.49189	5.21513	4.96360
28	6.53351	6.15204	5.80985	5.50160	5.22280	4.96967
29	6.55088	6.16555	5.82039	5.50983	5.22924	4.97472
30	6.56598	6.17720	5.82939	5.51681	5.23466	4.97894
31	6.57911	6.18724	5.83709	5.52272	5.23921	4.98245
32	6.59053	6.19590	5.84366	5.52773	5.24303	4.98537
33	6.60046	6.20336	5.84928	5.53197	5.24625	4.98781
34	6.60910	6.20979	5.85409	5.53557	5.24895	4.98984
35	6.61661	6.21534	5.85820	5.53862	5.25122	4.99154
36	6.62314	6.22012	5.86171	5.54120	5.25312	4.99295
37	6.62881	6.22424	5.86471	5.54339	5.25472	4.99412
38	6.63375	6.22779	5.86727	5.54525	5.25607	4.99510
39	6.63805	6.23086	5.86946	5.54682	5.25720	4.99592
40	6.64178	6.23350	5.87133	5.54815	5.25815	4.99660
41	6.64502	6.23577	5.87294	5.54928	5.25895	4.99717
42	6.64785	6.23774	5.87430	5.55024	5.25962	4.99764
43	6.65030	6.23943	5.87547	5.55105	5.26019	4.99803
44	6.65244	6.24089	5.87647	5.55174	5.26066	4.99836
45	6.65429	6.24214	5.87733	5.55232	5.26106	4.99863
46	6.65591	6.24323	5.87806	5.55281	5.26140	4.99886
47	6.65731	6.24416	5.87868	5.55323	5.26168	4.99905
48	6.65853	6.24497	5.87922	5.55359	5.26191	4.99921
49	6.65959	6.24566	5.87967	5.55389	5.26211	4.99934
50	6.66051	6.24626	5.88006	5.55414	5.26228	4.99945
51	6.66132	6.24678	5.88039	5.55436	5.26242	4.99954
52	6.66201	6.24722	5.88068	5.55454	5.26254	4.99962
53	6.66262	6.24760	5.88092	5.55469	5.26264	4.99968
54	6.66315	6.24793	5.88113	5.55483	5.26272	4.99974
55	6.66361	6.24822	5.88131	5.55494	5.26279	4.99978
56	6.66401	6.24846	5.88146	5.55503	5.26285	4.99982
57	6.66435	6.24868	5.88159	5.55511	5.26290	4.99985
58	6.66466	6.24886	5.88170	5.55518	5.26294	4.99987
59	6.66492	6.24902	5.88180	5.55524	5.26297	4.99989
60	6.66515	6.24915	5.88188	5.55529	5.26300	4.99991

TABLE VI   Capital-Recovery Factor $\dfrac{i(1+i)^n}{(1+i)^n-1}=\dfrac{1}{a_{\overline{n}|i}}$

				i		
n	0.01 (1%)	0.015 (1.5%)	0.02 (2%)	0.025 (2½%)	0.03 (3%)	0.035 (3½%)
1	1.01000	1.01500	1.02000	1.02500	1.03000	1.03500
2	0.50751	0.51128	0.51505	0.51883	0.52261	0.52640
3	0.34002	0.34338	0.34675	0.35014	0.35353	0.35693
4	0.25628	0.25944	0.26262	0.26582	0.26903	0.27225
5	0.20604	0.20909	0.21216	0.21525	0.21835	0.22148
6	0.17255	0.17553	0.17853	0.18155	0.18460	0.18767
7	0.14863	0.15156	0.15451	0.15750	0.16051	0.16354
8	0.13069	0.13358	0.13651	0.13947	0.14246	0.14548
9	0.11674	0.11961	0.12252	0.12546	0.12843	0.13145
10	0.10558	0.10843	0.11133	0.11426	0.11723	0.12024
11	0.09645	0.09929	0.10218	0.10511	0.10808	0.11109
12	0.08885	0.09168	0.09456	0.09749	0.10046	0.10348
13	0.08241	0.08524	0.08812	0.09105	0.09403	0.09706
14	0.07690	0.07972	0.08260	0.08554	0.08853	0.09157
15	0.07212	0.07494	0.07783	0.08077	0.08377	0.08683
16	0.06794	0.07077	0.07365	0.07660	0.07961	0.08268
17	0.06426	0.06708	0.06997	0.07293	0.07595	0.07904
18	0.06098	0.06381	0.06670	0.06967	0.07271	0.07582
19	0.05805	0.06088	0.06378	0.06676	0.06981	0.07294
20	0.05542	0.05825	0.06116	0.06415	0.06722	0.07036
21	0.05303	0.05587	0.05878	0.06179	0.06487	0.06804
22	0.05086	0.05370	0.05663	0.05965	0.06275	0.06593
23	0.04889	0.05173	0.05467	0.05770	0.06081	0.06402
24	0.04707	0.04992	0.05287	0.05591	0.05905	0.06227
25	0.04541	0.04826	0.05122	0.05428	0.05743	0.06067
26	0.04387	0.04673	0.04970	0.05277	0.05594	0.05921
27	0.04245	0.04532	0.04829	0.05138	0.05456	0.05785
28	0.04112	0.04400	0.04699	0.05009	0.05329	0.05660
29	0.03990	0.04278	0.04578	0.04889	0.05211	0.05545
30	0.03875	0.04164	0.04465	0.04778	0.05102	0.05437
31	0.03768	0.04057	0.04360	0.04674	0.05000	0.05337
32	0.03667	0.03958	0.04261	0.04577	0.04905	0.05244
33	0.03573	0.03864	0.04169	0.04486	0.04816	0.05157
34	0.03484	0.03776	0.04082	0.04401	0.04732	0.05076
35	0.03400	0.03693	0.04000	0.04321	0.04654	0.05000
36	0.03321	0.03615	0.03923	0.04245	0.04580	0.04928
37	0.03247	0.03541	0.03851	0.04174	0.04511	0.04861
38	0.03176	0.03472	0.03782	0.04107	0.04446	0.04798
39	0.03109	0.03405	0.03717	0.04044	0.04384	0.04739
40	0.03046	0.03343	0.03656	0.03984	0.04326	0.04683
41	0.02985	0.03283	0.03597	0.03927	0.04271	0.04630
42	0.02928	0.03226	0.03542	0.03873	0.04219	0.04580
43	0.02873	0.03172	0.03489	0.03822	0.04170	0.04533
44	0.02820	0.03121	0.03439	0.03773	0.04123	0.04488
45	0.02771	0.03072	0.03391	0.03727	0.04079	0.04445
46	0.02723	0.03025	0.03345	0.03683	0.04036	0.04405
47	0.02677	0.02980	0.03302	0.03641	0.03996	0.04367
48	0.02633	0.02937	0.03260	0.03601	0.03958	0.04331
49	0.02591	0.02896	0.03220	0.03562	0.03921	0.04296
50	0.02551	0.02857	0.03182	0.03526	0.03887	0.04263
51	0.02513	0.02819	0.03146	0.03491	0.03853	0.04232
52	0.02476	0.02783	0.03111	0.03457	0.03822	0.04202
53	0.02440	0.02749	0.03077	0.03425	0.03791	0.04174
54	0.02406	0.02715	0.03045	0.03395	0.03763	0.04147
55	0.02373	0.02683	0.03014	0.03365	0.03735	0.04121
56	0.02341	0.02652	0.02985	0.03337	0.03708	0.04097
57	0.02310	0.02622	0.02956	0.03310	0.03683	0.04073
58	0.02281	0.02594	0.02929	0.03284	0.03659	0.04051
59	0.02252	0.02566	0.02902	0.03259	0.03636	0.04029
60	0.02224	0.02539	0.02877	0.03235	0.03613	0.04009

TABLE VI Capital-Recovery Factor $\dfrac{i(1+i)^n}{(1+i)^n-1}=\dfrac{1}{a_{\overline{n}|i}}$ *(Continued)*

			i			
n	0.04 (4%)	0.045 (4½%)	0.05 (5%)	0.06 (6%)	0.07 (7%)	0.08 (8%)
1	1.04000	1.04500	1.05000	1.06000	1.07000	1.08000
2	0.53020	0.53400	0.53780	0.54544	0.55309	0.56077
3	0.36035	0.36377	0.36721	0.37411	0.38105	0.38803
4	0.27549	0.27874	0.28201	0.28859	0.29523	0.30192
5	0.22463	0.22779	0.23097	0.23740	0.24389	0.25046
6	0.19076	0.19388	0.19702	0.20336	0.20980	0.21632
7	0.16661	0.16970	0.17282	0.17914	0.18555	0.19207
8	0.14853	0.15161	0.15472	0.16104	0.16747	0.17401
9	0.13449	0.13757	0.14069	0.14702	0.15349	0.16008
10	0.12329	0.12638	0.12950	0.13587	0.14238	0.14903
11	0.11415	0.11725	0.12039	0.12679	0.13336	0.14008
12	0.10655	0.10967	0.11283	0.11928	0.12590	0.13270
13	0.10014	0.10328	0.10646	0.11296	0.11965	0.12652
14	0.09467	0.09782	0.10102	0.10758	0.11434	0.12130
15	0.08994	0.09311	0.09634	0.10296	0.10979	0.11683
16	0.08582	0.08902	0.09227	0.09895	0.10586	0.11298
17	0.08220	0.08542	0.08870	0.09544	0.10243	0.10963
18	0.07899	0.08224	0.08555	0.09236	0.09941	0.10670
19	0.07614	0.07941	0.08275	0.08962	0.09675	0.10413
20	0.07358	0.07688	0.08024	0.08718	0.09439	0.10185
21	0.07128	0.07460	0.07800	0.08500	0.09229	0.09983
22	0.06920	0.07255	0.07597	0.08305	0.09041	0.09803
23	0.06731	0.07068	0.07414	0.08128	0.08871	0.09642
24	0.06559	0.06899	0.07247	0.07968	0.08719	0.09498
25	0.06401	0.06744	0.07095	0.07823	0.08581	0.09368
26	0.06257	0.06602	0.06956	0.07690	0.08456	0.09251
27	0.06124	0.06472	0.06829	0.07570	0.08343	0.09145
28	0.06001	0.06352	0.06712	0.07459	0.08239	0.09049
29	0.05888	0.06241	0.06605	0.07358	0.08145	0.08962
30	0.05783	0.06139	0.06505	0.07265	0.08059	0.08883
31	0.05686	0.06044	0.06413	0.07179	0.07980	0.08811
32	0.05595	0.05956	0.06328	0.07100	0.07907	0.08745
33	0.05510	0.05874	0.06249	0.07027	0.07841	0.08685
34	0.05431	0.05798	0.06176	0.06960	0.07780	0.08630
35	0.05358	0.05727	0.06107	0.06897	0.07723	0.08580
36	0.05289	0.05661	0.06043	0.06839	0.07672	0.08534
37	0.05224	0.05598	0.05984	0.06786	0.07624	0.08492
38	0.05163	0.05540	0.05928	0.06736	0.07580	0.08454
39	0.05106	0.05486	0.05876	0.06689	0.07539	0.08419
40	0.05052	0.05434	0.05828	0.06646	0.07501	0.08386
41	0.05002	0.05386	0.05782	0.06606	0.07466	0.08356
42	0.04954	0.05341	0.05739	0.06568	0.07434	0.08329
43	0.04909	0.05298	0.05699	0.06533	0.07404	0.08303
44	0.04866	0.05258	0.05662	0.06501	0.07376	0.08280
45	0.04826	0.05220	0.05626	0.06470	0.07350	0.08259
46	0.04788	0.05184	0.05593	0.06441	0.07326	0.08239
47	0.04752	0.05151	0.05561	0.06415	0.07304	0.08221
48	0.04718	0.05119	0.05532	0.06390	0.07283	0.08204
49	0.04686	0.05089	0.05504	0.06366	0.07264	0.08189
50	0.04655	0.05060	0.05478	0.06344	0.07246	0.08174
51	0.04626	0.05033	0.05453	0.06324	0.07229	0.08161
52	0.04598	0.05008	0.05429	0.06305	0.07214	0.08149
53	0.04572	0.04983	0.05407	0.06287	0.07200	0.08138
54	0.04547	0.04961	0.05386	0.06270	0.07186	0.08127
55	0.04523	0.04939	0.05367	0.06254	0.07174	0.08118
56	0.04500	0.04918	0.05348	0.06239	0.07162	0.08109
57	0.04479	0.04899	0.05330	0.06225	0.07151	0.08101
58	0.04458	0.04880	0.05314	0.06212	0.07141	0.08093
59	0.04439	0.04862	0.05298	0.06199	0.07132	0.08086
60	0.04420	0.04845	0.05283	0.06188	0.07123	0.08080

TABLE VI **Capital-Recovery Factor** $\dfrac{i(1+i)^n}{(1+i)^n - 1} = \dfrac{1}{a_{\overline{n}|i}}$ **(Continued)**

				i		
n	0.09 (9%)	0.10 (10%)	0.11 (11%)	0.12 (12%)	0.13 (13%)	0.14 (14%)
1	1.09000	1.10000	1.11000	1.12000	1.13000	1.14000
2	0.56847	0.57619	0.58393	0.59170	0.59948	0.60729
3	0.39505	0.40211	0.40921	0.41635	0.42352	0.43073
4	0.30867	0.31547	0.32233	0.32923	0.33619	0.34320
5	0.25709	0.26380	0.27057	0.27741	0.28431	0.29128
6	0.22292	0.22961	0.23638	0.24323	0.25015	0.25716
7	0.19869	0.20541	0.21222	0.21912	0.22611	0.23319
8	0.18067	0.18744	0.19432	0.20130	0.20839	0.21557
9	0.16680	0.17364	0.18060	0.18768	0.19487	0.20217
10	0.15582	0.16275	0.16980	0.17698	0.18429	0.19171
11	0.14695	0.15396	0.16112	0.16842	0.17584	0.18339
12	0.13965	0.14676	0.15403	0.16144	0.16899	0.17667
13	0.13357	0.14078	0.14815	0.15568	0.16335	0.17116
14	0.12843	0.13575	0.14323	0.15087	0.15867	0.16661
15	0.12406	0.13147	0.13907	0.14682	0.15474	0.16281
16	0.12030	0.12782	0.13552	0.14339	0.15143	0.15962
17	0.11705	0.12466	0.13247	0.14046	0.14861	0.15692
18	0.11421	0.12193	0.12984	0.13794	0.14620	0.15462
19	0.11173	0.11955	0.12756	0.13576	0.14413	0.15266
20	0.10955	0.11746	0.12558	0.13388	0.14235	0.15099
21	0.10762	0.11562	0.12384	0.13224	0.14081	0.14954
22	0.10590	0.11401	0.12231	0.13081	0.13948	0.14830
23	0.10438	0.11257	0.12097	0.12956	0.13832	0.14723
24	0.10302	0.11130	0.11979	0.12846	0.13731	0.14630
25	0.10181	0.11017	0.11874	0.12750	0.13643	0.14550
26	0.10072	0.10916	0.11781	0.12665	0.13565	0.14480
27	0.09973	0.10826	0.11699	0.12590	0.13498	0.14419
28	0.09885	0.10745	0.11626	0.12524	0.13439	0.14366
29	0.09806	0.10673	0.11561	0.12466	0.13387	0.14320
30	0.09734	0.10608	0.11502	0.12414	0.13341	0.14280
31	0.09669	0.10550	0.11451	0.12369	0.13301	0.14245
32	0.09610	0.10497	0.11404	0.12328	0.13266	0.14215
33	0.09556	0.10450	0.11363	0.12292	0.13234	0.14188
34	0.09508	0.10407	0.11326	0.12260	0.13207	0.14165
35	0.09464	0.10369	0.11293	0.12232	0.13183	0.14144
36	0.09424	0.10334	0.11263	0.12206	0.13162	0.14126
37	0.09387	0.10303	0.11236	0.12184	0.13143	0.14111
38	0.09354	0.10275	0.11213	0.12164	0.13126	0.14097
39	0.09324	0.10249	0.11191	0.12146	0.13112	0.14085
40	0.09296	0.10226	0.11172	0.12130	0.13099	0.14075
41	0.09271	0.10205	0.11155	0.12116	0.13087	0.14065
42	0.09248	0.10186	0.11139	0.12104	0.13077	0.14057
43	0.09227	0.10169	0.11125	0.12092	0.13068	0.14050
44	0.09208	0.10153	0.11113	0.12083	0.13060	0.14044
45	0.09190	0.10139	0.11101	0.12074	0.13053	0.14039
46	0.09174	0.10126	0.11091	0.12066	0.13047	0.14034
47	0.09160	0.10115	0.11082	0.12059	0.13042	0.14030
48	0.09146	0.10104	0.11074	0.12052	0.13037	0.14026
49	0.09134	0.10095	0.11067	0.12047	0.13033	0.14023
50	0.09123	0.10086	0.11060	0.12042	0.13029	0.14020
51	0.09112	0.10078	0.11054	0.12037	0.13026	0.14018
52	0.09103	0.10071	0.11049	0.12033	0.13023	0.14015
53	0.09094	0.10064	0.11044	0.12030	0.13020	0.14014
54	0.09087	0.10059	0.11039	0.12026	0.13018	0.14012
55	0.09079	0.10053	0.11035	0.12024	0.13016	0.14010
56	0.09073	0.10048	0.11032	0.12021	0.13014	0.14009
57	0.09067	0.10044	0.11029	0.12019	0.13012	0.14008
58	0.09061	0.10040	0.11026	0.12017	0.13011	0.14007
59	0.09056	0.10036	0.11023	0.12015	0.13010	0.14006
60	0.09051	0.10033	0.11021	0.12013	0.13009	0.14005

TABLE VI   Capital-Recovery Factor $\dfrac{i(1+i)^n}{(1+i)^n - 1} = \dfrac{1}{a_{\overline{n}|i}}$ **(Continued)**

n	0.15 (15%)	0.16 (16%)	0.17 (17%)	0.18 (18%)	0.19 (19%)	0.20 (20%)
1	1.15000	1.16000	1.17000	1.18000	1.19000	1.20000
2	0.61512	0.62296	0.63083	0.63872	0.64662	0.65455
3	0.43798	0.44526	0.45257	0.45992	0.46731	0.47473
4	0.35027	0.35738	0.36453	0.37174	0.37899	0.38629
5	0.29832	0.30541	0.31256	0.31978	0.32705	0.33438
6	0.26424	0.27139	0.27861	0.28591	0.29327	0.30071
7	0.24036	0.24761	0.25495	0.26236	0.26985	0.27742
8	0.22285	0.23022	0.23769	0.24524	0.25289	0.26061
9	0.20957	0.21708	0.22469	0.23239	0.24019	0.24808
10	0.19925	0.20690	0.21466	0.22251	0.23047	0.23852
11	0.19107	0.19886	0.20676	0.21478	0.22289	0.23110
12	0.18448	0.19241	0.20047	0.20863	0.21690	0.22526
13	0.17911	0.18718	0.19538	0.20369	0.21210	0.22062
14	0.17469	0.18290	0.19123	0.19968	0.20823	0.21689
15	0.17102	0.17936	0.18782	0.19640	0.20509	0.21388
16	0.16795	0.17641	0.18500	0.19371	0.20252	0.21144
17	0.16537	0.17395	0.18266	0.19149	0.20041	0.20944
18	0.16319	0.17188	0.18071	0.18964	0.19868	0.20781
19	0.16134	0.17014	0.17907	0.18810	0.19724	0.20646
20	0.15976	0.16867	0.17769	0.18682	0.19605	0.20536
21	0.15842	0.16742	0.17653	0.18575	0.19505	0.20444
22	0.15727	0.16635	0.17555	0.18485	0.19423	0.20369
23	0.15628	0.16545	0.17472	0.18409	0.19354	0.20307
24	0.15543	0.16467	0.17402	0.18345	0.19297	0.20255
25	0.15470	0.16401	0.17342	0.18292	0.19249	0.20212
26	0.15407	0.16345	0.17292	0.18247	0.19209	0.20176
27	0.15353	0.16296	0.17249	0.18209	0.19175	0.20147
28	0.15306	0.16255	0.17212	0.18177	0.19147	0.20122
29	0.15265	0.16219	0.17181	0.18149	0.19123	0.20102
30	0.15230	0.16189	0.17154	0.18126	0.19103	0.20085
31	0.15200	0.16162	0.17132	0.18107	0.19087	0.20070
32	0.15173	0.16140	0.17113	0.18091	0.19073	0.20059
33	0.15150	0.16120	0.17096	0.18077	0.19061	0.20049
34	0.15131	0.16104	0.17082	0.18065	0.19051	0.20041
35	0.15113	0.16089	0.17070	0.18055	0.19043	0.20034
36	0.15099	0.16077	0.17060	0.18047	0.19036	0.20028
37	0.15086	0.16066	0.17051	0.18039	0.19030	0.20024
38	0.15074	0.16057	0.17044	0.18033	0.19026	0.20020
39	0.15065	0.16049	0.17037	0.18028	0.19022	0.20016
40	0.15056	0.16042	0.17032	0.18024	0.19018	0.20014
41	0.15049	0.16037	0.17027	0.18020	0.19015	0.20011
42	0.15042	0.16031	0.17023	0.18017	0.19013	0.20009
43	0.15037	0.16027	0.17020	0.18015	0.19011	0.20008
44	0.15032	0.16023	0.17017	0.18012	0.19009	0.20007
45	0.15028	0.16020	0.17015	0.18010	0.19008	0.20005
46	0.15024	0.16017	0.17012	0.18009	0.19006	0.20005
47	0.15021	0.16015	0.17011	0.18008	0.19005	0.20004
48	0.15018	0.16013	0.17009	0.18006	0.19004	0.20003
49	0.15016	0.16011	0.17008	0.18005	0.19004	0.20003
50	0.15014	0.16010	0.17007	0.18005	0.19003	0.20002
51	0.15012	0.16008	0.17006	0.18004	0.19003	0.20002
52	0.15010	0.16007	0.17005	0.18003	0.19002	0.20002
53	0.15009	0.16006	0.17004	0.18003	0.19002	0.20001
54	0.15008	0.16005	0.17004	0.18002	0.19002	0.20001
55	0.15007	0.16005	0.17003	0.18002	0.19001	0.20001
56	0.15006	0.16004	0.17003	0.18002	0.19001	0.20001
57	0.15005	0.16003	0.17002	0.18001	0.19001	0.20001
58	0.15005	0.16003	0.17002	0.18001	0.19001	0.20001
59	0.15004	0.16003	0.17002	0.18001	0.19001	0.20000
60	0.15003	0.16002	0.17001	0.18001	0.19001	0.20000

TABLE VII Monthly Payment per Dollar of Mortgage Loan

ANNUAL INTEREST RATE	Mortgage Period			
	15 YEARS (180 PAYMENTS)	20 YEARS (240 PAYMENTS)	25 YEARS (300 PAYMENTS)	30 YEARS (360 PAYMENTS)
7.50	0.00927012	0.00805593	0.00738991	0.00699214
7.75	0.00941276	0.00820948	0.00755329	0.00716412
8.00	0.00955652	0.00836440	0.00771816	0.00733764
8.25	0.00970140	0.00852065	0.00788450	0.00751266
8.50	0.00984740	0.00867823	0.00805227	0.00768913
8.75	0.00999448	0.00883710	0.00822143	0.00786700
9.00	0.01014266	0.00899725	0.00839196	0.00804622
9.25	0.01029192	0.00915866	0.00856381	0.00822675
9.50	0.01044224	0.00932131	0.00873696	0.00840854
9.75	0.01059362	0.00948516	0.00891137	0.00859154
10.00	0.01074605	0.00965021	0.00908700	0.00877572
10.25	0.01089951	0.00981643	0.00926383	0.00896101
10.50	0.01105399	0.00998380	0.00944182	0.00914739
10.75	0.01120948	0.01015229	0.00962093	0.00933481
11.00	0.01136597	0.01032188	0.00980113	0.00952323
11.25	0.01152345	0.01049256	0.00998240	0.00971261
11.50	0.01168190	0.01066430	0.01016469	0.00990291
11.75	0.01184131	0.01083707	0.01034798	0.01009410
12.00	0.01200168	0.01101086	0.01053224	0.01028613
12.25	0.01216299	0.01118565	0.01071744	0.01047896
12.50	0.01232522	0.01136140	0.01090354	0.01067258
12.75	0.01248837	0.01153817	0.01109052	0.01086693
13.00	0.01265242	0.01171576	0.01127835	0.01106200
13.25	0.01281736	0.01189431	0.01146700	0.01145412
13.50	0.01298319	0.01207375	0.01165645	0.01165113
13.75	0.01314987	0.01225405	0.01184666	0.01184872
14.00	0.01331741	0.01243521	0.01203761	0.01884872
14.25	0.01348580	0.01261719	0.01222928	0.01204687
14.50	0.01365501	0.01279998	0.01242163	0.01224556
14.75	0.01382504	0.01298355	0.01261465	0.01244476
15.00	0.01399587	0.01316790	0.01280831	0.01264444
15.25	0.01416750	0.01335299	0.01300258	0.01284459
15.50	0.01433990	0.01353881	0.01319745	0.01304517
15.75	0.01451308	0.01372534	0.01339290	0.01324617
16.00	0.01468701	0.01391256	0.01358889	0.01344757
16.25	0.01486168	0.01410046	0.01378541	0.01364935
16.50	0.01503709	0.01428901	0.01398245	0.01385148
16.75	0.01521321	0.01447820	0.01417998	0.01405396
17.00	0.01539004	0.01466801	0.01437797	0.01425675
17.25	0.01556757	0.01485842	0.01457641	0.01445986
17.50	0.01574578	0.01504942	0.01477530	0.01466325
17.75	0.01592467	0.01524099	0.01497460	0.01486692
18.00	0.01610421	0.01543312	0.01517430	0.01507085
18.25	0.01628440	0.01562578	0.01537439	0.01527503
18.50	0.01646523	0.01581897	0.01557484	0.01547945
18.75	0.01664669	0.01601266	0.01577565	0.01568408
19.00	0.01682876	0.01620685	0.01597680	0.01588892
19.25	0.01701143	0.01640152	0.01617827	0.01609397
19.50	0.01719470	0.01659665	0.01638006	0.01629920
19.75	0.01737855	0.01679223	0.01658215	0.01650461
20.00	0.01756297	0.01698825	0.01678452	0.01671019

A REVIEW OF ALGEBRA (OPTIONAL)

A.1 THE REAL NUMBER SYSTEM
A.2 POLYNOMIALS
A.3 FACTORING
A.4 FRACTIONS
A.5 EXPONENTS AND RADICALS

Algebra is the only prerequisite for using this text. This appendix provides a brief review of algebra. To guide you in your review of algebra it is suggested that you take the following self-correcting algebra test. Its purpose is to help you diagnose those areas which you need to review. The results of the test can guide you in your review of Secs. A.1 to A.5.

❏ ALGEBRA PRETEST

		CORRESPONDING SECTION IN APPENDIX
1	$|-10| =$	A.1
2	$x^3 \cdot x^4 =$	A.2
3	$[(x^3)^2]^3 =$	A.2
4	$x^5/x^3 =$	A.2
5	$(4x - 2y + z) - (-3x + 4y - 2z) =$	A.2
6	$\dfrac{2x^2(3x^3)}{(-2x^2)^2} =$	A.2
7	Factor $2a^3b^2c + 4a^2bc^2$.	A.3

8 Factor $x^2 - 4$.   A.3

9 Factor $x^2 - 5x + 4$.   A.3

10 $\frac{1}{5} + \frac{2}{15} - \frac{1}{6} =$ A.4

11 $\dfrac{2x^2}{3} \div \dfrac{4x^3}{9} =$ A.4

12 $x^{1/2}x^{4/3} =$ A.5

13 $\sqrt[3]{a^2b}\,\sqrt[3]{ab^2} =$ A.5

14 $3\sqrt{2} - 2\sqrt{8} =$ A.5

15 $\sqrt{\dfrac{4a^2}{9}} =$ A.5

16 Express $x^{2/3}$ in radical form.   A.5

❑ ANSWERS FOR ALGEBRA PRETEST

1 10	**2** x^7	**3** x^{18}	**4** x^2	**5** $7x - 6y + 3z$	**6** $3x/2$
7 $2a^2bc(ab + 2c)$		**8** $(x+2)(x-2)$		**9** $(x-4)(x-1)$		**10** $\frac{1}{6}$
11 $3/2x$	**12** $x^{11/6}$	**13** ab	**14** $-\sqrt{2}$	**15** $2a/3$	**16** $\sqrt[3]{x^2}$

A.1 THE REAL NUMBER SYSTEM

Real Numbers

In this book we are concerned with the mathematics of **_real numbers._** As indicated in Fig. A.1, the real number system consists of rational numbers and irrational numbers. **_Rational numbers_** are numbers which can be expressed as the **_ratio,_** or quotient, of two integers, with the divisor being a nonzero integer. Thus, a rational number is a number which can be expressed in the form a/b, where a and b are integers and b does not equal 0 (stated $b \neq 0$). The numbers $\frac{1}{5}$, $-\frac{2}{7}$, $\frac{23}{455}$, and $137/(-750)$ are all examples of rational numbers.

Because any integer a can be written in the form of the quotient $a/1$, all integers are also rational numbers. Examples include $-5 = -5/1$ and $54 = 54/1$. Zero is

Figure A.1
The real number system.

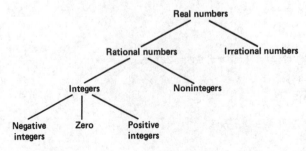

also considered to be an integer (neither negative nor positive), and it can be written in the quotient form $0/b = 0$, $b \neq 0$.

Irrational numbers are real numbers which cannot be expressed as the ratio of two integers. Numbers such as $\pi = 3.14159265 \ldots$ (which is the ratio of the circumference of a circle to its diameter), $\sqrt{2} = 1.4142 \ldots$, $\sqrt{3} = 1.7321 \ldots$, and $\sqrt{5} = 2.2361 \ldots$ are all examples of irrational numbers.

Figure A.2
Number line.

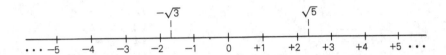

The set of real numbers can be represented using a *number line.* (See Fig. A.2.) The number line has a zero point, often called the *origin,* which is used to represent the real number 0. To each and every point on the number line there corresponds a real number. The correspondence is that the real number represented by a point equals the *directed distance* traveled in moving from the origin to that point. Movements from left to right along the number line are considered to be in a positive direction. Thus, points to the right of the origin correspond to positive real numbers whereas points to the left correspond to negative real numbers. Note that for each and every real number there corresponds a unique point on the number line.

The *inequality symbols* $>$ or $<$ are used to indicate that two numbers are not equal but they can be compared. When an inequality symbol is placed between two numbers, it "opens" in the direction of the larger number. Given two real numbers a and b, the notation $a > b$ is read "a is greater than b." The statement $a > b$ implies that on the real number line a is located to the right of b.

Absolute Value

The *absolute value* of a real number is the magnitude or size of the number without the sign. The notation $|a|$ denotes the absolute value of a.

DEFINITION: ABSOLUTE VALUE
For any real number a,

$$|a| = \begin{cases} a & \text{if } a \text{ is positive or zero} \\ -a & \text{if } a \text{ is negative} \end{cases}$$

EXAMPLE 1 The absolute value of the number $+5$ is $|+5| = 5$. The absolute value of -20 is $|-20| = 20$. The absolute value of 0 is $|0| = 0$.

The concept of absolute value is examined in greater detail in Chap. 1.

Section A.1 Follow-up Exercises

In Exercises 1–12, place an inequality symbol ($<$ or $>$) between the two given numbers to indicate the appropriate inequality relationship.

1	10____6	**2**	8____3
3	-1____-4	**4**	2____-3
5	20____10	**6**	-5____-2
7	-10____-15	**8**	5____0
9	-3____0	**10**	0____-2
11	1____-1	**12**	3____1
13	$|-5|=$	**14**	$|-3|=$
15	$|-5-10|=$	**16**	$|-10+5|=$
17	$|16|=$	**18**	$|2|=$
19	$|10-(4-3)|=$	**20**	$|-5-(-5+2)|=$

A.2 POLYNOMIALS

Positive Integer Exponents

When a real number a is multiplied times itself, we denote this product as $a \cdot a$ or aa. If the same number is multiplied times itself 5 times, the product is denoted by $aaaaa$. A shorthand notation which can be used to express these products is

$$aa = a^2$$

and

$$aaaaa = a^5$$

The number written above and to the right of a is called an *exponent*. The exponent indicates the number of times a occurs as a factor.

> **DEFINITION**
> If n is a positive integer and a is any real number,
> $$a^n = \underbrace{a \cdot a \cdot a \cdots a}_{n \text{ factors}}$$

The term a^n can be verbalized as "a raised to the nth power," where a is considered the **base** and n is the exponent or **power**.

EXAMPLE 2

(a) $(-2)(-2)(-2)(-2)(-2)(-2) = (-2)^6$
(b) $(5)(5)(5) = (5)^3$
(c) $aaaabbb = a^4 b^3$
(d) $aa/(bbbb) = a^2/b^4$

◻

DEFINITION

If n is a positive integer and $a \neq 0$,

$$a^{-n} = \frac{1}{a^n}$$

EXAMPLE 3

(a) $a^{-2} = 1/a^2$
(b) $(2)^{-3} = 1/(2)^3 = \frac{1}{8}$

$\square$

DEFINITION

If a is real and not equal to 0, $a^0 = 1$.

EXAMPLE 4

(a) $(10)^0 = 1$
(b) $(4x)^0 = 1, x \neq 0$
(c) $-5y^0 = -5(1) = -5, y \neq 0$

$\square$

The following laws of exponents apply when a and b are any real numbers and m and n are positive integers.

LAWS OF EXPONENTS

I $a^m \cdot a^n = a^{m+n}$

II $(a^m)^n = a^{mn}$

III $(ab)^n = a^n b^n$

IV $\dfrac{a^m}{a^n} = a^{m-n}$ *where $a \neq 0$*

V $\left(\dfrac{a}{b}\right)^n = \dfrac{a^n}{b^n}$ *where $b \neq 0$*

EXAMPLE 5

(a) $(b^5)(b) = b^{5+1} = b^6$
(b) $(-2)^3(-2)^2 = (-2)^{3+2} = (-2)^5$
(c) $(2)(2)^3(2)^{-2} = (2^{1+3})(2^{-2}) = (2^4)(2^{-2}) = 2^2 = 4$
(d) $(a^2)^3 = a^{2 \cdot 3} = a^6$
(e) $[(3)^2]^4 = (3)^{2 \cdot 4} = 3^8$
(f) $[(-1)^3]^5 = (-1)^{3 \cdot 5} = (-1)^{15} = -1$
(g) $(ab)^4 = a^4 b^4$
(h) $(2x)^3 = (2)^3(x)^3 = 8x^3$
(i) $\dfrac{a^6}{a^3} = a^{6-3} = a^3$

(j) $\dfrac{x^2}{x^4} = x^{2-4} = x^{-2} = \dfrac{1}{x^2}$

(k) $(2)^3/(2)^7 = (2)^{3-7} = (2)^{-4} = 1/(2)^4 = \frac{1}{16}$

(l) $(x/y)^5 = x^5/y^5$

(m) $(2a/5b^2)^3 = (2a)^3/(5b^2)^3 = 8a^3/125b^6$

(n) $x^5/x^5 = x^{5-5} = x^0 = 1$

❑

Polynomial Expressions

In this section we will discuss some important definitions and terminology. First, **constants** are quantities which do not change in value. A constant may be represented by a letter or by the real number which equals the constant. For example, 5 is a constant, as is the letter b if $b = -20$. **Variables** are quantities whose value may change. These are usually represented by letters. For example, the letter t may be used to represent the temperature each hour in a particular city measured on either the Fahrenheit or Celsius scale. The value of t is likely to be different each hour.

An **algebraic expression** is a collection of constants and variables connected by a series of additions, subtractions, multiplications, divisions, radical signs, and parentheses or other grouping symbols. For example,

$$5x^2y - 10x^3 + 75$$

is an algebraic expression. This algebraic expression consists of the three *terms* $5x^2y$, $10x^3$, and 75. A **term** consists of either a single number or the product of a number and powers of one or more variables. The term $5x^2y$ consists of the **factors** 5, x^2, and y. The constant factor 5 is referred to as the *coefficient* of the term. In this book, **coefficient** will always refer to a constant which is a factor in a term. For instance, 10 is the coefficient on the term $10x^3$. The term 75 in the algebraic expression contains no variables and is referred to as a **constant term.**

A **polynomial** is the sum of one or more terms, with the following restrictions:

❑ The terms of a polynomial consist of a number or the product of a number and *positive integer* powers of one or more variables. This definition excludes terms which have variables under a radical sign and any terms which have variables in the denominator.

❑ A polynomial consisting of one term is called a **monomial.** A polynomial consisting of two terms is called a **binomial.** A polynomial consisting of three terms is called a **trinomial.** Polynomials consisting of more than three terms are referred to simply as polynomials.

EXAMPLE 6

(a) The algebraic expression 25 is a polynomial having one term; thus it is called a monomial.

(b) The algebraic expression $5x^2 - 2x + 1$ is a polynomial consisting of three terms; thus it is referred to as a trinomial.

(c) The algebraic expression $2x^2y/z$ is not a polynomial because the variable z appears in the denominator of the term.

(d) The algebraic expression $\sqrt{x}$ is not a polynomial because the variable appears under a radical.

(e) The algebraic expression $x^5 - 2x^4 - x^3 + 2x^2 + x + 9$ is a polynomial consisting of six terms.

❑

The ***degree of a term*** is the sum of the exponents on the variables contained in the term. For a term involving one variable, the degree is simply the exponent of the variable. The degree of the term $5x^3$ is 3 since the exponent is 3. The degree of the term $5x^2y^3z$ is 6 since the sum of the exponents of x, y, and z equals 6. The degree of a nonzero constant term is 0. To illustrate, the term -20 can be written in the equivalent form $-20x^0$. Thus, the degree of the term equals 0.

In addition to the categorization of terms by degree, polynomials may be classified by their degree. The ***degree of a polynomial*** is defined as the degree of the highest-degree term in the polynomial.

EXAMPLE 7

(a) The polynomial $2x^3 - 4x^2 + x - 10$ has terms of degree 3, 2, 1, and 0, respectively. Therefore, the degree of the polynomial is 3.

(b) The polynomial $4x^2y^3 - 6xy^5 + 2xy$ has terms of degree 5, 6, and 2, respectively. Thus the degree of the polynomial is 6.

❑

Addition and Subtraction of Polynomials

In adding and subtracting polynomials, we combine like terms. ***Like terms*** are terms which involve the same variables raised to the same powers. The terms $3x$ and $-4x$ are considered to be like terms because each involves the variable x raised (implicitly) to the first power. The fact that their coefficients (3 and -4) are different has no bearing on whether the two terms are like terms. Any real constants are considered to be like terms. The constants -5 and 18 can be envisioned as having the form $-5x^0$ and $18x^0$, which qualifies them as like terms.

When polynomials are added or subtracted, like terms may be combined into a simpler form. For example, the like terms $4x$ and $3x$ may be added in the following manner:

$$4x + 3x = (4 + 3)x$$
$$= 7x$$

Similarly,

$$5xy^2 - 2xy^2 + 6xy^2 = [5 + (-2) + 6]xy^2$$
$$= 9xy^2$$

Terms which are not like terms cannot be combined into a simpler form (the old "apples and oranges" problem). The sum $5x + 2y$ cannot be written in a simpler form.

To add or subtract polynomials, like terms should be identified and combined. Unlike terms are added or subtracted as indicated. The following examples illustrate this process.

EXAMPLE 8

$$(2x^2 - 5x + 10) + (4x^2 + 3x - 5) = 2x^2 - 5x + 10 + 4x^2 + 3x - 5$$
$$= 2x^2 + 4x^2 - 5x + 3x + 10 - 5$$
$$= 6x^2 - 2x + 5$$

EXAMPLE 9

$$(5x^2y + 2xy^2 - 4y^3) - (-3x^2y + y^3 - 10) = 5x^2y + 2xy^2 - 4y^3 + 3x^2y - y^3 + 10$$
$$= 5x^2y + 3x^2y + 2xy^2 - 4y^3 - y^3 + 10$$
$$= 8x^2y + 2xy^2 - 5y^3 + 10$$

❑

Multiplication of Polynomials

All the rules and properties of multiplication for real numbers apply when polynomials are multiplied. We will discuss two different multiplication situations: (1) multiplication of two monomials and (2) multiplication of two polynomials.

> **MONOMIAL MULTIPLICATION**
> To multiply two monomials, multiply their coefficients and multiply the variable terms using the rules of exponents.

EXAMPLE 10

(a) $(2x)(3x) = (2)(3)xx = 6x^2$
(b) $(5x^2)(-2x^3) = (5)(-2)x^2x^3 = -10x^5$
(c) $(3ab^2)(6a^3b) = (3)(6)aa^3b^2b = 18a^4b^3$
(d) $(mn^2)(4m^2n^3)(-3m^3n) = -12m^6n^6$

❑

> **POLYNOMIAL MULTIPLICATION**
> To multiply two polynomials, multiply *each* term of one polynomial by *every* term of the other polynomial.

EXAMPLE 11

(a) $(2)(4x - 2y) = (2)(4x) + (2)(-2y) = 8x - 4y$
(b) $4x^2y(x^2 + 2x - 1) = 4x^2y(x^2) + (4x^2y)(2x) + (4x^2y)(-1)$
$$= 4x^4y + 8x^3y - 4x^2y$$
(c) $(2x - 6)(4x + 7) = (2x)(4x + 7) + (-6)(4x + 7)$
$$= 8x^2 + 14x - 24x - 42$$
$$= 8x^2 - 10x - 42$$
(d) $(5x^2 - 2x)(x^3 + 2x^2 - 5x) = (5x^2)(x^3 + 2x^2 - 5x) + (-2x)(x^3 + 2x^2 - 5x)$
$$= 5x^5 + 10x^4 - 25x^3 - 2x^4 - 4x^3 + 10x^2$$
$$= 5x^5 + 8x^4 - 29x^3 + 10x^2$$

❑

Division of Polynomials

The only type of polynomial division *explicitly* required in this book will be the division of a polynomial by a monomial. When the division of two polynomials is required, the quotient can be found by simplifying the factored forms of the two polynomials. Factoring of polynomials is reviewed in the next section.

> **DIVISION OF MONOMIALS**
> To divide a monomial by a monomial, divide the coefficients of each monomial and divide the variables using the appropriate rule(s) of exponents.

EXAMPLE 12

(a) $\dfrac{12x^5}{3x^2} = \left(\dfrac{12}{3}\right)\left(\dfrac{x^5}{x^2}\right) = 4x^{5-2} = 4x^3$

(b) $\dfrac{-8x^3y^2}{2xy^2} = \left(\dfrac{-8}{2}\right)\left(\dfrac{x^3}{x}\right)\left(\dfrac{y^2}{y^2}\right) = -4x^{3-1}y^{2-2} = -4x^2(1) = -4x^2$

> **POLYNOMIAL DIVISION BY A MONOMIAL**
> To divide a polynomial by a monomial, divide each term of the polynomial by the monomial and algebraically sum the individual quotients.

EXAMPLE 13

(a) $\dfrac{4x^3 - 8x^2 + 6x}{2x} = \dfrac{4x^3}{2x} - \dfrac{8x^2}{2x} + \dfrac{6x}{2x} = 2x^2 - 4x + 3$

(b) $\dfrac{24a^4b^5 + 18a^2b^3}{-3a^2b^4} = \dfrac{24a^4b^5}{-3a^2b^4} + \dfrac{18a^2b^3}{-3a^2b^4} = -8a^2b - \dfrac{6}{b}$

> **NOTE** You can always check your answer in division by multiplying the answer times the divisor. If your answer is correct, this product should equal the dividend (numerator).

Section A.2 Follow-up Exercises

In Exercises 1–12, express the indicated operations using exponents.

1 $(5)(5)(5)(5)$

2 $(-1)(-1)(-1)(-1)(-1)(-1)(-1)$

3 $(3)(3)(-2)(-2)(-2)$

4 $(7)(7)(7)/[(3)(3)]$

5 $(-x)(-x)(-x)$

6 $aaa/(bb)$

7 $aabbbcc$

8 $xxyyyy/(zzz)$

9 $xxxx/yyzzzz$

10 $ppqqq/rrrrss$

11 $(xy)(xy)(xy)(xy)$

12 $(abc)(abc)(abc)/(3)(3)(3)(3)(3)$

In Exercises 13–32, perform the indicated operations.

13 $(2)^3(2)^4$

14 $(3)^3(3)^2$

15 x^3x^5

16 yy^4y^3

17 $x^2y^3x^3y$

18 aa^3a^2a

19 $(x^2)^3$

20 $(a^2)^5$

21 $(x^3)^2(x^2)^4$

22 $a^3(a^3)^4$

23 $[(a^2)^3]^2$

24 $[(-1)^4]^3$

25 $(3x^2)^3$

26 $(5a^3)^2$

27 $(2m^3)^2$

28 $(4b^4)^3$

29 $12(a^2)^4(b)^3$

30 $2(2x^2)^3(3y^3)^2$

31 $[(2x^2)^3]^4$

32 $[2(3a^2)^3]^2$

In Exercises 33–40, rewrite the expression, using positive exponents.

33 a^{-4}

34 $(xy)^{-2}$

35 $(\frac{1}{2})^{-3}$

36 x^{-1}

37 $(\frac{1}{3})^{-4}$

38 $(abc)^{-3}$

39 $(xy)^{-5}$

40 $(4x)^{-2}$

In Exercises 41–60, perform the indicated operation.

41 x^3/x

42 m^7/m^4

43 $(2)^5/(2)^8$

44 x^6/x^6

45 $(3)^4/(3)^3$

46 $(2x^2)^2/2(x^2)$

47 $(xy)^0$

48 $-(25x^0)^2$

49 $(x/y)^3$

50 $(\frac{4}{5})^3$

51 $(x^2/y)^4$

52 $(xy/z)^3$

53 $(a^2b/c^3)^4$

54 $(2x^2/5yz^3)^3$

55 $(3xy^2/z^3)^3$

56 $[(x^2/y)^3]^2$

57 $-5[2(x^0)^5]^2$

58 $2a^2[a^3/4b]^2$

59 $(a^2/b^3)^2(b^2/a^3)^3$

60 $(ab/c)^3(c/ab)^3$

In Exercises 61–94, perform the indicated operations.

61 $10x + 3x$

62 $5x^2 - 4x^2 + 2x^2$

63 $(5y^3 - 2y^2 + y) + (4y^2 - 5y)$

64 $(2m^2 - 3m) + (4m^2 + 2m) - (m^2 + 6)$

65 $(40x^3y^2 - 25xy^3) - (15x^3y^2)$

66 $abc - cab - 4bac$

67 $(x - 2y) - (2x - 3y) + (x - y)$

68 $(-5x)(4x^2)$

69 $(7x^3)(3xy^2)$

70 $(3x^2)(2x)(-4x^3)$

71 $(a^2)(4a^5)(-2a^3)$

72 $5x(x - 10)$

73 $(-2x^2)(x^2 - y)$

74 $2a(a^2 - 2a + 5)$

75 $x^2y(x^2 - 2xy + y^2)$

76 $(x - 5)(x + 6)$

77 $(a + b)(a + b)$

78 $(2x - 3)(2x - 3)$

79 $(a - b)(a - b)$

80 $(x + 4)(x - 4)$

81 $(x - 2)(x^2 - 4x + 4)$

82 $21x^5/(3x)$

83 $16x^2y^3/(4xy^2)$

84 $10a^4b^2/(5ab^2)$

85 $-9xy^2/(3xy^3)$

86 $25a^2bc^3/(5ab^2c^4)$

87 $(15x^2 - 24x)/(3x)$

88 $(4x^3y - 2x^2y + 8xy)/(2x)$

89 $(12a^3 - 9a^2 + 6a)/(-3a)$

90 $(3x^2yz^3 - 4xy^2z)/(-xyz)$

91 $(4x^6 + 6x^3 - 8x^2)/(2x)$

92 $(8a^3b^2c - 4a^2b^3c^2)/4a^2bc$

93 $(48x^3y^2 - 16x^2y^4 + 24xy^3)/-4xy^2$

94 $(-12x^8y^6z^2 + 28x^5y^4z^5)/(-4x^3y)$

A.3 FACTORING

In this section we discuss *factoring of polynomials.* To factor a polynomial means to express it as the product of two or more other polynomials. The *distributive law of multiplication* is

$$a(b + c) = a \cdot b + a \cdot c$$

The binomial on the right of the equals sign can be expressed as the product of the polynomials a and $b + c$. These two polynomials are considered the factors of the expression $a \cdot b + a \cdot c$. With multiplication of polynomials, we are given the factors and must find the product. With factoring we are given the product and must determine the polynomials which, when multiplied, will yield the product.

Monomial Factors

The distributive law represents an example of *monomial factors.* That is,

$$ab + ac = a(b + c)$$

indicates that the two terms on the left side of the equals sign contain a common factor a. The common factor a may represent any monomial. For example, the polynomial $2x + 2y$ can be rewritten in the *factored form* $2(x + y)$ since each term has a common factor of 2.

EXAMPLE 14   (a) The terms of the polynomial $x^3 - x^2 + x$ have a common factor x. We can rewrite the polynomial as

$$x^3 - x^2 + x = x(x^2 - x + 1)$$

(b) The terms of the polynomial $6x^2y^3 - 10xy^2$ have a common factor $2xy^2$. Factoring $2xy^2$ from each term, we obtain

$$6x^2y^3 - 10xy^2 = 2xy^2(3xy - 5)$$

❑

We are usually interested in factoring polynomials *completely.* That means simply that the factors themselves cannot be factored any further. The right side of the equation

$$x^3y^2 + x^4y^3 = xy(x^2y + x^3y^2)$$

is not factored completely. The term x^2y can be factored from each term inside the parentheses. The polynomial is completely factored when it is written as $x^3y^2(1 + xy)$.

The goal in monomial factoring is usually to identify the *largest* common monomial factor. The largest common monomial factor is the one containing the largest common numerical factor and the highest powers of variables common to all terms.

Quadratic Polynomials

A second-degree polynomial is often referred to as a *quadratic* polynomial. We will see these types of polynomials frequently, and factoring them will be important. Specifically, we will be interested in expressing quadratic polynomials, if possible, as the product of two first-degree polynomials. The factoring process often involves trial and error. Sometimes it is easy; at other times it can be frustrating. The following cases will help you.

CASE 1

$$x^2 + (a+b)x + ab = (x+a)(x+b)$$

Consider the product $(x+a)(x+b)$. Multiplying these two binomials, we get

$$(x+a)(x+b) = x^2 + ax + bx + ab$$
$$= x^2 + (a+b)x + ab$$

The result of the multiplication is a trinomial having an x^2 term, an x term, and a constant term. Note the coefficients of each term of the trinomial. The x^2 term has a coefficient equal to 1; the x term has a coefficient $a + b$, which is equal to the sum of the constants contained in the ***binomial factors;*** and the constant term ab is the product of the two constants contained in the binomial factors. In factoring a trinomial of this form, the objective is to determine the values of a and b which generate the coefficient of x, or the middle term of the polynomial, and the constant term.

EXAMPLE 15   To find the factors of $x^2 - 5x + 6$, we seek values for a and b such that

$$(x+a)(x+b) = x^2 - 5x + 6$$

The coefficient of the middle term is -5. From our previous discussion, the values of a and b must be such that $a + b = -5$. And, the third term in the trinomial equals 6, suggesting that $ab = 6$. Using a trial-and-error approach, you should conclude that the values satisfying these two conditions are -2 and -3. It makes no difference which of these two values we assign to a and b. The binomial factors are $(x-2)(x-3)$ or $(x-3)(x-2)$.

EXAMPLE 16   To find the factors of $m^2 + 6m - 21$, we look for values of a and b such that

$$(m+a)(m+b) = m^2 + 6m - 21$$

Our earlier discussions suggest that the relationships between a and b are

$$a + b = 6$$

and

$$ab = -21$$

Verify that there are no real (integer) values for a and b which satisfy these equations. Thus, this quadratic expression cannot be factored.

❑

CASE 2

$$acx^2 + (ad + bc)x + bd = (ax + b)(cx + d)$$

Consider the product

$$(ax + b)(cx + d) = acx^2 + (ad + bc)x + bd$$

Assuming that a and c are integers, both of which are not equal to 1, the product is a trinomial which differs from case 1 in that the coefficient of the term x^2 equals an integer other than 1. When a trinomial has an integer coefficient other than 1 on the x^2 term, the binomial factors contain four constants which must be identified. The coefficient of the x^2 term equals the product of a and c, the coefficient of the x term equals $ad + bc$, and the constant term equals the product bd. Identifying the values of the four constants which satisfy these conditions can be difficult. Verify that when $a = 1$ and $c = 1$, case 2 is simply case 1.

EXAMPLE 17 To find the factors of $6x^2 - 25x + 25$, we seek values of a, b, c, and d such that

$$6x^2 - 25x + 25 = (ax + b)(cx + d)$$

The conditions which must be satisfied are

$$ac = 6$$

$$ad + bc = -25$$

and $$bd = 25$$

Verify that the values $a = 3$, $b = -5$, $c = 2$, and $d = -5$ satisfy the conditions. And

$$(3x - 5)(2x - 5) = 6x^2 - 25x + 25$$

EXAMPLE 18 The first step in factoring is to look for any common monomial factors. Given the trinomial $12x^2 - 27x + 6$, we can factor 3 from each term, or

$$12x^2 - 27x + 6 = 3(4x^2 - 9x + 2)$$

The next step is to determine if the trinomial *factor* can be factored. If so,

$$ac = 4$$

$$ad + bc = -9$$

and $$bd = 2$$

Values satisfying these conditions are $a = 1$, $b = -2$, $c = 4$, and $d = -1$. Thus,

$$12x^2 - 27x + 6 = 3(x - 2)(4x - 1)$$

❏

CASE 3

$$x^2 - a^2 = (x + a)(x - a)$$

This case involves factoring the ***difference between perfect squares.*** The binomial to be factored is the difference between the squares of two quantities, x and a. This binomial can be factored as the product of the *sum* and *difference* of x and a.

EXAMPLE 19
$$\begin{aligned} x^2 - 9 &= (x)^2 - (3)^2 \\ &= (x + 3)(x - 3) \end{aligned}$$

EXAMPLE 20
$$\begin{aligned} 16x^4 - 81 &= (4x^2)^2 - (9)^2 \\ &= (4x^2 + 9)(4x^2 - 9) \end{aligned}$$

However, the binomial $4x^2 - 9$ is the difference between two squares. Thus,

$$16x^4 - 81 = (4x^2 + 9)(2x + 3)(2x - 3)$$

❏

Other Special Forms
The following rules of factoring are used less frequently in the book.

CASE 4

$$a^3 - b^3 = (a - b)(a^2 + ab + b^2)$$

This case involves factoring the ***difference between two cubes.***

EXAMPLE 21
(a) $x^3 - 1 = (x)^3 - (1)^3 = (x - 1)(x^2 + x + 1)$
(b) $8x^3 - 64 = (2x)^3 - (4)^3 = (2x - 4)(4x^2 + 8x + 16)$
(c) $m^3 - n^3 = (m - n)(m^2 + mn + n^2)$

❏

CASE 5

$$a^3 + b^3 = (a + b)(a^2 - ab + b^2)$$

This case involves factoring the **sum of two cubes.**

EXAMPLE 22
(a) $x^3 + 8 = (x)^3 + (2)^3 = (x + 2)(x^2 - 2x + 4)$
(b) $27y^3 + 64 = (3y)^3 + (4)^3 = (3y + 4)(9y^2 - 12y + 16)$

❑

Section A.3 Follow-up Exercises

Completely factor (if possible) the polynomials in the following exercises. Do not forget to check your answers!

1 $2ax - 8a^3$

2 $21m^2 - 7mn$

3 $4x^3y - 6xy^3 + 8x^2y^2$

4 $65a^3b^2 - 13a^2b^3$

5 $9a^3 - 15a^2 - 27a$

6 $x^2 - 8x + 12$

7 $x^2 + x + 3$

8 $x^2 + 7x + 12$

9 $p^2 + 9p - 36$

10 $x^2 - 2x - 15$

11 $r^2 - 21r - 22$

12 $x^2 - 16x + 48$

13 $x^5 + y^5$

14 $9x^2 + 12x + 4$

15 $6m^2 - 19m + 3$

16 $2x^2 - 7x - 4$

17 $8x^2 - 2x - 3$

18 $2x^3 + 4x^2 - 42x$

19 $x^4 - 81$

20 $100x^2 - 225$

21 $81x^4 - 625$

22 $10x^2 + 13x - 3$

23 $x^2 + 4$

24 $27 - 8m^3$

25 $1 + 8x^3$

26 $a^3 - 125$

27 $x^4 - x^3 - 2x^2$

28 $4x^6 - 4x^2$

29 $x^3 - 3x^2 - 40x$

30 $8 - 6x + x^2$

31 $x^5 - 4x^4 - 21x^3$

32 $x^6 - x^5$

33 $a^5b - 81ab$

34 $a^2b^3y^4 - 625a^2b^3$

35 $162uv - 2u^5v$

36 $6x^4y^3 + x^3y^3 - 5x^2y^3$

A.4 FRACTIONS

Fractions, or **rational numbers,** constitute an important part of the real number system. This section discusses some useful rules for performing computations with fractions.

Addition and Subtraction of Fractions

RULE 1: COMMON DENOMINATORS
If two fractions have the same denominator, their sum (difference) is found by adding (subtracting) their numerators and placing the result over the common denominator.

EXAMPLE 23
(a) $\dfrac{3}{7} + \dfrac{2}{7} = \dfrac{3 + 2}{7} = \dfrac{5}{7}$

(b) $\dfrac{7}{8} - \dfrac{4}{8} = \dfrac{7 - 4}{8} = \dfrac{3}{8}$

❑

> **RULE 2: DIFFERENT DENOMINATORS**
> To add (subtract) two fractions which have different denominators, restate the fractions as equivalent fractions having the same denominator. The sum (difference) is then found by applying rule 1.

In applying rule 2, *any* common denominator may be identified when the equivalent fractions are found. However, the usual practice is to identify the **least common multiple** (lcm) of the denominators or the **least common denominator** (lcd).

The procedure for finding the least common denominator is as follows:

1   *Write each denominator in a completely factored form.*
2   *The lcd is a product of the factors. To form the lcd, each distinct factor is included the greatest number of times it appears in any one of the denominators.*

EXAMPLE 24   To find the lcd for the fractions $\frac{5}{8}$ and $\frac{3}{20}$, each denominator is factored completely:

$$8 = 8 \cdot 1 = 4 \cdot 2 \cdot 1 = 2 \cdot 2 \cdot 2 \cdot 1$$

$$20 = 20 \cdot 1 = 10 \cdot 2 \cdot 1 = 5 \cdot 2 \cdot 2 \cdot 1$$

These denominators are factored completely since each of the factors can be expressed only as the product of itself and 1 (assuming we are seeking integer-valued factors). Such factors are called **prime factors.**

In forming the lcd, *each distinct prime factor is included the greatest number of times it appears in any one denominator.* The distinct prime factors are 2, 5, and 1. Thus,

$$\text{lcd} = 2 \cdot 2 \cdot 2 \cdot 5 \cdot 1 = 40$$

EXAMPLE 25   Determine the sum $\frac{5}{8} + \frac{3}{20}$.

SOLUTION

Having identified the lcd in the previous example, we must restate each fraction with the common denominator 40. Restating the fractions and applying rule 1, we get

$$\frac{5}{8} + \frac{3}{20} = \frac{5 \cdot 5}{8 \cdot 5} + \frac{3 \cdot 2}{20 \cdot 2} = \frac{25}{40} + \frac{6}{40} = \frac{25 + 6}{40} = \frac{31}{40}$$

EXAMPLE 26   Determine the difference $3/(4x) - 5/(6x^2)$.

SOLUTION

Factoring each denominator, we obtain

$$4x = 4 \cdot x \cdot 1 = 2 \cdot 2 \cdot x \cdot 1$$

$$6x^2 = 6 \cdot x \cdot x \cdot 1 = 3 \cdot 2 \cdot x \cdot x \cdot 1$$

The distinct factors of these denominators are 2, 3, x, and 1, and

$$
\begin{aligned}
\text{lcd} &= 2 \cdot 2 \cdot 3 \cdot x \cdot x \cdot 1 \\
&= 12x^3
\end{aligned}
$$

The fractions, when restated in terms of the lcd, are subtracted, yielding

$$
\begin{aligned}
\frac{3}{4x} - \frac{5}{6x^2} &= \frac{3 \cdot 3x}{4x \cdot 3x} - \frac{5 \cdot 2}{6x^2 \cdot 2} \\
&= \frac{9x}{12x^2} - \frac{10}{12x^2} \\
&= \frac{9x - 10}{12x^2}
\end{aligned}
$$

EXAMPLE 27 To find the algebraic sum $3/(x-1) - 5x/(x+1) + x^2/(x^2-1)$, we first determine the least common denominator, or

$$
\text{lcd} = (x+1)(x-1) \cdot 1 = (x^2 - 1)
$$

The three fractions are restated using the lcd, yielding

$$
\begin{aligned}
\frac{3}{x-1} - \frac{5x}{x+1} + \frac{x^2}{x^2-1} &= \frac{3 \cdot (x+1)}{(x-1) \cdot (x+1)} - \frac{5x(x-1)}{(x+1)(x-1)} + \frac{x^2}{x^2-1} \\
&= \frac{3(x+1) - 5x(x-1) + x^2}{x^2-1} \\
&= \frac{3x + 3 - 5x^2 + 5x + x^2}{x^2-1} \\
&= \frac{-4x^2 + 8x + 3}{x^2-1}
\end{aligned}
$$

□

Multiplication and Division

> **RULE 3: MULTIPLICATION**
> The product of two or more fractions is found by dividing the product of their numerators by the product of their denominators. That is,
>
> $$
> \frac{a}{b} \frac{c}{d} = \frac{ac}{bd}
> $$

EXAMPLE 28 (a) $\dfrac{3}{5} \dfrac{2}{7} = \dfrac{(3)(2)}{(5)(7)} = \dfrac{6}{35}$

(b) $\dfrac{15}{x}\dfrac{x^2}{3} = \dfrac{15x^2}{3x} = \dfrac{5x}{1} = 5x$

(c) $\dfrac{x-1}{10}\dfrac{15}{x^2-1} = \dfrac{(15)(x-1)}{(10)(x-1)(x+1)} = \dfrac{3}{2(x+1)}$

RULE 4: DIVISION

The quotient of two simple fractions can be determined by inverting the divisor fraction and multiplying by the dividend fraction. That is,

$$\frac{a/b}{c/d} = \frac{a}{b}\frac{d}{c} = \frac{ad}{bc}$$

EXAMPLE 29

(a) $\dfrac{-\frac{5}{12}}{\frac{3}{4}} = \left(-\dfrac{5}{12}\right)\left(\dfrac{4}{3}\right) = -\dfrac{20}{36} = -\dfrac{5}{9}$

(b) $\dfrac{\frac{4}{10}}{2} = \dfrac{\frac{4}{10}}{2/1} = \left(\dfrac{4}{10}\right)\left(\dfrac{1}{2}\right) = \dfrac{4}{20} = \dfrac{1}{5}$

(c) $\dfrac{3x^2/4}{9x/2} = \dfrac{3x^2}{4}\dfrac{2}{9x} = \dfrac{6x^2}{36x} = \dfrac{x}{6}$

(d) $\dfrac{1-2/x}{4/x} = \dfrac{x/x-2/x}{4/x} = \dfrac{(x-2)/x}{4/x} = \dfrac{x-2}{x}\dfrac{x}{4} = \dfrac{x-2}{4}$

Section A.4 Follow-up Exercises

In Exercises 1–25, perform the indicated operations.

1 $\frac{1}{5} + \frac{5}{30}$

2 $\frac{2}{7} - \frac{4}{21}$

3 $\frac{1}{3} - \frac{5}{8} + \frac{5}{12}$

4 $\frac{4}{25} - \frac{3}{10} + \frac{7}{5}$

5 $\dfrac{1}{x} - \dfrac{2}{x^2}$

6 $\dfrac{5}{2a} + \dfrac{6}{a^3}$

7 $\dfrac{5x}{x^2-4} + \dfrac{x}{x-2}$

8 $\dfrac{5}{1} + \dfrac{1}{x}$

9 $\dfrac{10}{1} - \dfrac{2}{x^2}$

10 $\dfrac{4}{a} + \dfrac{3}{2ab}$

11 $\dfrac{3a}{a+1} - \dfrac{5}{a^2+2a+1}$

12 $\frac{3}{11}\cdot\frac{33}{6}$

13 $(\frac{1}{5})(\frac{10}{3})(-\frac{9}{5})$

14 $\left(\dfrac{1}{x}\right)\left(\dfrac{2x^3}{3}\right)\left(\dfrac{6}{5}\right)$

15 $\left(\dfrac{ab}{c}\right)\left(\dfrac{c^2}{3a^2b}\right)\left(\dfrac{1}{abc}\right)$

16 $\left(\dfrac{5}{x-4}\right)\left(\dfrac{x^2-16}{10}\right)\left(\dfrac{x+4}{2}\right)$

17 $\frac{7}{27} \div \frac{5}{9}$

18 $3x^2/5 \div x/5$

19 $a^2b/(5c) \div 3c^2/(10ab)$

20 $abc/8 \div 3a^2b/4$

21 $\dfrac{x-1}{x^2-5x-4} \div \dfrac{x-1}{x-4}$

22 $\dfrac{1-2/(3x)}{3/x+4}$

23 $\dfrac{x^2-1}{x^2} \div \dfrac{x-1}{x^3}$

24 $\dfrac{x^2+x-6}{x^2+x-2} \div \dfrac{x^2-9}{x^2-1}$

25 $\left[\dfrac{x^2+x-2}{x^2+x-6} \div \dfrac{x^2+7x+10}{x^2-x-2}\right] \cdot \dfrac{x^2+x-6}{x^2+3x+2}$

A.5 EXPONENTS AND RADICALS

In Sec. A.2 we discussed the following five laws of exponents:

$$\textbf{I}\quad a^m \cdot a^n = a^{m+n}$$
$$\textbf{II}\quad (a^m)^n = a^{mn}$$
$$\textbf{III}\quad (ab)^n + a^n b^n$$
$$\textbf{IV}\quad \dfrac{a^m}{a^n} = a^{m-n} \qquad a \neq 0$$
$$\textbf{V}\quad \left(\dfrac{a}{b}\right)^n = \dfrac{a^n}{b^n} \qquad b \neq 0$$

Recall that the exponents were restricted to integer values.

Fractional Exponents

Occasionally we will need to deal with fractional exponents. The laws of exponents are valid for any real values of m and n. The next example illustrates the application of the laws of exponents when the exponents are fractions.

EXAMPLE 30

(a) $x^{1/2} \cdot x^{1/2} = x^{1/2+1/2} = x$

(b) $x^{3/2} \cdot x^{1/3} = x^{3/2+1/3} = x^{9/6+2/6} = x^{11/6}$

(c) $(x^{1/2})^4 = x^{(1/2)(4)} = x^2$

(d) $(x^{2/3})^{-3} = x^{(2/3)(-3)} = x^{-2} = 1/x^2$

(e) $(2x^{1/4})^4 = (2)^4(x^{1/4})^4 = 16x$

(f) $x^{3/4}/x^{1/2} = x^{3/4-1/2} = x^{3/4-2/4} = x^{1/4}$

(g) $x^{5/8}/x^{3/4} = x^{5/8-3/4} = x^{5/8-6/8}$
$\qquad\qquad = x^{-1/8} = 1/x^{1/8}$

(h) $(x/y)^{1/2} = x^{1/2}/y^{1/2}$

☐

Radicals

Frequently we need to determine the value of x which satisfies an equation of the form

$$x^n = a$$

For example, what values of x satisfy these equations?

$$x^2 = 4 \qquad x^3 = 8 \qquad x^4 = 81$$

In the first equation, we want to determine the value x which, when multiplied times itself, yields a product equal to 4. You should conclude that values of $+2$ and

-2 satisfy the equation, i.e., make the left and right sides of the equation equal. Similarly, the second equation seeks the value of x which, when cubed, generates a product of 8. A value of $+2$ satisfies this equation. Verify that $+3$ and -3 satisfy the third equation.

DEFINITION
If $a^n = b$, a is called the nth root of b.

The nth root of b is denoted by $\sqrt[n]{b}$, where the symbol $\sqrt{}$ is the **radical sign,** n is the **index** on the radical sign, and b is the **radicand.** Thus, we can state:

$$\boxed{\text{If } a^n = b, \text{ then } a = \sqrt[n]{b}.}$$

Referring to the three previous equations,

$$\text{If } x^2 = 4 \qquad x = \sqrt[2]{4} = \sqrt{4}$$

where x is said to equal the **square root** of 4. *If no index appears with the radical sign, the index is implicitly equal to 2.*

For the second equation we can state

$$\text{If } \quad x^3 = 8 \qquad x = \sqrt[3]{8}$$

where x is said to equal the **cube root** of 8. And for the third equation,

$$\text{If } \quad x^4 = 81 \qquad x = \sqrt[4]{81}$$

where x is said to equal the **fourth root** of 81.

As we have seen with these equations, there may exist more than one nth root of a real number. We usually will be interested in just one of these roots — the **principal nth root.** Given $\sqrt[n]{b}$:

1   *The principal nth root is positive if b is positive.*
2   *The principal nth root is negative if b is negative and n is odd.*

The following examples indicate the principal nth root.

EXAMPLE 31   (a) $\sqrt{9} = 3$
(b) $\sqrt[3]{-27} = -3$
(c) $\sqrt[5]{32} = 2$
(d) $\sqrt[5]{-243} = -3$

❏

The following laws apply to computations involving radicals.

LAWS OF RADICALS

I $(\sqrt[n]{a})^n = a$

II $a\sqrt[n]{x} + b\sqrt[n]{x} = (a + b)\sqrt[n]{x}$

III $\sqrt[n]{ab} = \sqrt[n]{a}\sqrt[n]{b}$

IV $\sqrt[n]{\dfrac{a}{b}} = \dfrac{\sqrt[n]{a}}{\sqrt[n]{b}}$ for $b \neq 0$

V $b^{m/n} = (\sqrt[n]{b})^m = \sqrt[n]{b^m}$

EXAMPLE 32

(a) $(\sqrt{4})^2 = (2)^2 = 4$

(b) $(\sqrt[5]{36})^5 = 36$

(c) $(\sqrt[3]{-8})^3 = (-2)^3 = -8$

(d) $\sqrt[3]{a} - 3\sqrt[3]{a} + 5\sqrt[3]{a} = 3\sqrt[3]{a}$

(e) $\sqrt{x} + \sqrt[3]{x}$ cannot be simplified using the laws of radicals because the indices on the two radicals are different.

(f) $\sqrt[4]{x^3} + \sqrt[4]{x^2}$ cannot be simplified using the laws of radicals because the radicands are not equal.

(g) $\sqrt[3]{128} = \sqrt[3]{(64)(2)} = \sqrt[3]{64}\,\sqrt[3]{2} = 4\sqrt[3]{2}$

(h) $\sqrt{x^3} = \sqrt{x^2 \cdot x} = \sqrt{x^2}\,\sqrt{x} = x\sqrt{x}, \, x \geq 0$

(i) $\sqrt{\frac{4}{9}} = \sqrt{4}/\sqrt{9} = \frac{2}{3}$

(j) $\sqrt[3]{(-1)/125} = \sqrt[3]{-1}/\sqrt[3]{125} = -1/5$

(k) $x^{1/2} = \sqrt{x}$

(l) $x^{1/3} = \sqrt[3]{x}$

(m) $x^{1/n} = \sqrt[n]{x}$

(n) $(64)^{2/3} = \sqrt[3]{(64)^2} = (\sqrt[3]{64})^2 = 4^2 = 16$

(o) $(49)^{-1/2} = 1/(49)^{1/2} = 1/\sqrt{49} = \frac{1}{7}$

☐

Section A.5 Follow-up Exercises

In Exercises 1–14, perform the indicated operations.

1 $a^{3/2} \cdot a^{4/3}$

2 $b^{1/6} \cdot b^{1/4}$

3 $x^{1/3} \cdot x^{2/5} \cdot x^{3/10}$

4 $(x^{1/2})^{2/3}$

5 $(a^{3/2})^{5/6}$

6 $(2x^{3/4})^4$

7 $(-3x^{2/3})^3$

8 $x^{5/2}/x^{1/2}$

9 $a^{3/2}/a^{1/6}$

10 $(x^4y^2)^{1/2}$

11 $(x^{2/3} \cdot x^{4/5})^2$

12 $(a^6b^{15})^{1/3}$

13 $(x^{2/5} \cdot x^{1/3}) \div x^{3/5}$

14 $(2a^{2/3})^5 \div 4a^{1/3}$

In Exercises 15–28, determine the principal nth root.

15 $\sqrt{625}$

16 $\sqrt[4]{625}$

17 $\sqrt[3]{-a^3}$

18 $\sqrt[5]{-1}$

19 $\sqrt[3]{-8x^6}$

20 $\sqrt[3]{27a^9}$

21 $\sqrt{144x^6}$

22 $\sqrt[3]{-64x^3y^6}$

23 $\sqrt[4]{16a^8b^4}$

24 $\sqrt[3]{x^{12}y^{21}}$

25 $\sqrt[5]{a^{15}b^5c^{30}}$

26 $\sqrt[5]{-32x^{20}y^{40}}$

27 $\sqrt[4]{1296a^8}$

28 $\sqrt{900a^2b^6c^4}$

In Exercises 29–44, simplify the radical expressions.

29 $2\sqrt{7} + 3\sqrt{7}$

30 $5\sqrt{x} - 3\sqrt{x}$

31 $\sqrt{32} + 3\sqrt{2}$

32 $2\sqrt{45} - 2\sqrt{5}$

33 $4\sqrt{x} - \sqrt{x^3}$

34 $\sqrt{20} - 2\sqrt{5} + 3\sqrt{45}$

35 $\sqrt{2}\,\sqrt{8}$

36 $\sqrt[3]{5}\,\sqrt[3]{10}\,\sqrt[3]{5}$

37 $\sqrt{\frac{64}{9}}$

38 $\sqrt[3]{-\frac{1}{27}}$

39 $\sqrt{625x^2/(49y^4)}$

40 $\sqrt[4]{1/(81a^8)}$

41 $\sqrt[3]{\dfrac{64x^6}{27y^9}}$

42 $\sqrt[4]{10}\,\sqrt[4]{100}\,\sqrt[4]{10}$

43 $\dfrac{\sqrt{2}\,\sqrt{8}\,\sqrt{4}}{\sqrt[3]{32}\,\sqrt[3]{2}}$

44 $\dfrac{\sqrt{x^3y^5}\,\sqrt{xy^3}}{\sqrt[4]{x^7y^2}\,\sqrt[4]{xy^6}}$

In Exercises 45–56, express the term in radical form.

45 $x^{2/3}$

46 $x^{1/5}$

47 $(ab)^{3/5}$

48 $(xy)^{3/4}$

49 $x^{-1/2}$

50 $a^{-2/3}$

51 $(8)^{-1/3}$

52 $(32)^{-1/5}$

53 $a^{3/5}$

54 $(x+y)^{2/3}$

55 $(100-x)^{1/4}$

56 $(64x^{12}y^{24})^{1/6}$

In Exercises 57–68, express the term using fractional exponents.

57 $\sqrt{45x}$

58 $\sqrt[3]{a^2}$

59 $\sqrt[4]{x^3}$

60 $\sqrt{xy}$

61 $\sqrt[3]{x^5}$

62 $\sqrt[5]{(ab)^3}$

63 $\sqrt{x^4}$

64 $\sqrt[3]{(-1)^9}$

65 $\sqrt{x+y}$

66 $\sqrt[3]{(x-y)^2}$

67 $\sqrt[4]{(3-x)^3}$

68 $\sqrt[5]{(x-2x+y)^2}$

SUMMATION NOTATION

The Greek letter Σ (sigma) is the mathematical symbol which denotes the summation or addition operation. It provides a type of "shorthand" notation for representing addition. The expression

$$\sum_{j=l}^{u} f(j) \tag{B.1}$$

is read "summation of $f(j)$, where j goes from l to u." To the right of Σ is the general function or expression being added. The letter j beneath Σ is the summation index. The summation index increments one unit at a time from a lower limit l to an upper limit u. For each value of j, $f(j)$ is evaluated and added to the other values of $f(j)$.

Suppose that we wanted to add the positive integers 1 through 10. One way of denoting this is by the expression

$$\sum_{j=1}^{10} j$$

The longhand equivalent of this expression is

$$1 + 2 + 3 + 4 + 5 + 6 + 7 + 8 + 9 + 10$$

The following are other examples of summation notation.

$$\sum_{j=5}^{8} j^2 = (5)^2 + (6)^2 + (7)^2 + (8)^2 = 174$$

$$\sum_{i=1}^{4} (i^3 - 1) = [(1)^3 - 1] + [(2)^3 - 1] + [(3)^3 - 1] + [(4)^3 - 1] = 96$$

$$\sum_{i=1}^{3} (-3i) = (-3)(1) + (-3)(2) + (-3)(3) = -18$$

$$\sum_{j=1}^{5} x_j = x_1 + x_2 + x_3 + x_4 + x_5$$

Note that the name of the index is not restricted to j.

Summation notation can provide considerable efficiency in expressing the summation operation. It is a convenient way of representing systems of equations. And it has particular value when the computer can be used to perform computations.

SELECTED ANSWERS: Follow-up Exercises and Chapter Tests

CHAPTER 1

SEC. 1.1   **1** $x = 3$;   **3** $x = 4$;   **5** $x = -18$;   **7** $t = -1$;   **9** $t = 1$;
11 $t = 16$;   **13** $x = -1$;   **15** $t = 5$;   **17** no roots;   **19** $x = 3$;   **21** $x = 6$;
23 $x = 5$

SEC. 1.2   **1** $x = -3, 2$;   **3** $x = -1$;   **5** $x = -2, 5$;   **7** $t = -\frac{1}{2}, -4$;
9 $y = -\frac{1}{2}, 2$;   **11** $r = -4, 4$;   **13** cannot be factored (no real roots if solved using quadratic formula);   **15** $y = \frac{1}{2}, -5$;   **17** $x = -2, -6$;   **19** $r = -1$;
21 no roots;   **23** $x = -4, 1$;   **25** no roots;   **27** no roots;   **29** no roots;
31 no roots

SEC. 1.3   **1, 3, 5, 7, 9, 11, 13, 15** see figures (p. A-26);   **17** $x \geq -10$;
19 no solution;   **21** $x \leq 4$;   **23** $x \geq -6$;   **25** $-36 \leq x \leq -4$;
27 no solution;   **29** $-18 \leq x \leq 7$;   **31** $20 \leq x \leq 40$;   **33** $-4 \leq x \leq 4$;
35 $-6 \leq x \leq 3$;   **37** $x \leq -1$ or $x \geq 3$;   **39** $-\frac{1}{2} < x < 2$;   **41** $x < -5$ or $x > 3$;
43 $-5 < x < 5$

SEC. 1.4   **1** $x = -8, 8$;   **3** no solution;   **5** $x = 3, 9$;   **7** $x = -10, 4$;
9 $x = 2, 3$;   **11** $x = -\frac{1}{3}, -9$;   **13** $x = -2, \frac{12}{5}$;   **15** $-1 < x < 1$;
17 $x = $ any real number;   **19** $-2.5 \leq x \leq 2.5$;   **21** $-5 < x < 15$;
23 $x < -1$ or $x > 4$;   **25** no solution;   **27** $t \geq 24$ or $t \leq -24$;
29 $x = 0$, $x \leq -2$, or $x \geq 2$

SEC. 1.5   **1** $(1.5, 3.5)$;   **3** $(7.5, -1)$ **5** $(7.5, 15)$;   **7** $(-2, 15)$;   **9** $(6, -8)$;
11 $(7.5, -3)$;   **13** $(0, 0)$;   **15** $\sqrt{52} = 7.21$;   **17** 5;   **19** $\sqrt{85} = 9.219$;
21 $\sqrt{244} = 15.62$;   **23** $\sqrt{116} = 10.77$;   **25** 5;   **27** $\sqrt{41} = 6.40$;
29 $\sqrt{68} = 8.246$

CHAPTER TEST   **1** no roots;   **2** no roots;   **3** $x = 3, 4$;   **4** $x \geq 4$;
5 $-2 \leq x \leq -1$;   **6** $x = 8$;   **7** $-20 \leq x \leq -4$;   **8** (a) $(1, -2)$, (b) $\sqrt{500} = 22.36$

CHAPTER 2

SEC. 2.1   **1** linear (L);   **3** L;   **5** nonlinear (NL);   **7** L;   **9** L;   **11** L;
13 L;   **15** NL;   **17** L;   **19** (a) $a = 8$, $b = 0$, $c = 120$, (b) $(15, 10)$,
(c) no pair of values; x must equal 15, (d) $S = \{(x, y) | x = 15 \text{ and } y \text{ is any real number}\}$;

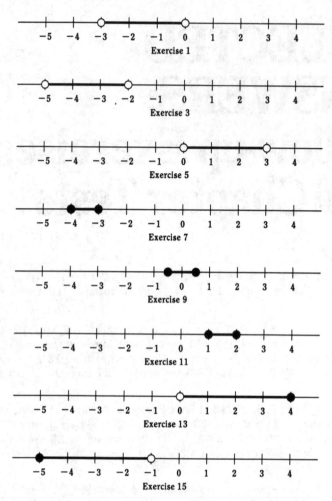

Exercise 1

Exercise 3

Exercise 5

Exercise 7

Exercise 9

Exercise 11

Exercise 13

Exercise 15

Sec. 1.3, Exercises 1, 3, 5, 7, 9, 11, 13, 15

21 (a) (2, 7, 1), (b) (0, 0, 0);   **23** (a) (4, 2, 20, 15), (b) no values, (c) (0, 0, 20, 0);
25 (a) $8x_1 + 24x_2 + 16x_3 = 120$, (b) 15 ounces of food 1, 5 ounces of food 2,
7.5 ounces of food 3;   **27** (a) $30x_1 + 60x_2 + 50x_3 + 80x_4 = 15,000$,
(b) $x_1 = 500$, $x_2 = 250$, $x_3 = 300$, $x_4 = 187.5$; (c) $x_1 = 500$ (both weight and volume),
$x_2 = 200$ (weight), $x_3 = 240$ (weight), $x_4 = 120$ (weight);
29 (a) $150,000x_1 + 180,000x_2 + 250,000x_3 = 100,000,000$, (b) 196 miles,
(c) 666.67 (or 666) buses, 555.55 (or 555) subway cars, or 400 miles of repaving

SEC. 2.2   1 (8, 0) and (0, −6);   **3** (−9, 0) and (0, 3);
5 (−3, 0), no y intercept;   **7** (0, 0) is both x and y intercept;
9 (2.5, 0) and (0, −4);   **11** (−18, 0) and (0, 6);   **13** no x intercept, (0, 6);
15 (0, t/b), (t/a, 0);   **17** no y intercept unless $q = 0$,
in which case there is an infinite number, (q/p, 0);
19 no x intercept unless $s = 0$ (then, infinite number), (0, −s/r);
21, 23, 25, 27, 29, 31, 33, and **35** see figures (pp. A-27 and A-28);

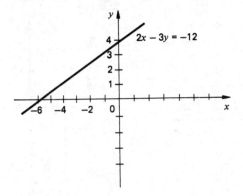

Sec. 2.2, Exercise 21

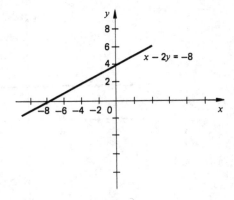

Sec. 2.2, Exercise 23

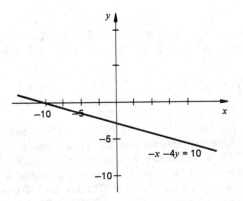

Sec. 2.2, Exercise 25

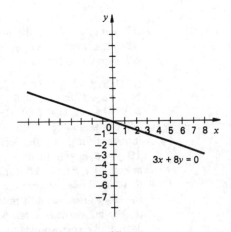

Sec. 2.2, Exercise 27

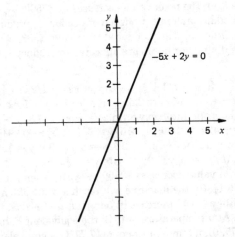

Sec. 2.2, Exercise 29

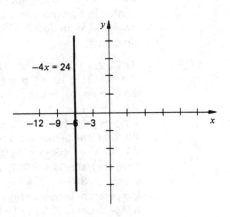

Sec. 2.2, Exercise 31

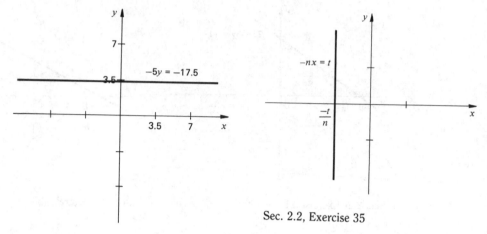

Sec. 2.2, Exercise 33

Sec. 2.2, Exercise 35

37 $y = 0$, $x = 0$;　　**39** $m = -3$;　　**41** $m = 3.5$;　　**43** $m = -2.5$;　　**45** $m = -3$;　　**47** undefined;　　**49** $m = 0$;　　**51** $m = 2$;　　**53** $m = b/a$;　　**55** $m = 1$;　　**57** $m = 0$;　　**59** $m = -1$

SEC. 2.3　**1** $y = \frac{3}{2}x - 7.5$, $m = \frac{3}{2}$, $(0, -7.5)$;　　**3** $y = \frac{4}{3}x - 6$, $m = \frac{4}{3}$, $(0, -6)$;　**5** $y = x + 8$, $m = 1$, $(0, 8)$;　　**7** $y = -x/2 - 6$, $m = -\frac{1}{2}$, $(0, -6)$;　**9** $y = \frac{3}{5}x + 4$, $m = \frac{3}{5}$, $(0, 4)$;　　**11** $y = -\frac{3}{2}x$, $m = -\frac{3}{2}$, $(0, 0)$;　**13** $y = \frac{4}{3}x$, $m = \frac{4}{3}$, $(0, 0)$;　　**15** $y = x/3 + \frac{5}{3}$, $m = \frac{1}{3}$, $(0, \frac{5}{3})$;　　**17** no slope-intercept form ($x = 0$), slope undefined, infinite number of y intercepts;　**19** $y = 3$, $m = 0$, $(0, 3)$;　　**21** $y = -(m/n)x + p/n$, $m = -m/n$, $(0, p/n)$;　　**23** $y = c/d$, $m = 0$, $(0, c/d)$;　　**25** (b) $m = 1.2$, $(0, 29.6)$, (c) Each year, the number of women (aged 35 to 44) in the labor force increases by 1.2 million; in 1981, there were 29.6 million women in the labor force, (d) 46.4 million in 1995, 52.4 million in 2000;　　**27** (a) $m = \frac{5}{9}$, $(0, -\frac{160}{9})$, (b) For an increase in temperature of $1°F$, the Celsius temperature increases by $\frac{5}{9}$ of a degree, $0°F$ corresponds to $-\frac{160}{9}°C$, (c) $F = \frac{9}{5}C + 32$;　　**29** (a) $(8, 0)$ and $(0, 60,000)$, (b) At age 8 years, the book value equals 0; when new, the book value equals $60,000, (c) For each year the machine ages, the book value decreases by $7500;　**31** $y = -\frac{2}{3}x + 33.33$, for each additional unit produced of product 1, production of product 2 must be decreased by $\frac{2}{3}$ unit; if no units are produced of product 1, 33.33 units can be produced of product 2; $(50, 0)$, if all labor hours are allocated to product 1, 50 units can be produced

SEC. 2.4　**1** $y = -2x + 10$;　　**3** $y = x/2 + \frac{3}{4}$;　　**5** $y = -rx - t/2$;　**7** $y = -3x + 10$;　　**9** $y = \frac{3}{2}x - \frac{1}{2}$;　　**11** $y = 2.5x + 10$;　　**13** $y = 5.6x - 18.24$;　**15** $y = wx + q - wp$;　　**17** no slope-intercept form, $x = -3$;　　**19** $y = v$;　**21** $y = -4x - 11$;　　**23** $y = -42x + 1080$;　　**25** $y = -1.042x + 20.994$;　**27** $y = [(d - b)/(c - a)]x + (bc - ad)/(c - a)$;　　**29** $y = b$;　　**31** $y = \frac{3}{4}x - \frac{11}{2}$;　**33** (a) slope undefined; $x = 7$, (b) $m = 0$, $y = 2$;　　**35** $y = -\frac{1}{4}x - 6$;　**37** (a) $V = -3500t + 18,000$, (b) value decreases $3500 per year; when the machine is new, the value is $18,000, (c) $(5.14, 0)$, the machine will reach a value of 0 after 5.14 years;　　**39** $F = 1.8C + 32$, For each $1°$ increase in temperature Celsius, the Fahrenheit temperature increases by $1.8°$. At a temperature of $0°C$, the equivalent Fahrenheit temperature is $32°$. For $(-17.77, 0)$, a temperature of $-17.77°C$ is equivalent to $0°F$　**41** (a) error $= 0.12$, (b) June 30, 1991: actual $= \$12.93$, forecast $= \$13.00$, error $= \$0.07$; March 31, 1992: actual $= \$13.43$, forecast $= \$13.72$, error $= \$0.29$

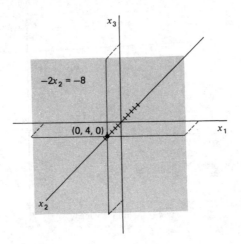

Sec. 2.5, Exercise 5

SEC. 2.5   1 $A(0, 4, 0)$, $B(3, 4, 0)$, $C(3, 0, 0)$, $D(-6, 0, 0)$, $E(-6, 0, 6)$, $F(-6, -2, 6)$, $G(0, -2, 6)$, $H(0, 0, 6)$, $I(0, -2, 0)$;   **3** $(7.5, 0, 0)$, $(0, -5, 0)$, $(0, 0, 15)$;
5 see figure;   **7** Planes which are parallel to the axis of the missing variable and which intersect the axes of the two included variables

SEC. 2.6   1 $0 \le x_1 \le 300$, $0 \le x_2 \le 200$, $0 \le x_3 \le 750$, $0 \le x_4 \le 1000$;
3 $0 \le x_1 \le 142{,}857$, $0 \le x_2 \le 83{,}333$, $0 \le x_3 \le 40{,}000$, $0 \le x_4 \le 50{,}000$, $0 \le x_5 \le 10{,}000$, $0 \le x_6 \le 20{,}000$;   **5** emergency airlift;   **7** $5x_1 + 3.5x_2 + 7.5x_3 = 240$, $x_1 = 48$, $x_2 = 68.57$, $x_3 = 32$;   **9** $25{,}000x_1 + 18{,}000x_2 + 15{,}000x_3 = 10{,}000{,}000$, only TV: \$400 (thousand), only radio: \$555.56 (thousand), only newspaper: \$666.67 (thousand).

CHAPTER 3

SEC. 3.1   1 infinite;   **3** unique;   **5** no solution;   **7** infinite;   **9** unique;
11, 13, 15, 17, 19 see figures (pp. A-29 and A-30);   **21** no solution;

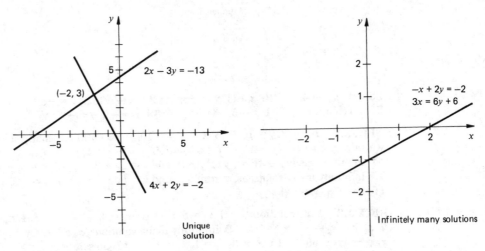

Sec. 3.1, Exercise 11   Sec. 3.1, Exercise 13

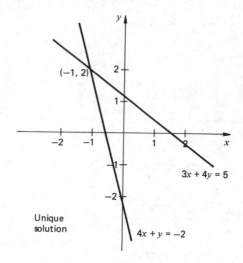

$(-1, 2)$

$3x + 4y = 5$

$4x + y = -2$

Unique
solution

Sec. 3.1, Exercise 15

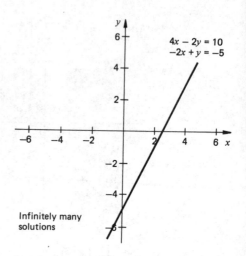

$4x - 2y = 10$
$-2x + y = -5$

Infinitely many
solutions

Sec. 3.1, Exercise 17

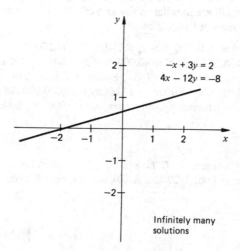

$-x + 3y = 2$
$4x - 12y = -8$

Infinitely many
solutions

Sec. 3.1, Exercise 19

23 $x = 0, y = 4$;   **25** $x = 3, y = -3$;   **27** $x = -10.5, y = -36$;
29 no solution;   **31** $x = 5, y = -2$;   **33** $x = -2, y = -3$

SEC. 3.2   1 $x = 3, y = 1$;   **3** $x = 5, y = 10$;   **5** infinitely many solutions,
y arbitrary and $x = 2y - 4$;   **7** no solution;   **9** $x = 1, y = -4$;
11 infinitely many solutions, y arbitrary and $x = 2y + 8$;   **13** $x = 0, y = -2$;
15 infinitely many solutions, y arbitrary and $x = \frac{1}{2}y + \frac{7}{4}$;   **17** $x = -3, y = -3$;
19 $x = -\frac{2}{11}, y = -\frac{21}{11}$

SEC. 3.3   1 no solution;   **3** $x_1 = 5, x_2 = 10, x_3 = 0$;   **5** $x_1 = -1, x_2 = 0, x_3 = 3$
7 $x_1 = 4, x_2 = -2, x_3 = 1$;   **9** infinitely many solutions, x_3 arbitrary, $x_1 = -\frac{14}{3}x_3 - 48$,
$x_2 = -\frac{13}{3}x_3 - 66$ **11** $x_1 = 3, x_2 = 1, x_3 = 2$;   **13** no solution;

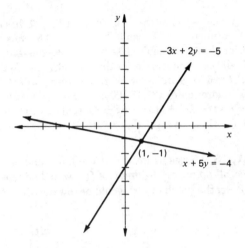

Chap. 3, Chapter Test, Problem 1

15 infinitely many solutions, x_2 arbitrary, x_3 arbitrary, $x_1 = 2x_2 - x_3 - 10$;
17 no solution;　**19** infinitely many solutions, x_2 arbitrary, x_3 arbitrary, and
$x_1 = \frac{1}{3}x_2 + \frac{2}{3}x_3 - 1$;　**21** (a), (c), (d), (e) unique, no solution, or infinitely many,
(b) no solution or infinitely many

SEC. 3.4　1 $x_1 = 200$, $x_2 = 350$, $x_3 = 100$;　**3** $x_1 = 100$, $x_2 = 200$, and $x_3 = 200$;
5 a blend consisting only of 60,000 gallons of component 2;　**7** 3 ounces of food 1,
2 of food 2, and 4 of food 3

CHAPTER TEST　1 see figure;　**2** $x = 3$, $y = -5$;　**3** (a) unique, no solution, or
infinitely many, (b) no solution or infinitely many;　**4** infinitely many solutions,
x_3 arbitrary, $x_1 = -\frac{3}{2}x_3 + 5$, $x_2 = -\frac{1}{4}x_3 - \frac{10}{4}$;　**5** (a) no solution, (b) $x_1 = 4$, $x_2 = -2$,
$x_3 = 1$, $x_4 = 3$;　**6** $x_1 = 10$, $x_2 = -5$, $x_3 = 10$, there is no combination of the three
products which would use the weekly capacities in all three departments

CHAPTER 4

SEC. 4.1　1 $-10, -20, 5a + 5b - 10$;　**3** $4, 6, -a - b + 4$;
5 $b, -2m + b, ma + mb + b$;　**7** $-9, -5, a^2 + 2ab + b^2 - 9$;
9 $-5, -3, a^2 + 2ab + b^2 + a + b - 5$;　**11** $-10, -18, a^3 + 3a^2b + 3ab^2 + b^3 - 10$;
13 $0, 16, (a + b)^4$;　**15** $4, 0, a^3 + 3a^2b + 3ab^2 + b^3 - 2a - 2b + 4$;
17 all real numbers;　**19** all real numbers;　**21** all real numbers;
23 all real numbers;　**25** $x \geq -4$;　**27** $t \leq -8$;　**29** all real numbers;
31 $x \neq 4$;　**33** $u \neq 1 + \sqrt{6}$ or $1 - \sqrt{6}$;　**35** $x \geq 8$;　**37** $h > 2, h \leq -2$ except
$h \neq -3$;　**39** $x \geq -3$ or $x \leq -5$;　**41** $0 \leq x \leq 50,000$, $80,000 \leq C(x) \leq 830,000$;
43 $0 \leq p \leq 6000$, $0 \leq q \leq 180,000$;　**45** $10 \leq x \leq 500$, $\$175 \leq p \leq \$1400$;
47 $200 \leq k \leq 1500$, $\$24 \leq c \leq \$147.50$;　**49** (a) $C = 6x + 250,000$, (b) 1,450,000,
(c) $0 \leq x \leq 300,000$, $\$250,000 \leq C \leq \$2,050,000$　**51** $p = \begin{cases} 200 & n \leq 60 \\ 200 - 2(n - 60) & n > 60 \end{cases}$;
53 (a) 0, (b) 8, (c) 1300, (d) $2x^2 + 5xy + y^3$;　**55** (a) 4, (b) 9, (c) 36;　**57** (a) -5,
(b) 7;　**59** (a) 110, (b) -16, (c) $ab - 5cd$;　**61** (a) $q_1 = 130$ (thousand),
$q_2 = 60$ (thousand), (b) $q_1 = 60$ (thousand), $q_2 = 70$ (thousand);
63 $S = 2.5x_1 + 4x_2 + 3x_3 + 60$, $\$295$

SEC. 4.2   1 can't classify (yet);   **3** linear;   **5** constant;   **7** linear;
9 can't classify (yet);   **11** rational;   **13** can't classify;   **15** constant;
17 can't classify;   **19** rational;   **21** can't classify;   **23** constant;
25 all real numbers;   **27** all real numbers *except* those resulting in $h(x) = 0$;
29 (*a*) linear, (*b*) \$11,000, (*c*) $t = 6.67$ years;   **31** (*a*) quadratic, (*b*) \$10,000,
(*c*) $p = \$0$ or \$30;   **33** (*a*) quadratic, (*b*) \$31,455,000;   **35** (*a*) quadratic, (*b*) 225,
(*c*) \$45;   **37** (*a*) $x^2 - 2x + 7$, (*b*) $-2x^3 + 10x^2 + 6x - 30$, (*c*) $(x^2 - 3)/(10 - 2x)$;
39 (*a*) $x^2 - 12x + 42$, (*b*) 70, (*c*) 31;   **41** (*a*) $x^6 + 2x^3$, (*b*) 0, (*c*) 80;   **43** (*a*) $(2)^{x+2}$,
(*b*) 32, (*c*) 1

SEC. 4.3   1, 3, 5, 7, 9, 11, 13, 15 see figures;   **17** (*c*), (*d*), and (*e*) are
functions;   **19** for $f(x) + c$, all points on the graph of $f(x)$ would be raised by c units
relative to the y axis; for $f(x) - c$, the graph of $f(x)$ would be lowered by c units

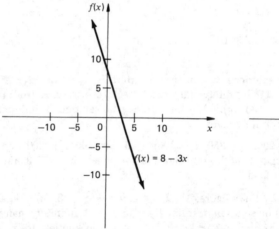

Sec. 4.3, Exercise 1

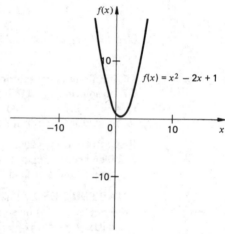

Sec. 4.3, Exercise 3

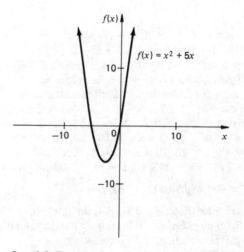

Sec. 4.3, Exercise 5

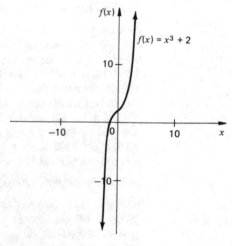

Sec. 4.3, Exercise 7

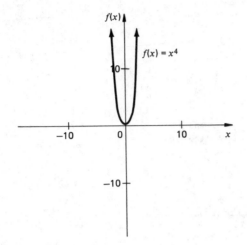

Sec. 4.3, Exercise 9

Sec. 4.3, Exercise 11

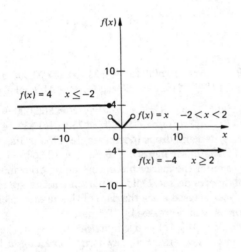

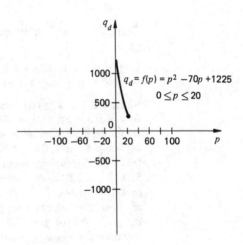

Sec. 4.3, Exercise 13

Sec. 4.3, Exercise 15

·**CHAPTER TEST** **1** $x \geq 2, x \leq -3$; **2** $0 \leq t \leq 8, 0 \leq V \leq 36{,}000$;

3 $p = \begin{cases} 450 & n \leq 150 \\ 450 - 2.5(n - 150) & n > 150 \end{cases}$; **4** (a) $(x^2 + 10x + 25)/(-2x - 9)$, (b) 0; ·

5 (a) can't classify, (b) rational, (c) quadratic, (d) linear; **6** see figure (p. A-34)

CHAPTER 5

SEC. 5.1 **1** $y = f(x_1, x_2, x_3, x_4, x_5) = a_1x_1 + a_2x_2 + a_3x_3 + a_4x_4 + a_5x_5 + a_0$;
3 $y = 5x_1 + 3x_2 + 25$ when $x_1 + x_2 \leq 80$, $y = 5x_1 + 3x_2 + 25 +$
$2.5(x_1 + x_2 - 80) = 7.5x_1 + 5.5x_2 - 175$ when $x_1 + x_2 > 80$;
5 (a) $R = 500x_1 + 1000x_2 + 1500x_3$, (b) $C = 275x_1 + 550x_2 + 975x_3 + 25{,}000{,}000$,
(c) $P = 225x_1 + 450x_2 + 525x_3 - 25{,}000{,}000$, (d) \$2,750,000; **7** (a) $R = 0.33x$,

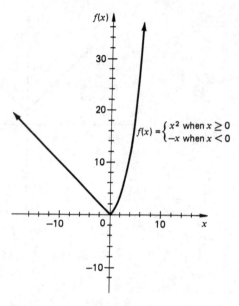

$$f(x) = \begin{cases} x^2 \text{ when } x \ge 0 \\ -x \text{ when } x < 0 \end{cases}$$

Chap. 4, Chapter Test, Problem 6

(b) $C = 0.18x + 10{,}500$, (c) $P = 0.15x - 10{,}500$, (d) $= -\$1500$, (e) 70,000 miles;
9 (a) $R = 1.299x_1 + 1.379x_2$, (b) $C = 1.219x_1 + 1.289x_2$, (c) $P = 0.08x_1 + 0.09x_2$, (d) \$11,600

SEC. 5.2   1 $V = f(t) = 80{,}000 - 13{,}333.33t$;   **3** $V = f(t) = 300{,}000 - 37{,}500t$;
5 $V = f(t) = 25{,}000 - 6466.67t$;   **7** (a) $q = 115{,}000 - 1750p$, (b) \$37.14,
(c) for each \$1 increase in price, demand drops (decreases) by 1750 units,
(d) $0 \le p \le 65.71$, $0 \le q \le 115{,}000$, (e) see figure;   **9** (a) $q = 10{,}800p - 15{,}200$, (b) \$5.57,
(c) $(1.407, 0)$, for any supply to enter the marketplace, the selling price must exceed
\$1.407;   **11** (a) At time of divorce decree, child support payments are made in 90
percent of cases, (b) for each year elapsed since the date of the divorce decree, the percent
of cases in which child support is paid decreases by 12.5 percent, (c) 27.5 percent; (d) see
figure.   **13** (a) $p = f(a) = -3a + 153$, (b) as age increases by 1 year, the probability of
marriage decreases by 3 percent; a never-married, newborn, woman has a 153 percent
chance of marrying (which is not meaningful), (c) p intercept not meaningful,
(d) $f(20) = 93$, $f(30) = 63$, $f(40) = 33$, $f(50) = 3$;   **15** (a) $E = f(t) = 79.327t + 1512.02$,

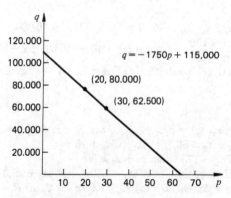

Sec. 5.2, Exercise 7e

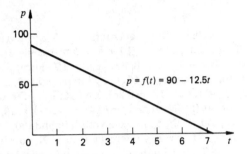

Sec. 5.2, Exercise 11d

(b) the expenditure per student has been increasing at a rate of approximately $79.327 per year; in 1955 the rate of expenditure was approximately $1512.02, (c) $5081.735;
17 (a) $V = f(t) = 1.267t + 8.133$, (b) the percent of vacant offices increases by 1.267 each year; in 1985, the estimated vacancy rate was 8.133 percent, (c) 20.803 percent;
19 $p_1 = 5, p_2 = 3, q_1 = 70, q_2 = 90$

SEC. 5.3 **1** 37,500; **3** $50; **5** (a) 11,111.11, (b) 37,777.78, (c) $9.07,
(d) $350,000; **7** (a) 8000, (b) 10,000, (c) $3,750 loss; **9** (a) 60,000,
(b) the more expensive machine ($520,000); **11** (a) 40,000 lines, (b) $75,000/$67,500,
(c) $1.25/line; **13** (a) 2000, (b) 8000 PacPerson, 6000 Astervoid, 3000 Haley's,
2000 Black Hole; **15** machine 1 if output $\leq$ 40,000; machine 2 if output between
40,000 and 51,428.57 units; machine 4 if output greater than 51,4287.57

CHAPTER TEST **1** (a) $P = f(x) = 40x - 200,000$, (b) 23,750;
2 (a) $n = f(t) = -6000t + 245,000$, (b) for each year, ridership decreases by approximately 6000 passengers, (c) 131,000, (d) $t = 10.833$ (between 1991 and 1992);
3 (a) $N = f(t) = 217.5t + 1720$, (b) average number of inmates increasing by approximately 217.5/year; in 1984, the average number of inmates was estimated to equal 1720; (d) 3025; **4** machine 1 for $x \leq 20,000$; machine 3 for $x \geq 20,000$, see figure;
5 (a) increases x_{BE}, (b) decreases x_{BE}, (c) decreases x_{BE}

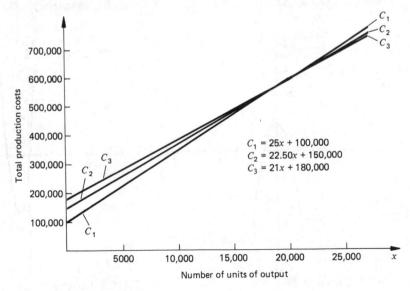

Chap. 5, Chapter Test, Question 4

CHAPTER 6

SEC. 6.1   **1** $a = 4, b = 0, c = -20$;   **3** not quadratic;   **5** $a = 1, b = -8, c = 16$;
7 $a = \frac{1}{3}, b = -\frac{4}{3}, c = \frac{5}{3}$;   **9** not quadratic;   **11** $a = 1, b = 0, c = -2$;
13 $a = 20, b = 4, c = -5$;   **15** $a = 1, b = -8, c = 16$;   **17** down, y intercept,
x intercept, and vertex at $(0, 0)$, see figure;   **19** up, y intercept at $(0, 2)$,
x intercepts at $(1, 0)$ and $(2, 0)$, vertex at $(1.5, -0.25)$, see figure;   **21** down,
y intercept at $(0, 5)$, x intercepts at $(2.236, 0)$ and $(-2.236, 0)$ vertex at $(0, 5)$,
see figure;   **23** up, y intercept at $(0, 0)$, x intercepts at $(0, 0)$ and $(20, 0)$,
vertex at $(10, -50)$, see figure;   **25** down, y intercept at $(0, -9)$, no x intercepts,
vertex at $(0, -9)$, see figure;   **27** up, y intercept at $(0, 4)$, no x intercepts,
vertex at $(0, 4)$, see figure;   **29** up, y intercept at $(0, -12)$, x intercepts at $(-\frac{3}{2}, 0)$ and
$(\frac{4}{3}, 0)$, vertex at $(-\frac{1}{12}, -\frac{289}{24})$, see figure;   **31** down, y intercept at $(0, 0)$,

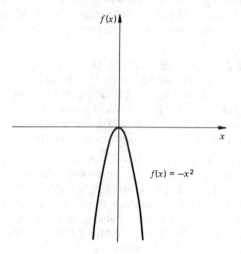

Sec. 6.1, Exercise 17

Sec. 6.1, Exercise 19

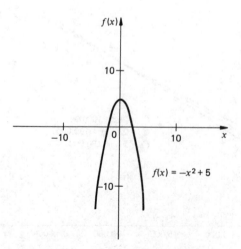

Sec. 6.1, Exercise 21

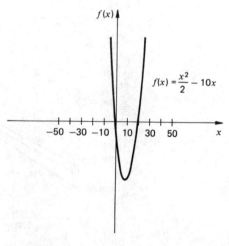

Sec. 6.1, Exercise 23

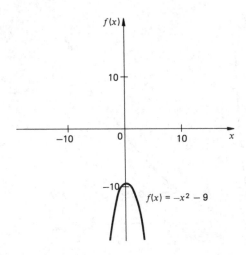

Sec. 6.1, Exercise 25

$f(x) = 2x^2 + 4$

Sec. 6.1, Exercise 27

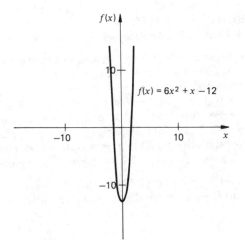

$f(x) = 6x^2 + x - 12$

Sec. 6.1, Exercise 29

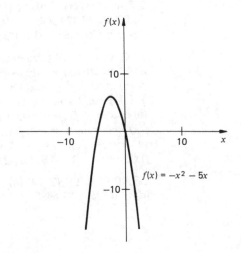

$f(x) = -x^2 - 5x$

Sec. 6.1, Exercise 31

x intercepts at $(0, 0)$ and $(-5, 0)$, vertex at $(-2.5, 6.25)$, see figure;   **33** see figure (p. A-38);   **35** $f(x) = -3x^2 + 2x$;   **37** $n = f(t) = 4t^2 - 41.167t + 125$, 32.831, beyond vertex of parabola; estimating function not valid

SEC. 6.2   1 $R = g(p) = 600{,}000p - 2500p^2$, concave down, $(0, 0)$, \$23,750,000, 475,000 units, \$120;   **3** $R = g(p) = 30{,}000p - 25p^2$, concave down, $(0, 0)$, \$1,710,000, 28,500 units, \$600;   **5** $R = f(q) = 160q - q^2/15$;   **7** (a) $q_s = f(p) = 3p^2 - 1200$, (b) $p \geq 20$, (c) $(20, 0)$ at prices of \$20 (or less), no units will be supplied, or, for units to be supplied, the price must be greater than \$20, (d) 9600;   **9** (a) $q_s = f(p) = 3p^2 - 4200$, (b) $p \geq 37.416$, (c) $(37.416, 0)$, for $p \leq 37.416$, no units will be supplied, (d) 25,800 units;   **11** $q_d = f(p) = 3p^2 - 240p + 4800$, 3675 units;   **13** \$75, 5225 units;   **15** (a) $S = f(x) = (x - 12)^2 + (x - 20)^2 + (x - 30)^2 = 3x^2 - 124x + 1444$, (b) $x = 20.67$;

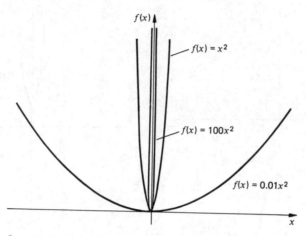

Sec. 6.1, Exercise 33

17 (a) $S = f(x) = 10,000 (10 - x)^2 + 6000 (50 - x)^2 + 18,000 (70 - x)^2 = 34,000x^2 - 3,320,000x + 104,200,000$, (b) $x = 48.82$;
19 $n = f(t) = 687.5t^2 + 125t + 2000$; 86,562.5 employees

SEC. 6.3   **1** (a) third degree, (b) as $x \to \infty, f(x) \to -\infty$; as $x \to -\infty, f(x) \to \infty$;
3 (a) eighth degree, (b) as $x \to \infty, f(x) \to \infty$; as $x \to -\infty, f(x) \to \infty$;
5 (a) seventh degree, (b) as $x \to \infty, f(x) \to \infty$; as $x \to -\infty, f(x) \to -\infty$;
7 (a) ninth degree, (b) as $x \to \infty, f(x) \to \infty$; as $x \to -\infty, f(x) \to -\infty$;   **9** (a) fifth degree,
(b) as $x \to \infty, f(x) \to -\infty$; as $x \to -\infty, f(x) \to \infty$;   **11** (a) eighth degree, (b) as $x \to \infty$,
$f(x) \to -\infty$; as $x \to -\infty, f(x) \to -\infty$;   **13** (a) fifth degree, (b) as $x \to \infty, f(x) \to -\infty$,
as $x \to -\infty, f(x) \to \infty$;   **15** (a) sixth degree, (b) as $x \to \infty; f(x) \to \infty$; as $x \to -\infty$,
$f(x) \to \infty$;   **17, 19, 21, 23, 25** see figures;   **27** see figure, 1082.5 persons

CHAPTER TEST   **1** (a) up, (b) $(0, -20)$, (c) $(1.696, 0)$ and $(-2.946, 0)$,
(d) $(-0.625, -21.5625)$;   **2** (a) $R = f(p) = 360,000p - 45p^2$, (b) \$4000;

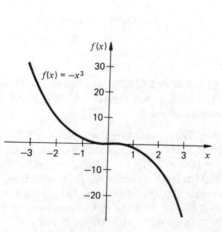

Sec. 6.3, Exercise 17

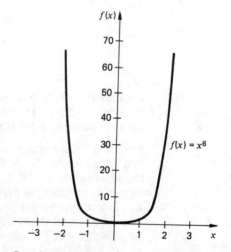

Sec. 6.3, Exercise 19

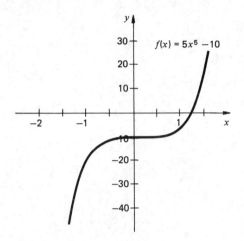

Sec. 6.3, Exercise 21

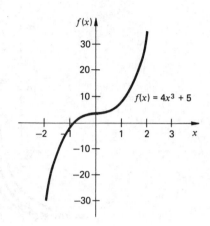

Sec. 6.3, Exercise 23

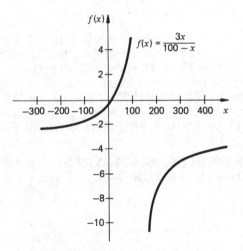

Sec. 6.3, Exercise 25

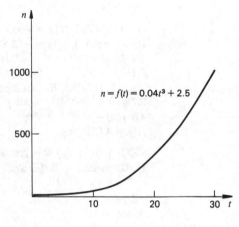

Sec. 6.3, Exercise 27

3 $I = f(t) = 0.144t^2 - 0.201t + 0.6$, \$29.985 billions; **4** (a) fifteenth degree, as $x \to \infty$, $f(x) \to -\infty$; as $x \to -\infty$, $f(x) \to \infty$; (b) twelfth degree, as $x \to \infty$, $f(x) \to -\infty$; as $x \to -\infty$, $f(x) \to -\infty$; **5** $V = f(x) = (24 - 2x)(16 - 2x)(x) = 384x - 80x^2 + 4x^3$

CHAPTER 7

SEC. 7.1 1 (a), (b), (c), (e), (g), and (h) are exponential functions **3** (a) see figure, (b) the larger the magnitude of a, the greater the rate of increase in $f(x)$ as x increases in value; **5** (a) I. domain is the set of real numbers, II. the graph of f lies entirely above the x axis, III. the graph of f is asymptotic to the line $y = c$, IV. the y intercept occurs at $(0, c + 1)$, (b) I. domain is set of real numbers, II. the graph of f lies above and below the x axis, III. the graph of f is asymptotic to the line $y = c$, IV. the y intercept occurs at $(0, c + 1)$; **7** $f(0) = 1, f(-3) = 1/\sqrt{8} = 0.3535, f(1) = \sqrt{2} = 1.414$; **9** $f(0) = 1$,

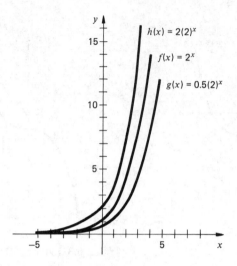

Sec. 7.1, Exercise 3a

$f(-3) = 4.482, f(1) = 0.6065;$ **11** $f(0) = 1, f(-3) = 90.017, f(1) = 1.649;$ **13** $f(0) = 0,$
$f(-3) = 9.975, f(1) = -63.89;$ **15** $f(0) = 2.5, f(-3) = 50.214, f(1) = 0.9197;$
17 $f(0) = 81, f(-3) = 0.0041, f(1) = 27;$ **19** $f(0) = 9, f(-3) = 11.99,$
$f(1) = -10.167;$ **21** $f(0) = -4, f(-3) = -0.199, f(1) = -6.873;$
23, 25, 27, 29, 31 see figures;   **33** $f(x) = e^{0.69x^2};$ **35** $f(x) = e^{0.405x};$
37 $f(t) = 10e^{-1.2t};$ **39** $f(t) = -2e^{4.5t}$ **41** $f(u) = 5e^{0.429u};$ **43** $f(z) = e^{3.91z};$
45 $f(x) = e^{1.0986x^2-0.3465x};$ **47** (a) $A = f(t) = 300(\frac{1}{2})^t$, (b) $f(2) = 75$ mg, $f(2.5) = 53.03$ mg,
$f(6) = 4.6875$ mg

SEC. 7.2   **1** (a) $V = 200,000e^{0.08t}$, (b) \$298,364.94, \$445,108.19;   **3** 8.66 years,
17.328 years;   **5** (a) \$2,339,646.80, (b) \$8,372,897.50, (c) 4.77 years;

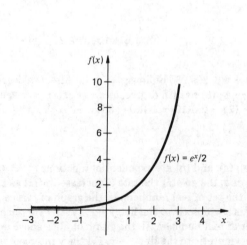

Sec. 7.1, Exercise 23

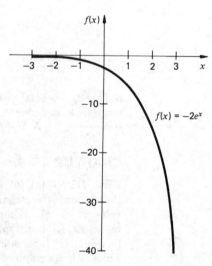

Sec. 7.1, Exercise 25

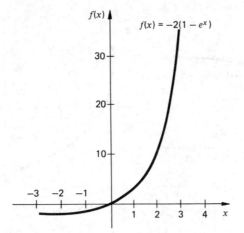

Sec. 7.1, Exercise 27

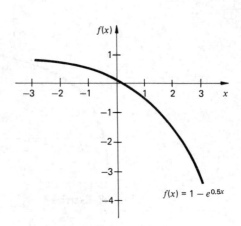

Sec. 7.1, Exercise 29

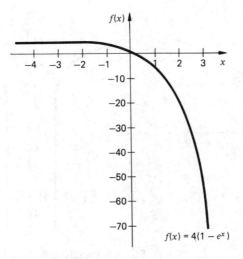

Sec. 7.1, Exercise 31

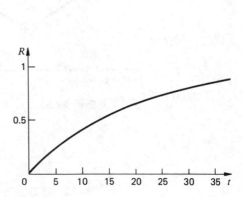

Sec. 7.2, Exercise 17b

7 (a) $P = f(t) = 40e^{0.025t}$, (b) 51.36 million, 74.729 million;　　**9** (a) 5563.85 tons, (b) in 5.875 years;　　**11** 23.1 years;　　**13** 17.328 years (or during 1997); **15** (a) 19.67 percent, 31.6 percent, (b) 0.5 (or 50 percent);　　**17** (a) 4.88 percent, 22.12 percent, 39.35 percent, 63.21 percent, (b) see figure;　　**19** $R = g(p) = 10,000\, pe^{-0.1p}$, $g(10) = \$36,787.94$, $f(10) = 3678.79$ units;　　**21** (a) $f(0) = pe^{-c}$, (b) p

SEC. 7.3　**1** $\log_5 25 = 2$;　　**3** $\log_4 64 = 3$;　　**5** $\log_7 343 = 3$;　　**7** $\log_8 4096 = 4$; **9** $\log_4 4096 = 6$;　　**11** $\log_5 0.04 = -2$;　　**13** $\log_{0.2} 625 = -4$; **15** $\log_{0.4} 15.625 = -3$;　　**17** $\log_2 0.0625 = -4$;　　**19** $\log_{0.2} 125 = -3$; **21** $2^7 = 128$;　　**23** $4^3 = 64$;　　**25** $3^6 = 729$;　　**27** $2^{-4} = 0.0625$;　　**29** $5^5 = 3125$; **31** $0.1^{-4} = 10,000$;　　**33** $0.2^{-2} = 25$;　　**35** $0.25^{-3} = 64$;　　**37** $e^{1.6094} = 5$; **39** $e^{4.6052} = 100$;　　**41** 6.3969;　　**43** 4.3820;　　**45** -2.3026;　　**47** -0.2877; **49** -4.6052;　　**51** 9.2103　　**53** 6.6201　　**55** 7.7832　　**57** 13.8155;

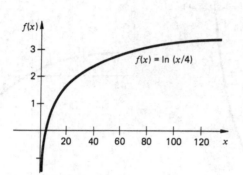

Sec. 7.3, Exercise 81

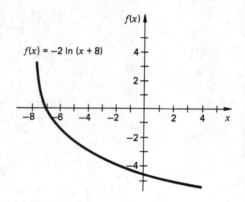

Sec. 7.3, Exercise 83

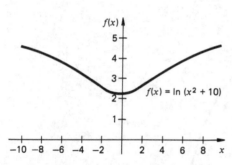

Sec. 7.3, Exercise 85

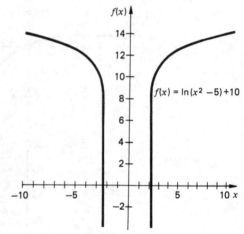

Sec. 7.3, Exercise 87

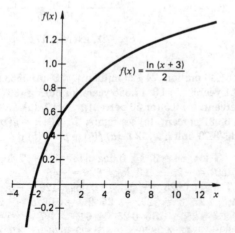

Sec. 7.3, Exercise 89

59 6.5147; **61** $x = 27.299$; **63** $x = 2, -2,$ or 1;
65 $x = \pm\sqrt{3/(e-1)} = \pm1.3214$; **67** $x = -1.8444$; **69** $x = -11.0904$;
71 $x = 1.4756$; **73** $x = -1.6094$; **75** $x = -3, 3, 1$; **77** no solution;
79 $x = -2.4079$; **81, 83, 85, 87, 89** see figures;
91 $t = 27.23$ (28th Super Bowl); **93** $R = f(t) = 71e^{0.0518t}$; **95** (a) ($e^{c/a} - b$, 0),
(b) (a, $a \ln b - c$); **97** (a) \$4.37, \$5.53; **99** (a) $P = 10^5 e^{0.6t}$, (b) 1.155 hours,
(c) 1.831 hours; **101** 1.98 hours, 0.822 hours; **103** $k = 0.0000347$;
105 (a) 90 percent, 57.81 percent, 43.95 percent

CHAPTER TEST **1** 0.0817; **2** \$276,091.48, \$256,091.48; **3** $\log_3 6561 = 8$;
4 $N = f(x) = 5,000,000 \, (1 - e^{-0.075x}) - 6000x$; **5** $x = 121.51$;
6 (a) $V = 281,158 e^{-0.1703t}$, (b) 10.14 years

CHAPTER 8

SEC. 8.1 **1** \$36,000; **3** \$10,500, \$420,000;

5

Semiannual Period	Principal (P)	Interest (I)	Compound Amount ($S = P + I$)
1	\$10,000	\$400	\$10,400.00
2	10,400	416	10,816.00
3	10,816	432.64	11,248.64
4	11,248.64	449.95	11,698.59

\$1,698.59 in interest; **7** (a) \$11,716.60, (b) quarterly by \$18.01

SEC. 8.2 **1** \$13,416.80, \$5416.80; **3** \$79,304.25, \$54,304.25;
5 \$1,342,530, \$842,530; **7** \$45,152.75, \$20,152.75; **9** (a) \$2.54035, \$1.54035,
(b) \$2.57508, \$1.57508, (c) \$2.58707, \$1.58707; **11** 21,159; **13** \$19,529.47;
15 \$14,849.40, \$5150.60; **17** \$27,644, \$22,356; **19** \$81,860, \$118,140;
21 between 16 and 17 semiannual periods, or just over 8 years; **23** between 37 and
38 quarters, or during the second quarter of the tenth year; **25** (a) 16.64 percent,
(b) 16.986 percent; **27** (a) 6.09 percent, (b) 6.136 percent;
29 just under 8 percent; **31** between 10 and 12 percent

SEC. 8.3 **1** (a) \$72,432.80, (b) \$22,432.80; **3** (a) \$58,198.73, (b) \$31,198.73;
5 (a) \$61,051.00, (b) \$62,889.45, (c) \$63,861.65; **7** \$14,832, \$25,840;
9 \$24,840, \$506,400; **11** \$745.60, \$1052.80

SEC. 8.4 **1** \$160,441.50; **3** \$140,939.40; **5** \$10,465.15; **7** \$112,387.60;
9 (a) \$490,907.50, (b) \$509,092.50; **11** \$44,245; **13** \$481,125; **15** \$236,200;
17 (a) \$78,022, (b) \$118,132; **19** (a) \$498.15, (b) \$2933.40; **21** \$772.02,
\$185,284.80, \$105,284.80 **23** \$786.33, \$235,899, \$145,899;
25 \$817.83, \$245,349, \$155,349; **27** \$1198.06, \$287,534.40, \$167,534.40;
29 \$53.73, \$12,895.20; **31** \$63.43, \$19,029; **33** \$64.27, \$19,281;
35 \$81.66, \$19,598.40; **37** \$79,433.84; **39** \$147,569.68

SEC. 8.5 **1** \$274,474, yes; **3** \$245,146, yes; **5** \$419,412, yes;
7 between 15 and 16 percent

CHAPTER TEST **1** \$101,118.50; **2** \$81,630; **3** (a) \$134,351.85,
(b) \$34,351.85; **4** \$6360, \$16,400; **5** \$21,225; **6** 12.551 percent;
7 \$125.46; **8** yes, \$4019.19

CHAPTER 9

SEC. 9.2 **1** (1×4), $\begin{pmatrix} 8 \\ -8 \\ 5 \\ 3 \end{pmatrix}$;   **3** (4×2), $\begin{pmatrix} 0 & 5 & -6 & -2 \\ 1 & 2 & 8 & 4 \end{pmatrix}$;

5 (3×3), $\begin{pmatrix} 1 & 0 & 0 \\ 0 & 1 & 0 \\ 0 & 0 & 1 \end{pmatrix}$;   **7** (4×1), $(1 \quad 2 \quad 3 \quad 4)$;

9 (5×3), $\begin{pmatrix} 1 & 6 & 0 & 4 & 5 \\ 3 & 4 & 1 & 6 & 1 \\ 5 & 2 & 2 & 3 & 2 \end{pmatrix}$;   **11** $\begin{pmatrix} 2 & 0 & 0 & 0 \\ 0 & 4 & 0 & 0 \end{pmatrix}$

SEC. 9.3 **1** $\begin{pmatrix} -7 & -10 \\ -13 & -12 \end{pmatrix}$;   **3** $\begin{pmatrix} 84 & 89 \\ -13 & -20 \end{pmatrix}$;   **5** $\begin{pmatrix} -70 & 125 \\ 70 & 30 \end{pmatrix}$;   **7** 4;
9 cannot perform;   **11** $ax + by$;   **13** -2;   **15** $ae + bf + cg + dh$;
17 $\begin{pmatrix} 8 & 24 \\ -67 & 60 \end{pmatrix}$;   **19** $(20 \quad -8)$;   **21** $\begin{pmatrix} 10 & -5 & 10 \\ 0 & 13 & 0 \end{pmatrix}$;   **23** cannot do;
25 $\begin{pmatrix} 8 & 27 & 26 \\ -3 & -16 & 6 \\ 2 & 2 & 29 \end{pmatrix}$;   **27** $\begin{pmatrix} a_{11}x_1 + a_{12}x_2 \\ a_{21}x_1 + a_{22}x_2 \end{pmatrix}$;   **29** cannot do;

31 $\begin{pmatrix} 8 & 4 & 6 \\ -2 & -10 & -3 \end{pmatrix}$;   **33** $\begin{pmatrix} -2 & -4 & 0 & 12 & 16 \\ 18 & 1 & 0 & 11 & 24 \\ -24 & -8 & 0 & 8 & 0 \\ -6 & -7 & 0 & 19 & 24 \end{pmatrix}$;

35 $\begin{pmatrix} 20 & -34 & 39 & -27 \\ 3 & -6 & 9 & -2 \\ -10 & 0 & 40 & -30 \end{pmatrix}$;   **37** $\begin{pmatrix} 1 & -3 \\ 2 & 3 \end{pmatrix} \begin{pmatrix} x \\ y \end{pmatrix} = \begin{pmatrix} 15 \\ -10 \end{pmatrix}$;

39 $\begin{pmatrix} 5 & -2 & 3 \\ 3 & -1 & -2 \end{pmatrix} \begin{pmatrix} x_1 \\ x_2 \\ x_3 \end{pmatrix} = \begin{pmatrix} 12 \\ 15 \end{pmatrix}$;   **41** $\begin{pmatrix} a & b \\ d & e \\ g & h \end{pmatrix} \begin{pmatrix} x_1 \\ x_2 \end{pmatrix} = \begin{pmatrix} c \\ f \\ i \end{pmatrix}$;

43 $\begin{pmatrix} a_1 & a_2 & a_3 \\ a_4 & a_5 & a_6 \end{pmatrix} \begin{pmatrix} x^2 \\ x \\ 1 \end{pmatrix} = \begin{pmatrix} b_1 \\ b_2 \end{pmatrix}$;   **45** $\begin{pmatrix} 5 & -2 & 1 \\ 3 & 0 & 0 \\ 0 & 5 & 0 \end{pmatrix} \begin{pmatrix} x^3 \\ x^2 \\ x \end{pmatrix} = \begin{pmatrix} 100 \\ -18 \\ 125 \end{pmatrix}$

SEC. 9.4 **1** $\Delta = -5$;   **3** $\Delta = -26$;   **5** $\Delta = 28$;   **7** $\Delta = 1$;   **9** $\Delta = 140$;
11 $\Delta = -120$;   **13** $\Delta = 188$;   **15** $\Delta = 80$;   **17** $\begin{pmatrix} -4 & -10 \\ 2 & 8 \end{pmatrix}$;   **19** $\begin{pmatrix} 1 & 0 \\ 0 & 1 \end{pmatrix}$;

21 $\begin{pmatrix} -12 & 10 & -6 \\ -18 & 2 & -22 \\ -16 & -4 & -8 \end{pmatrix}$;   **23** $\begin{pmatrix} -1 & 0 & -1 \\ 0 & 0 & 0 \\ -1 & 0 & -1 \end{pmatrix}$;   **25** $\begin{pmatrix} -6 & -2 \\ 8 & 4 \end{pmatrix}$

27 $\begin{pmatrix} 3 & 5 \\ -8 & 10 \end{pmatrix}$;   **29** $\begin{pmatrix} -24 & 6 & -42 \\ -24 & 0 & -48 \\ -56 & 4 & -92 \end{pmatrix}$;   **31** $\begin{pmatrix} -80 & -50 & -55 \\ -100 & -250 & 25 \\ -160 & -100 & 40 \end{pmatrix}$;

33 $\begin{pmatrix} -8 & 16 & 8 \\ 5 & -28 & -17 \\ 17 & -76 & -29 \end{pmatrix}$;   **35** $\Delta = -12$;   **37** $\Delta = 1$;   **39** $\Delta = -52$;
41 $\Delta = 0$;   **43** $\Delta = -8$;   **45** $\Delta = 70$;   **47** $\Delta = 48$;   **49** $\Delta = -1500$;
51 $\Delta = -48$;   **53** $\Delta = 1260$;   **55** $\Delta = -876$;   **57** $x_1 = -3, x_2 = 2$;
59 $x_1 = -10, x_2 = 15$;   **61** no solution;   **63** $x_1 = 1, x_2 = 4, x_3 = -2$;
65 $x_1 = -2, x_2 = 0, x_3 = 4$;   **67** no solution

SEC. 9.5 **1** $\begin{pmatrix} 3 & -1 \\ 2 & -1 \end{pmatrix}$; **3** no inverse exists; **5** $\begin{pmatrix} \frac{1}{2} & \frac{1}{2} \\ -\frac{1}{2} & \frac{1}{2} \end{pmatrix}$;

7 $\begin{pmatrix} -0.375 & 0.75 & 0.125 \\ 0.375 & 0.25 & -0.125 \\ -0.125 & -0.75 & 0.375 \end{pmatrix}$; **9** $\begin{pmatrix} 4 & -3 \\ -5 & 4 \end{pmatrix}$; **11** $\begin{pmatrix} 1.4 & 0.2 & -0.4 \\ 1.2 & -0.4 & -0.2 \\ -1.6 & 0.2 & 0.6 \end{pmatrix}$;

13 $\begin{pmatrix} 5 & -7 \\ -2 & 3 \end{pmatrix}$; **15** no inverse exists; **17** $\frac{1}{26}\begin{pmatrix} 2 & 5 \\ -4 & 3 \end{pmatrix}$;

19 no inverse exists; **21** $\begin{pmatrix} -1.6 & 0.6 & 0.8 \\ 3.8 & -0.8 & -1.4 \\ -1.2 & 0.2 & 0.6 \end{pmatrix}$ **23** $x_1 = 2, x_2 = 3$;

25 either no solution or infinitely many; **27** $x_1 = 24, x_2 = 35$;
29 $x_1 = 2, x_2 = 0, x_3 = 1$; **31** $x_1 = 2, x_2 = 3$; **33** $x_1 = 3, x_2 = -2, x_3 = 1$;
35 $x_1 = 6, x_2 = -3$; **37** no solution or infinitely many; **39** $x_1 = 4, x_2 = -2$;
41 no solution or infinitely many; **43** $x_1 = 2, x_2 = 0, x_3 = -4$;
45 $2x_1 + 3x_2 = 17$, $(4, 3)$; **47** $2x_1 + 2x_2 + 3x_3 = 3$, $(2, 7, -5)$
$\quad\quad\ x_1 + 2x_2 = 10$ $\quad\quad\quad\quad\quad x_2 + x_3 = 2$
$\quad\quad\quad\quad\quad\quad\quad\quad\quad\quad\quad x_1 + x_2 + x_3 = 4$

SEC. 9.6 **1** Democrat: 96,850, Republican: 98,850, Independent: 104,300;
3 T_1: Brand 3 will ultimately have 100 percent of the market, T_2: Brands 2 and 3 will each
have 50 percent of the market; **5** $P_n = 23.333$ millions, $P_s = 46.667$ million:

7 From
$\begin{array}{c} \\ A \\ B \\ C \\ D \end{array}$
$\begin{array}{cccc} \text{To} \\ A & B & C & D \\ \end{array}$
$\begin{pmatrix} 0 & 0 & 1 & 0 \\ 1 & 0 & 1 & 0 \\ 1 & 1 & 0 & 1 \\ 0 & 0 & 1 & 0 \end{pmatrix}$
From
$\begin{array}{cccc} \text{To} \\ A & B & C & D \\ \end{array}$
$\begin{pmatrix} 1 & 1 & 0 & 1 \\ 1 & 1 & 1 & 1 \\ 1 & 0 & 3 & 0 \\ 1 & 1 & 0 & 1 \end{pmatrix}$;
Adjacency
matrix

9 (a) industry 1 \$381,181,180, industry 2 \$322,122,120, industry 3 \$446,246,270,

(b) Supplier $\begin{array}{c} 1 \\ 2 \\ 3 \end{array}$
$\begin{array}{ccc} & \text{User} & \\ 1 & 2 & 3 \end{array}$
$\begin{pmatrix} 95.295 & 96.637 & 89.249 \\ 76.236 & 96.637 & 89.249 \\ 152.472 & 32.212 & 111.562 \end{pmatrix}$
(measured in
millions of dollars)

11 (a) $D = N^t \begin{pmatrix} 1 \\ 1 \\ 1 \\ 1 \\ 1 \end{pmatrix} = \begin{pmatrix} 53,000 \\ 120,000 \\ 115,000 \end{pmatrix}$ midsized
compact
subcompact

(b) $P = RD = \begin{pmatrix} 454,600 \\ 2,171,000 \\ 195,200 \\ 1,682,000 \end{pmatrix}$ fan belts
plugs
batteries
tires
(c) $T = CP = (\$67,031,050)$

(d) $P^t \begin{pmatrix} 1.25 & 0 & 0 & 0 \\ 0 & 0.80 & 0 & 0 \\ 0 & 0 & 30.00 & 0 \\ 0 & 0 & 0 & 35.00 \end{pmatrix} = (\$568,250 \quad \$1,736,800 \quad \$5,856,000 \quad \$58,870,000)$

or, C $\begin{pmatrix} 454{,}600 & 0 & 0 & 0 \\ 0 & 2{,}171{,}000 & 0 & 0 \\ 0 & 0 & 195{,}200 & 0 \\ 0 & 0 & 0 & 1{,}682{,}000 \end{pmatrix}$

$= (\$568{,}250 \quad \$1{,}736{,}800 \quad \$5{,}856{,}000 \quad \$58{,}870{,}000)$

CHAPTER TEST **1** $\begin{pmatrix} 4 & 0 & 6 \\ 1 & -3 & -2 \\ -6 & 10 & -4 \\ 8 & 5 & 19 \end{pmatrix}$;   **2** $ae + bf + cg + dh$;

3 (a) cannot multiply, (b) $\begin{pmatrix} -14 & -90 \\ 32 & 137 \\ -6 & -139 \end{pmatrix}$, (c) cannot multiply, (d) cannot multiply;

4 $\begin{pmatrix} 1 & 0 & 0 & -1 \\ 0 & 1 & 1 & 0 \\ 0 & 0 & 1 & 1 \\ 0 & 0 & 0 & 1 \end{pmatrix} \begin{pmatrix} x_1 \\ x_2 \\ x_3 \\ x_4 \end{pmatrix} = \begin{pmatrix} 20 \\ 15 \\ 18 \\ 9 \end{pmatrix}$;   **5** $\Delta = 504$;

6 no inverse exists;   **7** $3x_1 + 7x_2 = 15$
$\qquad\qquad\qquad\qquad\qquad\quad\;\; 2x_1 + 5x_2 = 11$

CHAPTER 10

SEC. 10.2 **1, 3, 5, 7, 9** see figures;   **11, 13, 15, 17, 19** see figures;
21 $z = 160$ when $x_1 = 0$ and $x_2 = 20$ (see figure);   **23** $z = 270$ when $x_1 = 3$ and $x_2 = 9$
(see figure);   **25** $z = 160$ when $x_1 = 0$ and $x_2 = 20$ (see figure);   **27** alternative
optimal solutions along the line segment connecting $(0, 12)$ and $(12, 6)$, $z = 96$ (see
figure);   **29** alternative optimal solutions along the line segment connecting $(\frac{4}{3}, \frac{10}{3})$
and $(2.5, 1)$, $z = 24$ (see figure);   **31** no feasible solution (see figure);
33 (a) Maximize $z = 3x_1 + 4x_2$ (b) $z = 85$, $x_1 = 15$, and $x_2 = 10$; (c) profit will be
$\qquad\quad$ subject to $2x_1 + 3x_2 \le 60$
$\qquad\qquad\qquad\qquad 4x_1 + 2x_2 \le 80$
$\qquad\qquad\qquad\qquad\quad x_1, x_2 \ge 0$
maximized at \$85/week if 15 units of product A and 10 units of product B are produced
and sold; the weekly labor capacities in both departments will be totally consumed;

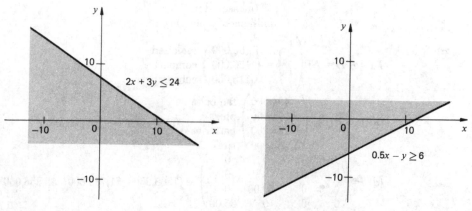

Sec. 10.2, Exercise 1   Sec. 10.2, Exercise 3

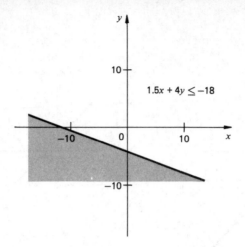

Sec. 10.2, Exercise 5

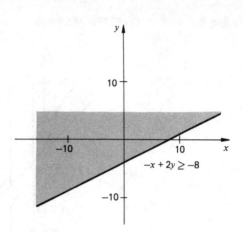

Sec. 10.2, Exercise 7

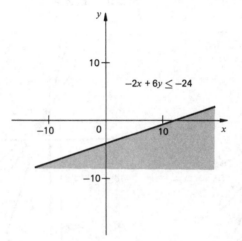

Sec. 10.2, Exercise 9

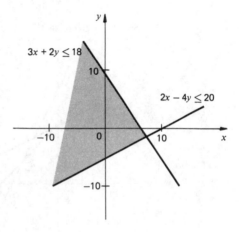

Sec. 10.2, Exercise 11

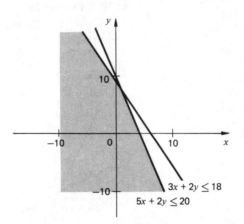

Sec. 10.2, Exercise 13

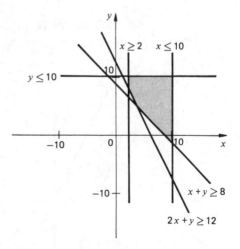

Sec. 10.2, Exercise 15

A-47

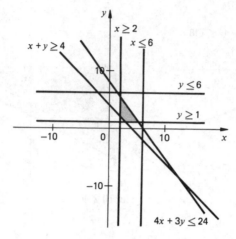

Sec. 10.2, Exercise 17

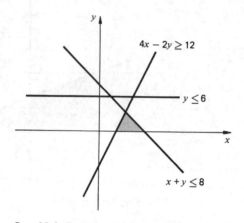

Sec. 10.2, Exercise 19

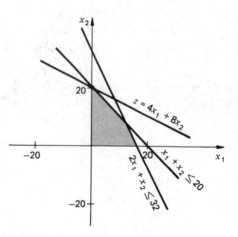

Sec. 10.2, Exercise 21

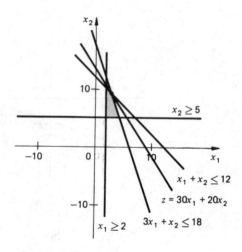

Sec. 10.2, Exercise 23

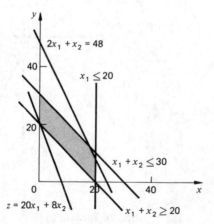

Sec. 10.2, Exercise 25

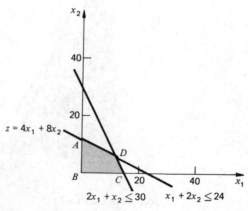

Sec. 10.2, Exercise 27

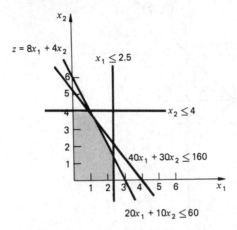

Sec. 10.2, Exercise 29

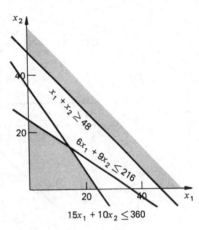

Sec. 10.2, Exercise 31

35 (a) Minimize $z = 0.12x_1 + 0.15x_2$ (b) Cost is minimized at $0.975 per meal if 3.75
  subject to $2x_1 + 3x_2 \geq 18$
$$4x_1 + 2x_2 \geq 22$$
$$x_1, x_2 \geq 0$$
ounces of food 1 and 3.5 ounces of food 2 are served; 100 percent of the MDR will be
realized for both vitamins

SEC. 10.3 1 (a) add the constraints $x_1 \geq 44$, $x_2 \geq 36$, $x_3 \geq 50$, $x_4 \geq 30$, $x_5 \geq 80$,
$x_6 \geq 24$, (b) $x_3 \geq x_4$, (c) $x_3 + x_6 \geq 30$, (d) $x_5 \geq 1.4x_6$, (e) $x_2 \leq 0.8x_1$;
3 x_j = number of ounces of food j
  Minimize $z = 0.15x_1 + 0.18x_2 + 0.22x_3$
  subject to $20x_1 + 40x_2 + 30x_3 \geq 240$
$$10x_1 + 25x_2 + 15x_3 \geq 120$$
$$20x_1 + 30x_2 + 25x_3 \geq 180$$
$$x_1 + x_2 + x_3 \geq 6$$
$$x_1, x_2, x_3 \geq 0$$
5 x_{ij} = thousands of gallons shipped from plant i to depot j
  Minimize $z = 50x_{11} + 40x_{12} + 35x_{13} + 20x_{14} + 30x_{21} + 45x_{22} + 40x_{23} + 60x_{24} + 60x_{31}$
$$+ 25x_{32} + 50x_{33} + 30x_{34}$$
subject to $x_{11} + x_{12} + x_{13} + x_{14}$ ≤ 1000
$$x_{21} + x_{22} + x_{23} + x_{24} \leq 1400$$
$$x_{31} + x_{32} + x_{33} + x_{34} \leq 1800$$
$$x_{11} \qquad + x_{21} \qquad + x_{31} \qquad = 800$$
$$x_{12} \qquad + x_{22} \qquad + x_{32} \qquad = 750$$
$$x_{13} \qquad + x_{23} \qquad + x_{33} \qquad = 650$$
$$x_{14} \qquad + x_{24} \qquad + x_{34} = 900$$
$$x_{ij} \geq 0 \qquad \text{for all } i, j$$

7 (a) $x_1 + x_2 + x_3 \geq 50$, (b) $x_1 \leq 2x_2$, (c) $x_2 = x_3$, (d) $x_2 \leq 0.5 (x_1 + x_2 + x_3)$;
9 x_{ij} = number of pounds of component i used in blend j
  Maximize $z = 0.6x_{11} + 0.75x_{12} + 1.15x_{13} + 0.45x_{21} + 0.6x_{22} + 1.00x_{23} + 0.35x_{31} + 0.5x_{32}$
$$+ 0.9x_{33} + 0.5x_{41} + 0.65x_{42} + 1.05x_{43}$$

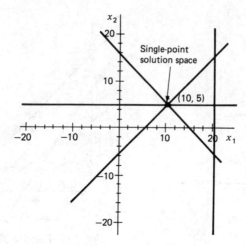

Chap. 10, Chapter Test, Problem 1

subject to
$$x_{11} + x_{12} + x_{13} \leq 80{,}000$$
$$x_{21} + x_{22} + x_{23} \leq 40{,}000$$
$$x_{31} + x_{32} + x_{33} \leq 30{,}000$$
$$x_{41} + x_{42} + x_{43} \leq 50{,}000$$
$$x_{13} + x_{23} + x_{33} + x_{43} \geq 50{,}000$$
$$x_{23} \geq 0.3\,(x_{13} + x_{23} + x_{33} + x_{43})$$
$$x_{21} \leq 0.2\,(x_{11} + x_{21} + x_{31} + x_{41})$$
$$x_{33} = 0.25\,(x_{13} + x_{23} + x_{33} + x_{43})$$
$$x_{41} \geq 0.4\,(x_{11} + x_{21} + x_{31} + x_{41})$$
$$x_{42} \leq 0.18\,(x_{12} + x_{22} + x_{32} + x_{42})$$
$$x_{ij} \geq 0 \qquad \text{for all } i, j$$

CHAPTER TEST   1 see figure;
2 Maximize $z = 20x_1 + 10x_2 + 50x_3 + 25x_4 + 2x_5$
   subject to $2x_1 + \quad x_2 + 0.5x_3 + 1.6x_4 + 0.75x_5 \leq 120$
$$1.5x_1 + 0.8x_2 + 1.5x_3 + 1.2x_4 + 2.25x_5 \leq 150$$
$$x_1 \qquad\qquad\qquad\qquad \geq 20$$
$$x_2 \qquad\qquad\qquad \geq 10$$
$$x_1 + \quad x_2 + \quad x_3 + \quad x_4 + \quad x_5 \leq 75$$
$$x_3 = x_5$$
$$x_1 + \quad x_2 \qquad\qquad\qquad \leq 0.5\,(x_3 + x_4 + x_5)$$
$$x_3 \geq x_1$$
$$x_1, x_2, x_3, x_4, x_5 \geq 0$$
3 z maximized at 30 when $x_1 = 8$ and $x_2 = 6$

CHAPTER 11

SEC. 11.1   1 $x_1 + \quad x_2 + \quad x_3 - E_1 + A_1 \qquad\qquad = 15$
$$-3x_1 \qquad\quad + 2x_3 - E_2 + A_2 \qquad = 5$$
$$4x_1 - 2x_2 + \quad x_3 \qquad\qquad\qquad + S_3 = 24$$
$$x_1, x_2, x_3, E_1, A_1, E_2, A_2, S_3 \geq 0;$$

3
$$
\begin{aligned}
x_1 + x_2 + x_3 + x_4 - E_1 + A_1 &= 25 \\
x_1 - 3x_2 \quad\quad + 2x_4 - E_2 + A_2 &= 20 \\
3x_1 \quad\quad\quad - 4x_4 \quad\quad + A_3 &= 10 \\
5x_1 - x_2 + 3x_3 + 8x_4 \quad\quad\quad\quad + S_4 &= 125 \\
x_1 \quad\quad\quad\quad\quad - E_5 + A_5 &= 5 \\
x_3 \quad\quad\quad\quad\quad\quad + S_6 &= 30
\end{aligned}
$$

$x_1, x_2, x_3, x_4, E_1, A_1, E_2, A_2, A_3, S_4, E_5, A_5, S_6 \geq 0;$

5
$$
\begin{aligned}
50x_1 + 30x_2 + 20x_3 - E_1 + A_1 &= 290 \\
20x_1 + 10x_2 + 30x_3 - E_2 + A_2 &= 200 \\
10x_1 + 50x_2 + 20x_3 - E_3 + A_3 &= 210 \\
x_1 + x_2 + x_3 - E_4 + A_4 &= 9
\end{aligned}
$$

$x_1, x_2, x_3, E_1, E_2, E_3, E_4, A_1, A_2, A_3, A_4 \geq 0;$

7
$$
\begin{aligned}
x_1 + x_2 + x_3 + x_4 + x_5 + x_6 \quad + S_1 \quad &= 1000 \\
x_1 \quad\quad\quad\quad\quad\quad + S_2 \quad &= 200 \\
x_2 \quad\quad\quad\quad\quad\quad + S_3 \quad &= 180 \\
x_3 \quad\quad\quad\quad\quad\quad + S_4 \quad &= 250 \\
x_4 \quad\quad\quad\quad\quad\quad + S_5 \quad &= 150 \\
x_5 \quad\quad\quad\quad\quad\quad + S_6 \quad &= 400 \\
x_6 \quad\quad\quad\quad\quad\quad + S_7 \quad &= 120 \\
x_5 \quad\quad\quad\quad - E_8 \quad + A_8 &= 200 \\
x_1 + x_2 \quad\quad\quad\quad - E_9 \quad + A_9 &= 300
\end{aligned}
$$

$x_1, \ldots, x_6, S_1, \ldots, S_7, E_8, E_9, A_8, A_9 \geq 0;$

9 67 total variables, 20 slack, 12 surplus, 20 artificial; **11** (a) constraints 1 and 2 are $\leq$, constraint 3 is $\geq$, (b) 2 slack, 1 surplus, and 1 artificial, (c) 3 basic, 3 nonbasic, (d) A: basic (x_2, S_1, S_2), nonbasic (x_1, E_3, A_3), B: basic (x_2, S_1, E_3), nonbasic (x_1, S_2, A_3), C: basic (x_1, x_2, E_3), nonbasic (S_1, S_2, A_3), D: basic (x_1, x_2, S_2), nonbasic (S_1, E_3, A_3); **13** (a) all $\geq$ constraints, (b) 4 surplus and 4 artificial variables, (c) 4 basic and 6 nonbasic variables, (d) A: basic (x_2, E_2, E_3, E_4), nonbasic $(x_1, E_1, A_1, A_2, A_3, A_4)$, C: basic (x_1, x_2, E_1, E_4), nonbasic $(E_2, E_3, A_1, A_2, A_3, A_4)$

SEC. 11.2 1 (a) $5x_1 + 4x_2 + S_1 = 48$
$$
\begin{aligned}
2x_1 + 5x_2 + S_2 &= 26 \\
x_1, x_2, S_1, S_2 &\geq 0
\end{aligned}
$$

(b)

Solution	Variable Set Equal to Zero	Value of Other Variables	z
1	x_1, x_2	$S_1 = 48, S_2 = 26$	0
2	x_1, S_1	$S_2 = -34, x_2 = 12$	
3	x_1, S_2	$S_1 = 27.2, x_2 = 5.2$	52
4	x_2, S_1	$x_1 = 9.6, S_2 = 6.8$	134.4
5	x_2, S_2	$x_1 = 13, S_1 = -17$	
6	S_1, S_2	$x_1 = 8, x_2 = 2$	132

(c) feasible solutions 1, 3, 4, and 6, (e) maximum of 134.4 reached when $x_1 = 9.6, S_2 = 6.8$, and $x_2 = S_1 = 0$;

3

Basic Variables	z	x_1	x_2	S_1	S_2	b_i	Row No.	r_i
	1	-4	-2	0	0	0	(0)	
S_1	0	1	1	1	0	50	(1)	50
S_2	0	⑥	0	0	1	240	(2)	40*
	1	0	-2	0	$\frac{2}{3}$	160	(0)	
S_1	0	0	①	1	$-\frac{1}{6}$	10	(1)	10*
x_1	0	1	0	0	$\frac{1}{6}$	40	(2)	
	1	0	0	2	$\frac{1}{3}$	180	(0)	
x_2	0	0	1	1	$-\frac{1}{6}$	10	(1)	
x_1	0	1	0	0	$\frac{1}{6}$	40	(2)	

z maximized at value of 180 when $x_1 = 40$ and $x_2 = 10$;

5

Basic Variables	z	x_1	x_2	S_1	S_2	S_3	b_i	Row No.	r_i
	1	-10	-12	0	0	0	0	(0)	
S_1	0	1	1	1	0	0	150	(1)	150
S_2	0	3	⑥	0	1	0	300	(2)	50*
S_3	0	4	2	0	0	1	160	(3)	80
	1	-4	0	0	2	0	600	(0)	
S_1	0	$\frac{1}{2}$	0	1	$-\frac{1}{6}$	0	100	(1)	200
x_2	0	$\frac{1}{2}$	1	0	$\frac{1}{6}$	0	50	(2)	100
S_3	0	③	0	0	$-\frac{1}{3}$	1	60	(3)	20*
	1	0	0	0	$\frac{14}{9}$	$\frac{4}{3}$	680	(0)	
S_1	0	0	0	1	$-\frac{1}{9}$	$-\frac{1}{6}$	90	(1)	
x_2	0	0	1	0	$\frac{2}{9}$	$-\frac{1}{6}$	40	(2)	
x_1	0	1	0	0	$-\frac{1}{9}$	$\frac{1}{3}$	20	(3)	

z maximized at value of 680 when $x_1 = 20$, $x_2 = 40$, and $S_1 = 90$;

7

Basic Variables	z	x_1	x_2	x_3	S_1	S_2	S_3	b_i	Row No.	r_i
	1	-10	-3	-4	0	0	0	0	(0)	
S_1	0	⑧	2	3	1	0	0	400	(1)	50*
S_2	0	4	3	0	0	1	0	200	(2)	50
S_3	0	0	0	1	0	0	1	40	(3)	
	1	0	$-\frac{1}{2}$	$-\frac{1}{4}$	$\frac{5}{4}$	0	0	500	(0)	
x_1	0	1	$\frac{1}{4}$	$\frac{3}{8}$	$\frac{1}{8}$	0	0	50	(1)	200
S_2	0	0	②	$-\frac{3}{2}$	$-\frac{1}{2}$	1	0	0	(2)	0*
S_3	0	0	0	1	0	0	1	40	(3)	
	1	0	0	$-\frac{5}{8}$	$\frac{9}{8}$	$\frac{1}{4}$	0	500	(0)	
x_1	0	1	0	$\frac{9}{16}$	$\frac{3}{16}$	$-\frac{1}{8}$	0	50	(1)	88.88
x_2	0	0	1	$-\frac{3}{4}$	$-\frac{1}{4}$	$\frac{1}{2}$	0	0	(2)	
S_3	0	0	0	①	0	0	1	40	(3)	40*
	1	0	0	0	$\frac{9}{8}$	$\frac{1}{4}$	$\frac{5}{8}$	525	(0)	
x_1	0	1	0	0	$\frac{3}{16}$	$-\frac{1}{8}$	$-\frac{9}{16}$	27.5	(1)	
x_2	0	0	1	0	$-\frac{1}{4}$	$\frac{1}{2}$	$\frac{3}{4}$	30	(2)	
x_3	0	0	0	1	0	0	1	40	(3)	

z maximized at value of 525 when $x_1 = 27.5$, $x_2 = 30$, and $x_3 = 40$;

9

	z	x_1	x_2	E_1	A_1	S_2	E_3	A_3	b_i	Row No.	r_i
	1	-3	-6	0	$-M$	0	0	$-M$	0	(0)	
A_1	0	4	1	-1	1	0	0	0	20	(1)	
S_2	0	1	1	0	0	1	0	0	20	(2)	
A_3	0	1	1	0	0	0	-1	-1	10	(3)	
	1	$5M-3$	$2M-6$	$-M$	0	0	$-M$	0	$30M$	(0)	
A_1	0	4	1	-1	1	0	0	0	20	(1)	5*
S_2	0	1	1	0	0	1	0	0	20	(2)	20
A_3	0	1	1	0	0	0	-1	1	10	(3)	10
	1	0	$\frac{3M-21}{4}$	$\frac{M-3}{4}$	$\frac{3-5M}{4}$	0	$-M$	0	$15+5M$	(0)	
x_1	0	1	$\frac{1}{4}$	$-\frac{1}{4}$	$\frac{1}{4}$	0	0	0	5	(1)	20
S_2	0	0	$\frac{3}{4}$	$\frac{1}{4}$	$-\frac{1}{4}$	1	0	0	15	(2)	20
A_3	0	0	$\frac{3}{4}$	$\frac{1}{4}$	$-\frac{1}{4}$	0	-1	1	5	(3)	$\frac{20}{3}$*
	1	0	0	1	$-M-1$	0	-7	$7-M$	50	(0)	
x_1	0	1	0	$-\frac{1}{3}$	$\frac{1}{3}$	0	$\frac{1}{3}$	$-\frac{1}{3}$	$\frac{10}{3}$	(1)	
S_2	0	0	0	0	0	1	1	-1	10	(2)	
x_2	0	0	1	$\frac{1}{3}$	$-\frac{1}{3}$	0	$-\frac{4}{3}$	$\frac{4}{3}$	$\frac{20}{3}$	(3)	20*
	1	0	-3	0	$-M$	0	-3	$3-M$	30	(0)	
x_1	0	1	1	0	0	0	-1	1	10	(1)	
S_2	0	0	0	0	0	1	1	-1	10	(2)	
E_1	0	0	3	1	-1	0	-4	4	20	(3)	

z minimized at value of 30 when $x_1 = 10$, $S_2 = 10$, and $E_1 = 20$;

SEC. 11.3 **1**

Basic Variables	z	x_1	x_2	S_1	S_2	b_i	Row No.	r_i
	1	−4	−2	0	0	0	(0)	
S_1	0	1	1	1	0	15	(1)	15
S_2	0	②	1	0	1	20	(2)	10*
	1	0	0	0	2	40	(0)	
S_1	0	0	①⁄₂	1	−½	5	(1)	10*
x_1	0	1	½	0	½	10	(2)	20

Optimal solution exists when $x_1 = 10$ and $S_1 = 5$, resulting in $z = 40$. However, row (0) coefficient of zero for x_2 indicates that an alternative optimal solution exists.

Basic Variables	z	x_1	x_2	S_1	S_2	b_i	Row No.	r_i
	1	0	0	0	2	40	(0)	
x_2	0	0	1	2	−1	10	(1)	
x_1	0	1	0	−1	1	5	(2)	

Alternative optimal solution occurs when $x_1 = 5$, $x_2 = 10$, and $z = 40$;
3 Alternative optimal solutions occur when $x_1 = 4$, $x_2 = 8$, and $z = 48$ or when $x_1 = 8$ and $S_1 = 12$;   **5** no feasible solution exists;   **7** optimal solution occurs when $x_1 = 200$, $x_2 = 133.33$, $S_1 = 333.33$, and $z = 11,666.67$

SEC. 11.4 **5** z maximized at 660 when $x_2 = 35$, $x_3 = 30$, and $S_3 = 5$. *Shadow Prices:* constraint 1, 6.333; constraint 2, 0.333; constraint 3, 0. *Sensitivity Analysis on Objective Function Coefficients:* nonbasic variable c_1, delta = 11, upper limit = 13.

Basic Variables	Lower Delta	Upper Delta	Lower Limit	Upper Limit
c_2	−14.67	4.00	−2.67	16.00
c_3	−2.00	no limit	6.00	no limit

Right-hand side constants:

Constraint	Lower Delta	Upper Delta	Lower Limit	Upper Limit
1	−84.00	1.818	16.00	101.818
2	−180.00	20.000	−100.00	100.000
3	−5.00	no limit	295.00	no limit

7 z minimized at $3925 when $x_{11} = 300$, $x_{13} = 500$, $x_{22} = 350$, and $x_{24} = 350$

SEC. 11.5 **1** Minimize $z = 45y_1 + 30y_2 + 50y_3$
subject to $y_1 + 4y_2 - y_3 \geq 3$
$y_1 + 5y_2 + 3y_3 \geq 4$
$y_1 - 3y_2 - 4y_3 \geq 2$
$y_1, y_2, y_3 \geq 0$

3 Minimize $z = 60y_1 + 25y_2 + 35y_3$
subject to
$$5y_1 + y_2 \qquad\ \ge 20$$
$$-3y_1 + y_2 - y_3 \ge 15$$
$$10y_1 + y_2 + 4y_3 \ge 18$$
$$4y_1 \qquad + 7y_3 = 10$$
$$y_1 \qquad\qquad \ge 0$$
$$y_2 \quad \text{unrestricted}$$
$$y_3 \le 0$$

5 Maximize $z = 45y_1 + 24y_2 + 20y_3$
subject to $y_1 + 3y_2 \qquad \le 4$
$$y_1 + 5y_2 \qquad \le 5$$
$$y_1 \qquad + 7y_3 = 2$$
$$y_1 - 2y_2 - 5y_3 \le 3$$
$$y_1 \qquad + 3y_3 = 1$$
$$y_1 \quad \text{unrestricted}$$
$$y_2 \le 0$$
$$y_3 \ge 0$$

7 (a) Minimize $z = 32y_1 + 24y_2$
subject to $2y_1 + 3y_2 \ge 5$
$$4y_1 + 2y_2 \ge 3$$
$$y_1, y_2 \ge 0$$

(b)

Basic Variables	z	x_1	x_2	S_1	S_2	b_i	Row No.	r_i
	1	−5	−3	0	0	0	(0)	
S_1	0	2	4	1	0	32	(1)	16
S_2	0	③	2	0	1	24	(2)	8*
	1	0	$\frac{1}{3}$	$\boxed{0}$	$\boxed{\frac{5}{3}}$	40	(0)	
S_1	0	0	$\frac{8}{3}$	1	−2	16	(1)	
x_1	0	1	$\frac{2}{3}$	0	$\frac{1}{3}$	8	(2)	

z maximized at value of 40 when $x_1 = 8$ and $S_1 = 16$;

(c) The two boxed values under S_1 and S_2 in row (0) of the optimal primal tableau indicate that in the optimal dual solution $y_1 = 0$ and $y_2 = \frac{5}{3}$. When substituted into the dual objective function, the minimum value for z is 40.

CHAPTER TEST 1 $\quad 4x_1 - 2x_2 + x_3 + x_3 + S_1 = 25$
$$-x_1 - 3x_2 + S_2 \qquad\qquad = 10$$
$$-2x_1 - 3x_3 + A_3 \qquad\ \ = 20$$
$$x_1, x_2, x_3, S_1, S_2, A_3 \ge 0$$

2 Maximum reached when $x_1 = 4$, $x_2 = 8$, and $z = 272$;

3 (a)

Basic Variables	z	x_1	x_2	E_1	A_1	A_2	b_i	Row No.
	1	$3M - 5$	−4	$-M$	0	0	$25M$	(0)
A_1	0	1	1	−1	1	0	10	(1)
A_2	0	2	−1	0	0	1	15	(2)

(b) A_2 will leave first, (c) x_1 will enter first;

5 Maximize $z = 25y_1 + 10y_2 + 48y_3 + 12y_4$
subject to $y_1 + 4y_2 + y_3 \qquad \le 8$
$$y_1 - 5y_2 - y_3 + y_4 \le 5$$
$$y_1 \qquad + 2y_3 \qquad = 6$$
$$y_1 \text{ unrestricted}$$
$$y_2 \ge 0$$
$$y_3, y_4 \le 0$$

CHAPTER 12

SEC. 12.2   **1** $z = \$3350$

Origin	Destination 1	Destination 2	Destination 3	Supply
1	5	10 (55)	10	55
2	20	30 (40)	20 (40)	80
3	10 (70)	20 (5)	30	75
Demand	70	100	40	210

3 (*a*) $z = \$2715$

Origin	Destination 1	Destination 2	Destination 3	Supply
1	8 (110)	6 (15)	10	125
2	4	9 (70)	8 (80)	150
3	7	6	5 (95)	95
Demand	110	85	175	370

(*b*) $z = \$2145$

Origin	Destination 1	Destination 2	Destination 3	Supply
1	8	6 (85)	10 (40)	125
2	4 (110)	9	8 (40)	150
3	7	6	5 (95)	95
Demand	110	85	175	370

5 $z = \$10,375$

Origin	Destination 1		2		3		Dummy		Supply
		20		30		10		0	
1					(150)				150
		30		40		25		0	
2	(150)		(25)		(75)		(50)		300
		35		15		20		0	
3			(100)						100
Demand	150		125		225		50		550

7 (a) Minimize $z = 20x_{11} + 35x_{12} + \cdots + 50x_{35}$

subject to $x_{11} + x_{12} + x_{13} + x_{14} + x_{15} \le 400$

$x_{21} + x_{22} + x_{23} + x_{24} + x_{25} \le 350$

$x_{31} + x_{32} + x_{33} + x_{34} + x_{35} \le 450$

$x_{11} + x_{21} + x_{31} = 150$

$x_{12} + x_{22} + x_{32} = 300$

$x_{13} + x_{23} + x_{33} = 200$

$x_{14} + x_{24} + x_{34} = 250$

$x_{15} + x_{25} + x_{35} = 175$

$x_{ij} \ge 0 \qquad \text{for all } i, j$

(c) z minimized at $\$34,350$ when $x_{11} = 150$, $x_{13} = 125$, $x_{15} = 125$, $x_{22} = 300$, $x_{25} = 50$, $x_{33} = 75$, $x_{34} = 250$;   **9** (a) 1500, (b) 80, (c) 1580;

11 (a) Minimize $z = 30x_{11} + 40x_{12} + 25x_{13} + 45x_{14} + 35x_{15} + 45x_{21} + 55x_{22} + 25x_{24}$
$+ 30x_{25} + 40x_{26} + 50x_{27} + 60x_{31} + \cdots + 45x_{58}$

subject to $x_{11} + x_{12} + x_{13} + x_{14} + x_{15} \le 300$

$x_{21} + x_{22} + x_{24} + x_{25} + x_{26} + x_{27} \le 350$

$x_{31} + x_{33} + x_{35} + x_{36} + x_{37} + x_{38} \le 325$

$x_{41} + x_{42} + x_{44} + x_{45} + x_{46} + x_{47} + x_{48} \le 250$

$x_{51} + x_{52} + x_{53} + x_{54} + x_{55} + x_{56} + x_{57} + x_{58} \le 400$

$x_{11} + x_{21} + x_{31} + x_{41} + x_{51} = 150$

$x_{12} + x_{22} + x_{42} + x_{52} = 100$

$x_{13} + x_{33} + x_{53} = 250$

$x_{14} + x_{24} + x_{44} + x_{54} = 175$

$x_{15} + x_{25} + x_{35} + x_{45} + x_{55} = 225$

$x_{26} + x_{36} + x_{46} + x_{56} = 200$

$x_{27} + x_{37} + x_{47} + x_{57} = 180$

$x_{38} + x_{48} + x_{58} = 220$

$x_{ij} \ge 0 \qquad \text{for all } i, j$

(c) z minimized at $\$41,225$ when $x_{11} = 150$, $x_{12} = 20$, $x_{13} = 80$, $x_{15} = 50$, $x_{24} = 175$, $x_{25} = 175$, $x_{36} = 200$, $x_{38} = 50$, $x_{42} = 80$, $x_{48} = 170$, $x_{53} = 170$, and $x_{57} = 180$

SEC. 12.3   **1** team 1 $\rightarrow$ Far West, team 2 $\rightarrow$ East, team 3 $\rightarrow$ Southwest, team 4 $\rightarrow$ Midwest, cost minimized at $\$27,000$;

3 (a) Maximize $z = 2500x_{11} + 1000x_{12} + 2800x_{13} + 3200x_{14} + 3500x_{15} + 1800x_{21} + \cdots$
$+ 2700x_{53} + 3000x_{54} + 2500x_{55}$

subject to $x_{11} + x_{12} + x_{13} + x_{14} + x_{15} = 1$
$$x_{21} + x_{22} + x_{23} + x_{24} + x_{25} = 1$$
$$x_{31} + x_{32} + x_{33} + x_{34} + x_{35} = 1$$
$$x_{41} + x_{42} + x_{43} + x_{44} + x_{45} = 1$$
$$x_{51} + x_{52} + x_{53} + x_{54} + x_{55} = 1$$
$$x_{11} + x_{21} + x_{31} + x_{41} + x_{51} = 1$$
$$x_{12} + x_{22} + x_{32} + x_{42} + x_{52} = 1$$
$$x_{13} + x_{23} + x_{33} + x_{43} + x_{53} = 1$$
$$x_{14} + x_{24} + x_{34} + x_{44} + x_{54} = 1$$
$$x_{15} + x_{25} + x_{35} + x_{45} + x_{55} = 1$$
$$x_{ij} = 0 \text{ or } 1 \quad \text{for all } i, j$$

(*b*) Profit maximized at $16,600 when following assignments made

Plane	Assigned to	Charter
1		4
2		3
3		5
4		1
5		2

CHAPTER TEST   **1** $x_{11} = 800$, $x_{12} = 400$, $x_{22} = 1500$, $x_{33} = 2000$, $x_{34} = 400$, $x_{44} = 1000$, $z = \$215,000$;   **2** $x_{12} = 100$, $x_{14} = 200$, $x_{21} = 350$, $x_{23} = 150$, $x_{32} = 300$, $x_{33} = 150$, $z = 22,650$;   **3** person 1 → job 1, person 2 → job 3, person 3 → job 2, person 4 → job 4, minimum number of days equals 48

CHAPTER 13

SEC. 13.1   **1** $A = \{a | a$ is a positive odd integer less than 20$\}$;   **3** $V = \{v | v$ is a vowel$\}$;
5 $C = \{c | c$ is the cube of a negative integer greater than $-5\}$;
7 $B = \{1, 2, 3, 4, 5, 6, 7\}$;   **9** $B = \{-3\}$;   **11** $B' = \{1, 3, 5, 7, 9, 10\}$;
13 $S = \{8, 11\}$;   **15** $A \subset \mathcal{U}, B \subset \mathcal{U}, C \subset \mathcal{U}, A \subset C, B \subset C$;   **17** see figure;
19 $A = B = C = D$, each set is a subset of all others;   **21** (*a*) $\emptyset$,
(*b*) $\{x | x$ is a positive integer greater than 10$\}$,
(*c*) $\{x | x$ equals 2, 4, 6, 8 or a positive integer greater than 10$\}$, (*d*) $\{1, 3, 5, 7\}$,
(*e*) $\{x | x$ equals 1, 3, 5, 7 or a positive integer greater than 10$\}$,
(*f*) $\{x | x$ is a positive integer greater than 8$\}$

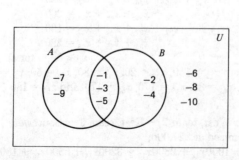

Sec. 13.1, Exercise 17

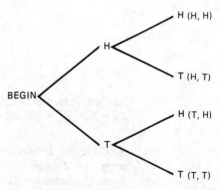

Sec. 13.2, Exercise 1

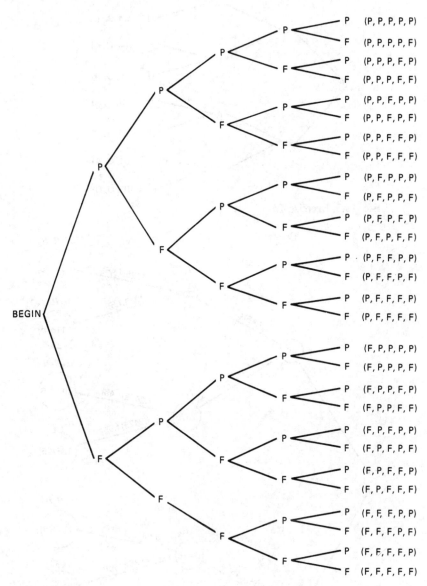

Sec. 13.2, Exercise 3

SEC. 13.2 **1** see figure; **3** see figure; **5** 676,000; **7** 2880; **9** 5040; **11** $1.3077 \times 10^{12} = 1,307,700,000,000$; **13** 210; **15** 6720; **17** 840; **19** 336; **21** 720; **23** 20,160; **25** 1; **27** 70; **29** 3,628,800; **31** 70; **33** 56; **35** 495; **37** 15; **39** 6.3501356×10^{11}; **41** 1,058,400

SEC. 13.3 **1** (a) $S = \{AAA, AAU, AUA, AUU, UAA, UAU, UUA, UUU\}$, (b) see figure (p. A-60), (c) AAU, UAA, AUA, (d) all outcomes except UUU; **3** (a) {MC, MH, MN}, (b) {MC, FC}; **5** (a) $510/1000 = 0.51$, (b) $310/1000 = 0.31$, (c) $40/1000 = 0.04$, (d) $275/1000 = 0.275$; **7** (a) not mutually exclusive, collectively exhaustive, (b) mutually exclusive, collectively exhaustive,

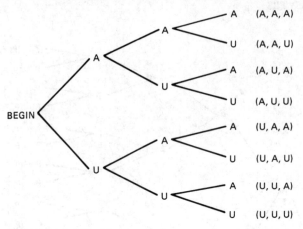

Sec. 13.3, Exercise 1*b*

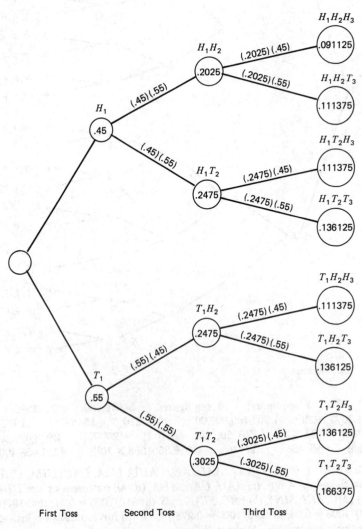

First Toss   Second Toss   Third Toss

Sec. 13.4, Exercise 3

(c) not mutually exclusive, not collectively exhaustive, (d) not mutually exclusive, not collectively exhaustive, (e) mutually exclusive;   **9** (a) 2,720,900/9,100,000 = 0.299, (b) 18,200/9,100,000 = 0.002, (c) 809,900/9,100,000 = 0.089;   **11** (a) 0.24, (b) 0.92, (c) 0.80;   **13** (a) 8/52, (b) 12/52, (c) 16/52, (d) 32/52;   **15** (a) 650/3000 = 0.2167, (b) 150/3000 = 0.05, (c) 0.25, (d) 2100/3000 = 0.70;   **17** (a) 0.7, (b) 0.3, (c) 0.4, (d) 0.9, (e) 0.1, (f) 1.0, (g) 1.0, (h) 0.6

SEC. 13.4   **1** (a) 0.7225, (b) 0.0921, (c) 0.4437;   **3** see figure, P(2 tails) = 0.408375, P(2 heads) = 0.334125;   **5** (a) 4950/8000 = 0.61875, (b) 3400/8000 = 0.4250, (c) 700/8000 = 0.0875, (d) 950/8000 = 0.11875, (e) 500/3050 = 0.1639, (f) 1500/2450, (g) 1500/2450 = 0.6122   **9** (a) 156/2652 = 0.0588, (b) 28,561/6,497,400 = 0.0044, (c) 24/132,600 = 0.00018, (d) 4,669,920/6,497,400 = 0.7187;   **11** (a) 1200/4000 = 0.30, (b) 1700/4000 = 0.425, (c) 760/4000 = 0.19, (d) 380/1700 = 0.2235, (e) 920/1200 = 0.7667;   **13** (a) 0.18, (b) 0.42;   **15** (a) 0.15, (b) 0.333, (c) 0.60

CHAPTER TEST   1 (a) {2, 4, 6, 8}, (b) {2}, (c) {−2, 0, 1, 2, 3};   **3** (a) 120, (b) 20; **4** 24/132,600 = 0.00018;   **5** (a) 12/26 = 0.4615, (b) 18/34 = 0.5294

CHAPTER 14

SEC. 14.1   1 (a) continuous, (b) discrete, (c) continuous, (d) continuous, (e) discrete, (f) continuous, (g) discrete, (h) continuous;   **3** (a)

X	$P(X)$
0	0.2083
1	0.2222
2	0.2139
3	0.1111
4	0.0778
5	0.0667
6	0.0556
7	0.0444

(b) see figure,

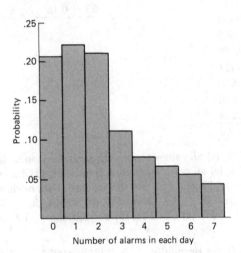

Sec. 14.1, Exercise 3*b*

(c) 0.7555, 0.3556;   **5** (a)

X	$P(X)$
0	0.0533
1	0.1333
2	0.1733
3	0.1867
4	0.1733
5	0.1067
6	0.0933
7	0.0667
8	0.0134

(b) see figure (p. A-62), (c) 0.9467, 0.2801

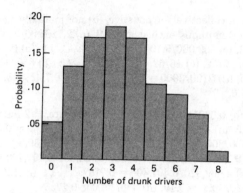

Sec. 14.1, Exercise 5*b*

7

Sum of Dots on Two Dice (x)	P(x)	
2	$\frac{1}{36} =$	.0278
3	$\frac{2}{36} =$	.0556
4	$\frac{3}{36} =$	.0833
5	$\frac{4}{36} =$	.1111
6	$\frac{5}{36} =$	.1389
7	$\frac{6}{36} =$	.1666
8	$\frac{5}{36} =$	.1389
9	$\frac{4}{36} =$	.1111
10	$\frac{3}{36} =$	.0833
11	$\frac{2}{36} =$	.0556
12	$\frac{1}{36} =$	.0278
		1.0000

9 (a)

X^2	$P(X^2)$
1	0.15
4	0.20
9	0.30
16	0.25
25	0.15

(b)

$X+1$	$P(X+1)$
2	0.15
3	0.20
4	0.30
5	0.25
6	0.15

SEC. 14.2 **1** (a) 110, (b) 110, (c) all values occur with same frequency, (d) 180, (e) 57.45; **3** (a) 29.5, (b) 29, (c) all occur with same frequency, (d) 42, (e) 12.796; **5** (a) 60, (b) 60, (c) 60, (d) 90, (e) 23.06; **7** $\bar{x} = 40$, median $= 40$, mode $= 50$; **9** $\bar{x} = 3.53$, median $= 3.5$, mode $= 0$; **11** (a) $\bar{x} = 31$, (b) $\sigma = 11.18$, **13** $\bar{x}_1 = 500$, $\bar{x}_2 = 500$, $\sigma_1 = 0$, $\sigma_2 = 284.60$; **15** (a) $\mu = 2.26$, (b) $\sigma = 2.0008$; **17** (a) $u = 3.40$, (b) $\sigma = 1.968$

SEC. 14.3 **1** (b), (d), (e), (f) not Bernoulli; **3** 0.1157, 0.9992;

5

X	P(X)
0	0.00032
1	0.00640
2	0.05120
3	0.20480
4	0.40960
5	0.32768

7 0.9872, yes, 0.9977; **9** 0.2508, 0.3020; **11** 0.8, 0.8579; **13** 0.1875, 0.1875; **15** 0.08192

SEC. 14.4 **1** (a) 0.75, (b) -1.00, (c) 2.00, (d) -1.75, (e) 3.125; **3** (a) 1.50, (b) -1.25, (c) -2.75, (d) 2.25, (e) -3.00; **5** (a) 0.0082, (b) 0.8849, (c) 0.21055, (d) 0.9867; **7** (a) 0.5987, (b) 0.3446, (c) 0.2075, (d) 0.5768; **9** (a) 0.8997, (b) 0.8686,

(c) 0.6546, (d) 0.0453;   **11** (a) 0.0122, (b) 0.8822, (c) 0.2012, (d) 0.3345;
13 0.0668, 0.1587;   **15** 0.8664, 0.0062;   **17** 0.9938, 0.9332;   **19** (a) 0.2420,
(b) 0.0668, (c) 0.3830, (d) 0.9938

CHAPTER TEST   **1** (a) 230.00, 16,248, (b) 0.78;   **2** $P(X = 0) = 0.2401$,
$P(X = 1) = 0.4116$, $P(X = 2) = 0.2646$, $P(X = 3) = 0.0756$, $P(X = 4) = 0.0081$;
3 32, 5.426;   **4** 0.234375;   **5** (a) 0.3264, (b) 0.0819, (c) 0.2743

CHAPTER 15

SEC. 15.1   **1** (a) 0, (b) 0, (c) 0, (d) 5, (e) -5, (f) no limit;   **3** (a) no limit,
(b) no limit, (c) no limit, (d) -2, (e) -2, (f) -2;   **5** 18;   **7** 10;
9 no limit exists;   **11** 54;   **13** -8;   **15** 16;   **17** -5;
19 no limit exists;   **21** 4/3;   **23** -14;   **25** -7;   **27** 3

SEC. 15.2   **1** 3;   **3** $-25\frac{1}{3}$;   **5** 11/14;   **7** 175;   **9** 6;   **11** 38;   **13** 0;
15 4200;   **17** 30;   **19** -8;   **21** $4c^3 - 5c^2 + 10$;
23 0, horizontal asymptote at $y = 0$, vertical asymptote at $x = 0$;
25 $-3/5$, horizontal asymptote at $y = -3/5$, vertical asymptote at $x = -20$;
27 no limit, no asymptotes;   **29** -2, horizontal asymptote at $y = -2$,
vertical asymptote at $x = -250$;   **31** 3, horizontal asymptote at $y = 3$,
vertical asymptote at $x = 0$;   **33** no discontinuities;   **35** no discontinuities;
37 discontinuity at $x = 4$;   **39** discontinuities at $x = -7$ and $x = 3$;
41 discontinuities at $x = 0$, $x = 3$, and $x = -2$;
43 discontinuities at $x = 5$ and $x = -2$;
45 discontinuities at $x = 2$, $x = -2$, and $x = 5$;   **47** discontinuities at $x = 0$ and $x = 3$

SEC. 15.3   **1** 3;   **3** -1;   **5** 1/9;   **7** 8;   **9** 2;   **11** -15;   **13** 5;
15 (a) 96 ft/s, 64 ft/s, 0 ft/s, (b) 8 s;   **17** \$4.5 million/yr, \$4.6 million/yr,
\$4.4 million/yr, \$5.55 million/yr;   **19** 0.9 million/yr, 1.25 million/yr,
1.067 million/yr;   **21** (a) 17 mi/h, 27 mi/h, (b) 32 mi/h;   **23** (a) $\Delta x + 2x + 3$,
(b) 7;   **25** (a) 0, (b) 0;   **27** (a) $10x + 5\,\Delta x + 20$, (b) 40;   **29** (a) $\frac{2}{3}x + \frac{1}{3}\Delta x + 5$,
(b) $6\frac{1}{3}$;   **31** (a) $2mx + m\,\Delta x$, (b) $4\,m$;   **33** (a) $-6x^2 - 6x\,\Delta x - 2\,\Delta x^2$, (b) -26;
35 (a) $-5/(x^2 + x\,\Delta x)$, (b) $-5/3$;   **37** (a) $-15x^2 - 15x\,\Delta x - 5\,\Delta x^2$, (b) -65;
39 (a) $6/(x^2 + x\,\Delta x)$, (b) 2

SEC. 15.4   **1** (a) $dy/dx = 4$, (b) 4, 4;   **3** (a) $dy/dx = 16x$, (b) 16, -32;
5 (a) $dy/dx = 6x - 5$, (b) 1, -17;   **7** (a) $dy/dx = 2x - 3$, (b) -1, -7;
9 (a) $dy/dx = -12x + 3$, (b) -9, 27;   **11** (a) $dy/dx = 40x$, (b) 40, -80;
13 (a) $dy/dx = a$, (b) a, a;   **15** (a) $dy/dx = 2/x^2$, (b) 2, $\frac{1}{2}$;   **17** (a) $dy/dx = -a/x^2$,
(b) $-a$, $-a/4$;   **19** (a) $dy/dx = 3/x^2$, (b) 3, 3/4;   **21** (a) $dy/dx = 3x^2/2$, (b) 3/2, 6;
23 (a) $dy/dx = 3x^2 + 6x$, (b) 9, 0

SEC. 15.5   **1** 0;   **3** 0;   **5** $3x^2 - 4$;   **7** $10x^4$;   **9** $3/5\sqrt[5]{x^2}$;   **11** $\frac{5}{2}\sqrt{x^3}$;
13 $10x^9$;   **15** $2x^5 - 2$;   **17** $3x^2/2$;   **19** $-10/x^3$;   **21** $40/x^5$;
23 $1 + 1/x^{3/2}$;   **25** $-2/3\sqrt[3]{x^4}$;   **27** $8x^7 + 30x^4 - 12x^5 - 36x^2$;
29 $9x^8 - 77x^6 + 18x^5 + 50x^4 - 120x^3$;   **31** $\frac{15}{2}x^4 - 2x^3 + 3x^2/4 + 60x - 10$;
33 $(1 + x^2)/(1 - x^2)^2$;   **35** $(x^2 - 20x - 2)/(x^2 + 2)^2$;
37 $(-20x^4 + 6x)/(4x^5 - 3x^2 + 1)^2$;   **39** (a) 10, (b) no values;   **41** (a) -2,
(b) ± 2.449;   **43** (a) $4a_2 + a_1$, (b) $-a_1/2a_2$;   **45** (a) 4/9, (b) 0;   **47** (a) 5, (b) ± 3

SEC. 15.6   **1** $-60x^2(1 - 4x^3)^4$;   **3** $4(3x^2 - 2)(x^3 - 2x + 5)^3$;
5 $-15x^2/2\sqrt{1 - 5x^3}$;   **7** $-x/\sqrt{(x^2 - 1)^3}$;   **9** e^x;   **11** $20xe^{x^2}$;
13 $15e^x(5e^x)^2 = 375e^{3x}$;   **15** $4e^{x^2} + 8x^2e^{x^2} = 4e^{x^2}(1 + 2x^2)$;   **17** $(2x - 2)e^{x^2 - 2x + 5}$;

19 $(xe^x - e^x)/x^2$;   **21** $3e^{3x}$;   **23** $1/x$;   **25** $2x/(x^2 - 3)$;   **27** $2x \ln x + x$;
29 $(10 \ln x - 10)/(\ln x)^2$;   **31** $[\ln 3x - 1 + 1/x]/(\ln 3x)^2$;   **33** $(1/x - 2x \ln x)/e^{x^2}$;
35 $\frac{1}{2}[(x - 5)^5(6x - 5)]^{-1/2}[5(x - 1)^4(6x - 5) + 6(x - 1)^5]$;
37 $6\sqrt{x^2 - 5x + 3} + (6x^2 - 17x + 5)/\sqrt{x^2 - 5x + 3}$;   **39** -15;   **41** $4x^3 - 4x$;
43 $(15 - 45x^2)(-x + x^3)^2$;   **45** $x/2\sqrt{x^2/2} = \sqrt{2}/2$;   **47** $(10x - 10)(x^2 - 2x - 3)^4$;
49 $(4x - 5)e^{2x^2 - 5x}$;   **51** $(60x^2 - 30x)/(20x^3 - 15x^2 - 3)$;   **53** (a) 12,000,000,
(b) $0, \pm\sqrt{2}$;   **55** (a) $\pm\frac{2}{5}$, (b) 0;   **57** (a) -0.1353, (b) no values;   **59** (a) -0.1353;
(b) 1;   **61** (a) $-\frac{1}{2}$, (b) 1;   **63** (a) $-3/2$, (b) 0.5

SEC. 15.7   **1** (a) 250 (hundred) ft/s, (b) 750 (hundred) ft/s, 3000 (hundred) ft/s;
3 (a) -48 ft/s, (b) -96 ft/s, (c) -128 ft/s;   **5** (a) 177.38 million, (b) 4.375 $e^{0.035t}$,
(c) 6.208 million/yr;   **7** (a) 50,274, (b) -1875 $e^{-0.025t}$, (c) -1256.85/yr

SEC. 15.8   **1** (a) 0, 0, (b) 0, 0;   **3** (a) $8x - 1$, 8, (b) 7, 8;   **5** (a) $15x^2$, $30x$,
(b) 15, 30;   **7** (a) $x^3 - x^2 + 10$, $3x^2 - 2x$, (b) 10, 1;   **9** (a) $-1/x^2$, $2/x^3$, (b) -1, 2;
11 (a)$15x^4 - 6x^2$, $60x^3 - 12x$, (b) 9, 48;   **13** (a) $-4/x^3$, $12/x^4$, (b) -4, 12;
15 (a) $4(x - 5)^3$, $12(x - 5)^2$, (b) -256, 192;   **17** (a) $1/2\sqrt{x + 1}$, $-1/4\sqrt{(x + 1)^3}$,
(b) 0.3535, -0.0884;   **19** (a) $2e^{2x}$, $4e^{2x}$, (b) 14.778, 29.556;   **21** (a) $2xe^{x^2}$, $2e^{x^2} + 4x^2e^{x^2}$,
(b) 5.4366, 16.3097;   **23** (a) $1/x$, $-1/x^2$, (b) 1, -1;   **25** (a) $2x/(x^2 - 5)$,
$(-2x^2 - 10)/(x^2 - 5)^2$, (b) $-1/2$, $-3/4$;   **27** (a) $(1 + x^2)/(1 - x^2)^2$,
$(6x + 2x^3)/(1 - x^2)^3$, (b) undefined, undefined;   **29** (a) $e^x \ln x + e^x/x$,
$e^x \ln x + e^x/x + (xe^x - e^x)/x^2$, (b) 2.718, 2.718

CHAPTER TEST   **1** (a) 5, (b) 250;   **2** no discontinuities;   **3** $-6x - 2$;
4 (a) $-24/5x^{9/5}$, (b) $42x^5 - 40x^4 + 9x^2$, (c) $(-8x^3 - 162x^2 + 8)/(x^3 + 8)^4$, (d) $2xe^{4x} + 4x^2e^{4x}$
(e) $15x^2 - 2x)/(5x^3 - x^2)$, (f) $4(e^x \ln x)^3[e^x \ln x + e^x/x]$;   **5** $x = 3$;
6 $f'(x) = 5x^3 - 3x^2 + 12x + 10$, $f''(x) = 15x^2 - 6x + 12$, $f'''(x) = 30x - 6$, $f''''(x) = 30$,
$f'''''(x) = 0$;   **7** $8x^7 + 240x^5 + 2400x^3 + 8006x$

CHAPTER 16

SEC. 16.1   **1** (a) decreasing, (b) no values, (c) all real values, (d) no values;
3 (a) decreasing, (b) $x > 1.5$, (c) $x < 1.5$, (d) $x = 1.5$;   **5** (a) increasing,
(b) $x > 0$ or $x < -1$, (c) $-1 < x < 0$, (d) $x = 0$ or -1;   **7** (a) increasing, (b) $x > 0$,
(c) $x < 0$, (d) $x = 0$;   **9** (a) increasing, (b) $x > -3$, (c) no values, (d) $x = -3$;
11 (a) increasing, (b) $x < 2$, (c) $x > 2$, (d) $x = 2$;   **13** (a) increasing, (b) $x > -4$,
(c) $x < -4$, (d) $x = -4$;   **15** (a) function undefined at $x = 1$, (b) $x > 4$, (c) no values,
(d) $x = 4$;   **17** (a) increasing, (b) all real values except $x = 5$, (c) no values,
(d) $x = 5$;   **19** (a) increasing, (b) $x \neq -5$, (c) no values, (d) $x = -5$;
21 down, down;   **23** up, up;   **25** up, up;   **27** down, up;   **29** down, up;
31 down, up;   **33** up, no conclusion;   **35** up, up;
37 function not defined at $x = -2$ or $x = 1$;   **39** up, up;   **41** down, down;
43 (a) all values of x, (b) no values, (c) linear function, no concavity,
(d) same as part c;   **45** (a) $x > -b/2a$, (b) $x < -b/2a$, (c) all values of x,
(d) no values of x;   **47** (a) $x \neq 0$, (b) none, (c) $x > 0$, (d) $x < 0$;   **49** $x = 3$;
51 $x = -3, 5$;   **53** $x = 5$;   **55** no inflection points;   **57** $x = -2$;
59 $x = -3, 2$;   **61** $x = \pm 1.732$;   **63** no inflection point;   **65** $x = 5$;
67 no inflection points;   **69** no inflection points;   **71** (a) $a \leq x < c$, $e < x < g$,
$i < x < k$, (b) $c < x < e$, $g < x < i$, (c) $x = c, e, g, i, k$;   **73** (a) $e < x < f$, $i < x < j$,
$a \leq x < b$, (b) $b < x < c$, $f < x < g$, $j < x < k$, (c) $d < x < e$, $h < x < i$, (d) $c < x < d$,
$g < x < h$;   **75** (a) $a_1 < x < a_2$, $a_3 < x < a_5$, $x > a_7$, (b) $a_2 < x < a_3$, $a_5 < x < a_7$,
(c) $x = a_2, a_3, a_5, a_7$

SEC. 16.2 **1** relative min at $(8, -92)$; **3** inflection point at $(0, 5)$;
5 relative max at $(1, 1.833)$, relative min at $(4, -2.667)$;
7 relative max at $(0, 0)$, relative min at $(-5, -468.75)$ and $(5, -468.75)$;
9 inflection point at $(0, -10)$; **11** relative max at $(6, 21)$;
13 relative min at $(0, 4)$; **15** relative max at $(-2, 12.67)$,
relative min at $(2.5, -17.708)$; **17** relative min at $(3, -777.6)$,
relative max at $(-3, 777.6)$; **19** relative max at $(-8, 42.67)$, relative min at $(0, 0)$;
21 relative max at $(0, 0)$, relative minima at $(-2, -4)$ and $(2, -4)$;
23 inflection point at $(0, 0)$, relative max at $(-2.5, 29.30)$, relative min at $(3, -58.05)$;
25 relative min at $(1, -0.8)$, relative max at $(-1, 0.8)$; **27** relative max at $(0, 2)$,
relative minima at $(-1, 5/3)$ and $(1, 5/3)$; **29** relative min at $(-10, 0)$;
31 inflection point at $(-\frac{1}{2}, 0)$; **33** relative min at $(0, -4096)$,
relative maxima at $(-2, 0)$ and $(2, 0)$; **35** no critical points;
37 relative min at $(0, 500)$; **39** no critical points;
41 relative max at $(1, 0.3679)$; **43** relative min at $(-21.97, -21.82)$;
45 no critical points; **47** inflection point at $(1, -0.307)$;
49 relative min at $(0.3679, -1.4716)$; **51** relative max at $(0.1, -1.6931)$;
53 relative min at $(0.5, 1.44)$; **55** relative min at $(-1, -0.5)$,
relative max at $(1, 0.5)$; **57** relative min at $(-b/2a, -b^2/4a + c)$

SEC. 16.3 **1, 3, 5, 7, 9, 11, 13, 15, 17, 19** see figures (pp. A-65 to A-67)

SEC. 16.4 **1** absolute max at $(8, 101)$, absolute min at $(2, 5)$;
3 absolute max at $(2, -40)$, absolute min at $(8, -256)$; **5** absolute max at $(8, 6520.6)$,
absolute min at $(3, 20.6)$; **7** absolute max at $(4, 298\frac{2}{3})$, absolute min at $(0, 0)$;
9 absolute max at $(0.75, -7.75)$, absolute min at $(10, -350)$;
11 absolute max at $(10, 2116)$, absolute min at $(5, 72.25)$;
13 absolute max at $(0, -10)$, absolute min at $(6, -1082.8)$;
15 absolute max at $(-5, 7291\frac{2}{3})$, absolute min at $(0, 0)$; **17** absolute max at $(16, 4)$,
absolute min at $(4, 2)$

CHAPTER TEST **1** (a) $x > 10$ or $x < -4$, (b) $-4 < x < 10$, (c) $x = -4$ or 10;
2 see figure (p. A-67); **3** (a) relative min at $(7, -125.66)$, relative max at $(-3, 41)$,

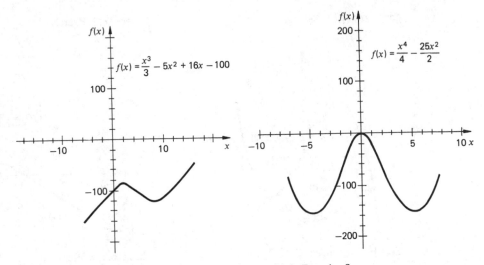

Sec. 16.3, Exercise 1   Sec. 16.3, Exercise 3

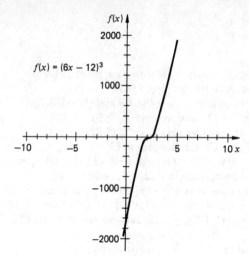

Sec. 16.3, Exercise 5

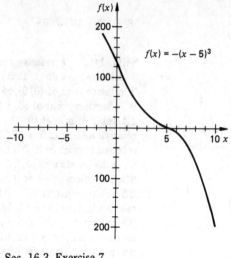

Sec. 16.3, Exercise 7

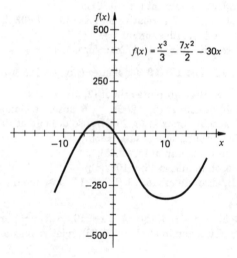

Sec. 16.3, Exercise 9

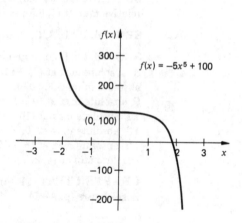

Sec. 16.3, Exercise 11

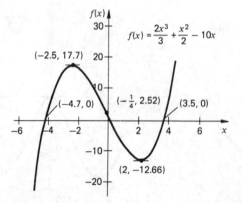

Sec. 16.3, Exercise 13

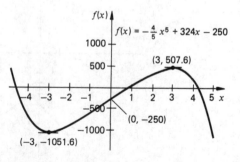

Sec. 16.3, Exercise 15

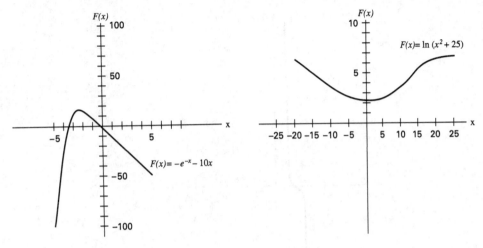

Sec. 16.3, Exercise 17   Sec. 16.3, Exercise 19

(b) relative max at (0, 20.09);   **4** inflection point when $x = -2$ and 4;
5 absolute max at $(1, -8.833)$, absolute min at $(3, -17.5)$;   **6** see figure

CHAPTER 17

SEC. 17.1   1 (a) $87.50, (b) $76,562.50;   **3** (a) 250,000 units, (b) 624,975,000;
5 (a) 3200 cents = $32.00, (b) $128 million, (c) 4,000,000 sets;   **7** (a) 20 tons,
(b) $754,160, (c) costs $112.28 higher;   **9** (a) 1200 dozens, (b) $2,000, 413.30,
(c) costs $3.40 higher;   **11** (a) 1183.216, (b) $8091.61, (c) $9,574,120;   **13** (a) $27.50,
(b) $32,812,500;   **15** (a) 50,000, (b) $575, (c) $9,650,000;
17 40,000, loss of $400,000;   **19** (a) $q = 400$, (b) $79,500;
21 (a) cannot determine using marginal approach, (b) $71,955,000;   **23** (a) 2500 units,
(b) $22,500

SEC. 17.2   1 38.73 ft by 38.73 ft;   **3** $x = 150, y = 75$, area = 11,250 m²;
5 $x = 900, y = 600$, area = 540,000 ft²;   **7** $x = 63.33$;   **9** (a) 76,376.26 mi,
(b) $0.4416/mi, (c) $11,043.56;   **11** (a) $p = 100 - x$, (b) $0 \leq x \leq 100$,
(c) $R = 5000 + 50x - x^2$, (d) 25, (e) 75, (f) $5625, (g) $75, (h) no;   **13** (a) 96.81 days,
(b) $611,370, (c) 91.11 percent;   **15** 40 officers, 352.84 crimes/day;   **17** (a) 6,
(b) $4.72;   **19** (a) $w = f(x) = 28.5 - 1.5x$, (b) 9.5 h, (c) $14.25, (d) $135.375,
(e) for units requiring 15 or more hours, wages = $90, which is $45.375 less than the

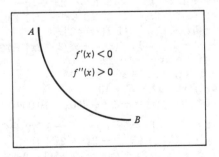

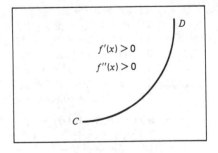

Chap. 16, Chapter Test, Problem 2

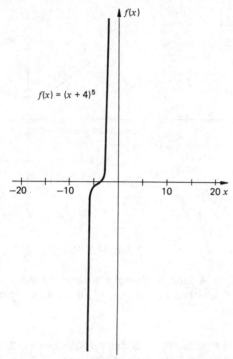

Chap. 16, Chapter Test, Problem 6

maximum;   **21** (a) $-60p/(1200-60p) = -p/(20-p)$, (b) $-\frac{1}{3}$ (inelastic),
-1 (unit elastic), -3 (elastic);   **23** (a) $-20p^2/(12,000-10p^2) = -2p^2/(1200-p^2)$,
(b) -0.18 (inelastic), -1 (unit elastic), -6 (elastic);   **25** \$15

CHAPTER TEST   **1** $R = g(p) = 60,000p - 7.5p^2$;
2 (a) $P = h(x) = -5x^2 + 450x - 5000$, (b) 45(100's) = 4500 units,
(c) \$5125(100's) = \$512,500;   **3** (a) 316.23 by 316.23, (b) \$25,298.40;
4 (a) 200 units, (b) \$80,200;   **5** (a) \$20, (b) \$257,515.61;   **6** (a) 7 persons, (b) \$4.29

CHAPTER 18

SEC. 18.1   **1** $80x + C$;   **3** $x/5 + C$;   **5** $3x^2/2 + C$;   **7** $x^3/6 + C$;
9 $x^5/5 + C$;   **11** $x^3/3 - 2x^2 + C$;   **13** $x^3/3 + 4x^2 + 10x + C$;
15 $3x^3 + 5x^2 + C$;   **17** $x^3 + 9x^2 + 12x + C$;   **19** $2x^4 - 2x^3 + C$;
21 $3x^4 - 3x^3 + 3x + C$;   **23** $x^5 - 3x^3 - 6x + C$;   **25** $6x^5 - x^4/2 + 4x^2 - 5x + C$;
27 $x^4/8 + C$;   **29** $-3x^6 + 3x^3 - 5x^2 + C$;   **31** $f(x) = 20x$;
33 $f(x) = 5x^2 - 10$;   **35** $f(x) = x^4 - 1$;   **37** $f(x) = -x^3/3 + 2x^2 + 36$;
39 $f(x) = 2x^3 + 4x^2 - 52$;   **41** $f(x) = 3x^3 + x^2 + 22$;
43 $f(x) = x^4 - x^3 + x^2 + 17$;   **45** $f(x) = 2x^4 - x^3 - 16$;
47 $f(x) = -x^4 - 6x^2 + 815$;   **49** $f(x) = x^6 - x^3 + 10$;
51 $R = f(x) = 40,000x - 2x^2$;   **53** $P = f(x) = -3x^2 + 450x - 10,000$

SEC. 18.2   **1** $60x + C$;   **3** $x/8 + C$;   **5** $4x^2 + C$;   **7** $-3x^2/2 + C$;
9 $3x^2/2 + 6x + C$;   **11** $x^2/6 - x/4 + C$;   **13** $x^3 - 2x^2 + 2x + C$;
15 $-6x^3 + x^2/2 - 5x + C$;   **17** $x^4 + 2x^3 - 3x + C$;   **19** $x^6/6 - 3x^4 + 3x + C$;
21 $5x^{6/5}/6 + C = 5x\sqrt[5]{x}/6 + C$;   **23** $-1/2x^2 + C$;   **25** $ax^5/5 + bx^3/3 + C$;
27 $(a/b)(x^{1-n})/(1-n) + C, n \neq 1$;   **29** $x^{1-n}/(1-n) + C, n \neq 1$;

31 $\frac{3}{2}x^{4/3} + C = 3x\sqrt[3]{x}/2 + C;$ **33** $5x^{8/5}/8 + C = 5x\sqrt[5]{x^3}/8 + C;$
35 $\frac{16}{15}x^{5/4} + C = \frac{16}{15}x\sqrt[4]{x} + C;$ **37** $-1/3x^3 + C;$ **39** $-16/x + C;$ **41** $2\sqrt{x} + C;$
43 $-\frac{75}{4}x^{4/5} + C;$ **45** $ax^5/5 + bx^4/4 + cx^3/3 + dx^2/2 + ex + C;$
47 $ax^4 + bx^3 + cx^2 + C;$ **49** $x^{1-n}/a(1-n) + C, n \neq 1$

SEC. 18.3   **1** $(x + 20)^6/6 + C;$ **3** $(8x + 20)^4/4 + C;$ **5** $[x/3 + 15]^3/3 + C;$
7 cannot be done;   **9** $(x^2 + 3)^5/5 + C;$ **11** $(x^2 + 5)^4/12 + C;$
13 $(2x^2 - 4x)^7/28 + C;$ **15** $2\sqrt{x^2 + 8} + C;$ **17** cannot do;   **19** cannot do;
21 $(2x^3 + 3)^{3/2}/9 + C;$ **23** $(2x^2 + 8x)/16 + C;$ **25** cannot do;   **27** cannot do;
29 $e^{3x}/3 + C;$ **31** $e^{ax}/a + C;$ **33** $-\frac{1}{2}\ln(x^2 + 5) + C;$ **35** $3\ln(6x + 5) + C;$
37 $(1/a)\ln(ax + b) + C;$ **39** $\ln(ax^2 + bx) + C$

SEC. 18.4   **1** $-xe^{-x} - e^{-x} + C;$ **3** $3x(x + 1)^{4/3}/4 - 9(x + 1)^{7/3}/28 + C;$
5 $x(e^{2x})/2 - e^{2x}/4 + C;$ **7** $(x^2/2 + 4x)\ln x - x^2/4 - 4x + C;$
9 $x(x + 2)^5/5 - (x + 2)^6/30 + C;$ **11** $2x(x - 3)^{1/2} - \frac{4}{3}(x - 3)^{3/2} + C;$
13 $-\ln x/x - 1/x + C;$ **15** $-8\ln(x + 3) + 7\ln(x + 2) + C;$
17 $-2\ln(x - 1) - 5/(x - 1) + C;$ **19** $-4\ln x + 4.75\ln(x + 4) + 4.25\ln(x - 4) + C;$
21 $6\ln x - 2\ln(x - 1) + C;$ **23** $(x^5/5)\ln(10x) - x^5/25 + C;$
25 $[(\ln x/-2) - \frac{1}{4}]/x^2 + C;$ **27** $x(\ln x)^2 - 2x\ln x + 2x + C;$
29 $x^5(\ln x/5 - \frac{1}{25}) + C;$ **31** $2e^{2.5x}/5 + C;$ **33** $(e^{5x}/25)(5x - 1) + C;$
35 $x/5 - \frac{1}{10}\ln(5 + 3e^{2x}) + C$

SEC. 18.5   **1** first order, first degree;   **3** second order, third degree;
5 first order, first degree;   **7** first order, third degree;
9 second order, third degree;   **11** $y = \frac{1}{5}x^5 + x^2 + 6x + C;$ **13** $y = \ln x + C;$
15 $y = \frac{1}{2}\ln(5x^2 + 10) + C;$ **17** $y = \frac{1}{6}x^3 + \frac{5}{2}x^2 + C_1x + C_2;$
19 $y = -\frac{1}{20}x^5 + \frac{1}{3}x^3 - 8x^2 + C_1x + C_2;$ **21** $y = e^x + C_1x + C_2;$
23 $y = e^x + \frac{5}{2}x^2 + C_1x + C_2;$ **25** $y = -\frac{1}{6}x^3 + \frac{5}{2}x^2 + C_1x + C_2;$
27 $y = \frac{1}{4}(x^2 + 15)^4 + C;$ **29** $y = x^2 + C, y = x^2 - 50;$
31 $y = x^3/3 + 3x^2/2 + 8x + C, y = x^3/3 + 3x^2/2 + 8x - \frac{7}{3};$
33 $y = x^3 + 9x^2 + C_1x + C_2, y = x^3 + 9x^2 - 175x + 336;$ **35** $y = e^{5x} + C_1x + C_2,$
$y = e^{5x} - x - 3;$ **37** $y = x^3 - \frac{9}{2}x^2 + C_1x + C_2, y = x^3 - \frac{9}{2}x^2 + 16x + 48;$
39 $P = f(t) = 500e^{0.458t};$ **41** $P = f(t) = 40,000e^{-0.0446t}$

CHAPTER TEST   **1** $f(x) = x^4 + x^2 - 20x - 350;$ **2** (a) $-\frac{3}{2}x^{-2/3} + C,$
(b) $\frac{1}{24}(x^4 + 10)^6 + C,$ (c) $-e^{-10x}/10 + C;$ **3** $y = x^3/6 + 3x^2 + C_1x + C_2,$
$y = x^3/6 + 3x^2 - 17.5x + 45\frac{2}{3};$ **4** $R = 220,000x - 9x^2;$ **5** $xe^{10x}/10 - e^{10x}/100 + C;$
6 $10\ln(x - 3) + 10\ln(x + 2) + C$

CHAPTER 19

SEC. 19.1   **1** 4.5;   **3** 4;   **5** 8;   **7** $10\frac{2}{3};$ **9** $228\frac{2}{3};$ **11** 0.4054;
13 $-8102.084;$ **15** 0;   **17** 0;   **19** 36;   **21** 60;   **23** $32\frac{2}{3};$ **25** 27;
27 $-57;$ **29** 53.598;   **31** 0.3465;   **33** 56;   **35** 22,025.466;   **37** $2645\frac{1}{3};$
39 8.3178;   **41** 0.8735;   **43** 3.1012;   **45** cannot evaluate;   **47** $\frac{3}{2}m - b;$
49 0.4373;   **51** $41\frac{1}{3};$ **53** $-218.667;$ **55** 22;   **57** 79.5;   **59** 133.599

SEC. 19.2   **1** 9;   **3** 168;   **5** 180;   **7** 93.33;   **9** $38\frac{1}{3};$ **11** 17.367;
13 $7\frac{1}{3};$ **15** $7666\frac{2}{3};$ **17** 4,443,027.9;   **19** 0.69316;   **21** (a) $\int_0^a g(x)\,dx,$

(b) $-\int_0^a f(x)\,dx,$ (c) $\int_b^c f(x)\,dx - \int_b^c g(x)\,dx,$ (d) $-\int_a^b g(x)\,dx;$

23 (a) $\int_c^b h(x)\,dx - \int_0^a g(x)\,dx - \int_a^b f(x)\,dx,$ (b) $\int_0^a g(x)\,dx,$ (c) $-\int_0^a f(x)\,dx,$

(d) $\int_a^b f(x)\,dx + \int_b^c h(x)\,dx$, (e) $-\int_a^c g(x)\,dx$,

(f) $\int_b^d f(x)\,dx - \int_b^c h(x)\,dx - \int_c^d g(x)\,dx$;   **25** (a) see figure, (b) 54;

27 (a) see figure, (b) 24;   **29** (a) see figure, (b) $333\frac{1}{3}$;   **31** 8;   **33** $21\frac{1}{3}$;

35 85.896

SEC. 19.3   **1** (a) 21, (b) $21\frac{1}{3}$, (c) underestimated by $\frac{1}{3}$;   **3** (a) 15.5, (b) 16,
(c) underestimated by 0.5;   **5** (a) 51.428, (b) 53.598, (c) underestimated by 2.170;
7 (a) 22.375, (b) $22\frac{2}{3}$, (c) underestimated by 0.2917;   **9** (a) 2,900,099.24, (b) 3,025,680,
(c) 125,580.76 (underestimated);   **11** (a) 52, (b) 52, (c) 0;   **13** (a) 31, (b) $30\frac{2}{3}$,
(c) overestimated by $\frac{1}{3}$;   **15** (a) 115.9839, (b) 107.1963, (c) overestimated by 8.7876;
17 (a) 4,022,336, (b) 3,029,184, (c) overestimated by 993,152;   **19** (a) 72.189,
(b) 53.598, (c) overestimated by 18.591;   **21** (a) 26, (b) 26, (c) 0;   **23** (a) $105\frac{1}{3}$,
(b) $105\frac{1}{3}$, (c) 0;   **25** (a) $30\frac{1}{12}$, (b) 30, (c) overestimated by $\frac{1}{12}$;   **27** (a) 269.3192,

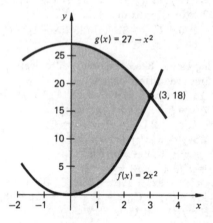

Sec. 19.2, Exercise 25

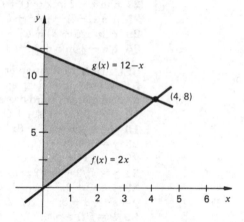

Sec. 19.2, Exercise 27

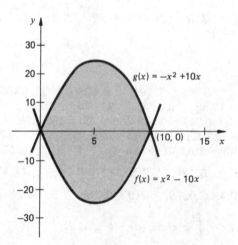

Sec. 19.2, Exercise 29

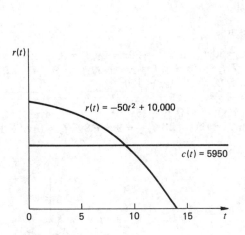

Sec. 19.4, Exercise 3

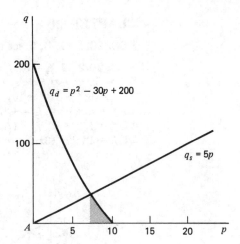

Sec. 19.4, Exercise 7

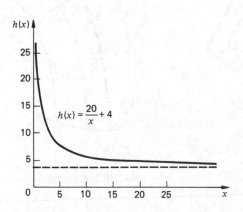

Sec. 19.4, Exercise 13

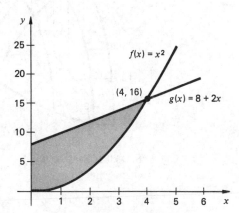

Chap. 19, Chapter Test, Problem 2a

(b) 267.9907, (c) overestimated by 1.3285;   **29** (a) 1,063,670,388, (b) 1,060,498,832, (c) overestimated by 3,171,556;   **31** (a) 20, (b) 20, 20, 20, (c) all are the same; **33** (a) $49\frac{1}{3}$, (b) 49, 50, $49\frac{1}{3}$, (c) Simpson's rule;   **35** (a) 16, (b) 15.5, 17, 16, (c) Simpson's rule;   **37** (a) 3,488.31, (b) 3,452.23, 3560.68, 3489.48, (c) Simpson's rule; **39** (a) 1,057,062,656, (b) 571,474,808.20; 2,177,820,000, 1,488,744,000, (c) Simpson's rule;   **41** 3.48453, 3.47474, 3.44781;   **43** 2.4008, 2.3090, 2.2536; **45** 0.1202, 0.1329, 0.1247;   **47** 3.8385, 3.8982, 3.8575

SEC. 19.4   1 (a) \$1200, (b) \$400;   **3** (a) see figure, (b) 9 days, (c) \$53,550, (d) \$24,300;   **5** (a) 195,082, (b) 9.4 hours;   **7** (a) see figure, (b) $p = \$7.192$, $q = 35.96$, (c) \$46.80;   **9** 10.986 days;   **11** (a) 0.642 billion tons/year; (b) 12.9744 billion tons;   **13** (a) 6 hours/unit, (b) 135.91 hours, (c) see figure, (d) 4 hours/unit;   **15** 407.15 cubic units

SEC. 19.5   1 0.444, 0.888;   **3** 0.3888

CHAPTER TEST   1 (a) $\frac{4}{3}$, (b) 9.3416;   **2** (a) see figure, (b) $26\frac{2}{3}$;   **3** (a) \$248/year, (b) \$432;   **4** 0.800;   **5** (a) 512, (b) 504, 528, 512, (c) Simpson's rule

CHAPTER 20

SEC. 20.1   1, 3, 5 see figures

SEC. 20.2   1 $f_x = 6x, f_y = -30y^2$;   **3** $f_x = 20x + 2y, f_y = 2x - 12y$;
5 $f_x = 3x^2y^5, f_y = 5x^3y^4$;   **7** $f_x = 12x - y, f_y = -x + 60y$;   **9** $f_x = 4/x^2y^2, f_y = 8/xy^3$;
11 $f_x = 6x^2 + 8xy^5 - 10y, f_y = 20x^2y^4 - 10x - 120y^5$;
13 $f_x = 4(x - y)^3, f_y = -4(x - y)^3$;   **15** $f_x = x/\sqrt{x^2 + y^2}, f_y = y/\sqrt{x^2 + y^2}$;
17 $f_x = 1/(x + y), f_y = 1/(x + y)$;   **19** $f_x = 15x^2e^{5x^3 - 2y^2}, f_y = -4ye^{5x^3 - 2y^2}$;
21 $f_x = 12x^3 - 16xy^3, f_y = -24x^2y^2$;   **23** $f_x = -9x^2 + 8y^2, f_y = 16xy + 3y^2$;

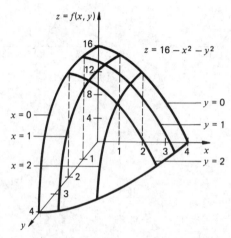

Sec. 20.1, Exercise 1

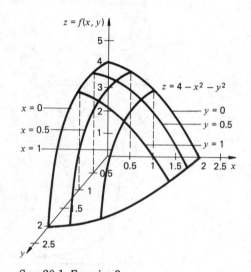

Sec. 20.1, Exercise 3

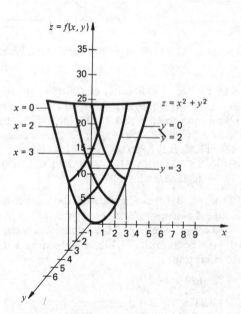

Sec. 20.1, Exercise 5

25 $f_x = 4x/3y^3, f_y = -2x^2/y^4$;   **27** $f_x = (-4x + 5y)/x^6, f_y = -1/x^5$;

29 $f_x = (-5x^2 - 10xy)/(x + y)^2, f_y = 5x^2/(x + y)^2$;

31 $f_x = 3(x - y)^2, f_y = -3(x - y)^2$;   **33** $f_x = 2x/3\sqrt[3]{(x^2 + 2y^2)^2}, f_y = 4y/3\sqrt[3]{(x^2 + 2y^2)^2}$;

35 $f_x = 2x/(x^2 - y^2), f_y = -2y/(x^2 - y^2)$;   **37** $f_x = 2xye^{x^2y}, f_y = x^2e^{x^2y}$;

39 $f_x = e^x \ln y, f_y = e^x/y$;   **41** $f_{xx} = 6, f_{xy} = 0, f_{yy} = 30y, f_{yx} = 0$;

43 $f_{xx} = 30x, f_{xy} = -3, f_{yy} = 6, f_{yx} = -3$;

45 $f_{xx} = 6xy^4, f_{xy} = 12x^2y^3, f_{yy} = 12x^3y^2, f_{yx} = 12x^2y^3$;   **47** $f_{xx} = f_{xy} = f_{yy} = f_{yx} = e^{x+y}$;

49 $f_{xx} = y^2e^{xy}, f_{xy} = e^{xy} + xye^{xy}, f_{yy} = x^2e^{xy}, f_{yx} = e^{xy} + xye^{xy}$;

51 $f_{xx} = 20(x - y)^3, f_{xy} = -20(x - y)^3, f_{yy} = 20(x - y)^3, f_{yx} = -20(x - y)^3$;

53 $f_{xx} = 12x^2y^2, f_{xy} = 8x^3y, f_{yy} = 2x^4, f_{yx} = 8x^3y$;

55 $f_{xx} = 0, f_{xy} = -2/y^3, f_{yy} = 6x/y^4, f_{yx} = -2/y^3$;

57 $f_{xx} = 27/\sqrt{6x - 8y}, f_{xy} = -36/\sqrt{6x - 8y}, f_{yy} = 48/\sqrt{6x - 8y}, f_{yx} = -36/\sqrt{6x - 8y}$;

59 $f_{xx} = -3/x^2, f_{xy} = 0, f_{yy} = -2/y^2, f_{yx} = 0$;   **61** (a) 88,000, (b) +1800,
(c) actual = +1900, (d) projected = +7900, actual = +8100;   **63** (a) 116,000,
(b) −1500 units, (c) +1900 units, (d) radio;   **65** (a) $f_{p_1} = p_1, f_{p_2} = 2p_2, f_{p_3} = 0.8p_3$,
(b) $f_{p_1} = -30, f_{p_2} = 20, f_{p_3} = 16$, (c) competing products

SEC. 20.3   **1** saddle point at $(-10, 10, -310)$;
3 saddle point at $(0, 2, -12)$, relative min at $(5, 2, -32.83)$;
5 relative min at $(1, 3.5, -20.75)$;   **7** relative min at $(1, -3, -1)$,
saddle point at $(-1, -3, 3)$;   **9** saddle point at $(0, 0, 0)$, relative min at $(1, 1, -1)$;
11 saddle point at $(1.414, -0.707, -10.15)$;   **13** relative min at $(20, 10, -300)$;
15 relative min at $(8, 4, -104)$;   **17** saddle point at $(0, 0, 0)$,
relative max at $(\frac{1}{6}, \frac{1}{12}, \frac{1}{432})$;   **19** relative max at $(-1, -4, 20)$;
21 saddle point at $(1, 2, 27)$, relative max at $(-1, 2, 31)$;
23 relative min at $(4, -2, -6)$;   **25** relative min at $(1, -3, -1)$,
saddle point at $(-1, -3, 3)$

SEC. 20.4   **1** (a) $2000 (1000's) or $2 million for TV and $2000 (1000's) or $2 million
for radio, (b) 100 million;   **3** (a) $p_1 = 10, p_2 = 10$,
(b) $q_1 = 60$ (thousand), $q_2 = 40$ (thousand), (c) $1,000,000;   **5** $x = -3.75, y = 10$;
7 $y = -3x + 8$

SEC. 20.5   **1** no conclusion about critical point located at $(-0.15, 0.25, 0.05, -0.075)$;
3 relative min at $(0, 0, 0, 2)$;   **5** relative max at $(0, 0, 0, 25)$;
7 (a) $p_1 = $9,666.67, p_2 = $7,500, p_3 = $9,833.34$,
(b) $q_1 = 2,000, q_2 = 3,000, q_3 = 2,500$, (c) $66,416,690

SEC. 20.6   **1** relative max at $(48, 52, 37,600)$, $\lambda = 752$;
3 relative max at $(33, 9, 1,926)$, $\lambda = +93$;   **5** relative max at $(9, 7, 528)$, $\lambda = 66$;
7 relative min at $(1.6, -2.24, -1.92, 12.6976)$, $\lambda = 0.96$;   **9** $q_1 = 50, q_2 = 150, (\lambda = 350)$

CHAPTER TEST   **1** (a) instantaneous change in $f(x, y)$ given a change in x, y assumed
to be held constant, (b) f_x represents a general expression for the tangent slope of the
family of traces which are parallel to the xz plane;   **2** given $z = f(x, y)$, a trace is the
graphical representation of $f(x, y)$ when one variable is held constant;
3 $f_x = 45x^2 + 10xy^3, f_y = -8y + 15x^2y^2$;
4 $f_{xx} = 160x^3 + 16y^3 + 12, f_{yy} = 48x^2y, f_{xy} = f_{yx} = 48xy^2$;
5 (a) relative min when $x = 1$ and $y = 3.5$, (b) $f(1, 3.5) = -20.75$;
6 200 acres of soybeans, 200 acres of corn, $400,000;
7 relative max at $(4, -2.5, 4, 278)$;
8 $L(x_1, x_2, \lambda) = -4x_1^3 + 3x_2^2 - 4x_1x_2 - \lambda(x_1 - 2x_2 - 20)$;

APPENDIX A

SEC. A.1   **1** $>$;   **3** $>$;   **5** $>$;   **7** $>$;   **9** $<$;   **11** $>$;   **13** 5;   **15** 15; **17** 16;   **19** 9

SEC. A.2   **1** 5^4;   **3** $(3)^2(-2)^3$;   **5** $(-x^3)$;   **7** $a^2b^3c^2$;   **9** x^4/y^2z^4; **11** x^4y^4;   **13** $2^7 = 128$;   **15** x^8;   **17** x^5y^4;   **19** x^6;   **21** x^{14};   **23** a^{12}; **25** $27x^6$;   **27** $4m^6$;   **29** $12a^8b^3$;   **31** $4096x^{24}$;   **33** $1/a^4$;   **35** 8; **37** 81;   **39** $1/x^5y^5$;   **41** x^2;   **43** $\frac{1}{8}$;   **45** 3;   **47** 1;   **49** x^3/y^3; **51** x^8/y^4;   **53** a^8b^4/c^{12};   **55** $27x^3y^6/z^9$;   **57** -20;   **59** $1/a^5$;   **61** $13x$; **63** $5y^3 + 2y^2 - 4y$;   **65** $25x^3y^2 - 25xy^3$;   **67** 0;   **69** $21x^4y^2$;   **71** $-8a^{10}$; **73** $-2x^4 + 2x^2y$;   **75** $x^4y - 2x^3y^2 + x^2y^3$;   **77** $a^2 + 2ab + b^2$; **79** $a^2 - 2ab + b^2$;   **81** $x^3 - 6x^2 + 12x - 8$;   **83** $4xy$;   **85** $-3/y$; **87** $5x - 8$;   **89** $-4a^2 + 3a - 2$;   **91** $2x^5 + 3x^2 - 4x$;   **93** $-12x^2 + 4xy^2 - 6y$

SEC. A.3   **1** $2a(x - 4a^2)$;   **3** $2xy(2x^2 - 3y^2 + 4xy)$;   **5** $3a(3a^2 - 5a - 9)$; **7** cannot be factored;   **9** $(p + 12)(p - 3)$;   **11** $(r + 1)(r - 22)$; **13** cannot be factored;   **15** $(6m - 1)(m - 3)$;   **17** $(2x + 1)(4x - 3)$; **19** $(x^2 + 9)(x + 3)(x - 3)$;   **21** $(9x^2 + 25)(3x + 5)(3x - 5)$; **23** cannot be factored;   **25** $(1 + 2x)(1 - 2x + 4x^2)$;   **27** $x^2(x - 2)(x + 1)$; **29** $x(x - 8)(x + 5)$;   **31** $x^3(x - 7)(x + 3)$;   **33** $ab(a^2 + 9)(a + 3)(a - 3)$; **35** $2uv(9 + u^2)(3 + u)(3 - u)$

SEC. A.4   **1** $\frac{11}{30}$;   **3** $\frac{1}{8}$;   **5** $(x - 2)/x^2$;   **7** $(x^2 + 7x)/(x^2 - 4)$; **9** $(10x^2 - 2)/x^2$;   **11** $(3a^2 + 3a - 5)/(a^2 + 2a + 1)$;   **13** -3;   **15** $1/3a^2b$; **17** $\frac{7}{15}$;   **19** $2a^3b^2/3c^3$;   **21** $(x - 4)/(x^2 - 5x - 4)$;   **23** $x(x + 1)$; **25** $(x^2 - 3x + 2)/(x^2 + 7x + 10)$

SEC. A.5   **1** $a^{17/6}$;   **3** $x^{31/30}$;   **5** $a^{5/4}$;   **7** $-27x^2$;   **9** $a^{4/3}$;   **11** $x^{44/15}$; **13** $x^{2/15}$;   **15** 25;   **17** $-a$;   **19** $-2x^2$;   **21** $12x^3$;   **23** $2a^2b$;   **25** a^3bc^6; **27** $6a^2$;   **29** $5\sqrt{7}$;   **31** $7\sqrt{2}$;   **33** $4\sqrt{x} - x\sqrt{x}$;   **35** 4;   **37** $\frac{2}{3}$; **39** $25x/7y^2$;   **41** $4x^2/3y^3$;   **43** 2;   **45** $\sqrt[3]{x^2}$;   **47** $\sqrt[5]{(ab)^3}$;   **49** $1/\sqrt{x}$; **51** $\frac{1}{2}$;   **53** $\sqrt[5]{a^3}$;   **55** $\sqrt[4]{(100 - x)}$;   **57** $(45x)^{1/2}$;   **59** $x^{3/4}$;   **61** $x^{5/3}$; **63** x^2;   **65** $(x + y)^{1/2}$;   **67** $(3 - x)^{3/4}$

INDEX

x	e^x	e^{-x}	x	e^x	e^{-x}	x	e^x	e^{-x}
0.880	2.4109	0.4148	3.700	40.447	0.0247	7.000	1096.6	0.0009
0.890	2.4351	0.4107	3.750	42.521	0.0235	7.050	1152.9	0.0009
0.900	2.4596	0.4066	3.800	44.701	0.0224	7.100	1212.0	0.0008
0.910	2.4843	0.4025	3.850	46.993	0.0213	7.150	1274.1	0.0008
0.920	2.5093	0.3985	3.900	49.402	0.0202	7.200	1339.4	0.0007
0.930	2.5345	0.3946	3.950	51.935	0.0193	7.250	1408.1	0.0007
0.940	2.5600	0.3906	4.000	54.598	0.0183	7.300	1480.3	0.0007
0.950	2.5857	0.3867	4.050	57.397	0.0174	7.350	1556.2	0.0006
0.960	2.6117	0.3829	4.100	60.340	0.0166	7.400	1636.0	0.0006
0.970	2.6379	0.3791	4.150	63.434	0.0158	7.450	1719.9	0.0006
0.980	2.6645	0.3753	4.200	66.686	0.0150	7.500	1808.0	0.0006
0.990	2.6912	0.3716	4.250	70.105	0.0143	7.550	1900.7	0.0005
1.000	2.7183	0.3679	4.300	73.700	0.0136	7.600	1998.8	0.0005
1.050	2.8576	0.3499	4.350	77.478	0.0129	7.650	2100.6	0.0005
1.100	3.0042	0.3329	4.400	81.451	0.0123	7.700	2208.3	0.0005
1.150	3.1582	0.3166	4.450	85.627	0.0117	7.750	2321.6	0.0004
1.200	3.3201	0.3012	4.500	90.017	0.0111	7.800	2440.6	0.0004
1.250	3.4903	0.2865	4.550	94.637	0.0106	7.850	2565.7	0.0004
1.300	3.6693	0.2725	4.600	99.484	0.0101	7.900	2697.3	0.0004
1.350	3.8574	0.2592	4.650	104.58	0.0096	7.950	2835.6	0.0004
1.400	4.0552	0.2466	4.700	109.95	0.0091	8.000	2981.0	0.0003
1.450	4.2631	0.2346	4.750	115.58	0.0087	8.050	3133.8	0.0003
1.500	4.4817	0.2231	4.800	121.51	0.0082	8.100	3294.5	0.0003
1.550	4.7114	0.2122	4.850	127.74	0.0078	8.150	3463.4	0.0003
1.600	4.9530	0.2019	4.900	134.29	0.0074	8.200	3641.0	0.0003
1.650	5.2069	0.1921	4.950	141.17	0.0071	8.250	3827.6	0.0003
1.700	5.4739	0.1827	5.000	148.41	0.0067	8.300	4023.9	0.0002
1.750	5.7545	0.1738	5.050	156.02	0.0064	8.350	4230.2	0.0002
1.800	6.0496	0.1653	5.100	164.02	0.0061	8.400	4447.1	0.0002
1.850	6.3597	0.1572	5.150	172.43	0.0058	8.450	4675.1	0.0002
1.900	6.6858	0.1496	5.200	181.27	0.0055	8.500	4914.8	0.0002
1.950	7.0286	0.1423	5.250	190.57	0.0052	8.550	5166.8	0.0002
2.000	7.3891	0.1353	5.300	200.34	0.0050	8.600	5431.7	0.0002
2.050	7.7678	0.1287	5.350	210.61	0.0047	8.650	5710.1	0.0002
2.100	8.1660	0.1225	5.400	221.41	0.0045	8.700	6002.9	0.0002
2.150	8.5847	0.1165	5.450	232.76	0.0043	8.750	6310.7	0.0002
2.200	9.0250	0.1108	5.500	244.69	0.0041	8.800	6634.2	0.0002
2.250	9.4875	0.1054	5.550	257.24	0.0039	8.850	6974.4	0.0001
2.300	9.9740	0.1003	5.600	270.43	0.0037	8.900	7332.0	0.0001
2.350	10.486	0.0954	5.650	284.29	0.0035	8.950	7707.9	0.0001
2.400	11.023	0.0907	5.700	298.87	0.0033	9.000	8103.1	0.0001
2.450	11.588	0.0863	5.750	314.19	0.0032	9.050	8518.5	0.0001
2.500	12.182	0.0821	5.800	330.30	0.0030	9.100	8955.3	0.0001
2.550	12.807	0.0781	5.850	347.23	0.0029	9.150	9414.4	0.0001
2.600	13.464	0.0743	5.900	365.04	0.0027	9.200	9897.1	0.0001
2.650	14.154	0.0707	5.950	383.75	0.0026	9.250	10405.	0.0001
2.700	14.880	0.0672	6.000	403.43	0.0025	9.300	10938.	0.0001
2.750	15.643	0.0639	6.050	424.11	0.0024	9.350	11499.	0.0001
2.800	16.445	0.0608	6.100	445.86	0.0022	9.400	12088.	0.0001
2.850	17.287	0.0578	6.150	468.72	0.0021	9.450	12708.	0.0001
2.900	18.174	0.0550	6.200	492.75	0.0020	9.500	13360.	0.0001
2.950	19.106	0.0523	6.250	518.01	0.0019	9.550	14045.	0.0001
3.000	20.086	0.0498	6.300	544.57	0.0018	9.600	14765.	0.0001
3.050	21.115	0.0474	6.350	572.49	0.0017	9.650	15522.	0.0001
3.100	22.198	0.0451	6.400	601.85	0.0017	9.700	16318.	0.0001
3.150	23.336	0.0429	6.450	632.70	0.0016	9.750	17154.	0.0001
3.200	24.533	0.0408	6.500	665.14	0.0015	9.800	18034.	0.0001
3.250	25.790	0.0388	6.550	699.24	0.0014	9.850	18958.	0.0001
3.300	27.113	0.0369	6.600	735.10	0.0014	9.900	19930.	0.0001
3.350	28.503	0.0351	6.650	772.78	0.0013	9.950	20952.	0.0000
3.400	29.964	0.0334	6.700	812.41	0.0012	10.000	22026.	0.0000
3.450	31.500	0.0317	6.750	854.06	0.0012			
3.500	33.115	0.0302	6.800	897.85	0.0011			
3.550	34.813	0.0287	6.850	943.88	0.0011			
3.600	36.598	0.0273	6.900	992.27	0.0010			
3.650	38.475	0.0260	6.950	1043.1	0.0010			

TABLE 2 NATURAL LOGARITHMS

x	ln x	x	ln x	x	ln x	x	ln x
0.050	− 2.9957	3.350	1.2089	6.650	1.8946	9.950	2.2976
0.100	− 2.3026	3.400	1.2238	6.700	1.9021	10.000	2.3026
0.150	− 1.8971	3.450	1.2384	6.750	1.9095	10.500	2.3514
0.200	− 1.6094	3.500	1.2528	6.800	1.9169	11.000	2.3979
0.250	− 1.3863	3.550	1.2669	6.850	1.9242	11.500	2.4423
0.300	− 1.2040	3.600	1.2809	6.900	1.9315	12.000	2.4849
0.350	− 1.0498	3.650	1.2947	6.950	1.9387	12.500	2.5257
0.400	− 0.9163	3.700	1.3083	7.000	1.9459	13.000	2.5649
0.450	− 0.7985	3.750	1.3217	7.050	1.9530	13.500	2.6027
0.500	− 0.6932	3.800	1.3350	7.100	1.9601	14.000	2.6391
0.550	− 0.5978	3.850	1.3481	7.150	1.9671	14.500	2.6741
0.600	− 0.5108	3.900	1.3610	7.200	1.9741	15.000	2.7080
0.650	− 0.4308	3.950	1.3737	7.250	1.9810	15.500	2.7408
0.700	− 0.3567	4.000	1.3863	7.300	1.9879	16.000	2.7726
0.750	− 0.2877	4.050	1.3987	7.350	1.9947	16.500	2.8034
0.800	− 0.2231	4.100	1.4110	7.400	2.0015	17.000	2.8332
0.850	− 0.1625	4.150	1.4231	7.450	2.0082	17.500	2.8622
0.900	− 0.1054	4.200	1.4351	7.500	2.0149	18.000	2.8904
0.950	− 0.0513	4.250	1.4469	7.550	2.0215	18.500	2.9178
1.000	− 0.0000	4.300	1.4586	7.600	2.0281	19.000	2.9444
1.050	0.0488	4.350	1.4702	7.650	2.0347	19.500	2.9704
1.100	0.0953	4.400	1.4816	7.700	2.0412	20.000	2.9957
1.150	0.1398	4.450	1.4929	7.750	2.0477	20.500	3.0204
1.200	0.1823	4.500	1.5041	7.800	2.0541	21.000	3.0445
1.250	0.2231	4.550	1.5151	7.850	2.0605	21.500	3.0681
1.300	0.2624	4.600	1.5260	7.900	2.0668	22.000	3.0910
1.350	0.3001	4.650	1.5369	7.950	2.0732	22.500	3.1135
1.400	0.3365	4.700	1.5476	8.000	2.0794	23.000	3.1355
1.450	0.3716	4.750	1.5581	8.050	2.0857	23.500	3.1570
1.500	0.4055	4.800	1.5686	8.100	2.0919	24.000	3.1781
1.550	0.4382	4.850	1.5790	8.150	2.0980	24.500	3.1987
1.600	0.4700	4.900	1.5892	8.200	2.1041	25.000	3.2189
1.650	0.5008	4.950	1.5994	8.250	2.1102	25.500	3.2387
1.700	0.5306	5.000	1.6094	8.300	2.1162	26.000	3.2581
1.750	0.5596	5.050	1.6194	8.350	2.1222	26.500	3.2771
1.800	0.5878	5.100	1.6292	8.400	2.1282	27.000	3.2958
1.850	0.6152	5.150	1.6390	8.450	2.1342	27.500	3.3142
1.900	0.6418	5.200	1.6486	8.500	2.1401	28.000	3.3322
1.950	0.6678	5.250	1.6582	8.550	2.1459	28.500	3.3499
2.000	0.6932	5.300	1.6677	8.600	2.1517	29.000	3.3673
2.050	0.7178	5.350	1.6771	8.650	2.1575	29.500	3.3844
2.100	0.7419	5.400	1.6864	8.700	2.1633	30.000	3.4012
2.150	0.7655	5.450	1.6956	8.750	2.1690	25.000	3.2189
2.200	0.7884	5.500	1.7047	8.800	2.1747	30.000	3.4012
2.250	0.8109	5.550	1.7138	8.850	2.1804	35.000	3.5553
2.300	0.8329	5.600	1.7228	8.900	2.1860	40.000	3.6889
2.350	0.8544	5.650	1.7316	8.950	2.1916	45.000	3.8067
2.400	0.8755	5.700	1.7405	9.000	2.1972	50.000	3.9120
2.450	0.8961	5.750	1.7492	9.050	2.2028	55.000	4.0073
2.500	0.9163	5.800	1.7578	9.100	2.2083	60.000	4.0943
2.550	0.9361	5.850	1.7664	9.150	2.2137	65.000	4.1744
2.600	0.9555	5.900	1.7749	9.200	2.2192	70.000	4.2485
2.650	0.9745	5.950	1.7834	9.250	2.2246	75.000	4.3175
2.700	0.9932	6.000	1.7918	9.300	2.2300	80.000	4.3820
2.750	1.0116	6.050	1.8000	9.350	2.2354	85.000	4.4427
2.800	1.0296	6.100	1.8083	9.400	2.2407	90.000	4.4998
2.850	1.0473	6.150	1.8164	9.450	2.2460	95.000	4.5539
2.900	1.0647	6.200	1.8245	9.500	2.2513	100.000	4.6052
2.950	1.0818	6.250	1.8326	9.550	2.2565	200.000	5.2983
3.000	1.0986	6.300	1.8405	9.600	2.2617	300.000	5.7038
3.050	1.1151	6.350	1.8484	9.650	2.2669	400.000	5.9915
3.100	1.1314	6.400	1.8563	9.700	2.2721	500.000	6.2146
3.150	1.1474	6.450	1.8641	9.750	2.2773	600.000	6.3969
3.200	1.1631	6.500	1.8718	9.800	2.2824		
3.250	1.1786	6.550	1.8795	9.850	2.2875		
3.300	1.1939	6.600	1.8871	9.900	2.2925		